1 MONTH OF
FREE
READING

at
www.ForgottenBooks.com

By purchasing this book you are eligible for one month membership to ForgottenBooks.com, giving you unlimited access to our entire collection of over 700,000 titles via our web site and mobile apps.

To claim your free month visit:
www.forgottenbooks.com/free892597

ISBN 978-0-265-80823-8
PIBN 10892597

DEPARTMENT OF THE INTERIOR

UNITED STATES GEOLOGICAL SURVEY

CHARLES D. WALCOTT, Director

WATER POWERS OF ALABAMA

WITH

AN APPENDIX ON STREAM MEASUREMENTS IN MISSISSIPPI

BY

BENJAMIN M. HALL

WASHINGTON

GOVERNMENT PRINTING OFFICE

1904

CONTENTS.

4 CONTENTS.

ILLUSTRATIONS.

5

LETTER OF TRANSMITTAL.

DEPARTMENT OF THE INTERIOR,
UNITED STATES GEOLOGICAL SURVEY,
HYDROGRAPHIC BRANCH,
Washington, D. C., February 27, 1904.

SIR: I have the honor to transmit herewith a manuscript entitled "The Water Powers of Alabama, with an Appendix on Stream Measurements in Mississippi," and to request that it be published as a water-supply paper of the Survey.

Very respectfully,

F. H. NEWELL,
Chief Engineer.

Hon. CHARLES D. WALCOTT,
Director United States Geological Survey.

LETTER OF TRANSMITTAL.

DEPARTMENT OF THE INTERIOR,
UNITED STATES GEOLOGICAL SURVEY,
HYDROGRAPHIC BRANCH,
Washington, D. C., February 27, 1904.

SIR: I have the honor to transmit herewith a manuscript entitled "The Water Powers of Alabama, with an Appendix on Stream Measurements in Mississippi," and to request that it be published as a water-supply paper of the Survey.

Very respectfully,

F. H. NEWELL,
Chief Engineer.

Hon. CHARLES D. WALCOTT,
Director United States Geological Survey.

WATER POWERS OF ALABAMA.

By Benjamin M. Hall.

INTRODUCTION.

Prior to 1896 no systematic investigation had ever been made of the water supply and water powers of Alabama. In 1885 Prof. Dwight Porter, in his report for the Tenth Census of the United States, Volume XVI, upon the water powers of the eastern Gulf slope, gave an excellent description of the most important water-power streams in the Mobile drainage basin, with such estimates of power as were possible from a reconnaissance and a study of rainfall, drainage areas, and the navigation surveys made by the United States Engineer Corps.

During the last seven years the systematic stream measurements by the United States Geological Survey, at stations maintained by the army engineers, the Weather Bureau, and the State geological survey, have furnished a reliable basis for water-power estimates. The United States Geological Survey has also made measurements of many smaller streams in order to compare their discharges with those at the regular stations and to ascertain the percentage of water furnished to the main rivers by different tributaries. It has also run levels along some of the streams that have not been surveyed by the army engineers.

In 1902 the writer prepared for Dr. E. A. Smith, State geologist, a report embodying all data obtainable at that time. This was published by the Alabama geological survey as Bulletin No. 7. This paper contains all of the material of Bulletin No. 7, and includes also the results of hydrographic work in 1902 and 1903. It gives an estimate and short description of the water powers in the crystalline and Paleozoic regions, and is intended to meet the demands of those who are interested in water powers and desire to have in one volume the records and results of all the hydrographic investigations from 1896 to the present.

Very recently two large water powers have been developed on Tallapoosa River; one of these is at Tallassee, Ala., and the other is 3 miles above Tallassee.

9

Some of the largest undeveloped powers in the State are mentioned below:

Power site No. 3, on Tallapoosa River, at Double Bridge Ferry, about 10 miles above Tallassee, where a dam 80 feet high is proposed by the Cherokee Development and Manufacturing Company.

At Black and Sanford Shoal, on Big Sandy Creek, near Dadeville, there is a fall of 80 feet.

Thirty-one locks on the Coosa River are capable of furnishing from 1,300 to 4,500 horsepower each, or an aggregate of 100,000 horsepower during the low season of an ordinary year like 1900.

Seven power sites on the Cahaba River are capable of furnishing from 500 to 1,100 horsepower each.

The following shoals are on Tennessee River:

Power sites on Tennessee River.

Shoal.	Fall.	Minimum horsepower.	
		Driest years.	Average years.
	Feet.		
Elk River	26	15,600	30,550
Mussel	85	51,000	99,875
Little Mussel	23	13,800	27,025
Colbert	21	12,600	24,675

These and other powers will be described more fully in the body of the work.

The water powers of Alabama are conveniently located for running cotton factories and other manufacturing plants, and also for generating electricity that can be transmitted to cities for power, light, etc. The larger powers are all close to water transportation, and are also on important railroads. These advantages naturally make them more valuable.

GENERAL TOPOGRAPHIC AND GEOLOGIC FEATURES.

The five principal drainage basins of the State are as follows:

The Apalachicola basin, draining to the Chattahoochee and Apalachicola River and entering the Gulf at Apalachicola, Fla.

The Choctawhatchee basin, draining to the Gulf through Choctawhatchee Bay.

The Pensacola basin, draining to Pensacola Bay and Perdido Bay, near Pensacola, Fla.

The Mobile basin, including the waters of Tallapoosa, Coosa, Cahaba, Alabama, Black Warrior, and Tombigbee rivers and draining into the Gulf at Mobile, Ala.

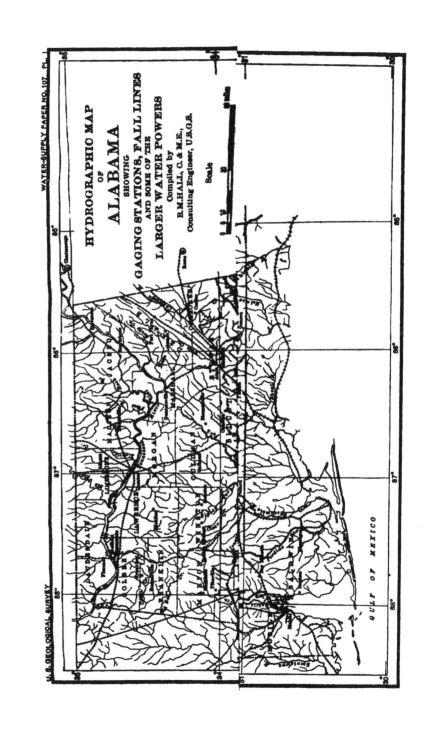

HYDROGRAPHIC MAP
OF
ALABAMA
SHOWING
GAGING STATIONS, FALL LINES
AND SOME OF THE
LARGER WATER POWERS
Compiled by
B. M. HALL, C. & M. E.,
Consulting Engineer, U.S.G.S.

Scale

GULF OF MEXICO

The Tennessee basin, draining into Tennessee River and thence through the Mississippi to the Gulf at New Orleans.

The water powers of the State are mainly in the Mobile and Tennessee basins, which practically cover the entire State, except a small area in the southeast corner.

Geologically the State may be divided into three areas, which differ very greatly from one another in the hardness and durability of their rocks and show a corresponding difference in the profiles of their streams: First, the crystalline area underlain by igneous and metamorphic rocks; second, the Paleozoic area of hard sedimentary rocks, sandstone, limestone, and shale; and third, the Coastal Plain formed by Mesozoic and later sediments, chiefly soft limestones, and unconsolidated sands and clays.

It may be said in a general way that the streams have their greatest falls in passing from an older to a younger geologic formation. Tallassee Falls, on the Tallapoosa, and Wetumka Falls, on the Coosa, are at the point where the streams flow from the crystalline to the Cretaceous rocks. The falls on Talladega Creek and other small streams entering Coosa River from the southeast in Talladega, Calhoun, and Cleburne counties are at the contact of the crystalline and the Paleozoic. The shoals above Centerville on the Cahaba, above Tuscaloosa on the Black Warrior, and near Tuscumbia on Tennessee River are at the point where the Paleozoic adjoins the Cretaceous. As Coosa River flows from the Paleozoic on to the crystalline near Talladega Springs, the shoals above this point reverse the general order by being made in passing from a younger to an older formation.

The crystalline area in Alabama is a triangle on the east side of the State, including Cleburne, Randolph, Chambers, Lee, Tallapoosa, Clay, Coosa, and parts of Elmore, Chilton, and Talladega counties. The "fall line" dividing the crystalline region from the Cretaceous and later formations of the Coastal Plain on the southwest runs from Columbus, Ga., crossing the Tallapoosa at Tallassee and the Coosa at Wetumka. The northwestern boundary between the crystalline rocks and the Paleozoic formations recrosses the Coosa near Marble Valley post-office in Coosa County, and runs in a northeasterly direction toward Cedartown, Ga., crossing the Alabama line near Warner.

The crystalline area is a plateau ranging in elevations from 500 to 2,000 feet above sea level. The rivers flow over bed rock in a succession of shoals and eddies between high hills, and present conditions most favorable to the development of water powers with high head.

Tallapoosa River and its tributaries drain the larger part of the area. On Tallapoosa River there is a fall of 64 feet utilized as one power at Tallassee, and also a 40-foot dam about 3 miles above Tallassee. Similar developments are contemplated at several different points above. The falls on Coosa River from Marble Valley to Wetumka, 235 feet in 45 miles, and on numerous eastern tributaries of the Coosa, are in this area; as are also the western tributaries of the Chattahoochee between West Point and Columbus, Ga.

The Paleozoic area includes the greater portion of northern Alabama, being bounded on the southeast by the crystalline area, and on the southwest by the Cretaceous and later formations of the Coastal Plain. The line of division between the Paleozoic area and the Coastal Plain begins near Strasburg, in Chilton County, and runs northwesterly through Centerville, Tuscaloosa, and Tuscumbia to a point on Tennessee River near Waterloo. The Paleozoic area is somewhat higher than the Coastal Plain, and slightly lower than the crystalline area. Its rivers have considerable fall, as is shown by the following figures:

Coosa River from Greensport to Marble Valley falls 132 feet in 97 miles.

Cahaba River above Centerville has a fall of 120 feet in 21 miles.

Black Warrior River above Tuscaloosa has a fall of 100 feet in 30 miles.

Tennessee River above Waterloo has a fall of 155 feet in 41 miles, 85 feet of which is in a distance of only 14 miles.

There are many important creeks and many large limestone springs in this region on which no measurements have yet been made. The area is rich in coal and iron, the most productive mines being in the drainage basins of Cahaba and Black Warrior rivers.

The Coastal Plain is a large area in southern and western Alabama, covering about two-thirds of the State, and is underlain by Cretaceous and younger formations. In the upper portion of this area the streams are not sluggish. Alabama River is navigable in the whole region, but the Cahaba falls 120 feet in a distance of 87 miles from Centerville to the mouth of the river; and the Black Warrior falls 59 feet in 132 miles from Tuscaloosa down to Demopolis.

There are many streams in this area that have a constant water supply and sufficient fall for the development of good water powers. One of these is Pea River, in the southeastern part of the State, where a dam is now being constructed to give a power head of 20 feet. No hydrographic work has yet been done on the streams of southern Alabama that rise in the Coastal Plain, but systematic work is to be taken up in this region at an early date.

As the investigations in this State have been confined so far mainly to the Mobile and Tennessee basins, only the streams of these basins will be considered. It is to be remembered that from West Point, Ga., southward the boundary between Georgia and Alabama is on the west bank of Chattahoochee River, along the line where ordinary vegetation ceases to grow. This leaves all of the water power of the main stream in Georgia. There are many creeks flowing into the river from Alabama, some of which have considerable fall, as they come from a high plateau. Holland Creek, opposite Columbus, Ga., furnishes the Columbus water supply by gravity, having a fall of 117 feet in less than 4 miles. No doubt many of the others have as much fall, but as they have not been examined, a report on them can not be made at present. However, a recent reconnaissance along the Chattahoochee gives the following estimate of power obtained from some of them for twelve hours each day.

Powers on tributaries of Chattahoochee River, per foot of fall.

	Horsepower.
Big Uchee Creek, Russell County	7
Ihagee Creek, Russell County	2
Hatchechubbee Creek, Russell County	7
Cowikee Creek, Barbour County	11
Yattayabba Creek, Henry County	9
Omussee Creek, Henry County	7

FIELD WORK.

Aside from certain surveys made to obtain maps and profiles of Tallapoosa River and Big Sandy Creek, the work done by the United States Geological Survey in this State deals exclusively with the amount of water flowing in the streams, and is intended to give a safe basis for calculation of low-water volumes at all seasons of the year, and for several consecutive years, in order to arrive at the value of the streams for water power, irrigation, municipal supply, mining, navigation, etc. In order to do this certain convenient stations have been established on important rivers. At each of these stations a gage rod is set to show the fluctuations of the streams; and a gage reader is employed to observe the height of the water every morning at the same hour, and to make a weekly report of the same. As far as possible the river stations of the United States Weather Bureau and the United States Engineer Corps have been utilized for this purpose. From time to time the hydrographer or one of his field assistants visits the station and makes an accurate meter measurement of the stream, noting the height of the water on the gage at the time the measurement is made. After a number of such discharge measurements have been made at different gage heights, a rating table is made from the data thus obtained. This table gives the amount of water flowing in the stream, at that station, for any gage height shown on the rod. Thus, by inspection of the table of daily gage heights, the flow of the stream is shown for every day in the year, or years, covered by the observation of gage height. At seasons of uniform low water, when the daily fluctuations of the rod are very slight for weeks at a time, discharge measurements are made of the stream at many points above and below the gage station in order to establish a relation between the discharge at these points and at the station. In like manner the principal tributaries are measured for the same purpose, where it is practicable to do so. In this way it is possible to arrive at a close estimate of the flow of all the streams of the watershed, and to make a rating table for each that will represent its flow under average conditions, not including the floods caused by local rains. Such tributaries as have not been measured can be estimated by watershed comparison with similar tributaries that have been measured.

The actual gage heights and discharge measurements are published in order to show the data upon which the conclusions are based. The regular gaging stations are as follows:

Gaging stations.

Station.	Stream.	Observer.
Milstead	Tallapoosa River	Seth Johnson.
Sturdevant	....do	B. F. Neighbors.
Dadeville	Big Sandy Creek	T. H. Finch.
Alexander	Hillabee Creek	J. H. Chisolm.
Nottingham	Talladega Creek	R. M. McClatchy.
Riverside	Coosa River	J. W. Foster.
Cordova	Black Warrior River	A. B. Logan.
Montgomery	Alabama River	United States Weather Bureau.
Selma	....do	Do.
Tuscaloosa	Black Warrior River	W. S. Wyman, jr.
Epes	Tombigbee River	J. C. Horton.
Rome, Ga	Coosa River	W. M. Towers.
Chattanooga, Tenn	Tennessee River	United States Weather Bureau.
Palos	Locust Fork of Black Warrior River.	United States Engineer Corps.
Centerville	Cahaba River	Clyde Lowrey.
Jenifer	Choccolocco Creek	W. J. Tolbert.
Columbus, Miss	Tombigbee River	J. J. Richards.

EXPLANATION OF TABLES.

GAGE HEIGHTS.

The "Table of gage heights" is a record of the height of water on a gage rod, graded to feet and hundredths of a foot, set into the river vertically, and fastened permanently to a convenient tree or pier. The rod is read every day in the year, at the same time of day, which is about 8 o'clock in the morning. Inches are not used in these records, as the daily height of water on the gage is written in feet and decimals of a foot.

DISCHARGE MEASUREMENTS.

These records show the date, the gage height at time of measurement. and the amount of water in cubic feet per second, or "second-feet." flowing in the river. A small stream filling a rectangular flume 1 foot wide and 1 foot deep has a stream whose sectional area is 1 square foot. The volume of this stream will vary in proportion to the speed with which the water flows through the flume. If the water is moving at a velocity of 1 foot per second, the flow or volume of water is 1 cubic foot per second, or 1 second-foot, and would fill a vessel 5 feet wide, 5 feet long, and 4 feet deep in just 100 seconds, as such a vessel would hold 100 cubic feet of water. If the water in the flume 1 foot wide and 1 foot deep flows with a velocity of 2 feet per second, the volume will be 2 cubic feet per second, or 2 second-feet, and so on for any other velocity. In the same way if the flume is 20 feet wide, and 5 feet deep, its sectional area will be 100 square feet, and if the average velocity is 3 feet per second, the volume will be 300 cubic feet per second, or 300 second-feet. In each of the discharge measurements here enumerated, a cross section of the stream is measured, and velocities taken with an electric current meter at many points of the cross section. Instead of multiplying the entire cross section by an average velocity, the area is divided up into a large number of small sections by soundings from 5 to 10 feet apart, and the area of each of the small sections multiplied by the velocity at the small section, thus giving the second-feet flowing in each small section. The sum of the discharges of all the small sections makes the total discharge of the stream.

RATING TABLE.

This is a table showing the discharges in second-feet (cubic feet per second) for all stages of water on the gage. Hence, when the gage heights are known, the corresponding discharges can be taken from the rating table and written opposite each daily gage height, thus giving the flow in second-feet on each day in the entire year.

ESTIMATED MONTHLY DISCHARGE.

This table gives in the first three columns the maximum, minimum, and mean discharge for each month in second-feet.

Columns 4 and 5 give the "run-off" from the drainage area. The run-off, like rainfall, is given in inches. For instance, a run-off of 2.23 inches from a given drainage area means that enough water ran off to have covered the entire drainage area or watershed to a depth of 2.23 inches.　This is convenient in estimating the proportion of the rainfall that can be stored for irrigation, city water supply, or other purposes.　The run-off in second-feet per square mile of drainage area is obtained by dividing the mean discharge by the number of square miles in the drainage area, and is useful in estimating the mean discharge of a tributary whose drainage area is known, and in comparing different drainage areas.　The "run-off" is not a fixed percentage of the rainfall, but is that part of the rainfall which is not lost by evaporation into the air or by percolation into subterranean outlets.　Being a remainder and not a percentage, it necessarily forms a much larger proportion of a heavy annual rainfall than it does of a small annual rainfall.　For instance, in the crystalline region of Georgia or Alabama, where the annual precipitation is 45 to 55 inches, the run-off is equal to fully one-half of the rainfall, while in regions having an annual precipitation of only 10 to 20 inches annually the run-off is frequently less than one-fifth of the rainfall.　Again, the geologic character of the watershed makes a vast difference in the run-off, even where the annual rainfall is the same and where practically the same conditions of climate, topography, forest area, and cultivation exist.　There will be a smaller run-off from the watershed where the rocks are permeable and a part of the rain water sinks into the ground and furnishes the supply to artesian wells in the lower country under which the same strata run.　In a comparison of two such watersheds, one in the crystalline region and the other in a regularly stratified formation, the difference of run-off should form a basis for estimating the artesian supply obtainable from the latter as a fountain head.

HORSEPOWER.

The data for the table headed "Net horsepower per foot of fall, with a turbine efficiency of 80 per cent for the minimum monthly discharge of Tallapoosa River at Milstead, Ala.," were obtained as follows:

The theoretical horsepower available at any point on a stream is the product of the fall by the weight of water falling in a given time.

It may be found by the formula $H. P. = \dfrac{Q \times h \times 62\frac{1}{2}}{550}$, where Q is discharge in second-feet and h is fall in feet.　To find the net horsepower developed by the water wheel the theoretical horsepower has to be reduced by a coefficient dependent upon the efficiency of the wheel. In the table this efficiency is assumed to be 80 per cent, and we have

$$\text{net } H. P. = 80 \text{ per cent} \times \frac{Qh \times 62.5}{550}.$$

Thus in the table for Tallapoosa River, pages 23–24, for June, 1899, the mean monthly discharge was 1,287 second-feet, so that for a fall of 1 foot the minimum net $H. P. = \dfrac{1,287 \times 1 \times 62.5}{550} \times 80 \text{ per cent} = 117.$

To find the minimum net horsepower available at a shoal on any stream, multiply the total fall of the shoal by the "net horsepower per foot of fall."

TALLAPOOSA RIVER AND TRIBUTARIES.

TALLAPOOSA RIVER AT MILSTEAD.

Tallapoosa River rises in west-central Georgia and flows southwesterly into Alabama. Six miles above Montgomery it joins the Coosa to form Alabama River. Its upper tributaries drain an area between the Chattahoochee and Coosa basins. At Tallassee, Ala., it crosses the southern fall line. The shoals at this place have a fall of 60 feet, forming an obstruction to navigation. The drainage area is largely wooded, with cultivated fields at short intervals.

A gaging station was established on August 7, 1897, at the bridge of the Tallassee and Montgomery Railway, about one-fourth of a mile from Milstead, Ala. The bridge is of iron, two spans of about 155 feet each, with short wooden trestles at each end. The initial point of measurement is the end of the iron bridge, on the left bank, downstream side. The rod of the wire gage is fastened to the outside of the guard rail on the downstream side of the bridge. The bench mark is the top of the second crossbeam from the left bank pier at the downstream end, and is 60 feet above datum. The channel is straight at the bridge and bends above and below. The current is sluggish at low water and obstructed by the center pier of the bridge. The banks are high, but overflow at extreme high water for several hundred feet on each side. The bed is fairly constant, and all the water is confined to the main channel by railroad embankments.

Milstead station has been discontinued on account of the large dams of the Tallassee Manufacturing Company and the Montgomery Power Company a short distance above. When these dams are storing water the flow is almost entirely cut off at Milstead and makes the record show a deficiency of water that does not exist. The dams were completed early in 1902, and as the estimated minimum flow for that year is a great deal less than at Sturdivant, 39 miles above, it is thought best not to publish the Milstead records for 1902 and 1903 and to discontinue the station. Sturdivant station is above both dams and gives the correct flow at that point.

The following discharge measurements were made by M. R. Hall, James R. Hall, and others:

Discharge measurements of Tallapoosa River at Milstead.

Date.	Gage height.	Discharge.	Date.	Gage height.	Discharge.
1897.	*Feet.*	*Second-feet.*	**1899.**	*Feet.*	*Second-feet.*
May 3	6. 20	7, 333	April 17	6. 34	7, 444
July 15	1. 95	1, 692	April 18	5. 63	6, 853
August 7............	2. 42	2, 292	May 17	2. 80	3, 000
September 4.........	1. 60	1, 271	June 26.............	2. 05	1, 847
November 23........	1. 20	677	September 9.........	1. 36	1, 016
December 16	3. 58	4, 210	November 8........	1. 25	972
1898.			December 18	2. 66	2, 844
January 19..........	2. 13	1, 889	**1900.**		
February 19.........	2. 20	2, 045	February 23........	9. 20	9, 956
March 18	2. 56	2, 646	March 5	6. 70	7, 088
April 26	5. 83	6, 648	December 3	2. 95	3, 031
May 17	1. 55	1, 059	**1901.**		
June 22.............	3. 05	3, 421	February 12........	10. 70	11, 759
July 7	1. 62	1, 262	March 13	5. 55	5, 644
August 5............	13. 67	15, 295	October 29	1. 70	1, 583
September 3	2. 76	3, 010			
November 29........	5. 16	5, 477			

Daily gage height, in feet, of Tallapoosa River at Milstead.

Day.	Jan.	Feb.	Mar.	April.	May.	June.	July.	Aug.	Sept.	Oct.	Nov.	Dec.
1897.												
1.........									1.70	0.80	0.90	1.50
2.........									1.80	.80	1.00	1.50
3.........									1.60	.80	1.10	1.50
4.........									1.60	.70	1.10	1.60
5.........									1.60	.70	1.10	1.80
6.........									1.40	.70	1.10	2.00
7.........								2.45	1.40	.70	1.10	2.10
8.........								1.90	1.30	.60	1.10	2.00
9.........								1.70	1.20	.70	1.20	1.90
10.........								1.50	1.20	.70	1.20	1.90
11.........								1.50	1.10	.70	1.80	1.90
12.........								2.70	1.10	.70	1.80	2.00
13.........								2.20	1.10	.70	1.80	1.90
14.........								2.00	1.10	.80	1.80	5.50
15.........								1.80	1.40	.80	1.20	4.70
16.........								1.60	1.20	.80	1.20	3.60
17.........								1.90	1.10	.80	1.20	2.80
18.........								2.20	1.00	.90	1.20	2.40
19.........								2.80	1.10	.90	1.20	2.10
20.........								9.70	1.10	.90	1.10	2.00
21.........								7.40	1.10	.80	1.20	1.90
22.........								8.50	1.00	.80	1.10	2.00
23.........								5.30	1.00	.90	1.10	2.40
24.........								3.40	1.00	.90	1.20	2.40
25.........								2.90	1.00	.80	1.20	2.40
26.........								2.80	1.00	.90	1.20	2.80
27.........								2.30	1.00	.90	1.20	2.60
28.........								2.00	.90	.90	1.60	2.50
29.........								1.80	.90	.90	1.50	2.30
30.........								1.70	.80	.90	1.50	2.20
31.........								1.70	....	.90	....	2.40

Daily gage height, in feet, of Tallapoosa River at Milstead—Continued.

Day.	Jan.	Feb.	Mar.	April.	May.	June.	July.	Aug.	Sept.	Oct.	Nov.	Dec.
1898.												
1	2.90	2.30	1.90	4.80	2.60	1.10	1.50	3.60	4.30	1.30	2.30	4.90
2	1.90	2.20	1.80	3.90	2.30	1.10	1.30	2.90	3.20	1.30	2.20	4.40
3	1.80	2.10	1.90	3.10	2.20	1.10	1.20	2.80	2.70	1.40	2.20	7.70
4	1.70	2.00	2.40	3.00	2.10	1.10	1.10	11.60	2.70	16.00	2.10	10.00
5	1.70	2.00	2.90	17.00	2.00	1.10	1.20	14.00	4.00	32.00	2.10	8.60
6	1.80	2.00	2.80	18.80	2.00	1.10	1.30	8.90	4.60	23.50	2.20	6.80
7	1.80	2.00	2.50	12.10	1.90	1.00	1.70	10.10	6.00	16.40	2.70	5.60
8	1.80	2.00	2.30	7.10	1.90	.90	2.90	8.50	5.60	22.80	4.20	4.80
9	1.90	2.00	2.20	5.20	1.80	.90	4.90	7.90	4.50	14.00	2.80	4.60
10	1.90	1.90	2.00	4.00	1.80	.90	8.60	5.30	3.50	7.90	3.00	6.70
11	1.90	1.90	2.00	3.40	1.70	.80	4.40	7.30	2.90	5.90	7.10	6.10
12	2.40	1.90	2.00	3.10	1.70	.90	2.00	22.60	2.60	4.90	5.10	5.50
13	2.40	1.90	1.90	3.00	1.70	1.20	1.70	10.10	2.50	3.80	5.00	5.00
14	2.60	1.90	1.90	2.90	1.60	1.20	2.70	7.70	2.40	3.40	5.90	4.40
15	2.40	1.80	2.70	2.80	1.60	1.10	3.50	6.10	2.20	3.00	5.50	4.00
16	2.40	1.80	2.60	2.60	1.60	1.20	5.10	4.20	2.00	2.70	5.40	3.80
17	2.40	1.80	2.50	2.40	1.60	2.10	2.40	3.40	1.80	2.60	6.20	3.60
18	2.30	1.90	2.60	2.40	1.60	1.80	2.80	2.80	1.70	4.40	6.10	3.40
19	2.20	2.20	2.50	2.20	1.50	1.40	3.05	2.60	1.60	5.00	14.40	4.30
20	2.20	2.30	2.40	2.40	1.50	1.40	2.30	2.80	1.60	4.40	12.20	7.60
21	3.10	2.30	2.20	2.80	1.40	1.40	1.90	2.50	1.50	3.70	10.00	7.00
22	3.10	2.10	2.00	2.70	1.40	3.00	1.60	2.30	1.50	3.30	8.00	5.20
23	3.00	2.00	1.90	2.80	1.40	2.40	1.60	2.20	1.50	3.10	10.00	4.40
24	2.90	2.00	1.90	14.50	1.30	2.20	1.50	2.00	1.60	2.70	8.80	6.40
25	2.60	1.90	1.80	11.60	1.20	2.10	1.50	2.10	1.80	2.60	7.90	6.00
26	2.90	1.80	1.80	5.90	1.20	1.80	2.50	8.10	2.00	2.50	5.30	4.60
27	3.60	1.90	1.80	4.30	1.20	1.50	2.60	10.20	1.80	2.40	4.60	4.20
28	3.90	1.90	1.70	3.30	1.30	1.80	4.10	8.40	1.60	2.40	4.10	3.90
29	3.10	......	2.20	2.95	1.20	2.70	2.80	7.00	1.50	2.40	4.90	3.80
30	2.65	......	4.20	2.80	1.20	1.90	2.90	5.20	1.50	2.40	5.40	3.70
31	2.40	......	5.30	......	1.10	......	3.80	5.10	......	2.30	......	4.00
1899.												
1	5.00	17.00	27.00	18.00	4.30	2.40	2.80	3.00	2.40	.70	1.50	2.40
2	4.60	11.60	19.00	12.20	4.00	2.90	2.30	2.70	2.30	.70	1.40	2.50
3	3.90	24.50	13.50	6.50	3.90	2.40	1.90	3.10	2.10	.80	1.40	2.60
4	3.90	20.00	9.60	6.40	3.80	2.40	1.80	3.10	2.00	.90	1.30	2.60
5	3.90	12.90	14.20	8.50	3.70	2.80	1.60	2.50	1.80	1.00	1.30	2.90
6	3.80	11.90	13.20	9.70	3.60	2.20	1.50	2.30	1.60	1.40	1.20	2.20
7	7.40	17.50	10.10	10.30	3.50	2.10	1.50	2.20	1.60	1.40	1.20	2.00
8	8.00	27.00	8.60	13.00	3.60	2.00	2.00	2.00	1.40	1.30	1.20	1.90
9	7.10	19.00	7.70	11.20	3.40	1.70	2.00	1.80	1.40	1.30	1.20	1.90
10	6.40	13.80	7.20	8.40	3.30	1.90	1.80	1.60	1.30	1.40	1.30	2.00
11	18.50	10.00	6.80	6.60	3.20	2.00	1.70	1.70	1.20	1.40	1.30	15.20
12	16.80	8.30	6.60	7.00	3.20	2.00	1.50	1.60	1.20	1.30	1.30	13.20
13	13.00	7.40	6.50	6.50	3.10	2.70	1.60	1.60	1.10	1.30	1.30	8.20
14	11.60	7.00	6.40	6.20	3.00	2.70	1.30	1.50	1.00	1.20	1.30	5.00
15	9.40	6.10	7.20	6.00	2.90	2.60	1.30	4.00	1.00	1.10	1.30	3.70
16	7.80	10.40	12.20	7.10	2.80	2.10	1.20	3.90	1.00	1.20	1.60	3.00
17	12.70	11.50	11.00	6.60	2.70	1.90	1.10	2.20	1.00	1.10	1.50	2.60
18	10.00	10.60	10.20	5.50	2.60	1.80	1.60	1.90	1.10	1.20	1.50	2.50
19	8.00	9.30	9.30	14.80	2.50	1.80	1.40	1.70	4.00	1.30	1.50	1.90
20	6.50	8.30	13.90	5.40	2.60	1.70	8.40	1.50	.90	1.40	1.40	2.70
21	5.70	8.30	10.40	5.20	2.70	1.50	16.75	1.60	1.00	1.50	1.40	2.70
22	5.30	8.40	8.30	5.00	2.60	1.50	14.00	2.00	1.00	1.60	1.60	2.60
23	5.10	7.60	8.10	4.90	3.30	1.50	16.95	2.00	.90	1.80	1.80	9.30
24	5.20	6.90	12.70	6.00	4.60	1.50	7.90	1.90	.90	1.50	2.20	9.40
25	5.20	6.40	8.70	10.00	8.30	2.00	6.70	1.80	.90	1.40	4.60	7.20
26	5.10	6.30	7.30	7.50	2.80	2.50	6.80	3.70	.90	1.30	6.20	5.00
27	4.80	25.00	6.90	6.60	2.60	2.60	8.40	2.80	.80	1.20	4.90	4.40
28	4.70	37.00	6.80	5.50	2.60	2.20	8.40	2.10	.80	1.50	8.60	3.50
29	5.10	......	9.00	4.90	2.50	2.20	10.10	2.10	.80	1.50	2.80	3.50
30	5.20	......	8.90	4.60	2.60	2.70	5.40	1.90	.80	1.50	......	3.10
31	6.50	......	13.85	......	2.50	......	4.40	2.30	......	1.60	......	2.90

Daily gage height, in feet, of Tallapoosa River at Milstead—Continued.

Day.	Jan.	Feb.	Mar.	April.	May.	June.	July.	Aug.	Sept.	Oct.	Nov.	Dec.
1900.												
1	2.70	2.50	13.20	5.50	6.30	2.60	9.00	6.50	2.70	1.80	2.50	3.30
2	2.60	2.40	13.10	5.30	5.90	2.70	9.10	4.50	8.00	1.60	3.30	3.30
3	2.40	2.30	10.70	5.00	5.40	3.00	10.60	3.50	8.10	1.50	6.10	3.30
4	2.30	4.50	8.00	5.30	4.90	3.50	7.10	2.80	4.40	1.80	6.50	3.50
5	2.30	4.20	6.80	4.80	4.70	3.30	8.00	2.70	8.00	3.30	7.10	3.90
6	2.30	4.30	6.10	4.70	3.90	3.50	5.20	2.50	2.40	6.40	5.50	3.90
7	2.20	4.40	5.60	4.60	3.80	3.80	4.00	2.30	2.20	8.40	4.10	3.80
8	2.20	3.80	10.90	4.60	3.70	5.90	4.00	2.20	2.00	3.20	3.50	3.80
9	2.20	5.20	13.80	4.50	3.60	9.70	3.50	2.10	1.80	2.60	3.00	3.10
10	2.20	8.90	12.70	4.60	3.50	6.90	7.60	2.00	1.70	2.60	2.90	3.10
11	2.60	19.00	10.00	5.50	3.30	5.60	5.50	1.90	1.70	8.00	2.70	2.80
12	7.30	30.00	7.90	10.60	3.20	4.60	4.10	1.90	1.60	2.80	2.60	2.80
13	6.00	43.25	6.60	11.50	3.10	5.30	4.70	1.80	1.50	8.40	2.60	2.80
14	4.50	42.00	5.90	9.00	3.00	4.00	6.70	1.90	1.50	8.30	2.50	7.70
15	4.00	31.90	5.40	6.60	2.90	3.80	6.00	2.20	14.00	8.30	2.50	9.10
16	3.40	22.80	7.00	5.30	2.90	3.50	5.10	2.70	25.60	2.70	2.50	7.30
17	3.00	13.50	7.20	4.70	2.80	4.50	3.80	3.10	18.00	2.70	2.50	5.20
18	2.90	8.90	5.60	13.90	2.80	5.90	3.80	2.50	11.00	2.20	2.50	4.20
19	3.70	7.00	5.40	17.00	2.80	5.60	3.60	4.00	5.30	2.10	2.50	4.00
20	9.50	6.10	6.00	15.00	2.90	5.90	3.10	4.20	3.60	2.00	2.50	8.60
21	7.50	6.50	11.40	16.90	2.80	5.90	3.00	2.50	3.00	1.90	2.50	17.00
22	5.90	9.80	10.50	13.30	8.00	5.50	2.80	2.20	2.50	2.00	4.00	13.50
23	4.60	9.50	7.60	10.30	3.10	5.40	2.70	2.00	2.30	4.80	4.00	10.40
24	3.90	8.90	15.50	13.20	8.40	20.00	2.60	2.30	2.20	12.10	3.90	11.30
25	3.50	8.40	15.20	12.50	3.50	25.04	2.50	3.30	2.00	10.50	3.50	8.70
26	3.10	8.00	16.00	9.40	5.00	20.00	2.50	3.40	2.00	9.00	10.50	6.60
27	3.00	7.00	13.70	7.50	4.50	16.00	2.80	3.60	2.00	6.00	5.80	5.60
28	2.80	6.00	11.20	6.40	3.20	18.00	2.90	2.70	2.00	4.20	6.70	5.00
29	2.70	......	8.70	6.20	2.70	13.80	3.20	2.60	2.00	2.20	4.50	4.60
30	2.60	......	7.10	8.10	8.00	9.00	8.10	2.10	1.90	2.90	3.70	4.30
31	2.50	......	6.20	......	2.50	......	10.60	3.20	......	2.70	......	11.00
1901.												
1	14.30	7.40	5.30	26.00	5.00	8.70	5.10	2.10	4.00	2.20	1.70	2.00
2	15.50	7.20	5.20	22.00	4.80	8.60	4.20	2.20	3.50	2.90	1.70	2.00
3	14.50	7.50	5.10	23.00	4.70	9.70	4.30	2.30	2.90	6.10	1.70	1.95
4	11.00	24.07	5.00	18.00	4.60	10.90	4.00	2.20	2.70	6.40	1.80	2.20
5	9.80	26.00	4.90	14.20	4.40	8.00	3.30	2.10	2.40	5.10	1.95	2.40
6	8.70	18.30	4.70	10.80	4.30	7.80	3.60	2.10	2.10	4.00	1.95	2.40
7	7.40	12.60	4.50	9.60	4.20	12.10	4.00	2.60	2.10	8.20	2.00	2.30
8	6.10	10.50	4.40	7.20	4.20	10.10	4.80	2.60	2.00	2.50	1.90	2.20
9	5.30	13.10	4.40	6.80	4.10	6.30	4.70	2.30	2.00	2.10	1.90	2.20
10	5.00	14.00	4.70	6.20	3.90	5.90	3.70	2.30	1.90	2.00	2.00	2.20
11	6.00	12.60	7.70	5.80	3.80	4.50	2.90	2.10	1.90	1.90	1.90	2.30
12	24.20	10.90	6.00	5.60	3.80	4.10	2.60	2.30	1.80	1.90	1.80	2.50
13	30.50	9.80	5.60	6.10	8.80	4.20	2.60	2.90	1.80	1.90	1.80	2.40
14	22.00	8.20	6.00	8.80	4.50	5.00	2.10	2.30	1.80	2.00	1.75	2.60
15	18.00	7.70	5.50	10.70	4.30	7.30	2.20	2.10	4.20	2.10	1.75	6.00
16	11.50	6.90	5.10	8.70	4.00	5.70	2.30	3.20	3.10	2.10	1.80	13.40
17	12.20	6.60	4.50	7.70	3.60	6.00	5.30	8.90	3.00	2.00	1.80	9.00
18	12.10	6.50	4.10	11.00	3.50	5.20	4.80	8.00	5.60	1.90	1.80	6.00
19	9.70	6.30	4.00	23.00	3.50	4.40	4.30	6.60	9.10	1.90	1.90	3.90
20	8.00	6.20	6.00	22.00	5.50	4.00	3.50	5.90	7.10	1.80	2.10	3.60
21	7.10	5.90	7.90	13.40	10.50	3.60	3.30	9.60	5.00	1.80	2.20	3.40
22	6.80	5.80	6.30	10.70	16.20	3.80	3.10	7.20	8.80	1.70	2.20	3.20
23	6.40	5.70	6.20	8.90	14.20	3.80	2.80	20.75	2.70	1.70	2.80	3.10
24	6.20	6.00	10.10	7.60	12.00	3.20	3.50	21.00	2.30	1.70	2.20	3.10
25	7.00	6.10	8.90	6.90	9.00	9.10	2.40	9.40	2.10	1.70	2.20	3.00
26	6.80	6.10	10.20	6.50	10.60	3.00	2.20	5.90	2.10	1.60	2.10	2.90
27	6.70	6.00	14.80	6.10	8.40	2.90	2.50	4.20	2.00	1.70	2.80	2.95
28	6.70	5.50	17.50	5.80	6.40	2.80	2.50	5.90	2.10	1.70	2.20	3.05
29	6.60	......	15.00	5.50	6.00	2.60	2.30	8.70	2.80	1.70	2.10	38.00
30	6.80	......	9.90	5.20	5.10	2.60	2.20	7.70	2.50	1.70	2.00	47.00
31	7.00	......	31.50	......	4.80	......	2.20	4.70	......	1.70	......	39.00

Rating table for Tallapoosa River at Milstead for 1897.

Gage height.	Discharge.	Gage height.	Discharge.	Gage height.	Discharge.	Gage height.	Discharge.
Feet.	*Second-feet.*	*Feet.*	*Second-feet.*	*Feet.*	*Second-feet.*	*Feet.*	*Second-feet.*
0.5	330	1.5	1,070	3.0	3,129	5.0	5,909
.6	350	1.6	1,200	3.2	3,407	5.2	6,187
.7	380	1.7	1,333	3.4	3,685	5.4	6,465
.8	420	1.8	1,467	3.6	3,963	5.6	6,743
.9	470	1.9	1,600	3.8	4,241	5.8	7,021
1.0	530	2.0	1,733	4.0	4,519	6.0	7,299
1.1	620	2.2	2,007	4.2	4,797	7.0	8,689
1.2	720	2.4	2,285	4.4	5,075	8.0	10,079
1.3	830	2.6	2,573	4.6	5,353	9.0	11,469
1.4	950	2.8	2,851	4.8	5,631		

Rating table for Tallapoosa River at Milstead for 1898.

Gage height.	Discharge.	Gage height.	Discharge.	Gage height.	Discharge.	Gage height.	Discharge.
Feet.	*Second-feet.*	*Feet.*	*Second-feet.*	*Feet.*	*Second-feet.*	*Feet.*	*Second-feet.*
0.8	540	2.0	1,920	4.0	4,220	10.0	11,120
.9	655	2.1	2,035	4.5	4,795	10.5	11,695
1.0	770	2.2	2,150	5.0	5,370	11.0	12,270
1.1	885	2.3	2,265	5.5	5,945	11.5	12,845
1.2	1,000	2.4	2,380	6.0	6,520	12.0	13,420
1.3	1,115	2.5	2,495	6.5	7,095	12.5	13,995
1.4	1,230	2.6	2,610	7.0	7,670	13.0	14,570
1.5	1,345	2.7	2,725	7.5	8,245	13.5	15,145
1.6	1,460	2.8	2,840	8.0	8,820	14.0	15,720
1.7	1,575	2.9	2,955	8.5	9,395	14.5	16,295
1.8	1,690	3.0	3,070	9.0	9,970	15.0	16,870
1.9	1,805	3.5	3,645	9.5	10,545		

Rating table for Tallapoosa River at Milstead for 1899.

Gage height.	Discharge.	Gage height.	Discharge.	Gage height.	Discharge.	Gage height.	Discharge.
Feet.	*Second-feet.*	*Feet.*	*Second-feet.*	*Feet.*	*Second-feet.*	*Feet.*	*Second-feet.*
0.7	320	4.0	4,362	12.5	14,817	21.0	25,272
.8	430	4.5	4,977	13.0	15,432	21.5	25,887
.9	550	5.0	5,592	13.5	16,047	22.0	26,502
1.0	672	5.5	6,207	14.0	16,662	22.5	27,117
1.1	795	6.0	6,822	14.5	17,277	23.0	27,732
1.2	918	6.5	7,437	15.0	17,892	23.5	28,347
1.3	1,041	7.0	8,052	15.5	18,507	24.0	28,962
1.4	1,164	7.5	8,667	16.0	19,122	24.5	29,577
1.5	1,287	8.0	9,282	16.5	19,737	25.0	30,192
1.6	1,410	8.5	9,897	17.0	20,352	25.5	30,807
1.7	1,533	9.0	10,512	17.5	20,977	26.0	31,422
1.8	1,656	9.5	11,127	18.0	21,582	26.5	32,037
1.9	1,779	10.0	11,742	18.5	22,179	27.1	32,652
2.0	1,902	10.5	12,357	19.0	22,812	27.5	33,287
2.5	2,517	11.0	12,972	19.5	23,427	27.9	33,779
3.0	3,132	11.5	13,587	20.0	24,042		
3.5	3,747	12.0	14,202	20.5	24,657		

Rating table for Tallapoosa River at Milstead for 1900 and 1901.

Gage height.	Discharge.	Gage height.	Discharge.	Gage height.	Discharge.	Gage height.	Discharge.
Feet.	*Second-feet.*	*Feet.*	*Second-feet.*	*Feet.*	*Second-feet.*	*Feet.*	*Second-feet.*
1.5	1,337	3.5	3,587	7.5	8,087	26.0	28,900
1.6	1,450	3.6	3,700	8.0	8,650	27.0	30,025
1.7	1,562	3.7	3,812	8.5	9,212	28.0	31,150
1.8	1,675	3.8	3,925	9.0	9,775	29.0	32,275
1.9	1,787	3.9	4,037	10.0	10,900	30.0	33,400
2.0	1,900	4.0	4,150	11.0	12,025	31.0	34,525
2.1	2,012	4.1	3,262	12.0	13,150	32.0	35,650
2.2	2,125	4.2	4,375	13.0	14,275	33.0	36,775
2.3	2,237	4.3	4,487	14.0	15,400	34.0	37,900
2.4	2,350	4.4	4,600	15.0	16,525	35.0	39,025
2.5	2,462	4.5	4,712	16.0	17,650	36.0	41,150
2.6	2,575	4.6	4,825	17.0	18,775	37.0	41,275
2.7	2,687	4.7	4,937	18.0	19,900	38.0	42,400
2.8	2,800	4.8	5,050	19.0	21,025	39.0	43,525
2.9	2,912	4.9	5,162	20.0	22,150	40.0	44,650
3.0	3,025	5.0	5,275	21.0	23,275	41.0	45,775
3.1	3,137	5.5	5,837	22.0	24,400	42.0	46,900
3.2	3,250	6.0	6,400	23.0	25,525	43.0	48,025
3.3	3,362	6.5	6,962	24.0	26,650		
3.4	3,475	7.0	7,525	25.0	27,775		

Estimated monthly discharge of Tallapoosa River at Milstead.

[Drainage area, 3,840 square miles.]

Month.	Discharge in second-feet.			Run-off.	
	Maximum.	Minimum.	Mean.	Second-feet per square mile.	Depth in inches.
1897.					
August 7–31	12,440	1,070	3,173	0.83	0.77
September	1,467	420	742	.19	.21
October	470	380	424	.11	.12
November	1,200	470	729	.19	.21
December	6,604	1,070	2,214	.58	.67
1898.					
January	4,105	1,575	2,426	.63	0.72
February	2,265	1,690	1,912	.50	.52
March	5,715	1,575	2,313	.60	.69
April	21,240	2,150	5,748	1.50	1.67
May	2,610	885	1,493	.39	.45
June	3,070	540	1,314	.34	.38
July	5,485	885	2,493	.65	.75
August	25,610	1,920	7,418	1.93	2.22
September	6,520	1,345	2,637	.69	.77
October	36,420	1,115	7,280	1.90	2.19
November	16,180	2,035	6,049	1.58	1.76
December	11,120	3,530	5,741	1.50	1.73
The year	36,420	540	3,902	1.02	13.85
1899.					
January	22,197	4,116	8,417	2.19	2.53
February	44,952	6,945	15,688	4.09	4.26
March	32,652	7,314	12,399	3.23	3.72
April	21,582	5,100	9,016	2.35	2.62
May	4,731	2,517	3,351	.87	1.00
June	2,999	1,287	2,040	.53	.59
July	20,290	795	4,985	1.30	1.50
August	4,362	1,287	2,222	.58	.67
September	2,394	430	984	.26	.29
October	1,656	320	1,014	.26	.30
November	7,068	918	1,787	.47	.53
December	18,138	1,656	4,728	1.23	1.42
The year	44,952	320	5,553	1.45	19.43

Estimated monthly discharge of Tallapoosa River at Milstead—Continued.

Month.	Discharge in second-feet.			Run-off.	
	Maximum.	Minimum.	Mean.	Second-feet per square mile.	Depth in inches.
1900.					
January	10, 335	2, 125	3, 728	.97	1. 12
February	48, 305	2, 237	12, 950	3. 37	3. 50
March	17, 650	5, 723	10, 208	2. 66	3. 07
April	18, 775	4, 712	9, 016	2. 35	2. 62
May	6, 736	2, 462	3, 718	.97	1. 12
June	27, 831	2, 575	8, 317	2. 17	2. 42
July	11, 572	2, 462	5, 405	1. 41	1. 63
August	6, 960	1, 675	2, 814	.73	.84
September	28, 447	1, 337	4, 975	1. 30	1. 45
October	13, 262	1, 337	3, 787	.99	1. 14
November	11, 460	2, 462	4, 224	1. 10	1. 23
December	18, 775	2, 800	6, 475	1. 69	1. 95
The year	48, 305	1, 337	6, 301	1. 64	22. 09
1901.					
January	33, 962	5, 275	11, 476	2. 99	3. 45
February	28, 900	5, 837	10, 440	2. 72	2. 83
March	35, 087	4, 150	8, 374	2. 18	2. 52
April	28, 900	5, 499	12, 020	3. 13	3. 49
May	17, 875	3, 587	6, 440	1. 68	1. 94
June	13, 262	2, 775	5, 976	1. 56	1. 74
July	5, 387	2, 012	3, 398	.88	1. 01
August	23, 275	2, 012	5, 904	1. 54	1. 78
September	9, 887	1, 675	3, 137	.82	.91
October	6, 849	1, 562	2, 364	.62	.71
November	2, 237	1, 562	1, 855	.48	.54
December	a 70, 000	1, 843	8, 282	2. 16	2. 49
The year	a 70, 000	1, 562	6, 639	1. 73	23. 41

a Approximate.

Net horsepower per foot of fall with a turbine efficiency of 80 per cent for the minimum monthly discharge of Tallapoosa River at Milstead.

Month.	1899.			1900.			1901.		
	Minimum discharge.	Minimum net horsepower per foot of fall.	Duration of minimum.	Minimum discharge.	Minimum net horsepower per foot of fall.	Duration of minimum.	Minimum discharge.	Minimum net horsepower per foot of fall.	Duration of minimum.
	Sec.-feet.		Days.	Sec.-feet.		Days.	Sec.-feet.		Days.
January	4,116	374	1	2,125	193	4	5,275	480	1
February	6,945	631	1	2,237	203	1	5,837	531	1
March	7,314	665	1	5,725	520	2	4,150	377	1
April	5,100	464	1	4,712	428	1	5,500	500	1
May	2,517	229	3	2,462	224	1	3,587	326	2
June............	1,287	117	4	2,575	234	1	2,575	234	2
July	795	72	1	2,462	224	2	2,012	183	1
August..........	1,287	117	3	1,675	152	1	2,012	183	5
September	430	39	3	1,337	122	2	1,675	152	2
October	320	29	2	1,337	122	1	1,562	142	10
November	918	83	5	2,462	244	9	1,562	142	3
December......	1,656	151	1	2,800	255	3	1,843	168	1

TALLAPOOSA RIVER NEAR SUSANNA.

This station was established July 27, 1900, by J. R. Hall. It is located at the mouth of Blue Creek, which is 10 feet above the east landing of McCartys Ferry, 13 miles southwest of Dadeville and 3 miles from Susanna, the nearest post-office. The rod is graduated to feet and tenths. It is 18 feet long and is nailed vertically to a tree overhanging the water on the south side of the creek at the junction of the creek and the river. The gage is referred to a bench mark on a white hickory tree about 40 feet from the rod on the south bank of the creek, and is 376.67 feet above tide water. Discharge measurements are made from a boat held in place by a wire stretched across the river, upon which the distances from the initial point are tagged. The section is an exceptionally good one, depth and current being almost uniform the entire width of the stream. This station was discontinued March 30, 1901.

During 1900 and 1901 the following discharge measurements were made by James R. Hall:

Discharge measurements of Tallapoosa River near Susanna.

Date.	Gage height.	Discharge.
1900.	*Feet.*	*Second-feet.*
July 27	1.80	2,309
August 9	1.55	1,900
September 28	1.50	1,809
November 24	2.40	3,629
1901.		
July 9	2.80	5,628
February 27	2.90	5,135

Daily gage height, in feet, of Tallapoosa River near Susanna.

Day.	Jan.	Feb.	Mar.	Apr.	May.	June.	July.	Aug.	Sept.	Oct.	Nov.	Dec.
1900.												
1								5.80	2.40	1.40	1.80	2.00
2								4.00	3.80	1.40	1.70	2.00
3								2.00	4.80	1.40	1.70	2.10
4								1.80	4.20	1.35	1.65	2.20
5								1.80	2.25	1.30	1.65	2.30
6								2.10	1.50	2.80	1.65	2.50
7								2.20	1.45	3.00	1.60	2.40
8								1.70	1.45	2.50	1.60	2.40
9								1.55	1.40	1.85	1.60	2.40
10								1.50	1.35	1.80	1.60	2.30
11								1.40	1.35	1.75	1.60	2.10
12								1.40	1.85	1.70	1.55	2.00
13								1.40	1.35	1.90	1.55	1.90
14								1.40	1.30	2.40	1.60	3.90
15								1.40	1.35	2.45	1.60	3.90
16								1.90	11.70	2.40	1.60	2.80
17								1.95	8.40	2.35	1.75	2.60
18								1.95	4.80	2.30	1.75	2.50
19								1.80	3.00	2.20	1.80	2.40
20								1.75	2.50	2.10	1.85	4.50
21								1.70	1.80	1.90	1.85	5.80
22								1.70	1.80	1.70	1.90	4.50
23								1.90	1.80	3.90	2.40	4.00
24								2.00	1.70	6.00	2.40	4.00
25								2.05	1.60	5.00	3.00	3.70
26								2.50	1.50	4.80	4.90	3.20
27							1.80	2.15	1.50	4.10	4.20	2.80
28							1.90	2.00	1.50	2.80	3.90	2.70
29							1.80	1.90	1.45	2.20	3.00	2.60
30							4.00	1.80	1.45	1.90	2.80	2.70
31							6.80	2.25		1.85		2.90

Daily gage height, in feet, of Tallapoosa River near Susanna—Continued.

Day.	Jan.	Feb.	Mar.	Apr.	May.	June.	July.	Aug.	Sept.	Oct.	Nov.	Dec.
1901.												
1	6.0	3.40	2.70									
2	6.0	3.45	2.60									
3	5.1	3.80	2.65									
4	4.5	11.50	2.70									
5	3.9	9.50	2.65									
6	3.5	6.50	2.60									
7	3.2	4.40	2.50									
8	3.0	4.30	2.40									
9	2.9	4.80	2.50									
10	2.8	4.90	2.70									
11	3.4	4.60	3.30									
12	13.5	4.00	3.10									
13	11.5	3.50	2.90									
14	8.0	3.04	2.60									
15	6.1	3.30	2.50									
16	4.5	3.20	2.40									
17	5.0	3.10	2.40									
18	4.5	3.10	2.40									
19	3.9	3.05	2.45									
20	3.4	3.00	3.00									
21	3.2	2.90	8.40									
22	3.1	2.85	2.90									
23	3.0	2.80	2.70									
24	3.1	2.90	3.80									
25	3.4	3.00	3.40									
26	3.1	3.00	8.60									
27	3.2	2.90	7.40									
28	3.1	2.80	6.90									
29	3.0		6.10									
30	3.2		4.10									
31	3.3											

Rating table for Tallapoosa River near Susanna for 1900 and 1901.

Gage height.	Discharge.	Gage height.	Discharge.	Gage height.	Discharge.	Gage height.	Discharge.
Feet.	*Second-feet.*	*Feet.*	*Second-feet.*	*Feet.*	*Second-feet.*	*Feet.*	*Second-feet.*
1.0	----------	2.6	4,730	4.2	11,930	7.0	24,530
1.2	----------	2.8	5,630	4.4	12,830	8.0	29,030
1.4	1,680	3.0	6,530	4.6	13,730	9.0	33,530
1.6	1,960	3.2	7,430	4.8	14,630	10.0	38,030
1.8	2,320	3.4	8,330	5.0	15,530	11.0	42,530
2.0	2,740	3.6	9,230	5.5	17,780	11.7	45,680
2.2	3,230	3.8	10,130	6.0	20,030		
2.4	3,850	4.0	11,030	6.5	22,280		

Estimated monthly discharge of Tallapoosa River near Susanna.

[Drainage area, 2,610 square miles.]

Month.	Discharge in second-feet.			Run-off.	
	Maximum.	Minimum.	Mean.	Second-feet per square mile.	Depth in inches.
1900.					
July 27 to 31....................	....	....	8,364	....	....
August.........................	19,130	1,680	3,258	1.25	1.44
September......................	45,680	1,570	6,083	2.33	2.60
October........................	20,030	1,570	4,776	1.83	2.11
November......................	15,080	1,885	3,676	1.41	1.57
December	19,130	2,520	6,288	2.41	2.78
1901.					
January........................	53,780	5,630	13,265	5.08	5.86
February.......................	44,780	5,630	11,303	4.33	4.51
March.........................	26,330	3,850	7,546	2.89	3.31

Net horsepower per foot of fall, with a turbine efficiency of 80 per cent for the minimum monthly discharge of Tallapoosa River near Susanna.

Month.	1900.			1901.		
	Minimum discharge.	Minimum net horsepower per foot of fall.	Duration of minimum.	Minimum discharge.	Minimum net horsepower per foot of fall.	Duration of minimum.
	Second-feet.		Days.	Second-feet.		Days.
January..............	....	....	....	5,630	512	1
February.............	....	....	....	5,630	512	2
March	....	....	....	3,850	350	4
July	2,320	211	2	....	....	....
August..............	1,680	153	5	....	....	....
September	1,570	143	1	....	....	....
October	1,570	143	1	....	....	....
November...........	1,885	171	2	....	....	....
December	2,520	229	1	....	....	....

TALLAPOOSA RIVER AT STURDEVANT.

This station was established July 19, 1900, by J. R. Hall. It is located at the Columbus and Western Railroad bridge, a fourth of a mile west of Sturdevant. This railroad belongs to the Central of Georgia Railway. The gage rod is 20 feet long, and is graduated to feet and tenths. It is in two sections, and is fastened vertically, the shorter section to a post at the edge of the water on the east bank about 20 feet below the bridge, and the longer section to the first stone pier from the east bank. The initial point of sounding is the east end of the bridge. The section is broken by three piers and by some large rocks below the bridge. The gage is referred to a bench mark consisting of a nail in the southwest corner of pier No. 2, on the east side of the river, 455.70 feet above tide water and 14.20 feet above the zero of the gage. The observer is B. F. Neighbors, farmer and postmaster at Sturdevant, who lives a fourth of a mile from the station. This station being above the big new dams at Tallassee, is intended to replace Milstead station.

The following discharge measurements were made by James R. Hall, W. E. Hall, M. R. Hall, and assistants:

Discharge measurements of Tallapoosa River at Sturdevant.

Date.	Gage height.	Discharge.	Date.	Gage height.	Discharge.
1900	*Feet.*	*Second-feet.*	**1903**	*Feet.*	*Second-feet.*
July 20	2.85	2,603	May 22	4.00	4,580
August 13	1.95	1,887	July 25	2.45	2,247
1901			August 22...........	2.20	1,837
March 8	3.40	3,774	August 24...........	1.88	1,485
			Do	1.86	1,616
1902			October 3	1.05	834
July 11	1.85	1,440	Do	1.05	835
September 17	.80	658	November 24.......	1.58	1,148
October 9	1.08	858			
November 12........	1.34	1,000			

Rating table for Tallapoosa River at Sturdevant for 1900, 1901, and 1902.

Gage height.	Discharge.	Gage height.	Discharge.	Gage height.	Discharge.	Gage height.	Discharge.
Feet.	*Second-feet.*	*Feet.*	*Second-feet.*	*Feet.*	*Second-feet.*	*Feet.*	*Second-feet.*
0.6	555	3.0	3,140	5.4	6,860	10.5	14,765
.8	660	3.2	3,450	5.6	7,170	11.0	15,540
1.0	775	3.4	3,760	5.8	7,480	11.5	16,315
1.2	910	3.6	4,070	6.0	7,790	12.0	17,090
1.4	1,055	3.8	4,380	6.5	8,565	12.5	17,865
1.6	1,220	4.0	4,690	7.0	9,340	13.0	18,640
1.8	1,410	4.2	5,000	7.5	10,115	13.5	19,415
2.0	1,640	4.4	5,310	8.0	10,890	14.0	20,190
2.2	1,910	4.6	5,620	8.5	11,665	14.5	20,965
2.4	2,210	4.8	5,930	9.0	12,440	15.0	21,740
2.6	2,520	5.0	6,240	10.0	13,215		
2.8	2,830	5.2	6,550	9.5	13,990		

Rating table for Tallapoosa River at Sturdevant for 1903.

Gage height.	Discharge.	Gage height.	Discharge.	Gage height.	Discharge.	Gage height.	Discharge.
Feet.	*Second-feet.*	*Feet.*	*Second-feet.*	*Feet.*	*Second-feet.*	*Feet.*	*Second-feet.*
0.90	765	2.20	1,885	4.60	5,470	7.00	9,070
.95	789	2.30	2,025	4.70	5,620	7.60	9,970
1.00	810	2.40	2,170	4.80	5,770	8.00	10,570
1.05	834	2.50	2,320	4.90	5,920	8.90	11,920
1.10	859	2.60	2,470	5.00	6,070	9.00	12,070
1.15	885	2.70	2,620	5.10	6,220	9.20	12,370
1.20	912	2.80	2,770	5.20	6,370	9.30	12,520
1.25	940	2.90	2,920	5.30	6,520	9.50	12,820
1.30	970	3.00	3,070	5.40	6,670	9.80	13,270
1.35	1,002	3.10	3,220	5.50	6,820	10.00	13,570
1.40	1,036	3.20	3,370	5.60	6,970	10.20	13,870
1.45	1,072	3.30	3,520	5.70	7,120	11.00	15,070
1.50	1,110	3.40	3,670	5.80	7,270	12.00	16,570
1.55	1,150	3.50	3,820	5.90	7,420	12.10	16,720
1.60	1,194	3.60	3,970	6.00	7,570	13.00	18,070
1.65	1,240	3.70	4,120	6.10	7,720	13.80	19,270
1.70	1,288	3.80	4,270	6.20	7,870	14.00	19,570
1.75	1,338	3.90	4,420	6.30	8,020	14.60	20,470
1.80	1,390	4.00	4,570	6.40	8,170	16.00	22,570
1.85	1,444	4.10	4,720	6.50	8,320	19.00	27,070
1.90	1,500	4.20	4,870	6.60	8,470	19.20	27,370
1.95	1,560	4.30	5,020	6.70	8,620		
2.00	1,620	4.40	5,170	6.80	8,770		
2.10	1,750	4.50	5,320	6.90	8,920		

Estimated monthly discharge of Tallapoosa River at Sturdevant.

[Drainage area, 2,500 square miles.]

Month.	Discharge in second-feet.			Run-off.	
	Maximum.	Minimum.	Mean.	Second-feet per square mile.	Depth in inches.
1900.					
July 19–31			4,002	1.60	0.77
August	5,155	1,520	2,533	1.01	1.16
September	17,090	1,220	3,602	1.44	1.61
October	9,805	1,220	3,398	1.36	1.57
November	7,635	1,910	3,275	1.31	1.46
December	10,115	2,520	4,330	1.73	1.99
1901.					
January	20,345	4,690	7,035	2.81	3.24
February	18,485	4,535	6,468	2.59	2.70
March	12,285	3,450	5,315	2.13	2.46
April	12,440	4,535	6,772	2.61	2.91
May	10,270	3,140	4,885	1.95	2.25
June	7,170	2,365	4,452	1.78	1.99
July	5,155	1,640	2,795	1.12	1.29
August	16,625	1,640	4,793	1.92	2.21
September	9,340	1,640	2,852	1.14	1.27
October	7,790	1,220	1,946	.78	.90
November	2,055	1,220	1,502	.70	.78
December	24,150	1,310	4,670	1.87	2.16
The year	24,150	1,220	4,457	1.78	24.16
1902.					
January	15,695	2,520	4,550	1.82	2.10
February	23,245	3,760	6,288	2.52	2.62
March	23,245	5,310	9,708	3.88	4.47
April	10,890	3,914	5,677	2.27	2.53
May	5,000	1,910	3,240	1.30	1.50
June	4,070	840	1,544	.62	.69
July	2,830	660	1,004	.40	.46
August	7,790	470	1,298	.52	.60
September	4,845	510	1,255	.50	.56
October	2,985	715	1,180	.47	.54
November	6,550	660	2,011	.80	.89
December	10,890	1,640	4,412	1.76	2.03
The year	23,245	470	3,514	1.40	18.99

Estimated monthly discharge of Tallapoosa River at Sturderant—Continued.

Month.	Discharge in second-feet.			Run-off.	
	Maximum.	Minimum.	Mean.	Second-feet per square mile.	Depth in inches.
1903.					
January	4,120	2,320	3,128	1.25	1.44
February	27,370	2,470	9,841	3.94	4.10
March	15,070	5,320	8,035	3.21	3.70
April..........................	14,470	4,420	6,988	2.79	3.11
May	19,120	3,370	5,688	2.27	2.62
June	10,720	2,770	4,845	1.94	2.16
July	7,270	1,500	3,204	1.28	1.48
August	6,970	1,110	2,771	1.11	1.28
September....................	3,370	765	1,271	.51	.57
October	2,620	765	939	.37	.43
November....................	1,750	912	1,285	.51	.57
December	2,920	1,036	1,410	.56	.65
The year	27,370	765	4,117	1,645	22.11

Net horsepower per foot of fall with a turbine efficiency of 80 per cent for the minimum monthly discharge of Tallapoosa River at Sturdevant.

Month.	1901.			1902.			1903.		
	Minimum discharge.	Minimum net horse-power per foot of fall.	Dura-tion of mini-mum.	Minimum discharge.	Minimum net horse-power per foot of fall.	Dura-tion of mini-mum.	Minimum discharge.	Minimum net horse-power per foot of fall.	Dura-tion of mini-mum.
	Sec.-feet.		*Days.*	*Sec.-feet.*		*Days.*	*Sec.-feet.*		*Days.*
January	4,690	426	3	2,520	229	2	2,320	211	7
February	4,535	412	2	3,760	342	2	2,470	225	1
March	3,450	314	2	5,310	483	1	5,320	484	2
April	4,535	412	3	3,914	356	1	4,420	402	3
May	3,140	285	2	1,910	174	1	3,370	306	1
June...........	2,365	215	2	840	76	1	2,770	252	2
July...........	1,640	149	2	660	60	2	1,500	136	1
August	1,640	149	2	470	43	1	1,110	101	2
September	1,640	149	7	510	46	2	765	70	2
October........	1,220	111	5	715	65	3	765	70	6
November	1,220	111	4	660	60	3	912	83	1
December......	1,310	119	2	1,640	15	1	1,036	94	2

CANAL, POWER HOUSE, AND MILLS OF TALLASSEE FALLS MANUFACTURING COMPANY AT TALLASSEE, ALA.

On Tallapoosa River; view from east bank.

DESCRIPTION OF TALLAPOOSA RIVER.

A survey of a part of Tallapoosa River in Alabama was made in June and July, 1900, under supervision of B. M. Hall, resident hydrographer, by Field Engineer James R. Hall, levelman and topographer.

The survey began at the hydrographic station on Tallapoosa River, at Milstead, Ala., and ran up the river 64 miles to the head of the shoal above Griffin Ferry. The elevations are above sea level.

Elevations of bench marks and water surface along Tallapoosa River from Milstead to Griffin Shoals.

Distance from Milstead.	Location.	Bench mark.	Water surface.
Miles.		*Feet.*	*Feet.*
6.0	Water surface of tail-water at Tallassee mills		206.3
6.2	Water surface above crest of Tallassee dam		269.9
8.5	Upper end of Tallassee pond		269.9
9.5	Water surface below Montgomery Power Company's dam		295.25
9.5	Crest of Montgomery Power Company's dam		335.25
15.7	Upper end of Montgomery Power Company's pond		335.25
16.5	Water surface at Double Bridge Ferry		351.46
16.8	Water surface at mouth of Wind Creek		352.45
16.8	Bench mark No. 7, bunch of mulberry trees at the mouth of Wind Creek	357.85	....
17.8	Bench mark No. 22, crooked willow on small branch at north end of Taylor's field	363.30	....
17.8	Water surface at bench mark No. 22		356.18
18.5	Water surface opposite mouth of Kowaliga Creek		357.16
18.75	Bench mark No. 33, mulberry 100 feet above old Baker field.	371.73	....
18.75	Water surface at bench mark No. 33		359.75
19.4	Bench mark No. 42, willow at Garnetts Ford	364.60	....
19.4	Water surface at Garnetts Ford		360.55
19.7	Bench mark No. 46, pine at mouth of High Falls Branch.	373.98	....
19.7	Water surface at "blue hole" at mouth of High Falls Branch		362.40
20.1	Water surface at "blue hole" at foot of Long Branch Shoals.		362.40
21.0	Bench mark No. 62, mulberry, 300 yards above mouth of Long Branch	382.45	....
21.0	Water surface at bench mark No. 62, top of Long Branch Shoals		367.23
21.3	Bench mark No. 70, white hickory at McCartys Ferry, mouth of Blue Creek	376.67	....
21.3	Water surface at McCartys Ferry, mouth of Blue Creek		367.80
23.0	Top of shoal opposite mouth of Peru Branch		372.55
23.8	Water surface at mouth of Gold Mine Branch		375.17
23.8	Bench mark No. 100, mulberry at mouth of Gold Mine Branch	386.00	....
24.4	Bench mark No. 110, water oak at Robinsons Ferry	404.40	....
24.4	Water surface at Robinsons Ferry		380.20
25.6	Water surface at top of upper Robinsons Shoals		389.10
25.6	Bench mark No. 124, small sycamore at mouth of small branch	395.10	....

Elevations of bench marks and water surface along Tallapoosa River from Milstead to Griffin Shoals—Continued.

Distance from Milstead.	Location.	Bench mark.	Water surface.
Miles.		*Feet.*	*Feet.*
27.7	Water surface at mouth of small branch in Pace's field....	....	390.09
28.7	Bench mark No. 140, water oak at foot of Hardy Shoals, in Pace's field...............................	414.30	....
29.5	Bench mark No. 150, dead stump 100 feet below the mouth of Big Sandy Creek............................	398.08	....
29.5	Water surface at mouth of Big Sandy Creek..............	....	....
30.0	Bench mark No. 165, big red oak at Youngs Ferry	413.50	....
30.0	Water surface at Youngs Ferry	....	394.00
31.0	Water surface at Cherokee Bluff........................	....	394.60
31.2	Bench mark No. 175, big walnut 200 yards above Monowa Creek	416.75	....
34.0	Bench mark No. 180, 10-inch pine tree at third bar of Seago Shoals......................................	424.72	....
34.0	Water surface at third bar of Seago Shoals, opposite bench mark No. 180..	....	399.92
35.8	Bench mark No. 190, large white oak at east landing at Walkers Ferry.............................	436.90	....
35.8	Water surface at Walkers Ferry.........................	....	429.65
37.0	Bench mark No. 210, leaning white oak at mouth of small branch at upper end of Upshaw place..................	438.60	....
37.4	Water surface at bench mark No. 210.....................	....	432.00
37.6	Water surface at top of fish trap.......................	....	436.47
38.3	Bench mark No. 215, 16-inch white oak on small branch at upper end of Locke's old field	448.90	....
38.3	Water surface at bench mark No. 215	....	438.00
39.3	Water surface under Central Railroad bridge at Sturdevant	....	444.25
39.3	Bench mark on top of rail over first pier of the east end of Central Railroad bridge.............................	505.90	....
41.2	Bench mark No. 240, large water oak at east landing of Dennis Ferry	457.15	....
41.2	Water surface at Dennis Ferry	....	445.85
42.2	Water surface at mouth of branch on left bank of river ..	....	448.20
45.3	Water surface 600 feet below mouth of Hillabee Creek ...	....	472.60
48.3	Bench mark No. 310 water oak at east landing of Welchs Ferry	504.15	....
48.3	Water surface at Welchs Ferry..........................	....	492.30
50.0	Bench mark No. 330, beech 150 feet above mouth of Freemans Branch	526.62	....
50.0	Water surface 150 feet above mouth of Freemans Branch ..	....	521.04
52.0	Water surface at Whaleys Ferry	....	529.48
52.0	Bench mark No. 340, birch at Whaleys Ferry............	539.38	....
55.4	Bench mark No. 350, 10-inch birch at Millers Ferry	552.16	....
55.4	Water surface at Millers Ferry	....	544.00
60.8	Water surface at Griffin Ferry	....	557.10
60.8	Bench mark No. 380, double ash tree on left bank at Griffin Ferry	564.76	....
62.0	Bench mark No. 390, 12-inch birch at head of Griffin Shoals.	573.87	....
62.0	Water surface at head of Griffin Shoals	....	570.30

DAM AND SHOALS OF TALLASSEE FALLS MANUFACTURING COMPANY AT TALLASSEE, ALA.

On Tallapoosa River, looking upstream.

Above Milstead the river flows on granitic bed rock, and has numerous bluffs along its banks, affording excellent sites for dams.

There are two large developed water powers on the river, the Tal-

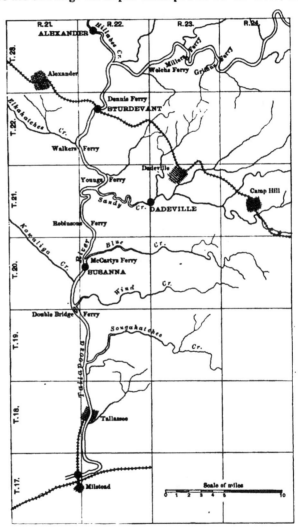

FIG. 1.—Map of Tallapoosa River from top of Griffin Shoals to Milstead.

lassee Falls plant and the Montgomery Power Company's plant, both of which are near the lower end of the area surveyed.

The Tallassee Falls dam and canal, which are 6 miles above Milstead, utilize a fall of 64 feet. The power and the large cotton manufacturing

plant were described in the Twentieth Annual Report of the United
States Geological Survey, Part IV, pages 192–193. In October, 1901,

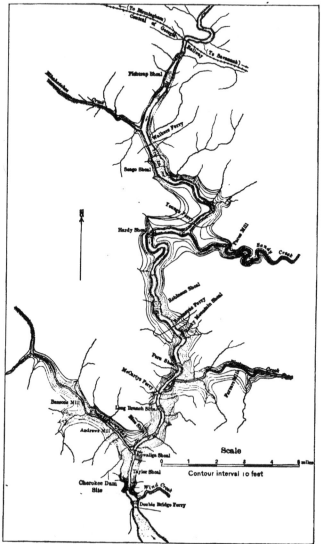

FIG. 2.—Map of Tallapoosa River between Double Bridge Ferry and Central of Georgia Railway,
showing reservoir above Cherokee dam site.

this power was capable of furnishing 8,900 net horsepower without
storage during low water. A break which occurred in the dam on

A CANAL AND DAM OF TALLASSEE FALLS MANUFACTURING COMPANY AT
TALLASSEE ALA.

On Tallapoosa River, view across canal from left bank

B. DAM AND NEW MILL OF TALLASSEE FALLS MANUFACTURING COMPANY
AT TALLASSEE, ALA., DURING CONSTRUCTION

On Tallapoosa River, view from right bank of river below old mill.

December 29, 1901, decreased the available head, but did not stop the machinery.

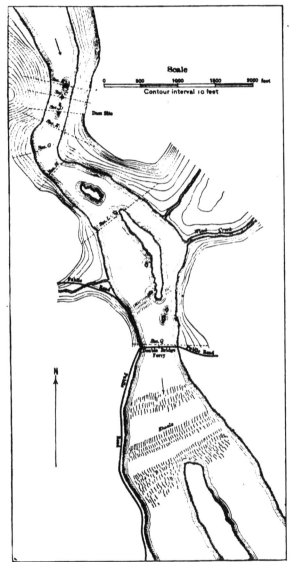

FIG. 3.—Topographic plan and location of cross sections of Tallapoosa River at Cherokee dam site.

The Montgomery Power Company has a 40-foot dam 9½ miles above Milstead. This dam backs the water 6½ miles up the river, and forms .

an immense storage basin. The power from this plant is transmitted electrically a distance of about 27 miles to Montgomery.

With river at stage of lowest water observed this plant will develop at the wheels 5,572 net horsepower from the run of the river without drawing on the storage.

The equalizing storage of this dam will add fully 25 per cent to this power and to the power at Tallassee for continuous running without materially lowering the head at either plant.

Surveys have been made for a large dam, 35 or 40 feet in height, at or near Double Bridge Ferry, to back the water beyond Robinsons Ferry, a distance of about 8 miles. There is an excellent site for a dam, and the project is entirely feasible. The horsepower in proportion to head would be the same as that available at the Montgomery Power Company's dam.

From the mouth of Big Sandy Creek to a point 1 mile above Griffin Ferry, a distance of 32 miles, the fall of Tallapoosa River is 176.5 feet. Nearly all of this fall can be utilized for power by developments similar to those which have been made. A study of the table of distances and elevations will give the distribution of the fall, and will show the distance to which dams of certain heights will back the water at the various shoals. The question of the best power sites and the proper plan of development, height, and location of dams, etc., can be determined only by special investigation and surveys. It will be safe, however, to assume that a practicable site for a dam up to 40 feet high can be found in the vicinity of any location which may be selected, and the power obtainable can be estimated by multiplying the volume of water, or its equivalent net horsepower per foot of fall, by the proposed head to be developed.

The water supply or discharge of Tallapoosa River at different points may be closely approximated from the foregoing records of the Milstead, Susanna, and Sturdevant hydrographic stations, and also from those at the Dadeville and Alexander stations on the tributaries.

PROPOSED DAM ON TALLAPOOSA RIVER NEAR DOUBLE BRIDGE FERRY, TALLAPOOSA COUNTY.

The following is an extract from a recent report by Mr. Henry C. Jones, of Montgomery, Ala., on a proposed development of the Cherokee and Seago water powers as a combined property:

Surveys recently completed for two 40-foot dams on Tallapoosa River, one at the Cherokee dam site and one at Seago Shoals, some miles above, both the property of the Cherokee Development and Manufacturing Company, a water-power corporation enjoying a comprehensive special charter, demonstrates conclusively that the erection of one dam of 80 feet at the lower site instead of 40 feet on the separate properties greater proportional output at less development cost per horsepower.

ection of one dam would also consolidate and cheapen construction and sub peration of the plants, but, most of all, such a dam would create a storage

A. DAM AND POWER HOUSE OF MONTGOMERY POWER COMPANY, ABOVE
TALLASSEE, ALA.

On Tallapoosa River, view from right bank below dam

B. INTERIOR OF POWER HOUSE OF MONTGOMERY POWER COMPANY, ABOVE
TALLASSEE, ALA.

reservoir 20 miles long, averaging more than one-half mile in width, from which upward of 5,000,000,000 cubic feet of water would be available without affecting speed of water-wheel machinery where proper size wheels are installed, as noted later on.

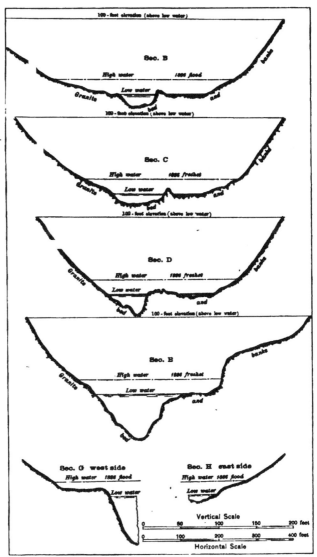

FIG. 4.—Cross section of Tallapoosa River at Cherokee dam site.

This vast quantity of stored water would be available to bridge over low-water periods, greatly increasing the power possible to derive from the actual flow of the river.

The combination of advantages at the Cherokee dam site is very interesting. The section for the dam is extremely narrow for so large a river, being less than 400 feet from anchorage to anchorage in solid granite hills. The contour of these hills permit

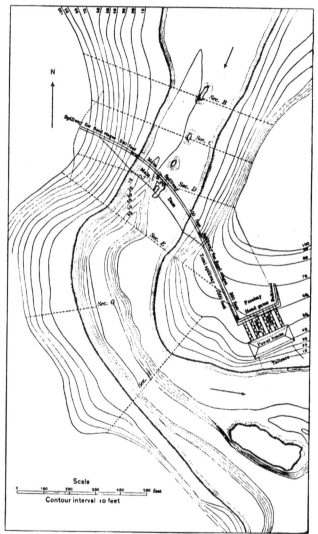

FIG. 5.—Plan of proposed development of Tallapoosa River at Cherokee power site.

the 80-foot elevation to continue, and thus increase the spillway at little cost, and sufficient for all flood stages of the river.

The bed of the stream at this section is free from bowlders or deposit, and offers a clean rock foundation to build upon.

The shape of the river at this point, with reference to its sharp bend, permits placing the power house well away from the spray, flood water, and other bad features of location at main dam, and slight excavation will form a forebay connecting main dam with power house.

This narrow section, so favorable for a short, high dam, widening of the river immediately below—to more than 2,000 feet—thereby releasing flood water, ideal location for a power house, and the vast storage created by erecting a high dam assures a power development of magnitude and at less cost per horse power per year than water is now sold in many localities simply as hydraulic power in the canal, where manufacturers install and operate their own head and race work, power house, water wheel lay out, and electrical machinery.

MATERIAL.

Adjacent to this water power there is abundant timber of first quality for coffer-damming, cribbing, and all timber work. The hills on both banks are a mass of granite, or rather gneiss, which is durable and workable and can be quarried high above the river for cheap and rapid handling by gravity tram cars, conveyors, cable ways, etc.

TRANSPORTATION.

The State of Alabama and this part especially is well covered by railroads. One of the main lines from the Birmingham district to the Atlantic Seaboard passes near this water-power property. This railroad can be reached by a spur track or by utilizing the reservoir for water transportation to connect with the main line at its river crossing.

LABOR.

There is labor of two kinds—negroes, adapted to all construction work, and native whites for work requiring intelligent, persevering application.

In building the dam and on all work of a general development nature, negro labor can be employed at low cost to do good work, under proper white direction. The native white comes from a line of early settlers—sturdy stock, who laid well the foundation of thrift, law and order.

HEALTH.

The location of this water power is in that part of the South noted for its health-fulness. It is on an upland formation generally known as the Piedmont Plateau, which is well drained, has a considerable elevation above sea level, produces well, and offers every condition desired for successful home building.

RIVER DATA.

Mean monthly discharge, in second-feet, of Tallapoosa River at Milstead and Sturderant.

Month.	At Milstead.				At Sturdevant.	
	1898.	1899.	1900.	1901.	1902.	1903.
January	2,426	8,417	3,728	11,476	4,550	3,140
February	1,912	15,688	12,950	10,440	6,288	9,500
March	2,313	12,399	10,208	8,374	9,708	8,000
April	5,784	9,016	9,016	12,020	5,677	7,170
May	1,493	3,351	3,718	6,440	3,240	5,750
June	1,314	2,040	8,317	5,976	1,544	4,800
July	2,493	4,985	5,405	3,398	1,004	3,140
August	7,418	2,222	2,814	5,904	1,298	2,830
September	2,637	984	4,975	3,137	1,255	1,100
October	7,280	1,014	3,787	2,364	1,180	910
November	6,049	1,787	4,224	1,855	2,011	1,300
December	5,741	4,728	6,475	8,282	4,412	1,410
The year	3,902	5,553	6,301	6,639	3,514	4,070

Several miles above the Milstead gaging station Tallapoosa River leaves a granite bed and enters a younger formation. Milstead is on this younger formation and the river there runs over a bowlder and gravel bed—débris washed from the mountain formation above.

Measurements and observations made during low-water periods demonstrate beyond question that the part of the river flowing over a granite bed shows a greater volume than on the bowlder gravel formation.

There is still another reason why the Milstead record is inaccurate, for low water, at present:

Where there are impounding dams above, the beginning of a day's operation could not bring the river to its normal flow and reach the Milstead gage, many miles below, by 8 a. m., the time prescribed for daily observation, the Milstead gage being read at the time that over-night impounding operations would be seriously felt. Especially is this the case since the completion of the two large dams in 1902, at and near Talassee, 7 and 9 miles, respectively, above Milstead.

A gaging station was located later at Sturdevant, upon the granite formation, which shows more water at low-water stages in 1902 and 1903 than the Milstead gage 40 miles below, having 1,340 square miles more drainage area than the Sturdivant station.

The United States Geological Survey has abandoned the Milstead gage as unreliable for low-water records. (See Water-Supply and Irrigation Paper No. 83, pp. 135 and 137).

Records at Milstead for four years are given in this report. They are correct, except as to low water, as explained. The Sturdevant records are given for the past two years. This new river station is a satisfactory location, but is above the Cherokee power site, and so far above, that the drainage area is materially reduced. The discharge for the Cherokee dam site can be estimated, approximately, by adding one-fifth to one-third, varying with river stage, to the discharge for Sturdivant.

Cost of developing the Cherokee and Seago water powers as a combined property.

Main dam, containing 60,000 cubic yards of rubble masonry, at $5 per cubic yard	$300,000
Other masonry, excavating, and tailrace..............................	50,000
Head-gates, feeders and draft tubes, machinery foundations, and power house	100,000
Water wheels, total capacity 42,000 horsepower, complete with governors, relief valves, and two exciter units, all set up to operate	175,000
(This difference between water wheel and generator capacity is a reserve required to maintain speed of wheels where storage is used with consequent reduction of head.)	
Electrical installation, total capacity 22,500 kilowatts, with exciters and switchboards, all set up to operate..................................	225,000
Legal and incidental..	50,000
Contingencies..	50,000
Total development cost...	950,000
Cost of Cherokee and Seago power sites, riparian and charter rights, and lands required for reservoir (approximately 15,000 acres), all controlled by the Cherokee Water Power Corporation	250,000
Actual development cost.................................	950,000
Total investment cost..	1,200,000

Total development cost per horsepower.

Electrical power, constant load at full-rated capacity twenty-four hours per day, three hundred and sixty-five days per year, river discharge supplemented by storage based on the records of the United States Geological Survey station at Milstead and Sturdivant.

Minimum water and storage, 12,500 horsepower............................	$76.00
Normal water and storage, 20,000 horsepower..............................	47.50
Ordinary river stages and storage, available the greater part of the year and generally whole years together (see discharge tables, a part of this report), 30,000 horsepower...	31.66

Total investment cost per horsepower.

Electrical power, constant load at full-rated capacity, twenty-four hours per day, three hundred and sixty-five days per year. River discharge supplemented by storage.

Minimum water and storage, 12,500 horsepower............................	$96.00
Normal water and storage, 20,000 horsepower	60.00
Ordinary river stages and storage, available the greater part of the year and generally whole years together (see discharge tables, a part of this report), 30,000 horsepower.......................................	40.00

Operating expense.

Interest at 5 per cent on $1,200,000	$60,000
Depreciation (estimated on proportion of investment subject thereto)......	40,000
Total fixed charges	100,000
Actual operating	25,000
Maintenance and miscellaneous	25,000
Total operating...	150,000

Total operating cost per horsepower.

Electrical power, constant load at full-rated capacity, twenty-four hours per day, three hundred and sixty-five days per year. River discharge supplemented by storage.

Minimum water and storage, 12,500 horsepower............................ $12. 00
Normal water and storage, 20,000 horsepower 7. 50
Ordinary river stages and storage, available the greater part of the year and generally whole years together (see discharge tables, a part of this report), 30,000 horsepower... 5. 00

The available power, as will be seen by the foregoing, is based on a constant full load twenty-four hours per day, three hundred and sixty-five days per annum, which conditions are not met with in practice.

Power for industrial purposes, such as cotton mills, would be used ten to twelve hours per day on working days, or three hundred and eight days per year. Some requirements, such as general lighting and traction work, would represent a fluctuating demand, while other power of large and small units would be intermittent in character.

This means that in supplying power on a large scale for commercial purposes, water required for propulsion would be at a less ratio than full load and for less than full time, and therefore there would be available for such fluctuating load water in excess of that required to maintain the maximum or peak of that load on a constant basis.

Therefore, under actual conditions of power supply, assuming maximum load from 5 p. m. to 11 p. m., minimum from 11 p. m. to 5 a. m., a morning peak from 5 a. m. to 8 a. m., and a normal power load the remainder of the day—8 a. m. to 5 p. m.—there would be available at the various river stages for such periods of demand the following net power:

Horsepower available at proposed Tallapoosa River dam.

	5 p. m. to 11 p. m.	11 p. m. to 5 a. m.	5 a. m. to 8 a. m.	8 a. m. to 5 p. m.	Total horsepower, hours.
Minimum water and storage	25, 000	3, 750	12, 500	10, 000	300, 000
Normal water and storage	40, 000	6, 000	20, 000	16, 000	480, 000
Ordinary river stages and storage, available the greater part of the year and generally whole years together (see river data, a part of this report)......................	60, 000	9, 000	30, 000	24, 000	720, 000

LOCATION OF POWER WITH RESPECT TO INDUSTRIES.

Alabama ranks high in extent of production and quality of its cotton, for cotton mills; in wealth of native timber, for paper making; in vast and varied mineral resources, for smelting, fusing, and refining; and for laws of extreme favor to capital and industry, promoting and protecting industrial effort.

The total output of this plant can conveniently and profitably be utilized for either of the three principal industries—cotton mills, carbide production, or manufacture of aluminum, or for such combination of the different lines offered as might give flexibility to the operation of a hydraulic plant and stability to a power company's income.

d it be deemed expedient to disregard the advantage so apparent in utilizing
rer where produced, in order to reach a present market, the total output can
mitted to the city of Birmingham and the Birmingham district, which can
every foot-pound of energy delivered.

BIG SANDY CREEK NEAR DADEVILLE.

s station, which was established by J. R. Hall, August 2, 1900,
is located about 4½ miles southwest of Dadeville, at the highway
bridge on the Dadeville-Susanna road. The gage, which is graduated
to feet and tenths, is 16 feet high, and is fastened vertically to the first
pier on the north side of the creek. The initial point of sounding is
at the gage rod. The section is good for ordinary or flood measurements, but is rather wide and shoaly for low-water measurements.
The latter can, however, be made a short distance from the gage. The
station was discontinued December 31, 1901. During 1900 the following measurements were made by James R. Hall:

> July 6: Gage height, 1.20 feet; discharge, 260 second-feet.
> August 8: Gage height, 1.00 foot; discharge, 110 second-feet.
> August 8: Gage height, 1.00 foot; discharge, 116 second-feet.
> August 25: Gage height, 1.35 feet; discharge, 281 second-feet.
> November 16: Gage height, 1.10 feet; discharge, 155 second-feet.
> December 31: Gage height, 2.00 feet; discharge, 870 second-feet.

The measurements of August 8 and November 16 were made a half
mile below Smith's bridge.

Daily gage height, in feet, of Big Sandy Creek near Dadeville.

Day.	Jan.	Feb.	Mar.	Apr.	May.	June.	July.	Aug.	Sept.	Oct.	Nov.	Dec.
1900.												
1								1.00	1.10	0.95	1.00	1.10
2								1.20	1.10	.90	1.30	1.15
3								1.10	1.40	.90	2.00	1.25
4								1.10	1.30	.90	1.80	1.40
5								1.10	1.20	3.50	1.40	1.35
6								1.05	1.05	1.80	1.20	1.30
7								1.05	1.00	1.25	1.20	1.25
8								1.00	1.00	1.20	1.20	1.20
9								1.05	.95	1.10	1.15	1.15
10								1.00	2.00	1.15	1.15	1.10
11								1.00	1.80	1.10	1.15	1.10
12								9.05	1.40	1.20	1.15	1.10
13								9.00	1.20	1.30	1.10	1.10
14								9.00	1.20	1.15	1.10	2.20
15								1.80	2.00	1.10	1.10	1.80
16								1.00	2.20	1.05	1.10	1.45
17								1.80	3.90	1.10	1.05	1.45
18								1.20	1.50	1.00	1.10	1.30
19								1.10	1.10	1.05	1.10	2.40
20								1.00	1.05	1.05	1.10	4.50
21								1.00	1.00	1.00	1.10	3.50
22								.90	1.00	1.50	1.80	1.70
23								.90	1.00	1.45	1.25	1.50
24								1.70	1.00	1.40	1.20	1.40
25								1.40	1.00	1.50	1.50	1.40
26								1.60	1.00	1.15	1.90	1.35
27								1.15	1.05	1.10	1.80	1.35
28								1.10	1.05	1.05	1.20	1.35
29								1.00	1.00	1.00	1.15	1.80
30								1.00	.90	1.00	1.10	1.75
31								1.80		1.05		2.00
1901.												
1	1.90	1.35	1.40	4.40	1.40	1.40	1.30	1.10	1.30	4.20	.85	.90
2	1.90	1.35	1.40	7.40	1.40	1.40	1.25	1.10	1.30	4.00	.85	.90
3	1.95	3.10	1.35	2.70	1.45	4.00	1.20	1.00	1.20	3.00	.85	1.40
4	1.70	6.00	1.35	2.00	1.35	5.60	1.20	1.05	1.10	3.50	.85	1.40
5	1.60	1.90	1.35	1.80	1.35	1.90	1.40	1.10	1.10	2.00	.85	1.40
6	1.50	1.75	1.40	1.60	1.35	3.50	1.50	1.05	1.10	1.80	.90	1.30
7	1.45	1.55	1.35	1.50	1.40	2.40	1.40	1.30	1.00	1.80	.90	1.10
8	1.40	2.10	1.35	1.50	1.35	1.90	1.25	1.20	1.00	1.50	1.00	1.00
9	1.40	3.50	1.40	1.50	1.35	1.50	1.20	1.10	1.00	1.40	1.00	1.10
10	1.40	2.20	1.45	1.50	1.30	1.60	1.20	1.10	.90	1.20	1.00	1.00
11	1.90	2.00	1.40	1.50	1.30	1.70	1.10	1.10	.90	1.20	1.00	1.00
12	1.70	1.70	1.35	1.55	1.35	2.00	1.10	1.10	.85	1.00	.90	1.00
13	2.50	1.50	1.35	1.65	1.35	1.80	1.10	1.15	.80	1.90	1.00	1.10
14	1.90	1.50	1.35	1.65	1.50	1.90	1.10	1.10	1.40	1.90	1.00	4.40
15	1.60	1.50	1.30	1.60	1.50	1.15	1.80	1.00	1.40	1.80	1.00	3.80
16	1.55	1.45	1.30	1.50	1.45	1.70	1.50	1.80	1.80	1.80	1.00	3.00
17	2.00	1.45	1.30	1.50	1.25	1.50	1.50	1.80	2.00	1.70	1.00	2.90
18	2.00	1.45	1.35	1.45	1.40	1.50	1.50	1.80	1.80	1.70	.90	2.50
19	1.60	1.50	1.35	6.00	1.40	1.45	1.30	1.40	1.80	1.70	.90	2.40
20	1.50	1.50	1.80	2.50	1.70	1.45	1.20	1.30	1.70	1.70	.80	2.40
21	1.50	1.45	1.60	2.40	7.00	1.40	1.15	1.20	1.40	1.60	.80	2.00
22	1.45	1.40	1.40	2.10	3.40	1.35	1.15	4.50	1.30	1.60	1.00	2.00
23	1.45	1.40	1.40	1.80	1.80	1.40	1.15	1.50	1.30	1.00	1.00	1.90
24	1.45	1.50	2.20	1.80	1.70	1.40	1.10	1.40	1.20	1.00	.85	1.80
25	1.50	1.50	2.10	1.70	1.50	1.35	1.10	1.40	1.20	.90	.85	1.50
26	1.55	1.45	1.70	1.60	2.70	1.30	1.10	1.20	1.10	.90	.90	3.00
27	1.45	1.45	1.70	1.45	1.80	1.30	1.15	1.20	1.00	a.70	.90	3.00
28	1.40	1.45	1.40	1.45	1.50	1.20	1.20	2.00	1.00	a.60	.90	21.00
29	1.40		1.40	1.45	1.50	1.15	1.15	1.50	1.80	a.70	.90	16.00
30	1.40		1.40	1.45	1.45	1.20	1.10	1.40	1.80	a.70	.90	8.00
31	1.40		2.30		1.40		1.10	1.40		.80		4.00

a Water was being held back by dams above in the morning when readings were made; 0.8 is assumed as minimum for October.

Rating table for Big Sandy Creek near Dadeville for 1900 and 1901.

Gage height.	Discharge.	Gage height.	Discharge.	Gage height.	Discharge.	Gage height.	Discharge.
Feet.	*Second-feet.*	*Feet.*	*Second-feet.*	*Feet.*	*Second-feet.*	*Feet.*	*Second-feet.*
0.8	67	4.4	1,868	8.0	3,740	11.6	5,612
.9	85	4.5	1,920	8.1	3,792	11.7	5,664
1.0	115	4.6	1,972	8.2	3,844	11.8	5,716
1.1	152	4.7	2,024	8.3	3,896	11.9	5,768
1.2	204	4.8	2,076	8.4	3,948	12.0	5,820
1.3	256	4.9	2,128	8.5	4,000	12.1	5,872
1.4	308	5.0	2,180	8.6	4,052	12.2	5,924
1.5	360	5.1	2,232	8.7	4,104	12.3	5,976
1.6	412	5.2	2,284	8.8	4,156	12.4	6,028
1.7	464	5.3	2,336	8.9	4,208	12.5	6,080
1.8	516	5.4	2,388	9.0	4,260	12.6	6,132
1.9	568	5.5	2,440	9.1	4,312	12.7	6,184
2.0	620	5.6	2,492	9.2	4,364	12.8	6,236
2.1	672	5.7	2,544	9.3	4,416	12.9	6,288
2.2	724	5.8	2,596	9.4	4,468	13.0	6,340
2.3	776	5.9	2,648	9.5	4,520	13.1	6,392
2.4	828	6.0	2,700	9.6	4,572	13.2	6,444
2.5	880	6.1	2,752	9.7	4,624	13.3	6,496
2.6	932	6.2	2,804	9.8	4,676	13.4	6,548
2.7	984	6.3	2,856	9.9	4,728	13.5	6,600
2.8	1,036	6.4	2,908	10.0	4,780	13.6	6,652
2.9	1,088	6.5	2,960	10.1	4,832	13.7	6,704
3.0	1,140	6.6	3,012	10.2	4,884	13.8	6,756
3.1	1,192	6.7	3,064	10.3	4,936	13.9	6,808
3.2	1,244	6.8	3,116	10.4	4,988	14.0	6,860
3.3	1,296	6.9	3,168	10.5	5,040	14.1	6,912
3.4	1,348	7.0	3,220	10.6	5,092	14.2	6,964
3.5	1,400	7.1	3,272	10.7	5,144	14.3	7,016
3.6	1,452	7.2	3,324	10.8	5,196	14.4	7,068
3.7	1,504	7.3	3,376	10.9	5,248	14.5	7,120
3.8	1,556	7.4	3,428	11.0	5,300	14.6	7,172
3.9	1,608	7.5	3,480	11.1	5,352	14.7	7,224
4.0	1,660	7.6	3,532	11.2	5,404	14.8	7,276
4.1	1,712	7.7	3,584	11.3	5,456	14.9	7,328
4.2	1,764	7.8	3,636	11.4	5,508	15.0	7,380
4.3	1,816	7.9	3,688	11.5	5,560		

Estimated monthly discharge of Big Sandy Creek near Dadeville.

[Drainage area, 195 square miles.]

Month.	Discharge in second-feet.			Run-off.	
	Maximum.	Minimum.	Mean.	Second-feet per square mile.	Depth in inches.
1900.					
August	655	80	207	1. 06	1. 22
September	3, 150	80	355	1. 82	2. 03
October	2, 670	80	264	1. 35	1. 56
November	870	110	261	1. 34	1. 50
December	3, 870	150	560	2. 87	3. 31
1901.					
January	880	308	425	2. 18	2. 51
February	2, 700	282	545	2. 78	2. 90
March	6, 392	256	552	2. 83	3. 26
April	3, 428	334	689	3. 53	3. 94
May	3, 220	230	480	2. 46	2. 84
June	2, 492	178	523	2. 68	2. 99
July	516	152	227	1. 16	1. 34
August	2, 180	115	369	1. 89	2. 18
September	620	67	257	1. 32	1. 47
October	1, 764	a 45	462	2. 37	2. 73
November	115	67	92	. 47	. 52
December	10, 500	85	1, 265	6. 49	7. 48
The year	10, 500	a 45	490	2. 51	34. 16

a See footnote under gage heights for 1901.

Net horsepower per foot of fall with a turbine efficiency of 80 per cent for the minimum monthly discharge of Big Sandy Creek near Dadeville.

	1900.			1901.		
	Minimum discharge.	Minimum net horse-power per foot of fall.	Duration of mini-mum.	Minimum discharge.	Minimum net horse-power per foot of fall.	Duration of mini-mum.
	Second-feet.		*Days.*	*Second-feet.*		*Days.*
January..............				308	28	7
February..............				282	26	2
March				256	23	3
April				334	30	5
May				230	21	1
June..............				178	16	2
July				152	14	9
August..............	80	7	4	115	10	2
September	80	7	1	67	6	1
October	80	7	3	a 45	a 4	1
November..........	110	10	1	67	6	3
December..........	150	14	5	85	8	2

a See footnote under gage heights for 1901.

POWERS ON BIG SANDY CREEK.

Elevations of bench marks and water surface along Big Sandy Creek between its mouth and the new bridge near Dadeville.

Dis-tance above mouth.	Location.	Bench marks.	Water sur-face.
Miles.		*Feet.*	*Feet.*
0.0	Bench mark No. 150, dead stump at mouth of creek......	398.08	....
.0	Water surface at mouth of Big Sandy Creek..................		393.80
.95	Water surface below Pace's dam		402.00
.95	Water surface above Pace's dam		412.10
1.06	Bench mark No. 160, big pine on north side, 175 feet above Pace's bridge	422.30	....
1.52	Creek surface.....................................		416.00
1.89	At point of Ivy Bend.....................		419.00
2.22	Bench mark No. 162, large walnut at Tucker's house	503.85	....
2.56	Bench mark No. 163, small oak at Tucker's fish trap......	432.85	....
2.56	Water surface above Tucker's fish trap		430.00
3.79	Bench mark No. 164, large sycamore at mouth of Lowry Branch.....................................	445.20	....
3.79	Water surface at mouth of Lowry Branch		436.10
5.02	Bench mark No. 166, oak post at north end of Smith's bridge...	463.95	....

Elevation of bench marks and water surface along Big Sandy Creek between its mouth and the new bridge near Dadeville—Continued.

Distance above mouth.	Location.	Bench mark.	Water surface.
Miles.		*Feet.*	*Feet.*
4.92	Water surface at Smith's bridge......................	....	441.70
5.02	Zero of United States Geological Survey gage at Dadeville.	440.50	....
6.34	Bench mark No. 167, wahoo tree at mouth of Young Branch.....................	559.58	....
6.34	Water surface at mouth of Young Branch...............	....	446.50
6.70	Water surface at Barnes basin.......................	....	452.30
7.18	Water surface at foot of Black Shoals................	....	465.00
7.44	Water surface at top of Black Shoals	....	496.30
7.78	Water surface at mouth of Buck Creek	....	497.30
7.78	Bench mark No. 168, small double oak at mouth of Buck Creek...............	503.65	....
7.98	Eddy water below Sanford's dam	....	500.00
8.06	Bench mark No. 169, hickory at Sanford's mill..........	522.10	....
8.06	Floor of Sanford's mill	514.00	....
8.29	Water surface at Sanford's bridge above dam	....	506.70
8.63	Water surface at head of Sanford's pond	....	506.70
9.47	Water surface at second shoal above Sanford's pond	....	512.50
9.91	Bench mark No. 173, large white oak near north end of Cook's bridge............	539.35	....
9.91	Water surface at Cook's bridge.......................	....	513.80
10.63	Water surface opposite mouth of Chattasofka Creek	....	520.60
10.63	Bench mark, 16-inch water oak on west bank of Chattasofka Creek, 50 feet above mouth...................	527.20	....
11.10	Water surface at top of old factory shoal..............	....	540.15
12.38	Water surface at new bridge	....	550.80
12.38	Bench mark on upstream end of sill on west end of new bridge...........	562.30	....
12.38	Bench mark No. 176, 6-inch maple at new bridge........	563.00	....

The best shoal on this creek is the Sanford and Black Shoal, near Dadeville, which has a fall of 85.8 feet in a distance of 5.2 miles. With a dam 54 feet high and a canal 1,370 feet long a practical working head of 80 feet can be developed, having 1 foot extra for grade of canal and 4 feet extra for storage at top of dam.

The city of Dadeville is now developing the most precipitous part of this shoal (December, 1903), known as the Sanford Shoal, and obtaining a power head of 40 feet, by a dam 4½ feet high and a canal about 600 feet long. The power is to be transmitted about 2 miles to Dadeville for electric lights and other purposes.

HILLABEE CREEK NEAR ALEXANDER, ALA.

This station, which was established August 20, 1900, by J. R. Hall, is located 6½ miles northeast of Alexander, on the road leading from that town to Newsite. The gage, which is graduated to feet and tenths and is placed vertically, is in two sections, the short section, which reads from 0 to 5.50 feet, being fastened to a post in the edge of the water on the north bank 20 feet from the upstream side of the bridge, the long section, which reads from 5.50 feet to 16 feet, being fastened to the upstream end of the first pier on the north bank, and arranged so that when water rises above the short section the readings are made from the long one, both sections being easily read from the north approach to the bridge. The initial point of sounding is on the south side of the first pier on the north bank. The gage is referred to a bench mark at the top of a chord on the downstream side of the bridge at the second pier from the north bank, and is 27.6 feet above the zero of the gage. The bridge is in three spans, having a total length of 276 feet, with a north approach of 116 feet and a south approach of 124 feet, making a total over all of 516 feet. The observer is J. H. Chisholm, a farmer; post-office address, Alexander, Ala.

The following measurements were made by James R. Hall, M. R. Hall, and others:

Discharge measurements of Hillabee Creek near Alexander.

Date.	Gage height.	Discharge.	Date.	Gage height.	Discharge.
1900.	*Feet.*	*Second-feet.*	1903.	*Feet.*	*Second-feet.*
August 29...........	1.40	184	May 21	2.65	766
November 28........	2.00	390	July 24	1.50	212
1901.			August 21	1.50	205
January 22..........	2.50	606	Do..............	1.50	213
September 12	1.00	139	October 5	.94	84
1902.			November 25	1.15	114
July 16	1.12	169			

Daily gage height, in feet, of Hillabee Creek near Alexander.

Day.	Jan.	Feb.	Mar.	Apr.	May	June.	July.	Aug.	Sept.	Oct.	Nov.	Dec.
1900.												
1									2.30	1.30	1.60	1.90
2									2.30	1.10	2.60	1.90
3									1.60	1.10	6.80	1.80
4									1.40	1.10	3.20	1.90
5									1.30	3.20	2.20	1.80
6									1.20	2.00	1.80	1.90
7									1.20	2.80	1.70	1.80
8									1.10	2.60	1.70	1.70
9									1.20	2.40	1.80	1.70
10									1.20	2.30	1.70	1.70
11									1.10	2.30	1.70	1.70
12									1.10	1.80	1.70	1.70
13									1.10	1.60	1.70	1.80
14									1.40	1.40	1.70	3.80
15									8.10	1.40	1.70	2.90
16									5.00	1.20	1.60	2.80
17									2.60	1.30	1.50	2.70
18									2.20	1.30	1.40	2.80
19									1.60	1.20	1.50	3.00
20									1.50	1.20	1.60	2.90
21									1.50	1.20	1.70	6.00
22									1.40	1.90	2.10	4.00
23									1.20	5.90	1.90	3.00
24									1.30	2.90	1.90	2.90
25									1.40	2.10	5.10	2.90
26									1.30	1.90	2.90	2.80
27									1.40	1.80	2.40	2.60
28									1.30	1.70	2.00	2.50
29								1.40	1.40	1.60	1.90	2.50
30								1.30	1.30	1.60	1.50	5.50
31								1.80		1.50		5.70
1901.												
1	5.00	2.40	2.30	5.20	2.50	2.50	2.50	1.80	2.60	1.20	1.00	1.00
2	4.90	2.40	2.30	4.40	2.50	2.40	2.50	1.40	2.00	1.20	1.00	1.00
3	4.90	4.60	2.40	4.80	2.50	3.00	2.40	1.40	1.10	1.10	1.00	2.00
4	4.70	9.40	2.40	3.10	2.40	2.50	2.30	1.20	1.20	1.10	1.00	1.60
5	3.90	6.00	2.30	2.90	2.40	2.50	2.00	1.10	1.10	1.10	1.30	1.40
6	3.60	4.10	2.20	2.80	2.40	2.60	1.80	1.10	1.10	1.20	1.10	1.20
7	3.60	3.60	2.10	2.70	2.40	2.50	1.70	1.10	1.10	1.40	1.10	1.20
8	3.40	3.10	2.10	2.60	2.40	2.50	2.10	1.00	1.10	1.20	1.00	1.20
9	2.60	3.60	2.10	2.60	2.30	2.40	1.80	1.20	1.10	1.20	1.00	1.20
10	2.50	3.00	2.40	2.50	2.30	2.40	1.80	1.20	1.10	1.20	1.00	1.90
11	8.00	2.90	2.50	2.50	2.80	2.30	1.60	1.10	1.00	1.60	1.00	1.50
12	7.60	2.90	2.30	2.40	2.20	2.00	1.60	1.10	1.00	2.00	1.00	1.40
13	7.00	2.80	2.20	3.10	2.50	2.10	1.50	1.60	2.50	2.90	1.00	1.40
14	5.90	3.00	2.20	2.90	2.30	2.00	1.40	1.60	2.50	1.60	1.00	2.00
15	4.50	3.10	2.30	2.80	2.50	2.00	1.40	4.40	2.40	1.40	1.00	3.00
16	4.30	2.60	2.40	2.70	2.20	1.90	1.40	2.40	2.40	1.20	1.00	3.00
17	4.00	2.60	2.20	2.60	2.20	2.20	1.30	2.00	2.00	1.00	1.00	2.80
18	3.50	2.40	2.20	2.60	2.10	1.90	3.40	2.00	1.80	1.00	1.00	2.70
19	3.40	2.60	2.10	10.00	2.10	1.80	2.20	2.00	1.60	1.00	1.30	2.60
20	3.00	2.70	2.10	3.20	2.90	1.80	1.80	4.00	1.40	1.00	1.30	2.00
21	2.90	2.60	3.00	3.10	3.80	1.70	1.70	4.10	1.20	1.00	1.30	1.50
22	2.90	2.40	2.40	3.00	2.90	1.70	1.90	3.40	1.40	1.00	1.30	1.50
23	2.80	2.60	2.30	2.90	2.50	1.60	1.80	3.10	1.20	1.00	1.30	1.50
24	2.60	2.80	2.80	2.80	2.50	1.60	1.80	2.90	1.10	1.00	1.30	1.60
25	2.50	2.50	2.70	2.70	2.40	1.70	1.70	2.20	1.10	1.00	1.20	1.60
26	2.60	2.40	3.50	2.60	3.90	1.70	1.70	2.20	1.20	1.00	1.20	1.70
27	2.60	2.40	3.00	2.50	2.80	1.80	1.80	2.10	1.00	1.00	1.10	1.70
28	2.50	2.40	2.70	2.50	2.60	1.80	1.60	3.80	1.60	1.00	1.10	1.80
29	2.40		2.60	2.50	2.50	1.70	1.40	3.60	1.40	1.00	1.10	11.00
30	2.40		2.70	2.50	2.80	2.60	1.40	3.40	1.20	1.00	1.10	4.90
31	2.40		5.20		2.90		2.00	3.00		1.00		3.90

Daily gage height, in feet, of Hillabee Creek near Alexander—Continued.

Day.	Jan.	Feb.	Mar.	Apr.	May.	June.	July.	Aug.	Sept.	Oct.	Nov.
1902											
1	3.70	5.30	4.10	3.40	1.90	1.60	0.80	0.80	1.40	3.20	0.90	1.
2	3.50	7.70	4.00	3.20	1.90	1.60	.80	.80	3.60	3.20	.80	6.
3	3.00	3.70	3.90	3.00	1.80	1.70	.80	.90	1.90	3.00	.70	4.
4	2.00	3.00	2.90	3.00	1.80	1.70	.80	1.20	1.50	3.00	2.60	3.
5	2.00	2.70	3.00	3.00	1.90	1.70	.70	1.10	1.40	3.00	3.60	2.
6	2.00	2.50	3.50	2.90	1.90	1.60	.70	1.10	1.30	2.80	2.60	1.
7	2.00	2.40	3.60	4.10	1.90	1.60	.70	1.00	1.20	2.60	1.10	1.
8	1.90	2.10	2.60	3.20	1.90	1.60	.70	1.00	1.10	2.60	1.00	1.
9	1.80	2.10	2.40	3.40	1.80	1.60	.80	.90	1.40	3.30	1.00	1.
10	1.70	2.00	2.30	2.80	1.80	1.50	1.00	.90	1.30	4.10	1.00	1.
11	1.70	2.00	2.30	2.60	1.80	1.50	1.10	.90	1.20	3.80	1.00	1.
12	1.70	2.00	2.00	2.60	1.70	1.40	1.30	.90	1.10	2.10	1.00	1.
13	1.70	1.90	2.50	2.60	1.70	1.40	1.30	.80	1.00	2.40	.90	1.
14	1.60	1.90	3.00	2.50	1.70	1.40	1.30	.80	1.00	2.60	.90	L.
15	1.60	2.00	3.70	2.40	1.60	1.40	1.20	.80	1.00	2.30	.80	6.
16	1.60	2.20	8.00	2.40	1.60	1.40	1.20	.80	1.00	2.20	.80	4.
17	1.60	2.10	3.90	2.40	1.80	1.40	1.10	1.00	2.00	2.00	1.00	2.
18	1.60	2.10	3.40	2.40	2.00	1.30	.90	.90	1.00	1.60	1.00	2.
19	1.90	2.00	3.10	2.30	1.90	1.30	.90	.70	1.00	1.60	.90	1.
20	2.80	2.10	3.00	2.40	1.90	1.20	.90	.70	.90	2.60	.90	1.
21	2.10	2.00	3.90	2.30	1.80	1.20	.80	.70	.80	2.40	.90	1.
22	2.00	2.00	3.50	2.20	1.80	1.20	.80	.70	.80	2.00	.80	2.
23	2.00	2.00	4.00	2.20	1.80	1.20	.80	.70	.80	1.40	.70	2.
24	2.00	2.00	3.40	2.10	1.70	1.10	.70	.70	.80	1.40	.70	1.
25	2.00	2.20	3.00	2.10	1.70	1.00	.70	.70	2.10	1.10	1.30	1.
26	1.90	2.00	3.00	2.00	1.70	.90	.80	.70	2.00	1.00	1.40	1.
27	1.90	7.00	7.00	2.00	1.70	.90	.80	2.00	1.80	.90	1.40	1.
28	2.90	12.50	14.00	2.00	1.60	.90	.80	1.70	1.80	.90	1.50	3.
29	3.00	------	8.00	2.00	1.60	.80	.70	8.60	2.10	.90	1.60	2.
30	2.70	------	4.10	------	1.60	.90	.70	1.90	3.40	.90	1.70	1.
31	2.80	------	3.90	------	1.60	------	.70	1.50	------	1.00	------	1.
1903											
1	1.90	1.60	------	2.20	2.20	2.30	2.40	2.40	2.70	1.10	1.00	1.
2	4.00	1.60	------	2.10	2.30	2.30	2.20	4.00	2.60	1.10	1.00	1.
3	3.60	1.50	------	2.20	2.30	2.70	3.30	4.00	2.40	1.10	1.50	1.
4	3.00	3.00	------	2.10	2.30	2.70	3.30	1.80	2.40	1.00	1.60	1.
5	2.50	2.00	------	2.70	2.30	3.30	2.10	1.80	2.40	1.00	1.60	1.(
6	1.80	1.90	------	2.30	2.90	3.50	2.00	2.00	2.80	1.10	1.30	1.(
7	1.80	4.00	------	2.60	3.10	2.80	2.90	1.70	2.60	1.10	1.00	1.(
8	1.90	16.50	------	3.10	2.10	2.80	3.40	2.00	2.00	1.00	1.00	1.(
9	1.90	------	------	2.60	2.20	2.90	2.40	2.00	1.90	1.00	1.10	1.:
10	1.90	------	------	2.50	2.50	2.80	2.40	1.80	1.70	1.00	1.20	1.:
11	3.00	------	------	2.20	2.90	3.50	2.70	1.80	1.50	1.00	1.90	1.(
12	2.10	------	------	2.40	2.90	2.10	2.60	1.90	1.30	1.00	1.20	1.(
13	2.00	------	------	2.80	4.20	2.80	2.00	1.60	1.30	1.00	1.50	1.:
14	1.70	------	------	2.70	12.08	1.80	2.00	1.60	1.30	1.00	1.30	1.:
15	1.70	------	------	2.60	4.60	1.70	1.90	1.60	1.40	1.10	1.30	1.:
16	1.60	------	------	2.60	3.90	1.70	1.90	1.60	1.30	1.20	1.70	1.:
17	1.60	------	------	2.70	2.90	1.70	1.90	1.60	1.30	1.10	1.70	1.:
18	1.50	------	------	2.70	4.80	1.60	2.20	1.70	1.30	1.00	1.50	1.:
19	1.70	------	------	11.00	2.80	1.60	2.00	1.70	1.20	1.00	1.20	1.:
20	1.70	------	------	2.80	2.70	2.20	2.00	1.70	1.20	1.00	1.20	1.:
21	1.80	------	3.10	2.80	2.60	5.90	2.00	1.70	1.20	1.00	1.50	1.:
22	1.90	------	3.60	2.70	2.40	1.90	1.90	2.00	1.20	1.00	1.50	1.:
23	1.80	------	3.30	2.50	2.00	1.80	1.80	2.00	1.20	1.00	1.50	1.:
24	1.80	------	3.10	2.20	1.90	1.80	2.00	1.50	1.10	1.00	1.50	1.:
25	1.80	------	3.00	2.90	1.80	3.80	1.80	1.50	1.00	1.00	1.50	1.:
26	1.90	------	2.90	2.70	1.80	4.00	1.80	1.50	1.00	1.00	1.50	1.:
27	3.00	------	2.80	2.70	1.80	3.80	1.90	1.40	1.00	1.00	1.50	1.:
28	2.70	------	3.10	2.50	1.90	8.60	1.90	1.40	1.10	1.00	1.20	1.:
29	2.50	------	3.60	2.50	1.90	8.00	1.80	2.40	1.10	1.00	------	1.:
30	1.90	------	4.10	------	1.80	------	3.80	2.80	------	1.00	------	1.:
31	------	------	------	------	------	------	------	------	------	------	------	

Rating table for Hillabee Creek near Alexander for 1900 and 1901.

Gage height.	Discharge.	Gage height.	Discharge.	Gage height.	Discharge.	Gage height.	Discharge.
Feet.	*Second-feet.*	*Feet.*	*Second-feet.*	*Feet.*	*Second-feet.*	*Feet.*	*Second-feet.*
1.0	138	2.1	434	3.2	918	4.3	1,402
1.1	146	2.2	478	3.3	962	4.4	1,446
1.2	156	2.3	522	3.4	1,006	4.5	1,490
1.3	169	2.4	566	3.5	1,050	4.6	1,534
1.4	184	2.5	610	3.6	1,094	4.7	1,578
1.5	204	2.6	564	3.7	1,138	4.8	1,622
1.6	230	2.7	698	3.8	1,182	4.9	1,666
1.7	263	2.8	742	3.9	1,226	5.0	1,710
1.8	303	2.9	786	4.0	1,270		
1.9	346	3.0	830	4.1	1,314		
2.0	390	3.1	874	4.2	1,358		

Rating table for Hillabee Creek near Alexander for 1902.

Gage height.	Discharge.	Gage height.	Discharge.	Gage height.	Discharge.	Gage height.	Discharge.
Feet.	*Second-feet.*	*Feet.*	*Second-feet.*	*Feet.*	*Second-feet.*	*Feet.*	*Second-feet.*
0.8	125	2.2	478	3.6	1,094	5.0	1,710
1.0	138	2.4	566	3.8	1,182	6.0	2,150
1.2	156	2.6	654	4.0	1,270	7.0	2,590
1.4	184	2.8	742	4.2	1,358	8.0	3,030
1.6	230	3.0	830	4.4	1,446	9.0	3,470
1.8	303	3.2	918	4.6	1,534	10.0	3,910
2.0	390	3.4	1,006	4.8	1,622	11.0	4,350

Rating table for Hillabee Creek near Alexander for 1903.

Gage height.	Discharge.	Gage height.	Discharge.	Gage height.	Discharge.	Gage height.	Discharge.
Feet.	*Second-feet.*	*Feet.*	*Second-feet.*	*Feet.*	*Second-feet.*	*Feet.*	*Second-feet.*
0.90	80	2.00	425	3.10	920	4.20	1,415
1.00	90	2.10	470	3.20	965	4.30	1,460
1.10	108	2.20	515	3.30	1,010	4.40	1,505
1.20	126	2.30	560	3.40	1,055	4.50	1,550
1.30	150	2.40	605	3.50	1,100	4.60	1,595
1.40	178	2.50	650	3.60	1,145	4.70	1,640
1.50	210	2.60	695	3.70	1,190	4.80	1,685
1.60	248	2.70	740	3.80	1,235	4.90	1,730
1.70	290	2.80	785	3.90	1,280		
1.80	335	2.90	830	4.00	1,325		
1.90	380	3.00	875	4.10	1,370		

Estimated monthly discharge of Hillabee Creek near Alexander.

[Drainage area, 214 square miles.]

Month.	Discharge in second-feet.			Run-off.	
	Maximum.	Minimum.	Mean.	Second-feet per square mile.	Depth in inches.
1900.					
September	3,074	146	370	1.73	1.93
October	2,106	146	387	1.81	2.09
November	2,502	184	471	2.20	2.45
December	2,150	263	716	3.35	3.86
1901.					
January	3,030	566	1,198	5.60	6.46
February	3,646	566	920	4.30	4.48
March	1,798	434	617	2.88	3.32
April	3,910	566	911	4.26	4.75
May	1,226	434	624	2.92	3.37
June	830	230	439	2.05	2.29
July	1,006	169	357	1.67	1.93
August	1,446	138	535	2.50	2.89
September	654	138	249	1.16	1.29
October	786	138	181	.85	.98
November	169	138	148	.69	.77
December	4,350	138	526	2.46	2.84
The year	4,350	138	559	2.61	35.37
1902.					
January	1,138	230	459	2.14	2.47
February	5,010	346	854	3.99	4.15
March	5,670	390	1,284	6.00	6.92
April	1,314	390	647	3.02	3.37
May	390	230	294	1.37	1.58
June	263	125	186	.87	.97
July	169	120	137	.64	.74
August	1,094	120	182	.85	.98
September	1,094	125	254	1.19	1.33
October	1,314	131	525	2.45	2.82
November	1,094	120	213	1.00	1.12
December	2,150	230	569	2.66	3.07
The year	5,670	120	467	2.18	29.52

Estimated monthly discharge of Hillabee Creek near Alexander—Continued.

Month.	Discharge in second-feet.			Run-off.	
	Maximum.	Minimum.	Mean.	Second-feet per square mile.	Depth in inches.
1903.					
January....................	1,325	210	479	2.24	2.58
February 1 to 8..............			a 1,333	a 6.23	a 1.85
March 22 to 31..............			b 992	b 4.64	b 1.72
April......................	4,475	470	796	3.72	4.15
May	4,925	335	865	4.04	4.66
June......................	2,180	248	739	3.45	3.85
July	1,235	335	547	2.56	2.95
August....................	1,325	178	393	1.84	2.12
September.................	785	90	298	1.39	1.55
October	126	90	95	.44	.51
November.................	380	90	173	.81	.90
December	380	90	133	.62	.71
The year					

a 8-day period.   b 10-day period.

Net horsepower per foot of fall with a turbine efficiency of 80 per cent for the minimum monthly discharge of Hillabee Creek near Alexander.

Month.	1901.			1902.			1903.		
	Minimum discharge.	Minimum net horsepower per foot of fall.	Duration of minimum.	Minimum discharge.	Minimum net horsepower per foot of fall.	Duration of minimum.	Minimum discharge.	Minimum net horsepower per foot of fall.	Duration of minimum.
	Sec.-ft.		*Days.*	*Sec.-ft.*		*Days.*	*Sec.-ft.*		*Days.*
January	566	51	3	230	21	5	210	19	1
February	566	51	7	346	31	2	a 210	19	1
March	434	39	5	390	35	1	b 785	71	1
April	566	51	1	390	35	5	470	43	2
May	434	39	2	230	21	6	335	30	1
June..........	230	21	2	125	11	1	248	23	2
July	169	15	2	120	11	9	335	30	3
August........	138	13	1	120	11	8	178	16	2
September	138	13	2	125	11	4	90	8	3
October........	138	13	15	131	12	4	90	8	22
November	138	13	15	120	11	3	90	8	5
December......	138	13	2	230	21	3	90	8	7

a Record for only 7 days in month.   b Record for only 10 days in month.

TRIBUTARIES OF TALLAPOOSA RIVER ABOVE MILSTEAD.

Side.	Stream.	Point on stream.	Drainage area.	Estimated discharge low water 1900-1901.	Net horse-power per foot of fall on 80 per cent tur-bine.
			Sq. miles.	*Sec.-ft.*	
Left	Uphapee Creek	Mouth of creek	450	45	4.1
	....do	Chehaw, Ala	360	40	3.6
	Sougahatchee Creek ...	Mouth of creek	240	48	4.3
Right ..	Cedar Creek	....do	55	14	1.3
Left	Wind Creek	....do	25	8	.7
Right ..	Kowaliga Creek.........	....do	135	40	3.6
	....do	Kowaliga, Ala	115	35	3.2
Left	Blue Creek	Mouth of creek	60	20	1.8
	Big Sandy Creek	....do	200	70	6.3
	....do	Smith's bridge.......	195	67	6.1
Right ..	Elkhatchee Creek......	Mouth of creek	75	37	3.3
	Hillabee Creek.........	....do	220	141	12.8
	....do	Chisholme's bridge...	214	138	12.6
	Emucklaw Creek......	Mouth of creek	78	46	4.2
Left	Cohoasanocsa Creek...	....do	70	42	3.8
	High Pine Creek	....do	82	49	4.4
Right ..	Hurricane Creek	....do	14	8	.7
Left	Cornhouse Creek.......	....do	72	43	3.9
Right ..	Crooked Creek.........	....do	95	57	5.2
	Fox Creek.............	....do	37	22	2.0
Left	Little Tallapoosa River.	Mouth of river	590	354	32.2
	....do	Alabama-Georgia State line.	311	186	16.9
	Tallapoosa River	Above Little Talla-poosa River.	767	460	41.8
Right ..	Ketchepedrakee Creek.	Mouth of creek	49	29	2.6
	Cane Creek	....do	55	33	3.0
	Muscadine Creek......	....do	36	21	1.9
	Tallapoosa River	Alabama-Georgia State line.	302	181	16.4

All of these tributaries to Tallapoosa River are in the crystalline region, and have fine shoals all along their courses.

No State or Government surveys have ever been made to determine their profiles, and it is, therefore, impossible at present to make a detailed statement of the water powers. The tabulated statement given above shows discharge at certain places during low season of ordinary years, like 1900 and 1901.

This flow at any point, multiplied by the total practical fall in feet that can be brought upon a water wheel on the given stream at that point and divided by 11, gives the net available horsepower at that point during low season of a year, like 1900 or 1901.

From the discharge and drainage area at a given point can be computed the discharge at other points on the same stream if drainage area is known.

Discharge measurements have been made on these streams at various points, as is shown by the following list. As the dates of these measurements are given, the stage of water as related to minimum for 1900–1901 can be approximated by noting the stage at regular stations on the same dates.

Miscellaneous discharge measurements of tributaries of Tallapoosa River.

Date.	Stream.	Locality.	Discharge.
1900.			*Sec.-ft.*
Aug. 2	Sougahatchee Creek..........	Meader's bridge	125
3	Blue Creek	Susanna	34
28	Elkhatchee Creek............	Island Home	184
30	Timbercut Creek.............	Near Welche's ferry.............	18
Dec. 12	Chattasofka Creek	New bridge, near Dadeville......	35
1901.			
Feb. 11	Wind Creek	Starr's bridge, near Meltons Mill.	66
11	Sougahatchee Creek..........	Lovelady's bridge, near Thaddeus.	453
13	Blue Creek	Farrow's mill, Susanna	117
13	Channahatchee Creek	Freeman's mill, Channahatchee..	80
27	Kowaliga Creek...............	Benson's bridge, Kowaliga.......	154
Mar. 5	Emuckfaw Creek	Hamlett's mill, Zana	113
11	Moores Creek.................	Near Dudleyville................	29
12	Chattahaspa Creek	Scott's mill, near Tiller Crossroads.	203
12	Cohoasanocsa Creek..........	Leverett's mill, near Milltown ...	122
12	High Pine Creek..............	Lile's gin, Happy Land..........	89
12	Beaverdam Creek.............	Near Louina	30
13	Cornhouse Creek.............	Swann's store, near Levelroad ...	31
13	Wild Cat Creek	Murphy's mill, near Gay	32
13	Tallapoosa River.............	Below mouth of Little Tallapoosa River, near Goldburg.	2,400
13	Crooked Creek...............	Near Goldburg...................	183
13	Hurricane Creek	Near Almond	29

A SHOALS, NEAR HIGH SHOALS, ALA.

On tributary of Tallapoosa River

B. GILES MILL, NEAR OFELIA, ALA.

On tributary of Tallapoosa River

The following discharge measurement was made on Tallapoosa River near Tallapoosa, Ga., in the year 1902, by M. R. Hall:

June 2: Gage height, 1.30 feet; discharge, 114 second-feet.

COOSA RIVER AND TRIBUTARIES.

Coosa River is formed by the junction of Etowah and Oostanaula rivers at Rome, Ga. The drainage area is 4,006 square miles. Both of the tributary rivers rise in the northern part of Georgia and flow for the most part through a hilly, broken country, well wooded, about one-fourth of the land being under cultivation. Coosa River flows in a southwesterly direction into Alabama and joins the Tallapoosa 6 miles above Montgomery, Ala., to form Alabama River.

The regular stations at which measurements have been made are Riverside, Ala., and Rome, Ga., on Coosa River, and Nottingham, Ala., on Talladega Creek. Numerous miscellaneous discharge measurements have been made at other points.

COOSA RIVER AT RIVERSIDE.

This station is at Riverside, at the bridge of the Southern Railway, Georgia Pacific division, across Coosa River. The river here flows in a southerly direction, the railroad running from east to west. The town of Riverside is on the right, or west, bank of the river, and the railroad station is about 1,000 feet west of the bridge, which is of iron and about 30 feet above low water. Beginning at the left bank, there are two spans of 154 feet each; then a drawbridge 220 feet, revolving on a large center pier; then a stationary span 80 feet in length, to west, or right, bank abutment. There is no running water at low stages under the last-named span.

At low water the flowing river is 480 feet wide, including three piers, and is from 4 to 10 feet deep. Very little of the current is too slow to turn any meter. The channel is somewhat irregular, as there are shoals and some old cribs just above the bridge, but for all stages it is probably the best station that can be found on the river at a bridge and easy of access.

On September 8, 1896, a discharge measurement was made by B. M. Hall, and two bench marks were established. On September 22, 1896, another discharge measurement was made, a wire gage was put in, and Mr. J. W. Foster, sawyer at a large sawmill about 300 feet distant, on right bank of river, below the bridge, was employed as observer.

The initial point is top of left abutment at the edge toward the river, on the downstream side of the bridge, from which side soundings and meter measurements are made. The rod of wire gage is nailed to outside guard rail, downstream side, next to the last panel of stationary bridge before reaching the pier at end of draw span. The rod is 14 feet long and divided to feet and tenths. The bench mark is the top of the capstone on the large circular center pier of turn span. It is 26.80 feet above datum of gage at downstream side of pier.

The drainage area is 6,850 square miles, and is mapped on the following atlas sheets of the United States Geological Survey: Springville, Anniston, Gadsden, Fort Payne, Rome, Tallapoosa, Marietta, Cartersville, Suwanee, Ellijay, Dalton, Cleveland, Ringgold, and Stevenson.

The following discharge measurements were made by B. M. Hall, M. R. Hall, and others.

Discharge measurements of Coosa River at Riverside.

Date.	Gage height.	Discharge.	Date.	Gage height.	Discharge.
1896.	*Feet.*	*Second-feet.*	**1899.**	*Feet.*	*Second-feet.*
September 8	0.70	1,630	November 7	0.85	2,271
September 25	.50	1,403	December 9........	1.20	2,727
October 30..........	.88	1,986	**1900.**		
December 21........	1.57	3,272	February 10	5.03	13,493
1897.			March 21..........	12.50	43,759
March 31...........	4.53	12,515	May 5.............	4.15	11,196
June 17	1.54	3,747	August 21	2.32	5,609
July 21............	5.55	16,925	December 28.......	4.25	11,335
August 20	2.58	6,174	**1901.**		
November 29	.80	1,854	January 8	3.85	9,572
1898.			March 18..........	3.70	9,333
January 27	10.00	30,359	August 24..........	12.95	44,554
March 9	1.60	3,538	November 14	1.70	4,039
May 3.............	3.22	7,758	**1902.**		
May 25.............	1.39	3,172	April 8	7.30	21,138
August 3	3.92	9,524	June 3	2.00	4,720
September 7	11.05	37,811	October 17.........	2.30	5,128
October 19.........	6.80	14,484	**1903.**		
November 22	5.85	16,384	March 16..........	10.75	40,072
1899.			April 11	8.80	30,710
April 26	9.00	29,069	June 20...........	2.87	7,374
May 3.............	4.05	10,592	July 22...........	2.30	5,549
May 20............	2.70	6,276	August 26	1.64	4,001
June 14...........	2.20	5,010	October 1..........	1.05	2,687
August 26..........	1.42	3,791	November 13	1.37	3,136
September 23	1.00	2,457			

Daily gage height, in feet, of Coosa River at Riverside.

Day.	Jan.	Feb.	Mar.	Apr.	May.	June.	July.	Aug.	Sept.	Oct.	Nov.	Dec.
1896.												
1										0.60	1.10	1.30
2										1.75	1.40	2.10
3										8.10	1.20	4.88
4										2.75	1.10	8.80
5										2.00	1.05	8.20
6										1.50	1.10	2.50
7										1.20	1.20	2.20
8										.85	2.55	1.90
9										.70	2.30	1.70
10										.60	1.90	1.60
11										.65	1.80	1.55
12										.60	1.60	1.55
13										.60	2.25	1.60
14										.60	2.70	1.60
15										.56	4.00	1.80
16										.55	5.20	2.00
17										.55	4.70	2.10
18										.65	4.20	2.20
19										.80	8.20	2.00
20										.85	2.30	1.80
21										.75	1.50	1.70
22										.70	1.40	1.50
23										.60	1.35	1.45
24										.55	1.30	1.40
25										.60	1.25	1.35
26									0.45	.70	1.20	1.30
27									.45	.80	1.15	1.25
28									.45	.85	1.15	1.20
29									.45	.90	1.10	1.10
30									.50	.95	1.20	1.10
31										.85		1.10
1897.												
1	1.10	2.00	5.00	4.30	3.40	1.90	1.45	2.00	1.20	.50	.70	.80
2	1.10	2.50	4.50	4.45	3.10	1.90	1.45	1.80	1.10	.50	.70	.90
3	1.10	5.35	4.25	5.20	3.05	1.90	1.40	1.60	1.30	.50	.70	.95
4	1.10	7.35	3.90	7.00	3.60	1.85	1.40	1.50	1.60	.50	.65	1.20
5	1.10	7.70	4.20	8.60	3.20	1.85	1.45	1.45	1.30	.45	.80	2.50
6	1.20	7.90	5.80	9.50	3.00	1.90	1.50	1.40	1.20	.45	1.05	3.00
7	1.25	9.00	11.40	10.50	3.80	1.90	1.50	1.50	1.10	.45	1.15	2.90
8	1.30	7.70	13.30	11.15	3.70	2.15	2.40	1.50	1.00	.45	1.10	2.40
9	1.35	6.40	12.55	12.15	2.65	2.10	2.30	1.60	.90	.45	1.00	2.15
10	1.35	5.90	12.65	11.90	2.60	1.90	2.05	1.70	.85	.45	.95	2.00
11	1.35	5.20	12.70	10.70	2.50	1.90	2.50	1.70	.80	.40	.85	1.70
12	1.30	7.35	12.80	9.10	2.55	2.00	2.70	2.00	.80	.40	.85	1.60
13	1.40	8.30	13.45	7.80	2.65	1.90	2.50	2.50	.75	.45	.85	1.60
14	2.00	8.20	14.80	6.05	2.10	1.85	2.00	2.30	.75	.45	.85	2.00
15	3.50	7.50	14.60	5.60	8.90	1.60	1.80	2.00	.80	1.15	.80	2.50
16	4.00	6.60	14.80	5.30	4.00	1.70	1.80	1.80	.85	1.65	.75	3.00
17	4.90	5.70	14.70	5.60	4.00	1.50	1.80	1.60	.85	1.40	.70	3.30
18	5.35	5.00	14.70	5.40	3.60	1.60	1.95	1.50	.80	1.35	.70	3.15
19	5.00	4.50	14.50	5.00	3.20	1.90	2.00	1.90	.80	1.20	.70	2.65
20	4.80	4.00	15.30	4.60	3.00	2.00	3.00	2.60	.75	1.00	.65	2.10
21	6.50	4.60	14.90	4.80	2.70	1.80	5.20	2.00	.75	.90	.65	2.20
22	7.00	4.65	14.70	4.00	2.35	1.70	6.40	1.70	.70	.85	.65	2.80
23	7.35	6.00	14.50	8.80	2.30	1.60	8.00	1.60	.70	.80	.65	4.20
24	7.00	7.90	13.70	3.60	2.25	1.55	6.20	1.70	.70	.70	.65	4.85
25	5.40	9.00	12.20	8.40	2.25	1.50	4.50	1.75	.65	.80	.65	4.95
26	4.70	9.00	10.60	3.30	2.15	1.45	4.00	1.60	.65	.75	.65	4.55
27	3.80	8.00	8.50	3.25	2.05	1.45	3.00	1.60	.60	.60	.65	8.85
28	3.00	6.20	6.50	2.20	2.00	1.40	2.60	1.60	.55	.65	.65	3.20
29	2.70		5.30	8.10	2.00	1.45	2.50	1.40	.55	.80	.70	2.95
30	2.50		4.90	3.20	1.95	1.45	3.00	1.35	.55	.75	.75	2.85
31	2.20		4.60		1.90		2.60	1.30		.70		2.50

Daily gage height, in feet, of Coosa River at Riverside—Continued.

Day.	Jan.	Feb.	Mar.	Apr.	May.	June.	July.	Aug.	Sept.	Oct.	Nov.	Dec.
1896.												
1	2.30	6.00	1.65	6.80	3.90	1.30	1.15	4.25	2.20	2.20	2.60	3.70
2	2.15	5.25	1.70	7.50	3.60	1.25	1.05	4.00	1.80	1.80	2.50	4.00
3	2.10	4.00	1.80	6.80	3.30	1.20	1.00	4.10	1.75	1.70	2.50	3.90
4	2.00	3.25	1.80	5.50	3.05	1.20	.95	4.00	5.80	2.00	2.45	3.70
5	1.90	3.00	1.80	5.80	2.90	1.30	.95	3.30	9.30	6.80	2.40	3.70
6	1.85	2.80	1.75	9.30	2.70	1.25	.90	4.00	10.20	11.20	2.40	3.90
7	.1.75	2.75	1.70	10.50	2.55	1.20	.95	5.50	11.00	11.90	2.40	4.00
8	1.70	2.70	1.70	10.80	2.40	1.10	1.00	5.30	11.30	15.80	2.40	4.00
9	1.65	2.60	1.65	10.40	2.30	1.05	1.15	4.50	11.60	14.70	2.45	3.70
10	1.60	2.50	1.65	8.90	2.20	1.05	1.65	4.30	10.80	12.50	2.55	3.30
11	1.60	2.40	1.60	7.50	2.15	1.00	2.15	4.50	8.70	12.00	2.75	3.25
12	1.65	2.30	1.60	6.00	2.10	1.48	2.15	4.70	5.80	11.20	2.75	3.10
13	1.80	2.30	1.70	5.00	2.00	1.10	2.10	6.70	4.75	8.80	2.70	3.00
14	2.00	2.20	1.80	4.40	1.95	1.25	2.15	5.90	4.10	5.50	·2.65	2.90
15	3.10	2.10	2.00	4.00	1.90	1.15	2.05	4.70	3.40	4.40	2.70	2.80
16	3.00	2.00	2.25	3.70	1.85	1.00	2.30	3.70	3.00	3.60	2.80	2.65
17	2.80	1.95	3.00	3.50	1.80	1.65	3.10	3.00	2.70	3.00	3.10	2.60
18	2.60	1.90	.4.75	4.00	1.75	1.70	3.05	2.75	2.50	3.50	3.15	2.60
19	2.80	1.85	5.50	5.10	1.70	1.65	2.50	2.55	2.25	5.40	3.25	2.70
20	3.00	1.80	4.70	5.80	1.60	1.55	1.90	2.45	2.20	6.40	3.70	2.90
21	4.10	1.80	4.00	5.50	1.70	1.95	1.65	2.40	2.15	6.30	4.20	4.00
22	5.80	1.80	3.25	4.00	1.65	2.10	1.50	2.40	2.20	6.00	5.15	3.80
23	6.05	1.75	3.00	4.50	1.55	2.30	1.35	2.35	2.55	5.80	7.00	3.40
24	6.50	1.75	2.75	5.75	1.50	2.50	1.25	2.30	8.55	5.00	5.90	3.00
25	7.20	1.70	2.30	7.10	1.45	2.05	1.20	2.15	4.30	4.35	5.20	2.75
26	9.00	1.70	2.15	7.80	1.40	1.75	1.40	2.00	3.90	4.00	4.90	2.70
27	10.20	1.70	2.00	7.45	1.55	1.50	2.35	1.80	3.40	3.75	4.60	2.50
28	10.65	1.65	2.00	6.45	1.70	1.60	3.15	2.00	3.15	3.30	4.20	2.40
29	10.45	......	2.30	5.50	1.70	1.45	3.10	2.50	3.00	8.00	4.00	2.30
30	9.45	......	3.00	4.75	1.45	1.30	3.40	3.00	2.75	2.75	3.80	2.30
31	7.55	......	4.50	......	1.35	......	4.00	2.60	......	2.70	......	2.40
1899												
1	2.80	5.90	12.10	10.30	5.00	2.70	1.70	3.95	2.30	0.90	0.95	2.60
2	2.75	6.30	12.20	10.20	4.50	2.80	1.70	3.00	2.40	.90	.95	2.00
3	2.75	7.50	12.30	10.00	4.20	2.80	1.65	2.30	2.80	.90	.95	1.60
4	2.80	7.40	12.10	9.80	3.95	2.60	1.60	2.00	2.40	.90	.95	1.50
5	2.80	9.10	10.10	8.90	3.80	2.30	1.50	1.90	2.10	.90	.95	1.45
6	3.00	12.10	9.00	8.75	3.75	2.10	1.50	1.75	2.00	.90	.95	1.40
7	3.50	14.10	8.00	9.00	3.60	2.00	1.45	1.65	1.85	.90	.95	1.35
8	4.30	14.30	7.50	10.00	4.00	2.00	1.40	1.50	1.40	.90	.95	1.30
9	4.20	14.30	7.25	12.30	3.85	1.95	1.70	1.50	1.30	.95	.95	1.30
10	4.40	14.10	7.00	12.00	3.60	1.95	1.60	1.55	1.25	1.20	.95	1.35
11	5.20	13.80	6.15	11.70	3.45	1.95	1.50	1.55	1.25	1.30	.95	1.40
12	5.90	13.00	5.20	10.00	3.30	1.95	1.50	1.60	2.20	1.40	.95	5.40
13	5.60	12.00	5.50	8.90	3.20	2.00	1.40	1.50	2.50	1.80	.95	8.25
14	5.00	10.90	7.50	7.90	3.00	2.15	1.40	1.45	2.65	1.20	1.00	8.00
15	4.70	8.70	8.30	6.30	3.00	2.80	1.30	1.50	4.30	1.20	1.00	6.00
16	4.90	7.90	16.00	5.55	2.95	8.20	1.30	1.50	1.40	1.20	1.00	4.50
17	5.00	7.60	17.40	5.25	2.95	2.95	1.80	1.50	1.25	1.10	1.00	3.75
18	4.90	7.80	17.00	5.10	2.80	2.50	1.20	1.55	1.20	1.00	1.00	3.60
19	4.70	8.10	16.50	5.00	2.80	2.25	1.30	1.50	1.20	.95	1.00	3.40
20	4.60	8.20	16.30	4.80	2.75	2.00	1.50	1.45	1.10	.95	1.00	3.00
21	4.20	8.00	16.35	4.60	2.75	1.80	1.50	1.45	1.00	.90	.95	2.85
22	4.00	7.65	16.20	4.30	2.70	1.70	1.90	1.35	.95	.90	.95	2.75
23	3.90	8.00	15.90	4.75	2.70	1.70	3.20	1.30	1.00	.95	.95	8.00
24	3.90	8.10	15.70	5.65	2.65	1.60	4.70	1.30	1.00	1.20	.95	5.40
25	4.00	7.30	15.50	8.90	2.60	1.75	3.60	1.30	1.00	1.15	1.00	6.40
26	4.25	7.00	14.90	9.00	2.50	1.60	8.20	1.60	1.00	1.00	2.15	7.10
27	4.15	8.30	13.25	8.90	2.45	1.60	3.00	1.20	.95	1.00	2.90	7.00
28	4.00	11.00	11.00	8.30	2.35	1.65	8.60	1.10	.95	.95	3.00	6.60
29	3.90	......	8.00	6.90	2.30	1.70	4.20	1.50	.90	.95	3.00	6.00
30	3.75	......	7.90	5.45	2.20	1.65	5.20	2.10	.90	.90	2.75	4.85
31	8.70	......	8.50	......	2.70	......	4.75	2.10	......	.95	......	4.00

Daily gage height, in feet, of Coosa River at Riverside—Continued.

Day.	Jan.	Feb.	Mar.	Apr.	May.	June.	July.	Aug.	Sept.	Oct.	Nov.	Dec.
1900.												
1	3.50	2.70	6.90	6.65	5.00	2.75	11.60	4.70	1.85	1.55	2.30	5.75
2	3.00	2.65	7.55	6.25	5.30	2.80	10.10	3.90	2.00	1.55	2.25	4.35
3	2.50	2.60	6.90	6.00	5.30	2.90	8.90	3.00	2.20	1.50	2.50	3.75
4	2.40	2.50	6.25	5.60	4.75	2.60	8.20	2.75	2.10	1.50	2.50	3.40
5	2.15	2.60	5.40	5.10	4.30	2.70	7.50	2.55	2.00	1.50	2.50	3.30
6	2.05	2.80	5.00	4.90	4.20	3.45	6.45	2.40	1.80	1.45	2.40	4.35
7	1.95	2.95	4.90	4.75	4.00	3.90	5.50	2.25	1.70	1.40	2.40	6.05
8	1.95	3.00	6.00	4.40	3.65	4.20	4.70	2.15	1.60	1.50	2.35	5.40
9	2.00	3.75	8.75	4.35	3.40	7.05	5.00	2.10	1.50	2.20	2.30	4.80
10	2.00	4.25	10.00	4.30	3.30	8.30	4.30	2.00	1.45	2.35	2.30	4.00
11	2.10	5.80	10.55	6.50	3.15	8.00	4.20	2.00	1.35	3.85	2.15	3.60
12	3.50	6.50	10.05	12.40	2.95	7.70	4.10	1.90	1.30	3.60	2.10	3.15
13	6.00	13.30	8.75	12.90	2.70	6.70	5.65	2.25	1.25	3.80	2.10	2.95
14	7.40	15.30	7.50	11.70	2.70	4.30	4.65	2.00	1.20	2.80	2.00	2.80
15	7.00	15.20	5.60	9.50	2.65	4.50	3.75	1.90	3.35	3.00	1.90	2.70
16	6.40	14.50	6.00	7.20	2.65	4.70	3.60	1.85	6.00	2.90	1.80	2.65
17	5.10	14.00	6.30	12.40	2.60	5.00	3.50	2.00	7.00	2.80	1.75	2.60
18	4.00	13.25	6.00	18.10	2.60	4.90	3.35	2.00	7.50	2.65	1.70	2.55
19	4.25	12.80	6.50	17.55	2.60	6.90	3.10	2.20	6.00	2.50	1.80	2.55
20	8.00	12.10	10.00	16.65	2.60	6.90	3.00	2.10	4.35	2.40	2.00	3.00
21	9.70	9.00	12.20	12.95	2.65	6.45	2.90	2.20	3.20	2.80	2.50	3.30
22	10.00	7.80	12.85	13.15	3.20	6.10	2.70	2.00	2.50	2.20	4.00	5.20
23	9.40	6.80	12.60	12.65	2.90	7.00	2.45	2.00	2.00	2.15	3.80	7.00
24	8.75	7.20	11.80	12.20	3.00	11.35	2.50	1.95	1.90	3.00	3.20	7.30
25	7.75	6.90	10.60	10.80	3.25	12.50	2.60	1.90	1.85	5.25	3.10	6.90
26	6.00	6.50	10.30	9.15	3.10	14.10	2.50	2.20	1.80	7.50	3.00	6.85
27	4.10	5.25	10.20	7.90	3.20	14.42	2.60	2.10	1.80	5.00	4.35	5.90
28	3.60	5.00	9.85	6.50	3.00	14.60	3.70	2.00	1.65	3.80	6.40	4.90
29	3.80	......	9.50	5.70	2.80	18.80	5.70	1.95	1.60	3.00	9.20	4.30
30	3.00	......	8.50	5.35	2.70	12.80	5.30	1.90	1.55	2.65	8.20	4.00
31	2.70	......	7.20	......	2.60	......	5.90	1.90	......	2.50	......	6.50
1901.												
1	7.30	6.00	3.55	14.35	4.80	5.20	3.00	2.15	5.60	2.70	1.90	1.85
2	7.35	6.50	3.50	14.40	4.60	8.20	2.90	2.10	5.00	4.00	1.85	1.90
3	6.50	9.00	3.50	13.30	4.80	8.25	2.85	2.10	4.50	3.20	1.85	1.95
4	6.00	10.60	3.50	11.90	4.15	7.20	2.80	2.05	4.30	3.00	1.85	1.95
5	5.50	11.00	3.40	10.40	4.00	5.75	2.80	2.00	3.65	2.90	1.80	2.20
6	5.00	12.00	3.30	10.30	3.95	5.00	2.75	2.20	3.30	2.75	1.80	2.10
7	4.25	11.90	3.20	9.00	3.90	4.80	2.75	4.10	3.05	2.60	1.80	2.00
8	3.95	11.50	3.10	7.25	3.75	4.30	2.70	4.80	3.00	2.60	1.85	2.00
9	3.70	9.90	3.10	6.50	3.60	4.20	3.50	4.25	2.90	2.30	1.85	2.10
10	3.60	9.60	3.40	5.60	3.45	6.20	4.00	3.50	2.85	2.80	1.80	2.10
11	6.50	9.90	5.50	5.00	3.25	6.00	3.50	2.90	2.75	2.25	1.80	2.05
12	14.10	9.60	6.50	4.80	3.15	5.00	2.90	2.75	2.70	2.30	1.80	2.00
13	15.70	8.90	7.50	4.80	3.10	4.25	2.50	2.60	2.65	3.00	1.80	2.00
14	15.40	7.60	7.00	6.50	3.10	3.75	2.40	3.20	2.65	2.80	1.75	2.90
15	15.10	6.50	6.30	7.50	3.00	3.75	2.35	3.20	2.60	2.70	1.75	11.50
16	14.90	5.80	5.25	8.30	2.95	3.80	2.20	3.40	2.85	2.50	1.80	12.00
17	14.30	5.45	4.25	7.80	2.95	3.80	3.15	5.30	3.90	2.45	1.80	12.50
18	13.90	5.00	3.75	6.50	2.90	4.90	2.90	9.60	5.60	2.60	1.80	12.40
19	13.30	4.80	3.55	10.50	2.90	5.00	2.90	9.65	6.30	2.40	1.80	11.00
20	11.30	4.60	3.55	14.00	2.85	4.50	3.20	10.00	7.50	2.35	1.75	10.20
21	8.30	4.35	8.75	14.50	4.00	3.90	3.00	11.50	7.20	2.30	1.75	9.20
22	6.25	4.20	8.70	14.80	8.20	3.60	2.95	11.00	6.00	2.30	1.75	8.00
23	5.30	4.00	8.50	13.50	9.90	8.50	2.90	11.50	4.50	2.25	1.75	4.50
24	4.60	3.90	4.30	12.50	10.75	3.25	2.75	12.90	3.30	2.20	1.85	3.60
25	5.30	3.85	13.90	11.30	11.85	3.15	2.70	12.90	3.00	2.10	1.90	4.10
26	5.80	3.65	15.90	9.90	11.90	3.00	2.60	12.40	2.70	2.10	2.10	4.60
27	6.30	3.65	15.20	7.90	11.80	2.90	2.40	12.00	2.60	2.10	2.00	5.10
28	6.40	3.60	14.70	6.80	11.00	2.90	2.40	11.50	2.50	2.05	2.00	6.50
29	5.85	......	14.25	5.90	9.60	2.90	2.30	9.95	2.65	2.05	1.85	11.50
30	5.50	......	14.60	5.20	6.20	3.10	2.40	8.35	2.60	2.00	1.85	15.60
31	5.80	......	15.50	......	4.60	......	2.20	6.85	......	1.95	......	16.00

Daily gage height, in feet, of Coosa River at Riverside—Continued.

Day.	Jan.	Feb.	Mar.	Apr.	May.	June.	July.	Aug.	Sept.	Oct.	Nov.	Dec.
1902.												
1	15.70	10.40	15.00	15.50	3.35	2.15	1.60	1.40	2.40	2.50	1.20	2.80
2	13.30	13.35	15.40	14.80	3.25	2.10	1.50	1.60	2.00	2.25	1.20	2.40
3	15.10	14.60	15.20	14.60	3.20	2.05	1.50	2.00	1.50	2.00	1.20	4.50
4	15.30	14.40	14.90	14.40	3.80	2.00	1.45	1.60	1.40	1.90	1.15	6.00
5	15.10	14.00	15.60	12.70	3.40	2.00	1.45	1.40	2.00	1.80	1.20	7.20
6	14.50	13.40	15.00	8.90	3.20	1.90	1.40	1.45	1.80	1.75	1.60	6.80
7	13.10	13.20	14.90	6.40	3.10	1.90	1.40	1.90	1.40	1.70	1.80	5.40
8	9.30	11.25	14.10	7.20	8.00	1.85	1.40	1.85	1.50	2.00	1.90	4.80
9	6.20	9.10	18.00	7.60	2.90	1.80	1.35	1.80	1.50	1.75	1.85	4.20
10	4.80	6.00	9.90	7.70	2.75	1.80	1.35	1.80	1.40	1.60	1.70	3.90
11	4.60	5.20	8.10	7.00	2.70	1.80	1.40	2.00	1.40	2.40	1.60	3.50
12	2.90	5.00	6.90	6.20	2.65	2.20	1.40	1.80	1.45	4.00	1.50	2.90
13	2.85	4.90	6.65	5.70	2.75	2.20	1.45	1.70	1.40	8.80	1.30	2.40
14	4.50	4.50	6.40	5.30	2.80	1.95	1.45	1.60	1.40	3.25	1.25	2.45
15	5.25	5.00	6.20	4.85	2.85	1.85	1.50	1.60	1.30	2.65	1.20	2.50
16	7.20	5.80	7.20	4.70	2.65	1.80	1.60	1.35	1.25	2.50	1.20	3.60
17	6.70	5.20	9.00	4.60	2.60	1.80	2.25	1.25	2.00	2.40	1.15	5.30
18	6.20	5.00	9.90	4.90	3.30	1.85	2.85	1.20	1.65	2.10	1.20	5.60
19	2.90	4.90	10.20	4.95	2.90	2.20	2.80	1.20	1.60	1.90	1.25	5.40
20	2.85	4.85	9.40	5.35	2.60	2.00	2.20	1.20	1.20	1.75	1.40	4.90
21	4.50	4.60	8.20	5.00	2.50	1.95	1.95	1.25	1.10	1.60	1.60	5.00
22	5.25	4.90	6.90	4.90	2.45	1.90	1.90	1.25	1.15	1.65	2.00	6.90
23	7.20	4.90	5.90	4.50	2.40	1.90	1.80	1.40	1.25	1.45	1.80	7.80
24	6.70	5.00	5.40	4.30	2.40	1.85	1.75	1.60	1.25	1.35	1.40	7.40
25	6.20	5.35	5.45	3.90	2.35	1.85	1.45	1.70	1.20	1.30	1.60	6.35
26	5.20	5.75	5.30	3.70	2.30	2.00	1.40	1.70	1.35	1.30	2.20	5.10
27	4.60	6.35	6.20	3.70	2.25	1.90	1.50	1.65	2.00	1.30	2.90	4.50
28	5.10	12.50	8.90	3.60	2.25	1.85	2.30	1.65	4.00	1.25	3.45	3.50
29	6.50	......	13.30	3.50	2.20	1.80	1.60	1.60	3.50	1.25	3.90	3.40
30	7.00	......	17.30	3.40	2.20	1.75	1.50	2.20	3.40	1.25	3.50	3.35
31	7.40	......	16.50	......	2.20	......	1.40	2.40	......	1.20	......	3.25
1903.												
1	3.20	4.20	12.90	18.40	4.20	6.10	4.00	3.00	1.25	1.10	1.10	1.20
2	4.00	3.90	14.30	13.60	4.00	7.80	8.80	3.45	1.25	1.05	1.15	1.20
3	4.50	3.85	14.60	13.75	3.85	7.60	3.00	3.20	1.30	1.10	1.25	1.15
4	5.50	5.40	14.60	13.65	3.80	6.90	2.80	3.00	1.40	1.10	1.25	1.15
5	5.55	6.30	14.65	13.50	3.60	7.00	2.70	2.65	1.30	1.05	1.30	1.10
6	5.60	9.40	15.15	12.80	3.50	9.30	2.65	3.00	1.25	1.00	1.35	1.05
7	5.35	11.30	15.00	10.50	3.45	8.60	2.65	3.75	1.25	1.00	1.60	1.05
8	5.10	a15.20	14.50	7.75	3.40	10.05	2.60	4.95	1.20	1.30	1.80	1.05
9	4.50	14.80	13.40	8.90	4.00	9.00	2.60	3.50	1.20	1.10	1.90	1.10
10	3.50	14.20	10.90	9.70	4.10	7.30	2.60	2.45	1.20	1.00	1.80	1.15
11	4.20	15.10	9.30	8.80	4.00	5.60	2.55	2.20	1.15	1.20	1.60	1.20
12	5.00	16.00	9.60	7.40	3.65	4.80	2.65	2.00	1.15	1.40	1.40	1.15
13	5.90	15.75	9.70	6.60	3.35	4.40	2.70	1.80	1.10	1.50	1.35	1.15
14	6.60	14.80	11.60	8.90	3.60	5.00	3.30	1.10	1.10	1.80	1.40	1.15
15	6.30	14.10	11.55	10.10	b8.70	4.40	6.00	1.50	1.20	1.20	1.40	1.20
16	5.30	14.00	10.80	10.35	7.90	3.90	6.60	1.50	1.25	1.10	1.35	1.20
17	4.45	17.25	9.30	10.00	6.30	3.60	5.50	1.80	1.10	1.20	1.30	1.20
18	3.90	16.40	7.90	8.10	5.25	3.40	3.90	2.00	2.00	1.20	1.25	1.15
19	3.60	15.50	6.85	6.70	4.05	8.10	3.15	3.25	2.05	1.20	1.25	1.15
20	3.40	14.40	6.20	6.30	3.90	2.90	2.75	2.70	1.80	1.30	1.30	1.20
21	8.25	14.20	5.70	5.60	3.70	2.75	2.60	2.35	1.50	1.20	1.25	1.20
22	2.90	14.00	7.10	5.10	8.20	2.65	2.35	2.30	1.35	1.20	1.20	1.25
23	2.85	13.80	9.10	4.80	3.10	2.60	2.20	2.05	1.25	1.25	1.30	1.25
24	2.85	12.20	10.20	4.75	3.00	2.60	2.20	1.90	1.20	1.30	1.30	1.30
25	2.65	8.20	11.00	4.65	2.90	2.65	2.30	1.75	1.20	1.25	1.35	1.40
26	2.70	5.80	11.65	4.55	2.90	2.80	2.20	1.65	1.15	1.20	1.30	1.35
27	2.70	5.40	11.90	4.50	2.90	2.70	2.10	1.50	1.15	1.15	1.30	1.30
28	2.90	8.40	11.55	4.70	2.65	2.10	2.10	1.40	1.20	1.05	1.40	1.30
29	3.40	......	9.40	4.40	2.60	2.80	2.20	1.40	1.30	1.00	1.35	1.30
30	3.60	......	10.00	4.30	3.00	5.10	2.15	1.35	1.20	1.05	1.25	1.20
31	3.90	......	12.35	......	3.90	......	2.20	1.30	......	1.10	......	1.25

a 15.30 maximum.   b 9.09 maximum.

Rating table for Coosa River at Riverside for 1896.

Gage height.	Discharge.	Gage height.	Discharge.	Gage height.	Discharge.	Gage height.	Discharge.
Feet.	*Second-feet.*	*Feet.*	*Second-feet.*	*Feet.*	*Second-feet.*	*Feet.*	*Second-feet.*
0.5	1,400	1.7	3,560	2.9	7,010	4.1	11,030
.6	1,500	1.8	3,820	3.0	7,320	4.2	11,390
.7	1,630	1.9	4,080	3.1	7,640	4.3	11,750
.8	1,780	2.0	4,360	3.2	7,970	4.4	12,110
.9	1,930	2.1	4,630	3.3	8,300	4.5	12,470
1.0	2,100	2.2	4,920	3.4	8,630	4.6	12,840
1.1	2,280	2.3	5,200	3.5	8,960	4.7	13,210
1.2	2,480	2.4	5,500	3.6	9,300	4.8	13,580
1.3	2,680	2.5	5,800	3.7	9,640	4.9	13,950
1.4	2,880	2.6	6,100	3.8	9,980	5.0	14,330
1.5	3,090	2.7	6,400	3.9	10,330	5.1	14,710
1.6	3,320	2.8	6,708	4.0	10,680	5.2	15,100

Rating table for Coosa River at Riverside for 1897.

Gage height.	Discharge.	Gage height.	Discharge.	Gage height.	Discharge.	Gage height.	Discharge.
Feet.	*Second-feet.*	*Feet.*	*Second-feet.*	*Feet.*	*Second-feet.*	*Feet.*	*Second-feet*
0.4	1,350	1.4	3,070	3.0	7,530	7.0	20,566
.5	1,400	1.6	3,540	3.2	8,178	8.0	23,826
.6	1,500	1.8	4,020	3.4	8,830	9.0	27,086
.7	1,650	2.0	4,520	3.6	9,482	10.0	30,346
.8	1,820	2.2	5,100	3.8	10,134	11.0	33,606
.9	2,010	2.4	5,700	4.0	10,786	12.0	36,866
1.0	2,210	2.6	6,300	5.0	14,046	13.0	40,126
1.2	2,630	2.8	6,910	6.0	17,306	14.0	43,386

Rating table for Coosa River at Riverside for 1898.

Gage height.	Discharge.	Gage height.	Discharge.	Gage height.	Discharge.	Gage height.	Discharge.
Feet.	*Second-feet.*	*Feet.*	*Second-feet.*	*Feet.*	*Second-feet.*	*Feet.*	*Second-feet.*
0.9	2,140	4.7	12,301	8.5	25,335	12.3	46,960
1.0	2,320	4.8	12,644	8.6	25,678	12.4	47,680
1.1	2,520	4.9	12,987	8.7	26,021	12.5	48,400
1.2	2,720	5.0	13,330	8.8	26,364	12.6	49,120
1.3	2,925	5.1	13,673	8.9	26,707	12.7	49,840
1.4	3,130	5.2	14,016	9.0	27,050	12.8	50,560
1.5	3,340	5.3	14,359	9.1	27,433	12.9	51,280
1.6	3,550	5.4	14,702	9.2	27,800	13.0	52,000
1.7	3,760	5.5	15,045	9.3	28,175	13.1	52,720
1.8	3,970	5.6	15,388	9.4	28,550	13.2	53,440
1.9	4,185	5.7	15,731	9.5	28,965	13.3	54,160
2.0	4,400	5.8	16,074	9.6	29,380	13.4	54,885
2.1	4,620	5.9	16,417	9.7	29,815	13.5	55,600
2.2	4,840	6.0	16,760	9.8	30,250	13.6	56,320
2.3	5,070	6.1	17,103	9.9	30,725	13.7	57,040
2.4	5,300	6.2	17,446	10.0	31,200	13.8	57,760
2.5	5,540	6.3	17,789	10.1	31,725	13.9	58,480
2.6	5,780	6.4	18,132	10.2	32,250	14.0	59,200
2.7	6,030	6.5	18,475	10.3	32,825	14.1	59,920
2.8	6,280	6.6	18,818	10.4	33,400	14.2	60,640
2.9	6,540	6.7	19,161	10.5	34,067	14.3	61,360
3.0	6,800	6.8	19,540	10.6	34,725	14.4	62,080
3.1	7,080	6.9	19,847	10.7	35,442	14.5	62,800
3.2	7,360	7.0	20,190	10.8	36,160	14.6	63,520
3.3	7,655	7.1	20,533	10.9	36,880	14.7	64,240
3.4	7,950	7.2	20,876	11.0	37,600	14.8	64,960
3.5	8,260	7.3	21,219	11.1	38,320	14.9	65,680
3.6	8,570	7.4	21,562	11.2	39,040	15.0	66,400
3.7	8,895	7.5	21,905	11.3	39,760	15.1	67,120
3.8	9,220	7.6	22,248	11.4	40,480	15.2	67,840
3.9	9,560	7.7	22,591	11.5	41,200	15.3	68,560
4.0	9,900	7.8	22,934	11.6	41,920	15.4	69,280
4.1	12,243	7.9	23,277	11.7	42,640	15.5	70,000
4.2	10,586	8.0	23,620	11.8	43,360	15.6	70,720
4.3	10,929	8.1	23,963	11.9	44,080	15.7	71,440
4.4	11,272	8.2	24,306	12.0	44,800	15.8	72,160
4.5	11,615	8.3	24,649	12.1	45,520	15.9	72,880
4.6	11,958	8.4	24,992	12.2	46,240	16.0	73,600

Rating table for Coosa River at Riverside for 1899.

Gage height.	Discharge.	Gage height.	Discharge.	Gage height.	Discharge.	Gage height.	Discharge.
Feet.	*Second-feet.*	*Feet.*	*Second-feet.*	*Feet.*	*Second-feet.*	*Feet.*	*Second-feet.*
0.9	2,330	5.2	14,740	9.5	30,650	13.8	46,580
1.0	2,460	5.3	15,110	9.6	31,020	13.9	46,930
1.1	2,600	5.4	15,480	9.7	31,390	14.0	47,300
1.2	2,760	5.5	15,850	9.8	31,760	14.1	47,670
1.3	2,920	5.6	16,220	9.9	32,130	14.2	48,040
1.4	3,100	5.7	16,590	10.0	32,500	14.3	48,410
1.5	3,300	5.8	16,960	10.1	32,870	14.4	48,780
1.6	3,500	5.9	17,330	10.2	33,240	14.5	49,150
1.7	3,720	6.0	17,700	10.3	33,610	14.6	49,520
1.8	3,940	6.1	18,070	10.4	33,980	14.7	49,890
1.9	4,160	6.2	18,440	10.5	34,350	14.8	50,260
2.0	4,400	6.3	18,810	10.6	34,720	14.9	50,630
2.1	4,600	6.4	19,180	10.7	35,090	15.0	51,000
2.2	4,900	6.5	19,550	10.8	35,460	15.1	51,370
2.3	5,160	6.6	19,920	10.9	35,830	15.2	51,740
2.4	5,430	6.7	20,290	11.0	36,200	15.3	52,110
2.5	5,700	6.8	20,660	11.1	36,570	15.4	52,480
2.6	5,970	6.9	21,030	11.2	36,940	15.5	52,850
2.7	6,250	7.0	21,400	11.3	37,310	15.6	53,220
2.8	6,530	7.1	21,770	11.4	37,680	15.7	53,590
2.9	6,810	7.2	22,140	11.5	38,050	15.8	53,960
3.0	7,100	7.3	22,510	11.6	38,420	15.9	54,330
3.1	7,400	7.4	22,880	11.7	38,790	16.0	54,700
3.2	7,700	7.5	23,250	11.8	39,160	16.1	55,070
3.3	8,010	7.6	23,620	11.9	39,530	16.2	55,440
3.4	8,330	7.7	23,990	12.0	39,900	16.3	55,810
3.5	8,650	7.8	24,360	12.1	40,270	16.4	56,280
3.6	8,970	7.9	24,730	12.2	40,640	16.5	56,650
3.7	9,290	8.0	25,100	12.3	41,010	16.6	57,020
3.8	9,620	8.1	25,470	12.4	41,380	16.7	57,390
3.9	9,950	8.2	25,840	12.5	41,750	16.8	57,760
4.0	10,300	8.3	26,210	12.6	42,120	16.9	58,130
4.1	10,670	8.4	26,580	12.7	42,490	17.0	58,400
4.2	11,040	8.5	26,950	12.8	42,860	17.1	58,770
4.3	11,410	8.6	27,320	12.9	43,230	17.2	59,140
4.4	11,780	8.7	27,690	13.0	43,600	17.3	59,510
4.5	12,150	8.8	28,060	13.1	43,970	17.4	59,880
4.6	12,520	8.9	28,430	13.2	44,340	17.5	60,250
4.7	12,890	9.0	28,800	13.3	44,710	17.6	60,620
4.8	13,260	9.1	29,170	13.4	45,080	17.7	60,990
4.9	13,630	9.2	29,540	13.5	45,450	17.8	61,360
5.0	14,000	9.3	29,910	13.6	45,720	17.9	61,730
5.1	14,370	9.4	30,280	13.7	46,190	18.0	62,100

Rating table for Coosa River at Riverside for 1900 and 1901.

Gage height.	Discharge.	Gage height.	Discharge.	Gage height.	Discharge.	Gage height.	Discharge.
Feet.	*Second-feet.*	*Feet.*	*Second-feet.*	*Feet.*	*Second-feet.*	*Feet.*	*Second-feet.*
1.0	2,460	5.6	15,900	10.2	33,900	14.8	52,300
1.1	2,610	5.7	16,250	10.3	34,300	14.9	52,700
1.2	2,760	5.8	16,600	10.4	34,700	15.0	53,100
1.3	2,930	5.9	16,950	10.5	35,100	15.1	53,500
1.4	3,100	6.0	17,300	10.6	35,500	15.2	53,900
1.5	3,300	6.1	17,680	10.7	35,900	15.3	54,300
1.6	3,500	6.2	18,060	10.8	36,300	15.4	54,700
1.7	3,720	6.3	18,440	10.9	36,700	15.5	55,100
1.8	3,940	6.4	18,820	11.0	37,100	15.6	55,500
1.9	4,170	6.5	19,200	11.1	37,500	15.7	55,900
2.0	4,400	6.6	19,580	11.2	37,900	15.8	56,300
2.1	4,650	6.7	19,960	11.3	38,300	15.9	56,700
2.2	4,900	6.8	20,340	11.4	38,700	16.0	57,100
2.3	5,165	6.9	20,720	11.5	39,100	16.1	57,500
2.4	5,430	7.0	21,100	11.6	39,500	16.2	57,900
2.5	5,700	7.1	21,500	11.7	39,900	16.3	58,300
2.6	5,970	7.2	21,900	11.8	40,300	16.4	58,700
2.7	6,250	7.3	22,300	11.9	40,700	16.5	59,100
2.8	6,530	7.4	22,700	12.0	41,100	16.6	59,500
2.9	6,845	7.5	23,100	12.1	41,500	16.7	59,900
3.0	7,100	7.6	23,500	12.2	41,900	16.8	60,300
3.1	7,400	7.7	23,900	12.3	42,300	16.9	60,700
3.2	7,700	7.8	24,300	12.4	42,700	17.0	61,100
3.3	8,015	7.9	24,700	12.5	43,100	17.1	61,500
3.4	8,330	8.0	25,100	12.6	43,500	17.2	61,900
3.5	8,650	8.1	25,500	12.7	43,900	17.3	62,300
3.6	8,970	8.2	25,900	12.8	44,300	17.4	62,700
3.7	9,295	8.3	26,300	12.9	44,700	17.5	63,100
3.8	9,620	8.4	26,700	13.0	45,100	17.6	63,500
3.9	9,960	8.5	27,100	13.1	45,500	17.7	63,900
4.0	10,300	8.6	27,500	13.2	45,900	17.8	64,300
4.1	10,650	8.7	27,900	13.3	46,300	17.9	64,700
4.2	11,000	8.8	28,300	13.4	46,700	18.0	65,100
4.3	11,350	8.9	28,700	13.5	47,100	18.1	65,500
4.4	11,700	9.0	29,100	13.6	47,500	18.2	65,900
4.5	12,050	9.1	29,500	13.7	47,900	18.3	66,300
4.6	12,400	9.2	29,900	13.8	48,300	18.4	66,700
4.7	12,750	9.3	30,300	13.9	48,700	18.5	67,100
4.8	13,100	9.4	30,700	14.0	49,100	18.6	67,500
4.9	13,450	9.5	31,100	14.1	49,500	18.7	67,900
5.0	13,800	9.6	31,500	14.2	49,900	18.8	68,300
5.1	14,150	9.7	31,900	14.3	50,300	18.9	68,700
5.2	14,500	9.8	32,300	14.4	50,700	19.0	69,100
5.3	14,850	9.9	32,700	14.5	51,100		
5.4	15,200	10.0	33,100	14.6	51,500		
5.5	15,550	10.1	33,500	14.7	51,900		

Rating table for Coosa River at Riverside for 1902.

Gage height.	Discharge.	Gage height.	Discharge.	Gage height.	Discharge.	Gage height.	Discharge.
Feet.	*Second-feet.*	*Feet.*	*Second-feet.*	*Feet.*	*Second-feet.*	*Feet.*	*Second-feet.*
1.0	2,470	4.0	10,300	7.0	21,100	13.0	45,100
1.2	2,760	4.2	11,000	7.2	21,900	13.5	47,100
1.4	3,100	4.4	11,700	7.4	22,700	14.0	49,100
1.6	3,500	4.6	12,400	7.6	23,500	14.5	51,100
1.8	3,940	4.8	13,100	7.8	24,300	15.0	53,100
2.0	4,400	5.0	13,800	8.0	25,100	15.5	55,100
2.2	4,900	5.2	14,500	8.5	27,100	16.0	57,100
2.4	5,430	5.4	15,200	9.0	29,100	16.5	59,100
2.6	5,970	5.6	15,900	9.5	31,100	17.0	61,100
2.8	6,530	5.8	16,600	10.0	33,100	17.5	63,100
3.0	7,100	6.0	17,300	10.5	35,100	18.0	65,100
3.2	7,700	6.2	18,060	11.0	37,100	18.5	67,100
3.4	8,330	6.4	18,820	11.5	39,100	19.0	69,100
3.6	8,970	6.6	19,580	12.0	41,100		
3.8	9,620	6.8	20,340	12.5	43,100		

Rating table for Coosa River at Riverside for 1903.

Gage height.	Discharge.	Gage height.	Discharge.	Gage height.	Discharge.	Gage height.	Discharge.
Feet.	*Second-feet.*	*Feet.*	*Second-feet.*	*Feet.*	*Second-feet.*	*Feet.*	*Second-feet.*
1.00	2,620	4.80	14,300	8.60	30,310	12.40	46,650
1.10	2,800	4.90	14,690	8.70	30,740	12.50	47,080
1.20	2,990	5.00	15,090	8.80	31,170	12.60	47,510
1.30	3,180	5.10	15,490	8.90	31,600	12.70	47,940
1.40	3,380	5.20	15,890	9.00	32,030	12.80	48,370
1.50	3,590	5.30	16,290	9.10	32,460	12.90	48,800
1.60	3,810	5.40	16,700	9.20	32,890	13.00	49,230
1.70	4,040	5.50	17,110	9.30	33,320	13.10	49,660
1.80	4,270	5.60	17,520	9.40	33,750	13.20	50,090
1.90	4,510	5.70	17,930	9.50	34,180	13.30	50,520
2.00	4,750	5.80	18,340	9.60	34,610	13.40	50,950
2.10	5,010	5.90	18,760	9.70	35,040	13.50	51,380
2.20	5,290	6.00	19,180	9.80	35,470	13.60	51,810
2.30	5,580	6.10	19,600	9.90	35,900	13.70	52,240
2.40	5,870	6.20	20,020	10.00	36,330	13.80	52,670
2.50	6,160	6.30	20,440	10.10	36,760	13.90	53,100
2.60	6,460	6.40	20,860	10.20	37,190	14.00	53,530
2.70	6,770	6.50	21,280	10.30	37,620	14.10	53,960
2.80	7,090	6.60	21,710	10.40	38,050	14.20	54,390
2.90	7,420	6.70	22,140	10.50	38,480	14.30	54,820
3.00	7,750	6.80	22,570	10.60	38,910	14.40	55,250
3.10	8,080	6.90	23,000	10.70	39,340	14.50	55,680
3.20	8,410	7.00	23,430	10.80	39,770	14.60	56,110
3.30	8,750	7.10	23,860	10.90	40,200	14.70	56,540
3.40	9,090	7.20	24,290	11.00	40,630	14.80	56,970
3.50	9,440	7.30	24,720	11.10	41,060	14.90	57,4~0
3.60	9,790	7.40	25,150	11.20	41,490	15.00	57,830
3.70	10,140	7.50	25,580	11.30	41,920	15.10	58,260
3.80	10,500	7.60	26,010	11.40	42,350	15.20	58,690
3.90	10,870	7.70	26,440	11.50	42,780	15.50	59,980
4.00	11,240	7.80	26,870	11.60	43,210	15.70	60,840
4.10	11,610	7.90	27,300	11.70	43,640	15.80	61,270
4.20	11,990	8.00	27,730	11.80	44,070	16.00	62,130
4.30	12,370	8.10	28,160	11.90	44,500	16.40	63,850
4.40	12,750	8.20	28,590	12.00	44,930	17.00	66,430
4.50	13,130	8.30	29,020	12.10	45,360	17.20	67,290
4.60	13,520	8.40	29,450	12.20	45,790	17.30	67,720
4.70	13,910	8.50	29,880	12.30	46,220		

Estimated monthly discharge of Coosa River at Riverside.

[Drainage area, 6,850 square miles.]

Month.	Discharge in second-feet.			Run-off.	
	Maximum.	Minimum.	Mean.	Second-feet per square mile.	Depth in inches.
1896.					
September 27 to 30	1,400	1,350	1,363	0.20	0.03
October	7,640	1,450	2,218	.32	.37
November	15,100	2,190	4,637	.68	.75
December	12,110	2,280	4,125	.60	.69
1897.					
January	21,707	2,420	8,434	1.23	1.42
February	27,086	4,520	18,658	2.72	2.83
March	47,624	10,460	32,481	4.74	5.47
April	37,355	5,100	17,698	2.58	2.87
May	10,786	4,270	7,040	1.03	1.19
June	4,950	3,070	3,915	.57	.63
July	23,826	3,070	7,142	1.04	1.20
August	6,300	2,850	3,870	.56	.64
September	3,540	1,440	1,976	.29	.32
October	3,660	1,350	1,819	.27	.31
November	2,525	1,570	1,786	.26	.29
December	13,883	1,820	6,566	.96	1.10
The year	47,624	1,350	9,282	1.35	18.27
1898.					
January	35,084	3,550	11,572	1.69	1.95
February	16,760	3,655	5,763	.84	.87
March	15,045	3,550	5,852	.59	.68
April	36,160	8,260	18,133	2.65	2.95
May	9,560	3,028	4,684	.68	.78
June	5,540	2,320	3,281	.48	.54
July	9,900	2,140	4,289	.63	.72
August	19,161	3,970	8,758	1.28	1.48
September	41,920	3,865	13,927	2.03	2.26
October	72,160	3,760	19,936	2.91	3.36
November	20,190	5,300	8,375	1.22	1.36
December	9,900	5,070	7,376	1.08	1.25
The year	72,160	2,140	9,329	1.34	18.20

Estimated monthly discharge of Coosa River at Riverside—Continued.

Month.	Discharge in second-feet.			Run-off.	
	Maximum.	Minimum.	Mean.	Second-feet per square mile.	Depth in inches.
1899.					
January	17,330	6,390	10,865	1.54	1.78
February	48,410	17,330	30,974	4.38	4.56
March	60,880	14,740	38,094	5.39	6.21
April	41,010	11,410	24,915	3.53	3.94
May	14,000	4,900	7,742	1.10	1.27
June	7,700	3,500	4,771	.68	.75
July	14,740	2,760	5,318	.75	.86
August	10,125	2,600	3,806	.54	.62
September	6,530	2,330	3,555	.50	.56
October	3,100	2,330	2,510	.36	.41
November	7,100	2,395	3,086	.44	.49
December	26,025	2,920	10,631	1.50	1.73
The year	60,880	2,330	12,189	1.73	23.18
1900.					
January	33,100	4,280	13,344	1.89	2.18
February	54,300	5,700	23,487	3.32	3.45
March	44,500	13,450	26,822	3.80	4.38
April	65,500	11,350	29,813	4.22	4.71
May	14,850	5,970	8,198	1.16	1.34
June	51,500	5,970	22,216	3.14	3.51
July	39,500	5,565	13,610	1.93	2.23
August	12,750	4,050	5,147	.73	.84
September	23,100	2,760	6,483	.92	1.03
October	23,100	3,100	6,910	.98	1.13
November	29,900	3,720	7,673	1.09	1.22
December	22,300	5,835	11,773	1.67	1.93
The year	65,500	2,760	14,623	2.07	27.95

Estimated monthly discharge of Coosa River at Riverside—Continued.

Month.	Discharge in second-feet.			Run-off.	
	Maximum.	Minimum.	Mean.	Second-feet per square mile.	Depth in inches.
1901.					
January	55,900	8,970	26,089	3.69	4.25
February	41,100	8,970	21,784	3.08	3.21
March	56,700	7,400	20,613	2.92	3.37
April	51,100	a 14,500	30,616	4.33	4.83
May	40,700	6,670	16,195	2.29	2.64
June	26,100	6,810	12,335	1.75	1.95
July	10,300	4,900	6,535	.93	1.07
August	44,700	4,400	20,370	2.88	3.32
September	23,100	5,700	9,977	1.41	1.57
October	10,300	4,280	5,694	.81	.93
November	4,650	3,830	4,016	.57	.64
December	57,100	4,050	18,885	2.67	3.08
The year	57,100	3,830	16,092	2.28	30.86
1902.					
January	55,900	6,670	23,804	3.37	3.89
February	51,500	12,050	24,839	3.52	3.67
March	62,300	14,850	34,762	4.92	5.67
April	55,100	8,330	20,872	2.95	3.29
May	8,330	4,900	6,375	.90	1.04
June	4,900	3,830	4,247	.60	.67
July	6,670	3,015	3,718	.53	.61
August	5,430	2,760	3,577	.51	.59
September	10,300	2,610	3,938	.56	.62
October	10,300	2,760	4,576	.65	.75
November	9,950	2,685	3,994	.57	.64
December	24,300	5,430	12,719	1.80	2.08
The year	62,300	2,610	12,285	1.74	23.52

a Should have been 13,100.

Estimated monthly discharge of Coosa River at Riverside—Continued.

Month.	Discharge in second-feet.			Run-off.	
	Maximum.	Minimum.	Mean.	Second-feet per square mile.	Depth in inches.
1903.					
January	2,170	6,610	12,066	1.71	1.97
February	67,500	10,680	43,155	6.11	6.36
March	58,470	17,930	40,682	5.76	6.64
April	52,450	12,370	28,983	4.10	4.57
May	30,740	6,460	11,294	1.60	1.84
June	36,540	6,460	15,654	2.22	2.48
July	21,710	5,010	7,994	1.13	1.30
August	14,890	3,180	5,910	.84	.97
September	4,880	2,800	3,211	.45	.50
October	3,590	2,620	2,922	.41	.47
November	4,510	2,800	3,334	.47	.52
December	3,380	2,710	2,983	.42	.48
The year	67,500	2,620	14,849	2.10	28.10

Net horsepower per foot of fall with a turbine efficiency of 80 per cent for the minimum monthly discharge of Coosa River at Riverside.

Month.	1899.			1900.			1901.		
	Minimum discharge.	Minimum net horsepower per foot of fall.	Duration of minimum.	Minimum discharge.	Minimum net horsepower per foot of fall.	Duration of minimum.	Minimum discharge.	Minimum net horsepower per foot of fall.	Duration of minimum.
	Sec-feet.		Days	Sec.-feet.		Days.	Sec.-feet.		Days.
January	6,390	581	2	4,285	390	2	8,970	815	1
February	17,330	1,575	1	5,700	518	1	8,970	815	1
March	14,740	1,340	1	13,450	1,223	1	7,400	673	2
April	11,410	1,037	1	11,350	1,032	1	14,500	1,310	2
May	4,900	445	1	5,970	543	5	6,687	608	1
June	3,500	318	3	5,970	543	1	6,810	619	2
July	2,760	251	1	5,565	506	1	4,900	445	3
August	2,600	236	1	4,050	369	1	4,400	400	1
September	2,330	212	2	2,760	251	1	5,700	518	1
October	2,330	212	11	3,100	282	1	4,285	390	1
November	2,395	218	17	3,720	338	1	3,830	348	6
December	2,920	265	2	5,835	530	2	4,050	368	1

COOSA RIVER AT ROME, GA.

The measurements at Rome are made on the Oostanaula and Etowah just above their junction. Etowah River is measured at Second Avenue Bridge and the Oostanaula at Fifth Avenue Bridge in Rome, and the result added to give the flow of Coosa River. The gage height is taken from the United States Weather Bureau gage at Fifth Avenue Bridge, on the Oostanaula. There is practically no fall on Oostanaula River from Fifth Avenue Bridge to the junction, hence the gage is used as Coosa River gage and gives the fluctuations of Coosa River. This gage is a 4 by 6 inch timber, graduated to feet and tenths and fastened to the downstream left-hand corner of the first pier from the left bank. The zero of the gage is 575.79 feet above sea level. The United States Weather Bureau has maintained the station here for many years. It is now maintained only as a half-year station, from November 1 to April 30, inclusive, but W. M. Towers, the river observer, kindly reads the gage and furnishes the Survey with monthly reports of the daily gage heights for the entire year without charge. Mr. Towers has kept the records for many years and has predicted floods with great precision. The channel of the Etowah is straight, current swift and unobstructed, but the Oostanaula is rather sluggish and somewhat obstructed by piers. The banks are high, but liable to overflow in times of high water.

The following discharge measurements were made by M. R. Hall and others:

Discharge measurements of Coosa River at Rome, Ga.

Date.	Gage height.	Discharge.	Date.	Gage height.	Discharge.
1896.	*Feet.*	*Second-feet.*	**1900.**	*Feet.*	*Second-feet.*
September 29........	0.20	1,209	September 13	0.90	1,992
1897.			December 8	3.73	6,066
May 7	2.75	4,646	**1901.**		
October 5	.15	990	January 23.........	3.60	6,454
1898.			April 5	9.90	16,692
May 11	1.90	2,946	June 22...........	3.70	6,030
September 17	2.60	3,913	October 15	3.15	5,388
October 11	5.05	8,324			
October 22	4.10	6,489	**1902.**		
November 30........	3.90	6,039	June 24............	1.30	2,483
1899.			October 8	.80	1,800
January 25..........	3.80	6,540	November 8........	1.10	2,332
Do	3.60	5,932	**1903.**		
May 19	2.75	4,394	March 14	9.70	16,146
June 16.............	2.40	3,352	June 5.............	12.55	25,008
August 4............	1.45	2,835	July 1	2.80	5,305
October 13	.60	1,769	July 3	2.70	4,653
1900.			July 18	2.85	4,403
February 21........	4.80	8,115	September 4	.90	2,211
May 19	2.30	4,496	November 28.......	.75	1,892

Daily gage height, in feet, of Coosa River at Rome, Ga.

Day.	Jan.	Feb.	Mar.	Apr.	May.	June.	July.	Aug.	Sept.	Oct.	Nov	Dec.
1897.												
1	1.0	2.8	3.3	7.1	4.1	1.8	1.7	0.8	1.0	0.0	0.5	1.1
2	1.0	9.7	3.2	7.5	4.0	2.3	1.9	.7	.5	.0	.9	1.0
3	1.0	11.5	3.1	8.2	3.5	2.0	1.0	1.2	.3	.0	1.0	1.2
4	1.0	9.6	3.3	9.4	8.3	3.0	.9	1.0	.5	.1	1.0	2.3
5	1.0	8.2	3.5	14.8	3.0	2.4	2.0	.8	.4	.1	1.0	3.2
6	1.3	5.2	7.6	18.9	3.0	2.0	1.9	.8	.3	.1	.8	3.7
7	1.1	5.0	19.7	17.0	3.0	2.0	1.9	1.9	.3	.1	.8	3.2
8	1.1	4.3	18.9	14.7	2.8	2.0	3.0	2.0	.2	.1	.8	2.2
9	1.0	5.0	15.4	12.1	2.6	2.0	2.1	2.0	.1	.1	.8	1.9
10	1.0	4.4	13.5	9.6	2.6	1.9	1.9	1.6	.0	.1	.7	1.7
11	.9	4.5	12.0	7.2	2.6	1.9	2.5	2.4	.0	.1	.7	1.5
12	.9	7.4	11.5	6.2	3.0	1.9	2.8	1.8	.0	1.1	.7	1.4
13	.9	8.7	18.6	5.8	3.4	1.8	2.0	1.3	.0	1.6	.7	1.3
14	2.8	7.2	21.3	5.0	4.0	1.7	1.6	.8	.0	1.3	.6	2.2
15	6.2	5.5	23.8	6.0	5.0	1.7	1.3	.6	.0	1.0	.6	4.0
16	5.0	4.5	23.4	7.4	4.0	2.0	1.0	.6	.0	.8	.6	3.5
17	3.5	4.0	22.6	7.0	3.3	2.8	5.2	2.1	.0	.7	.6	2.5
18	3.9	3.7	21.4	5.0	2.8	2.3	4.2	3.2	.1	.6	.6	2.2
19	5.0	3.4	19.7	4.5	2.7	2.0	4.8	2.4	.2	.6	.6	1.8
20	3.5	3.0	18.9	4.0	2.6	1.8	8.8	1.4	.2	.6	.6	1.7
21	8.7	4.0	17.7	3.8	2.5	1.6	12.8	1.3	.2	1.5	.6	3.2
22	9.5	3.9	15.3	3.7	2.4	1.5	7.3	1.5	.2	1.3	.5	4.1
23	5.7	5.6	13.7	3.5	2.4	1.5	4.4	1.5	.2	1.0	.5	5.8
24	4.0	11.7	12.9	3.5	2.4	1.4	3.9	1.5	.2	.8	.5	5.3
25	3.5	8.6	9.1	3.5	2.3	1.3	2.6	1.1	.3	.8	.5	3.7
26	3.0	6.7	6.0	3.5	2.2	1.2	2.6	.8	.3	.7	.5	2.8
27	2.5	4.7	5.2	3.4	2.1	1.2	3.8	.5	.4	.7	.5	3.0
28	2.5	3.5	4.8	3.4	2.0	1.0	3.0	.4	.4	.7	.9	2.8
29	2.5	......	4.5	3.4	2.0	1.1	2.4	.4	.4	.6	1.1	2.3
30	2.8	......	4.2	3.2	1.9	2.0	1.4	.4	.4	.5	1.1	2.0
31	2.2	......	4.0	......	1.9	......	1.2	.5	......	.5	......	2.0
1898.												
1	1.8	3.6	1.2	9.0	2.8	1.4	1.2	4.8	2.0	2.0	2.2	4.2
2	1.8	3.1	1.2	6.1	2.6	1.4	1.0	4.4	7.8	2.0	2.2	4.0
3	1.7	2.8	1.2	4.2	2.4	1.4	1.0	3.2	21.7	2.0	2.2	3.8
4	1.7	2.6	1.2	3.6	2.3	1.4	1.0	4.4	24.3	4.9	2.2	3.8
5	1.6	2.4	1.2	9.9	2.2	1.3	1.0	8.0	22.2	22.0	2.0	4.3
6	1.6	2.2	1.2	17.2	2.1	1.3	1.3	5.6	20.0	23.8	2.2	5.0
7	1.3	2.0	1.2	14.5	2.0	1.3	2.0	4.4	17.6	19.0	2.6	4.3
8	1.3	1.8	1.2	10.9	2.0	1.3	2.8	4.4	16.4	18.4	2.4	4.0
9	1.3	1.8	1.2	7.0	2.0	1.3	3.2	3.4	9.7	16.6	2.3	3.7
10	1.3	1.7	1.2	4.1	2.0	1.3	1.7	3.0	5.0	14.0	2.1	3.4
11	1.4	1.5	1.2	4.0	2.0	1.8	2.8	9.9	5.4	5.6	2.0	3.3
12	2.0	1.5	1.2	3.8	1.9	1.2	2.0	7.2	4.6	4.2	2.0	3.3
13	4.0	1.3	1.2	3.6	1.8	1.4	1.8	4.2	3.8	3.8	2.0	3.2
14	4.0	1.3	1.3	3.5	1.8	1.8	1.6	8.4	3.2	3.7	2.3	3.0
15	3.8	1.3	1.6	3.5	1.7	1.8	3.7	3.0	3.0	3.5	2.3	3.0
16	3.6	1.3	3.7	3.4	1.6	1.7	3.7	2.5	2.9	3.2	2.9	2.8
17	3.6	1.2	7.3	3.0	1.5	1.8	2.2	2.0	2.7	3.1	2.9	2.7
18	3.2	1.2	5.8	3.0	1.5	1.8	1.9	2.2	2.5	6.5	4.0	2.6
19	2.8	1.2	3.7	8.0	1.5	2.2	1.7	2.2	2.3	9.0	5.0	2.6
20	4.4	1.2	8.0	3.6	1.4	3.6	1.6	3.2	2.2	6.0	4.5	2.6
21	6.5	1.2	2.5	3.6	1.4	3.2	1.5	2.8	2.2	4.2	5.0	2.8
22	6.4	1.2	2.5	8.2	1.4	3.0	1.4	3.9	2.3	3.9	4.0	2.9
23	5.0	1.2	2.3	3.0	1.4	2.8	1.3	2.2	2.6	4.0	5.0	3.2
24	4.5	1.2	2.2	7.2	1.4	2.6	1.8	2.2	4.1	5.9	7.0	3.6
25	7.0	1.2	2.1	8.2	1.4	2.0	3.7	1.9	3.1	3.5	4.7	3.0
26	14.0	1.2	2.0	6.0	1.4	1.8	3.8	2.7	3.0	3.3	3.9	2.9
27	14.6	1.2	1.9	4.6	1.4	1.8	2.9	4.0	2.7	3.1	4.5	2.7
28	11.6	1.2	1.8	4.0	1.4	1.8	3.7	4.4	2.5	8.0	4.3	2.6
29	8.6	......	2.0	3.7	1.4	1.6	4.2	3.4	2.3	2.8	4.3	2.5
30	4.6	......	8.5	3.2	1.4	1.4	4.1	2.0	2.1	2.6	3.9	2.4
31	3.9	......	11.4	......	1.4	......	4.2	2.8	......	2.4	......	2.4

Daily gage height, in feet, of Coosa River at Rome, Ga.—Continued.

Day.	Jan.	Feb.	Mar.	Apr.	May	June.	July.	Aug.	Sept.	Oct.	Nov.	Dec.
1899.												
1	3.0	6.9	19.7	13.2	4.0	3.0	1.7	2.2	3.4	.4	.7	1.1
2	3.4	7.8	15.0	10.6	3.7	2.6	1.5	1.9	2.0	.3	.6	1.1
3	3.0	6.0	8.6	7.9	3.7	2.6	1.0	1.7	1.6	.3	.5	1.5
4	2.7	9.2	6.6	7.2	3.5	2.0	.9	1.5	1.4	.3	.5	1.3
5	2.6	15.3	7.8	9.5	3.5	2.0	2.0	1.4	1.3	.3	.1	1.1
6	2.6	18.2	9.0	8.2	3.5	2.0	1.9	1.5	1.3	.5	.4	1.0
7	3.6	27.8	8.0	8.2	3.7	2.0	1.9	1.6	1.2	.7	.3	
8	5.9	24.0	6.8	15.0	3.7	1.9	3.0	1.5	1.0	.7	.3	.8
9	5.9	22.4	5.7	13.4	3.6	1.8	2.1	1.8	1.0	.8	.3	.8
10	4.9	21.0	5.4	11.2	3.5	1.8	1.9	1.6	1.0	.6	.3	.8
11	4.0	19.0	5.2	9.5	3.3	1.8	2.5	1.4	2.9	1.0	.3	.8
12	4.5	16.5	4.9	7.0	3.1	2.2	2.8	1.4	2.3	.9	.3	2.8
13	4.0	7.0	4.5	6.4	3.1	3.8	2.0	1.2	1.5	.7	.3	6.1
14	3.8	5.0	6.0	5.9	3.0	4.0	1.6	1.1	1.0	.7	.4	5.0
15	3.6	5.0	16.6	5.6	3.0	3.5	1.3	1.3	.9	.6	.4	3.2
16	3.6	5.5	27.7	5.4	2.9	2.5	1.8	1.9	.8	.6	.5	2.0
17	4.0	8.9	29.2	5.2	2.8	2.1	5.2	1.6	.6	.6	.9	1.8
18	4.2	9.5	25.8	4.8	2.8	2.0	4.2	1.4	.6	.6	.7	1.7
19	4.0	8.5	24.9	4.7	2.8	2.0	4.8	1.1	.6	.6	.5	1.3
20	3.7	7.7	26.2	4.6	2.8	2.0	8.8	.9	.7	.6	.5	1.6
21	3.3	6.8	24.6	4.3	2.6	1.8	12.8	.9	.7	.7	.5	2.0
22	3.2	6.9	23.0	4.1	2.6	2.2	7.9	.8	.6	.7	.4	2.0
23	3.1	7.3	22.6	4.0	2.4	1.7	4.8	.8	.6	.6	1.0	1.8
24	3.5	6.6	21.9	5.4	2.6	1.7	3.9	.7	.5	.5	2.1	7.2
25	3.8	5.8	18.0	7.4	2.5	1.7	2.6	.7	.5	.4	1.5	7.5
26	3.8	5.5	10.5	9.1	2.4	1.7	2.6	.7	.5	.4	2.5	5.0
27	3.3	19.1	7.7	6.7	2.2	2.1	3.8	2.5	.5	.4	3.0	3.5
28	3.0	23.4	6.8	5.5	2.2	1.9	3.0	2.5	.6	.4	2.2	3.0
29	3.0	......	8.8	4.8	2.0	1.9	2.4	2.5	.5	.4	1.9	3.0
30	2.9		9.3	4.2	2.0	1.8	1.4	2.0	.4	.5	1.4	3.4
31	4.4		10.2	......	3.3	......	1.2	2.5		.8	......	2.0
1900.												
1	2.0	2.0	4.2	4.4	6.2	2.4	10.5	3.4	1.5	1.2	2.1	3.2
2	1.6	1.8	5.8	4.2	4.8	2.5	8.0	8.2	1.5	1.2	2.2	2.8
3	1.5	1.6	5.6	4.0	4.0	2.8	8.0	3.0	1.7	1.0	2.0	2.6
4	1.5	2.0	4.4	4.0	4.0	4.2	7.0	2.8	1.7	1.0	2.3	3.5
5	1.5	3.0	4.1	4.0	3.8	4.2	5.5	2.6	1.5	.9	2.3	7.4
6	1.5	3.8	3.8	4.0	3.7	4.2	4.2	2.5	1.5	.9	2.1	6.8
7	1.5	2.8	5.0	3.8	3.6	4.8	3.8	2.2	1.4	.9	2.1	5.2
8	1.5	2.4	8.2	3.8	3.4	13.0	3.8	2.2	1.0	3.8	2.0	3.8
9	1.5	4.0	15.0	3.6	3.0	12.6	4.0	2.0	1.0	5.9	1.9	3.6
10	1.5	6.9	13.4	3.5	3.0	8.0	4.9	2.0	.8	2.6	1.8	3.2
11	2.0	7.0	10.3	6.0	3.0	5.9	3.8	1.8	.8	2.0	1.8	2.8
12	7.0	6.4	7.5	11.0	3.0	5.0	3.4	1.8	.8	1.8	1.6	2.6
13	9.0	22.6	5.5	7.4	2.5	5.2	3.8	1.8	.8	2.5	1.5	2.6
14	7.2	27.2	4.8	5.5	2.4	5.3	3.4	1.7	.8	3.2	1.5	2.4
15	5.5	25.3	4.2	4.5	2.4	4.2	3.4	2.0	6.5	3.0	1.5	2.2
16	3.5	21.2	5.3	5.6	2.4	3.8	3.3	1.7	11.1	2.0	1.5	2.2
17	3.0	18.0	5.6	6.2	2.4	4.8	3.1	1.6	7.0	1.6	1.5	2.2
18	2.9	10.7	4.5	11.0	2.4	6.0	3.0	1.8	3.2	1.5	1.4	2.0
19	5.0	5.0	5.2	11.1	2.9	6.5	2.8	2.2	2.3	1.5	1.4	2.0
20	11.3	4.0	15.9	11.4	3.0	7.2	2.6	2.0	2.0	1.4	1.6	2.8
21	10.6	4.1	17.5	13.6	2.6	4.2	2.5	1.6	1.8	1.8	1.8	6.7
22	8.5	6.8	14.6	12.7	2.5	3.6	2.4	1.6	1.8	1.8	2.1	8.0
23	5.8	7.6	10.4	10.5	2.3	5.5	2.4	1.6	1.6	1.6	2.1	7.0
24	4.0	6.0	7.2	8.6	2.9	14.2	2.4	1.9	1.6	1.6	2.0	6.6
25	3.4	5.8	8.8	8.5	3.2	18.2	3.6	2.4	1.5	1.5	5.0	6.6
26	3.1	5.2	13.0	6.5	2.7	17.0	2.8	2.0	1.5	1.5	11.0	5.6
27	2.8	4.6	12.1	5.3	2.6	15.5	6.2	1.8	1.4	1.4	11.5	4.0
28	2.6	4.0	8.9	4.8	2.5	15.6	6.8	1.6	1.4	2.2	8.6	3.8
29	2.4	......	5.8	4.3	2.4	14.2	6.2	1.5	1.3	2.2	7.0	3.6
30	2.1		5.7	6.0	2.9	10.0	4.5	1.5	1.3	2.1	4.0	3.5
31	2.0		5.3	......	3.0	......	4.0	1.5		2.1		5.6

Daily gage height, in feet, of Coosa River at Rome, Ga.—Continued.

Day.	Jan.	Feb.	Mar.	Apr.	May.	June.	July.	Aug.	Sept.	Oct.	Nov.	Dec.
1901.												
1	7.4	6.4	3.0	8.8	4.0	10.6	3.6	1.8	6.4	2.6	1.2	1.3
2	6.4	5.8	8.0	8.6	3.8	7.6	3.0	1.8	5.8	2.6	1.2	1.3
3	5.2	5.5	3.0	13.0	8.8	5.6	3.0	1.8	3.2	3.2	1.2	1.3
4	4.2	15.8	3.0	13.0	8.8	6.4	2.6	1.6	3.4	3.0	1.2	1.5
5	4.0	18.5	3.0	10.0	8.6	5.0	2.4	1.6	3.0	2.8	1.2	2.0
6	3.8	13.8	3.0	7.9	3.5	4.0	2.2	2.6	2.9	2.2	1.2	1.8
7	3.5	9.5	3.0	6.4	3.5	7.0	5.2	5.3	2.6	2.0	1.2	1.8
8	3.2	6.5	3.0	5.6	3.4	7.6	4.8	5.9	2.2	1.9	1.1	1.8
9	3.0	9.6	2.8	5.2	3.3	5.4	3.3	3.0	2.0	1.9	1.1	1.8
10	2.8	12.5	5.5	4.5	3.1	4.3	2.6	2.6	2.0	1.8	1.1	1.8
11	8.8	10.5	7.8	4.3	8.0	4.0	2.4	2.5	2.0	1.7	1.0	2.6
12	23.5	7.6	8.0	4.2	2.9	3.8	2.3	8.4	2.0	1.6	1.0	2.6
13	27.0	6.5	6.7	4.3	2.8	3.8	2.0	3.0	1.8	1.8	1.0	2.1
14	23.8	5.6	4.8	10.4	2.8	4.0	2.0	2.3	2.0	2.0	1.0	2.2
15	21.4	5.0	4.0	10.1	2.7	4.3	2.0	4.5	3.0	3.2	1.0	16.4
16	19.8	4.8	3.6	7.7	2.6	6.9	1.9	7.2	2.4	2.6	1.0	17.6
17	17.4	4.2	3.2	5.8	2.6	6.0	1.7	10.5	2.4	2.4	1.0	14.7
18	8.9	4.2	3.0	5.2	2.5	5.0	5.5	9.8	11.2	2.4	1.0	14.0
19	5.0	4.2	3.0	9.0	2.5	4.8	3.0	10.8	11.1	2.0	1.0	13.0
20	4.0	4.0	3.0	18.6	3.0	4.0	3.0	12.5	7.0	1.8	1.0	5.6
21	3.8	3.8	3.0	17.2	10.0	3.8	2.4	10.8	8.9	1.8	1.6	3.0
22	3.8	3.7	3.0	15.5	23.6	3.6	2.4	14.5	8.7	1.6	1.2	2.0
23	3.8	3.6	3.0	14.6	26.4	3.6	2.4	20.8	8.3	1.6	1.4	2.0
24	3.8	3.6	3.6	12.7	21.8	3.6	2.0	23.2	2.8	1.6	1.4	3.6
25	6.7	3.5	3.6	6.8	18.9	2.7	2.0	18.3	2.6	1.6	1.3	4.0
26	6.6	3.2	22.0	5.6	16.5	2.7	1.9	13.1	2.5	1.4	1.3	8.7
27	5.4	8.2	27.0	4.8	11.1	3.2	1.7	6.6	2.3	1.3	1.3	5.7
28	5.2	3.0	24.5	4.4	5.5	3.0	2.8	8.8	2.0	1.3	1.3	6.0
29	5.0		21.3	4.2	4.9	3.6	1.9	7.5	2.0	1.3	1.3	21.5
30	4.6		19.2	4.1	4.7	3.6	1.9	6.2	2.5	1.3	1.3	29.8
31	6.3		16.1		5.4		1.6	5.6		1.2		32.6
1902.												
1	28.00	11.80	28.50	21.30	2.90	1.80	1.00	1.20	1.00	1.60	.20	1.90
2	24.60	22.00	27.60	14.80	2.90	1.70	1.00	1.00	.80	1.60	.20	2.20
3	21.90	24.00	24.10	7.20	3.70	1.70	1.10	.80	1.20	1.60	.20	6.20
4	17.60	20.00	21.60	5.80	3.40	1.70	1.00	.80	1.00	1.50	.20	6.80
5	6.60	15.60	19.20	5.70	8.20	1.70	.90	1.50	1.00	1.00	.20	5.80
6	5.60	9.80	14.00	5.20	3.00	1.70	.90	1.40	1.00	2.00	.40	5.00
7	4.60	6.30	10.10	6.70	2.90	1.70	1.00	1.20	.80	1.40	2.00	4.00
8	4.20	5.70	8.70	6.70	2.80	1.70	.90	1.20	.60	.90	1.60	2.70
9	4.00	5.00	7.20	7.70	2.60	2.00	.90	.60	.90	.80	1.00	2.50
10	3.90	4.60	7.00	6.60	2.60	1.90	.90	.40	1.00	.70	.80	1.90
11	3.60	4.40	6.60	5.60	2.50	1.90	.90	.30	.90	2.30	.80	1.60
12	3.50	4.00	6.00	5.00	2.40	1.80	1.40	.30	.70	2.20	.70	1.40
13	3.40	3.90	5.60	4.70	2.30	1.70	1.50	1.00	.60	2.20	.60	1.20
14	3.20	3.80	6.50	4.50	2.30	1.70	1.80	.60	2.50	2.00	.60	1.20
15	3.00	3.80	5.00	4.40	2.30	1.60	1.40	.60	1.90	1.80	.60	1.10
16	2.80	4.00	7.00	4.30	2.40	1.60	2.00	.80	1.00	1.60	.60	2.40
17	2.70	4.50	14.00	4.50	2.50	1.60	1.40	.60	.70	1.50	.60	4.30
18	2.70	4.70	11.60	5.70	2.30	1.50	1.20	.50	.40	1.30	.90	4.50
19	3.00	4.50	8.60	5.00	2.30	1.50	1.00	.40	.40	1.10	1.50	3.80
20	3.20	4.00	6.20	4.00	2.40	1.90	1.00	1.00	.40	.90	1.30	3.00
21	3.60	4.00	5.60	3.90	2.40	1.60	.90	1.30	1.00	.80	1.10	4.90
22	4.20	5.00	5.10	3.80	2.80	1.40	.80	1.00	.80	.70	1.10	5.80
23	5.00	5.60	5.00	3.70	2.30	1.50	.80	.90	.60	.60	1.00	5.80
24	4.00	5.30	4.70	3.50	2.10	1.30	.70	.70	.30	.50	.90	5.20
25	3.90	5.00	4.70	3.30	2.00	1.30	.60	.50	2.20	.40	1.60	3.80
26	3.60	5.50	4.60	3.20	2.00	1.20	.50	.50	4.00	.30	3.10	3.00
27	3.60	5.50	4.20	3.20	2.00	1.20	.50	.60	3.30	.20	4.80	2.60
28	4.20	22.70	4.20	3.00	1.80	1.10	.50	1.20	3.30	.40	3.80	2.20
29	6.50		20.80	3.00	1.80	1.00	.70	1.70	2.00	.40	2.40	2.20
30	5.80		28.90	2.90	1.80	1.00	2.00	1.00	1.80	.30	2.20	2.60
31	5.40		26.80		1.80		1.60	1.00		.80		4.00

Daily gage height, in feet, of Coosa River at Rome, Ga.—Continued.

Day.	Jan.	Feb.	Mar.	Apr.	May.	June.	July.	Aug.	Sept.	Oct.	Nov.	Dec.
1903												
1	3.0	2.8	28.6	24.9	3.6	6.0	3.0	2.8	1.0	.4	.7	.9
2	3.9	2.8	27.1	22.0	3.6	7.7	3.2	2.6	.9	.4	.7	.9
3	3.9	4.2	24.0	19.5	3.6	9.0	2.9	3.0	.9	.4	.8	.8
4	5.0	8.4	22.3	14.6	3.5	7.5	2.7	3.3	.9	.3	1.5	.7
5	4.3	13.2	20.5	8.0	3.5	11.7	2.6	3.2	.8	.3	1.5	.7
6	4.0	13.4	15.4	7.0	3.3	17.1	2.6	6.2	.7	.3	2.0	.7
7	3.7	9.7	9.9	6.4	3.4	11.7	2.6	4.6	.7	.2	2.0	.7
8	3.0	18.7	7.7	7.7	3.5	6.9	2.4	3.6	.6	.5	1.6	.7
9	2.8	21.6	6.6	10.5	3.5	5.3	2.6	3.4	.6	1.4	1.4	.7
10	2.4	16.5	10.5	8.4	3.4	4.4	2.6	3.0	.6	1.6	1.3	.7
11	3.0	15.1	11.6	6.7	3.3	4.8	2.5	2.6	.6	1.5	1.2	.7
12	5.0	21.8	16.0	6.3	3.3	6.0	2.9	2.2	.6	1.4	1.2	.6
13	5.0	19.6	14.0	5.6	3.0	4.6	5.9	2.0	.6	1.3	1.2	.6
14	4.8	14.5	10.1	14.3	3.0	4.2	11.1	2.0	.6	1.1	1.2	.6
15	3.9	11.1	9.4	12.0	3.1	4.0	6.7	1.6	.6	.9	1.0	.6
16	2.5	8.7	7.8	9.5	3.3	3.8	3.7	4.3	.9	.8	1.0	.6
17	2.4	24.7	6.8	7.4	3.0	3.4	3.5	3.2	1.7	1.2	1.0	.6
18	2.4	28.7	6.3	6.2	3.0	3.0	3.3	2.8	2.0	1.3	1.0	.6
19	2.4	25.5	5.7	5.3	2.8	2.4	3.0	2.2	1.5	1.2	1.5	.6
20	2.4	21.0	5.3	5.7	2.5	2.7	2.9	2.1	1.2	1.0	1.4	.6
21	2.2	15.2	11.0	5.8	2.5	2.5	2.5	2.0	.9	.8	1.3	.7
22	2.0	7.1	11.6	5.5	2.6	2.4	2.4	2.0	.8	.8	1.2	1.3
23	2.0	5.8	16.9	4.9	2.6	2.2	3.0	1.9	.7	.7	1.1	1.3
24	2.0	5.1	22.6	4.7	2.5	2.2	3.0	1.8	.5	.7	1.0	1.0
25	2.0	4.8	20.6	4.4	2.5	2.2	2.6	1.6	.5	.7	1.0	.9
26	2.0	4.5	16.0	4.9	2.3	2.2	2.4	1.5	.5	.6	1.0	.9
27	2.0	4.0	9.7	4.7	2.1	2.4	2.1	1.4	.5	.6	1.0	.9
28	2.4	23.1	6.9	4.0	2.0	6.8	2.0	1.4	.5	.5	.9	.9
29	3.2	....	7.8	3.9	2.0	6.8	1.8	1.3	.4	.4	.9	.9
30	4.2	....	22.5	3.7	2.4	4.0	2.3	1.3	.4	.4	.9	.9
31	3.8	....	27.6	....	6.7	....	2.6	1.2	....	.7	....	.9

Rating table for Coosa River at Rome, Ga., for 1897 and 1898.

Gage height.	Discharge.	Gage height.	Discharge.	Gage height.	Discharge.	Gage height.	Discharge.
Feet.	*Second-feet.*	*Feet.*	*Second-feet.*	*Feet.*	*Second-feet.*	*Feet.*	*Second-feet.*
—0.15	990	2.5	3,760	5.1	8,445	7.7	13,515
.0	1,070	2.6	3,910	5.2	8,640	7.8	13,710
.1	1,140	2.7	4,060	5.3	8,835	7.9	13,905
.2	1,210	2.8	4,220	5.4	9,030	8.0	14,100
.3	1,280	2.9	4,380	5.5	9,225	8.1	14,2.5
.4	1,360	3.0	4,540	5.6	9,420	8.2	14,490
.5	1,440	3.1	4,700	5.7	9,615	8.3	14,685
.6	1,520	3.2	4,860	5.8	9,810	8.4	14,880
.7	1,610	3.3	5,020	5.9	10,005	8.5	15,075
.8	1,700	3.4	5,180	6.0	10,200	8.6	15,270
.9	1,800	3.5	5,340	6.1	10,395	8.7	15,465
1.0	1,900	3.6	5,520	6.2	10,590	8.8	15,660
1.1	2,000	3.7	5,715	6.3	10,785	8.9	15,855
1.2	2,110	3.8	5,910	6.4	10,980	9.0	16,050
1.3	2,220	3.9	6,105	6.5	11,175	10.0	18,000
1.4	2,330	4.0	6,300	6.6	11,370	11.0	19,950
1.5	2,450	4.1	6,495	6.7	11,565	12.0	21,900
1.6	2,570	4.2	6,690	6.8	11,760	13.0	23,850
1.7	2,690	4.3	6,885	6.9	11,955	14.0	25,800
1.8	2,810	4.4	7,080	7.0	12,150	15.0	27,750
1.9	2,930	4.5	7,275	7.1	12,345	16.0	29,700
2.0	3,060	4.6	7,470	7.2	12,540	17.0	31,650
2.1	3,190	4.7	7,665	7.3	12,735	18.0	33,600
2.2	3,320	4.8	7,860	7.4	12,930	20.0	37,500
2.3	3,460	4.9	8,055	7.5	13,125	22.0	41,400
2.4	3,610	5.0	8,250	7.6	13,320	24.0	45,300

Rating table for Coosa River at Rome, Ga., for 1899.

Gage height.	Discharge.	Gage height.	Discharge.	Gage height.	Discharge.	Gage height.	Discharge.
Feet.	*Second-feet.*	*Feet.*	*Second-feet.*	*Feet.*	*Second-feet.*	*Feet.*	*Second-feet.*
1.0	2,030	5.0	8,710	9.7	18,157	13.7	26,197
1.1	2,124	5.1	8,911	9.8	18,358	13.8	26,398
1.2	2,218	5.2	9,112	9.9	18,559	13.9	26,599
1.3	2,312	5.3	9,313	10.0	18,760	14.0	26,800
1.4	2,406	5.4	9,514	10.1	18,961	14.1	27,001
1.5	2,500	5.5	9,715	10.2	19,162	14.2	27,202
1.6	2,620	5.6	9,916	10.3	19,363	14.3	27,403
1.7	2,740	5.7	10,117	10.4	19,564	14.4	27,604
1.8	2,860	5.8	10,318	10.5	19,765	14.5	27,805
1.9	2,980	5.9	10,519	10.6	19,966	14.6	28,006
2.0	3,100	6.0	10,720	10.7	20,167	14.7	28,207
2.1	3,260	6.1	10,921	10.8	20,368	14.8	28,408
2.2	3,420	6.2	11,122	10.9	20,569	14.9	28,609
2.3	3,580	6.3	11,323	11.0	20,770	15.0	28,810
2.4	3,740	6.4	11,524	11.1	20,971	15.1	29,011
2.5	3,900	6.5	11,725	11.2	21,172	15.2	29,212
2.6	4,060	6.6	11,926	11.3	21,373	15.3	29,413
2.7	4,220	6.7	12,127	11.4	21,574	15.4	29,614
2.8	4,380	6.8	12,328	11.5	21,775	15.5	29,815
2.9	4,540	6.9	12,529	11.6	21,976	15.6	30,016
3.0	4,700	7.0	12,730	11.7	22,177	15.7	30,217
3.1	4,900	7.1	12,931	11.8	22,378	15.8	30,418
3.2	5,100	7.2	13,132	11.9	22,579	15.9	30,619
3.3	5,300	7.3	13,333	12.0	22,780	16.0	30,820
3.4	5,500	7.4	13,534	12.1	22,981	16.1	31,021
3.5	5,700	7.5	13,735	12.2	23,182	16.2	31,222
3.6	5,900	7.6	13,936	12.3	23,383	16.3	31,423
3.7	6,100	7.7	14,137	12.4	23,584	16.4	31,624
3.8	6,300	7.8	14,338	12.5	23,785	16.5	31,825
3.9	6,500	7.9	14,539	12.6	23,986	16.6	32,026
4.0	6,700	8.0	14,740	12.7	24,187	16.7	32,227
4.1	6,901	8.1	14,941	12.8	24,388	16.8	32,428
4.2	7,102	8.2	15,142	12.9	24,589	16.9	32,629
4.3	7,303	9.0	16,750	13.0	24,790	17.0	32,830
4.4	7,504	9.1	16,951	13.1	24,991	17.1	33,031
4.5	7,705	9.2	17,152	13.2	25,192	17.2	33,232
4.6	7,906	9.3	17,353	13.3	25,393	17.3	33,433
4.7	8,107	9.4	17,554	13.4	25,594	17.4	33,634
4.8	8,308	9.5	17,755	13.5	25,795	17.5	33,835
4.9	8,509	9.6	17,956	13.6	25,996	17.6	34,036

Rating table for Coosa River at Rome, Ga., for 1899—Continued.

Gage height.	Discharge.	Gage height.	Discharge.	Gage height.	Discharge.	Gage height.	Discharge.
Feet.	*Second-feet.*	*Feet.*	*Second-feet.*	*Feet.*	*Second-feet.*	*Feet.*	*Second-feet.*
17.7	34,237	20.8	40,468	23.9	46,699	27.0	52,930
17.8	34,438	20.9	40,669	24.0	46,900	27.1	53,131
17.9	34,639	21.0	40,870	24.1	47,101	27.2	53,332
18.0	34,840	21.1	41,071	24.2	47,302	27.3	53,533
18.1	35,041	21.2	41,272	24.3	47,503	27.4	53,734
18.2	35,242	21.3	41,473	24.4	47,704	27.5	53,935
18.3	35,443	21.4	41,674	24.5	47,905	27.6	54,136
18.4	35,644	21.5	41,875	24.6	48,106	27.7	54,337
18.5	35,845	21.6	42,076	24.7	48,307	27.8	54,538
18.6	36,046	21.7	42,277	24.8	48,508	27.9	54,739
18.7	36,247	21.8	42,478	24.9	48,709	28.0	54,940
18.8	36,448	21.9	42,679	25.0	48,910	28.1	55,141
18.9	36,649	22.0	42,880	25.1	49,111	28.2	55,342
19.0	36,850	22.1	43,081	25.2	49,312	28.3	55,543
19.1	37,051	22.2	43,282	25.3	49,513	28.4	55,744
19.2	37,252	22.3	43,483	25.4	49,714	28.5	55,945
19.3	37,453	22.4	43,684	25.5	49,915	28.6	56,146
19.4	37,654	22.5	43,885	25.6	50,116	28.7	56,347
19.5	37,855	22.6	44,086	25.7	50,317	28.8	56,548
19.6	38,056	22.7	44,287	25.8	50,518	28.9	56,749
19.7	38,257	22.8	44,488	25.9	50,719	29.0	56,950
19.8	38,458	22.9	44,689	26.0	50,920	29.1	57,151
19.9	38,659	23.0	44,890	26.1	51,121	29.2	57,352
20.0	38,860	23.1	45,091	26.2	51,322	29.3	57,553
20.1	39,061	23.2	45,292	26.3	51,523	29.4	57,754
20.2	39,262	23.3	45,493	26.4	51,724	29.5	57,955
20.3	39,463	23.4	45,694	26.5	51,925	29.6	58,156
20.4	39,664	23.5	45,895	26.6	52,126	29.7	58,357
20.5	39,865	23.6	46,096	26.7	52,327	29.8	58,558
20.6	40,066	23.7	46,297	26.8	52,528	29.9	58,759
20.7	40,267	23.8	46,498	26.9	52,729	30.0	58,960

Rating table for Coosa River at Rome, Ga., for 1900 and 1901.

Gage height.	Discharge.	Gage height.	Discharge.	Gage height.	Discharge.	Gage height.	Discharge.
Feet.	*Second-feet.*	*Feet.*	*Second-feet.*	*Feet.*	*Second-feet.*	*Feet.*	*Second-feet.*
0.8	1,930	1.6	2,850	2.4	4,000	3.2	5,230
.9	2,020	1.7	2,985	2.5	4,150	3.3	5,405
1.0	2,110	1.8	3,120	2.6	4,300	3.4	5,580
1.1	2,230	1.9	3,260	2.7	4,450	3.5	5,755
1.2	2,350	2.0	3,400	2.8	4,600	a3.6	5,930
1.3	2,475	2.1	3,550	2.9	4,750		
1.4	2,600	2.2	3,700	3.0	4,900		
1.5	2,725	2.3	3,850	3.1	5,065		

a Above 3.6 feet gage height the rating for 1900–1901 is the same as for 1899.

Rating table for Coosa River at Rome, Ga., for 1902.

Gage height.	Discharge.	Gage height.	Discharge.	Gage height.	Discharge.	Gage height.	Discharge.
Feet.	*Second-feet.*	*Feet.*	*Second-feet.*	*Feet.*	*Second-feet.*	*Feet.*	*Second-feet.*
0.2	1,410	4.2	7,102	8.2	15,142	15.5	29,815
.4	1,555	4.4	7,504	8.4	15,544	16.0	30,820
.6	1,720	4.6	7,906	8.6	15,946	17.0	32,830
.8	1,905	4.8	8,308	8.8	16,348	18.0	34,840
1.0	2,110	5.0	8,710	9.0	16,750	19.0	36,850
1.2	2,350	5.2	9,112	9.2	17,152	20.0	38,860
1.4	2,600	5.4	9,514	9.4	17,554	21.0	40,870
1.6	2,850	5.6	9,916	9.6	17,956	22.0	42,880
1.8	3,120	5.8	10,318	9.8	18,358	23.0	44,890
2.0	3,400	6.0	10,720	10.0	18,760	24.0	46,900
2.2	3,700	6.2	11,122	10.5	19,765	25.0	48,910
2.4	4,000	6.4	11,524	11.0	20,770	26.0	50,920
2.6	4,300	6.6	11,926	11.5	21,775	27.0	52,930
2.8	4,600	6.8	12,328	12.0	22,780	28.0	54,940
3.0	4,900	7.0	12,730	12.5	23,785	29.0	56,950
3.2	5,230	7.2	13,132	13.0	24,790	30.0	58,960
3.4	5,580	7.4	13,534	13.5	25,795	31.0	60,970
3.6	5,930	7.6	13,936	14.0	26,800	32.0	62,980
3.8	6,300	7.8	14,338	14.5	27,805	33.0	64,990
4.0	6,700	8.0	14,740	15.0	28,810	34.0	67,000

Rating table for Coosa River at Rome, Ga., for 1903.

Gage height.	Discharge.	Gage height.	Discharge.	Gage height.	Discharge.	Gage height.	Discharge.
Feet.	*Second-feet.*	*Feet.*	*Second-feet.*	*Feet.*	*Second-feet.*	*Feet.*	*Second-feet.*
0.20	1,280	1.50	2,920	2.80	4,795	4.10	6,950
.30	1,390	1.60	3,060	2.90	4,945	4.20	7,140
.40	1,510	1.70	3,200	3.00	5,095	4.30	7,330
.50	1,630	1.80	3,340	3.10	5,250	4.40	7,520
.60	1,750	1.90	3,480	3.20	5,405	4.50	7,710
.70	1,880	2.00	3,620	3.30	5,565	4.60	7,910
.80	2,010	2.10	3,765	3.40	5,725	4.70	8,110
.90	2,140	2.20	3,910	3.50	5,890	4.80	8,310
1.00	2,270	2.30	4,055	3.60	6,055	4.90	8,510
1.10	2,400	2.40	4,200	3.70	6,225	*a*5.00	8,710
1.20	2,530	2.50	4,345	3.80	6,400		
1.30	2,660	2.60	4,495	3.90	6,580		
1.40	2,790	2.70	4,645	4.00	6,760		

a Use 1902 table above 5 feet.

Estimated monthly discharge of Coosa River at Rome, Ga.

[Drainage area, 4,006 square miles.]

Month.	Discharge in second-feet.			Run-off.	
	Maximum.	Minimum.	Mean.	Second-feet per square mile.	Depth in inches.
1897.					
January	17,025	1,800	4,820	1.20	1.38
February	20,925	4,220	10,100	2.52	2.62
March	44,910	4,700	22,537	5.63	6.49
April	35,150	4,860	12,304	3.07	3.43
May	8,250	2,930	4,421	1.10	1.27
June	4,540	1,900	2,884	.72	.80
July	23,460	1,800	5,184	1.30	1.50
August	4,860	1,360	2,256	.56	.64
September	1,900	900	1,106	.28	.31
October	2,570	1,010	1,518	.38	.44
November	2,000	1,440	1,626	.41	.46
December	9,810	1,900	4,806	1.02	1.18
The year	44,910	900	6,070	1.52	20.52

Estimated monthly discharge of Coosa River at Rome, Ga.—Continued.

Month.	Discharge in second-feet.			Run-off.	
	Maximum.	Minimum.	Mean.	Second-feet per square mile.	Depth in inches.
1898.					
January	26,970	2,220	7,272	1.82	2.10
February	5,520	2,110	2,705	.68	.71
March	20,730	2,110	4,384	1.10	1.27
April	32,040	4,540	9,430	2.36	2.63
May	4,220	2,330	2,778	.69	.79
June	5,520	2,110	2,866	.72	.80
July	6,690	1,900	3,670	9.17	10.59
August	17,805	2,930	6,079	1.52	1.75
September	45,885	3,060	12,114	3.03	2.26
October	44,910	3,060	11,830	2.96	3.41
November	12,150	3,060	5,213	1.30	1.45
December	8,250	3,610	4,996	1.25	1.44
The year	45,885	1,900	6,111	2.22	29.20
1899.					
January	10,519	4,060	6,092	1.52	1.75
February	54,538	8,710	22,536	5.62	5.85
March	57,352	7,705	26,314	6.57	7.57
April	28,810	6,700	13,333	3.33	3.72
May	6,700	3,100	4,783	1.19	1.37
June	6,700	2,740	3,489	.87	.97
July	24,388	1,950	5,499	1.37	1.58
August	3,900	1,790	2,595	.65	.75
September	5,500	1,550	2,219	.55	.61
October	2,030	1,470	1,684	.42	.48
November	4,700	1,470	2,009	.50	.56
December	13,735	1,870	4,314	1.08	1.25
The year	57,352	1,470	7,906	1.97	26.46

Estimated monthly discharge of Coosa River at Rome, Ga.—Continued.

Month.	Discharge in second-feet.			Run-off.	
	Maximum.	Minimum.	Mean.	Second-feet per square mile.	Depth in inches.
1900.					
January	21,373	2,725	6,854	1.71	1.97
February	53,332	2,850	14,736	3.68	3.83
March	33,835	6,300	14,714	3.67	4.33
April	25,996	5,755	12,050	3.01	3.36
May	11,122	3,850	5,129	1.28	1.48
June	35,242	4,000	14,154	3.53	3.94
July	19,765	4,000	7,589	1.89	2.18
August	5,580	2,725	3,488	.87	1.00
September	20,971	1,930	3,960	.99	1.10
October	10,519	2,010	3,408	.85	.98
November	21,775	2,600	5,438	1.36	1.52
December	14,740	3,400	7,096	1.77	2.04
The year	53,332	1,930	8,218	2.05	27.73
1901.					
January	52,930	4,600	15,450	3.86	4.45
February	35,845	4,900	12,186	3.04	3.17
March	52,930	4,600	13,406	3.34	3.85
April	36,046	6,901	15,578	3.88	4.33
May	51,724	4,150	12,533	3.12	3.60
June	19,966	4,450	8,316	2.08	2.32
July	9,715	2,850	4,441	1.10	1.27
August	45,292	2,850	13,780	3.44	3.97
September	21,172	3,120	6,389	1.59	1.77
October	5,230	2,350	3,414	.85	.98
November	2,850	2,110	2,316	.58	.65
December	64,186	2,475	13,428	3.35	3.86
The year	64,186	2,110	10,103	2.52	34.22

Estimated monthly discharge of Coosa River at Rome, Ga.—Continued.

Month.	Discharge in second-feet.			Run-off.	
	Maximum.	Minimum.	Mean.	Second-feet per square mile.	Depth in inches.
1902.					
January	54,940	4,450	11,816	2.95	3.40
February	46,900	6,300	14,812	3.70	3.85
March	56,749	7,102	21,944	5.48	6.32
April	41,473	4,750	10,015	2.50	2.79
May	6,115	3,120	4,089	1.02	1.18
June	3,400	2,110	2,836	.71	.79
July	3,400	1,635	2,214	.55	.63
August	2,985	1,480	1,998	.50	.58
September	6,700	1,480	2,505	.63	.70
October	3,850	1,410	2,346	.59	.68
November	8,308	1,410	2,572	.64	.71
December	12,328	2,225	5,885	1.47	1.69
The year	56,749	1,410	6,920	1.73	23.32
1903.					
January	8,710	3,620	5,442	1.36	1.57
February	56,347	4,795	25,376	6.34	6.60
March	56,146	9,313	27,111	6.78	7.82
April	48,709	6,225	15,788	3.95	4.41
May	12,127	3,620	5,278	1.32	1.52
June	33,031	3,910	9,594	2.40	2.68
July	20,971	3,340	5,616	1.40	1.61
August	11,122	2,530	4,472	1.12	1.29
September	3,620	1,510	2,002	.50	.56
October	3,060	1,280	2,002	.50	.58
November	3,620	1,880	2,512	.63	.70
December	2,660	1,750	1,985	.50	.58
The year	56,347	1,280	8,932	2.23	29.92

Net horsepower per foot of fall with a turbine efficiency of 80 per cent for the minimum monthly discharge of Coosa River at Rome, Ga.

Month.	1899.			1900.			1901.		
	Minimum discharge.	Minimum net horsepower per foot of fall.	Duration of minimum.	Minimum discharge.	Minimum net horsepower per foot of fall.	Duration of minimum.	Minimum discharge.	Minimum net horsepower per foot of fall.	Duration of minimum.
	Sec.-ft.		*Days.*	*Sec.-ft.*		*Days.*	*Sec.-ft.*		*Days.*
January	4,060	369	2	2,725	248	8	4,600	418	1
February	8,710	792	2	2,850	259	1	4,900	445	1
March	7,705	700	1	6,300	573	1	4,600	418	1
April	6,700	609	1	5,755	523	1	6,901	627	1
May	3,100	282	2	3,850	350	1	4,150	377	2
June...........	2,740	249	4	4,000	364	1	4,450	405	2
July	1,950	177	1	4,000	364	3	2,850	259	1
August.........	1,790	163	3	2,725	248	3	2,850	259	2
September	1,550	141	1	1,930	175	5	3,120	284	1
October........	1,470	134	4	2,010	183	3	2,350	214	1
November......	1,470	134	7	2,600	236	2	2,110	192	10
December......	1,870	170	4	3,400	309	2	2,475	225	3

SURVEY OF COOSA RIVER.

Coosa River has its beginning at the junction of Etowah and Oostanaula rivers, at Rome, Ga., a short distance east of the Alabama line.

From Rome down to Greensport, Ala., a distance of about 180 miles by river, navigation has been carried on for many years. The total fall in this section is only about 55 feet, and is so well distributed that it has not been necessary to construct locks at any point, though improvements have been made by the United States Government in the way of deepening channels, blasting out reefs, building wing dams, etc.

This part of the river will therefore not be considered as having any water-power value.

Below Greensport, Ala., the river has a large amount of fall, and although it is proposed to make the whole distance navigable by the construction of locks, there are many fine water powers which can be developed in connection with the river improvements without interfering with navigation.

A complete survey has been made of this portion of the river by the United States engineers, and a system of locks planned.

The level notes herein presented are reproduced from that survey, and show the river profile and the location of the shoals.

The total distance between Greensport and Wetumka, Ala., is 142
miles, and the number of locks proposed is 31, varying in lift from
5.83 feet to 15 feet. Of these, only three have been completed—
Nos. 1, 2, and 3. No. 4 is in process of construction.

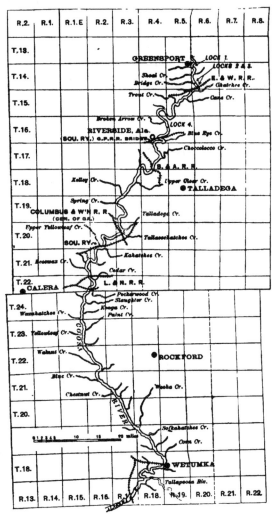

FIG. 6.—Map of Coosa River from Greensport to Wetumka.

The following table shows the lift or fall at each lock, the discharge
of the river in second-feet for the minimum low stages of water in 1897
and in 1900, and the equivalent net horsepower for the fall shown.

The minimum low water is based on the exceptionally low stages of 1896 and 1897, the lowest of which there is any record; the minimum for the year 1900 represents lowest water for average years.

In estimating the amount of horsepower that will be available it will be necessary to deduct the amount of water which will be necessary for lockage. This will depend upon the amount of traffic on the river, but will probably in no case amount to more than 10 per cent of the discharge.

At most of these locks and proposed locks, reservations have been made by the original owners of the river front of the privilege of utilizing for power the water not needed for lockage. By constructing a plant at the opposite end of the Government dam from the lock, the surplus water can be used for power without interfering with navigation.

Discharge and net horsepower at proposed locks on Coosa River at lowest water of 1897 and 1900.

[80 per cent of theoretical horsepower.]

Distance from Wetumka.	Lock No.	Elevation of top of lock.	Lift or fall at lock.	1897.		1900.	
				Discharge.	Net horsepower.	Discharge.	Net horsepower.
Miles.		*Feet.*	*Feet.*	*Sec.-ft.*		*Sec.-ft.*	
141.5	1	521.30	5.33	1,320	640	2,700	1,308
138.5	2	515.97	5.57	1,320	668	2,700	1,367
137.0	3	510.40	12.00	1,320	1,440	2,700	2,945
116.2	4	492.30	10.00	1,350	1,227	2,760	2,510
105.8	5	482.30	12.00	1,350	1,472	2,760	3,012
92.0	6	455.32	10.00	1,440	1,310	2,940	2,673
88.3	7	445.32	10.00	1,450	1,317	2,960	2,690
81.3	8	435.32	12.00	1,490	1,625	3,040	3,317
56.2	9	420.00	8.00	1,580	1,149	3,220	2,342
53.5	10	412.00	12.00	1,585	1,728	3,230	3,523
46.7	11	399.64	10.00	1,585	1,440	3,230	2,936
44.9	12	389.64	10.00	1,600	1,454	3,260	2,964
43.0	13	379.64	12.00	1,600	1,745	3,260	3,557
41.9	14	367.64	12.00	1,600	1,745	3,260	3,557
40.2	15	355.64	10.00	1,605	1,460	3,270	2,973
37.5	16	345.64	14.00	1,605	2,044	3,270	4,162
36.1	17	331.64	15.00	1,605	2,190	3,270	4,460
34.8	18	316.64	13.00	1,610	1,903	3,280	3,877
33.8	19	303.64	12.00	1,610	1,757	3,280	3,578
31.5	20	291.64	10.00	1,610	1,464	3,280	2,982
25.5	21	281.33	10.00	1,700	1,545	3,460	3,145
21.4	22	270.80	12.00	1,700	1,854	3,460	3,774

A. LOCK NO 4 ON COOSA RIVER, NEAR LINCOLN, ALA, DURING CONSTRUCTION.

B. PRATT COTTON GIN MANUFACTURING COMPANYS PLANT, ON AUGUSTA
CREEK AT PRATTVILLE, ALA

Snowing dam with flash boards to raise as flood gate

Discharge and net horsepower at proposed locks on Coosa River at lowest water of 1897 and 1900—Continued.

Distance from Wetumka.	Lock No.	Elevation of top of lock.	Lift or fall at lock.	1897.		1900.	
				Discharge.	Net horsepower.	Discharge.	Net horsepower.
Miles.		*Feet.*	*Feet.*	*Sec.-feet.*		*Sec.-feet.*	
18.5	23	258.80	14.00	1,710	2,175	3,480	4,430
16.3	24	244.80	10.00	1,710	1,554	3,480	3,164
12.9	25	234.80	10.00	1,710	1,554	3,480	3,164
11.7	26	224.80	12.00	1,720	1,877	3,500	3,818
8.8	27	212.80	14.00	1,720	2,190	3,500	4,455
7.4	28	198.80	12.00	1,720	1,877	3,500	3,818
4.6	29	186.37	8.00	1,740	1,266	3,540	2,574
2.0	30	178.37	10.00	1,740	1,582	3,540	3,218
0.0	31	168.37	14.00	1,740	2,215	3,540	4,505

Total net horsepower, 1897, 49,467; 1900, 100,798.

Locks and proposed locks on Coosa River are located as follows:

Lock No. 1 is 1 mile south of Greensport and 5 miles north of Singleton, a station on the East and West Railroad of Alabama. Lock No. 1 is 3 miles above lock No. 2.

Lock No. 2 is 1½ miles above lock No. 3 at the head of Ten Island Shoal canal. It is located at the head of Woods Island, and is 2 miles northeast of Singleton, a station on the East and West Railroad.

Lock No. 3 is 1½ miles below lock No. 2, near the foot of Woods Island, and on Ten Island Shoal canal. It is 1 mile east of Singleton and 20.8 miles above lock No. 4.

Lock No. 4 is 3½ miles above the United States Geological Survey gage at Riverside and 3 miles northwest of Lincoln. Lincoln and Riverside are on the Georgia Pacific division of the Southern Railway. Lock No. 4 has a lift of 12 feet, and is three-fourths of a mile below Densons Island, and 10 miles above proposed lock No. 5.

Proposed lock No. 5 is to be at the head of Ogletree Island, 1 mile above the mouth of Choccolocco Creek, and 5 miles northeast of Hamilton, on the Talladega and Coosa Valley Railroad. It has a lift of 10 feet.

Proposed lock No. 6 is to be located one-fourth of a mile above the mouth of Upper Clear Creek, 1½ miles above Grissom Ferry, and 9 miles northeast of Vincent, a station on the Columbus and Western division of the Central of Georgia Railroad.

Proposed lock No. 7 is to be located 2 miles above Kelly Creek, and 5½ miles northeast of Vincent.

Proposed lock No. 8 is to be located at Myers Ferry, at the mouth of Clear Creek, 6 miles east of Harpersville, and 3 miles northeast of Creswell, a station on the Columbus and Western division of the Central of Georgia Railroad.

Proposed lock No. 9 is to be located at the mouth of Kelly Branch, at Fort Williams Shoals. It is to be 13¼ miles east of Columbiana, and 8 miles east of Shelby.

Lock No. 10 is to be located one-half mile above Peckerwood Creek, at the foot of Peckerwood Shoals, and is 8 miles east of Shelby, and 2 miles west of Talladega Springs.

Lock No. 11 is to be located at the foot of Weduska Shoals, immediately above the narrows, 2 miles above Waxahatchee Creek, and 6 miles southeast of Shelby, a station on the Shelby Iron Works Railroad, which connects with the Southern Railway at Columbiana.

Lock No. 12 is to be located 1.8 miles below lock No. 11, immediately below the mouth of Waxahatchee Creek, and 8 miles southeast of Shelby.

Lock No. 13 is to be located 1.9 miles below lock No. 12, at a place known as Devils Race, 3 miles above the mouth of Yellowleaf Creek, and 16 miles northeast of Clanton, on the Louisville and Nashville Railroad.

Lock No. 14 is to be located 1 mile below lock No. 13, 2 miles above Yellowleaf Creek, and 14 miles northeast of Clanton.

Lock No. 15 is to be located 1.7 miles below lock No. 14, three-tenths of a mile above Yellowleaf Creek, and 12 miles northeast of Clanton, on the Louisville and Nashville Railroad.

Lock No. 16 is to be located 2.7 miles below lock No. 15, at Butting Ram Shoals, which is 11 miles northeast of Clanton.

Lock No. 17 is to be located 1.4 miles below lock No. 16, and is 10¼ miles northeast of Clanton.

Lock No. 18 is to be located 1.3 miles below lock No. 17, and 11 miles east of Clanton.

Lock No. 19 is to be located 1 mile below lock No. 18, about 11 miles east of Clanton.

Lock No. 20 is to be 31.5 miles above Wetumka, one-fourth mile above Zimmermans Ferry, 1.2 miles above the mouth of Hatchet Creek.

Lock No. 21 is to be 25.5 miles above Wetumka, 1.6 miles below mouth of Blue Creek, 7 miles east of Cooper, on the Louisville and Nashville Railroad.

Lock No. 22 is to be 21.4 miles above Wetumka, three-fourths of a mile below the mouth of Proctor Creek, and 1.1 miles above the mouth of Pinchoulee Creek, and 7 miles east of Verbena, on the Louisville and Nashville Railroad.

Lock No. 23 is to be 18.5 miles from Wetumka, 1.5 miles below the mouth of Pinchoulee Creek.

Lock No. 24 is to be 16 miles above Wetumka, 0.4 mile below the mouth of Welcree Creek, and 7¼ miles east of Mountain Creek station, on the Louisville and Nashville Railroad.

Lock No. 25 is to be 12.9 miles above Wetumka, 0.1 mile above the mouth of Shoal Creek, and about 8 miles east of Wadsworth, on the Louisville and Nashville Railroad.

Lock No. 26 is to be 11.7 miles above Wetumka, at Staircase Falls, just above the mouth of Wewoka Creek.

Lock No. 27 is to be 8.8 miles above Wetumka, 0.6 mile above the mouth of Sofkahatchee Creek, and about 9 miles east of Deatsville, on the Louisville and Nashville Railroad.

Lock No. 28 is to be 7.4 miles above Wetumka.

Lock No. 29 is to be 4.6 miles above Wetumka.

Lock No. 30 is to be 2 miles above Wetumka.

Lock No. 31 is to be at Wetumka.

Elevations of water surface on Coosa River between Wetumka and Greensport.

[Determined in 1889 by Charles Firth, assistant engineer under Capt. Philip M. Price, Corps of Engineers, U. S. Army.]

Distance from Wetumka.	Location.	Elevation above sea level.
Miles.		*Feet.*
0.0	Wetumka Bridge	154.37
0.4	Mouth of Valley Brook	164.37
2.0	Mouth of Corn Creek	168.37
4.0	Mouth of creek, top of Moccasin Reef	177.00
4.7	Foot of Grays Island Shoals	178.37
6.0	Big Eddy, top of Grays Island Shoals	186.37
8.3	Mouth of Sofkahatchee Creek	190.80
8.8	Top of Gunns Island Shoals	196.80
11.7	Mouth of Weoka, or Wenone, Creek, at Staircase Falls	212.80
12.9	Mouth of Shoal, or Mill, Creek	225.00
15.3	Greys Ferry	234.00
15.8	Mouth of Town Creek	234.80
16.8	Welona Reef, mouth of Welona Creek	241.00
20.3	Mouth of Pinchoulee Creek	254.00
21.4	Foot of Hells Gap Shoal	256.00
22.0	Mouth of Proctor Creek	265.00
22.6	Mouth of Chestnut Creek	267.00
24.0	Knights Ferry	270.80
25.6	Foot of Duncans Riffle	270.80
27.2	Mouth of Blue Creek	279.30
28.7	Mouth of Cargal Creek, Smiths Ferry	280.00
30.3	Mouth of Hatchet Creek	281.90

Elevations of water surface on Coosa River between Wetumka and Greensport—Continued.

Distance from Wetumka	Location.	Elevation above sea level.
Miles.		*Feet.*
31.3	Zimmermans Ferry	281.33
34.0	Foot of Tuckaleague Shoal	292.00
37.0	Head of Tuckaleague Shoal	331.64
40.0	Mouth of Yellowleaf Creek, at Adam Ferry....................	343.64
41.3	Lower mouth of Paint Creek.................................	360.00
45.0	Mouth of Waxahatchee Creek	380.00
46.0	Mouth of Kooga Creek	388.00
47.5	Spring Branch, head of Weduska Shoal.......................	399.64
48.6	Mouth of Slaughter Creek..................................	399.64
52.0	Merrills Ferry	399.64
53.0	Mouth of Peckerwood Creek, at foot of Peckerwood Shoal	399.64
54.3	Mouth of Sulphur Branch from Talladega Springs	402.80
55.0	Mouth of Cedar Creek, near Fort William Ferry..............	403.00
56.2	Mouth of Kelly Branch, foot of Fort William Shoals...........	407.80
56.9	Mouth of Flat Branch	413.00
59.0	Mouth of Sally Branch	413.00
60.2	Mouth of Beeswax Creek	413.00
60.8	McRaes Ferry	413.00
63.9	Mouth of Hay's spring branch	415.00
65.7	McGowans Ferry	421.00
66.6	Mouth of Upper Yellowleaf Creek	421.00
68.0	Southern Railway bridge, at mouth of Kahatchee Creek	421.00
74.1	Chancellors Ferry	421.00
75.2	Columbus and Western Railway bridge (Central of Georgia Railway)	422.00
75.7	Mouth of Tallaseehatchee Creek............................	422.00
76.7	Mouth of Talladega Creek	422.00
81.1	Mouth of Clear Creek, at Meyers Ferry......................	423.00
84.7	Mouth of Spring Creek, at Glovers Ferry	429.00
86.2	Mouth of Kelly Creek	431.00
90.4	Grissom Ferry	439.00
91.8	Mouth of Upper Clear Creek, at Howell's mill shoals...........	444.00
95.9	Griffiths Ferry, head of Drake's mill shoals..................	456.00
101.2	Collin Ferry and bridge of the Birmingham and Atlantic Railroad..	461.00
102.0	Foot of Choccolocco Shoals	461.00
105.0	Mouth of Choccolocco Creek	469.00
106.8	Head of Choccolocco Shoals	475.00
108.0	Truss Ferry..	475.00
115.3	Mouth of Blue Eye Creek..................................	476.00

Elevations of water surface on Coosa River between Wetumka and Greensport—Continued.

Distance from Wetumka.	Location.	Elevation above sea level.
Miles.		*Feet.*
112. 6	Bridge of Southern Railway, Riverside	476. 00
114. 8	Mouth of Blue Spring Branch	477. 00
116. 2	Foot of Lock 4	477. 30
116. 2	Zero of gage below Lock 4	477. 30
116. 2	Zero of gage above Lock 4	489. 30
117. 7	Mouth of Broken Arrow Creek	483. 00
125. 5	Wood Ferry	492. 30
127. 8	Mouth of Alligator Creek	493. 00
129. 0	Mouth of Trout Creek	493. 00
130. 0	Mouth of Bruner Creek, at Mays Ferry	494. 00
131. 4	Mouth of Cane Creek	495. 00
136. 0	Mouth of Ohatchee Creek, at Harts Ferry	496. 00
136. 4	Below East and West Railroad bridge	496. 00
136. 4	Above East and West Railroad bridge	498. 00
137. 0	Below Lock 3	498. 40
137. 0	Zero of gage below Lock 3 (U. S. A. Engineers)	494. 40
137. 0	Zero of gage above Lock 3 (U. S. A. Engineers)	505. 25
137. 0	Present (1903) water level above Lock 3	510. 40
138. 4	Present (1903) water level below Lock 2	510. 40
138. 4	Present (1903) water level above Lock 2	515. 97
138. 4	Zero of gage below Lock 2	505. 20
138. 4	Zero of gage above Lock 2	510. 57
141. 7	Present (1903) water level below Lock 1	515. 97
141. 7	Present (1903) water level above Lock 1	521. 30
141. 7	Zero of gage below Lock 1	512. 72
141. 7	Zero of gage above Lock 1	517. 89

TALLADEGA CREEK AT NOTTINGHAM.

This station is located on the Southern Railroad bridge, a fourth of a mile from the station at Nottingham and 1 mile north of Alpine. The gage, which is graduated to feet and tenths and is 20 feet long, is fastened vertically to a tree on right bank about 50 feet above the bridge. The initial point of sounding is the end of the iron bridge, right bank, upstream. The bench mark is the top rail on the upstream side of the bridge, and is 24.13 feet above gage datum. The station is a good one, and is free from piers. The observer is R. M. McClatchy, station agent at Nottingham. The following measurements were made by James R. Hall, M. R. Hall, and others:

Discharge measurements of Talladega Creek at Nottingham.

Date.	Gage height.	Discharge.	Date.	Gage height.	Discharge.
1900.	*Feet.*	*Second-feet.*	1902.	*Feet.*	*Second-feet.*
August 16	1.10	102	October 18	0.80	78
November 29	1.70	240	November 13	.70	63
1901.			1903.		
April 5	3.00	526	May 25	2.05	243
October 22	1.00	90	July 27	1.37	111
1902.			August 20	1.30	116
January 16	1.30	155	October 2	1.00	57
July 17	.90	97	November 14	1.21	80

Daily gage height, in feet, of Talladega Creek at Nottingham.

Day.	Jan.	Feb.	Mar.	Apr.	May.	June.	July.	Aug.	Sept.	Oct.	Nov.	Dec.
1900.												
1									1.2	1.0	1.3	1.6
2									2.0	1.0	1.3	1.4
3									1.3	1.0	2.3	1.4
4									1.1	1.0	2.1	1.5
5									1.0	1.2	1.7	1.7
6									1.0	1.3	1.5	1.6
7									1.0	1.3	1.4	1.5
8									.9	1.4	1.4	1.5
9									11.0	1.3	1.3	1.4
10									8.0	1.3	1.3	1.3
11									10.3	1.3	1.3	1.3
12									8.0	1.5	1.3	1.3
13									8.3	1.4	1.2	1.6
14				.A.					9.3	1.3	1.2	1.8
15									9.3	1.2	1.2	1.7
16								1.1	8.9	1.2	1.2	1.6
17								1.2	2.3	1.1	1.2	1.5
18								1.2	1.7	1.1	1.2	1.5
19								1.1	1.6	1.1	1.2	1.4
20								1.1	1.5	1.0	1.5	2.7
21								1.1	1.3	1.2	2.2	3.0
22								1.1	1.2	1.4	1.9	2.6
23								1.1	1.2	1.6	1.8	4.0
24								1.0	1.2	2.9	1.6	2.5
25								.9	1.2	3.5	3.9	2.1
26								1.0	1.2	3.0	3.8	2.0
27								1.1	1.2	2.6	3.4	1.8
28								1.2	1.2	1.4	2.0	1.7
29								1.0	1.1	1.3	1.8	1.6
30								1.0	1.1	1.3	1.7	1.9
31								1.6		1.3		5.1

Daily gage height, in feet, of Talladega Creek at Nottingham—Continued.

Day.	Jan.	Feb.	Mar.	Apr.	May.	June.	July.	Aug.	Sept.	Oct.	Nov.	Dec.
1901.												
1	3.2	2.3	2.0	3.4	2.2	2.0	1.5	1.1	1.2	1.1	1.0	1.0
2	2.8	2.1	2.0	5.9	2.2	2.0	1.4	1.0	1.2	2.8	1.0	1.4
3	3.2	3.8	2.0	4.5	2.1	2.0	1.3	1.0	1.2	2.5	1.0	1.3
4	2.7	8.0	2.0	3.4	2.0	1.8	1.3	1.0	1.1	1.5	1.0	1.2
5	2.4	3.9	1.8	3.0	2.0	1.7	1.3	1.0	1.1	1.3	1.0	1.2
6	2.2	3.2	1.8	2.9	1.9	1.7	1.4	1.0	1.1	1.2	1.0	1.3
7	2.1	2.8	1.8	2.6	1.9	3.0	1.4	1.0	1.1	1.1	1.0	1.1
8	2.1	2.8	1.8	2.5	1.9	1.9	1.8	1.0	1.1	1.2	1.0	1.0
9	2.0	3.3	1.8	2.3	1.9	1.7	1.3	1.0	1.0	1.1	1.0	1.0
10	1.9	2.8	2.0	2.2	1.9	1.6	1.2	1.0	1.0	1.0	1.0	1.1
11	5.8	2.6	1.9	2.1	1.9	1.5	1.2	1.0	1.0	1.0	1.0	1.1
12	8.8	2.7	1.8	2.1	1.9	1.5	1.2	1.1	1.0	1.0	1.0	1.0
13	4.7	2.5	1.8	2.7	2.0	1.6	1.2	1.0	1.0	1.1	1.0	1.0
14	3.4	2.3	1.8	3.7	1.9	1.8	1.2	1.0	1.0	1.0	1.0	2.4
15	2.9	2.2	1.8	2.7	1.9	1.7	1.2	1.0	1.0	1.0	1.0	3.5
16	2.6	2.2	1.7	2.5	1.9	1.6	1.2	2.3	1.0	1.0	1.0	1.7
17	2.7	2.2	1.7	2.8	1.8	1.5	1.2	2.2	3.4	1.0	1.0	1.5
18	2.4	2.1	1.7	2.4	1.7	1.4	2.0	1.9	2.8	1.0	1.0	1.3
19	2.2	2.1	1.7	11.2	1.9	1.4	1.5	1.7	1.6	1.0	1.0	1.2
20	2.2	2.0	1.8	6.3	2.5	1.4	1.3	1.5	1.5	1.0	1.0	1.1
21	2.1	2.0	2.1	4.1	3.9	1.4	1.3	1.4	1.4	1.0	1.0	1.1
22	2.2	2.0	2.0	3.4	2.5	1.3	1.2	1.3	1.3	1.0	1.1	1.1
23	2.2	2.0	1.8	3.1	2.0	1.8	1.2	1.2	1.3	1.0	1.2	1.1
24	2.2	2.1	2.6	2.8	1.8	1.3	1.2	1.2	1.2	1.0	1.3	1.1
25	2.2	2.1	2.8	2.7	1.7	1.8	1.2	1.5	1.2	1.0	1.2	1.3
26	2.3	2.1	8.9	2.6	1.7	1.3	1.2	1.3	1.1	1.0	1.1	1.2
27	2.2	2.0	4.6	2.5	1.7	1.3	1.2	1.4	1.1	1.0	1.1	1.1
28	2.2	2.0	3.3	2.4	1.7	1.3	1.2	1.5	1.1	1.0	1.0	1.2
29	2.2	....	2.8	2.3	1.7	1.3	1.2	1.4	1.1	1.0	1.0	8.4
30	2.6	....	2.9	2.2	1.7	1.4	1.2	1.3	1.1	1.0	1.0	7.5
31	2.5	....	5.5	....	2.2	....	1.2	1.2	....	1.0	....	3.5
1902.												
1	2.40	5.60	5.00	3.50	1.80	2.60	.90	.90	1.00	1.40	.70	.80
2	2.20	8.90	3.70	3.10	1.70	1.50	.90	.80	.90	1.10	.70	2.40
3	2.00	4.40	3.10	2.90	1.60	1.30	.90	.80	.70	.90	.70	2.70
4	1.80	3.80	2.90	2.90	1.50	1.30	.90	.80	.70	.80	.70	1.60
5	1.70	2.70	4.60	2.70	1.50	1.20	1.00	.80	.70	.70	.70	1.40
6	1.60	2.50	8.30	3.30	1.50	1.10	1.00	.80	.70	.70	1.90	1.20
7	1.50	2.30	2.90	3.60	1.60	1.10	1.00	.80	.70	.70	1.10	1.30
8	1.50	2.10	2.60	3.20	1.90	1.10	.90	.80	.70	.70	.90	1.20
9	1.50	2.10	2.50	3.00	1.70	1.10	1.00	.80	.70	.70	.80	1.10
10	1.50	2.00	2.40	2.70	1.60	1.10	1.50	1.00	.70	.70	.80	1.00
11	1.40	1.90	2.80	2.50	1.60	1.10	2.00	1.20	.70	4.10	.70	.90
12	1.40	1.90	2.20	2.30	1.50	1.10	1.50	1.50	.70	1.90	.70	.80
13	1.80	1.90	2.20	2.30	1.40	1.10	1.20	1.10	.70	1.80	.70	.80
14	1.30	1.90	2.10	2.30	1.40	1.10	1.00	1.00	.70	1.60	.70	.80
15	1.80	2.30	2.60	2.20	1.40	1.10	1.00	.90	.70	.80	.70	.80
16	1.80	2.10	7.60	2.10	1.40	1.10	1.00	.80	.70	.80	.70	6.30
17	1.30	1.90	3.80	2.50	1.50	1.10	.90	.80	.70	.70	.70	2.50
18	1.30	1.80	2.70	2.30	1.50	1.00	.90	.80	.70	.70	1.20	1.80
19	1.40	1.80	2.60	2.20	1.50	1.00	.90	.80	.70	.80	1.00	1.40
20	1.90	1.90	2.40	2.10	1.40	1.00	.90	.80	.70	.80	.90	1.20
21	2.10	1.80	2.40	2.00	1.30	1.00	1.50	.80	.60	.80	.80	2.30
22	1.70	2.00	2.30	1.90	1.30	1.10	1.10	.70	.60	.80	.80	2.60
23	1.60	1.90	2.20	1.90	1.30	1.10	1.00	.70	.60	.80	.80	2.20
24	1.50	1.80	3.20	1.90	1.80	.90	1.00	.70	.60	.80	.80	1.50
25	1.80	1.90	3.20	1.80	1.20	.90	1.00	.70	.70	.80	1.90	1.30
26	1.60	1.80	2.50	1.80	1.20	.90	1.00	.70	.70	.80	1.40	1.10
27	1.80	3.60	2.50	1.80	1.20	.90	1.00	.70	.60	.80	1.10	1.10
28	3.80	12.20	12.50	1.80	1.10	.90	1.00	1.00	1.10	.80	.90	1.10
29	4.10	....	10.70	1.80	1.10	.90	1.00	5.10	1.10	.70	.80	1.30
30	2.80	....	5.20	1.80	1.10	.90	1.00	3.10	1.70	.70	.70	2.40
31	3.50	....	3.90	....	1.10	....	1.00	1.00	....	.70	....	1.80

Daily gage height, in feet, of Talladega Creek at Nottingham—Continued.

Day.	Jan.	Feb.	Mar.	Apr.	May.	June.	July.	Aug.	Sept.	Oct.	Nov.	Dec.
1903.												
1	1.7	1.3	6.4	3.8	2.0	2.5	1.9	1.7	1.0	1.0	1.0	1.1
2	2.1	1.3	4.2	3.4	1.9	2.3	1.8	1.8	1.0	1.0	1.0	1.1
3	2.2	1.3	4.0	3.2	1.8	2.0	1.7	1.6	.9	1.0	1.0	1.1
4	2.1	2.3	3.8	3.0	1.8	1.9	2.0	1.6	.9	1.0	1.0	1.1
5	1.9	2.1	3.5	2.8	1.8	1.8	2.5	1.4	.9	1.0	1.0	1.1
6	1.8	1.7	4.2	2.6	1.8	2.2	2.0	1.5	.9	1.0	1.0	1.2
7	1.6	6.8	3.4	2.5	2.8	2.0	2.0	1.5	.9	1.0	1.0	1.1
8	1.5	13.0	3.2	2.4	2.3	1.8	1.7	1.4	.9	1.6	1.1	1.1
9	1.3	5.5	3.2	2.2	2.0	1.8	2.0	1.4	.9	1.5	1.1	1.1
10	1.2	3.6	3.5	2.0	1.9	1.7	1.8	1.3	.9	1.3	1.1	1.1
11	1.5	a9.2	3.6	2.2	1.8	3.6	1.8	1.3	.9	1.1	1.1	1.1
12	2.7	6.5	3.5	3.0	1.8	2.2	1.8	1.3	.9	1.0	1.5	1.1
13	1.8	3.8	3.3	3.2	2.8	2.0	1.7	1.3	.9	1.0	1.5	1.1
14	1.6	3.2	3.6	3.0	4.1	1.8	1.6	1.3	.9	1.0	1.2	1.1
15	1.5	2.8	3.3	2.8	10.8	1.8	1.5	1.8	.9	1.1	1.2	1.1
16	1.4	b7.5	3.4	2.6	5.2	1.8	1.5	1.5	1.0	1.1	1.2	1.1
17	1.4	10.0	3.2	2.5	8.5	1.8	1.5	1.3	1.5	1.2	1.2	1.1
18	1.5	4.4	2.9	2.5	2.9	1.7	1.5	1.3	1.2	1.2	1.2	1.1
19	1.3	3.5	2.8	2.4	2.7	1.6	1.5	1.4	1.2	1.2	1.2	1.1
20	1.3	3.1	2.7	4.7	2.6	1.5	1.8	1.4	1.2	1.2	1.2	1.1
21	1.2	2.8	3.0	3.0	2.4	1.5	1.4	1.3	1.1	1.2	1.2	1.1
22	1.2	2.6	3.5	2.5	2.2	2.5	1.4	1.2	1.1	1.2	1.2	1.1
23	1.2	2.3	5.5	2.4	2.0	1.8	1.4	1.2	1.1	1.2	1.1	1.1
24	1.2	2.2	4.0	2.3	2.0	1.6	1.4	1.1	1.1	1.2	1.1	1.1
25	1.2	2.2	3.5	2.2	2.0	1.5	1.4	1.1	1.0	1.1	1.1	1.1
26	1.2	2.2	3.1	2.2	2.0	1.7	1.2	1.0	1.0	1.1	1.1	1.1
27	1.2	2.3	2.9	2.1	2.0	2.5	1.3	1.0	1.0	1.1	1.1	1.1
28	1.6	16.0	2.7	2.0	1.9	6.0	1.5	1.0	1.0	1.0	1.1	1.1
29	1.5	......	2.9	2.0	1.9	2.5	1.4	1.0	1.0	1.0	1.1	1.1
30	1.4	......	8.7	2.0	2.8	2.0	1.4	1.0	1.0	1.1	1.1	1.1
31	1.3	......	4.8	....	2.3	....	1.6	1.0	....	1.1	....	1.1

a Maximum, 10.2.   b Maximum, 9.8.

Rating table for Talladega Creek at Nottingham for 1900 and 1901.

Gage height.	Discharge.	Gage height.	Discharge.	Gage height.	Discharge.	Gage height.	Discharge.
Feet.	Second-feet.	Feet.	Second-feet.	Feet.	Second-feet.	Feet.	Second-feet.
1.0	90	2.1	328	3.2	570	4.3	812
1.1	109	2.2	350	3.3	592	4.4	834
1.2	130	2.3	372	3.4	614	4.5	856
1.3	152	2.4	394	3.5	636	4.6	878
1.4	174	2.5	416	3.6	658	4.7	900
1.5	196	2.6	438	3.7	680	4.8	922
1.6	218	2.7	460	3.8	702	4.9	944
1.7	240	2.8	482	3.9	724	5.0	966
1.8	262	2.9	504	4.0	746		
1.9	284	3.0	526	4.1	768		
2.0	306	3.1	548	4.2	790		

Rating table for Talladega Creek at Nottingham for 1902.

Gage height.	Discharge.	Gage height.	Discharge.	Gage height.	Discharge.	Gage height.	Discharge.
Feet.	Second-feet.	Feet.	Second-feet.	Feet.	Second-feet.	Feet.	Second-feet.
0.6	51	2.8	482	5.0	966	7.2	1,450
.8	78	3.0	526	5.2	1,010	7.4	1,494
1.0	109	3.2	570	5.4	1,054	7.6	1,538
1.2	143	3.4	.614	5.6	1,098	7.8	1,582
1.4	179	3.6	658	5.8	1,142	8.0	1,626
1.6	219	3.8	702	6.0	1,186	8.5	1,736
1.8	262	4.0	746	6.2	1,230	9.0	1,846
2.0	306	4.2	790	6.4	1,274	9.5	1,956
2.2	350	4.4	834	6.6	1,318	10.0	2,066
2.4	394	4.6	878	6.8	1,362	10.5	2,176
2.6	438	4.8	922	7.0	1,406	11.0	2,286

Rating table for Talladega Creek at Nottingham for 1903.

Gage height.	Discharge.	Gage height.	Discharge.	Gage height.	Discharge.	Gage height.	Discharge.
Feet.	Second-feet.	Feet.	Second-feet.	Feet.	Second-feet.	Feet.	Second-feet.
0.90	46	2.70	400	4.50	832	6.30	1,264
1.00	57	2.80	424	4.60	856	6.40	1,288
1.10	69	2.90	448	4.70	880	6.50	1,312
1.20	83	3.00	472	4.80	904	6.60	1,336
1.30	99	3.10	496	4.90	928	6.70	1,360
1.40	116	3.20	520	5.00	952	6.80	1,384
1.50	133	3.30	544	5.10	976	6.90	1,408
1.60	151	3.40	568	5.20	1,000	7.00	1,432
1.70	170	3.50	592	5.30	1,024	7.50	1,552
1.80	190	3.60	616	5.40	1,048	8.00	1,672
1.90	211	3.70	640	5.50	1,072	8.70	1,840
2.00	233	3.80	664	5.60	1,096	8.80	1,864
2.10	256	3.90	688	5.70	1,120	9.00	1,912
2.20	280	4.00	712	5.80	1,144	9.20	1,960
2.30	304	4.10	736	5.90	1,168	10.00	2,152
2.40	328	4.20	760	6.00	1,192	13.00	2,872
2.50	352	4.30	784	6.10	1,216	16.00	3,592
2.60	376	4.40	808	6.20	1,240		

Estimated monthly discharge of Talladega Creek at Nottingham.

Month.	Discharge in second-feet.			Run-off.	
	Maximum.	Minimum.	Mean.	Second-feet per square mile.	Depth in inches.
1900.					
August 16–31	....	....	113	0. 72	0. 43
September...................	2, 286	74	575	3. 69	4. 12
October	636	90	190	1. 22	1. 41
November...................	724	130	249	1. 60	1. 79
December	746	152	291	1. 87	2. 16
1901.					
January	1, 802	284	485	3. 11	3. 59
February...................	1, 626	306	449	2. 88	3. 00
March....................	1, 824	240	405	2. 60	3. 00
April....................	2, 330	328	591	3. 79	4. 23
May	724	240	306	1. 96	2. 26
June	526	152	218	1. 40	1. 56
July	196	130	149	. 96	1. 11
August	372	90	148	. 95	1. 10
September...................	614	90	148	. 95	1. 06
October	482	90	123	. 79	. 91
November...................	152	90	97	. 62	. 69
December	1, 714	90	264	1. 69	1. 95
The year	2, 330	90	282	1. 81	24. 46
1902.					
January...................	768	161	276	1. 77	2. 04
February...................	2, 550	262	513	3. 29	3. 43
March....................	2, 616	328	676	4. 33	4. 99
April....................	658	262	390	2. 50	2. 79
May	284	126	187	1. 20	1. 38
June	438	93	128	. 82	. 91
July	306	93	121	. 78	. 90
August	988	64	135	. 87	1. 00
September...................	240	51	74	. 47	. 52
October	768	64	111	. 71	. 82
November...................	284	64	97	. 62	. 69
December	1, 252	78	238	1. 53	1. 76
The year	2, 616	51	246	1. 57	21. 23

Estimated monthly discharge of Talladega Creek at Nottingham—Continued.

Month.	Discharge in second-feet.			Run-off.	
	Maximum.	Minimum.	Mean.	Second-feet per square mile.	Depth in inches.
1903.					
January	400	83	144	.92	1.06
February	3,592	99	832	5.33	5.55
March	1,840	400	637	4.08	4.70
April	880	233	388	2.49	2.78
May	2,344	190	387	2.48	2.86
June	1,192	133	265	1.70	1.90
July	352	83	161	1.03	1.19
August	190	57	105	.67	.77
September	133	46	59	.38	.42
October	151	57	74	.47	.54
November	133	57	75	.48	.54
December	83	69	68	.44	.51
The year	3,592	46	266	1.71	22.82

Net horsepower per foot of fall with a turbine efficiency of 80 per cent for the minimum monthly discharge of Talladega Creek at Nottingham.

Month.	1901.			1902.			1903.		
	Minimum discharge.	Minimum net horsepower per foot of fall.	Duration of minimum.	Minimum discharge.	Minimum net horsepower per foot of fall.	Duration of minimum.	Minimum discharge.	Minimum net horsepower per foot of fall.	Duration of minimum.
	Sec.-feet.		Days.	Sec.-feet.		Days.	Sec.-feet.		Days.
January	284	26.0	1	161	15.0	6	83	7.5	8
February	306	28.0	6	262	24.0	5	99	9.0	3
March	240	22.0	4	328	30.0	1	400	36.0	2
April	328	30.0	2	262	24.0	6	233	21.0	4
May	240	22.0	7	126	11.0	4	190	17.0	6
June	152	14.0	8	93	8.0	7	133	12.0	3
July	130	12.0	18	93	8.0	9	83	7.5	1
August	90	8.0	13	64	6.0	6	57	5.0	7
September	90	8.0	8	51	4.6	5	46	4.0	13
October	90	8.0	21	64	6.0	10	57	5.0	12
November	90	8.0	24	64	6.0	13	57	5.0	7
December	90	8.0	5	78	7.0	4	69	6.0	31

CHOCCOLOCCO CREEK AT JENIFER.

Choccolocco Creek rises in Calhoun and Cleburne counties in a mountainous crystalline region and flows in a, southwesterly course past Anniston, Oxford, and Jenifer, and enters Coosa River near Eureka. The Choccolocco is a very rapid stream and drains a mountainous, well-wooded country.

The gaging station is located at the Louisville and Nashville Railroad bridge, 1¼ miles north of Jenifer. Discharge measurements were made at this point April, 1901, October, 1902, and July, 1903, but the gage was not put in until August 20, 1903.

The bridge is a single-span iron through bridge, 150 feet long, having a trestle approach of 428 feet at the right bank. Measurements are made from the upstream side. The initial point is the end of the bridge, left bank. Distances are marked with white paint on the upstream guard rail. The channel is straight for a quarter of a mile above and a quarter of a mile below, and the current is swift, as there are shoals both above and below the bridge. The bed of the stream is rocky and unchangeable and presents a fair section for accurate measurements. The gage is a 1 by 4 inch rod painted white and graded to feet and tenths. It is nailed to a 3 by 8 inch pine timber, which is spiked to a birch tree on the left bank, 20 feet above the bridge.

Bench mark No. 1: Top of cross beam at 50 feet from initial point, 23 feet above zero of gage. Bench mark No. 2: Copper plug in upstream wing of abutment, 14.19 feet above zero of gage.

The observer is Mr. W. J. Tolbert, a farmer, living one-fourth of a mile from the gage.

The drainage area above this station is 280 square miles.

The following discharge measurements were made by M. R. Hall and others:

Discharge measurements of Choccolocco Creek at Jenifer.

Date.	Gage height.	Discharge	Date.	Gage height.	Discharge.
1901.	*Feet.*	*Second-feet.*	**1903.**	*Feet.*	*Second-feet.*
April 5..............	4.00	1,170	August 20..........	2.26	183
1902.			August 25	2.12	146
October 18	1.90	220	October 2	1.82	90
1903.			November 14.......	2.08	130
July 23	2.25	186			

Daily gage height of Choccolocco Creek at Jenifer.

Day.	Aug.	Sept.	Oct.	Nov.	Dec.	Day.	Aug.	Sept.	Oct.	Nov.	Dec.
1903.						1903.					
1		2.0	1.8	1.8	1.9	17		2.1	1.8	1.9	1.9
2		2.0	1.8	1.8	1.8	18		2.1	1.9	1.9	1.9
3		2.0	1.8	1.8	1.8	19		2.0	1.9	1.9	1.9
4		2.0	1.8	1.8	1.8	20	2.2	2.0	1.9	1.9	1.9
5		2.0	1.8	1.9	1.9	21	2.2	1.9	1.9	1.9	1.9
6		2.0	1.8	2.1	1.9	22	2.1	1.9	1.9	1.9	1.9
7		2.0	1.8	2.0	1.9	23	2.1	1.9	1.8	1.9	2.1
8		2.0	2.6	1.9	1.9	24	2.1	1.9	1.8	1.9	2.0
9		1.9	2.1	1.9	1.9	25	2.1	1.9	1.8	1.9	1.9
10		1.9	2.0	1.9	1.9	26	2.1	1.9	1.8	1.9	2.0
11		1.9	2.0	1.8	1.9	27	2.1	1.9	1.8	1.9	1.9
12		1.9	1.9	2.0	1.9	28	2.0	1.9	1.8	1.9	1.9
13		1.9	1.9	2.0	1.9	29	2.1	1.8	1.8	1.9	1.9
14		1.9	1.8	2.0	1.9	30	2.0	1.8	1.8	1.9	1.9
15		1.9	1.8	2.0	1.9	31	2.0		1.8		1.9
16		2.0	1.8	1.9	1.9						

Rating table for Choccolocco Creek at Jenifer for 1903.

Gage height.	Discharge.	Gage height.	Discharge.	Gage height.	Discharge.	Gage height.	Discharge.
Feet.	Second-feet.	Feet.	Second-feet.	Feet.	Second-feet.	Feet.	Second-feet.
1.80	88	2.10	138	2.40	249	2.70	400
1.85	94	2.15	150	2.45	273	3.00	580
1.90	101	2.20	164	2.50	297	3.50	880
1.95	109	2.25	181	2.55	321	4.00	1,180
2.00	117	2.30	202	2.60	345		
2.05	127	2.35	225	2.65	372		

Estimated monthly discharge of Choccolocco Creek at Jenifer for 1903.

Room.	Discharge in second-feet.		
	Maximum.	Minimum.	Mean.
August 20 to 31 a	164	117	b 137
September	138	88	108
October	345	88	103
November	138	88	103
December	138	88	102

a 12 days.   b Mean for 12 days.

TRIBUTARIES OF COOSA RIVER.

Miscellaneous discharge measurements of tributaries of Coosa River were made by B. M. Hall, M. R. Hall, and assistants, as follows:

Miscellaneous discharge measurements of tributaries of Coosa River.

Date.	Stream.	Location.	Dis-charge.	Remarks.
1898.		●	*Sec.-ft.*	
May 26	Choccolocco Creek	Eureka	171	Low water.
1900.				
Aug. 16	Talladega Creek.........	Kymulga post-office.....	107	Medium.
15	Tallaseehatchee Creek...	Childersburg	102	
17	Hatchet Creek	Goodwater	84	
1901.				
Oct. 16	Wills Creek............	Wesson's mill, 2 miles north of Attalla.	107	Low water.
1903.				
July 2	Little River............	Cedar Bluff.............	123.	
2	Chattooga River........	Gaylesville	325	

Tributaries of Coosa River above Wetumka.

Side.	Stream.	Point.	Drainage area.	Estimated discharge low water 1900–1901.	Net horse-power per foot of fall on 80 per cent turbine.
			Sq. miles	*Sec.-feet.*	
Left....	Sofkahatchee Creek....	Mouth of creek.......	40	12	1.1
	Weoka Creek...........	.....do	85	28	2.5
Right ..	Chestnut Creek.........	.....do	90	30	2.7
Left....	Hatchet Creek..........	.....do	500	165	15.0
	.....do................	Goodwater	105	40	3.6
	Pinthlocco Creek	Mouth of creek.......	60	24	2.2
Right ..	Weogufka Creek........	.....do	120	48	4.3
	Waxahatchee Creek ...	.....do	196	75	6.8
	Yellowleaf Creek	.....do	192	75	6.8
Left....	Tallaseehatchee Creek..	.....do	172	70	6.3
	Talladega Creek	.....do	188	75	6.8
	.....do................	Nottingham	156	66	6.0
Right ..	Kelly Creek...........	Mouth of creek.......	218	88	8.0
Left. ..	Choccolocco Creek.....	.....do	510	153	13.9
	.....do................	Jenifer	273	95	8.6
	Blue Eye Creek........	Mouth of creek.......	26	7	.6

Tributaries of Coosa River above Wetumka—Continued.

Side.	Stream.	Point.	Drainage area.	Estimated discharge low water 1900–1901.	Net horsepower per foot of fall on 80 per cent turbine.
			Sq. miles.	*Sec.-feet.*	
Right ..	Broken Arrow Creek...	Mouth of creek.......	49	18	1. 6
	Trout Creek..............	.....do	23	10	.9
Left....	Cane Creek..............	.....do	94	35	3. 2
	Ohatchee Creek	.....do	217	85	7. 7
	.....do..............	Above Tallasee-hatchee Creek.	86	35	3. 2
	Tallaseehatchee Creek..	Mouth of creek.......	125	50	4. 5
Right ..	Shoal Creek	.....do	31	12	1. 1
	Beaver Creek...........	.....do	33	12	1. 1
	Big Canoe Creek.......	.....do	248	90	8. 2
	.....do..............	Above Little Canoe Creek.	165	65	5. 9
	Little Canoe Creek.....	Mouth of creek.......	34	14	1. 3
	Wills Creek............	.....do	354	160	14. 4
	.....do..............	Above Little Wills Creek.	249	115	10. 4
	.....do..............	Above Wesson's mill .	200	107	9. 7
Left....	Black Creek............	Mouth of creek.......	59	25	2. 3
Right ..	Little Wills Creek	.....do	30	14	1. 3
Left....	Ballplay Creek	.....do	33	15	1. 4
	Terrapin Creek.........	.....do	282	130	11. 8
Right ..	Chattooga River.......	Above Little River ...	384	170	15. 4
	.....do..............	Alabama-Georgia State line.	246	121	11. 0
	Little River	Mouth of river.......	280	130	11. 8
	Coosa River...........	Alabama-Georgia State line.	4, 340	2, 000	181. 8

On the above-named tributaries there are many important water powers, very few of which have been surveyed. The above list, giving the drainage area, the discharge for low season, 1900–1901, and the corresponding net horsepower per foot fall for each of the streams will be very useful in estimating the horsepower available on any shoal, the fall of which may hereafter be surveyed by the owners or by parties contemplating development.

Talladega Creek, in the vicinity of Taylor's mill, has a fall of 73 feet in 1 mile. During the low water of 1900 and 1901 this 73 feet of fall would have produced 438 feet net horsepower without storage. This 73 feet is probably the most precipitous shoal on the large creek, but above it for 4 or 5 miles the creek has a number of rapids and shoals that will admit of good development.

The headwaters of this stream in the neighborhood of the pyrites mines in Clay County have high falls on them.

Choccolocco Creek is a very large and constant stream, and has many rapids where good power could be developed by dams. During a season such as low water of 1900 or 1901 a 10-foot dam near Jenifer would develop 86 net horsepower. A 10-foot dam at any point near the mouth of the creek would develop 140 net horsepower during the given season.

Wills Creek, at the old Wesson mill, two miles north of Attalla, offers a good site for a 25-foot dam. The flow at this point on October 16, 1901, was 107 second-feet, which, with a fall of 25 feet, will give 242 net horsepower. The fall on other tributaries named has not been ascertained.

ALABAMA RIVER AND TRIBUTARIES.

ALABAMA RIVER AT SELMA.

This station was originally established by the United States Engineer Corps; readings are now taken by the United States Weather Bureau. The gage, which is attached to the iron highway bridge, the floor of which is about 60 feet above low water, is in two sections. The lower section, which reads from −0.3 feet to +2.30 feet, is secured to the pile on the lower side of the cofferdam on the draw pier; the upper section, which reads from .2.30 feet to 48 feet, is spiked to the highway bridge. The bench mark, which is an iron bolt driven into the face of a rock bluff 182.3 feet from the first bridge pier, on the road ascending to the city, is 26 feet above the zero of the gage and 87.30 feet above mean sea level. The top of the coping stone of the pivot pier at the highway bridge to which the gage is attached is 56 feet above the zero of the gage, and 117.30 feet above mean sea level. Graduations extend from −3.0 feet to +48 feet. No measurements of discharge were made here during 1899.

The following measurements were made by M. R. Hall and others:

Discharge measurements of Alabama River at Selma.

	Gage height.	Discharge.		Gage height.	Discharge.
1900.	*Feet.*	*Second-feet.*	**1901.**	*Feet.*	*Second-feet.*
April 14	23.60	66,607	August 9	4.35	12,519
May 26	6.10	17,049	October 30	1.10	7,710
August 24	3.10	9,879	**1903.**		
1901.			April 10	22.35	59,101
March 14	14.20	35,518	June 19	6.45	18,815
April 25	34.00	90,332	November 11	1.00	8,290

Daily gage height, in feet, of Alabama River at Selma.

Day.	Jan.	Feb.	Mar.	Apr.	May.	June.	July.	Aug.	Sept.	Oct.	Nov.	Dec.
1899.												
1	6.2	10.8	35.8	23.8	13.9	4.5	2.5	11.1	3.7	−1.2	−0.2	4.8
2	6.2	17.0	36.8	24.3	11.5	4.0	2.5	9.9	4.3	−1.3	−.3	3.7
3	6.5	20.2	38.8	24.9	9.9	4.0	2.6	9.0	4.6	−1.3	−.5	3.0
4	6.2	24.0	37.7	24.1	9.8	4.0	2.6	6.7	4.8	1.3	−.6	3.0
5	6.2	26.8	35.3	22.3	8.5	3.8	2.1	6.5	4.4	−1.1	−.6	2.8
6	5.8	27.2	32.6	20.9	8.1	3.9	1.6	5.8	4.4	1.0	−.7	2.8
7	6.6	27.2	30.5	20.0	7.6	3.8	1.6	5.3	4.1	−1.6	−.8	1.7
8	8.3	29.8	27.5	19.8	7.4	3.7	1.5	4.7	3.9	−.5	−.8	1.7
9	9.7	32.2	23.4	22.3	7.8	3.8	1.4	4.5	3.6	−.4	−.8	1.3
10	11.6	33.9	19.7	25.6	7.3	3.1	1.3	3.7	.6	.4	.9	1.3
11	13.9	34.4	16.9	26.9	7.5	2.8	1.3	3.5	.4	.5	.9	1.4
12	14.8	33.9	15.0	26.6	7.2	2.3	1.2	3.0	.3	−.6	−1.0	4.0
13	21.2	32.0	13.9	25.1	6.8	2.6	1.5	2.9	.3	−.6	1.1	10.4
14	21.9	30.0	16.2	22.6	6.5	2.4	1.2	2.7	.2	−.2	−1.2	16.6
15	19.8	28.0	16.8	19.5	6.3	2.4	1.1	2.6	.2	−.2	−1.2	17.8
16	18.0	26.5	19.4	16.3	6.2	2.3	1.0	2.6	.1	−.2	−1.3	16.3
17	17.5	26.8	21.4	14.0	6.1	2.8	1.0	2.6	.6	−.3	−1.3	13.4
18	17.2	24.0	27.7	13.0	5.6	3.4	.7	3.6	1.3	−.5	−1.3	9.9
19	17.0	22.3	31.6	12.0	5.4	3.9	.6	4.1	1.0	−.7	−1.2	8.4
20	15.8	19.9	33.5	11.2	5.0	3.9	.6	4.3	.9	−.6	−1.0	3.8
21	14.2	19.9	34.7	10.5	4.8	3.1	1.0	3.6	.6	−.7	−1.0	3.8
22	12.6	19.5	34.8	10.4	4.5	2.8	1.6	3.5	−.8	−.8	−.6	3.4
23	10.2	18.8	34.2	10.2	4.4	1.6	5.5	3.4	.7	.2	−.6	3.4
24	10.3	17.8	33.4	10.1	4.8	1.4	10.7	3.1	.8	.0	−.3	3.8
25	9.4	17.3	31.1	12.4	5.0	1.4	14.8	3.6	−1.0	.0	.1	7.6
26	9.0	16.4	32.6	13.5	6.0	1.6	17.0	4.2	1.0	−.5	.2	12.2
27	8.9	20.3	31.8	16.3	6.1	1.7	17.0	3.9	−1.0	−.5	.9	13.5
28	8.6	31.2	30.5	17.9	5.2	2.0	14.9	3.6	−1.0	−.6	3.2	13.7
29	8.6	......	29.3	17.7	4.5	2.2	13.1	4.3	−1.1	−.6	4.6	12.3
30	9.0	......	27.8	16.1	4.3	2.4	12.8	4.6	−1.1	−.6	4.8	11.4
31	9.6	......	26.3	......	4.2	......	11.9	3.8	......	.4	......	8.3

Daily gage height, in feet, of Alabama River at Selma—Continued.

Day.	Jan.	Feb.	Mar.	Apr.	May.	June.	July.	Aug.	Sept.	Oct.	Nov.	Dec.
1900.												
1	7.2	4.8	17.2	19.8	15.4	4.8	34.8	14.0	3.9	.8	2.0	16.0
2	6.6	3.6	19.7	16.8	13.9	4.6	33.0	13.0	3.9	.7	2.0	14.0
3	4.3	3.5	22.2	14.0	13.0	4.4	29.8	11.0	5.8	.6	4.0	11.0
4	3.3	3.9	22.0	12.0	12.0	4.2	26.5	9.0	4.5	.6	9.0	9.0
5	3.0	4.7	20.6	10.9	11.8	5.6	23.5	7.0	5.4	.5	14.0	8.0
6	3.0	6.2	17.8	8.0	10.2	5.1	20.2	6.0	5.0	1.0	13.5	7.5
7	3.0	8.2	15.0	7.8	9.6	4.6	17.0	5.5	4.5	1.8	9.4	7.0
8	3.0	8.4	13.9	8.9	8.8	4.2	14.0	5.0	4.1	2.5	6.3	7.0
9	2.7	8.5	14.9	9.6	8.0	6.8	11.5	4.8	3.3	4.0	4.0	9.0
10	2.7	10.7	18.8	9.5	7.8	11.6	10.0	4.5	1.8	4.2	2.0	9.4
11	3.3	16.0	20.9	9.8	7.5	13.5	9.8	3.2	1.0	2.0	2.0	6.0
12	7.7	22.2	22.2	12.0	7.3	14.0	10.2	3.0	.7	2.0	1.9	5.0
13	12.4	29.9	22.0	17.7	7.0	13.9	10.0	2.8	.6	2.5	1.6	2.0
14	13.5	38.6	19.9	23.4	6.6	12.8	9.9	2.5	1.0	4.3	1.6	3.2
15	14.7	44.0	19.0	25.5	6.4	11.0	9.9	2.5	1.6	6.7	1.0	9.0
16	14.0	47.0	16.9	25.0	6.0	9.0	10.0	2.4	11.0	6.0	1.0	11.0
17	13.2	48.0	15.3	22.5	5.7	8.9	9.9	2.8	18.0	5.2	1.0	11.0
18	12.0	47.9	13.9	23.5	5.5	8.8	9.0	2.7	19.0	2.5	1.0	10.0
19	11.1	47.0	14.3	29.0	5.2	8.6	7.0	2.7	19.4	1.0	1.0	6.0
20	11.1	44.1	14.6	34.8	5.1	10.0	7.0	2.5	16.0	1.0	1.0	5.1
21	13.4	41.6	18.8	39.0	5.0	10.9	6.5	2.3	12.5	.9	1.0	9.0
22	16.9	36.9	23.0	39.8	5.0	12.0	6.5	2.6	10.0	1.0	1.6	14.5
23	18.5	33.2	26.5	41.0	4.8	12.9	6.3	3.6	6.0	1.5	6.0	17.0
24	18.3	22.6	29.0	40.0	5.5	14.0	6.0	3.8	3.0	6.0	9.0	17.2
25	17.0	22.6	30.2	38.5	6.1	17.6	5.8	4.0	1.9	11.5	9.8	17.6
26	14.7	21.1	32.7	35.8	6.2	24.5	5.0	3.5	1.6	12.0	9.9	18.0
27	13.0	19.0	33.3	32.7	6.6	29.0	4.5	3.5	1.0	11.5	13.0	17.0
28	11.2	16.9	32.5	28.5	6.8	32.0	4.4	3.4	1.0	12.3	16.0	14.5
29	8.4		30.5	23.0	6.0	33.5	7.5	3.5	.9	13.0	16.8	12.9
30	6.5		27.7	18.0	5.5	35.0	8.0	4.0	.8	11.0	17.0	11.2
31	4.8		24.4		5.0		11.8	4.2		5.0		11.0
1901.												
1	16.0	13.0	11.3	35.6	12.0	19.0	1.0	2.8	17.0	4.3	1.4	1.8
2	21.0	13.0	9.8	36.5	10.4	17.0	2.0	2.6	13.8	4.3	1.4	1.8
3	24.0	13.6	9.5	37.4	10.0	16.5	2.4	2.6	10.6	5.0	1.4	1.9
4	24.6	17.0	9.3	38.5	9.6	18.5	6.6	3.0	8.8	7.4	1.5	2.2
5	24.0	24.9	9.6	38.4	8.2	19.0	6.0	3.0	8.0	7.9	1.5	2.3
6	23.0	30.1	9.6	37.2	7.0	19.8	5.6	2.8	7.4	6.4	1.4	2.3
7	18.0	33.0	9.4	35.5	7.0	18.5	5.6	6.6	6.0	5.8	1.6	2.4
8	15.0	35.1	8.0	38.0	6.8	17.4	5.6	5.6	5.2	5.0	1.6	2.1
9	13.3	35.6	7.7	28.0	6.4	16.1	5.5	4.2	4.4	4.3	1.5	2.0
10	10.0	35.7	7.9	22.6	6.0	14.8	5.5	4.4	4.4	3.4	1.5	2.2
11	8.1	33.0	9.0	17.4	5.0	12.0	5.2	4.0	3.6	2.9	1.5	2.4
12	16.5	31.4	10.2	14.0	5.0	11.0	5.2	5.3	3.4	2.8	1.5	2.4
13	28.0	31.2	12.0	12.0	4.8	9.5	5.1	6.2	3.2	2.6	1.4	2.4
14	34.0	27.0	14.1	11.8	4.5	8.0	4.4	4.4	3.7	2.5	1.4	2.6
15	38.0	26.0	15.4	12.0	4.0	7.6	3.7	3.8	4.3	2.2	1.4	5.0
16	39.5	20.6	15.0	15.0	3.8	7.0	3.7	6.0	4.0	2.5	1.3	10.0
17	40.0	16.9	14.8	16.5	3.4	6.1	3.5	7.4	5.0	2.5	1.3	18.0
18	39.0	14.6	12.0	17.3	8.2	6.0	4.8	11.0	5.0	2.4	1.3	21.6
19	37.5	13.1	11.1	22.0	3.0	5.0	6.3	12.0	5.5	2.0	1.4	22.0
20	35.0	12.6	10.5	28.6	3.0	4.1	7.3	16.0	9.5	2.3	1.6	21.5
21	32.4	12.0	11.0	35.0	3.5	3.4	6.0	17.6	11.4	2.5	1.8	18.7
22	29.0	11.8	11.9	38.0	4.7	3.0	5.5	18.8	11.5	2.5	1.8	15.0
23	24.0	11.7	12.2	39.0	9.4	3.0	5.4	20.0	11.2	2.2	1.8	14.2
24	22.0	11.6	13.0	38.0	17.0	2.6	5.4	20.9	10.0	2.0	1.8	12.0
25	14.0	11.2	14.7	35.8	19.0	2.2	4.4	22.8	7.5	2.0	1.9	11.0
26	12.8	11.5	17.0	31.9	20.0	2.0	4.0	24.6	6.0	2.0	1.9	7.1
27	12.8	11.4	22.5	28.0	20.9	1.5	4.0	24.8	4.4	1.8	2.0	6.2
28	12.7	11.3		24.2	22.0	1.3	8.6	22.9	4.0	1.6	2.0	6.0
29	12.7		31.0	19.5	21.8	1.2	2.9	21.0	4.0	1.5	2.0	11.0
30	13.0		33.0	15.0	20.7	1.2	2.9	20.6	4.2	1.3	1.8	35.0
31	13.0		34.5		19.5		2.8	19.5		1.3		35.0

Daily gage height, in feet, of Alabama River at Selma—Continued.

Day.	Jan.	Feb.	Mar.	Apr.	May.	June.	July.	Aug.	Sept.	Oct.	Nov.	Dec.
1902.												
1	41.8	16.8	34.3	50.1	8.4	3.6	1.5	.4	3.5	3.7	1.4	6.0
2	45.0	23.0	41.5	50.7	8.7	3.5	1.2	.4	3.7	5.2	.8	6.3
3	46.6	29.8	45.2	50.0	8.4	3.5	1.0	.7	3.0	5.0	.7	8.8
4	46.3	35.0	47.1	48.6	7.8	4.0	1.0	.8	3.8	4.0	.5	10.5
5	41.4	37.5	47.1	46.0	7.4	4.0	.9	.9	3.0	3.6	.2	13.6
6	37.0	37.9	46.2	42.4	7.0	3.8	.8	1.1	3.0	3.0	.4	14.0
7	35.0	37.9	44.4	39.0	7.0	3.5	.7	1.4	2.5	2.8	.8	14.2
8	33.0	37.0	43.0	35.6	7.0	3.2	.6	1.3	2.2	2.5	3.9	18.7
9	30.0	34.0	41.8	32.0	6.9	2.9	.6	1.1	2.0	2.5	4.8	12.0
10	25.4	30.0	39.0	28.9	6.8	2.9	.6	1.1	1.0	2.0	4.0	9.8
11	17.0	24.0	35.1	25.6	6.7	2.9	1.0	1.5	.9	2.0	2.8	7.8
12	14.9	17.5	30.0	22.0	6.5	2.8	1.1	1.5	.8	2.0	1.9	7.2
13	10.1	14.0	25.5	19.0	6.3	2.6	1.1	1.1	.7	4.0	1.5	5.3
14	9.0	11.7	21.0	16.2	6.0	2.5	1.8	.9	.6	6.0	1.4	4.2
15	8.1	11.6	19.2	14.6	5.8	2.5	1.9	.8	.2	5.8	.9	8.8
16	7.5	12.0	19.8	14.2	6.0	2.8	1.9	.8	.0	5.0	.7	7.6
17	7.0	12.3	23.0	13.0	7.4	2.6	1.9	.7	.1	4.6	.5	13.8
18	6.5	13.0	28.0	12.2	8.7	2.4	1.9	.6	.2	4.5	.5	21.5
19	6.5	12.0	32.0	12.8	9.5	2.3	1.9	.1	.3	4.0	.7	25.5
20	6.5	11.8	33.6	13.2	9.0	2.2	2.0	.0	.3	2.0	.9	25.6
21	8.0	11.8	33.8	13.0	8.6	2.1	2.1	-.1	.0	1.8	1.4	22.0
22	8.1	11.9	29.8	13.0	7.8	2.0	2.5	.2	.0	1.4	1.5	17.0
23	9.0	12.0	26.0	11.4	6.6	2.0	2.4	-.3	-.1	1.0	1.5	15.0
24	12.0	12.2	23.0	10.6	5.9	2.0	1.8	.2	-.2	1.0	1.3	15.0
25	13.8	12.0	22.8	10.0	5.4	1.9	1.0	.4	-.2	.9	1.7	15.0
26	13.0	12.0	24.0	9.8	5.4	1.9	.8	-.3	-.2	.6	2.6	14.0
27	11.5	14.0	24.9	9.4	5.0	1.8	.6	.4	.2	.2	4.8	12.8
28	11.4	23.6	30.0	9.0	4.8	1.8	.5	-.2	.2	.0	7.6	10.9
29	12.0	......	38.0	8.6	4.4	1.8	.4	.1	.5	-.1	6.9	9.0
30	14.6	......	45.1	8.5	4.0	1.7	.5	.2	.9	1.2	5.8	7.0
31	16.0	......	48.9	......	3.8	......	.5	.4	......	1.8	......	6.4
1903.												
1	6.8	7.0	33.5	28.0	9.6	7.0	10.0	3.5	1.5	.1	-.2	.4
2	8.0	6.9	38.0	30.2	9.4	8.0	9.8	4.8	1.6	.0	.2	.5
3	8.9	6.8	41.0	31.4	8.8	9.6	9.0	6.8	1.6	.0	-.1	.5
4	11.0	7.0	42.6	31.4	8.7	12.7	8.0	6.8	1.4	.2	.0	.4
5	11.9	7.0	42.8	30.2	8.7	13.5	7.0	6.1	1.4	.2	.1	.5
6	13.0	9.6	42.0	29.0	8.4	13.2	6.8	5.0	1.2	-.2	.1	.3
7	11.0	14.0	40.2	27.8	7.8	14.5	6.4	5.5	1.0	-.3	.3	.4
8	10.0	23.5	38.5	26.3	7.8	16.0	6.4	5.0	.7	-.3	.2	.4
9	9.5	23.0	36.7	24.5	8.0	17.0	7.6	5.0	.5	-.3	.5	.5
10	8.2	39.0	35.0	22.4	10.0	17.0	8.8	5.4	.4	.0	.9	.5
11	7.0	44.3	34.0	22.2	9.8	17.0	9.0	6.6	.4	.4	1.0	.5
12	7.9	48.0	31.3	20.7	9.0	16.0	8.0	6.1	.3	.5	1.1	.6
13	10.6	49.5	29.8	21.5	11.2	14.0	6.8	5.0	.2	.6	1.0	.8
14	12.8	50.2	27.5	20.8	16.0	11.0	6.6	4.6	.1	.4	1.0	.9
15	13.0	50.6	27.0	20.0	21.0	9.5	5.8	3.3	.1	.2	1.0	.9
16	12.0	49.9	27.0	20.3	25.0	8.6	5.8	8.0	.0	.4	.9	.9
17	11.8	49.0	27.0	20.8	30.0	8.0	6.5	2.8	.6	.4	.9	.9
18	10.0	47.7	26.3	20.9	31.2	7.4	8.8	4.0	1.3	.6	.8	.8
19	9.0	47.3	24.0	19.0	29.7	6.8	8.6	5.0	2.2	.6	.7	.6
20	7.9	47.8	21.3	17.6	24.0	6.5	7.0	6.7	2.5	.5	.7	.6
21	7.0	47.9	18.0	15.4	16.5	5.9	6.0	7.7	1.7	.5	.8	.5
22	6.4	47.0	16.0	17.2	11.8	5.9	5.0	7.4	1.5	.4	.8	.6
23	6.0	45.0	16.0	17.5	9.8	6.4	4.5	5.0	1.5	.4	.8	1.0
24	5.8	42.0	19.0	14.8	8.7	7.4	3.8	4.0	1.3	.4	.7	1.0
25	5.4	38.0	23.0	12.8	7.9	7.5	3.6	3.1	.8	.5	.8	1.0
26	5.2	33.8	25.1	11.7	7.6	7.5	3.5	3.0	.6	.6	.9	1.1
27	5.2	28.0	26.3	11.0	7.0	7.0	3.7	2.6	.6	.5	1.0	2.0
28	5.8	28.9	26.7	10.8	6.8	6.8	4.0	2.0	.6	.2	.8	8.0
29	6.0	......	25.4	10.0	6.4	9.5	4.0	1.7	.6	.0	.5	3.1
30	7.0	......	25.2	9.8	6.7	11.4	3.5	1.6	.5	.0	.5	2.6
31	7.2	......	26.2	......	6.5	......	3.5	1.5	......	.0	......	1.8

Rating table for Alabama River at Selma for 1900 and 1901.

Gage height.	Discharge.	Gage height.	Discharge.	Gage height.	Discharge.	Gage height.	Discharge.
Feet.	*Second feet.*	*Feet.*	*Second feet.*	*Feet.*	*Second feet.*	*Feet.*	*Second feet.*
0.0	6,700	4.0	11,820	8.0	22,180	12.0	32,820
.1	6,770	4.1	12,015	8.1	22,446	12.1	33,086
.2	6,845	4.2	12,220	8.2	22,712	12.2	33,352
.3	6,925	4.3	12,435	8.3	22,978	12.3	33,618
.4	7,010	4.4	12,660	8.4	23,244	12.4	33,884
.5	7,100	4.5	12,900	8.5	23,510	12.5	34,150
.6	7,184	4.6	13,150	8.6	23,776	12.6	34,416
.7	7,282	4.7	13,405	8.7	24,042	12.7	34,682
.8	7,384	4.8	13,668	8.8	24,308	12.8	34,948
.9	7,488	4.9	13,934	8.9	24,574	12.9	35,214
1.0	7,596	5.0	14,200	9.0	24,840	13.0	35,480
1.1	7,706	5.1	14,466	9.1	25,106	13.1	35,746
1.2	7,818	5.2	14,732	9.2	25,372	13.2	36,012
1.3	7,931	5.3	14,998	9.3	25,638	13.3	36,278
1.4	8,045	5.4	15,264	9.4	25,904	13.4	36,544
1.5	8,160	5.5	15,530	9.5	26,170	13.5	36,810
1.6	8,270	5.6	15,796	9.6	26,436	13.6	37,076
1.7	8,393	5.7	16,062	9.7	26,702	13.7	37,342
1.8	8,511	5.8	16,328	9.8	26,968	13.8	37,608
1.9	8,630	5.9	16,594	9.9	27,234	13.9	37,874
2.0	8,750	6.0	16,860	10.0	27,500	14.0	38,140
2.1	8,872	6.1	17,126	10.1	27,760	14.1	38,406
2.2	8,996	6.2	17,392	10.2	28,032	14.2	38,672
2.3	9,124	6.3	17,658	10.3	28,290	14.3	38,838
2.4	9,256	6.4	17,924	10.4	28,564	14.4	39,104
2.5	9,392	6.5	18,190	10.5	28,830	14.5	39,370
2.6	9,532	6.6	18,456	10.6	29,096	14.6	39,676
2.7	9,676	6.7	18,722	10.7	29,362	14.7	40,002
2.8	9,822	6.8	18,988	10.8	29,628	14.8	40,268
2.9	9,970	6.9	19,254	10.9	29,894	14.9	40,534
3.0	10,120	7.0	19,520	11.0	30,160	15.0	40,800
3.1	10,272	7.1	19,786	11.1	30,426	15.1	41,066
3.2	10,428	7.2	20,052	11.2	30,692	15.2	41,332
3.3	10,588	7.3	20,318	11.3	30,958	15.3	41,598
3.4	10,752	7.4	20,584	11.4	31,224	15.4	41,864
3.5	10,920	7.5	20,850	11.5	31,490	15.5	42,130
3.6	11,092	7.6	21,116	11.6	31,756	15.6	42,396
3.7	11,268	7.7	21,382	11.7	32,022	15.7	42,662
3.8	11,448	7.8	21,648	11.8	32,228	15.8	42,928
3.9	11,632	7.9	21,914	11.9	32,556	15.9	43,194

Rating table for Alabama River at Selma for 1900 and 1901—Continued.

Gage height.	Discharge.	Gage height.	Discharge.	Gage height.	Discharge.	Gage height.	Discharge.
Feet.	*Second-feet.*	*Feet.*	*Second-feet.*	*Feet.*	*Second-feet.*	*Feet.*	*Second-feet.*
16.0	43,460	19.6	53,036	23.2	62,612	26.8	72,188
16.1	43,726	19.7	53,302	23.3	62,878	26.9	72,454
16.2	43,992	19.8	53,568	23.4	63,144	27.0	72,720
16.3	44,258	19.9	53,884	23.5	63,410	27.1	72,986
16.4	44,524	20.0	54,100	23.6	63,676	27.2	73,252
16.5	44,790	20.1	54,366	23.7	63,942	27.3	73,518
16.6	45,056	20.2	54,632	23.8	64,208	27.4	73,784
16.7	45,322	20.3	54,898	23.9	64,474	27.5	74,050
16.8	45,588	20.4	55,164	24.0	64,740	27.6	74,316
16.9	45,854	20.5	55,430	24.1	65,006	27.7	74,582
17.0	46,120	20.6	55,696	24.2	65,272	27.8	74,848
17.1	46,386	20.7	55,962	24.3	65,538	27.9	75,114
17.2	46,652	20.8	56,228	24.4	65,804	28.0	75,380
17.3	46,918	20.9	56,494	24.5	66,070	28.1	75,646
17.4	47,184	21.0	56,760	24.6	66,336	28.2	75,912
17.5	47,450	21.1	57,026	24.7	66,602	28.3	76,178
17.6	47,716	21.2	57,292	24.8	66,868	28.4	76,444
17.7	47,982	21.3	57,558	24.9	67,134	28.5	76,710
17.8	48,248	21.4	57,824	25.0	67,400	28.6	76,976
17.9	48,514	21.5	58,090	25.1	67,666	28.7	77,242
18.0	48,780	21.6	58,356	25.2	67,932	28.8	77,508
18.1	49,046	21.7	58,622	25.3	68,198	28.9	77,744
18.2	49,312	21.8	58,888	25.4	68,464	29.0	78,040
18.3	49,578	21.9	59,154	25.5	68,730	29.1	78,306
18.4	49,844	22.0	59,420	25.6	68,996	29.2	78,572
18.5	50,110	22.1	59,686	25.7	69,262	29.3	78,838
18.6	50,376	22.2	59,952	25.8	69,528	29.4	79,104
18.7	50,642	22.3	60,218	25.9	69,794	29.5	79,370
18.8	50,908	22.4	60,484	26.0	70,060	29.6	79,636
18.9	51,174	22.5	60,750	26.1	70,326	29.7	79,902
19.0	51,440	22.6	61,016	26.2	70,592	29.8	80,168
19.1	51,706	22.7	61,282	26.3	70,858	29.9	80,434
19.2	51,972	22.8	61,548	26.4	71,124	30.0	80,700
19.3	52,238	22.9	61,814	26.5	71,390		
19.4	52,504	23.0	62,080	26.6	71,656		
19.5	52,770	23.1	62,346	26.7	71,922		

Rating table for Alabama River at Selma for 1902.

Gage height.	Discharge.	Gage height.	Discharge.	Gage height.	Discharge.	Gage height.	Discharge.
Feet.	*Second-feet.*	*Feet.*	*Second-feet.*	*Feet.*	*Second-feet.*	*Feet.*	*Second-feet.*
—0.4	6,470	3.6	11,092	11.5	31,490	28.0	75,380
—.2	6,575	3.8	11,448	12.0	32,820	29.0	78,040
.0	6,700	4.0	11,820	12.5	34,150	30.0	80,700
.2	6,845	4.2	12,220	13.0	35,480	31.0	83,360
.4	7,009	4.4	12,660	13.5	36,810	32.0	86,020
.6	7,189	4.6	13,150	14.0	38,140	33.0	88,680
.8	7,384	4.8	13,668	14.5	39,370	34.0	91,340
1.0	7,596	5.0	14,200	15.0	40,800	35.0	94,000
1.2	7,818	5.5	15,530	1€.0	43,460	36.0	96,660
1.4	8,045	6.0	16,860	17.0	46,120	37.0	99,320
1.6	8,276	6.5	18,190	18.0	48,780	38.0	101,980
1.8	8,511	7.0	19,520	19.0	51,440	39.0	104,640
2.0	8,750	7.5	20,850	20.0	54,100	40.0	107,300
2.2	8,996	8.0	22,180	21.0	56,760	41.0	109,960
2.4	9,256	8.5	23,510	22.0	59,420	42.0	112,620
2.6	9,532	9.0	24,840	23.0	62,080	43.0	115,280
2.8	9,822	9.5	26,170	24.0	64,740	44.0	117,940
3.0	10,120	10.0	27,500	25.0	67,400	45.0	120,600
3.2	10,428	10.5	28,830	26.0	70,060	46.0	123,260
3.4	10,752	11.0	30,160	27.0	72,720	47.0	125,920

Rating table for Alabama River at Selma for 1903.

Gage height.	Discharge.	Gage height.	Discharge.	Gage height.	Discharge.	Gage height.	Discharge.
Feet.	*Second-feet.*	*Feet.*	*Second-feet.*	*Feet.*	*Second-feet.*	*Feet.*	*Second-feet.*
−0.30	6,262	4.30	13,570	8.90	23,540	13.50	34,410
−.20	6,364	4.40	13,770	9.00	23,760	13.60	34,670
−.10	6,470	4.50	13,980	9.10	23,990	13.70	34,940
.00	6,580	4.60	14,190	9.20	24,220	13.80	35,210
.10	6,692	4.70	14,400	9.30	24,450	13.90	35,480
.20	6,806	4.80	14,610	9.40	24,680	14.00	35,750
.30	6,922	4.90	14,820	9.50	24,910	15.00	38,450
.40	7,040	5.00	15,030	9.60	25,140	16.00	41,200
.50	7,160	5.10	15,240	9.70	25,370	17.00	43,950
.60	7,282	5.20	15,450	9.80	25,600	18.00	46,700
.70	7,406	5.30	15,660	9.90	25,830	19.00	49,450
.80	7,532	5.40	15,870	10.00	26,060	20.00	52,200
.90	7,660	5.50	16,080	10.10	26,290	21.00	54,950
1.00	7,790	5.60	16,290	10.20	26,520	22.00	57,700
1.10	7,920	5.70	16,500	10.30	26,750	23.00	60,450
1.20	8,060	5.80	16,720	10.40	26,980	24.00	63,200
1.30	8,200	5.90	16,940	10.50	27,210	25.00	65,950
1.40	8,340	6.00	17,160	10.60	27,440	26.00	68,700
1.50	8,480	6.10	17,380	10.70	27,670	27.00	71,450
1.60	8,620	6.20	17,600	10.80	27,900	28.00	74,200
1.70	8,770	6.30	17,820	10.90	28,130	29.00	76,950
1.80	8,920	6.40	18,040	11.00	28,360	30.00	79,700
1.90	9,070	6.50	18,260	11.10	28,590	31.00	82,450
2.00	9,230	6.60	18,480	11.20	28,820	32.00	85,200
2.10	9,390	6.70	18,700	11.30	29,050	33.00	87,950
2.20	9,550	6.80	18,920	11.40	29,280	34.00	90,700
2.30	9,710	6.90	19,140	11.50	29,510	35.00	93,450
2.40	9,880	7.00	19,360	11.60	29,740	36.00	96,200
2.50	10,050	7.10	19,580	11.70	29,970	37.00	98,950
2.60	10,220	7.20	19,800	11.80	30,200	38.00	101,700
2.70	10,400	7.30	20,020	11.90	30,430	39.00	104,450
2.80	10,590	7.40	20,240	12.00	30,660	40.00	107,200
2.90	10,780	7.50	20,460	12.10	30,900	41.00	109,950
3.00	10,970	7.60	20,680	12.20	31,140	42.00	112,700
3.10	11,170	7.70	20,900	12.30	31,380	43.00	115,450
3.20	11,370	7.80	21,120	12.40	31,620	44.00	118,200
3.30	11,570	7.90	21,340	12.50	31,860	45.00	120,950
3.40	11,770	8.00	21,560	12.60	32,110	46.00	123,700
3.50	11,970	8.10	21,780	12.70	32,360	47.00	126,450
3.60	12,170	8.20	22,000	12.80	32,610	48.00	129,200
3.70	12,360	8.30	22,220	12.90	32,860	49.00	131,950
3.80	12,570	8.40	22,440	13.00	33,110	50.00	134,700
3.90	12,770	8.50	22,660	13.10	33,370	51.00	137,450
4.00	12,970	8.60	22,880	13.20	33,630		
4.10	13,170	8.70	23,100	13.30	33,890		
4.20	13,370	8.80	23,320	13.40	34,150		

Estimated monthly discharge of Alabama River at Selma.

[Drainage area, 13,500 square miles.]

Month.	Discharge in second-feet.			Run-off.	
	Maximum.	Minimum.	Mean.	Second-feet per square mile.	Depth in inches.
1900.					
January	50,110	9,676	26,495	1.96	2.26
February	128,540	10,920	63,763	4.72	4.91
March	89,478	37,874	58,272	4.32	4.98
April	109,960	21,648	60,909	4.51	5.03
May	41,864	13,668	21,090	1.56	1.80
June	94,000	12,220	35,288	2.61	2.91
July	93,468	12,660	33,964	2.52	2.90
August	38,140	9,124	14,156	1.05	1.21
September	52,504	7,189	17,366	1.29	1.44
October	35,480	7,097	14,492	1.07	1.23
November	46,120	7,596	18,506	1.37	1.53
December	48,780	8,750	28,989	2.15	2.48
The year	128,540	7,097	33,772	2.34	32.68
1901.					
January	107,300	22,446	61,213	4.53	5.22
February	95,862	30,692	55,037	4.08	4.25
March	92,670	21,382	39,017	2.89	3.33
April	104,640	32,288	73,048	5.41	6.04
May	59,420	10,120	26,966	2.00	2.31
June	53,568	7,818	26,030	1.93	2.15
July	21,318	7,596	13,536	1.00	1.15
August	66,868	9,532	30,853	2.29	2.64
September	46,120	10,428	19,394	1.44	1.61
October	21,914	7,931	11,022	.82	.95
November	8,750	7,931	8,266	.61	.68
December	94,000	8,511	26,638	1.97	2.27
The year	107,300	7,596	32,585	2.47	32.60

*Estimated monthly discharge of Alabama River at Selma—*Continued.

Month.	Discharge in second-feet.			Run-off.	
	Maximum.	Minimum.	Mean.	Second-feet per square mile.	Depth in inches.
1902.					
January	124,856	18,190	52,655	3.42	3.94
February	101,714	31,756	54,898	3.56	3.71
March	130,974	51,972	90,404	5.87	6.77
April	135,762	23,510	62,017	4.03	4.49
May	26,170	11,448	18,859	1.22	1.41
June	11,820	8,393	9,682	.63	.70
July	9,392	7,009	7,897	.51	.59
August	8,160	6,470	7,176	.47	.54
September	11,448	6,520	7,871	.51	.57
October	16,860	6,635	10,184	.66	.76
November	21,116	6,845	9,557	.62	.69
December	68,996	11,448	33,122	2.15	2.48
The year	135,762	6,470	30,360	1.97	26.65
1903.					
January	33,100	15,450	23,039	1.50	1.73
February	136,350	18,920	90,958	5.91	6.15
March	114,900	41,200	78,139	5.07	5.85
April	83,560	25,600	53,852	3.50	3.90
May	83,000	18,040	33,338	2.16	2.49
June	43,950	16,950	27,142	1.76	1.96
July	26,060	11,970	18,122	1.18	1.36
August	20,900	8,480	14,205	.92	1.06
September	10,050	6,580	7,781	.51	.57
October	7,282	6,262	6,835	.44	.51
November	7,920	6,364	7,310	.47	.52
December	11,170	6,922	7,801	.51	.59
The year	136,350	6,262	30,710	1.99	26.69

Net horsepower per foot of fall with a turbine efficiency of 80 per cent for the minimum monthly discharge of Alabama River at Selma.

Month.	1899.			1900.			1901.		
	Minimum discharge.	Minimum net horsepower per foot of fall.	Duration of minimum.	Minimum discharge.	Minimum net horsepower per foot of fall.	Duration of minimum.	Minimum discharge.	Minimum net horsepower per foot of fall.	Duration of minimum.
	Sec.-ft.		*Days.*	*Sec.-ft.*		*Days.*	*Sec.-ft.*		*Days.*
January	16,328	1,484	1	9,676	880	2	22,446	2,041	1
February	29,628	2,693	1	10,920	993	1	30,692	2,790	1
March	37,874	3,443	1	37,874	3,443	2	21,382	1,944	1
April	27,760	2,524	1	21,648	1,968	1	32,288	2,935	1
May	12,220	1,111	1	13,668	1,243	1	10,120	920	2
June	8,045	731	2	12,220	1,111	2	7,818	711	2
July	7,184	653.	2	12,660	1,151	1	7,596	691	1
August	9,532	867	3	9,124	829	1	9,532	867	2
September	5,800	527	2	7,189	653	1	10,428	948	1
October	5,400	491	1	7,100	645	1	7,931	721	2
November	5,700	518	3	7,596	691	7	7,931	721	3
December	7,931	721	2	8,750	795	1	8,511	774	2

CAHABA RIVER AND TRIBUTARIES.

CAHABA RIVER AT CENTERVILLE.

This station is at the Bibb County highway bridge, one-fourth of a mile west of the court-house at Centerville. The bridge is a single-span iron through bridge. The length of the span is about 175 feet. The floor of the bridge is 41½ feet above low water, and the stream is 130 feet wide at low water.

The initial point of sounding is at the end of the iron bridge, left bank, downstream. The gage is of wire, with rod fastened to the outside of the downstream guard rail, and graded to feet and tenths. The gage pulley is at station 100. Bench mark No. 1, at the downstream end of the top of the iron crossbeam under the bridge floor at station 100 from initial point, is 42.85 above gage datum. Bench mark No. 2, at the top of the bottom flange of the same crossbeam, directly under bench mark No. 1, is 41.40 above datum of gage. The banks are high but overflow at time of high water. The section is swift and tolerably uniform, and the bottom appears to be rock.

The river observer is Mr. Clyde Lowrey, who lives about a third of a mile from the gage.

The following discharge measurements have been made on Cahaba River at Centerville by M. R. Hall and others:

Discharge measurements of Cahaba River at Centerville.

Date.	Gage height.	Discharge.	Date.	Gage height.	Discharge.
1901.	*Feet.*	*Second-feet.*	**1903.**	*Feet.*	*Second-feet.*
April 25	5.50	1,925	June 17.............	2.05	416
August 7.............	1.30	399	June 18.............	2.00	394
1902.			July 20	3.23	757
January 25...........	5.15	1,707	July 21	2.36	516
April 7	8.60	2,823	September 28	1.31	212
July 9	2.40	251	September 29	1.30	218
1903.			November 9........	1.32	203
April 8	5.15	1,637	November 10.......	1.40	223
April 9	6.65	2,225			

Daily gage height, in feet, of Cahaba River at Centerville.

Day.	Jan.	Feb.	Mar.	April.	May.	June.	July.	Aug.	Sept.	Oct.	Nov.	Dec.
1901.												
1									2.2	2.1	1.2	1.4
2									2.0	2.0	1.2	1.3
3									1.9	1.8	1.2	1.6
4									1.8	1.4	1.2	1.9
5									1.6	2.3	1.2	1.8
6									1.5	4.6	1.1	1.6
7								1.3	1.4	3.1	1.1	1.7
8								1.3	1.4	2.6	1.1	1.6
9								1.2	1.4	2.3	1.2	1.7
10								1.2	1.4	1.9	1.2	2.1
11								1.2	1.3	1.6	1.2	2.0
12								1.2	1.3	1.4	1.3	1.8
13								1.3	1.3	1.4	1.4	1.9
14								1.5	2.6	1.4	1.3	2.9
15								2.1	2.4	1.3	1.3	19.0
16								7.9	2.0	1.3	1.3	15.0
17								7.7	2.1	1.3	1.3	12.3
18								8.1	2.6	1.3	1.3	4.4
19								9.1	3.9	1.3	1.7	3.6
20								10.6	4.1	1.3	1.6	3.1
21								14.7	2.0	1.3	1.6	2.9
22								10.3	1.9	1.3	1.5	2.6
23								7.9	1.8	1.3	1.6	2.5
24								5.6	1.6	1.2	1.6	2.6
25								4.8	1.5	1.2	1.5	2.7
26								4.1	1.4	1.2	1.5	2.7
27								3.9	1.4	1.2	1.5	2.7
28								3.4	1.4	1.2	1.4	6.0
29								3.0	2.6	1.3	1.4	24.0
30								2.8	2.2	1.3	1.4	24.0
31								2.5		1.3		21.0

Daily gage height, in feet, of Cahaba River at Centerville—Continued.

Day.	Jan.	Feb.	Mar.	April.	May.	June.	July.	Aug.	Sept.	Oct.	Nov.	Dec.
1902.												
1	13.6	16.8	24.0	15.4	4.9	2.8	2.3	2.2	-2.3	2.4	2.0	3.0
2	9.0	24.6	18.2	12.1	4.5	2.8	2.3	2.8	2.3	2.6	2.0	2.5
3	6.9	21.2	13.3	11.4	4.0	2.7	2.2	2.5	3.3	2.6	2.0	2.9
4	5.7	16.6	9.8	10.9	3.7	2.6	2.2	2.5	3.2	2.5	2.0	3.1
5	4.9	10.6	12.0	10.4	3.6	2.6	2.2	2.4	2.9	2.4	2.0	5.6
6	4.5	8.3	12.1	8.6	3.5	2.5	2.2	2.4	2.4	2.2	2.0	5.1
7	4.1	7.3	8.7	8.2	3.6	2.5	2.2	2.3	2.3	2.2	2.9	4.9
8	3.9	6.7	8.3	13.6	3.5	2.4	2.2	2.3	2.3	2.2	2.4	4.2
9	8.7	5.8	7.9	12.2	3.4	2.8	2.3	2.8	2.3	2.7	2.4	3.9
10	8.4	5.4	7.6	9.6	3.4	2.8	2.5	2.8	2.2	2.8	2.8	3.6
11	8.2	5.1	6.3	8.0	3.8	2.4	2.9	2.8	2.2	10.8	2.8	8.2
12	3.0	4.8	6.1	7.2	3.3	2.4	3.5	2.2	2.1	7.1	2.2	3.0
13	2.9	4.3	5.9	6.9	8.4	2.3	3.3	2.1	2.1	5.4	2.2	8.0
14	2.7	5.7	5.2	6.5	3.5	2.3	3.2	2.1	2.0	8.4	2.2	3.0
15	2.6	5.9	6.2	6.1	4.1	2.3	2.6	2.1	2.0	8.2	2.1	3.9
16	2.4	6.3	22.3	5.6	8.4	2.8	2.4	2.1	2.0	8.0	2.2	15.3
17	2.3	5.6	22.2	5.9	8.4	2.8	2.2	2.1	2.0	2.9	2.2	13.5
18	8.1	5.3	17.3	5.7	5.3	2.2	2.2	2.1	2.0	2.9	2.3	12.7
19	3.6	4.9	11.1	5.6	4.8	2.3	2.2	2.1	2.0	2.6	2.4	9.9
20	3.9	4.8	8.6	5.2	4.2	2.5	2.3	2.1	2.1	2.5	2.2	7.5
21	4.6	5.1	8.9	4.9	4.2	2.4	2.4	2.1	2.1	2.5	2.2	6.3
22	6.9	5.1	9.1	4.6	3.9	2.6	2.3	2.1	2.0	2.4	2.2	7.8
23	5.8	5.0	7.3	4.5	3.6	2.5	2.2	2.1	2.0	2.3	2.2	8.2
24	5.1	5.1	7.8	4.5	3.5	2.4	2.2	2.1	2.0	2.2	2.2	10.3
25	5.0	5.9	13.6	4.4	3.3	2.4	2.2	2.1	2.0	2.2	2.2	7.3
26	4.9	6.5	11.9	4.3	3.2	2.4	2.1	2.1	2.0	2.1	2.9	6.5
27	6.2	8.1	28.8	4.2	3.0	2.4	2.1	2.2	2.1	2.1	3.4	5.1
28	7.0	27.6	35.0	4.1	2.9	2.3	2.1	2.3	2.3	2.0	4.2	4.9
29	8.3		29.2	4.3	2.8	2.3	2.1	2.5	2.2	2.0	3.5	4.5
30	8.1		25.2	4.9	2.8	2.3	2.2	2.6	3.0	2.0	3.2	4.1
31	11.6		24.0		2.9		2.2	2.5		2.0		4.9
1903.												
1	5.9	5.8	23.8	8.2	3.0	2.8	2.4	4.8	1.5	1.3	1.6	1.4
2	6.4	4.9	18.1	7.0	3.1	2.7	2.2	6.1	1.5	1.2	1.6	1.4
3	7.7	6.4	12.6	6.3	3.0	2.7	2.1	4.6	1.5	1.2	1.5	1.4
4	7.4	8.8	12.0	5.8	2.9	2.8	2.1	3.0	1.5	1.3	1.5	1.4
5	6.1	12.9	13.3	5.1	2.8	2.9	2.0	2.9	1.5	1.3	1.5	1.4
6	5.6	12.0	11.1	4.9	2.7	2.9	1.9	2.4	1.5	1.3	1.5	1.5
7	4.9	20.6	9.8	4.7	2.8	3.1	3.6	2.8	1.5	1.3	1.4	1.6
8	4.5	31.6	12.3	4.7	3.1	3.2	3.4	2.9	1.4	2.8	1.4	1.6
9	4.3	26.8	15.1	6.6	3.0	8.4	2.8	2.8	1.4	2.1	1.3	1.5
10	4.1	20.5	12.4	6.3	2.8	2.8	2.9	3.7	1.4	1.8	1.3	1.5
11	10.6	22.0	16.3	5.6	2.7	2.9	2.8	2.1	1.4	1.6	1.4	1.5
12	14.7	20.5	15.8	5.2	3.3	2.8	2.9	1.9	1.4	1.5	1.6	1.5
13	11.3	16.9	15.2	4.6	5.0	2.5	2.8	1.9	1.4	1.5	1.7	1.5
14	7.9	11.7	14.6	5.8	12.9	2.4	2.8	1.8	1.6	1.5	1.6	1.6
15	6.5	9.6	13.9	5.7	17.8	2.3	2.3	2.7	1.7	1.4	1.5	1.6
16	5.8	22.9	12.2	4.8	17.7	2.2	2.1	8.5	1.7	1.4	1.3	1.6
17	5.3	26.4	10.9	4.4	9.4	2.1	2.0	5.1	1.6	1.4	1.3	1.6
18	4.9	24.5	8.2	4.2	7.2	2.0	2.0	4.7	1.6	1.4	1.3	1.5
19	4.6	16.6	7.3	4.0	5.8	2.0	2.8	3.8	1.5	1.3	1.4	1.5
20	4.3	12.4	6.8	3.9	4.9	1.9	3.8	3.4	1.5	1.3	1.4	1.5
21	4.2	9.9	6.3	5.5	4.4	1.9	2.3	2.7	1.6	1.3	1.4	1.5
22	4.0	8.1	7.0	5.1	4.0	2.7	1.9	2.4	1.6	1.3	1.3	1.6
23	3.9	7.3	6.8	4.3	3.7	8.4	1.7	2.2	1.5	1.3	1.3	1.6
24	3.8	6.7	6.3	3.9	3.5	2.6	1.6	2.1	1.5	1.3	1.4	1.6
25	4.0	5.8	5.9	3.7	3.2	2.8	2.4	2.0	1.5	1.3	1.4	1.7
26	3.8	5.2	5.4	3.6	3.1	2.9	1.7	1.9	1.5	1.3	1.4	2.0
27	3.8	5.9	5.1	3.5	3.0	8.4	1.7	1.8	1.5	1.3	1.4	1.9
28	6.0	27.2	5.1	3.3	2.9	3.3	1.6	1.8	1.4	1.8	1.4	1.7
29	6.4		6.3	3.2	2.8	8.1	1.6	1.6	1.3	1.3	1.4	1.5
30	6.2		7.7	3.0	2.7	2.6	1.5	1.5	1.3	1.3	1.4	1.5
31	5.9		9.2		2.9		5.7	1.5		1.5		1.5

Rating table for Cahaba River at Centerville for 1901.

Gage height.	Discharge.	Gage height.	Discharge.	Gage height.	Discharge.	Gage height.	Discharge.
Feet.	Second-feet.	Feet.	Second-feet.	Feet.	Second-feet.	Feet.	Second-feet.
1.1	326	4.9	1,694	8.7	3,062	12.5	4,430
1.2	362	5.0	1,730	8.8	3,098	12.6	4,466
1.3	398	5.1	1,766	8.9	3,134	12.7	4,502
1.4	434	5.2	1,802	9.0	3,170	12.8	4,538
1.5	470	5.3	1,838	9.1	3,206	12.9	4,574
1.6	506	5.4	1,874	9.2	3,242	13.0	4,610
1.7	542	5.5	1,910	9.3	3,278	13.1	4,646
1.8	578	5.6	1,946	9.4	3,314	13.2	4,682
1.9	614	5.7	1,982	9.5	3,350	13.3	4,718
2.0	650	5.8	2,018	9.6	3,386	13.4	4,754
2.1	686	5.9	2,054	9.7	3,422	13.5	4,790
2.2	722	6.0	2,090	9.8	3,458	13.6	4,826
2.3	758	6.1	2,126	9.9	3,494	13.7	4,862
2.4	794	6.2	2,162	10.0	3,530	13.8	4,898
2.5	830	6.3	2,198	10.1	3,566	13.9	4,934
2.6	866	6.4	2,234	10.2	3,602	14.0	4,970
2.7	902	6.5	2,270	10.3	3,638	14.1	5,006
2.8	938	6.6	2,306	10.4	3,674	14.2	5,042
2.9	974	6.7	2,342	10.5	3,710	14.3	5,078
3.0	1,010	6.8	2,378	10.6	3,746	14.4	5,114
3.1	1,046	6.9	2,414	10.7	3,782	14.5	5,150
3.2	1,082	7.0	2,450	10.8	3,818	14.6	5,186
3.3	1,118	7.1	2,486	10.9	3,856	14.7	5,222
3.4	1,154	7.2	2,522	11.0	3,890	14.8	5,258
3.5	1,190	7.3	2,558	11.1	3,926	14.9	5,294
3.6	1,226	7.4	2,594	11.2	3,962	15.0	5,330
3.7	1,262	7.5	2,630	11.3	3,998	15.1	5,366
3.8	1,298	7.6	3,666	11.4	4,034	15.2	5,402
3.9	1,334	7.7	2,702	11.5	4,070	15.3	5,438
4.0	1,370	7.8	2,738	11.6	4,106	15.4	5,474
4.1	1,406	7.9	2,774	11.7	4,142	15.5	5,510
4.2	1,442	8.0	2,810	11.8	4,178	15.6	5,546
4.3	1,478	8.1	2,846	11.9	4,214	15.7	5,582
4.4	1,514	8.2	2,882	12.0	4,250	15.8	5,618
4.5	1,550	8.3	2,918	12.1	4,286	15.9	5,654
4.6	1,586	8.4	2,954	12.2	4,322	16.0	5,690
4.7	1,622	8.5	2,996	12.3	4,358		

Rating table for Cahaba River at Centerville for 1902.

Gage height.	Discharge.	Gage height.	Discharge.	Gage height.	Discharge.	Gage height.	Discharge.
Feet.	*Second-feet.*	*Feet.*	*Second-feet.*	*Feet.*	*Second-feet.*	*Feet.*	*Second-feet.*
1.0	245	4.6	1,510	11.5	3,925	20.5	7,075
1.2	320	4.8	1,580	12.0	4,100	21.0	7,250
1.4	390	5.0	1,650	12.5	4,275	22.0	7,600
1.6	460	5.2	1,720	13.0	4,450	23.0	7,950
1.8	530	5.4	1,790	13.5	4,625	24.0	8,300
2.0	600	5.6	1,860	14.0	4,800	25.0	8,650
2.2	670	5.8	1,930	14.5	4,975	26.0	9,000
2.4	740	6.0	2,000	15.0	5,150	27.0	9,350
2.6	810	6.5	2,175	15.5	5,325	28.0	9,700
2.8	880	7.0	2,350	16.0	5,500	29.0	10,050
3.0	950	7.5	2,525	16.5	5,675	30.0	10,400
3.2	1,020	8.0	2,700	17.0	5,850	31.0	10,750
3.4	1,090	8.5	2,875	17.5	6,025	32.0	11,100
3.6	1,160	9.0	3,050	18.0	6,200	33.0	11,450
3.8	1,230	9.5	3,225	18.5	6,375	34.0	11,800
4.0	1,300	10.0	3,400	19.0	6,550	35.0	12,150
4.2	1,370	10.5	3,575	19.5	6,725		
4.4	1,440	11.0	3,750	20.0	6,900		

Rating table for Cahaba River at Centerville for 1903.

Gage height.	Discharge.	Gage height.	Discharge.	Gage height.	Discharge.	Gage height.	Discharge
Feet.	Second-feet.	Feet.	Second-feet.	Feet.	Second-feet.	Feet.	Second-feet.
1.00	143	4.10	1,195	7.20	2,435	13.00	4,755
1.10	163	4.20	1,235	7.30	2,475	13.30	4,875
1.20	184	4.30	1,275	7.40	2,515	13.90	5,115
1.30	206	4.40	1,315	7.50	2,555	14.00	5,155
1.40	229	4.50	1,355	7.60	2,595	14.60	5,295
1.50	253	4.60	1,395	7.70	2,635	14.70	5,435
1.60	279	4.70	1,435	7.80	2,675	15.00	5,555
1.70	306	4.80	1,475	7.90	2,715	15.10	5,595
1.80	335	4.90	1,515	8.00	2,755	15.20	5,635
1.90	365	5.00	1,555	8.10	2,795	15.80	5,875
2.00	397	5.10	1,595	8.20	2,835	16.00	5,959
2.10	430	5.20	1,635	8.30	2,875	16.30	6,075
2.20	463	5.30	1,675	8.80	3,075	16.60	6,195
2.30	497	5.40	1,715	9.00	3,155	16.90	6,315
2.40	532	5.50	1,755	9.20	3,235	18.00	6,755
2.50	568	5.60	1,795	9.60	3,395	18.10	6,795
2.60	604	5.70	1,835	9.80	3,475	20.00	7,555
2.70	640	5.80	1,875	9.90	3,515	20.50	7,755
2.80	676	5.90	1,915	10.00	3,555	20.60	7,795
2.90	715	6.00	1,955	10.60	3,795	22.00	8,355
3.00	755	6.10	1,995	10.90	3,915	22.90	8,715
3.10	795	6.20	2,035	11.00	3,955	23.00	8,755
3.20	835	6.30	2,075	11.10	3,995	23.80	9,075
3.30	875	6.40	2,115	11.30	4,075	24.00	9,155
3.40	915	6.50	2,155	11.70	4,235	24.50	9,355
3.50	955	6.60	2,195	12.00	4,355	26.00	9,955
3.60	995	6.70	2,235	12.20	4,435	26.40	10,155
3.70	1,035	6.80	2,275	12.30	4,475	26.80	10,275
3.80	1,075	6.90	2,315	12.40	4,515	31.00	11,955
3.90	1,115	7.00	2,355	12.60	4,595	31.60	12,195
4.00	1,155	7.10	2,395	12.90	4,715		

Estimated monthly discharge of Cahaba River at Centerville.

[Drainage area, 1,040 square miles.]

Month.	Discharge in second-feet.			Run-off.	
	Maximum.	Minimum.	Mean.	Second-feet per square mile.	Depth in inches.
1902.					
January	4,660	705	1,739	1.67	1.93
February	9,560	1,405	2,955	2.84	2.96
March	12,150	1,720	4,799	4.61	5.31
April	5,290	1,335	2,464	2.37	2.64
May	2,840	880	1,282	1.23	1.42
June	880	670	749	.72	.80
July	1,125	635	730	.70	.81
August	880	635	690	.66	.76
September	1,055	600	689	.66	.74
October	3,680	600	939	.90	1.04
November	1,370	600	746	.72	.80
December	5,255	775	1,974	1.90	2.19
The year	12,150	600	1,646	1.58	21.70
1903.					
January	5,435	1,075	1,940	1.87	2.16
February	12,195	1,515	5,412	5.20	5.41
March	9,075	1,595	3,849	3.70	4.27
April	2,835	755	1,514	1.46	1.63
May	6,675	640	1,518	1.44	1.68
June	915	365	648	.62	.69
July	1,835	253	560	.54	.62
August	1,995	253	721	.69	.80
September	306	206	252	.24	.27
October	676	184	243	.23	.27
November	306	206	237	.23	.26
December	397	229	268	.26	.30
The year	12,195	184	1,430	1.38	18.36

Net horsepower per foot of fall with a turbine efficiency of 80 per cent for the minimum monthly discharge of Cahaba River at Centerville.

Month.	1901.			1902.			1903.		
	Minimum discharge.	Minimum net horsepower per foot of fall.	Duration of minimum.	Minimum discharge.	Minimum net horsepower per foot of fall.	Duration of minimum.	Minimum discharge.	Minimum net horsepower per foot of fall.	Duration of minimum.
	Sec.-ft.		*Days.*	*Sec.-ft.*		*Days.*	*Sec.-ft.*		*Days.*
January				705	64	1	1,075	98	3
February				1,405	128	1	1,515	138	1
March				1,720	156	1	1,595	145	2
April				1,335	121	1	755	69	1
May				880	80	2	640	58	2
June				670	61	1	365	33	2
July				635	58	4	253	23	1
August	362	33	4	635	58	14	253	23	2
September	398	36	3	600	55	11	206	19	3
October	362	33	5	600	55	4	184	17	2
November	326	30	3	600	55	6	206	19	7
December	398	36	1	775	70	1	229	21	5

WATER POWERS ON CAHABA RIVER.

Cahaba River rises near Birmingham, and, flowing in a southerly direction, enters Alabama River just below Selma.

The Corps of Engineers, United States Army, made a survey of this stream, beginning at the southwestern boundary of Shelby County,

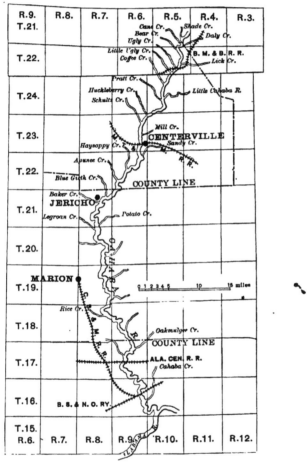

FIG. 7.—Map of portion of Cahaba River surveyed by Corps of Engineers, United States Army.

and running down the river 110 miles to its mouth, in which distance there is a fall of 256 feet. The level notes of this survey are given on the following pages. The map (fig. 7) shows the location of points referred to.

Elevations of water surface of Cahaba River in Alabama from its mouth up to the line of Shelby County.

[Survey by United States Engineer Corps in 1874.]

Distance above mouth.	Location.	Elevation above mean low water of Alabama River at mouth.
Miles.		*Feet.*
0.0	Alabama River at mouth of Cahaba............................	0.00
13.2	Below Tallys Defeat Island	22.40
13.3	Above Tallys Defeat Island	23.66
16.0	Foot of Log Creek Shoals	25.45
20.0	Head of Log Creek Shoals	31.85
24.6	Line between Dallas and Perry counties	37.50
25.2	Mouth of Oakmulgee Creek...................................	38.00
41.8	Fikes Ferry	62.47
42.0	Mouth of Waters Creek......................................	64.15
44.0	Shoal..	67.50
44.3	.....do	69.00
46.7	.....do	73.62
49.0	Foot of Burras Island Shoal................................	77.50
49.5	On Burras Island Shoal.....................................	80.92
50.0	Head of Burras Island Shoal	90.11
53.0	Shoal..	92.14
53.8	.....do	95.08
55.1	Blocks Cut-off	97.90
55.3	.....do	98.50
55.7	.....do	99.70
56.8	Shoals...	100.39
57.0	Below Potato Creek Shoals..................................	101.82
58.1	Above Potato Creek Shoals	107.00
61.6	Below Jericho Island.......................................	109.30
63.2	Above Jericho Island.......................................	111.20
65.3	Cluster of islands, mouth of Blue Girth Creek..............	113.55
65.7	Cluster of islands above mouth of Blue Girth Creek	114.65
67.6	Below small shoal..	115.50
67.7	Above small shoal..	116.30
69.2	Small shoal..	116.50
70.2	Line between Perry and Bibb counties	118.00
74.8	Foot of shoal..	123.40
75.5	Head of shoal..	125.30
81.6	Foot of shoal..	131.80
82.6	Head of shoal..	132.80

Elevations of water surface of Cahaba River in Alabama from its mouth up to the line of Shelby County—Continued.

Distance above mouth.	Location.	Elevation above mean low water of Alabama River at mouth.
Miles.		*Feet.*
85.7	Maberrys Island	133.70
86.7	Ferry, Centerville, Ala., below shoal	134.40
86.7	Ferry, Centerville, Ala., above shoal	136.40
87.5	Mouth of Mill Creek	139.60
88.4	Top of Centerville Shoal (in township 23)	148.00
90.0	Logans Ferry, mouth of Schultz Creek	148.00
91.0	Crossing of range lines 9 and 10	149.00
92.0	East and west line townships 23 and 24	149.00
92.5	Small shoal	149.88
93.0	Mouth of Buckhalter or Huckleberry Creek	151.00
93.2	Jones Ferry	151.00
94.0	Point Lookout	151.90
94.2	Mouth of Rocky Creek	152.00
94.5	Mouth of Palmetto Creek	152.40
94.8	Pratts Ferry, in township 24 N., R. 10 E.	152.61
96.0	Below Little Cahaba Shoals	155.00
97.1	Mouth of Little Cahaba River	158.00
97.2	Below small shoal	158.00
97.3	Above small shoal	159.40
97.8	Mouth of Pratts Creek	161.50
98.1	Mouth of Lewis Branch	163.00
98.5	Mouth of Coalbed Branch	172.00
99.1	Opposite Lyman coal shaft at Lick Branch	184.26
99.5	Below small shoal	185.20
99.6	Above small shoal	189.60
99.7	Mouth of Coffee Creek	189.60
100.1	Foot of Bailey Reach Rapids	192.00
100.6	Head of Bailey Reach Rapids	204.80
101.4	Mouth of Little Ugly Creek	205.00
102.1	Mouth of Ugly Creek, in township 22 S., R. 5 W	205.00
102.7	Below small shoal	205.00
102.8	Above small shoal	208.50
103.6	Foot of Lily Shoals	213.00
103.7	Lily Shoals Ford	216.60
104.4	Mouth of Lick Creek	220.00
104.6	Mouth of Bear Branch	220.00
105.7	Mouth of Daly Creek, east and west line sections 11 and 14	221.80

Elevations of water surface of Cahaba River in Alabama from its mouth up to the line of Shelby County—Continued.

Distance above mouth.	Location.	Elevation above mean low water of Alabama River at mouth.
Miles.		*Feet.*
105.8	Above small shoal	224.50
106.3	Small shoal	226.60
106.5	.....do	228.40
107.0	Foot of Long Island Shoal	231.30
107.4	Mouth of Cane Creek	243.50
107.9	Head of Halfmile Rapids	252.00
109.2	Locke Ford	254.00
110.3	Mouth of Shades Creek, Shelby County line	255.88

Miscellaneous discharge measurements of Cahaba River and tributaries.

Date.	Stream.	Location.	Discharge.
1901.			*Second-feet.*
Jan. 28	Hawkins Spring	Birmingham	15.8
29	Cahaba River	Sydenton	549
Mar. 28	Valley Creek	Adgers Station	378
28	Blocton Creek	Blocton	107
29	Cahaba River	Sydenton	1,117
Apr. 25	.....do	Harrall	6,560
1903.			
Sept. 30	East Cahaba River	Near Bridgeton, below Dishazo's mill.	30
30	.....do	.....do	29
Nov. 12	.....do	Bridgeton	28
12	.....do	Bridgeton, above Dishazo's mill.	21
12	.....do	Near Pledger's mill, Shelby County.	13
12	.....do	Near Leeds	(

The foregoing measurements give a fair idea of the river flow at all seasons. It may be safely assumed that the flow at different points will bear the same proportion to drainage area as that at Centerville.

In the foregoing level notes the stations are 1 mile apart, and are numbered from zero, at the mouth of the river, up to 110, at the Shelby County line. In the following description of powers that can be developed these mile stations will be referred to as stations:

Power No. 1.—From the head of "Halfmile Rapids," at station 108, there is a succession of shoals, known as Halfmile, Long Island, Fish-trap, Ford, Reach, and Dry Creek shoals, in which the aggregate fall is 30 feet in 2¼ miles. There is also a fall of about 4 feet from the Shelby County line down to the head of Halfmile Shoal, making a total fall of 34 feet in 4 miles. This can be developed either by building a dam 34 feet high at the mouth of Dry Creek and backing the water to the Shelby County line, or by building a low dam near the head of the shoals and a canal from it to a point opposite the mouth of Dry Creek. Such a development will give about 500 net horsepower, with an 80 per cent turbine at ordinary low season. This power would be near Blocton.

Power No. 2.—By building a 15-foot dam at the head of "Bailey Reach Rapids," near station 101 and near the mouth of Ugly Creek, to back the water to the mouth of Persimmon Branch, near station 104, and constructing from this point a canal along the river bank about 4 miles long, to a point opposite station 97, at the mouth of Little Cahaba River, a practical head of 54 feet can be developed. This allows 8 feet for storage and grade, as the total fall is 62 feet. A 54-foot fall would produce about 800 net horsepower.

The same power can be developed by building a high dam lower down the river and having the canal shorter; or the power can be divided into two separate powers. This power site is between River Bend and Cadle, in Bibb County.

Power No. 3.—From the mouth of Little Cahaba down to station 88½, at the top of Centerville Shoals, there is a fall of 10 feet in 8½ miles, and from the top of Centerville Shoals down to the foot of Centerville Shoals, at Centerville, there is a fall of 13.6 feet in about 1¾ miles. This power can be developed by a 10-foot dam at top of Centerville Shoals and a canal from there to Centerville, 1½ miles long. Allowing 2.6 feet for storage and canal grade, a head of 21 feet can be obtained, which will give 650 net horsepower.

It is probable that a much better method of development will be to erect a dam at Centerville 23.6 feet high to back the water to the mouth of the Little Cahaba. This will produce 732 net horsepower, with storage. The incidental storage of such a dam would add largely to the amount and efficiency of the power. A plant running only twelve hours a day and storing the water at night could utilize 1,440 net horsepower.

This power site is at Centerville, on the Mobile and Ohio Railroad.

Power No. 4.—A 16-foot dam can be built at shoal No. 9, station 69½, in Perry County, just below the Bibb County line. This dam would back the water for 12 miles to shoal No. 2, 4½ miles below Centerville. A 16-foot head will produce 670 horsepower without storage, or 1,340 horsepower by storing the water at night and running only twelve hours a day. This dam site is about 17 miles below Centerville by river.

Power No. 5.—A 15-foot dam at "Blocks Cut-off," near station 55, will back the water 10 miles to the mouth of Taylors Creek, and will produce 750 continuous, or 1,500 twelve-hour horsepower.

Power No. 6.—At shoal No. 24, station 50, there is a fall of 9 feet in less than half a mile. A 14-foot dam at foot of this shoal, or a 5-foot dam at its head, and a short canal will develop a head of 14 feet and realize 720 continuous, or 1,400 twelve-hour horsepower.

This site is just above Burras Island, 8 or 10 miles northeast of Marion, Ala.

Power No. 7.—From Burras Island to Fikes Ferry there is a fall of 22 feet in a distance of 7 miles, 20 feet of which could probably be utilized by a dam at Fikes Ferry, producing 1,100 continuous, or 2,200 twelve-hour horsepower. Fikes Ferry is near Marion, Ala.

In making the above statement of powers that can be developed, it has been assumed that there are suitable banks for dam sites. The system proposed, or some other system approximating to it, would not interfere with navigation improvements, as locks could be constructed at the dams.

BLACK WARRIOR RIVER AND TRIBUTARIES.

Black Warrior River is formed by the junction of the Mulberry and Sipsey forks of Black Warrior at old Warriortown in Walker County, and runs in a southwesterly direction past Tuscaloosa to Demopolis, Ala., at which point it enters Tombigbee River. Above Tuscaloosa it is known as Black Warrior River and below Tuscaloosa as Warrior River.

BLACK WARRIOR RIVER AT TUSCALOOSA.

A gage at Tuscaloosa was placed in position by the United States Corps of Engineers in 1888. It is about three-fourths of a mile from the business center of Tuscaloosa, and is reached by passing down Bridge street to the river, thence down the east bank 1,800 feet to the gage. It consists of an inclined timber 2 by 6 inches, supported on posts and graduated by means of notches placed 1 foot vertically apart. The observer is W. S. Wyman, jr., Tuscaloosa. Observations are taken daily at 7 a. m. The drainage area draining at this point is 4,900 square miles.

The bench marks are fixed, one on a willow 10 feet west of gage, 97.84 feet above Mobile datum, the other on a small hackberry 30 feet south of the upper end of the gage and 139.36 feet above Mobile datum. The current here is rather sluggish, being almost imperceptible at low stages. Both banks are of earth and subject to overflow. Observations of gage heights have been obtained through the courtesy of Mr. R. C. McCalla, jr., of the United States engineers in charge of Black Warrior River, from the time the gage was established until December 31, 1896. A measurement made by Mr. McCalla September 14, 1896, showed a gage height of −0.60 foot area, 1,022 square feet; mean velocity, 0.16; discharge, 164 second-feet.

Measurements at the same place have been furnished by Mr. Horace Harding, United States assistant engineer. Velocities were obtained by means of rod floats reaching from the water surface to near the bottom. The highest flood occurred on April 8, 1892. The gage height was 62.5, the sectional area 33,600 square feet, and the estimated mean velocity 4.5 feet per second. This gave a discharge of 151,200 second-feet. From this estimate and the following list of measurements a curve has been platted and a rating table constructed, and this rating table applied to all gage heights observed. The estimates of discharge thus obtained are shown in diagrammatic form in Pl. V. The highest discharges are merely approximations, but the discharges shown by the diagrams serve as a basis for comparison of the state of the river during the various years.

Discharge measurements have been made as follows:

Discharge measurements of Black Warrior River at Tuscaloosa.

Date.	Gage height.	Discharge.	Remarks.
1895.	*Feet.*	*Second-feet.*	
December 17.........................	1.10	617	Stationary.
December 21.........................	2.61	1,344	Do.
December 24.........................	3.60	1,733	Rising slowly.
1896.			
January 30	9.99	5,073	Falling 0.05 per hour.
January 31	8.65	4,363	Do.
February 26	8.25	4,360	Falling 0.01 per hour.
February 28	7.27	3,657	Falling 0.02 per hour.
February 29	6.92	3,522	Stationary.
March 2	7.67	4,211	Do.
March 3	7.28	3,632	Falling 0.03 per hour.
March 6	6.94	4,558	Rising 0.15 per hour.
March 24	24.85	13,550	Falling 0.12 per hour.
April 10	9.71	5,331	Falling.
April 11	8.89	4,755	Do.
April 14	8.25	4,675	Rising.
April 20	7.55	3,862	Falling.
April 21	6.65	3,388	Do.
April 22	5.96	2,940	Do.
April 23	5.46	2,704	Do.
April 24	5.88	3,158	Rising.
April 27	5.68	3,049	Do.
1897.			
January 12	1.70	829	

Discharge measurements of Black Warrior River at Tuscaloosa—Continued.

Date.	Gage height.	Discharge.	Date.	Gage height.	Discharge.
1899.	*Feet.*	*Second-feet.*	1899.	*Feet.*	*Second-feet.*
February 21..........	19.36	12,855	March 18	56.40	86,410
Do	19.25	12,640	March 23	40.30	23,911
February 24.........	22.85	16,216	1901.		
February 28........:.	39.47	48,010	February 1.........	15.10	9,300
March 1	35.50	24,988	March 15	18.72	9,461
March 2	30.35	18,052	June 27.............	1.77	828
March 4	23.70	12,609	1903.		
March 14	31.18	36,653	July 20	5.45	862
Do	34.37	40,331	July 21	5.44	719
March 17	59.50	119,533			

Daily gage height, in feet, of Black Warrior River at Tuscaloosa.

Day.	Jan.	Feb.	Mar.	Apr.	May.	June.	July.	Aug.	Sept.	Oct.	Nov.	Dec.
1889.												
1............	15.00	23.50	18.80	8.50	8.80	2.30	2.40	3.70	0.75	2.40	0.90	6.00
2............	19.50	20.50	16.50	9.80	14.50	2.90	3.10	3.60	7.30	1.96	1.05	5.65
3............	18.00	17.50	25.00	10.30	13.20	2.75	4.40	3.00	6.40	1.55	1.85	5.30
4............	16.00	14.80	31.50	11.60	11.00	2.50	5.10	2.95	5.30	1.30	1.90	4.80
5............	25.00	13.00	29.40	10.50	8.50	2.80	6.25	2.90	6.20	1.15	3.85	4.50
6............	33.00	11.50	26.50	9.50	7.00	2.10	6.85	3.70	9.30	1.05	3.65	4.20
7............	33.50	10.40	23.00	8.40	5.80	1.90	6.45	3.45	14.90	.90	3.30	4.00
8............	29.50	9.30	20.10	7.30	4.60	1.70	5.70	2.95	19.00	.70	3.00	3.80
9............	26.80	8.30	17.50	6.80	4.30	2.00	4.75	2.60	14.00	.50	2.75	3.60
10............	28.40	8.00	15.00	6.40	4.00	2.60	3.85	2.00	9.90	.40	2.55	3.40
11............	29.00	7.80	12.80	6.30	3.80	2.85	3.05	1.75	5.50	.30	3.65	3.20
12............	25.50	7.60	11.00	6.00	3.20	4.20	2.85	1.55	5.25	.30	4.30	2.95
13............	22.50	7.80	10.00	5.60	3.00	3.40	2.45	3.85	4.15	.20	4.15	2.75
14............	19.00	7.00	9.20	5.50	3.10	4.00	2.15	3.60	3.80	.15	3.75	2.70
15............	16.00	7.30	8.80	8.00	3.20	3.45	1.96	8.35	2.70	.15	3.50	2.65
16............	13.60	27.50	8.10	16.80	3.00	3.25	4.20	8.15	2.25	.15	4.20	2.50
17............	29.00	49.00	7.20	16.70	2.80	3.05	7.35	8.80	2.00	.10	5.20	2.30
18............	40.50	56.40	7.10	14.00	2.50	2.80	11.35	3.50	1.55	.10	12.45	2.20
19............	38.50	56.60	11.00	12.00	2.35	2.65	11.50	3.20	1.40	.06	18.90	2.10
20............	34.00	53.00	12.50	10.00	2.25	2.45	8.40	2.90	1.80	.05	16.90	2.05
21............	30.30	47.00	12.30	8.90	2.05	2.25	4.60	2.40	1.20	.06	14.10	2.00
22............	28.10	41.50	11.80	8.00	1.85	3.25	4.55	1.90	1.10	.00	11.70	2.00
23............	26.00	36.50	10.80	7.00	1.65	3.45	4.50	1.50	1.00	— .10	9.70	2.00
24............	23.10	32.50	10.00	6.40	1.50	3.20	3.70	1.20	1.00	— .30	7.90	2.00
25............	21.00	28.50	9.80	6.30	1.30	3.00	3.15	.90	.95	— .15	6.60	1.95
26............	20.40	26.50	11.20	6.80	1.25	2.40	3.10	.75	2.25	+ .10	6.30	1.95
27............	24.00	23.50	11.80	6.50	1.15	1.95	4.55	1.00	3.70	.35	6.10	1.95
28............	33.50	21.40	12.30	6.20	1.05	1.55	4.05	.96	3.50	.35	5.90	1.90
29............	33.80	....	11.00	5.80	.95	1.25	4.00	.90	3.25	.40	5.80	1.90
30............	30.00	....	10.50	5.50	1.15	1.10	4.00	1.10	2.60	.90	6.30	8.00
31............	27.00	....	9.80	....	1.10	....	3.80	1.05	....	1.00	....	4.85

Daily gage height, in feet, of Black Warrior River at Tuscaloosa—Continued.

Day.	Jan.	Feb.	Mar.	Apr.	May.	June.	July.	Aug.	Sept.	Oct.	Nov.	Dec.
1890.												
1	7.70	30.20	58.90	24.50	16.70	7.75	1.80	5.00	8.00	9.30	3.75	1.00
2	7.20	26.20	57.40	34.50	13.70	6.50	1.65	4.05	5.80	7.40	3.35	1.00
3	6.20	22.50	52.40	34.10	11.95	5.65	1.45	3.25	4.35	6.10	3.10	1.00
4	5.20	19.05	45.80	43.60	12.75	5.00	1.25	2.85	3.40	5.10	2.85	1.00
5	5.20	16.40	40.85	45.90	14.65	4.40	1.10	3.15	3.10	4.40	2.55	1.00
6	4.80	13.90	37.15	44.50	16.50	4.00	.80	3.35	3.85	3.80	2.40	1.00
7	4.50	12.10	35.40	38.70	15.00	5.20	.60	3.75	4.35	3.50	2.25	1.20
8	4.55	44.30	32.60	34.00	13.50	5.60	.50	4.55	4.80	3.45	2.15	10.05
9	5.10	53.95	30.50	29.95	11.95	5.40	.45	4.30	2.85	4.50	2.05	10.10
10	5.30	52.90	27.50	26.70	10.10	5.25	.30	4.30	3.20	4.45	2.00	8.95
11	5.60	47.50	25.00	23.45	9.30	5.00	.20	6.10	2.90	4.10	1.95	7.10
12	5.45	42.20	23.00	20.45	8.35	6.00	.15	6.40	3.45	3.80	1.75	5.80
13	5.30	37.20	20.75	17.45	8.35	6.20	.05	5.30	3.95	3.50	1.65	4.60
14	5.20	32.65	22.50	14.50	9.00	6.20	1.05	4.50	5.80	3.20	1.60	3.70
15	5.60	29.45	38.20	12.15	8.75	5.60	2.55	3.20	5.55	2.80	1.60	3.30
16	11.20	26.95	38.00	10.65	8.65	5.05	3.10	2.80	4.50	2.73	1.55	2.95
17	19.30	23.95	35.80	10.85	9.50	4.35	3.65	2.50	3.53	2.60	1.65	2.80
18	21.00	21.45	32.30	12.95	8.90	4.30	3.65	2.70	2.90	2.40	1.60	2.75
19	18.40	18.45	29.00	18.20	8.15	4.20	4.15	2.55	2.45	2.30	1.55	2.20
20	15.40	15.65	27.20	18.95	7.95	3.70	4.75	2.40	2.10	2.10	1.50	2.00
21	13.30	18.45	32.40	12.15	7.95	3.20	4.20	2.20	1.75	2.00	1.50	2.00
22	11.60	12.00	34.45	10.45	9.00	2.80	3.00	1.80	1.70	2.45	1.45	1.90
23	10.10	10.40	41.25	9.30	8.40	2.50	2.10	1.40	3.05	4.90	1.35	1.90
24	10.70	9.40	40.25	10.65	7.65	2.25	1.75	1.10	3.10	8.45	1.30	1.80
25	12.50	9.50	36.75	15.45	6.95	2.25	2.25	1.00	13.05	10.55	1.25	1.80
26	13.50	13.50	33.35	26.50	6.85	2.15	12.65	.95	23.90	9.60	1.20	2.30
27	12.80	35.20	29.70	28.45	9.60	2.55	13.40	.95	22.65	8.05	1.15	6.80
28	11.80	53.10	27.45	26.20	13.20	2.25	10.50	5.10	18.00	6.45	1.10	9.70
29	10.60	......	24.55	23.35	13.55	2.05	9.30	9.55	14.70	5.30	1.05	10.20
30	17.60	......	21.95	19.80	11.95	1.90	7.45	9.15	11.90	4.90	1.00	9.10
31	31.70	......	18.80	......	9.25	......	6.50	10.45	......	4.15	......	7.30
1891.												
1	6.10	33.20	24.00	34.00	5.70	1.80	2.10	6.50	1.00	−0.40	−0.80	3.40
2	8.60	39.60	23.50	39.40	5.50	1.90	2.00	13.00	1.00	−.30	−.80	3.00
3	14.30	40.90	20.50	36.80	5.30	1.80	1.90	14.80	1.00	−.20	−.80	2.40
4	17.90	40.20	19.60	32.50	4.90	1.70	1.80	16.00	.80	−.20	−.80	6.20
5	15.20	39.30	17.20	28.50	4.50	1.50	1.50	13.00	.70	−.20	−.80	6.80
6	13.10	36.30	29.00	25.00	4.50	1.40	1.30	9.50	.70	−.20	−.80	21.50
7	11.10	37.00	53.00	22.20	4.20	1.40	1.40	7.40	.60	−.40	−.80	19.50
8	10.00	51.50	58.00	19.00	4.00	1.50	2.10	5.80	.60	−.50	−.80	20.00
9	9.50	51.50	60.40	16.00	3.90	2.10	7.00	5.40	.60	−.60	−.80	20.50
10	17.50	52.20	58.00	14.20	3.50	2.80	10.40	3.70	.60	−.70	+.70	17.00
11	20.90	53.50	54.00	17.20	3.20	3.50	8.20	3.10	.70	−.80	2.10	14.00
12	26.30	50.50	48.00	20.00	3.20	10.20	6.00	2.70	1.00	−.80	2.80	11.50
13	30.10	47.60	43.00	26.00	3.10	10.50	4.80	2.50	1.20	−.80	4.40	9.00
14	25.50	51.40	40.00	22.50	3.00	9.80	3.20	3.40	1.20	−.80	3.80	6.20
15	21.00	49.50	36.50	19.50	2.80	8.00	2.30	3.40	1.00	−.60	2.50	7.00
16	18.50	46.50	33.20	17.20	2.60	6.50	2.60	4.00	1.00	−.70	2.00	9.00
17	17.10	44.30	30.00	15.60	2.50	5.60	2.60	3.00	.60	−.70	1.80	11.00
18	17.70	41.00	28.00	16.40	2.50	6.30	2.90	2.60	.20	−.70	1.80	11.20
19	17.10	37.50	26.40	14.50	2.50	7.20	2.50	2.20	.10	−.60	1.50	10.00
20	15.60	35.00	24.00	13.40	2.50	7.00	2.20	2.00	.10	−.70	1.30	8.50
21	12.60	33.50	21.00	12.00	2.50	6.00	1.80	1.50	.10	−.70	1.40	7.00
22	20.20	39.50	19.90	11.00	2.70	5.50	1.60	1.50	.10	−.60	1.30	6.00
23	31.60	41.00	17.50	10.00	3.00	5.20	1.40	1.40	.10	−.60	8.00	6.00
24	31.80	39.00	15.00	9.00	3.00	4.80	1.40	1.30	.10	−.50	12.80	6.20
25	30.10	36.50	12.50	8.20	2.80	4.50	1.40	1.20	.00	−.50	13.30	7.50
26	30.70	33.00	12.10	7.60	2.60	4.20	1.60	1.10	.00	−.50	10.80	13.20
27	29.00	29.00	15.00	8.40	2.50	4.10	2.00	1.10	−.20	−.70	7.20	31.00
28	26.80	26.50	18.20	8.00	2.40	3.70	2.20	1.00	−.20	−.80	5.80	38.90
29	23.70	......	19.40	7.30	2.30	2.90	2.00	1.00	−.40	−.80	4.20	31.20
30	32.70	......	18.00	6.50	2.10	2.50	2.10	1.00	−.50	−.80	3.80	27.00
31	33.00	......	17.00	......	2.00	......	2.40	1.00	......	−.80	......	22.10

Daily gage height, in feet, of Black Warrior River at Tuscaloosa—Continued.

Day.	Jan.	Feb.	Mar.	Apr.	May.	June.	July.	Aug.	Sept.	Oct.	Nov.	Dec.
1892.												
1	17.90	21.20	10.70	21.80	12.00	3.40	4.60	4.80	9.30	3.90	0.40	5.90
2	15.90	10.50	10.30	18.30	11.20	3.20	3.90	3.20	7.60	3.50	.40	5.60
3	18.50	9.40	9.70	15.50	9.90	3.50	3.40	7.00	6.50	3.30	1.20	5.30
4	18.70	8.50	8.80	15.20	8.80	3.70	3.00	7.40	5.60	3.10	1.60	4.90
5	16.40	7.80	8.00	12.00	7.20	3.50	2.90	6.60	4.90	2.90	1.60	6.30
6	14.60	7.40	7.80	11.60	6.50	4.70	3.50	5.40	4.90	2.50	1.60	7.50
7	13.00	7.80	8.00	56.90	5.90	5.00	5.20	4.80	4.80	2.40	2.00	7.50
8	11.60	7.50	16.50	63.20	5.30	4.90	11.00	3.90	6.40	2.20	3.90	12.20
9	10.80	9.00	26.80	62.20	5.20	4.10	26.70	3.60	6.20	2.10	4.60	15.00
10	10.50	16.00	28.50	58.00	4.90	4.00	43.50	3.50	5.40	2.00	8.90	18.20
11	25.70	13.00	26.70	52.80	4.90	3.80	46.20	3.00	5.10	2.00	10.90	11.00
12	34.80	11.30	22.00	45.40	4.80	3.50	41.40	3.00	3.90	2.00	8.30	9.40
13	53.00	10.80	18.00	40.70	4.70	3.30	38.80	3.30	3.40	2.00	7.60	8.00
14	57.40	9.50	15.80	36.50	4.60	2.80	37.50	3.10	3.80	1.90	5.80	7.40
15	55.90	9.00	13.80	32.80	4.50	2.50	34.80	3.90	4.00	1.90	5.00	8.60
16	51.70	10.00	13.30	29.50	4.00	2.40	32.80	3.50	5.10	1.90	4.50	10.80
17	45.00	11.80	11.20	27.00	3.70	2.00	41.40	3.50	4.90	1.80	4.00	18.10
18	40.10	11.00	12.90	24.50	3.60	1.80	38.00	6.00	4.30	1.70	4.20	26.70
19	36.60	9.90	24.00	22.40	4.30	2.10	33.00	9.50	6.00	1.00	4.40	28.40
20	41.50	10.50	25.50	20.20	4.60	3.90	29.00	11.50	18.30	1.00	4.70	28.40
21	41.00	13.50	22.90	18.10	6.50	7.40	28.30	9.50	23.90	.90	5.00	36.50
22	36.80	18.00	20.00	15.90	6.30	10.80	29.30	7.40	20.90	.90	4.80	35.80
23	34.40	23.90	18.30	13.80	6.00	10.70	25.10	10.20	17.00	.90	4.50	31.50
24	31.00	21.20	22.00	12.30	6.00	8.40	21.50	12.30	13.30	1.00	4.20	27.00
25	28.50	18.50	29.00	11.30	4.80	7.80	18.80	14.00	10.50	1.00	3.80	23.00
26	26.00	16.00	32.00	10.50	3.50	7.90	15.80	14.20	8.00	.90	3.50	19.50
27	23.80	14.00	35.80	8.80	4.70	7.20	13.00	13.20	6.10	.80	3.30	16.00
28	21.50	12.50	34.00	7.90	4.00	6.10	10.50	11.50	5.00	.70	3.20	13.60
29	19.00	11.50	30.50	7.70	3.80	5.20	7.80	12.00	4.90	.70	3.90	11.40
30	16.80	......	26.80	10.50	3.50	5.40	6.20	11.50	4.80	.50	5.90	9.80
31	14.00	......	23.90	......	3.50	......	5.50	10.50	......	.50	......	8.50
1893.												
1	8.20	18.10	23.00	9.40	24.50	12.50	2.50	0.60	0.30	1.10	0.40	1.50
2	8.60	15.90	21.70	9.30	21.20	33.50	2.30	.70	.20	1.00	.40	1.50
3	9.80	14.00	19.70	9.00	30.00	49.60	2.80	1.00	.10	1.20	.40	1.80
4	10.70	12.80	20.50	8.70	51.20	46.00	3.60	1.20	.00	1.40	.40	3.50
5	10.10	12.50	23.00	8.00	52.20	39.00	3.60	1.30	− .10	1.30	.40	3.50
6	9.00	11.80	24.00	8.70	48.00	37.70	3.20	1.20	− .10	1.20	.40	4.20
7	8.40	11.20	22.60	9.30	42.90	39.90	2.70	1.30	.00	1.20	.40	4.10
8	7.90	10.80	20.10	9.00	40.40	39.30	2.40	1.20	+ .20	1.20	.40	3.50
9	7.00	10.60	20.00	8.30	37.40	34.00	2.10	1.00	1.40	1.00	.40	3.00
10	6.70	10.20	22.00	7.60	34.30	29.20	1.80	.90	2.00	.90	.60	2.80
11	6.20	12.00	21.80	7.00	30.80	25.10	1.60	.90	2.10	.80	.60	2.30
12	6.50	23.90	20.50	6.40	27.50	21.20	1.60	.90	2.20	.80	.50	1.90
13	7.80	28.30	19.30	6.10	24.40	17.40	2.20	.90	3.20	.80	.50	1.70
14	9.00	27.00	18.10	5.80	21.90	14.70	2.60	3.60	4.30	.70	.60	1.60
15	9.60	25.90	16.80	23.00	19.10	11.60	2.40	4.90	4.30	.60	.60	1.50
16	11.40	52.20	14.70	27.10	16.20	9.60	2.00	5.10	3.90	.50	.60	1.70
17	12.20	55.60	13.20	24.00	14.00	7.60	1.70	4.70	3.30	.50	.60	2.30
18	11.80	54.70	12.20	20.00	11.90	6.70	1.40	3.70	2.50	.50	.60	2.40
19	12.00	51.40	11.30	16.40	11.00	6.50	1.30	2.80	2.10	.50	.60	2.30
20	12.40	46.50	10.70	13.90	8.40	6.90	1.20	2.20	1.80	.50	.60	2.10
21	11.80	41.80	9.90	12.00	6.90	6.90	1.20	1.70	1.40	.50	.70	2.50
22	10.90	37.90	9.30	11.20	6.20	6.60	1.30	1.50	1.30	.50	1.00	2.40
23	11.20	34.50	8.80	10.70	5.60	6.40	1.60	1.30	1.20	.40	1.00	2.30
24	12.90	31.30	12.30	9.90	5.10	5.70	1.50	1.10	1.00	.40	1.10	2.10
25	15.90	28.40	22.20	8.90	4.80	4.90	1.50	1.00	1.30	.40	1.00	1.90
26	19.20	25.90	22.50	7.00	4.50	4.40	1.30	.90	1.20	.40	.90	1.70
27	22.00	23.80	20.00	10.50	4.30	4.00	1.20	.80	1.20	.40	.60	1.60
28	23.30	23.00	17.00	32.50	5.10	3.50	.90	.80	1.10	.40	1.70	1.50
29	23.10	......	14.40	33.90	6.10	3.10	.90	.60	1.10	.40	1.70	1.40
30	22.10	......	12.40	29.00	12.90	2.80	.90	.60	1.10	.40	1.50	3.70
31	20.40	......	10.90	......	14.10	......	.80	.20	......	.40	......	7.60

Daily gage height, in feet, of Black Warrior River at Tuscaloosa—Continued.

Day.	Jan.	Feb.	Mar.	Apr.	May.	June.	July.	Aug.	Sept.	Oct.	Nov.	Dec.
1894												
1	10.30	8.50	27.00	11.40	6.50	1.70	2.50	1.20	5.80	0.55	—0.05	—0.05
2	9.60	7.80	25.60	14.10	5.90	1.60	1.80	1.30	4.60	.45	— .10	— .05
3	7.90	7.30	23.60	23.50	5.90	1.50	1.50	1.40	3.60	.40	— .20	.00
4	6.20	7.90	21.00	24.00	5.80	1.35	1.10	1.95	2.95	.32	— .25	.00
5	5.40	16.90	18.20	22.20	5.50	1.33	.80	2.30	2.45	.25	— .30	.00
6	6.40	22.60	15.90	20.60	5.20	1.20	.60	2.40	2.05	.20	— .30	.00
7	16.80	20.40	14.80	18.90	5.10	1.20	1.00	2.00	1.85	.15	— .30	.00
8	22.60	17.80	14.40	16.30	4.90	1.10	1.30	1.60	1.75	.00	— .30	+ .05
9	19.70	17.30	13.80	14.70	4.50	.95	.70	1.80	1.65	— .10	— .30	.05
10	26.90	26.00	12.60	15.70	4.10	.90	.70	1.10	1.60	— .20	— .30	.50
11	34.50	27.80	11.60	25.00	3.70	.80	.60	.80	1.85	— .25	— .30	2.10
12	35.80	25.40	11.50	25.00	3.90	.80	.60	.70	3.40	— .30	— .30	5.70
13	30.90	29.90	16.50	22.70	4.30	.70	.50	.50	5.60	— .30	— .30	10.40
14	25.80	32.10	17.30	19.50	7.00	.60	.50	.50	7.50	— .30	— .15	11.20
15	22.20	29.70	15.80	16.30	6.20	.60	.60	.85	5.70	— .30	.00	8.90
16	23.60	26.00	15.50	15.50	5.70	.60	.70	.60	4.20	— .30	+ .05	6.40
17	24.80	22.30	33.10	16.00	6.10	.55	.75	.55	3.25	— .30	.05	4.90
18	22.40	19.10	36.70	17.00	5.00	.50	.70	.45	3.50	— .30	.05	3.90
19	19.30	16.80	33.70	25.30	4.70	.60	.70	.40	3.25	— .30	.00	3.20
20	15.90	16.00	29.50	24.80	4.20	.70	.60	.55	3.25	— .30	— .10	2.90
21	14.10	16.90	27.80	24.00	3.60	.85	.50	'1.00	4.10	— .30	— .10	2.60
22	18.60	17.90	29.30	20.40	3.10	.80	.95	1.35	3.60	— .30	— .10	2.20
23	21.30	17.00	29.60	17.30	2.80	.90	1.30	3.80	2.90	— .30	— .10	2.00
24	19.90	16.10	28.00	14.20	2.60	1.10	1.35	5.30	2.20	— .35	.00	1.80
25	17.60	17.30	27.80	12.20	2.40	1.00	1.60	9.20	1.75	— .40	.00	1.70
26	15.50	29.30	25.30	10.50	2.30	1.05	1.50	16.00	1.45	— .40	.00	2.00
27	14.00	31.00	22.10	9.80	2.10	1.40	1.40	20.40	1.20	— .45	— .05	7.70
28	12.40	28.80	19.00	8.70	2.10	1.70	1.30	16.00	1.05	— .45	— .05.	9.70
29	11.10		16.10	7.90	2.00	2.35	.20	11.80	.90	— .45	— .05	8.40
30	10.00		14.30	7.00	1.90	2.60	1.10	8.80	.70	— .20	— .05	7.20
31	9.10		12.70		1.90		1.20	6.80		— .05	— .05	6.50
1895												
1	6.10	24.10	8.60	16.20	15.20	7.00	4.80	1.80	1.90	0.10	—0.04	0.75
2	5.40	21.50	13.20	14.50	13.00	5.90	4.50	1.50	1.80	.00	+ .07	.95
3	6.80	21.40	35.30	12.60	11.00	5.10	5.40	1.30	1.80	— .07	.12	1.05
4	9.00	21.40	36.70	11.00	9.40	4.60	7.60	1.20	1.70	— .10	.09	1.00
5	8.80	19.90	32.40	10.20	8.30	4.20	17.40	1.20	1.90	— .10	.11	.90
6	8.00	17.90	27.80	8.40	7.40	6.70	17.70	1.20	2.60	— .10	.25	1.30
7	7.40	16.50	23.80	7.90	7.00	5.70	18.40	1.10	3.05	— .10	.31	1.50
8	36.00	16.80	20.60	20.50	8.20	5.10	15.70	1.00	3.70	.00	.30	1.45
9	50.60	16.90	20.10	24.00	13.50	4.30	13.10	.90	3.40	— .03	.28	1.90
10	49.30	15.80	19.80	21.20	19.70	3.80	11.10	.97	2.90	— .07	.70	1.80
11	45.10	14.60	17.50	18.00	23.30	3.10	9.80	1.40	2.40	— .10	1.20	1.45
12	40.10	14.10	17.20	15.10	22.00	2.70	7.90	1.50	2.20	— .14	1.10	1.40
13	35.00	13.90	18.40	12.60	18.50	2.40	6.50	1.40	2.10	— .14	1.18	1.40
14	29.80	12.80	24.90	11.00	15.20	2.20	5.30	1.30	2.00	+ .06	1.35	1.37
15	25.70	11.60	37.50	9.50	12.20	2.00	5.40	1.10	1.80	.04	1.35	1.33
16	23.40	10.50	47.40	8.60	9.70	2.00	5.60	1.30	1.60	— .08	1.28	1.23
17	31.20	9.40	52.00	10.00	8.00	2.70	5.90	1.40	1.40	— .18	1.20	1.12
18	32.90	9.00	47.30	15.90	6.90	3.30	5.80	2.00	1.40	— .27	1.05	1.05
19	29.20	9.00	42.10	15.80	6.30	4.20	5.10	4.00	1.80	— .31	.88	1.02
20	25.80	9.50	38.80	14.00	5.90	4.90	4.90	4.30	1.00	— .37	.82	1.57
21	23.10	10.20	48.70	12.00	5.30	4.70	4.70	8.60	1.20	— .42	.70	2.55
22	21.60	10.90	51.30	10.40	4.80	4.30	4.30	8.60	1.50	— .45	.60	2.60
23	21.60	11.40	47.60	8.90	4.30	4.10	4.10	3.40	1.30	- .50	.50	3.03
24	19.80	11.30	42.10	8.00	4.10	3.70	3.30	4.80	1.10	— .35	.75	3.65
25	17.40	10.90	37.30	7.20	4.00	3.40	3.60	4.90	.80	— .38	.60	3.31
26	16.70	10.30	32.80	7.00	4.30	3.00	3.70	4.00	.60	— .53	.38	4.50
27	20.00	9.70	29.10	7.10	5.60	4.40	3.30	8.20	.85	— .70	.64	10.40
28	21.20	9.00	26.50	12.50	11.00	2.90	3.20	2.40	.10	— .68	.65	21.02
29	22.00		23.60	15.00	13.30	6.00	2.40	2.30	.50	— .67	.65	16.88
30	27.40		21.10	16.40	11.20	5.40	2.30	2.00	.30	— .50	.60	13.10
31	27.10		18.60		8.90		2.10	2.00		— .32		11.67

Daily gage height, in feet, of Black Warrior River at Tuscaloosa—Continued.

Day.	Jan.	Feb.	Mar.	Apr.	May.	June.	July.	Aug.	Sept.	Oct.	Nov.	Dec.
1896												
1	11.41	7.92	6.91	10.61	18.24	4.92	1.94	0.53	0.38	.52	-0.78	7.70
2	10.26	8.51	7.66	14.78	38.30	4.28	1.71	.80	.25	- .62	- .77	6.82
3	8.90	21.98	7.41	23.00	37.18	3.50	1.44	1.34	.15	- .66	- .42	4.63
4	7.60	33.12	6.74	23.50	30.88	5.44	1.48	.80	.05	- .20	- .39	3.40
5	6.55	30.02	6.14	19.85	26.40	5.24	1.30	.54	- .06	+ .78	- .33	2.60
6	5.74	30.75	6.64	16.70	21.98	4.90	1.15	.42	- .15	.74	- .19	2.05
7	5.09	35.92	10.18	14.26	17.88	3.62	1.32	.50	- .23	.68	- .19	1.70
8	5.24	35.08	12.04	12.55	13.95	3.27	2.20	.60	- .30	.40	- .17	1.47
9	5.76	36.21	14.89	11.20	10.80	6.60	8.40	.68	- .37	.09	- .18	1.21
10	6.46	36.52	13.69	10.02	8.60	8.18	6.96	.64	- .44	- .15	- .21	1.10
11	6.58	33.65	12.45	9.04	7.11	18.39	5.37	.55	- .44	- .28	- .30	.89
12	6.25	29.45	13.97	8.83	6.09	15.13	3.95	.45	- .46	- .28	- .20	.84
13	5.79	25.97	16.15	7.62	5.34	10.77	2.99	.38	- .50	- .48	- .22	.80
14	5.26	27.41	15.85	7.86	4.77	7.66	2.45	.85	- .60	- .49	- .11	.75
15	4.82	33.25	13.56	11.65	6.13	5.60	2.16	.31	- .59	- .46	- .10	1.20
16	4.87	31.02	13.86	13.90	5.45	4.33	2.13	.26	- .59	- .43	- .01	1.25
17	6.03	27.80	22.30	12.36	4.65	4.74	2.15	.13	- .60	- .60	+ .26	1.25
18	8.61	23.65	27.75	10.57	3.94	4.20	2.42	.40	- .60	- .63	.46	1.60
19	9.14	20.09	29.70	9.03	3.44	6.18	2.46	.50	- .60	- .78	.52	1.90
20	8.58	17.00	37.68	7.81	3.05	6.00	2.24	.40	- .61	- .82	.45	1.70
21	7.87	14.45	37.92	6.88	2.91	5.82	2.16	.34	- .64	- .82	.39	1.50
22	8.85	12.20	33.55	6.15	2.77	4.91	1.90	.29	- .64	- .84	.88	1.33
23	22.47	10.32	29.12	5.56	2.87	4.46	1.71	.22	- .36	- .84	.38	1.20
24	29.26	9.18	25.85	5.67	2.90	4.19	2.06	1.06	- .45	- .78	.87	.98
25	26.52	8.60	23.54	5.76	2.87	4.04	2.58	.95	- .61	- .80	.87	.92
26	22.44	8.35	21.28	5.80	2.97	3.87	2.17	.79	- .64	- .82	.37	.75
27	18.55	7.90	18.72	5.26	8.66	2.85	1.70	.77	- .71	- .82	.96	.71
28	14.92	7.40	16.35	10.47	3.59	2.69	1.28	1.35	- .76	- .78	.09	.59
29	12.14	7.00	14.30	16.06	6.52	2.54	.94	1.30	- .67	- .80	.83	.52
30	10.18		12.69	14.18	6.50	2.21	.76	.95	- .55	- .80	1.00	.48
31	8.88		11.40		5.78		.64	.60		- .78		.48
1897.												
1	0.34	3.90	11.14	18.90	9.51	1.83	-0.15	1.36	0.60	-1.65	-1.39	-1.28
2	.47	6.00	9.57	15.28	9.95	1.75	- .18	1.08	.96	-1.71	-1.30	-1.12
3	.40	11.50	8.58	22.20	9.22	1.70	- .20	.87	1.02	-1.72	-1.29	- .48
4	.90	12.60	8.72	21.11	7.66	2.01	- .18	.62	.92	-1.75	-1.31	+1.29
5	1.24	11.70	10.14	22.00	6.23	1.98	- .05	.50	.76	-1.79	-1.33	13.10
6	1.11	12.37	16.33	26.95	5.36	2.65	1.63	.40	.62	-1.86	-1.28	14.24
7	1.40	16.24	51.42	25.82	4.67	3.44	3.40	.25	.51	-1.79	-1.27	10.72
8	2.60	18.70	54.77	22.10	4.20	2.87	3.90	.24	.47	-1.85	-1.27	7.39
9	2.66	21.04	51.59	21.80	3.87	2.25	3.71	1.10	.42	-1.88	-1.17	5.12
10	2.30	19.80	44.69	29.27	3.50	1.85	3.05	2.10	.36	-1.90	-1.18	3.72
11	2.08	17.90	40.54	29.57	3.20	1.60	2.58	3.26	.29	-1.88	-1.14	3.05
12	1.76	28.42	42.58	25.48	3.64	1.39	2.42	8.22	.23	-1.88	-1.14	2.70
13	1.52	25.90	48.70	21.60	11.40	1.24	2.16	2.73	.14	-1.89	-1.17	2.56
14	1.33	23.84	50.96	18.10	20.36	1.11	1.86	2.27	.11	-1.92	-1.25	3.06
15	1.23	20.30	48.57	16.32	20.46	.95	1.46	1.63	.10	-1.92	-1.25	8.54
16	8.23	16.96	45.20	18.43	16.59	.85	1.16	1.28	.06	-1.92	-1.27	4.12
17	9.70	14.04	47.21	18.33	12.68	.75	.97	1.00	.01	-1.90	-1.33	4.10
18	13.10	11.72	46.72	15.92	9.77	.83	1.50	.73	- .05	-1.90	-1.35	3.78
19	19.35	9.97	42.90	18.66	7.73	1.30	3.50	.52	- .86	-1.90	-1.36	3.50
20	18.70	8.77	42.57	11.86	6.35	1.45	12.50	1.58	- .75	-1.88	-1.86	3.82
21	17.48	8.08	44.54	10.45	5.37	1.11	14.50	1.78	- .95	-1.84	-1.86	6.70
22	18.64	8.32	41.50	9.24	4.66	.80	11.30	1.58	-1.07	-1.77	-1.37	19.58
23	16.52	11.00	37.70	8.15	4.13	.55	8.42	1.33	-1.17	-1.64	-1.42	29.96
24	13.30	20.20	35.66	7.37	3.70	.33	6.64	1.08	-1.26	-1.58	-1.44	24.08
25	10.60	21.24	32.40	6.80	3.45	.23	4.77	1.27	-1.24	-1.63	-1.40	18.97
26	8.60	18.97	28.86	6.37	3.13	.16	3.46	.97	-1.36	-1.63	-1.36	13.10
27	7.20	16.18	25.60	5.90	2.90	.12	2.72	.80	-1.41	-1.61	-1.36	15.67
28	6.00	13.31	23.15	5.49	2.60	.03	2.92	.61	-1.44	-1.64	-1.28	13.10
29	5.10		20.33	5.10	2.28	.00	2.15	.54	-1.50	-1.61	-1.22	11.00
30	4.42		17.52	5.90	2.09	- .08	1.80	.46	-1.55	-1.63	-1.24	9.38
31	3.80		14.98		2.00		1.62	.31		-1.63		8.00

Daily gage height, in feet, of Black Warrior River at Tuscaloosa—Continued.

Day.	Jan.	Feb.	Mar.	Apr.	May.	June.	July.	Aug.	Sept.	Oct.	Nov.	Dec.
1898.												
1	7.10	21.68	3.67	25.50	11.34	0.23	0.66	2.45	1.20	−0.60	0.90	5.40
2	6.28	18.30	3.78	22.40	9.82	.08	.43	1.95	.82	− .70	.70	5.50
3	5.52	14.90	4.03	18.50	8.67	.23	.30	2.45	.70	− .80	.60	5.40
4	4.96	12.12	4.05	15.31	7.68	.27	.53	2.80	.30	− .90	.40	5.00
5	4.41	10.15	3.98	20.82	6.88	.27	.16	2.96	.00	−1.00	.00	4.70
6	4.17	9.42	8.90	38.70	6.20	.15	.10	3.90	− .10	− .90	.30	4.50
7	4.13	8.90	3.57	38.55	5.59	.02	− .18	3.40	− .30	− .90	.50	5.00
8	4.18	8.43	3.33	32.83	5.00	− .10	− .04	3.10	− .20	− .70	.40	5.00
9	4.17	7.84	3.12	27.70	4.50	− .29	− .04	3.98	.00	− .80	.40	4.70
10	3.97	7.30	3.00	23.04	4.20	− .41	+ .10	6.20	.20	+2.00	.50	4.20
11	4.00	6.68	2.90	19.50	3.91	− .54	.08	12.20	.40	3.30	.50	3.70
12	4.18	6.30	2.83	17.03	3.50	− .62	− .07	14.10	.20	2.60	.40	3.30
13	6.70	6.20	2.80	15.01	3.21	− .71	− .10	10.60	.00	2.00	.50	3.00
14	9.70	6.08	3.00	12.93	2.88	− .62	− .07	7.30	− .30	1.40	.80	2.80
15	11.97	5.86	4.80	11.65	2.58	− .53	− .11	5.00	− .30	1.00	1.10	2.60
16	11.71	5.40	8.00	10.78	2.30	− .38	− .07	3.50	− .40	.80	1.30	2.50
17	15.00	5.00	15.40	9.60	2.08	− .48	− .14	2.62	− .60	.40	1.50	2.40
18	15.53	4.90	14.10	8.63	1.97	− .56	.00	1.91	− .70	.80	1.60	2.10
19	14.32	4.83	11.92	8.15	1.64	− .30	1.24	1.50	− .70	.80	1.90	4.40
20	24.50	4.84	10.08	27.80	1.43	− .21	1.30	1.46	− .80	1.20	2.20	18.80
21	33.54	4.70	8.82	33.11	1.23	.00	1.12	1.12	− .90	3.10	2.50	23.90
22	31.42	4.48	7.80	28.45	1.07	.07	.88	1.26	− .50	3.80	4.00	21.30
23	28.50	4.30	6.96	23.77	1.00	.18	.47	1.20	− .60	4.10	8.60	17.40
24	30.38	4.00	6.30	22.72	.93	.30	1.08	.90	− .70	4.80	11.70	13.30
25	30.12	3.83	6.00	24.04	.78	.17	1.95	.50	− .80	4.30	11.60	10.50
26	42.50	3.64	5.90	22.62	.57	.06	2.50	1.10	− .90	3.90	9.50	8.60
27	43.48	3.56	5.53	19.67	.43	1.30	2.00	1.02	− .90	2.90	7.40	7.30
28	39.41	3.72	5.12	17.12	.43	1.10	1.80	.98	− .80	2.30	5.90	6.20
29	33.80	....	5.23	15.05	.52	.46	2.03	1.30	− .80	2.00	5.50	5.60
30	28.50	....	13.40	13.03	.44	.68	2.55	1.52	− .60	1.40	5.30	5.00
31	24.90	....	25.68	....	.30	....	2.86	1.56	....	1.10	....	4.70
1899.												
1	4.40	26.10	37.70	30.30	7.90	2.00	.10	6.50	1.20	−1.50	−1.03	4.83
2	4.50	29.50	32.00	29.20	7.00	1.50	.80	4.90	.70	−1.61	−1.02	3.50
3	4.60	27.00	27.70	24.80	6.30	1.20	.30	3.40	.62	−1.80	−1.03	2.77
4	4.70	29.00	24.10	22.00	5.80	1.00	− .70	2.70	.23	−1.79	− .99	2.03
5	4.70	45.50	21.00	23.20	5.30	1.00	− .10	2.00	− .02	−1.51	−1.00	1.50
6	11.20	50.60	20.80	22.60	4.80	.90	− .30	1.90	− .17	−1.48	−1.02	1.22
7	42.50	51.40	19.50	23.30	4.40	1.00	− .20	1.80	− .38	−1.46	−1.03	.98
8	49.30	51.70	16.90	33.90	4.40	.80	− .30	1.30	− .40	−1.34	−1.03	.88
9	46.60	48.60	14.40	34.00	7.60	.40	− .40	1.20	− .46	−1.30	−1.03	.58
10	40.40	43.10	12.90	30.80	6.50	.40	− .40	1.10	− .68	− .96	−1.01	.60
11	33.70	37.80	11.90	27.00	5.10	.50	− .40	2.00	− .73	− .78	−1.01	2.20
12	31.90	32.80	11.30	23.60	4.30	.50	− .50	1.70	− .79	− .72	− .98	23.50
13	28.00	28.80	10.00	20.10	4.20	.50	− .60	1.30	− .71	− .70	− .96	39.53
14	25.00	25.70	28.80	17.20	4.30	.50	− .60	.90	− .73	.88	− .98	35.71
15	22.20	22.90	44.50	14.90	4.70	.50	− .70	.60	− .78	− .94	− .99	26.50
16	20.00	21.60	59.30	13.00	4.80	.70	− .70	.40	− .83	−1.03	−1.00	20.63
17	19.60	19.90	60.30	11.60	3.70	.60	− .70	.30	− .86	−1.07	− .97	15.21
18	20.10	20.10	57.70	10.50	3.30	.40	− .70	.30	− .88	−1.12	− .97	10.83
19	18.60	20.80	52.40	9.80	2.80	.30	− .70	.60	− .90	−1.16	− .95	8.02
20	16.10	20.60	49.30	9.60	2.70	.20	− .70	.60	− .92	−1.12	− .96	8.09
21	14.00	19.60	46.80	9.50	3.70	.10	− .60	.70	− .94	−1.10	− .92	9.63
22	12.20	18.50	41.60	8.70	3.80	.40	− .50	1.50	− .99	− .84	− .89	10.40
23	11.00	22.70	36.80	8.60	3.30	− .10	− .20	2.60	−1.03	− .73	− .33	10.63
24	11.00	23.10	33.00	11.30	3.10	.10	+ .60	2.60	−1.05	− .71	+ .17	22.01
25	20.30	20.90	29.50	13.60	2.70	− .60	4.90	2.60	−1.04	.88	.88	29.04
26	29.30	18.50	26.50	13.20	2.30	+ .20	7.60	2.30	−1.05	− .94	2.60	25.91
27	26.20	23.50	24.20	12.60	2.00	.20	7.40	2.10	−1.04	−1.02	4.50	20.98
28	22.50	89.00	22.30	11.40	1.75	.20	7.90	1.90	1.18	− .60	10.48	17.09
29	18.60	....	21.10	10.00	1.50	.20	9.50	1.60	−1.28	− .73	9.49	15.80
30	15.70	....	19.30	8.90	1.60	.10	9.30	1.50	−1.38	− .96	6.67	14.62
31	14.80	....	18.10	....	2.50	....	8.50	1.10	....	−1.03	....	12.51

Daily gage height, in feet, of Black Warrior River at Tuscaloosa—Continued.

Day.	Jan.	Feb.	Mar.	Apr.	May.	June.	July.	Aug.	Sept.	Oct.	Nov.	Dec.
1900.												
1	10.90	6.92	18.14	21.55	21.23	4.24	41.00	9.85	5.50	0.75	4.20	12.50
2	9.48	6.40	25.44	18.60	18.95	5.30	33.88	8.00	5.00	.80	5.10	9.65
3	8.32	5.00	26.00	15.80	16.33	13.95	32.27	6.50	4.07	.60	5.40	8.50
4	7.10	5.71	23.30	13.60	13.82	15.16	32.13	5.15	4.30	.25	5.55	7.90
5	6.18	6.40	20.29	12.27	11.60	20.98	30.96	4.40	3.58	.40	5.05	8.20
6	5.53	9.58	17.43	12.17	9.60	21.50	27.31	3.80	2.80	.40	4.45	10.20
7	5.12	11.50	15.34	11.44	8.20	20.92	23.70	3.35	2.30	.80	3.90	10.65
8	4.84	10.53	27.58	10.29	7.28	21.75	20.80	3.00	1.90	1.80	3.50	9.65
9	4.60	12.23	39.00	9.97	6.80	38.52	18.45	2.50	1.50	3.90	3.10	8.75
10	4.50	20.60	36.03	10.58	6.50	31.70	15.96	2.30	1.40	6.05	2.90	8.15
11	7.12	23.00	31.63	26.35	6.12	24.73	12.96	2.10	1.10	6.10	2.70	7.50
12	31.63	20.64	27.34	52.79	5.90	19.60	11.05	2.00	.90	14.30	2.50	6.90
13	31.80	41.37	23.78	53.40	5.48	15.95	10.60	1.75	.80	22.50	2.40	6.45
14	28.18	47.96	20.43	48.69	4.98	15.60	12.90	1.50	1.40	21.85	2.25	6.40
15	24.09	45.73	17.40	42.30	4.50	19.23	10.35	1.40	5.45	16.40	2.20	6.65
16	20.12	40.23	16.63	37.10	3.85	29.15	8.50	1.50	10.65	11.60	2.10	6.45
17	16.50	34.74	21.18	63.00	3.65	28.89	7.00	1.48	8.95	8.20	2.05	6.10
18	14.28	29.75	20.28	64.06	3.50	25.33	6.00	1.40	6.34	6.13	2.00	5.75
19	15.80	25.88	18.80	62.17	3.52	24.51	5.60	3.00	4.44	5.10	1.95	5.30
20	25.00	22.43	45.10	59.35	3.39	30.10	6.35	4.40	3.35	4.30	3.90	5.45
21	32.60	19.90	51.00	56.10	3.65	27.80	8.50	3.75	2.60	3.60	8.40	6.30
22	29.44	23.58	47.98	51.71	3.50	25.38	8.00	3.00	2.20	3.15	10.25	8.35
23	24.54	26.50	42.40	46.20	3.35	24.55	8.10	2.50	1.87	4.60	12.50	11.90
24	20.43	24.20	38.24	41.88	4.40	50.00	8.70*	2.55	1.65	7.30	11.00	17.00
25	17.12	22.38	35.41	37.94	7.55	58.35	7.40	2.10	1.60	10.40	9.45	18.70
26	14.12	21.00	36.78	33.94	7.68	56.35	6.80	4.48	1.45	8.55	15.90	17.00
27	12.14	18.63	35.33	30.89	6.70	52.90	9.20	4.30	1.30	6.75	22.20	14.60
28	10.47	16.52	31.80	28.15	5.30	49.05	16.20	3.25	1.05	5.40	21.00	12.50
29	9.10	....	28.85	25.73	4.25	48.35	13.50	2.40	.90	4.70	17.35	10.95
30	8.27	....	26.25	23.50	3.60	46.85	11.83	2.10	.75	4.10	14.00	10.15
31	7.60	....	24.20	....	3.02	....	11.05	2.50	....	3.70	....	14.20
1901.												
1	19.50	15.10	7.50	28.00	9.40	6.50	1.70	0.70	6.11	5.40	0.85	1.60
2	19.00	14.80	7.30	26.50	8.50	12.25	1.73	.60	5.10	6.91	.80	1.50
3	17.90	15.00	8.40	32.60	7.90	16.65	2.00	.85	4.05	9.00	.75	1.40
4	14.70	35.20	9.65	35.10	7.30	14.80	2.30	1.00	3.60	6.72	1.00	1.40
5	13.00	42.00	9.70	31.50	7.00	12.00	2.50	.80	3.05	5.30	1.10	1.31
6	11.50	38.35	9.50	27.50	6.35	11.20	2.30	.75	2.81	4.50	1.10	1.95
7	10.25	32.15	9.00	24.30	6.15	11.10	2.50	.60	2.35	3.45	1.05	1.90
8	9.50	29.15	8.50	20.90	6.00	11.60	2.40	.50	2.04	3.02	1.00	2.58
9	8.75	28.05	8.05	17.90	5.70	9.40	2.25	.35	1.90	2.50	.95	2.60
10	8.30	30.97	8.70	14.50	5.60	7.60	2.10	.25	1.70	2.31	.90	3.90
11	17.40	29.75	29.50	12.50	5.00	6.50	2.00	.20	1.50	2.10	.90	5.00
12	52.70	27.46	34.00	11.00	4.60	6.20	1.70	.65	1.41	1.95	.85	5.80
13	56.50	25.15	28.50	10.00	5.00	6.40	1.30	.65	1.30	3.41	1.00	5.70
14	53.25	22.50	23.60	11.35	6.60	5.50	.90	.80	2.52	3.30	1.05	7.40
15	47.25	19.60	19.70	12.80	8.70	5.05	.70	1.20	6.70	3.97	1.00	31.00
16	41.45	17.45	16.00	13.00	8.30	4.80	.60	5.70	8.00	3.80	.98	40.75
17	36.30	15.70	13.00	11.70	6.80	4.50	.43	17.00	6.10	3.51	.90	35.00
18	31.85	13.70	11.05	12.80	5.50	4.10	2.60	26.80	12.21	3.00	.93	27.00
19	28.15	12.60	10.90	25.70	4.70	3.80	4.00	22.70	16.00	2.70	1.10	21.50
20	25.15	11.90	9.60	39.80	4.90	3.35	5.70	25.70	14.50	2.30	1.30	16.41
21	22.50	10.90	13.00	42.60	8.30	3.00	7.33	32.10	10.60	2.00	1.40	12.10
22	19.85	9.95	16.50	38.00	17.30	2.75	6.00	32.50	7.50	1.81	1.70	9.95
23	17.55	9.20	15.70	32.80	19.80	2.30	4.50	31.97	5.70	1.61	2.00	7.00
24	15.35	8.75	14.00	28.41	17.50	2.20	3.20	26.80	4.51	1.45	2.05	6.80
25	14.60	8.65	13.80	24.40	14.10	2.00	2.50	22.40	3.90	1.31	2.00	7.30
26	18.00	8.50	28.50	21.00	11.30	1.97	2.00	18.30	3.40	1.20	1.95	8.34
27	17.20	8.20	37.25	17.90	9.30	1.85	1.50	14.10	2.91	1.05	1.95	9.00
28	15.50	7.90	34.50	14.90	8.15	1.70	1.05	9.85	2.60	.81	2.00	10.96
29	14.80	....	29.00	12.50	7.00	1.65	.90	7.90	2.91	.55	1.95	36.30
30	15.10	....	24.30	10.90	6.90	1.65	.80	7.41	4.80	.90	1.91	49.00
31	14.60	....	24.85	....	6.35	....	.60	7.10	....	.90	....	49.00

Daily gage height, in feet, of Black Warrior River at Tuscaloosa—Continued.

Day.	Jan.	Feb.	Mar.	Apr.	May.	June.	July.	Aug.	Sept.	Oct.	Nov.	Dec.
1902.												
1	44.00	37.80	49.87	52.88	10.85	1.52	0.35	0.04	2.52	5.61	4.90	8.50
2	38.80	47.50	48.10	45.50	9.90	1.48	.30	.05	2.20	5.50	4.89	9.60
3	34.10	48.40	44.00	39.70	8.32	1.40	.15	.08	2.85	5.00	4.89	13.55
4	28.50	45.00	37.50	35.00	6.51	1.52	.00	.15	4.87	5.50	4.90	17.10
5	24.80	40.00	33.50	32.00	5.50	1.40	.00	1.80	4.65	5.52	4.91	17.92
6	21.15	36.00	35.78	29.00	5.00	1.80	.03	2.30	3.50	5.31	4.94	16.65
7	15.50	31.00	35.00	26.91	4.75	1.25	.05	1.90	2.52	5.15	5.00	14.90
8	14.00	28.15	31.61	32.70	5.11	1.15	.05	1.35	2.00	5.00	4.94	13.40
9	12.50	24.38	28.50	35.75	5.15	1.10	.15	.81	1.60	4.95	4.94	11.75
10	11.00	21.50	25.51	32.81	4.75	1.05	.20	.69	1.32	4.90	4.89	10.25
11	9.30	18.00	23.10	29.05	4.35	1.00	.25	.49	1.12	9.65	4.89	9.10
12	8.50	15.40	20.40	26.00	4.05	.91	.25	.48	.85	15.40	4.90	8.50
13	8.00	13.00	17.70	28.32	3.75	.80	.15	.39	.71	14.90	4.90	8.05
14	7.30	11.00	16.52	20.61	3.52	.72	.15	.25	.50	11.75	4.91	7.85
15	6.70	11.50	15.00	18.00	3.15	.70	.10	.18	.40	8.98	4.90	7.45
16	6.11	19.85	17.50	15.82	3.50	.65	.10	.15	.25	7.45	4.89	15.00
17	5.65	21.81	31.80	14.41	6.01	.53	.05	.14	.15	6.40	4.90	29.65
18	5.00	19.70	29.60	12.93	6.52	.50	.05	.10	.12	6.00	5.25	28.85
19	6.05	17.55	25.00	11.85	6.10	.80	.04	.08	.75	5.83	5.50	23.99
20	7.81	16.00	21.30	10.57	5.31	1.00	.03	.05	2.21	5.75	5.54	19.15
21	11.00	14.51	18.50	9.61	4.55	1.60	.03	.01	3.51	5.80	5.50	16.30
22	16.15	14.00	17.65	8.80	3.90	1.70	.04	.10	3.92	5.60	5.50	18.00
23	20.81	18.60	15.21	8.05	3.60	1.72	.04	.12	4.30	5.52	5.45	20.30
24	20.81	13.05	13.80	7.72	3.05	1.80	.10	.15	4.60	5.45	5.42	18.10
25	18.50	12.50	18.30	7.30	2.82	1.85	.11	.10	4.75	5.35	5.69	15.41
26	16.85	13.00	19.00	7.05	2.75	1.50	.10	.08	4.83	5.29	6.75	13.50
27	15.50	14.00	35.50	6.72	2.35	1.10	.11	.15	4.90	5.00	13.90	12.00
28	20.00	41.61	60.35	6.41	2.03	.65	.07	.61	5.20	4.90	13.40	10.95
29	28.00	......	60.60	5.90	1.95	.40	.05	1.45	5.08	4.91	10.42	10.00
30	27.87	......	58.30	5.75	1.81	.40	.05	1.68	5.65	4.90	9.65	11.10
31	28.50	......	57.40	......	1.65	......	.04	1.98	......	4.92	......	12.51
1903.												
1	18.20	13.35	54.30	16.70	8.50	9.60	6.43	6.95	5.10	3.98	5.03	4.71
2	13.75	12.55	52.41	16.65	8.15	12.50	5.90	7.40	6.15	3.95	5.17	4.72
3	16.58	13.75	47.25	15.30	7.90	14.21	5.65	8.55	5.00	3.80	5.68	4.74
4	23.50	18.55	41.42	14.12	7.75	12.35	5.44	7.65	4.91	8.75	5.55	4.74
5	22.52	35.00	36.90	13.10	7.45	12.40	5.80	6.95	4.80	3.65	5.21	4.85
6	19.81	37.85	33.11	12.30	7.15	18.35	6.35	6.50	4.73	8.98	5.04	4.87
7	17.10	36.00	37.82	11.50	6.92	16.83	6.15	6.80	4.55	4.20	4.98	4.88
8	15.11	56.35	36.50	11.15	6.89	16.50	5.90	6.72	4.45	4.70	4.90	4.86
9	13.31	55.85	39.15	12.30	7.00	12.20	5.75	6.55	4.40	4.75	4.88	4.88
10	12.10	51.50	39.20	22.35	6.90	10.50	6.50	6.10	4.42	4.90	4.84	4.89
11	12.53	52.00	37.75	20.10	6.75	10.85	6.25	6.35	4.31	4.97	4.80	4.90
12	23.95	53.75	35.12	16.50	7.05	11.00	6.35	6.15	4.25	5.10	4.87	4.92
13	28.41	51.10	33.80	14.35	8.65	10.35	6.52	6.10	4.21	5.15	4.91	4.87
14	25.75	46.15	36.15	21.20	14.35	9.46	6.70	6.00	4.19	4.95	4.87	4.87
15	21.90	41.11	36.00	32.12	36.85	8.30	6.55	5.96	4.13	4.96	4.92	4.87
16	19.01	40.50	33.84	28.20	43.40	7.65	6.25	6.25	4.10	4.93	4.97	4.86
17	16.11	56.65	31.12	23.40	43.10	7.08	5.90	6.60	4.00	4.98	4.98	4.86
18	14.50	a56.75	28.65	18.35	37.10	6.70	5.81	6.85	4.03	5.01	4.91	4.84
19	12.51	52.95	26.15	16.00	30.65	6.45	5.70	7.82	4.08	4.99	4.90	4.80
20	11.85	46.75	23.50	14.50	25.80	6.20	5.70	7.60	4.03	4.98	4.88	4.92
21	11.05	41.35	20.85	14.90	21.55	6.23	5.10	6.91	3.98	4.90	4.86	5.05
22	10.84	36.50	18.80	17.95	17.60	7.20	5.05	6.75	3.94	4.85	4.80	5.28
23	9.98	32.30	17.55	15.90	18.85	6.95	4.75	6.10	3.90	4.76	4.78	5.15
24	9.51	28.85	16.15	13.81	11.10	6.75	4.70	5.65	3.89	4.65	4.75	5.18
25	8.65	26.10	14.80	12.30	9.80	6.83	4.83	5.45	3.86	4.60	4.73	5.60
26	8.25	23.95	13.48	11.30	9.00	7.80	4.90	5.35	3.80	4.50	4.75	5.51
27	8.05	22.00	12.40	10.50	8.35	7.52	4.80	5.15	3.75	4.50	4.74	5.39
28	9.85	45.42	11.85	10.00	7.80	7.15	5.05	5.70	3.70	4.60	4.75	5.15
29	12.35	......	11.62	9.45	7.45	6.91	7.85	4.95	3.81	4.68	4.73	5.12
30	14.72	......	11.78	8.95	7.15	6.55	4.75	5.35	3.95	4.76	4.70	5.09
31	14.55	......	13.50	......	7.32	......	6.50	5.15	......	4.81	......	5.07

a Maximum 57.10.

Rating table for Black Warrior River at Tuscaloosa from 1895 to 1901.

Gage height.	Discharge.	Gage height.	Discharge.	Gage height.	Discharge.	Gage height.	Discharge.
Feet.	Second-feet.	Feet.	Second-feet.	Feet.	Second-feet.	Feet.	Second-feet.
−2.0	88	3.0	1,470	7.0	3,665	11.0	5,885
−1.8	92	3.1	1,520	7.1	3,715	11.1	5,935
−1.6	100	3.2	1,570	7.2	3,765	11.2	5,985
−1.4	110	3.3	1,620	7.3	3,815	11.3	6,035
−1.2	120	3.4	1,670	7.4	3,865	11.4	6,085
−1.0	130	3.5	1,725	7.5	3,925	11.5	6,145
− .8	150	3.6	1,780	7.6	3,985	11.6	6,205
− .6	175	3.7	1,835	7.7	4,045	11.7	6,265
− .4	205	3.8	1,890	7.8	4,105	11.8	6,325
− .2	240	3.9	1,945	7.9	4,165	11.9	6,385
.0	280	4.0	2,000	8.0	4,220	12.0	6,440
.1	310	4.1	2,055	8.1	4,270	12.1	6,490
.2	340	4.2	2,111	8.2	4,320	12.2	6,540
.3	370	4.3	2,166	8.3	4,370	12.3	6,590
.4	400	4.4	2,222	8.4	4,420	12.4	6,640
.5	430	4.5	2,277	8.5	4,480	12.5	6,700
.6	460	4.6	2,333	8.6	4,540	12.6	6,760
.7	490	4.7	2,388	8.7	4,600	12.7	6,820
.8	530	4.8	2,444	8.8	4,660	12.8	6,880
.9	565	4.9	2,500	8.9	4,720	12.9	6,940
1.0	600	5.0	2,555	9.0	4,775	13.0	6,995
1.1	635	5.1	2,610	9.1	4,825	13.1	7,045
1.2	670	5.2	2,666	9.2	4,875	13.2	7,095
1.3	710	5.3	2,721	9.3	4,925	13.3	7,145
1.4	750	5.4	2,777	9.4	4,975	13.4	7,195
1.5	790	5.5	2,832	9.5	5,035	13.5	7,255
1.6	830	5.6	2,888	9.6	5,095	13.6	7,315
1.7	870	5.7	2,943	9.7	5,155	13.7	7,375
1.8	910	5.8	3,000	9.8	5,215	13.8	7,435
1.9	955	5.9	3,054	9.9	5,275	13.9	7,495
2.0	1,000	6.0	3,110	10.0	5,330	14.0	7,550
2.1	1,045	6.1	3,160	10.1	5,380	14.1	7,600
2.2	1,090	6.2	3,210	10.2	5,430	14.2	7,650
2.3	1,135	6.3	3,260	10.3	5,480	14.3	7,700
2.4	1,180	6.4	3,310	10.4	5,530	14.4	7,750
2.5	1,225	6.5	3,370	10.5	5,590	14.5	7,810
2.6	1,270	6.6	3,430	10.6	5,650	14.6	7,870
2.7	1,320	6.5	3,490	10.7	5,710	14.7	7,930
2.8	1,370	6.8	3,550	10.8	5,770	14.8	7,990
2.9	1,420	6.9	3,610	10.9	5,830	14.9	8,050

Rating table for Black Warrior River at Tuscaloosa from 1895 to 1901—Continued.

Gage height.	Discharge.	Gage height.	Discharge.	Gage height.	Discharge.	Gage height.	Discharge.
Feet.	*Second-feet.*	*Feet.*	*Second-feet.*	*Feet.*	*Second-feet.*	*Feet.*	*Second-feet.*
15.0	8,105	18.0	9,770	21.0	11,600	24.0	14,700
15.1	8,155	18.1	9,820	21.1	11,690	24.1	14,830
15.2	8,205	18.2	9,870	21.2	11,780	24.2	14,960
15.3	8,255	18.3	9,920	21.3	11,870	24.3	15,090
15.4	8,305	18.4	9,970	21.4	11,960	24.4	15,200
15.5	8,365	18.5	10,030	21.5	12,050	24.5	15,350
15.6	8,425	18.6	10,090	21.6	12,140	24.6	15,480
15.7	8,485	18.7	10,150	21.7	12,230	24.7	15,610
15.8	8,545	18.8	10,210	21.8	12,320	24.8	15,740
15.9	8,605	18.9	10,270	21.9	12,410	24.9	15,870
16.0	8,660	19.0	10,325	22.0	12,500	25.0	16,000
16.1	8,710	19.1	10,375	22.1	12,600	26.0	17,600
16.2	8,760	19.2	10,425	22.2	12,700	28.0	21,500
16.3	8,810	19.3	10,475	22.3	12,800	30.0	26,500
16.4	8,860	19.4	10,525	22.4	12,900	32.0	31,700
16.5	8,920	19.5	10,585	22.5	13,000	34.0	38,000
16.6	8,980	19.6	10,645	22.6	13,100	36.0	45,000
16.7	9,040	19.7	10,705	22.7	13,200	38.0	53,000
16.8	9,100	19.8	10,765	22.8	13,300	40.0	61,000
16.9	9,160	19.9	10,825	22.9	13,400	42.0	69,000
17.0	9,215	20.0	10,880	23.0	13,500	44.0	77,000
17.1	9,265	20.1	10,950	23.1	13,620	46.0	85,000
17.2	9,315	20.2	11,020	23.2	13,740	48.0	93,000
17.3	9,365	20.3	11,090	23.3	13,860	50.0	101,000
17.4	9,415	20.4	11,160	23.4	13,980	52.0	109,000
17.5	9,475	20.5	11,230	23.5	14,100	54.0	117,000
17.6	9,535	20.6	11,305	23.6	14,220	55.0	121,000
17.7	9,595	20.7	11,380	23.7	14,340		
17.8	9,655	20.8	11,455	23.8	14,460		
17.9	9,715	20.9	11,530	23.9	14,580		

Estimated monthly discharge of Black Warrior River at Tuscaloosa.

[Drainage area, 4,900 square miles.]

Month.	Discharge in second-feet.			Run-off.	
	Maximum.	Minimum.	Mean.	Second-feet per square mile.	Depth in inches.
1895.					
January	103,400	2,777	25,464	5.20	6.00
February	14,830	4,775	7,603	1.55	1.61
March	109,000	4,540	39,977	8.16	9.42
April	14,700	3,665	6,895	1.41	1.57
May	13,860	2,000	5,511	1.12	1.29
June	3,665	1,000	2,133	.44	.49
July	9,970	1,000	3,581	.73	.84
August	2,500	565	1,098	.22	.25
September	1,835	310	883	.18	.20
October	310	140	233	.05	.06
November	750	280	488	.10	.11
December	11,600	530	2,021	.41	.47
The year	109,000	140	7,991	1.63	22.31
1896.					
January	24,610	2,444	5,981	1.22	1.41
February	47,000	3,665	19,161	3.91	4.22
March	52,600	3,160	12,996	2.65	3.06
April	14,100	2,721	6,072	1.24	1.38
May	49,800	1,370	7,420	1.51	1.74
June	9,970	1,090	2,910	.59	.65
July	4,420	460	1,232	.25	.29
August	750	310	478	.10	.12
September	400	120	201	.04	.04
October	260	120	157	.03	.03
November	600	120	307	.06	.07
December	4,045	430	955	.19	.22
The year	52,600	120	4,822	.98	13.23

Estimated monthly discharge of Black Warrior River at Tuscaloosa—Continued.

Month.	Discharge in second-feet.			Run-off.	
	Maximum.	Minimum.	Mean.	Second-feet per square mile.	Depth in inches.
1897.					
January	10,500	385	3,493	0.71	0.82
February	17,440	1,945	8,409	1.72	1.79
March	120,080	4,540	52,883	10.79	12.44
April	25,285	2,610	9,657	1.97	2.20
May	11,195	1,000	3,600	.73	.84
June	1,697	260	715	.15	.17
July	7,810	240	1,809	.37	.43
August	1,595	355	701	.14	.16
September	600	102	295	.06	.07
October	102	90	93	.02	.02
November	125	107	115	.02	.03
December	29,000	115	5,549	1.13	1.30
The year	120,080	90	7,277	1.48	20.27
1898.					
January	75,000	1,972	16,577	3.38	3.90
February	12,230	1,752	3,902	.80	.83
March	17,120	1,370	3,626	.74	.85
April	55,800	4,295	15,620	3.19	3.56
May	6,060	370	1,766	.36	.41
June	710	160	303	.06	.07
July	1,395	250	549	.11	.13
August	7,600	430	1,785	.36	.41
September	670	140	252	.05	.06
October	2,444	130	880	.18	.21
November	6,265	280	1,626	.33	.37
December	14,580	1,045	3,763	.77	.89
The year	75,000	130	4,221	.86	11.69

Estimated monthly discharge of Black Warrior River at Tuscaloosa—Continued.

Month.	Discharge in second-feet.			Run-off.	
	Maximum.	Minimum.	Mean.	Second-feet per square mile.	Depth in inches.
1899.					
January	81, 375	2, 222	18, 118	3. 70	4. 27
February	90, 375	10, 030	30, 923	6. 31	6. 57
March.............................	122, 625	5, 330	35, 308	7. 21	8. 31
April..............................	32, 800	4, 540	11, 901	2. 43	2. 71
May	4, 165	790	2, 092	. 43	. 49
June	1, 000	175	448	. 09	. 10
July	5, 035	160	1, 111	. 23	. 26
August	3, 370	370	963	. 20	. 23
September.........................	670	110	200	. 04	. 04
October...........................	175	92	130	. 03	. 03
November	5, 590	127	721	. 15	. 17
December	47, 650	460	8, 880	1. 81	2. 09
The year	122, 625	92	9, 233	1. 89	25. 27
1900.					
January	29, 760	2, 277	9, 857	2. 01	2. 32
February	76, 312	2, 555	18, 356	3. 75	3. 90
March.............................	87, 750	8, 280	27, 105	5. 53	6. 37
April..............................	136, 687	5, 302	48, 426	9. 88	11. 02
May...............................	11, 825	1, 645	3, 702	. 76	. 88
June	115, 312	2, 138	32, 614	6. 66	7. 43
July...............................	52, 000	2, 888	10, 952	2. 24	2. 59
August	5, 245	750	1, 674	. 34	. 39
September.........................	5, 680	512	1, 580	. 32	. 36
October	13, 000	355	3, 382	. 69	. 80
November	12, 700	977	3, 701	. 76	. 85
December	10, 150	2, 721	5, 119	1. 05	1. 21
The year	136, 687	355	13, 872	2. 83	38. 12

Daily gage height, in feet, of Black Warrior River at Tuscaloosa—Continued.

Day.	Jan.	Feb.	Mar.	Apr.	May.	June.	July.	Aug.	Sept.	Oct.	Nov.	Dec.
1894												
1	10.30	8.50	27.00	11.40	6.50	1.70	2.50	1.20	5.80	0.55	-0.05	-0.05
2	9.60	7.80	25.60	14.10	5.90	1.60	1.80	1.30	4.60	.45	-.10	-.05
3	7.90	7.30	23.60	23.50	5.90	1.50	1.35	1.40	3.60	.40	-.20	.00
4	6.20	7.90	21.00	24.00	5.80	1.35	1.10	1.95	2.95	.32	-.25	.00
5	5.40	16.90	18.20	22.20	5.50	1.33	.80	2.30	2.45	.25	-.30	.00
6	6.40	22.60	15.90	20.60	5.20	1.20	.60	2.40	2.05	.20	-.30	.00
7	16.80	20.40	14.80	18.90	5.10	1.20	1.00	2.00	1.85	.15	-.30	.00
8	22.60	17.80	14.40	16.30	4.90	1.10	1.30	1.60	1.75	.00	-.30	+.05
9	19.70	17.30	13.80	14.70	4.50	.95	.70	1.80	1.65	-.10	-.30	.05
10	26.90	26.00	12.60	15.70	4.10	.90	.70	1.10	1.60	-.20	-.30	.50
11	34.50	27.80	11.60	25.00	3.70	.80	.60	.80	1.85	-.25	-.30	2.10
12	35.80	25.40	11.50	25.60	3.90	.80	.60	.70	3.40	-.30	-.30	5.70
13	30.90	29.90	16.50	22.70	4.80	.70	.50	.60	5.60	-.30	-.30	10.40
14	25.80	32.10	17.30	19.50	7.00	.60	.50	.50	7.50	-.30	-.15	11.20
15	22.20	29.70	15.80	16.30	6.20	.60	.60	.35	5.70	-.30	.00	8.90
16	23.60	26.00	15.60	15.50	5.70	.60	.70	.60	4.20	-.30	+.05	6.40
17	24.80	22.30	33.10	16.00	5.10	.55	.75	.55	3.25	-.30	.05	4.90
18	22.40	19.10	36.70	17.00	5.00	.50	.70	.45	3.50	-.30	.05	3.90
19	19.30	16.80	33.70	25.30	4.70	.60	.70	.40	3.25	-.30	.00	3.20
20	15.90	16.00	29.50	24.80	4.20	.70	.60	.55	3.25	-.30	-.10	2.90
21	14.10	16.90	27.80	24.00	3.60	.85	.50	1.00	4.10	-.30	-.10	2.60
22	18.60	17.90	29.30	20.40	3.10	.80	.95	1.35	3.60	-.30	-.10	2.20
23	21.80	17.00	29.60	17.30	2.80	.95	1.30	3.80	2.90	-.30	-.10	2.00
24	19.90	16.10	28.00	14.20	2.60	1.10	1.35	5.30	2.20	-.35	.00	1.80
25	17.60	17.30	27.80	12.20	2.40	1.00	1.60	9.20	1.75	-.40	.00	1.70
26	15.50	29.80	25.30	10.50	2.30	1.05	1.50	16.00	1.45	-.40	.00	2.00
27	14.00	31.00	22.10	9.80	2.10	1.40	1.40	20.40	1.20	-.45	-.05	7.70
28	12.40	28.80	19.00	8.70	2.10	1.70	1.30	16.00	1.05	-.45	.05.	9.70
29	11.10	....	16.10	7.90	2 00	2.35	.20	11.80	.90	-.45	.05	8.40
30	10.00	....	14.80	7.00	1.90	2.60	1.10	8.30	.70	-.20	.05	7.20
31	9.10	....	12.70	....	1.90	....	1.20	6.80	....	-.05	.05	6.50
1895												
1	6.10	24.10	8.60	16.20	15.20	7.00	4.80	1.80	1.90	0.10	-0.04	0.75
2	5.40	21.50	13.20	14.50	13.00	5.90	4.50	1.50	1.80	.00	+.07	.95
3	6.80	21.40	35.80	12.60	11.00	5.10	5.40	1.30	1.80	-.07	.12	1.05
4	9.00	21.40	36.70	11.00	9.40	4.60	7.60	1.20	1.70	-.10	.09	1.00
5	8.80	19.90	32.40	10.20	8.30	4.20	17.40	1.20	1.90	-.10	.11	.90
6	8.00	17.90	27.80	8.40	7.40	6.70	17.70	1.20	2.60	-.10	.25	1.20
7	7.40	16.50	23.80	7.90	7.00	5.70	18.40	1.10	3.05	-.10	.31	1.50
8	35.00	16.80	20.60	20.50	8.20	5.10	15.70	1.00	3.70	.00	.30	1.45
9	50.60	16.90	20.10	24.00	13.50	4.30	13.10	.90	3.40	-.03	.28	1.90
10	49.30	15.80	19.30	21.20	19.70	3.80	11.10	.97	2.90	-.07	.70	1.80
11	45.10	14.60	17.50	18.00	23.30	3.10	9.80	1.40	2.40	-.10	1.20	1.45
12	40.10	14.10	17.20	15.10	22.00	2.70	7.90	1.50	2.20	-.14	1.10	1.40
13	35.00	13.90	18.40	12.60	18.50	2.40	6.50	1.40	2.10	-.14	1.18	1.40
14	29.80	12.80	24.90	11.00	15.20	2.20	5.30	1.30	2.00	+.06	1.35	1.37
15	25.70	11.60	37.50	9.50	12.20	2.00	5.40	1.10	1.80	.04	1.35	1.33
16	23.40	10.50	47.40	8.60	9.70	2.00	5.60	1.30	1.60	-.06	1.28	1.28
17	31.20	9.40	52.00	10.00	8.00	2.70	5.90	1.40	1.40	-.18	1.20	1.12
18	32.90	9.00	47.30	15.90	6.90	3.30	5.80	2.00	1.40	-.27	1.05	1.05
19	29.20	9.00	42.10	15.80	6.30	4.20	5.10	4.00	1.30	-.31	.88	1.02
20	25.80	9.50	38.80	14.00	5.90	4.90	4.90	4.30	1.00	-.37	.82	1.57
21	23.10	10.20	48.70	12.00	5.30	4.70	4.70	4.00	1.20	-.42	.70	2.55
22	21.60	10.90	51.30	10.40	4.80	4.30	3.60	2.80	1.50	-.45	.60	2.60
23	21.60	11.40	47.60	8.90	4.30	4.10	3.40	2.40	1.30	-.50	.50	3.03
24	19.80	11.30	42.10	8.00	4.10	3.70	3.80	4.80	1.10	-.35	.75	3.65
25	17.40	10.90	37.30	7.20	4.00	3.40	3.60	4.90	.90	-.33	.60	3.31
26	16.70	10.80	32.80	7.00	4.30	3.00	8.70	4.00	.60	-.53	.38	4.50
27	20.00	9.70	29.10	7.10	5.60	4.40	8.30	3.20	.35	-.70	.64	10.40
28	21.20	9.00	26.10	12.50	11.00	5.40	8.30	2.30	.10	-.68	.65	21.02
29	22.00	....	23.60	15.00	13.30	6.00	2.40	2.30	.50	-.67	.65	16.88
30	27.40	....	21.10	16.40	11.20	5.40	2.00	2.10	.30	-.50	.60	13.10
31	27.10	....	18.60	....	8.90	....	2.10	2.00	....	-.32	....	11.67

Daily gage height, in feet, of Black Warrior River at Tuscaloosa—Continued.

Day.	Jan.	Feb.	Mar.	Apr.	May.	June.	July.	Aug.	Sept.	Oct.	Nov.	Dec.
1896												
1	11.41	7.92	6.91	10.61	18.24	4.92	1.94	0.53	0.38	-0.52	-0.78	7.70
2	10.26	8.51	7.66	14.78	33.30	4.28	1.71	.90	.25	-.62	-.77	6.82
3	8.90	21.98	7.41	23.00	37.18	3.50	1.44	1.34	.15	-.66	-.42	4.68
4	7.60	33.12	6.74	23.50	30.88	5.44	1.48	.89	.06	-.20	-.39	3.40
5	6.55	30.02	6.14	19.85	26.40	5.24	1.30	.54	-.06	+.78	-.33	2.60
6	5.74	30.75	6.64	16.70	21.93	4.30	1.15	.42	-.15	.74	-.19	2.05
7	5.09	35.92	10.18	14.26	17.88	3.62	1.32	.50	-.23	.68	-.19	1.70
8	5.24	35.08	12.04	12.55	18.95	3.27	2.20	.60	-.30	.40	-.17	1.47
9	5.76	36.21	14.89	11.20	10.80	6.60	8.40	.68	-.37	.09	-.18	1.21
10	6.46	36.52	13.69	10.02	8.60	8.18	6.98	.64	-.44	.15	-.21	1.10
11	6.58	33.65	12.45	9.04	7.11	18.39	5.37	.55	-.44	-.28	-.30	.89
12	6.25	29.45	13.97	8.83	6.09	15.13	3.95	.45	-.46	-.28	-.20	.84
13	5.79	25.97	16.15	7.62	5.34	10.77	2.99	.38	-.50	-.48	-.22	.80
14	5.26	27.41	15.85	7.86	4.77	7.65	2.45	.85	-.60	-.49	-.11	.75
15	4.82	33.25	13.56	11.65	6.13	5.60	2.16	.81	-.59	-.46	-.10	1.20
16	4.87	31.02	18.86	13.90	5.45	4.33	2.13	.26	-.59	-.43	-.01	1.25
17	6.08	27.80	22.30	12.36	4.65	4.74	2.15	.13	-.60	-.60	+.26	1.25
18	8.61	23.65	27.75	10.57	3.94	4.20	2.42	.40	-.60	-.63	.46	1.60
19	9.14	20.09	29.70	9.03	3.44	6.18	2.46	.50	-.60	-.78	.52	1.90
20	8.58	17.00	37.68	7.81	3.05	6.00	2.24	.40	-.61	-.82	.45	1.70
21	7.87	14.45	37.92	6.88	2.91	5.32	2.16	.34	-.64	-.82	.39	1.50
22	8.85	12.20	33.55	6.15	2.77	4.91	1.90	.29	-.64	-.84	.38	1.33
23	22.47	10.32	29.12	5.56	2.87	4.46	1.71	.22	-.36	-.84	.38	1.20
24	29.26	9.18	25.85	5.67	2.90	4.19	2.06	1.06	-.45	-.78	.37	.96
25	26.52	8.60	23.54	5.76	2.87	4.04	2.58	.95	-.61	-.80	.37	.92
26	22.44	8.85	21.28	5.80	2.97	3.37	2.17	.79	-.64	-.82	.37	.77
27	18.55	7.90	18.72	5.26	3.66	2.85	1.70	.77	-.71	-.82	.98	.71
28	14.92	7.40	16.35	10.47	3.59	2.69	1.28	1.35	-.76	-.78	.09	.59
29	12.14	7.00	14.30	16.06	6.52	2.54	.94	1.30	-.67	-.80	.83	.52
30	10.18	......	12.69	14.18	6.50	2.21	.76	.95	-.55	-.80	1.00	.48
31	8.88	......	11.40	......	5.78	......	.64	.60	......	-.78	......	.48
1897.												
1	0.34	3.90	11.14	13.90	9.51	1.83	-0.15	1.36	0.60	-1.65	-1.39	-1.28
2	.47	6.00	9.57	15.28	9.96	1.75	-.18	1.08	.96	-1.71	-1.29	-1.12
3	.40	11.50	8.58	22.20	9.22	1.70	-.20	.87	1.02	-1.72	-1.29	-.48
4	.90	12.60	8.72	21.11	7.66	2.01	-.18	.62	.92	-1.75	-1.31	+1.29
5	1.24	11.70	10.14	22.00	6.23	1.98	-.05	.50	.76	-1.79	-1.33	13.10
6	1.11	12.87	16.83	26.95	5.86	2.65	1.63	.40	.62	-1.86	-1.28	14.24
7	1.40	16.24	51.42	25.32	4.67	3.44	3.40	.25	.51	-1.79	-1.27	10.72
8	2.60	18.70	54.77	22.10	4.20	2.87	3.90	.24	.47	-1.85	-1.27	7.89
9	2.66	21.04	51.59	21.30	3.87	2.25	3.71	1.10	.42	-1.88	-1.17	5.12
10	2.30	19.80	44.69	29.27	3.50	1.85	3.05	2.10	.36	-1.90	-1.13	3.72
11	2.03	17.90	40.54	29.57	3.20	1.60	2.58	3.26	.29	-1.88	-1.10	3.05
12	1.76	23.42	42.58	25.48	3.64	1.39	2.42	3.22	.23	-1.88	-1.14	2.70
13	1.52	25.90	48.70	21.60	11.40	1.24	2.16	2.73	.14	-1.89	-1.17	2.56
14	1.33	23.84	50.96	18.10	20.36	1.11	1.86	2.27	.11	-1.92	-1.25	3.05
15	1.23	20.30	48.57	16.32	20.46	.95	1.46	1.63	.10	-1.92	-1.25	8.54
16	8.23	16.96	45.20	18.43	16.59	.85	1.16	1.28	.06	-1.92	-1.27	4.12
17	9.70	14.04	47.21	18.33	12.68	.75	.97	1.00	.01	-1.90	-1.33	4.10
18	13.10	11.72	46.72	15.92	9.77	.63	1.50	.73	-.05	-1.90	-1.35	3.78
19	19.35	9.97	42.90	13.66	7.73	1.30	3.50	.52	-.36	-1.90	-1.36	3.50
20	18.70	8.77	42.57	11.86	6.35	1.45	12.50	1.53	-.75	-1.88	-1.36	3.82
21	17.43	8.08	44.54	10.45	5.37	1.11	14.50	1.78	-.95	-1.88	-1.36	6.70
22	18.64	8.82	41.50	9.24	4.66	.80	11.30	1.58	-1.07	-1.77	-1.37	19.58
23	16.52	11.00	37.70	8.15	4.13	.80	8.42	1.33	-1.17	-1.64	-1.42	31.00
24	13.30	20.20	35.66	7.87	3.70	.33	6.64	1.08	-1.26	-1.58	-1.44	29.96
25	10.60	21.24	32.40	6.80	3.45	.22	4.77	1.27	-1.24	-1.68	-1.40	24.08
26	8.60	18.97	28.86	6.37	3.13	.16	8.46	.97	-1.36	-1.63	-1.36	18.97
27	7.20	16.18	25.60	5.90	2.90	.12	2.72	.80	-1.41	-1.61	-1.36	15.67
28	6.00	13.81	23.15	5.49	2.60	.03	2.92	.61	-1.44	-1.64	-1.28	13.10
29	5.10	......	20.33	5.10	2.28	.00	2.15	.54	-1.50	-1.61	-1.22	11.00
30	4.42	......	17.52	5.90	2.09	-.08	1.80	.46	-1.55	-1.63	-1.24	9.38
31	3.80	......	14.96	......	2.00	......	1.62	.31	......	-1.63	......	8.00

Daily gage height, in feet, of Black Warrior River at Tuscaloosa—Continued.

Day.	Jan.	Feb.	Mar.	Apr.	May.	June.	July.	Aug.	Sept.	Oct.	Nov.	Dec.
1898.												
1	7.10	21.68	3.67	25.50	11.34	0.23	0.66	2.45	1.20	-0.60	0.90	5.40
2	6.28	18.30	3.78	22.40	9.82	.08	.43	1.95	.82	-.70	.70	5.50
3	5.52	14.90	4.03	18.50	8.67	.23	.30	2.45	.70	-.80	.60	5.40
4	4.96	12.12	4.05	15.31	7.68	.27	.53	2.80	.30	-.90	.40	5.00
5	4.41	10.15	3.98	20.82	6.88	.27	.16	2.96	.00	-1.00	.00	4.70
6	4.17	9.42	3.90	38.70	6.20	.15	.10	3.90	.10	-.90	.30	4.50
7	4.13	8.90	3.57	38.55	5.59	.02	-.18	3.40	-.30	-.90	.50	5.00
8	4.18	8.43	3.33	32.83	5.00	-.10	-.04	3.10	-.20	-.70	.40	5.00
9	4.17	7.84	3.12	27.70	4.50	-.29	-.04	3.98	.00	-.30	.40	4.70
10	3.97	7.30	3.00	23.04	4.20	-.41	+.10	6.20	.20	+2.00	.50	4.20
11	4.00	6.68	2.90	19.50	3.91	-.54	.08	12.20	.40	3.30	.50	3.70
12	4.18	6.30	2.83	17.03	3.50	-.62	-.07	14.10	.20	2.60	.40	3.30
13	6.70	6.20	2.80	15.01	3.21	-.71	-.10	10.60	.00	2.00	.50	3.00
14	9.70	6.08	3.00	12.93	2.88	-.62	-.07	7.30	.30	1.40	.80	2.80
15	11.97	5.86	4.80	11.65	2.58	-.53	-.11	5.00	.30	1.00	1.10	2.60
16	11.71	5.40	8.00	10.78	2.30	-.38	-.07	3.50	.40	.80	1.30	2.50
17	15.00	5.00	15.40	9.60	2.08	-.48	-.14	2.62	.60	.40	1.50	2.40
18	15.53	4.90	14.10	8.63	1.97	-.56	.00	1.91	.70	.80	1.60	2.10
19	14.32	4.83	11.92	8.15	1.64	-.30	1.24	1.50	.70	.80	1.90	4.40
20	24.50	4.84	10.08	27.80	1.43	-.21	1.30	1.46	.80	1.20	2.20	18.80
21	33.54	4.70	8.82	33.11	1.23	.00	1.12	1.12	.90	3.10	2.50	23.90
22	31.42	4.48	7.80	28.45	1.07	.07	.88	1.26	.50	3.80	4.00	21.30
23	28.50	4.30	6.96	23.77	1.00	.18	.47	1.20	.60	4.10	8.60	17.40
24	30.38	4.00	6.30	22.72	.93	.30	1.08	.90	.70	4.80	11.70	13.30
25	30.12	3.83	6.00	24.04	.78	.17	1.95	.50	.80	4.30	11.60	10.50
26	42.50	3.64	5.90	22.62	.57	.06	2.50	1.10	.90	3.90	9.50	8.60
27	43.48	3.56	5.53	19.67	.43	1.80	2.00	1.02	.90	2.90	7.40	7.30
28	39.41	3.72	5.12	17.12	.43	1.10	1.80	.98	.80	2.30	5.90	6.20
29	33.80	......	5.23	15.05	.52	.46	2.03	1.30	.80	2.00	5.50	5.60
30	28.50	......	13.40	13.03	.44	.68	2.55	1.52	.60	1.40	5.30	5.00
31	24.90	......	25.68	......	.30	......	2.86	1.56	......	1.10	......	4.70
1899.												
1	4.40	26.10	37.70	30.30	7.90	2.00	.10	6.50	1.20	-1.50	-1.03	4.83
2	4.50	29.50	32.00	29.20	7.00	1.50	.80	4.90	.70	-1.61	-1.02	3.50
3	4.60	27.00	27.70	24.80	6.30	1.20	.30	3.40	.62	-1.80	-1.03	2.77
4	4.70	29.00	24.10	22.00	5.80	1.00	-.70	2.70	.23	-1.79	-.99	2.03
5	4.70	45.50	21.00	23.20	5.30	1.00	-.10	2.00	-.02	-1.51	-1.00	1.50
6	11.20	50.60	20.80	22.60	4.80	.90	-.30	1.90	-.17	-1.48	-1.02	1.22
7	42.50	51.40	19.50	23.30	4.40	1.00	-.20	1.80	-.38	-1.46	-1.03	.98
8	49.30	51.70	16.90	33.90	4.40	.80	-.30	1.80	-.40	-1.34	-1.03	.88
9	46.60	48.60	14.40	34.00	7.60	.40	-.40	1.20	-.46	-1.30	-1.03	.58
10	40.40	43.10	12.90	30.80	6.50	.40	-.40	1.10	-.68	-.96	-1.01	.60
11	33.70	37.80	11.90	27.00	5.10	.50	-.40	2.00	-.73	-.78	-1.01	2.20
12	31.90	32.80	11.30	23.60	4.30	.50	-.50	1.70	-.79	-.79	-.98	23.50
13	28.00	28.80	10.00	20.10	4.20	.50	-.60	1.30	-.71	-.70	-.96	39.53
14	25.00	25.70	28.80	17.20	4.30	.50	-.60	.90	-.73	-.88	-.98	35.71
15	22.20	22.90	44.50	14.90	4.70	.50	-.70	.60	-.78	-.94	-.99	26.50
16	20.70	21.60	59.30	13.00	4.80	.70	-.70	.40	-.83	-1.03	-1.00	20.63
17	19.60	19.90	60.30	11.60	3.70	.60	-.70	.30	-.86	-1.07	-.97	15.21
18	20.10	20.10	57.70	10.50	3.30	.40	-.70	.30	-.88	-1.12	-.97	10.83
19	18.60	20.80	52.40	9.80	2.80	.30	-.70	.60	-.90	-1.16	-.95	8.02
20	16.10	20.60	49.30	9.60	2.70	.20	-.70	.60	-.92	-1.12	-.96	8.09
21	14.00	19.60	46.80	9.50	3.70	.10	-.60	.70	-.94	-1.10	-.92	9.63
22	12.20	18.50	41.60	8.70	3.80	.40	-.50	.70	-.94	-.84	-.89	10.80
23	11.00	22.70	36.80	8.60	8.30	-.10	-.20	2.60	-1.03	-.73	-.33	10.63
24	11.00	23.10	33.00	11.30	3.10	.10	+.60	2.60	-1.05	-.71	+.17	22.01
25	20.30	20.90	29.50	13.60	2.70	-.60	4.90	2.60	-1.04	.88	.88	29.04
26	29.30	18.50	26.50	13.20	2.30	+.20	7.60	2.10	-1.05	.94	2.60	25.91
27	26.20	23.50	24.20	12.60	2.00	.20	7.40	2.10	-1.04	1.02	4.50	20.98
28	22.50	39.00	22.30	11.40	1.75	.20	7.90	1.80	-1.18	-.60	10.48	17.09
29	18.60	......	21.10	10.00	1.50	.20	9.50	1.60	-1.28	-.73	9.49	15.80
30	15.70	......	19.30	8.90	1.60	.10	9.30	1.50	-1.38	.96	6.67	14.62
31	14.80	......	18.10	......	2.50	......	8.50	1.10	......	-1.03	......	12.51

Daily gage height, in feet, of Black Warrior River at Tuscaloosa—Continued.

Day.	Jan.	Feb.	Mar.	Apr.	May.	June.	July.	Aug.	Sept.	Oct.	Nov.	Dec.
1900.												
1	10.90	6.92	18.14	21.55	21.23	4.24	41.00	9.85	5.50	0.75	4.20	12.50
2	9.48	6.40	18.60	18.95	18.95	5.30	33.88	8.00	5.00	.80	5.10	9.65
3	8.32	5.00	26.00	15.80	16.83	13.95	32.27	6.50	4.07	.60	5.40	8.50
4	7.10	5.71	23.30	13.60	13.82	15.16	32.13	5.15	4.30	.25	5.55	7.90
5	6.18	6.40	20.29	12.27	11.60	20.98	30.96	4.40	3.58	.40	5.05	8.20
6	5.53	9.58	17.43	12.17	9.60	21.50	27.31	3.80	2.80	.40	4.45	10.20
7	5.12	11.50	15.34	11.44	8.20	20.92	23.70	3.35	2.30	.80	3.90	10.65
8	4.84	10.53	27.58	10.29	7.28	21.75	20.80	3.00	1.90	1.80	8.50	9.65
9	4.60	12.23	39.00	9.97	6.80	38.52	18.45	2.50	1.50	3.90	3.10	8.75
10	4.50	20.60	36.03	10.58	6.50	31.70	15.96	2.30	1.40	6.05	2.90	8.15
11	7.12	23.00	31.63	26.35	6.12	24.73	12.96	2.10	1.10	6.10	2.70	7.50
12	31.63	20.64	27.34	52.79	5.90	19.60	11.05	2.00	.90	14.30	2.50	6.90
13	31.80	41.37	23.78	53.40	5.48	15.95	10.60	1.75	.80	22.50	2.40	6.45
14	28.18	47.96	20.43	48.69	4.98	15.60	12.90	1.50	1.40	21.85	2.25	6.40
15	24.09	45.73	17.40	42.30	4.50	19.23	10.35	1.40	5.45	16.40	2.20	6.65
16	20.12	40.23	16.63	37.10	3.85	29.15	8.50	1.50	10.65	11.60	2.10	6.45
17	16.50	34.74	21.18	63.00	3.65	28.89	7.00	1.48	8.95	8.20	2.05	6.10
18	14.28	29.75	20.28	64.05	3.50	25.33	6.00	1.40	6.34	6.13	2.00	5.75
19	15.80	25.88	18.80	62.17	3.52	24.51	5.60	3.00	4.44	5.10	1.95	5.30
20	25.00	22.43	45.10	59.35	3.39	30.10	6.35	4.40	3.35	4.80	3.90	5.45
21	32.60	19.90	51.00	56.10	3.65	27.80	8.50	3.75	2.60	3.60	8.40	6.30
22	29.44	23.58	47.98	51.71	3.50	25.38	8.00	3.00	2.20	3.15	10.25	8.35
23	24.54	26.50	42.40	46.20	3.35	24.55	8.10	2.60	1.87	4.60	12.50	11.90
24	20.43	24.20	38.24	41.88	4.40	50.00	8.70	1.65	1.65	7.30	11.00	17.00
25	17.12	22.88	35.41	37.94	7.55	58.35	7.40	2.10	1.60	10.40	9.45	18.70
26	14.12	21.00	36.78	33.94	7.68	56.35	6.80	4.48	1.45	8.55	15.98	17.00
27	12.14	18.63	35.33	30.89	6.70	52.90	9.20	4.30	1.30	6.75	22.20	14.60
28	10.47	16.52	31.80	28.15	5.30	49.06	16.20	3.25	1.05	5.40	21.00	12.50
29	9.10		28.85	25.73	4.25	48.35	13.50	2.40	.90	4.70	17.35	10.95
30	8.27		26.25	23.50	3.60	46.35	11.83	2.10	.75	4.10	14.00	10.15
31	7.60		24.20		3.62		11.05	2.50		3.70		14.20
1901.												
1	19.50	15.10	7.50	28.00	9.40	6.50	1.70	0.70	6.11	5.40	0.85	1.60
2	19.00	14.80	7.30	26.50	8.50	12.25	1.73	.60	5.10	6.91	.80	1.50
3	17.90	15.00	8.40	32.60	7.90	16.65	2.00	.85	4.05	9.00	.75	1.40
4	14.70	35.20	9.65	35.10	7.30	14.80	2.30	1.00	3.60	6.72	1.00	1.40
5	13.00	42.00	9.70	31.50	7.00	12.00	2.50	.90	3.05	5.30	1.10	1.31
6	11.50	38.35	9.50	27.50	6.35	11.20	2.30	.75	2.81	4.50	1.10	1.95
7	10.25	32.15	9.00	24.30	6.15	11.10	2.50	.60	2.35	3.45	1.05	1.90
8	9.50	29.15	8.50	20.90	6.00	11.60	2.40	.50	2.04	3.02	1.00	2.58
9	8.75	28.05	8.05	17.90	5.70	9.40	2.25	.35	1.90	2.50	.95	2.60
10	8.30	30.97	8.70	14.50	5.60	7.60	2.10	.25	1.70	2.31	.90	3.80
11	17.40	29.75	29.50	12.50	5.00	6.50	2.00	.20	1.50	2.10	.90	5.00
12	52.70	27.46	34.00	11.00	4.60	6.20	1.70	.65	1.41	1.95	.85	5.80
13	56.50	25.15	28.50	10.00	5.00	6.40	1.30	.65	1.30	3.41	1.00	5.70
14	53.25	22.50	23.50	11.35	6.60	5.50	.90	.80	2.52	3.30	1.05	7.40
15	47.25	19.60	19.70	12.80	8.70	5.05	.70	1.20	6.70	3.97	1.00	31.00
16	41.45	17.45	16.00	13.00	8.30	4.80	.60	5.70	8.00	3.80	.98	40.75
17	36.30	15.70	13.00	11.70	6.80	4.50	.43	17.00	6.10	3.51	.90	35.00
18	31.85	13.70	11.05	12.80	5.50	4.10	2.60	26.80	12.21	8.00	.93	27.00
19	28.16	12.60	10.00	25.70	4.79	8.80	4.00	22.70	16.00	2.70	1.10	21.50
20	25.15	11.90	9.80	39.80	4.90	3.35	5.70	25.70	14.50	2.30	1.30	16.41
21	22.50	10.90	13.00	42.60	8.30	3.00	7.33	32.10	10.60	2.00	1.40	12.10
22	19.85	9.95	16.50	38.00	17.30	2.75	6.00	32.50	7.50	1.81	1.70	9.95
23	17.55	9.20	15.70	32.80	19.80	2.30	4.50	31.97	5.70	1.61	2.00	7.00
24	15.35	8.75	14.00	28.41	17.50	2.20	3.20	26.80	4.51	1.45	2.05	6.80
25	14.60	8.65	13.80	24.40	14.10	2.00	2.50	22.40	3.90	1.31	2.00	7.30
26	18.00	8.50	28.50	21.00	11.30	1.97	2.00	18.30	3.40	1.20	1.95	8.34
27	17.20	8.20	37.25	17.90	9.30	1.85	1.50	14.10	2.91	1.05	1.95	9.00
28	15.50	7.90	34.50	14.90	8.15	1.70	1.05	8.15	2.60	.81	2.00	10.96
29	14.80		29.00	12.50	7.00	1.65	.90	7.90	2.91	.55	1.95	36.30
30	15.10		24.30	10.90	6.90	1.65	.80	7.41	4.80	.90	1.91	49.00
31	14.60		24.85		6.35		.60	7.10		.90		49.00

Daily gage height, in feet, of Black Warrior River at Tuscaloosa—Continued.

Day.	Jan.	Feb.	Mar.	Apr.	May.	June.	July.	Aug.	Sept.	Oct.	Nov.	Dec.
1902.												
1	44.00	37.80	49.87	52.88	10.85	1.52	0.35	0.04	2.52	5.61	4.90	8.50
2	38.80	47.50	48.10	45.50	9.90	1.48	.30	.05	2.20	5.50	4.89	9.50
3	34.10	48.40	44.00	39.70	8.32	1.40	.15	.08	2.85	5.00	4.89	13.55
4	28.50	45.00	37.50	35.00	6.51	1.52	.00	.15	4.87	5.50	4.90	17.10
5	24.80	40.00	33.50	32.00	5.50	1.40	.00	1.80	4.65	5.52	4.91	17.92
6	21.15	36.00	35.78	29.00	5.00	1.80	.03	2.30	3.50	5.31	4.94	15.65
7	15.50	31.00	35.00	26.91	4.75	1.25	.05	1.90	2.52	5.15	5.00	14.90
8	14.00	28.15	31.61	32.70	5.11	1.15	.05	1.35	2.00	5.00	4.94	13.40
9	12.50	24.88	28.50	35.75	5.15	1.10	.15	.81	1.60	4.95	4.94	11.75
10	11.00	21.50	25.51	32.81	4.75	1.05	.20	.69	1.32	4.90	4.89	10.25
11	9.30	18.00	23.10	29.05	4.35	1.00	.25	.49	1.12	9.65	4.89	9.10
12	8.50	15.40	20.40	26.00	4.05	.91	.25	.48	.85	15.40	4.90	8.50
13	8.00	13.00	17.70	23.82	3.75	.80	.15	.39	.71	14.90	4.90	8.05
14	7.30	11.00	15.52	20.61	3.52	.72	.15	.25	.50	11.75	4.91	7.85
15	6.70	11.50	15.00	18.00	3.15	.70	.10	.18	.40	8.98	4.90	7.45
16	6.11	19.85	17.50	15.82	3.50	.65	.10	.15	.25	7.45	4.89	15.00
17	5.65	21.81	31.80	14.41	6.01	.53	.05	.14	.15	6.40	4.90	29.65
18	5.00	19.70	29.60	12.93	6.52	.50	.05	.10	.12	6.00	5.25	28.85
19	6.05	17.55	25.00	11.85	6.10	.80	.04	.08	.75	5.83	5.50	23.90
20	7.81	16.00	21.30	10.57	5.31	1.00	.03	.05	2.21	5.75	5.54	19.15
21	11.00	14.51	18.50	9.61	4.55	1.60	.03	.04	3.51	5.60	5.50	16.30
22	16.15	14.00	17.65	8.80	3.90	1.70	.04	.10	3.92	5.60	5.50	18.00
23	20.81	13.60	15.21	8.05	3.50	1.72	.04	.12	4.80	5.52	5.45	20.30
24	20.81	13.05	13.80	7.72	3.05	1.80	.10	.15	4.50	5.45	5.42	18.10
25	18.50	12.50	18.30	7.30	2.82	1.85	.11	.10	4.75	5.35	5.69	15.41
26	16.85	13.00	19.00	7.05	2.75	1.50	.10	.08	4.83	5.29	6.75	13.50
27	15.50	14.00	35.50	6.72	2.35	1.10	.11	.15	4.90	5.00	13.90	12.00
28	20.00	41.61	60.35	6.41	2.03	.65	.07	.61	5.20	4.90	13.40	10.95
29	28.00	......	60.60	5.90	1.95	.40	.05	1.45	5.08	4.91	10.42	10.00
30	27.87	......	58.30	5.75	1.81	.40	.05	1.68	5.65	4.90	9.65	11.10
31	28.50	......	57.40	......	1.65	......	.04	1.98	......	4.92	......	12.51
1903.												
1	13.20	13.35	54.30	16.70	8.50	9.60	6.43	6.95	5.10	3.98	5.03	4.71
2	13.75	12.55	52.41	16.65	8.15	12.50	5.90	7.40	5.15	3.95	5.17	4.72
3	16.53	13.75	47.25	15.30	7.90	14.21	5.65	8.55	5.00	3.80	5.68	4.78
4	23.50	18.55	41.42	14.12	7.75	12.35	5.80	7.65	4.91	3.75	5.56	4.84
5	22.52	35.00	36.90	13.10	7.45	12.40	5.80	6.95	4.80	3.65	5.21	4.85
6	19.81	37.85	33.11	12.30	7.15	13.35	6.35	6.80	4.73	3.98	5.04	4.87
7	17.10	36.00	37.82	11.50	6.92	16.88	6.15	6.80	4.55	4.20	4.98	4.88
8	15.11	56.35	36.50	11.15	6.89	16.50	5.90	6.62	4.70	4.90	4.90	4.86
9	13.31	55.85	39.15	12.30	7.00	12.20	5.75	6.55	4.40	4.75	4.88	4.88
10	12.10	51.50	39.20	22.35	6.90	10.50	6.50	6.10	4.42	4.90	4.84	4.89
11	12.53	52.00	37.75	20.10	6.75	10.85	6.25	6.35	4.31	4.97	4.80	4.90
12	23.96	53.75	35.12	16.50	7.05	11.00	6.15	6.15	4.25	5.10	4.87	4.92
13	28.41	51.10	33.80	14.35	8.65	10.35	6.52	6.10	4.21	5.15	4.91	4.87
14	25.75	46.15	36.15	21.20	14.35	9.45	6.70	6.00	4.19	4.95	4.87	4.85
15	21.90	41.11	36.00	32.12	36.85	8.30	6.55	5.96	4.13	4.96	4.92	4.87
16	19.01	40.50	33.84	28.20	43.40	7.65	6.25	6.25	4.10	4.93	4.97	4.86
17	16.11	56.65	31.12	23.40	43.10	7.08	5.90	6.60	4.00	4.98	4.98	4.86
18	14.50	a56.75	28.65	18.35	37.10	6.70	5.81	6.85	4.08	5.01	4.91	4.84
19	12.51	52.95	26.15	16.00	30.65	6.45	5.70	7.82	4.08	4.99	4.90	4.80
20	11.85	46.75	23.50	14.50	25.80	6.20	5.45	7.60	4.03	4.98	4.88	4.92
21	11.05	41.35	20.85	14.90	21.55	6.23	5.10	6.91	3.98	4.90	4.86	5.05
22	10.84	36.50	18.80	17.95	17.60	7.20	5.05	6.55	3.94	4.85	4.80	5.23
23	9.98	32.30	17.55	15.90	18.85	6.95	4.75	6.10	3.90	4.76	4.78	5.15
24	9.51	28.85	16.15	13.81	11.10	6.75	4.70	5.65	3.89	4.65	4.75	5.48
25	8.65	26.10	14.80	12.30	9.80	6.88	4.83	5.45	3.86	4.60	4.73	5.60
26	8.25	23.95	13.48	11.30	9.00	7.80	4.80	5.15	3.80	4.50	4.75	5.51
27	8.05	22.00	12.40	10.50	8.35	7.52	4.80	5.15	3.75	4.50	4.74	5.39
28	9.85	45.42	11.85	10.00	7.80	7.15	4.80	5.15	3.70	4.60	4.75	5.15
29	12.35	......	11.62	9.45	7.45	6.91	7.85	4.95	3.81	4.68	4.73	5.12
30	14.72	......	11.78	8.95	7.15	6.55	4.75	5.35	3.95	4.76	4.70	5.09
31	14.55	......	13.50	......	7.32	......	6.50	5.15	......	4.81	......	5.07

a Maximum 57.10.

Rating table for Black Warrior River at Tuscaloosa from 1895 to 1901.

Gage height.	Discharge.	Gage height.	Discharge.	Gage height.	Discharge.	Gage height.	Discharge.
Feet.	*Second-feet.*	*Feet.*	*Second-feet.*	*Feet.*	*Second-feet.*	*Feet.*	*Second-feet.*
—2.0	88	3.0	1,470	7.0	3,665	11.0	5,885
—1.8	92	3.1	1,520	7.1	3,715	11.1	5,935
—1.6	100	3.2	1,570	7.2	3,765	11.2	5,985
—1.4	110	3.3	1,620	7.3	3,815	11.3	6,035
—1.2	120	3.4	1,670	7.4	3,865	11.4	6,085
—1.0	130	3.5	1,725	7.5	3,925	11.5	6,145
— .8	150	3.6	1,780	7.6	3,985	11.6	6,205
— .6	175	3.7	1,835	7.7	4,045	11.7	6,265
— .4	205	3.8	1,890	7.8	4,105	11.8	6,325
— .2	240	3.9	1,945	7.9	4,165	11.9	6,385
.0	280	4.0	2,000	8.0	4,220	12.0	6,440
.1	310	4.1	2,055	8.1	4,270	12.1	6,490
.2	340	4.2	2,111	8.2	4,320	12.2	6,540
.3	370	4.3	2,166	8.3	4,370	12.3	6,590
.4	400	4.4	2,222	8.4	4,420	12.4	6,640
.5	430	4.5	2,277	8.5	4,480	12.5	6,700
.6	460	4.6	2,333	8.6	4,540	12.6	6,760
.7	490	4.7	2,388	8.7	4,600	12.7	6,820
.8	530	4.8	2,444	8.8	4,660	12.8	6,880
.9	565	4.9	2,500	8.9	4,720	12.9	6,940
1.0	600	5.0	2,555	9.0	4,775	13.0	6,995
1.1	635	5.1	2,610	9.1	4,825	13.1	7,045
1.2	670	5.2	2,666	9.2	4,875	13.2	7,095
1.3	710	5.3	2,721	9.3	4,925	13.3	7,145
1.4	750	5.4	2,777	9.4	4,975	13.4	7,195
1.5	790	5.5	2,832	9.5	5,035	13.5	7,255
1.6	830	5.6	2,888	9.6	5,095	13.6	7,315
1.7	870	5.7	2,943	9.7	5,155	13.7	7,375
1.8	910	5.8	3,000	9.8	5,215	13.8	7,435
1.9	955	5.9	3,054	9.9	5,275	13.9	7,495
2.0	1,000	6.0	3,110	10.0	5,330	14.0	7,550
2.1	1,045	6.1	3,160	10.1	5,380	14.1	7,600
2.2	1,090	6.2	3,210	10.2	5,430	14.2	7,650
2.3	1,135	6.3	3,260	10.3	5,480	14.3	7,700
2.4	1,180	6.4	3,310	10.4	5,530	14.4	7,750
2.5	1,225	6.5	3,370	10.5	5,590	14.5	7,810
2.6	1,270	6.6	3,430	10.6	5,650	14.6	7,870
2.7	1,320	6.5	3,490	10.7	5,710	14.7	7,930
2.8	1,370	6.8	3,550	10.8	5,770	14.8	7,990
2.9	1,420	6.9	3,610	10.9	5,830	14.9	8,050

Rating table for Black Warrior River near Cordova for 1903.

Gage height.	Discharge.	Gage height.	Discharge.	Gage height.	Discharge.	Gage height.	Discharge.
Feet.	*Second-feet.*	*Feet.*	*Second-feet.*	*Feet.*	*Second-feet.*	*Feet.*	*Second-feet.*
−0.90	78	1.60	1,740	5.00	6,200	12.20	17,000
− .85	89	1.70	1,840	5.10	6,350	13.00	18,200
− .80	100	1.80	1,940	5.20	6,500	13.50	18,950
− .75	112	1.90	2,040	5.30	6,650	13.70	19,250
− .70	126	2.00	2,140	5.40	6,800	14.00	19,700
− .65	142	2.10	2,250	5.50	6,950	14.30	20,150
− .60	159	2.20	2,360	5.60	7,100	15.00	21,200
− .55	178	2.30	2,475	5.70	7,250	15.70	22,250
− .50	198	2.40	2,590	5.80	7,400	16.00	22,709
− .45	221	2.50	2,705	5.90	7,550	16.10	22,850
− .40	247	2.60	2,820	6.00	7,700	17.00	24,200
− .35	276	2.70	2,935	6.20	8,000	17.80	25,400
− .30	307	2.80	3,050	6.60	8,600	18.00	25,700
− .25	340	2.90	3,175	6.70	8,750	19.00	27,200
− .20	374	3.00	3,300	6.80	8,900	19.40	27,800
− .15	408	3.10	3,430	7.00	9,200	20.00	28,700
− .10	442	3.20	3,560	7.20	9,500	20.10	28,850
− .05	476	3.30	3,695	7.80	10,400	21.00	30,200
.00	510	3.40	3,830	7.90	10,550	21.20	30,500
.10	580	3.50	3,970	8.00	10,700	21.60	31,100
.20	650	3.60	4,110	8.20	11,000	21.90	31,550
.30	725	3.70	4,255	8.60	11,600	25.00	36,200
.40	800	3.80	4,400	8.70	11,750	25.30	36,650
.50	875	3.90	4,550	9.00	12,200	25.40	36,800
.60	950	4.00	4,700	10.00	13,700	26.00	37,700
.70	1,025	4.10	4,850	10.10	13,850	26.10	37,850
.80	1,100	4.20	5,000	10.20	14,000	27.00	39,200
.90	1,175	4.30	5,150	10.50	14,450	27.30	39,650
1.00	1,250	4.40	5,300	11.00	15,200	27.60	40,100
1.10	1,325	4.50	5,450	11.10	15,350	29.00	42,200
1.20	1,400	4.60	5,600	11.20	15,500	29.50	42,950
1.30	1,480	4.70	5,750	11.40	15,800		
1.40	1,560	4.80	5,900	11.90	16,550		
1.50	1,650	4.90	6,050	12.00	16,700		

Estimated monthly discharge of Black Warrior River near Cordova.

[Drainage area, 237 square miles.]

Month.	Discharge in second-feet.			Run-off.	
	Maximum.	Minimum.	Mean.	Second-feet per square mile.	Depth in inches.
1900.					
June	49,284	1,134	16,185	8.52	9.51
July	11,248	810	2,975	1.57	1.81
August	2,144	660	1,016	.53	.61
September	1,951	205	556	.29	.32
October	13,098	60	1,732	.91	1.06
November	6,808	384	1,487	.78	.87
December	6,660	968	2,154	1.13	1.30
1901.					
January	51,800	1,762	8,713	4.59	5.29
February	31,820	1,486	6,616	3.48	3.62
March	29,600	1,486	6,637	3.49	4.02
April	23,088	2,242	6,967	3.67	4.10
May	10,064	968	2,539	1.34	1.54
June	5,920	320	1,582	.83	.93
July	3,404	205	631	.33	.38
August	17,168	205	4,155	2.19	2.53
September	6,660	320	1,415	.74	.83
October	1,669	320	687	.36	.42
November	660	320	468	.25	.28
December	31,820	518	4,923	2.59	2.99
The year	51,800	205	3,778	1.99	26.93

Estimated monthly discharge of Black Warrior River near Cordova—Continued.

Month.	Discharge in second-feet.			Run-off.	
	Maximum.	Minimum.	Mean.	Second-feet per square mile.	Depth in inches.
1902.					
January	17,020	1,396	5,692	3.00	3.46
February	32,856	2,652	8,599	4.53	4.72
March	56,980	3,258	13,764	7.24	8.35
April	26,048	1,050	5,366	2.82	3.15
May	1,307	205	630	.33	.38
June	518	125	193	.10	.11
July	125	74	99	.05	.06
August	660	74	170	.09	.10
September	260	74	105	.06	.07
October	3,552	160	536	.28	.32
November	11,100	160	880	.46	.51
December	15,984	1,050	4,395	2.31	2.66
The year	56,980	74	3,369	1.77	23.89
1903.					
January	13,700	1,480	4,306	2.27	2.62
February	42,950	2,475	17,941	9.44	9.83
March	40,100	1,840	10,203	5.37	6.19
April	17,000	1,250	4,773	2.51	2.80
May	28,850	800	4,176	2.20	2.54
June	5,750	510	1,661	.87	.97
July	510	159	317	.17	.20
August	875	126	306	.16	.18
September	126	78	97	.05	.06
October	307	78	148	.08	.09
November	247	159	198	.10	.11
December	875	159	252	.13	.15
The year	42,950	78	3,698	1.95	25.74

Net horsepower per foot of fall with a turbine efficiency of 80 per cent for the minimum monthly discharge of Black Warrior River near Cordova.

Month.	1901.			1902.			1908.		
	Minimum discharge.	Minimum net horsepower per foot of fall.	Duration of minimum.	Minimum discharge.	Minimum net horsepower per foot of fall.	Duration of minimum.	Minimum discharge.	Minimum net horsepower per foot of fall.	Duration of minimum.
	Sec.-feet.		*Days.*	*Sec.-feet.*		*Days.*	*Sec.-feet.*		*Days.*
January	1,762	160	1	1,396	127	1	1,480	134	1
February	1,486	135	1	2,652	241	1	3,300	300	1
March	1,486	135	1	3,258	296	1	1,840	167	1
April	2,242	204	2	1,050	95	1	1,250	114	1
May	968	88	1	205	19	1	800	73	1
June	320	29	1	125	11	6	510	46	3
July	205	19	3	74	7	8	159	14	4
August..........	205	19	3	74	7	13	126	11	3
September	320	29	1	74	7	18	78	7	13
October.........	320	29	3	160	14	7	78	7	7
November	320	29	2	160	14	6	159	14	9
December	518	47	2	1,050	95	1	159	14	12

LOCUST FORK OF BLACK WARRIOR RIVER AT PALOS.

Locust Fork of Black Warrior River rises in Blount County, and, flowing in a southwesterly course, enters Black Warrior River a short distance above Wilmington. Its drainage basin is hilly, and about half its area is under cultivation. Palos station was established November 26, 1901, by R. C. McCalla, United States assistant engineer, who furnishes the daily gage heights to the Geological Survey. It is maintained by the United States Engineer Corps. The gage is a 4 by 8 inch timber on the right bank of Locust Fork of Black Warrior River just below the Kansas City, Memphis and Birmingham Railroad bridge. One section follows the slope of the bank from low water to a tree on top of the bank, and from there up a vertical section is fastened to the tree. The slope is 17 feet in elevation, measured vertically, and the vertical section of the rod is 15 feet. The rod is graduated to feet and tenths, with copper figures at the 5-foot points and round-head tacks at intermediate foot marks. The total height is 32 feet. The plane of reference (about 251.71 feet above Mobile datum) is supposed to be extreme low water. High water April, 1900, was about 37 feet above the plane of reference.

Measurements are made from the Drennan bridge, which is about a quarter of a mile below the Kansas City, Memphis and Birmingham

Railroad bridge. The Drennan bridge is the property of the Drennan Coal Mining Company. It is a mining railroad bridge, having width for a double-track tramway of 3 feet gage. One track is laid and in operation. The bridge has two iron spans of 100 feet each and trestle approaches at both ends.

The initial point for measurements is the left-bank end of the iron bridge on the downstream side.

The bench mark is the top of the iron crossbeam at station 80 from the initial point, and is 44.3 feet above the water when the gage reads 2.50 feet.

This bridge is lower and the section is better than that at the upper bridge, from which some of the measurements have been made; but there is a raft against the center pier that should be removed. At minimum stages the water at both bridges is too sluggish for accurate measurement.

Discharge measurements of Locust Fork of Black Warrior River at Palos.

Date.	Gage height.	Discharge.	Date.	Gage height.	Discharge.
1902.	*Feet.*	*Second-feet.*	1903.	*Feet.*	*Second-feet.*
January 18..........	0.85	849	May 20	1.75	2,148
April 5	2.50	3,224	June 16.............	.62	404
Do	2.50	3,292	Do	.62	431
September 20	.00	a 117	July 18	.40	228
October 15	.60	473	August 28..........	.13	84
1903.			Do	.13	93
March 7	4.75	7,450	September 26	.02	47
March 14	3.55	5,342			

a Estimated from float measurement.

Daily gage height, in feet, of Locust Fork of Black Warrior River at Palos.

Day.	Jan.	Feb.	Mar.	Apr.	May.	June.	July.	Aug.	Sept.	Oct.	Nov.	Dec.
1902.												
1	5.50	8.30	13.30	4.90	0.90	0.20	0.10	0.40	0.10	0.10	0.00	1.10
2	3.40	12.80	6.70	3.50	.80	.20	.10	.80	.00	.10	.00	1.10
3	2.70	10.20	4.40	2.70	.70	.20	.10	.80	.00	.10	.00	1.90
4	2.20	5.70	3.00	2.60	.70	.20	.10	.80	.00	.10	.00	1.90
5	2.00	3.90	5.00	2.50	.60	.20	.10	.80	.00	.10	.00	1.70
6	1.70	3.00	5.30	2.30	.60	.20	.00	.20	.00	.10	.00	1.60
7	1.60	2.40	3.70	2.10	.50	.20	.00	.10	.00	.00	.00	1.50
8	1.50	2.00	3.00	5.80	.50	.20	.00	.10	.00	.00	.00	1.20
9	1.30	1.80	2.50	5.00	.50	.20	.10	.10	.00	.00	.00	1.00
10	1.30	1.70	2.10	3.40	.50	.20	.10	.00	.00	.00	.00	.90
11	1.20	1.60	1.90	2.60	.40	.10	.10	.00	.00	1.70	.00	.80
12	1.10	1.50	1.70	2.20	.40	.10	.10	.00	.00	2.00	.00	.70
13	1.10	1.40	1.60	2.00	.40	.10	.10	.00	.00	1.30	.00	.70
14	1.00	1.30	1.60	1.80	.40	.10	.10	.00	.00	.80	.00	.60
15	.90	2.20	1.50	1.70	.40	.10	.10	.00	.00	.60	.00	3.10
16	.90	2.80	3.40	1.60	.30	.10	.00	.00	.00	.40	.00	3.30
17	.90	2.20	4.50	1.50	.60	.10	.00	.00	.00	.40	.20	2.40
18	.90	2.00	8.00	1.50	.50	.10	.00	.00	.00	.30	.20	1.90
19	1.00	1.80	2.30	1.40	.50	.10	.00	.00	.00	.20	.20	1.50
20	1.10	1.70	2.00	1.30	.40	.10	.00	.00	.00	.20	.20	1.60
21	1.60	1.60	1.80	1.10	.40	.10	.10	.00	.00	.10	.20	1.60
22	2.50	1.60	1.70	1.00	.40	.10	.10	.00	.00	.10	.20	2.50
23	2.40	1.50	1.60	1.00	.40	.10	.00	.00	.00	.10	.20	2.20
24	2.30	1.50	1.50	.90	.30	.10	.00	.00	.00	.00	.10	1.80
25	2.40	1.80	1.90	.90	.30	.10	.00	.00	.00	.00	.20	1.50
26	1.90	1.60	1.70	.90	.30	.10	.00	.00	.00	.00	1.10	1.30
27	1.80	2.20	6.30	.80	.30	.10	.00	.00	.00	.00	1.40	1.10
28	2.10	14.90	24.00	.80	.20	.10	.10	.00	.10	.00	1.10	1.00
29	3.10	......	27.00	.70	.20	.20	.10	.10	.00	.00	.80	1.00
30	3.30	......	25.00	1.20	.20	.20	.20	.10	.10	.00	.60	1.20
31	3.60	......	17.90	......	.20	......	.50	.10	......	.00	......	1.40
1903.												
1	1.20	1.10	17.00	2.60	.90	1.20	.4	.70	.40	.00	.20	.10
2	1.30	1.10	7.40	2.20	.90	2.10	.4	1.00	.80	.00	.20	.10
3	1.50	1.60	4.20	1.90	.80	1.40	.4	.70	.20	.00	.20	.10
4	1.50	8.30	3.20	1.70	.80	1.20	.8	.90	.10	.00	.10	.10
5	1.60	7.60	2.80	1.60	.80	1.10	.8	.80	.10	.00	.10	.10
6	1.50	4.80	4.60	1.50	.70	1.60	.7	.60	.10	.00	.10	.10
7	1.40	4.30	5.20	1.40	.70	1.40	.6	.50	.10	.00	.10	.10
8	1.30	18.10	8.80	1.50	.70	1.20	.6	.40	.10	.00	.00	.10
9	1.20	12.20	4.30	4.00	.60	1.10	.5	.40	.00	.40	.00	.10
10	1.00	16.40	3.50	3.70	.60	1.10	.5	.40	.00	.30	.00	.10
11	1.50	12.60	3.80	2.20	.60	1.50	.5	.40	.00	.20	.20	.10
12	3.20	12.50	2.80	1.80	.60	1.10	.4	.30	.00	.20	.30	.10
13	2.70	6.60	2.50	1.80	.70	1.00	.5	.30	.00	.10	.30	.10
14	2.00	4.40	3.60	10.20	1.00	.80	.8	.30	.00	.10	.30	.10
15	1.70	4.20	3.70	7.20	5.80	.70	.9	.60	.00	.00	.20	.10
16	1.60	7.50	3.10	8.80	11.90	.60	.7	.80	.00	.00	.10	.10
17	1.40	23.00	2.60	2.60	5.70	.60	.5	.70	.00	.00	.10	.10
18	1.40	18.10	2.30	2.10	3.00	.60	.4	.70	.00	.00	.10	.10
19	1.20	7.40	2.00	1.90	2.10	.60	.3	.70	.00	.00	.10	.20
20	1.10	4.20	1.80	1.90	1.80	.50	.3	.60	.00	.00	.10	.20
21	1.10	3.10	1.70	2.50	1.50	.50	.2	.60	.00	.00	.10	.10
22	1.00	2.60	1.70	2.00	1.30	.40	.2	.60	.00	.00	.10	.10
23	1.00	2.20	1.80	1.70	1.20	.50	.2	.40	.00	.00	.10	.10
24	.90	2.00	1.80	1.50	1.10	.40	.2	.30	.00	.00	.10	.10
25	.90	1.80	1.60	1.40	1.00	.40	.2	.30	.00	.00	.10	.10
26	.90	1.70	1.50	1.30	.90	.40	.2	.20	.00	.00	.10	.20
27	.80	1.60	1.40	1.20	.80	.50	.2	.20	.00	.00	.10	.20
28	.80	20.00	1.30	1.10	.70	.50	.1	.10	.00	.00	.10	.20
29	1.25	......	1.30	1.00	.70	.50	.1	.10	.00	.00	.10	.20
30	1.30	......	1.40	1.00	.80	.50	.1	.10	.00	.00	.10	.20
31	1.20	......	2.70	......	.90	......	.3	.10	......	.30	......	.20

Rating table for Locust Fork of Black Warrior River at Palos for 1902.

Gage height.	Discharge.	Gage height.	Discharge	Gage height.	Discharge.	Gage height.	Discharge.
Feet.	Second-feet.	Feet.	Second-feet.	Feet.	Second-feet.	Feet.	Second-feet.
0.0	100	0.9	900	1.8	2,250	2.7	3,600
.1	140	1.0	1,050	1.9	2,400	2.8	3,750
.2	190	1.1	1,200	2.0	2,550	2.9	3,900
.3	245	1.2	1,350	2.1	2,700	3.0	4,050
.4	310	1.3	1,500	2.2	2,850	3.1	4,200
.5	390	1.4	1,650	2.3	3,000	3.2	4,350
.6	490	1.5	1,800	2.4	3,150	3.3	4,500
.7	610	1.6	1,950	2.5	3,300	3.4	4,650
.8	750	1.7	2,100	2.6	3,450	3.5	4,800

Rating table for Locust Fork of Black Warrior River at Palos for 1903.

Gage height.	Discharge.	Gage height.	Discharge.	Gage height.	Discharge.	Gage height.	Discharge.
Feet.	Second-feet.	Feet.	Second-feet.	Feet.	Second-feet.	Feet.	Second-feet.
0.00	46	1.50	1,590	3.70	5,550	5.90	9,510
.05	61	1.60	1,770	3.80	5,730	6.00	9,690
.10	77	1.70	1,950	3.90	5,910	6.40	10,410
.15	95	1.80	2,130	4.00	6,090	6.60	10,770
.20	115	1.90	2,310	4.10	6,270	7.00	11,490
.25	139	2.00	2,490	4.20	6,450	7.20	11,850
.30	167	2.10	2,670	4.30	6,630	7.40	12,210
.35	197	2.20	2,850	4.40	6,810	7.50	12,390
.40	230	2.30	3,030	4.50	6,990	7.60	12,570
.45	266	2.40	3,210	4.60	7,170	10.00	16,890
.50	306	2.50	3,390	4.70	7,350	10.20	17,250
.55	350	2.60	3,570	4.80	7,530	11.90	20,310
.60	398	2.70	3,750	4.90	7,710	12.00	20,490
.65	448	2.80	3,930	5.00	7,890	12.20	20,850
.70	500	2.90	4,110	5.10	8,070	12.50	21,390
.80	605	3.00	4,290	5.20	8,250	12.60	21,570
.90	720	3.10	4,470	5.30	8,430	17.00	29,490
1.00	845	3.20	4,650	5.40	8,610	18.00	31,290
1.10	980	3.30	4,830	5.50	8,790	18.10	31,470
1.20	1,120	3.40	5,010	5.60	8,970	20.00	34,890
1.30	1,265	3.50	5,190	5.70	9,150	23.00	40,290
1.40	1,420	3.60	5,370	5.80	9,330		

Estimated monthly discharge of Locust Fork of Black Warrior River at Palos.

[Drainage area, 1,020 square miles.]

Month.	Discharge in second-feet.			Run-off.	
	Maximum.	Minimum.	Mean.	Second-feet per square mile.	Depth in inches.
1902.					
January	7,800	900	2,468	2.42	2.79
February	21,900	1,500	4,746	4.65	4.84
March	40,050	1,800	8,400	8.24	9.50
April	8,250	610	2,635	2.58	2.88
May	900	190	374	.37	.43
June	190	140	160	.16	.18
July	390	100	133	.13	.15
August	900	100	157	.15	.17
September	140	100	104	.10	.11
October	2,550	100	359	.35	.40
November	1,650	100	282	.28	.31
December	4,500	490	1,784	1.75	2.02
The year	40,050	100	1,800	1.76	23.78
1903.					
January	4,650	605	1,428	1.40	1.61
February	40,290	980	11,498	11.27	11.74
March	29,490	1,265	4,833	4.74	5.46
April	17,250	845	3,246	3.18	3.55
May	20,310	398	2,097	2.06	2.37
June	2,670	230	758	.74	.83
July	720	77	276	.27	.31
August	845	77	326	.32	.37
September	230	46	64	.06	.07
October	230	46	66	.06	.07
November	167	46	89	.09	.10
December	115	77	86	.08	.09
The year	40,290	46	2,064	2.02	26.57

Net horsepower per foot of fall with a turbine efficiency of 80 per cent for the minimum monthly discharge of Locust Fork of Black Warrior River at Palos.

Month.	1902.			1903.		
	Minimum discharge.	Minimum net horse-power per foot of fall.	Duration of mini-mum.	Minimum discharge.	Minimum net horse-power per foot of fall.	Duration of mini-mum.
	Second-feet.		*Days.*	*Second-feet.*		*Days.*
January	900	82	4	605	55	2
February	1,650	150	1	980	89	2
March	1,800	164	2	1,265	115	2
April	610	55	1	845	77	2
May	190	17	4	398	36	4
June	140	13	18	230	21	4
July	100	9	13	77	7	3
August	100	9	19	77	7	4
September	100	9	27	46	4	22
October	100	9	11	46	4	24
November	100	9	17	46	4	3
December	490	45	2	77	7	24

SURVEY OF BLACK WARRIOR RIVER.

A great deal of work is being done by the Government on this river in order to make it navigable as an outlet to important coal fields above.

In the 92 miles from old Warriortown to Tuscaloosa, there is a

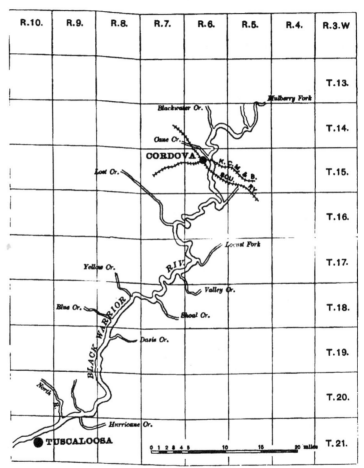

FIG. 8.—Map of part of Black Warrior River surveyed by Corps of Engineers, U. S. A.

fall of 158 feet. The distribution of this fall is shown by the following table, giving distances in miles above Tuscaloosa and elevations of water surface above sea level.

Elevations of locks and gages and water surface on Black Warrior River in Alabama from Tuscaloosa up to mouth of Mulberry Fork.

[Survey by United States Engineer Corps.]

Distance from Tuscaloosa gage.	Location.	Locks and gages.	Water surface.
Miles.		*Feet.*	*Feet.*
0.0	Zero of Tuscaloosa gage	86.86	
.0	Water surface, mean low water, original river		86.86
.0	Back water from lock below Tuscaloosa	91.30	
.9	Back water below lock No. 1	91.30	
.9	Zero of gage at lock No. 1	84.36	
.9	Crest of dam at lock No. 1	101.16	
.9	Water surface original below lock No. 1		90.00
.9	Water surface original above lock No. 1		91.00
1.2	Water surface original below lock No. 2		91.00
1.2	Water surface original above lock No. 2		96.00
1.2	Zero of gage at lock No. 2	94.36	
1.2	Crest of dam at lock No. 2	109.66	
2.0	Zero of gage at lock No. 3	102.86	
2.0	Water surface original below lock No. 3		106.00
2.0	Water surface original above lock No. 3		107.00
2.0	Crest of dam at lock No. 3	120.16	
5.0	Water surface original at mouth of North River		111.50
7.9	Water surface original at mouth of Hurricane Creek		113.00
8.5	Water surface original at mouth of Yellow Creek		113.50
8.8	Back water from lock No. 3 at foot of lock No. 4	120.16	
8.8	Water surface original below lock No. 4		113.50
8.8	Water surface original above lock No. 4		114.00
8.8	Zero of gage at lock No. 4	113.36	
8.8	Crest of dam at lock No. 4	132.30	
16.0	Water surface original at head of Mossy Shoal		131.30
19.2	Water surface original at foot of Rose Shoals		132.30
19.7	Water surface at head of Rose Shoals		136.50
21.8	Water surface at foot of Fair Shoals		139.80
24.2	Water surface at head of Crowder Shoals		151.20
25.6	Water surface at mouth of Blue Creek		151.30
25.8	Water surface at foot of Squaw Shoals		151.30
26.0	Water surface on Squaw Shoals		158.00
26.8	.....do		169.20
27.4	.....do		169.20
28.1	.....do		183.80
28.6	.....do		184.30
28.8	.....do		187.20

Elevations of locks and gages and water surface on Black Warrior River in Alabama from Tuscaloosa up to mouth of Mulberry Fork—Continued.

Distance from Tuscaloosa gage.	Location.	Locks and gages.	Water surface.
Miles.		*Feet.*	*Feet.*
29. 4	Water surface at head of Squaw Shoals		192. 80
37. 8	Water surface below Black Rock		193. 10
38. 0	Water surface above Black Rock		202. 20
43. 3	Water surface below Knight's mill dam		202. 50
43. 3	Water surface above Knight's mill dam		206. 10
44. 7	Water surface at mouth of Valley Creek		206. 10
47. 5	Water surface at foot of Fork Shoals		206. 10
47. 9	Water surface at mouth of Locust Fork		215. 10
48. 2	Water surface at head of Fork Shoals		215. 50
44. 8	Water surface at mouth of Lost Creek		216. 60
56. 9	Water surface at foot of Franklins Shoals		216. 70
59. 0	Water surface at head of Franklins Shoals		219. 10
63. 0	Water surface at foot of Copelands Shoals		220. 00
65. 4	Water surface at foot of Lanes Shoals		223. 50
68. 0	Water surface at foot of Tuggle Shoals		225. 60
69. 0	Water surface at head of Tuggle Shoals		230. 00
74. 4	Water surface at foot of Bee Shoals		231. 80
75. 5	Water surface at head of Bee Shoals		236. 90
76. 1	Water surface below Payne's mill dam		237. 10
76. 1	Water surface above Payne's mill dam		240. 80
79. 6	Water surface at mouth of Cane Creek, at Cordova, Ala.		242. 10
84. 6	Water surface at foot of Sanders Shoals		242. 30
85. 3	Water surface at head of Sanders Shoals, mouth of Black-water Creek		248. 80
92. 4	Water surface at mouth of Mulberry Fork		249. 75

Systematic discharge measurements have been made at Tuscaloosa and Cordova, the results of which are given in the foregoing pages. Comparative measurements at the two stations at the same stage in November, 1901, shows a discharge of 825 second-feet at Tuscaloosa and 285 second-feet at Cordova. At minimum stage of dry years the water gets considerably lower, as is shown by the records referred to, but the figures named are safe for low season in all ordinary years, and will be used in this discussion for determining the power available at different sites along the river.

The locks and proposed locks on this section of the river begin with No. 1, at Tuscaloosa, and are numbered up the river. Locks 1, 2, 3, and 4 are about completed, and others are projected, but the locations of the latter in the following list are approximated. However, the exact location of each is immaterial in showing the power available. The following is a table showing positions of locks and lock sites in miles above Tuscaloosa, the sea-level elevation of water below each, the lift at each, and the net horsepower that can be developed at each day on an 80 per cent turbine during the dry season in ordinary years, like 1900, after deducting 100 second-feet for lockage:

Powers on Black Warrior River.

No. of lock or site.	Distance from Tuscaloosa.	Elevation of water below lock.	Lift.	Discharge after deducting lockage.	Net horsepower on 80 per cent turbine without storage.	Location.
	Miles.	Feet.	Feet.	Sec.-ft.		
1	0.7	91.30	9.86	725	650	Bottom University Shoal, Tuscaloosa.
2	1.3	101.16	8.50	725	560	On University Shoal, Tuscaloosa.
3	2.0	109.66	10.50	725	690	Do.
4	8.7	120.16	12.14	704	777	Near mouth of Yellow Creek.
5	19.3	132.30	10.00	660	600	Foot of Rose Shoals.
6	21.7	142.30	9.00	660	540	Foot of Fair Shoals.
7	25.8	151.30	14.00	660	840	Foot of Squaw Shoals.
8	26.3	165.30	14.00	660	840	On Squaw Shoals.
9	27.8	179.30	14.00	660	840	Do.
10	37.7	193.3	14.00	550	700	Below Black Rock.
11	47.6	207.3	14.00	550	700	Mouth of Little Warrior River, or Locust Fork.
12	63.4	221.3	14.00	374	476	
13	75.0	235.3	14.00	285	364	

The best power on the river is at Squaw Shoals, 26 miles above Tuscaloosa, covered on the above table by locks Nos. 7, 8, and 9, each having a lift of 14 feet, and making a total fall on Squaw Shoals of 42 feet. This can be developed to best advantage by constructing a canal from the top of proposed dam at lock No. 9, along the river bank, 2 miles in length, to a point opposite the foot of Squaw Shoals, below lock No. 7. This canal, taking the river water not needed for lockage and allowing 2 feet for grade and storage, will utilize a net head of 40 feet, and produce 2,400 net horsepower continuously, or 4,800 net horsepower for a twelve-hour run per day, storing the water above lock No. 9 during the twelve idle hours.

It is to be remembered that the above estimates of power are for low season during ordinary years. There will be exceptional periods

A. DAM AND LOCK NO 1 ON BLACK WARRIOR RIVER AT TUSCALOOSA, ALA.

View from east bank.

B. DAM AND LOCK NO 2 ON BLACK WARRIOR RIVER AT TUSCALOOSA, ALA.

View from west bank.

of minimum water in extremely dry years in which the entire flow of the river will be as low as 100 second-feet, and will, therefore, barely suffice for lockage during a busy season of boating on the river. (See Nineteenth Annual Report, United States Geological Survey, Part IV, p. 251.) But such seasons are rare, and the facilities for water transportation should compensate for them to a great extent. It is admitted that the cheapness of coal along this river would naturally make the water powers less valuable, but the cheapness of development in connection with Government dams would partly offset the cheapness of coal. It is believed that the proposed development at Squaw Shoals could be made at a very moderate cost, and that such an investment would pay handsomely.

The following additional information concerning the Warrior and Black Warrior River is from Mr. R. C. McCalla, United States assistant engineer, Tuscaloosa, Ala., who is in charge of the improvements on that river:

Tuscaloosa is 361 miles by river above Mobile. Above Tuscaloosa the river is called the Black Warrior, and below it is called the Warrior. The locks on the two parts of the stream are numbered as two separate systems, the lowest lock in each system being No. 1, and the numbers running upstream. The following table gives the lift and location of the locks in both systems:

No. of lock.	Lift in feet.	Miles above Mobile.	
1 a	10.00	230.5	0.5 miles below mouth of Warrior; located but not begun.
2 a	10.00	246.2	Located but not begun.
3 a	10.00	266.7	Do.
4 a	10.00	282.3	Under construction.
5 a	10.00	298.3	Do.
6 a	10.00	315.2	Do.
1 b	9.86	361.9	In operation.
2 b	8.50	362.3	Do.
3 b	10.50	363.1	Do.
4 b	12.14	370.1	Under construction.

a Below Tuscaloosa.　　　　b Above Tuscaloosa.

Between lock No. 4 and the junction of Mulberry and Locust forks, 407.8 miles above Mobile, there are projected seven locks at 14 feet lift each, but none of these are yet located. The following table gives the location, etc., of gages now established and read daily at 7 a. m.:

Name of gage.	Number of gages.	Miles above Mobile.	Elevation of zero above mean low tide, Mobile.	Remarks.
Demopolis ^a	1	229.7	28.07	Zero about 1½ feet above mean low water.
Millwood ^a	1	259.8	45.97	Zero about mean low water.
Lock 4 ^a........	1	282.3	54.50	Zero top of lower miter sill.
A. G. S. bridge ^a.	1	288.0	61.26	Zero about mean low water.
Lock 5 ^a........	1	298.3	64.50	Zero top of lower miter sill.
Lock 6 ^a........	1	315.2	74.50	Do.
Grays Landing ^a.	1	319.5	80.41	Zero about mean low water.
Tuscaloosa......	1	361.1	86.86	Zero about 1 foot above mean low water.
Lock 1 ^b........	2	361.9	84.36	Zero top of lower miter sill.
Lock 2 ^b........	2	362.3	94.36	Do.
Lock 3 ^b........	2	363.1	102.86	Do.
Lock 4 ^b........	1	370.1	113.36	Do.
Cordova ^b.......	1	445.0	237.85	Zero about mean low water.

a Below Tuscaloosa.　　　　b Above Tuscaloosa.

TRIBUTARIES OF BLACK WARRIOR RIVER.

At Clear Creek Falls, in Winston County, within a distance of half a mile, there is a fall of over 100 feet, distributed as follows:

Fall on Clear Creek, Winston County.

	Feet.
Rapids above Upper Falls in 100 yards	6
Upper Falls, about	45
Still pool for 275 yards..	00
Lower Falls..	27
Rapids below Lower Falls	30

No discharge measurements have ever been made on this stream. It is thought best not to attempt to approximate its flow, as the stream originates from big springs.

TOMBIGBEE RIVER AND TRIBUTARIES.

TOMBIGBEE RIVER AT COLUMBUS, MISS.

This stream enters Alabama a short distance below Columbus.

The station is located about 1,000 feet below the highway bridge, 1¼ miles from the Southern Railway station at Columbus. The rod, which is in three sections, is fastened vertically to the rock bluff on the left bank. It is a 3 by 10 inch pine timber 45 feet long, marked with brass figures and copper nails, the graduation extending from −5 feet to +40 feet. The initial point of sounding is the end of the iron bridge, right bank, downstream side. Bench mark No. 1 is 250 feet from the initial point of sounding. The bridge floor is 40.85 feet above the zero of the rod, and the top of the iron girder under the floor timbers is 39.85 feet above the zero. Bench mark No. 2 is the top of the rail at the station of the Southern Railway, and is 55.2 feet above gage datum and 190.9 feet above mean sea level. The width of the river at low water is 160 feet. The maximum record height of the river was, on April 8, 1892, when the gage registered, 42 feet. The lowest recorded height was on October 26, 1893, when the gage reading was −3.9 feet. The danger line is at 33 feet. No measurements of discharge were made during 1900.

The following discharge measurements were made by K. T. Thomas, M. R. Hall, and others.

Discharge measurements of Tombigbee River at Columbus, Miss.

Date.	Gage height.	Discharge.	Date.	Gage height.	Discharge.
1901.	*Feet.*	*Second-feet.*	1903.	*Feet.*	*Second-feet.*
March 11	12.33	19,425	March 9	15.50	26,452
April 16	1.10	3,926	March 12	17.30	29,015
June 25	− 2.50	698	May 18	12.50	17,804
October 30	− 3.00	657	July 16	− 1.70	1,340
1902.			July 17	− 1.80	1,278
April 3	27.60	45,214	September 22	− 3.70	252
July 11	− 3.40	697	September 25	− 3.70	263
July 15	− 2.70	1,058			
September 23	− 3.25	545			
September 26	− 3.30	493			

Daily gage height, in feet, of Tombigbee River at Columbus, Miss.

Day.	Jan.	Feb.	Mar.	April.	May.	June.	July.	Aug.	Sept.	Oct.	Nov.	Dec.
1900.												
1	3.5	0.4	7.1	4.5	8.0	1.0	19.7	5.8	-1.2	-2.7	+2.6	+ 3.4
2	2.2	.3	6.8	3.8	6.5	8.0	18.3	4.0	-1.0	-2.8	+4.9	+ 2.5
3	1.8	.2	5.6	2.7	5.9	10.0	17.6	2.8	-1.0	-3.0	+5.5	+ 1.9
4	1.6	.4	4.4	1.9	4.5	13.4	16.2	1.9	- .7	-3.1	+5.1	+ 1.0
5	1.4	1.8	3.8	1.5	2.8	15.3	15.9	1.0	- .6	-3.3	+4.4	+ .1
6	1.2	3.5	3.4	1.3	1.5	17.0	15.4	.0	- .9	-3.5	+3.1	+ .2
7	1.1	3.5	7.6	1.3	.8	20.7	14.5	.5	-1.3	-3.5	+1.9	+ .3
8	1.0	3.3	14.4	1.1	.5	23.6	13.5	.9	-1.9	-2.2	+ .9	+ .3
9	.9	4.2	15.1	.7	1.4	25.5	10.0	-1.4	-2.1	-1.0	-- .1	.0
10	.9	8.4	18.8	.5	2.3	25.0	6.8	-1.8	-2.3	- .6	-- .5	.3
11	2.0	7.8	11.3	11.7	2.3	23.6	5.5	-2.2	-2.4	- .4	- .9	.5
12	6.6	7.6	9.9	16.2	3.6	21.6	4.0	-2.6	-2.5	+2.6	-1.1	+ .6
13	8.3	10.2	6.7	17.4	2.8	20.0	3.5	-2.6	-2.6	+4.8	-1.3	+ .9
14	7.1	9.8	4.3	19.3	2.2	18.5	2.0	-2.7	-2.7	+5.2	-1.3	+ .9
15	5.6	8.1	4.8	20.8	1.6	17.8	1.9	-2.2	-2.8	+5.6	-1.4	+ .9
16	4.6	5.8	5.6	20.9	+ .7	17.1	1.5	-2.2	-2.8	+5.4	-1.5	+ .9
17	2.8	4.6	5.2	22.9	.9	17.3	1.0	-2.3	-2.9	+2.4	-1.5	+ .9
18	2.4	3.8	4.6	26.9	- .4	17.8	.5	-2.3	-3.0	+ .4	· 1.6	- 1.0
19	2.2	3.2	9.4	27.6	-1.0	18.0	.0	-1.8	-3.0	- .4	· 1.6	- 1.0
20	2.1	2.8	15.6	27.5	- .8	16.8	.4	-1.4	-3.1	- .9	-1.6	- 1.0
21	1.9	3.5	18.2	27.1	- .5	15.2	1.3	-1.1	-8.0	-1.1	· 1.6	- 1.0
22	1.8	5.6	19.0	25.5	- .5	13.5	2.0	-1.6	-2.9	-1.3	+1.2	+ 3.0
23	1.6	5.4	19.2	23.3	+ .1	13.8	1.9	-1.9	-2.7	-1.0	.0	- 5.8
24	1.4	4.5	18.1	21.3	.2	18.5	1.4	-2.0	-1.3	-- .2	+1.7	+ 6.0
25	1.2	5.0	15.2	19.4	.3	21.5	.7	- .1	-1.7	+ .8	+2.3	+ 6.0
26	1.1	5.0	11.4	17.3	2.3	24.1	.1	+1.2	-2.0	+1.0	+2.8	+ 6.8
27	.9	3.9	7.8	14.8	2.4	25.0	1.8	+ .6	-2.2	+5.1	+3.3	+ 6.1
28	.7	4.2	4.6	11.8	2.2	24.8	5.0	- .3	-2.4	+5.6	+3.8	+ 5.0
29	.6	......	3.9	9.5	1.6	23.5	7.2	-1.1	......	+3.0	......	+ 4.0
30	.5	......	3.1	9.4	1.0	21.7	5.9	-1.0	-2.5	+4.6	+3.9	+ 4.2
31	.4	......	3.3	......	.8	......	6.7	-1.4	-2.5	+4.0	+3.8	+ 4.8
1901.												
1	4.0	6.0	0.9	2.6	0.9	1.3	- 1.6	- 3.2	-0.4	-2.3	-3.0	- 2.2
2	3.7	5.9	.8	6.4	.5	3.1	- 2.4	- 3.4	- .9	-2.0	-3.0	- 2.2
3	3.2	6.8	2.2	8.0	.2	8.8	- 2.4	- 3.5	-1.3	-2.2	-3.0	- 2.3
4	2.6	12.3	2.4	7.1	.0	3.9	- 2.1	- 3.4	-1.6	-2.2	-2.9	- 2.3
5	1.8	13.2	2.1	6.8	- .3	3.5	- 2.4	- 3.8	-1.6	-2.5	-2.9	- 2.2
6	1.0	13.9	1.8	6.2	- .5	3.1	- 1.6	- 3.2	-1.9	-2.5	-2.9	- 2.2
7	.7	15.1	1.4	5.3	- .6	2.8	- 2.6	- 3.0	-2.0	-2.6	-3.0	- 2.2
8	.1	15.9	1.2	4.0	- .9	1.9	- 2.6	- 2.8	-2.2	-2.7	-2.8	- 2.1
9	.0	15.6	1.0	2.9	-1.0	1.4	- 2.7	- 2.8	-2.3	-2.7	-2.8	- 2.0
10	.3	14.5	8.8	2.0	-1.1	1.0	- 2.8	- 2.9	-2.4	-2.8	-2.8	- 1.8
11	10.9	18.0	12.1	1.4	-1.2	- .1	- 2.8	- 2.9	-2.5	-2.8	-2.7	- 1.1
12	16.9	12.5	14.0	.9	-1.2	- .7	- 2.9	- 3.1	-2.5	-2.8	-2.8	- .5
13	19.4	12.4	17.8	.7	- .4	- 1.0	- 3.0	- 3.0	-2.4	· 2.6	-2.8	+ .1
14	21.7	11.0	19.4	.7	+3.3	- .9	- 3.1	- 3.2	-2.1	-2.5	-2.8	4.5
15	22.7	8.9	19.0	.9	4.4	- .8	- 3.1	- 1.5	-1.8	-2.2	-2.5	9.5
16	22.3	6.0	17.1	1.2	4.4	- .8	- 3.2	+ 4.0	-1.0	-2.0	-2.7	9.8
17	20.9	4.4	13.8	1.5	3.6	- .8	- 3.2	11.5	-- .2	-2.0	-2.7	9.2
18	18.8	3.3	10.8	6.0	2.7	- 1.0	- 3.0	12.4	+2.5	-2.2	-2.7	9.8
19	16.0	2.7	8.0	11.8	2.1	- 1.2	- 3.0	12.4	3.4	-2.3	-2.6	10.0
20	13.6	2.2	6.3	12.4	2.1	- 1.7	- 3.0	14.0	3.5	-2.4	· 2.5	8.8
21	9.4	1.8	4.6	12.7	6.3	- 2.0	- 2.0	15.6	2.6	·-2.5	--2.4	5.9
22	6.2	1.5	3.5	13.5	6.7	- 2.2	- 1.9	15.9	1.4	· 2.6	-2.2	4.8
23	3.8	1.2	3.2	13.5	6.3	- 2.4	- 2.3	14.8	.4	-2.7	-2.3	2.4
24	3.0	1.0	3.1	11.8	5.3	- 2.5	- 2.5	12.1	-- .6	· 2.8	-2.1	1.4
25	5.9	1.0	2.8	8.2	4.0	- 2.5	- 2.6	8.9	-1.2	-2.9	-2.0	.8
26	6.5	1.1	2.5	5.1	2.2	- 2.6	- 2.9	6.5	-1.6	-2.9	- 1.9	.7
27	6.3	1.2	2.2	8.4	1.5	- 2.6	- 3.0	4.1	-1.9	-2.9	- 1.8	1.1
28	6.0	1.0	1.9	2.5	1.8	- 2.6	- 3.1	2.0	-2.1	· 2.9	-1.8	1.0
29	5.8	......	1.6	1.8	1.7	- 2.7	- 3.2	1.6	-2.2	-3.0	2.0	9.0
30	5.7	......	1.4	1.3	1.2	- 2.6	- 3.3	1.1	-2.3	-3.0	--2.1	11.0
31	5.5	......	2.3	......	.9	......	- 3.4	.3	......	-3.0	......	9.6

Daily gage height, in feet, of Tombigbee River at Columbus, Miss.—Continued.

Day.	Jan.	Feb.	Mar.	Apr.	May.	June.	July.	Aug.	Sept.	Oct.	Nov.	Dec.
1902.												
1	9.20	14.50	9.50	30.30	0.00	-2.60	-3.20	-2.60	-2.40	-2.20	-3.30	3.80
2	8.90	16.80	9.30	29.40	-.50	-2.50	-3.20	-1.30	2.70	-1.80	3.30	3.90
3	7.70	17.90	9.30	28.20	-1.50	-2.50	-3.20	-.50	2.40	-.90	3.30	4.90
4	5.80	18.50	9.20	26.80	-.90	-2.60	-3.20	-.20	2.20	.60	-3.30	5.10
5	4.00	19.00	9.80	23.70	-.20	-2.60	-3.20	-.10	-2.50	1.00	3.30	5.70
6	2.50	17.60	9.20	20.50	-.10	-2.70	3.30	-.40	2.20	-1.30	3.20	5.70
7	1.50	16.00	8.20	18.20	.40	-2.70	3.30	1.00	2.20	1.60	3.10	4.90
8	.90	13.50	7.80	18.90	-.60	2.80	-3.30	-1.60	2.10	-2.00	3.10	4.60
9	.50	10.70	7.00	18.30	-.40	-2.80	-3.40	2.00	2.20	2.30	-3.00	3.90
10	.10	7.60	6.00	17.80	-.20	2.90	-3.40	2.40	2.40	2.50	2.90	3.00
11	-.20	5.60	6.00	17.40	-.50	-2.80	3.20	2.70	2.70	-2.90	-3.00	2.20
12	-.20	3.00	4.20	16.00	-.80	-2.90	3.40	2.90	3.00	2.60	-3.00	1.50
13	.50	2.10	4.30	13.00	-1.00	-2.80	3.00	2.90	-3.20	2.70	3.10	.80
14	-.60	1.40	4.30	9.00	-1.30	-2.90	2.50	-3.10	3.30	2.50	3.10	.60
15	-.90	4.00	4.80	5.60	-1.50	-3.00	2.70	3.20	3.40	-2.20	-3.10	.60
16	-1.00	5.40	6.00	4.20	1.60	-3.10	-3.00	3.30	3.50	-2.20	-3.00	1.20
17	-1.00	5.60	7.40	3.50	-1.10	-3.20	3.10	3.30	3.50	-2.00	3.00	1.90
18	-1.00	5.90	7.60	3.00	-1.30	-3.20	3.20	3.30	3.50	2.50	3.00	11.10
19	-.60	5.60	7.50	2.70	-1.10	3.20	-3.30	3.40	3.50	2.50	3.00	11.80
20	+.50	4.60	6.79	2.20	-1.10	3.20	3.30	3.40	3.50	2.60	3.00	13.00
21	3.10	4.40	5.70	2.00	-1.00	-3.10	2.70	3.40	3.40	-2.90	-2.90	13.20
22	5.20	5.30	4.70	1.80	-1.20	-3.00	3.00	-3.50	3.30	3.00	2.90	14.80
23	5.60	5.60	3.50	1.60	-1.50	-2.90	-3.00	-3.50	-3.30	3.00	2.90	15.80
24	5.80	5.30	4.00	1.10	-1.70	-2.80	2.90	-3.50	3.30	3.00	-2.90	12.10
25	5.50	6.40	4.30	.90	1.90	-3.00	2.90	-3.60	3.40	3.10	1.40	10.00
26	4.60	5.80	6.10	.70	2.10	3.00	-2.80	3.60	3.40	-3.10	.30	10.40
27	4.00	5.70	13.40	.60	2.20	3.00	2.90	-3.60	3.30	3.10	1.20	10.60
28	6.10	8.10	21.00	.40	-2.30	-3.00	-2.90	-3.10	3.20	8.20	2.50	10.80
29	7.40	....	28.00	.30	-2.40	3.10	2.60	-3.00	-3.10	3.30	2.90	11.30
30	10.00	....	30.50	.30	2.50	3.20	-3.00	-2.50	-2.90	-3.30	3.50	10.40
31	12.40	....	30.60	....	-2.50	....	-3.00	-2.40	....	3.30	....	9.50
1903.												
1	8.8	9.0	10.4	8.5	-0.4	3.1	0.0	-2.2	-3.4	-3.7	3.2	-3.0
2	8.4	8.2	10.8	3.4	-.6	2.9	-.8	-2.1	3.4	-3.7	2.8	-2.6
3	11.8	7.7	11.5	3.0	-.9	3.0	-1.3	-2.2	-3.5	-3.7	2.0	2.6
4	12.2	8.5	11.7	2.5	-1.1	3.0	-1.4	-2.4	-3.5	-3.7	1.8	2.8
5	11.8	8.8	11.4	2.0	1.1	2.9	-1.6	-1.6	-3.5	3.7	1.7	3.0
6	12.2	9.0	11.5	1.7	-1.2	2.3	-1.7	-2.8	-3.6	3.7	1.6	3.0
7	12.0	12.0	12.4	1.3	-1.3	2.3	-2.0	-2.9	3.6	3.7	1.4	3.0
8	11.0	18.6	13.6	1.0	-1.4	2.3	-1.8	-3.0	-3.6	-3.6	1.4	3.0
9	8.8	19.8	15.3	.8	-1.4	2.1	-2.0	-2.4	3.6	-3.7	1.5	3.0
10	8.5	20.5	16.0	1.6	-1.4	1.9	-2.0	-2.3	3.6	-3.7	1.5	3.0
11	8.9	22.3	17.0	2.6	-1.4	1.3	-2.0	-1.5	3.6	3.7	1.5	-3.0
12	9.5	23.9	17.3	2.4	-1.1	1.2	-2.0	-1.0	-3.7	-3.6	1.5	2.9
13	9.8	23.8	17.5	2.2	1.0	.8	-2.1	-1.0	3.7	3.6	1.4	-2.8
14	9.9	22.7	18.5	1.9	8.0	.1	-1.7	-.8	-3.7	-3.6	2.3	2.6
15	10.4	21.5	19.8	2.7	10.6	-.4	-1.8	-1.2	-3.7	-3.6	2.6	2.6
16	10.4	20.9	20.0	3.6	12.0	-.8	-1.9	-1.2	-3.7	-3.6	3.0	2.9
17	9.4	22.5	19.8	3.6	12.0	-1.0	-1.9	-1.5	-3.7	-3.5	3.0	-3.0
18	7.6	23.0	19.2	3.3	12.5	-1.6	2.1	-1.7	-3.7	-3.5	3.2	-3.0
19	5.6	23.1	18.3	2.8	11.9	-1.8	-2.3	-1.9	-3.7	-3.6	2.7	-2.8
20	4.6	23.6	16.7	2.2	10.1	-1.9	-2.4	-1.0	-3.7	-3.6	2.7	-2.8
21	5.5	23.3	14.0	1.5	7.6	-1.9	-2.5	-1.7	-3.7	-3.5	2.7	-2.5
22	3.2	21.7	12.1	1.0	5.7	2.0	-2.6	-1.4	-3.7	-3.5	2.8	-2.6
23	2.7	19.8	10.8	1.6	8.0	2.0	-2.6	-2.0	-3.8	-3.5	2.8	-2.6
24	2.3	17.0	9.8	2.0	1.2	-1.7	-3.0	-2.3	-3.8	-3.6	2.8	-2.9
25	2.0	15.8	9.5	2.0	.4	1.3	-3.0	2.5	-3.7	-3.6	2.6	-2.9
26	1.9	14.6	8.8	1.6	-.2	-1.6	-3.0	2.6	-3.7	3.5	2.7	2.7
27	1.9	8.7	6.9	1.4	-.6	.8	-3.0	2.9	-3.7	3.5	2.7	2.1
28	5.6	9.2	5.1	.6	-1.0	1.9	3.0	2.9	-3.7	3.5	2.8	-1.7
29	7.6	....	4.0	.1	-1.0	1.5	-3.1	3.0	-3.7	3.5	....	-1.6
30	8.0	....	3.4	-.2	-1.1	.8	-3.1	3.1	-3.7	3.6	....	1.6
31	8.8	....	3.5	....	.1	....	-3.1	3.3	....	3.5	....	-1.8

Rating table for Tombigbee River at Columbus, Miss., for 1900 and 1901.

Gage height.	Discharge.	Gage height.	Discharge.	Gage height.	Discharge.	Gage height.	Discharge.
Feet.	Second-feet.	Feet.	Second-feet.	Feet.	Second-feet.	Feet.	Second-feet.
−3.0	650	1.0	3,790	5.0	9,310	9.0	14,830
−2.9	668	1.1	3,928	5.1	9,448	9.1	14,968
−2.8	688	1.2	4,066	5.2	9,586	9.2	15,106
−2.7	712	1.3	4,204	5.3	9,724	9.3	15,244
−2.6	736	1.4	4,342	5.4	9,862	9.4	15,382
−2.5	752	1.5	4,480	5.5	10,000	9.5	15,520
−2.4	780	1.6	4,618	5.6	10,138	9.6	15,658
−2.3	810	1.7	4,756	5.7	10,276	9.7	15,796
−2.2	842	1.8	4,894	5.8	10,414	9.8	15,934
−2.1	877	1.9	5,032	5.9	10,552	9.9	16,072
−2.0	915	2.0	5,170	6.0	10,690	10.0	16,210
−1.9	956	2.1	5,308	6.1	10,828	10.1	16,348
−1.8	1,000	2.2	5,446	6.2	10,966	10.2	16,486
−1.7	1,047	2.3	5,584	6.3	11,104	10.3	16,624
−1.6	1,097	2.4	5,722	6.4	11,242	10.4	16,762
−1.5	1,150	2.5	5,860	6.5	11,380	10.5	16,900
−1.4	1,206	2.6	5,998	6.6	11,518	10.6	17,038
−1.3	1,265	2.7	6,136	6.7	11,656	10.7	17,176
−1.2	1,328	2.8	6,274	6.8	11,794	10.8	17,314
−1.1	1,394	2.9	6,412	6.9	11,932	10.9	17,452
−1.0	1,464	3.0	6,550	7.0	12,070	11.0	17,590
−.9	1,537	3.1	6,688	7.1	12,208	11.5	18,280
−.8	1,613	3.2	6,826	7.2	12,346	12.0	18,970
−.7	1,692	3.3	6,964	7.3	12,484	12.5	19,660
−.6	1,775	3.4	7,102	7.4	12,622	13.0	20,350
−.5	1,863	3.5	7,240	7.5	12,760	13.5	21,040
−.4	1,957	3.6	7,378	7.6	12,898	14.0	21,730
−.3	2,057	3.7	7,516	7.7	13,036	14.5	22,420
−.2	2,165	3.8	7,654	7.8	13,174	15.0	23,110
−.1	2,283	3.9	7,792	7.9	13,312	15.5	23,800
.0	2,410	4.0	7,930	8.0	13,450	16.0	24,490
.1	2,548	4.1	8,068	8.1	13,588	16.5	25,180
.2	2,686	4.2	8,206	8.2	13,726	17.0	25,870
.3	2,824	4.3	8,344	8.3	13,864	17.5	26,560
.4	2,962	4.4	8,482	8.4	14,002	18.0	27,250
.5	3,100	4.5	8,620	8.5	14,140	18.5	27,940
.6	3,238	4.6	8,758	8.6	14,278	19.0	28,630
.7	3,376	4.7	8,896	8.7	14,416	19.5	29,320
.8	3,514	4.8	9,034	8.8	14,554	20.0	30,010
.9	3,652	4.9	9,172	8.9	14,692		

Rating table for Tombigbee River at Columbus, Miss., for 1902.

Gage height.	Discharge.	Gage height.	Discharge.	Gage height.	Discharge.	Gage height.	Discharge
Feet.	*Second-feet.*	*Feet.*	*Second-feet.*	*Feet.*	*Second-feet.*	*Feet.*	*Second-feet.*
−3.4	500	0.2	3,230	6.5	10,400	15.5	24,750
−3.2	630	.4	3,410	7.0	11,100	16.0	25,600
−3.0	760	.6	3,590	7.5	11,815	17.0	27,300
−2.8	890	.8	3,780	8.0	12,540	18.0	29,000
−2.6	1,030	1.0	3,970	8.5	13,290	19.0	30,700
−2.4	1,170	1.2	4,160	9.0	14,040	20.0	32,400
−2.2	1,310	1.4	4,350	9.5	14,815	21.0	34,100
−2.0	1,450	1.6	4,550	10.0	15,590	22.0	35,800
−1.8	1,590	1.8	4,750	10.5	16,370	23.0	37,500
−1.6	1,740	2.0	4,950	11.0	17,170	24.0	39,200
−1.4	1,890	2.5	5,475	11.5	17,975	25.0	40,900
−1.2	2,040	3.0	6,020	12.0	18,800	26.0	42,600
−1.0	2,190	3.5	6,585	12.5	19,650	27.0	44,300
−.8	2,350	4.0	7,170	13.0	20,500	28.0	46,000
−.6	2,520	4.5	7,775	13.5	21,350	29.0	47,700
−.4	2,690	5.0	8,400	14.0	22,200	30.0	49,400
−.2	2,870	5.5	9,050	14.5	23,050		
.0	3,050	6.0	9,720	15.0	23,900		

Rating table for Tombigbee River at Columbus, Miss., for 1903.

Gage height.	Discharge.	Gage height.	Discharge.	Gage height.	Discharge.	Gage height.	Discharge.
Feet.	*Second-feet.*	*Feet.*	*Second-feet.*	*Feet.*	*Second-feet.*	*Feet.*	*Second-feet.*
—3.80	220	—1.30	1,660	1.20	3,940	3.70	6,740
-3.70	260	—1.20	1,740	1.30	4,050	3.80	6,860
-3.60	340	—1.10	1,820	1.40	4,160	3.90	6,985
-3.50	340	—1.00	1,900	1.50	4,270	4.00	7,110
-3.40	380	— .90	1,980	1.60	4,380	4.10	7,235
-3.30	425	— .80	2,060	1.70	4,490	4.20	7,360
-3.20	470	— .70	2,150	1.80	4,600	4.30	7,485
-3.10	520	— .60	2,220	1.90	4,710	4.40	7,610
—3.00	570	— .50	2,310	2.00	4,820	4.50	7,735
-2.90	620	— .40	2,400	2.10	4,930	4.60	7,860
-2.80	675	— .30	2,490	2.20	5,040	4.70	7,990
-2.70	730	— .20	2,580	2.30	5,150	4.80	8,120
—2.60	785	— .10	2,670	2.40	5,260	4.90	8,250
-2.50	840	.00	2,760	2.50	5,370	5.00	8,380
—2.40	895	.10	2,850	2.60	5,480	5.10	8,510
-2.30	950	.20	2,940	2.70	5,590	5.20	8,640
-2.20	1,010	.30	3,040	2.80	5,700	5.30	8,775
-2.10	1,070	.40	3,140	2.90	5,810	5.40	8,910
—2.00	1,135	.50	3,240	3.00	5,920	5.50	9,045
—1.90	1,205	.60	3,340	3.10	6,030	5.60	9,180
-1.80	1,275	.70	3,440	3.20	6,140	5.70	9,315
-1.70	1,345	.80	3,540	3.30	6,260	5.80	9,450
—1.60	1,420	.90	3,640	3.40	6,380	5.90	9,585
-1.50	1,500	1.00	3,740	3.50	6,500	a 6.00	9,720
—1.40	1,580	1.10	3,840	3.60	6,620		

a Above 6 feet use 1902 table.

Estimated monthly discharge of Tombigbee River at Columbus, Miss.

[Drainage area, 4,440 square miles.]

Month.	Discharge in second-feet.			Run-off.	
	Maximum.	Minimum.	Mean.	Second-feet per square mile.	Depth in inches.
1900.					
January	13,864	2,962	5,588	1.26	1.45
February	16,486	2,686	8,659	1.95	2.03
March	23,938	6,688	15,285	3.42	3.85
April	40,498	3,100	21,265	4.79	5.34
May	13,450	1,464	4,944	1.11	1.28
June	37,600	3,790	27,692	6.24	6.96
July	29,596	2,410	11,411	2.57	2.97
August	10,414	707	2,257	.51	.59
September	1,775	632	950	.21	.23
October	10,138	566	3,989	.90	1.04
November	10,000	1,097	4,304	.97	1.08
December	11,794	1,464	5,239	1.18	1.36
The year	40,498	566	9,299	2.09	28.18
1901.					
January	33,736	2,410	14,193	3.20	3.69
February	24,352	3,790	12,533	2.83	2.95
March	29,182	3,514	10,884	2.45	2.33
April	21,040	3,376	9,890	2.23	2.49
May	11,656	1,328	4,949	1.11	1.28
June	7,792	707	2,767	.62	.69
July	1,097	582	730	.16	.18
August	24,352	582	7,673	1.73	1.99
September	7,240	753	2,608	.45	.50
October	915	650	748	.17	.20
November	1,000	650	756	.17	.19
December	17,590	810	6,730	1.52	1.75
The year	33,736	582	6,155	1.39	18.74

Estimated monthly discharge of Tombigbee River at Columbus, Miss.—Continued.

Month.	Discharge in second-feet.			Run-off.	
	Maximum.	Mimimum.	Mean.	Second-feet per square mile.	Depth in inches.
1902.					
January	19,480	2,190	7,009	1.58	1.82
February	30,700	4,350	14,148	3.19	3.32
March	50,420	6,585	15,583	3.51	4.05
April	49,910	3,320	18,180	4.09	4.56
May	3,050	1,100	2,060	.46	.53
June	1,100	630	825	.19	.21
July	1,100	500	717	.16	.18
August	2,960	370	1,095	.25	.29
September	1,380	435	790	.18	.20
October	2,520	565	1,165	.26	.30
November	6,585	565	1,482	.33	.37
December	23,560	3,590	11,730	2.64	3.04
The year	50,420	370	6,232	1.40	18.87
1903.					
January	19,140	4,710	12,300	2.77	3.19
February	39,030	12,105	27,631	6.22	6.48
March	32,400	6,380	20,465	4.61	5.32
April	6,620	2,580	4,828	1.09	1.22
May	19,650	1,580	6,250	1.41	1.63
June	6,030	1,135	3,439	.77	.86
July	2,760	520	1,084	.24	.28
August	2,060	425	1,139	.26	.30
September	380	220	281	.06	.07
October	340	260	300	.07	.08
November	1,580	470	961	.22	.25
December	1,420	570	770	.17	.20
The year	39,030	220	6,621	1.49	19.88

Net horsepower per foot of fall with a turbine efficiency of 80 per cent for the minimum monthly discharge of Tombigbee River at Columbus, Miss.

Month.	1900.			1901.			1902.		
	Minimum discharge.	Minimum net horsepower per foot of fall.	Duration of minimum.	Minimum discharge.	Minimum net horsepower per foot of fall.	Duration of minimum.	Minimum discharge.	Minimum net horsepower per foot of fall.	Duration of minimum.
	Sec.-ft.		*Days.*	*Sec.-ft.*		*Days.*	*Sec.-ft.*		*Days.*
January	2,962	269	1	2,410	219	1	2,190	191	3
February	2,686	244	1	3,790	345	3	4,350	395	1
March	6,688	608	1	3,514	319	1	6,585	599	1
April	3,100	282	1	3,376	307	2	3,320	302	1
May	1,464	133	1	1,328	121	2	1,100	100	2
June	3,790	345	1	707	65	1	630	57	5
July	2,410	219	1	582	53	1	500	45	2
August	707	65	1	582	53	1	370	34	3
September	632	57	1	753	68	2	435	40	5
October........	566	51	2	650	59	3	565	51	3
November	1,097	100	4	650	59	4	565	51	5
December......	1,464	133	4	810	74	2	3,590	326	2

TOMBIGBEE RIVER NEAR EPES.

· A record of gage heights has been kept at this station for the last ten years by the Alabama Great Southern Railway Company. The gage is painted on the center brick pier of the railway bridge of that company across the Tombigbee, a half mile east of Epes, and is referred to two bench marks. The first, the top of the iron girder at the third crossbeam at the station, 80 feet from the right-bank end of the iron bridge, is 64.70 feet above datum of gage; the second, the top of the cross-tie or the base of the rail at the station, 80 feet from the right-bank end of the iron bridge, is 65.50 feet above datum of gage. The west bank of the river is a solid wall of limestone; the east bank is flat and is subject to overflow. The trestle at the east end of the bridge is seven-eighths of a mile long. The section is good, though the water is very deep and rather swift.

The following discharge measurements were made during 1901 by K. T. Thomas:

    January 31: Gage height, 12.70 feet; discharge, 13,738 second-feet.
    March 14: Gage height, 21.10 feet; discharge, 23,824 second-feet.
    June 28: Gage height, 1 foot; discharge, 1,496 second-feet.
    November 13: Gage height, 0.70 foot; discharge, 1,290 second-feet.

Daily gage height, in feet, of Tombigbee River near Epes.

Day.	Jan.	Feb.	Mar.	Apr.	May.	June.	July.	Aug.	Sept.	Oct.	Nov.	Dec.
1900.												
1	8.5	3.0	18.0	24.5	43.0	6.0	44.5	14.5	2.0	0.5	7.0	8.0
2	7.5	3.0	19.0	19.5	41.0	8.0	44.5	12.0	2.5	.5	8.0	7.5
3	6.5	3.0	18.0	16.0	39.0	15.0	44.5	10.5	2.0	.5	9.0	7.0
4	6.0	3.0	17.0	12.0	33.5	21.0	44.0	8.0	2.0	.5	8.5	6.0
5	5.5	5.0	15.0	10.0	28.0	24.5	42.5	6.0	2.0	.5	8.5	5.0
6	5.0	6.0	14.0	8.0	20.5	27.0	42.0	5.0	2.0	.5	7.5	4.5
7	4.5	7.0	13.0	7.5	16.0	29.5	41.0	5.0	2.0	.5	7.0	4.0
8	4.0	7.5	18.0	7.5	10.0	32.0	40.0	4.0	2.0	.5	5.5	4.0
9	3.5	11.5	21.0	7.0	7.0	34.5	39.0	3.0	2.0	.5	4.5	4.0
10	3.5	13.5	23.0	6.5	7.0	37.0	38.0	2.0	2.0	2 0	4.0	3.5
11	13.0	15.0	24.0	20.5	7.0	38.5	34.0	2.0	1.5	4.0	4.0	3.0
12	20.0	20.5	24.0	26.0	8.0	39.5	26.0	2.0	1.5	6.0	3.0	3.0
13	23.0	26.0	23.0	29.0	8.0	40.5	23.0	2.0	1.5	7.5	3.0	8.5
14	23.5	28.0	20.0	30.0	7.0	41.0	15.5	2.0	2.0	8.5	2.5	3.5
15	22.0	28.0	17.5	31.0	6.0	41.5	13.0	2.0	1.5	10.0	2.0	3.5
16	21.0	26.0	17.0	38.0	6.0	42.0	8.0	1.5	1.0	10.5	2.0	3.0
17	18.5	24.0	18.0	46.0	5.0	42.0	7.0	1.5	1.0	10.0	2.0	3.0
18	15.0	22.0	18.0	48.5	4.0	42.0	6.0	1.5	1.0	7.5	2.0	3.0
19	11.5	18.5	18.5	51.0	3.5	41.5	5.0	1.5	1.0	5.0	2.6	3.0
20	10.0	16.0	26.0	51.5	3.5	41.5	6.5	1.5	.5	4.0	2.5	6.0
21	8.5	14.0	30.0	52.0	3.5	41.5	7.0	1.0	.5	3.5	3.0	5.0
22	8.0	15.0	32.0	52.0	3.5	41.0	8.0	1.0	.5	3.0	3.5	5.0
23	8.0	17.5	34.0	51.5	3.5	41.0	9.0	2.0	.5	2.0	3.5	8.5
24	7.0	18.0	35.5	51.0	4.0	42.5	8.0	1.5	.5	2.0	3.5	10.0
25	6.5	17.5	37.5	49.5	4.5	42.5	6.0	1.5	.5	3.0	6.5	11.5
26	6.0	17.0	38.0	47.5	5.0	43.5	5.5	1.5	.5	4.0	8.5	11.5
27	5.0	16.0	39.0	47.0	6.0	43.5	5.5	5.0	.5	4.5	7.5	12.5
28	3.0	17.0	38.5	46.5	7.0	43.5	5.5	4.5	.5	7.5	7.5	12.0
29	3.0	......	35.0	46.0	7.0	44.0	14.0	4.0	.5	9.0	8.0	10.0
30	3.0	......	33.0	44.5	6.0	44.5	14.5	3.0	.5	8.0	8.5	10.0
31	3.0	......	30.0	......	5.0	......	14.5	2.0	......	7.5	......	10.5
1901.												
1	10.0	12.0	6.5	13.0	7.5	5.5	1.0	......	5.5	1.5	0.7	1.7
2	10.0	10.5	6.5	13.5	7.0	6.0	1.0	......	4.0	6.0	.7	1.7
3	9.0	16.8	9.0	15.5	6.0	7.0	1.0	......	3.5	2.5	.7	1.7
4	8.0	21.5	9.0	18.0	5.0	9.0	1.0	......	2.5	1.5	.0	8.5
5	7.0	25.5	8.5	17.0	4.5	9.0	1.0	......	2.0	1.5	.0	2.0
6	6.0	26.5	7.0	16.0	5.0	11.0	1.0	......	2.0	1.5	.0	2.0
7	5.5	27.5	6.5	15.0	4.0	12.0	1.0	......	1.5	1.5	.0	2.0
8	5.0	29.0	6.5	14.0	3.5	10.0	1.0	......	1.5	1.0	.0	3.5
9	4.5	30.5	6.5	11.0	3.5	8.0	1.0	......	1.5	1.0	.7	5.9
10	4.0	31.0	13.0	9.5	3.0	6.5	1.0	......	1.5	1.0	.7	6.5
11	18.0	31.0	18.0	8.0	3.0	5.5	.5	......	1.5	1.0	.7	5.5
12	29.5	31.0	19.5	7.5	3.5	4.5	.5	......	1.5	1.0	.7	4.7
13	33.0	30.0	20.0	9.0	4.0	4.0	.5	......	1.0	1.0	.7	4.2
14	35.0	29.5	24.5	8.0	7.0	3.5	.5	......	1.0	1.0	.7	13.0
15	36.0	28.5	25.5	7.0	8.0	3.5	.5	......	1.0	1.0	.7	19.5
16	38.0	26.0	26.5	7.0	8.5	3.5	.5	8.5	1.5	1.0	.7	22.0
17	39.0	23.0	26.5	6.5	8.0	3.0	.5	15.0	2.0	1.0	1.0	23.0
18	39.5	16.0	26.0	20.0	7.0	3.0	.5	20.0	2.5	1.0	1.7	22.5
19	40.0	12.0	25.0	28.0	6.5	2.5	.5	22.5	6.0	1.0	1.7	22.0
20	40.5	10.0	23.0	29.5	9.0	2.0	1.0	23.0	......	1.0	1.5	20.0
21	39.0	8.5	20.0	30.0	12.0	2.0	1.0	24.0	7.5	1.0	1.5	18.0
22	38.0	8.0	17.0	29.5	12.0	2.0	1.0	24.5	7.5	1.0	1.7	14.0
23	34.5	7.5	12.5	28.5	11.5	1.5	1.0	26.0	7.0	1.0	1.7	12.0
24	29.0	7.0	11.0	28.0	11.0	1.2	1.0	26.5	5.5	.5	1.8	9.0
25	24.5	6.5	12.0	27.0	10.0	1.0	1.0	26.0	4.0	.5	1.8	8.0
26	20.0	6.5	13.5	24.0	8.0	1.0	1.0	25.0	3.0	.5	2.0	7.0
27	16.0	6.5	13.5	18.0	6.5	1.0	.5	23.0	2.0	.5	2.0	7.0
28	15.0	6.5	11.5	13.0	6.5	1.0	.5	19.0	1.5	.5	2.0	12.0
29	14.0	......	10.0	11.0	6.5	1.0	− .2	12.0	1.5	.5	1.9	20.0
30	13.0	......	9.0	8.0	6.5	1.0	− .2	6.5	1.5	.5	1.8	26.0
31	12.7	......	13.5	......	7.0	......	− .2	5.5	......	.5	......	27.0

Rating table for Tombigbee River near Epes for 1900 and 1901.

Gage height.	Discharge.	Gage height.	Discharge.	Gage height.	Discharge.	Gage height.	Discharge.
Feet.	Second-feet.	Feet.	Second-feet.	Feet.	Second-feet.	Feet.	Second-feet.
−0.2	810	3.8	3,698	7.8	7,806	11.8	12,660
−.1	840	3.9	3,788	7.9	7,980	11.9	12,780
.0	880	4.0	3,878	8.0	8,100	12.0	12,900
.1	830	4.1	3,969	8.1	8,220	12.1	13,020
.2	985	4.2	4,060	8.2	8,340	12.2	13,140
.3	1,043	4.3	4,152	8.3	8,460	12.3	13,260
.4	1,103	4.4	4,245	8.4	8,580	12.4	13,380
.5	1,164	4.5	4,338	8.5	8,700	12.5	13,500
.6	1,226	4.6	4,432	8.6	8,820	12.6	13,620
.7	1,289	4.7	4,527	8.7	8,940	12.7	13,740
.8	1,353	4.8	4,622	8.8	9,060	12.8	13,860
.9	1,418	4.9	4,718	8.9	9,180	12.9	13,980
1.0	1,484	5.0	4,815	9.0	9,300	13.0	14,100
1.1	1,551	5.1	4,912	9.1	9,420	13.1	14,220
1.2	1,619	5.2	5,010	9.2	9,540	13.2	14,340
1.3	1,688	5.3	5,109	9.3	9,660	13.2	14,460
1.4	1,758	5.4	5,208	9.4	9,780	13.4	14,580
1.5	1,829	5.5	5,308	9.5	9,900	13.5	14,700
1.6	1,903	5.6	5,409	9.6	10,020	13.6	14,820
1.7	1,976	5.7	5,511	9.7	10,140	13.7	14,940
1.8	2,050	5.8	5,613	9.8	10,260	13.8	15,060
1.9	2,125	5.9	5,716	9.9	10,380	13.9	15,180
2.0	2,200	6.0	5,820	10.0	10,500	14.0	15,300
2.1	2,276	6.1	5,925	10.1	10,620	14.1	15,420
2.2	2,353	6.2	6,030	10.2	10,740	14.2	15,540
2.3	2,431	6.3	6,136	10.3	10,860	14.3	15,660
2.4	2,510	6.4	6,243	10.4	10,980	14.4	15,780
2.5	2,590	6.5	6,350	10.5	11,100	14.5	15,900
2.6	2,671	6.6	6,458	10.6	11,220	14.6	16,020
2.7	2,753	6.7	6,566	10.7	11,340	14.7	16,140
2.8	2,835	6.8	6,675	10.8	11,460	14.8	16,260
2.9	2,918	6.9	6,785	10.9	11,580	14.9	16,380
3.0	3,002	7.0	6,900	11.0	11,700	15.0	16,500
3.1	3,087	7.1	7,020	11.1	11,820	15.1	16,620
3.2	3,172	7.2	7,140	11.2	11,940	15.2	16,740
3.3	3,258	7.3	7,260	11.3	12,060	15.3	16,860
3.4	3,345	7.4	7,380	11.4	12,180	15.4	16,980
3.5	3,432	7.5	7,500	11.5	12,300	15.5	17,100
3.6	3,520	7.6	7,620	11.6	12,420	15.6	17,220
3.7	3,609	7.7	7,740	11.7	12,540	15.7	17,340

Rating table for Tombigbee River near Epes for 1900 and 1901—Continued.

Gage height.	Discharge	Gage. height.	Discharge.	Gage height.	Discharge.	Gage height.	Discharge.
Feet.	*Second-feet.*	*Feet.*	*Second-feet.*	*Feet.*	*Second-feet.*	*Feet.*	*Second-feet.*
15.8	17,460	17.4	19,380	19.0	21,300	26.6	23,220
15.9	17,580	17.5	19,500	19.1	21,420	20.7	23,340
16.0	17,700	17.6	19,620	19.2	21,540	20.8	23,460
16.1	17,820	17.7	19,740	19.3	21,660	20.9	23,580
16.2	17,940	17.8	19,860	19.4	21,780	21.0	23,700
16.3	18,060	17.9	19,980	19.5	21,900	21.1	23,820
16.4	18,180	18.0	20,100	19.6	22,020	21.2	23,940
16.5	18,300	18.1	20,220	19.7	22,140	21.3	24,060
16.6	18,420	18.2	20,340	19.8	22,260	21.4	24,180
16.7	18,540	18.3	20,460	19.9	22,380	21.5	24,300
16.8	18,660	18.4	20,580	20.0	22,500	21.6	24,420
16.9	18,780	18.5	20,700	20.1	22,620	21.7	24,540
17.0	18,900	18.6	20,820	20.2	22,740	21.8	24,660
17.1	19,020	18.7	20,940	20.3	22,860	21.9	24,780
17.2	19,140	18.8	21,060	20.4	22,980	22.0	24,900
17.3	19,260	18.9	21,180	20.5	23,100		

Estimated monthly discharge of Tombigbee River near Epes.

[Drainage area, 8,830 square miles.]

Month.	Discharge in second-feet.			Run-off.	
	Maximum.	Minimum.	Mean.	Second-feet per square mile.	Depth in inches.
1901.					
January	47,100	3,878	25,579	2.90	3.34
February	35,700	6,350	20,999	2.38	2.48
March	30,300	6,350	16,198	1.83	2.11
April	34,500	6,350	18,102	2.05	2.29
May	12,900	3,002	6,880	.78	.90
June	12,900	1,484	4,585	.52	.58
July	1,484	810	1,295	.15	.17
August 16–31			21,541	2.44	1.41
September	7,500	1,484	3,205	.36	.40
October	5,820	1,164	1,633	.18	.21
November	2,200	880	1,550	.18	.20
December	30,900	1,960	12,249	1.39	1.60

Net horsepower per foot of fall with a turbine efficiency of 80 per cent for the minimum monthly discharge of Tombigbee River near Epes.

Month.	1900.			1901.		
	Minimum discharge.	Minimum net horse-power per foot of fall.	Duration of mini-mum.	Minimum discharge.	Minimum net horse-power per foot of fall.	Duration of mini-mum.
	Second-feet.		Days.	Second-feet.		Days.
January	3,002	273	4	3,878	353	1
February	3,002	273	4	6,350	577	4
March	14,100	1,282	1	6,350	577	5
April	6,350	577	1	6,350	577	1
May	3,432	312	5	3,003	273	2
June	5,820	529	1	1,484	135	..
July	4,815	438	2	810	74	
August	1,484	135	2	....	....	....
September	1,164	106	11	1,484	135	3
October.................	1,164	106	9	1,164	106	8
November	2,200	200	4	880	80	5
December...............	3,002	273	5	1,960	178	3

TRIBUTARIES OF TOMBIGBEE RIVER.

There are several large creeks in Marion and Lamar counties that flow into Mississippi and enter Tombigbee River near Columbus. One of these, Buttahatchee River, in Marion County, has numerous rapids, especially near the crossing of the military road.

Luxapallila Creek, in Lamar County, has two prongs that are both good power streams. They come together before the creek enters Mississippi, making Big Luxapallila Creek, which enters the Tombigbee at Columbus, Miss.

The following discharge measurements were made on Big Luxapal-lila Creek, at Water Works, near Columbus, Miss., by M. R. Hall and assistants:

Discharge measurements of Big Luxapallila Creek near Columbus, Miss.

Date.	Gage height.	Discharge.	Date.	Gage height.	Discharge.
1901.	*Feet.*	*Second-feet.*	**1902.**	*Feet.*	*Second-feet.*
February 18.........	4. 95	957	April 4	12. 40	3, 864
March 11	8. 20	2, 459	July 11	1. 60	141
April 16	4. 45	873	September 23	1. 70	322
June 26.............	1. 90	109	**1903.**		
October 31	2. 00	126	July 16	....	272

TENNESSEE RIVER AND TRIBUTARIES.

TENNESSEE RIVER AT CHATTANOOGA, TENN.

This river, after passing Chattanooga, enters Alabama. It then makes a bend to the west and later to the north, returning to Tennessee. Flowing through this State and Kentucky, it empties into the Ohio 50 miles above Cairo. In 1879 a gage was established at Chattanooga, Tenn., at the foot of Lookout street, just below Chattanooga Island, by the Signal Corps of the United States Army. This gage has been in charge of the Weather Bureau since July 1, 1891. The drainage area above this station is 21,382 square miles, and is mapped on the following atlas sheets of the United States Geological Survey: Morristown, Greenville, Roan Mountain, London, Knoxville, Mount Guyot, Asheville, Murphy, Briceville, Standingstone, Wartburg, Pikeville, Maynardville, Cumberland Gap, Jonesville, Estillville, Bristol, Whitesburg, Grundy, Abington, Tazewell, Pocahontas, Wytheville, Cranberry, Morganton, Mount Mitchell, Saluda, Pisgah, Como, Nantahala, Walhalla, Dahlonega, Ellijay, Dalton, Cleveland, Ringgold, Kingston, and Chattanooga. The gage is on an inclined railroad iron for about 20 feet of its lower portion. Above this it is a vertical rod bolted to the rock bluff forming the river bank. The zero of the gage is 630.4 feet above sea level. Measurements are made from the Hamilton County steel highway bridge at the foot of Walnut street, a short distance below the gage. Gage heights are obtained from L. M. Pindell, United States Weather Bureau observer. In 1900 a new gage on the same datum was established. It is a vertical rod bolted to the south side of the third stone pier from the south end of the bridge.

Discharge measurements of Tennessee River at Chattanooga, Tenn.

Date.	Gage height.	Discharge.	Date.	Gage height.	Discharge.
1893.	*Feet.*	*Second-feet.*	1898.	*Feet.*	*Second-feet.*
March 15	10.3	63,039	November 29	4.70	31,340
March 16	9.2	58,310	1899.		
April 3	5.1	32,628	May 3..............	6.71	37,770
April 4	5.1	32,643	May 26.....	4.76	25,526
May 5...............	26.0	156,187	June 21.............	4.15	21,391
May 8..............	26.0	151,660	September 15......	1.90	10,819
May 9..............	16.0	98,979	October 27.........	.80	6,566
May 17.............	9.9	65,867	1900.		
May 18.............	10.4	67,883	March 13...........	11.25	66,012
1897.			July 27.............	3.45	18,470
May 8...............	7.07	44,187	1901.		
May 28.............	4.52	25,892	January 24........	5.60	30,317
June 29.............	5.76	32,943	April 4.............	24.20	155,457
July 13.............	4.59	26,884	July 31.............	2.80	15,393
September 7........	1.67	10,313	August 18...........	31.70	198,718
October 6...........	.48	5,969	1902.		
November 16	.83	5,552	June 25.............	3.80	17,773
December 23........	10.30	67,000	October 9...........	2.00	10,678
1898.			November 14......	1.55	9,282
May 10.............	4.14	22,066	1903.		
July 29.............	5.30	29,693	March 26....	28.85	190,279
August 19	6.37	38,671	July 21.............	3.85	20,936
October 6...........	17.60	120,359	September 5.......	1.60	10,472
October 28..........	6.00	35,953	October 21.........	1.10	8,063
November 29	4.75	29,569			

Daily gage height, in feet, of Tennessee River at Chattanooga, Tenn.

Day.	Jan.	Feb.	Mar.	Apr.	May.	June.	July.	Aug.	Sept	Oct.	Nov.	Dec.
1890.												
1	4.9	8.0	40.2	10.0	7.6	6.1	3.2	5.8	7.6	9.4	6.8	2.4
2	5.2	7.3	42.5	9.8	7.1	6.6	2.9	5.2	5.6	8.1	6 6	2.4
3	5.1	7.3	41.0	9.4	6.7	5.3	2.9	4.7	5.7	7.7	5.7	2.3
4	5.0	7.4	34.4	9.7	6.8	5.3	3.0	4.7	4.0	7.2	5.3	2.3
5	5.0	7.2	23.0	12.2	7.0	5.2	3.0	4.8	3.6	6.4	5.0	2.3
6	4.8	7.8	15.1	14.0	7.5	5.3	3.0	4.5	3.2	5.7	4.6	2.5
7	4.7	8.3	14.2	13.6	8.9	5.2	3.0	5.2	3.6	5.7	4.6	2.5
8	4.6	11.5	11.4	13.4	9.1	4.9	2.9	5.7	3.2	5.2	4.3	2.9
9	4.7	19.3	12.8	11.9	8.7	4.7	2.8	6.6	2.8	5.5	4.0	7.1
10	4.9	20.4	11.2	10.5	8.7	4.5	2.7	7.5	2.8	5.3	3.9	8.1
11	4.7	17.8	10.0	9.6	8.1	4.5	2.7	7.2	3.0	4.8	3.8	8.2
12	4.6	14.7	9.2	8.7	7.7	4.4	3.0	6.3*	3.4	4.6	3.6	7.7
13	4.6	12.0	8.6	8.0	7.2	4.0	2.7	5.8	3.6	4.4	3.5	7.4
14	4.9	10.0	8.7	7.5	6.7	3.9	2.5	5.2	4.6	4.2	3.4	6.4
15	4.6	9.8	9.7	7.1	6.6	4.0	2.3	4.8	4.0	3.9	3.3	3.7
16	5.4	9.6	13.7	6.9	7.8	4.1	2.1	4.2	4.0	3.7	3.2	4.2
17	7.2	9.0	15.1	7.1	9.3	4.0	2.0	3.8	3.7	3.7	3.2	3.9
18	9.2	9.3	14.9	9.4	8.6	3.9	2.2	3.7	4.0	3.7	3.0	3.9
19	8.2	8.5	13.0	16.6	7.9	3.7	4.7	8.3	5.3	4.2	3.2	4.0
20	7.5	7.8	11.7	20.4	8.8	3.6	4.7	3.1	3.7	4.0	3.1	4.0
21	7.0	7.3	12.4	18.2	10.7	3.7	4.1	2.6	3.7	4.0	3.1	3.9
22	9.6	7.1	14.0	14.3	11.9	3.8	3.5	2.5	5.8	3.8	3.1	3.9
23	13.0	7.2	20.0	11.3	11.9	3.7	3.2	3.8	3.5	5.3	2.9	3.9
24	12.3	7.4	25.5	9.6	11.6	3.8	3.2	4.0	3.3	7.2	2.8	3.9
25	11.7	12.1	27.2	8.7	9.2	4.0	4.1	3.3	3.4	8.8	2.8	4.3
26	10.0	18.7	26.0	8.4	7.8	4.0	5.9	3.8	4.0	9.2	2.7	4.6
27	8.3	26.4	21.4	8.4	7.4	3.9	7.5	3.6	4.3	9.5	2.6	9.4
28	7.3	34.8	15.4	8.5	8.0	3.9	7.7	3.0	4.0	8.6	2.6	12.5
29	6.6	....	13.0	8.4	8.1	3.5	7.0	8.8	3.8	7.6	2.5	12.9
30	7.7	....	11.9	8.1	7.4	3.1	6.3	5.4	5.3	7.1	2.4	12.4
31	8.3		10.7		6.8		6.2	6.5		7.0		9.3
1891.												
1	7.7	9.8	18.6	15.4	5.9	5.8	3.9	8.3	5.4	1.8	1.3	3.7
2	7.8	13.2	17.5	16.3	5.7	5.6	3.8	10.3	4.7	1.9	1.3	3.2
3	9.9	16.1	15.6	16.3	5.6	5.3	3.6	15.1	4.3	1.9	1.3	2.9
4	14.1	19.8	13.3	15.7	5.5	4.8	3.6	16.4	4.2	1.8	1.2	2.8
5	15.5	22.6	15.4	15.1	5.2	4.4	3.6	12.0	5.1	1.8	1.2	5.6
6	15.2	21.6	20.0	12.6	5.1	4.1	3.6	8.7	5.1	1.7	1.2	6.1
7	10.4	18.3	23.6	11.6	4.9	3.9	3.4	6.9	5.2	1.7	1.2	6.6
8	8.2	16.9	29.1	10.8	4.7	4.1	3.4	5.8	5.2	1.6	1.2	8.6
9	7.1	14.5	34.5	9.8	4.6	4.6	3.3	5.1	5.8	1.7	1.2	10.8
10	6.3	21.0	37.5	9.6	4.5	4.7	4.5	4.6	4.9	1.7	1.2	10.9
11	6.5	27.8	38.9	9.8	4.4	5.5	5.1	4.4	4.4	1.8	1.5	10.2
12	8.9	34.3	37.6	9.9	4.3	7.0	4.4	4.0	3.9	1.8	1.7	8.5
13	10.7	36.5	33.5	10.6	4.2	6.5	3.9	4.0	3.6	1.7	2.7	6.8
14	10.0	37.5	27.0	11.3	4.1	5.7	3.5	3.9	3.6	1.7	3.6	5.7
15	9.2	35.5	22.2	12.2	4.2	5.5	8.1	3.5	3.5	1.7	4.1	5.1
16	7.3	29.0	19.8	10.8	4.2	5.7	2.9	3.6	3.5	1.6	3.5	5.0
17	7.8	21.1	18.1	9.4	4.5	5.8	2.8	3.5	3.5	1.6	2.8	5.2
18	7.5	19.7	15.3	8.4	4.7	6.1	2.7	3.4	3.2	1.5	2.5	5.3
19	7.5	18.2	13.5	8.2	4.7	6.8	4.1	3.0	2.9	1.5	2.4	1.8
20	7.6	16.5	12.3	7.9	4.6	7.3	5.0	3.0	2.7	1.5	2.5	4.5
21	7.3	15.5	11.3	7.9	4.5	6.8	4.5	3.4	2.6	1.5	2.3	4.2
22	8.2	18.8	10.8	7.6	4.3	6.8	4.0	4.0	2.5	1.5	2.1	4.0
23	12.5	24.0	10.7	7.4	4.1	6.5	3.8	4.6	2.4	1.5	3.0	3.8
24	15.3	27.7	10.8	7.4	4.0	7.1	3.6	5.5	2.3	1.5	4.6	3.7
25	14.0	29.0	10.6	7.5	3.8	7.4	3.5	5.6	2.2	1.5	6.2	4.1
26	13.6	26.7	10.4	7.5	3.9	7.6	3.5	7.7	2.2	1.5	6.7	1.9
27	11.2	20.6	10.5	7.4	4.0	6.2	3.5	8.2	2.1	1.5	6.3	8.1
28	9.7	19.0	11.1	7.2	4.1	4.9	3.6	8.1	2.0	1.5	5.6	10.2
29	7.9	....	13.6	6.5	4.1	4.3	3.7	7.0	1.9	1.4	1.7	9.6
30	7.9	....	13.0	6.2	4.7	4.1	3.8	6.4	1.9	1.4	4.0	8.1
31	8.9		13.1		5.3		5.7	6.1		1.1		7.8

Daily gage height, in feet, of Tennessee River at Chattanooga, Tenn.—Continued.

Day.	Jan.	Feb.	Mar.	Apr.	May.	June.	July.	Aug.	Sept.	Oct.	Nov.	Dec.
1892.												
1	6.6	6.5	5.7	9.1	8.7	5.6	6.6	4.2	2.1	2.2	1.1	4.1
2	6.6	6.2	5.6	8.3	8.2	5.4	5.6	4.1	1.8	2.0	1.2	4.0
3	8.1	6.0	5.5	7.4	7.6	5.5	5.5	4.2	1.7	1.9	1.2	3.8
4	8.8	5.8	5.3	6.8	7.3	5.8	5.4	4.0	1.5	1.8	1.6	3.7
5	8.7	5.5	5.1	6.5	7.4	8.8	6.6	4.1	1.3	1.7	2.0	3.6
6	8.4	5.4	5.0	8.5	7.0	9.2	8.9	4.9	2.2	1.6	2.3	3.6
7	9.0	5.3	4.9	21.7	6.6	9.3	11.2	4.2	2.1	1.6	2.4	3.3
8	9.8	5.8	5.0	31.6	6.3	8.7	11.8	3.8	2.0	1.5	2.4	3.2
9	10.0	8.1	6.0	34.2	6.2	8.6	10.1	3.5	2.0	1.5	2.9	3.2
10	9.3	11.5	7.1	34.3	6.9	8.3	8.6	3.3	1.9	1.5	4.4	3.0
11	8.1	11.3	8.0	31.0	5.5	7.8	9.0	3.3	2.0	1.5	5.9	2.9
12	8.3	10.5	7.9	26.6	5.7	8.0	9.5	3.1	1.8	1.5	6.6	2.7
13	11.2	8.9	7.6	18.0	5.7	8.1	9.4	3.5	2.1	1.4	7.0	2.6
14	22.9	7.7	7.6	12.9	5.5	7.8	8.9	3.6	2.1	1.4	4.1	2.6
15	32.9	7.2	6.8	11.7	5.3	6.0	8.7	3.7	2.1	1.4	1.2	3.3
16	37.1	7.4	6.2	10.9	5.2	5.3	8.5	3.5	3.1	1.4	1.8	3.4
17	37.9	8.0	5.9	10.0	5.1	4.8	8.4	3.1	4.5	1.4	5.6	5.3
18	35.2	7.9	6.5	9.4	4.8	4.4	8.1	2.9	4.1	1.3	6.2	5.1
19	26.3	7.5	7.5	8.8	5.1	4.4	7.5	2.8	3.5	1.3	6.4	8.7
20	18.7	7.7	8.2	12.3	5.5	4.3	7.2	2.8	3.0	1.2	6.2	8.4
21	19.0	7.9	8.2	16.2	6.1	7.0	6.9	3.0	2.8	1.2	5.5	9.1
22	19.0	8.9	7.8	16.3	6.5	7.8	6.2	2.9	3.0	1.2	4.8	9.3
23	17.4	8.9	7.6	15.5	6.9	7.4	5.6	2.9	2.6	1.2	4.4	8.9
24	14.9	8.4	8.4	14.8	7.1	7.1	5.4	3.0	2.3	1.2	1.2	7.8
25	12.2	7.9	9.6	18.5	7.6	6.9	5.2	2.7	2.3	1.2	3.9	6.7
26	10.5	7.5	9.5	13.7	7.8	6.8	5.0	2.8	3.6	1.1	3.4	5.8
27	9.7	6.7	10.0	13.6	6.7	6.7	4.6	3.2	3.6	1.1	3.1	5.2
28	8.5	6.4	10.6	10.4	6.2	7.3	4.5	3.6	3.1	1.1	3.0	4.8
29	7.7	5.9	10.3	9.1	5.8	7.5	4.3	1.4	2.6	1.1	3.0	4.3
30	6.9	......	9.7	8.8	5.7	7.2	4.1	2.8	2.4	1.1	3.6	3.9
31	6.8		9.4		5.6		3.8	3.3		1.1		3.3
1893.												
1	3.4	10.4	8.1	5.5	10.2	7.4	3.9	2.6	8.4	2.6	3.1	2.5
2	3.8	12.1	9.1	5.3	9.6	8.0	3.8	2.5	6.8	2.5	3.0	3.9
3	4.7	10.6	8.9	5.1	11.0	10.0	4.2	3.1	5.0	2.4	2.8	3.7
4	5.7	8.6	8.9	5.1	18.4	8.1	5.2	3.5	6.2	2.4	2.7	4.1
5	5.6	8.0	8.7	5.1	24.5	6.6	4.8	3.3	6.0	2.6	2.6	4.3
6	5.3	7.7	9.0	5.2	28.2	8.3	3.9	3.2	4.9	2.5	2.5	4.7
7	5.2	7.0	9.1	5.4	30.0	16.0	3.8	1.9	4.3	2.9	2.4	4.7
8	4.7	6.5	8.8	5.4	28.2	20.7	3.5	5.0	3.5	3.1	2.1	4.5
9	4.0	6.1	8.8	5.1	18.0	19.1	3.4	4.1	3.2	2.9	2.3	4.1
10	3.8	6.2	9.4	5.1	12.8	15.2	3.6	3.8	2.8	2.9	2.4	4.0
11	3.4	8.5	11.1	5.1	11.7	11.8	3.4	3.3	2.8	2.6	2.5	3.7
12	2.9	14.7	11.7	5.0	10.4	8.9	3.4	3.0	3.7	2.5	4.8	3.2
13	2.9	21.8	11.5	4.8	9.4	7.3	3.3	2.7	5.5	2.4	3.8	3.1
14	(a)	23.6	12.0	10.2	8.8	6.8	3.2	2.5	10.9	2.5	3.6	3.0
15	(a)	22.6	10.6	12.1	8.1	6.5	3.0	2.8	12.7	2.0	3.5	2.9
16	(a)	21.3	9.5	10.4	7.8	6.2	2.8	2.9	9.6	1.7	3.0	3.0
17	(a)	23.6	8.4	8.6	9.4	5.6	2.7	3.8	8.0	9.6	2.8	3.2
18	(1)	29.4	7.6	7.4	10.4	5.4	2.8	5.2	7.0	6.1	2.7	3.5
19	(a)	32.4	7.0	6.5	8.9	5.3	3.0	4.9	6.1	5.7	2.5	3.8
20	(a)	33.4	6.7	6.4	7.7	5.2	3.2	4.0	5.1	5.2	2.6	3.9
21	(a)	32.0	6.3	7.2	7.4	5.4	3.5	2.9	4.2	1.1	2.6	3.9
22	(a)	29.5	6.0	7.2	6.7	5.6	3.6	2.6	3.6	3.1	2.6	3.5
23	2.9	18.2	5.8	7.1	6.1	5.7	4.6	2.4	3.4	3.1	2.5	3.3
24	3.1	12.3	5.7	6.8	5.7	5.4	5.2	2.3	3.2	2.8	2.5	3.1
25	3.1	10.4	6.4	6.7	5.4	5.4	5.5	2.2	3.0	3.1	2.4	2.9
26	3.4	9.3	6.8	6.0	5.2	5.6	3.7	1.9	2.9	3.3	2.5	2.8
27	3.7	8.4	6.8	5.7	5.0	5.2	1.8	2.7	4.9	2.6	2.7
28	3.8	8.2	6.3	7.0	4.6	5.1	2.9	1.6	2.6	1.6	2.8	2.6
29	4.4		5.9	9.5	5.4	4.7	3.7	1.6	2.4	4.0	2.7	2.5
30	5.3	......	5.8	10.4	6.5	4.1	2.6	1.7	2.5	3.5	2.5	2.7
31	7.1		5.7		7.4		2.6	1.6		3.2		3.1

a Frozen at gage.

Daily gage height, in feet, of Tennessee River at Chattanooga, Tenn.—Continued.

Day.	Jan.	Feb.	Mar.	Apr.	May.	June.	July.	Aug.	Sept.	Oct.	Nov.	Dec.
1894.												
1	2.9	5.1	7.7	5.0	3.9	3.8	4.4	2.9	4.0	0.9	1.4	1.0
2	3.4	5.0	8.2	4.8	3.8	3.6	4.0	2.9	3.8	.9	1.7	1.0
3	3.8	4.9	9.4	5.4	3.7	3.5	3.7	2.9	3.0	1.0	2.3	.9
4	3.9	5.5	9.7	5.3	3.7	3.4	4.4	2.8	2.6	1.5	1.6	.9
5	3.5	21.9	9.5	6.8	8.6	3.3	4.2	2.9	2.1	1.8	1.4	.9
6	3.1	25.5	9.3	6.9	3.5	3.2	3.7	3.0	2.0	1.8	1.5	.9
7	4.9	23.9	8.5	7.2	3.4	2.9	3.2	2.9	1.8	1.5	1.5	.9
8	6.1	19.7	8.2	7.4	3.3	2.8	3.3	2.9	1.7	1.3	1.3	.9
9	9.3	16.1	7.9	6.6	3.4	2.6	3.1	3.0	1.5	1.1	1.2	1.1
10	9.0	16.0	7.2	5.7	3.3	2.5	3.3	2.6	1.4	1.0	1.1	1.2
11	8.5	16.7	6.9	5.9	3.2	2.5	3.7	2.3	1.4	.8	1.0	1.6
12	7.9	15.4	6.6	7.2	4.7	2.4	3.3	2.1	1.4	.9	1.0	3.8
13	8.3	15.2	6.7	8.5	5.1	2.3	2.7	1.9	1.3	1.2	.9	8.6
14	8.0	14.1	7.2	7.8	4.8	2.3	2.4	1.8	1.2	1.9	.8	11.1
15	7.8	12.2	7.0	7.2	4.3	2.2	2.1	2.0	1.5	2.4	.8	11.2
16	7.8	10.3	6.9	6.9	4.0	2.1	1.9	3.6	1.8	2.1	.8	10.8
17	7.1	9.5	6.8	6.3	4.1	2.0	1.8	4.6	1.8	1.7	.8	8.6
18	7.2	8.6	7.3	5.5	5.2	2.1	2.8	3.5	2.0	1.4	.7	6.6
19	6.3	8.4	7.4	5.0	5.0	2.4	2.4	3.0	2.0	1.1	.8	4.7
20	6.0	8.3	7.7	5.1	5.4	2.5	2.4	3.1	1.6	1.0	.9	4.2
21	5.3	8.5	7.1	4.9	5.6	2.6	3.3	3.6	1.5	.9	.9	3.6
22	5.0	8.7	8.8	4.8	6.2	2.	3.7	3.7	1.5	.8	.9	3.2
23	5.0	8.8	8.7	4.7	6.8	2.3	3.8	4.0	1.8	.8	1.0	2.8
24	5.2	8.2	8.1	4.6	6.9	2.2	3.4	3.6	1.8	.8	1.0	2.7
25	5.8	7.9	7.7	4.5	7.1	2.2	4.0	3.0	1.6	.8	1.1	2.5
26	5.2	7.0	7.3	4.3	6.7	2.5	1.4	2.6	1.3	.8	1.2	2.4
27	5.4	7.7	7.0	4.2	6.0	2.6	3.9	2.2	1.1	.7	1.2	4.2
28	5.4	7.7	6.5	4.1	5.6	2.7	3.8	2.7	1.0	.7	1.1	6.9
29	5.1		6.6	4.0	5.1	2.9	3.6	2.4	.9	.7	1.1	8.4
30	5.0		5.7	4.0	4.7	4.3	3.3	2.7	.8	1.0	1.1	7.9
31	4.9		5.2		4.2		3.8	4.8		1.1		5.8
1895.												
1	4.7	7.6	6.8	7.8	6.0	5.6	3.2	4.4	3.3	.9	1.1	1.4
2	3.9	7.2	7.3	7.4	5.8	5.2	3.4	4.1	3.4	.8	1.2	1.5
3	3.8	7.3	12.1	6.8	5.5	4.8	3.8	3.8	3.2	.8	1.3	1.4
4	3.2	7.5	18.2	6.5	5.4	4.5	4.0	3.5	3.1	.8	1.6	1.4
5	3.1	7.6	19.9	6.3	5.7	4.2	4.5	3.4	2.9	.8	1.6	1.4
6	3.1	7.4	18.2	6.2	6.0	4.4	5.0	3.0	2.8	.8	1.3	1.6
7	3.3	6.9	13.4	6.0	6.5	4.6	5.0	3.2	2.8	.8	1.3	1.5
8	4.0	6.5	10.5	9.6	7.0	5.2	5.1	3.8	2.8	.9	1.2	1.4
9	10.9	6.4	9.2	10.7	8.2	5.1	5.5	3.3	2.5	.8	1.1	1.3
10	20.5	5.0	8.6	11.4	8.6	4.6	5.7	3.0	2.4	.9	1.2	1.3
11	28.5	4.0	8.1	13.0	9.0	4.2	5.1	3.0	2.3	1.0	1.3	1.4
12	32.1	3.3	7.5	12.5	8.8	3.8	4.4	2.9	2.4	1.0	1.7	1.5
13	31.2	4.2	7.8	10.4	8.9	3.6	3.8	3.2	2.5	1.0	2.1	1.8
14	28.3	(a)	8.0	8.8	9.5	3.5	3.4	2.9	2.4	1.0	2.4	1.9
15	19.5	4.3	8.7	7.9	9.0	3.4	3.2	2.8	2.3	.9	2.2	2.0
16	12.3	4.7	9.4	7.4	8.2	3.4	3.2	2.7	2.2	1.0	2.0	1.9
17	10.9	4.7	9.2	7.0	7.7	3.8	3.7	3.1	2.1	1.0	1.9	1.8
18	10.0	4.2	9.6	9.0	7.1	3.7	3.6	4.3	2.3	1.0	1.8	1.7
19	9.7	4.7	9.4	11.8	7.0	3.5	3.3	4.9	2.2	.9	1.7	1.6
20	9.1	4.6	8.9	11.8	7.2	3.2	3.0	5.7	2.0	.9	1.5	1.5
21	9.6	5.1	14.3	9.9	7.1	3.1	2.7	5.3	2.1	.8	1.3	1.6
22	10.2	5.6	20.6	8.6	6.7	3.2	2.7	6.1	1.9	.8	1.1	2.1
23	9.9	6.1	22.7	7.7	6.5	3.1	2.5	5.3	1.6	.7	1.3	3.3
24	9.1	6.7	22.0	7.1	5.8	3.0	2.4	4.8	1.4	.7	1.4	4.2
25	10.8	6.8	18.2	6.7	5.6	2.9	2.4	4.6	1.3	.7	1.3	4.6
26	10.8	6.5	18.0	6.3	5.6	2.9	2.9	4.7	1.3	.7	1.3	4.3
27	10.0	6.3	11.3	6.0	7.0	2.6	3.8	4.2	1.2	.7	1.3	4.2
28	9.3	6.3	10.6	6.0	7.5	2.5	10.2	3.7	1.1	.7	1.3	4.5
29	8.8		9.5	5.9	7.4	2.5	10.4	3.5	1.0	.7	1.3	5.2
30	8.6		8.9	5.9	6.7	2.6	7.4	3.7	0.9	.7	1.5	4.7
31	8.4		8.4		6.0		5.3	3.4		1.0		4.7

a Frozen

Daily gage height, in feet, of Tennessee River at Chattanooga, Tenn.—Continued.

Day.	Jan.	Feb.	Mar.	Apr.	May.	June.	July.	Aug.	Sept.	Oct.	Nov.	Dec.
1896.												
1	4.9	4.4	4.1	14.8	3.4	2.6	3.3	5.5	2.4	2.5	1.2	2.4
2	5.0	6.2	3.9	27.7	3.4	3.0	3.2	5.2	2.1	2.3	1.2	2.5
3	4.9	10.0	3.8	34.4	3.4	4.4	3.1	4.8	1.9	2.6	1.3	2.5
4	4.9	11.6	3.7	38.8	3.5	5.7	3.2	4.6	1.8	3.0	1.5	2.6
5	4.7	10.5	3.6	40.5	4.0	5.2	3.3	4.5	1.6	2.7	1.5	2.6
6	4.3	9.3	3.5	36.9	4.6	4.7	3.2	4.9	1.6	2.6	1.6	2.7
7	3.6	11.8	3.4	23.3	4.6	4.1	3.6	5.0	1.5	2.1	2.3	2.9
8	3.3	14.0	3.5	11.6	4.3	3.5	5.0	4.2	2.0	1.7	3.5	3.0
9	3.2	13.8	3.5	9.0	4.0	3.5	3.9	3.8	2.8	1.6	4.2	2.8
10	3.2	13.2	3.4	8.0	3.7	4.5	14.2	3.4	2.7	1.4	4.1	2.8
11	3.1	12.8	3.6	7.2	3.4	7.0	21.1	3.3	2.4	1.2	3.3	2.7
12	3.1	11.4	3.6	6.7	3.1	6.3	21.6	3.4	2.0	1.2	3.2	2.6
13	2.9	10.1	3.8	6.2	2.9	5.1	15.6	3.2	1.8	1.2	5.8	2.4
14	2.7	11.1	3.8	5.8	2.8	4.3	11.5	3.2	1.6	1.2	7.3	2.6
15	2.6	12.8	8.7	5.5	2.7	3.6	11.2	3.1	1.6	1.5	6.5	4.1
16	2.4	13.6	3.8	5.2	2.6	3.2	11.4	3.0	1.5	1.7	5.5	6.5
17	2.3	12.5	5.5	5.0	2.5	3.0	11.0	3.0	1.3	1.6	4.9	6.6
18	2.3	11.0	10.1	4.8	2.4	2.8	13.9	2.9	1.4	1.6	4.3	6.3
19	2.3	9.0	13.1	4.2	2.4	2.9	12.5	2.7	1.4	1.7	3.8	6.4
20	2.3	7.6	15.7	4.4	2.2	3.1	9.6	2.6	1.3	1.6	3.4	6.8
21	2.3	6.7	13.8	4.2	2.1	3.7	7.6	2.4	1.2	1.6	3.0	7.0
22	2.5	6.0	11.2	4.1	2.1	3.5	6.5	2.4	1.2	1.4	2.8	7.2
23	3.1	5.4	9.5	4.1	2.5	3.5	8.5	2.2	1.3	1.2	2.5	7.3
24	5.0	4.9	8.4	4.0	3.2	3.3	8.8	2.2	1.4	1.2	2.4	7.0
25	6.5	4.7	7.9	4.0	3.6	3.1	8.6	2.8	1.6	1.2	2.3	6.6
26	8.2	4.6	7.5	3.8	3.8	2.9	7.8	2.6	2.0	1.2	2.2	5.9
27	8.0	4.5	7.2	3.8	3.2	2.6	11.1	2.7	1.7	1.2	2.1	5.3
28	7.0	4.4	6.7	3.8	3.1	2.6	12.2	3.2	1.5	1.2	2.2	4.8
29	6.0	4.2	6.2	3.6	2.8	2.8	9.3	4.0	1.5	1.1	5.3	4.4
30	5.3	....	5.8	3.6	2.7	3.0	7.2	3.6	2.7	1.1	9.4	3.7
31	4.8	....	7.7	......	2.5	......	6.2	2.8	......	1.3	....	3.0
1897.												
1	2.4	3.0	12.5	8.7	5.9	4.3	5.0	4.4	2.1	.8	.8	1.0
2	2.5	7.0	9.6	12.2	6.3	4.2	4.6	3.9	2.2	.9	.9	1.2
3	2.8	10.1	8.6	15.0	7.4	4.1	3.8	3.8	1.9	.7	1.0	1.3
4	2.6	10.5	9.0	16.0	9.6	4.1	3.4	3.6	1.8	.6	1.2	2.0
5	2.6	9.4	9.5	26.0	9.6	4.1	3.4	3.8	1.8	.5	1.2	3.3
6	2.7	8.3	12.1	30.4	8.5	4.1	4.0	3.5	1.7	.5	1.3	3.8
7	2.9	8.8	19.2	29.7	7.7	4.4	3.8	4.4	1.7	.5	1.4	3.9
8	3.0	10.7	25.1	25.4	7.2	4.4	3.8	4.2	1.6	.4	1.2	3.5
9	2.8	14.1	24.2	20.0	6.6	4.0	4.4	4.2	1.6	.4	1.2	2.9
10	2.8	15.5	21.3	16.0	6.2	5.2	4.0	5.6	1.4	.4	1.2	2.6
11	2.7	13.2	22.3	14.0	6.0	5.0	4.1	5.2	1.3	.5	1.1	2.4
12	2.6	10.8	28.4	26.0	6.2	5.7	4.5	4.6	1.2	.6	1.0	2.1
13	2.4	9.9	34.9	11.4	7.8	5.0	4.6	4.1	1.2	.9	1.0	1.8
14	2.6	10.0	37.9	10.3	18.4	4.3	4.2	3.5	1.1	1.4	.9	1.8
15	4.1	10.5	37.9	9.7	22.4	3.9	3.8	3.1	1.0	1.1	.8	2.5
16	6.5	10.7	37.0	9.8	20.3	3.6	3.6	2.8	1.0	1.2	.8	2.7
17	6.6	9.8	36.0	10.2	16.5	3.7	4.5	2.8	.9	1.2	.8	2.5
18	6.3	8.6	33.8	9.9	11.9	3.6	6.3	3.0	.9	1.2	.8	2.5
19	6.4	7.6	29.6	9.3	9.1	3.3	6.1	3.4	.8	1.1	.8	2.6
20	6.8	7.0	29.6	8.8	7.7	3.3	5.6	3.0	.8	1.4	.7	3.4
21	7.0	7.0	32.4	8.1	6.9	4.1	6.7	3.0	.9	2.0	.7	4.5
22	7.2	8.3	33.3	7.5	6.4	5.0	6.1	3.4	.9	1.0	.7	7.1
23	7.3	13.2	30.9	7.0	5.9	4.8	5.8	3.1	.8	1.6	.7	10.2
24	7.0	25.2	25.0	6.7	5.6	5.3	6.0	3.8	.9	1.4	.7	9.3
25	6.6	31.6	18.1	6.4	5.3	5.5	5.5	3.4	.8	1.6	.7	7.7
26	5.9	34.8	14.2	6.2	5.1	6.2	9.7	2.9	.8	1.4	.7	6.4
27	5.3	33.8	12.2	6.0	4.8	5.4	13.3	2.8	.7	1.2	.7	5.6
28	4.8	23.6	10.8	6.1	4.6	5.5	8.7	2.8	.7	1.0	.7	5.0
29	4.4	......	9.8	6.2	4.4	6.2	6.7	2.5	.7	.9	.7	4.5
30	3.7	......	9.1	5.8	4.2	5.2	5.6	2.2	.8	.8	.9	4.0
31	3.0	......	8.6	......	4.2	......	5.0	2.1	......	.8	......	3.3

Daily gage height, in feet, of Tennessee River at Chattanooga, Tenn.—Continued.

Day.	Jan.	Feb.	Mar.	Apr.	May.	June.	July.	Aug.	Sept.	Oct.	Nov.	Dec.
1898.												
1	3.45	7.55	3.30	17.45	6.40	3.35	2.45	8.15	3.55	3.65	4.60	5.10
2	3.25	6.70	3.15	17.80	5.80	3.30	2.35	7.55	3.95	3.30	4.30	5.05
3	3.05	6.15	3.00	15.00	5.40	3.30	2.25	6.45	9.15	3.20	4.25	4.90
4	2.90	5.40	2.95	11.45	5.05	3.30	2.05	5.35	18.50	3.90	4.05	5.00
5	2.75	5.00	3.30	10.85	4.70	2.85	2.35	6.25	25.00	8.90	3.90	5.05
6	2.65	4.55	3.45	12.15	4.45	2.55	2.10	11.85	22.15	16.90	3.90	5.00
7	2.70	4.45	3.50	11.60	4.35	2.35	2.10	14.65	15.70	16.50	4.25	5.95
8	2.80	4.85	3.40	10.30	4.10	2.20	2.15	12.55	11.25	10.75	4.45	5.90
9	3.05	4.30	3.25	9.30	4.20	2.05	2.60	10.15	9.50	8.80	4.50	5.85
10	3.25	4.15	3.05	8.30	4.15	1.96	3.05	8.50	8.60	8.40	4.45	5.55
11	3.25	4.00	2.90	8.60	4.45	1.80	3.50	9.05	7.45	7.55	4.65	5.10
12	5.60	3.90	2.85	9.50	4.65	1.95	3.40	12.30	6.45	6.56	5.05	4.75
13	13.20	3.80	2.80	9.40	4.40	1.80	3.20	14.85	5.70	6.00	5.30	4.55
14	14.40	3.80	2.85	9.00	4.15	1.75	2.80	15.85	5.20	5.70	4.95	4.30
15	12.25	3.80	3.15	9.05	3.95	1.65	2.85	14.95	4.80	5.35	4.55	4.00
16	12.20	3.70	5.10	9.15	3.90	1.75	3.50	11.60	4.45	4.90	4.40	3.85
17	12.35	3.55	5.05	8.60	3.80	2.00	4.95	8.90	4.25	4.70	4.55	2.70
18	10.00	3.50	5.20	8.20	3.70	2.35	5.35	7.10	3.95	5.25	4.75	3.00
19	9.20	3.30	5.50	8.00	3.70	3.50	4.60	6.40	3.75	6.70	4.95	3.75
20	11.70	3.30	6.10	7.95	3.65	4.05	4.15	6.05	3.55	7.75	5.30	5.20
21	13.80	3.20	5.70	7.40	3.60	5.35	4.00	5.95	3.45	9.30	5.85	5.85
22	13.40	3.25	5.45	7.05	3.60	5.55	3.30	5.65	3.55	8.80	6.05	6.00
23	12.55	3.40	5.15	6.60	3.50	5.05	3.30	5.40	5.00	7.65	6.55	5.85
24	12.35	3.50	4.65	6.85	3.40	4.55	3.40	4.75	5.05	7.25	6.85	5.35
25	12.85	3.60	4.30	6.65	3.45	3.70	3.55	4.30	6.40	7.20	6.55	5.20
26	16.05	3.80	4.15	6.60	3.70	3.40	3.55	4.05	7.20	7.60	6.10	5.40
27	18.20	3.50	4.45	6.45	4.95	2.90	4.55	4.00	6.15	6.90	5.65	5.90
28	16.70	3.35	4.45	7.10	5.60	2.80	5.56	4.10	5.00	6.20	5.15	5.70
29	14.15	......	4.65	7.05	4.90	2.80	5.35	4.30	4.30	5.65	4.80	5.15
30	11.20	......	5.55	6.75	4.20	2.55	5.90	4.10	3.85	5.15	4.95	4.80
31	8.95	......	13.25	......	3.70	......	7.90	3.85	......	4.85	......	4.50
1899.												
1	4.75	5.70	19.25	22.80	7.60	4.15	3.45	4.20	2.20	1.20	1.10	1.70
2	4.95	5.65	17.60	19.50	7.10	4.25	3.30	3.55	2.35	1.10	1.10	1.70
3	5.30	5.60	15.15	14.90	6.70	4.40	3.05	3.05	2.80	1.05	1.05	1.70
4	5.80	10.70	14.15	12.95	6.30	4.85	2.80	2.75	3.05	.95	1.10	1.70
5	5.95	23.10	17.95	13.25	6.15	4.65	2.60	2.45	2.65	.90	1.50	1.80
6	7.25	30.45	24.50	14.70	7.10	4.25	2.60	2.45	2.25	.80	1.50	1.70
7	18.80	34.30	26.55	15.70	8.50	4.05	2.65	2.40	1.95	.85	1.50	1.60
8	18.40	36.95	27.60	18.05	9.35	3.75	8.05	2.40	1.80	1.00	1.45	1.50
9	17.85	38.25	27.70	17.75	10.00	3.55	2.90	2.25	1.60	1.15	1.35	1.40
10	17.15	36.75	16.15	15.70	10.70	3.40	2.60	2.10	1.80	1.60	1.20	1.40
11	13.85	30.30	11.85	14.20	11.15	3.90	2.55	2.10	1.70	1.80	1.15	1.60
12	10.50	19.35	10.60	12.90	10.40	4.30	2.45	2.00	2.00	1.85	1.10	5.20
13	9.15	12.15	9.55	11.65	9.60	5.25	2.30	2.00	1.80	1.70	1.00	6.45
14	8.10	9.50	11.20	10.70	9.30	5.80	2.20	2.00	1.65	1.00	1.00	7.10
15	7.55	8.50	21.55	10.00	9.55	6.45	2.15	2.65	1.85	1.40	1.00	5.15
16	7.30	7.55	34.25	9.40	9.20	6.10	1.95	2.65	1.65	1.25	1.00	6.20
17	7.40	7.95	36.90	8.75	8.70	6.40	1.90	2.40	1.45	1.15	1.00	5.20
18	7.45	9.55	36.15	8.40	7.75	6.20	1.80	1.35	1.35	1.15	1.00	4.25
19	7.25	11.30	35.85	8.00	6.90	5.25	1.90	2.15	1.20	1.10	1.00	3.85
20	7.00	12.65	37.05	7.55	6.40	4.70	2.05	1.90	1.05	1.10	.95	4.25
21	6.80	11.50	39.20	7.35	5.90	4.20	2.05	1.70	1.00	1.10	.85	4.10
22	6.45	10.65	40.00	7.05	5.60	3.75	2.10	1.60	1.05	1.10	.85	4.10
23	5.90	10.10	38.70	7.85	5.35	3.50	2.70	1.45	1.30	1.05	1.00	4.15
24	5.65	9.75	32.70	9.65	5.30	3.25	3.50	1.30	1.50	1.00	1.15	5.65
25	6.05	9.50	23.15	9.35	5.05	3.15	3.40	1.20	1.50	.95	1.30	6.15
26	6.35	9.20	16.80	10.75	4.80	3.06	3.00	1.20	1.45	.85	1.70	6.30
27	5.85	13.20	13.65	10.30	4.65	3.25	3.05	1.20	1.30	.80	1.80	5.85
28	5.75	18.45	13.95	9.20	4.40	3.65	4.15	1.25	1.20	.80	1.85	5.55
29	5.55	......	17.30	8.35	4.30	3.50	4.25	1.50	1.25	.90	1.80	5.10
30	5.30	......	21.20	7.75	4.20	3.30	4.25	1.85	1.30	1.00	1.75	4.65
31	5.30	......	22.80	......	4.25	......	5.15	1.75	......	1.05	......	3.85

Daily gage height, in feet, of Tennessee River at Chattanooga, Tenn.—Continued.

Day.	Jan.	Feb.	Mar.	Apr.	May.	June.	July.	Aug.	Sept.	Oct.	Nov.	Dec.
1900.												
1	3.05	3.25	8.05	7.85	6.20	2.85	8.85	6.20	2.10	2.00	2.90	8.70
2	2.95	(a)	8.70	7.20	5.65	2.80	8.15	5.40	2.00	1.80	2.60	6.50
3	(a)	2.60	10.90	6.85	5.35	3.00	6.95	4.70	2.30	1.70	2.50	5.60
4	(a)	2.50	12.50	7.25	5.15	3.20	6.30	4.20	2.50	1.50	2.90	5.10
5	(a)	2.90	12.75	8.05	4.95	3.20	5.80	3.60	2.30	1.40	3.50	5.60
6	2.10	3.50	10.65	8.55	4.80	3.50	6.40	3.20	2.00	1.30	3.70	6.90
7	2.20	3.95	10.00	7.85	4.65	5.65	5.00	2.90	1.70	1.30	4.20	8.30
8	2.30	3.90	11.65	7.05	4.45	6.65	4.50	2.60	1.60	1.60	1.20	9.20
9	2.85	5.35	14.55	6.60	4.45	6.15	4.20	2.10	1.40	1.80	3.70	8.50
10	2.45	8.40	16.50	6.10	4.35	5.30	4.20	2.30	1.30	2.10	3.20	7.00
11	3.35	9.40	16.15	6.50	4.30	5.00	4.30	2.10	1.20	2.10	3.00	6.10
12	6.05	8.95	14.25	7.50	4.30	4.90	3.80	2.00	1.10	2.50	2.70	5.40
13	8.15	13.90	11.65	7.40	4.15	4.50	3.40	1.90	1.00	3.00	2.50	4.90
14	8.70	21.55	9.85	7.00	4.00	5.20	3.30	1.90	1.10	2.50	2.30	1.50
15	8.45	24.00	8.65	6.50	3.85	5.30	3.30	2.10	1.80	1.90	2.20	4.70
16	7.80	21.40	8.00	6.30	3.75	5.25	3.30	2.20	3.10	1.80	2.10	1.20
17	6.35	17.00	7.80	8.75	3.60	5.45	3.20	2.30	4.00	1.60	2.00	4.00
18	5.50	12.05	7.55	10.65	3.50	6.15	3.10	2.30	4.10	1.50	2.00	3.60
19	5.80	9.25	7.55	9.75	3.40	8.85	3.00	2.30	4.60	1.40	1.90	3.40
20	5.50	7.70	8.55	9.40	3.40	9.20	2.20	2.20	4.70	1.30	1.90	3.30
21	9.40	7.10	11.80	11.70	3.25	8.90	2.70	1.90	3.90	1.20	2.10	3.30
22	8.85	7.70	14.95	12.00	3.15	7.65	2.50	1.80	3.00	1.20	2.20	4.00
23	7.95	8.50	17.40	11.35	3.05	6.40	2.50	1.70	2.60	1.40	2.30	4.20
24	7.20	8.55	16.45	10.70	3.00	6.25	2.80	1.80	2.40	2.20	2.80	4.70
25	6.15	8.55	12.65	9.75	3.15	7.15	3.00	1.90	2.70	4.10	3.20	5.20
26	5.50	9.50	11.15	8.50	3.20	7.60	3.10	2.50	2.70	7.00	7.80	5.40
27	5.00	9.30	10.90	7.80	3.35	8.05	3.30	3.10	2.60	7.50	13.90	5.20
28	4.65	8.45	10.70	7.45	3.60	8.20	4.60	2.70	2.40	6.00	15.60	4.10
29	4.20	......	10.20	7.05	3.60	8.60	8.00	2.50	2.30	4.90	15.60	4.50
30	3.90	......	9.35	6.60	3.35	8.70	8.20	2.30	2.20	3.70	13.20	4.20
31	3.55	......	8.50	......	3.06	......	7.30	2.20	......	3.40	......	4.50
1901.												
1	5.2	6.5	8.7	12.4	10.8	12.0	6.0	2.8	9.9	4.2	2.6	2.3
2	5.7	6.7	8.7	13.2	9.3	11.1	5.9	2.8	9.8	4.5	2.6	2.2
3	5.8	7.2	8.7	19.7	8.5	9.8	6.3	2.9	9.7	4.6	2.5	2.3
4	5.6	8.7	3.7	24.1	7.6	8.5	6.4	2.8	10.3	4.5	2.5	2.5
5	5.1	10.1	3.8	23.9	7.0	7.7	6.0	2.6	9.4	4.1	2.5	2.5
6	4.7	10.0	4.0	22.4	6.7	6.9	5.2	2.6	7.9	4.8	2.5	2.5
7	4.4	9.4	4.1	18.9	6.4	6.9	5.1	3.2	6.9	4.5	2.4	3.0
8	4.1	8.9	4.1	14.2	6.2	6.9	5.4	9.1	6.4	4.1	2.4	3.2
9	3.9	8.5	4.0	11.8	5.9	6.5	5.6	12.2	5.9	3.1	2.4	3.2
10	3.8	7.7	7.0	10.3	5.6	6.9	6.3	9.9	5.5	3.7	2.4	3.5
11	6.1	7.6	9.8	9.2	5.6	8.2	6.6	7.3	5.3	3.1	2.4	3.5
12	15.4	7.0	11.2	8.4	5.4	7.4	5.6	5.8	5.1	3.4	2.4	4.0
13	26.6	7.1	9.7	7.9	5.6	6.4	5.0	5.3	5.7	3.5	2.5	4.1
14	28.1	7.2	8.2	6.8	5.5	6.1	4.4	6.5	5.9	4.0	2.5	4.7
15	25.3	7.3	7.3	9.8	5.5	6.4	4.1	14.0	6.0	4.3	2.5	17.9
16	19.5	6.4	6.4	10.3	5.4	7.5	3.6	27.3	6.1	4.1	2.5	26.8
17	12.7	5.8	5.8	10.2	5.2	8.9	3.7	32.8	6.3	4.1	2.4	28.8
18	9.7	5.3	5.4	9.6	4.9	9.8	3.9	32.6	8.8	4.0	2.4	26.7
19	8.1	5.1	5.0	10.8	5.3	9.3	3.7	28.6	9.9	3.7	2.3	19.9
20	7.2	5.0	4.7	21.1	8.0	8.9	3.7	23.4	9.3	3.5	2.3	11.4
21	6.4	4.9	4.7	26.5	10.2	8.4	4.2	18.6	8.3	8.1	2.2	8.3
22	5.9	4.7	4.8	24.7	20.2	7.7	3.9	17.0	7.4	3.1	2.1	6.6
23	5.4	4.5	5.2	23.0	26.5	10.1	3.7	16.5	6.4	3.1	2.2	5.7
24	5.6	4.4	5.0	22.2	29.7	9.5	3.5	18.5	5.6	3.0	2.5	5.8
25	5.8	4.2	5.0	19.0	32.4	7.6	8.1	16.5	5.2	3.0	2.5	6.9
26	5.8	4.1	5.7	18.5	32.5	9.6	3.0	13.1	4.9	2.9	2.5	7.9
27	5.4	3.8	15.9	14.9	23.5	9.8	2.9	11.0	4.6	2.8	2.6	10.2
28	5.2	3.7	22.3	14.9	13.5	8.4	2.9	10.3	4.4	2.7	2.5	16.0
29	5.2	......	21.7	14.5	12.1	7.2	2.8	10.7	4.4	2.6	2.5	24.0
30	5.2	......	18.4	13.8	11.9	6.4	2.8	10.0	4.3	2.5	2.4	32.0
31	5.5	......	14.7	......	12.3	......	2.8	9.8	......	2.5	......	37.4

a Frozen at gage.

Daily gage height, in feet, of Tennessee River at Chattanooga, Tenn.—Continued.

Day.	Jan.	Feb.	Mar.	Apr.	May.	June.	July.	Aug.	Sept.	Oct.	Nov.	Dec.
1902.												
1	40.1	20.1	24.0	30.9	5.6	4.0	9.8	2.2	1.4	3.9	1.3	5.2
2	40.8	21.8	31.9	27.0	8.5	3.8	10.2	2.0	1.2	3.7	1.3	5.1
3	37.6	23.2	35.8	18.0	9.3	3.6	8.8	2.1	1.2	3.9	1.4	6.7
4	26.8	21.7	38.0	12.3	8.0	3.5	7.5	2.4	1.5	3.8	1.5	7.4
5	15.0	18.0	35.9	10.7	6.8	3.4	6.5	2.4	1.3	3.0	1.5	7.8
6	10.9	14.5	30.3	10.0	6.0	3.4	5.5	2.8	1.4	2.8	1.5	7.5
7	9.7	11.7	25.5	9.5	5.6	3.2	4.5	2.5	1.5	2.4	1.4	7.5
8	8.9	10.0	20.7	9.8	5.6	3.2	4.0	2.1	1.5	2.1	1.8	6.6
9	8.0	8.8	17.9	9.9	5.5	3.3	3.7	1.9	1.5	2.0	2.2	6.0
10	7.7	8.1	15.6	9.5	5.2	3.4	3.5	2.2	1.5	1.9	2.1	5.3
11	7.3	7.5	14.2	8.9	5.0	3.2	3.4	2.8	1.4	2.2	1.9	4.7
12	6.9	6.9	12.9	8.4	4.8	3.4	3.4	2.7	1.5	2.9	1.8	4.2
13	6.5	6.4	12.1	8.0	4.6	3.5	3.6	2.4	1.9	2.7	1.8	3.9
14	6.2	6.0	11.2	7.5	4.5	3.3	4.8	2.0	2.1	2.7	1.6	3.6
15	5.8	6.0	10.5	7.3	4.5	3.1	5.0	1.7	2.0	3.3	1.5	3.4
16	5.5	6.1	10.0	7.1	4.6	3.0	4.2	1.6	1.9	3.4	1.4	3.5
17	5.3	6.0	12.2	6.9	4.7	3.0	3.8	1.5	1.8	3.5	1.4	4.8
18	5.1	5.8	14.5	6.8	4.5	3.0	3.3	1.5	1.5	3.0	1.4	7.4
19	5.1	5.6	14.9	6.7	4.5	4.0	3.1	1.5	1.4	2.8	1.7	7.8
20	5.2	5.5	14.1	6.6	4.4	4.5	2.9	1.4	1.4	2.6	1.9	7.2
21	5.4	5.2	12.3	6.5	4.4	4.6	2.6	1.5	1.4	2.2	2.1	7.0
22	6.2	5.7	10.6	6.3	4.5	4.4	2.7	1.6	2.0	2.0	1.9	8.7
23	6.5	6.8	9.5	6.1	4.8	4.0	2.9	1.6	2.6	1.8	1.9	9.6
24	6.7	7.9	8.9	6.0	4.6	3.9	2.6	1.5	2.6	1.8	2.0	8.7
25	6.6	8.1	8.8	5.9	4.6	3.7	2.3	1.4	2.8	1.6	2.1	7.2
26	6.1	8.5	7.8	5.7	4.5	3.5	2.0	1.4	3.3	1.5	3.5	6.5
27	6.3	8.6	7.5	5.5	4.2	4.0	2.0	1.5	3.6	1.4	6.1	5.7
28	8.5	13.3	7.2	5.4	4.0	3.9	1.9	1.5	3.6	1.4	6.3	5.1
29	12.6	......	12.9	5.2	4.4	4.0	1.9	1.6	3.6	1.4	5.8	4.5
30	15.8	......	26.5	5.2	4.5	5.0	2.0	1.9	3.5	1.2	5.4	4.3
31	18.9	......	31.0	......	4.3	......	2.3	1.7	......	1.2	......	4.6
1903.												
1	4.7	4.8	26.5	17.4	9.0	6.5	4.1	2.8	1.7	0.6	0.8	1.1
2	4.5	4.9	31.0	16.6	8.3	8.2	3.8	2.6	1.6	.6	.9	1.1
3	5.3	5.1	29.2	14.8	7.8	11.6	3.7	2.9	1.4	.6	1.1	1.1
4	6.1	7.6	23.6	13.0	7.3	11.5	3.8	3.8	1.4	.6	1.2	1.0
5	6.9	15.4	16.5	12.3	7.0	10.1	3.6	3.8	1.3	.6	1.8	1.0
6	6.8	19.6	13.1	12.4	6.7	9.3	3.8	4.2	1.3	.6	1.8	.9
7	8.2	17.5	12.3	11.9	6.6	9.8	3.9	4.9	1.3	.6	1.5	1.0
8	7.6	18.0	14.6	15.5	6.3	11.2	3.8	4.2	1.3	.8	1.4	.9
9	6.7	17.2	20.7	24.5	6.0	10.5	4.0	3.6	1.3	1.0	1.3	1.0
10	6.0	15.3	24.4	30.3	5.8	9.0	4.0	3.0	1.2	1.5	1.3	.9
11	5.5	14.4	28.9	31.8	5.6	8.7	3.9	2.6	1.2	1.4	1.3	.9
12	6.2	16.0	21.0	28.0	5.4	7.8	3.8	2.7	1.1	1.3	1.3	.9
13	7.1	16.3	18.3	19.7	5.2	8.0	4.1	3.3	1.1	1.2	1.3	1.0
14	7.2	14.8	16.2	17.5	5.0	7.1	5.7	2.9	1.1	1.2	1.3	.9
15	6.5	13.6	14.9	20.4	4.9	6.3	5.5	3.2	1.2	1.0	1.2	.9
16	6.0	12.3	13.6	21.9	4.8	5.6	5.4	3.8	1.1	.9	1.2	1.0
17	5.8	18.4	11.9	21.2	4.7	5.1	5.6	4.1	1.0	1.0	1.5	1.2
18	5.4	25.9	10.7	18.8	4.6	4.7	5.1	3.8	1.2	1.0	3.0	1.1
19	5.1	29.3	9.8	16.1	4.5	4.4	4.7	3.4	1.3	1.0	6.1	1.0
20	4.8	29.0	9.0	14.2	4.3	4.3	4.1	3.8	1.3	.9	5.8	1.4
21	4.5	24.4	9.0	13.2	4.2	4.3	3.8	3.3	1.1	.9	4.6	3.0
22	4.4	15.4	9.0	13.0	4.1	4.3	4.0	3.0	1.2	.9	3.8	4.7
23	4.2	11.3	10.0	13.0	4.0	4.3	3.6	2.9	1.2	.9	3.1	4.4
24	4.0	9.9	16.7	11.8	4.0	4.3	3.2	2.5	1.1	.8	2.5	3.7
25	4.0	8.8	25.8	10.8	3.8	4.2	2.9	2.2	1.0	.8	2.1	3.7
26	4.2	8.0	28.8	10.1	3.8	4.1	2.7	2.0	1.0	.7	1.8	3.4
27	4.3	7.5	27.3	9.6	3.6	4.1	2.6	1.8	.8	.7	1.6	3.1
28	4.4	12.7	20.1	9.2	3.5	4.5	2.4	1.7	.7	.6	1.5	3.7
29	4.6	......	13.8	9.2	3.5	4.7	2.3	1.6	.7	.6	1.3	3.8
30	4.7	......	13.1	6.6	3.8	4.5	2.3	1.6	.6	.6	1.2	3.8
31	4.8	......	16.0	......	4.7	......	2.2	1.8	......	.7	......	3.7

Rating table for Tennessee River at Chattanooga, Tenn., from 1890 to 1895.

Gage height.	Discharge.	Gage height.	Discharge.	Gage height.	Discharge.	Gage height.	Discharge.
Feet.	*Second-feet.*	*Feet.*	*Second-feet.*	*Feet.*	*Second-feet.*	*Feet.*	*Second-feet.*
0.7	16,360	11.0	66,850	22.5	133,665	34.0	293,820
.8	16,560	11.5	69,755	23.0	136,570	34.5	302,720
.9	16,780	12.0	72,660	23.5	139,475	35.0	311,620
1.0	17,000	12.5	75,565	24.0	142,380	35.5	320,520
1.5	18,160	13.0	78,470	24.5	145,285	36.0	329,420
2.0	19,500	13.5	81,375	25.0	148,190	36.5	338,320
2.5	21,100	14.0	84,280	25.5	151,095	37.0	347,220
3.0	23,000	14.5	87,185	26.0	154,000	37.5	356,120
3.5	25,090	15.0	90,090	26.5	162,500	38.0	365,020
4.0	27,300	15.5	92,995	27.0	171,000	38.5	373,920
4.5	29,660	16.0	95,900	27.5	179,900	39.0	382,820
5.0	32,200	16.5	98,805	28.0	188,800	39.5	391,720
5.5	34,895	17.0	101,710	28.5	197,700	40.0	400,620
6.0	37,800	17.5	104,615	29.0	206,600	40.5	409,520
6.5	40,705	18.0	107,520	29.5	215,500	41.0	418,420
7.0	43,610	18.5	110,425	30.0	224,400	41.5	427,320
7.5	46,515	19.0	113,330	30.5	233,300	42.0	436,220
8.0	49,420	19.5	116,235	31.0	242,200	42.5	445,120
8.5	52,325	20.0	119,140	31.5	251,100	43.0	454,020
9.0	55,230	20.5	122,045	32.0	260,000	43.5	462,920
9.5	58,135	21.0	124,950	32.5	268,900	44.0	471,820
10.0	61,040	21.5	127,855	33.0	276,020	44.5	480,720
10.5	63,945	22.0	130,760	33.5	284,920		

Rating table for Tennessee River at Chattanooga, Tenn., for 1896 and 1897.

Gage height.	Discharge.	Gage height.	Discharge.	Gage height.	Discharge.	Gage height.	Discharge.
Feet.	*Second-feet.*	*Feet.*	*Second-feet.*	*Feet.*	*Second-feet.*	*Feet.*	*Second-feet.*
0.2	3,080	1.4	10,208	3.4	22,088	12.0	73,172
.3	3,674	1.6	11,396	3.6	23,276	13.0	79,112
.4	4,268	1.8	12,584	3.8	24,464	14.0	85,052
.5	4,862	2.0	13,772	4.0	25,652	15.0	90,992
.6	5,456	2.2	14,960	4.4	28,028	16.0	96,932
.7	6,050	2.4	16,148	4.8	30,404	18.0	108,812
.8	6,644	2.6	17,336	6.0	37,532	20.0	120,690
.9	7,238	2.8	18,524	8.0	49,412	22.0	132,570
1.0	7,832	3.0	19,712	10.0	61,292	24.0	144,450
1.2	9,020	3.2	20,900	11.0	67,232	a 26.0	156,330

a Above 26 feet use above table as applicable to 1890–1895.

Rating table for Tennessee River at Chattanooga, Tenn., for 1898.

Gage height.	Discharge.	Gage height.	Discharge.	Gage height.	Discharge.	Gage height.	Discharge.
Feet.	*Second-feet.*	*Feet.*	*Second-feet.*	*Feet.*	*Second-feet.*	*Feet.*	*Second-feet.*
0.5	5,900	3.7	19,550	6.9	42,370	10.5	68,650
.6	6,266	3.8	20,120	7.0	43,100	11.0	72,300
.7	6,634	3.9	20,700	7.1	43,830	11.5	75,950
.8	7,004	4.0	21,320	7.2	44,560	12.0	79,600
.9	7,376	4.1	21,950	7.3	45,290	12.5	83,250
1.0	7,750	4.2	22,580	7.4	46,020	13.0	86,900
1.1	8,126	4.3	23,350	7.5	46,750	13.5	90,550
1.2	8,504	4.4	24,120	7.6	47,480	14.0	94,200
1.3	8,884	4.5	24,850	7.7	48,210	14.5	97,850
1.4	9,266	4.6	25,580	7.8	48,940	15.0	101,500
1.5	9,650	4.7	26,310	7.9	49,670	15.5	105,150
1.6	10,046	4.8	27,040	8.0	50,400	16.0	108,800
1.7	10,444	4.9	27,770	8.1	51,130	16.5	112,450
1.8	10,844	5.0	28,500	8.2	51,860	17.0	116,100
1.9	11,246	5.1	29,230	8.3	52,590	17.5	119,750
2.0	11,650	5.2	29,960	8.4	53,320	18.0	123,400
2.1	12,056	5.3	30,690	8.5	54,050	18.5	127,050
2.2	12,464	5.4	31,420	8.6	54,780	19.0	130,700
2.3	12,874	5.5	32,150	8.7	55,510	19.5	134,350
2.4	13,286	5.6	32,880	8.8	56,240	20.0	138,000
2.5	13,700	5.7	33,610	8.9	56,970	20.5	141,650
2.6	14,126	5.8	34,340	9.0	57,700	21.0	145,300
2.7	14,562	5.9	35,070	9.1	58,430	21.5	148,950
2.8	15,008	6.0	35,800	9.2	59,160	22.0	152,600
2.9	15,464	6.1	36,530	9.3	59,890	22.5	156,250
3.0	15,930	6.2	37,260	9.4	60,620	23.0	159,900
3.1	16,410	6.3	37,990	9.5	61,350	23.5	163,550
3.2	16,900	6.4	38,720	9.6	62,080	24.0	167,200
3.3	17,400	6.5	39,450	9.7	62,810	24.6	171.680
3.4	17,920	6.6	40,180	9.8	63,540		
3.5	18,460	6.7	40,910	9.9	64,270		
3.6	19,000	6.8	41,640	10.0	65,000		

Rating table for Tennessee River at Chattanooga, Tenn., for 1899 and 1900.

Gage height.	Discharge.	Gage height.	Discharge.	Gage height.	Discharge.	Gage height.	Discharge.
Feet.	*Second-feet.*	*Feet.*	*Second-feet.*	*Feet.*	*Second-feet.*	*Feet.*	*Second-feet.*
0.8	6,600	4.8	25,760	8.8	50,560	12.8	75,380
.9	6,950	4.9	26,380	8.9	51,180	12.9	75,980
1.0	7,300	5.0	27,000	9.0	51,800	13.0	76,600
1.1	7,670	5.1	27,620	9.1	52,420	13.1	77,220
1.2	8,040	5.2	28,240	9.2	53,040	13.2	77,840
1.3	8,430	5.3	28,860	9.3	53,660	13.3	78,460
1.4	8,820	5.4	29,480	9.4	54,280	13.4	79,080
1.5	9,220	5.5	30,100	9.5	54,900	13.5	79,700
1.6	9,620	5.6	30,720	9.6	55,520	13.6	80,320
1.7	10,020	5.7	31,340	9.7	56,140	13.7	80,940
1.8	10,430	5.8	31,960	9.8	56,760	13.8	81,560
1.9	10,840	5.9	32,580	9.9	57,380	13.9	82,180
2.0	11,250	6.0	33,200	10.0	58,000	14.0	82,800
2.1	11,660	6.1	33,820	10.1	58,620	14.1	83,420
2.2	12,080	6.2	34,440	10.2	59,240	14.2	84,040
2.3	12,500	6.3	35,060	10.3	59,860	14.3	84,660
2.4	12,930	6.4	35,680	10.4	60,480	14.4	85,280
2.5	13,360	6.5	36,300	10.5	61,100	14.5	85,900
2.6	13,800	6.6	36,920	10.6	61,720	14.6	86,520
2.7	14,240	6.7	37,540	10.7	62,340	14.7	87,140
2.8	14,680	6.8	38,160	10.8	62,960	14.8	87,760
2.9	15,140	6.9	38,780	10.9	63,580	14.9	88,380
3.0	15,600	7.0	39,400	11.0	64,200	15.0	89,000
3.1	16,080	7.1	40,020	11.1	64,820	15.1	89,620
3.2	16,550	7.2	40,640	11.2	65,440	15.2	90,240
3.3	17,050	7.3	41,260	11.3	66,060	15.3	90,860
3.4	17,550	7.4	41,880	11.4	66,680	15.4	91,480
3.5	18,050	7.5	42,500	11.5	67,300	15.5	92,100
3.6	18,550	7.6	43,120	11.6	67,920	15.6	92,720
3.7	19,050	7.7	43,740	11.7	68,540	15.7	93,340
3.8	19,600	7.8	44,360	11.8	69,160	15.8	93,960
3.9	20,200	7.9	44,980	11.9	69,780	15.9	94,580
4.0	20,800	8.0	45,600	12.0	70,400	16.0	95,200
4.1	21,420	8.1	46,220	12.1	71,020	16.1	95,820
4.2	22,040	8.2	46,840	12.2	71,640	16.2	96,440
4.3	22,660	8.3	47,460	12.3	72,260	16.3	97,060
4.4	23,280	8.4	48,080	12.4	72,880	16.4	97,680
4.5	23,900	8.5	48,700	12.5	73,500	16.5	98,300
4.6	24,520	8.6	49,320	12.6	74,120	16.6	98,920
4.7	25,140	8.7	49,940	12.7	74,740	16.7	99,540

Rating table for Tennessee River at Chattanooga Tenn., for 1899 and 1900—Continued.

Gage height.	Discharge.	Gage height.	Discharge	Gage height.	Discharge.	Gage height.	Discharge.
Feet.	*Second-feet.*	*Feet.*	*Second-feet.*	*Feet.*	*Second-feet.*	*Feet.*	*Second-feet.*
16.8	100,160	20.8	124,960	24.8	149,760	28.8	174,560
16.9	100,780	20.9	125,580	24.9	150,380	28.9	175,180
17.0	101,400	21.0	126,200	25.0	151,000	29.0	175,800
17.1	102,020	21.1	126,820	25.1	151,620	29.1	176.420
17.2	102,640	21.2	127,440	25.2	152,240	29.2	177,040
17.3	103,260	21.3	128,060	25.3	152,860	29.3	177,6 0
17.4	103,880	21.4	128,680	25.4	153,480	29.4	178,280
17.5	104,500	21.5	129,300	25.5	154,100	29.5	178,900
17.6	105,120	21.6	129,920	25.6	154,720	29.6	179,520
17.7	105,740	21.7	130,540	25.7	155,340	29.7	180,140
17.8	106,360	21.8	131,160	25.8	155,960	29.8	180,760
17.9	106,980	21.9	131,780	25.9	156,580	29.9	181,380
18.0	107,600	22.0	132,400	26.0	157,200	30.0	183,000
18.1	108,220	22.1	133,020	26.1	157,820	30.1	182,620
18.2	108,840	22.2	133,640	26.2	158,440	30.2	183,240
18.3	109,460	22.3	134,260	26.3	159,060	30.3	183,8?0
18.4	110,080	22.4	134,880	26.4	159,680	30.4	184,480
18.5	110,700	22.5	135,500	26.5	160,300	30.5	185,100
18.6	111,320	22.6	136,120	26.6	160,920	30.6	185,720
18.7	111,940	22.7	136,740	26.7	161,540	30.7	186,340
18.8	112,560	22.8	137,360	26.8	162,160	30.8	186,960
18.9	113,180	22.9	137,980	26.9	162,780	30.9	187,580
19.0	113,800	23.0	138,600	27.0	163,400	31.0	188,200
19.1	114,420	23.1	139,220	27.1	164,020	31.1	188,820
19.2	115,040	23.2	139,840	27.2	164,640	31.2	189,440
19.3	115,660	23.3	140,460	27.3	165,260	31.3	190,060
19.4	116,280	23.4	141,080	27.4	165,880	31.4	190,680
19.5	116,900	23.5	141,700	27.5	166,500	31.5	191,300
19.6	117,520	23.6	142,320	27.6	167,120	31.6	191,920
19.7	118,140	23.7	142,940	27.7	167,740	31.7	192,540
19.8	118,760	23.8	143,560	27.8	168,360	31.8	193,160
19.9	119,380	23.9	144,180	27.9	168,980	31.9	193,780
20.0	120,000	24.0	144,800	28.0	169,600	32.0	194,400
20.1	120,620	24.1	145,420	28.1	170,220	32.1	195,020
20.2	121,240	24.2	146,040	28.2	170,840	32.2	195,640
20.3	121,860	24.3	146,660	28.3	171,460	32.3	196,260
20.4	122,480	24.4	147,280	28.4	172,080	32.4	196,880
20.5	123,100	24.5	147,900	28.5	172,700	32.5	197,500
20.6	123,720	24.6	148,520	28.6	173,320	32.6	198,120
20.7	124,340	24.7	149,140	28.7	173,940	32.7	198,740

Rating table for Tennessee River at Chattanooga, Tenn., for 1899 and 1900—Continued.

Gage height.	Discharge.	Gage height.	Discharge.	Gage height.	Discharge.	Gage height.	Discharge.
Feet.	*Second-feet.*	*Feet.*	*Second-feet.*	*Feet.*	*Second-feet.*	*Feet.*	*Second-feet.*
32.8	199,360	34.7	211,140	36.6	222,920	38.5	234,700
32.9	199,980	34.8	211,760	36.7	223,540	38.6	235,320
33.0	200,600	34.9	212,380	36.8	224,160	38.7	235,940
33.1	201,220	35.0	213,000	36.9	224,780	38.8	236,560
33.2	201,840	35.1	213,620	37.0	225,400	38.9	237,180
33.3	202,460	35.2	214,240	37.1	226,020	39.0	237,800
33.4	203,080	35.3	214,860	37.2	226,640	39.1	238,420
33.5	203,700	35.4	215,480	37.3	227,260	39.2	239,040
33.6	204,320	35.5	216,100	37.4	227,880	39.3	239,660
33.7	204,940	35.6	216,720	37.5	228,500	39.4	240,280
33.8	205,560	35.7	217,340	37.6	229,120	39.5	240,900
33.9	206,180	35.8	217,960	37.7	229,740	39.6	241,520
34.0	206,800	35.9	218,580	37.8	230,360	39.7	242,140
34.1	207,420	36.0	219,200	37.9	230,980	39.8	242,760
34.2	208,040	36.1	219,820	38.0	231,600	39.9	243,380
34.3	208,660	36.2	220,440	38.1	232,220	40.0	244,000
34.4	209,280	36.3	221,060	38.2	232,840		
34.5	209,900	36.4	221,680	38.3	233,460		
34.6	210,520	36.5	222,300	38.4	234,080		

Rating table for Tennessee River at Chattanooga, Tenn., for 1901.

Gage height.	Discharge.	Gage height.	Discharge.	Gage height.	Discharge.	Gage height.	Discharge.
Feet.	*Second-feet.*	*Feet.*	*Second-feet.*	*Feet.*	*Second-feet.*	*Feet.*	*Second-feet.*
2.0	11,250	3.4	17,550	4.8	25,760	6.2	34,500
2.1	11,660	3.5	18,050	4.9	26,380	6.3	35,150
2.2	12,080	3.6	18,550	5.0	27,000	6.4	35,800
2.3	12,500	3.7	19,050	5.1	27,620	6.5	36,450
2.4	12,930	3.8	19,600	5.2	28,240	6.6	37,100
2.5	13,360	3.9	20,200	5.3	28,860	6.7	37,750
2.6	13,800	4.0	20,800	5.4	29,480	6.8	38,400
2.7	14,240	4.1	21,420	5.5	30,100	6.9	39,050
2.8	14,680	4.2	22,040	5.6	30,720	7.0	39,700
2.9	15,140	4.3	22,660	5.7	31,340	7.1	40,350
3.0	15,600	4.4	23,280	5.8	31,960	7.2	41,000
3.1	16,080	4.5	23,900	5.9	32,580	7.3	41,650
3.2	16,550	4.6	24,520	6.0	33,200	7.4	42,300
3.3	17,050	4.7	25,140	6.1	33,850	7.5	42,950

Rating table for Tennessee River at Chattanooga, Tenn., for 1901—Continued.

Gage height.	Discharge.	Gage height.	Discharge.	Gage height.	Discharge.	Gage height.	Discharge.
Feet.	Second-feet.	Feet.	Second-feet.	Feet.	Second-feet.	Feet.	Second-feet.
7.6	43,600	11.7	70,250	15.8	96,900	19.9	123,550
7.7	44,250	11.8	70,900	15.9	97,550	20.0	124,200
7.8	44,900	11.9	71,550	16.0	98,200	20.1	124,850
7.9	45,550	12.0	72,200	16.1	98,850	20.2	125,500
8.0	46,200	12.1	72,850	16.2	99,500	20.3	126,150
8.1	46,850	12.2	73,500	16.3	100,150	20.4	126,800
8.2	47,500	12.3	74,150	16.4	100,800	20.5	127,450
8.3	48,150	12.4	74,800	16.5	101,450	20.6	128,100
8.4	48,800	12.5	75,450	16.6	102,100	20.7	128,750
8.5	49,450	12.6	76,100	16.7	102,750	20.8	129,400
8.6	50,100	12.7	76,750	16.8	103,400	20.9	130,050
8.7	50,750	12.8	77,400	16.9	104,050	21.0	130,700
8.8	51,400	12.9	78,050	17.0	104,700	21.1	131,350
8.9	52,050	13.0	78,700	17.1	105,350	21.2	132,000
9.0	52,700	13.1	79,350	17.2	106,000	21.3	132,650
9.1	53,350	13.2	80,000	17.3	106,650	21.4	133,300
9.2	54,000	13.3	80,650	17.4	107,300	21.5	133,950
9.3	54,650	13.4	81,300	17.5	107,950	21.6	134,600
9.4	55,300	13.5	81,950	17.6	108,600	21.7	135,250
9.5	55,950	13.6	82,600	17.7	109,250	21.8	135,900
9.6	56,600	13.7	83,250	17.8	109,900	21.9	136,550
9.7	57,250	13.8	83,900	17.9	110,550	22.0	137,200
9.8	57,900	13.9	84,550	18.0	111,200	22.1	137,850
9.9	58,550	14.0	85,200	18.1	111,850	22.2	138,500
10.0	59,200	14.1	85,850	18.2	112,500	22.3	139,150
10.1	59,850	14.2	86,500	18.3	113,150	22.4	139,800
10.2	60,500	14.3	87,150	18.4	113,800	22.5	140,450
10.3	61,150	14.4	87,800	18.5	114,450	22.6	141,100
10.4	61,800	14.5	88,450	18.6	115,100	22.7	141,750
10.5	62,450	14.6	89,100	18.7	115,750	22.8	142,400
10.6	63,100	14.7	89,750	18.8	116,400	22.9	143,050
10.7	63,750	14.8	90,400	18.9	117,050	23.0	143,700
10.8	64,400	14.9	91,050	19.0	117,700	23.1	144,350
10.9	65,050	15.0	91,700	19.1	118,350	23.2	145,000
11.0	65,700	15.1	92,350	19.2	119,000	23.3	145,650
11.1	66,350	15.2	93,000	19.3	119,650	23.4	146,300
11.2	67,000	15.3	93,650	19.4	120,300	23.5	146,950
11.3	67,650	15.4	94,300	19.5	120,950	23.6	147,600
11.4	68,300	15.5	94,950	19.6	121,600	23.7	148,250
11.5	68,950	15.6	95,600	19.7	122,250	23.8	148,900
11.6	69,600	15.7	96,250	19.8	122,900	23.9	149,550

*Rating table for Tennessee River at Chattanooga, Tenn., for 1901—*Continued.

Gage height.	Discharge.	Gage height.	Discharge.	Gage height.	Discharge.	Gage height.	Discharge.
Feet.	*Second-feet.*	*Feet.*	*Second-feet.*	*Feet.*	*Second-feet.*	*Feet.*	*Second-feet.*
24.0	150,200	28.1	176,850	32.2	203,500	36.3	230,150
24.1	150,850	28.2	177,500	32.3	204,150	36.4	230,800
24.2	151,500	28.3	178,150	32.4	204,800	36.5	231,450
24.3	152,150	28.4	178,800	32.5	205,450	36.6	232,100
24.4	152,800	28.5	179,450	32.6	206,100	36.7	232,750
24.5	153,450	28.6	180,100	32.7	206,750	36.8	233,400
24.6	154,100	28.7	180,750	32.8	207,400	36.9	234,050
24.7	154,750	28.8	181,400	32.9	208,050	37.0	234,700
24.8	155,400	28.9	182,050	33.0	208,700	37.1	235,350
24.9	156,050	29.0	182,700	33.1	209,350	37.2	236,000
25.0	156,700	29.1	183,350	33.2	210,000	37.3	236,650
25.1	157,350	29.2	184,000	33.3	210,650	37.4	237,300
25.2	158,000	29.3	184,650	33.4	211,300	37.5	237,950
25.3	158,650	29.4	185,300	33.5	211,950	37.6	238,600
25.4	159,300	29.5	185,950	33.6	212,600	37.7	239,250
25.5	159,950	29.6	186,600	33.7	213,250	37.8	239,900
25.6	160,600	29.7	187,250	33.8	213,900	37.9	240,550
25.7	161,250	29.8	187,900	33.9	214,550	38.0	241,200
25.8	161,900	29.9	188,550	34.0	215,200	38.1	241,850
25.9	162,550	30.0	189,200	34.1	215,850	38.2	242,500
26.0	163,200	30.1	189,850	34.2	216,500	38.3	243,150
26.1	163,850	30.2	190,500	34.3	217,150	38.4	243,800
26.2	164,500	30.3	191,150	34.4	217,800	38.5	244,450
26.3	165,150	30.4	191,800	34.5	218,450	38.6	245,100
26.4	165,800	30.5	192,450	34.6	219,100	38.7	245,750
26.5	166,450	30.6	193,100	34.7	219,750	38.8	246,400
26.6	167,100	30.7	193,750	34.8	220,400	38.9	247,050
26.7	167,750	30.8	194,400	34.9	221,050	39.0	247,700
26.8	168,400	30.9	195,050	35.0	221,700	39.1	248,350
26.9	169,050	31.0	195,700	35.1	222,350	39.2	249,000
27.0	169,700	31.1	196,350	35.2	223,000	39.3	249,650
27.1	170,350	31.2	197,000	35.3	223,650	39.4	250,300
27.2	171,000	31.3	197,650	35.4	224,300	39.5	250,950
27.3	171,650	31.4	198,300	35.5	224,950	39.6	251,600
27.4	172,300	31.5	198,950	35.6	225,600	39.7	252,250
27.5	172,950	31.6	199,600	35.7	226,250	39.8	252,900
27.6	173,600	31.7	200,250	35.8	226,900	39.9	253,550
27.7	174,250	31.8	200,900	35.9	227,550	40.0	254,200
27.8	174,900	31.9	201,550	36.0	228,200		
27.9	175,550	32.0	202,200	36.1	228,850		
28.0	176,200	32.1	202,850	36.2	229,500		

Rating table for Tennessee River at Chattanooga, Tenn., for 1902.

Gage height.	Discharge.	Gage height.	Discharge.	Gage height.	Discharge.	Gage height.	Discharge.
Feet.	*Second-feet.*	*Feet.*	*Second-feet.*	*Feet.*	*Second-feet.*	*Feet.*	*Second-feet.*
1.2	8,040	5.2	28,240	14.0	85,200	24.0	150,200
1.4	8,820	5.4	29,480	14.5	88,450	24.5	153,450
1.6	9,620	5.6	30,720	15.0	91,700	25.0	156,700
1.8	10,430	5.8	31,960	15.5	94,950	25.5	159,950
2.0	11,250	6.0	33,200	16.0	98,200	26.0	163,200
2.2	12,080	6.5	36,450	16.5	101,450	27.0	169,700
2.4	12,930	7.0	39,700	17.0	104,700	28.0	176,200
2.6	13,800	7.5	42,950	17.5	107,950	29.0	182,700
2.8	14,680	8.0	46,200	18.0	111,200	30.0	189,200
3.0	15,600	8.5	49,450	18.5	114,450	31.0	195,700
3.2	16,550	9.0	52,700	19.0	117,700	32.0	202,200
3.4	17,550	9.5	55,950	19.5	120,950	33.0	208,700
3.6	18,550	10.0	59,200	20.0	124,200	34.0	215,200
3.8	19,600	10.5	62,450	20.5	127,450	35.0	221,700
4.0	20,800	11.0	65,700	21.0	130,700	36.0	228,200
4.2	22,040	11.5	68,950	21.5	133,950	37.0	234,700
4.4	23,280	12.0	72,200	22.0	137,200	38.0	241,200
4.6	24,520	12.5	75,450	22.5	140,450	39.0	247,700
4.8	25,760	13.0	78,700	23.0	143,700	40.0	254,200
5.0	27,000	13.5	81,950	23.5	146,950		

Rating table for Tennessee River at Chattanooga, Tenn., for 1903.

Gage height.	Discharge.	Gage height.	Discharge.	Gage height.	Discharge.	Gage height.	Discharge.
Feet.	*Second-feet.*	*Feet.*	*Second-feet.*	*Feet.*	*Second-feet.*	*Feet.*	*Second-feet.*
0.60	6,100	2.20	12,790	3.80	20,560	5.40	29,600
.70	6,490	2.30	13,240	3.90	21,090	5.50	30,210
.80	6,880	2.40	13,700	4.00	21,620	5.60	30,820
.90	7,280	2.50	14,160	4.10	22,160	5.70	31,430
1.00	7,680	2.60	14,620	4.20	22,700	5.80	32,050
1.10	8,080	2.70	15,090	4.30	23,250	5.90	32,670
1.20	8,490	2.80	15,560	4.40	23,800	6.00	33,290
1.30	8,900	2.90	16,040	4.50	24,360	6.10	33,920
1.40	9,320	3.00	16,520	4.60	24,920	6.20	34,550
1.50	9,740	3.10	17,010	4.70	25,490	6.30	35,180
1.60	10,160	3.20	17,500	4.80	26,060	6.40	35,820
1.70	10,590	3.30	18,000	4.90	26,640	6.50	36,460
1.80	11,020	3.40	18,500	5.00	27,220	6.60	37,100
1.90	11,460	3.50	19,010	5.10	27,810	6.70	37,750
2.00	11,900	3.60	19,520	5.20	28,400		
2.10	12,340	3.70	20,040	5.30	29,000		

Estimated monthly discharge of Tennessee River at Chattanooga, Tenn.

[Drainage area, 21,382 square miles.]

Month.	Discharge in second-feet.			Run-off.	
	Maximum.	Minimum.	Mean.	Second-feet per square mile.	Depth in inches.
1890.					
January	78,470	30,150	42,749	2.00	2.31
February	308,060	44,191	76,081	3.56	3.72
March	445,120	52,906	129,093	6.03	6.96
April	121,464	43,029	64,855	3.03	3.38
May	72,079	41,286	51,200	2.39	2.76
June	41,286	23,400	29,102	1.36	1.52
July	47,677	19,500	27,036	1.26	1.45
August	46,515	21,100	30,881	1.44	1.66
September	47,096	22,200	27,843	1.30	1.45
October	58,135	25,950	37,982	1.77	2.04
November	42,448	20,700	26,394	1.23	1.37
December	77,889	20,400	36,088	1.69	1.95
The year	445,120	19,500	48,275	2.26	30.57
1891.					
January	92,995	39,543	59,484	2.78	3.21
February	356,120	59,878	154,822	7.23	7.53
March	381,040	63,364	135,160	6.32	7.30
April	97,643	38,962	61,873	2.89	3.22
May	37,219	26,380	30,215	1.41	1.63
June	47,096	26,840	36,276	1.70	1.90
July	36,057	21,800	26,429	1.24	1.43
August	98,224	23,000	40,402	1.89	2.18
September	36,638	19,200	25,777	1.20	1.34
October	19,200	17,910	18,461	.86	.99
November	41,867	17,440	23,510	1.10	1.23
December	66,269	22,200	39,299	1.84	2.12
The year	381,040	17,440	54,309	2.54	34.08

Estimated monthly discharge of Tennessee River at Chattanooga, Tenn.—Continued.

Month.	Discharge in second-feet.			Run-off.	
	Maximum.	Minimum.	Mean.	Second-feet per square mile.	Depth in inches.
1892.					
January	363,240	41,286	103,453	4.83	5.57
February	69,755	33,733	46,755	2.25	2.43
March	64,526	31,680	45,769	2.14	2.47
April	299,160	40,705	101,287	4.73	5.27
May	53,487	31,160	39,772	1.86	2.14
June	56,972	28,680	43,265	2.02	2.25
July	71,498	26,380	44,520	2.08	2.40
August	31,680	21,800	25,121	1.17	1.35
September	29,660	17,660	21,403	1.00	1.11
October	19,800	17,220	17,952	.84	.97
November	43,610	17,220	27,924	1.30	1.45
December	56,972	21,450	32,793	1.53	1.76
The year	363,240	17,220	45,835	2.15	29.17
1893.					
January	44,191	22,600	26,812	1.25	1.44
February	283,140	38,381	105,921	4.95	5.15
March	72,660	36,057	50,320	2.35	2.71
April	73,241	31,160	42,137	1.97	2.20
May	224,400	30,150	71,525	3.34	3.85
June	123,207	27,760	49,679	2.32	2.59
July	34,895	21,450	25,741	1.20	1.38
August	33,152	18,410	23,477	1.10	1.27
September	76,727	20,750	33,933	1.59	1.77
October	58,716	18,660	25,550	1.19	1.37
November	31,160	20,400	22,263	1.04	1.16
December	30,640	21,100	24,970	1.17	1.35
The year	283,140	18,410	41,861	1.96	26.24

Estimated monthly discharge of Tennessee River at Chattanooga, Tenn.—Continued.

Month.	Discharge in second-feet.			Run-off.	
	Maximum.	Minimum.	Mean.	Second-feet per square mile.	Depth in inches.
1894.					
January	56,972	22,600	37,389	1.75	2.02
February	151,095	31,680	70,893	3.31	3.45
March	59,297	33,152	46,796	2.19	2.53
April	52,325	27,300	36,287	1.70	1.90
May	44,191	23,800	31,137	1.43	1.68
June	28,680	19,500	21,983	1.03	1.15
July	29,170	18,910	24,486	1.14	1.31
August	30,150	18,910	22,971	1.07	1.23
September	27,300	16,560	19,160	.90	1.00
October	20,750	16,360	17,445	.82	.94
November	20,400	16,360	17,330	.81	.90
December	68,012	16,780	30,862	1.44	1.66
The year	151,095	16,360	31,395	1.53	19.77
1895.					
January	261,780	23,400	76,446	3.57	4.12
February	47,096	24,200	35,787	1.67	1.74
March	134,827	42,448	72,341	3.38	3.90
April	78,470	37,219	51,047	2.39	2.67
May	58,135	34,314	43,929	2.05	2.37
June	35,476	21,100	26,417	1.23	1.37
July	63,364	20,750	29,638	1.39	1.60
August	38,381	21,800	26,927	1.26	1.45
September	24,660	16,780	20,316	.95	1.05
October	17,000	16,360	16,665	.78	.90
November	20,750	17,220	18,162	.85	.94
December	33,152	17,660	21,561	1.01	1.16
The year	261,780	16,360	36,603	1.71	23.27

Estimated monthly discharge of Tennessee River at Chattanooga, Tenn.—Continued.

Month.	Discharge in second-feet.			Run-off.	
	Maximum.	Minimum.	Mean.	Second-feet per square mile.	Depth in inches.
1896.					
January	50,600	15,554	26,169	1.22	1.41
February	85,052	26,840	55,577	2.59	2.79
March	95,150	22,088	39,257	1.83	2.11
April	409,520	23,276	87,649	4.09	4.56
May	29,216	14,366	20,574	.96	1.10
June	43,472	17,336	24,365	1.14	1.27
July	130,196	20,306	55,390	2.59	2.99
August	34,562	14,960	22,433	1.05	1.21
September	18,524	9,020	12,346	.58	.64
October	19,712	8,426	11,588	.54	.62
November	57,728	9,020	22,603	1.06	1.18
December	45,254	16,148	27,951	1.31	1.51
The year	409,520	8,426	33,825	1.58	21.39
1897.					
January	45,254	16,148	27,932	1.30	1.50
February	308,060	19,712	89,962	4.20	4.37
March	363,240	52,976	165,448	7.72	8.90
April	231,520	36,344	81,056	3.78	4.22
May	134,948	26,543	50,124	2.34	2.70
June	38,126	21,494	29,107	1.36	1.52
July	74,657	21,791	34,428	1.61	1.86
August	34,562	14,366	25,847	1.21	1.39
September	14,960	6,050	8,951	.42	.47
October	13,772	4,268	7,842	.37	.43
November	9,614	6,050	7,330	.34	.38
December	62,183	8,129	24,627	1.15	1.33
The year	363,240	4,268	46,055	2.15	29.07

Estimated monthly discharge of Tennessee River at Chattanooga, Tenn.—Continued.

Month.	Discharge in second-feet.			Run-off.	
	Maximum.	Minimum.	Mean.	Second-feet per square mile.	Depth in inches.
1898.					
January	124,860	14,344	59,509	2.77	3.20
February	47,115	16,900	22,994	1.07	1.11
March	88,725	15,008	24,774	1.11	1.28
April	121,940	39,085	60,048	2.80	3.12
May	38,720	17,920	23,701	1.11	1.28
June	32,515	10,245	16,395	.77	.85
July	49,670	11,853	20,063	.94	1.08
August	107,705	20,410	50,638	2.36	2.72
September	174,500	18,190	47,349	2.21	2.46
October	115,370	16,900	44,215	2.06	2.38
November	42,005	20,700	28,415	1.33	1.48
December	35,800	19,000	28,909	1.35	1.56
The year	174,500	10,245	35,584	1.66	22.52
1899.					
January	112,560	25,450	47,250	2.21	2.55
February	233,150	30,720	95,554	4.46	4.64
March	244,000	55,210	142,700	6.66	7.68
April	137,360	39,710	69,286	3.23	3.59
May	65,130	22,040	40,450	1.89	2.18
June	35,990	15,600	23,088	1.08	1.20
July	27,930	10,430	15,053	.70	.81
August	22,040	8,040	11,900	.56	.64
September	15,840	7,300	10,118	.47	.53
October	10,635	6,600	7,851	.37	.43
November	10,635	6,775	8,216	.38	.43
December	41,880	8,820	22,061	1.03	1.19
The year	244,000	6,600	41,127	1.09	25.87

Estimated monthly discharge of Tennessee River at Chattanooga, Tenn.—Continued.

Month.	Discharge in second-feet.			Run-off.	
	Maximum.	Minimum.	Mean.	Second-feet per square mile.	Depth in inches.
1900.					
January	54,280	11,660	30,807	1.44	1.66
February	144,800	13,360	52,077	2.43	2.53
March	103,880	42,810	66,020	3.08	3.55
April	70,400	32,820	46,819	2.19	2.44
May	34,440	15,600	21,086	.98	1.13
June	53,040	14,680	33,295	1.55	1.73
July	50,870	13,360	24,674	1.15	1.33
August	34,440	10,020	14,602	.68	.78
September	25,140	7,300	13,393	.63	.70
October	42,500	8,040	14,230	.66	.76
November	92,720	10,840	25,138	1.17	1.31
December	53,040	17,050	29,001	1.35	1.56
The year	144,800	7,300	30,928	1.44	19.48
1901.					
January	189,200	19,600	50,641	2.36	2.72
February	59,850	19,050	36,516	1.70	1.77
March	139,150	19,050	44,952	2.10	2.42
April	166,450	38,400	95,080	4.44	4.95
May	205,450	26,380	68,736	3.21	3.70
June	72,200	33,850	47,673	2.23	2.49
July	37,100	14,680	23,932	1.12	1.29
August	207,400	13,800	75,761	3.54	4.08
September	61,150	22,660	38,859	1.81	2.02
October	25,760	13,360	18,979	.89	1.03
November	13,800	11,660	13,076	.61	.68
December	237,300	12,080	65,509	3.06	3.53
The year	237,200	11,660	48,310	2.26	30.68

Estimated monthly discharge of Tennessee River at Chattanooga, Tenn.—Continued.

Month.	Discharge in second-feet.			Run-off.	
	Maximum.	Minimum.	Mean.	Second-feet per square mile.	Depth in inches.
1902.					
January	259,400	27,620	70,567	3.29	3.79
February	145,000	28,240	60,106	2.81	2.93
March	241,200	41,000	108,411	5.06	5.83
April	195,050	28,240	54,811	2.56	2.86
May	54,650	20,800	28,283	1.32	1.52
June	27,000	15,600	19,043	.89	.99
July	60,500	10,840	22,414	1.05	1.21
August	14,680	8,820	10,892	.51	.59
September	18,550	8,040	11,376	.53	.59
October	20,200	8,040	13,260	.62	.71
November	35,150	8,430	13,298	.62	.69
December	56,600	17,550	33,763	1.58	1.82
The year	259,400	8,040	37,185	1.74	23.53
1903.					
January	47,500	21,620	30,435	1.42	1.64
February	184,650	26,060	90,229	4.21	4.38
March	195,700	52,700	109,690	5.12	5.90
April	200,900	54,000	99,890	4.66	5.20
May	52,700	19,010	29,098	1.36	1.57
June	69,600	22,160	38,661	1.81	2.02
July	31,430	12,790	20,855	.97	1.12
August	26,640	10,160	16,818	.79	.91
September	10,590	6,100	8,337	.39	.44
October	9,740	6,100	7,124	.33	.38
November	37,100	6,880	12,597	.59	.66
December	25,490	7,280	12,176	.57	.66
The year	200,900	6,100	39,659	1.85	24.88

Net horsepower per foot of fall with a turbine efficiency of 80 per cent for the minimum monthly discharge of Tennessee River at Chattanooga, Tenn.

Month.	1899.			1900.			1901.		
	Minimum discharge.	Minimum net horsepower per foot of fall.	Duration of minimum.	Minimum discharge.	Minimum net horsepower per foot of fall.	Duration of minimum.	Minimum discharge.	Minimum net horsepower per foot of fall.	Duration of minimum.
	Sec.-ft.		*Days.*	*Sec.-ft.*		*Days.*	*Sec.-ft.*		*Days.*
January	25,450	2,314	1	11,660	1,060	1	19,600	1,782	1
February	30,720	2,793	1	13,360	1,215	1	19,050	1,732	1
March	55,210	5,019	1	42,810	3,892	2	19,050	1,732	4
April	39,710	3,610	1	32,820	3,075	1	38,400	3,491	1
May	22,040	2,004	1	15,600	1,418	1	26,380	2,398	1
June	15,600	1,418	1	14,680	1,335	1	33,850	3,077	1
July	10,430	948	1	13,360	1,215	2	14,680	1,335	3
August	8,040	731	3	10,020	911	1	13,800	1,255	2
September	7,300	664	1	7,300	664	1	22,660	2,060	1
October	6,600	600	3	8,040	731	2	13,360	1,215	2
November	6,775	616	2	10,840	985	2	11,660	1,060	1
December	8,820	802	2	17,050	1,550	2	12,080	1,098	1

SHOALS IN TENNESSEE RIVER NEAR FLORENCE.

In Tennessee River, in the vicinity of Florence, Ala., are several shoals capable of the development of power. The compiler has brought together the data regarding these, his intention being not to discuss the manner in which the immense water power of these shoals can be developed, but to give some idea of its magnitude and the possibility of its utilization.

The shoals are a succession of cascades amid many islands, in a river bed varying in width from a half mile to 3 miles. The numerous channels thus formed are very irregular in fall and direction. The difference between high and low water is only 5 or 6 feet, corresponding to a rise of 50 feet at Chattanooga. Beginning at Browns Ferry, 12 miles below Decatur, Ala., the river has the following falls:

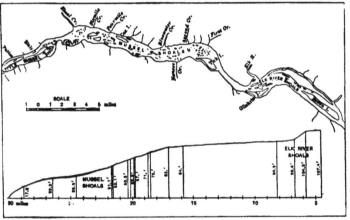

FIG. 9.—Map showing shoals in Tennessee River near Florence.

From Browns Ferry to the mouth of Elk River the fall is 26 feet in 11 miles. This is known as Elk River Shoals. Its most precipitous part is at the lower end, where there is a fall of 16.5 feet in about 4 miles.

From the mouth of Elk River to the head of Mussel Shoals, a distance of 5 miles, there is a fall of only 2 feet.

From the head of Mussel Shoals to Bainbridge the fall is 85 feet in 17 miles, and is known as Mussel Shoals.

From Bainbridge to Florence the fall is 23 feet in 7 miles, and is known as Little Mussel Shoals.

From Florence to the head of Colbert Shoals the fall is 3 feet in 11 miles.

From the head of the Colbert Shoals to Waterloo the fall is 21 feet in 6 miles.

The total fall from Browns Ferry to Waterloo is, therefore, 160 feet in a distance of 57 miles. Sixteen miles of the distance, however, has a fall of only 5 feet, leaving a fall of 155 feet in the 41 miles that cover the four shoals mentioned. The shoals are really more precipitous than the foregoing figures would indicate. For instance, 84.6 feet of the fall at Mussel Shoals is in a distance of 14 miles.

The bed rock at Elk River Shoals is Carboniferous limestone; that of Mussel Shoals is a hard siliceous rock of dark color and flinty structure.

Assuming that tributaries entering the river below Chattanooga will safely supply all of the water needed for lockage, the discharge at Chattanooga can be used in estimating the water power of these shoals, which are about 200 miles below Chattanooga, by river, and of which the drainage area is more than 7,000 square miles greater than that at Chattanooga.

Estimated minimum net horsepower of Tennessee River in Alabama on turbines realizing 80 per cent of the theoretical power.

Locality.	Fall.	Minimum net power in driest years.	Minimum net power in average years.
	Feet.		
Elk River Shoals..	26	15,600	30,550
Mussel Shoals..	85	51,000	99,875
Little Mussel Shoals..	23	13,800	27,025
Colbert Shoals	21	12,600	24,675
Total...	155	93,000	182,125

The foregoing table assumes that the total fall can in each case be utilized. While this assumption is not correct, it stands as an offset to the assumption that the water supply available will be as low as the minimum discharge at Chattanooga, 200 miles above. The drainage area above Chattanooga is 21,418 square miles, while the drainage area above the shoals under consideration is about 29,000 square miles. It may therefore safely be assumed that the actual power available for development at the shoals is greater than that shown by the table.

The foregoing statements of fall and distance are from a report by Mr. William B. Gaw, chief assistant engineer, U. S. Army, 1868, and the map and profile are from drawings prepared under the direction of Lieut. Col. J. W. Barlow, United States Engineers, 1890.

BIG SPRING AT HUNTSVILLE, ALA.

TRIBUTARIES OF TENNESSEE RIVER.

Paint Rock Creek, Elk River, Shoal Creek, Flint Creek, Nance Creek, Town Creek, and Big Bear Creek are all large streams, and most of them have fine undeveloped water powers. But no surveys have been made of them, and no measurements of discharge so far. There are also many large bold springs in this basin that are said to have a pure and unfailing water supply, but no report can be made on them at this time.

Miscellaneous discharge measurements on streams that enter Tennessee River in Alabama.

Date.	Stream.	Locality.	Discharge.
1903.			*Second-feet.*
July 20	Big Spring	Tuscumbia	177
1904.			
Feb. 6	Big Spring	Tuscumbia	33
Do	Creek from Big Spring	do	43
Feb. 4	Paint Rock Creek	Paint Rock	159
Feb. 5	Flint Creek	Brownsboro	209
Feb. 8	Elk River	Prospect, Tenn	5,296
Feb. 9	Sequatchie River	2 miles from Jasper, Tenn	1,916
Do	do	2 miles northeast of Jasper, Tenn.	1,829
Do	Battle Creek	Near South Pittsburg, Tenn	482

UTILIZED WATER POWERS.

The following is a list by counties of the water powers that are utilized. The most of these powers are small, but they make a large aggregate, and they represent only an insignificant part of the power that is capable of development.

Water powers in Alabama, by counties.

AUTAUGA COUNTY.

Name.	Post-office.	Industry.	Horse-power.
Charity P. Carter *a*	Billingsley	Flour and grist mill	15
Montgomery's mill *a*	Prattville *b*	do	30
Public gristmill *a*	Billingsley	do	9
Parker's mill *a*	Milton	do	20
Dawson's mill *a*	Netezen	Lumber and timber mill.	20
Ellis's mill *a*	Jones Switch	do	4
Long Leaf Yellow Pine sawmill.*a*	Autaugaville	do	15
Ray's sawmill *a*	Jones Switch	do	10
Swift Creek Mill Co. (Swift Creek).*a*	Autaugaville	do	70
John H. Herod *a*	Netezen	do	6
Prattville Cotton Mills and Banking Co. (Autauga Creek).*a*	Prattville	Cotton goods	200
Continental Gin Co. (Autauga Creek). *c*	do	Cotton gin	
Prattville ice factory (Autauga Creek). *c*	do	Ice factory	
Doster ginnery (Autauga Creek). *c*	do	Cotton gin	
G. H. Roy *c*	Vine Hill	do	

a From U. S. Census, 1900.

b The water power at Prattville was first developed about 1830, when it was used by a man named May to operate a small sawmill. About 1833 this water power and the adjacent lands were purchased by Mr. Daniel Pratt, who then erected a cotton-gin factory, which was driven by the water power. The dam at that time was about 8 feet high. A number of years after the purchase of this property by Mr. Pratt he increased the dam so that it now has a height of 16 feet, and is built of brick. At present it is used jointly by the Prattville Cotton Mills and Banking Company and the Continental Gin Company, the former using about 255 horsepower and the latter 100 horsepower. About half a mile below the dam above referred to is another dam affording about 8 feet head and owned by the M. E. Pratt estate. This power operates a gristmill, cotton ginnery, and ice factory, and the water wheel at that point has a rated capacity of 54 horsepower. About 1 mile above the dam of the Cotton Mill and Gin Company there was formerly another dam 12 feet high, which afforded power for a cotton mill. This mill, however, was burned a number of years ago, and the dam has been allowed to go to ruin. It would probably afford 200 horsepower, or possibly a little more, should it be rebuilt.

There is also a dam about 2 miles below Prattville, known as the Montgomery mill property. This dam is about 12 feet high and affords power for a gristmill and ginnery. Only a small portion of the available power is used. It could afford easily 250 horsepower, if the proper wheels were installed.

c From report of probate judge.

Water powers in Alabama, by counties—Continued.

BARBOUR COUNTY.[a]

Name.	Post-office.	Industry.	Horse-power.
Hagler's mill	Louisville	Flour and grist mill	17
Carpenter's mill	....do	....do	15
Hoffman's mill	Clayton	....do	50
Hartman's mill	....do	....do	10
Zorn mills	Lodi	....do	8
William M. Wood	Bush	....do	12
Will Stewart	Whiteoak Springs	....do	12
Winn's mill	Clayton	....do	12
John White	Spivey	....do	10
Weston's mill	Louisville	....do	8
H. J. Turner	Whiteoak Springs	....do	10
Spencer's mill	Clayton	....do	10
Perkin's mill	Elamville	....do	12
Angus McSwain	Whiteoak Springs	....do	12
William Johnson	Clayton	....do	10
John M. Jenkins	Starhill	....do	10
Solomon's mills	Solomons Mills	....do	25
Danner mill	Elamville	....do	12
William H. Chambers	Oateston	....do	12
Wilson Deshazo	Cottonhill	....do	16

BIBB COUNTY.

Name.	Post-office.	Industry.	Horse-power.
Scottsville flour and grist mill.[a]	Scottsville	Flour and grist mill	30
Palmetto flouring and grist mill.[a]	Brierfield	....do	30
Williams's gristmill[a]	Blocton	....do	10
William S. Mathews[a]	Data	....do	8
Sixmile custom mill[a]	Sixmile	....do	15
Mayfield Bros.[a]	Mertz	Lumber and timber mill.	29
Scottsville Wool Carder[b]	Scottsville	Woolen goods	20
J. M. Battle (Sixmile Creek).[b]	Sixmile	Flour and grist mill	50
W. C. Trott (Sixmile Creek).[b]	....do	Cotton gin and grist mill.	50
W. H. Thomas (Sixmile Creek).[b]	Ashley	Lumber and grist mill.	35
Dock Mahan (Mahans Creek).[b]	Brierfield	Wool carder and grist-mill.	40

a From U. S. Census, 1900.   b From report of probate judge.

Water powers in Alabama, by counties—Continued.

BIBB COUNTY—Continued.

Name.	Post-office.	Industry.	Horse-power.
Bessemer Land and Improvement Co. (Schultz Creek).*a*	Lopez	Wood carder, gristmill, and cotton gin.	100
R. R. McCally (Hills Creek)*a*	Blocton	Gin, lumber, and grist..	30
E. M. Timbro (Schultz Creek).*a*	Centerville	Gristmill	30
F. H. James (Haysoppy Creek).*a*	.....do	.....do	20
A. L. Elam (Affonee Creek)*a*	Affonee	.....do	15

BLOUNT COUNTY.*b*

Name.	Post-office.	Industry.	Horse-power.
Logan Snead	Snead	Flour and grist mill	10
E. B. Head	Gumspring	.....do	16
E. R. Wood	Wynnville..........	.....do	8
Hendrick's mill............	Swansea...........	.....do	30
Jones M. Burns..........	Clarence	.....do	15
Wilson Adcock	Tidmore...........	.....do	10
G. M. D. Tidwell & Sons ..	Tidwell	.....do	20
Alldridge & Bro...........	Liberty...........	.....do	10
Brittain's mill	Summit	.....do	20
Rufus F. Wyatt...........	Bangor	.....do	10
Sam Mardis...............	Blountsville	.....do	60
Jno. H. Donahoo & Geo. W. Darden.	Rosa	Lumber and timber	20

BULLOCK COUNTY.

Name.	Post-office.	Industry.	Horse-power.
Brooks's mill	Mascotte..........	Flour and grist mill	6
Union Springs waste mill..	Union Springs	.....do	15
Chappell's gristmill	.....do	.....do	10
D. H. Mason (McBrides Creek).	Indian	Lumber, gin, and grist..	20
Chas. Radford (Conecuh Creek).*a*	Union Springs	Gristmill	10

BUTLER COUNTY.*b*

Name.	Post-office.	Industry.	Horse-power.
John W. Halso	Pigeoncreek	Flour and grist mill	10
Glen Graham.............	Pontus............	.....do	6
The Four-mile mill........	Greenville.........	.....do	10
The N. M. Rhodes Mill and Mercantile Co.	Shell	Flour, grist, and lumber mill.	50
Mrs. M. E. Crane..........	Monterey..........	Flour and grist mill	15
Rouse & Whiddon	Greenville.........	.....do	50

a From report of probate judge.　　*b* From U. S. Census, 1900.

Water powers in Alabama, by counties—Continued.

CALHOUN COUNTY.

Name.	Post-office.	Industry.	Horse-power.
Joseph Francis [a]	Cane Creek	Flour and grist mill	50
Richey's mill [a]	Jacksonville	....do	20
Canada gristmill [a]	Womack	....do	16
Coldwater mills [a]	Coldwater	....do	20
Read's mill [a]	Reads	....do	60
Luther Barton [a]	Piedmont	....do	20
W. F. McCulley [a]	Oxford	....do	20
A. McCurdy [a]	Whiteplains	....do	34
Morris's gristmill [a]	Morrisville	....do	18
Nisbet's mill [a]	Jacksonville	....do	30
James A. Weatherly [a]	De Armanville	....do	8
Wood Milling Co. [a]	Ohatchee	....do	26
Davis & Henderson [a]	Piedmont	....do	24
Hendon's gristmill [a]	Ironcity	....do	10
Hughes's sawmill [a]	Oxford	Lumber and timber mill	28
F. M. Whiteside (Choccolocco Creek). [b]	Whiteplains		25 or 30
Downing & Morris (Choccolocco Creek). [b]	Choccolocco		50
J. T. De Arman (Choccolocco Creek). [b]	Anniston		15
W. E. Mellon (Choccolocco Creek). [b]	Oxford		40
Lee's mill (Choccolocco Creek). [b]	....do		30
T. G. Slaughter (Choccolocco Creek). [b]	....do		15
J. H. Savage (Terrapin Creek). [b]	Anniston		20
Do	....do		20
Frank Aderhold (Nances Creek). [b]	Ladiga		20
John Ramagnand (Champion Creek). [b]	Jacksonville		15
James Crook (Tallaseehatchee Creek). [b]	....do		10
W. J. Edmondson (Tallaseehatchee Creek). [b]	Anniston		30
W. A. Prickett (Tallaseehatchee Creek). [b]	Alexandria		10
Beaty Estate (Tallaseehatchee Creek). [b]	....do		30
Peter Heifner (Tallaseehatchee Creek). [b]	....do		15

[a] From U. S. Census, 1900.   [b] From report of probate judge.

Water powers in Alabama, by counties—Continued.

CALHOUN COUNTY—Continued.

Name.	Post-office.	Industry.	Horse-power.
James Aderhold (Ohatchee Creek).[b]	Reads		20
Pleas. Martin (Ohatchee Creek).[b]	Peekshill		25
C. J. Wood (Ohatchee Creek).[b]	Jacksonville		30
Wm. Thompson (Ohatchee Creek).[b]	Peekshill		8
R. L. Treadway (Tallasee-hatchee Creek).[b]	Anniston		10
J. H. Francis (Tallasee-hatchee Creek).[b]			25
R. H. Cobb (Tallasee-hatchee Creek).[b]	Anniston		20
G. W. S. Loyd (Cane Creek).[b]	Peaceburg		10
Mrs. Loyd (Cane Creek).[b]	...do	Gin	6
Morris Manufacturing Co. (Cane Creek).[b]	Morrisville	Shops	30
E. G. Morris (Cane Creek).[b]	...do		30
P. H. Brothers (Cane Creek).[b]	Zula		30
J. H. Francis (Cane Creek).[b]			50

CHAMBERS COUNTY.[a]

Name.	Post-office.	Industry.	Horse-power.
D. E. M. Smith	Barber	Flour and grist mill	24
Cumbee's gristmill	Stroud	...do	20
Thomas H. Fuller	Lafayette	...do	10
R. T. Humphrey	Westpoint, Ga	...do	42
J. T. Hudson	Hickoryflat	...do	4
Wyche Robinson	Lafayette	...do	16
Stephens's mill	Driver	...do	8
Ripville mills	Wise	...do	20
Charles F. Higgins	Finley	...do	20
J. E. Dixon	Lafayette	...do	10
Ratchford & Tucker	...do	...do	10
Benjamin F. Knight	...do	...do	10
Wooddy & Beall	Moorefield	...do	6
Leverett & Abernathy's Mill.	Milltown	...do	4
G. L. Leverett	Lafayette	...do	16
Westpoint Manufacturing Co.	Westpoint	Cotton goods	1,100

[a] From U. S. Census, 1900.   [b] From report of L. J. Morris.

Water powers in Alabama, by counties—Continued.

CHEROKEE COUNTY.

Name.	Post-office.	Industry.	Horse-power.
Shamblin & Toles mill [a] ...	Broomtown.........	Flour and grist mill	8
Chandler & Stinson [a]	Center	....do	20
Shamblin & Toles mill [a] ...	Broomtown.........	....do	20
J. A. Lumpkin [a]	Forney	....do	13
Hurley's mill [a]	Hurley	....do	12
Tyre G. Craig [a]	Grover	....do	12
Rush mill [a]	Lawrence.........	....do	10
E. W. Ragdale [a]	Spring Garden	....do	30
W. F. Timmerman [a]	Round Mountain...	....do	8
M. E. Cobia [a].............	Cedar Bluff........	....do	24
M. J. Abernathy [a]	Pleasant Gap	Lumber and timber mill.	15
Hurricane Creek Manufacturing and Mining Co. [a]	Spring Garden	Cotton goods...........	65
W. A. Stinson (Terrapin Creek). [b]	Center	Gin, flour, and grist mill.	60
J. J. Scroggin (Terrapin Creek). [b]	Coloma............	......do	60
T. F. Stewart (Terrapin Creek). [b]	Spring Garden	Flour and grist mill	60
J. M. Adderhold (Mill Creek). [b]	Piedmont..........	Flour, grist, and gin mill.	40
M. L. Braswell (Hurricane Creek). [b]	Pleasant Gap	Flour and grist mill	40
B. F. Newberry (Yellow Creek). [b]	Round Mountain...	Flour, grist, and gin mill.	40
E. Cobia (Chattooga River) [b]	Cedar Bluff........	......do	60
R. A. Russell & Co. (Chattooga River). [b]	Gaylesville	....do	60
W. F. Henderson (Mill Creek). [b]	Fullerton..........	......do	40
Rush & Rinehart (Chattooga River). [b]	.....do	......do	60
J. G. Toles (Mill Creek) [b]..	Broomtown.........	Grist and gin mill......	40
Elliott Bros. (North Spring Creek). [b]	Grassland	......do	40
J. T. Webb & Bros. (Spring Creek). [b]	Hurley	......do	40
J. D. Jordan (South Spring Creek). [b]	Noah..............	......do	20

[a] From U. S. Census, 1900.   [b] From report of L. J. Morris.

Water powers in Alabama, by counties—Continued.

CHILTON COUNTY.a

Name.	Post-office.	Industry.	Horse-power.
James Dorming	Jemison	Flour and grist mill	10
Mahan's mill	Clanton	....do	20
W. W. Sansome	Adams	....do	12
Honeycutt mill	Jemison	Flour, grist, lumber, and timber mill.	20

CHOCTAW COUNTY.a

Pink Blackwell	Hinton	Flour and grist mill	12
Aquilla mills	Aquilla	Lumber and timber mill.	16

CLARKE COUNTY.a

Gate's mill	Vashti	Flour and grist mill	30
Fleming's gristmill	Nealton	....do	10
Dacy's mill	Whatley	....do	5

CLAY COUNTY.a

Henry F. Smedley	Mellow Valley	Flour and grist mill	15
Hezakiah Ingram	Hatchett Creek	....do	10
Allen P. Jenkins	Delta	....do	14
Knight's mill	Wesobulga	....do	14
F. M. Munroe	Millerville	....do	40
John R. Gilbert	Pinckneyville	....do	8
Hodnett & Co	Hat	....do	10
Moses R. Watts	Dean	....do	6
Thomas J. Watts	Shinbone	....do	8
Bishop, Carpenter & Co	Cherry	....do	10
Cockrell & Mitchell	Goldburg	....do	14
McRairie, Gladney & Co	Cherry	....do	20
Virginia Whellen	Coleta	....do	6
Stephens & East	Delta	....do	4
Deberry & Griffin	Flatrock	....do	15
Child's mill	Swann	....do	5
James B. Brown	Pinckneyville	....do	6
James J. Bachus	Fishhead	....do	24
Brooks & Handley	Hatchett Creek	....do	8
Columbus Bell	Lineville	Lumber and timber mill.	10
J. C. Kennedy	Fishhead	....do	14
William M. Patterson	Meadow	....do	30
Ward & Ford	Lineville	....do	15

a From U. S. Census, 1900.

Water powers in Alabama, by counties—Continued.

CLEBURNE COUNTY.ᵃ

Name.	Post-office.	Industry.	Horse-power.
J. T. & E. W. Beason	Beasons Mill	Flour and grist mill	10
W. M. Evans	Edwardsville	do	20
Robert mill	Oaklevel	do	16
Teague & Co	Eudora	do	13
H. F. Alsabrook	Borden Springs	do	30
Buttram's mill	Bucham	do	20
John A. Brown	Bell Mills	do	16
John I. Burgess	Edwardsville	do	20
Wade H. Barnes	Muscadine	do	4
J. W. Connor	Chulafinnee	do	6
Lyon & Killebrue		do	34
W. G. Miligan	Oakfuskee	do	8
James McMahan	Edwardsville	do	12
E. W. Pitchford	Oaklevel	do	15
William J. Thrash	Oakfuskee	do	6
Wade H. Barnes	Muscadine	do	30
W. H. Tumlin & D. S. Baber.	Ai	do	16

COFFEE COUNTY.ᵃ

Name.	Post-office.	Industry.	Horse-power.
Levy Wise	Ino	Flour and grist mill	5
Bell mill	Dot	do	8
Lenora F. Hildreth	Enterprise	do	17
Harper flour mills	Brockton	do	4
F. M. Prestwood	Fresco	do	20
McIntosh mill	Eta	do	8
Wise's lower mill	Elba	do	12
Wise's upper mill	do	do	10
Buck & Co	Penn	Lumber and timber mill.	50

COLBERT COUNTY.ᵃ

Name.	Post-office.	Industry.	Horse-power.
George Martin	Allsboro	Flour and grist mill	8
James Burns	Mand	do	4
Tuscumbia mill	Tuscumbia	do	40
C. C. Hester	do	do	40
Chambee's gristmill	do	do	8
Dillard's mills	Russellville	Lumber and timber	12
Steenson's mills	Sheffield	do	30

ᵃ From U. S. Census, 1900.

Water powers in Alabama, by counties—Continued.

CONECUH COUNTY. a

Name.	Post-office.	Industry.	Horse-power.
George Stenson	Bonnette	Flour and grist mill	12
James B. Pate	Brooklyn	....do	5
William M. Robinson	....do	....do	5
Jimson C. Cox	Gem	....do	5
John N. Varner & Chas. M. Varner.	Herbert	....do	10
James E. Wilson	Mount Union	....do	20
Ransom H. Finley	Zern	....do	8
G. G. Broker	Bowles	Lumber and timber mill.	10
Cary & Johnston	Brooklyn	....do	15
T. N. Piggott	Gravella	....do	40
Robinson Bros	Brooklyn	....do	30
H. J. Robinson	Burnt Corn	....do	40
Henry Wills	Finklet	....do	30

COOSA COUNTY. a

Name.	Post-office.	Industry.	Horse-power.
Miller's mill	Bentleyville	Flour and grist mill	20
Nolen's mill	Darden	....do	15
J. T. M. Hodnett & O. P. Hodnett.	Equality	....do	12
W. N. Neighbors	Goodwater	....do	23
Smith's mill	Nixburg	....do	10
George P. Waits	Rockford	....do	8
Crawford mill	....do	....do	4
Lawson grist and sawmill	....do	Lumber and timber mill.	36

COVINGTON COUNTY. a

Name.	Post-office.	Industry.	Horse-power.
A. J. Fletcher	Andalusia	Flour and grist mill	10
Uatu gristmill	....do	....do	10
William Sharp	Ealums	....do	10
Davis B. Gantt	Gantt	....do	12
C. E. Rawls	....do	....do	10
Dorsey's mill	Glasiasko	....do	10
James Aplin	Green Bay	....do	20
William Watkins	Liberty Hill	....do	8
Kearsey's mill	Redlevel	....do	5
Ephram F. Lassiter	Rosehill	....do	10
Thomas sawmill	Redlevel	Lumber and timber mill.	25

a From U. S. Census, 1900.

Water powers in Alabama, by counties—Continued.

COVINGTON COUNTY *a*—Continued.

Name.	Post-office.	Industry.	Horse-power.
Simmons mill	Beck	Lumber and timber mill.	40
J. A. Prestwood, jr	Andalusia	.....do	40
George W. Lee	Rat	.....do	20
Buck Creek mill	River Falls	.....do	80
J. F. Guthrie	Veracruz	.....do	25
Gunter's mill	Andalusia	.....do	40
Gunter's sawmill	Gantt	.....do	15
Gantt's mill	River Falls	.....do	70
Pollard Gantt	Searight	.....do	35
Davis B. Gantt	Gantt	.....do	40
N. B. Dixon	Mason	.....do	60
Bartlett & Barker		.....do	60

CRENSHAW COUNTY. *a*

Name.	Post-office.	Industry.	Horse-power.
E. P. Lasseter	Bullock	Flour and grist mill	8
G. B. Morgan	.....do	.....do	15
Folmar's mill	Goshen	.....do	8
N. Skipper	Honoraville	.....do	10
Daniel & Co	Lapine	.....do	30
John S. Marsh	Rutledge	.....do	20
G. B. Sasser	Luverne	.....do	15

CULLMAN COUNTY. *a*

Name.	Post-office.	Industry.	Horse-power.
Joseph W. Hyatt	Baileyton	Flour and grist mill	10
Miles Humphries	.....do	.....do	4
D. H. Laney	Battleground	.....do	6
Robert J. Waldrop	Cranehill	.....do	20
Andrew J. Miller	Summit	.....do	6

DALE COUNTY. *a*

Name.	Post-office.	Industry.	Horse-power.
Archer McCall	Candy	Flour and grist mill	10
Floyd mill	Dothan	.....do	10
Lewis mill	Clopton	.....do	15
Murphy mill	Dothan	.....do	5
Maunds corn mill	Ewells	.....do	10
Pope's mill	Grimes	.....do	60

a From U. S. Census, 1900.

Water powers in Alabama, by counties—Continued.

DALE COUNTY—Continued.

Name.	Post-office.	Industry.	Horse-power.
Charles Thrower	Kleg	Flour and grist mill	16
Daniel McSwean	Ozark	do	20
Preston's mill	Peach	do	20
The Kelley gristmill	Pinckard	do	150
Atkinson's sawmill	Newton	Lumber and timber mill.	16
J. F. Bell	Daleville	do	22

DALLAS COUNTY.a

Calhoun's mill	Carlowville	Flour and grist mill	10
Ivey & Williams	Morrowville	do	8

DEKALB COUNTY.

L. D. Wooten a	Blake	Flour and grist mill	8
J. D. Hall a	Chavies	do	10
J. S. Ward a	Chumley	do	12
Kean & Warren a	Cordell	do	20
Swindell's mill a	Cotnam	do	12
Griffin's mill a	do	do	12
Emeline Clayton a	Crossville	do	6
Swader's mill a	Dekalb	do	15
James Clark a	Eula	do	15
David J. Harper a	Floy	do	3
Elrod's gristmill a	do	do	4
Davis mill a	Fort Payne	do	16
Thomas F. Everett a	Luna	do	8
Elrod's mill a	Geraldine	do	30
Pruitt's mill a	Skirum	do	12
Lebanon flour and grist mill.a	Lebanon	do	36
Robert F. Ellison a	Mentone	do	25
Ellic Ellsworth a	Ophir	do	6
Warren's gristmill a	Portersville	do	12
John F. Williams a	Rains	do	8
Edward W. Williams a	do	do	6
McGee's mill a	Sandrock	do	5
Charles G. Matheny a	Sauty Mills	do	20
Dixie Mills a	Sulphur Springs	do	10
Phillips's mill a	Valleyhead	do	4

a From U. S. Census, 1900.

Water powers in Alabama, by counties—Continued.

DEKALB COUNTY—Continued.

Name.	Post-office.	Industry.	Horse-power.
The Roberts Mill Co.[a]	Collinsville	Flour and grist mill	25
W. E. Brown & Son[a]	Sulphur Springs	Lumber and timber mill.	15
James M. Durham[a]	Chavies	....do	16
William C. Hill & Co.[a]	Blanche	....do	40
D. D. Hughes[a]	Hughes	....do	15
Ward, Pickens & Co.[a]	Dawson	....do	15
John A. Davis (Wills Creek)[b]	Fort Payne	Grist mill and gin	
M. S. Brown and W. C. Thomas (Lookout Creek).[b]	Sulphur Springs	Flouring mill	
D. D. Hughes (Wills Creek)[b]	Hughes	Flour and grist mill	
P. M. Frazier (Wills Creek)[b]	Lebanon	....do	
S. D. Warren (Wills Creek)[b]	....do	....do	
Grif. Elrod (Town Creek)[b]	South Hill	....do	
Durham & Co. (Town Creek).[b]	Chavies	Flour, grist, and saw mill	

ELMORE COUNTY.[a]

Name.	Post-office.	Industry.	Horse-power.
E. & H. T. Andrews	Channahatchee	Flour and grist mill	25
Benjamin Spigener	Elmore	....do	5
Sykes mill	Sykes Mills	....do	16
John C. Birt (Lancaster old mill).	Tallassee	....do	24
Freeman's gristmill	....do	....do	5
J. J. Benson	Kowaliga	Lumber and timber mill.	20
J. T. Rogers	Spigners	....do	36
Tallassee Falls Manufacturing Co. (Tallapoosa River).[c]	Tallassee	Cotton and woolen goods	8,900
Montgomery Power Co. (Tallapoosa River).	....do	Electric transmission to Montgomery, Ala.	5,600

ESCAMBIA COUNTY.[a]

Name.	Post-office.	Industry.	Horse-power.
Bradley mill		Flour and grist mill	10
S. S. Overstreet	Roberts	....do	20
James F. Douglas	Mason	Lumber and timber	25

[a] From U. S. Census, 1900.
[b] From report of probate judge.
[c] This is the same company that is now organized under the name of the Mount Vernon-Woodberry Cotton Duck Company, with office at Montgomery, Ala.

Water powers in Alabama, by counties—Continued.

ETOWAH COUNTY.

Name.	Post-office.	Industry.	Horse-power.
Wesson mills [a]	Attalla	Flour and grist mill	25
Cox & Bro. [a]	Avery	....do	6
B. H. Rogers [a]	Etowahton	....do	40
T. G. Ewing [a]	Ewings	....do	60
John C. Rollins [a]	Fenton	....do	8
Reese mill [a]	Hill	....do	10
John H. Helms [a]	Ballplay	....do	6
Ford & Sibert's mill [a]	Hokesbluff	....do	30
Morgan & Cochran [a]	Keener	....do	8
W. J. Harris [a]	Nix	....do	12
John B. Burns [a]	Seaborn	....do	8
A. B. Stephens [a]	....do	....do	8
W. H. Cobb [a]	Steels Depot	....do	20
P. C. Turner [a]	Walnutgrove	....do	30
Do. [a]	....do	Woolen goods	13
W. M. Brothers & Son [a]	Gallant	....do	8
Gadsden Times-News [a]	Gadsden	Printing and publishing.	4
J. M. Morague (Wills Creek). [b]	....do	Gristmill	100
Wm. McClendon (Wills Creek). [b]	Attalla	....do	40
—— Griffith (Wills Creek). [b]	Keener	....do	35
Bob Rigers (Canoe Creek). [b]	Gadsden	....do	75
Tom Ewing (Cane Creek) [b]	....do	....do	40

FAYETTE COUNTY.

Name.	Post-office.	Industry.	Horse-power.
Rodolphus Cotton [a]	Bankston	Flour and grist mill	20
D. G. Hester [a]	Covin	....do	12
John W. Anthony [a]	Glenallen	....do	30
Landon Miles [a]	Hester	....do	13
Bishop Emick [a]	Rena	Lumber and timber mill.	40
Phillip N. Fortenberry [a]	Bankston	....do	8
W. L. Caine (Sipsey River) [b]	Fayette	Saw and gristmill	40
T. E. Newton & Bro. (Sipsey River). [b]	....do	....do	40
Licurgas Ray (Luxapallila Creek). [b]	Montcalm	....do	30
John Barnes (Luxapallila Creek). [b]	Covin	Gin and grist mill	30

a From U. S. Census, 1900.   *b* From report of probate judge.

Water powers in Alabama, by counties—Continued.

FAYETTE COUNTY—Continued.

Name.	Post-office.	Industry.	Horse-power.
E. Bishop (Luxapallila Creek).[a]	Rainy	Saw, gin, and grist mill.	30
John Williams (Luxapallila Creek).[b]	Covin	Gin and grist mill	30
Washington Hubbert (Shirley Creek).[a]		.....do	10
Gilpin & Jones (Shirley Creek).[a]		Saw, gin, and grist mill.	16
Jones & Jones (Shirley Creek).[a]	Hugent	.....do	20
P. N. Fortenberry (Davis Creek).[a]	Bankston	.....do	o
G. H. White (Davis Creek)[a]	Davis Creek	.....do	16
J. W. Blackburn (Davis Creek).[a]	.....do	.....do	18
M. I. Barnette (Davis Creek).[a]	Ridge	.....do	20
Dolphus Cotton (Clear Creek).[a]	Bankston	.....do	16
M. Miller (Clear Creek) [a]	.....do	.....do	....
John G. Kizer (North River).[a]	Berry station	.....do	40·
Marshall Jones (Bear Creek).[a]	Bear	.....do	20
R. G. Walker (Bear Creek)[a]	.....do	.....do	24
Landon Miles (Stewart Creek).[a]	Hester	Gristmill	12
J. T. McCaleb (Mountain Creek).[a]	Newriver	.....do	18
W. A. Ayers (Beaver Creek)[a]	Fayette	Gin and grist mill	12
G. W. Gray (Boxes Creek)[a]	Stough	Gristmill	16
Miles Whitson (Clear Creek).[a]	Handy	.....do	12
Bud Wade (Hollingsworth Creek).[a]	Newriver	.....do	12

FRANKLIN COUNTY.[b]

Helm's mill	Belgreen	Flour and grist mill	6
M. J. Height	Baggett	.....do	10
James McNair	Kirby	.....do	20
Andrew Posey	Igoburg	.....do	24
Thomas Watson	Phil Campbell	.....do	20
S. T. Bonds	Pleasant Site	.....do	80

a From report of probate judge.   b From U. S. Census, 1900.

Water powers in Alabama, by counties—Continued.

FRANKLIN COUNTY—Continued.

Name.	Post-office.	Industry.	Horse-power.
Jes. S. Scott	Russellville	Flour and grist mill	10
Sparks mill	Underwood	...do	10
John T. McAlister	Phil Campbell	Lumber and timber mill.	10

GENEVA COUNTY.a

Name.	Post-office.	Industry.	Horse-power.
Avant's mill	Geneva	Flour and grist mill	15
Lowry's mill	...do	...do	10
Bell's mill	Fadette	...do	15
W. J. Keith and R. Y. Daniels.	Geneva	...do	15
Clark's gristmill	Highnote	...do	4
Underwood's gristmill	Sanders	...do	20
Condry's gristmill	Whitaker	...do	15
John T. Coleman	Geneva	Lumber and timber mill.	30
Clark Bros. & Co.	Wicksburg	...do	10
Wilson Deshoga	Dundee	Lumber and timber	15
Nathan Hall	Dotham	...do	20

HALE COUNTY.

Name.	Post-office.	Industry.	Horse-power.
William Steward a	Fivemile	Flour and grist mill	8
William A. Avery (Five-mile Creek).a	...do	...do	10
J .H. Payne & Co.a	Ingram	...do	10
M. M. Avery a	Havana	...do	15
Pickens mill a	Greensboro	Lumber and timber mill.	15
Greensboro Carriage and Wagon Shops.a	...do	Carriages and wagons...	6
Richardson's mills (Five-mile Creek).b	Fivemile	Grist mill and gin	20
J. H. Payne's mill (Five-mile Creek).b	Havana	...do	20
Avery's mill (Five-mile Creek).b	...do	...do	25
J. A. Stephenson (Prairie Creek).b	Newbern	...do	20
Irwin & Martin (Big Creek).b	Greensboro	...do	25

a From U. S. Census, 1900.   b From report of probate judge.

Water powers in Alabama, by counties—Continued.

HENRY COUNTY.*a*

Name.	Post-office.	Industry.	Horse-power.
Kennedy's mill	Shorterville	Flour and grist mill	8
Joshua A. Hart	Granger	......do	15
Jeffcoat mill	Gordon	......do	8
Blacksheer & Saunders	Haleburg	......do	25
Cumming's mill	Bush	......do	20
Joe Baker	Headland	......do	27
Badiford gristmill	Little Rock	......do	15
John L. Smith	Ashford	......do	13
Mark Shelley	Balkum	......do	6
Singleterry's water mill	Kinsey	Lumber and timber	27
J. P. Williams & Co	Columbia	......do	25

JACKSON COUNTY.*a*

Name.	Post-office.	Industry.	Horse-power.
Moody's flouring mill	Kyles	Flour and grist mill	40
George W. Brown	Kosh	......do	8
J. F. Bell	Maxwell	......do	4
Coffey's mill	Scottsboro	......do	8
Gross mill	Parks Store	......do	10
Hackworth's mills	Boliver	......do	8
John S. Henegar	Rosalie	......do	20
Bort Harrison	Section	......do	6
W. A. Howell	Hollytree	......do	5
Mathew's gristmill	Carns	......do	10
Page's mill	Woodville	......do	6
Paint Rock Milling Co	Paintrock	......do	8
Reid & Prince	Estillfork	......do	20
David H. Starkey	Kosh	......do	8
Shork mills	Hollywood	......do	60
Cagle mill	Oakley	......do	12
John Thomas	Pisgah	......do	20
Martin Walker	Trenton	......do	40
James P. Williams	....do	......do	20
John V. Wheeler	Pisgah	......do	20
Charles W. Brown	Glenzaida	Lumber and timber mill	25
J. N. Gonce	Anderson	......do	20
Melton Morris	Daugherty	......do	12
David M. Starkey	Kosh	......do	20
Tomon shingle mill	Culver	......do	10

a From U. S. Census, 1900.

Water powers in Alabama, by counties—Continued.

JEFFERSON COUNTY.ᵃ

Name.	Post-office.	Industry.	Horse-power.
J. M. Landrum............	Pinson	Flour and grist mill	20
John Lowery mill.........	Gary	.....do	12
Hendon's corn mill	Trussville	.....do	10
Posey's mill	Morris	.....do	20
James W. Raney..........	Ezra	.....do	35
William B. Rogers	Toadvine	.....do	32
G. W. Underwood	Argo	.....do	15
William J. Wedgworth....	Cardiff	.....do	10
W. W. Woodruff	Adamsville	.....do	8
W. M. Self.................	Oneonto.............	.....do	15
William M. Phillip........	Greene	.....do	40
Morris's mill..............	Ensley	.....do	10
Hurst & Johnson	Pinson	Lumber and timber mill.	18
James W. Raney	Ezra	Woolen goods..........	35

LAMAR COUNTY.

Name.	Post-office.	Industry.	Horse-power.
John H. Cantrell ᵃ	Pharos	Flour and grist mill	15
Claborn E. Carterᵃ........	Detroit..............	.....do	12
Kirk's mill(Yellow Creek)ᵃ	Sizemore	.....do	8
Mote's mill (Beaver Creek)ᵃ	Guin	.....do	6
John T. Moore (Yellow Creek).ᵃ	Vernon..............	.....do	35
W. H. Miller (Luxapallila Creek).ᵃ	Millport.............	.....do	20
Stanford mills ᵃ	Detroit	.....do	12
S. B. Thomas ᵃ.............	Arcola	.....do	10
Lafayette J. Hayes ᵃ.......	Molloy	Lumber and timber mill.	15
Hiram Hollis ᵃ.............	Vernon	.....do	35
Dr. Wm. H. Kennedy ᵃ....	Kennedy	.....do	50
S. B. Thomas ᵃ.............	Arcola	.....do	15
J. O. Kennedy ᵇ............	Kennedy	Mill and gin............	....
J. W. Thomas, jr. (Hills Creek).ᵇ	Alfred..............	Gin, saw, and grist mill.	....
W. M. Thomas (Hills Creek).ᵇ	.....do	.....do	....
Osborn & Hill (Yellow Creek).ᵇ	Blowhorn	.....do	....
D. M. Hollis (Beaver Creek).ᵇ	Beaverton	.....do	....

ᵃ From U. S. Census, 1900.   ᵇ From report of probate judge.

Water powers in Alabama, by counties—Continued.

LAMAR COUNTY—Continued.

Name.	Post-office.	Industry.	Horse-power.
B. G. Boman (Yellow Creek).*a*	Vernon	Gin, saw, and grist mill	
A. A. Mathews (Yellow Creek).*a*	Arcola	do	
W. L. Morton (Yellow Creek).*a*	Vernon	do	
Penning Bros*a*	Baxter	do	

LAUDERDALE COUNTY.*b*

Name.	Post-office.	Industry.	Horse-power.
William M. Thornton	Rogersville	Flour and grist mill	20
James A. Bevis	Threet	do	8
Jessie J. Bevis	Kendell	do	6
George M. Bretherick	Hines	do	24
Isa B. Eastep	Eastep	do	8
Ingram Brothers	Anderson	do	8
Thomas D. Pruitt	Pruitton	do	24
Sharpe's mill	Florence	do	40
Nancy Williams	Lexington	do	20
H. N. Call	Reserve	do	18
Chandler & Chittam	Oliver	do	20

LAWRENCE COUNTY.

Name.	Post-office.	Industry.	Horse-power.
Burrell & Casteel *b*	Progress	Flour and grist mill	10
George's mill *b*	Leighton	do	18,
Jones's estate *b*	Kinlock	do	10
Kerby's mill *b*	Avoca	do	16
Thomas Oliver *b*	Hatton	do	16
John S. Stephenson & Co. *b*	Kinlock	do	27
Wesley L. Stover *b*	Crow	do	15
Terry & Terry *b*	Courtland	do	20
Wallace mill *b*	Avoca	do	10
W. M. Willingham *b*	Camp Spring	Lumber and timber mill	1
H. C. McClannaher (Town Creek).*a*	Mount Hope	Gristmill	
John S. Stephenson (Sipsey River).*a*	Moulton	Flour and gristmill	
Ben F. Masterson (Nances Creek).*a*	do	Gristmill	

a From report of probate judge.   *b* From U. S. Census, 1900.

Water powers in Alabama, by counties—Continued.

LAWRENCE COUNTY—Continued.

Name.	Post-office.	Industry.	Horse-power.
W. G. Hamilton (Nances Creek).[a]	Pitt	Gristmill	
J. M. Key (Brushey Creek)[a]	Pool	do	
W. L. Stover (Flint Creek)[a]	Oakville	Flour and grist mill	
B. A. Casteel (Flint Creek)[a]	Sewickley	do	

LEE COUNTY.

Name	Post-office	Industry	Horse-power
Shelton's mill[b]	Opelika	Flour and grist mill	40
Floyd mill[b]	do	do	10
George W. McKinnon[b]	Yale	do	24
Vaugh mill[b]	Loachapoka	do	20
N. G. Macon (Reed Creek)[b]	do	do	30
W. O. Moore[b]	Auburn	do	40
W. K. Meadows (Halawochee Creek)[b].	Hattie	do	36
James Crosby[b]	Osanippa	do	15
Benjamin F. Stripling[b]	Yale	Lumber and timber	20
W. W. Wright (Chewacla Creek).[a]	Auburn	Not in use now	
W. W. Wright & Geo. P. Harrison (Saugahatchee Creek).[a]	Opelika		
H. J. Spratling (Frazer Creek).[a]	do	Gristmill	25
B. F. Meadows (Halawochee Creek).[a]	do	do	40

LIMESTONE COUNTY.

Name	Post-office	Industry	Horse-power
Weatherford Bros.[b]	Elkmont	Flour and grist mill	6
Carter's mill[b]	Athens	do	16
Dupree & Stepp[b]	Mount Rozell	do	25
Haye's gristmill[b]	Mooresville	do	15
T. M. Holmes[b]	Elkmont	do	12
John M. Head[b]	Pettusville	do	8
Nancy Haney[b]	Legg	do	20
Edward G. Hampleton[b]	Goodsprings	do	15
Thomas D. Hastings[b]	Elkmont	do	5
James L. Lamar[b]	Goodsprings	do	8
Eugene Parham (Piney Creek).[b]	Athens	do	8

[a] From report of probate judge.   [b] From U. S. Census, 1900.

Water powers in Alabama, by counties—Continued.

LIMESTONE COUNTY—Continued.

Name.	Post-office.	Industry.	Horse-power.
M. A. Phillips [a]	Shoalford	Flour and grist mill	12
Ripley's mill [a]	Ripley	....do	15
George Vassar [a]	Lax	....do	8
Witty's Mill (Birds Branch). [a]	Athens	....do	15
William J. Woodfin [a]	Pettusville	....do	15
Pioneer mill [a]	Mount Rozell	....do	20
A. P. Andrews [a]	Elkmont	....do	8
William N. Webb [a]	Elkriver Mills	....do	12
Baker's mill [a]	....do	....do	8
Allison Miller [a]	Rowland	....do	10
Grisham Bros. [a]	Elkriver Mills	Lumber and timber	40
Do	....do	Carriages and wagons	40
L. C. Hightower (Big Creek). [b]	....do	Saw, flour, and grist mill	
Wm. Bailey (Big Creek) [b]	Quidnunc	Flour and grist mill	
J. W. Carter (Big Creek) [b]	O'Neal	Gin, flour, and grist mill	
M. J. Witty (Birds Branch) [b]	Athens	Flour and grist mill	
J. C. Vaughn (Sulphur Creek). [b]	Elkmont	Gin, flour, and grist mill	
R. B. Malone (Sulphur Creek). [b]	Athens	....do	
Wm. Woodfin (Ragsdale Creek). [b]	Elkmont	....do	
J. W. Carter (Panther Creek). [b]	Carter	....do	
John Carroll (Leslie Creek). [b]	Centerhill	....do	
Wm. Davidson (Limestone Creek). [b]	Lax	....do	
R. M. Clem (Piney Creek) [b]	Fairmount	....do	
Eugene Parkam (Piney Creek). [b]	Athens	....do	
W. M. Hayes (Limestone Creek). [b]	Mooresville	....do	
W. H. Roberts (Sugar Creek). [b]	Athens	....do	
W. H. Marbut [b]	Goodsprings	....do	

LOWNDES COUNTY.

Name.	Post-office.	Industry.	Horse-power.
G. B. Holley [a]	Lowndesboro	Flour and grist mill	10
W. N. Bozeman [b]	Benton	Gin and mill	

[a] From U. S. Census, 1900.   [b] From report of probate judge.

Water powers in Alabama, by counties—Continued.

MADISON COUNTY.*a*

Name.	Post-office.	Industry.	Horse-power.
Fannie J. Ridley	Haden	Flour and grist mill	8
D. L. Middleton water mill	Gurley	do	20
Delop's mill	Dan	do	8
Hardy Keel water mill	Gurley	do	15
Annie M. Taylor	Hazelgreen	do	8
Bell Factory mill	Huntsville	do	25
Key's mill	Keysmill	do	28
William S. Russell	Madison Station	do	12
Chas. F. Rountree	Maysville	do	15
William S. Garvin	Monrovia	do	15
A. D. & W. E. Rogers	Newmarket	do	60
B tler Mill Co	Poplarridge	do	30
Payne & Miller	Huntsville	do	30
Martin's gristmill	do	do	15
H. C. Turner	Dan	Lumber and timber	16
Daily Mercury	Huntsville	Printing and publishing	6

MACON COUNTY.*b*

H. H. Robinson	Loachapoka	Flour and grist mill	4
M. W. Glass	Societyhill	do	8
J. O. H. Perry	Tuskegee	do	20

MARION COUNTY.

The Carter mill *a*	Ur	Flour and grist mill	5
Bexar Mercantile Co.*a*	Bexar	do	8
Eads & Fowler *a*	Glenallen	do	12
The Boatwright mill *a*	Inez	do	12
Samuel A. & Wm. V. Read *a*	Eldridge	do	20
Jasper N. Green & Sons *a*	Brilliant	do	20
Elishu Vickery *a*	Winfield	do	16
The Shirley mill *a*	Ur	do	10
Jesse G. Poe *a*	Bearcreek	do	6
Bull, Atkins & Donaldson *a*	Haleysville	do	52
Buttahatchee Mill Co.*a*	do	Lumber and timber	52
John Cumens *a*	do	do	12
Kelly sawmill *a*	do	do	15
John R. Phillips *a*	Bearcreek	do	50

a From U. S. Census, 1900.　　　　*b* From U. S. Census.

Water powers in Alabama, by counties—Continued.

MARION COUNTY—Continued.

Name.	Post-office.	Industry.	Horse-power.
Simon W. Moss[a]	Winfield	Lumber and timber	36
The Powell Mill and Wool Carder.[a]	Duffey	Woolen goods	50
Albert J. Hamilton (Williams Creek).[b]	Hamilton	Flour and grist mill	
W. C. Gann (Sipsey Creek)[b]	Bexar	do	
Q. Northington (Sipsey Creek).[b]	Hamilton	do	
Crane & Riggs (Sipsey Creek).[b]	Delhi	do	
T. L. Shotts (Bull Mountain Creek).[b]	Shottsville	do	
I. J. Loyd (Bull Mountain Creek).[b]	Bull Mountain	do	
D. F. Ballard (Williams Creek).[b]	Hamilton	do	
James P. Pearce (Buttahatchee River).[b]	Pearces Mills	do	
James P. Pearce (New River).[b]	Texas	do	
J. C. Carter (Woods Creek)[b]	Elmira	do	
James Young (Cantrell Mill Creek).[b]	Hamilton	do	
W. J. Wright (Barnesville Mill Creek).[b]	Barnesville	do	
Henry Guin[b]	Guin	do	
Tucker Moss (Luxapallila Creek).[b]	Winfield	do	
D. G. Morrow (Woods Creek).[b]	Elmira	do	

MARSHALL COUNTY[c]

Name.	Post-office.	Industry.	Horse-power.
J. M. Ellison	Preston	Flour and grist mill	4
Mathis mill	Albertville	do	10
James B. Powell	Columbus City	do	4
James F. Prentice	Arab	do	7
P. C. Ragsdale	Uniongrove	do	10
James P. Smith	Warrenton	do	10
Scott's mill	Friendship	do	8
John D. Sumers	Boaz	do	15
Lakey mill	Bartlett	do	10
George E. Whisnant & Son	Oleander	do	10

a From U. S. Census, 1900.   *b* From report of probate judge   *c* From U. S. Census

Water powers in Alabama, by counties—Continued.

MARSHALL COUNTY—Continued.

Name.	Post-office.	Industry.	Horse-power.
I. G. Gross	Columbus City.....	Flour and grist mill	12
Walker & Fowler mills....	Friendship	.....do	20
William J. Copelan	Diamond	.....do	5
James Wm. Barclay.......	Woodville..........	.....do	10
The Winston mill.........	Meltonsville	.....do	12
W. G. Smith Estate.......	Sidney	.....do	10
Jas. M. Selvage	Grant	.....do	4

MARENGO COUNTY.a

Name.	Post-office.	Industry.	Horse-power.
Rhodes mill	Sweetwater........	Flour and grist mill	12

MOBILE COUNTY.a

Name.	Post-office.	Industry.	Horse-power.
N. Q. Thompson..........	Citronelle	Flour and grist mill	10
H. Brannan & Son	Pierce............	Lumber and timber	30
T. A. Hatter & Son........	Creola	.....do	75
Littleton Lee	Pierce............	.....do	60

MONROE COUNTY.a

Name.	Post-office.	Industry.	Horse-power.
J. B. Solomon	Manistee	Flour and grist mill	15
James H. Simpson	Mexia............	.....do	10
Benjamin Johnson	Hollinger..........	.....do	15
Andrew Bohanon..........	Franklin	.....do	15
David J. Hatter & Son.....	Wait	Lumber and timber	60
Do	.....do	.....do ......:.......	20
C. C. Yarbrough	Monroeville	.....do	20

MONTGOMERY COUNTY.a

Name.	Post-office.	Industry.	Horse-power.
Daniel's mill..............	Sellers............	Flour and grist mill	25
Montgomery cotton mill...	Montgomery.......	Cotton goods	35

MORGAN COUNTY.a

Name.	Post-office.	Industry.	Horse-power.
Sarah M. McCutcheon.....	Briscoe............	Flour and grist mill	10

a From U. S. Census, 1900.

Water powers in Alabama, by counties—Continued.

PERRY COUNTY.

Name.	Post-office.	Industry.	Horse-power.
Henry C. Nichols (Dobynes Creek).[a]	Theo	Flour and grist mill	20
Mary G. Wallace[a]	Marion	do	4
Hodger's mill[a]	Newbern	do	15
W. F. Moore[a]	Marion	do	4
Downey's sawmill[a]	Greensboro	Lumber and timber	15
Stevenson's saw and water mills.[a]	Newbern	do	20
Lucinda Washburn (Taylors Creek).[a]	Jericho	do	18
W. T. Downey (Limestone Creek).[b]	Folsom	Gristmill	6
James Wallace (Legroane Creek).[b]	Jericho	do	o
Dr. J. B. Tucker (Taylors Creek).[b]	do	do	6
Lucinda Washburn (Taylors Creek).[b]	do	do	8
S. M. Bolling (Branch of Oakmulgee Creek).[b]	Pinetucky	do	8
C. C. Cosby (Oakmulgee Creek).[b]	Perryville	do	8
Thomas J. Fountain (Little Creek).[b]	Oakmulgee	Gin, saw, and grist mill.	8
Pann Patterson (Little Creek).[b]	do	do	8
Sarah Fountain (Little Creek).[b]	do	do	o
Thaddeus Smith (Little Creek).[b]	Active	Gristmill	8
W. M. Eiland (Fords Mill Creek).[b]	Marion	do	20
J. F. Morton (Potato Patch Creek).[b]	Levert	do	6
Elijah Smith (Beaver Creek).[b]	Bliss	do	6
Noah Coker (Beaver Dam Creek).[b]	Bethlehem	do	6
W. A. Fountain (Oakmulgee Creek).[b]	Oakmulgee	Rice mill	10

a From U. S. Census, 1900.   b From report of probate judge.

Water powers in Alabama, by counties—Continued.

PICKENS COUNTY.

Name.	Post-office.	Industry.	Horse-power.
Richardson & Prichards ...	Coalfire	Flour and grist mill	25
James Mullenix...........	Gordo...............	....do	6
H. B. & A. W. Latham....	Carrollton	....do	12
Slaughter's mill...........	Raleigh	....do	16
W. A. Kerr...............	Reform.............	Lumber and timber	10

PIKE COUNTY.a

Name.	Post-office.	Industry.	Horse-power.
M. J. Youngblood	Youngblood	Flour and grist mill....	110
William F. Ingram........	Josie	....do	20
Nancy Cotton (Cotton's mill).	Milo	....do	12
Ely Dees & J. D. Murphee.	Pronto	....do	20
George W. King	Goshen.............	....do	30
The Lewis mill............	Rodney	....do	24
McQuaggis mill...........	Ansley	....do	15
George F. Williams	Tatum	....do	4
Slatting's gristmill	Henderson	....do	25
P. A. Motia...............	Wingard	....do	8
Bowden & Daughtry	Tennille...........	....do	16
William E. Brown	Josie	....do	10
G. B. Howard	Goshen.............	....do	20

RANDOLPH COUNTY.a

Name.	Post-office.	Industry.	Horse-power.
W. W. Dobson............	Wedowee...........	Flour and grist mill....	20
J. H. White & Z. N. Lipham.	Clack	....do	11
Mrs. Georgia Gibbs	Wedowee...........	....do	10
Giles mill	Ofelia	....do	10
Eppie M. White	Bernice............	....do	5
Larkin & M. B. Taylor	Lamar.............	....do	8
Joseph B. Taylor	Roanoke...........	....do	24
Owins mill	Potash	....do	15
Rogers mill...............	Ofelia	....do	8
C. A. Prescott............	Wedowee...........	....do	20
H. A. Merrill.............	Lamar	....do	6
Elizabeth H. Merrill	Micaville	....do	12
J. E. McCosh & Co........	Lime..............	....do	40
William S. McCarley	Graham	....do	20
John H. Landers...........	Lofty	....do	8

a From U. S. Census, 1900.

Water powers in Alabama, by counties—Continued.

RANDOLPH COUNTY a—Continued.

Name.	Post-office.	Industry.	Horse-power.
Edward Lavoorn		Flour and grist mill	8
Thomas J. Lavoorn	Hawk	do	16
Thomas J. Lavoorn, sr	Newell	do	8
James L. & John T. Kaylor	Kaylor	do	60
Henry C. Jordon	Clack	do	6
J. B. Hammond	Sewell	do	8
T. M. Halaway	Tolbut	do	15
Robert H. Harris	Louina	do	15
Dock Huckaby	Almond	do	10
Holley's mill	Rock Mills	do	30
F. C. Heaton	Hawk	do	10
William N. Gladney	Roanoke	do	12
A. B. East	Christiana	do	2
Adamson & Edwards's mills	Ofelia	do	25
Bailey Mill	Haywood	do	12
F. P. Parker	Foresters Chapel	do	10
John C. Murphy	Gay	do	2
E. L. Pool	Happyland	do	20
James M. Kitchens	Rockdale	do	8
James H. Wright	Jeptha	do	12
Adamson & Edwards	Ofelia	Lumber and timber	40
William W. Brooks	Lofty	do	15
William A. Camp	Almond	do	10
James L. & John T. Kaylor	Kaylor	do	20
H. H. Stephens	Pencil	do	20
Samuel H. Striplin	Roanoke	Leather, tanned, curried, and finished.	6
Wehadkee cotton mills	Rock Mills	Cotton goods	108

RUSSELL COUNTY.

Name.	Post-office.	Industry.	Horse-power.
Davis's mill a	Crawford	Flour and grist mill	20
H. R. Dudley a	Seale	Lumber and timber	40
E. M. Anderson (Watermelon Creek). b	do	Grist mill and gin	20

a From U. S. Census, 1900.   b From report of probate judge.

Water powers in Alabama, by counties—Continued.

SHELBY COUNTY.ᵃ

Name.	Post-office.	Industry.	Horse-power.
W. C. Denson	Pelham	Flour and grist mill	12
William H. Schrader	Shelby	....do	20
William H. Pledger	Pelham	....do	40
Hendrick & Alverson	Vincent	....do	40
David A. Whitfield	Vandiver	....do	10
Brownings mill	Columbiana	Lumber and timber	30

ST. CLAIR COUNTY.ᵃ

Name	Post-office	Industry	Horse-power
The Yarbrough mill	Ashville	Flour and grist mill	8
Hare's mill	....do	....do	8
John R. Dyke	Wolfcreek	....do	30
Perry E. Wyatt	Coal City	....do	10
Henry A. Palmer	Partlow	....do	10
J. M. McLaughlin	Springville	....do	25
The Machen mill	Partlow	....do	10
The Lindsey mill	Ashville	....do	10
Hill & Foreman	Springville	....do	28
Henderson's mill	Ragland	....do	5
Helm & Truss	Helms	....do	20
Grout's mills	Wolfcreek	....do	10
The Gilchrist mill	Ashville	....do	5
The Cox mill	....do	....do	10
Rufus W. Beason	Whitney	....do	11
Rock Bridge mill	Gallant	Lumber and timber mill	20

SUMTER COUNTY.ᵃ

Name	Post-office	Industry	Horse-power
E. B. Hearn (Kinterbish Creek).	Gaston		40
R. H. Stephens (Kinterbish Creek).	Alamuchee		20
R. D. Simmons (Toomsooba Creek).	Bells Station		30
R. W. Shaw	Cuba		10
W. H. Walker (Silver Creek).	Alamuchee		20
J. U. Gillespie (Coatopa Creek).	Coatopa		10

ᵃ From U. S. Census, 1900.

Water powers in Alabama, by counties—Continued.

TALLADEGA COUNTY.

Name.	Post-office.	Industry.	Horse-power.
Jefferson Roberson [a]	Fayetteville	Flour and grist mill	10
J. C. Brock [a]	Eastaboga	do	12
Riser & Bro. [a]	Talladega	do	40
Shock E. Jemison [a]	Sunnyside	do	15
Vincent's mill [a]	Talladega	do	25
O. F. Luttrell [a]	do	do	40
Riddle mills [b]	Waldo	do	16
J. F. Smith [a]	Eastaboga	do	40
John W. Thweatt [a]	McFall	do	12
J. B. Turner [a]	do	do	15
Allison's mill [a]	Talladega	do	60
J. F. Smith [a]	Eastaboga	Lumber and timber	40
Cragdale mill [a]	Talladega	do	40
J. B. Turner [a]	McFall	do	20
Priebes mill (Choccolocco Creek). [a]	Jenifer	Gristmill	200
J. F. Smith's mill (Choccolocco Creek). [a]	Oxford	do	225
B. Schmidt's mill (Choccolocco Creek). [a]	Lincoln	do	200
Craig's mill (Choccolocco Creek). [a]	Oxford	do	150
Wilson's mill (Choccolocco Creek). [a]	Jenifer	do	150
Eureka mills (Choccolocco Creek). [a]	Eureka	do	150
Turner's mill (Chehawhaw Creek). [a]	McFall	do	150
Kants mill (Talladega Creek). [a]	Chandler Springs	do	50
Riddle's mill (Talladega Creek). [a]	Waldo	do	75
Taylor's mill (Talladega Creek). [a]	Talladega	do	150
Reynold's mill (Talladega Creek). [a]	Nottingham	do	150
Allison's mill (Talladega Creek). [a]	Talladega	do	75
Duncan's mill (Talladega Creek). [a]	Alpine	do	75
Baker's mill (Talladega Creek). [a]	Kymulga	do	100
Vincent's mill (Crooked Creek). [a]	Sylacauga	do	50

[a] From report of probate judge.

Water powers in Alabama, by counties—Continued.

TALLEDEGA COUNTY—Continued. .

Name.	Post-office.	Industry.	Horse-power.
Oden's mill (Short Creek).[a]	Sylacauga	Gristmill	75
Jemison's mill (Kelly Creek).[a]	Sunnyside	....do	50
Camp & Sons' mill (Salt Creek).[a]	Hopeful	....do	50
Robinson's mill (Cedar Creek).[a]	Fayetteville	....do	50
Lackey's mill (Horse Creek).[a]	Ironaton	....do	25
Talladega Company (Choccolocco Creek).[a]	Talladega	Organized for electric transmission.	....

TALLAPOOSA COUNTY.[b]

Name.	Post-office.	Industry.	Horse-power.
George Stewart	Thaddeus	Flour and grist mill	12
John W. Britt	Jacksons Gap	....do	20
Benjamin F. Jarvis	Yates	....do	12
T. J. Hamlet	Hamlet	....do	15
T. W. Whitman	Dadeville	....do	20
Sanford Milling and Manufacturing Co.	....do	....do	25
John W. Hay	Camphill	....do	15
John B. Calhoun	Camphill	....do	8
Hammond's mill	Dadeville	....do	20
Hodnett grist and flour mill.	Acme	....do	16
Thomas L. Bulger	Dadeville	....do	15
Vines mills	Easton	....do	40
A. T. & H. C. Vickers	Newsite	....do	20
J. C. Street	Anniston	....do	25
Shephard Bros. & Co	Tohopeka	....do	10
G. W. Stewart	Thaddeus	....do	25
Albert J. Hollaway	Alexander	....do	20
Mrs. Milliner	Mary	....do	25
Jno. L. Patterson	Hackneyville	....do	12
Thomas B. Griffin	Matilda	....do	10
Daviston mill	Daviston	....do	8
Lamberth & Dewberry	Logpit	....do	20
Silver Shoals mill	Buttston	....do	80
M. R. Hays & Bro	Notasulga	....do	40

a From report of probate judge.   b From U. S Census, 1900.

Water powers in Alabama, by counties—Continued.

TALLAPOOSA COUNTY—Continued.

Name.	Post-office.	Industry.	Horse-power.
Farrows flour and grist mill.	Susanna	Flour and grist mill	60
J. H. Yarbrough	Hackneyville	do	12
T. F. Garnett	Tallassee	Lumber and timber	20
G. W. Stewart	Thaddeus	do	20

WINSTON COUNTY.

Name.	Post-office.	Industry.	Horse-power.
Richard H. Blake[a]	Houston	Flour and grist mill	8
Thomas O. Partridge[a]	Elk	do	10
Wm. D. Shadix (Sandy Creek).[a]	Double Springs	do	4
George D. Wilson[a]	Haleysville	do	8
Manna A. Posey[a]	Motes	do	10
Martin A. & Martha Peak[a]	Peaks Mill	do	10
Miligan mill[a]	Double Springs	do	10
James Cantrell[a]	Addison	do	4
Burks mill[a]	Cranal	do	10
Nauvoo mill (Blackwater Creek).[b]	Nauvoo	Grist mill and gin	
Anderson Ward mill (Clear Creek).[b]	Haleysville	Flour and grist mill	
J. Calvin Cagle (Clear Creek).[b]	Double Springs	Saw, flour, and grist mill and gin	
Jonathan Barton mill (Clear Creek).[b]	Deer	Gristmill	
Hadder mill (Clear Creek)[b].	Double Springs	do	
Posey mill (Clear Creek)[b].	Motes	Gristmill, saw, and gin	
S. D. Spain (Clear Creek)[b].	Malta	do	
Gus Posey mill (Clear Creek).[b]	Elk	do	
Wm. Dodd (Splunge Creek).[b]	Natural Bridge	do	
Kelley mill (Blackwater Creek).[b]	Lynn	do	
Peaks mill (Grindstone Creek).[b]	Peaks Mill	do	
Jack Curtis (Sandy Creek)[b].	Double Springs	do	
Manley Payne (Beech Creek).[b]	Gumpond	do	
Christian mill (Christian Creek).[b]	Peaks Mill	Grist and saw mill	

a From U. S. Census, 1900.   b From report of probate judge.

APPENDIX.

STREAM MEASUREMENTS IN MISSISSIPPI.

In the foregoing sections (pp. 167–175, 179–180) have been included discharge measurements, gage heights, and estimates of flow at Columbus, Miss., on Tombigbee River, which flows into Alabama, and Luxapallila Creek, which flows out of Alabama into Mississippi.

The following is a statement of the hydrographic work on the remaining important streams of Mississippi, exclusive of Mississippi River:

PEARL RIVER AT JACKSON, MISS.

This station was established June 24, 1901, and is situated at the highway bridge 2 miles from the union station at Jackson and one-eighth mile above the Alabama and Vicksburg Railway bridge.

The gage is a wire gage fastened to the guard timber on the downstream side of the bridge. The rod is 10 feet long, graded to feet and tenths, and is marked with staples and brass figures. Above 10 feet the guard timber is marked. The bench mark is the downstream end of the top of the iron crossbeam 120 feet from the right-bank end of the bridge, which latter is the initial point for soundings. The elevation of the bench mark is 39 feet above datum. The bridge floor at the same point is 40.15 feet above datum.

The observer is James Hurst. The following discharge measurements were made by K. T. Thomas, M. R. Hall, and others:

Discharge measurements of Pearl River at Jackson, Miss.

Date.	Gage height.	Discharge.	Date.	Gage height.	Discharge.
1901.	*Feet.*	*Second-feet.*	1903.	*Feet.*	*Second-feet.*
June 24	2.10	430	March 10	24.00	16,050
August 15	5.85	1,880	July 13	4.15	1,348
October 28	1.55	262	July 14	5.37	1,988
1902.			September 24	.80	128
July 14	1.10	193	Do	.78	142
September 25	.90	290			

Daily gage height, in feet, of Pearl River at Jackson, Miss.

Day.	Jan.	Feb.	Mar.	Apr.	May.	June.	July.	Aug.	Sept.	Oct.	Nov.	Dec.
1901.												
1							1.90	1.50	5.90	1.90	1.40	3.50
2							1.70	1.50	5.70	1.80	1.30	3.30
3							1.80	1.50	5.60	1.60	1.30	3.40
4							1.80	1.40	5.40	1.40	1.40	3.20
5							1.70	1.30	5.20	1.50	1.20	3.30
6							1.80	1.80	4.70	1.60	1.00	3.30
7							1.80	1.20	3.90	1.50	1.30	3.20
8							1.80	1.20	3.60	1.50	1.50	3.40
9							1.80	1.20	3.20	1.40	1.80	3.90
10							1.80	1.20	3.00	1.10	1.80	4.20
11							1.80	1.20	2.00	1.30	1.60	5.40
12							1.80	1.20	2.60	1.40	1.60	5.80
13							1.80	1.30	2.50	2.00	1.60	6.40
14							1.60	1.40	2.30	3.00	1.60	10.00
15							1.80	1.60	2.30	3.40	1.50	14.90
16							1.70	1.80	2.50	3.40	1.50	15.00
17							1.70	2.10	2.80	3.00	1.50	15.20
18							1.70	2.60	3.40	2.10	1.50	15.90
19							1.70	3.70	3.40	2.30	1.50	15.40
20							1.60	4.50	3.60	2.10	1.60	15.00
21							1.70	6.00	3.80	2.00	1.60	14.60
22							1.70	6.30	3.70	1.90	1.80	14.30
23							1.70	7.40	3.50	1.80	1.90	13.20
24						2.10	1.80	7.40	3.10	1.80	2.10	12.60
25						2.10	2.20	7.30	2.20	1.70	2.40	12.70
26						2.00	2.00	7.30	2.40	1.70	2.70	11.90
27						2.00	1.90	7.40	2.50	1.60	3.80	11.30
28						1.90	1.90	7.60	2.50	1.60	3.50	10.60
29						2.00	1.80	6.90	2.30	1.60	3.60	13.60
30						2.10	1.80	6.90	2.10	1.50	3.70	14.60
31							1.60	6.70		1.40		15.00
1902.												
1	15.00	13.90	13.00	37.20	6.00	2.20	1.30	1.30	7.90	1.90	1.00	3.10
2	15.20	16.60	13.00	36.20	5.90	2.20	1.30	1.20	4.80	2.00	1.00	4.60
3	15.10	17.10	13.00	31.60	5.60	2.10	1.30	1.70	3.60	2.40	1.00	4.10
4	15.00	18.30	13.20	33.10	5.10	2.00	1.20	1.70	3.40	2.80	1.10	5.30
5	14.80	18.60	13.40	32.00	4.90	2.00	1.20	1.60	3.20	3.00	1.10	5.80
6	14.50	18.40	13.70	31.00	4.80	1.90	1.20	1.60	3.60	8.60	1.00	6.30
7	14.20	18.30	14.00	30.10	4.60	1.90	1.30	1.40	3.90	3.20	1.00	6.80
8	13.90	18.20	13.80	29.10	4.30	1.80	1.30	1.40	3.80	3.10	1.00	6.40
9	13.40	17.00	13.60	28.10	4.00	1.80	1.30	1.40	3.60	2.90	1.20	6.00
10	12.60	17.70	13.20	27.10	3.90	1.80	3.00	1.40	3.50	2.30	1.20	5.90
11	11.80	17.60	12.00	26.30	3.50	1.80	3.50	1.40	3.60	2.30	1.30	5.70
12	10.80	17.40	12.20	25.60	3.20	1.70	3.00	1.40	3.50	2.60	1.30	5.40
13	10.00	16.90	11.70	24.80	3.20	1.70	1.20	1.40	3.50	2.60	1.30	5.20
14	9.10	15.80	11.90	24.40	3.10	1.70	1.20	1.30	3.90	2.80	1.40	5.70
15	8.10	14.30	10.30	24.00	3.00	1.60	1.20	1.30	3.70	3.00	1.40	5.10
16	8.00	12.60	10.60	23.50	2.10	1.60	1.30	1.30	3.60	3.20	1.40	10.60
17	7.90	10.90	10.60	23.00	2.80	1.50	1.80	1.20	3.40	3.30	1.40	10.70
18	6.70	9.70	10.60	22.30	2.60	1.50	1.20	1.20	3.60	3.60	1.30	10.50
19	5.40	9.00	10.50	21.20	3.00	1.50	1.10	1.10	2.90	3.50	1.30	11.10
20	5.10	8.60	10.00	19.50	2.90	1.40	1.10	1.10	2.60	3.30	1.20	11.30
21	5.40	8.30	9.90	17.20	2.80	1.30	1.10	1.10	2.40	3.20	1.20	11.60
22	5.60	8.20	9.90	14.00	2.60	1.80	1.10	1.10	1.20	2.00	1.20	12.20
23	5.80	8.30	9.90	11.10	2.80	1.30	1.10	.90	1.20	1.90	1.10	12.20
24	6.40	9.00	10.80	9.30	3.00	1.40	1.10	1.00	1.10	1.80	1.80	11.90
25	6.90	10.70	12.90	8.10	3.00	1.40	1.20	.90	1.10	1.50	1.70	11.70
26	6.20	11.40	13.80	7.20	3.00	1.30	1.40	.80	1.30	1.30	1.80	10.80
27	7.00	11.60	22.70	7.10	2.70	1.30	1.30	1.00	1.30	1.10	1.70	10.80
28	7.10	12.30	28.55	6.50	2.40	1.30	1.40	2.20	1.60	1.30	1.60	10.00
29	7.20		30.40	6.40	2.50	1.30	1.40	4.50	1.60	1.20	1.60	10.90
30	8.50		32.10	6.20	2.40	1.30	1.40	7.20	1.80	2.60	2.60	8.80
31	10.10		36.50		2.20		1.40	7.40		1.10		9.90

Daily gage height, in feet, of Pearl River at Jackson, Miss.—Continued.

Day.	Jan.	Feb.	Mar.	Apr.	May.	June.	July.	Aug.	Sept.	Oct.	Nov.	Dec.
1903.												
1	10.90	6.7	26.0	12.6	2.8	2.3	1.7	2.0	2.2	0.7	0.3	0.3
2	13.80	7.4	25.1	10.9	2.7	2.3	1.8	2.0	2.0	.7	.3	.3
3	14.90	12.2	24.2	9.8	2.6	2.2	1.8	1.8	1.9	.7	.3	.3
4	15.20	13.4	23.0	9.2	2.6	2.2	1.9	1.7	1.8	.7	.3	.3
5	15.80	14.6	22.4	8.5	2.5	2.1	2.0	1.5	1.6	.7	.3	.4
6	16.60	15.8	21.8	7.8	2.5	2.2	2.1	1.6	1.6	.7	.3	.5
7	16.00	19.0	21.6	7.1	2.4	2.2	2.2	2.5	1.5	.7	.3	.6
8	16.00	23.0	21.6	6.7	2.4	2.3	2.3	2.6	1.3	.6	.3	.8
9	15.90	23.4	23.5	6.1	2.5	3.5	2.4	2.8	1.2	.6	.3	.9
10	15.60	24.6	23.9	5.7	2.5	3.4	2.5	2.9	1.2	.6	.3	.9
11	15.60	27.5	24.7	5.2	2.4	2.5	2.7	3.5	1.1	.6	.3	1.0
12	15.40	30.4	24.9	5.0	2.4	2.4	2.8	4.4	1.1	.6	.3	1.0
13	14.00	32.5	25.2	5.8	2.5	2.3	2.7	4.5	1.0	.6	.3	1.0
14	14.60	33.7	25.4	4.6	2.6	2.3	5.3	4.6	1.0	.6	.3	1.0
15	13.80	33.5	25.7	4.4	2.8	2.4	4.6	4.6	1.0	.5	.3	1.2
16	13.40	33.2	25.9	4.3	3.0	2.5	4.3	4.5	1.0	.5	.3	1.2
17	13.10	33.1	25.9	4.5	3.1	2.3	3.8	4.6		.5	.3	1.3
18	13.20	32.8	25.8	4.7	3.5	2.2	2.7	4.0	.9	.5	.3	1.3
19	13.10	32.3	25.8	4.6	4.0	2.1	2.5	4.6	.9	.5	.3	1.3
20	13.00	31.5	24.6	4.5	4.4	2.0	2.4	4.8	.8	.5	.3	1.3
21	12.50	30.9	23.7	4.8	4.5	1.9	2.2	5.1	.8	.6	.8	1.3
22	11.40	30.9	23.7	4.2	4.2	1.8	2.0	4.8	.8	.6	.3	1.3
23	10.40	30.1	23.4	4.0	3.9	1.7	1.9	4.6	.8	.6	.3	1.3
24	9.50	30.0	22.8	3.8	3.7	1.6	1.7	4.3	.8	.6	.3	1.4
25	8.40	29.8	21.9	3.6	3.5	1.7	1.6	4.1	.8	.5	.3	1.4
26	8.90	28.7	20.9	3.4	3.4	1.8	1.5	3.8	.8	.5	.3	1.4
27	7.80	27.6	19.6	3.2	3.3	1.8	1.5	3.4	.8	.5	.3	1.4
28	7.30	26.7	18.0	3.0	3.1	1.7	1.4	3.0	.8	.5	.3	1.4
29	6.80		16.5	2.9	3.5	1.7	1.3	2.7	.9	.4	.3	1.4
30	6.50		15.3	2.8	2.7	1.8	1.2	2.5	.8	.3	.3	1.3
31	6.60		13.8		2.5		1.2	2.3		.3		1.3

Rating table for Pearl River at Jackson, Miss., for 1901, 1902, and 1903.

Gage height.	Discharge.	Gage height.	Discharge.	Gage height.	Discharge.	Gage height.	Discharge.
Feet.	*Second-feet.*	*Feet.*	*Second-feet.*	*Feet.*	*Second-feet.*	*Feet.*	*Second-feet.*
0.30	98	4.90	1,730	9.50	4,860	14.10	8,310
.40	103	5.00	1,790	9.60	4,935	14.20	8,385
.50	109	5.10	1,850	9.70	5,010	14.30	8,460
.60	117	5.20	1,910	9.80	5,085	14.40	8,535
.70	126	5.30	1,970	9.90	5,160	14.50	8,610
.80	136	5.40	2,030	10.00	5,235	14.60	8,685
.90	148	5.50	2,090	10.10	5,310	14.70	8,760
1.00	162	5.60	2,150	10.20	5,385	14.80	8,835
1.10	177	5.70	2,210	10.30	5,460	14.90	8,910
1.20	194	5.80	2,270	10.40	5,535	15.00	8,983
1.30	214	5.90	2,330	10.50	5,610	15.10	9,060
1.40	235	6.00	2,390	10.60	5,685	15.20	9,135
1.50	257	6.10	2,450	10.70	5,760	15.30	9,210
1.60	280	6.20	2,510	10.80	5,835	15.40	9,285
1.70	305	6.30	2,570	10.90	5,910	15.50	9,360
1.80	333	6.40	2,630	11.00	5,985	15.60	9,435
1.90	365	6.50	2,695	11.10	6,060	15.70	9,510
2.00	398	6.60	2,760	11.20	6,135	15.80	9,585
2.10	432	6.70	2,825	11.30	6,210	15.90	9,660
2.20	467	6.80	2,890	11.40	6,285	16.00	9,735
2.30	503	6.90	2,955	11.50	6,360	16.50	10,135
2.40	540	7.00	3,020	11.60	6,435	17.00	10,535
2.50	580	7.10	3,090	11.70	6,510	17.50	10,935
2.60	620	7.20	3,160	11.80	6,585	18.00	11,355
2.70	660	7.30	3,230	11.90	6,660	18.50	11,735
2.80	700	7.40	3,300	12.00	6,735	19.00	12,135
2.90	740	7.50	3,370	12.10	6,810	19.50	12,535
3.00	785	7.60	3,440	12.20	6,885	19.80	12,935
3.10	830	7.70	3,510	12.30	6,960	19.90	13,335
3.20	875	7.80	3,585	12.40	7,035	21.00	13,735
3.30	920	7.90	3,660	12.50	7,110	22.00	14,535
3.40	965	8.00	3,735	12.60	7,185	23.00	15,335
3.50	1,010	8.10	3,810	12.70	7,260	24.00	16,135
3.60	1,055	8.20	3,885	12.80	7,335	25.00	16,935
3.70	1,100	8.30	3,960	12.90	7,410	26.00	17,735
3.80	1,150	8.40	4,035	13.00	7,485	27.00	18,535
3.90	1,200	8.50	4,110	13.10	7,560	28.00	19,335
4.00	1,250	8.60	4,185	13.20	7,635	29.00	20,135
4.10	1,300	8.70	4,260	13.30	7,710	30.00	20,935
4.20	1,350	8.80	4,335	13.40	7,785	31.00	21,735
4.30	1,400	8.90	4,410	13.50	7,860	32.00	22,535
4.40	1,455	9.00	4,485	13.60	7,935	33.00	23,335
4.50	1,510	9.10	4,560	13.70	8,010	34.00	24,135
4.60	1,565	9.20	4,635	13.80	8,085		
4.70	1,620	9.30	4,710	13.90	8,160		
4.80	1,675	9.40	4,785	14.00	8,235		

Estimated monthly discharge of Pearl River at Jackson, Miss.

Month.	Discharge in second-feet.		
	Maximum.	Minimun.	Mean.
1903.			
January	10,215	2,695	7,082
February	23,895	2,825	17,421
March	17,735	8,085	15,162
April	7,185	700	2,265
May	1,510	540	821
June	1,010	280	468
July	1,970	194	558
August	1,850	257	1,023
September	467	136	197
October	126	98	115
November	98	98	98
December	235	98	178
The year	23,895	98	3,782

Net horsepower per foot of fall with a turbine efficiency of 80 per cent for the minimum monthly discharge of Pearl River at Jackson, Miss.

Month.	1901.			1902.			1903.		
	Minimum discharge.	Minimum net horsepower per foot of fall.	Duration of minimum.	Minimum discharge.	Minimum net horsepower per foot of fall.	Duration of minimum.	Minimum discharge.	Minimum net horsepower per foot of fall.	Duration of minimum.
	Sec.-feet.		Days.	Sec.-feet.		Days.	Sec.-feet.		Days.
January				1,850	168	1	2,695	245	1
February				3,885	353	1	2,825	257	1
March				5,160	469	4	8,085	735	1
April				2,510	228	1	700	64	1
May				432	39	1	540	49	4
June	398	36	3	214	19	9	280	25	1
July	280	25	3	177	16	6	194	18	2
August	194	18	6	136	12	1	257	23	1
September	398	36	1	162	15	1	136	12	12
October	235	21	5	177	16	2	98	9	2
November	194	18	1	162	15	7	98	9	30
December	875	80	2	830	75	1	98	9	4

YAZOO RIVER AT YAZOO CITY, MISS.

A gage has been maintained at this point by the Engineer Corps of the Army. It was replaced in 1901 by a new gage rod in three sections, marked with brass figures and brass tacks, the sections being placed as follows: The lowest, marked from −3 to +4.5 feet, is attached to the protecting work of the bridge; the middle section, marked from 4.5 to 18.5 feet, is attached to the piling that protects the bridge pier; the uppermost section, continuing the graduation up to 32.3 feet, is on a post under the approach to the bridge. The highest known water occurred in 1882, reaching a gage height of 36.5 feet; the lowest occurred on October 15 to 17 and 20 to 22, 1896, with a gage height of −2.8 feet. The danger line is at 25 feet. A bench mark was established on the top of the upstream cylinder of the second pier from the left bank, at a distance of 85 feet from the initial point for soundings, which is on the downstream end of iron bridge on the left bank. The elevation of the mark is 35.85 feet above the zero of the gage. Other important bench marks in Yazoo City are the following: P. B. M. 12, Yazoo City, is a copper bolt in stone underground, surmounted by an iron pipe and cap, in the north corner of the county court-house yard. It is 44.1 feet above the zero of the gage and 116.2 feet above mean sea level. P. B. M. 13, Yazoo City, is a copper bolt in a stone underground, surmounted by an iron pipe and cap, in the north corner of the public school yard, near Washington and Main streets. It is 29.2 feet above the zero of the gage, and 101.3 feet above mean sea level. Discharge measurements are made by the United States Geological Survey from the city toll bridge, one-half mile northwest from the Illinois Central station. The observer is P. C. Battaille. Daily gage heights are furnished by the Weather Bureau.

The following measurements were made by K. T. Thomas, M. R. Hall, and others:

Discharge measurements of Yazoo River at Yazoo City, Miss.

Date.	Gage height.	Discharge.	Date.	Gage height.	Discharge.
1901.	*Feet.*	*Second-feet.*	1902.	*Feet.*	*Second-feet.*
March 9	11.80	11,618	September 24........	−1.0	2,108
April 12.............	13.40	11,779	September 25........	−1.0	2,048
June 22.............	2.92	3,935	1903.		
1902.			July 13.............	5.80	4,755
July 12..............	3.40	2,887	September 23........	−2.00	1,623
July 14..............	4.40	3,672			

Daily gage height, in feet, of Yazoo River at Yazoo City, Miss.

Day.	Jan.	Feb.	Mar.	Apr.	May.	June.	July.	Aug.	Sept.	Oct.	Nov.	Dec.
1900.												
1	5.0	1.0	15.0	19.7	22.9	8.0	21.1	18.7	0.6	1.6	9.6	9.6
2	4.6	.7	15.0	19.5	22.9	9.5	21.2	18.2	.6	1.5	9.6	9.8
3	4.3	.4	15.0	19.2	22.9	8.1	21.3	17.8	.6	1.4	9.6	10.0
4	4.0	.4	15.3	18.8	22.9	8.3	21.4	17.4	.7	1.2	9.5	10.3
5	3.5	.4	15.4	18.5	22.8	8.6	21.5	17.0	.7	.9	5.3	10.5
6	3.0	.3	15.4	18.1	22.7	9.4	21.6	16.5	.6	.6	9.2	10.8
7	2.4	.2	15.5	17.6	22.6	10.3	21.7	16.0	.4	.2	9.0	11.1
8	1.7	.2	16.0	17.0	22.4	11.0	21.9	15.5	.1	.1	8.8	11.5
9	1.2	1.1	16.1	16.4	22.3	11.5	22.0	14.7	.9	.3	8.5	11.7
10	.9	1.5	16.2	15.6	22.0	12.0	22.1	13.7	.9	.5	8.1	11.9
11	1.3	2.0	16.3	17.6	21.7	12.6	22.1	12.9	.8	.7	7.7	12.0
12	1.3	3.2	16.4	17.2	21.4	13.1	22.1	11.7	.8	.7	7.3	12.1
13	1.0	4.9	16.5	17.0	21.0	13.9	22.1	10.5	.7	.6	6.8	12.2
14	1.0	6.0	16.5	17.0	20.7	17.6	22.1	9.2	.6	.5	6.2	12.2
15	1.6	6.9	16.5	17.2	20.3	17.8	22.1	7.8	.5	.3	5.5	12.0
16	2.1	7.5	17.0	17.6	20.0	17.6	21.9	6.5	.4	.1	4.7	11.7
17	2.5	8.0	17.2	21.5	19.6	17.7	21.9	5.3	.3	.8	3.9	11.5
18	2.8	8.3	17.3	24.4	19.2	18.1	21.8	4.3	.2	.8	3.2	11.1
19	3.0	8.5	18.8	21.8	18.6	18.8	21.6	3.5	.1	.9	2.5	10.7
20	3.3	8.8	19.5	22.0	18.0	19.1	22.3	2.8	.1	1.0	2.1	11.8
21	3.5	9.9	19.5	22.0	17.4	19.8	22.0	2.4	.6	.9	1.6	11.1
22	3.6	10.5	19.8	22.0	16.6	20.0	21.5	2.0	.5	1.3	1.4	10.7
23	8.7	11.0	20.0	22.2	16.0	20.0	21.2	1.7	.5	1.4	2.2	10.5
24	3.7	11.4	20.2	22.2	15.0	20.2	20.0	1.4	.7	1.2	3.6	10.5
25	3.6	12.1	20.5	22.2	13.8	20.3	20.6	1.2	.0	1.9	5.0	10.5
26	3.4	12.4	20.5	22.2	12.7	20.5	20.5	1.0	.4	3.9	6.2	10.5
27	8.1	12.5	20.5	22.3	11.6	20.8	20.4	.9	.9	6.0	7.0	10.2
28	2.7	14.4	20.5	22.3	10.4	20.8	20.2	.8	1.3	7.1	8.0	10.8
29	2.3	......	20.8	22.9	9.2	20.9	19.9	.7	1.5	8.0	8.6	11.1
30	1.8	......	20.2	23.0	8.0	21.0	19.4	.6	1.6	8.7	9.1	11.4
31	1.4		20.0		7.1		19.0	.6		9.1		11.7
1901.												
1	11.8	16.8	16.4	16.5	16.6	9.0	−.1	−1.4	9.1	7.4	−1.5	.1
2	11.9	16.7	16.0	16.4	16.6	8.1	−.2	−1.4	8.9	7.3	−1.5	.1
3	12.1	17.7	15.6	16.2	16.5	7.6	−.3	−1.4	8.4	7.0	−1.5	.0
4	12.2	18.2	15.1	16.0	16.3	7.5	−.3	−1.3	7.8	6.9	−1.0	.4
5	12.3	17.8	14.7	15.6	16.0	7.4	−.2	−1.0	7.0	6.8	−1.3	.5
6	12.4	17.6	14.1	15.3	15.8	7.2	−.1	−.8	6.2	6.5	−1.4	.5
7	12.4	17.6	13.4	14.9	15.6	7.5	+.1	−.8	5.3	6.0	−1.5	.6
8	12.4	18.0	12.6	14.5	15.5	7.0	.3	−1.0	4.3	5.6	−1.5	.7
9	12.4	17.7	11.8	14.2	15.4	6.5	.4	−1.2	3.4	4.8	−1.5	+2.1
10	12.4	17.8	12.0	13.9	15.2	6.1	.4	−1.3	2.5	4.0	−1.6	2.5
11	18.5	17.7	12.3	13.7	15.2	5.7	.3	−1.4	1.8	3.1	−1.5	2.3
12	18.2	17.9	12.4	18.6	15.2	5.4	.1	−1.4	1.1	2.3	−1.6	2.8
13	17.6	17.9	12.6	13.4	15.2	5.0	−.3	−.3	1.0	1.7	−1.6	3.5
14	17.6	17.8	12.9	13.3	15.5	4.7	−.6	−.7	2.5	1.1	−1.7	6.7
15	17.8	17.7	13.0	13.1	15.5	4.5	−.9	+1.2	2.5	.7	−1.7	6.9
16	17.8	17.7	13.1	12.9	15.4	4.2	−1.0	2.0	3.0	.4	−1.8	7.3
17	17.8	17.7	13.3	12.8	15.4	4.1	−1.2	1.2	3.5	.2	−1.8	8.0
18	17.8	17.6	13.5	14.4	15.4	3.9	−1.2	1.8	4.7	.0	−1.8	8.7
19	17.7	17.5	13.7	14.6	15.1	3.7	−1.2	2.6	5.9	−.3	−1.7	9.3
20	17.6	17.5	14.0	14.8	14.7	3.5	−1.2	3.8	7.0	−.5	−1.7	9.5
21	17.5	17.4	14.3	15.2	14.2	3.3	−.5	4.9	7.7	−.7	−1.7	9.7
22	17.4	17.3	14.7	15.5	13.2	3.0	−.3	5.6	8.2	−.7	−1.4	9.8
23	17.3	17.2	15.0	15.8	11.9	2.6	−.7	6.1	8.3	−.8	−.4	9.8
24	17.4	17.1	15.5	16.0	10.7	2.2	−.7	6.9	8.3	−.8	−.1	9.8
25	17.4	17.0	15.9	16.1	11.7	1.8	−.8	6.9	8.3	−.8	−.1	9.8
26	17.3	16.8	16.2	16.3	12.2	1.4	−.9	7.3	8.1	−.9	−.1	9.7
27	17.3	16.5	16.2	16.5	11.3	1.0	−1.0	8.4	8.0	−1.0	+.5	9.7
28	17.3	16.2	16.2	16.6	10.7	.6	−1.1	8.3	7.8	−1.1	.5	10.0
29	17.2	......	16.3	16.6	10.2	.3	−1.2	8.8	7.7	−1.1	.6	11.6
30	17.1	......	16.3	16.6	9.5	.1	1.3	9.0	7.5	−1.3	.5	11.5
31	17.0		16.6		9.2		−1.4	9.1		−1.4		11.1

Daily gage height, in feet, of Yazoo River at Yazoo City, Miss.—Continued.

Day.	Jan.	Feb.	Mar.	Apr.	May.	June.	July.	Aug.	Sept.	Oct.	Nov.	Dec.
1902.												
1	11.40	11.30	16.70	23.90	25.30	2.30	.10	3.20	-.70	-1.00	-2.10	4.90
2	11.50	12.40	16.60	23.80	25.00	1.90	.20	3.00	-.90	-.90	2.10	5.30
3	11.40	12.80	16.60	23.80	24.70	1.80	.20	3.00	-.50	-.40	2.10	5.80
4	11.00	13.20	16.60	23.80	21.40	1.60	.00	3.20	-.50	.00	2.10	5.90
5	10.90	13.40	16.90	23.80	24.10	1.50	.00	3.60	-.50	.60	2.10	6.00
6	10.80	13.70	16.90	23.80	23.70	1.50	-.10	4.00	-.30	1.00	-2.10	6.30
7	10.50	13.90	16.80	25.10	23.30	1.70	-.20	4.40	.00	1.30	-2.10	6.50
8	10.30	14.10	16.80	25.10	22.90	2.00	.00	4.50	.20	1.50	-2.10	6.90
9	10.00	14.30	16.80	25.00	22.40	2.20	.70	4.70	.30	1.50	2.10	7.00
10	9.70	14.60	16.70	25.00	21.80	2.40	1.50	4.70	.00	1.20	2.10	7.20
11	9.30	14.90	16.70	25.10	20.90	2.50	2.40	4.70	-.20	.90	-2.10	7.30
12	8.90	15.20	16.90	25.30	20.10	2.40	3.40	4.70	-.50	.70	-2.10	7.40
13	8.40	15.50	17.40	25.50	19.20	2.10	4.00	4.60	-.70	.50	2.10	7.40
14	7.90	15.80	17.80	25.60	18.00	1.80	4.50	4.60	-1.10	.50	2.10	7.50
15	7.30	16.00	17.30	25.80	16.90	1.50	4.90	4.50	1.20	.50	2.10	7.50
16	6.50	16.30	17.90	26.00	15.50	1.20	5.00	4.50	-1.30	.20	2.10	9.80
17	5.70	16.40	17.80	26.20	14.20	.90	5.00	4.40	-1.40	.20	2.10	9.80
18	4.90	16.40	17.70	26.30	13.00	.50	4.70	4.40	-1.50	.50	2.10	10.20
19	4.20	16.40	17.70	26.40	11.90	.40	4.40	4.30	-1.50	-1.00	2.10	10.90
20	3.50	16.30	17.80	26.50	10.80	.30	4.00	4.20	-1.50	-1.20	-2.10	11.50
21	3.20	16.20	18.00	26.60	9.70	.00	3.90	4.00	-1.50	-1.30	-2.10	12.00
22	3.50	16.10	18.10	26.60	8.50	-.10	3.50	3.80	-1.50	-1.50	2.10	12.70
23	3.00	16.00	18.10	26.50	7.70	-.20	3.40	3.30	-1.50	-1.60	-2.10	13.20
24	3.50	16.00	19.00	26.50	6.70	-.30	3.00	2.60	-1.10	-1.60	-2.10	13.80
25	4.00	16.00	19.00	26.50	5.80	-.40	2.90	1.90	-1.10	-1.70	-1.10	14.30
26	4.70	16.10	18.90	26.40	5.00	-.40	2.90	1.30	-1.10	-1.70	.80	14.80
27	5.40	16.20	21.00	26.20	4.40	-.40	2.90	2.90	.50	-1.00	.80	15.20
28	6.40	16.60	24.80	26.00	3.90	-.10	2.90	.10	-1.00	-1.80	1.50	15.40
29	7.40	......	24.80	25.80	3.30	-.20	2.90	.80	-.10	-1.00	3.00	15.80
30	9.40	......	24.30	25.60	3.00	.00	2.70	-.30	-1.00	-2.10	4.40	16.00
31	10.30	......	24.00	......	2.50	......	2.60	-.60	......	-2.10	......	16.20
1903.												
1	16.4	17.3	24.1	28.6	22.0	6.7	16.9	-.6	.5	-2.2	-2.5	-2.2
2	18.0	17.2	24.2	28.6	21.8	6.5	16.8	-.6	.4	-2.2	-2.5	-2.1
3	18.0	19.0	24.4	28.6	21.4	6.7	16.5	-.5	.2	-2.2	-2.5	-2.1
4	18.1	19.0	24.6	28.6	21.0	7.4	16.0	-.5	.4	-2.2	-2.5	-2.1
5	18.4	18.5	24.7	28.7	20.7	8.3	15.5	-.5	-.6	-2.3	-2.5	-2.2
6	18.7	18.2	24.8	28.7	20.3	9.2	14.8	-.4	-.9	-2.3	-2.5	-2.3
7	18.9	19.7	24.9	28.7	20.0	10.0	13.9	-.3	-1.0	-2.3	-2.5	-2.3
8	19.0	22.4	25.2	28.7	19.8	10.9	12.5	-.1	-1.2	-2.3	-2.5	-2.3
9	19.0	21.9	25.3	28.6	19.5	11.5	11.0	-.1	-1.3	-2.3	-2.5	-2.3
10	19.0	21.8	25.4	28.5	19.2	12.4	9.6	-.1	-1.4	-2.3	-2.5	-2.3
11	19.4	22.4	25.8	28.4	19.0	13.0	8.2	1.0	-1.5	-2.3	-2.5	-2.4
12	19.4	22.0	25.9	28.3	18.4	13.5	6.9	1.0	-1.6	-2.4	-2.4	-2.4
13	19.3	21.9	26.0	28.2	18.5	14.1	6.0	.9	-1.7	-2.4	-2.3	-2.3
14	19.3	21.9	26.2	28.0	18.0	14.7	5.0	.9	-1.7	-2.4	-2.2	-2.2
15	19.3	22.0	26.4	27.8	17.5	15.0	4.0	1.4	-1.7	-2.4	-2.1	-.2
16	19.2	23.1	26.5	27.5	16.9	15.5	3.5	1.6	-1.8	-2.4	-2.0	-2.2
17	19.0	23.4	26.6	27.2	16.2	15.8	2.8	1.8	-1.8	-2.4	-1.9	-2.1
18	19.0	23.3	26.8	26.9	15.5	16.0	2.4	1.8	-1.8	-2.4	-1.9	-2.1
19	18.9	23.3	26.9	26.6	14.5	16.2	2.0	2.0	-1.8	-2.4	-2.1	-2.0
20	18.8	23.3	27.0	26.3	13.4	16.3	1.8	2.5	-2.0	-2.4	-2.1	-1.8
21	18.6	23.3	27.2	25.9	12.0	16.4	1.6	1.8	-2.0	-2.4	-2.2	-1.6
22	18.4	23.4	27.5	25.5	10.8	16.5	1.2	1.5	-2.1	-2.4	-2.3	-1.4
23	18.1	23.5	27.6	25.0	9.5	16.6	1.0	1.5	-2.2	-2.4	-2.3	-1.2
24	18.0	23.6	27.8	24.7	8.8	16.7	.8	1.5	-2.2	-2.4	-2.3	-.2
25	17.8	23.6	27.9	24.3	8.2	16.9	.5	1.5	-2.2	-2.4	-2.3	-.8
26	17.6	23.7	28.0	23.9	7.8	17.0	.5	1.5	2.2	-2.4	-2.3	1.6
27	17.5	23.8	28.2	23.5	7.6	17.1	.4	1.4	-2.2	-2.4	-2.3	1.9
28	17.8	24.1	28.3	23.0	7.5	17.1	.2	1.3	-2.2	-2.4	-2.3	2.1
29	17.6	......	28.4	22.8	7.4	17.1	-.2	1.2	-2.2	-2.4	-2.3	2.3
30	17.5	......	28.5	22.5	7.2	17.1	-.4	1.0	-2.2	-2.4	-2.3	2.5
31	17.4	......	28.6	......	6.9	......	-.5	.8	......	-2.4	......	2.5

Rating table for Yazoo River at Yazoo City, Miss., for 1901, 1902, and 1903.

Gage height.	Discharge.	Gage height.	Discharge.	Gage height.	Discharge.	Gage height.	Discharge.
Feet.	*Second-feet.*	*Feet.*	*Second-feet.*	*Feet.*	*Second-feet.*	*Feet.*	*Second-feet.*
—2.0	1,830	1.0	2,620	4.0	3,870	11.5	10,500
—1.8	1,870	1.2	2,690	4.5	4,145	12.0	11,000
—1.6	1,915	1.4	2,760	5.0	4,440	12.5	11,500
—1.4	1,965	1.6	2,830	5.5	4,790	13.0	12,000
—1.2	2,015	1.8	2,900	6.0	5,150	13.5	12,500
—1.0	2,065	2.0	2,970	6.5	5,550	14.0	13,000
— .8	2,115	2.2	3,050	7.0	6,000	14.5	13,500
— .6	2,165	2.4	3,130	7.5	6,500	15.0	14,000
— .4	2,215	2.6	3,210	8.0	7,000	15.5	14,500
— .2	2,265	2.8	3,300	8.5	7,500	16.0	15,000
.0	2,320	3.0	3,390	9.0	8,000	16.5	15,500
.2	2,380	3.2	3,480	9.5	8,500	17.0	16,000
.4	2,440	3.4	3,570	10.0	9,000	17.5	16,500
.6	2,500	3.6	3,660	10.5	9,500	18.0	17,000
.8	2,560	3.8	3,760	11.0	10,000		

Estimated monthly discharge of Yazoo River at Yazoo City, Miss.

Month.	Discharge in second-feet.		
	Maximum.	Minimum.	Mean.
1901.			
January	17,300	10,800	14,829
February	17,200	15,200	16,443
March	15,600	10,800	13,377
April	15,600	11,800	14,043
May	15,600	8,200	13,094
June	8,000	2,350	4,478
July	2,440	1,965	2,181
August	8,100	1,965	3,805
September	8,100	2,620	5,456
October	6,400	1,965	3,327
November	2,500	1,870	2,046
December	10,600	2,115	5,978
The year	17,300	1,965	8,255

Estimated monthly discharge of Yazoo River at Yazoo City, Miss.—Continued.

Month.	Discharge in second-feet.		
	Maximum.	Minimum.	Mean.
1902.			
January	10,500	3,390	8,493
February	15,600	10,300	14,082
March	23,800	15,600	17,448
April	25,600	22,800	24,477
May	24,300	3,170	14,004
June	3,170	2,215	2,651
July	4,440	2,265	3,281
August	4,260	2,165	3,578
September	2,410	1,940	2,104
October	2,795	1,810	2,223
November	4,090	1,810	2,023
December	15,200	4,380	9,065
The year	25,600	1,810	8,619
1903.			
January	18,400	15,400	17,432
February	23,100	16,200	20,807
March	27,600	23,100	25,442
April	27,700	21,500	25,753
May	21,000	5,910	14,075
June	16,100	5,550	12,610
July	15,900	2,190	6,549
August	3,170	2,165	2,593
September	2,470	1,790	1,953
October	1,790	1,750	1,760
November	1,850	1,730	1,767
December	3,170	1,750	2,089
The year	27,700	1,730	11,068

*Net horsepower per foot of fall, with a turbine efficiency of 80 per cent, for the minimum
monthly discharge of Yazoo River at Yazoo City, Miss.*

Month.	1901.			1902.			1903.		
	Minimum discharge.	Minimum net horsepower per foot of fall.	Duration of minimum.	Minimum discharge.	Minimum net horsepower per foot of fall.	Duration of minimum.	Minimum discharge.	Minimum net horsepower per foot of fall.	Duration of minimum.
	Sec.-ft.		*Days.*	*Sec.-ft.*		*Days.*	*Sec.-ft.*		*Days.*
January	10,800	982	1	3,390	308	1	15,400	1,400	1
February	15,200	1,382	1	10,300	936	1	16,200	1,473	1
March	10,800	982	1	15,600	1,418	3	23,100	2,100	1
April	11,800	1,073	1	22,800	2,073	5	21,500	1,955	1
May	8,200	745	1	3,170	288	1	5,910	537	1
June..........	2,350	214	1	2,215	201	4	5,550	505	1
July	1,965	179	1	2,265	206	1	2,190	199	1
August........	1,965	179	4	2,165	197	1	2,165	197	2
September	2,620	238	1	1,940	176	6	1,790	163	8
October........	1,965	179	1	1,810	164	2	1,750	159	20
November	1,870	170	3	1,810	164	24	1,730	157	11
December......	2,115	192	1	4,380	398	1	1,750	159	1

MISCELLANEOUS MEASUREMENTS.

The following miscellaneous measurements have been made:

Miscellaneous measurements in Mississippi.

Date.	Stream.	Locality.	Discharge.
1901.			*Second-feet.*
Aug. 15..	Sakatonchee River	Mhoon Valley	559
Aug. 16..	Tombigbee River....................	Waverly...............	6,726
Mar. 7...	Yalobusha River....................	Grenada..........	1,336
Apr. 15..	Big Black River....................	Goodman..............	614
1903.			
July 14 ..	Big Black River....................	Morey.................	491
July 15 ..	.....do	Way	1,397

INDEX.

249

O

[Mount each slip upon a separate card, placing the subject at the top of the second slip. The name of the series should not be repeated on the series card, but the additional numbers should be added, as received, to the first entry.]

Hall, Benjamin M[ortimer] 1853-

. . . Water powers of Alabama, with an appendix on stream measurements in Mississippi; by Benjamin M. Hall. Washington, Gov't print. off., 1904.

253 p., 1 l. illus., 9 pl. (incl. map) 23ᶜᵐ. (U. S. Geological survey. Water-supply and irrigation paper no. 107)
Subject series: N, Water power, 8.

1. Water power—Alabama. 2. Stream measurements—Mississippi.   .

Hall, Benjamin M[ortimer] 1853-

. . . Water powers of Alabama, with an appendix on stream measurements in Mississippi; by Benjamin M. Hall. Washington, Gov't print. off., 1904.

253 p., 1 l. illus., 9 pl. (incl. map) 23ᶜᵐ. (U. S. Geological survey. Water-supply and irrigation paper no. 107)
Subject series: N, Water power, 8.

1. Water power—Alabama. 2. Stream measurements—Mississippi.

U. S. Geological survey.

Water-supply and irrigation papers.

no. 107. Hall, B. M. Water powers of Alabama, with an appendix on stream measurements in Mississippi. 1904.

U. S. Dept. of the Interior.

see also

U. S. Geological survey.

Water-Supply and Irrigation Paper No. 108.   Series L, Quality of Water, 7

DEPARTMENT OF THE INTERIOR
UNITED STATES GEOLOGICAL SURVEY
CHARLES D. WALCOTT, DIRECTOR

QUALITY OF WATER

IN THE

SUSQUEHANNA RIVER DRAINAGE BASIN

BY

MARSHALL ORA LEIGHTON

WITH

AN INTRODUCTORY CHAPTER ON PHYSIOGRAPHIC FEATURES

BY

GEORGE BUELL HOLLISTER

WASHINGTON
GOVERNMENT PRINTING OFFICE
1904

CONTENTS.

ILLUSTRATIONS.

5

LETTER OF TRANSMITTAL.

DEPARTMENT OF THE INTERIOR,
UNITED STATES GEOLOGICAL SURVEY,
HYDROGRAPHIC BRANCH,
Washington, D. C., March 10, 1904.

SIR: I have the honor to transmit herewith a manuscript entitled "Quality of Water in Susquehanna River Drainage Basin," by Marshall Ora Leighton, and to request that it be published as one of the series of Water-Supply and Irrigation Papers.

In this paper is presented a brief introductory chapter on the physiographic features of the Susquehanna basin, by George B. Hollister, which is followed by a detailed discussion of the population and industries in New York. Numerous analytical reports are presented which show the character of the unpolluted waters in the main stream and its various tributaries, and the effects which have been produced by industrial and domestic pollution. Of special importance are the statements concerning the effect of mine wastes. It is shown that such wastes are not without their beneficial effects, especially in those parts of the river which have been set aside as areas for sewage disposal. The discussion of this matter, together with the consideration of the amount of mine wastes discharged into Susquehanna River, is one of the important features of the paper.

It is intended that this paper shall be followed soon by another (No. 109), entitled "Hydrography of Susquehanna River Drainage Basin," by John C. Hoyt and Robert H. Anderson. The two papers will make available a large amount of valuable information with reference to the resources of the Susquehanna River system.

Very respectfully,

F. H. NEWELL,
Chief Engineer.

Hon. CHARLES D. WALCOTT,
Director U. S. Geological Survey.

QUALITY OF WATER IN THE SUSQUEHANNA RIVER DRAINAGE BASIN.

By M. O. Leighton.

PHYSIOGRAPHIC FEATURES OF SUSQUEHANNA BASIN.

By G. B. Hollister.

The Susquehanna is the largest river of the Atlantic slope, its drainage area covering approximately 27,400 square miles. There are various speculations regarding the origin and development of the present river system, but the evidences on which they are based are too meager to permit the acceptance of the conclusions as final. Disregarding conjectures as to the conditions of drainage existing during Permian time in what is now eastern Pennsylvania, it is safe to assume that during and after the regional depression when the Triassic beds of the eastern portion of the continent were laid down there were a number of streams heading in what is now central Pennsylvania that found their way to the Atlantic coast. It is also probable that this depression gave new vigor to the eastward-flowing streams, which, becoming more active, gradually worked back and succeeded in capturing tributaries of adjacent systems less advantageously situated, so that by the end of this epoch the more important streams became firmly established, and continued in spite of the subsequent uplift. In taking this course, all of these streams were obliged to cross the truncated series of resistant sandstones then practically base-leveled, which have since remained as the Allegheny ridges. So firmly established, however, do the streams appear to have become that these adverse conditions were not sufficient in the main to change their direction and they have persisted in their eastward courses, cutting notches and gaps in the ridges even in spite of the regional uplift which followed the Triassic deposition.

One of these streams corresponded in part to the present Schuylkill and was comparatively large, owing to its capture of parts of a former north-flowing stream.[a] According to Davis another and possi-

---

[a] Davis, W. M., Rivers and valleys of Pennsylvania: Nat. Geog. Mag., vol. 1, p. 206.

bly a smaller stream was the parent of the Susquehanna. Its headwaters lay in the mountain region of the central portion of the State
and it flowed across the Allegheny ridges to the southeast, approxi-

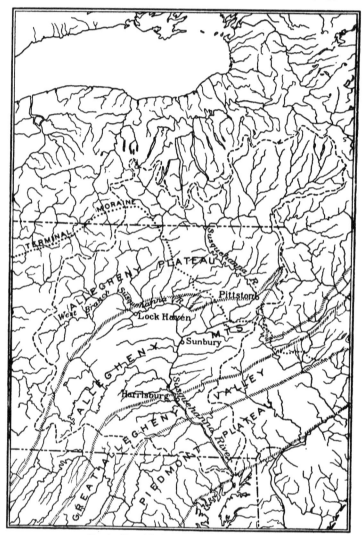

Fig. 1.—Map of drainage area of Susquehanna River.

mately in the position of the lower portion of the present Susquehanna. Various causes combined to render this stream more vigorous in its action than the one previously referred to, and in the

course of time it succeeded in capturing many of the branches of the Schuylkill and even in tapping and capturing its entire upper waters. In this manner the Schuylkill was left with a mere remnant of its former volume. The Susquehanna was also strengthened by the capture of the Juniata and other streams on the west, and gradually assumed its prominence as the master stream of the region. This outline of its previous history must be considered merely as a suggestion rather than as demonstrated fact.

The watershed of the Susquehanna embraces portions of four great physiographic regions of the eastern part of the United States—the Allegheny Plateau, the Allegheny Mountains, the great Allegheny Valley, and the Piedmont Plateau. Its distribution among these provinces is approximately as follows:

Physiographic divisions of Susquehanna basin.

	Square miles.	Per cent.
Allegheny Plateau	15,400	56
Allegheny Mountains	8,500	31
Allegheny Valley	1,700	6
Piedmont Plateau	1,800	7

ALLEGHENY PLATEAU.

More than half the Susquehanna drainage area, approximately 56 per per cent, is included in the Allegheny Plateau in New York and Pennsylvania. This region, dissected by the stream and its branches into a succession of high hills and deep valleys, is the remnant of an extended plain. The plain is not confined to the Susquehanna watershed, but may be traced eastward into the Catskill region and southward to Alabama. Its eastward face is usually recognizable in a pronounced escarpment, while westward it merges gradually into the great plain of the Mississippi Valley.

Geologically this plain is composed of Paleozoic sandstones, shales, and limestones, which in New York lie nearly level, with a slight dip toward the south. In Pennsylvania the folds, which are so evident in the Appalachian Mountain region, gradually die out westward in the Allegheny Plateau. The physiographic evidence leads to the belief that the region has been base-leveled and, in common with the Allegheny Mountain region, reduced to a well-defined peneplain. The region was then elevated, and the streams, thus given new energy, have eroded the surface of the plain, which now presents the appearance of a very hilly country. However, from the summits of the hills between the lesser watersheds evidences of the former plain are seen in the remaining hilltops, which stand at approximately even heights as far as the eye can see. The hills, as a rule, rise from 500 to 800 feet above the valleys, with comparatively steep but beautifully symmetrical slopes.

A large proportion of that part of the Susquehanna drainage area within the Allegheny Plateau has been glaciated. The former presence of the ice sheet is recognized by the drift which covers the surface, particularly in the valleys, where immense accumulations are found. Well borings show the thickness of the valley deposits to be from 50 to over 1,000 feet. Owing to the action of ice and the subsequent outflow of water from the ice fields, combined with the distribution of the drift, many changes of drainage seem to have resulted, and not a few streams which previously flowed northward into the Great Lakes have apparently been diverted and now form part of the Susquehanna system.

The effect of the ice on the excavation of the north-south valleys has not yet been fully determined, but from the evidence offered in the Finger Lake region and elsewhere in central New York it may be fairly surmised that many of the valleys were more or less deepened in this manner.

The principal streams in the Susquehanna basin that drain the Allegheny Plateau are the Susquehanna River and the West Branch. Susquehanna River rises in Lake Otsego, Otsego County, N. Y. It flows generally southwestward through the southern tier of counties of New York and enters Pennsylvania in Bradford County near Sayre. Thence its course is generally southeastward to Pittston, where it leaves the plateau region. Its most important tributaries to this point are Chenango and Chemung rivers, both in New York.

West Branch rises in the highlands of Cambria County, Pa., and with its various tributaries drains a number of the central and north-central counties of that State. It leaves the Allegheny Plateau near Lock Haven, about 95 miles from its source. Its chief tributaries on the plateau are Moshannon, Sinnamahoning, Kettle, Pine, Lycoming, and Loyalsock creeks. In the region they traverse the strata are bent in broad, simple folds. The present aspect of the country is extremely rugged, though its plateau character here, as in New York State, is distinctly recognized from the tops of the hills. These rise to an elevation of 1,500 to 2,000 feet and more, and contain a large amount of forest lands.

The general topographic features of this portion of the plateau are distinct from those in the area drained by the Susquehanna in New York. There the slopes are usually gentle and symmetrical, the weathering process having been applied to strata lying approximately horizontal, while here the topography presents more rugged features and more uneven slopes. The rocks are tilted to a greater extent, and the more resistant layers, the sandstones and grits, outlast the more yielding shales and limestones and stand out as ridges. The steeper slope of a hill of this character is usually on the side which exposes the upturned faces of the strata, while the slope parallel to the dip of the strata is gentler and longer.

Lake Otsego, which may be considered the source of the Susquehanna, lies at an elevation of 1,193 feet above sea level. The altitude of the Susquehanna at its junction with the Chenango at Binghamton is 822 feet; at the junction of the Chemung at Athens it is 744 feet; and near Pittston, 232 miles from its source, where the river leaves the Allegheny Plateau and enters the Appalachian belt, it is 536 feet, giving a total fall of 657 feet, or an average fall on the Allegheny Plateau of 2.8 feet per mile.

ALLEGHENY MOUNTAIN REGION.

After leaving the Allegheny Plateau Susquehanna River enters the Alleghany Mountain region. Like the Allegheny Plateau, this physiographic region is not confined to the Susquehanna drainage basin, but extends as a well-defined feature from Alabama to Canada. The ridges generally run northeast and southwest, sweeping off at the north in broad curves to a more easterly direction.

Geologically this is a region of alternating hard and soft sedimentary beds, bent by enormous lateral compression into folds or waves technically known as "anticlines" and "synclines." These folds have a general northeast-southwest trend, and in some limited districts a slight pitch to the southwest. After the rocks had been folded the whole country was base-leveled by erosion; or, in other words, all the layers, hard and soft, were planed down to an approximately uniform surface. Then followed a general uplift of the region, which gave the streams renewed vigor and inaugurated another period of denudation, which has been continued to the present time. As a result of this last cycle of erosion the softer rocks have been gradually worn down and carried away and the more resistant layers stand out as ridges.

The effect of the southward pitch of the folds upon the topography is particularly interesting. Instead of being merely approximately parallel ridges, as would be the case if there were no longitudinal tilting of the layers, the general planation and subsequent denudation have left a peculiar series of canoe-shaped valleys. Where the synclinal folds are eroded the mountains surrounding the valleys gradually converge to form what would represent the prow. Where the anticlinal folds are truncated a series of hemi-cigar-shaped mountains results. This peculiar system of ridges surrounding blind valleys and inclosing narrow valleys within valleys has had a decided influence upon the region in a number of ways. It has greatly increased the difficulty of travel across it, and thus retarded development, and has also increased the expense of railroad construction. Its influence upon the drainage system as a whole has been marked, and the original drainage systems have suffered many very radical changes.

In the Allegheny Mountain region is included the drainage of the lower portion of the Susquehanna, of the Juniata and its tributaries, and of almost the entire West Branch below Lock Haven, with the exception of its northerly and westerly tributaries—in brief, the portion of the Susquehanna system between Pittston and Harrisburg, with the exception of the tributaries of West Branch. The area of this part of the watershed is approximately 8,500 square miles, or about 31 per cent of the entire drainage basin. The slopes of the main stream and its principal branches are as follows: From Pittston to the junction with West Branch, at Sunbury, a distance of 68 miles, the fall is 114 feet—an average of 1.6 feet per mile. From Sunbury to Harrisburg, 53 miles, the fall is 124 feet—an average of 2.3 feet per mile. The portion of West Branch in this region is 65 miles long, and falls 110 feet at an average rate of 1.6 feet per mile from Lock Haven to its mouth. The Juniata has an average fall of 3.1 feet per mile.

The slopes in this region differ greatly from those in the Allegheny Plateau. There the streams have etched the plateau, leaving symmetrical and well-rounded hills. The elevations of the Allegheny Mountain region are long, straight, or slightly curved ridges, with level tops, but comparatively steep sides. Some of them may be traced for long distances as continuous mountains, which are broken only where they have been cut by streams. A notable example of such a range is Kittatinny Mountain, which forms the southeastern-most ridge of the group. This ridge runs out of Pennsylvania through northwestern New Jersey and southern New York to Hudson River, interrupted only by notches and water gaps.

One of the most striking features of the river in this part of its course is its bold persistence across the trend of the Allegheny ridges. Just above Pittston it flows into the fertile Wyoming Valley, which lies parallel to the mountain chain. It follows this valley until it reaches Nanticoke, where it bends gradually southward across the Lee-Penobscot Mountain and again resumes its southwestward course, which it holds until reaching Sunbury. Here it turns southward again, crossing Mahantango, Berry, Peters, Second, and Blue mountains, and emerges into Allegheny Valley near Harrisburg.

ALLEGHENY VALLEY.

After leaving the Allegheny Mountain region the river crosses the greater Allegheny Valley, which forms a striking contrast in topographic feature to the other portions of the region. Instead of being composed of level-crested parallel ridges, as is the region lying immediately west, Allegheny Valley is, as the term indicates, a wide depression having true valley characteristics. It is in reality a strip of low country extending from St. Lawrence Valley southward, embracing a portion of Hudson River Valley and all of Wallkill Valley

in New York, the valleys of the Wallkill and Paulins Kill in New
Jersey, the Cumberland and Hagerstown in Pennsylvania and Maryland, and the Shenandoah in West Virginia. The Susquehanna
enters it near Harrisburg and leaves it where it crosses the western
edge of the crystalline and Triassic areas. It is somewhat difficult to
define with precision the limits of Allegheny Valley in this section.
The river, however, may be said to cross the dividing line between
the valley and the Allegheny Mountain region near Rockville, a few
miles north of Harrisburg, and to intersect its southern edge near
Highspire, where it meets the Piedmont Plateau. The distance
between these points is approximately 11 miles, in traversing which
the river falls 19 feet, or an average of 1.7 feet per mile. Allegheny
Valley is not a large portion of the watershed, containing in all but
1,700 square miles, or approximately 6 per cent of the total drainage area, but its features are distinct and its general characteristics
strikingly different from those of the other physiographic divisions
under consideration. The valley is characterized geologically by
limestones, which are much more easily eroded and have undergone
greater denudation than the shales to the west. The tributaries in
this portion of the watershed are small, being Conedogwinet and
Yellow creeks on the west and Swatara Creek on the east. The fall
of the main stream and of its tributaries in the Allegheny Valley
system is slight. This is particularly noticeable in the case of Conedogwinet Creek, which flows in wide meanders in the last 15 or 20
miles of its course.

PIEDMONT PLATEAU.

After leaving Allegheny Valley the river flows through the Piedmont Plateau, whose northern limits are not clearly defined in this
portion of the State. The Piedmont Plateau extends from New England to Alabama. It receives its name from the fact that it lies at the
foot of the Allegheny Mountain ridges and borders them for their
entire length on the southeast. It consists mainly of crystalline and
metamorphic rocks highly altered and disturbed, and also includes
Triassic shales and sandstones with their trap intrusives. Susquehanna River may be said to enter the region 6 to 10 miles below Harrisburg.

The topography is characterized by comparatively low, well-rounded
hills on the divides and by rather steep-sided valleys along the streams,
especially in their lower portions. South Mountain, the continuation
of the Highlands, has a somewhat greater elevation. This division as
a whole is a farming country, with a small proportion of forested areas.
It has a heavy soil, resulting from the disintegration of the metamorphic series underlying it. The side streams flow with rather steep
grades on bed rock, and the Susquehanna passes over a number of
reefs and shoals suitable for the development of power. On the east-

ern edge of the Piedmont Plateau the river finds its mouth at the head of Chesapeake Bay, its lower course in particular being confined in a steep-sided valley.

In crossing the Piedmont Plateau the river flows approximately 59 miles. In this distance it falls 286 feet, or an average of about 4.5 feet per mile.

SUMMARY OF SLOPES.

For the sake of comparison the slopes of the main stream and its principal branch on the different portions of the watershed are here given.

Slopes of Susquehanna River and the West Branch.

ALLEGHENY PLATEAU.

Stream.	Length.	Fall.	Average per mile.
	Miles.	*Feet.*	*Feet.*
Susquehanna	232	657	2.8
West Branch	95	578	5.9

ALLEGHENY MOUNTAIN REGION.

Susquehanna (Pittston to Rockville)	116	234	2
West Branch (Lock Haven to Rockville)	113	231	2

ALLEGHENY VALLEY.

Susquehanna (Rockville to Highspire)	11	19	1.7

PIEDMONT PLATEAU.

Susquehanna (Highspire, near Harrisburg, to mouth)	63	286	4.5

The striking feature observable from the above table is the relatively sharp drop of the stream on the Piedmont Plateau, which, it will be remembered, is traversed by the river in the last 63 miles of its course. This condition is unusual, for streams in their erosive cycles normally reach base-level along their lower courses first, their gradients becoming relatively steeper as the headwaters are approached. This does not seem to be the case with the Susquehanna, which shows three distinct conditions of grade:

First, in the Allegheny Plateau the fall appears to be normal. To be sure, the figures show that the main stream drops 2.8 feet per mile while West Branch falls 5.9 feet per mile, but the two streams fall approximately through the same distance and the difference in average gradient seems to be entirely due to the difference in their length, which is marked, the shorter stream having relatively the steeper gradient, as would be expected.

Second, in the Allegheny Mountain region the fall is also normal. The lengths of the main stream and of West Branch are approximately the same, and their average fall is remarkably uniform. The difference in average fall per mile between portions of the stream on the Allegheny Plateau and in the Allegheny Mountain region is also normal, as it is to be expected that the fall on the upper portions of the system will be greater than on any portion lower down.

Third, the stream, which has thus far followed the laws of all well-developed river systems, begins its last stage—namely, that upon the Piedmont Plateau—with an entire reversal of the natural conditions; for here, instead of entering smoothly upon the last part of its slope curve, which with steadily decreasing fall should approach flatness at the river's mouth, it abruptly changes its grade, and the average fall suddenly increases from 2 feet to 4.5 feet per mile. The waters of the stream from this point pass rapidly over a series of shoals and successive drops which reach almost to sea level, the fall in about 60 miles being approximately 300 feet. This is a fall more than twice as great as that of the river in its previous stage across the Allegheny Mountain region and nearly double its fall in crossing the Allegheny Plateau.

This unusual fall in the Susquehanna on the Piedmont Plateau is a matter of considerable interest, and an investigation of the causes producing it is worthy of more attention than can be given in the limits of this chapter. It may be of interest, however, to call attention to a few of the facts which may throw some light upon the problem. First, the narrow, gorge-like valley of the Susquehanna upon the Piedmont Plateau, and also of its tributaries on the same area, indicates a comparatively recent elevation of the entire region. It is an extremely suggestive fact that indications of this uplift appear on the lower Potomac and to a less extent on the lower Delaware during the passage of each of these streams across the Piedmont Plateau, where this same condition of gorge-like valley is also found. On the other hand, the extensive embayment of the mouths of the Delaware, Hudson, Susquehanna, and Potomac rivers gives evidence of a wide regional depression which must have involved the entire portion of the seaboard region in question.

It is also evident that the changes producing the present peculiar conditions on the lower Susquehanna were not confined to the limits of its watershed, as the same sudden increase in fall is found on Delaware and Potomac rivers as they cross the Piedmont Plateau. Hence it seems that the causes for these phenomena must be found in a study of the regional changes that are known to have taken place along the whole portion of the Atlantic seaboard drained by these systems. Studies of the sand and gravel deposits near Washington, D. C., and in Maryland and New Jersey, and recent investigations on Long Island, indicate quite clearly that between the late Tertiary and the present time the Atlantic coast has been subjected to a series of regional uplifts and depressions. It was during these uplifts that the profound degradation of the Coastal Plain was accomplished and its present topographic features produced. The graded streams flowing across the Piedmont Plateau must have been considerably accelerated by these uplifts, and must have begun at once to alter their grades to conform to the new conditions.

According to this idea the sudden drop of the Susquehanna from Harrisburg to its mouth may be considered as a temporary stage in the river's attempt to again reach a normal slope curve. In the accomplishment of this object the slope has been affected only as far back as the limits of the Piedmont Plateau, but given sufficient time the effects may be expected to extend even to the headwaters.

The unusual occurrence of such conditions of grade at the lower end of a large stream like the Susquehanna offers opportunities for the development of water power. At a number of places on the lower course of the river it seems possible to erect power plants of considerable magnitude, and it is understood that capitalists are now studying the conditions at some of these with a view to future developments.

POPULATION AND INDUSTRIAL DEVELOPMENT IN NEW YORK.

Industrial development in the drainage basin of Susquehanna River has been important, especially in Pennsylvania, where the enormous mineral wealth has attracted a large number of settlers and encouraged industries.

In this basin in New York are the important cities of Binghamton, Elmira, and Corning, together with smaller ones like Hornellsville, Bath, Owego, Oneonta, Norwich, and Cortland. In this portion of the basin dairying and agriculture are the principal industries.

The run-off water from the Susquehanna drainage area in New York is of excellent character. There are, as a rule, not sufficient dissolved alkaline earth constituents to render the water very hard, and except at times of extraordinary floods little suspended material is carried in the streams. The water is comparatively free from color, and for all purposes is probably as good as any that can be found in the United States, with the possible exception of New England. Dairying is the principal industry in a large part of the region. The farms are famous for the excellent character of the water, which is available at small cost.

The following table shows the population of the cities and towns in the New York portion of the Susquehanna drainage area:

Population in drainage area of Susquehanna River in New York, by counties and towns.

ALLEGANY COUNTY:
Alfred (one-half)	808
Almond	1,436
Birdsall (one-third)	211
West Almond	601

BROOME COUNTY:
Barker	1,072
Binghamton	40,494
Chenango	1,872
Colesville	2,773
Conklin	946
Kirkwood	918
Lisle	1,710
Maine	1,534
Nanticoke	666
Triangle	1,727
Union	5,707
Vestal	1,850
Windsor	2,967

CHEMUNG COUNTY:
Ashland	954
Baldwin	664
Big Flats	1,705
Catlin	1,109
Chemung	1,500
Elmira	35,672
Erin	996
Horseheads	4,944
Southport	2,201
Van Etten	1,406
Veteran (one-third)	551

CHENANGO COUNTY:
Afton	1,920
Bainbridge	1,991
Columbus	997
Coventry	987
German	423
Greene	3,152
Guilford	2,208
Lincklaen	646
McDonough	907
New Berlin	2,525
Norwich	7,004
Otselic	1,234
Oxford	3,545
Pharsalia	780
Pitcher	751

CHENANGO COUNTY—Continued.
Plymouth	1,026
Preston	662
Sherburne	2,614
Smithville	1,105
Smyrna	1,290

CORTLAND COUNTY:
Cincinnatus	912
Cortland	9,014
Cortlandville	2,907
Cuyler	991
Freetown	610
Harford	753
Homer	3,864
Lapeer	538
Marathon	1,664
Preble	857
Scott	852
Solon	622
Taylor	762
Truxton	1,217
Virgil	1,326
Willett	687

DELAWARE COUNTY:
Davenport	1,620
Franklin	2,529
Harpersfield (two-thirds)	814
Kortright (one-half)	737
Masonville	1,245
Meredith	1,508
Sidney	4,023

HERKIMER COUNTY:
Columbia	1,268
Litchfield	931
Warren	1,240
Winfield	1,475

LIVINGSTON COUNTY:
Springwater (one-third)	672

MADISON COUNTY:
Brookfield	2,726
De Ruyta	1,410
Georgetown	998
Hamilton	3,744
Lebanon	1,243

ONONDAGA COUNTY:
Fabius	1.686
Tully (one-half)	733

OTSEGO COUNTY:

Burlington	1,263
Butternuts	1,698
Cherry Valley	1,802
Decatur	559
Edmeston	1,767
Exeter	1,087
Hartwick	1,800
Laurens	1,488
Maryland	1,998
Middlefield	2,100
Milford	2,007
Morris	1,689
New Lisbon	1,225
Oneonta	8,910
Otego	1,817
Otsego	4,497
Pittsfield	1,101
Plainfield	1,101
Richfield	2,526
Roseboon	1,031
Springfield	1,762
Unadilla	2,601
Westford	910
Worcester	2,409

SCHUYLER COUNTY:

Catharine (one-half)	693
Cayuta	459
Orange	1,391
Tyrone	1,586

SCHOHARIE COUNTY:

Jefferson (one-half)	704
Summit (one-half)	608

STEUBEN COUNTY:

Addison	2,637
Avoca	2,125
Bath	8,437
Bradford	771
Cameron	1,353

STEUBEN COUNTY—Continued.

Campbell	1,467
Canisteo	3,432
Cohocton	8,197
Erwin	1,851
Fremont	1,033
Greenwood	1,129
Hartsville	787
Hornellsville	1,833
Howard	1,704
Jasper	1,430
Lindley	1,306
Prattsburg	2,197
Rathbone	1,059
Thurston	1,017
Troupsburg	2,015
Tuscarora	1,301
Wayne (one-half)	419
West Union	1,025
Wheeler	1,188
Woodhull	1,787

TIOGA COUNTY:

Barton	6,381
Berkshire	1,011
Candor	3,330
Newark Valley	2,164
Nichols	1,564
Owego	8,378
Richford	1,112
Spencer	1,868
Tioga	2,113

TOMPKINS COUNTY:

Caroline	1,938
Danby (one-third)	483
Newfield (one-half)	951

YATES COUNTY:

Italy Hill	1,094
Total	345,850

Area of Susquehanna basin in New York _____ square miles __ 6,267
Population per square mile _____ 55

CHARACTER OF SURFACE WATERS.

SUSQUEHANNA RIVER IN NEW YORK.

The character of run-off water from the drainage area of this river is well shown by the analyses given below, which were made by George C. Whipple, of Mount Prospect Laboratory, Brooklyn.

Analyses of water from Susquehanna River above Binghamton, N. Y.

[Parts per million.]

Date.	Turbidity.	Color.	Odor.	Nitrogen as—				Chlorine.	Total residue.	Hardness.		Iron.	Gage heights, Binghamton, N. Y.
				Albuminoid ammonia.	Free ammonia.	Nitrites.	Nitrates.			Alkalinity.	Normal hardness.		
													Feet
Jan. 3, 1901...		10	None.	0.045	0 005	0.004	1.540	3.500	75.0	50	0	......	......
Mar 24, 1902	9	12	2V	.086	.004	.000	.400	1.400	74.0	35	9.5	0.60	6.32
June 17, 1902	8	15	3V+1M	.104	.024	.002	.040	1 600	109.0	47	8.0	.15	3.28
Sept. 16, 1902.	2	10	3V	.098	.044	.002	.030	2.100	101.0	50	0	.20	2.85
Jan. 7, 1903...	31	21	3V	.102	.012	.001	.200	1 000	96.0	33	3.5	.65	3 83
Aug. 18, 1903.	2	13	3V	.114	.016	.001	.020	1.500	86.5	54	6.0	.10	2.30

The analyses in the above table are of a water which is typical of a large drainage area which has received a small amount of sewage. The condition of the organic matter as shown by the analyses would not condemn the water for domestic purposes. On the contrary, the report is somewhat reassuring. Only in the high and unsteady chlorine content is there any indication of pollution, although the table gives no information concerning the time at which the polluting ingredients entered the stream. So far as the analysis is concerned the water might be a highly desirable potable beverage. It is known, however, from observation that such is not the case.

Below Binghamton Susquehanna River runs southwesterly, and a short distance below the State line it is joined by the Chemung. This stream drains a basin of practically the same character as that drained by the Susquehanna, and the conditions with respect to settlement are analogous. The following determinations were made by James M. Caird, Troy, N. Y.:

Analyses of water from Chemung River at Elmira, N. Y.

[Parts per million.]

Date.	Turbidity.	Color.	Odor.	Nitrogen as—				Chlorine.	Total residue.	Hardness.	Iron.	Bacteria per cubic centimeter.	Number of analyses made.
				Albuminoid ammonia.	Free ammonia.	Nitrites.	Nitrates.						
Mar. 28–31, 1900	16	20								65		7,000	6
May 18–24, 1900	18	26									114		1,800	7
Aug. 2–8, 1900	26	32									160		1,300	7
Aug. 20–31, 1900	10	95											600	12
Sept. 1–26, 1900	7	95											525	26
Sept. 27–30, 1900	18	25									68		950	4
Oct. 1–3, 1900	13	23									61		1,250	3
Dec. 18–20, 1900	30	55									39		26,500	3
Apr. 27, 1901	0			0.048	0.024	0	2.55	4.0	75.7		0	500	1
Jan. 6–11, 1902	25	25									52		11,500	6
Mar. 9, 1902	30	35									55		4,200	1
Mar. 10, 1902	98	40									20		26,000	1
Oct. 5, 1902	40	45									68		7,200	1
Oct. 6–10, 1902	25	29									65		11,100	5

The above table contains only four determinations, but they are probably the most valuable of all for practical purposes in this case. The record shows that the water of Chemung River is of good physical quality, but is polluted by wastes. Compared with Susquehanna River at Binghamton the water here shows, on the average, somewhat higher turbidity and color, while it contains considerably more hardening constituents. From a bacteriologic standpoint the record for Chemung River is unsteady, showing that usually the water contains little impurity, and only now and then betrays the excess of organic matter which is poured into the stream from the cities on its banks.

SUSQUEHANNA RIVER JUST ABOVE NORTHERN ANTHRACITE COAL BASIN.

In the table below are given a few results of analysis of samples taken from the river just above the northern anthracite coal basin.

Analyses of water from Susquehanna River just above northern anthracite coal basin.

[Parts per million.]

| Date. | Turbid-ity. | Color. | Odor. | Nitrogen as— | | | | Chlorine. | Total residue. | Hardness. | | | Bacteria per cubic centimeter, 1,000. | Gage height. Wilkesbarre. |
				Albuminoid ammonia.	Free ammonia.	Nitrites.	Nitrates.			Alkalinity.	Normal hardness.		
Feb. 11, 1899	0	10	0	0.072	0.004	0.000	0.250	1.850	127		6.2	......	....
Mar. 16, 1899	Very turbid.	20	Faint.	.156	.088	.000	.200	1.786	168		......	......	....
Apr. 7, 1899	Cons.	15	Faint.	.202	.014	.000	.000	1.607	161		......	......	7.4
May 12, 1899	0	12	Very faint.	.102	.020	.000	.080	1.965	142.5		......	......	5.1
Oct. 27, 1901	Slight.	10	Very faint.	.082	.242	.000	.000	6.000	166		107.5	......	3.4
Dec. —, 1902	0	10	0	.058	.018	.000	.000	4.090	130		81.4	......	....
Feb. 15, 1903	40	17	0	.118	.030	.000	.006	2.000	112.5	3.9	32.5	1,940	10.8
May 22, 1903	5	8	Very faint.	.132	.036	.001	.000	3.200	132.5	20.1	69.9	2,750	3.5
July 20, 1903	10	20	0	.158	.082	.000	.000	3.000	126	12.7	47.3	2,450	4.6
Sept. 15, 1903	8	17	0	.106	.044	.000	.002	2.400	129.5	4.3	57.1	2,000	5.2

The analyses in the above table denote a water which, except for its high chlorine content, appears to be excellent, and if the history of the stream above this point were not known it might be maintained that the water could be used with safety for domestic purposes. In its course from Binghamton and Elmira across the sparsely settled country the water has become somewhat purified and shows a distinct improvement. It is clear, almost colorless, and comparatively soft, but is unfit for domestic uses unless purified.

SUSQUEHANNA RIVER IN NORTHERN ANTHRACITE COAL BASIN.

SEWAGE POLLUTION.

In the northern anthracite coal basin the conditions along the Susquehanna are entirely different from those previously encountered. At Pittston the river is joined by the Lackawanna, which flows near the median line of the coal basin and draws most of its water from it. The Lackawanna is not a large river, and of all streams in the United States it is probably the least attractive in appearance. It contains sewage from Carbondale, Archbald, Jermyn, and Scranton.

Carbondale, situated in Lackawanna County, upon Lackawanna River, has a population, according to the Twelfth Census, of 13,536. The public works consist of a sewerage system and municipal water supply, the latter being under the control of a private corporation. The municipal sewage is poured directly into Lackawanna River, but there are no manufacturing plants which produce any important amount of polluting material. Above the city the river presents a fairly acceptable appearance. It has a sluggish flow at low water, but rises quickly after heavy precipitation. Considerable discomfort is experienced at times by persons living along the river below the city.

Carbondale is supplied with an impounded water conserved in Crystal Lake, a few miles west. An artificial distributing reservoir is situated east of the city on high ground. The daily consumption is 3,500,000 gallons, equivalent to 260 gallons per capita. Crystal Lake is a body of water 195 acres in extent, while the artificial reservoir covers 35 acres. Its surface is about 200 feet above the average distributing reservoir.

Scranton, situated in Lackawanna County, along Lackawanna River, has a population, according to the Twelfth Census, of 102,026. It is provided with a sewerage system, the total cost of which up to August, 1902, was $720,500. The river which flows past the city is usually a black, foul, ropy mass of fluid, due to the enormous amounts of culm and municipal waste turned into it. There are no industrial plants in Scranton from which damaging waste is turned into the sewers.

Scranton is provided with an impounded water supply, the system being owned by the Scranton Gas and Water Company. A description of this system will be found in subsequent pages.

COAL-MINE WASTE.

In addition to the large amounts of sewage there are discharged into Lackawanna River enormous amounts of coal-mine waste. This is a peculiar solution containing a high percentage of calcium sulphate, free sulphuric acid, and iron. Concerning this waste, Dr. Charles B. Dudley, of Altoona, Pa., makes the following statement:

Much trouble is experienced in Pennsylvania, Ohio, Indiana, and Illinois from the corrosive condition of the water that must necessarily be used in steam boilers. The sources of supply, especially the rivers and small streams in those States, receive large amounts of drainage from coal mines, and this drainage is extremely corrosive. To such an extent is this the case that at the Homestead Steel Works, in order to neutralize the corrosive characteristic of Monongahela River water, many barrels of soda ash are used every month.

The corrosive material in water contaminated with mine drainage may be two-fold, namely, it may be free sulphuric acid or it may be sulphates of iron and alumina. The origin of the sulphuric acid and sulphates is undoubtedly from the coal. It is believed that coal may contain sulphur in three forms, namely, as sulphate of lime, as sulphide of iron or iron pyrites, and possibly as organic sulphur. Just exactly how this sulphur is converted into either sulphuric acid or sulphates of iron and alumina has possibly not yet been determined, but samples of coal have been obtained from certain mines, which on being coarsely crushed and put in a funnel and washed with distilled water actually yield free sulphuric acid in the filtrate. More commonly, however, the drainage from the mines comes out of the mines in the form of protosulphate of iron, $FeSO_4$. On exposure to the air this salt breaks up, yielding hydrated oxide or free basic sulphate of iron, which precipitates and gives a yellowish color to many of the waters of the small streams into which the mine drainage runs, and sesquisulphate or ferric sulphate, which remains in solution. This ferric sulphate when brought in contact with metallic iron or steel, such as the inside of a boiler, attacks iron freely, the metal being converted thereby into protosalt again. Fresh water containing oxygen being added to the boiler as feed water, apparently the protosalt again breaks up into hydrated oxide or basic sulphate, which is thrown out of solution, and ferric sulphate. the same operation being repeated over and over again. resulting. as is readily seen, in very rapid destruction of boilers. The separated hydrate or basic sulphate of iron is washed out usually as a reddish powder. Waters containing mine drainage do not usually form scale in the boilers, and as long as any carbonates of any kind are present, even if the water contains mine drainage, there is no corrosion. Accordingly where mine-drainage waters must be used it is customary to add soda ash in certain amounts sufficient to a little more than react with the ferric salts that may be in the water, and if perchance mixed waters are used, some of which contain mine drainage, and some contain sufficient carbonate of lime in solution as bicarbonate to react with the iron salts, there is no corrosion.

Results of analysis of three representative samples of this mine waste are given below. The analyses were made for the Geological Survey by Mr. George C. Whipple, of Mount Prospect Laboratory, Brooklyn, N. Y.

Analyses of three samples of mine waste from northern anthracite coal basin, Pennsylvania.

[Parts per million.]

	1.	2.	3.
Turbidity	130	210	190
Color	26	23	6
Odor	0	0	0
Albuminoid ammonia in solution	.082	.094	.034
Albuminoid ammonia in suspension	.094	.070	.022
Albuminoid ammonia total	.176	.164	.056
Free ammonia	.840	2.400	.282
Nitrites	.004	.007	.000
Nitrates	.000	.000	.000
Total residue on evaporation	1,977	3,003	792
Loss on ignition	279	280	.64
Fixed solids	1,698	2,723	728
Calcium and magnesium sulphates	1,036	1,424	477
Sulphuric acid	164	150	39
Oxide of iron	143	393	78.6
Sodium chloride	3.3	6	.7
Ammonium sulphate	4	11.3	1.3
Silica and alumina	347.7	738.7	131.4

1. From Mineral Railroad and Mining Company, collected from pump No. 1. sump, Cameron colliery, Shamokin, Pa.

2. From slope No. 4, Susquehanna Coal Company, Nanticoke, Pa.

3. From pump delivery at Short Mountain colliery, Lykens Valley Coal Company, Lykens, Pa.

The three samples above described are representative types in the anthracite region.

The above analyses do not show the most important facts desirable in this discussion. The determinations of turbidity, color, odor, and the four nitrogens are without special consequence, although they retain a measure of interest because the amounts of albuminoid and free ammonia are very large in comparison with the nitrites and nitrates. This would seem to indicate that the organic matter in the mine water is the result of recent pollution. This, however, is probably incorrect. The organic matter has a remote source and has remained available in the first stages of oxidation because of the lack of oxygen.

The most important part of the analytical statement is that which begins with the results of the determination of total residue on evaporation. The residue is extremely large, especially in Nos. 1 and 2. The loss on ignition is extensive in all three samples, and without doubt there was driven off a large proportion of matter, the nature and amount of which should have appeared in the list of inorganic ingredients. For example, the amounts of sulphuric acid which appear are entirely insufficient to combine with all the iron and alumina, and it is probable that a large part of the matter lost on ignition was sulphuric acid. These wastes are known to be highly acid, and there is probably an excess of sulphuric acid over that required to combine with the iron and alumina in the solution, although it is apparent that the evidences of free acid which appear by reason of the corrosive properties of the water may be due to the ferric sulphate.

It is unfortunate, too, that a separate determination was not made of the amount of alumina, for, as will be shown in the following pages, this mine waste has remarkable coagulating and precipitating properties, which are believed to be due to aluminum sulphate.

The analyses show unmistakably that the water is highly corrosive and probably scale-forming, although in the presence of acid the calcium sulphate, of which the water contains so high a proportion, would not be precipitated as a scale, but in the form of a powder.

CULM WASTE.

The third waste which is turned into Lackawanna River is very fine coal, or culm. This material is discharged into the streams of this region in enormous quantities from the coal washeries. A washery is a plant erected for the purpose of working over the old waste piles (see Pl. I, A) to secure coal which in the earlier times and cruder methods of handling passed out as waste. The fine material is separated from the coarse by a sort of decantation process, the supernatant water loaded with coal dust being turned into the streams. Wherever it is turned into a small brook, where the amount of water is not suf-

A. RECOVERY OF COAL FROM CULM PILE.

B. SUSQUEHANNA RIVER IN WYOMING VALLEY.

ficient to transport the culm, the stream bed soon becomes filled, and in some cases no channel is left, the water running at will over the lands adjoining.

This black combination of sewage, acid mine waste, and culm slowly makes its way down the valley. Here the sediment has filled a depression and there it has been piled high on one side of the stream. Along the Lackawanna near its confluence with the Susquehanna the level of the bottom land has been raised and the stream winds through a broad tract, changing its course through the loose, shifting piles at every influx of storm water.

The Susquehanna at its confluence with the Lackawanna undergoes a complete change. It enters the coal basin a bright, clear stream. Below the mouth of the Lackawanna the inky black trail extends downstream along the eastern shore and may be seen for a long distance, while on the opposite side the water remains clear. As progress is made downstream the black swath becomes wider and wider until finally it covers the entire breadth of the channel.

From the mouth of Lackawanna River southward the Susquehanna traverses the heart of the northern anthracite coal basin through what is known as Wyoming Valley. (See Pl. I, B.) In this beautiful basin are situated the cities of Pittston, West Pittston, Wilkesbarre, Kingston, Plymouth, and Nanticoke, which combined have a population of 99,734.

EFFECT OF MINE WASTES.

The sewage from these towns and enormous amounts of culm and mine waste are poured into the Susquehanna. The river along this portion has, however, only a slight grade, and at the lower end of the valley there is a dam. This structure, about 7 feet in height, backs up the water to Wilkesbarre and forms a slack-water basin which serves as an excellent sedimentation reservoir in all except high stages of the river.

Before discussing in detail the effect of mine waste a brief description of the northern anthracite basin will be given.

This basin is divided into four districts: Carbondale, Scranton, Pittston, and Wilkesbarre. In that part of the coal basin above Nanticoke dam there were at the time of the report of the Pennsylvania geological survey, in 1883, about 160 shafts or slopes. Without doubt some of the collieries reported at that time have been "worked out," but it is equally probable that as many more have been opened. Therefore the total number is at least as great as in 1883, although precise information upon this subject has not been obtained.

It is of course impossible to determine with any degree of accuracy the amount of mine waste that is pumped from shafts or allowed to run out from slopes into Susquehanna River and its tributaries. At only a few mines are continuous records kept. At some others the results of a few days' run are noted, while at others records are made occasionally.   As the amount of water which it is necessary to pump varies widely, it would be imperative to have a continuous record at each shaft for a term of years to secure data of any value.   Even then there would be no record of the water which runs out of the slopes requiring no pumping.   Below are given some data which show the amount pumped at the shafts represented.   These figures are only approximately accurate, but are in nearly all cases less than the actual amount pumped.

Mine waste pumped from collieries in northern anthracite coal basin, Pennsylvania.

Company.	Colliery.	Cubic feet per second.	Total.
Scranton Coal Co	Johnson No. 1	7.04	
	Johnson No. 2	.90	
	Richmond No. 3	1.29	
	West Ridge	.26	
	Richmond No. 4	.89	
	Raymond	3.56	
	Ontario openings	3.10	
	Capouse	4.08	
	Pine Brook	3.05	
	Mount Pleasant	1.39	
	Pancoast	.26	
			25.82
Hillside Coal Co	Forest City	2.74	
	Clifford	2.83	
	Erie	6.01	
	Glenwood	6.35	
	Consolidated	2.01	
	Elmwood	4.01	
			23.95
Pennsylvania Coal Co	Gypsy Grove	.79	
	No. 1	.81	
	No. 5	.49	
	Old Forge	2.59	
	Central	6.51	
	Barnum	.28	
			11.47
Lehigh and Wilkesbarre Coal Co.	Hollenbach	1.22	
	Empire	1.10	
	Stanton	1.30	
	South Wilkesbarre	.18	
	Sugar Notch	.41	
	Lance	.41	
	Nottingham	2.25	

Mine waste pumped from collieries in northern anthracite coal basin, Pennsylvania—Continued.

Company.	Colliery.	Cubic feet per second.	Total.
Lehigh and Wilkesbarre Coal Co.	Reynolds	.04	
	Wanamie	.93	
	Maxwell	1.07	
			8.91
Parish Coal Co	Plymouth	1.32	
	Buttonwood	.59	
			1.91
Delaware and Hudson Co	Conyngham (in April, 1902)	a 11.8	
	Conyngham (in May 1902)	a 10.2	b 11.00
	Plymouth No. 1 (in April, 1902)	a 31.22	
			31.22
Plymouth Coal Co	Dodson	}	1.44
	Gaylord c		
Delaware, Lackawanna and Western R. R.	Archbald	1.34	
	Avondale	13.25	
	Bellevue	6.14	
	Bliss	2.76	
	Brisbin	2.73	
	Cayuga	2.91	
	Continental	1.58	
	Diamond	4.64	
	Dodge	5.44	
	Halstead	17.82	
	Hampton	1.67	
	Holden	.49	
	Hyde Park	1.78	
	Manville	.54	
	Pettebone	1.57	
	Pyne	2.24	
	Sloan	.75	
	Storrs	2.64	
	Taylor	4.53	
	Woodward	2.24	
	Central	4.64	
	Central Air Shaft	1.00	
			82.75
Temple Coal and Iron Co	Lackawanna	6.68	
	Edgerton	1.11	
	Mount Lookout	3.12	

a Accurate records.

b Average for Conyngham.

c Gaylord belongs to another company. Shut down at time of measurement, and water was pumped through Dodson colliery.

Mine waste pumped from collieries in northern anthracite coal basin, Pennsylvania—Continued.

Company.	Colliery.	Cubic feet per second.	Total.
Temple Coal and Iron Co	Sterrick Creek	4.46	
	Babylon	2.67	
	Forty Fort	1.78	
	Harry E	1.34	
			21.16
Susquehanna Coal Co., Nanticoke district.	No. 1 a	2.47	
	No. 2 a	2.09	
	No. 2 Slope a	.61	
	No. 4 a	1.30	
	Stearns a	.80	
	Glen Lyon a	2.03	
			9.30
Total			227.49

a Actual pumpage, one day of twenty-four hours, June, 1902.

The daily records of pumping at the Luke Fiddler colliery by the Susquehanna Coal Company, contributed to this report by R. V. Norris, chief engineer, show that the average daily pumpage was 1.27 cubic feet per second in 1899; 1.33 cubic feet per second in 1900, and 1.29 cubic feet per second in 1901. These records are from a mine outside of the immediate area under discussion, but are added for the purpose of showing the average yearly pumpage and its general uniformity over long periods.

There are enumerated in the above table 74 collieries, from which are pumped 227.49 cubic feet of acid mine water per second. If from the remainder of the collieries in the Susquehanna basin above Nanticoke dam there is pumped a proportionately equivalent amount— and this assumption is believed to be fairly safe—there is an average of 491.9 cubic feet of waste per second turned into the river.

The flow of Susquehanna River at Wilkesbarre on September, 1902, was 2,100 second-feet. At Nanticoke dam (see Pl. II, A), 7 miles below Wilkesbarre, the flow at that time was somewhat more, probably 2,500 second-feet. Assuming that the pumpage of 491.9 second-feet of mine water is sufficiently close for practical purposes, about one-fifth of the water flowing in Susquehanna River through Wyoming Valley was acid mine waste artificially turned into the stream. Under such conditions any effects which this mine water might have would be most pronounced.

The appearance of a small stream into which coal-mine waters are

A. SUSQUEHANNA RIVER AT NANTICOKE DAM.

B. ELMHURST RESERVOIR, SCRANTON, PA.

discharged is peculiar. The bottom of the channel is colored a light yellow, and there appear no signs of vegetation of any kind. All fish life in a stream is immediately destroyed at the first appearance of coal-mine waste. Where culm as well as acid mine waste is dumped into the channel the appearance is well-nigh beyond description. Many of the small brooks emptying into Susquehanna River in Wyoming Valley have no permanent channel; the old channel has been filled by deposits of culm, and the stream takes a new course whenever freshets arise, often covering fertile fields with culm and doing great damage.

The important question to be considered in this connection is the effect of the acid mine waste upon the water of Susquehanna River. It has been shown that the run-off from the Lackawanna basin is befouled with sewage, impregnated with acid, and blackened by culm. Susquehanna River below this point is generally of the same character, as it receives the pumpage waters from all the mines and the sewage from the cities in Wyoming Valley. Below this great influx of putrescible matter one would confidently expect to find a water of high organic content, supporting enormous numbers of bacteria. The remarkable fact is that a series of chemical analyses shows that the water is actually more free from organic matter at the lower end of Wyoming Valley than at the upper. This effect is traceable to nothing else than the large amounts of mine waste which are turned into the stream. The analyses in the table below show results which are altogether unique. The fact that the enormous quantity of fine culm turned into Susquehanna River a short distance above Nanticoke is not apparent is due to the coagulating property of the combination which causes the large quantities of fine coal and the organic matter from city sewers to precipitate on the bottom of the stream. It is really a somewhat crude application of the coagulating process used in connection with mechanical filtration. The bottom of the channel of Susquehanna River shows that precipitation occurs rapidly, for there are places at which the bottom has been raised from 8 to 12 feet, and, in fact, this has been a contributory cause of recent damaging floods in the city of Wilkesbarre.

Analyses of water from Susquehanna River at Nanticoke dam.

[Parts per million.]

Date.	Turbidity.	Color.	Odor.	Nitrogen as—				Chlorine.	Total residue.	Hardness.		Gage height, Wilkesbarre, Pa.	Bacteria per cubic centimeter.
				Albuminoid ammonia.	Free ammonia.	Nitrites.	Nitrates.			Alkalinity.	Normal hardness.		
1902.												*Feet.*	
Nov. 22	12	2.00	0	0.078	0.070	0.000	0.0000	3.8	134.5	0.0	80.0	4.2	1,278
Dec. 20	60	.22	1E	.110	.046	.000	.0000	2.4	109.5	.0	32.5	11.3	2,816
1903.													
Jan. 10	25	10.00	0	.090	.050	.000	.0000	2.8	116.0	2.9	47.1	6.8	524
Feb. 14	40	17.00	0	.088	.030	.000	.0080	2.4	119.5	1.3	40.3	11.0	1,520
Mar. 7	50	18.00	1E	.104	.032	.000	.0010	1.8	122.0	5.2	27.3	12.1	....
Apr. 11	35	18.00	0	.080	.030	.000	.0000	1.8	93.0	4.3	51.4	9.8	980
May 8	10	15.00	0	.104	.088	.000	.0004	3.0	134.5	5.7	82.9	4.0	840
July 14	20	2.00	0	.114	.102	.000	.0000	3.2	139.5	15.7	70.0	3.8	7,000
Aug. 15	15	.20	......	.110	.086	.000	.0080	2.4	121.5	5.7	74.3	4.9	1,110
Sept. 14	7	17.00	0	.098	.062	.000	.0020	2.8	115.5	10.0	70.0	5.6	1,270
Oct. 28	15	26.00	......	.070	.072	.000	.0050	2.8	122.0	5.7	84.8	6.5	490
Nov. 28	15	.23	0	.062	.042	.000	.0000	2.0	90.0	2.8	68.6	5.8	542
Dec. 21	25	.21	......	.082	.028	.000	.0000	2.2	94.5	4.2	52.9	8.9	1,380

IMPORTANT INGREDIENTS AND THEIR VARIATION.

The monthly analyses of the water in Susquehanna River at the head and foot of Wyoming Valley afford an opportunity to show the variation in character of water according to the stage of the river. The United States Geological Survey gage at Wilkesbarre, which is read twice daily, is situated about midway between the sampling points, and therefore the readings express accurately the relative conditions with regard to the river stage at those points. Sufficient determinations are not at hand to enable one to make a positive statement regarding the relation of the character of water to the amount flowing in the channel. However, the material available shows the relative changes and the general variation in the ingredients according as the river rises or falls. Figs. 2 and 3 show the relation of turbidity, chlorine, normal hardness, and alkalinity to quantity of water in Susquehanna River water at opposite ends of Wyoming Valley. In these figures the character of the water has been used as ordinates and the gage heights as abscissæ. In fig. 2, which represents the determinations made at the upper end of the valley, the curve does not vary widely from a straight line drawn between the determinations representing the highest and lowest gage heights. The curves show that with an increased flow in the stream the turbidity increases, while the chlorine, alkalinity, and normal hardness decrease. This is a very natural variation often exemplified in rivers, and it would indicate that the ground waters contribute the carbonates, sulphates, and chlorides, while the waters that run off directly after a storm carry large amounts of suspended matter, with an extremely small proportion of carbonates, sulphates, and chlorides, and serve to dilute the inorganic constituents generally.

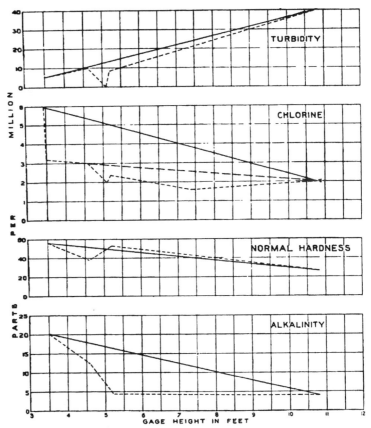

FIG. 2.—Diagram showing relation of character and quantity of water in Susquehanna River at upper end of Wyoming Valley.

Turning to the curves in fig. 3, which show the relations at the lower end of Wyoming Valley, there is a somewhat wider variation in the amounts as the water in the river increases, this being especially true in the case of turbidity and alkalinity. This is readily accounted for by the fact that at the head of Wyoming Valley the water in the

IRR 108—04——3

stream is the result of natural drainage and has not been polluted to
any degree since passing Binghamton, a long distance above, while at
Nanticoke dam the water has been highly charged with sewage and

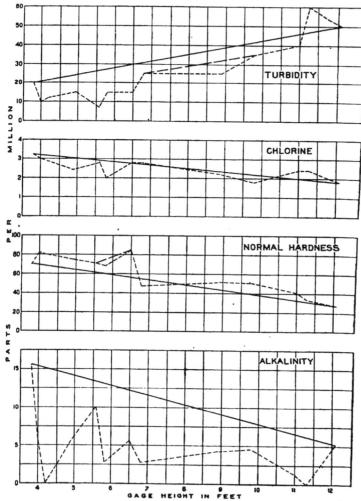

FIG. 3.—Diagram showing relation of character and quantity of water in Susquehanna River
at lower end of Wyoming Valley.

acid coal-mine waste. The amounts of sewage and mine waste, and
consequently the character of the water, vary from day to day. It
may be expected, then, that the artificial conditions which exist to a
large extent just above the sampling point will cause the relations
between characteristics and gage height to be less uniform than above

the entrance of this unusual pollution. However, in fig. 3 the general trend of the curves is the same as in fig. 2, and even the pronounced effect of the acid mine waste and sewage in Wyoming Valley does not make an exception to the general statement that there is increase in turbidity and decrease in chlorine, normal hardness, and alkalinity with the rise of water in Susquehanna River.

PUBLIC WATER SUPPLIES.

Before leaving the consideration of that part of the Susquehanna drainage area included in the northern anthracite coal basin it will be of interest to discuss the domestic water supply in that region.

Along the heights bordering the valley of Lackawanna River and in the mountains forming the perimeter of Wyoming Valley (Pl. I, *B*) are exceptional sites for storage systems. The water-supply systems of Scranton and Wilkesbarre and of neighboring towns have been developed in the highland regions in a manner worthy of note. Naturally these highlands are ideal for water-supply purposes. They are excellent examples of that class of lands which in the opinion of hydrographic experts should be forever set aside for the raising of timber and the conservation of water. Indeed, they are good for nothing else. Agriculture is practically an impossibility, for the soil is both poor and shallow. A plowed field in this highland region after a rain has washed the loose earth from the small stones appears like a waste heap of some quarry. Only along the bottoms of the Susquehanna is there any good, arable land.

The entire region has been deforested. Save for scrub thicket and immature trees, which may at some distant date produce timber, the hills are bare. There are few places in the east where the natural beauties of mountain scenery and the natural resources of timber lands have been destroyed to the extent that has taken place in northern Pennsylvania. One may journey for miles without encountering a mature tree. The situation is an excellent one in which to consider the value and possibilities of intelligent forestation. The water resources of the country are great, but this very greatness only serves to indicate what might have been if the timber had not been destroyed. Judicious cutting would have been beneficial; in any event, it would not have been practical devastation.

UNPOLLUTED SOURCES.

Although the locality here dealt with is generally populous, settlement along the ridges is sparse. Water of great purity is available almost anywhere. In the following table are recorded analyses of waters from various small tributaries of the Susquehanna. Some of these waters are normal, and the whole table indicates extremely well the general character of the run-off in this region. The records were contributed by the Spring Brook Water Supply Company, of Wilkesbarre, through its chemist, W. H. Dean.

Analyses of waters from unpolluted tributaries of Susquehanna River in northern anthracite coal basin.

[Parts per million.]

HUNTSVILLE RESERVOIR.

Date.	Turbidity.	Color.	Odor.	Nitrogen as—				Chlorine.	Total residue.	Total hardness.	Location.
				Albuminoid ammonia.	Free ammonia.	Nitrites.	Nitrates.				
1899.											
May 3	Slight.	1.5	Strong.	0.290	0.040	0.000	0.000	1.072	65.0	......	
June 9	Cons.	2.2	Strong.	.236	.224	.000	.000	1.070	105.5	......	
July 12	Slight.	3.5	Strong.	.378	.272	.000	.000	1.432	70.0	......	
Oct. 21	Cons.	2.8	Strong.	.910	.390	.000	.000	1.280	98.5	......	
1900.											
June 16	Cons.	......	Strong.	.820	.020	.000	.000	......	92.5	......	
Sept. 17	Slight.	3.5	Strong.	.254	.238	.000	.000	1.600	70.0	......	
Oct. 19	Slight.	2.0	Faint.	.296	.026	.000	.000	1.600	68.5	......	
1901.											
Aug. 16	Cons.	5.0	Rank.	.252	.288	.000	.000	2.000	101.0	......	
1902.											
Mar. 29	Cons.	2.0	Faint.	.136	.084	.000	.000	2.400	117.5	......	

GARDNERS CREEK.

Date.	Turbidity.	Color.	Odor.	Albuminoid ammonia.	Free ammonia.	Nitrites.	Nitrates.	Chlorine.	Total residue.	Total hardness.	Location.
1899.											
Aug. 10	0	1.8	0	0.110	0.056	0.000	0.000	......	89.5	30.0	
Dec. 27	0	1.8	0	.028	.010	.000	.000	1.200	74.0	18.2	
1900.											
Feb. 22	Cons.	2.5	Faint.	.116	.032	.000	.000	1.000	166.0	......	
June 15	0	0	Very faint.	.0054	.012	.000	.000	.800	68.5	......	

MILL CREEK.

Date.	Turbidity.	Color.	Odor.	Albuminoid ammonia.	Free ammonia.	Nitrites.	Nitrates.	Chlorine.	Total residue.	Total hardness.	Location.
1899.											
Mar. 4	Slight.	1.0	0	0.046	0.020	0.000	0.000	1.428	72.5	......	Intake waterworks.
Apr. 22	0	.5	0	.038	.012	.000	.000	1.072	70.0	......	Do.
Apr. 26	0	.8	0	.044	.020	.000	.000	.714	76.5	......	Storage reservoir.
July 19	Cons.	3.0	Strong.	.088	.186	.000	.000	.714	75.0	12.7	Bottom reservoir.
July 19	Cons.	2.0	Moldy.	.254	.096	.000	.000	.714	86.5	11.1	Top reservoir.
July 19	Slight.	2.0	Foul.	.080	.022	.000	.040	.714	75.0	12.7	Intake waterworks.
July 20	Cons.	1.5	Moldy.	.330	.246	.000	.000	......	......	......	Top reservoir.
July 28	Slight.	2.0	Sw'mpy.	.140	.270	.000	.000	......	......	......	Bottom reservoir.
Aug. 10	0	1.2	Foul.	.064	.042	.000	.210	.750	65.0	26.0	Intake waterworks.
Dec. 27	0	2.0	0	.076	.002	.000	.000	1.200	60.5	11.1	Reservoir.
1900.											
Feb. 22	Cons.	2.0	Foul.	.062	.028	.000	.000	1.200	62.0	......	Intake waterworks.
June 15	0	1.5	Very foul.	.056	.010	.000	.000	.800	64.5	......	Do.

Analyses of waters from unpolluted tributaries of Susquehanna River in northern anthracite coal basin—Continued.

MILL CREEK—Continued.

| Date. | Turbidity. | Color. | Odor. | Nitrogen as— | | | | Chlorine. | Total residue. | Total hardness. | Location. |
				Albuminoid ammonia.	Free ammonia.	Nitrites.	Nitrates.				
1900.											
June 15	0	1.5	Very foul.	.046	.090	.000	.000	1.200	65.0	......	Bottom reservoir.
June 30	0	2.0	Faint.	.070	.054	.000	.000	.714	59.0	11.1	Storage reservoir.
Sept.14	Slight.	2.1	Moldy.	.114	.164	.000	.000	1.200	73.5	......	Bottom reservoir.
1901.											
Feb. 8	Slight.	1.2	0	.028	.008	.000	.000	1.200	52.5	......	Stream opposite dam.
May 15	0	1.2	0	.028	.020	.000	.000	1.200	35.0	......	Bottom reservoir.
July 30	0	2.0	0	.050	.026	.000	.000	1.600	81.5	9.5	Intake waterworks.
July 30	0	2.5	W.	.042	.064	.000	.000	1.200	66.5	15.6	Bottom reservoir.
Sept. 9	0	.8	0	.026	.008	.000	.000	1.400	78.0	......	Intake waterworks.

LAUREL RUN.

Date.	Turbidity.	Color.	Odor.	Albuminoid ammonia.	Free ammonia.	Nitrites.	Nitrates.	Chlorine.	Total residue.	Total hardness.	Location.
1899.											
Mar. 4	0	0	0	0.032	0.008	0.000	0.000	0.357	84.0	......	Laurel Run.
Apr. 19	0	0	0	.026	.008	.000	.000	.893	98.5	......	No.1 reservoir.
Apr. 22	0	.5	0	.098	.010	.000	.000	1.250	64.0	......	Kelleys Run.
Apr. 22	Slight.	0	0	.048	.012	.000	.350	.536	74.0	......	Gin Creek.
Apr. 22	0	.5	0	.060	.022	.000	.000	1.070	65.0	......	Above Gin Creek.
1900.											
Mar. 29	0	.5	0	.006	.006	.000	.000	1.000	118.5	......	Kelleys Run.
Mar. 29	Slight.	.5	0	.020	.008	.000	.300	1.400	39.0	......	Laurel Run.
Apr. 20	0	.5	0	.022	.002	.000	.400	1.600	57.5	......	Do.
May 4	0	.5	0	.022	.006	.000	.350	.800	68.5	......	Do.
May 12	0	.8	0	.008	.012	.000	.000	.800	......		Below Gin Creek.
May 30	0	0	0	.012	.084	.000	.350	.800	57.5	......	Dynamite works.
May 30	0	0	0	.008	......	.000	.060	1.200	41.5	......	Above dynamite works.
July 7	0	.5	0	.082	.006	.000	......	3.200	60.0	......	Below dynamite works.
Oct. 29	0	.5	0	.024	.016	.000	......	1.600	79.0	......	
Nov. 29	Cons.	1.0	0	.012	.028	.000	.800	2.200	......	......	Gin Creek.
1901.											
July 1	0	.5	0	.016	.002	.000	.200	1.000	95.0	......	
Sept.20	0	.7	0	.014	.008	.000	.500	1.400	103.0	......	
Oct. 31	0	0	0	.084	.006	.000	.500	1.200	69.5	......	No 2 dam.
1902.											
Feb. 6	0	.5	0	.020	.020	.000	.500	2.000	55.5	......	Do.
Mar. 28	0	.2	0	.028	.016	......	......	......	53.0	......	Do.

SCRANTON SYSTEM.

The water supply of the city of Scranton is owned by the Scranton Gas and Water Supply Company, of which W. M. Marple is chief engineer. The water is obtained from upland brooks from which all apparent dangerous contamination has been eliminated. The following facts with reference to the system were contributed by Mr. Marple. The principal impounding reservoirs are known as Scranton Lake and Elmhurst reservoir. (Pl. II, *B*, and Pl. III, *B*.)

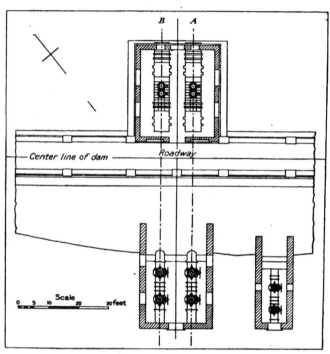

Fig. 4.—Plan of valve houses at Scranton Lake dam.

Scranton Lake has an area of about 225 acres, a capacity of 2,513,972,000 gallons, and a spillway elevation of 1,282.28 feet. It is situated in the basin of Stafford Meadow Brook and has a direct contributing area of 8 square miles. In addition to this, however, 37 square miles are indirectly contributory through a 36 by 30 inch cast-iron pipe, which conducts water from Elmhurst reservoir. The latter is situated in Roaring Brook basin and has an area of 195 acres and a capacity of 1,393,459,000 gallons. The spillway is at an elevation of 1.425.5 feet, while the business portion of the city of Scranton is but 743 feet above sea level.

Reservoirs in Scranton water-supply system.

Reservoir.	Drainage area.	Area of water surface	Capacity	Extent of contributing drainage area.	Elevation of spillway above tide.
		Acres.	*Million gallons.*	*Square miles.*	*Feet.*
Scranton Lake [a]	Stafford Meadow Brook.	225	2,514	45	1,282.8
Williams Bridge	....do	42.6	343	5½	1,360.6
No. 5 distributing [b]	....do	10	32	11½	922.2
Elmhurst	Roaring Brook	195	1,393	34½	1,425.5
Oak Run [c]	....do	75	418	2½	1,467.8
Lake Henry [d]	....do	69	205	....	1,905.8
No. 7 distributing [e]	....do	19	100	13	1,058.5
Dunmore system	Little Roaring Brook.	22.5	78	5½	1,212.62
Providence system: [f]					
Griffin	Leggitts Creek	105	579	2 to 8	1,354.4
Summit Lake	....do	54	212	¼	1,878.4
3 small distributing reservoirs (combined area).	....do	4	9	....	....

[a] Immediate drainage, 8 square miles. Supplementary by 36 by 30 inch pipe from Elmhurst reservoir, 37 square miles.

[b] No. 5 distributing reservoir is 3 miles below Scranton Lake. It is kept full by water of Scranton Lake.

[c] Oak Run reservoir is on a small creek of same name and is included in drainage area of Elmhurst reservoir.

[d] Lake Henry is principally fed by springs.

[e] No. 7 distributing reservoir is on Roaring Brook, nearly 7 miles below Elmhurst (in borough of Dunmore). There is 18 square miles of drainage area between it and Elmhurst.

[f] Undetermined.

Scranton Lake dam is about 325 feet long (Pl. III, *A*). At the
south end is a core wall of hydraulic masonry 265 feet long, with earth
embankments, which slope 3½ to 1 on the inside and 3 to 1 on the outside.   The spillway is situated about one-fourth mile from the dam
in a natural depression 100 feet wide. Details of the dam construction are shown in figs. 4, 5, and 6. There are two gate houses, one at
the dam and the other at the opposite end of the reservoir. The
former has sluice gates at different levels and there are four 36-inch

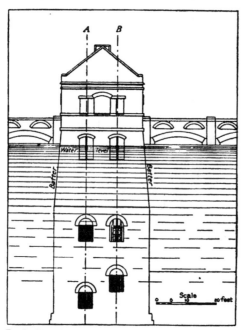

Fig. 5.—Rear elevation of valve house at Scranton Lake dam.

valves. The masonry structure, which is surmounted by an arch-
bridge driveway, is very substantial and represents an excellent type
of the municipal water-supply dam common in the northeastern part
of the United States.

WYOMING VALLEY SYSTEM.

The water-supply system of Wyoming Valley, which includes the
municipalities of Pittston, West Pittston, Wilkesbarre, Kingston,
Plymouth, Nanticoke, and smaller places, is owned by the Spring

A. SCRANTON LAKE DAM, SCRANTON, PA.

B. ELMHURST RESERVOIR DAM, SCRANTON, PA.

Brook Water Supply Company. The system is an extensive one. A portion was constructed for the various municipalities by separate companies which were afterwards consolidated, and other parts have been constructed since. Some of the original plant has been abandoned, notably those systems which involved the pumping of water from Susquehanna River.

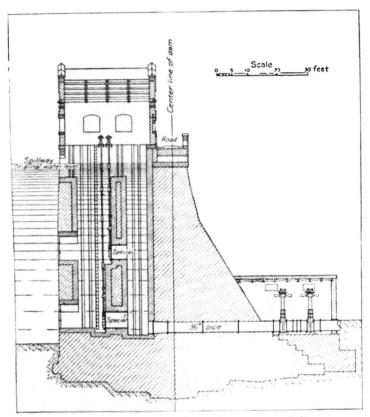

FIG 6 —Cross section of valve house and dam at Scranton Lake

Thirty-three reservoirs and intakes are scattered over the region. As many of these basins are connected by an intricate system of pipes and gate valves, the water from almost any of them can be delivered to any part of the valley, according to the temporary conditions in the various drainage areas.

The following list of reservoirs comprising the Wyoming Valley system is published by courtesy of O. M. Lance, general manager for Spring Brook Water Supply Company.

Reservoirs of water supply system controlled by Spring Brook Water Supply Company, of Wilkesbarre, Pa.

	Reservoir.	Stream.	Drainage area.	Capacity.	
			Sq. miles.	*Gallons.*	*Cubic feet.*
1	Maple Lake	Spring Brook	1.07	300,000,000	40,000,000
2	Nesbitt	do	15.74	75,000,000	1,000,000
3	Round Hole	do	36.00	1,322,300,000	173,306,000
4	Intake	do	53.00	50,000,000	6,666,600
5	Covey	Covey	.47	10,000,000	1,333,300
6	Trout and Monument.	Reservoir	7.52	10,000	1,330
7	Intake	Gardners Creek	2.13	70,400,000	9,386,600
8	Laflin	do	2.88	10,000	1,330
9	No. 1	Mill Creek	5.07	616,700,000	82,226,600
10	Intake	do	10.00	4,000,000	533,300
11	do	Deep Hollow	.87	1,000	130
12	No. 2	Laurel Run	8.35	10,000,000	1,333,300
13	No. 1	do	8.95	6,000,000	800,000
14	Crystal Lake		2.95	1,310,000,000	174,666,600
15	Intake	Statlers Creek	.17	10,000	1,330
16	No. 4	Crystal Spring	2.96	10,000	1,330
17	No. 1	do	4.00	1,000,000	133,300
18	No. 6	do	.44	10,000	1,330
19	No. 5	do	2.18	10,000	1,330
20	Red Mill	do	7.38	20,000	2,660
21	Intake	Sugar Notch	1.35	20,000	2,660
22	Storage	do	.04	5,000,000	666,600
23	do	Hanover	.46	4,000,000	533,300
24	Intake	do	.54	50,000	6,600
25	Storage	Wanamie	.77	10,000,000	1,333,300
26	Intake	do	.91	50,000	6,600
27	do	Coal Creek	4.07	1,000,000	133,300
28	No. 2	do	3.47	2,500,000	333,300
29	No. 3	do	2.50	2,500,000	333,300
30	No. 4	do	1.45	6,000,000	800,000
31	Storage	Huntsville	8.44	1,750,000,000	233,333,300
32	Intake 2	do	14.84	2,000,000	266,600
33	Intake 1		14.84	3,000,000	400,000
	Total		115.91	5,541,600,000	

A.

B.

ROUND HOLE DAM, MOOSIC, PA.

A. Buttress walls and sluice openings

B Dam and spillway

The largest reservoir noted in the above table is that at Round Hole on Spring Brook. As this reservoir is of recent construction it is of considerable interest in that locality and of general interest to the engineering profession. The construction did not involve any unusual engineering difficulties, although it presented some peculiar features. The work was done by the Water Supply Company with its own laborers working by the day. The project was planned and executed by Mr. John Lance, engineer of the company. The stone was quarried upon the drainage area and brought to the site by a specially constructed switch-back road, which was also built by the company. The whole plant was completed at exceedingly small cost.

Round Hole reservoir is situated 6 miles from Moosic station on the Delaware and Hudson Railroad. It has a length of 1.5 miles, an area of 118.7 acres, and a capacity of 1,322,000,000 gallons. The drainage area above the dam is 36 square miles. At the dam the water is 87 feet deep and at the spillway is at an elevation of 1,155.13 feet above sea level. The dam, of which two views are shown on Pl. IV, is 500 feet long; 280 feet of this is masonry. The remainder is an earthen embankment with masonry core wall. Two-foot wing walls (Pl. IV, A) retain the embankment at its junction with the masonry. On the west end of the masonry the structure joins the natural rock, which rises in a precipitous bluff to a height of about 20 feet above the dam crest. The dam foots in the bed rock, which was excavated to a depth of 15 feet. The height from foundation to capstones is 104 feet.

The cross section of the dam (see fig. 7) is based upon a right triangle having its base three-fourths of its altitude. The crest is 9 feet thick and the lower edge along the spillway is appropriately curved. Below this, on the lower face of the dam, is a batter of 9 inches per foot. The determination of stress showed the necessity of a batter of one-half inch per foot on the back face of the dam, commencing 30 feet below the overflow. On the foot of the spillway a curve of 20 feet radius was introduced for horizontal deflection of falling water. The spillway channel below the dam is heavily paved with large cut blocks for about 100 yards.

There are 37,710 cubic yards of masonry in the dam; 25,085 barrels of cement were used, which is equivalent to about 1 barrel to 1.26 cubic yards of masonry. The mortar used in construction was made of Portland cement and sand in proportions, by volume, of 1 cement to 3 sand; that used for pointing and filling exposed joints was mixed in proportions of 1 cement to 1 sand. The cements used were Lehigh, Saylor's, and Atlas. The stone used in building the dam was Pottsville conglomerate, a tough, gray, heavy stone, which dresses roughly and gives an admirable adhering surface for the cement. This conglomerate has a specific gravity of 2.742 and weighs 165 pounds per cubic foot.   The average size of the stones was 0.54 cubic yards and the largest laid contained 3 cubic yards.

The embankment at the end of the masonry section of the dam rises to a height equal to that of the masonry.   Its front and back slope is 2½ to 1, and it is paved on the back face to a depth of 18 feet below the spillway elevation.   The core wall supporting this embank-

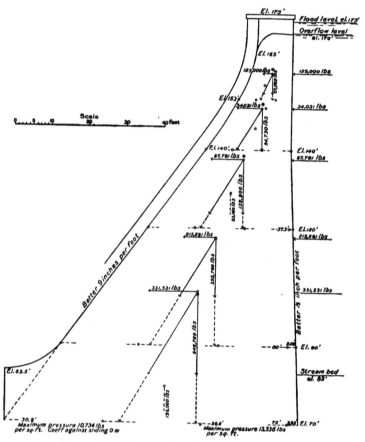

FIG. 7.—Cross section of Round Hole dam.

ment is 3 feet wide at the top, extending to a height of 3 feet above the spillway.   The core is also reenforced by buttresses on front and back faces sloping 1 in 6.

The discharge pipes are 2 in number, 30 inches in diameter and 36 inches apart.   They were carefully laid on wedges to a true line and a grade of 1 per cent, and then mortar was worked under and

around them. When this had not sufficiently the wedges were
removed and large blocks of stone, of such a height as to reach about
3 inches above the pipes, were set on either side between them, the
space between these blocks and the pipes being carefully filled with
mortar. Other large blocks, having a thickness of at least 15 inches
and a length at least 5½ feet, were laid on full beds of mortar across
the tops of pipes, but not resting upon them. Although it was impossible to break joints the mortar was so worked in about the pipe that
no moisture was noticeable at the downstream face of the dam. Two
gates were set on each line below the dam, separated by a piece of
pipe 30 inches long, and the lower gate on each line was anchored
back to the masonry to a distance of 15 feet by iron bars, which were
bent over a length of 12 inches at the ends. The capacity of these
pipes is about 350,000,000 gallons per day.

In the construction of this work it was necessary to lay 2.3 miles of
railroad on a graded roadbed of an old lumber road, at a cost of $7,235.
A branch from the construction road to the quarry, 2 miles long,
erected on a steep side hill with about 1,500 feet of rock work, cost
$4,600, exclusive of rails. Workable stone was obtained with the
first shot at the quarry, and thereafter cost of quarrying was about
$1.59 per cubic yard.

The average number of men employed on the construction was 220,
and the time occupied was from August 16, 1899, to December 14,
1901, leaving out about 100 days each winter.

The total cost of the dam, including everything but the railroad,
was $240,547.93. The masonry alone cost $5.98 per cubic yard.

SUSQUEHANNA RIVER BELOW NORTHERN ANTHRACITE COAL BASIN.

After passing Nanticoke dam the Susquehanna emerges from the
northern anthracite coal basin through a deep gorge, and for some distance flows in a southwesterly direction along the edge of the coal
field. It makes an abrupt turn at Shickshinny and flows south for
some distance, and then takes a southwesterly direction to its junction with West Branch at Sunbury.

The country tributary to Susquehanna River between the northern
anthracite coal basin and Sunbury is sparsely settled, and save
for a comparatively small amount of contamination from a few
towns along its banks and considerable mine-water drainage from
Nescopeck Creek, Catawissa Creek, and Roaring Creek, which enter
from the south, draining the middle coal basin, the stream is not
damaged to any extent. The municipalities of Nescopeck, Berwick,
Mifflinville, Bloomsburg, Catawissa, and Danville, containing a total
population of 22,204, contribute a small amount of pollution to the
stream; but from the physical appearance of the water and the general

survey of the entire region it is apparent that the stream at Sunbury is in a better condition than at Nanticoke.

WEST BRANCH OF SUSQUEHANNA RIVER.

West Branch of Susquehanna River, which joins the main stream at Sunbury, drains a country in northern Pennsylvania which has not been so thickly settled and upon which there are fewer municipalities than along the main stream. The principal towns located along this branch are Milton, Lewisburg, Muncy, Lock Haven, Renovo, Driftwood, and Jersey Shore. The population in the drainage area, by counties, is as follows:

Population in drainage area of West Branch of Susquehanna River.

Cambria County 17,516
Cameron County 7,048
Clearfield County 74,754
Center County 19,498
Columbia County 512
Elk County 5,416
Lycoming County 52,658
Montour County 2,574
Northumberland County 6,959
Potter County 10,642
Sullivan County 12,134
Tioga County 16,897
Union County 15,492

The character of the water in West Branch with respect to pollution is not different from that of the main stream above Binghamton, shown in the analyses on page 21. Along its lower reaches the water is certainly not fit for domestic supply in its raw state, yet it is probably as good or better than that in any large stream in Pennsylvania. The following analyses of water from the drainage area of West Branch are here reproduced by courtesy of Dr. Charles B. Dudley, of the Pennsylvania Railroad:

Analyses of waters from various sources in drainage area of West Branch of Susquehanna River.

[Parts per million.]

Location and source.	Date.	Turbidity.	Color.	Odor.	Nitrogen as — Albuminoid ammonia.	Free ammonia.	Nitrites.	Nitrates.	Chlorine.	Total residue.	Iron.	Total bacteria per c. c.	B. Coli Com.
West Branch at Williamsport, Pa.	May 26, 1897	Trace.			0.028	0.028	Trace.	0.080	21.00	207.9			0
Water supply, Williamsport, Pa.	Apr. 11, 1901	0	0		.047	.080	.000	1.000	8.00	25.7			
West Branch at railway bridge 168, Watsontown,	Feb. 21, 1898	0			.018	.082	Trace.	2.460	7.00	168.8			
West Branch at bridge, Northumberland, Pa.	Nov. 28, 1902	0			.084	.082	Small amount	.350	4.40	79.0	0	840	0
West Branch, 2,500 feet west of bridge, Northumberland, Pa.	...do...	0			.104	.040	Small amount	.400	4.40	79.1	0	420	Few.
Paddy Run above dam, Renovo, Pa.	July 6, 1898	Slight.			.080	.000	.000	.200	Trace.	20.0	Very little.		
Reservoir, Renovo, Pa.	Nov. 29, 1898	Slight.			.600	Trace.	.000	.200	8.50	24.5	0	8,800	Some.
Drury Run, Renovo, Pa.	Jan. 31, 1903	0			.049	.086	.000	.240	2.90	24.0	0	8,350	0
Do.	...do...	0			.049	.088	.000	.240	2.90	24.0		1,070	Some.
Do.	Apr. 7, 1903	Trace.			.088	.088	.000	.200	2.90	22.9		2,800	Some.
Do.		Slight.			.061	.020	.000	.200	5.90	31.0			
Stream, Ralston, Pa.	Jan. 9, 1896				.164	.006	.000	.880	5.90	186.5			
City supply sample No. 1, Osceola, Pa.	July 29, 1899				.016	Trace.	.000	.900	13.80	44.0	0		
Delaware Run, Dewart, Pa.	Feb. 21, 1898				.078	.034	.000	.900	7.00	116.7			
Stream, Benezett, Pa.	Jan. 25, 1901	0			.016	.006	.000	1.700	4.50	29.6	0		
Stream, Trout Run, Pa.	Apr. 25, 1901	0			.022	Trace.	.000	.110	2.90	19.8	Trace.	29,000	0
Pond, Penhryn, Pa.	Apr. 27, 1901	0			.154	.088	.000	.375	5.00	59.9	Trace.	468	
Water station, Bellefonte, Pa.	May 25, 1901	0			.088	.088	.000	2.300	5.00	145.4	Trace.	64,000	
Water station, Lewisburg, Pa.	...do...	0			.142	.072	.000	.880	11.00	47.9	Trace.	7,900	Some.
Stream, Caledonia, Pa.	Jan. 22, 1902	Slight.			.088	.016	.000	.100	3.90	24.9	0	1,684	
Stream, Tyler, Pa.	Jan. 28, 1902				.090	.040	.000	.220	6.00	66.5	0	8,600	
Stream, Driftwood, Pa.	Jan. 27, 1902				.016	.041	.000	.250	5.00	39.2	0	680	
Stream, Spangler, Pa.	May 8, 1902	Marked.			.070	.022	.000	.400	1.60	34.0	0	10,600	
Water supply, Spangler, Pa.	May 29, 1902	Slight.			.102	.030	.000	.190	2.90	67.9	0	168	
Cushaua reservoir, Lock Haven, Pa.	Mar. 18, 1903	0			.043	.030	.000	.120	2.50	23.8	0	24,400	

SUSQUEHANNA RIVER BELOW WEST BRANCH.

The city of Sunbury is located immediately below the confluence of Susquehanna River and West Branch. At this point the slack water caused by the Shamokin dam affords a basin which is quite effectual in disposing of such sewage as is discharged from Sunbury, and from all evidences at hand it is practically certain that the city sewage has little effect upon the chemical characteristics of the river water below. The country traversed by the Susquehanna immediately below West Branch is sparsely settled and receives little or no sewage contamination. The tributaries entering from the east, however, carry large quantities of coal-mine waste of the character described in the preceding pages, while those entering from the west are in many cases highly alkaline.

In the table below are given results of analyses of water from Susquehanna River at Clarks Ferry, made by Prof. W. B. Lindsay, of Dickinson College, Carlisle, Pa.

Analyses of water from Susquehanna River at Clarks Ferry, Pa.

[Parts per million.]

| Date. | Turbidity. | Color. | Odor. | Nitrogen as— | | | | Chlorine. | Total residue. | Hardness. | | Bacteria per cubic centimeter. | Gage heights at Harrisburg, Pa. | Character of sample. |
				Albuminoid ammonia.	Free ammonia.	Nitrites.	Nitrates.			Alkalinity.	Normal hardness.			
1902.													*Feet.*	
Nov. 4	60	2	None	0.101	0.182	0.0	0.15	5.35	170	42.99	87.84	875	4.00	Composite.
Nov. 25	5	10	None	.235	.078	.0	.25	4.37	252.5	15.89	154.19	475	1.66	Do.
Dec. 16	5	5	None	.1071	.017	.0	1.25	2.3	155	1.87	75.69	850	4.00	Do.
1903.														
Jan. 27	10	2.5	None	.112	.104	.0	1.2	1.68	155	.0	88.88	12,000	3.50	Do.
Feb. 24	10	5	None	.066	.073	.0	1.2	.43	115	2.63	90.05	2,400	4.83	Do.
Mar. 17	55	5	None	.149	.069	.0	1	1.38	163	.0	78.67	850	7.83	Do.
Apr. 14	35	3	None	.128	.063	.001	.6	.46	190	.0	76.85	300	6.50	Do.
May 12	— 7	3	None	.032	.125	.0	.4	1.38	185	.0	149.51	200	2.16	Do.
June 1	5	3	1 M	.237	.209	.0	.2	2.16	413	.0	165.65	600	1.00	Do.

Below Harrisburg the character of the water in Susquehanna River is well shown by the following results of analyses made by Prof. W. B. Lindsay, of Dickinson College, Carlisle, Pa.

Analyses of water from Susquehanna River at Steelton and New Cumberland, Pa.

[Parts per million.]

Date.	Turbidity.	Color.	Odor.	Nitrogen as—				Chlorine.	Total residue.	Alkalinity.	Hardness. Normal hardness.	Bacteria per cubic centimeter.	Gage height at Harrisburg. Pa.	Character of sample.
				Albuminoid ammonia.	Free ammonia.	Nitrites.	Nitrates.							
1912.													*Feet*	
Nov. 4	55	1	1 M	0.110	0.157	0.004	1.5	6.34	148.3	41.12	74.46	6,600	4.00	Composite.
Nov. 25	10	7	1 M	.101	.185	.007	.55	4.27	147.5	58.88	71.96	3,500	1.66	Do.
Dec. 16	20	15	2 M	.198	.181	.002	2.55	3.68	179	12.47	75.49	81,000	4.00	Do.
1903.														
Jan. 27	15	15	2 M	.136	.115	.012	2.80	2.53	200	53.04	80.28	2,500	3.50	Do.
Feb. 24	15	10	2 M	.166	.155	.005	4.80	1.72	160	66.54	53.03	186,000	4.33	Do.
Mar. 17	20	17	2 M	.176	.114	.0008	1.40	1.96	110	6.67	53.34	2,700	7.83	Do.
Apr. 14	110	30	1 M	.268	.247	.005	1.20	3.22	200	26.29	44.58	1,500	6.50	Do.
May 12	10	20	1 M	.155	.207	.008	1.96	3.90	161	44.45	81.83	Enormous.	2.16	Do.
June 1	10	25	1 M	.207	.190	.008	1.30	3.20	170	21.83	90.13	21,000	1.50	Do.

JUNIATA RIVER.

Juniata River enters the Susquehanna from the west a few miles above Harrisburg. In the entire basin drained by this tributary there are no important cities except Altoona, near the extreme western divide. Below Altoona there is without doubt a considerable amount of organic pollution discharging into the headwaters of the river, but long before it reaches the Susquehanna all chemical evidence at least is destroyed.

Population in drainage area of Juniata River.

Bedford County ... 39,468
Blair County ... 85,099
Cumberland County ... 21,645
Franklin County ... 25,213
Fulton County ... 1,894
Huntingdon County ... 34,650
Juniata County ... 12,181
Mifflin County ... 23,160
Perry County ... 9,119

In the table below are set forth results of analyses of waters taken from several points in the Juniata basin. The analyses were made in the laboratory of the Pennsylvania Railroad at Altoona under the direction of Dr. Charles B. Dudley, chief chemist.

Analyses of waters from various points in Juniata River basin.

[Parts per million.]

Source.	Date.	Turbidity.	Color.	Odor.	Nitrogen as—				Chlorine.	Total residue.	Iron.	Bacteria per cubic centimeter.
					Albuminoid ammonia.	Free ammonia.	Nitrites.	Nitrates.				
City supply, Altoona, Pa	July 21, 1897	0			0.036	0.094	0.00	0.200	7.00			
Pottsgrove Reservoir No. 1, Altoona, Pa	Aug. 10, 1887				.152	.160	Slight trace.	.360				
Pottsgrove Reservoir No. 2, Altoona, Pa	..do..				.128	.066	Slight trace.	Trace.				
Pottsgrove Reservoir No. 3, Altoona, Pa	..do..				.246	.022	Slight trace.	Small amt.				
Well at car shops, Altoona, Pa	Oct. 3, 1901			0	.072	.080	.000	.790	5.00	21.8		22,800
Mountain Reservoir, near Altoona, Pa	Apr. 1, 1903	Trace.	0	0	.052	.024	.000	.170	2.50	29.5		71
Do	..do..	Trace.	0	0	.090	.016	.000	.120	2.10	76.5		1,780
City supply, Altoona, Pa	..do..	Trace.			.029	.016	.000	.140	5.00	42.5		60
Stream above Fountain Inn, near Altoona, Pa	July 20, 1903	0			.051	.014	.000	.280	3.00	41.0		1,260
The same, below Fountain Inn	..do..	0	0	0	.076	.028	.000	.380	3.00			1,380
Bell Mill dam, Altoona, Pa	Oct. 14, 1901	0	0	0	.156	.454	Some.	.560	9.10			15,600
Water station, Blair Furnace, Pa	July 29, 1897			Slight.	.170	.040	.000	.140	Trace.			
Stream at Barre, Pa	Mar. 23, 1903	Cons.	0		.040	.012	.000	.140	2.80	21.4		180
City supply, Kittanning Point, Pa	Aug. 11, 1893				.066	.014	Trace.	.400	Trace.			
Do	Nov. 8, 1895				.018	.022	.000	.440	6.50	77.4	Some.	
Dam at Elizabeth Furnace, Pa	..do..				.006	.048	.000	1.040	5.00			15,000
Reservoir, Martinsburg, Pa	Oct. 8, 1900				.104	.062	.000	.000	Trace.		Little.	
Water supply, Tyrone, Pa	July 24, 1895				.012	Trace.	Trace.	.000	Trace.	98.1		
Reservoir, Lewistown, Pa	Jan. 20, 1897				Trace.	Trace.	.000	.900	8.50	88.4		
Do	..do..				.010	Trace.	.000	.000	7.00	211.9		
Kishacoquillas Creek, Lewistown, Pa	..do..	Marked.			.256	.080	.000	1.800	8.50	581.90	Trace.	

In the table immediately below are results of analyses of waters from Juniata River near its confluence with the Susquehanna, made by Prof. W. B. Lindsay, of Dickinson College, Carlisle, Pa. The results show that there is little or no chemical evidence of the sewage pollution which is turned into the river at Altoona.

Analyses of water from Juniata River above its confluence with the Susquehanna.

[Parts per million.]

Date.	Turbidity.	Color.	Odor.	Nitrogen as—				Chlorine.	Total residue.	Hardness.			Bacteria per cubic centimeter.
				Albuminold ammonia.	Free ammonia.	Nitrites.	Nitrates.			Alkalinity.	Normal hardness.		
1903.													
Jan. 27	10	17.5	0	0.1155	0.008	0.0	1.10	2.183	125	85.84	64.51		4,200
Feb. 24	12	17.5	0	.057	.005	.0005	1.80	1.075	125	81.98	49.34		6,100
Mar. —	15	15	0	.084	.010	Trace.	1.40	1.840	123	40	....		900
Apr. 14	15	23	0	.111	.014	.004	1.00	1.840	190	43.43	36.58		1,700
May 12	— 7	.15	0	.105	.012	.001	1.00	2.070	117	75.77	44.45		300
June 1	— 7	.15	0	.098	.017	.005	1.30	1.470	160	46.85	61.70		700

INORGANIC INGREDIENTS OF UNDERGROUND AND SURFACE WATERS.

From the analytical records of the Pennsylvania and the New York Central and Hudson River railroads there has been collected a large amount of data with reference to the inorganic ingredients of the waters of the Susquehanna River basin. These analyses are supplied through the courtesy of Dr. Charles B. Dudley and Mr. R. H. Mahon, chemists of the respective roads. The analyses give in definite terms the determinations of total solids, total chlorines, and incrusting and nonincrusting matter; the other constituents are stated qualitatively. The four determinations above mentioned will therefore be dealt with in this discussion.

CHEMUNG BASIN.

UNDERGROUND WATERS.

The table on page 52 shows that the ground waters in various parts of the Chemung basin contain only a fairly large amount of incrusting constituents. In fact, none of the analyses indicate that the water would be especially objectionable for use in boilers, and it is clearly a water that is treatable if it should be necessary to reduce the small amount of incrusting constituents contained.

It will be noted that the sources from which the samples were taken are all shallow wells and that there is no evidence with reference to the character of the water from greater depths. It is apparent, however, that the supply from shallow wells is sufficient for all needs that have so far been apparent in this region, and therefore the supply, so far as its incrusting constituents are concerned, may be considered unobjectionable.

Inorganic ingredients of ground waters in Chemung drainage basin.

[Parts per million.]

Source	Depth.	Location.	Total solids.	Incrusting solids.	Nonincrusting solids.	Total chlorine.
	Feet.					
Driven well	15	Snediker, Pa	123.462	101.061	22.401	77.866
Well	64	....do	205.200	184.680	20.520	14.022
Do	64	....do	225.720	145.350	80.370	6.498
Do		....do	297.198	173.394	123.804	30.438
Driven well	15	Beaver Dams, N. Y.	205.200	129.447	75.753	5.472
Well	12	....do	183.483	155.61	27.873	5.985
Do	45	Horseheads, N. Y.	226.233	176.814	49.419	7.011
Do		Southport, N. Y	209.988	142.956	67.032	13.680
Do	22	Elmira, N. Y	193.401	122.949	70.452	8.892

SURFACE WATERS.

From the boiler standpoint the river water is even better than the water from shallow wells. The analytical statements in the table below include waters from widely distributed points in the Chemung basin, as well as from the main river itself. It is apparent upon examination of this table that the water from any of the streams may be accepted for use in boilers without prejudice.

Inorganic ingredients of surface waters in Chemung drainage area.

[Parts per million]

Stream.	Location.	Mineral solids.	Incrusting solids.	Nonincrusting solids.	Total chlorine.
Crooked River	Hammond, Pa	102.70	75.75	27.873	9.92
Tioga River	C. V. Junction, Pa	125.57	87.723	39.843	9.92
Do	Lawrence, Pa	102.70	69.768	33.858	4.959
Cowanesque River.	Academy Corners, Pa.	83.79	61.902	21.888	3.933
Tioga River	Presho, N. Y	119.7	101.75	35.06	3.933
Poster Creek	Beaver Dams, N. Y	114.57	78.66	35.91	3.591
Do	Townleys, N. Y	143.64	119.70	23.94	2.907
Chemung River	Corning, N. Y	253.08	199.386	53.694	16.929
Do	Elmira, N. Y.	75.411	48.051	27.360	3.933

SUSQUEHANNA BASIN NORTH OF SUNBURY, CHEMUNG BASIN EXCLUDED.

There are available only a few boiler analyses of the waters of that part of the Susquehanna drainage area outside of the Chemung basin and most of the waters analyzed are from the main stream. The table below gives the results of determinations of minerals and incrusting solids in waters taken from this basin. It will be readily noted that none of these samples contain an amount of dissolved solids of incrusting constituents sufficient to be objectionable for boiler use. In fact, the surface waters and most of the ground waters in the basin of Susquehanna River are generally unobjectionable from this standpoint.

Inorganic ingredients of surface waters of Susquehanna basin, exclusive of Chemung basin.

[Parts per million.]

Source.	Location.	Mineral solids.	Incrusting solids.	Nonincrusting solids.	Total chlorine.
Susquehanna River....	Susquehanna, Pa..	63.954	44.118	19.836	2.548
Chenango River	Binghamton, N.Y.	99.522	63.099	36.423	15.048
Brook................	Canton, Pa........	59.850	36.765	23.085	8.933
City water	Wilkesbarre, Pa....	20.007	15.732	4.275	3.420
Do..............	....do	91.827	63.783	28.044	10.260
Do..............	Nanticoke, Pa.....	39.843	29.925	9.918	3.420
Harvey Creek	....do	67.887	23.940	43.947	6.840
Boyds Run	Pond Hill, Pa . ..	48.906	36.936	11.970	3.420
Susquehanna River...	Nescopeck .-......	29.925	23.940	5.985	Trace.
Do..............	Danville	92.511	69.255	23.256	8.550
Do..............	....do	134.577	77.976	56.601	12.483

WEST BRANCH OF SUSQUEHANNA RIVER BASIN.

UNDERGROUND WATERS.

At Barnesboro there are three wells, 18, 55, and 230 feet deep, respectively, which are characterized by water containing traces of iron and much calcium carbonate and magnesium sulphate. Boiler water high in solids and incrustants, but containing moderate amounts of chlorine is obtained from the 55-foot well, while the two deeper wells supply drinking water. The chlorine in the 55-foot well is low and that in the 230-foot well is moderate in quantity. One of the samples from the 230-foot well was very turbid and carried a large amount of undissolved limestone, clay, and some oxide of iron in suspension.

At Gallitzin there is a flowing well with water that contains moderate amounts of chlorine and much magnesium carbonate, and is used for boilers. Water secured from a test hole at the same place contains very little chlorine. The city water is neutral in reaction, low in solids, contains moderate amounts of chlorine, and but little magnesium and calcium sulphate, the former predominating.

A 300-foot well at Osceola supplies boiler water carrying moderate amounts of solids, incrustants, and chlorine, and much calcium carbonate and magnesium sulphate.

At Viaduct eight samples were collected from two wells, one known as the 240-foot and the other as the drilled well. The 240-foot well supplies a natural water that is almost colorless, but carries some sediment containing oxides, moderate amounts of solids and incrustants, and low chlorine contents. The analyses indicate the presence of small amounts of calcium carbonate and sulphate, magnesium sulphate, and sodium chloride. The water in the drilled well is slightly alkaline, is colored because of suspended sediment, and contains moderate amounts of chlorine, solids, and incrustants. Of five analyses each shows traces of oxides, calcium carbonates, calcium sulphate, magnesium carbonate, magnesium sulphate, magnesium chloride, alkali carbonates, alkali sulphates, and alkali chlorides.

The 900-foot well at Sizerville, in Cameron County, supplies a slightly alkaline and turbid boiler water containing very considerable amounts of chlorine, solids, and nonincrustants, and only a small quantity of incrustants. The nonincrustants are composed of sodium chloride chiefly, with a small amount of magnesium chloride and some calcium chloride, while the incrustants include some calcium carbonate, calcium sulphate, and a small amount of clay. Drinking water taken from a spring at Shippen was found to be low in solids, incrustants, nonincrustants, and chlorine. The water from two springs at Rock Run is similar in composition.

The analysis of boiler water taken from an 18-foot well at Rock Run shows very large amounts of solids, incrustants, nonincrustants, and chlorine. In its normal condition the water is slightly alkaline and clear, and contains a large quantity of calcium carbonate and magnesium chloride, with traces of oxides, magnesium carbonate, and alkali carbonates and chlorides.

A spring at Hecklin, near Gillintown, furnishes a neutral and clear water, very low in both solids and chlorine.

At Lock Haven the water drawn from the Castenea reservoir is neutral and low in solids and chlorine. It contains a little clay sediment and some calcium and magnesium sulphate.

The 24-foot McKinney well at Oakgrove supplies water with a moderate quantity of solids, incrustants, and chlorine. This is also true of a well at Stokesdale Junction which furnishes a slightly alkaline water containing some calcium carbonate and traces of alkali carbonate and magnesium chloride.

Springs at Leaches and Jersey Shore contain clear and neutral water, low in solids and chlorine, and suitable for domestic purposes.

Two wells at Ralston, 19 feet and 234 feet deep, furnish a boiler water showing moderately low amounts of solids and incrustants and moderately high amounts of chlorine.

There were a number of wells at Williamsport whose waters were analyzed. Boiler and drinking water from one of the wells contains so large a quantity of chlorine and total solids, chiefly calcium carbonate, that it is worthy of further investigation. Waters from a well 140 feet deep and from an artesian well sunk to the depth of 166 feet show moderately large amounts of solids and incrustants, and small quantities of chlorine. The above artesian well furnishes a supply of both drinking and boiler water carrying much calcium carbonate and magnesium sulphate in solution. The Williamsport city water is very low in total solids, moderately low in chlorine, and contains but little calcium and magnesium carbonate.

Analyses of underground waters obtained in the drainage basin of West Branch of Susquehanna River.

[Parts per million.]

Location.	Sources of supply.	Total residue.	Incrustants.	Nonincrustants.	Chlorine.
Barnesboro	230-foot well	446.139	400.311	45.828	6.327
Do	18-foot well	397.746	289.161	108.585	6.840
Do	55-foot well	169.461	119.700	49.761	1.026
Do	230-foot well	151.506	119.529	31.977	3.420
Gallitzin	Test hole	174.078	113.373	60.705	1.026
Do	Flowing well	162.450	109.440	53.010	8.550
Do	City water	59.850	43.947	15.903	3.420
Do	Water supply	43.605	29.925	13.680	6.840
Osceola	300-foot well	125.514	84.645	40.869	6.840
Munson		30.007	25.992	4.015	1.881
Viaduct	240-foot well	176.301	125.001	51.300	2.907
Do	do	170.316	120.555	49.761	2.907
Do	do	165.870	127.908	37.962	2.907
Do	Drilled well	170.487	114.912	55.575	6.498
Do	do	118.845	88.920	29.925	6.498
Do	do	107.901	71.991	35.910	6.498
Do	do	92.169	56.601	35.568	6.498
Do	do	78.489	22.572	55.917	6.498
Sizerville	900-foot well	1,246.077	137.313	1,108.764	616.284
Shippen	Spring	35.568	30.267	5.301	3.078
Rock Run	do	32.832	23.940	8.892	2.052
Do	18-foot well	497.097	324.216	172.881	53.523
Gillintown	Spring at Hecklin	31.977	5.903	16.074	2.052

Analyses of underground waters obtained in the drainage basin of West Branch of Susquehanna River—Continued.

Location.	Sources of supply.	Total residue.	Incrust-ants.	Nonin-crustants.	Chlorine.
Forks	Spring	61.902	47.880	14.022	2.907
Do	Big spring	21.888	16.416	5.472	1.881
Do	do	20.007	14.022	5.985	2.907
Lock Haven	Castanea reservoir.	30.780	20.178	10.602	2.394
Do	do	22.743	17.271	5.472	2.394
Oakgrove	24-foot McKinney well.	226.746	148.428	78.318	5.643
Stokesdale Junction	do	199.386	167.580	31.806	7.011
Leaches	Spring	39.843	34.884	4.959	2.052
Jersey Shore	do	33.858	21.888	11.970	3.933
Ralston	234-foot well	175.104	121.410	53.694	13.509
Do	Well	160.911	106.704	54.207	7.524
Do	19-foot well	91.143	57.114	34.029	8.892
Williamsport	Well	521.379	287.280	234.099	97.812
Do	300-foot well	218.880	140.220	78.660	....
Do	140-foot well	217.854	143.469	74.385	2.907
Do		207.423	173.394	34.029	20.862
Do	Artesian well 166 feet.	201.096	126.198	74.898	3.591
Do	City water	25.650	19.665	5.985	2.993
Lewisburg	Tank	99.693	79.857	19.836	Little.

SURFACE WATERS.

Near the headwaters of West Branch of Susquehanna River, at Spangler, Barnesboro, and Cush Creek Junction, the waters analyzed are rather low in solids, incrustants, and chlorine. This holds true especially of the water used for drinking and boiler purposes at Spangler, which is neutral in reaction and contains a small amount of calcium carbonate and magnesium sulphate, besides a little clay sediment.

At Mahaffey West Branch is joined by Chest Creek, at the head of small branches of which are the towns of Carrolltown and Hastings. Near the former place is located Peter Campbell's pond, containing turbid water with rather large amounts of oxides and a medium quantity of chlorine. At Hastings a tributary of Chest Creek evidently receives a large amount of material that considerably raises its contents in solids and incrustants (chiefly calcium sulphate), while the chlorine remains nearly stationary. As a result West Branch at Mahaffey is moderately high in solids and incrustants. The sample tested was found to be high in chlorine, this being possibly due to pollution by sewage from the town, as the water in the neighboring run was found to be very low in chlorine.

At Bower the water of West Branch is evidently of good quality, neutral in reaction, and low in solids and chlorine. Curwensville receives from Anderson Creek a water supply that is free from carbonates, but contains some little calcium and magnesium sulphates.

Clearfield Creek empties into West Branch of Susquehanna River below Clearfield. The water of Clearfield Creek near Smoke Run is low in solids and incrustants and shows only a trace of chlorine. It is used for both drinking and boiler purposes, and is of excellent quality. At the creek dam, on the other hand, the water used for steam-generating purposes was found to be high in solids and incrustants, and showed a moderate amount of chlorine, together with much carbonate and sulphate of lime, magnesia, and iron. At Belsena the water of Clearfield Creek varies within rather large limits, and contains from a very low to a moderately large quantity of solids (sulphates of lime, magnesia, and iron), incrustants, and chlorine. Pine Run at Belsena is very low in all of these constituents.

At Clearfield Bridge Clearfield Creek is joined by Little Clearfield Creek, on which are located the towns of Kerrmoor and Porters. At these towns the water is neutral in reaction. It contains very moderate amounts of chlorine, solids, and incrustants at Kerrmoor, and small amounts at Porters, the difference being probably largely due to suspended sediment.

At and near Clearfield a large number of water samples were collected from various places on West Branch, and from Clearfield, Moose, and Montgomery creeks. The town supply of Clearfield and the water from Moose and Montgomery creeks are very low in solids, incrustants, and chlorine, thus indicating a very good quality of potable water. There were found present in the water, which has a neutral reaction, only traces of the following constituents: Iron and aluminum oxides, calcium and magnesium sulphates, magnesium carbonate, and the chlorides and sulphates of the alkalies. Directly at the inlet on West Branch of Susquehanna River the solids and incrustants were found to be large in quantity and the chlorine content extremely high, indicating a condition worthy of serious investigation. The river at the time of sampling was very low, however, and the water was turbid in appearance. It is significant that the analyses show a considerable quantity of magnesium chloride and sulphate besides a large amount of the alkaline chlorides.

Nearly all the numerous other samples of waters from Clearfield Creek and West Branch of Susquehanna River, especially at the first and second dams near Clearfield, show moderate amounts of solids, such as lime and alkali sulphates, and a little calcium and magnesium carbonate, a neutral reaction, and moderately large amounts of chlorine.

Moshannon Creek rises above Osceola, flows past Philipsburg, Munson, and Viaduct, and empties into West Branch of Susquehanna River above the town of Karthaus.

At Osceola the water of Moshannon Creek used in boilers is very high in chlorine and solids, the sulphates of lime, iron, and magnesia constituting the incrustants. The drinking water of Osceola is obtained from Mountain Branch, is low in solids, and contains only moderate amounts of chlorine. At Philipsburg the city water and the water of Cold Stream are low in solids, incrustants, and chlorine, and are evidently neutral waters of good quality, while the water obtained from Moshannon Creek was found to be moderately high in solids and chlorine and to contain large amounts of sulphates, probably due to the inflow of acid mine waters. At Viaduct the water of Moshannon Creek has an acid reaction, is high in solids, and contains a large amount of calcium and iron sulphates.

The water of Upper Three Run below Karthaus is low in solids, incrustants, and chlorine.

Bennett Branch and Driftwood Branch flow into Sinnamahoning Creek, which, in turn, empties into West Branch of Susquehanna River at Keating. The towns of Tyler, Caledonia, and Benezett, located on Bennett Branch, all secure good potable water from this stream. The samples tested gave a neutral reaction, were low in solids and incrustants, and moderately low in chlorine. This is also true of the water supply of Emporium, on Driftwood Branch, and Driftwood, on Sinnamahoning Creek.

At Keating the water of Jewels Run is clear, neutral in reaction, free from carbonates, and contains only traces of oxides and sulphates of the alkalies, calcium and magnesium. It is very low in solids and moderately low in chlorine.

Near Renovo, Drury Run and Paddy Run empty into West Branch. Every sample of the waters of the streams analyzed shows very low contents in solids, incrustants, and chlorine, has a neutral reaction, and contains but a small quantity of sulphates and sediment. The water of these streams is excellent for drinking and steam-generating purposes.

Beech Creek passes the towns of Forks and Beech Creek, and near the latter place empties in Bald Eagle Creek, which in turn empties into West Branch of Susquehanna River below Lock Haven.

The waters of Beech, Bald Eagle, and Canoe creeks at or near Beech Creek show moderate to rather large quantities of solids and incrustants and moderate contents in chlorine. At Beech Creek, when the stream was low, it was found to be strongly acid, possibly due to acid waters from the wood-alcohol works in this neighborhood, and to contain calcium and magnesium sulphate, but no carbonates. Bald

Eagle Creek, on the other hand, was alkaline in reaction and contained calcium and magnesium carbonates, while Canoe Creek carried a neutral water, even at a low stage. At Lock Haven, Bald Eagle Creek, while not much higher in solids (principally calcium carbonates) than at Beech Creek, showed a considerable increase in chlorine.

At Oak Hall station, near the head of one of the smaller branches of Bald Eagle Creek, the water is rather high in solids and moderately low in chlorine. At Bellefonte the city and station water supply is moderately high in chlorine and solids, principally carbonate of lime, carbonate of iron, and sulphate of magnesium.

Pine Creek rises in Potter County and flows east and south past Cedar Run, Cammal, and Ramsey station into West Branch of Susquehanna River near Jersey Shore Junction. Near Ansonia it is joined by Marsh Creek, which rises near Wellsboro. Marsh Creek at Wellsboro is moderately low in solids, incrustants, and chlorine, and is generally characterized by a neutral reaction and clear appearance. The water of the run at Stokesdale Junction and of Marsh Creek at Marshcreek has a slightly alkaline reaction, is dark, owing to coloration by organic matter, and contains also a moderately high quantity of chlorine and carbonate of lime. Darling Run, near Ansonia, is clear in appearance, neutral in reaction, and low in chlorine and solids. The stream at Antrim is neutral and low in solids and chlorine, while the water taken from the pond at the same place, which receives drainage from a charcoal iron furnace and wood-alcohol distillery, is naturally considerably higher in all of these constituents, excepting chlorine. The latter constituent is only increased by a moderate amount. The water in the pond carries much calcium carbonate and magnesium sulphate in solution. The water of Jacob Run at the town of Cedar Run and the water of Ramsey Run at the station of the same name are clear, neutral in reaction, and low in solids and chlorine, while samples of water taken from the stream at Cammal are similar in composition, except higher contents in chlorine. Both West Branch and Pine Creek near Jersey Shore Junction are clear, neutral in reaction, moderately low in solids, and contain moderate amounts of chlorine.

Lycoming Creek starts at Penhryn and flows in a generally southerly direction by Ralston and Trout Run into West Branch of Susquehanna River above Williamsport. Red Run at Ralston contains moderate amounts of solids and chlorine, while Rock Run at the same place is somewhat lower in these constituents. The water of Red Run is peculiar in that it carries a considerable quantity of a red sediment in suspension, is acid in reaction, and contains hydrogen sulphide besides considerable sulphate of lime and magnesia. Lycoming Creek at Trout Run furnishes an excellent potable water, which is neutral in reaction, free from all turbidity and sediment, and low in dissolved solids and chlorine.

Below Lock Haven are the towns and cities of Youngdale, Oakgrove, Jersey Shore, Newberry Junction, Williamsport, Dewart, Watsontown, Lewisburg, and Northumberland, the surface waters of which will be discussed in the order given.

The water of McElhattan Creek at Youngdale is neutral, low in solids, and moderately low in chlorine. At Oakgrove the waters of West Branch of Susquehanna River and Pine Creek contain moderate amounts of solids, but rather high chlorine contents, and the city supply is low in solids and chlorine and possesses a slightly alkaline reaction.

At Newberry Junction the water supplied to the city, while neutral in reaction and not carrying more than moderate quantities of solids, is high in chlorine. This water posssibly needs further investigation to ascertain if it is not contaminated by sewage.

At Dewart the waters of Delaware Run and of the stream 2 miles west of the town are very much like those of West Branch of Susquehanna River at Watsontown, and contain exactly the same quantity of chlorine, which is present in moderately large amounts. The solids and incrustants are not very large in quantity. While some of the samples are free from carbonates and contain only sulphates, they usually carry but small amounts of the latter and much combined carbonic acid.

At Northumberland, finally, the water of West Branch of Susquehanna River, used for municipal purposes by the city and as a source of boiler water, is neutral in reaction, carries a little sediment, and contains some calcium carbonate and magnesium sulphate, besides a moderate amount of chlorine.

Analyses of surface waters obtained in the drainage basin of West Branch of Susquehanna River.

[Parts per million.]

· Location.	Source of supply.	Total residue.	Incrustants.	Nonincrustants.	Chlorine.
Spangler	Tank	116.622	85.500	31.122	6.840
Do	Water supply	67.716	53.865	13.851	2.565
Do	Stream	83.790	68.724	15.066	1.539
Cherry Tree	Killen Run	76.095	55.746	20.349	2.565
Carroltown	Peter Campbell's pond	161.595	139.536	22.059	8.037
Cush Creek Junction.		65.151	52.497	12.654	3.078
Hastings		537.966	343.539	194.427	3.591
Mahaffey	West Branch Susquehanna River.	159.543	99.693	59.850	20.862
Do	Run	33.858	25.992	7.866	2.907
Do	do	23.940	14.022	9.918	1.881

Analyses of surface waters obtained in the drainage basin of West Branch of Susquehanna River—Continued.

Location.	Source of supply.	Total residue.	Incrustants.	Nonincrustants.	Chlorine.
Bower	West Branch Susquehanna River.	74.727	55.917	18.810	3.933
Do.		55.917	45.828	10.089	3.933
Do..	Hazletts Run	35.055	23.940	11.115	2.052
Kerrmoor.	Little Clearfield Creek.	155.610	105.678	49.932	7.011
Porters		33.858	23.940	9.918	2.907
Curwensville	Anderson Creek	39.672	29.070	10.602	6.840
Clearfield	West Branch, directly at inlet.	481.707	116.793	364.914	196.137
Do	Clearfield Creek	187.416	111.663	75.753	5.985
Do	do	186.903	137.826	49.077	10.944
Do.	River at dam	107.730	69.768	37.962	11.970
Do	Tank	119.700	79.857	39.843	11.970
Do	West Branch Susquehanna River.	112.689	77.805	34.884	14.022
Do	Second dam breast	103.626	67.203	36.423	15.903
Do	River	95.760	61.902	33.858	11.970
Do	Moose Creek, 2,000 feet from mouth.	41.382	21.888	19.494	4.617
Do	City supply	17.100	11.970	5.130	2.907
Do	Moose and Montgomery creeks.	16.416	13.851	2.565	2.394
Do	Town supply	15.732	13.509	2.223	2.052
Clearfield Junction.	Creek	167.580	108.756	58.824	7.011
Woodland		61.902	47.880	14.022	4.959
Summit	Little Spring Branch	79.857	23.940	55.917	34.884
Belsena	Clearfield Creek	169.461	112.518	56.943	13.680
Do	Pine Run	29.925	20.007	9.918	Trace.
Do	Clearfield Creek	49.932	39.843	10.089	Trace.
Smoke Run	Creek dam	771.728	612.698	159.030	6.840
Do	Stream	67.716	59.850	7.866	Trace.
Osceola	Moshannon Creek	364.059	234.954	129.105	29.070
Do	Small Creek	60.192	43.947	16.245	6.840
Do	Mountain Branch Creek	43.947	24.795	19.152	6.840
Philipsburg	Moshannon Creek	279.243	233.415	45.828	13.680
Do	do	187.759	110.808	76.951	4.788
Do	City water (Water Co).	37.962	26.505	11.457	3.933
Do	Reservoir of Citizens' Water Co.	29.925	19.494	10.431	3.933
Do	Cold Stream	33.858	27.873	5.985	2.052
Viaduct	Moshannon Creek	339.093	193.401	145.692	9.918
Do	Run	29.925	23.940	5.985	2.052

Analyses of surface waters obtained in the drainage basin of West Branch of Susquehanna River—Continued.

Location.	Source of supply.	Total residue.	Incrustants.	Nonincrustants.	Chlorine.
Tyler, Clearfield County.	....	66.177	57.627	8.550	4.959
Caledonia	....	24.795	15.732	9.063	4.959
Benezett	....	29.583	20.007	9.576	4.446
Emporium	....	45.144	34.713	10.431	4.959
Driftwood........	....	88.988	27.531	11.457	4.959
Keating.........	Jewels Run	17.955	11.970	5.985	5.130
Renovo	Drury Run..........	30.609	25.479	5.130	2.052
Do............	.....do	23.940	16.074	7.866	2.052
Do............	.....do	23.940	15.903	8.037	2.052
Do............	.....do	22.743	18.810	3.933	2.052
Do............	Paddy Run..........	39.843	31.977	7.866	6.840
Do............	.....do	23.940	17.955	5.985	Trace.
Do............	.....do	20.007	14.022	5.985	Slight.
Do...........	Reservoir	24.282	17.955	6.327	3.420
Do...........	Pennsylvania R. R. reservoir.	23.940	15.903	8.037	Trace.
Rock Run	....	17.955	13.851	4.104	3.933
Karthaus........	Upper Three Run	34.029	24.111	9.918	3.984
Saltlick, near Karthaus.	Run	29.925	25.992	3.933	2.052
Beech Creek.....	Beech Creek	237.348	129.618	107.730	3.933
Do..........	Bald Eagle Creek......	144.495	119.700	24.795	3.933
Do..........	Canoe Creek	55.917	43.947	11.970	3.933
Oak Hall........	Water station	254.106	218.880	35.226	4.446
Do	Stream........	199.386	171.855	27.531	Very little.
Bellefonte	Water station	145.179	115.083	30.096	0.944
Do	City water	131.499	108.243	23.256	11.457
Howards	West Creek	82.935	55.062	27.873	4.959
Lock Haven	Bald Eagle Creek......	186.903	147.060	39.843	17.100
Youngdale.......	McElhattan Creek.....	29.925	20.007	9.918	3.933
Oakgrove........	West Branch Susquehanna River.	89.262	62.928	26.334	6.840
Do	Pine Creek	73.872	42.921	30.951	10.944
Do	City supply..........	39.843	26.163	13.680	1.026
Wellsboro	Stream...............	79.857	65.835	14.022	2.907
Do	.....do	67.887	47.880	20.007	4.959
Stokesdale Junction.	Run	121.581	79.857	41.724	9.918
Do	Marsh Creek	101.574	81.738	19.836	6.327
Ansonia	Darling Run	71.820	55.917	15.903	2.907

Analyses of surface waters obtained in the drainage basin of West Branch of Susquehanna River—Continued.

Location.	Source of supply.	Total residue.	Incrustants.	Nonincrustants.	Chlorine.
Blackwell	Run	37.962	31.977	5.985	2.907
Antrim	Pond receiving drainage from charcoal iron furnace and wood-alcohol distillery.	205.371	155.610	49.761	3.420
Do	Stream	39.843	23.940	15.903	2.907
Cedar Run	Jacob Run	31.977	23.940	8.037	2.052
Cammal	Stream	79.857	69.768	10.089	7.011
Ramsey	Ramsey Run	22.914	17.955	4.959	2.052
Jersey Shore Junction.	Pine Creek	61.902	35.910	25.992	11.970
Do	West Branch Susquehanna River.	58.824	41.895	16.929	7.011
Newberry Junction.	City	73.872	53.865	20.007	14.364
Penhryn	Pond	59.679	44.631	15.048	Not given.
Ralston	Red Run	116.622	62.244	54.378	5.180
Do	....do	83.790	34.200	49.590	4.446
Do	Rock Run	30.780	18.126	12.654	4.446
Trout Run		19.665	13.338	6.327	2.907
Dewart	Delaware Run	116.109	87.552	28.557	7.011
Do	Stream 2 miles west	48.393	29.925	18.468	7.011
Watsontown	Stream at Ridge	153.387	123.291	30.096	7.011
Do	Reservoir	128.763	101.403	27.360	7.011
Do	West Branch Susquehanna River.	80.028	52.668	27.360	7.011
Northumberland	....do	78.831	54.720	24.111	4.446
Do	....do	76.950	53.523	23.427	4.446

JUNIATA BASIN.

UNDERGROUND WATERS.

At Martinsburg, located near the headwaters of the Piney and Clover Creek branches of Juniata River, is a 12-foot well from which boiler water is obtained showing rather large amounts of dissolved solids, incrustants, and chlorine. The water, which is not acid in reaction, contains considerable calcium and magnesium sulphate and some little iron.

At Altoona water for both drinking and boiler purposes is obtained from a large number of wells sunk to various depths up to 2,000 feet. The best water, to judge from the analytical data submitted, is obtained from a 2,000-foot artesian well, which contains but a trace of chlorine and some calcium carbonate in solution, with a little magnesium carbonate and sulphate, and from the well of the American Brewing Company, which contains in solution but a small amount of salt and total solids, of which the greater part is magnesium sulphate.

There is a well 187 feet deep at the Altamont Hotel, Altoona, from which two samples of boiler and drinking water were collected, the first of which shows a slightly acid water, comparatively low total solids, incrustants, and chlorine, while a second sample, obtained but a week later, is very high in all of these constituents and might be considered dangerous for drinking purposes and to warrant investigation of the condition of this well. The second sample of water showed a slightly alkaline reaction, was slightly turbid, and contained magnesium chloride and calcium carbonate.

The analyses of the waters of other wells at Altoona, ranging in depth from 24 to 85 feet, show from moderate to rather high contents in solids, incrustants, and chlorine, the salts found being chiefly calcium carbonate and to a lesser extent magnesium sulphate. Some of these wells furnish drinking as well as boiler water, while the rest are used as a source of water supply for locomotives.

An 8-foot well at Tyrone shows water containing rather large amounts of solids and chlorine in solution, with much calcium and magnesium sulphates and a little free carbonic acid. It is used for both drinking and boiler purposes.

At Lewistown Junction there is a well from which both drinking and boiler water is obtained. The water is low in chlorine, alkaline in reaction, and moderately high in solids—chiefly magnesium sulphate and calcium carbonate.

A consideration of the analyses of samples of water from various sources of supply that can not be classified accurately or separately under the above heads of streams, wells, and springs leads to the following conclusions:

The city water of Altoona is slightly acid in reaction, contains a small amount of total solids and chlorine, with some iron, calcium, and magnesium sulphate. A like quality is shown by the water at Kittanning Point, which is drawn from the Altoona water supply.

The city water of Tyrone and boiler water at the Pennsylvania Railroad station at that place is of good quality and contains but a small amount of calcium and less of magnesium sulphate and no carbonates.

The boiler water obtained in East Tyrone is higher in solids than the former and contains some calcium carbonate and a small amount of sediment.

The city water supply of Martinsburg, Pa., is apparently excellent in quality. The water does not carry more than a trace of chlorine, is very low in solids, slightly acid (probably due to a trace of sulphates), and shows a little iron in solution. The boiler water secured at the Martinsburg railroad station is higher in all of the above constituents except the iron.

Analyses of water of the Lewistown reservoir and of the railroad water stations at Henrietta and Roaring Spring show the samples to be of the average composition of Pennsylvania waters, sulphates and carbonates of the alkaline earths constituting the principal incrustants.

SURFACE WATERS.

The spring furnishing water used in boilers by the Pennsylvania Railroad at Kittanning Point does not contain more than a very small quantity of salts in solution, the latter being calcium and magnesium sulphate and sodium chloride. This water is evidently of very good quality for the purpose for which it is employed and is also adapted for municipal use generally.

The analysis of spring water obtained at Altoona shows small amounts of total solids and probable incrustants with a medium amount of chlorine. Both calcium sulphate and magnesium chloride are found in this water, which is used in boilers by the Pennsylvania Railroad.

Winnemier's spring, at Sprucecreek, furnishes boiler water showing rather small amounts of solids and incrustants in solution and a medium quantity of saline matter. Calcium sulphate is the chief incrustant present.

At Huntingdon, on Juniata River, several springs, large and small, are found near the reformatory. Upon analysis they average from a small to a moderately large quantity of residue and incrustants, and, with one exception, show only traces of chlorine. Small quantities of calcium and iron carbonates and sulphates are present, but not in sufficent amount to prevent the use of the above water for drinking as well as for generating steam.

At Hollidaysburg, on the headwaters of Little Juniata River, one of the branches of Juniata River, the water from the Little Juniata shows medium amounts of residue after evaporation and of incrustants and chlorine, and in general composition would be classified with the average stream water found in Pennsylvania. Although somewhat high in calcium sulphate, apparently, and also slightly acid, it is regularly used in boilers by the Pennsylvania Railroad.

At Altoona, judging by samples taken both above and below Fountain Inn, the river water is soft, leaves little residue upon evaporation, and is low in incrusting compounds and chlorine. It is neutral in reaction, only slightly turbid, and is advantageously used for both drinking and steam-generating purposes.

Analyses of water from Little Juniata River at Tyrone, Pa., show medium to rather large amounts of total residue and incrustants and somewhat high average contents in chlorine.   In most of the samples tested the water is neutral in reaction, although occasionally alkaline, in which case it is also found to be high in calcium carbonate. Frequently calcium and magnesium sulphates are present in large amounts. This water is evidently not considered potable and is only used in boilers.

Farther down Little Juniata River, at Barree, on the Pennsylvania Railroad, very good soft water is again found, the analysis of the stream at this town showing very small amounts of total solids, probable incrustants, and chlorine.   This water is employed for both drinking and steam-generating purposes, for which it is certainly well adapted.

At Lewistown, on Juniata River, the water of Kishacoquillas Creek was found to be hard and to contain large amounts of residue upon evaporation, together with large quantities of probable incrustants, but only small amounts of chlorine.   The incrustants comprise chiefly clayey matter and calcium carbonate, while the principal nonincrustant appears to be magnesium sulphate. The above water is used for drinking purposes as well as in boilers.

Analyses of water from the drainage basin of Juniata River.

[Parts per million.]

Source of supply.	Total residue.	Incrustants.	Nonincrustants.	Chlorine.
STREAMS.				
Lewistown (Kishacoquillas Creek) ---	580.374	536.427	43.947	3.420
Barree ---	21.204	14.706	6.498	2.907
Tyrone (Little Juniata River) ---	127.566	76.779	50.787	Considerable.
Tyrone (Juniata River) ---	379.962	285.912	94.050	31.635
Tyrone (stream at water station) ---	143.127	111.321	31.806	5.985
Tyrone (river at water station) ---	146.547	114.228	32.319	11.970
Altoona (stream above Fountain Inn) ---	42.237	84.371	7.866	2.907
Altoona (stream below Fountain Inn) ---	40.869	31.464	9.405	2.907
Hollidaysburg (Little Juniata River) ---	129.618	93.708	35.910	6.840
SPRINGS.				
Huntingdon (springs at reformatory) ---	29.925	27.189	2.736	Trace.
Huntingdon (spring at reformatory) ---	169.461	159.543	9.918	3.420
Huntingdon (large springs at reformatory) ---	29.925	27.873	2.052	Trace.
Huntingdon (small springs at reformatory) ---	41.895	33.858	8.037	Trace.
Altoona ---	49.932	30.267	9.665	7.011
Sprucecreek (Winnemier's spring) ---	63.954	30.951	33.003	6.840
Kittanning Point ---	24.966	11.970	12.996	8.420

Analyses of water from the drainage basin of Juniata River—Continued.

Source of supply.	Total residue.	Incrustants.	Nonincrustants.	Chlorine.
WELLS.				
Altoona (American Brewing Co.)	42.408	28.899	13.509	3.420
Altoona (Mackey House)	503.424	250.344	253.080	71.820
Altoona (well at Eighteenth street) ...	227.772	162.621	65.151	20.520
Altoona (well at Nineteenth street) ...	325.071	239.400	85.671	27.873
Altoona (well at car shop, 24 feet)	276.938	237.177	39.761	17.442
Altoona (well, 84 feet)	236.322	178.524	57.798	11.970
Altoona (artesian well, 85 feet, Ninth avenue)	207.423	174.420	33.003	13.680
Altoona (well, 187 feet, Altoona Hotel).	692.892	406.638	286.254	112.689
Do	75.411	45.144	30.267	5.472
Altoona (artesian well, 2,000 feet)...	179.550	119.700	59.850	Trace.
Lewistown Junction	389.880	241.110	148.770	3.078
Tyrone (well, 8 feet)	346.617	220.932	125.685	42.750
Martinsburg (12-foot well)	414.504	161.424	253.080	46.512
NOT CLASSIFIED.				
Altoona (city water)	76.266	43.263	33.003	4.959
Kittanning Point (Altoona water supply)	84.987	49.077	35.910	6.840
Tyrone (city water)	38.133	27.873	10.260	3.420
Tyrone (water station)	42.408	31.635	10.773	5.985
East Tyrone (water station)	114.570	82.080	32.490	29.925
Martinsburg (city supply, mountain reservoir)	25.821	15.903	9.918	Trace.
Martinsburg (water station)	249.318	198.360	50.958	2.394
Lewistown (reservoir)	211.356	179.550	31.806	7.011
Henrietta (water station)	56.772	40.869	15.903	6.840
Roaring Spring (water station)	109.611	103.113	6.496	2.394

SUSQUEHANNA RIVER BELOW SUNBURY.

At Sunbury, Pa., located near the junction of West Branch and Susquehanna River, a soft water is used for drinking and other purposes. It contains but a trace of sediment, chiefly clay and iron, and a little calcium sulphate and sodium chloride.

Below Sunbury the first town is Millersburg, the water of which is about the same in quality as that of Sunbury. This is also true of some samples of water at Clarks Ferry, although others taken at the same place are moderately high in total solids and in chlorines, with a neutral or slightly acid reaction.

Rockville water is soft in character and low in solids, incrustants, and chlorine. At West Fairview, both on Susquehanna River and at Conedogwinet Creek, the drinking and boiler water used is appreciably higher in solids, incrustants, and chlorine, and contains a small amount of sediment.

The water of Susquehanna River at Enola is low in total solids, incrustants, and chlorine, is neutral in reaction, with a trace of sediment, and contains some carbonates and sulphates of lime and magnesia. In general it is of fairly good quality.

At Harrisburg the city water shows but very little chlorine upon careful analysis, slight turbidity, little magnesium sulphate, more calcium carbonate, and a little calcium sulphate and iron. At the roundhouse tap the water is moderately high in chlorine and in dissolved carbonates and sulphates of lime, magnesia, and iron. It reacts slightly alkaline after being boiled.

At New Cumberland the water is moderately high in dissolved solids and chlorines and contains but a very small amount of nonincrustants. Similar conditions hold true with the waters of Shenks Ferry, which are neutral in reaction.

The water in the canal at Harrisburg is decidedly acid in reaction, due to the presence of considerable sulphuric acid or sulphates, probably combined with some lime and a little magnesia and iron. The total solids are moderate in quantity and the chlorine contents moderately high.

At Lucknow, near Harrisburg, are some swamp springs which furnish neutral water for boiler and drinking purposes and are of average composition, excepting the moderately high contents in chlorine. Some magnesium sulphate and calcium carbonate with traces of organic matter were detected in this water.

Analyses of water from the drainage basin of Susquehanna River below Sunbury.

[Parts per million.]

Source of supply.	Total residue.	Incrustants.	Nonincrustants.	Chlorine.
STREAMS.				
Sunbury (city water)	55.404	45.315	10.089	3.933
Millersburg	62.415	47.709	14.706	3.933
Clarks Ferry	42.066	32.490	9.576	4.959
Do	251.712	150.309	101.403	5.985
Do	213.066	170.316	42.750	12.825
Rockville	53.010	44.460	8.550	3.420
West Fairview	101.916	73.359	28.557	7.011
West Fairview (Conedogwinet)	188.100	148.428	39.672	4.446
Enola	82.935	63.441	19.494	3.933
Harrisburg (city water)	115.767	89.775	25.992	(a)
Harrisburg (roundhouse tap)	174.762	139.878	34.884	5.985
Harrisburg	207.423	162.108	45.315	8.379
New Cumberland (city water)	131.670	127.737	3.933	7.011
New Cumberland (Kauffman's store)	135.774	126.711	9.063	4.446
Columbia (Pennsylvania R. R. reservoir)	123.633	94.221	29.412	6.840
Shenks Ferry	98.667	90.288	8.379	3.933
CANAL.				
Harrisburg	159.543	117.990	41.553	6.840
SWAMP SPRINGS.				
Lucknow (near Harrisburg)	135.090	108.585	26.505	8.208

a Very little.

SUMMARY.

The value of Susquehanna River as a resource to the inhabitants of the important area which it drains is without question. It has been noted in previous pages that the character of the water in the main stream is markedly variable, while as a whole that of the tributaries is quite satisfactory. The part of the stream which flows through New York is extensively used as a source of water supply, but as it leaves the northern anthracite coal basin in Pennsylvania its usefulness in this line of development is somewhat limited. It has been demonstrated by the water department of the city of Harrisburg that by reason of its peculiar character the water is exceedingly difficult of satisfactory purification.

It has been noted that the pollution of rivers by acid mine refuse is not without certain advantages. This does not mean, however, that such pollution is in the end beneficial, yet from the standpoint of public-water supply it is not as dangerous as the pollution which is freely poured into the stream from city sewers. Acid mine waste is certainly harmful to the resources of any stream, and it is doubtful if a stream so polluted could be used as a domestic water supply with satisfaction. Susquehanna River water could not, however, be used in its raw state for household purposes if no mine drainage was turned into it. This being the case the bad effects of mine drainage which are readily apparent in the small tributaries running from the coal regions, are not important in the main stream, for as shown on previous pages there the effects are altogether beneficial. Stated in another way, it may be said that acid mine drainage does great damage to a water-supply stream, but is of undoubted benefit in a stream which has been converted into a depository for sewage. These benefits are, however, somewhat limited. The precipitation of immense quantities of sewage matter and coal dust will eventually fill up the bed of the river along certain reaches and in the end be troublesome. In fact, the effects are already apparent in some sections.

It is probable that Susquehanna River, especially that portion below West Branch, is of more value as an agent of sewage disposal than it would be if attempts were made to purify it. It would be impossible, with the present large population upon the drainage area, to render the great stream pure; this being the case, it might as well receive city sewage up to a certain limit. The size of the stream and the purifying effect of acid mine waste make it extremely unlikely that the population along the river will increase enough to render the river a nuisance and a damage to realty values. Almost the only serious problem, with reference to the main stream, which presents itself at the present time is that of the disposal of culm from the coal washeries. Although this material has been turned into the stream during a comparatively short period only, its damaging effects are already apparent; therefore it is not difficult to foresee what will transpire in the future if steps are not taken to dispose of it in some other manner. Not the least important feature in this consideration is the enormous waste of material which may at some time in the future be a marketable product.

INDEX.

71

O

Leighton, Marshall Ora, 1874–

. . . Quality of water in the Susquehanna River drainage basin, by Marshall Ora Leighton; with an introductory chapter on physiographic features, by George Buell Hollister. Washington, Gov't print. off., 1904.

76 p., 1 l. 4 pl., diagrs. 23ᶜᵐ. (U. S. Geological survey. Water-supply and irrigation paper no. 108.)

Subject series: L, Quality of water, 7.

1. Water, Quality of—Susquehanna River basin. 2. Susquehanna River basin—Physiography. I. Hollister, George Buell, 1865–

Author.

Leighton, Marshall Ora, 1874–

. . . Quality of water in the Susquehanna River drainage basin, by Marshall Ora Leighton; with an introductory chapter on physiographic features, by George Buell Hollister. Washington, Gov't print. off., 1904.

76 p., 1 l. 4 pl., diagrs. 23ᶜᵐ. (U. S. Geological survey. Water-supply and irrigation paper no. 108.)

Subject series: L, Quality of water, 7.

1. Water, Quality of—Susquehanna River basin. 2. Susquehanna River basin—Physiography. I. Hollister, George Buell, 1865–

Subject.

U. S. Geological survey.

Water-supply and irrigation papers

no. 108. Leighton, M. O. Quality of water in the Susquehanna River drainage basin. 1904.

Series.

U. S. Dept. of the Interior.

see also

U. S. Geological survey.

Reference.

Water-Supply and Irrigation Paper No. 109   Series { M, General Hydrographic Investigations. 13   N, Water Power, 9

DEPARTMENT OF THE INTERIOR

UNITED STATES GEOLOGICAL SURVEY

CHARLES D. WALCOTT, DIRECTOR

HYDROGRAPHY

OF THE

SUSQUEHANNA RIVER DRAINAGE BASIN

BY

JOHN C. HOYT AND ROBERT H. ANDERSON

WASHINGTON
GOVERNMENT PRINTING OFFICE
1905

CONTENTS.

ILLUSTRATIONS.

5

6   ILLUSTRATIONS.

LETTER OF TRANSMITTAL.

DEPARTMENT OF THE INTERIOR,
UNITED STATES GEOLOGICAL SURVEY,
HYDROGRAPHIC BRANCH,
Washington, D. C., May 5, 1904.

SIR: I have the honor to transmit herewith a manuscript by John C. Hoyt and Robert H. Anderson, relating to the hydrography of the Susquehanna River drainage basin, and recommend its publication in the series of Water-Supply and Irrigation Papers.

In this paper has been brought together, in such form as to be of use to both the general and the engineering public, all the available hydrographic information in regard to this important area.

It is intended that this paper shall be published in sequence with another (No. 108) entitled "Quality of Water in the Susquehanna River Drainage Basin, by Marshall Ora Leighton, with an Introductory Chapter on Physiographic Features, by George Buell Hollister." The combination of the two papers will make available a large amount of valuable information with reference to the resources of this important river system.

Very respectfully,

F. H. NEWELL, *Chief Engineer.*

Hon. CHARLES D. WALCOTT,
Director United States Geological Survey.

HYDROGRAPHY OF THE SUSQUEHANNA RIVER BASIN.

By JOHN C. HOYT and ROBERT H. ANDERSON.

INTRODUCTION.

A detailed study of the hydrographic features of the Susquehanna River drainage basin has revealed the existence of a large amount of interesting data. These, however, are widely distributed in various publications and manuscripts which are in most cases inaccessible. This paper has been prepared to meet the constant demand for this information from both the general and the engineering public. The general deductions are intended to give the general reader a comprehensive review of the principal conditions which exist in this area, while the base data have been given for the use of the engineer, so that he may make his own deductions and have sufficient data for estimates in hydraulic investigations.

ACKNOWLEDGMENTS.

The records and reports of the United States Geological Survey have been the chief sources from which the data on flow have been obtained. These records have been carefully revised and in many cases recomputed. New rating tables based on all the discharge measurements to date have been prepared and the tables of estimated discharge have been revised to agree with these rating tables. These recomputations will account for the differences between the figures herein presented and many of those in the previous reports, as the latter were prepared from year to year with such information as was available. Special acknowledgment is due to E. G. Paul, resident hydrographer for Pennsylvania, who established the gaging stations and under whose direction the discharge measurements in this State have been made. The stations in New York were established and have been maintained under the direction of R. E. Horton, resident hydrographer for that State.

The base data from which the precipitation tables have been prepared were taken from the published reports of the United States Weather Bureau.

The tables showing the utilized horsepower in 1900 are from manuscript schedules furnished by the manufactures division of the Twelfth Census.

In the preparation of descriptive portions of the paper Vol. XVI of the reports of the Tenth Census (Water Powers, Part I), Rogers's Geology of Pennsylvania, and the Army Engineers' reports have been largely drawn upon.

The annual reports and original records of the Chief of Engineers, United States Army, have furnished valuable information in regard to declivity, and the profiles herewith given are largely based upon them.

The data for McCalls Ferry have been furnished through the kindness of Dr. Cary T. Hutchinson, of New York City, who is interested in the power development at that point and had charge of extensive surveys and studies there in 1902 and 1903. Special mention is due Boyd Ehle and R. H. Anderson, who established and carried on the measurements at the McCalls Ferry gaging station.

Acknowledgment is also due to Frank H. Brundage, H. J. Saunders, L. R. Stockman, and other members of the hydro-computing section of the United States Geological Survey for assistance given in the computations and in other work connected with the preparation of the many tables.

DESCRIPTION OF DRAINAGE AREA.

GENERAL FEATURES.

The Susquehanna River basin is the largest and most important drainage area commercially in the North Atlantic States, although it is not the most important as regards water power. The headwaters of this river system are on the elevated plateau which separates the waters which flow south and east into the Atlantic streams from those flowing north and west into the Mississippi, St. Lawrence, and Great Lakes.

Geologically, this watershed lies in four physiographic divisions: the Allegheny Plateau, the Allegheny Mountains, the Great Allegheny Valley, and the Piedmont Plateau. Its distribution among these provinces is approximately as follows: Allegheny Plateau, 56 per cent; Allegheny Mountains, 31 per cent; Great Allegheny Valley, 6 per cent; Piedmont Plateau, 7 per cent.

As the physical features of the foregoing divisions and the early history of the formation of this basin, as well as the quality of the water, have been fully discussed by Messrs. G. B. Hollister and M. O. Leighton in Water-Supply Paper No. 108, further discussion here is omitted.

The Susquehanna drainage basin, as shown in fig. 1, has a total area of 27,400 square miles. It comprises 21,060 square miles in Pennsylvania, or about 47 per cent of the area of the State; 6,080 square miles in New York, or 13 per cent of the area of the State; 260 square miles in Maryland, or about 2 per cent of the area of the State. It includes all or a portion of the counties in New York and Pennsylvania listed in the table below:

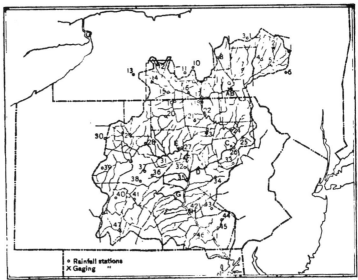

FIG. 1.—Map showing drainage area and location of gaging and rainfall stations.

Counties in New York and Pennsylvania drained wholly or in part by Susquehanna River and its tributaries.

New York:
    Madison.
    Cortland.
    Otsego.
    Chenango.
    Delaware.
    Broome.
    Tioga.
    Tompkins.
    Schuyler.
    Chemung.
    Steuben.
Pennsylvania:
    Potter.
    Tioga.
    Bradford.
    Susquehanna.
    Elk.
    Cameron.
    Clinton.
    Lycoming.
    Sullivan.
    Wyoming.
    Lackawanna.
    Luzerne.
    Columbia.

Pennsylvania—Continued.
    Montour.
    Northumberland.
    Union.
    Center.
    Clearfield.
    Indiana.
    Cambria.
    Blair.
    Huntingdon.
    Mifflin.
    Juniata.
    Snyder.
    Perry.
    Cumberland.
    York.
    Adams.
    Franklin.
    Fulton.
    Bedford.
    Somerset.
    Dauphin.
    Schuylkill.
    Lebanon.
    Lancaster.

In order to simplify the descriptive matter which follows, the following division has been made of the Susquehanna River system: Susquehanna River and its tributaries below mouth of West Branch; Susquehanna River and its tributaries above mouth of West Branch; West Branch of Susquehanna River and its tributaries. The principal streams in each division are shown by the following diagrams:

Tributaries of Susquehanna River below West Branch.

Shamokin Creek.
Penn Creek.
Middle Creek.
Mahanoy Creek.
Mahantango Creek.
Burgess Creek.
Wiconisco Creek.
Armstrong Creek.

Juniata River.

Frankstown Branch.
Sugar Creek.
Canoe Creek.
Piney Creek.
Clover Creek.
Little Juniata. Spruce Creek. Bald Eagle Creek.
Shavers Creek.
Standing Stone Creek.

Raystown Branch.
Buffalo Creek.
Dunnings Creek.
Cove Creek.
Shavers Creek.
Brush Creek.
Yellow Creek.
Great Trough Creek.

Aughwick Creek.
Kishacoquillas Creek.
Jacks Creek.
Lost Creek.
Tuscarora Creek.
Cocolanus Creek.
Buffalo Creek.

Powell Creek.
Shermans Creek.
Clark Creek.
Stoney Creek.
Fishing Creek No. 1.
Conedoguinet Creek.
Paxton Creek.
Yellows Breeches Creek.
Swatara Creek.
Conewago Creek.
Codorus Creek.
Conestoga Creek.
Pequea Creek.
Otter Creek.
Muddy Creek.

Tributaries of Susquehanna River below West Branch—Continued.

Fishing Creek No. 2.
Broad Creek.
Conowingo Creek.
Octoraro Creek.
Deer Creek.

Tributaries of Susquehanna River above West Branch.

Otsego Lake.
Oak Creek, Schuyler Lake.
Cherry Valley Creek.
Schenevus Creek.
Charlotte River.
Otsego Creek.
Ouleout Creek.
Carrs Creek.
Unadilla River. { Butternut Creek.
              { Wharton Creek.
Bennetts Creek.
Starucca Creek.
Salt Lick Creek.
Snake Creek.

Chenango River. {
  Castle Creek.
  Genegantslet Creek.
  Canaswacta Creek.
  Tioughnioga River. { Eastern branch Tioughnioga.
                 { Western branch Tioughniogo.
                 { Otselic River.

Choconut Creek.
Nanticoke Creek.
Apalachin Creek.
Owego Creek. { Cottalong Creek.
           { East Creek.
Wappasening Creek.
Cayuta Creek.

Chemung River. {
  Ten Mile Creek.
  Twelve Mile Creek.
  Five Mile Creek.
  Canisteo River. {
    Carr Valley Creek.
    Crosby Creek.
    Purdy Creek.
    Bennetts Creek.
    Tuscorora Creek.
    Tioga River. { Mill Creek.
               { Crooked Creek.
               { Cowanesque Creek.
  Hammond Creek.
  Bucks Creek.

Sugar Creek.
Towanda Creek.
Wysox Creek.
Wyalusing Creek.
Tuscarora Creek.
Meshoppen Creek.
Mehoopany Creek.

*Tributaries of Susquehanna River above West Branch—*Continued.

Tunkhannock Creek.
Buttermilk Creek.
Coray Creek.
Gardner Creek.
Abraham Creek.
Mill Creek.
Toby Creek.
Buttonwood Creek.
Warrior Creek.
Newport Creek.
Harvey Creek.
Hunlock Creek.
Shickshinny Creek.
Little Wapwallopen Creek.
Wapwallopen Creek.
Nescopec Creek.
Briar Creek.

Fishing Creek. {Little Fishing Creek. / Green Creek. / Huntington Creek.

Catawissa Creek.
Roaring Creek.
Mahoning Creek.

Tributaries of West Branch of Susquehanna River.

Anderson Creek.
Clearfield Creek.
Moshannon Creek.
Mosquito Creek.

Sinnemahoning Creek. {West Creek. / Bennetts Brook. / East Fork.

Kettle Creek.
Youngwomans Creek.

Bald Eagle Creek. {Spring Creek. / Beach Creek. / Fishing Creek.

Pine Creek. {Marsh Creek. / Babbs Creek. / Little Pine Creek.

Big Larrys Creek.
Lycoming Creek.
Loyalsock Creek.
Muncy Creek.
White Deer Hole Creek.
White Deer Creek.
Buffalo Creek.
Chillisquaque Creek.

The following table, compiled from Vol. XVI of the reports of the Tenth Census and from the publications of the United States Geological Survey, shows the drainage area at different points on Susquehanna River and its tributaries.

Drainage areas of Susquehanna River and its tributaries.

Stream.	Tributary to	Point of measurement.	Drainage area.
			Sq. miles.
Susquehanna River	Chesapeake Bay	Outlet of Otsego Lake.	*a* 81
Do	do	Oak Creek	97
Do	do	Below and including Oak Creek.	212
Do	do	Oneonta	*a* 686
Do	do	Below and including Charlotte River.	718
Do	do	Unadilla River	*a* 914
Do	do	Below and including Unadilla River.	*a* 1,480
Do	do	Nineveh	1,790
Do	do	Susquehanna	2,024
Do	do	Binghamton	*a* 2,400
Do	do	Below and including Chenango River.	*a* 3,960
Do	do	Chemung River	4,940
Do	do	Below and including Chemung River.	*a* 7,460
Do	do	Wilkesbarre	*a* 9,810
Do	do	Danville	*a* 11,070
Do	do	Mouth of west branch	*a* 11,140
Do	do	Sunbury	*a* 18,170
Do	do	Harrisburg	*a* 24,030
Do	do	McCalls Ferry	*a* 26,770
Do	do	Mouth	*a* 27,400
Shamokin Creek	Susquehanna River	do	165
Penn Creek	do	do	361
Middle Creek	do	do	147
Mahanoy Creek	do	do	133
Mahantango Creek	do	do	166
Wiconisco Creek	do	do	83
Clark Creek	do	do	47
Yellow Breeches Creek	do	do	247
Conedogwinit Creek	do	do	450
Swatara Creek	do	do	536
Conewago Creek	do	do	560
Shermans Creek	do	do	232
Peques Creek	do	do	148

a Measured by United States Geological Survey.

Drainage areas of Susquehanna River and its tributaries—Continued.

Steam.	Tributary to—	Point of measurement.	Drainage area.
			Sq. miles.
Conestoga Creek	Susquehanna River	Lancaster	332
Do	...do	Mouth	474
Conowingo Creek	...do	...do	31
Octorara Creek	...do	...do	178
Deer Creek	...do	...do	128
Oak Creek	...do	...do	115
Cherry Valley Creek	...do	...do	121
Schenevus Creek	...do	...do	127
Charlotte River	...do	...do	178
Otego Creek	...do	...do	106
Oaliout Creek	...do	...do	115
Unadilla River	...do	...do	561
Butternut Creek	Unadilla River	...do	123
Wharton Creek	...do	...do	92
Bennetts Creek	Susquehanna River	...do	47
Chenango River	...do	Canasawacta Creek	2*
Do	...do	Tioughnioga River	a7
Do	...do	Below and including Tioughnioga River.	a1,
Do	...do	Mouth	a1,
Canasawacta Creek	Chenango River	...do	
Genegantslet Creek	...do	...do	1.
Tioughnioga River	...do	Otselic River	a2*
Do	...do	Mouth	780
West Branch Tioughnioga River.	Tioughnioga River	...do	103
East Branch Tioughnioga River.	...do	...do	164
Otselic River	...do	...do	259
Starucca Creek	Susquehanna River	...do	75
Owego Creek	...do	...do	391
Caynta or Shepards Creek.	...do	...do	145
Chemung River	...do	Elmira	110
Do	...do	Mouth	2,520
Tioga River	Chemung River	...do	1,830
Do	...do	Cowanesque Creek	438
Do	...do	Canisteo River	776
Canisteo River	...do	Mouth	545
Tuscarora Creek	...do	...do	130
Cowanesque Creek	Tioga River	...do	288

a Measured by United States Geological Survey.

Drainage areas of Susquehanna River and its tributaries—Continued.

Steam.	Tributary to—	Point of measurement.	Drainage area.
			Sq. miles.
Sugar Creek	Susquehanna River	Mouth	177
Towanda Creek	do	do	220
Wysox Creek	do	do	90
Wyalusing Creek	do	do	204
Tunkhannock Creek	do	do	409
Lackawanna Creek	do	do	323
Little Wapwallopen Creek.	do	do	38
Big Wapwallopen Creek.	do	do	68
Nescopec Creek	do	do	145
Catawissa Creek	do	do	131
Fishing Creek	do	do	353
West Branch Susquehanna River.	do	Clearfield Creek	476
Do	do	Sinnemahoning Creek	1,440
Do	do	Queens Run	3,030
Do	do	Lock Haven	3,040
Do	do	Williamsport	*a*5,640
Do	do	Allenswood	*a*6,540
Do	do	Mouth	*a*7,030
Clearfield Creek	West Branch Susquehanna River.	do	342
Moshannon Creek	do	do	233
Mosquito Creek	do	do	54
Sinnemahoning Creek	do	Benezette	163
Do	do	Driftwood	334
Do	do	Mouth	962
Trout Run	Sinnemahoning Creek	do	48
Driftwood Branch	do	do	314
First Fork	do	do	240
Kettle Creek	West Branch Susquehanna River.	do	215
Bald Eagle Creek	do	do	726
Beach Creek	Bald Eagle Creek	do	157
Fishing Creek	do	do	169
Spring Creek	do	do	148
Pine Creek	West Branch Susquehanna River.	do	930
Big Larrys Creek	do	do	85
Lycoming Creek	do	do	261

a Measured by United States Geological Survey.

Drainage areas of Susquehanna River and its tributaries—Continued.

Stream.	Tributary to—	Point of measurement.	Drainage area.
			Sq. miles.
Loyalsock Creek	West Branch Susquehanna River.	Mouth	494
Muncy Creek	do	do	185
White Deer Creek	do	do	40
Chillisquaque Creek	do	do	119
Juniata River	Susquehanna River	Junction of and including its two branches.	1,842
Do	do	Newton Hamilton	2,270
Do	do	Lewistown dam	2,550
Do	do	Newport	a 3,480
Do	do	Mouth	a 3,530
Raystown Branch	Juniata River	Hopewell	588
Do	do	Mouth	909
Frankstown Branch	do	Holidaysburg	129
Do	do	Crooked dam	249
Do	do	Threemile dam	273
Do	do	Williamsburg	279
Do	do	Mud dam	333
Do	do	Smokers dam	333
Do	do	Donnellys dam	342
Do	do	Willow dam	347
Do	do	Water Street dam	356
Do	do	Alexandria	360
Do	do	Little Juniata	374
Do	do	Pipers dam	750
Do	do	Huntingdon dam	759
Do	do	Mouth	933
Standingstone Creek	Frankstown Branch	do	129
Shavers Creek	do	do	45
Little Juniata River	do	Tyrone (including Bald Eagle Creek).	154
Do	do	Barree	325
Do	do	Mouth	327
Spruce Creek	Little Juniata River	do	94
Bald Eagle Creek	do	do	54
Great Aughwick	Juniata River	do	316
Kishacoquillas Creek	do	do	174
Jacks Creek	do	do	55
Tuscarora Creek	do	do	252

a Measured by United States Geological Survey.

A   TYPICAL VIEW ON SUSQUEHANNA RIVER NEAR CATAWISSA, PA

B   BED OF SUSQUEHANNA RIVER AT McCALLS FERRY CABLE STATION, DURING
LOW WATER.

SUSQUEHANNA RIVER BELOW WEST BRANCH.

Susquehanna River is joined by the West Branch at Sunbury, Northumberland County. Below this point the river drains an area of 9,230 square miles. It flows nearly south, between Northumberland, Dauphin, and Lancaster counties on the east and Snyder, Juniata, Perry, Cumberland, and York counties on the west, passing then into Maryland, where it flows between Cecil County on the east and Harford County on the west, and empties into Chesapeake Bay at its northern extremity.

Below the mouth of the West Branch the fall becomes more irregular than above, and there are rapids where the stream flows over a rocky bottom. In the lower part of its course from Marietta to Havre de Grace the river occupies a deep valley, varying in width from a few hundred yards to more than 2 miles, and on either shore it is for the most part bounded by rocky bluffs surmounted by a table-land 100 to 500 feet above the stream. The channel is in many places filled with small rocky islands, some of which are cultivated. Pls. I, B, and VIII show typical views of this part of the river.

The fall of the main river is rapid. Its elevation at the mouth of the West Branch is about 400 feet above mean sea level at Havre de Grace. The distance between this point and Havre de Grace is about 125 miles, hence the mean slope of the main river is nearly $3\frac{1}{2}$ feet per mile. The slope is, however, extremely variable, being over 5 feet per mile in the lower 40 miles and about $2\frac{1}{2}$ feet per mile in the upper 40 miles. The change in slope takes place as the river passes from the Allegheny Mountain and the Allegheny Valley regions to the Piedmont Plateau region.

The tables on pages 207–210 give the elevation of the river and its branches at various points, and Pls. XXVIII and XXIX show their profiles.

This part of the river is described by Prof. H. D. Rogers as follows:[a]

Between Northumberland and the Kittatinny Valley the river leads us through many striking scenes. It is studded with many little islands, most of which are covered with trees or bushes to the water's edge, and it is here a wide and majestic river, flowing alternately for long reaches across highly cultivated belts of country and past the ends of steep and rugged mountains. Passing out from the mountains it traverses a beautiful country in the Kittatinny Valley, dividing Dauphin from Cumberland County. Quitting the limestone valley the river next traverses the red-shale belt, between the villages of Highspire and Bainbridge, crossing a rather monotonous country, except at the Conewago Falls, or rapids, where numerous hard trap dikes impede its course and cause it to rush in wild tumult, by deep and dangerous sluices, for a long distance between black and jutting reefs. At Chickies Ridge, 1 mile above Columbia, the river leaves the smoother country and passes between a range of high and picturesque crags. With two or three intermissions, caused by the softer limestone valleys which it next crosses, it runs the whole way thence to the vicinity of Port Deposit, or nearly to the Chesapeake Bay, between steep naked and half naked hillsides, rising

---

from 200 to 400 feet above its channel. In some parts of this long reach, as at Washington Borough, the river is greatly dilated and is filled with rocky islands and projecting reefs. In other localities its rugged banks approach. and the river rushes with tremendous force, especially during freshets, through these deeper gorges. The traveler. who finds only a rough and very toilsome path along its eastern shore from Turkey Hill to Port Deposit, a distance of more than 30 miles. will choose to descend it by its right bank along the towpath of the canal. He will pass an almost unbroken succession of interesting rocky scenes. affording much geological instruction, and he will witness many beautiful bits of river perspective, but he will find himself pent in all the way between the bold river hills.

The principal tributary below the West Branch is the Juniata, which has its source in Bedford, Blair, and Somerset counties, Pa., at an elevation of about 2,000 feet above sea level. The divide between its waters and those of the Ohio attains in places a height of nearly 2,800 feet. The valley of the stream is narrow and the banks are generally high. The stream has a number of both large and small tributaries. Doctor Rogers describes the Juniata as follows:[a]

This second great tributary of the Susquehanna has two chief upper divisions, the Frankstown and the Raystown branches, both of which. like the main stream below their junction, traverse much beautiful scenery. We will trace the Frankstown Branch as that which is most accessible. After gathering its headwaters from the eastern slope and the foothills of the Allegheny Mountains it begins to assume the volume of a small river near Frankstown. Below this point it first passes the cove of the Lock Mountain. a curious district of conical hills, in structure like the Muncy Hills of the West Branch. Its course is now by a wild and rocky gorge through the Lock or Canoe Mountain into Canoe Valley. Winding northeastward through this valley it next goes through Tusey Mountain into Hartslog Valley by an interesting curving pass of the form of the letter S. The mountain, which consists of two ridges, is trenched along its center for the passage of the river, and the western ridge is, moreover, breached at Water street by a lateral notch, which gives passage to a small tributary stream and heightens much the picturesqueness of the place, which is further enhanced by a great stone slide covering the ends of the mountain. Crossing Hartslog Valley it next traverses Warrior Ridge, passing by the Pulpit Rocks. Emerging from the Warrior Ridge and deflecting more toward the east it crosses the Huntingdon Valley and passes by the northern end or knob of Terrace Mountain or Slideling Hill, receiving first the Raystown Branch, which nearly doubles the volume of its waters. Here, bending southward, it follows a picturesque gap through Stone Ridge, and turning more eastward it presently enters the deep cleft in Jacks Mountain called "Jacks Narrows," upon the western side of which the mountain is covered with a great stone slide or field of naked angular blocks of sandstone, which imparts a most desolate aspect to the pass, especially when the forest is not in leaf.

On emerging from Jacks Narrows the river crosses a succession of open valleys divided by narrow ridges until it meets the base of Blue Ridge in Sugar Valley. There it makes a great loop, turning in an oxbow backward till it reaches Newton Hamilton, where it flows with many large sinuosities longitudinally through the Juniata or Lewistown Valley to the deep synclinal ravine called the "Long Narrows," formed by the near approach of the Blue and Shade mountains. The Long Narrows of the Juniata is a narrow trough between mountain ridges, deeply trenched on their flanks and thickly clothed with timber on their lower slopes and

---

a Geol. Pennsylvania, p. 50.

at their base, and overspread nearer their summits with extensive sloping sheets
of dark-gray angular blocks. The pass is 7 miles long and is one of the wildest and
most impressive within the mountains. At the eastern end of the Long Narrows
the river turns southeastward and winds between hills and valleys across the
country to the base of the Tuscarora Mountain, passing Mifflintown, Mexico, and
other villages. Below New Mexico it sweeps the base of the Tuscarora Mountain
for several miles, until it turns abruptly across its eastern end a mile northwest of
Millerstown. Below Millerstown the river crosses the Wildcat and Buffalo valleys, washing the end of the Buffalo Mountain. Pursuing its course, the Juniata,
after making two or three bends, flows through a belt of hills called the "Half-
Fall Mountain," where, as at nearly all its passes through the larger sandstone
ridges, it is impeded by ledges of hard strata and thrown into ripples or rapids.
From the Half-Fall Rapids it flows between steep but low cliffs and hills for about
4 miles farther, to its entrance into the main Susquehanna at Duncans Island,
having followed a winding course entirely across the central zone of the Appalachian chain through a distance of nearly 200 miles.

SUSQUEHANNA RIVER ABOVE WEST BRANCH.

This portion of the stream and its tributaries drain an area of about
11,140 square miles, of which 6,080 are in New York and 5,060 in
Pennsylvania. It rises in Otsego Lake, in Otsego County, N. Y.,
which is about 7½ miles long and 1½ miles wide, and has an
elevation of about 1,193 feet above sea level. It flows in a southwesterly direction through Otsego, Chenango, and Broome counties,
N. Y., into Susquehanna County, Pa. It then flows in 'a westerly-
northwesterly direction through this county and again enters New
York and takes a westerly course through Broome and Tioga counties to near the western boundary of Tioga County, where it turns
south and enters Pennsylvania. Before leaving New York its volume
is rapidly swelled by many large tributaries. After entering Pennsylvania the second time it flows through Bradford, Wyoming, Luzerne,
Columbia, Montour, and Northumberland counties to its junction with
the West Branch, above Sunbury.

This portion of the drainage basin is varied in character. In New
York it is a rolling and sometimes rather broken country, forming the
plateau bounding the mountain region on the north. The stream has
a very uniform declivity in this part of its course and offers comparatively little power. Its bed is gravel or sand, with an occasional
rocky ledge. Its banks are moderately high, shelving, and are subject to overflow only in extreme freshets.

After it enters Pennsylvania it flows through the mountain regions,
and its course is in many places tortuous as it winds along the parallel ranges of hills. In general, however, its fall is gradual, its bed
being composed mostly of drift materials—gravel, sand, and bowlders. The banks, as in New York, are generally high and are seldom
overflowed, although the river has an extreme rise of as much as 30
feet.

In this portion of the drainage area is located the great Lackawanna
and Wyoming coal basin, and J. H. Dager reported upon this, in sub-

stance, as follows: [a] This basin extends from Nanticoke on the south-west, where the river emerges from the Coal Measures, to Carbondale on the northeast.   It is about 50 miles in length and averages 3½ miles in width.   It is surrounded by the Allegheny Mountains, which are composed of the Catskill formation and rocks of the Carboniferous system.

In this vicinity there are several workable seams of coal, ranging from 3 to 14 feet in thickness and at depths varying from nothing to 800 feet.   These seams are from 10 to 200 feet apart vertically, and are underlain by sandstone and fire clay.

From the outcrop of the Coal Measures just above Pittston to the New York State line the country is traversed by long, narrow, parallel ranges of mountains whose axes are nearly at right angles to the general direction of the river.   At bends on the convex side there rise from the shore abrupt cliffs from 200 to 400 feet in height, opposite which, with one or two exceptions, are gently sloping cultivated lands.

Professor Rogers refers to this portion of the river as follows: [b]

That portion of the Susquehanna River which flows near the northern boundary of the State passes from its sharp elbow, called the "Great Bend," to the mouth of its affluent, the Chemung River, through a charming, broad valley, bounded by soft slopes terminating in wide, table-shaped hills. It is a fertile and very beautiful district, and with its westward extension, the plain of the Chemung River, is rapidly becoming one of the most attractive agricultural districts of New York.   From the mouth of the Chemung River to Pittston, where the river suddenly turns at a right angle on entering the Wyoming coal field, it flows, with many bendings, along a deep and picturesque valley, almost identical in its features with that o˙ the corresponding stretch of the Delaware, the main difference being that the bed of the valley is wider and the hillsides confining it less mountainous.   From the mouth of the Lackawanna, at Pittston, where it enters, to Nanticoke, where it leaves the beautiful Wyoming Valley, the scenery along the river is wholly different.   It flows through a broad and almost perfectly level, smooth plain—the Wyoming and Kingston flats—composed of a deep bed of diluvium or drift.   On either side of this plain rise the rolling hills of the coal basin, and behind these the long, gentle slopes of the high mountain barriers, which frame in the whole scene.   At Nanticoke the river turns abruptly northward out of the coal basin, through its steep barrier, by a highly picturesque pass, and then sweeps again as suddenly westward to run for several miles in a closely confined trench between the outer and the inner ridges of the basin.   It does not, however, run round the western end of this, but at the ravine of the Shickshinny turns suddenly southward and cuts across its point, leaving a high, isolated hill of the coal strata on its western or right-hand side.   Disengaging itself by a fine pass from the southern barrier of the coal basin, it passes out into an open valley and makes another rectangular bend, to run once more toward the west, parallel with the Nescopeck Mountain, which it follows to the neighborhood of Catawissa. Beyond this point it maintains its general course westward, somewhat south, parallel with the southern base of Montour Ridge, all the way to Northumberland, where it is joined by its great tributary, the West Branch.   In some portions of this long reach of the river the scenery adjoining it is uncommonly rich and pleasing.   A remarkably fine view up the river is presented from the hills on its west bank, a little below the mouth of Fishing Creek.

[a] Ann Rept. Chief of Engineers, U. S Army, 1884, pt. 1, p. 873.   [b] Geol. Pennsylvania, p. 48.

WEST BRANCH OF SUSQUEHANNA RIVER.

The drainage basin of the West Branch has an area of approximately 7,030 square miles, all of which is in Pennsylvania. The West Branch has its sources in the mountains of Cambria County at an elevation of not less than 2,000 feet above sea level. It flows first in a northward direction, receiving some tributaries from Indiana County on the west, into Clearfield County. Gradually bending to the right, it flows northeast between Center and Clinton counties, east through Clinton and Lycoming counties, and south between Union and Northumberland to join the main stream above Sunbury, Pa.

The watershed of this stream occupies the high table-lands of the north-central part of Pennsylvania. The crest of the watershed has an elevation of from 500 to 1,200 feet above sea level in the vicinity of the junction of the West Branch and the main stream, increasing to about 2,200 feet at its southwestern part; thence along its western side it maintains this latter elevation to its northern line, where, in the northern part of the Pine Creek basin, it attains an elevation of over 2,600 feet. Along the remainder of the northern crest the height quickly falls to about 1,200 feet, but rises again to about 2,000 feet along the eastern crest of the divide. The highest points in the State are along the crest of this watershed.

As far up as Queens Run the fall of this branch is comparatively small, while above that point, in the mountain region, it is much greater. Furthermore, the banks of both the stream and its tributaries above Queens Run are generally high, and there are few low grounds subject to overflow. Below Queens Run the river traverses a wide, fertile valley, without, however, overflowing its banks to any considerable extent. The bed of the river is generally gravel and sand, with a rocky ledge at places. In former years this portion of the drainage was largely used by lumbermen for floating logs. On most of the streams splash dams were built, sometimes flooding considerable areas, and serving to hold the logs which were sent down until a sufficient number were collected. The gates in the dam were then raised, letting the water out suddenly, so that the logs were carried down on the swell or wave to the next dam or to the main river, where the natural current would be sufficient to carry them along. As the forest areas are now largely cut off, but very little logging is done either on this or other portions of the river.

Professor Rogers describes this branch of the river as follows:[a]

The upper part of the West Branch of the Susquehanna, and also its tributaries, the Sinnemahoning, Kettle Creek, Pine Creek, etc., draining the high plateau northwest of the Allegheny Mountains, flow through deep trenches in the horizontal strata, very analogous in their features to those which give passage to the Delaware and the Main or North Susquehanna, in the northeastern part of the State. From the mouth of the Sinnemahoning out into the Bald Eagle Valley,

---

[a] Geol Pennsylvania. p 49.

the river hills are very high and steep, and admit extremely narrow strips of ground between their feet and the river, except near the openings of the lateral streams. The trough through which the lower half of Pine Creek flows is equally profound. Entering the valley between the Allegheny Mountains and the Bald Eagle ridge, the river pursues a beautiful winding course the whole way from Lockhaven to the neighborhood of Muncy, alternately sweeping toward the middle of the cultivated valley and back again, close in to the base of the steep and wood-covered ridge. Near Muncy it turns with a broad majestic curve round the end of the Bald Eagle Mountains, and in a few miles deflects from a southwest to a west course, through a highly fertile, richly cultivated open country, till it strikes the base of the Blue Hill, or range of red sandstone cliffs above Northumberland. Southwest of Muncy the river crosses a singular belt of deeply eroded country, full of conical hills.

NAVIGATION.

Information in regard to navigation along Susquehanna River and its tributaries is now only of historical interest. The official records of Pennsylvania and other papers published during the early part of the century show that from the first settlement Susquehanna River and its tributaries were regarded as a possible means of navigation.

In this relation the following quotation from Dager's report is of interest:[a]

General Sullivan, to punish the Six Nations. late in August. 1779. organized a force of 3,000 men and moved north from Wyoming, the artillery and stores being drawn up the North Branch in 150 boats. At Tioga he was joined by General Clinton with 1,000 New York troops. The latter had marched from Albany to Otsego Lake, where, finding the water too low to flo.t his bateaux. he built a dam across the stream, by which the lake was raised several feet, and when the dam was cut away the discharge wave floated his boats down to Tioga.

The Indians fled in dismay at the sight of a flood in the midst of the summer drought. believing it a signal of the displeasure of the Great Spirit. From this might be inferred that Otsego Lake could be made a reservoir to pay tribute to the river when there was an insufficient flow.

On March 9, 1777, an act was passed declaring Susquehanna River a public highway as far down as Wrights Ferry, and later on, March 31, 1785, the whole river through Pennsylvania was declared a public highway. An appropriation of £6,290 was made as early as April 11, 1791, for the improvement of the navigation of Susquehanna River. Other appropriations were made from time to time and active canals were maintained from Havre de Grace to the New York State line, on the West Branch from Northumberland to Lock Haven, and on the Juniata from Juniata Junction to Holidaysburg.

Between 1800 and 1830 several plans were proposed for connecting Susquehanna River with the Great Lakes and with Mississippi River. Nothing, however, came of any of these projects, and with the coming of the railroads the canals were gradually abandoned, being in most cases bought by the railroad companies. The North Branch extension, from the New York State line to Pittston, was abandoned in 1868 or 1869. The canal from Pittston down was used more or less

---

a Ann. Rept. Chief of Engineers. U S Army. 1884. pt. 1, p. 876.

until the fall of 1874, but the high floods of the spring of 1875 caused so much damage that no boats were run after that date above Wilkesbarre. The Lackawanna Canal served as a feeder for the Wilkesbarre Branch until the spring of 1882, when it was abandoned to the Nanticoke dam. The canals below Sunbury were abandoned about 1890.

MEASUREMENTS OF FLOW.

The records of the measurements of flow in the Susquehanna drainage have been divided into two classes: First, those at regular stations, where systematic observations have been carried on over a series of years; second, those at miscellaneous stations, which consist of short or broken series of observations. There have been nine regular stations maintained, as given in the following list:

Gaging stations in the Susquehanna drainage basin.

Stream.	Location.	Date established.	Established by—
A. Susquehanna..	Binghamton, N. Y....	Aug. 1, 1901	United States Geological Survey.
B. Chenango.....	.....do....	.do	Do.
C. Susquehanna..	Wilkesbarre, Pa	Mar. 30, 1899	Do.
D ....do	Danville, Pa	Mar. 25, 1899	Do.
E. West Branch..	Williamsport, Pa	Mar. 4, 1895	City engineer.
F. .....do	Allenwood, Pa	Mar. 25, 1899	United States Geological Survey.
G. Juniata.......	Newport, Pa	Mar. 21, 1899	Do.
H. Susquehanna..	Harrisburg, Pa........	Mar. 21, 1890	Water board.
I...|.....do	McCalls Ferry, Pa.....	May 17, 1902	Cary T. Hutchinson.

The locations of these stations are shown on fig. 1 (p. 11) by the letters in column 1 of the above table.

Miscellaneous records have been collected at the following points:

Chemung River at Chemung, N. Y.
Tioughnioga River at Chenango Forks, N. Y.
Cayuta Creek at Waverly, N. Y.
Chenango River at Oxford, N. Y.
Eaton and Madison creeks.
Diversions from Chenango River drainage.

The following pages give the data which have been collected at both regular and miscellaneous stations, also the results of the computations based upon these data.

SUSQUEHANNA RIVER AT BINGHAMTON, N. Y.

This gaging station was established by R. E. Horton July 31, 1901. The gage is located on the upstream side of the left span of the Washington street bridge. The bench mark is a chiseled draft on the corner of the left abutment on the upstream side. Its elevation is 23.71 feet above gage datum. This bridge is located about 800 feet upstream from the junction of Chenango and Susquehanna rivers. A rift extends diagonally across the stream underneath the bridge. The gage is above a stretch of smooth water extending from the crest of the rift to the dam 2,800 feet upstream, and the gage readings are not affected by backwater from Chenango River at ordinary stages. On account of unfavorable conditions of Washington Street Bridge discharge measurements are made at Exchange Street Bridge, which is 1,900 feet upstream. At this place the channel is about 300 feet wide at low water and about 450 feet wide at high water, and is straight

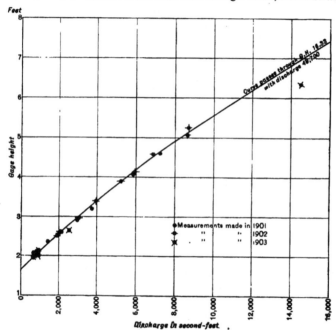

FIG. 2.—Rating curve for Susquehanna River at Binghamton, N. Y.

for about 500 feet above and below the bridge. The bed is naturally gravel and small stones. Formerly a wooden footbridge was located at this point, and the channel was divided into three parts by two piers. Large stones were piled around the piers. At present a steel bridge occupies this site, and there is but one pier, above which are two rows of short piles and a quantity of small stones. The upper parts of the old piers have been removed, but the stone filling around them remains, leaving the river bed irregular and rough.

The velocity is good at low water and swift at high water. The lowest observed mean velocity is 0.72 foot per second.

Within the time for which this record has been kept, the gage height has ranged between 1.84 and 19.22 feet, and the estimated discharge between 400 and 60,300 cubic feet per second. The gage is read twice daily by E. F. Weeks.

Discharge measurements of Susquehanna River at Binghamton, N. Y., 1901-4.

Date.	Hydrographer.	Area.	Mean velocity.	Gage height.	Discharge.
		Square feet.	*Feet per second.*	*Feet.*	*Second-feet.*
1901.					
July 3	E. C. Murphy	891	1.06	2.12	947
July 10	do	1,020	1.40	2.35	1,425
July 30	do	847	.72	1.99	608
August 20	do	909	1.04	2.05	942
August 20	do	923	1.03	2.06	952
August 21	do	1,989	3.65	4.60	7,244
August 22	do	1,439	2.61	3.19	3,752
August 22	do	1,324	2.25	2.90	2,983
August 23	do	1,189	1.83	2.60	2,176
1902.					
July 2	E. C. Murphy	1,790	3.26	4.08	5,839
July 4	do	1,717	3.28	3.90	5,230
July 14	do	1,320	2.32	2.96	3,064
August 3	do	2,187	3.95	5.08	8,633
August 4	do	1,952	3.53	4.59	6,902
August 15	do	1,140	1.85	2.61	2,105
August 16	do	1,103	1.74	2.50	1,920
1903.					
April 7	E. C. Murphy	1,773	3.35	4.18	5,946
May 15	do	794	.96	2.05	763
May 19	do	746	.86	1.96	640
June 13	C. C. Covert	2,293	3.80	5.25	8,726
August 22	do	1,241	2.07	2.65	2,572
September 3	do	544	1.81	2.00	948
October 1	H. H. Halsey	889	1.08	2.14	962
October 11	C. C. Covert	6,446	7.71	16.32	49,707
October 13	do	2,944	4.94	6.35	14,566
1904.					
March 8	C. C. Covert	3,975	3.58	*a* 11.24	14,254
March 12	do	2,846	2.60	*a* 7.90	7,400
April 8	R. E. Horton	2,524	4.50	6.94	11,118
July 13	C. C. Covert	786	1.07	2.04	786
September 10	do	825	1.29	2.13	1,061

a Ice gorge 3 miles below.

Mean daily gage height, in feet, of Susquehanna River at Binghamton, N. Y.,
1901-1904.

Day.	Jan.	Feb.	Mar.	Apr.	May.	June.	July.	Aug.	Sept.	Oct.	Nov.	Dec.
1901.												
1								1.84	2.21	2.19	2.04	2.49
2								1.96	2.16	2.19	2.02	2.49
3								1.91	2.16	2.16	1.94	2.64
4								1.86	2.21	2.16	1.94	2.56
5								1.86	2.18	2.14	1.96	2.64
6								1.86	2.16	2.06	1.94	2.44
7								1.86	2.06	2.04	1.94	2.32
8								1.91	2.04	1.99	1.94	2.34
9								1.91	2.04	2.04	1.92	2.44
10								1.86	1.96	1.99	1.94	5.21
11								1.94	1.98	2.02	1.92	6.12
12								1.94	2.06	1.96	1.96	5.32
13								1.91	2.04	1.99	2.49	----
14								1.96	2.01	2.06	2.96	4.62
15								1.94	2.08	2.14	2.79	14.86
16								1.94	2.16	2.32	2.54	13.74
17								1.96	2.21	2.30	2.44	9.24
18								2.11	2.36		2.42	5.66
19								2.16	2.36	2.26	2.36	4.29
20								2.06	2.34	2.24	2.39	3.46
21								3.66	2.24	2.24	2.39	2.96
22								2.98	2.16	2.26	2.32	2.76
23								2.61	2.06	2.24	2.29	3.74
24								4.51	2.06	2.19	2.71	4.68
25								8.86	2.06	2.14	3.42	3.96
26								3.21	2.04	2.09	2.94	3.32
27								2.78	2.00	2.06	2.52	3.28
28								2.46		2.06	2.24	2.86
29								2.36	2.02	2.04	2.34	2.44
30								2.26	2.04	2.04	2.39	3.69
31							1.91	2.31		2.06		4.06
1902.												
1	3.22	2.56	15.59	5.20	2.85	2.35	5.10	4.90	2.13	4.57	4.60	2.75
2	3.89	2.54	19.22	5.10	2.85	2.37	4.23	5.94	2.13	4.25	4.07	2.70
3	3.22	2.56	17.69	4.87	2.75	2.30	3.60	5.27	2.13	3.67	3.70	2.85
4	3.56	3.24	13.79	4.55	2.65	2.63	3.87	4.51	2.15	3.85	3.47	3.10
5	3.22	2.96	9.19	4.20	2.65	3.07	3.48	8.77	2.13	2.90	3.27	3.33
6	3.14	2.66	6.36		2.67	2.85	3.97	3.45	2.07	2.98	3.13	3.18
7	3.02	2.72	5.59	3.90	2.57	2.63	4.48	3.37	2.05	2.83	3.07	2.93
8	2.82	2.74	5.34	3.88	2.53	2.57	4.35	3.10	2.10	2.77	3.00	2.77
9	2.66	2.79	5.04	4.75	2.45	2.65	4.00	2.97	2.07	2.75	2.83	2.73
10	2.54	2.72	5.74	5.40	2.45	2.60	4.03	2.88	2.25	2.67	2.77	2.85
11	2.52	2.84	5.59	5.70	2.35	2.47	4.77	2.73	2.25	2.55	2.70	2.83
12	2.46	2.64	7.81	5.45	2.38	2.47	4.37	2.75	2.25	2.67	2.65	2.95
13	2.57	2.42	11.19	5.03	2.30	2.57	3.43	2.80	2.23	2.77	2.65	2.83
14	2.46	2.34	11.94	4.70	2.30	2.57	3.08	2.75	2.15	2.90	2.75	2.67
15	2.34	2.24	10.61	4.35	2.27	2.65	2.75	2.59	2.15	2.90	2.67	2.75
16	2.82	2.26	8.42	3.97	2.25	2.65	2.70	2.49	2.10	2.87	2.55	2.83
17	2.24	2.19	11.82	3.70	2.25	2.53	2.63	2.40	2.05	2.75	2.56	7.13
18	2.22	2.14	11.87	3.58	2.15	2.55	2.65	2.35	2.05	2.60	2.58	7.65
19	2.42	2.16	9.47	3.37	2.15	2.50	2.65	2.30	2.05	2.50	2.47	6.70
20	2.64	2.16	6.82	8.17	2.15	2.50	7.27	2.30	2.00	2.60	2.50	5.87
21	2.14	2.19	5.72	3.07	2.25	2.45	10.90	2.85	1.96	2.75	2.45	6.70
22	2.56	2.12	5.49	2.97	2.33	2.47	11.35	2.35	1.96	2.73	2.45	9.45
23	4.76	2.24	5.61	2.85	2.35	2.57	10.00	2.27	2.00	2.57	2.47	10.62
24	5.16	2.29	5.76	2.67	2.25	2.50	8.90	2.27	1.97	2.53	2.45	8.20
25	4.22	2.16	5.44	2.68	2.20	2.37	8.10	2.25	2.00	2.58	2.47	6.20
26	3.99	2.12	4.92	2.55	2.37	2.27	6.37	2.25	2.85	2.50	2.53	5.65
27	3.14	2.42	4.44	2.50	2.55	2.37	5.40	2.20	2.63	2.45	2.70	4.96
28	3.67	5.46	4.44	2.53	2.53	2.37	7.51	2.25	2.55	5.95	2.85	4.48
29	4.02		5.30	2.47	2.85		6.07	2.27	5.00	8.30	2.90	4.93
30	3.34		5.70	2.60	2.40	6.98	5.00	2.30	3.80	7.15	2.83	3.81
31	2.86		5.53		2.35		5.55	2.17		5.53		3.65

Mean daily gage height, in feet, of Susquehanna River at Binghamton, N. Y.,
1901–1904—Continued.

Day.	Jan.	Feb.	Mar.	Apr.	May.	June.	July.	Aug.	Sept.	Oct.	Nov.	Dec.
1903.												
1	3.40	8.60	12.98	6.65	2.88	1.85	3.35	2.55	6.55	2.07	3.26	2.68
2	3.30	7.90	10.82	5.85	2.27	1.87	3.00	2.43	5.17	2.13	3.07	2.69
3	3.70	7.23	7.75	5.15	2.25	1.85	2.73	2.80	4.80	2.15	2.95	2.65
4	5.15	8.27	6.17	5.05	2.25	1.80	2.57	2.25	3.70	2.10	2.85	2.65
5	5.33	9.60	5.68	4.80	2.23	1.83	2.45	2.50	......	2.13	2.85	2.62
6	4.63	7.95	6.43	4.83	2.20	1.80	2.36	3.17	......	2.25	3.05	2.47
7	3.83	6.35	6.30	4.17	2.17	1.77	2.27	3.25	......	2.36	3.17	2.52
8	3.75	5.00	6.35	4.96	2.15	1.85	2.25	3.06	2.70	2.70	2.97	2.52
9	3.45	4.65	10.75	5.63	2.15	1.80	2.20	2.80	2.67	7.97	2.85	2.57
10	6.05	4.83	10.55	5.05	2.10	1.80	2.17	2.63	2.55	15.49	2.75	2.29
11	5.55	4.20	11.55	4.70	2.05	1.80	2.13	2.73	2.65	16.35	2.72	2.45
12	5.93	5.47	11.47	4.40	2.05	2.77	2.10	2.83	2.67	12.12	2.67	2.55
13	6.00	6.95	9.57	4.08	2.05	3.35	2.10	2.70	2.60	8.17	2.62	2.65
14	6.07	6.07	7.75	3.73	2.05	3.45	2.07	2.60	2.50	5.99	2.59	3.17
15	5.85	4.97	6.65	4.05	2.05	3.03	2.13	2.55	2.37	5.09	2.52	3.22
16	5.80	4.40	6.03	3.97	2.00	2.63	2.07	2.43	2.30	4.49	2.52	3.12
17	5.53	3.65	5.55	3.73	2.00	2.50	2.05	2.38	2.37	4.22	5.70	2.97
18	5.10	3.13	5.45	3.47	2.00	2.45	2.10	2.30	2.50	7.55	6.89	2.85
19	4.60	3.27	5.13	3.23	2.00	2.35	2.17	2.27	2.45	7.89	5.45	2.79
20	4.15	3.57	4.75	3.07	1.95	2.30	2.15	2.27	2.45	6.55	4.25	2.62
21	4.30	3.75	4.50	2.90	1.95	2.53	2.23	2.45	2.35	5.47	3.67	4.37
22	6.53	3.53	5.60	2.77	1.95	3.77	2.25	2.65	3.27	4.83	3.35	5.39
23	6.63	3.55	7.57	2.70	1.95	4.45	3.50	2.40	2.90	4.25	3.29	4.97
24	5.63	3.25	12.11	2.65	1.87	5.08	4.65	2.30	2.23	4.08	3.39	4.25
25	4.90	3.20	11.48	2.60	1.85	4.43	3.43	2.25	2.20	3.92	3.32	4.05
26	4.53	3.15	9.20	2.57	1.85	3.97	2.80	2.70	2.15	3.67	3.05	3.79
27	4.23	2.95	7.15	2.50	1.87	3.40	2.60	4.13	2.10	3.52	2.87	3.73
28	4.20	6.80	6.07	2.45	1.90	2.95	2.45	3.57	2.10	3.45	2.79	3.45
29	5.35	......	5.70	2.40	1.90	3.08	2.35	10.63	2.10	3.45	2.85	3.57
30	9.68	......	5.30	2.35	1.87	3.65	2.47	10.58	2.07	3.42	2.85	3.65
31	10.23	......	6.20		1.85	......	2.70	8.57	......	3.35	......	3.75
1904.												
1	3.28	3.67	3.57	7.72	5.06	2.46	2.02	2.40	2.28	4.12	3.08	2.98
2	3.35	3.40	3.29	9.02	4.53	2.43	1.99	2.35	2.25	3.85	3.00	2.82
3	a3.42	3.59	3.92	......	4.08	2.38	2.14	2.94	2.28	2.90	2.92	2.85
4	3.88	3.67	6.65	6.95	3.68	2.36	2.14	2.95	2.28	2.80	2.88	2.70
5	3.52	3.55	8.48	6.20	3.51	2.38	2.06	2.60	2.20	2.64	2.82	2.85
6	3.58	3.15	7.68	6.15	3.83	2.41	2.09	3.52	2.22	2.62	2.80	2.68
7	3.90	4.42	7.52	6.35	3.13	2.46	2.04	3.40	2.28	2.62	2.92	2.68
8	3.28	10.49	11.40	6.98	2.98	2.57	2.04	2.72	2.22	2.52	2.90	2.60
9	3.15	11.92	13.62	7.14	2.86	3.67	2.04	2.50	2.22	2.45	2.80	2.60
10	3.20	10.86	12.25	8.74	2.80	4.23	2.04	2.38	2.20	2.42	2.75	2.68
11	3.10	8.62	9.80	8.24	2.69	3.43	2.04	2.50	2.18	2.40	2.75	2.58
12	2.78	7.15	8.02	6.94	2.65	2.93	1.99	2.45	2.18	2.88	2.75	2.98
13	2.85	6.09	6.88	6.09	2.65	2.65	2.04	2.30	2.20	5.60	2.70	2.50
14	2.72	5.27	6.08	5.51	2.49	2.50	2.03	2.22	2.15	4.64	2.70	2.58
15	2.85	4.77	5.30	4.97	2.50	2.43	1.95	2.20	3.00	3.65	2.68	2.58
16	3.05	6.12	4.75	4.61	3.22	2.45	1.92	2.28	3.10	3.45	2.70	2.58
17	2.85	b6.85	4.28	4.49	3.45	2.33	2.05	2.22	2.82	2.95	2.78	2.60
18	3.00	6.07	3.85	4.39	3.17	2.33	2.24	2.18	2.55	2.80	2.75	2.48
19	2.98	5.87	3.55	4.49	2.92	2.23	2.10	2.18	2.42	2.70	2.65	2.60
20	3.08	5.22	3.92	4.37	3.22	2.17	2.05	2.22	2.35	2.62	2.65	2.40
21	8.80	4.72	4.45	4.17	3.05	2.20	1.98	2.90	2.30	5.95	2.82	2.45
22	2.78	4.52	4.30	3.97	2.75	2.13	2.00	8.18	2.30	7.48	3.58	2.58
23	7.02	4.92	7.42	3.97	2.67	2.24	1.98	4.55	2.28	6.95	3.72	2.40
24	7.82	5.72	11.40	3.77	2.59	2.09	2.00	4.20	2.18	5.32	3.55	2.55
25	c8.27	5.52	12.12	3.79	2.62	2.05	2.02	3.88	3.52	4.40	3.38	3.08
26	6.85	4.67	15.98	3.96	2.52	2.02	2.02	2.92	3.25	4.40	3.32	3.15
27	5.95	4.19	15.70	3.95	2.49	1.99	2.05	2.78	3.22	4.35	3.18	3.40
28	5.25	3.75	12.62	5.83	2.45	1.99	2.52	......	2.85	3.92	2.90	8.80
29	4.42	3.67	8.50	6.36	2.36	2.04	2.48	2.48	2.65	3.65	2.78	9.60
30	4.27	......	6.90	5.63	2.36	1.99	3.12	2.38	2.80	3.42	2.88	7.05
31	3.80	......	6.72		2.36	......	2.65	2.35	......	3.18	......	5.25

a Anchor ice. January 6 river frozen nearly across.
b Heavy anchor ice. River frozen over 2,000 feet downstream from junction of the two rivers.
Ice gorge causes backwater March 4–15.
c Current of stream very sluggish.

Rating table for Susquehanna River at Binghamton, N. Y., for 1901 to 1904, inclusive.

Gage height.	Discharge.	Gage height.	Discharge.	Gage height.	Discharge.	Gage height.	Discharge.
Feet.	*Second-feet.*	*Feet.*	*Second-feet.*	*Feet.*	*Second-feet.*	*Feet.*	*Second-feet.*
1.75	210	3.9	5,255	7.2	15,260	11.6	30,860
1.8	315	4.0	5,510	7.4	15,920	11.8	31,580
1.9	525	4.1	5,770	7.6	16,590	12.0	32,300
2.0	740	4.2	6,030	7.8	17,270	12.2	33,020
2.1	960	4.3	6,300	8.0	17,950	12.4	33,740
2.2	1,180	4.4	6,570	8.2	18,650	12.6	34,470
2.3	1,400	4.5	6,845	8.4	19,350	12.8	35,210
2.4	1,625	4.6	7,125	8.6	20,060	13.0	35,950
2.5	1,855	4.7	7,405	8.8	20,780	13.5	37,820
2.6	2,085	4.8	7,690	9.0	21,500	14.0	39,720
2.7	2,315	4.9	7,980	9.2	22,220	14.5	41,650
2.8	2,545	5.0	8,280	9.4	22,940	15.0	43,600
2.9	2,785	5.2	8,880	9.6	23,660	15.5	45,550
3.0	3,025	5.4	9,495	9.8	24,380	16.0	47,500
3.1	3,265	5.6	10,120	10.0	25,100	16.5	49,500
3.2	3,505	5.8	10,760	10.2	25,820	17.0	51,500
3.3	3,755	6.0	11,400	10.4	26,540	17.5	53,500
3.4	4,005	6.2	12,040	10.6	27,260	18.0	55,500
3.5	4,255	6.4	12,680	10.8	27,980	18.5	57,500
3.6	4,505	6.6	13,320	11.0	28,700	19.0	59,500
3.7	4,755	6.8	13,960	11.2	29,420	19.5	61,500
3.8	5,005	7.0	14,600	11.4	30,140	20.0	63,500

Mean daily discharge, in second-feet, of Susquehanna River at Binghamton, N. Y., 1901-1904.

Day.	Jan.	Feb.	Mar.	Apr.	May.	June.	July.	Aug.	Sept.	Oct.	Nov.	Dec.
1901.												
1							399	1,180	1,180	850		1,855
2							652	1,070	1,180	784		1,855
3							546	1,070	1,070	609		2,200
4							441	1,180	1,070	609		1,970
5							441	1,136	1,070	652		2,200
6							441	1,070	860	609		1,740
7							441	850	850	609		1,444
8							546	860	718	609		1,510
9							546	850	850	567		1,740
10							441	652	718	609		8,880
11							609	696	784	567		11,720
12							609	850	652	652		9,185
13							546	850	718		1,855	8,655
14							652	740	860		2,905	7,125
15							609	916	1,070		2,545	48,210
16							609	1,070	1,444		1,970	38,580
17							652	1,180	1,625		1,740	22,220
18							968	1,510	1,458		1,671	10,280
19							1,092	1,510	1,290		1,510	6,300
20							872	1,510	1,290		1,685	4,130
21							4,630	1,290	1,290		1,625	2,905
22							2,977	1,070	1,290		1,444	2,430

Mean daily discharge, in second-feet, of Susquehanna River at Binghamton, N. Y., 1901-1904—Continued.

Day.	Jan.	Feb.	Mar.	Apr	May	June.	July	Aug.	Sept.	Oct.	Nov.	Dec
1901.												
23								2,085	850	1,290	1,400	4,880
24								6,845	850	1,180	2,315	7,265
25								5,130	850	1,070	4,055	5,380
26								8,505	850	980	2,905	3,805
27								2,499	740	850	1,901	3,630
28								1,740	762	850	1,290	2,665
29								1,570	784	850	1,510	2,785
30								1,290	850	850	1,625	4,755
31								1,400		850		5,640
1902.												
1	3,555	1,970	45,940	8,880	2,665	1,510	8,580	7,980	1,036	6,985	7,125	2,430
2	4,005	1,970	60,300	8,580	2,665	1,554	6,165	11,240	1,026	6,165	5,640	2,384
3	3,555	1,970	54,300	7,835	2,430	1,400	4,505	9,080	1,026	4,680	4,755	2,665
4	4,380	3,630	38,980	6,945	2,200	2,154	5,180	6,845	1,070	3,880	4,180	3,265
5	3,555	2,905	22,220	6,090	2,200	3,195	4,040	4,980	1,026	2,745	3,640	3,830
6	3,386	2,200	12,520	5,640	2,246	2,665	5,455	4,130	894	2,857	3,337	3,480
7	3,075	2,361	10,120	5,255	2,016	2,154	6,705	3,930	850	2,617	8,193	2,857
8	2,568	2,430	9,840	5,130	1,924	2,016	6,435	3,265	980	2,476	3,025	2,476
9	2,200	2,545	8,430	7,545	1,740	2,200	5,510	2,953	894	2,430	2,617	2,384
10	1,970	2,361	10,600	9,495	1,740	2,085	5,640	2,617	1,290	2,246	2,476	2,665
11	1,901	2,665	10,120	10,440	1,510	1,798	7,545	2,384	1,290	1,970	2,315	2,617
12	1,740	2,200	17,270	9,650	1,466	1,798	6,435	2,430	1,290	2,246	2,200	2,905
13	2,016	1,671	29,420	8,430	1,400	2,016	4,040	2,545	1,246	2,476	2,200	2,617
14	1,740	1,510	31,940	7,405	1,400	2,016	3,100	2,430	1,070	2,785	2,430	2,246
15	1,510	1,290	27,260	6,435	1,834	2,200	2,430	2,085	1,070	2,785	2,248	2,430
16	1,444	1,290	19,350	5,380	1,290	2,200	2,315	1,855	980	2,713	1,970	2,857
17	1,290	1,180	31,580	4,755	1,290	1,924	2,154	1,625	850	2,430	1,970	14,930
18	1,224	1,070	31,940	4,330	1,070	1,970	2,200	1,510	850	2,085	1,924	16,760
19	1,671	1,070	23,300	3,980	1,070	1,855	2,200	1,400	850	1,855	1,798	13,640
20	2,200	1,070	13,980	3,435	1,070	1,855	15,590	1,400	740	2,085	1,855	10,920
21	1,070	1,180	10,440	3,196	1,290	1,740	28,340	1,510	680	2,430	1,740	9,185
22	1,970	1,004	9,805	2,953	1,466	1,798	29,960	1,510	630	2,384	1,740	23,120
23	7,545	1,290	10,120	2,665	1,510	2,016	25,100	1,334	740	2,016	1,786	27,260
24	8,730	1,400	10,600	2,246	1,290	1,855	21,140	1,334	674	1,924	1,740	18,650
25	6,080	1,070	9,650	2,154	1,180	1,554	18,300	1,290	740	1,924	1,786	12,360
26	4,005	1,004	7,980	1,970	1,554	1,334	12,520	1,290	1,510	1,855	1,924	10,240
27	3,386	1,671	6,985	1,855	1,970	1,554	9,495	1,180	2,154	1,740	2,315	8,280
28	4,680	9,650	6,705	1,924	1,924	1,554	16,250	1,290	1,970	11,240	2,665	6,845
29	5,510		9,185	1,924	1,786	2,246	8,280	1,334	8,280	19,000	2,785	5,380
30	3,880		10,440	2,085	1,625	14,600	8,280	1,400	5,005	15,095	2,617	5,005
31	2,665		9,980		1,510		9,980	1,114		9,980		4,580
1903.												
1	4,005	20,060	35,580	13,480	1,466	420	3,880	1,970	18,160	894	3,680	2,131
2	8,755	15,260	27,980	10,920	1,384	462	3,025	1,694	8,730	1,038	8,198	2,315
3	4,755	15,260	17,100	8,730	1,290	420	2,384	1,400	6,300	1,070	2,905	2,200
4	8,730	19,000	11,880	8,430	1,290	315	2,016	1,290	4,755	980	2,665	2,200
5	9,340	28,650	10,280	7,690	1,246	378	1,740	1,855	3,535	1,026	2,665	2,131
6	7,285	17,780	12,840	6,435	1,180	315	1,510	3,435	3,585	1,290	3,145	1,786
7	5,080	12,520	12,380	5,900	1,114	252	1,334	3,630	3,535	1,510	3,435	1,901
8	4,880	8,280	1x,520	8,130	1,070	420	1,290	3,097	2,315	2,315	2,953	1,901
9	4,180	7,285	27,800	10,280	1,070	315	1,180	2,545	2,246	17,950	2,665	2,016
10	11,580	6,435	27,080	8,430	980	315	1,114	2,154	1,970	45,550	2,430	1,400
11	9,980	6,080	30,080	7,405	850	315	1,026	2,384	2,200	48,900	2,361	1,740
12	11,400	14,440	23,660	6,570	850	2,476	960	2,617	2,246	32,660	2,246	1,970
13	11,580	11,560	17,100	5,640	850	9,340	960	2,315	2,085	18,650	2,131	2,200
14	11,580	11,560	17,100	4,830	850	4,130	894	2,085	1,855	11,400	2,085	3,435
15	10,980	8,130	13,480	5,640	850	3,097	1,026	1,970	1,556	8,580	1,901	3,555
16	10,760	6,570	11,560	5,430	740	2,154	894	1,697	1,40)	6,845	1,901	3,313
17	9,980	4,680	9,980	4,890	740	1,855	850	1,466	1,556	6,030	10,440	2,953
18	8,480	3,340	9,650	4,180	740	1,740	980	1,400	1,855	16,420	14,280	2,665
19	7,125	3,680	8,730	3,580	740	1,510	1,114	1,334	1,740	17,610	9,650	2,545
20	5,900	4,430	7,545	3,193	630	1,400	1,070	1,334	1,740	13,160	6,165	2,131
21	6,300	4,880	6,845	2,785	630	1,924	1,246	1,740	1,510	9,650	4,680	6,435
22	13,160	4,330	10,120	2,476	630	4,930	1,290	2,200	1,334	7,690	3,880	9,495
23	12,480	4,380	16,580	2,315	630	6,705	4,255	1,625	1,180	6,165	3,755	8,130
24	10,280	3,630	32,660	2,200	462	8,430	7,265	1,400	1,246	5,510	4,005	6,165
25	7,690	8,505	30,500	2,085	420	6,705	4,080	1,290	1,180	5,305	3,805	5,640
26	6,985	8,385	22,220	2,016	420	5,430	2,545	1,070	980	4,680	3,145	5,005
27	6,165	2,905	15,095	1,855	462	4,005	2,085	5,900	980	4,805	2,713	4,805
28	6,080	18,980	11,560	1,740	525	2,905	1,740	4,430	980	4,130	2,545	4,180
29	9,340		10,440	1,625	525	3,097	1,510	27,260	980	4,130	2,665	4,430
30	24,020		9,185	1,510	462	4,630	1,786	26,900	894	4,055	2,685	4,630
31	25,830		12,040		420		2,315	20,060		8,880		4,880

Mean daily discharge, in second-feet, of Susquehanna River at Binghamton.
N. Y., 1901–1904—Continued.

Day.	Jan.	Feb.	Mar.	Apr.	May.	June.	July.	Aug.	Sept.	Oct.	Nov.	Dec.
1904.												
1	3,705	4,680	4,430	16,980	8,430	1,763	784	1,625	1,356	5,770	3,217	2,977
2	3,880	4,005	3,780	21,500	6,985	1,809	718	1,510	1,290	3,880	3,025	2,593
3	4,055	4,480	5,305	17,950	5,770	1,579	1,048	2,977	1,356	2,785	2,833	2,665
4	5,205	4,680	6,740	14,440	4,705	1,582	1,048	2,905	1,356	2,545	2,737	2,315
5	4,305	4,380	9,815	12,040	4,280	1,579	872	2,085	1,180	2,269	2,593	2,665
6	4,455	3,385	8,410	11,880	3,830	1,648	968	4,305	1,224	2,131	2,545	2,269
7	3,755	6,624	10,100	12,580	3,337	1,763	828	4,005	1,356	2,131	2,833	2,269
8	3,705	26,864	15,070	14,600	2,977	2,016	828	2,361	1,224	1,901	2,785	2,085
9	3,385	32,012	19,100	16,095	2,689	4,680	828	1,855	1,224	1,740	2,545	2,085
10	3,505	28,160	16,600	20,600	2,545	6,165	828	1,579	1,180	1,671	2,785	2,269
11	3,265	20,182	12,190	18,825	2,232	4,080	828	1,855	1,136	1,625	2,430	2,089
12	2,977	15,095	8,970	14,440	2,200	2,857	718	1,740	1,136	2,737	2,430	2,977
13	2,499	11,688	7,140	11,720	2,200	2,200	828	1,400	1,180	10,120	2,315	1,855
14	2,361	9,092	5,860	9,806	1,832	1,855	784	1,224	1,070	7,405	2,315	2,089
15	2,665	7,608	5,000	8,180	2,062	1,694	630	1,180	3,025	4,630	2,269	2,089
16	3,145	11,784	7,545	7,125	3,555	1,740	567	1,356	3,265	4,180	2,315	2,039
17	2,665	14,120	6,300	6,845	4,130	1,468	850	1,224	2,593	2,905	2,499	2,085
18	3,025	11,624	5,130	6,570	3,433	1,466	1,856	1,188	1,970	2,545	2,430	1,809
19	2,977	10,344	4,380	6,445	2,853	1,246	960	1,136	1,671	2,315	2,200	2,085
20	3,217	8,940	5,305	6,435	3,555	1,114	850	1,224	1,510	2,131	2,200	1,625
21	5,005	7,461	6,705	5,900	3,145	1,180	696	2,785	1,400	11,240	2,598	1,740
22	2,499	6,901	6,300	5,432	2,430	1,026	740	3,457	1,400	16,250	4,455	2,089
23	14,666	8,040	15,920	5,432	2,246	1,048	696	6,985	1,356	14,440	4,805	1,625
24	17,338	10,504	30,140	4,930	2,062	968	740	6,090	1,136	9,185	4,880	1,970
25	18,895	9,867	32,660	4,980	2,131	872	784	3,955	4,305	6,570	3,955	3,217
26	14,120	7,321	47,110	5,406	1,901	784	784	2,833	3,630	6,570	3,805	3,385
27	11,240	6,004	46,330	5,930	1,832	718	850	2,499	3,555	6,435	3,457	4,005
28	9,080	4,880	34,470	5,080	1,740	718	1,901	2,154	2,665	5,305	2,785	20,780
29	6,624	4,680	19,700	12,520	1,532	828	2,039	1,809	2,200	4,630	2,499	23,660
30	6,219		14,290	10,290	1,532	718	2,313	1,198	2,545	4,055	2,737	14,765
31	5,230		13,640		1,532		2,200	1,510		3,457		9,084

Estimated monthly discharge of Susquehanna River at Binghamton, N. Y.,
1901–1904.

[Drainage area, 2,400 square miles.]

Month.	Discharge in second-feet.			Run-off.	
	Maximum	Minimum	Mean.	Second-feet per square mile.	Depth in inches.
1901.					
August	6,845	399	1,475	0.61	0.70
September	1,510	652	988	.41	.46
October	1,625	652	1,034	.43	.50
November	4,055	567	1,454	.61	.68
December	43,210	1,444	7,514	3.13	3.61
1902.					
January	8,730	1,070	3,177	1.32	1.52
February	9,650	1,004	2,058	.86	.89
March	60,300	6,705	19,701	8.21	9.48
April	10,440	1,855	5,285	2.20	2.45
May	2,665	1,070	1,672	.70	.81
June	14,600	1,334	2,373	.99	1.10
July	29,960	2,154	9,587	4.00	4.61
August	11,240	1,114	2,941	1.23	1.42
September	8,280	630	1,420	.59	.66

*Estimated monthly discharge of Susquehanna River at Binghamton, N. Y.,
1901–1904—Continued.*

Month.	Discharge in second-feet.			Run-off.	
	Maximum.	Minimum.	Mean.	Second-feet per square mile.	Depth in inches.
1902.					
October	19,000	1,740	4,197	1.75	2.02
November	7,125	1,740	2,784	1.14	1.27
December	27,260	2,246	7,461	3.11	3.59
The year	60,300	630	5,217	2.18	29.82
1903.					
January	25,820	3,755	9,360	3.90	4.50
February	23,660	2,905	9,248	3.85	4.01
March	35,580	6,845	17,275	7.19	8.29
April	13,480	1,510	5,344	2.23	2.49
May	1,466	420	821	.34	.39
June	9,340	252	2,680	1.12	1.25
July	7,265	850	1,914	.80	.92
August	27,260	1,290	4,413	1.84	2.12
September	13,160	894	2,654	1.11	1.24
October	48,900	894	10,108	4.21	4.85
November	14,280	1,901	3,890	1.62	1.81
December	9,495	1,400	3,556	1.48	1.71
The year	48,900	252	5,980	2.47	33.58
1904.					
January	18,895	2,361	5,794	2.41	2.78
February	32,012	3,385	10,530	4.39	4.73
March	47,110	3,730	14,010	5.84	6.73
April	21,500	4,930	10,650	4.44	4.95
May	8,430	1,532	3,088	1.29	1.49
June	6,165	718	1,769	.737	.822
July	3,313	567	1,027	.428	.493
August	6,985	1,136	2,396	.996	1.151
September	4,305	1,070	1,850	.770	.859
October	16,250	1,625	5,016	2.09	2.41
November	4,805	2,200	2,881	1.20	1.34
December	23,660	1,625	4,226	1.76	2.03
The year	47,110	567	5,270	2.20	29.78

CHENANGO RIVER AT BINGHAMTON, N. Y.

This station was established by R. E. Horton July 31, 1901. The gage is located on the upstream side of the first span from the right bank of Court Street Bridge, Binghamton. It is a boxed wire gage secured to the vertical supports of the hand railing. The bench mark is a circular chisel draft on the upstream corner of the bridge seat on the left abutment. Its elevation is 34.02 feet above gage datum. Court Street Bridge stands squarely across the stream, which has a nearly horizontal bed of gravel and small cobblestones, affording a smooth, uniform current for gaging. The channel is obstructed by three masonry piers supporting the four spans of the bridge, 79 feet clear width each, the bridge having a total length of 337 feet between abutments. The bridge is situated 2,500 feet above the confluence of Chenango and Susquehanna rivers. A small rift below the bridge cuts off backwater from the Susquehanna at ordinary stages of the rivers. For periods during freshets or at times when there is an abnormal rise on one stream, accompanied by a similar rise in the other stream, either the Chenango or Susquehanna River record at Binghamton may be affected by backwater, indicating a too great discharge. For freshets of considerable duration the flow of the two streams will be more nearly equalized. Gage readings on Chenango River, as well as those on Susquehanna River at Binghamton, are taken by E. F. Weeks. In estimating run-off of Chenango River the area directly tributary to storage reservoirs from which diversion is made to supply Erie Canal has been deducted from the total area naturally tributary to Chenango River.

In estimating the run-off of Chenango River the area directly tributary to storage reservoirs, from which diversion is made to supply Erie Canal, has been deducted from the total area naturally tributary to Chenango River, as follows:

	Square miles.
Natural tributary area [a]	1,580
Diversion area, 6 reservoirs at head of Chenango River, whose overflow is turned into Erie Canal through Oriskany Creek	30
Diversion area, De Ruyter reservoir, at head of Tioughnioga River; outflow turned into Erie Canal through Limestone Creek	18
	48
Net area used for Chenango basin	1,532

Above estimate of diversion area is approximate. No allowance for direct inflow to feeder channels from additional areas nor fo.' waste into original stream. Gross area, from which more or less run-off is diverted, is about 105 square miles.

[a] From Bien's Atlas of New York State. Areas tributary to reservoirs are from New York Barge Canal Report, 1900.

Discharge measurements of Chenango River at Binghamton, N. Y., 1901-1904.

Date.	Hydrographer.	Area.	Mean velocity.	Gage height.	Discharge.
1901.		*Square feet.*	*Feet per second.*	*Feet.*	*Second-feet.*
July 2	E. C. Murphy	689	1.28	5.64	848
July 8	do	764	1.46	5.78	1,119
July 9	do	617	1.58	5.71	942
July 29	do	602	.61	5.21	405
Do	do	469	.90	5.21	425
August 19	do	547	1.04	5.48	566
Do	do	681	.85	5.49	577
October 19	do	646	1.58	5.81	987
Do	do	775	1.20	5.82	927
1902.					
March 27	E. C. Murphy	1,384	3.04	8.15	4,201
March 28	do	1,489	2.94	8.21	4,377
March 29	do	1,590	3.27	8.75	5,205
June 6 *a*	R. E. Horton	956	2.52	7.00	2,407
July 1	E. C. Murphy	1,534	3,14	8.49	4,815
July 3	do	1,155	2.33	7.24	2,688
July 15	do	995	2.13	6.64	2,098
August 3	do	1,775	3.12	9.16	5,548
August 14	do	877	1.83	6.32	1,605
August 15	do	841	1.48	6.20	1,341
September 3	C. C. Covert	675	.80	5.56	546
1903.					
April 6	E. C. Murphy	1,359	2.71	7.72	3,695
May 15	do	646	.83	5.49	538
June 13	C. C. Covert	1,490	1.93	8.06	2,877
August 19	J. C. Hoyt	621	- .97	5.62	601
August 21	C. C. Covert	1,006	2.23	6.72	2,243
October 1	H. H. Halsey	650	1.09	5.51	709
October 10	C. C. Covert	5,411	5.23	19.81	28,300
1904.					
March 8	C. C. Covert	3,702	3.45	*b* 14.90	9,104
April 8	R. E. Horton	2,459	5.42	10.86	11,632
July 12	C. C. Covert	595	.87	5.42	516
September 10	do	467	1.15	5.55	539
November 22	H. R. Beebe	1,022	2.45	6.86	2,505

a Rough measurement.   *b* Backwater, caused by ice jam.

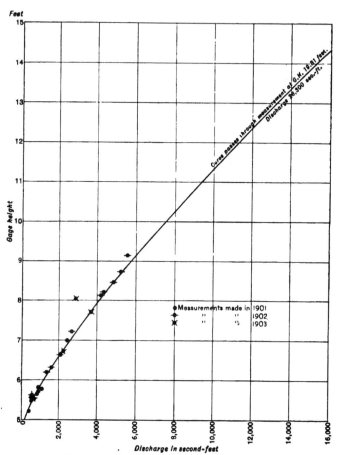

Fig. 3.—Rating curve for Chenango River at Binghamton. N. Y.

Mean daily gage height, in feet, of Chenango River at Binghamton, N. Y., 1901-1904.

Day.	Jan.	Feb.	Mar.	Apr.	May.	June.	July	Aug.	Sept.	Oct.	Nov.	Dec.
1901.												
1								5.18	5.58	5.70	5.46	6.12
2								5.12	5.75	5.50	5.30	6.88
3								5.10	5.58	5.51	5.25	6.60
4								5.10	5.50	5.68	5.24	6.52
5								5.05	5.42	5.54	5.26	6.19
6								5.20	5.24	5.50	5.25	5.95
7								5.05	5.22	5.46	5.25	5.90
8								5.10	5.20	5.47	5.22	6.02
9								5.20	5.18	5.40	5.23	6.08
10								5.20	5.15	5.37	5.21	8.14
11								5.22	5.15	5.34	5.13	10.00
12								5.20	5.18	5.33	5.26	8.82
13								5.18	5.30	5.42	6.85	
14								5.12	5.48	6.47	6.46	8.48
15								5.15	5.35	6.40	6.19	19.54
16								5.85	5.42	6.08	6.11	17.67
17								5.80	5.55	5.89	6.10	12.61
18								5.60	5.62	5.85	6.08	9.41
19								5.48	5.55	5.80	6.06	8.11
20								5.40	5.45	5.82	6.00	7.80
21								5.55	5.45	5.78	5.95	6.84
22								5.58	5.80	5.75	5.95	6.66
23								5.48	5.22	5.70	5.94	7.26
24								6.70	5.20	5.66	6.71	8.18
25								6.20	5.25	5.57	7.78	7.41
26								5.65	5.24	5.48	7.18	6.88
27								5.58	5.25	5.45	6.63	6.83
28								5.30		5.89	6.05	6.50
29								5.25	5.15	5.40	6.20	6.52
30								5.20	5.88	5.35	6.92	7.20
31								5.20		5.39		7.96
1902.												
1	6.62	6.31	18.75	8.65	6.54	6.25	8.58	8.46	5.58	7.28	8.04	6.54
2	6.64	6.25	22.75	8.61	6.82	6.13	7.88	9.46	5.54	7.26	7.56	6.48
3	6.74	6.13	21.65	8.45	6.22	6.00	7.30	8.47	5.56	6.68	7.26	6.68
4	6.91	6.34	17.35	8.10	6.22	6.27	7.43	7.82	5.48	6.24	6.98	7.24
5	6.64	6.20	12.80	7.82	6.22	7.00	7.13	7.82	5.46	6.04	6.84	7.14
6	6.61	6.19	9.94		6.12	6.63	7.46	7.00	5.44	6.24	6.74	6.74
7	6.52	6.16	9.25	7.60	6.12	6.35	8.20	7.02	5.48	6.56	6.71	6.61
8	6.30	6.20	9.02	7.58	6.12	6.35	8.00	6.87	5.46	6.44	7.58	6.51
9	4.22	6.21	8.68	8.12	6.12	6.37	7.80	6.80	5.48	6.46	6.44	6.26
10	6.12	6.08	9.45	8.50	6.00	6.35	7.88	6.57	5.89	6.31	6.34	6.18
11	6.14	6.10	9.28	8.98	5.97	6.20	9.23	6.00	6.08	6.14	6.28	6.56
12	6.02	5.98	11.60	8.78	5.92	6.87	8.40	6.77	5.81	6.16	6.24	6.54
13	5.87	5.90	15.08	8.48	5.87	6.30	7.40	6.72	5.86	6.16	6.48	6.24
14	5.88	5.84	15.78	8.22	5.82	6.35	6.96	8.40	5.61	6.36	6.41	6.01
15	5.89	5.77	14.18	7.80	5.77	6.25	6.56	6.24	5.56	6.64	6.31	6.11
16	5.91	5.86	11.98	7.42	5.72	6.23	6.56	6.22	5.48	6.31	6.16	6.04
17	5.88	5.76	15.86	7.18	5.74	6.25	6.50	5.41	5.41	6.11	6.11	10.58
18	5.76	5.78	15.72	7.05	5.72	6.15	6.48	6.04	5.86	6.01	6.08	10.94
19	5.78	5.74	18.10	6.90	5.72	6.05	6.60	6.00	5.36	5.96	6.08	9.91
20	5.78	5.71	10.48	6.80	5.77	6.05	11.36	6.00	5.31	6.81	6.11	9.08
21	5.66	5.74	9.40	6.72	6.05	6.03	15.02	5.71	5.28	6.86	6.06	8.51
22	6.02	5.87	9.20	6.64	5.83	6.28	15.02	6.00	5.26	6.51	6.08	12.84
23	8.24	5.68	9.82	6.52	5.83	6.33	13.52	5.91	5.31	6.34	6.08	14.03
24	8.66	5.66	9.38	6.40	5.77	6.16	12.34	5.88	5.28	6.31	6.16	11.28
25	7.02	5.68	8.95	6.32	8.00	6.00	11.47	5.84	5.31	6.31	6.21	9.31
26	6.86	5.73	8.48	6.20	6.35	6.06	9.62	5.81	5.54	6.21	6.28	8.71
27	6.86	6.08	8.15	6.20	6.63	6.18	8.62	5.71	5.76	6.78	6.78	8.24
28	7.28	8.92	8.15	6.14	6.35	6.16	11.62	5.78	5.66	9.30	7.06	7.64
29	7.39		8.95	6.14	6.25	6.79	9.70	5.74	7.64	11.71	6.78	7.24
30	6.85		9.28	6.30	6.23	10.56	8.02	5.74	6.44	10.41	6.61	7.28
31	6.40		8.98		6.20		9.30	5.66		8.96		6.98

Mean daily gage height, in feet, of Chenango River at Binghamton, N. Y., 1901-1904—Continued.

Day.	Jan.	Feb.	Mar.	Apr.	May.	June.	July.	Aug	Sept.	Oct.	Nov.	Dec.
1903.												
1	6.76	12.04	16.49	10.14	5.81	5.06	6.59	5.57	10.01	5.51	6.88	6.22
2	6.64	10.61	14.84	9.14	5.76	5.11	6.42	5.52	8.66	5.56	6.72	6.04
3	7.21	10.56	11.18	8.56	5.74	5.11	6.12	5.45	7.83	5.81	6.68	6.20
4	8.78	11.94	9.71	8.51	5.71	5.06	5.97	5.45	7.29	5.83	6.55	6.20
5	8.64	13.38	9.06	8.24	5.68	5.06	5.89	6.57	6.96	5.83	6.50	6.15
6	7.96	11.46	10.14	7.78	5.64	5.06	5.82	6.49	6.66	7.33	6.92	6.12
7	7.34	9.81	10.01	7.66	5.61	4.94	5.85	8.47	6.46	6.71	6.82	6.08
8	7.24	8.54	10.08	8.76	5.58	5.24	5.77	6.32	6.54	6.66	6.58	6.05
9	6.98	8.16	14.68	9.11	5.56	5.11	5.67	5.99	6.19	11.94	6.48	6.05
10	9.56	7.84	14.28	8.54	5.56	5.06	5.57	6.22	6.13	19.06	6.40	5.90
11	9.84	7.76	15.26	8.18	5.56	5.11	5.47	6.52	6.36	19.91	6.40	5.90
12	9.36	9.01	15.24	7.81	5.51	5.06	5.47	6.27	6.49	15.48	6.32	6.00
13	9.26	10.24	13.16	7.51	5.51	7.97	5.42	6.27	6.19	11.42	6.25	5.90
14	9.08	9.28	11.31	7.26	5.48	6.62	5.42	6.07	6.03	9.45	6.22	5.90
15	9.14	8.21	10.26	7.48	5.46	6.29	5.42	5.95	5.93	8.58	6.15	6.15
16	9.14	7.84	9.56	7.41	5.46	6.12	5.49	5.79	5.89	7.95	6.15	6.10
17	8.96	7.24	9.08	7.21	5.46	5.92	5.57	5.69	5.86	7.78	9.08	6.10
18	8.54	6.44	9.14	6.98	5.41	5.77	5.69	5.89	6.21	11.55	10.10	6.10
19	7.86	6.68	8.78	6.76	5.38	5.75	5.72	5.59	6.23	11.72	8.50	5.92
20	7.38	6.71	8.36	6.56	5.38	5.82	5.89	7.07	5.99	10.20	7.42	5.98
21	7.74	6.88	8.16	6.44	5.38	5.62	5.79	6.80	5.89	9.08	6.92	7.35
22	9.84	6.81	9.48	6.36	5.34	8.67	5.47	6.29	5.81	8.40	6.75	8.35
23	9.86	6.91	11.38	6.31	5.26	8.19	6.67	5.99	5.71	7.88	6.72	8.10
24	8.71	6.76	15.73	6.24	5.26	8.99	7.15	5.79	5.69	7.72	6.85	7.48
25	7.98	6.68	14.96	6.11	5.26	8.32	6.09	5.79	5.66	7.55	6.78	7.35
26	7.96	6.64	12.56	6.11	5.21	7.87	5.77	7.63	5.61	7.25	6.40	7.18
27	7.66	6.56	10.54	6.04	5.21	7.27	5.65	7.59	5.56	7.15	6.32	6.92
28	7.71	9.96	9.54	5.96	5.21	6.77	5.57	6.89	5.61	7.10	6.35	6.48
29	8.74		9.16	5.88	5.24	6.69	5.57	14.61	5.59	7.20	6.18	6.48
30	13.31		8.61	5.86	5.21	6.89	5.65	14.36	5.59	7.18	6.20	6.50
31	13.74		9.78		5.16		5.59	12.11		7.10		6.45
1904.												
1	6.42	7.32	7.60	11.30	8.72	7.14	5.59	6.10	5.70	7.69	6.22	6.15
2	6.55	7.20	7.40	12.90	8.19	6.79	5.73	7.08	5.72	6.85	6.20	5.95
3	6.42	7.18	7.88	a11.70	7.79	6.56	5.63	7.35	5.70	6.41	6.12	5.80
4	6.45	7.20	10.38	10.50	7.42	6.42	5.61	6.88	5.72	6.21	6.07	5.75
5	6.68	7.05	11.22	9.45	7.19	6.64	5.51	6.32	5.65	6.11	6.04	5.65
6	6.82	6.75	11.08	10.08	6.99	6.59	5.49	6.72	5.65	6.01	6.17	5.72
7	6.64	8.12	10.95	10.30	6.82	6.34	5.51	6.65	5.60	6.01	6.23	5.80
8	6.60	13.92	14.78	10.88	6.67	6.25	5.58	6.28	5.52	5.96	6.16	5.75
9	6.58	15.30	16.90	11.01	6.55	6.88	5.44	6.10	5.50	5.88	6.11	5.62
10	6.48	14.28	15.65	12.97	6.44	7.98	5.40	6.02	5.50	5.80	6.11	5.55
11	6.38	12.05	18.70	12.42	6.34	6.93	5.30	5.98	5.40	6.05	6.06	5.54
12	6.30	10.60	11.40	10.84	6.26	6.48	5.50	5.92	5.31	7.60	6.06	5.62
13	6.25	9.50	10.30	9.91	6.18	6.18	5.72	5.85	5.31	8.95	6.01	5.70
14	6.20	8.70	9.52	9.29	6.14	6.15	5.35	5.75	5.34	7.85	6.02	5.55
15	6.15	8.20	8.75	8.74	6.26	6.08	5.40	5.72	6.09	7.03	5.95	5.65
16	6.15	9.38	8.20	8.49	7.36	6.53	5.60	5.65	5.91	6.40	6.08	5.65
17	6.12	10.18	7.65	8.39	7.36	6.11	5.65	5.70	5.67	6.42	6.10	5.65
18	6.15	10.05	7.42	8.39	6.84	5.94	6.68	5.62	5.54	6.26	5.95	5.65
19	6.30	9.52	7.22	8.40	6.64	5.84	6.55	5.55	5.40	6.16	5.92	5.60
20	6.45	8.94	7.48	8.23	7.30	5.84	6.08	5.78	5.36	6.12	5.90	5.60
21	6.30	8.02	7.88	7.98	7.10	5.84	5.84	6.82	5.46	5.79	6.08	5.60
22	6.30	8.35	7.78	7.98	6.70	5.82	5.82	6.50	5.68	10.79	6.80	5.60
23	10.36	8.62	11.30	8.00	6.47	5.72	5.65	8.25	5.66	9.76	6.68	5.65
24	11.18	9.25	15.15	7.93	6.73	5.60	6.10	7.55	5.56	8.15	6.50	5.92
25	11.60	9.38	15.90	8.13	6.47	5.54	6.10	6.65	6.70	7.38	6.38	6.50
26	10.20	8.70	19.82	8.43	6.40	5.54	5.92	6.32	6.42	7.41	6.32	6.25
27	9.35	8.25	19.90	8.13	6.50	5.47	6.20	6.29	6.27	7.23	6.18	6.72
28	8.65	7.95	16.15	10.13	6.50	5.46	6.22	a6.05	6.15	6.92	5.98	12.75
29	8.10	7.88	12.08	10.18	6.40	5.46	6.65	5.90	5.95	6.68	5.80	13.28
30	7.88		10.62	9.39	6.26	5.49	6.90	5.80	6.92	6.53	6.20	10.15
31	7.60		10.58		6.76		6.32	5.72		6.82		5.25

a Interpolated.

Rating table for Chenango River at Binghamton, N. Y., for 1901 to 1904, inclusive.

Gage height.	Discharge.	Gage height.	Discharge.	Gage height.	Discharge.	Gage height.	Discharge.
Feet.	*Second-feet.*	*Feet.*	*Second-feet*	*Feet.*	*Second-feet.*	*Feet.*	*Second-feet.*
5.0	160	7.4	3,200	10.6	8,590	15.4	18,240
5.1	256	7.5	3,350	10.8	8,970	15.6	18,660
5.2	352	7.6	3,500	11.0	9,350	15.8	19,080
5.3	450	7.7	3,650	11.2	9,730	16.0	19,500
5.4	550	7.8	3,800	11.4	10,110	16.2	19,940
5.5	650	7.9	3,950	11.6	10,490	16.4	20,380
5.6	760	8.0	4,100	11.8	10,870	16.6	20,820
5.7	875	8.1	4,250	12.0	11,250	16.8	21,260
5.8	995	8.2	4,400	12.2	11,650	17.0	21,700
5.9	1,115	8.3	4,550	12.4	12,050	17.2	22,140
6.0	1,235	8.4	4,700	12.6	12,450	17.4	22,580
6.1	1,365	8.5	4,850	12.8	12,850	17.6	23,030
6.2	1,495	8.6	5,020	13.0	13,250	17.8	23,490
6.3	1,625	8.7	5,190	13.2	13,650	18.0	23,950
6.4	1,755	8.8	5,360	13.4	14,050	18.2	24,410
6.5	1,885	8.9	5,530	13.6	14,460	18.4	24,870
6.6	2,025	9.0	5,700	13.8	14,880	18.6	25,340
6.7	2,165	9.2	6,060	14.0	15,300	18.8	25,820
6.8	2,305	9.4	6,420	14.2	15,720	19.0	26,300
6.9	2,450	9.6	6,780	14.4	16,140	19.2	26,780
7.0	2,600	9.8	7,140	14.6	16,560	19.4	27,260
7.1	2,750	10.0	7,500	14.8	16,980	19.6	27,760
7.2	2,900	10.2	7,860	15.0	17,400	19.8	28,280
7.3	3,050	10.4	8,220	15.2	17,820		

Remarks: Tangent at 19.5 feet. Differences above this point 260 per tenth.

Mean daily discharge, in second-feet, of Chenango River at Binghamton, N. Y., 1901-1904.

Day.	Jan.	Feb.	Mar.	Apr	May.	June.	July.	Aug.	Sept.	Oct.	Nov.	Dec.
1901.												
1								333	738	875	610	1,391
2								275	935	650	450	1,664
3								256	788	661	400	2,025
4								256	650	851	430	1,913
5								208	570	694	410	1,482
6								352	430	650	400	1,175
7								208	371	610	400	1,115
8								256	352	620	371	1,261
9								352	833	560	381	1,359
10								352	304	520	361	4,325
11								371	304	490	285	7,500
12								352	333	480	410	5,380
13								333	450	570	2,375	5,105
14								275	630	1,846	1,833	4,850
15								304	500	1,755	1,482	27,630
16								1,690	570	1,839	1,378	23,145
17								1,115	705	1,102	1,366	12,450
18								760	782	1,055	1,313	6,420
19								630	705	995	1,313	4,250
20								550	700	1,019	1,235	3,200
21								705	700	971	1,175	2,361
22								738	450	985	1,175	2,109
23								630	371	875	1,163	2,975
24								2,165	352	827	2,180	4,400
25								1,495	400	727	3,800	3,200
26								815	381	630	2,900	2,420
27								530	400	600	2,067	2,347
28								450	352	540	1,300	1,885
29								400	334	550	1,495	1,913
30								352	1,001	500	1,651	2,900
31								352		540		3,125
1902.												
1	2,053	1,638	25,700	5,105	1,941	1,560	5,020	4,775	778	3,050	4,175	1,941
2	2,061	1,580	35,950	5,020	1,651	1,404	3,950	6,510	804	2,975	3,425	1,859
3	2,221	1,404	33,090	4,775	1,521	1,235	3,200	4,775	716	2,137	2,975	2,137
4	2,465	1,677	22,470	4,250	1,521	1,586	3,275	3,800	683	1,599	2,570	2,975
5	2,081	1,495	12,850	3,800	1,521	2,600	2,825	3,050	610	1,287	2,361	2,825
6	2,089	1,482	7,500	3,650	1,391	2,067	3,275	2,600	500	1,599	2,221	2,221
7	1,913	1,443	6,150	3,500	1,391	1,690	4,400	2,630	610	1,969	2,179	2,825
8	1,625	1,495	5,700	3,500	1,391	1,690	4,100	2,405	610	1,807	3,500	1,809
9	1,521	1,508	5,190	4,250	1,391	1,716	3,800	2,305	620	1,833	1,807	1,573
10	1,391	1,339	6,510	4,850	1,235	1,690	3,950	1,983	1,067	1,638	1,677	1,469
11	1,417	1,365	6,240	5,700	1,199	1,495	6,150	1,913	1,323	1,417	1,599	1,989
12	1,261	1,211	10,490	5,380	1,189	1,716	4,700	2,263	1,067	1,448	1,547	1,941
13	1,079	1,115	17,610	4,850	1,079	1,625	3,200	2,193	827	1,448	1,859	1,547
14	1,091	1,043	19,080	4,400	1,019	1,690	2,540	1,755	771	1,703	1,768	1,248
15	1,108	959	15,720	3,800	959	1,560	2,137	1,547	716	2,081	1,638	1,378
16	1,127	1,087	11,250	3,200	899	1,534	1,969	1,521	610	1,638	1,443	1,287
17	1,091	947	19,185	2,900	923	1,560	1,969	1,365	560	1,378	1,378	8,495
18	947	971	18,870	2,675	899	1,430	1,859	1,287	510	1,248	1,339	9,255
19	971	923	13,450	2,450	782	1,300	2,305	1,235	510	1,187	1,313	7,320
20	971	887	8,400	2,305	959	1,300	10,015	1,235	460	2,319	1,378	5,840
21	827	804	6,420	2,198	1,300	1,274	17,400	887	431	2,300	1,313	4,850
22	1,261	839	6,060	2,081	1,151	1,599	17,400	1,235	410	1,809	1,339	12,950
23	4,475	851	6,240	1,913	1,031	1,664	14,250	1,127	472	1,677	1,339	15,405
24	5,105	827	6,420	1,756	959	1,443	11,950	1,091	430	1,638	1,443	9,920
25	3,500	851	5,615	1,651	1,235	1,235	10,205	1,043	431	1,638	1,508	6,240
26	2,390	911	4,850	1,495	1,600	1,313	6,790	1,007	634	1,508	1,599	5,190
27	2,390	1,339	4,325	1,495	2,067	1,469	5,020	887	947	1,443	2,277	4,475
28	3,050	5,530	4,325	1,417	1,690	1,443	10,490	971	827	6,240	2,675	3,575
29	3,200		5,615	1,417	1,590	1,590	6,980	923	923	10,680	2,277	2,975
30	2,375		6,240	1,625	1,534	8,495	5,020	923	1,807	8,220	2,089	3,050
31	1,755		5,700		1,495		6,240	827		5,615		2,570

Mean daily discharge, in second-feet, of Chenango River at Binghamton, N. Y., 1901-1904—Continued.

Days.	Jan.	Feb.	Mar.	Apr.	May.	June.	July.	Aug.	Sept.	Oct.	Nov.	Dec.
1903.												
1	2,249	11,350	20,600	7,770	1,007	217	2,011	727	7,500	661	2,420	1,521
2	2,081	8,590	16,085	5,970	947	285	1,781	672	5,105	716	2,198	1,389
3	2,900	8,495	9,730	4,985	923	285	1,391	600	3,875	1,007	2,137	1,495
4	5,360	11,155	6,960	4,850	887	217	1,199	600	3,050	1,081	1,955	1,495
5	5,105	14,050	5,790	4,475	851	217	1,103	1,983	2,540	1,081	1,895	1,430
6	4,025	10,205	7,770	3,800	804	217	1,019	1,872	2,109	3,125	2,480	1,391
7	3,125	7,140	7,500	3,575	771	108	1,055	1,846	1,883	2,179	2,333	1,339
8	2,975	4,995	7,680	5,275	738	390	959	1,651	1,638	2,109	1,997	1,300
9	2,570	4,325	16,770	5,880	716	285	839	1,223	1,482	11,155	1,859	1,300
10	6,690	3,875	15,960	4,985	716	217	727	1,521	1,404	26,420	1,755	1,115
11	6,330	3,725	17,925	4,400	716	285	620	1,913	1,703	24,540	1,755	1,115
12	6,330	5,700	17,925	3,800	661	847	620	1,546	1,872	18,450	1,651	1,235
13	6,150	7,950	13,550	3,350	661	4,025	570	1,546	1,482	10,110	1,560	1,625
14	5,880	6,240	9,920	2,975	630	2,053	570	1,326	1,274	6,510	1,521	1,690
15	5,970	4,400	7,950	3,350	610	1,612	570	1,175	1,151	5,080	1,430	1,430
16	5,970	3,875	6,690	3,200	610	1,391	640	983	1,103	4,025	1,430	1,385
17	5,615	2,975	5,880	2,900	610	1,139	727	863	1,067	3,800	5,790	1,385
18	4,985	1,807	5,970	2,570	560	899	640	863	1,508	10,395	7,680	1,300
19	3,875	2,137	5,360	2,249	530	935	899	749	1,534	10,680	4,850	1,139
20	3,200	2,179	4,625	1,969	510	1,103	1,103	2,675	1,223	7,860	3,200	1,211
21	3,725	2,420	4,325	1,807	510	2,053	983	2,390	1,103	5,880	2,480	3,125
22	7,230	2,319	6,610	1,703	480	5,105	1,079	1,612	1,007	4,700	2,235	4,625
23	7,230	2,465	10,110	1,638	410	4,400	2,123	1,223	887	3,950	2,193	4,250
24	5,190	2,249	18,975	1,547	410	5,700	2,825	983	863	3,650	2,375	3,350
25	4,100	2,137	17,295	1,378	410	4,550	1,352	983	827	3,425	2,277	3,125
26	4,025	2,081	12,350	1,378	382	3,875	959	8,575	772	2,975	1,755	2,900
27	3,575	1,969	8,495	1,287	382	2,975	815	3,500	716	2,825	1,651	2,480
28	3,650	7,410	6,690	1,187	382	2,263	727	2,435	772	2,750	1,729	1,859
29	5,275		5,970	1,091	390	2,151	727	16,560	749	2,900	1,469	1,859
30	13,850		5,080	1,067	382	2,435	815	16,035	749	2,900	1,495	1,885
31	14,775		7,140		314		749	11,450		2,750		1,820
1904.												
1	1,781	3,050	3,500	9,920	5,190	2,825	749	1,365	875	3,650	1,521	1,430
2	1,955	2,900	3,200	13,051	4,400	2,291	911	2,750	899	2,375	1,495	1,175
3	1,781	2,900	3,950	10,680	3,800	1,969	793	3,125	875	1,768	1,391	995
4	1,880	2,900	5,790	8,400	3,200	1,781	771	2,420	899	1,508	1,326	935
5	2,137	2,675	9,000	6,510	2,900	2,081	661	1,651	815	1,378	1,287	818
6	2,333	2,235	8,500	7,680	2,585	2,011	640	2,193	815	1,248	1,456	899
7	2,137	4,250	8,300	8,040	2,333	1,677	661	2,095	760	1,248	1,534	995
8	2,028	15,060	8,985	9,180	2,123	1,560	738	1,599	672	1,187	1,443	935
9	1,997	18,060	11,400	9,350	1,955	2,420	630	1,365	650	1,091	1,378	783
10	1,859	15,930	10,700	13,150	1,807	4,100	550	1,261	650	995	1,378	705
11	1,729	11,350	8,950	12,050	1,677	2,495	450	1,211	550	1,300	1,313	738
12	1,625	8,590	6,670	9,065	1,573	1,859	650	1,139	460	3,500	1,313	783
13	1,560	6,610	5,700	7,320	1,469	1,560	705	1,055	460	5,615	1,248	875
14	1,495	5,190	4,950	6,240	1,417	1,430	500	935	490	3,875	1,261	705
15	1,430	4,400	4,170	5,275	1,573	1,339	550	899	1,352	2,675	1,175	818
16	1,430	6,420	3,600	4,850	3,125	1,927	760	815	1,127	1,755	1,859	818
17	1,391	7,980	3,080	4,700	2,361	1,378	815	875	839	1,781	1,985	818
18	1,430	7,590	2,800	4,700	2,361	1,163	2,137	782	694	1,573	1,175	818
19	1,625	6,610	2,680	4,700	2,081	1,043	1,955	705	550	1,443	1,139	760
20	1,880	5,190	3,015	4,475	3,050	1,043	1,330	971	510	1,391	1,115	760
21	1,625	5,080	3,555	4,100	2,750	1,043	1,091	2,333	610	983	1,859	760
22	1,625	4,625	3,350	4,100	2,165	1,019	1,019	1,885	851	8,970	2,305	760
23	8,190	5,080	9,920	4,100	1,846	899	815	4,475	827	7,050	2,187	818
24	9,730	6,330	17,715	4,025	2,207	760	1,385	3,425	716	4,325	1,885	1,139
25	10,490	6,420	19,290	4,325	1,846	694	1,261	2,096	2,165	3,200	1,729	1,885
26	7,860	5,190	28,280	4,775	1,755	694	1,139	1,651	1,781	3,200	1,651	1,560
27	6,330	4,475	28,540	4,325	1,895	620	1,495	1,495	1,612	2,975	1,469	2,193
28	5,105	4,125	19,830	7,770	1,895	610	1,521	1,300	1,430	2,480	1,211	12,750
29	4,250	3,950	11,450	7,860	1,775	610	2,095	1,115	1,175	2,137	995	13,810
30	3,950		8,590	6,420	1,573	640	2,450	995	2,480	1,927	1,495	7,770
31	3,500		8,590		2,249		1,651	899		1,651		401

The daily discharge during January, February, and March is only approximate, owing to the ice conditions. From March 4 to 22, 1904, the discharge was estimated from the measurement of March 8, which was approximately 50 per cent of normal conditions. This was due to an ice gorge.

Estimated monthly discharge of Chenango River at Binghamton, N. Y.,
1901–1904.

[Drainage area 1,530 square miles.]

Month.	Discharge in second-feet.			Run-off.			Rainfall in inches.
	Maximum.	Minimum.	Mean.	Second-feet per square mile.	Depth in inches.	Per cent of rainfall.	
1901.							
August	2,165	208	576	0.38	0.44	9	4.50
September	1,091	304	524	.34	.38	12	3.12
October	1,846	480	807	.58	.61	31	1.88
November	3,800	285	1,204	.78	.87	31	2.70
December	27,630	1,115	4,750	3.10	3.57	65	5.34
1902.							
January	5,105	827	1,960	1.28	1.48	108	1.83
February	5,530	804	1,339	.87	.91	29	2.99
March	35,950	4,325	11,717	7.64	8.81	241	3.56
April	5,700	1,417	3,246	2.12	2.37	136	1.68
May	2,067	782	1,307	.85	.98	36	2.64
June	8,495	1,235	1,820	1.19	1.38	22	5.87
July	17,400	1,859	6,011	3.92	4.52	54	8.07
August	6,510	827	2,002	1.30	1.50	48	3.07
September	3,575	410	809	.53	.59	17	3.28
October	10,680	1,187	2,539	1.66	1.91	47	3.92
November	4,175	1,313	1,999	1.30	1.43	117	1.21
December	15,405	1,248	4,273	2.79	3.22	71	4.36
The year	35,950	410	3,252	2.12	29.07	67	41.97
1903.							
January	14,775	2,081	5,289	3.44	3.99	145	2.67
February	14,050	1,807	5,291	3.44	3.58	142	2.45
March	20,600	4,325	10,114	6.59	7.40	147	5.03
April	7,770	1,067	3,210	2.09	2.33	140	1.61
May	1,007	314	608	.40	.46	142	.31
June	5,700	103	1,737	1.13	1.26	19	6.62
July	2,825	570	1,039	.68	.78	20	3.79
August	16,560	600	2,812	1.83	2.11	31	6.72
September	7,500	716	1,763	1.15	1.28	81	1.55
October	28,540	661	6,243	4.07	4.69	60	7.64
November	7,680	1,430	2,385	1.55	1.73	79	2.12
December	4,625	1,115	1,886	1.23	1.42	55	2.50
The year	28,540	103	3,532	2.30	31.21	71	43.00

*Estimated monthly discharge of Chenango River at Binghamton, N. Y.,
1901-1904—Continued.*

Month.	Discharge in second-feet.			Run-off.	
	Maximum.	Minimum.	Mean.	Second-feet per square mile.	Depth in inches.
1904.					
January	10,490	1,391	3,160	2.06	2.37
February	18,030	2,235	6,390	4.17	4.50
March	28,540	2,680	8,966	5.84	6.73
April	13,150	4,025	7,037	4.59	5.12
May	5,190	1,417	2,876	1.55	1.79
June	4,100	610	1,518	.990	1.105
July	2,450	450	1,060	.691	.807
August	4,475	705	1,641	1.07	1.23
September	2,480	460	953	.621	.693
October	8,970	983	2,587	1.69	1.95
November	2,305	995	1,429	.932	1.04
December	13,810	401	1,981	1.29	1.49
The year	28,540	401	3,258	2.12	28.82

SUSQUEHANNA RIVER AT WILKESBARRE, PA.

The Wilkesbarre station was established by E. G. Paul on March
30, 1899.

The standard chain gage is located on the upstream side of the
Market Street Bridge. The length of the chain from the end of the
weight to the marker is 40.83 feet. The gage is read once each day
by W. S. Bennett, the bridge keeper. When this gage was established, there was found to be a gage painted on the bridge pier,
being a portion of one established by the Weather Bureau. The
lower part of this gage, erected in January, 1898, originally consisted
of heavy cast-brass plates graduated to feet and tenths. The gage
plates were made in 4-foot sections and bolted to the stone bridge
pier. The two lower sections of the brass plates had been torn away
by ice, so that there was no graduation below the 8-foot mark, but
readings were made by the figures painted on the stone pier. The
zero of this old gage is at the base of the dressed-stone portion of the
pier and is reported to be 535 feet above sea level. During low
stages of the river the water recedes from the pier, rendering it
impracticable to read the gage. So far as could be ascertained, this

has not been connected with the city datum. On account of the low water, which in 1897 had gone below the city datum, it was decided to put the zero of the new gage 4 feet below the zero of the old Weather Bureau gage, so as to obviate minus readings. In order, therefore, to compare with former records, it is necessary to add 4 feet to the old figures. The danger mark of this Weather Bureau gage is at 14 feet, or 18 feet of new gage, as at this elevation the west bank of the river is under water in places. River reports from this locality were furnished as early as 1888. During low water measurements were made by wading at a better cross section, at Retreat, 10 miles below Wilkesbarre. The elevation of the Market Street toll bridge above the river bed requires 65 feet of cable to sound across the section.

Observations of fluctuations of Susquehanna River are made by the Weather Bureau, above Wilkesbarre, at Towanda, Pa., where the drainage area is estimated to be 8,000 square miles. The river gage, made of iron 1 foot wide and one-half inch thick, is on the east side of the road bridge over Susquehanna River, and is securely bolted to the masonry of the pier. The graduation is from 0 to 25 feet. The highest water was 29 feet in March, 1869, and the lowest, −0.1 foot, in October, 1895. The danger line is at 16 feet. The elevation of the zero is 633.7 feet.

Discharge measurements are made from the downstream side of the bridge, which has a total span of 700 feet between abutments. The initial point for soundings is the end of the iron handrail on the left bank, downstream side. The channel is straight for about one-fourth mile above and below the station. There is a bar across the river about one-half mile above the station, and another at about the same distance below, with deep water between these two points. This makes a sluggish current at low stages. The right bank is low and overflows at a gage height of about 20 feet. The left bank is above ordinary floods. The bed of the stream is composed of sand and gravel and is somewhat shifting. There is but one channel, broken by 3 bridge piers. There are a few willows growing under the right span. The bench mark is the extreme west end of the stone doorsill of the north entrance to the Coal Exchange Building. Its elevation is 32.99 feet above gage datum.

Discharge measurements of Susquehanna River at Wilkesbarre, Pa., 1899–1904.

Date.	Hydrographer.	Gage height.	Area of section.	Mean velocity.	Discharge.
1899.		*Feet.*	*Sq. ft.*	*Ft. per sec.*	*Sec.-ft.*
Mar. 30	E. G. Paul	9.00	6,846	3.62	24,800
June 6	.....do	4.30	3,064	1.20	3,668
July 26*a*	.....do	2.80	1,223	1.57	1,924
July 27	.....do	2.80	1,508	.90	1,357
Sept. 17	.....do	2.30	2,193	.38	851
Sept. 18*a*	...do	2.30	1,115	.98	1,096
Oct. 16	.....do	2.35	1,054	1.06	1,114
1900.					
May 20	E. G. Paul	5.60	3,599	1.88	6,772
Sept. 26*a*	.....do	2.20	1,023	.93	961
1901.					
Aug. 20	E. G. Paul	3.10	3,154	.69	2,170
1902.					
Sept. 20	E. G. Paul	3.10	3,154	.69	2,170
1903.					
Mar. 4	E. C. Murphy	13.50	9,996	4.61	46,112
Apr. 8	.....do	8.86	6,920	3.37	23,247
Aug. 4	John C. Hoyt	4.00	3,489	1.35	4,718
Oct. 10	W. C. Sawyer	19.00	13,168	6.57	86,500
1904.					
July 20	N. C. Grover	4.05	3,864	1.13	4,382
July 21*b*	.....do	4.20	4,077	1.15	4,680
Sept. 15	John C. Hoyt	3.70	3,670	.96	3,540
Oct. 1	...do	4.75	4,220	1.44	6,090
Nov. 5	H. D. Comstock	4.61	4,218	1.47	6,189
Nov. 7	...do	4.49	4,057	1.39	5,660

a Measured at Retreat.　　　　　*b* Measured at Pittston.

Mean daily gage height, in feet, of Susquehanna River at Wilkesbarre, Pa., 1899–1904.

Day.	Jan.	Feb.	Mar.	Apr.	May.	June.	July.	Aug.	Sept.	Oct.	Nov.	Dec.
1899.												
1				8.40	6.40	4.50	3.60	2.70	8.10	2.50	2.50	3.40
2				8.10	6.20	5.50	3.30	2.60	2.90	2.50	8.00	3.40
3				7.70	6.80	5.30	3.80	2.70	2.80	2.50	8.30	3.40
4				7.20	6.30	5.10	8.20	2.60	2.80	2.50	6.70	3.40
5				6.90	6.40	4.60	8.00	3.20	2.50	2.50	7.30	3.50
6				6.90	6.10	4.30	8.00	3.00	2.50	2.60	6.60	3.50
7				7.40	5.70	3.60	2.80	2.80	2.50	2.60	6.90	3.50
8				10.35	5.60	3.50	2.90	2.50	2.40	2.50	5.30	3.70
9				14.10	5.40	3.50	2.80	2.50	2.40	2.50	5.00	3.60
10				14.20	5.30	3.50	2.80	2.50	2.40	2.50	4.50	3.50
11				12.80	5.10	3.30	2.80	2.50	2.40	2.50	4.20	3.50
12				11.10	5.20	3.20	2.90	2.50	2.50	2.50	4.80	3.60
13				11.30	5.10	3.20	2.90	2.70	2.50	2.40	4.90	7.70
14				14.00	5.00	8.20	3.00	2.80	2.50	2.40	4.70	9.00
15				14.80	5.00	8.00	3.20	2.80	2.40	2.40	4.60	9.60
16				13.90	4.80	8.10	3.80	2.80	2.40	2.40	4.50	8.50
17				13.40	4.80	3.20	3.10	2.90	2.30	2.80	5.20	7.70
18				12.50	4.70	3.20	3.00	2.70	2.30	2.80	5.20	7.30
19				11.30	4.90	3.00	3.00	2.40	2.30	2.30	5.30	6.50
20				10.50	4.90	3.00	3.00	2.30	2.30	2.80	5.00	6.50
21				9.90	5.40	3.10	3.10	2.30	2.30	2.80	4.70	8.30
22				9.40	5.90	3.00	3.00	2.60	2.30	2.80	4.60	8.40
23				9.00	5.80	8.00	3.00	2.50	2.30	2.30	4.30	7.40
24				8.50	5.70	2.90	2.90	2.50	2.30	2.80	4.20	6.60
25				8.00	5.50	2.90	2.80	2.40	2.20	2.30	4.00	8.40
26				7.40	5.40	3.10	2.80	2.40	2.50	2.20	3.80	8.00
27				7.60	5.10	3.10	2.80	2.40	2.40	2.30	3.80	7.40
28				7.40	4.90	3.30	2.80	2.40	2.50	2.30	3.70	6.30
29				7.10	4.80	3.80	2.80	4.60	2.50	2.50	3.60	9.10
30			9.00	6.60	4.80	4.00		2.60	4.10	2.60	2.50	7.90
31			8.70		4.70		2.60	3.40		2.50		7.70
1900.												
1	6.80	7.40	10.40	6.90	6.10	3.80	3.00	3.20	3.10	2.80	2.70	10.50
2	6.20	6.80	17.75	7.50	5.80	3.70	2.80	3.20	3.00	2.80	2.60	9.20
3	6.40	6.30	14.55	9.80	5.50	4.20	2.70	3.00	3.10	2.80	2.60	8.10
4	6.80	6.50	11.80	11.40	5.30	3.90	2.90	2.90	3.00	2.80	2.50	7.40
5	7.00	8.40	9.90	11.10	5.20	3.70	2.90	2.90	2.90	2.80	2.70	9.20
6	7.00	8.50	8.40	9.40	5.00	3.80	3.40	2.90	2.80	2.80	2.80	11.90
7	6.90	7.90	8.20	9.60	4.80	3.70	2.90	2.90	2.70	2.70	3.00	11.30
8	6.80	7.80	8.10	11.70	4.70	3.60	3.60	2.90	2.70	2.70	2.90	9.20
9	6.50	14.45	7.70	12.20	4.60	3.60	3.40	2.90	2.60	2.60	2.90	8.50
10	6.10	9.20	8.40	10.90	4.50	3.80	3.20	2.80	2.60	2.60	2.90	8.90
11	5.80	9.80	9.00	9.20	4.50	3.90	3.10	2.80	2.70	2.70	3.00	7.60
12	5.90	9.20	7.80	7.90	4.80	4.80	2.90	2.70	2.70	2.80	3.10	6.60
13	5.60	9.20	6.80	7.30	4.90	4.80	2.70	2.70	2.80	2.80	3.30	6.20
14	5.90	12.10	6.30	7.70	4.80	4.80	3.00	2.60	2.50	2.80	3.50	6.10
15	5.60	13.65	5.70	8.10	4.70	4.00	3.00	2.40	2.20	2.80	8.50	a10.30
16	5.50	11.80	5.70	7.80	4.70	4.00	3.00	2.50	2.60	2.80	3.40	9.80
17	5.50	9.20	9.00	7.60	4.90	3.80	2.90	2.40	2.40	2.40	3.30	9.20
18	5.20	7.70	8.10	10.03	5.00	3.60	2.90	2.50	2.30	2.40	3.20	8.70
19	5.10	8.90	8.80	12.45	5.10	3.50	2.80	2.40	2.20	2.50	3.20	9.20
20	5.80	10.70	8.50	12.40	5.60	8.40	8.10	2.50	2.20	2.70	3.10	9.60
21	14.65	9.80	10.85	11.10	5.20	3.30	3.20	2.50	2.10	2.60	3.10	9.40
22	16.85	11.40	9.70	10.00	5.00	3.20	3.10	2.50	2.20	2.60	3.20	9.00
23	13.50	16.10	9.20	9.50	4.80	3.50	8.00	2.80	2.20	2.70	3.60	8.80
24	10.30	14.75	8.40	11.20	4.60	3.30	2.90	3.00	2.20	2.90	4.00	9.20
25	8.50	11.00	9.90	10.70	4.50	3.80	2.90	2.90	2.20	2.80	4.80	8.80
26	7.80	8.80	8.70	9.50	4.30	3.20	4.00	2.80	2.60	2.80	4.70	12.80
27	7.90	7.00	8.10	10.70	4.10	3.20	3.70	2.70	2.80	2.70	16.75	14.20
28	6.20	8.50	7.10	8.40	4.00	3.10	3.20	2.80	2.80	2.70	20.75	12.90
29	9.20		7.00	6.90	3.90	3.10	3.20	2.80	3.10	2.70	14.65	12.40
30	9.00		6.80	6.50	3.80	3.10	3.10	3.10	2.80	2.70	11.80	11.40
31	8.70		6.50		3.70		3.30	3.10		2.60		11.40

a Ice backed water at gage.

*Mean daily gage height, in feet, of Susquehanna River at Wilkesbarre, Pa.,
1899–1904—Continued.*

Day.	Jan.	Feb.	Mar.	Apr.	May	June.	July	Aug	Sept	Oct	Nov	Dec.
1901.												
1	10.60	8.60	6.20	9.70	7.80	14.55	4.50	3.40	5.80	3.90	3.20	7.30
2	10.60	8.40	6.10	8.80	7.20	11.70	4.00	3.20	5.60	3.70	3.20	6.70
3	9.50	8.30	6.10	8.30	8.70	11.00	3.50	3.00	5.70	4.00	3.10	6.80
4	8.70	8.40	6.10	9.30	8.90	10.60	3.60	3.00	5.30	4.30	3.10	a9.30
5	8.50	8.00	6.20	10.80	8.10	9.20	3.60	3.00	5.00	4.00	3.00	9.90
6	7.20	7.80	6.00	11.90	7.50	8.10	3.60	3.00	4.50	3.90	3.00	9.40
7	7.10	7.80	5.90	16.20	6.80	8.10	4.30	3.00	4.20	3.70	3.00	9.00
8	7.00	7.70	5.80	18.05	6.30	9.00	4.00	3.30	3.80	3.60	3.00	8.80
9	7.90	7.70	5.70	16.90	5.90	9.30	4.00	3.20	3.70	3.40	8.00	8.70
10	7.90	7.50	6.50	14.70	5.80	8.90	3.90	3.10	3.50	3.30	3.00	11.70
11	7.80	7.60	8.40	13.20	6.40	8.00	3.80	3.20	3.30	3.20	2.90	12.10
12	7.80	7.60	18.80	11.80	7.80	7.20	3.80	3.30	3.30	3.20	3.00	11.70
13	8.10	7.40	12.80	10.70	9.50	6.50	3.60	3.10	3.30	3.30	3.00	10.10
14	9.00	6.90	9.70	10.10	9.80	6.10	3.50	8.10	3.80	3.50	8.80	8.80
15	12.00	7.00	8.90	9.60	9.10	5.90	8.40	8.20	8.20	4.10	4.00	20.40
16	14.50	7.10	9.10	9.30	8.00	5.70	8.20	8.30	8.30	4.80	4.70	26.75
17	14.00	7.30	8.80	8.90	7.10	5.50	3.20	3.70	8.50	4.40	4.50	22.80
18	13.60	7.30	8.30	8.50	6.70	5.30	8.60	8.15	3.80	4.30	4.20	15.80
19	12.50	7.20	8.00	8.10	6.80	4.90	3.40	5.60	4.00	4.20	4.10	11.00
20	11.50	6.90	10.10	7.90	7.00	4.70	3.70	4.80	4.20	4.00	4.00	8.20
21	9.40	6.90	12.15	11.05	7.10	4.60	8.10	4.60	4.10	3.90	4.00	7.80
22	10.50	6.70	14.80	18.10	6.50	4.40	8.00	6.95	3.90	3.80	3.90	9.50
23	11.00	6.80	14.50	17.10	6.40	4.50	3.10	6.90	3.70	3.70	3.80	11.20
24	11.00	6.40	12.90	14.80	7.90	5.60	3.10	6.50	3.50	3.70	3.80	11.70
25	11.70	6.40	12.90	14.70	9.00	5.70	8.00	10.50	3.40	3.60	6.00	13.70
26	11.00	6.30	13.80	13.60	8.30	5.70	3.00	8.20	3.40	3.40	9.10	13.50
27	10.50	6.20	17.15	12.30	7.80	5.00	2.90	7.10	3.20	3.40	7.80	13.30
28	10.00	6.30	21.40	11.00	7.40	4.20	2.90	6.10	3.10	3.40	6.20	12.80
29	9.50		19.45	9.60	10.60	4.50	3.00	5.80	3.30	3.20	5.50	13.10
30	9.30		15.50	8.60	16.85	4.20	8.30	4.90	8.80	8.10	5.70	13.10
31	9.10		12.90		17.55		8.60	4.90		8.10		13.50
1902.												
1	14.00	12.70	29.57	9.70	5.00	4.10	10.60	8.80	8.60	9.60	9.50	5.10
2	13.00	11.40	30.75	9.20	4.90	4.00	10.50	9.50	8.50	10.80	8.20	5.00
3	12.10	10.80	30.05	9.00	5.10	3.90	11.10		8.40	10.60	7.40	5.00
4	10.90	10.70	25.25	8.50	5.10	3.90	7.80	9.60	8.40	8.50	6.80	5.20
5	9.60	8.50	20.30	8.10	4.80	3.80	8.80	8.80	8.20	7.30	6.40	5.50
6	9.90	7.00	14.65	7.90	4.80	3.80	8.26	7.50	8.20	7.10	6.00	5.90
7	9.80	9.10	11.65	7.60	4.70	4.80	12.70	8.60	3.20	6.90	5.80	5.80
8	9.60	9.80	10.70	7.70	4.70	4.50	14.20	6.50	3.20	6.70	5.50	5.50
9	9.70	9.00	10.30	11.85	4.50	4.40	13.15	6.20	3.20	6.20	5.60	5.20
10	9.40	9.40	11.00	15.80	4.40	4.20	8.75	5.80	3.20	5.80	5.70	5.90
11	9.20	9.00	12.50	15.45	4.80	4.20	9.60	5.60	3.60	5.50	5.00	7.20
12	9.00	9.00	14.80	12.80	4.20	4.20	9.70	5.50	3.50	5.80	4.70	8.00
13	8.20	9.00	18.00	14.40	4.10	4.10	9.50	5.40	3.60	6.50	4.70	9.85
14	7.20	8.80	19.60	10.30	4.00	4.20	7.40	5.40	3.50	6.00	4.70	10.20
15	6.40	8.00	18.20	9.40	3.90	4.20	6.30	5.20	3.60	5.80	4.70	9.20
16	6.80	8.20	15.80	8.60	3.80	4.20	5.80	5.00	3.40	5.90	4.60	10.70
17	7.20	7.80	18.50	8.00	3.80	5.00	5.40	4.40	3.30	5.90	4.50	13.45
18	7.00	7.70	20.20	7.40	3.70	4.70	5.20	4.40	3.30	5.60	4.40	12.70
19	6.70	7.20	17.45	7.00	3.70	4.40	5.10	4.10	8.20	5.30	4.30	12.40
20	6.10	6.60	14.30	6.70	3.60	4.60	5.40	4.10	3.10	4.90	4.20	11.30
21	6.20	6.60	11.60	6.40	3.50	4.30	12.10	4.00		4.80	4.20	10.00
22	10.60	6.50	10.20	6.20	3.50	4.30	15.90	4.00	3.00	4.90	4.20	15.60
23	16.70	6.40	9.60	6.00	3.50	4.20	13.90	4.00	3.00	5.20	4.10	17.65
24	12.20	7.20	9.60	5.70	3.70	4.20	13.45	3.90	3.00	5.00	4.10	16.35
25	10.70	7.70	9.50	5.50	3.70	4.20	13.85	3.90	3.00	4.70	4.10	13.70
26	9.70	7.70	9.00	5.20	3.70	4.20	14.90	3.80	4.20	4.70	4.10	11.00
27	8.90	8.80	8.50	5.00	3.80	4.70	9.70	3.70	7.10	4.80	4.50	9.70
28	8.20	14.08	8.00	4.80	3.90	3.90	9.70	3.60	6.00	7.62	4.70	8.50
29	7.70		8.00	4.70	4.60	3.90	10.80	3.60		11.05	5.00	8.00
30	7.60		10.40	4.90	4.60	5.10	10.60	3.60	10.70	12.05	5.20	7.00
31	13.80		9.80		4.20		9.30	3.60		11.10		6.80

a River frozen over.

Mean daily gage height, in feet, of Susquehanna River at Wilkesbarre, Pa., 1899–1904—Continued.

Day.	Jan.	Feb.	Mar.	Apr.	May.	June.	July.	Aug.	Sept.	Oct.	Nov.	Dec.
1903.												
1	8.50	15.30	20.40	11.20	4.80	3.00	6.90	4.60	13.80	3.60	5.60	7.20
2	11.00	13.10	19.94	12.00	4.60	3.00	6.80	4.60	11.90	3.60	5.60	7.30
3	12.80	13.00	16.28	10.70	4.40	3.00	6.10	4.30	9.90	3.60	5.40	7.50
4	13.00	14.65	13.60	9.70	4.30	2.90	5.50	4.00	8.40	3.60	5.20	6.70
5	13.50	18.78	11.30	9.80	4.20	2.90	5.40	4.80	7.40	3.60	5.00	5.20
6	9.70	16.50	10.50	9.90	4.10	2.90	5.50	6.70	6.70	3.70	5.00	4.50
7	8.10	13.90	12.10	8.70	4.00	2.90	7.30	7.90	6.20	3.80	5.00	4.50
8	7.90	11.30	11.60	8.80	4.00	2.90	7.30	7.60	5.80	4.70	5.20	4.29
9	6.90	10.00	16.20	10.60	4.00	3.00	9.40	6.80	5.53	10.70	5.30	4.30
10	6.80	8.60	18.60	10.80	8.80	3.00	4.80	6.00	5.30	19.20	5.00	4.10
11	10.70	8.00	17.94	9.80	8.70	2.90	4.40	5.70	5.20	21.25	4.90	4.00
12	10.00	8.50	18.91	9.00	3.70	3.60	4.30	5.40	5.30	21.15	4.70	3.70
13	9.50	9.10	17.80	8.90	8.60	6.60	4.00	5.50	6.00	18.15	4.60	3.90
14	9.10	11.00	15.70	8.30	8.50	5.00	3.80	5.20	6.60	13.70	4.50	4.70
15	9.10	10.80	13.20	12.20	3.50	7.50	3.70	4.90	5.20	10.50	4.40	4.80
16	10.00	9.80	11.70	14.20	3.50	6.40	8.60	4.70	4.80	9.00	4.20	5.80
17	10.50	8.40	10.60	12.30	3.40	5.80	3.60	4.50	4.80	8.00	7.90	6.60
18	10.40	7.40	9.90	10.50	3.40	5.20	3.40	4.30	5.20	8.30	12.90	6.80
19	9.60	10.00	9.60	9.00	3.30	5.00	4.30	3.90	4.80	12.50	13.70	6.30
20	8.70	9.20	9.20	8.00	3.30	4.80	4.60	3.70	5.00	12.40	10.80	5.80
21	8.60	9.40	8.70	7.90	3.30	4.70	4.80	3.80	4.80	10.90	8.70	8.90
22	9.40	10.00	8.30	6.80	3.50	6.80	5.10	5.60	4.70	9.40	7.10	9.00
23	9.80	10.50	13.92	6.40	3.30	8.00	5.30	4.40	4.20	8.30	6.70	8.40
24	10.40	10.90	20.88	6.10	3.30	8.93	4.40	5.00	4.20	7.50	6.20	8.00
25	10.00	11.20	21.16	5.90	3.10	9.45	4.40	4.60	4.00	7.00	6.20	7.50
26	9.60	10.40	18.00	5.70	3.10	10.40	6.10	4.40	3.90	6.80	6.10	7.10
27	8.70	9.60	15.40	5.50	3.10	10.20	5.20	4.30	3.90	6.50	5.80	7.20
28	8.20	10.20	12.60	5.80	3.10	8.00	4.50	5.40	3.80	6.20	5.50	10.40
29	8.20		10.70	5.10	3.00	8.50	4.10	9.15	3.70	6.00	6.00	9.70
30	14.54		9.90	4.90	3.00	7.60	4.20	19.40	3.60	5.80	7.70	9.20
31	17.60		9.80		3.00		4.70	16.83		5.60		8.40
1904.												
1	9.00	14.00	10.80	12.00	11.50	5.70	3.50	4.80	3.70	4.80	5.30	4.20
2	8.90	13.00	10.90	15.10	10.50	7.40	3.50	4.40	3.60	5.40	5.10	4.10
3	8.50	12.30	11.15	15.80	9.40	7.00	3.50	4.20	3.50	5.90	4.90	4.20
4	7.20	11.80	16.50	14.00	8.40	6.40	3.50	4.80	3.40	5.20	4.80	4.20
5	6.50	11.00	a18.20	12.00	7.60	6.00	3.50	5.80	3.40	4.70	4.60	3.60
6	6.70	b10.90	17.20	10.70	7.00	9.10	3.50	5.00	3.30	4.50	4.50	8.30
7	7.20	11.60	17.90	10.20	6.70	7.40	3.60	4.40	3.30	4.30	4.50	3.50
8	7.20	c21.70	25.20	10.50	6.30	6.40	3.70	4.60	3.30	4.00	4.50	3.60
9	7.30	25.30	d30.60	11.00	6.00	6.60	4.20	5.00	3.50	4.00	4.50	3.30
10	7.40	24.60	26.60	11.70	5.70	11.60	3.80	4.40	3.50	3.90	4.50	8.20
11	7.80	23.80	24.00	16.20	5.50	10.90	3.70	4.60	3.30	3.80	4.40	3.10
12	7.10	22.00	e22.00	14.80	5.20	8.50	4.10	4.00	3.30	3.80	4.40	3.30
13	7.00	20.30	e19.30	12.10	5.00	7.10	4.50	3.90	3.20	8.90	4.30	3.20
14	7.00	f18.00	e17.40	10.80	4.80	6.20	4.20	3.80	3.10	7.00	4.20	3.30
15	6.70	17.00	e15.90	9.70	4.80	5.60	3.90	3.60	3.60	8.30	4.30	3.20
16	6.40	15.70	e14.90	8.90	6.10	5.20	3.80	5.50		6.90	4.90	3.30
17	6.20	14.70	e14.00	8.30	8.00	5.10	3.60	3.40	4.30	6.00	4.30	3.30
18	6.00	12.90	e13.00	8.00	7.90	5.60	3.90	3.30	4.80	5.50	4.40	3.30
19	g5.90	12.60	e12.50	7.90	7.10	4.80	3.60	3.30	4.40	5.10	4.30	3.30
20	5.60	h12.90	12.80	7.90	7.90	11.20	4.50	3.70	3.20	4.80	4.30	3.40
21	i5.60	12.70	13.60	7.80	7.80	10.20	4.30	4.20	3.20	5.00	4.30	3.40
22	6.00	12.90	10.50	7.40	8.50	4.10	4.30	3.80	8.30	4.60	4.60	8.40
23	12.70	13.70	9.70	7.10	7.30	4.30	3.50	8.70	8.40	10.20	4.60	8.30
24	j18.10	12.80	16.90	7.10	6.50	6.40	4.90	4.40	3.40	10.20	5.30	3.50
25	13.50	12.70	16.90	7.00	6.50	8.90	3.30	6.40	8.40	8.80	5.50	3.60
26	k11.60	12.60	20.40	6.90	6.70	6.70	3.80	5.80	4.00	7.40	5.20	3.30
27	k10.10	12.00	22.90	7.20	6.50	3.70	3.70	5.30	5.40	6.90	5.00	3.50
28	k9.00	12.00	22.70	7.90	5.90	6.50	4.60	5.30	5.80	6.70	4.80	10.00
29	k8.20	11.50	18.40	12.40	6.00	3.50	3.60	4.30	5.20	6.40	4.20	13.85
30	k9.20		14.20	12.80	5.50	8.40	3.80	4.10	4.70	6.00	4.20	10.00
31	18.90		11.70		5.30		4.10	3.90		5.90		10.80

a Ice still unbroken.
b Closed with anchor ice as far up as Ransom.
c Ice started at 5 15 p. m.; moved until February 10, 12. m. Gorged below city.
d Highest gage reading 30.6.
e Still gorged
f Ice blocked as far as Tunkhannock, Pa.
g Ice started at Pittston at 1.30 p. m., at Wilkesbarre, 2 p. m. River closed December 10 to 28, inclusive.
h Ice blocked as far as Laceyville. Pa.
i 12 midnight ice still running; stream nearly full.
j River full of running ice all day; 10 p. m. very little ice running.
k Anchor ice.

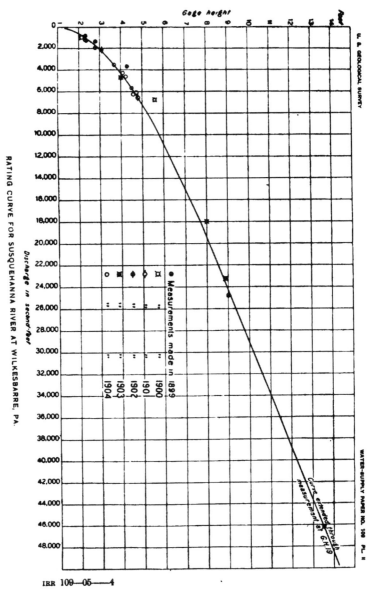

RATING CURVE FOR SUSQUEHANNA RIVER AT WILKESBARRE, PA.

Rating table for Susquehanna River at Wilkesbarre, Pa., from March 30, 1899, to December 31, 1904.

Gage height.	Discharge.	Gage height.	Discharge.	Gage height.	Discharge	Gage height.	Discharge.
Feet.	*Second-feet.*	*Feet.*	*Second-feet*	*Feet.*	*Second-feet.*	*Feet.*	*Second-feet.*
2.0	620	4.3	5,070	6.6	13,170	9.8	28,200
2.1	720	4.4	5,340	6.7	13,590	10.0	29,200
2.2	820	4.5	5,620	6.8	14,010	10.2	30,100
2.3	930	4.6	5,9:0	6.9	14,440	10.4	31,100
2.4	1,050	4.7	6,210	7.0	14,870	10.6	32,100
2.5	1,180	4.8	6,520	7.1	15,300	10.8	33,000
2.6	1,320	4.9	6,830	7.2	15,730	11.0	34,000
2.7	1,470	5.0	7,150	7.3	16,160	11.2	35,000
2.8	1,630	5.1	7,470	7.4	16,600	11.4	36,000
2.9	1,810	5.2	7,800	7.5	17,040	11.6	37,000
3.0	2,000	5.3	8,140	7.6	17,490	11.8	37,900
3.1	2,200	5.4	8,490	7.7	17,950	12.0	38,900
3.2	2,410	5.5	8,850	7.8	18,420	12.2	39,900
3.3	2,620	5.6	9,210	7.9	18,900	12.4	40,800
3.4	2,840	5.7	9,580	8.0	19,380	12.6	41,800
3.5	3,070	5.8	9,950	8.2	20,360	12.8	42,800
3.6	3,300	5.9	10,330	8.4	21,340	13.0	43,700
3.7	3,540	6.0	10,720	8.6	22,320	13.2	44,700
3.8	3,780	6.1	11,120	8.8	23,300	13.4	45,700
3.9	4,030	6.2	11,520	9.0	24,300	13.8	47,600
4.0	4,280	6.3	11,930	9.2	25,300	14.0	48,600
4.1	4,540	6.4	12,340	9.4	26,200		
4.2	4,800	6.5	12,750	9.6	27,200		

Table based on discharge measurements of 1899, 1900, 1901, 1902, 1903, and 1904. Well defined between 2 feet gage height and 19 feet gage height. Tangent at 9.80 feet gage height with a difference of 500 per tenth. Table applied to tenths.

*Mean daily discharge, in second-feet, of Susquehanna River at Wilkesbarre, Pa..
1899–1904.*

Day.	Jan.	Feb.	Mar.	Apr.	May.	June.	July	Aug.	Sept.	Oct.	Nov.	Dec.
1899.												
1				21,340	12,340	5,620	3,300	1,470	2,200	1,180	1,180	2,840
2				19,870	11,520	8,850	2,620	1,320	1,810	1,180	2,000	2,840
3				17,950	11,980	8,140	2,620	1,470	1,320	1,180	20,850	2,840
4				15,730	11,980	7,470	2,410	1,320	1,320	1,180	13,590	2,840
5				14,440	12,340	5,910	2,000	2,410	1,180	1,180	16,160	3,070
6				14,440	11,120	5,070	2,000	2,000	1,180	1,320	18,170	3,070
7				16,600	9,580	3,300	1,630	1,630	1,180	1,320	14,440	3,070
8				30,850	9,210	3,070	1,810	1,180	1,050	1,180	8,140	3,540
9				49,100	8,490	3,070	1,630	1,180	1,050	1,180	7,150	3,300
10				49,600	8,140	3,070	1,630	1,180	1,050	1,180	5,620	3,070
11				42,800	7,470	2,620	1,630	1,180	1,050	1,180	4,800	3,070
12				34,500	7,800	2,410	1,810	1,180	1,180	1,180	5,070	3,300
13				35,500	7,470	2,410	1,810	1,470	1,180	1,050	6,830	17,950
14				48,600	7,150	2,410	2,000	1,630	1,180	1,050	6,210	27,200
15				50,100	7,150	2,000	2,410	1,630	1,050	1,050	5,910	27,200
16				48,100	6,520	2,200	2,620	1,630	1,050	1,050	5,620	21,830
17				45,700	6,520	2,410	2,200	1,810	930	930	7,800	17,950
18				41,300	6,210	2,410	2,000	1,470	930	930	7,800	16,160
19				35,500	6,830	2,000	2,000	1,050	930	980	8,140	12,750
20				31,600	6,830	2,000	2,000	980	930	980	7,150	12,750
21				28,700	8,490	2,200	2,200	980	930	930	6,710	20,850
22				26,200	10,330	2,000	2,000	1,320	930	930	5,910	21,340
23				24,300	9,950	2,000	2,000	1,180	930	930	5,070	16,600
24				21,830	9,580	1,810	1,810	1,180	930	930	4,800	13,170
25				19,870	8,850	1,810	1,630	1,050	820	930	4,290	21,340
26				16,600	8,490	2,200	1,630	1,050	1,180	820	3,780	19,390
27				17,490	7,470	2,200	1,630	1,050	1,050	930	3,780	16,600
28				16,600	6,830	2,620	1,630	1,050	1,180	930	3,540	11,930
29				15,300	6,520	3,780	1,630	5,910	1,180	1,180	3,300	24,800
30				13,170	6,520	4,290	1,630	4,540	1,320	1,180	3,070	18,900
31					6,210		1,020	2,840		1,180		17,950
1900.												
1	14,010	16,600	31,100	14,440	11,120	3,780	2,000	2,410	2,200	930	1,470	31,600
2	11,520	14,010	75,900	17,040	9,950	3,540	1,630	2,410	2,000	930	1,810	25,300
3	12,340	11,930	52,200	28,200	8,850	4,800	1,470	2,000	2,200	930	1,320	19,870
4	14,010	12,750	37,900	36,000	8,140	4,000	1,810	1,810	2,000	930	1,180	16,600
5	14,870	21,340	28,700	34,500	7,800	3,540	1,810	1,810	1,810	930	1,470	25,300
6	14,870	21,830	21,340	26,200	7,150	3,780	2,840	1,810	1,630	820	1,630	38,400
7	14,440	18,900	20,360	27,200	6,520	3,540	1,810	1,810	1,470	720	2,000	35,500
8	14,010	18,420	19,870	37,400	6,210	3,300	1,810	1,810	1,470	720	1,810	28,700
9	11,930	51,600	17,950	30,900	5,910	3,300	2,410	1,810	1,320	820	1,810	23,800
10	11,120	25,300	21,340	31,600	5,620	3,780	2,410	1,630	1,320	820	1,810	20,360
11	9,950	28,200	24,300	25,300	5,620	4,000	1,630	1,470	1,470	820	2,200	13,170
12	10,330	25,300	18,420	18,900	6,520	5,070	1,810	1,470	1,470	820	2,620	11,520
13	9,210	25,300	14,010	16,160	6,830	5,070	2,000	1,470	1,470	820	3,070	11,120
14	10,330	39,400	11,930	17,950	6,520	6,520	2,000	1,320	1,180	820	3,070	11,120
15	9,210	46,900	9,580	18,420	6,210	5,070	2,000	1,320	1,050	820	2,840	28,200
16	8,850	37,900	9,580	18,420	6,210	4,290	2,000	1,320	1,180	930	2,840	28,200
17	8,850	25,300	24,300	6,830	3,780	1,180	1,050	2,620	25,300			
18	7,800	17,950	19,870	29,100	7,150	3,300	1,810	1,180	930	1,050	2,410	22,810
19	7,470	23,800	20,850	41,000	7,470	3,070	1,050	1,050	820	1,180	2,410	25,300
20	9,950	32,600	21,830	40,800	9,210	2,840	2,200	1,180	820	1,470	2,200	27,200
21	52,900	28,200	33,200	34,500	7,800	2,620	2,200	1,180	720	1,320	2,200	26,200
22	68,800	36,000	27,700	29,200	7,150	2,410	2,200	1,180	1,320	2,410	24,300	
23	46,200	63,200	25,300	26,700	6,520	2,000	2,000	1,630	930	1,470	4,290	23,300
24	30,600	53,600	21,340	35,500	5,910	2,620	1,810	2,000	820	1,810	4,280	25,300
25	21,830	34,000	28,700	32,100	5,620	2,620	1,810	1,810	930	1,630	5,070	23,300
26	18,420	23,300	22,810	26,700	5,070	2,410	4,290	1,320	820	1,630	6,210	42,800
27	18,900	14,870	19,870	21,340	4,540	3,540	1,470	930	1,470	68,000	49,600	
28	11,520	21,830	15,300	17,040	4,290	2,200	2,840	1,630	820	1,470	102,200	43,300
29	25,300		14,870	14,440	4,000	2,200	2,410	1,630	820	1,470	32,900	40,800
30	24,300		14,010	12,750	3,780	2,200	2,410	2,200	930	1,470	37,900	36,000
31	22,810		12,750		3,540		2,620	2,200		1,470		36,000

Mean daily discharge, in second-feet, of Susquehanna River at Wilkesbarre, Pa., 1899–1904—Continued.

Day.	Jan.	Feb.	Mar.	Apr.	May.	June.	July.	Aug.	Sept.	Oct.	Nov.	Dec.
1901.												
1	32,100	22,820	11,520	27,700	18,420	51,300	5,620	2,840	9,960	4,080	2,410	16,160
2	32,100	21,340	11,120	23,300	15,730	37,400	4,290	2,410	9,210	3,540	2,410	13,590
3	26,700	20,850	11,120	20,850	22,810	34,000	3,070	2,000	9,540	4,240	2,280	14,010
4	22,810	21,340	11,120	25,800	23,800	32,100	3,300	2,000	8,140	5,070	2,280	25,800
5	21,830	19,380	11,520	33,000	19,870	25,300	3,300	2,000	7,150	4,280	2,000	28,700
6	16,730	18,420	10,720	38,400	17,040	19,870	3,300	2,000	5,620	4,080	2,000	26,200
7	15,300	18,420	10,330	63,900	14,010	19,870	5,070	2,000	4,800	3,540	2,000	24,300
8	14,870	17,950	9,960	78,400	11,980	24,300	4,280	2,620	3,780	3,300	2,000	20,850
9	18,900	17,950	9,540	69,200	10,330	25,800	4,280	2,410	3,540	2,840	2,000	22,810
10	18,900	17,040	12,750	53,300	9,960	23,800	4,080	2,280	3,070	2,620	2,000	37,400
11	18,420	17,490	21,340	44,700	12,340	19,380	3,780	2,410	2,680	2,410	1,810	39,400
12	18,420	17,490	84,700	37,900	18,420	15,730	3,780	2,620	2,620	2,410	2,000	37,400
13	19,870	16,600	39,900	32,600	26,700	12,750	3,300	2,200	2,620	2,620	2,000	29,600
14	24,300	14,440	27,700	29,600	28,200	11,120	3,070	2,200	2,620	3,070	3,070	23,300
15	38,400	14,870	23,800	27,200	24,800	10,330	2,840	2,410	2,410	4,540	4,240	98,900
16	52,000	15,300	24,800	25,800	19,380	9,540	2,410	3,300	2,620	5,070	6,210	166,300
17	48,600	16,160	23,300	23,800	15,300	8,850	2,410	3,540	3,070	5,340	5,620	122,300
18	43,700	16,160	20,850	21,830	13,590	8,140	3,300	20,110	3,780	5,070	4,800	59,500
19	41,300	15,730	19,380	19,870	14,010	6,830	2,840	9,210	4,280	4,800	4,540	34,000
20	36,500	14,440	29,600	18,890	14,870	6,210	2,620	6,520	4,800	4,280	4,240	20,880
21	26,200	14,440	39,600	34,200	15,300	5,910	2,200	5,910	4,540	4,080	4,240	18,420
22	31,600	13,590	54,000	78,800	12,750	5,340	2,000	14,655	4,080	3,780	4,080	26,700
23	34,000	14,010	52,000	70,800	12,340	5,620	2,200	14,440	3,540	3,540	3,780	35,000
24	34,000	12,340	43,300	54,000	18,900	9,210	2,200	12,750	3,070	3,540	3,780	37,400
25	37,400	12,340	43,300	53,300	24,300	9,540	2,000	31,600	2,840	3,300	10,720	47,100
26	34,000	11,930	47,600	46,700	20,850	9,540	2,000	25,300	2,410	2,840	24,800	46,200
27	31,600	11,520	71,100	40,300	17,490	7,150	1,810	15,300	2,410	2,840	17,490	45,200
28	29,300	11,980	108,400	34,000	16,600	4,800	1,810	11,120	2,200	2,840	11,520	42,800
29	26,700		90,300	27,300	32,100	5,620	2,000	8,140	2,620	2,410	8,850	44,200
30	25,800		58,800	22,820	68,900	4,800	2,620	6,520	3,780	2,200	9,540	46,200
31	24,800		43,300		74,300		3,300	6,830		2,200		
1902.												
1	48,600	42,300	201,800	27,700	7,150	4,540	32,100	23,300	3,300	27,200	26,700	7,470
2	43,700	36,000	217,700	25,800	6,880	4,280	31,600	26,700	3,070	33,000	20,380	7,150
3	32,400	33,000	208,200	24,300	7,470	4,080	20,850	34,500	2,840	32,100	16,600	7,150
4	33,500	32,600	148,800	21,830	7,470	4,080	18,420	27,200	2,840	21,830	14,010	7,800
5	27,200	21,830	97,100	19,870	6,520	3,780	21,830	23,300	2,410	16,160	12,340	8,850
6	28,700	14,870	52,900	18,900	6,520	3,780	20,400	17,040	2,410	15,300	10,720	10,330
7	24,300	24,800	37,200	17,490	6,210	6,520	42,300	14,010	2,410	14,440	9,950	9,950
8	27,200	28,200	32,600	17,950	6,210	5,620	49,900	18,750	2,410	13,590	8,850	8,850
9	27,700	27,200	30,600	38,100	5,620	5,340	44,400	11,520	2,410	11,520	9,210	7,800
10	26,200	26,200	34,000	61,000	5,340	4,800	23,050	9,950	2,410	9,950	9,540	10,330
11	25,300	24,300	41,300	58,400	5,070	4,800	24,050	9,210	3,300	8,850	7,150	15,730
12	24,300	24,300	42,800	47,600	4,800	4,800	27,700	8,850	3,070	9,950	6,210	19,380
13	20,380	24,300	78,000	51,300	4,540	4,540	21,830	8,490	3,300	12,750	6,210	28,400
14	15,730	20,850	91,700	30,600	4,240	4,800	16,600	8,490	3,070	10,720	6,210	30,100
15	12,340	19,380	79,600	26,200	4,080	4,800	11,980	7,800	3,070	9,950	6,210	25,300
16	14,010	20,380	61,000	22,820	3,780	4,800	9,950	7,150	2,840	10,330	5,910	32,600
17	15,730	18,420	82,100	19,380	3,780	7,150	8,490	5,910	2,620	10,330	5,620	46,000
18	14,870	17,950	97,100	16,600	3,540	6,210	7,470	5,340	2,620	9,210	5,340	42,300
19	13,590	15,730	73,500	14,870	3,540	5,340	7,470	4,800	2,410	8,140	5,070	40,800
20	11,120	13,170	50,600	13,590	3,300	5,910	8,490	4,540	2,200	6,830	4,800	35,500
21	11,520	18,170	87,000	12,340	3,070	5,070	39,400	4,240	2,200	6,520	5,070	29,200
22	32,100	12,750	80,100	11,520	3,070	5,070	57,000	4,240	2,000	6,830	4,800	59,500
23	67,700	12,340	27,700	10,720	3,070	4,800	48,100	4,280	2,000	7,800	4,540	75,100
24	39,900	15,730	27,200	9,540	3,540	4,800	45,000	4,080	2,000	7,150	4,540	65,000
25	32,600	15,730	26,700	8,850	3,540	4,800	47,000	4,080	2,000	6,210	4,540	47,100
26	27,700	17,950	24,300	7,800	3,540	4,800	50,700	8,540		6,210	4,540	34,000
27	23,900	23,300	21,830	7,150	3,780	4,540	37,400	8,540	15,300	5,910	5,620	27,700
28	20,380	48,800	19,380	6,520	4,080	4,030	27,700	8,540	10,720	17,540	6,210	21,830
29	17,950		19,380	6,520	5,910	3,780	33,000	3,300	18,900	34,200	7,150	19,380
30	17,490		31,100	6,830	5,910	7,470	32,100	7,800	32,600	39,200	7,800	14,870
31	45,200		28,200		4,800		25,800	8,300		34,500		14,010

Mean daily discharge, in second-feet, of Susquehanna River at Wilkesbarre, Pa.,
1899–1904—Continued.

Day.	Jan.	Feb.	Mar.	Apr.	May.	June	July.	Aug.	Sept	Oct	Nov.	Dec.
1903.												
1	21,830	57,400	98,900	35,000	6,520	2,000	14,440	5,910	47,600	3,300	9,210	15,770
2	34,000	44,200	94,700	38,900	5,910	2,000	14,010	5,910	38,400	3,300	9,210	16,160
3	42,800	43,700	64,500	32,600	5,340	2,000	11,120	5,070	28,700	3,300	8,490	17,040
4	43,700	53,000	46,700	27,700	5,070	1,810	8,850	4,280	21,340	3,300	7,800	13,580
5	46,200	84,500	35,500	28,200	4,800	1,810	8,490	6,520	16,600	3,300	7,150	7,800
6	27,700	66,100	31,600	28,700	4,540	1,810	8,850	13,500	13,500	3,540	7,150	6,520
7	19,870	48,100	39,400	22,810	4,280	1,810	16,160	18,900	11,520	3,780	7,150	5,910
8	18,900	33,500	37,000	23,300	4,280	1,810	16,160	17,490	9,950	6,210	7,800	4,800
9	14,440	29,200	63,900	32,100	4,280	2,000	26,200	14,010	8,900	32,600	8,140	5,070
10	14,010	22,320	83,000	33,000	3,780	2,000	6,520	10,720	8,140	88,100	7,150	4,540
11	32,800	19,380	77,300	28,200	3,540	1,810	5,340	9,580	7,810	106,900	6,830	4,280
12	29,200	21,830	85,600	24,300	3,540	3,300	5,070	8,490	8,140	106,000	6,210	3,540
13	26,700	24,800	76,300	23,800	3,300	13,170	4,280	8,850	10,720	79,200	5,910	4,030
14	24,800	34,000	60,300	20,850	3,070	7,150	3,780	7,800	9,210	47,100	5,620	6,210
15	24,800	33,000	44,700	39,900	3,070	17,040	3,540	6,830	7,800	31,600	5,340	6,520
16	29,200	25,800	37,400	49,900	3,070	12,340	3,300	6,210	6,520	24,300	4,800	9,950
17	31,600	21,340	32,100	40,300	2,840	9,950	3,300	5,620	5,910	19,380	18,900	13,170
18	31,100	16,600	28,700	31,600	2,840	7,800	2,840	5,070	7,800	20,850	43,300	14,010
19	27,200	29,200	27,200	24,300	2,620	7,150	5,070	4,030	6,520	41,300	47,100	11,930
20	22,810	25,300	25,300	19,380	2,620	6,520	5,910	3,540	7,150	40,800	33,000	9,950
21	22,320	26,200	22,810	16,160	2,620	6,210	6,520	3,780	6,520	33,500	22,810	23,800
22	26,200	29,200	20,850	14,010	3,070	14,010	7,470	9,210	6,210	28,200	15,300	24,300
23	28,200	31,600	48,100	12,340	2,620	19,380	6,210	8,140	5,340	20,850	13,590	21,340
24	31,100	33,500	103,400	11,120	2,620	24,050	5,340	7,150	4,800	17,040	11,520	19,380
25	29,200	35,000	106,100	10,330	2,220	26,500	5,340	5,910	4,280	14,780	11,520	17,040
26	27,200	25,800	78,100	9,580	2,200	31,100	11,120	5,340	4,030	14,010	11,120	15,300
27	22,810	27,200	58,100	8,850	2,200	30,100	7,800	5,070	3,780	12,750	9,950	15,310
28	20,360	30,100	41,800	8,140	2,200	19,880	5,620	8,490	3,780	11,520	8,850	31,100
29	20,360		32,600	7,470	2,000	14,440	4,540	25,000	3,540	10,720	10,720	27,700
30	51,300		28,700	6,830	2,000	17,490	4,800	90,000	3,300	9,950	17,950	25,300
31	66,100		28,200		2,000		6,210	68,700		9,210		21,340
1904.												
1	24,300	48,000	16,600	38,900	36,500	9,580	3,070	6,520	3,540	6,520	8,140	4,800
2	23,800	43,700	16,900	56,000	31,600	16,600	3,070	5,340	3,300	8,490	7,470	4,540
3	21,830	40,300	18,350	61,000	26,200	14,870	3,070	4,800	3,070	10,330	6,890	4,800
4	15,780	37,000	33,300	48,600	21,340	12,340	3,070	5,070	2,840	7,800	6,520	4,800
5	12,750	34,000	40,100	38,900	17,490	10,720	3,070	8,140	2,840	6,210	5,910	8,300
6	13,590	31,500	36,100	32,600	14,870	24,800	3,070	7,150	2,620	5,620	5,620	2,620
7	15,780	37,000	38,900	30,100	18,590	16,600	3,300	5,340	2,620	5,070	5,620	3,070
8	15,780	55,000	74,720	31,600	11,930	12,330	5,340	5,910	2,620	4,280	5,620	3,300
9	16,160	75,100	103,700	34,000	10,720	13,170	4,800	7,150	3,070	4,280	5,620	2,620
10	16,600	71,300	82,900	37,400	9,580	87,000	3,540	5,340	3,070	4,080	5,620	2,410
11	16,160	67,000	68,000	63,900	8,850	83,500	3,540	5,910	2,620	8,780	5,840	2,200
12	15,300	57,000	57,600	50,600	7,800	21,830	4,540	4,280	2,620	8,780	5,340	2,620
13	14,870	40,400	44,900	39,400	7,150	15,300	5,620	4,030	2,410	6,080	5,070	2,410
14	14,870	30,300	36,800	33,000	6,520	11,520	3,780	2,500	14,870	4,800	2,620	
15	13,590	35,300	31,100	27,700	6,520	9,210	4,030	3,300	3,300	20,850	5,070	2,410
16	12,340	30,350	27,500	23,800	11,120	7,470	3,780	3,070	8,850	14,440	5,070	2,620
17	11,520	20,800	31,000	20,850	19,380	7,470	3,300	2,840	5,070	10,720	5,070	2,620
18	10,720	21,830	30,000	19,380	18,900	12,340	4,030	2,620	11,520	8,850	5,340	2,620
19	10,330	21,070	35,500	18,900	15,300	6,520	3,300	2,620	5,340	7,470	5,070	2,620
20	9,210	21,830	42,800	18,900	35,000	5,070	3,540	2,410	4,540	6,520	5,070	2,840
21	9,210	21,350	46,700	18,420	30,100	5,070	4,800	2,410	3,780	7,150	5,070	2,840
22	10,720	21,830	31,600	16,600	21,830	4,540	3,780	2,620	13,300	22,320	5,910	2,840
23	42,300	20,700	27,700	15,300	16,160	5,070	3,070	3,540	2,840	30,100	5,910	2,620
24	79,600	21,550	69,200	14,870	12,750	4,030	2,620	12,340	2,840	23,300	8,850	3,300
25	46,200	21,350	69,200	14,870	12,750	4,030	2,620	12,340	2,840	23,300	8,850	3,300
26	37,000	21,500	96,900	14,440	13,590	3,540	2,840	9,950	4,280	30,100	7,800	2,620
27	29,600	10,600	123,400	15,730	12,750	3,540	3,540	8,140	8,490	14,440	7,150	3,070
28	24,300	19,000	121,300	18,900	10,330	3,540	3,300	5,910	8,140	31,600	6,520	29,200
29	20,360	18,350	81,300	40,800	10,720	3,070	3,300	5,070	7,800	12,340	4,800	47,850
30	25,300		49,900	42,800	8,850	2,840	3,780	4,540	6,210	10,720	4,800	45,300
31	48,100		37,400		8,140		4,540	4,080		10,330		33,000

From February 8 to March 19, 1904, discharges reduced 50 per cent on account of ice gorge.

Estimated monthly discharge of Susquehanna River at Wilkesbarre, Pa., 1899–1904.

[Drainage area, 9,810 square miles.]

Month.	Discharge in second-feet.			Run-off.	
	Maximum.	Minimum.	Mean.	Second-feet per square mile.	Depth in inches.
1899.					
April	50,100	13,170	28,773	2.93	3.27
May	12,340	6,210	8,574	.87	1.00
June	8,850	1,810	3,378	.34	.38
July	8,300	1,320	1,965	.20	.23
August	5,910	980	1,653	.17	.20
September	2,200	820	1,140	.12	.13
October	1,320	820	1,072	.11	.13
November	20,850	1,180	7,046	.72	.80
December	27,200	2,840	12,694	1.29	1.49
1900.					
January	68,800	7,470	18,279	1.86	2.14
February	63,200	11,980	28,226	2.88	3.00
March	75,900	9,580	23,780	2.42	2.79
April	41,000	12,750	26,348	2.69	3.00
May	11,120	3,540	6,583	.67	.77
June	6,520	2,200	3,506	.36	.40
July	4,280	1,470	2,820	.24	.28
August	2,410	1,050	1,685	.17	.20
September	2,200	720	1,289	.13	.15
October	1,810	720	1,120	.11	.13
November	102,200	1,180	10,858	1.11	1.24
December	49,600	11,120	27,874	2.79	3.22
The year	102,200	720	12,606	1.29	17.32

Estimated monthly discharge of Susquehanna River at Wilkesbarre, Pa..
1899–1904—Continued.

Month.	Discharge in second-feet.			Run-off.	
	Maximum.	Minimum.	Mean.	Second-feet per square mile.	Depth in inches.
1901.					
January	52,000	14,870	29,018	2.96	3.41
February	22,320	11,520	16,278	1.66	1.73
March	108,400	9,580	34,736	3.54	4.08
April	78,800	18,890	39,255	4.00	4.46
May	74,300	9,950	21,462	2.19	2.52
June	51,300	4,800	15,676	1.60	1.79
July	5,620	1,810	3,065	.31	.36
August	31,600	2,000	7,405	.75	.86
September	9,950	2,200	4,257	.43	.48
October	5,340	2,200	3,570	.36	.42
November	24,800	1,810	5,289	.54	.60
December *a*	166,300	13,590	41,752	4.26	4.91
The year	166,300	1,810	18,480	1.88	25.62
1902.					
January	67,700	11,120	26,905	2.74	3.16
February	48,800	12,340	28,055	2.35	2.45
March	217,700	19,380	66,697	6.80	7.84
April	61,000	6,210	21,867	2.23	2.49
May	7,470	3,070	4,847	.49	.56
June	7,470	3,780	4,968	.51	.57
July	57,800	7,470	29,013	2.96	3.41
August	34,500	3,300	10,073	.10	.12
September	32,600	2,000	4,918	.50	.56
October	39,200	5,910	14,976	1.53	1.76
November	26,700	4,540	8,395	.86	.96
December	75,100	7,150	26,112	2.66	3.07
The year	217,700	2,000	20,152	1.96	26.95

a Frozen December 4 to 31. Rating table assumed to apply correctly.

Estimated monthly discharge of Susquehanna River at Wilkesbarre, Pa., 1899–1904—Continued.

Month.	Discharge in second-feet.			Run-off.	
	Maximum.	Minimum.	Mean.	Second feet per square mile.	Depth in inches.
1903.					
January	66,100	14,010	29,310	2.99	3.45
February	84,500	16,600	34,970	3.56	3.71
March	106,100	20,850	53,502	5.45	6.28
April	49,900	6,830	23,656	2.41	2.69
May	6,520	2,000	3,388	.35	.40
June	31,100	1,810	10,265	1.05	1.17
July	26,200	2,840	7,877	.80	.92
August	90,000	3,540	13,071	1.33	1.53
September	47,600	3,300	10,932	1.11	1.24
October	106,900	3,300	27,377	2.79	3.22
November	47,100	4,800	12,986	1.32	1.47
December	31,100	3,540	13,583	1.38	1.59
The year	106,900	1,810	20,076	2.04	27.67
1904.					
January	79,600	9,210	21,860	2.23	2.57
February	75,100	18,350	35,720	3.64	3.92
March	123,400	16,600	52,530	5.34	6.16
April	63,900	14,440	31,290	3.19	3.56
May	36,500	6,520	15,750	1.61	1.86
June	37,000	2,840	11,180	1.14	1.27
July	5,620	2,620	3,636	.371	.428
August	12,340	2,410	5,194	.529	.610
September	8,850	2,200	4,119	.420	.469
October	30,100	3,780	11,260	1.15	1.33
November	8,850	4,800	5,972	.609	.679
December	47,850	2,200	7,660	.781	.900
The year	123,400	2,200	17,180	1.75	23.76

SUSQUEHANNA RIVER AT DANVILLE, PA.

This station, 52 miles below Wilkesbarre and 11 miles above the mouth of the West Branch, was established on March 25, 1899, by E. G. Paul. It is located at the Mill Street Bridge, 600 feet south of the public square, Danville, Pa., near the Pennsylvania Railroad station at South Danville. The box of the standard chain gage is bolted to the hand rail on the lower side of the bridge 200 feet from the right bank. The length from the end of the weight to the marker is 42.85 feet. The gage is read once each day by E. F. Bell. Discharge measurements were made from the lower side of the Mill street covered wooden highway bridge. This bridge was carried away by the ice on March 9, 1904. From that time until the water dropped below gage height, 5 feet, its stage was observed on the Weather Bureau gage. After the water fell below 5 feet its stage was measured approximately, until September 30, 1904, by means of temporary gages set by the gage reader. This bridge had a total span of about 1,300 feet. The initial point for soundings was at the end of the wooden hand rail on the left bank, downstream side. The channel is straight for about one-half mile above and below the station. The right bank is low and liable to overflow. The left bank is high and is not subject to overflow. The bed of the stream is rocky, with some gravel, and is permanent. There is but one channel, broken by the six bridge piers, which do not obstruct the flow to any considerable extent. The current is moderately rapid, except at very low stages, when it becomes sluggish. The bench mark is the extreme south end of the stone doorsill at the east entrance to the city filter plant. Its elevation is 31.7 feet above gage datum.

Discharge measurements of Susquehanna River at Danville, Pa., 1899-1903.

Date.	Hydrographer.	Gage height.	Area of section.	Mean velocity	Discharge.
1899.		*Feet.*	*Sq. feet.*	*Feet per second.*	*Second-feet.*
Mar. 25	E. G. Paul	10.00	10,971	4.34	47,646
June 8 ... do .		3.00	2,235	1.76	3,927
July 27 ....do		2.40	1,607	1.41	2,272
Sept. 16 .....do		2.00	1,265	1.13	1,427
Oct. 17 .....do .		1.90	1,123	1.03	1,163
1900.					
May 20	E. G. Paul	4.60	3,799	2.76	10,515
Sept. 25 !... .do ..		1.60	798	1.03	822
1901.					
Aug. 19	E. G. Paul... ...	7.50	7,681	3.63	27,714
Oct. 27 .... do		3.10	2,051	2.20	4,510
1902.					
Apr. 22	E. G. Paul................. ..	5.20	4,541	3.17	14,898
Sept. 19 .....do		2.75	1,993	1.56	3,115
1903.					
Mar. 5	E. C. Murphy................	9.82	10,418	3.72	39,600
Apr. 9 .....do		8.60	8,848	3.66	33,000
May 9 .....do		3.44	2,688	1.85	4,963
Oct. 8	W. C. Sawyer..:....	3.46	2,845	2.01	5,728

Rating table for Susquehanna River at Wilkesbarre, Pa., from March 30, 1899, to December 31, 1904.

Gage height.	Discharge.	Gage height.	Discharge.	Gage height.	Discharge.	Gage height.	Discharge.
Feet.	Second-feet.	Feet.	Second-feet.	Feet.	Second-feet.	Feet.	Second-feet.
2.0	620	4.3	5,070	6.6	13,170	9.8	28,200
2.1	720	4.4	5,340	6.7	13,590	10.0	29,200
2.2	820	4.5	5,620	6.8	14,010	10.2	30,100
2.3	930	4.6	5,920	6.9	14,440	10.4	31,100
2.4	1,050	4.7	6,210	7.0	14,870	10.6	32,100
2.5	1,180	4.8	6,520	7.1	15,300	10.8	33,000
2.6	1,320	4.9	6,830	7.2	15,730	11.0	34,000
2.7	1,470	5.0	7,150	7.3	16,160	11.2	35,000
2.8	1,630	5.1	7,470	7.4	16,600	11.4	36,000
2.9	1,810	5.2	7,800	7.5	17,040	11.6	37,000
3.0	2,000	5.3	8,140	7.6	17,490	11.8	37,900
3.1	2,200	5.4	8,490	7.7	17,950	12.0	38,900
3.2	2,410	5.5	8,850	7.8	18,420	12.2	39,900
3.3	2,620	5.6	9,210	7.9	18,900	12.4	40,800
3.4	2,840	5.7	9,580	8.0	19,380	12.6	41,800
3.5	3,070	5.8	9,950	8.2	20,360	12.8	42,800
3.6	3,300	5.9	10,330	8.4	21,340	13.0	43,700
3.7	3,540	6.0	10,720	8.6	22,320	13.2	44,700
3.8	3,780	6.1	11,120	8.8	23,300	13.4	45,700
3.9	4,030	6.2	11,520	9.0	24,300	13.8	47,600
4.0	4,280	6.3	11,930	9.2	25,300	14.0	48,600
4.1	4,540	6.4	12,340	9.4	26,200		
4.2	4,800	6.5	12,750	9.6	27,200		

Table based on discharge measurements of 1899, 1900, 1901, 1902, 1903, and 1904. Well defined between 2 feet gage height and 19 feet gage height. Tangent at 8.80 feet gage height with a difference of 500 per tenth. Table applied to tenths.

Mean daily discharge, in second-feet, of Susquehanna River at Wilkesbarre, Pa.,
1899–1904.

Day.	Jan.	Feb.	Mar.	Apr.	May.	June.	July.	Aug.	Sept.	Oct.	Nov.	Dec.
1899.												
1				21,340	12,340	5,620	3,300	1,470	2,200	1,180	1,180	2,840
2				19,870	11,520	8,850	2,620	1,320	1,810	1,180	2,000	2,840
3				17,950	11,980	8,140	2,620	1,470	1,320	1,180	20,850	2,840
4				15,730	11,980	7,470	2,410	1,320	1,320	1,180	13,590	2,840
5				14,440	12,340	5,910	2,000	2,410	1,180	1,180	16,180	3,070
6				14,440	11,120	5,070	2,000	2,000	1,180	1,320	18,170	3,070
7				16,600	9,580	3,300	1,630	1,630	1,180	1,320	14,440	3,070
8				30,850	9,210	3,070	1,810	1,180	1,050	1,180	8,140	3,540
9				49,100	8,490	3,070	1,630	1,180	1,050	1,180	7,150	3,300
10				49,800	8,140	3,070	1,630	1,180	1,050	1,180	5,620	3,070
11				42,800	7,470	2,620	1,630	1,180	1,050	1,180	4,800	3,070
12				34,500	7,800	2,410	1,810	1,180	1,180	1,180	5,070	3,300
13				35,500	7,470	2,410	1,810	1,470	1,180	1,050	6,820	17,960
14				48,600	7,150	2,410	2,000	1,630	1,180	1,050	6,210	27,200
15				50,100	7,150	2,000	2,410	1,630	1,050	1,050	5,910	27,200
16				48,100	6,520	2,200	2,620	1,630	1,050	1,050	5,620	21,830
17				45,700	6,520	2,410	2,200	1,810	950	950	7,800	17,950
18				41,300	6,210	2,410	2,000	1,470	950	950	7,800	16,180
19				35,500	6,830	2,000	2,000	1,050	950	950	8,140	12,750
20				31,600	6,830	2,000	2,000	950	950	950	7,150	12,750
21				28,700	8,490	2,300	2,200	950	950	950	6,710	20,850
22				26,300	10,330	2,000	2,000	1,320	950	950	5,910	21,340
23				24,300	9,950	2,000	2,000	1,180	950	950	5,070	16,600
24				21,830	9,580	1,810	1,810	1,180	950	950	4,800	13,170
25				19,380	8,850	1,810	1,630	1,050	820	950	4,280	21,340
26				16,600	8,490	2,200	1,630	1,050	1,180	820	3,780	19,380
27				17,490	7,470	2,200	1,630	1,050	1,050	950	3,780	16,600
28				16,600	6,830	2,620	1,630	1,050	1,180	950	3,540	11,930
29				15,300	6,520	3,780	1,630	5,910	1,180	1,180	3,300	24,800
30				13,170	6,520	4,280	1,620	4,540	1,320	1,180	3,070	18,900
31					6,210		1,620	2,840		1,180		17,950
1900.												
1	14,010	16,600	31,100	14,440	11,120	3,780	2,000	2,410	2,200	950	1,470	31,600
2	11,520	14,010	75,900	17,040	9,950	3,540	1,630	2,410	2,000	950	1,320	25,310
3	12,340	11,680	52,200	28,200	8,850	4,800	1,470	2,000	2,000	950	1,320	19,870
4	14,010	12,750	37,900	36,000	8,140	4,080	1,810	1,810	2,000	950	1,180	16,600
5	14,870	21,340	28,700	34,500	7,800	3,540	1,810	1,810	1,810	950	1,470	38,400
6	14,870	21,830	21,340	26,300	7,150	3,780	2,840	1,810	1,630	820	1,630	38,400
7	14,440	18,900	20,360	27,200	6,520	3,300	4,080	1,810	1,470	720	1,810	35,500
8	14,010	18,420	19,870	37,400	6,210	3,300	3,300	1,810	1,470	720	1,810	28,700
9	11,560	51,600	17,950	39,900	5,910	3,300	2,840	1,810	1,320	820	1,810	28,700
10	11,120	25,300	21,340	31,500	5,620	3,780	2,410	1,630	1,320	820	1,810	20,360
11	9,950	28,200	24,300	25,300	5,620	4,080	2,200	1,630	1,470	820	2,400	17,040
12	10,330	25,300	18,420	18,900	6,520	5,070	1,810	1,470	1,470	820	2,200	13,170
13	9,210	25,300	14,010	16,160	6,830	5,070	2,000	1,470	1,470	820	2,620	11,120
14	10,330	30,400	11,930	17,950	6,520	6,520	2,000	1,320	1,180	820	3,070	11,120
15	9,210	46,900	9,580	18,420	6,210	5,070	2,000	1,320	1,050	820	2,840	30,600
16	8,850	37,900	9,580	18,420	6,210	4,280	2,000	1,320	1,180	950	2,840	28,200
17	8,850	25,300	24,300	17,400	6,830	4,080	1,810	1,180	1,050	1,050	2,680	25,300
18	7,800	17,950	19,870	29,400	7,150	3,300	1,810	1,180	950	1,050	2,410	22,810
19	7,470	24,800	20,850	41,000	7,470	3,300	1,630	1,050	820	1,180	2,410	25,300
20	9,950	32,000	21,830	40,800	9,210	2,840	2,200	1,180	820	1,470	2,200	27,200
21	52,900	28,200	33,200	34,500	7,800	2,410	2,410	1,140	720	1,320	2,410	26,300
22	64,800	36,000	27,700	29,200	7,150	2,410	2,200	1,140	820	1,320	2,410	24,300
23	46,900	63,200	25,300	34,500	6,520	3,070	2,000	1,050	820	1,470	3,300	23,300
24	30,600	53,600	21,340	35,500	5,910	1,620	1,810	2,000	820	1,810	4,280	25,310
25	21,830	34,000	28,700	32,600	5,620	1,620	1,810	1,810	820	1,630	5,070	23,300
26	18,420	28,200	22,810	26,700	5,070	2,410	4,240	1,320	950	1,630	6,210	42,800
27	18,900	34,870	19,870	21,340	4,540	2,840	3,540	1,470	950	1,470	68,000	49,600
28	11,520	21,830	15,300	17,040	4,240	2,200	2,840	1,630	820	1,470	102,200	43,300
29	25,300		14,870	14,440	4,080	2,200	2,410	1,630	820	1,470	52,900	40,800
30	24,300		14,010	12,750	3,780	2,840	2,410	1,320	950	1,470	37,900	38,000
31	22,810		12,750		3,540		2,620	2,200		1,470		38,000

Mean daily discharge, in second-feet, of Susquehanna River at Wilkesbarre, Pa., 1899–1904—Continued.

Day.	Jan.	Feb.	Mar.	Apr.	May.	June	July.	Aug.	Sept.	Oct.	Nov.	Dec.
1901.												
1	32,100	22,320	11,520	27,700	18,420	51,300	5,620	2,840	9,950	4,080	2,410	16,180
2	32,100	21,340	11,120	23,300	15,730	37,400	4,220	2,410	9,210	3,540	2,410	13,590
3	26,700	20,850	11,120	20,850	22,810	34,000	3,070	2,000	9,540	4,240	2,240	14,010
4	22,810	21,340	11,120	25,800	23,800	32,100	3,300	2,000	8,140	5,070	2,300	25,800
5	21,830	19,390	11,520	33,000	19,870	25,300	3,300	2,000	7,150	4,240	2,000	28,700
6	15,730	18,420	10,720	38,400	17,040	19,870	3,300	2,000	5,620	4,080	2,000	26,200
7	15,300	18,420	10,330	64,900	14,010	19,870	5,070	2,000	4,800	3,540	2,000	24,300
8	14,870	17,950	9,950	78,400	11,980	24,300	4,240	2,620	3,780	3,300	2,000	20,850
9	18,900	17,950	9,580	69,200	10,330	25,800	4,240	2,410	3,540	2,840	2,000	22,810
10	18,900	17,040	12,750	53,300	9,950	23,800	4,030	2,300	3,070	2,620	2,000	37,400
11	18,420	17,490	21,340	44,700	12,340	19,390	3,780	2,410	2,620	2,410	1,810	39,400
12	18,420	17,490	84,700	37,900	18,420	15,730	3,780	2,620	2,620	2,410	2,000	37,400
13	19,870	16,600	39,900	32,600	26,700	12,750	3,300	2,300	2,620	2,620	2,000	29,600
14	24,300	14,440	27,700	29,600	28,200	11,120	3,070	2,300	2,620	3,070	3,070	23,300
15	38,900	14,870	23,800	27,300	24,800	10,330	2,840	2,410	4,540	4,240	98,900	
16	52,000	15,300	24,800	25,800	19,380	9,580	2,410	3,300	2,620	5,070	6,210	166,300
17	48,600	16,160	23,300	23,800	15,300	8,850	2,410	8,540	3,070	5,340	5,620	122,300
18	43,700	16,160	20,850	21,830	13,590	8,140	3,300	20,110	3,780	5,070	4,800	59,500
19	41,300	15,730	19,380	19,870	14,010	6,830	2,840	9,210	4,240	4,900	4,540	34,000
20	36,500	14,440	29,600	18,890	14,870	6,210	2,620	6,520	4,800	4,240	4,240	20,880
21	26,200	14,440	39,600	34,300	15,300	5,910	2,200	5,910	4,540	4,030	4,240	18,420
22	31,600	13,590	54,000	78,800	12,750	5,340	2,000	14,655	4,030	3,780	4,030	26,700
23	34,000	14,010	52,000	70,800	12,340	5,620	2,200	14,440	3,540	3,540	3,780	35,000
24	34,000	12,340	43,300	54,000	18,900	9,210	2,200	12,750	3,070	3,540	3,780	37,400
25	37,400	12,340	43,300	53,300	24,300	9,580	2,000	31,600	2,840	3,300	10,720	47,100
26	34,000	11,930	47,600	46,700	20,850	9,580	2,000	25,300	2,410	2,840	24,800	46,200
27	31,600	11,520	71,100	40,300	17,490	7,150	1,810	15,300	2,410	2,840	17,490	45,200
28	29,200	11,930	108,400	34,000	16,600	4,800	1,810	11,120	2,200	2,840	11,520	42,800
29	26,700		90,300	27,300	32,100	5,620	2,000	8,140	2,620	2,410	8,850	44,200
30	25,800		58,800	22,320	68,900	4,800	2,620	6,520	3,780	2,300	9,580	46,200
31	24,800		43,300		74,300		3,300	6,830		2,200		
1902.												
1	44,600	42,300	201,800	27,700	7,150	4,540	32,100	23,300	3,300	27,300	26,700	7,470
2	44,700	86,000	217,700	25,800	6,880	4,240	31,600	26,700	3,070	33,000	20,380	7,150
3	39,400	33,000	208,200	24,300	7,470	4,030	20,850	34,500	2,840	32,100	16,600	7,150
4	33,500	82,600	148,800	21,830	7,470	4,080	18,420	27,300	2,840	21,830	14,010	7,800
5	27,300	21,830	97,100	19,870	6,520	3,780	21,830	23,	2,410	15,300	12,340	8,850
6	28,700	14,870	52,900	18,900	6,520	3,780	20,600	17,	2,410	15,300	10,720	10,330
7	28,200	24,800	37,200	17,480	6,210	6,520	42,300	14,010	2,410	14,440	9,950	9,950
8	27,300	27,200	32,600	17,950	6,210	5,620	49,900	12,	2,410	13,590	8,850	8,850
9	27,700	27,200	30,600	38,100	5,620	5,340	44,400	11,	2,410	11,520	9,210	7,800
10	26,200	26,200	34,000	61,000	5,340	4,800	23,060	9,	2,410	9,950	9,540	10,330
11	25,300	24,300	41,300	58,400	5,070	4,800	24,300	9,880	3,300	8,850	7,150	15,730
12	24,300	24,300	54,000	42,800	4,800	4,800	27,700	8,490	3,070	9,950	6,210	19,380
13	20,380	24,300	78,000	51,300	4,540	4,540	21,830	8,490	3,300	12,750	6,210	28,400
14	15,730	20,850	91,700	30,600	4,240	4,800	16,600	8,490	3,070	10,720	6,210	30,100
15	12,340	19,380	79,600	26,200	4,030	4,800	11,980	7,800	3,070	9,950	6,210	25,300
16	14,010	20,380	61,000	22,320	3,780	4,800	9,950	7,150	2,840	10,330	5,910	46,000
17	15,730	18,420	92,100	19,380	3,780	7,150	8,490	5,910	2,620	10,330	5,620	46,000
18	14,870	17,950	97,100	16,600	8,540	6,210	7,470	5,340	2,620	9,210	5,340	42,800
19	13,590	15,730	73,500	14,870	3,540	5,340	7,470	4,800	2,410	8,140	5,070	40,800
20	11,120	18,170	50,600	13,590	3,300	5,910	8,490	4,800	2,200	6,830	4,900	35,500
21	11,520	18,170	57,000	12,340	3,070	5,070	39,400	4,240	2,200	6,520	5,070	29,200
22	32,100	12,750	30,100	11,520	3,070	5,070	57,800	4,240	2,000	6,830	4,900	59,500
23	67,700	12,340	27,700	10,720	3,070	4,800	48,100	4,240	2,000	7,800	4,540	65,000
24	39,900	15,730	27,200	9,540	3,540	4,800	45,900	4,030	2,000	7,150	4,540	47,100
25	32,600	15,730	26,700	8,850	3,540	4,800	47,400	4,080	2,000	6,210	4,540	34,000
26	27,700	17,950	24,300	7,800	3,540	4,800	54,700	3,780	4,540	6,210	4,540	27,700
27	23,800	23,300	21,830	7,150	4,540	4,540	37,400	3,300	15,300	5,910	5,620	27,700
28	20,380	48,800	19,380	6,520	4,080	4,030	27,700	3,300	10,720	17,580	6,210	21,830
29	17,950		24,300	6,210	5,910	3,780	33,000	3,300	18,900	84,200	7,150	19,380
30	17,490		31,100	6,830	5,910	7,470	32,100	3,300	32,600	39,200	7,800	14,870
31	45,200		28,200		4,800		25,800	3,300		84,500		14,010

Mean daily discharge, in second-feet, of Susquehanna River at Wilkesbarre, Pa., 1899–1904—Continued.

Day.	Jan.	Feb.	Mar.	Apr.	May.	June.	July.	Aug.	Sept	Oct.	Nov.	Dec.
1903.												
1	21,830	57,400	98,900	35,000	6,520	2,000	14,440	5,910	47,000	3,300	9,210	15,730
2	34,000	44,200	94,700	38,900	5,910	2,000	14,010	5,916	38,400	3,300	9,210	16,160
3	42,900	43,700	64,500	32,600	5,340	2,000	11,120	5,070	28,700	3,300	8,490	17,040
4	43,700	53,000	46,700	27,700	5,070	1,810	8,850	4,280	21,340	3,300	7,800	13,590
5	46,300	34,500	35,500	28,200	4,800	1,810	8,490	6,520	16,000	3,300	7,150	7,800
6	27,700	65,100	31,690	28,700	4,540	1,810	8,850	13,500	13,500	3,540	7,150	5,620
7	19,870	48,100	39,400	22,810	4,280	1,810	16,160	18,900	11,520	3,780	7,150	4,800
8	18,900	30,500	37,000	23,300	4,290	1,810	16,160	17,490	9,050	6,210	7,800	4,800
9	14,440	39,200	63,930	32,100	4,280	2,000	26,200	14,010	8,900	32,600	8,140	5,070
10	14,010	22,320	83,000	33,000	3,780	2,000	6,520	10,720	8,140	88,100	7,150	4,540
11	32,000	19,380	77,300	28,200	3,540	1,810	5,340	9,580	7,800	106,900	6,830	4,280
12	29,200	21,850	85,600	24,700	3,540	3,300	5,070	8,480	8,140	106,000	6,210	3,540
13	26,700	24,800	76,300	23,800	3,300	13,170	4,280	8,850	10,720	79,200	5,910	4,030
14	24,800	34,000	60,300	20,850	3,070	7,150	3,780	7,800	9,210	47,100	5,620	6,210
15	24,800	33,000	44,700	30,000	3,070	17,040	3,540	6,830	7,800	37,600	5,340	6,520
16	29,200	25,800	37,400	30,000	3,070	12,340	3,300	6,210	6,520	24,300	4,800	9,950
17	31,000	21,340	32,100	40,000	2,840	9,950	3,300	5,620	5,910	19,380	18,900	13,170
18	31,100	16,600	28,700	31,000	2,840	7,800	5,910	5,070	7,800	20,850	43,300	14,010
19	27,200	29,200	27,200	24,300	2,620	7,150	5,070	4,030	6,520	11,300	47,100	11,930
20	22,810	26,200	25,300	19,380	2,620	6,520	5,910	3,540	7,170	40,800	33,000	9,950
21	22,320	26,200	22,810	16,160	2,620	6,210	6,520	3,780	6,520	33,500	22,810	23,800
22	26,200	29,200	20,850	14,010	3,070	14,010	6,210	9,210	6,210	26,200	15,300	24,300
23	28,200	31,000	48,100	12,340	2,620	19,380	6,210	8,140	5,340	20,850	13,500	21,340
24	31,100	33,500	103,400	11,120	2,620	24,000	5,340	7,150	4,800	17,040	11,520	19,380
25	29,200	35,000	106,100	10,550	2,520	26,500	5,340	5,910	4,280	14,780	11,520	17,040
26	27,200	31,100	78,100	34,000	2,290	31,500	11,120	5,340	4,030	14,010	11,120	15,300
27	22,810	27,200	58,100	8,850	2,290	30,100	7,800	5,070	3,780	12,750	9,950	15,300
28	20,300	30,100	41,800	8,140	2,390	21,890	6,620	6,400	3,780	11,520	8,850	31,100
29	20,300	.....	32,600	7,470	2,000	14,440	4,540	25,000	3,540	10,720	10,720	27,700
30	51,300	.....	28,700	6,830	2,000	17,040	4,800	90,000	3,300	9,950	17,950	25,300
31	66,100	.....	28,300	.....	2,000	.....	6,210	68,700	.....	9,210	.....	21,340
1904.												
1	24,300	48,600	16,600	38,900	36,500	3,540	6,520	3,540	6,520	8,140	4,800	
2	23,800	43,700	16,900	56,000	31,000	16,600	3,070	5,340	3,300	8,490	7,470	4,540
3	21,850	40,300	18,350	61,000	26,200	14,870	3,070	4,800	3,070	10,330	6,830	4,800
4	15,730	37,000	53,300	48,600	21,340	12,340	3,070	5,070	2,840	7,800	6,520	4,800
5	12,750	34,000	40,100	38,900	17,490	10,720	3,070	8,140	2,840	6,210	5,910	5,300
6	13,500	33,500	36,100	32,000	14,870	24,800	3,070	7,150	2,620	5,620	5,620	2,620
7	15,730	37,000	38,900	30,100	13,500	16,600	3,300	5,340	2,620	5,070	5,620	3,070
8	15,730	55,900	74,500	31,000	11,930	12,340	3,540	6,910	2,620	4,280	5,620	3,300
9	16,160	75,100	53,700	34,000	10,720	13,170	4,800	7,150	3,070	4,280	5,620	2,620
10	16,600	71,300	32,600	37,400	9,580	37,000	3,780	5,340	3,070	4,280	5,620	2,620
11	16,160	67,000	68,000	63,000	8,850	33,500	3,540	5,910	3,070	4,030	5,620	2,410
12	15,300	57,600	57,600	50,000	7,800	21,890	4,540	4,280	2,620	3,780	5,840	2,200
13	14,870	49,400	44,500	30,900	6,520	15,300	4,540	3,780	2,620	3,780	5,840	2,620
14	14,870	42,300	36,800	33,000	7,150	15,300	5,620	4,030	2,410	4,030	5,070	2,410
15	13,500	35,000	31,100	27,700	6,520	11,520	4,800	3,780	3,300	20,850	5,070	2,620
16	12,340	30,050	27,500	23,800	11,120	9,210	4,030	3,300	8,850	14,440	5,070	2,410
17	11,520	26,800	31,000	20,850	19,380	7,470	3,780	3,070	5,070	10,720	5,070	2,620
18	10,720	21,850	30,000	19,380	18,900	6,520	3,300	2,840	6,520	8,850	5,340	2,620
19	10,330	21,050	35,500	18,100	15,300	6,520	3,300	2,620	5,340	7,470	5,070	2,620
20	9,210	21,850	42,800	18,100	35,000	5,620	3,540	2,410	4,540	6,520	5,070	2,840
21	9,210	21,350	45,100	18,420	30,100	5,070	4,800	2,410	3,780	7,150	5,070	2,840
22	10,720	21,850	40,800	18,100	21,850	5,070	3,780	2,620	13,900	22,320	5,910	2,840
23	42,300	25,700	27,700	15,300	16,160	5,070	4,030	3,540	2,840	30,100	5,910	2,620
24	79,600	21,350	62,200	14,870	12,750	4,280	2,620	12,340	2,840	30,100	8,140	3,070
25	46,300	21,350	62,200	14,870	12,750	4,030	2,620	12,340	2,840	23,300	8,850	3,900
26	37,000	21,500	48,600	14,440	13,500	3,780	2,840	8,140	8,490	14,440	7,800	2,620
27	29,600	19,600	123,400	15,730	12,750	3,540	3,540	8,140	8,490	14,440	7,150	3,070
28	24,300	19,600	116,100	14,440	10,550	3,540	3,300	5,910	8,490	14,440	6,520	29,200
29	20,300	18,350	83,300	40,800	10,720	3,070	3,300	5,070	7,800	12,340	4,800	47,850
30	25,300	.....	49,400	42,400	8,850	2,840	3,780	4,540	6,210	10,720	4,800	45,200
31	48,100	.....	37,400	.....	8,140	.....	4,540	4,030	.....	10,330	.....	33,000

From February 8 to March 19, 1904, discharges reduced 50 per cent on account of ice gorge.

Estimated monthly discharge of Susquehanna River at Wilkesbarre, Pa.,
1899–1904.

[Drainage area, 9,810 square miles.]

Month.	Discharge in second-feet.			Run-off.	
	Maximum.	Minimum.	Mean.	Second-feet per square mile.	Depth in inches.
1899.					
April	50,100	13,170	28,773	2.93	3.27
May	12,340	6,210	8,574	.87	1.00
June	8,850	1,810	3,378	.34	.38
July	8,300	1,820	1,965	.20	.23
August	5,910	930	1,653	.17	.20
September	2,200	820	1,140	.12	.13
October	1,320	820	1,072	.11	.13
November	20,850	1,180	7,046	.72	.80
December	27,200	2,840	12,694	1.29	1.49
1900.					
January	68,800	7,470	18,279	1.86	2.14
February	63,200	11,980	28,226	2.88	3.00
March	75,900	9,580	23,780	2.42	2.79
April	41,000	12,750	26,848	2.69	3.00
May	11,120	3,540	6,583	.67	.77
June	6,520	2,200	3,506	.36	.40
July	4,280	1,470	2,320	.24	.28
August	2,410	1,050	1,685	.17	.20
September	2,200	720	1,289	.13	.15
October	1,810	720	1,120	.11	.13
November	102,200	1,180	10,858	1.11	1.24
December	49,600	11,120	27,874	2.79	3.22
The year	102,200	720	12,606	1.29	17.32

Estimated monthly discharge of Susquehanna River at Wilkesbarre, Pa.,
1899–1904—Continued.

Month.	Discharge in second-feet.			Run-off.	
	Maximum.	Minimum.	Mean.	Second-feet per square mile.	Depth in inches.
1901.					
January	52,000	14,870	29,018	2.96	3.41
February	22,320	11,520	16,278	1.66	1.73
March	108,400	9,580	34,736	3.54	4.08
April	78,800	18,890	39,255	4.00	4.46
May	74,300	9,950	21,462	2.19	2.52
June	51,300	4,800	15,676	1.60	1.79
July	5,620	1,810	3,065	.31	.36
August	31,600	2,000	7,405	.75	.86
September	9,950	2,200	4,257	.43	.48
October	5,340	2,200	3,570	.36	.42
November	24,800	1,810	5,289	.54	.60
December *a*	166,300	13,590	41,752	4.26	4.91
The year	166,300	1,810	18,480	1.88	25.62
1902.					
January	67,700	11,120	26,905	2.74	3.16
February	48,800	12,340	23,055	2.35	2.45
March	217,700	19,380	66,697	6.80	7.84
April	61,000	6,210	21,867	2.23	2.49
May	7,470	3,070	4,847	.49	.56
June	7,470	3,780	4,968	.51	.57
July	57,800	7,470	29,013	2.96	3.41
August	34,500	3,300	10,073	.10	.12
September	32,600	2,000	4,918	.50	.56
October	39,200	5,910	14,976	1.53	1.76
November	26,700	4,540	8,395	.86	.96
December	75,100	7,150	26,112	2.66	3.07
The year	217,700	2,000	20,152	1.98	26.95

a Frozen December 4 to 31. Rating table assumed to apply correctly.

*Estimated monthly discharge of Susquehanna River at Wilkesbarre, Pa.,
1899-1904—Continued.*

Month.	Discharge in second-feet.			Run-off.	
	Maximum.	Minimum.	Mean.	Second feet per square mile.	Depth in inches.
1903.					
January	66,100	14,010	29,310	2.99	3.45
February	84,500	16,600	34,970	3.56	3.71
March	106,100	20,850	53,502	5.45	6.28
April	49,900	6,830	23,656	2.41	2.69
May	6,520	2,000	3,388	.35	.40
June	31,100	1,810	10,265	1.05	1.17
July	26,200	2,840	7,877	.80	.92
August	90,000	3,540	13,071	1.33	1.53
September	47,600	3,300	10,932	1.11	1.24
October	106,900	3,300	27,377	2.79	3.22
November	47,100	4,800	12,986	1.32	1.47
December	31,100	3,540	13,583	1.38	1.59
The year	106,900	1,810	20,076	2.04	27.67
1904.					
January	79,600	9,210	21,860	2.23	2.57
February	75,100	18,350	35,720	3.64	3.92
March	123,400	16,600	52,530	5.34	6.16
April	63,900	14,440	31,290	3.19	3.56
May	36,500	6,520	15,750	1.61	1.86
June	37,000	2,840	11,180	1.14	1.27
July	5,620	2,620	3,636	.371	.428
August	12,340	2,410	5,194	.529	.610
September	8,850	2,200	4,119	.420	.469
October	30,100	3,780	11,260	1.15	1.33
November	8,850	4,800	5,972	.609	.679
December	47,850	2,200	7,660	.781	.900
The year	123,400	2,200	17,180	1.75	23.76

SUSQUEHANNA RIVER AT DANVILLE, PA.

This station, 52 miles below Wilkesbarre and 11 miles above the mouth of the West Branch, was established on March 25, 1899, by E. G. Paul. It is located at the Mill Street Bridge, 600 feet south of the public square, Danville, Pa., near the Pennsylvania Railroad station at South Danville. The box of the standard chain gage is bolted to the hand rail on the lower side of the bridge 200 feet from the right bank. The length from the end of the weight to the marker is 42.85 feet. The gage is read once each day by E. F. Bell. Discharge measurements were made from the lower side of the Mill street covered wooden highway bridge. This bridge was carried away by the ice on March 9, 1904. From that time until the water dropped below gage height, 5 feet, its stage was observed on the Weather Bureau gage. After the water fell below 5 feet its stage was measured approximately, until September 30, 1904, by means of temporary gages set by the gage reader. This bridge had a total span of about 1,300 feet. The initial point for soundings was at the end of the wooden hand rail on the left bank, downstream side. The channel is straight for about one-half mile above and below the station. The right bank is low and liable to overflow. The left bank is high and is not subject to overflow. The bed of the stream is rocky, with some gravel, and is permanent. There is but one channel, broken by the six bridge piers, which do not obstruct the flow to any considerable extent. The current is moderately rapid, except at very low stages, when it becomes sluggish. The bench mark is the extreme south end of the stone doorsill at the east entrance to the city filter plant. Its elevation is 31.7 feet above gage datum.

Discharge measurements of Susquehanna River at Danville, Pa., 1899–1903.

Date.	Hydrographer.	Gage height.	Area of section.	Mean velocity.	Discharge.
1899.		*Feet.*	*Sq. feet.*	*Feet per second.*	*Second-feet.*
Mar. 25	E. G. Paul	10.00	10,971	4.34	47,646
June 8	....do .	3.00	2,235	1.76	3,927
July 27	....do	2.40	1,607	1.41	2,272
Sept. 16	....do	2.00	1,265	1.13	1,427
Oct. 17	.....do .	1.90	1,123	1.03	1,163
1900.					
May 20	E. G. Paul	4.60	3,799	2.76	10,515
Sept. 25	.....do	1.60	798	1.03	822
1901.					
Aug. 19	E. G. Paul	7.50	7,631	3.63	27,714
Oct. 27	.... do	3.10	2,051	2.20	4,510
1902.					
Apr. 22	E. G. Paul.	5.20	4,541	3.17	14,393
Sept. 19	...do	2.75	1,998	1.56	3,115
1903.					
Mar. 5	E. C. Murphy.................	9.83	10,413	3.72	39,600
Apr. 9	.....do	8.60	8,848	3.66	33,000
May 9	.....do	3.44	2,688	1.85	4,963
Oct. 8	W. C. Sawyer.'....	3.46	2,845	2.01	5,728

Mean daily gage height, in feet, of Susquehanna River at Danville, Pa., 1899–1904.

Day.	Jan.	Feb.	Mar.	Apr.	May.	June.	July.	Aug.	Sept.	Oct.	Nov.	Dec.
1899.												
1				6.95	4.80	3.30	3.20	2.20	2.80	2.10	2.10	3.10
2				6.80	4.65	3.40	3.00	2.20	2.60	2.10	2.40	3.00
3				6.35	4.60	3.70	2.80	2.60	2.50	2.10	2.60	3.00
4				6.00	4.60	3.60	2.70	2.30	2.50	2.10	6.10	3.00
5				5.65	4.60	3.50	2.60	2.20	2.30	2.00	5.40	3.00
6				5.50	4.55	3.30	2.60	2.20	2.20	3.00	5.70	2.90
7				5.65	4.35	3.20	2.60	2.50	2.20	3.00	5.20	3.10
8				6.90	4.15	3.00	2.50	2.30	2.10	3.00	4.70	3.10
9				10.50	3.80	3.00	2.50	2.20	2.40	2.10	4.80	3.10
10				11.60	3.70	2.90	2.50	2.20	2.20	2.10	3.90	3.00
11				10.45	3.70	2.90	2.50	2.60	2.10	2.00	7.30	3.00
12				9.15	3.75	2.90	2.40	2.30	2.20	2.00	3.90	3.10
13				8.95	3.80	2.70	2.60	2.40	2.10	2.00	3.70	4.20
14				10.75	3.70	2.70	2.60	2.30	2.10	1.90	4.00	6.80
15				11.55	3.70	2.60	2.60	2.30	2.10	1.90	3.90	7.80
16				11.40	3.60	2.60	2.80	2.30	2.00	1.90	3.80	7.60
17				10.85	3.60	2.60	2.80	2.90	1.90	1.90	3.90	6.10
18				10.05	3.70	2.60	2.70	2.30	1.90	1.90	4.30	6.10
19				9.05	3.60	2.60	2.70	2.30	1.80	1.90	4.40	5.70
20				8.25	3.60	2.50	2.50	2.20	1.80	1.90	4.30	5.70
21				7.75	3.60	2.50	2.50	2.10	1.90	1.90	4.10	5.40
22				7.35	3.80	2.50	2.50	2.10	1.90	1.90	3.80	5.60
23				7.05	3.80	2.50	2.50	2.10	1.80	1.90	3.90	6.90
24				6.65	3.80	2.50	2.50	2.10	1.80	1.90	3.60	6.30
25			10.00	6.20	3.80	2.50	2.50	2.00	1.80	1.90	3.40	6.50
26			9.25	5.85	3.70	2.70	2.40	1.90	1.90	1.90	3.40	7.10
27			8.10	5.70	3.60	2.60	2.40	2.00	1.90	1.90	3.30	6.90
28			7.85	5.65	3.50	2.60	2.40	2.30	1.80	1.80	3.20	6.40
29			7.30	5.35	3.30	2.90	2.40	2.20	1.90	1.80	3.10	5.80
30			7.55	5.10	3.20	3.20	2.40	3.50	2.10	1.90	3.10	5.00
31			7.45		3.30		2.30	3.20		1.90		
1900.												
1	(a)	(a)	7.55	5.00	5.35	3.00	2.30	2.40	2.20	1.70	2.00	8.75
2	(a)	(a)	15.25	5.80	5.05	2.90	2.30	2.40	2.20	1.70	2.00	7.15
3	(a)	(a)	13.10	6.75	4.80	2.90	2.30	2.30	2.20	1.70	2.00	5.90
4	(a)	(a)	10.65	8.40	4.55	3.50	2.20	2.20	2.20	1.70	2.00	5.50
5	(a)	(a)	9.25	9.30	4.40	3.30	2.20	2.20	2.20	1.70	2.00	7.10
6	(a)	(a)	7.10	8.45	4.25	3.10	2.30	2.10	2.10	1.70	2.00	8.80
7	(a)	(a)	7.10	7.40	4.15	3.00	2.70	2.50	2.10	1.70	2.00	9.05
8	(a)	(a)	7.30	8.70	4.05	2.90	2.30	2.20	2.00	1.70	2.00	8.55
9	(a)	9.70	6.85	9.75	4.00	2.90	2.90	2.10	1.80	1.70	2.00	7.50
10	(a)	9.90	6.75	9.45	3.95	2.90	2.70	2.10	1.90	1.70	2.00	6.85
11	(a)	7.60	7.50	9.25	3.85	3.10	2.50	2.00	1.80	1.70	2.10	6.30
12	(a)	7.80	7.20	7.10	3.90	3.10	2.50	2.00	1.80	1.70	2.10	5.55
13	(a)	9.40	6.40	6.30	4.10	3.30	2.40	1.90	1.80	1.70	2.20	5.20
14	(a)	9.60	5.65	6.10	4.20	3.30	2.30	2.00	1.80	1.70	2.40	5.10
15	(a)	11.20	5.20	6.30	4.00	3.90	2.30	2.00	1.80	1.80	2.40	5.00
16	(a)	10.40	4.90	6.65	4.00	3.50	2.30	1.90	1.80	1.80	2.60	6.80
17	(a)	8.30	4.70	6.35	3.80	3.20	2.80	1.90	1.70	1.80	2.50	(a)
18	(a)	7.30	4.90	7.00	3.90	3.00	2.30	1.90	1.70	1.80	2.50	(a)
19	(a)	5.70	5.05	9.75	3.90	3.00	2.30	1.80	1.70	1.80	2.50	(a)
20	(a)	5.00	5.10	10.55	4.40	2.90	2.30	1.80	1.70	1.80	2.50	(a)
21	9.40	4.70	7.95	9.15	4.40	2.80	2.20	1.90	1.70	1.70	2.50	(a)
22	12.70	5.95	8.80	8.95	4.10	2.70	2.40	1.90	1.70	1.70	2.50	(a)
23	11.95	12.15	7.95	8.10	3.90	2.60	2.30	1.80	1.60	1.90	2.60	(a)
24	9.70	13.50	7.40	8.35	3.70	2.60	2.20	1.80	1.60	2.10	2.70	(a)
25	7.80	11.05	7.40	9.30	3.60	2.70	2.10	2.30	1.60	2.30	2.90	(a)
26	6.80	8.95	6.95	8.40	3.60	2.60	2.30	2.10	1.70	2.20	8.90	7.05
27	6.45	6.85	6.95	7.40	3.40	2.50	3.00	2.20	1.70	2.10	8.45	8.60
28	6.30	5.45	6.50	8.20		2.50	2.90	2.10	1.70	2.10	16.60	7.55
29	5.80		5.85	6.10	3.20	2.40	2.60	2.00	1.70	2.10	12.65	6.95
30	5.80		5.90	5.65	3.10	2.40	2.40	2.00	2.00	2.00	10.20	6.55
31	(a)		5.65		8.00		2.40	2.00		2.00		4.80

a River frozen.

Mean daily gage height, in feet, of Susquehanna River at Danville, Pa., 1899-1904—Continued.

Days.	Jan.	Feb.	Mar.	Apr.	May.	June.	July	Aug.	Sept.	Oct.	Nov.	Dec.
1901.												
1	5.70	(a)	(a)	8.50	6.65	13.60	3.70	3.30	4.60	3.50	2.90	4.10
2	8.60	(a)	(a)	7.65	6.10	9.05	3.70	3.00	5.10	3.50	2.90	3.90
3	(a)	(a)	(a)	7.90	6.50	9.65	3.40	2.70	4.90	3.85	2.80	3.90
4	(a)	(a)	(a)	7.60	7.60	9.15	3.20	2.60	4.90	4.05	2.80	4.55
5	(a)	(a)	(a)	8.65	7.35	8.30	3.10	2.60	4.80	3.85	2.80	5.95
6	(a)	(a)	(a)	9.40	6.65	7.30	3.10	2.60	4.25	3.70	2.70	7.90
7	(a)	(a)	(a)	8.60	6.05	6.80	3.10	2.70	3.95	3.50	2.70	b8.30
8	(a)	(a)	(a)	8.55	6.35	7.30	3.60	2.90	3.70	3.30	2.60	b8.70
9	(a)	(a)	(a)	7.80	5.30	7.60	3.50	3.10	3.50	3.20	2.60	b9.10
10	(a)	(a)	(a)	7.45	6.50	7.55	3.40	2.90	3.35	3.20	2.60	9.55
11	(a)	(a)	(a)	7.10	5.00	7.00	3.30	3.50	3.25	3.10	2.60	9.80
12	(a)	(a)	12.00	6.75	5.70	6.40	3.30	3.30	3.10	3.00	2.60	10.05
13	(a)	(a)	11.15	6.50	6.60	5.60	3.20	3.00	3.00	2.90	2.90	8.90
14	(a)	(a)	8.50	8.60	7.95	5.20	3.10	2.90	3.05	3.90	2.90	7.90
15	(a)	(a)	7.60	8.15	7.85	5.00	3.00	2.80	3.00	3.85	3.00	14.65
16	(a)	(a)	7.30	7.80	7.05	4.95	2.90	2.80	3.10	3.90	3.45	22.57
17	(a)	(a)	7.40	7.45	6.30	4.60	2.90	2.80	3.10	3.90	3.90	20.05
18	(a)	(a)	6.90	7.10	5.80	4.60	3.00	6.60	3.40	8.90	3.90	13.85
19	(a)	(a)	6.60	6.75	5.80	4.45	3.10	7.85	3.50	3.80	3.60	10.25
20	(a)	(a)	6.60	6.50	5.70	4.10	3.00	5.60	3.50	3.90	3.50	8.30
21	(a)	(a)	9.25	6.90	5.95	4.00	2.90	4.55	3.60	3.60	3.50	7.10
22	(a)	(a)	11.85	12.60	5.75	3.90	2.80	4.75	3.60	3.50	3.50	5.90
23	(a)	(a)	12.70	15.25	5.35	3.90	2.70	6.30	3.40	3.40	3.40	5.10
24	(a)	(a)	11.35	12.75	5.40	4.25	2.60	8.10	3.40	3.30	3.40	4.90
25	(a)	(a)	11.25	13.05	6.55	5.35	2.60	11.02	3.10	3.30	3.70	4.75
26	(a)	(a)	11.15	11.70	7.40	4.70	2.60	9.25	3.00	3.20	6.17	4.95
27	(a)	(a)	13.35	10.65	6.90	4.45	2.60	7.55	2.90	3.10	7.00	5.10
28	(a)	(a)	17.00	8.90	6.40	4.10	2.60	6.15	2.90	3.10	5.85	5.00
29	(a)	......	16.85	8.25	8.00	8.85	2.50	5.35	2.90	3.00	4.95	5.20
30	(a)	......	13.35	7.85	12.70	3.80	2.70	4.70	3.20	3.00	4.85	7.15
31	(a)	......	10.45	......	14.95	......	2.90	4.40	......	2.90	......	6.80
1902.												
1	6.60	4.85	20.67	7.85	4.40	3.50	6.10	7.70	3.10	8.95	7.05	4.30
2	6.20	5.05	24.43	7.60	4.30	3.40	8.95	7.75	8.00	9.15	6.80	4.20
3	5.40	(c)	26.07	7.40	4.20	3.30	7.40	8.70	8.00	9.05	5.80	4.20
4	5.50	......	22.25	7.10	4.20	3.30	6.90	8.20	2.90	7.65	5.45	4.60
5	6.70	......	18.20	6.65	4.20	3.30	6.90	7.20	2.90	6.75	5.20	4.70
6	(c)	......	14.50	6.45	4.10	3.20	6.90	6.75	2.80	6.80	5.00	4.90
7	......	......	10.75	6.80	4.00	3.20	6.90	5.85	2.80	6.50	4.85	4.90
8	......	......	8.55	6.50	4.00	3.90	11.90	5.45	2.90	6.10	4.70	4.80
9	10.80	......	8.35	7.30	3.90	3.70	10.45	5.20	2.70	5.60	4.70	4.70
10	9.45	......	9.10	11.90	3.80	3.50	7.85	5.00	2.90	5.20	4.50	4.30
11	9.10	......	10.25	13.10	3.80	3.50	7.25	4.70	8.00	4.90	4.30	4.20
12	9.30	......	11.55	11.20	3.70	3.60	7.40	4.60	3.10	5.40	4.20	4.30
13	(c)	......	14.15	9.75	3.60	3.50	7.90	4.50	3.00	6.00	4.10	4.40
14	......	......	16.15	8.65	3.50	3.50	7.20	4.50	3.10	5.60	4.00	5.00
15	......	......	15.55	7.70	3.50	3.60	6.55	4.40	3.10	5.25	3.90	6.50
16	......	......	13.95	7.05	3.40	3.60	6.15	4.30	3.00	5.10	3.90	7.80
17	......	......	14.25	6.60	3.30	3.70	4.85	4.10	2.90	5.10	3.90	9.40
18	......	......	16.60	6.35	3.30	4.10	4.60	3.80	2.90	5.00	8.80	10.30
19	......	......	15.60	6.15	3.20	4.00	4.40	3.60	2.70	4.70	8.70	10.60
20	......	......	12.80	5.90	3.10	3.80	4.40	3.60	2.60	4.50	3.60	9.40
21	......	......	10.95	5.45	3.10	3.10	5.30	3.50	2.60	4.20	3.60	8.80
22	......	......	8.90	5.30	3.10	3.70	11.90	3.50	2.60	4.10	3.60	12.70
23	8.10	......	8.00	5.10	3.00	3.60	12.00	3.40	2.60	4.20	3.50	14.80
24	9.45	......	6.40	4.90	3.00	3.50	11.30	3.40	2.50	4.30	3.50	14.40
25	8.50	......	7.20	4.70	3.20	3.50	10.90	3.40	2.60	4.20	3.50	11.80
26	7.40	......	7.10	4.50	3.30	3.60	11.90	3.30	4.75	4.00	3.60	9.75
27	6.90	......	7.05	4.80	3.20	3.70	10.20	3.20	6.85	5.60	3.80	8.40
28	6.75	13.75	6.65	4.10	3.80	3.70	8.30	3.20	6.20	8.90	3.90	7.60
29	6.40	......	6.75	4.00	3.80	3.60	8.10	3.10	6.05	9.70	4.00	6.80
30	6.20	......	8.15	4.30	3.80	4.20	9.30	3.10	7.95	9.35	4.20	6.30
31	5.55	......	8.30	......	3.70	......	8.20	3.10	......	8.20	......	5.70

a Ice
b Estimated
c Frozen from January 6 to 8, 13 to 21, February 3 to 27.

Mean daily gage height, in feet, of Susquehanna River at Danville, Pa., 1899–1904—Cont'd.

1903.

Day.	Jan.	Feb.	Mar.	Apr.	May.	June.	July.	Aug.	Sept.	Oct.	Nov.	Dec.
1	5.20	a13.80	16.40	8.50	4.10	2.70	6.50	4.10	11.60	3.00	4.80	4.00
2	5.20	a12.40	17.60	9.80	4.00	2.60	6.00	3.90	9.85	2.90	4.60	4.60
3	6.10	a10.20	14.40	8.90	3.80	2.60	5.55	3.90	8.00	2.90	4.50	5.10
4	6.60	a11.20	11.60	7.80	3.70	2.60	5.30	3.70	6.90	2.90	4.40	4.60
5	7.30	a14.00	9.60	7.60	3.70	2.50	5.00	4.15	6.05	2.90	4.30	4.00
6	8.20	a15.20	8.70	8.30	3.60	2.50	4.70	4.85	5.60	3.00	4.10	3.90
7	7.40	a11.80	9.20	7.60	3.50	2.50	4.70	6.70	5.10	3.00	4.10	4.40
8	6.60	a9.70	9.60	7.20	3.50	2.80	6.50	6.45	4.90	3.40	4.20	4.50
9	6.00	a7.80	10.40	8.80	3.40	2.90	5.30	6.00	4.50	4.70	4.30	4.00
10	5.70	a7.00	15.00	9.30	3.30	2.80	4.60	5.60	4.30	3.40	4.30	4.40
11	a9.40	a7.20	14.50	8.80	3.30	2.70	4.00	5.00	4.20	16.60	4.10	5.10
12	(b)	a7.10	15.00	7.90	3.20	2.80	3.40	4.90	4.30	17.00	4.00	(c)
13	(b)	a7.40	14.80	7.70	3.10	3.10	8.90	4.60	4.30	15.40	3.80	(c)
14	(b)	a8.50	12.80	7.30	3.10	5.00	3.60	4.70	4.70	11.60	3.80	(c)
15	(b)	a8.80	11.40	8.10	3.10	5.90	3.50	4.30	4.30	8.95	3.70	(c)
16	(b)	a8.10	9.60	11.35	3.10	5.65	3.40	4.30	4.00	7.60	3.90	(c)
17	(b)	a7.00	8.70	11.05	3.00	5.00	3.30	4.10	3.80	6.80	3.90	(c)
18	(b)	a6.60	7.60	9.05	3.00	4.60	3.20	3.90	4.30	7.50	7.75	(c)
19	(b)	a5.70	7.60	7.30	3.00	4.25	3.90	8.70	4.30	9.00	10.10	(c)
20	(b)	a6.00	7.40	7.10	2.90	4.15	4.50	8.50	3.90	10.20	7.80	(c)
21	(b)	(b)	7.00	6.40	2.90	4.00	4.40	3.50	4.10	9.40	7.50	(c)
22	(b)	(b)	6.80	5.90	2.90	4.30	4.50	3.85	3.90	8.20	6.80	(c)
23	(b)	(b)	8.00	5.50	3.00	6.40	4.40	4.50	3.80	7.20	5.80	(c)
24	(b)	(b)	15.85	5.30	3.00	6.95	4.10	4.20	3.70	6.50	5.50	(c)
25	(b)	(b)	18.05	5.00	2.90	7.75	3.80	3.90	3.40	6.00	5.20	(c)
26	(b)	(b)	15.25	4.80	2.90	7.80	5.30	3.70	3.80	5.60	5.00	(c)
27	(b)	(b)	12.80	4.70	2.80	8.55	4.90	3.50	3.20	5.40	4.80	(c)
28	(b)	10.85	10.70	4.50	2.70	6.90	4.10	3.70	3.20	5.20	4.70	(c)
29	(b)		9.30	4.40	2.80	6.80	3.80	5.15	3.20	5.00	4.30	(c)
30	(b)		8.30	4.20	2.80	7.30	3.80	10.73	3.00	4.80	4.20	(c)
31	a14.80		7.80		2.70		3.80	14.65		4.80		(c)

1904.k

Day.	Jan.	Feb.	Mar.	Apr.	May.	June.	July.	Aug.	Sept.	Oct.	Nov.	Dec.
1	(c)	14.70	11.40	11.05	8.10	4.00	2.00	2.40	1.90			
2	(c)	14.10	11.30	10.85	8.00	4.20	2.00	2.50	1.90			
3	(c)	13.80	11.80	10.60	7.50	4.70	1.90	2.50	1.80			
4	(c)	12.70	f12.90	10.40	6.40	4.20	1.90	2.70	1.70			
5	(c)	12.10	13.80	10.40	5.30	4.70	1.80	2.90	1.70			
6	(c)	11.70	16.00	9.70	4.20	5.10	2.40	2.50	1.60			
7	(c)	11.50	17.25	9.30	3.70	5.50	1.80	2.40	1.50			
8	(c)	13.10	19.95	8.80	3.60	4.70	2.10	2.70	1.50			
9	(c)	f20.00	f24.00	8.20	3.60	4.30	2.10	2.90	1.40			
10	(c)	w23.86		7.90	3.40	4.90	2.00	2.40	1.40			
11	(c)	21.25		7.40	3.30	7.10	1.90	1.90	1.40			
12	(c)	19.50		6.80	3.30	6.20	1.90	1.70	1.30			
13	(c)	18.05		6.30	3.20	4.80	2.00	1.70	1.30			
14	(c)	16.90		6.10	3.10	4.70	2.40	1.80	1.00			
15	(c)	15.40		5.80	2.90	4.50	2.60	1.80	1.90			
16	(c)	h13.90		5.40	2.70	4.30	2.20	1.50	2.20			
17	(c)	13.00		5.00	3.90	4.00	1.90	1.40	1.90			
18	(c)	12.40		4.70	4.50	3.70	1.80	1.70	1.70			
19	(c)	11.00		4.30	6.30	3.30	1.80	1.60	1.60			
20	(c)	10.60		4.10	6.90	3.00	1.70	1.50	1.50			
21	(c)	11.20		4.00	7.20	2.90	1.70	1.50	1.50			
22	(c)	12.30		3.70	6.90	2.60	1.60	1.40	1.90			
23	(c)	12.30		3.50	4.90	2.60	1.90	1.40	2.40			
24	d19.85	12.40		3.90	4.40	2.50	2.00	1.80	2.90			
25	d24.00	12.00		3.90	4.10	2.50	1.80	2.40	2.30			
26	23.25	11.70		3.20	4.70	2.30	1.70	2.90	2.00			
27	19.85	11.70	14.25	3.00	4.40	2.20	1.50	2.00	2.20			
28	17.90	11.40	13.80	4.20	3.90	2.30	1.80	2.50	2.70			
29	16.00	11.10	13.35	5.30	3.70	2.10	1.80	2.30	2.10			
30	15.55		12.55	6.90	3.70	2.10	2.00	2.00	2.40			
31	15.05		11.75		3.90		2.20	1.90				

a Water backed up by ice.
b River frozen.
c River frozen.
d The ice started at 11.30 a. m.
e The ice gorged 1 p. m.
f The river is still frozen over.
g The ice broke and gorged and left an open place by the bridge.
h The ice is still gorged in the river.
i The ice gorge is still in the river above and below town.
j The ice started at 4 o'clock and the water backed up to 29 feet
k The gage heights for 1904 are somewhat uncertain. therefore no estimates of flow have been made.

U. S. GEOLOGICAL SURVEY

WATER-SUPPLY PAPER NO. 109 PL. III

RATING CURVE FOR SUSQUEHANNA RIVER AT DANVILLE, PA.

Rating table for Susquehanna River at Danville, Pa., for 1899 to 1904.

Gage height.	Discharge.	Gage height.	Discharge.	Gage height.	Discharge.	Gage height.	Discharge.
Feet.	*Second-feet.*	*Feet.*	*Second-feet.*	*Feet.*	*Second-feet.*	*Feet.*	*Second-feet.*
1.5	700	3.8	6,880	6.1	19,230	9.8	42,900
1.6	830	3.9	7,330	6.2	19,800	10.0	44,800
1.7	970	4.0	7,780	6.3	20,370	10.2	46,700
1.8	1,120	4.1	8,230	6.4	20,940	10.4	48,600
1.9	1,270	4.2	8,690	6.5	21,510	10.6	50,400
2.0	1,440	4.3	9,160	6.6	22,080	10.8	52,300
2.1	1,620	4.4	9,660	6.7	22,660	11.0	54,300
2.2	1,810	4.5	10,170	6.8	23,240	11.2	56,300
2.3	2,010	4.6	10,700	6.9	23,820	11.4	58,300
2.4	2,230	4.7	11,250	7.0	24,400	11.6	60,400
2.5	2,470	4.8	11,820	7.2	25,600	11.8	62,500
2.6	2,720	4.9	12,390	7.4	26,800	12.0	64,600
2.7	3,000	5.0	12,960	7.6	28,000	12.2	66,700
2.8	3,280	5.1	13,530	7.8	29,100	12.4	68,900
2.9	3,580	5.2	14,100	8.0	30,300	12.6	71,200
3.0	3,900	5.3	14,670	8.2	31,600	12.8	73,500
3.1	4,230	5.4	15,240	8.4	32,800	13.0	75,800
3.2	4,570	5.5	15,810	8.6	34,100	13.5	81,800
3.3	4,920	5.6	16,380	8.8	35,400	14.0	87,800
3.4	5,280	5.7	16,950	9.0	36,700	14.5	94,300
3.5	5,650	5.8	17,520	9.2	38,000	15.0	101,000
3.6	6,040	5.9	18,090	9.4	39,500		
3.7	6,450	6.0	18,660	9.6	41,100		

Mean daily discharge, in second-feet, of Susquehanna River at Danville, Pa., 1899–1903.

Day.	Jan.	Feb.	Mar.	Apr.	May.	June.	July.	Aug.	Sept.	Oct.	Nov.	Dec.
1899.												
1				24,110	11,820	4,920	4,570	1,810	3,280	1,620	1,620	4,230
2				23,240	10,920	5,280	3,900	1,810	2,720	1,620	2,720	3,900
3				20,680	10,700	6,450	3,280	2,720	2,470	1,620	2,720	3,900
4				18,660	10,700	6,040	3,000	2,010	2,470	1,620	19,230	3,900
5				16,660	10,700	5,650	2,720	1,810	2,010	1,440	15,240	3,900
6				15,810	10,440	4,920	2,720	1,810	1,810	1,440	16,950	3,580
7				16,690	9,410	4,570	2,720	2,470	1,810	1,440	14,100	4,230
8				23,820	8,460	3,900	2,470	2,010	1,620	1,440	11,260	4,230
9				49,500	6,880	3,900	2,470	1,810	2,280	1,620	9,160	4,230
10				60,400	6,450	3,580	2,470	1,810	1,810	1,620	7,930	3,900
11				49,000	6,450	3,580	2,470	2,720	1,620	1,440	6,450	3,900
12				37,000	6,660	3,580	2,230	2,010	1,810	1,440	7,880	4,230
13				36,400	6,880	3,000	2,720	2,230	1,620	1,440	6,450	8,690
14				51,800	6,450	3,000	3,000	2,010	1,620	1,270	7,780	23,240
15				50,800	6,450	2,720	2,720	2,010	1,620	1,270	6,880	29,100
16				58,300	6,040	2,720	3,280	2,010	1,440	1,270	7,330	28,000
17				52,880	6,040	2,720	3,280	2,010	1,270	1,270	7,330	22,660
18				45,250	6,450	2,720	3,000	2,010	1,270	1,270	9,160	19,230
19				37,000	6,040	2,720	3,000	2,010	1,120	1,270	9,690	16,950
20				31,900	6,040	2,470	2,470	1,810	1,120	1,270	9,160	15,240
21				28,800	6,040	2,470	2,470	1,620	1,270	1,270	8,230	16,380
22				26,500	6,880	2,470	2,470	1,620	1,270	1,270	6,880	23,820
23				24,700	6,880	2,470	2,470	1,620	1,120	1,270	7,380	20,370
24				22,370	6,880	2,470	2,470	1,620	1,120	1,270	6,040	20,370
25			44,800	19,800	6,880	2,470	2,470	1,440	1,120	1,270	5,280	21,510
26			38,350	17,800	6,450	3,000	2,230	1,440	1,270	1,270	5,280	25,000
27			31,000	16,950	6,040	2,720	2,230	1,440	1,270	1,270	4,920	23,820
28			26,500	16,990	5,650	2,720	2,230	2,010	1,120	1,270	4,570	20,940
29			26,250	14,950	4,920	3,580	2,230	1,810	1,270	1,270	4,230	17,520
30			27,700	13,530	4,570	4,570	2,230	5,050	1,620	1,270	4,230	12,960
31			27,100		4,920		2,010	4,570		1,270		
1900.												
1			27,700	16,380	14,950	3,900	2,010	2,230	1,810	970	1,440	35,000
2			104,300	17,520	13,240	3,580	2,010	2,230	1,810	970	1,440	25,300
3			77,000	22,940	11,820	3,580	2,010	2,010	1,810	970	1,440	18,660
4			50,800	22,820	10,440	5,650	1,810	1,810	1,810	970	1,440	15,810
5			38,350	38,700	9,660	4,920	1,810	1,810	1,810	970	1,440	25,000
6			25,000	33,100	8,920	4,920	2,010	1,620	1,620	970	1,440	35,400
7			25,000	26,800	8,460	3,900	2,000	2,470	1,620	970	1,440	41,600
8			26,250	34,700	8,000	3,580	2,580	1,810	1,440	970	1,440	33,800
9		42,000	23,530	42,400	7,780	3,580	1,810	1,620	1,120	970	1,440	27,400
10		43,800	22,940	39,900	7,550	3,580	3,000	1,620	1,270	970	1,440	23,530
11		28,000	27,400	31,000	7,100	4,920	2,470	1,440	1,120	970	1,620	20,370
12		99,160	25,600	25,000	7,330	4,920	2,470	1,440	1,120	970	1,620	16,380
13		39,500	20,940	20,370	8,230	4,920	2,230	1,270	1,120	970	1,810	14,100
14		41,100	16,660	19,230	8,690	4,920	2,010	1,440	1,120	970	2,230	12,960
15		56,300	14,100	20,370	7,780	7,330	2,010	1,440	1,120	1,120	2,230	12,960
16		48,000	12,360	22,370	7,780	5,650	2,010	1,270	1,120	1,120	2,720	23,240
17		32,390	11,550	20,400	6,880	4,570	2,010	1,270	970	1,120	2,470	
18		26,390	12,360	24,400	7,330	3,900	2,010	1,120	970	1,120	2,470	
19		16,950	13,230	42,400	7,330	3,900	2,010	1,120	970	1,120	2,470	
20		12,960	13,530	50,000	9,660	3,580	2,010	1,120	970	1,120	2,470	
21	39,500	11,250	30,000	43,400	9,680	3,280	1,810	1,270	970	970	2,470	
22	72,300	18,350	25,400	36,400	8,230	3,000	2,230	1,270	970	970	2,470	
23	64,000	66,350	30,000	31,000	7,330	2,010	1,810	1,120	840	1,250	2,720	
24	42,000	81,840	26,800	32,560	6,450	2,230	1,810	1,120	840	1,620	3,000	
25	29,100	54,800	26,800	32,560	6,040	2,230	1,620	2,010	840	2,010	3,580	
26	23,240	36,400	28,350	32,840	6,040	2,720	2,010	1,620	970	1,810	7,330	24,700
27	21,230	32,560	22,110	26,800	5,280	2,470	3,300	1,810	970	1,620	33,100	34,100
28	20,370	15,530	21,510	22,370	4,570	2,470	3,280	1,620	970	1,620	121,600	27,700
29	17,520		18,050	16,950	4,570	2,720	2,720	1,440	970	1,620	71,600	24,110
30	17,520		18,050	16,950	4,230	2,230	2,230	1,440	970	1,440	46,700	21,800
31			16,660		3,900		2,230	1,440		1,440		20,370

Mean daily discharge, in second-feet, of Susquehanna River at Danville, Pa., 1899–1903—Continued.

Day.	Jan.	Feb.	Mar.	Apr.	May.	June.	July.	Aug.	Sept.	Oct.	Nov.	Dec.
1901.												
1	16,950			38,400	22,370	83,000	6,450	4,920	10,700	5,650	3,580	8,230
2	34,100			28,200	19,230	37,000	6,450	3,900	13,530	5,650	3,580	7,380
3				25,600	21,510	41,600	5,280	3,000	12,390	7,100	3,280	7,380
4				28,000	28,000	37,600	4,570	2,720	12,390	8,000	3,280	10,440
5				34,400	26,500	32,200	4,230	2,720	9,160	7,100	3,280	18,370
6				39,500	22,370	26,200	4,230	2,720	8,920	6,450	3,000	29,700
7				34,100	18,940	23,240	4,230	3,000	7,550	5,650	8,000	32,200
8				33,800	14,950	26,200	6,040	3,580	6,450	4,920	2,720	34,700
9				29,100	14,670	28,000	5,650	4,230	5,650	4,570	2,720	37,300
10				27,100	21,510	27,700	5,280	3,580	5,100	4,570	2,720	40,700
11				25,000	12,980	24,400	4,920	5,650	4,790	4,230	2,720	42,900
12				64,600	22,940	16,950	20,940	4,920	4,230	3,900	2,720	45,200
13				55,800	22,510	22,080	16,380	4,570	3,900	3,900	3,580	3,280	36,000
14				33,400	34,100	30,000	14,100	4,230	3,580	4,090	7,330	3,580	29,700
15				28,000	31,300	29,400	12,980	3,900	3,280	3,900	7,100	3,900	96,300
16				26,200	29,100	24,700	12,670	3,580	3,280	4,230	7,330	5,460	228,400
17				26,500	27,100	20,370	10,700	3,580	3,280	4,230	7,330	7,830	180,300
18				23,820	25,000	17,620	10,700	3,900	22,080	5,280	7,330	7,330	86,000
19				22,080	22,940	17,520	9,920	4,230	29,400	5,650	6,880	6,040	47,200
20				22,080	21,510	16,950	8,230	16,380	5,650	7,330	5,650	32,200
21				38,350	24,820	18,370	7,780	3,580	10,440	6,040	6,040	5,650	25,000
22				63,000	71,200	17,230	7,330	3,280	11,540	6,040	5,650	5,650	18,080
23				72,300	104,400	14,950	7,330	3,000	20,370	5,280	5,240	5,280	18,530
24				57,800	72,900	15,240	8,920	2,720	31,400	5,280	4,920	5,240	12,390
25				56,800	65,100	21,300	14,950	2,720	54,300	4,230	4,920	6,450	11,540
26				55,800	61,400	26,400	11,250	2,720	34,300	3,900	4,570	19,520	12,670
27				80,000	50,800	23,820	9,920	2,720	27,700	3,580	4,230	24,400	13,530
28				120,400	36,000	20,940	8,230	2,720	19,530	3,900	4,230	17,800	12,980
29				127,300	31,900	30,300	7,100	2,470	14,950	3,580	3,900	12,670	14,100
30				80,000	26,500	72,900	6,880	3,000	11,250	4,570	3,900	9,410	25,300
31				49,000		100,300		3,580	9,000		3,580		23,240
1902.												
1	22,080	12,100	191,400	20,400	5,650	19,230	28,500	4,230	36,400	24,700	9,160
2	19,800	13,240	267,400	28,000	5,280	36,400	25,000	3,900	37,600	20,370	8,690
3	15,240		204,800	20,000	4,920	26,500	34,700	3,000	37,000	17,520	8,690
4	15,810		222,000	25,000	4,920	23,820	31,400	3,580	28,200	15,520	10,700
5	22,660		148,500	21,230	4,570	23,820	22,940	3,580	22,940	14,100	11,250
6			94,300	21,230	4,570	23,820	22,940	3,280	23,240	12,980	12,380
7			51,800	20,370	7,330	63,500	15,520	3,280	21,510	12,100	12,380
8			33,800	21,510	6,450	49,000	14,100	3,000	19,230	11,250	11,820
9	50,400		32,500	26,200	6,450	29,400	12,100	3,580	16,080	11,250	9,160
10	39,900		37,300	63,500	5,650	29,400	12,100	3,580	14,100	10,170	9,160
11	37,800		47,200	77,000	5,650	25,900	11,250	3,900	12,390	9,160	9,160
12	38,700		59,800	56,300	6,450	29,100	10,700	4,230	15,240	8,690	9,160
13			80,600	42,400	5,650	29,700	10,170	3,900	18,620	8,230	9,160
14			117,600	34,400	5,020	5,650	25,600	10,170	4,230	16,380	7,780	12,980
15			108,400	28,700	5,020	6,040	16,200	9,000	4,230	14,380	7,330	21,510
16			87,300	24,700	5,280	6,040	13,810	9,160	3,900	13,530	7,330	29,100
17			91,000	22,660	4,920	6,450	12,100	8,230	3,580	13,530	7,330	39,500
18			124,000	20,600	4,230	8,230	10,700	6,580	3,280	12,390	6,880	47,600
19			109,100	19,520	4,570	7,780	9,660	6,450	6,450	11,250	6,450	60,400
20			73,500	16,080	4,230	6,880	9,660	6,040	2,720	10,170	6,040	39,500
21			55,800	15,520	4,230	6,880	14,670	6,040	2,720	8,690	6,040	35,400
22			36,000	14,670	4,230	6,450	63,500	5,650	2,720	8,230	6,040	72,300
23	31,000		33,400	13,530	3,580	6,040	64,800	5,240	2,720	8,230	5,650	98,300
24	39,900		20,940	12,380	3,580	5,650	57,300	5,240	2,470	9,160	5,650	93,000
25	33,400		25,600	11,250	4,570	5,650	53,300	4,920	3,580	8,690	5,650	62,500
26	26,800		20,370	10,170	4,230	6,040	63,500	4,920	11,540	7,780	6,040	42,400
27	23,820		24,700	9,160	4,570	6,450	46,700	4,570	16,380	6,880	32,900
28	22,940	84,800	22,370	8,230	4,570	6,450	32,200	4,570	19,800	36,000	7,330	28,000
29	20,940		22,940	7,780	5,020	5,650	30,240	4,570	18,940	42,000	7,780	23,240
30	19,800		21,230	9,160	5,880	38,700	4,230	30,000	39,100	8,690	20,370
31	16,200		22,200		6,450		31,600	4,230		31,600		16,950

Mean daily discharge, in second-feet, of Susquehanna River at Danville, Pa., 1899–1903—Continued.

Day.	Jan.	Feb.	Mar.	Apr.	May.	June.	July.	Aug.	Sept.	Oct.	Nov.	Dec.
1903.												
1	85,400	120,600	33,400		8,230	3,000	21,510	8,230	60,400	3,900	11,820	7,780
2	68,900	138,500	42,900		7,780	2,720	18,660	7,330	43,400	3,580	10,700	10,700
3	46,700	95,000	36,000		6,880	2,720	16,200	7,330	30,300	3,580	10,170	13,530
4	22,040	56,300	60,400	29,100	6,450	2,720	14,670	6,450	23,820	3,580	9,680	10,700
5	26,200	87,800	41,100	28,000	6,450	2,470	12,960	8,460	18,940	3,580	9,160	7,780
6	31,600	108,600	34,700	32,200	6,040	2,470	11,250	12,100	16,380	3,900	8,230	
7	26,800	62,500	38,000	28,000	5,650	2,470	11,250	22,660	13,530	3,900	8,230	
8	22,090	42,000	41,100	25,600	5,650	3,280	21,510	21,220	12,380	5,280	8,690	
9	18,660	29,100	48,600	35,400	5,280	3,580	14,670	18,660	10,170	11,250	9,160	
10	16,950	24,400	101,000	38,700	4,920	3,280	10,700	16,380	9,160	70,000	9,160	
11	39,500	25,600	94,300	35,400	4,920	3,000	7,780	12,980	8,600	123,600	8,230	
12		25,000	101,000	29,700	4,570	5,280	7,780	12,890	9,160	129,600	7,780	
13		26,800	95,300	28,500	4,230	8,230	7,330	10,700	9,160	106,300	6,880	
14		33,400	73,500	26,200	4,230	12,960	6,040	11,250	11,250	60,400	6,880	
15		35,400	58,300	31,000	4,230	18,090	5,650	9,160	9,100	36,400	6,450	
16		31,000	41,100	57,800	4,230	16,660	5,280	9,160	7,780	27,000	7,330	
17		24,400	34,700	54,800	3,900	12,960	4,920	8,230	6,880	23,240	7,330	
18		22,090	28,000	37,000	3,900	10,700	4,570	7,330	9,160	27,400	28,800	
19		16,950	28,000	26,200	3,900	8,920	7,330	6,450	9,160	36,700	45,700	
20	18,090	26,800	26,800	27,000	3,580	8,460	10,170	5,650	7,330	46,700	29,100	
21		24,400	20,940		3,580	7,780	9,660	5,650	8,230	39,500	27,400	
22		23,240	18,090		3,580	7,330	10,170	7,100	7,330	31,600	23,240	
23		30,300	15,810		3,900	20,940	9,660	10,170	6,880	25,000	17,520	
24		112,700	14,670		3,580	24,110	8,690	8,690	6,450	21,510	15,810	
25		146,100	12,980		3,580	28,800	6,880	7,330	5,280	18,660	14,100	
26		104,300	11,820		3,580	29,100	14,670	6,450	4,920	16,380	12,960	
27		73,500	11,250		3,280	33,800	12,390	5,650	4,570	15,240	11,820	
28	52	51,800	10,170		3,000	23,820	8,230	6,450	4,570	14,100	11,250	
29		38,700	9,090		3,280	23,240	6,880	13,810	4,570	12,980	9,160	
30		32,200	8,690		3,280	28,200	0,880	51,800	3,900	11,820	8,690	
31	96,300		29,100		3,000		0,880	96,300		11,820		

Estimated monthly discharge of Susquehanna River at Danville, Pa., 1899–1903.

[Drainage area, 11,070 square miles.]

Month.	Discharge in second-feet.			Run-off.	
	Maximum.	Minimum.	Mean.	Second-feet per square mile.	Depth in inches.
1899.					
March (25–31)	44,800	26,200	31,663	2.860	0.744
April	60,400	13,530	31,048	2.804	3.128
May	11,820	4,570	7,293	.659	.760
June	6,450	2,470	3,579	.323	.360
July	4,570	2,010	2,710	.245	.282
August	5,650	1,440	2,121	.192	.221
September	3,280	1,120	1,940	.175	.195
October	1,620	1,120	1,371	.124	.143
November	19,230	1,620	7,828	.707	.789
December (1–30)	29,100	3,580	13,798	1.246	1.390
The period	60,400	1,120	10,335	.934	8.012

Estimated monthly discharge of Susquehanna River at Danville, Pa., 1899-1903.

Month.	Discharge in second-feet.			Run-off.	
	Maximum.	Minimum.	Mean.	Second-feet per square mile.	Depth in inches.
1900.					
January (21-31)[a]	72,300	17,520	34,677	3.182	1.165
February (9-28)[a]	81,800	11,250	36,229	3.273	2.434
March	104,300	11,250	27,861	2.517	2.902
April	50,000	16.380	29,393	2.655	2.962
May	14,950	3,900	7,911	.715	.824
June	7,830	2,230	3,819	.345	.385
July	3,900	1,620	2,320	.210	.242
August	2,470	1,120	1,564	.141	.162
September	1,810	830	1,200	.108	.120
October	2,010	970	1,184	.107	.123
November	123,600	1,440	11,109	1.004	1.120
December (1-16 and 26-31)[a]	41,600	12,960	24,252	2.191	1.793
The year	123,600	830	15,127	1.366	13.989
1901.					
January (1-2)[a]	34,100	16,950	25,525	2.306	0.172
February[a]					
March (12-31)[a]	129,600	22,080	55,636	5.026	3.735
April	104,300	21,510	37,287	3.368	3.758
May	100,800	12,960	25,179	2.274	2.622
June	83,000	6,880	19,781	1.787	1.994
July	6,450	2,470	4,085	.369	.425
August	54,300	2,720	12,232	1.105	1.274
September	13,530	3,280	6,118	.553	.617
October	8,000	3,580	5,588	.505	.582
November	24,400	2,720	6,376	.576	.643
December	228,400	7,330	39,769	3.592	4.141
The year	228,400	2,470	19,798	1.788	19.963

a River frozen, for days not included.

66   HYDROGRAPHY OF SUSQUEHANNA BASIN.   [NO. 109.

Estimated monthly discharge of Susquehanna River at Danville, Pa., 1899–1903—Continued.

Month.	Discharge in second-feet.			Run-off.	
	Maximum.	Minimum.	Mean.	Second-feet per square mile.	Depth in inches.
1902.					
January (1–5, 9–12, 23–31)[a]	50,400	15,240	27,594	2.493	1.669
February (1–2, 28)[a]	84,800	12,100	36,713	3.316	.370
March	304,800	20,940	84,379	7.622	8.787
April	77,000	7,780	24,663	2.228	2.486
May	9,660	3,900	6,184	.559	.644
June	8,690	4,570	6,087	.550	.614
July	64,600	9,660	32,516	2.937	3.386
August	34,700	4,230	12,112	1.094	1.261
September	30,000	2,470	6,325	.571	.637
October	42,000	7,780	19,723	1.782	2.054
November	24,700	5,650	9,697	.876	.977
December	98,300	8,690	28,995	2.619	3.019
The year	304,800	2,470	24,582	2.221	25.904
1903.					
January (4–11, 31)[a]	98,300	16,950	33,574	3.033	1.015
February (1–20, 28)[a]	103,600	16,950	43,752	3.952	3.086
March	146,100	23,240	63,459	5.782	6.608
April	57,800	8,690	27,165	2.454	2.738
May	8,230	3,000	4,612	.417	.481
June	33,800	2,470	12,031	1.087	1.213
July	21,510	4,570	10,347	.935	1.081
August	96,300	5,650	14,242	1.286	1.483
September	60,400	3,900	12,764	1.153	1.286
October	129,600	3,580	30,648	2.768	3.191
November	45,700	6,450	13,380	1.209	1.349
December (1–5)	18,500	7,780	10,098	.912	.170
The year	146,100	2,470	23,006	2.078	23.701

[a] River frozen, for days not included.

WEST BRANCH OF SUSQUEHANNA RIVER AT WILLIAMSPORT, PA.

This station was established March 1, 1895, by George D. Snyder, who was at that time city engineer. On August 16, 1901, a standard chain gage was installed on the upper side of the Market Street Bridge. It is read once each day by Henry H. Guise, who is employed in the city engineer's office. The length of the chain from the end of the weight to the marker is 40.29 feet. Discharge measurements are made from the lower side of the Market street iron highway bridge. The initial point for soundings is the face of the abutment on the left bank. The channel is straight for several hundred feet above and below the station, is broken by four bridge piers, and is about 1,000 feet wide at the station. There is a dam about one-half mile above the station. Both banks are high and rocky. The bed of the stream is composed of gravel and silt, and will probably change to some extent in the shore spans. The current velocity is sufficient for accurate measurement, except at extreme low stages. The bench mark is a cut in the face of the left abutment 10.07 feet above gage datum.

Discharge measurements of West Branch of Susquehanna River at Williamsport, Pa., 1901–1904.

Date.	Hydrographer.	Gage height.	Area of section.	Mean velocity.	Discharge.
1901.		*Feet.*	*Sq. feet.*	*Ft.per sec.*	*Sec.-feet.*
Aug. 16	E. G. Paul	0.90	2,851	0.68	1,932
Oct. 25	____do	.66	2,510	.72	1,807
1902.					
Apr. 20	E. G. Paul	3.90	5,188	1.80	9,318
Sept. 18	____do	.41	1,997	.54	1,006
1903.					
Mar. 6	E. C. Murphy	7.12	8,629	2.80	24,138
Apr. 3	____do	5.24	6,840	2.14	14,675
June 4	J. C. Hoyt	.85	2,769	.70	1,954
June 27	E. D. Walker	6.40	9,130	2.22	20,400
Oct. 7	W. C. Sawyer	1.77	3,270	1.08	3,525
1904.					
July 19	R. J. Taylor	2.07	3,874	1.09	4,220
Sept. 14	J. C. Hoyt	0.52	2,550	0.53	1,340
Sept. 30	____do	1.10	3,040	0.67	2,060

Mean daily gage height, in feet, of West Branch of Susquehanna River at Williamsport, Pa., 1895-1904.

Day.	Jan.	Feb.	Mar.	Apr.	May.	June.	July.	Aug.	Sept.	Oct.	Nov.	Dec.
1895.												
1			8.0	6.0	2.1	2.4	4.5	0.3	0.4	0.1	-0.1	1.5
2			9.0	6.0	1.9	2.1	3.7	.2	.4	.1	-.1	1.6
3			10.5	7.2	1.9	1.9	3.0	.2	.3	.2	.0	1.5
4			9.5	6.5	1.9	1.8	2.3	.1	.3	.3	.0	1.4
5			9.0	5.8	1.8	1.5	1.7	.1	.2	.3	-.1	1.4
6			6.5	5.4	1.8	1.5	1.5	.0	.2	.2	-.1	1.3
7			4.5	6.0	1.6	1.4	1.5	.0	.1	.1	.0	1.0
8			4.5	7.0	2.2	1.2	1.3	.3	.0	.2	+.1	1.1
9			5.0	11.0	2.9	.8	1.2	.3	.0	.2	.1	1.1
10			5.2	12.0	3.2	.5	1.6	.3	.3	.1	.3	1.0
11			5.3	11.0	2.8	.4	1.5	.4	1.6	.1	.3	1.0
12			5.5	7.9	2.7	.2	1.5	.7	1.8	.2	.4	.9
13			5.5	6.5	2.8	.2	1.6	1.8	.9	.2	.4	.8
14			5.2	8.0	4.3	.4	1.5	1.5	.7	.2	.3	.6
15			6.0	10.5	3.8	.8	1.4	.5	.5	.2	.3	.4
16			6.5	8.5	3.3	.8	1.3	.6	.4	.3	.2	.3
17			5.5	6.0	3.0	.7	1.2	.7	.6	.2	.2	.3
18			5.0	5.3	2.8	.7	1.1	.9	.1	.2	.2	.3
19			4.7	5.8	2.6	.6	1.0	1.1	.2	.2	.2	.2
20			4.5	5.8	2.8	.6	.8	1.1	.2	.2	.3	.2
21			4.2	4.5	2.2	.6	.7	1.1	.1	.2	.3	.2
22			4.5	3.6	2.0	.4	.9	1.2	.0	.2	.3	1.6
23			5.0	3.4	1.9	1.0	.8	1.3	-.1	.2	.2	2.4
24			5.5	3.2	1.8	1.4	.8	1.4	-.2	.1	.2	2.6
25			6.0	2.9	1.7	1.7	.9	1.4	-.2	.0	.4	2.4
26			8.7	2.6	1.7	1.3	1.0	1.5	-.2	-.1	.5	2.2
27			9.2	2.6	2.0	1.7	-.2	1.8	-.1	-.1	2.9	2.4
28			7.7	2.5	3.5	6.2	.0	1.3	-.1	-.1	3.1	7.0
29			6.7	2.5	3.6	4.9	+.1	1.3	.0	-.1	2.8	6.5
30			6.5	2.2	3.2	4.0	.1	1.4	+.2	-.1	2.1	4.5
31			6.3		3.0		.4	1.4		.2		5.4
1896.												
1	6.8	1.9	6.5	18.0	3.5	1.8	3.1	6.5	.5	6.8	2.3	4.0
2	4.5	2.0	6.6	11.0	3.4	2.0	2.7	6.7	.4	6.8	2.3	3.8
3	4.1	2.4	6.1	10.0	3.1	1.7	2.3	6.9	.4	5.8	2.8	3.3
4	3.8	4.1	4.7	8.5	3.0	1.4	2.0	5.9	.4	4.5	2.1	3.1
5	3.5	4.1	3.9	7.1	2.8	1.3	2.3	4.8	.4	3.2	2.5	2.9
6	3.8	3.9	4.1	6.1	2.6	1.2	2.4	4.0	.5	1.7	7.5	2.4
7	3.1	10.8	4.0	5.8	2.4	1.4	2.3	3.5	.5	1.5	6.9	2.0
8	2.9	9.2	3.9	5.6	2.3	1.6	2.1	3.7	.6	1.3	6.2	1.7
9	2.4	6.8	3.9	5.1	2.2	2.6	2.0	3.2	.7	.9	5.4	3.2
10	2.4	6.1	3.8	4.7	2.1	5.3	3.3	2.9	.6	.9	4.5	4.4
11	2.3	5.3	3.6	4.8	1.9	4.3	3.0	2.7	.4	.8	4.3	5.0
12	2.1	4.3	3.1	5.3	1.9	3.4	2.6	2.5	.4	.7	3.9	4.2
13	2.0	4.1	2.4	5.7	1.6	2.9	2.2	2.2	.4	9.8	4.0	4.0
14	2.0	3.7	2.7	7.8	1.6	2.5	1.9	2.5	.4	10.8	4.0	3.5
15	1.8	3.7	2.4	8.3	1.5	2.2	1.7	2.3	.5	9.8	3.6	3.3
16	1.7	4.6	2.0	7.5	1.5	2.2	1.8	2.1	.6	8.2	3.1	3.2
17	1.5	4.8	2.4	6.8	1.5	2.1	1.9	1.8	.6	6.5	3.1	2.9
18	1.4	3.6	2.5	6.1	1.6	4.1	2.2	1.5	.9	6.1	3.0	2.9
19	1.8	3.2	2.4	5.7	1.4	4.0	2.3	1.4	.6	5.4	2.8	2.7
20	1.8	1.7	3.6	5.2	1.3	3.5	1.8	1.2	1.5	4.7	2.7	2.5
21	1.4	1.5	3.8	4.7	1.4	3.0	1.6	1.0	2.0	4.0	2.7	2.2
22	1.4	2.2	3.8	4.7	1.3	2.6	1.7	.9	1.6	3.7	2.7	2.2
23	1.4	1.9	4.5	4.5	1.3	2.4	1.8	.9	.7	3.6	2.6	2.1
24	1.6	2.3	4.2	4.4	1.1	2.1	2.0	.8	.6	3.5	2.6	2.4
25	2.5	3.2	4.1	4.1	1.0	3.5	2.5	1.0	.3	3.4	2.8	2.2
26	2.7	3.1	4.2	4.2	1.1	7.0	3.1	1.0	.3	3.3	2.8	2.0
27	2.9	2.8	4.8	4.1	1.1	5.1	3.8	.9	.4	3.2	2.8	1.8
28	2.9	2.6	5.6	8.8	1.2	5.1	3.9	.6	.5	3.0	2.8	1.5
29	2.5	4.0	7.1	8.7	1.2	4.4	5.0	.6	.6	2.7	3.3	1.3
30	2.2		10.8	8.7	1.0	3.8	5.8	.6	1.3	2.5	4.0	1.6
31	2.0		13.9		1.5		6.8	.5		2.5		1.8

Mean daily gage height, in feet, of West Branch of Susquehanna River at Williamsport, Pa., 1895-1904—Continued.

Day.	Jan.	Feb.	Mar.	Apr.	May.	June.	July.	Aug.	Sept.	Oct.	Nov.	Dec.
1897.												
1	1.9	1.6	4.0	4.8	2.6	1.9	1.0	3.1	0.7	0.9	0.4	4.4
2	2.0	1.5	3.5	4.0	3.9	1.8	1.0	3.0	.7	.9	1.0	3.8
3	2.1	1.5	3.1	3.7	5.2	1.8	1.0	2.5	.6	.8	4.8	3.4
4	2.1	1.5	5.1	8.4	8.8	2.4	.9	2.2	.6	.7	4.1	3.1
5	2.8	1.5	7.0	8.2	8.5	2.3	.9	2.1	.5	.6	3.1	4.0
6	3.9	1.5	7.4	8.3	7.9	2.0	.7	2.0	.4	.5	2.7	4.5
7	3.5	3.7	10.4	3.6	7.2	1.7	.7	2.4	.3	.5	2.3	5.0
8	3.0	4.1	9.1	8.8	6.1	1.6	.7	2.2	.3	.4	1.9	4.7
9	3.0	3.9	7.6	4.0	5.5	1.6	.7	2.1	.2	.3	1.8	4.1
10	3.0	3.7	6.9	8.0	4.9	1.8	.8	1.7	.2	.3	2.0	3.8
11	3.0	3.5	7.8	8.8	4.6	1.8	.8	1.6	.0	.3	2.1	3.6
12	3.2	3.6	8.6	7.8	4.5	1.7	1.0	2.0	.0	.3	2.9	3.8
13	2.9	3.3	8.8	6.7	4.4	1.5	.9	1.9	.1	.4	2.6	4.0
14	1.8	3.0	8.6	5.9	6.5	1.4	.9	1.7	.1	.5	2.4	4.1
15	1.7	2.7	7.7	5.6	7.4	1.3	.9	1.5	.1	.5	2.2	4.8
16	2.2	2.7	6.6	6.6	7.1	1.2	1.0	1.3	.2	.5	2.1	7.4
17	2.2	2.7	6.1	7.8	6.9	1.1	1.0	1.1	.3	.4	2.3	7.7
18	2.2	2.8	5.1	6.9	5.4	1.1	1.1	1.0	.4	.3	4.9	6.7
19	2.5	3.6	5.3	6.1	4.8	1.1	1.1	.8	.5	.8	4.5	6.3
20	2.2	3.6	5.4	5.4	4.8	1.8	1.1	1.0	.5	.3	3.8	5.9
21	1.4	3.7	8.3	4.9	4.6	1.5	1.1	1.1	.5	.4	3.4	5.3
22	1.6	3.9	8.8	4.4	4.4	1.2	1.2	1.0	.6	.5	3.0	4.9
23	2.0	5.1	8.5	4.0	3.8	1.1	1.2	.9	.7	.6	2.7	4.6
24	2.2	8.8	8.8	3.7	3.2	1.1	2.0	3.5	.8	.7	2.5	3.8
25	2.4	7.8	11.3	3.4	3.0	1.1	2.3	2.8	2.4	.6	2.3	3.6
26	2.2	6.3	10.2	3.1	2.8	1.2	2.5	2.2	2.3	.6	2.0	3.3
27	2.3	5.2	8.4	3.1	2.7	1.2	2.0	1.5	2.2	.6	2.5	3.0
28	1.5	4.3	7.1	3.0	2.6	1.2	3.1	1.2	2.0	.5	3.5	3.1
29	1.8		6.2	2.9	2.4	1.2	4.6	1.0	1.7	.5	5.7	2.4
30	1.9		5.3	2.7	2.2	1.0	4.0	.8	1.1	.4	5.0	2.0
31	1.8		4.7		2.0		3.8	.8		.4		2.2
1898.												
1	2.0	2.9	3.5	3.2	4.6	3.5	2.0	1.0	1.1	.6	3.3	1.8
2	1.9	2.6	3.2	6.9	4.1	3.1	1.6	.9	1.0	.5	3.0	1.9
3	1.7	2.5	3.2	6.1	4.0	2.8	1.4	1.0	.9	.5	2.7	1.9
4	1.7	2.1	3.1	5.3	3.8	2.5	1.3	1.3	.9	.5	2.4	2.1
5	1.8	2.8	3.0	4.8	3.4	3.2	1.1	2.8	.8	.6	2.2	2.3
6	2.0	2.9	2.9	4.4	3.5	2.0	1.0	2.9	.8	1.1	2.0	2.6
7	2.1	3.1	2.8	4.0	3.8	1.8	.9	2.0	.7	1.0	2.0	2.6
8	2.1	3.0	3.0	3.7	3.8	1.8	.8	1.5	.7	1.3	1.8	2.5
9	2.1	2.9	3.1	3.5	3.9	1.6	.8	1.3	.9	1.2	1.8	2.2
10	2.1	2.9	3.8	3.3	4.1	1.4	.8	1.2	1.0	1.2	1.8	2.0
11	2.5	3.1	4.7	3.2	3.7	1.6	.7	1.1	.8	1.1	4.8	1.8
12	2.6	3.8	6.3	3.0	3.4	1.8	.7	1.0	.6	1.1	9.4	1.6
13	2.9	8.4	9.0	2.9	3.2	2.0	.7	1.0	.7	1.2	7.3	1.8
14	9.6	8.0	9.4	2.7	3.0	2.5	.6	1.7	.6	1.3	6.3	1.6
15	8.7	7.1	9.4	2.7	3.0	3.4	.6	1.4	.6	1.5	5.3	1.5
16	7.5	6.3	7.2	3.5	2.9	3.1	.6	1.2	.6	1.4	4.9	1.4
17	8.2	4.7	6.2	3.5	3.0	2.5	.5	1.0	.5	1.3	4.1	1.3
18	7.2	4.6	5.8	3.5	3.0	2.0	.5	1.0	.5	1.2	3.6	1.4
19	6.1	4.3	5.4	3.2	4.0	1.9	.5	1.4	.4	1.3	3.4	1.5
20	5.3	4.8	9.0	3.1	3.9	1.8	.5	6.8	.5	2.3	3.2	1.7
21	5.6	5.3	10.8	3.0	5.1	1.7	.7	4.8	.4	2.7	3.0	2.0
22	6.2	6.4	10.2	3.0	4.8	1.6	.8	3.9	.5	4.2	2.8	2.6
23	7.0	6.0	14.9	2.9	5.1	1.6	.8	3.0	.4	9.0	2.7	5.3
24	9.9	5.3	21.0	4.0	5.1	1.4	.7	2.5	.4	8.9	2.6	8.3
25	9.3	5.0	14.8	7.7	6.0	1.3	.7	2.1	.4	7.0	2.4	7.3
26	7.6	4.6	10.4	8.7	6.3	1.2	.7	2.1	.5	5.0	2.3	6.8
27	6.8	4.2	9.6	8.2	5.6	1.1	1.9	1.9	.5	4.7	2.1	5.8
28	6.0	8.8	7.1	6.4	5.3	1.0	1.9	1.8	.6	5.0	1.9	4.7
29	5.3		6.3	5.7	4.8	2.1	1.3	1.7	.5	4.7	1.8	4.3
30	4.7		9.9	5.1	4.8	2.7	1.0	1.6	.6	4.2	1.8	4.1
31	4.1		10.1		3.9		1.0	1.5		3.6		3.9

Mean daily gage height, in feet, of West Branch of Susquehanna River at Williamsport, Pa., 1895–1904—Continued.

Day.	Jan.	Feb.	Mar.	Apr.	May.	June.	July.	Aug.	Sept.	Oct.	Nov.	Dec.
1899.												
1	3.9	3.0	7.8	6.8	2.9	2.4	1.2	0.4	1.4	0.4	0.4	1.5
2	3.9	2.8	7.3	6.4	2.8	2.4	1.1	.8	1.5	.5	3.8	1.5
3	3.9	2.6	7.3	6.0	2.7	2.3	1.0	.8	1.5	.5	3.8	1.6
4	3.8	2.5	7.8	5.3	2.7	2.2	.9	.1	1.5	.4	3.8	1.6
5	4.8	2.6	11.8	4.5	2.9	2.1	.8	.2	1.5	.4	3.4	1.9
6	7.0	2.8	18.1	4.3	2.5	1.9	.7	.1	1.4	.4	2.9	1.6
7	8.0	2.8	11.3	4.3	2.3	1.7	.7	.0	1.3	.4	2.4	1.5
8	6.3	2.9	9.1	6.8	2.2	1.5	.7	.0	1.3	.4	2.1	1.7
9	5.3	2.9	7.3	7.8	2.3	1.3	.6	.1	1.3	.4	1.9	1.6
10	4.3	2.8	6.3	7.8	2.4	1.3	.6	.1	1.2	.4	2.0	1.7
11	4.0	2.7	5.4	6.8	2.4	1.2	.6	.2	1.1	.4	2.1	1.7
12	3.9	2.6	6.3	6.3	2.7	1.2	.6	.2	1.0	.4	2.2	1.9
13	3.8	2.4	7.3	6.8	2.5	1.1	.6	.8	.9	.4	2.3	7.0
14	4.3	2.8	7.8	7.3	2.4	1.0	.6	.6	.8	.3	2.6	7.5
15	4.8	2.3	7.1	7.8	2.3	1.0	.6	.4	.7	.3	2.9	6.3
16	5.3	2.4	6.1	6.8	2.2	1.0	.6	.2	.7	.3	3.1	5.5
17	5.8	2.5	5.8	6.3	2.0	.9	.6	.8	.6	.3	3.2	4.7
18	5.8	2.6	5.8	5.1	2.7	.9	.8	.3	.5	.3	3.6	4.0
19	5.6	2.8	7.5	4.9	6.8	.8	1.1	.2	.5	.3	3.5	3.9
20	4.5	3.2	9.3	4.6	7.3	.8	1.4	.1	.4	.3	3.5	3.8
21	3.9	3.3	8.8	4.4	6.1	.7	1.7	.0	.4	.3	3.2	4.3
22	3.9	4.2	7.6	4.2	4.9	.5	1.2	.2	.3	.3	2.9	4.9
23	3.8	5.3	6.8	4.0	4.1	.6	1.0	.1	.3	.3	2.6	4.3
24	3.8	6.8	7.0	3.7	3.6	.5	.8	.1	.4	.3	2.4	4.5
25	4.0	7.3	5.8	8.5	3.1	1.3	.7	.1	.3	.3	2.3	4.8
26	4.2	6.8	5.8	8.3	2.9	1.0	.6	.2	.3	.2	2.2	5.0
27	3.6	5.3	5.8	8.7	2.7	1.3	1.4	.3	.3	.2	2.1	4.5
28	3.5	8.8	5.6	8.6	2.5	1.2	.4	2.5	.3	.2	1.9	4.3
29	3.4		6.5	8.8	2.4	1.3	.3	2.0	.4	.2	1.9	3.8
30	3.2		8.3	8.1	2.4	1.3	.4	1.7	.4	.1	1.7	3.7
31	3.0		7.8		2.5		.4	1.5		.1		3.5
1900.												
1	3.3	2.9	4.0	8.9	8.8	3.8	1.8	.6	.8	.1	1.0	5.8
2	3.2	2.8	9.0	8.8	3.1	2.9	1.5	.6	.7	.1	1.0	5.0
3	3.1	2.8	8.2	4.2	2.9	3.2	1.8	.6	.6	.1	.9	4.8
4	3.0	2.9	7.1	4.5	2.7	3.5	1.0	.5	.5	.2	.9	4.3
5	2.9	2.9	6.0	4.8	2.6	3.5	.9	.5	.5	.2	.9	6.8
6	2.8	3.3	5.2	4.5	2.5	3.0	1.0	.4	.4	.2	.9	7.2
7	2.6	3.0	5.3	5.0	2.3	2.7	1.1	.4	.3	.3	.8	5.8
8	2.5	3.0	7.1	6.5	2.2	2.5	1.1	.3	.3	.3	.8	5.7
9	2.6	4.5	6.5	6.8	2.0	2.4	1.0	.3	.3	.4	.8	4.8
10	2.6	6.0	6.2	6.1	2.0	2.2	.9	.2	.3	.9	.8	4.5
11	2.6	5.5	7.0	5.5	2.0	2.0	1.0	.2	.3	1.1	.8	4.2
12	2.7	5.0	6.3	4.8	2.0	1.9	1.1	.1	.2	1.0	.9	3.5
13	2.8	5.0	5.1	4.5	2.0	1.8	1.0	.1	.2	1.0	.9	3.0
14	2.9	3.7	4.5	4.3	2.0	1.6	1.7	.2	.2	1.0	.9	2.9
15	3.0	3.5	4.1	4.1	2.3	1.7	1.3	.1	.1	.9	.9	2.8
16	8.0	6.5	3.5	3.9	2.3	1.9	1.1	.1	.1	1.1	.8	2.3
17	8.0	5.5	2.8	3.9	2.0	1.7	.9	.2	.1	1.2	.8	1.9
18	3.3	4.7	2.7	5.1	2.0	1.6	.8	.2	.2	1.1	.7	1.8
19	3.8	3.8	2.5	6.9	2.0	1.5	.8	.2	.2	.9	.7	2.1
20	4.5	3.6	8.1	6.8	2.5	1.4	.7	.2	.2	.8	.7	2.0
21	18.0	3.5	7.0	6.2	2.5	1.3	.7	.3	.2	.8	.8	2.0
22	13.0	5.5	6.1	5.5	2.3	1.2	.6	.3	.2	.7	1.0	1.9
23	10.0	9.8	5.0	5.5	2.0	1.2	.6	1.0	.1	.7	1.4	1.9
24	8.0	7.4	5.5	5.9	1.8	1.1	.6	.9	.1	.7	1.5	1.8
25	6.5	5.4	6.0	5.7	1.7	1.0	.6	.9	.1	.9	2.7	1.9
26	5.8	5.2	5.2	5.2	1.8	1.1	.5	.9	.1	1.8	4.8	2.1
27	5.0	3.2	4.9	4.7	2.0	1.0	.7	1.0	.1	1.5	17.0	2.4
28	4.5	3.9	4.5	4.2	1.9	0.9	1.0	.9	.1	1.4	12.0	2.3
29	4.0		4.5	3.8	1.9	0.8	.9	1.0	.1	1.8	8.0	2.3
30	4.1		4.4	8.6	4.0	0.8	.9	1.0		1.1	5.5	2.3
31	3.3		4.1		8.6		.7	.9		1.1		2.2

Mean daily gauge height, in feet, of West Branch of Susquehanna River at Williamsport, Pa., 1895–1904—Continued.

Day.	Jan.	Feb.	Mar.	Apr.	May.	June.	July.	Aug.	Sept.	Oct.	Nov.	Dec.
1901.												
1	2.80	1.60	.90	5.50	4.00	9.80	3.10	1.20	3.00	1.80	0.70	3.00
2	2.30	1.40	1.00	4.80	3.80	7.20	2.60	1.10	3.50	1.50	.60	2.80
3	1.90	1.40	1.30	4.50	4.00	7.00	2.30	1.00	6.80	2.00	.60	3.00
4	1.10	1.40	1.40	6.00	4.60	6.50	2.20	.80	5.70	1.50	.70	2.60
5	1.00	1.60	2.10	6.20	4.20	5.70	2.10	.70	4.60	1.40	.60	2.30
6	1.00	1.80	3.00	7.00	4.00	5.10	2.00	.70	4.10	1.30	.60	2.00
7	1.10	2.50	3.70	9.50	3.70	5.00	1.90	.80	3.30	1.20	.60	1.80
8	1.10	1.90	3.00	11.50	3.50	5.50	1.80	1.80	2.70	.90	.60	1.80
9	1.40	1.40	2.60	11.20	3.10	5.30	1.60	1.90	2.40	.80	.50	1.80
10	1.50	1.30	3.00	9.50	3.80	5.00	1.50	1.70	2.20	.90	.50	2.80
11	1.80	1.30	7.00	8.20	3.40	4.50	1.40	1.50	2.00	.90	.40	6.90
12	2.10	1.90	10.50	7.20	3.40	4.10	1.30	1.40	2.00	.90	.50	6.10
13	3.60	2.40	9.20	6.20	3.40	3.90	1.10	1.10	2.00	1.00	.60	5.50
14	4.50	2.10	7.50	5.80	3.60	3.30	1.10	.90	2.30	1.30	.60	5.20
15	4.20	1.50	6.50	5.50	3.60	3.00	1.10	.80	2.50	1.10	1.50	20.17
16	4.00	1.80	6.80	5.30	3.50	3.60	1.00	.90	2.80	.80	1.20	18.20
17	3.70	1.80	6.00	4.80	3.30	3.40	1.00	3.30	2.70	1.10	1.30	12.00
18	3.50	1.80	5.50	4.20	3.50	2.90	1.20	3.30	3.00	1.00	1.50	8.80
19	2.90	1.20	5.00	4.20	3.40	2.70	1.20	4.50	3.00	.90	1.50	7.00
20	2.40	1.30	6.20	4.00	3.20	2.60	1.10	4.20	2.80	.80	1.10	5.50
21	2.00	1.40	7.50	12.00	3.00	2.90	1.00	4.00	2.50	.80	1.00	5.00
22	1.90	1.30	9.50	15.20	1.80	4.00	.90	4.60	2.30	.70	.90	4.40
23	2.20	1.20	8.50	12.50	5.80	4.50	.70	4.00	2.00	.60	.80	3.70
24	2.60	1.00	7.50	9.70	5.50	4.40	.70	5.40	1.90	.60	1.60	3.60
25	2.40	.90	6.50	8.50	5.50	4.30	.70	7.80	1.80	.60	5.60	3.60
26	2.50	.90	7.80	7.50	5.00	3.80	.80	6.80	1.50	.60	6.70	3.70
27	2.60	1.00	10.50	6.50	5.00	3.50	.90	5.20	1.40	.60	5.70	8.90
28	2.60	1.00	11.20	5.50	7.60	3.60	1.00	4.30	1.20	.60	4.40	3.40
29	2.70		9.30	5.00	11.50	3.70	1.10	3.50	1.50	.70	3.60	3.20
30	2.60		7.80	4.50	14.00	8.50	1.20	3.00	1.90	.80	3.50	3.00
31	1.70		6.20		12.80		1.20	2.70		.70		3.40
1902.												
1	3.20	4.20	20.38	6.00	2.50	1.30	8.30	5.00	.50	2.70	1.90	1.00
2	2.90	4.20	21.10	5.70	2.50	1.20	7.40	4.90	.60	4.10	1.70	1.30
3	2.60	5.00	16.45	5.30	2.40	1.20	6.40	4.60	.50	3.10	1.60	1.50
4	2.50	4.70	13.00	4.90	2.70	1.10	9.70	4.30	.50	2.50	1.50	2.20
5	2.40	4.50	10.00	4.50	2.70	1.10	10.80	3.80	.50	2.40	1.40	2.40
6	2.30	4.00	8.10	4.30	2.90	1.80	8.60	3.30	.40	2.30	1.40	2.50
7	2.30	8.90	6.80	4.50	2.90	1.20	8.80	3.10	.40	2.20	1.30	2.30
8	2.30	3.70	5.90	4.70	3.20	1.90	7.30	3.00	.40	2.20	1.40	2.30
9	2.40	3.60	5.30	13.80	3.40	1.10	6.30	2.80	.40	2.00	1.40	2.80
10	2.40	3.40	5.50	16.60	3.20	1.00	6.00	2.60	.60	1.80	1.30	1.90
11	2.40	2.90	6.30	12.90	3.00	1.10	7.70	2.40	.50	1.60	1.30	2.00
12	2.40	3.00	7.10	10.80	2.80	1.10	7.20	2.20	.50	1.40	1.20	2.80
13	2.40	2.90	9.60	8.40	2.60	1.30	6.50	2.00	.50	1.20	1.20	3.10
14	2.30	3.00	12.20	7.30	2.50	1.40	5.00	2.10	.50	1.00	1.10	4.40
15	2.10	2.60	10.80	6.30	2.40	1.60	4.20	1.90	.40	1.20	1.00	3.60
16	2.10	2.80	8.40	5.50	2.30	1.80	3.60	1.80	.40	1.30	1.00	3.00
17	2.00	2.10	13.80	5.00	2.00	1.90	3.10	1.60	.40	1.60	.90	5.80
18	2.00	2.10	12.70	4.70	1.90	2.00	3.90	1.50	.40	1.50	.90	8.10
19	1.80	2.50	10.00	4.30	1.80	2.00	3.70	1.40	.30	1.40	.90	6.40
20	1.60	2.20	8.10	3.90	1.70	1.80	4.40	1.30	.20	1.30	.80	5.30
21	2.00	1.90	6.80	4.40	1.70	1.80	6.80	1.30	.20	1.30	.90	5.10
22	5.30	2.20	6.00	3.50	1.70	1.70	6.80	1.40	.20	1.20	.90	8.00
23	6.73	1.90	5.40	3.20	1.60	1.50	6.30	1.30	.20	1.10	.90	10.70
24	4.50	1.80	5.00	2.90	1.60	1.40	5.70	1.10	.20	1.00	.90	9.10
25	4.50	1.80	4.50	2.80	1.60	*.60	5.60	1.00	.50	1.00	1.00	7.20
26	4.00	2.00	4.20	3.30	1.60	1.50	5.80	.90	.90	.90	1.10	6.00
27	4.10	3.10	3.90	2.50	1.70	1.90	6.10	.90	2.30	.90	1.10	5.40
28	4.00	10.89	3.70	2.40	1.80	2.80	5.50	.40	2.60	1.30	1.10	4.10
29	3.90		3.90	2.30	1.60	2.60	5.20	.50	2.80	1.20	1.00	4.40
30	4.10		5.60	2.50	1.50	4.30	4.50	.60	2.80	1.50	1.00	3.60
31	4.00		6.20		1.40		5.20	.50		1.70		2.50

a Splash on dam.

Mean daily gage height, in feet, of West Branch of Susquehanna River at Williamsport, Pa., 1895–1904—Continued.

Day.	Jan.	Feb.	Mar.	Apr.	May.	June.	July.	Aug.	Sept.	Oct.	Nov.	Dec.
1903.												
1	3.00	9.80	17.07	5.80	2.40	1.00	4.60	2.70	5.80	1.00	1.90	2.10
2	2.50	6.00	14.30	5.60	2.30	.50	4.20	2.30	5.30	1.00	1.70	2.00
3	3.00	7.50	10.20	5.30	2.20	.90	4.00	2.00	4.80	.90	1.70	2.00
4	4.30	a10.60	8.80	5.00	2.00	.60	4.40	1.80	4.00	.90	1.00	1.90
5	4.90	15.50	7.20	5.10	2.00	.60	4.00	2.50	3.50	1.00	1.50	1.70
6	5.30	13.20	7.10	4.70	2.00	.60	3.70	3.60	3.10	1.40	1.60	1.70
7	5.00	10.10	7.20	4.50	2.00	.70	6.00	3.80	2.70	1.70	1.50	1.70
8	4.40	7.80	7.60	4.50	1.70	1.00	5.30	4.00	2.50	2.00	1.50	1.70
9	3.70	6.70	b12.20	5.00	1.60	1.40	4.20	3.50	2.40	5.90	1.50	1.60
10	2.30	5.80	12.70	5.40	1.40	1.40	3.50	2.90	2.60	7.80	1.40	1.70
11	2.20	5.10	11.00	5.60	1.40	2.10	3.00	2.60	2.80	6.20	1.40	1.70
12	4.30	5.20	11.10	5.30	1.40	2.20	3.00	2.40	3.40	5.50	1.40	1.40
13	4.20	6.20	10.60	5.80	1.30	3.70	3.00	2.20	3.00	4.60	1.30	1.30
14	4.20	6.70	8.90	6.10	1.30	3.40	2.70	2.00	2.50	4.00	1.20	1.90
15	4.20	6.40	7.80	9.60	1.30	3.60	2.40	1.70	2.30	3.60	1.20	1.50
16	4.20	6.30	6.90	11.70	1.20	4.00	2.20	1.60	2.00	3.30	1.20	1.00
17	4.10	6.30	6.20	10.70	1.20	3.70	2.00	1.80	1.90	3.00	2.80	1.00
18	3.70	6.00	5.70	9.10	1.20	3.40	1.80	2.50	4.20	12.00	1.00	
19	3.70	4.40	5.80	7.60	1.20	2.90	5.40	1.60	2.60	5.30	9.20	1.00
20	3.60	4.00	4.70	6.50	1.20	2.60	8.00	1.50	2.50	5.00	7.40	1.00
21	3.70	4.10	4.40	5.70	1.20	2.50	6.20	1.60	2.30	4.60	5.50	1.80
22	3.40	4.00	4.60	5.10	1.10	2.50	5.50	2.20	2.00	4.00	4.90	2.10
23	3.20	4.50	5.30	4.60	1.10	2.60	5.00	1.80	1.80	3.60	4.40	2.00
24	3.10	3.80	13.30	4.20	1.00	4.10	4.20	1.50	1.60	3.20	4.10	1.90
25	3.10	3.90	12.20	4.00	.90	6.10	3.60	1.50	1.50	3.00	3.80	1.80
26	3.00	3.80	9.50	3.50	.90	9.20	3.30	1.40	1.50	2.80	3.40	1.70
27	3.00	3.60	7.70	3.40	1.00	7.00	2.70	1.50	1.30	2.60	3.00	2.00
28	3.00	c9.85	6.50	8.10	1.00	5.40	2.30	1.80	1.20	2.30	2.00	2.00
29	2.80		5.60	2.90	1.10	4.50	2.10	3.90	1.20	2.20	2.10	2.40
30	3.00		5.00	2.70	1.10	5.20	2.50	7.20	1.10	2.00	1.70	2.30
31	11.00		4.90		1.10		2.80	6.50		2.00		2.40
1904.												
1	2.2	3.8	2.7	6.2	7.8	3.6	2.3	1.0	.4	.8	1.0	0.6
2	2.2	3.4	7.0	16.8	7.0	3.7	2.1	.9	.4	1.0	1.0	.5
3	2.0	3.0	7.5	18.6	6.2	3.7	1.9	.9	.4	1.0	1.0	.5
4	2.0	2.8	19.0	9.8	5.5	3.5	1.7	.9	.3	.9	1.0	.4
5	1.8	3.0	16.5	8.0	5.0	6.1	1.5	.8	.3	.8	.9	g.4
6	1.8	2.4	9.2	6.8	4.5	4.5	1.5	.7	.3	.8	.9	.4
7	1.7	2.6	7.4	6.4	4.2	3.7	1.7	.7	.2	.7	.8	.4
8	1.7	d5.0	17.4	6.0	3.9	3.4	1.8	.6	.2	.6	.7	.4
9	1.7	e10.5	18.5	6.0	3.6	3.2	3.4	.6	.2	.6	.6	.4
10	1.7	e7.6	9.8	8.8	3.3	3.2	4.4	.5	.2	.6	.6	.4
11	1.7	c6.0	7.6	9.2	3.2	3.3	8.1	.5	.2	.6	.6	.4
12	1.7	c5.2	6.5	7.9	3.0	3.3	6.7	.4	.5	.5	.7	.4
13	1.7	c4.3	5.8	7.2	2.8	2.9	5.4	.4	.6		.7	.3
14	1.7	3.8	5.3	6.6	2.5	2.7	4.6	.5	.5	1.2	.8	.3
15	1.6	4.0	5.0	5.8	3.0	2.4	3.8	.5	.5	1.5	.8	.3
16	1.6	f3.8	4.4	5.2	8.4	2.6	8.4	.5	.6	1.4	.7	.3
17	1.5	f3.6	4.1	5.2	3.3	3.1	3.0	.3	.5	1.3	.7	.3
18	1.5	f3.5	3.8	5.1	3.2	2.8	2.5	.3	.5	1.2	.7	.2
19	1.5	3.3	4.0	5.0	4.7	2.6	2.1	.3	.4	1.1	.7	.2
20	1.5	f3.0	4.5	4.5	7.7	2.3	2.0	.4	.3	1.0	.7	.2
21	1.4	2.9	6.5	4.2	7.2	2.3	1.7	.5	.3	1.1	.6	.2
22	1.5	2.8	6.7	8.9	6.0	3.0	1.5	.5	.2	1.5	.6	.2
23	7.7	2.7	6.6	8.6	5.2	3.7	1.3	.9	.2	1.7	.6	.2
24	18.3	3.7	h9.9	8.8	4.7	4.0	1.8	1.0	.2	1.6	.7	.3
25	9.8	4.2	10.3	8.2	4.4	3.2	1.2	1.2	.2	1.5	.6	.3
26	7.0	3.8	11.3	3.6	4.2	2.8	1.1	1.0	.3	1.5	.6	.8
27	5.4	3.0	12.6	4.3	4.0	2.3	1.1	1.1	.4	1.4	.6	.4
28	4.9	2.7	10.6	5.1	3.8	2.1	1.1	.7	1.0	1.3	.6	1.8
29	3.5	2.5	8.0	6.8	8.5	1.9	1.1	.6	1.0	1.3	.5	5.4
30	3.2		6.9	8.4	3.3	1.7	1.0	.6	1.1	1.2	.5	5.5
31	3.6		6.0		3.3		1.0	.5		1.1		4.4

a 16.00, 11 p. m..
b 13.2, 11 p. m.
c 15.00, 12 p. m., rising 1 foot in 2 hours.
d Ice running

e Slush ice running.
f Anchor ice running.
g River frozen December 5 to 28, 1904.
h 18 feet at noon.

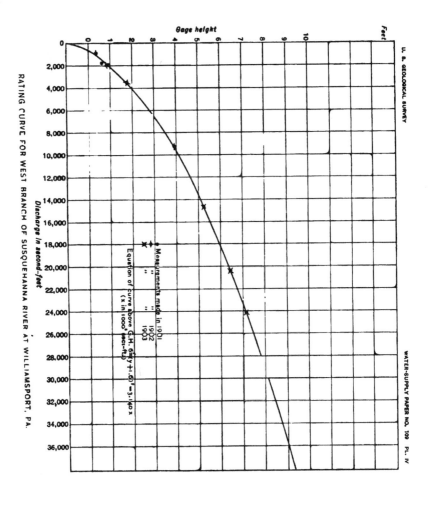

RATING CURVE FOR WEST BRANCH OF SUSQUEHANNA RIVER AT WILLIAMSPORT, PA.

Rating table for West Branch of Susquehanna River at Williamsport, Pa., for 1895 to 1904.

Gage height.	Discharge.	Gage height.	Discharge.	Gage height.	Discharge.	Gage height.	Discharge.
Feet.	*Second-feet.*	*Feet.*	*Second-feet.*	*Feet.*	*Second-feet.*	*Feet.*	*Second-feet.*
—0.2	410	2.2	4,530	6.0	18,330	10.6	47,400
.0	600	2.3	4,770	6.2	19,330	10.8	49,000
.1	710	2.4	5,010	6.4	20,340	11.0	50,600
.2	830	2.5	5,250	6.6	21,360	11.2	52,200
.3	970	2.6	5,500	6.8	22,380	11.4	53,800
.4	1,120	2.7	5,760	7.0	23,400	11.6	55,500
.5	1,280	2.8	6,020	7.2	24,600	11.8	57,200
.6	1,440	2.9	6,300	7.4	25,700	12.0	58,900
.7	1,610	3.0	6,580	7.6	26,900	12.2	60,700
.8	1,780	3.2	7,170	7.8	28,100	12.4	62,500
.9	1,960	3.4	7,780	8.0	29,300	12.6	64,300
1.0	2,140	3.6	8,400	8.2	30,500	12.8	66,100
1.1	2,320	3.8	9,030	8.4	31,800	13.0	67,900
1.2	2,510	4.0	9,690	8.6	33,100	13.2	69,800
1.3	2,700	4.2	10,400	8.8	34,400	13.4	71,700
1.4	2,890	4.4	11,150	9.0	35,800	13.6	73,600
1.5	3,080	4.6	11,940	9.2	37,200	13.8	75,500
1.6	3,270	4.8	12,750	9.4	38,600	14.0	77,500
1.7	3,460	5.0	13,600	9.6	40,000	14.5	82,600
1.8	3,660	5.2	14,500	9.8	41,400	15.0	87,800
1.9	3,860	5.4	15,420	10.0	42,800		
2.0	4,070	5.6	16,370	10.2	44,300		
2.1	4,300	5.8	17,340	10.4	45,800		

Mean daily discharge, in second-feet, of West Branch of Susquehanna River at Williamsport, Pa., 1895-1904.

Day.	Jan.	Feb.	Mar.	Apr.	May.	June.	July.	Aug.	Sept.	Oct.	Nov.	Dec.
1895.												
1			29,300	18,330	4,300	5,010	11,540	970	1,120	710	500	3,040
2			35,800	18,330	3,900	4,300	8,710	830	1,120	710	500	3,270
3			46,600	24,600	3,860	3,860	6,580	890	970	830	600	3,040
4			39,300	20,850	3,460	3,660	4,770	710	970	970	600	2,890
5			35,800	17,340	3,660	3,080	3,460	710	830	970	500	2,890
6			20,850	15,420	3,660	3,080	3,080	600	830	830	500	2,700
7			11,540	18,330	3,270	2,890	3,080	600	710	710	600	2,140
8			11,540	23,400	4,530	2,510	2,700	600	830	830	710	2,320
9			13,600	50,600	6,300	1,780	2,510	970	600	830	710	2,320
10			14,500	58,900	7,170	1,280	3,270	970	600	710	970	2,140
11			14,500	50,600	6,020	1,120	3,080	1,120	3,270	710	970	2,140
12			16,860	24,700	5,760	830	3,080	1,610	3,660	830	1,120	1,960
13			15,890	20,850	6,020	830	3,270	3,660	1,960	830	1,120	1,780
14			14,500	29,300	10,770	1,120	3,080	3,080	1,610	830	970	1,440
15			18,330	36,600	9,030	1,780	2,890	1,280	1,280	830	970	1,120
16			20,850	32,400	7,470	1,780	2,700	1,440	1,120	970	830	970
17			15,890	18,330	6,580	1,610	2,510	1,610	1,440	830	830	970
18			13,600	14,500	6,020	1,610	2,320	1,960	710	830	830	970
19			12,340	14,500	5,500	1,440	2,140	2,320	830	830	830	830
20			11,540	14,500	6,020	1,440	1,780	2,320	830	830	970	830
21			10,400	11,540	4,530	1,440	1,610	2,320	710	830	970	830
22			11,540	8,400	4,070	1,120	1,960	2,510	600	830	970	3,270
23			13,600	7,780	3,860	2,140	1,780	2,700	500	830	830	5,010
24			15,890	7,170	3,660	2,890	1,780	2,700	410	710	830	5,500
25			18,330	6,300	3,460	3,460	1,960	2,890	410	600	1,120	5,010
26			33,700	5,500	3,460	2,700	2,140	2,890	410	500	1,280	4,530
27			37,200	5,500	4,070	3,460	410	2,700	500	500	6,300	5,010
28			27,500	5,250	8,000	19,330	600	2,700	500	500	6,870	23,400
29			21,870	5,250	8,400	13,170	710	2,700	600	500	4,770	20,850
30			20,850	4,530	7,170	9,690	710	2,890	830	500	4,300	11,540
31			19,830		6,580		1,120	2,890		410		15,420
1896.												
1	22,380	3,860	20,850	87,900	8,000	3,660	6,870	20,850	1,280	22,380	4,770	9,690
2	11,540	4,070	21,900	50,600	7,780	4,070	5,760	21,870	1,120	22,380	4,770	9,030
3	10,040	5,010	18,830	42,900	6,870	3,860	4,770	22,890	1,120	17,340	4,770	7,470
4	9,030	10,040	12,340	32,400	6,580	2,890	4,070	17,830	1,120	11,540	4,300	6,870
5	8,090	10,040	9,390	24,000	6,020	2,510	4,770	12,750	1,120	7,170	5,250	6,300
6	7,470	9,390	10,040	18,830	5,500	2,510	5,010	9,690	1,280	3,460	26,300	5,010
7	6,870	49,000	9,690	17,340	5,010	2,890	4,770	9,030	1,280	3,080	22,890	4,070
8	6,300	37,300	9,390	16,370	4,770	3,270	4,300	8,710	1,440	2,700	19,330	3,460
9	5,010	22,380	9,390	14,050	4,530	5,500	4,770	7,170	1,610	1,980	15,420	7,170
10	5,010	18,830	9,030	12,340	4,300	14,500	7,470	6,300	1,440	1,980	11,540	11,150
11	4,770	14,500	8,400	12,750	3,860	10,770	6,580	5,760	1,120	1,780	10,770	13,600
12	4,300	10,770	6,870	14,950	3,800	7,780	5,500	5,250	1,120	1,610	9,390	10,400
13	4,070	10,040	5,010	16,870	3,270	6,300	5,010	5,250	1,120	41,400	9,090	8,090
14	4,070	8,710	5,760	28,100	3,270	5,250	31,800	5,250	1,120	49,000	8,400	7,470
15	3,660	8,710	5,010	31,100	3,080	4,530	4,770	4,770	1,280	41,400	8,400	7,470
16	3,460	11,940	4,070	26,300	3,080	4,530	3,660	4,300	1,440	30,500	6,870	7,170
17	3,080	10,770	5,010	22,380	3,080	3,800	3,860	3,660	1,440	20,850	6,870	6,300
18	2,890	8,400	5,250	18,830	3,270	10,040	4,530	3,660	1,980	18,830	6,580	6,300
19	2,700	7,170	5,010	16,370	2,890	8,000	3,660	2,890	1,440	15,420	6,020	5,760
20	2,700	3,460	8,400	14,500	2,700	8,090	3,660	2,510	3,080	12,340	5,760	5,250
21	2,890	3,080	9,030	12,340	2,890	6,580	3,460	3,270	4,070	9,690	5,760	4,530
22	2,890	4,530	9,030	12,340	2,700	5,500	3,460	1,960	3,270	8,710	5,760	4,530
23	2,890	3,860	11,540	11,540	2,700	5,010	3,460	1,960	1,610	8,400	5,500	4,300
24	3,270	4,770	10,400	11,150	2,520	4,300	4,070	1,780	1,440	22,380	5,500	5,010
25	5,250	7,170	10,040	10,040	2,140	4,300	4,530	2,140	970	7,780	6,020	4,530
26	5,760	6,870	10,400	10,400	2,320	23,400	6,870	2,140	970	7,470	6,020	4,070
27	6,300	4,770	12,750	10,040	2,320	19,330	5,010	1,960	1,120	7,170	6,020	3,660
28	6,300	5,500	16,370	9,690	2,510	14,050	5,500	1,440	1,280	6,540	6,020	3,460
29	5,250	9,690	24,000	8,710	2,320	13,600	4,530	1,440	1,440	5,760	7,470	2,700
30	4,530		49,000	8,710	2,140	9,690	17,340	1,440	2,700	5,250	9,690	3,270
31	4,070		76,500		3,080		22,380	1,280		5,250		3,660

Mean daily discharge, in second-feet, of West Branch of Susquehanna River at Williamsport, Pa., 1895–1904—Continued.

Day.	Jan.	Feb.	Mar.	Apr.	May.	June.	July.	Aug.	Sept.	Oct.	Nov.	Dec.
1897.												
1	3,800	3,270	9,680	10,770	5,500	3,890	2,140	6,870	1,610	1,980	1,120	11,150
2	4,070	3,080	8,080	9,690	9,360	3,660	2,140	6,580	1,610	1,980	2,140	9,080
3	4,300	3,080	6,870	8,710	14,500	3,660	2,140	5,250	1,440	1,780	12,750	7,780
4	4,300	3,080	14,050	7,780	34,400	5,010	1,980	4,530	1,440	1,610	10,040	6,870
5	6,020	3,080	23,400	7,170	32,400	4,770	1,980	4,300	1,280	1,440	6,870	9,690
6	9,360	3,080	25,700	7,470	28,700	4,070	1,610	4,070	1,120	1,280	5,760	11,540
7	8,080	8,710	45,800	8,400	24,600	3,460	1,610	5,010	970	1,280	4,770	13,600
8	6,580	10,040	36,500	9,080	18,830	3,270	1,610	4,530	970	1,120	3,860	12,340
9	6,580	9,360	26,300	9,690	15,880	3,270	1,610	4,300	830	970	3,660	10,040
10	6,580	8,710	22,890	29,300	13,170	3,660	1,780	3,830	830	970	4,070	9,080
11	6,580	8,080	28,100	34,400	11,940	3,660	1,780	3,270	600	970	4,300	8,400
12	7,170	8,400	31,100	28,100	11,540	3,460	2,140	4,070	600	970	6,300	9,080
13	6,300	7,470	34,400	21,870	11,150	3,080	1,980	3,860	710	1,120	5,500	9,690
14	3,660	6,580	31,100	17,830	20,850	2,890	1,980	3,460	710	1,280	5,010	10,040
15	3,460	5,760	27,500	16,370	25,700	2,700	1,980	3,080	710	1,280	4,530	12,750
16	4,530	5,760	21,870	21,860	24,000	2,510	2,140	2,700	830	1,280	4,300	25,700
17	4,530	5,760	18,830	28,100	22,890	2,325	2,140	2,325	970	1,120	4,770	27,500
18	4,530	6,020	14,050	22,890	15,420	2,320	2,320	2,700	1,120	970	18,170	21,870
19	5,250	8,400	14,980	18,830	12,750	2,320	2,320	1,780	1,280	970	11,540	19,830
20	4,530	8,400	15,420	15,420	12,750	2,700	2,320	2,320	1,280	960	9,080	17,830
21	2,890	8,710	31,100	13,170	11,940	3,080	2,320	2,320	1,280	1,120	7,780	14,980
22	3,270	9,360	34,400	11,150	11,150	2,510	2,510	2,320	1,440	1,280	6,580	18,170
23	4,070	14,050	32,400	9,680	9,000	2,320	2,510	1,980	1,610	1,440	5,760	11,940
24	4,530	34,400	34,400	8,710	7,170	2,320	4,070	8,080	1,780	1,610	5,250	9,080
25	5,010	28,100	53,100	7,780	6,580	2,325	4,770	6,020	5,010	1,440	4,770	8,400
26	4,530	19,830	44,300	6,870	6,020	2,510	5,250	4,530	4,770	1,440	4,070	7,470
27	4,770	14,500	31,800	6,870	5,760	2,510	4,070	3,080	4,530	1,440	5,250	6,580
28	3,080	10,770	24,000	6,580	5,500	2,510	6,870	2,510	4,070	1,280	8,090	6,870
29	3,660		19,830	6,300	5,010	2,510	11,940	2,140	3,860	1,280	16,850	5,010
30	3,860		14,980	5,760	4,770	2,140	12,750	1,780	2,320	1,120	18,600	4,070
31	3,680		12,340		4,070		9,080	1,780		1,120		4,530
1898.												
1	4,070	6,300	8,080	30,500	11,940	8,090	4,070	2,140	2,520	1,440	7,470	3,660
2	3,860	5,500	7,170	22,890	10,040	6,870	3,270	1,980	2,140	1,280	6,580	3,860
3	3,460	5,760	7,170	18,830	9,690	5,250	2,890	2,700	1,980	1,280	5,010	3,860
4	3,460	4,900	6,870	14,980	9,080	5,250	2,700	2,700	1,980	1,280	4,530	4,300
5	3,660	6,020	6,300	11,150	8,090	4,530	2,320	6,300	1,780	2,320	4,070	4,770
6	4,300	6,870	6,300	9,690	9,080	3,660	2,140	4,070	1,610	2,140	4,070	5,500
7	4,900	6,580	5,580	8,710	9,030	3,660	1,960	4,070	1,610	2,700	3,680	5,250
8	4,300	6,300	5,870	8,090	9,360	3,270	1,780	3,080	1,610	2,700	3,680	4,580
9	4,300	6,300	9,030	7,470	10,040	2,890	1,870	2,700	2,140	2,510	3,680	4,070
10	5,250	6,870	12,340	7,170	8,710	3,270	1,610	2,510	2,140	2,320	12,570	3,680
11	5,500	9,080	19,830	6,580	7,780	3,660	1,610	2,140	1,440	2,325	38,400	3,270
12	6,300	31,800	35,800	6,300	7,170	4,070	1,610	2,140	1,440	2,510	25,100	3,660
13	40,000	29,300	38,600	5,760	6,580	5,250	1,440	3,460	1,440	2,700	19,830	3,270
14	38,700	24,000	38,600	5,760	6,580	7,780	1,440	2,890	1,440	3,080	14,980	3,080
15	26,300	19,830	24,600	8,090	6,300	6,870	1,440	2,510	1,440	2,890	13,170	2,890
16	30,500	12,340	19,330	8,090	6,580	5,250	1,280	2,140	1,290	2,700	10,040	2,700
17	24,600	11,940	17,340	8,090	6,580	4,070	1,280	2,140	1,280	2,510	8,400	2,890
18	18,830	10,770	15,420	7,170	9,690	3,660	1,280	2,890	1,120	2,700	7,780	3,080
19	14,980	12,750	35,800	6,870	9,360	3,660	1,280	22,380	1,280	4,770	7,170	3,460
20	16,370	14,000	49,000	6,580	14,050	3,460	1,610	12,750	1,120	5,760	6,580	4,070
21	19,330	20,340	44,300	6,580	12,750	3,270	1,780	9,360	1,280	10,400	6,020	5,500
22	23,400	18,330	36,800	6,300	14,050	3,270	1,780	6,580	1,120	35,100	5,760	14,980
23	100	14,980	92,900	9,690	14,050	2,890	1,610	5,250	1,120	35,100	5,500	31,100
24	900	13,600	85,800	27,500	18,330	2,700	1,610	4,300	1,120	23,400	5,010	25,100
25	900	11,940	45,800	34,700	19,830	2,510	1,610	4,300	1,280	18,600	4,770	19,830
26	580	10,400	40,000	30,500	16,370	2,325	3,860	3,660	1,290	12,340	4,000	12,340
27	330	9,080	24,000	20,340	16,370	2,140	8,880	3,660	1,280	13,600	3,680	10,770
28	960		19,830	16,850	12,750	4,300	2,700	3,460	1,290	12,340	3,680	10,770
29	2,340		42,100	14,050	10,770	5,760	2,140	3,270	1,440	18,600	3,680	10,440
30	4,040		43,500	9,360		2,140	3,080	8,400		9,360
31							2,140	3,080		8,400		9,360

Mean daily discharge, in second-feet, of West Branch of Susquehanna River at Williamsport, Pa., 1895–1904—Continued.

Day.	Jan.	Feb.	Mar.	Apr.	May.	June.	July.	Aug.	Sept.	Oct.	Nov.	Dec.
1899.												
1	9,360	6,580	28,100	22,380	6,300	5,010	2,510	1,120	2,800	1,120	1,120	3,080
2	9,360	6,020	25,100	20,340	6,020	5,010	2,320	970	3,080	1,280	1,280	3,040
3	9,360	5,500	25,100	18,830	5,760	4,770	2,140	970	3,080	1,280	9,030	3,270
4	9,090	5,250	28,100	14,960	5,760	4,530	1,960	710	3,080	1,120	9,030	3,270
5	12,750	5,500	57,200	11,540	6,300	4,300	1,780	830	3,080	1,120	7,780	3,880
6	23,400	6,020	68,800	10,770	5,250	3,880	1,610	710	2,890	1,120	6,300	3,270
7	29,300	6,020	53,000	10,770	4,770	3,460	1,610	600	2,700	1,120	5,010	3,080
8	19,830	6,300	36,500	22,380	4,530	3,080	1,610	600	2,700	1,120	4,300	3,460
9	14,960	6,300	25,100	28,100	4,770	2,700	1,440	710	2,510	1,120	3,860	3,270
10	10,770	6,020	19,830	28,100	5,010	2,700	1,440	710	2,510	1,120	4,070	3,460
11	9,690	5,760	15,420	22,380	5,010	2,510	1,440	830	2,320	1,120	4,300	3,460
12	9,360	5,500	19,830	19,830	5,760	2,510	1,440	830	2,140	1,120	4,530	3,860
13	9,030	5,010	25,100	22,380	5,250	2,320	1,440	1,780	1,960	1,120	4,770	23,400
14	10,770	4,770	28,100	25,100	5,010	2,140	1,440	1,440	1,780	970	5,500	26,300
15	12,750	4,770	24,000	25,100	4,770	2,140	1,440	1,120	1,610	970	6,300	19,830
16	14,960	5,010	18,830	22,380	4,530	2,140	1,440	830	1,610	970	6,870	15,890
17	17,340	5,250	17,340	19,830	4,070	1,960	1,440	970	1,440	970	7,170	12,340
18	17,340	5,500	17,340	14,960	5,760	1,960	1,440	970	1,280	970	8,400	9,360
19	16,370	6,020	26,300	13,170	22,380	1,780	2,320	830	1,280	970	8,050	9,360
20	11,540	7,170	37,900	14,960	5,760	1,780	2,880	710	1,280	970	8,050	9,690
21	9,360	7,470	34,400	11,150	18,830	1,610	3,460	600	1,280	970	7,170	10,770
22	9,360	10,400	26,300	10,400	13,170	1,280	2,510	830	970	970	6,300	10,770
23	9,030	14,960	22,380	9,690	10,040	1,440	2,140	710	970	970	5,500	10,170
24	9,030	22,380	23,400	8,710	8,400	1,780	710	710	1,120	970	5,010	11,540
25	9,690	25,100	17,340	8,050	6,870	2,700	1,610	710	970	970	4,770	12,750
26	10,400	19,830	17,340	7,470	6,300	2,140	1,440	970	970	830	4,300	11,540
27	8,400	14,960	17,340	8,710	5,760	2,700	1,440	2,800	970	830	4,300	11,540
28	8,090	31,100	16,370	8,400	5,250	2,140	1,120	5,250	970	830	3,860	10,770
29	7,780		20,850	7,470	5,010	2,700	970	4,070	1,120	830	3,860	9,030
30	7,170		31,100	6,870	5,010	2,700	1,280	3,460	1,120	710	3,460	8,710
31	6,580		28,100		5,250		1,120	1,120		710		8,090
1900.												
1	7,470	6,300	9,030	9,360	7,470	7,470	2,700	1,440	1,780	710	2,140	17,340
2	7,170	6,020	35,800	9,030	6,870	6,300	3,080	1,440	1,610	710	2,140	13,600
3	6,870	6,020	30,500	10,400	6,300	7,170	2,700	1,440	1,440	710	1,960	12,750
4	6,580	6,300	24,000	11,540	5,760	8,090	2,140	1,280	1,280	830	1,960	10,770
5	6,300	6,300	18,330	12,750	5,500	6,580	2,140	1,280	1,280	830	1,960	22,380
6	6,020	7,470	14,500	11,540	5,250	6,580	2,140	1,120	1,120	830	1,960	24,600
7	5,500	6,580	14,960	14,960	4,770	5,500	2,320	1,120	970	830	1,780	17,340
8	5,250	6,580	24,000	20,850	4,530	5,250	2,320	970	970	970	1,780	16,850
9	5,500	11,540	22,380	22,380	4,070	5,010	2,140	970	970	1,780	1,780	12,750
10	5,500	18,830	19,830	18,830	4,010	4,530	1,960	830	970	1,960	1,780	11,540
11	5,500	15,890	28,100	15,890	4,070	5,010	2,140	830	970	2,320	1,780	10,400
12	5,760	13,600	19,830	12,750	4,070	3,880	2,320	710	830	2,140	1,960	8,090
13	6,020	13,600	14,960	11,540	4,070	3,880	2,140	710	830	2,140	1,960	6,300
14	6,300	33,700	11,540	10,770	4,070	3,270	2,140	710	830	2,140	1,960	6,300
15	6,580	22,400	10,040	10,040	4,530	3,460	1,960	710	830	1,780	1,960	6,020
16	6,580	20,850	8,090	9,360	4,770	3,080	1,780	710	710	2,320	1,780	4,770
17	6,580	15,890	6,870	9,360	4,070	3,080	1,960	830	710	2,510	1,780	3,690
18	7,470	12,340	5,760	11,540	4,070	3,270	1,780	830	830	2,320	1,610	4,300
19	9,030	9,030	5,760	22,380	5,250	2,880	1,610	830	880	1,780	1,610	4,070
20	11,540	8,400	6,870	22,380	5,250	2,880	1,610	830	830	1,780	1,610	4,070
21	67,900	8,090	23,400	19,830	5,250	2,510	1,610	970	830	1,610	1,780	3,880
22	67,900	15,890	18,830	15,890	4,770	2,510	1,440	1,610	830	1,610	2,880	3,860
23	42,800	41,400	19,830	17,830	3,660	2,320	1,440	2,140	710	1,610	2,880	3,690
24	29,300	25,700	15,890	17,830	3,660	2,320	1,440	1,960	710	1,610	3,080	3,880
25	20,850	15,420	18,330	15,890	3,100	2,140	1,440	1,960	710	1,960	4,530	5,760
26	17,340	14,500	14,500	14,500	3,460	2,320	1,280	1,960	710	3,460	12,750	4,300
27	18,600	7,170	13,170	9,360	4,070	2,140	1,610	2,140	710	3,080	110,100	5,010
28	11,540	9,860	11,540	10,400	3,860	1,960	2,140	1,960	710	2,890	58,300	4,770
29	9,690		11,540	9,360	3,660	1,960	2,140	1,960	710	2,700	20,300	4,770
30	10,040		11,150	8,400	3,660	1,780	1,780	2,140	710	2,320	15,890	4,770
31	7,470		10,040		8,400		1,610	1,960		2,320		4,530

Mean daily discharge, in second-feet, of West Branch of Susquehanna River at Williamsport, Pa., 1895–1904—Continued.

Day.	Jan.	Feb.	Mar.	Apr.	May.	June.	July.	Aug.	Sept.	Oct.	Nov.	Dec.
1901.												
1	4,770	3,270	1,990	15,890	9,690	41,400	6,870	2,510	6,580	3,660	1,610	6,580
2	4,770	2,890	2,140	12,750	9,030	24,600	5,500	2,320	8,090	3,080	1,440	6,020
3	3,860	2,890	2,700	11,540	9,690	23,400	4,770	2,140	22,380	4,070	1,440	6,580
4	2,320	2,890	2,890	18,830	11,940	20,850	4,530	1,780	16,850	3,080	1,610	5,500
5	2,140	3,270	4,300	19,830	10,400	16,850	4,300	1,610	11,940	2,890	1,440	4,770
6	2,140	3,660	6,580	23,400	9,690	14,050	4,070	1,610	10,040	2,700	1,440	4,070
7	2,320	5,250	8,710	69,300	8,710	13,600	3,860	1,780	7,470	2,510	1,440	3,660
8	2,320	3,860	6,580	54,600	8,090	15,890	3,660	3,660	5,760	1,980	1,440	3,660
9	2,890	2,890	5,500	52,300	6,870	14,980	3,270	3,860	5,010	1,780	1,280	3,660
10	3,080	2,700	6,580	89,300	7,470	13,600	3,080	3,860	4,530	1,980	1,280	6,020
11	3,660	2,700	23,400	30,500	7,780	11,540	2,890	2,890	4,070	1,980	1,120	22,890
12	4,300	3,860	46,600	24,600	7,780	10,040	2,700	2,320	4,070	1,980	1,280	18,890
13	8,400	5,010	37,200	19,830	7,780	9,380	2,700	2,320	4,070	2,140	1,440	15,890
14	11,540	4,300	26,300	17,340	8,400	7,470	2,510	1,980	4,770	2,700	1,440	14,500
15	10,400	8,090	20,850	15,890	8,400	6,580	2,320	1,780	5,250	2,320	1,080	150,900
16	9,690	2,890	22,380	14,980	8,090	8,400	2,140	1,980	6,020	1,780	2,510	124,900
17	8,710	2,700	18,330	12,750	7,470	7,780	2,140	2,140	5,760	2,320	2,700	58,900
18	8,090	2,510	15,890	10,400	8,090	6,300	2,510	2,140	6,580	2,140	1,980	34,400
19	6,300	2,510	13,600	10,400	7,780	5,760	2,510	11,540	6,580	1,980	2,700	23,400
20	5,010	2,700	19,330	9,690	7,170	5,540	2,320	10,400	6,020	1,780	2,700	15,890
21	4,070	2,890	26,300	58,900	6,580	6,300	2,140	9,380	5,250	1,780	2,140	13,600
22	3,860	2,700	39,300	89,300	3,660	9,680	11,940	11,940	4,770	1,610	2,250	11,150
23	4,530	2,510	32,400	63,400	17,340	11,540	1,610	9,680	4,070	1,440	1,780	8,710
24	5,500	2,140	26,300	40,710	15,890	11,150	1,610	9,680	3,800	1,440	3,250	8,400
25	5,010	1,980	20,850	32,400	15,890	10,400	1,610	24,100	3,660	1,440	16,350	8,400
26	5,250	1,980	28,100	26,300	13,600	9,680	1,780	22,380	8,090	1,440	21,570	8,710
27	5,500	2,140	46,600	20,850	13,600	8,090	1,980	14,500	2,890	1,440	16,850	9,380
28	5,500	2,140	52,300	15,890	26,300	8,400	2,320	10,770	2,510	1,440	11,150	7,780
29	5,760	......	37,200	18,600	54,600	8,710	2,320	8,400	3,080	1,610	8,400	7,170
30	5,500	......	28,100	11,540	77,500	8,080	2,510	6,580	3,860	1,780	8,090	6,580
31	3,460	......	19,330	......	61,600	......	2,510	5,760	......	1,610	......	7,780
1902.												
1	7,170	10,770	154,100	18,330	5,250	2,700	31,100	13,600	1,280	5,760	3,840	2,140
2	6,300	10,400	164,100	16,850	5,250	2,510	25,700	13,170	1,440	10,040	3,460	2,700
3	5,500	13,600	103,770	14,980	5,010	2,320	20,340	11,940	1,280	6,870	3,250	4,070
4	5,250	12,340	67,000	13,170	5,760	2,320	40,700	10,770	1,280	5,250	3,080	4,530
5	5,010	11,540	42,800	11,540	5,760	2,320	49,000	9,680	1,280	5,010	2,890	5,010
6	4,770	9,680	29,900	10,770	6,300	2,700	33,100	7,470	1,120	4,770	2,890	5,250
7	4,770	9,380	22,380	11,540	6,300	2,510	34,400	6,870	1,120	4,530	2,700	4,770
8	4,770	8,710	17,830	12,340	7,170	2,510	25,100	6,580	1,120	4,530	2,890	4,770
9	5,010	8,400	14,980	9,680	7,780	2,320	19,830	6,300	1,120	4,070	2,890	6,020
10	5,010	7,780	15,890	105,500	7,170	2,140	18,330	5,760	1,440	3,660	2,700	3,860
11	5,010	7,470	19,830	6,700	6,580	2,320	27,500	19,830	1,280	3,270	2,700	4,770
12	5,010	6,580	24,600	45,000	6,020	2,320	24,600	4,530	1,280	2,890	2,510	4,770
13	5,010	6,300	40,600	31,800	5,500	2,700	19,830	4,300	1,440	2,510	2,320	11,150
14	4,770	6,580	60,700	25,100	5,250	2,890	13,600	4,080	1,280	2,510	2,140	8,400
15	4,300	5,500	49,000	19,830	5,010	3,270	10,400	3,660	1,120	2,700	2,140	6,580
16	4,070	4,770	31,800	15,890	4,530	3,660	8,400	3,660	1,120	3,270	1,980	17,340
17	4,070	4,300	75,500	11,540	4,070	3,860	8,710	3,270	1,120	3,080	1,980	24,600
18	4,070	4,300	65,300	12,340	3,860	4,070	7,470	3,080	970	2,890	1,980	20,340
19	3,660	5,250	42,800	10,770	3,460	3,660	11,150	2,700	830	2,700	1,740	14,980
20	3,270	4,530	29,900	......	3,460	3,660	14,050	2,700	830	2,700	1,960	14,050
21	4,070	3,860	22,380	......	3,460	3,460	22,380	2,890	830	2,510	1,960	29,300
22	14,980	4,530	18,330	8,090	3,460	3,460	16,850	2,700	830	2,320	1,940	48,200
23	22,130	3,860	15,430	7,170	3,270	2,890	16,850	2,320	830	2,140	1,960	36,500
24	11,540	3,660	13,600	6,300	3,270	1,440	17,850	2,140	1,280	2,140	2,140	24,600
25	11,540	3,660	11,540	6,020	3,270	3,080	17,340	1,960	1,960	1,980	2,140	18,330
26	9,690	4,070	10,400	5,250	3,460	3,460	17,340	1,780	4,770	1,980	2,890	15,420
27	10,040	6,870	9,380	5,250	3,660	6,020	15,890	1,120	5,500	2,700	2,520	10,040
28	9,680	19,800	8,710	5,010	3,270	5,500	14,500	6,020	2,510	2,140	11,150
29	9,380	......	9,380	4,770	3,080	10,770	11,540	1,440	4,770	3,080	8,400
30	10,040	......	16,350	5,250	2,890	10,770	14,500	1,280	4,770	3,460	8,400
31	9,690	......	19,330	......	2,890	......	14,500	1,280	......	3,460	......	5,250

Mean daily discharge, in second-feet, of West Branch of Susquehanna River at Williamsport, Pa., 1895–1904—Continued.

Day.	Jan.	Feb.	Mar.	Apr.	May.	June.	July.	Aug.	Sept.	Oct.	Nov.	Dec.
1903.												
1	6,580	41,400	110,700	17,340	5,010	2,140	11,940	5,760	17,340	2,140	3,890	4,070
2	5,250	18,330	80,500	16,370	4,770	1,290	10,400	4,770	14,960	2,140	3,460	4,070
3	6,580	26,300	44,300	14,980	4,530	1,960	9,690	4,070	11,940	1,960	3,460	4,070
4	10,770	47,400	31,100	13,600	4,070	1,440	11,150	3,660	9,690	1,960	3,270	3,890
5	13,170	98,100	24,600	14,050	4,070	1,440	9,690	5,250	8,090	2,140	3,080	3,660
6	14,960	69,800	24,000	12,340	4,070	1,440	8,710	8,400	6,870	2,890	3,270	3,660
7	13,600	43,500	24,600	11,540	4,070	1,610	18,330	9,690	5,760	3,460	3,080	3,460
8	11,150	28,100	26,900	11,540	3,460	2,140	14,960	9,690	5,250	4,070	3,080	3,460
9	8,710	21,870	60,700	13,600	3,270	2,890	10,400	8,090	5,010	17,830	3,080	3,270
10	4,530	17,340	65,200	15,420	2,890	8,060	8,060	6,300	5,500	28,100	2,890	3,460
11	4,530	14,050	50,600	16,370	2,890	4,300	6,580	5,500	6,020	19,330	2,800	3,460
12	10,770	14,500	51,400	14,960	2,890	4,530	6,580	5,010	7,780	15,890	2,800	2,890
13	10,400	19,330	47,400	17,340	2,700	8,710	6,580	4,530	6,580	11,940	2,700	2,700
14	10,400	21,870	35,100	18,830	2,700	7,780	5,760	4,070	5,250	9,690	2,510	3,660
15	10,400	20,340	28,100	40,060	2,700	8,400	5,010	3,460	4,770	8,400	2,510	3,080
16	10,040	19,830	22,890	56,300	2,510	9,690	4,530	3,270	4,070	7,470	2,700	2,140
17	10,400	19,830	19,330	48,300	2,510	8,710	4,070	3,660	3,860	6,580	6,020	2,140
18	8,710	18,330	16,850	36,500	2,510	7,780	4,070	3,660	5,250	10,400	58,900	2,140
19	8,710	11,150	14,960	28,900	2,510	6,300	15,420	3,270	5,500	14,960	37,200	2,140
20	8,400	9,690	12,340	20,850	2,510	5,500	29,300	3,080	5,250	13,600	25,700	2,140
21	8,710	10,040	11,150	16,850	2,510	5,250	19,330	3,270	4,770	11,940	15,890	3,270
22	7,780	9,690	11,940	14,050	2,320	5,250	15,890	4,530	4,070	9,690	13,170	4,300
23	7,170	11,540	14,960	11,940	2,320	5,500	13,600	3,660	3,660	8,400	11,150	4,070
24	6,870	9,690	10,700	10,400	2,140	10,040	10,400	4,080	3,270	7,170	10,040	3,880
25	6,870	9,360	60,700	9,690	1,960	18,830	8,400	3,080	3,080	6,580	9,690	3,660
26	6,580	9,100	59,300	8,090	1,960	37,200	7,170	2,890	3,080	6,020	7,780	3,460
27	6,580	8,400	27,500	7,780	2,140	23,460	5,790	3,080	2,700	5,500	6,580	4,070
28	6,580	41,700	20,850	6,870	2,140	15,420	4,770	3,660	2,510	4,770	5,250	4,070
29	6,020		16,370	6,300	2,320	11,540	4,300	9,360	2,510	4,530	4,300	5,010
30	6,580		13,600	5,760	2,320	14,500	5,250	24,600	2,320	4,070	3,460	4,770
31	50,600		13,170		2,320		6,020	20,850		4,070		5,010
1904.												
1	4,530	9,690	5,760	19,330	28,100	8,400	4,770	2,140	1,120	1,780	2,140	1,440
2	4,530	7,780	23,460	107,800	23,460	8,710	4,300	1,960	1,120	2,140	2,140	1,280
3	4,070	6,580	20,300	73,690	19,330	8,710	3,460	1,960	1,120	2,140	2,140	1,280
4	4,070	6,020	135,100	41,400	15,890	8,090	3,460	1,960	970	1,960	2,140	1,120
5	3,690	6,580	104,300	29,300	13,600	18,830	3,080	1,780	970	1,780	1,960	1,120
6	3,660	5,010	37,200	22,340	11,540	11,540	3,080	1,610	970	1,780	1,960	1,120
7	3,460	5,500	25,700	20,340	10,400	8,710	3,460	1,610	830	1,610	1,780	1,120
8	3,460	13,600	115,000	18,330	9,360	7,780	3,660	1,610	830	1,610	1,610	1,120
9	3,460	46,600	72,600	18,330	8,400	7,170	7,780	1,440	830	1,440	1,440	1,120
10	3,460	26,900	41,400	34,400	7,470	7,170	11,150	1,280	830	1,440	1,440	1,120
11	3,460	18,330	26,900	37,200	7,170	7,470	29,950	1,280	830	1,440	1,610	1,120
12	3,460	11,540	20,850	28,700	6,580	7,470	21,870	1,120	1,280	1,280	1,610	1,120
13	3,460	10,770	17,340	24,600	6,020	6,300	15,420	1,280	1,440	1,610	1,610	970
14	3,660	9,690	14,960	21,390	5,250	5,760	11,940	1,280	1,280	2,510	1,780	970
15	3,270	9,690	16,280	17,340	6,580	5,760	6,580	1,280	3,080	1,780	1,780	970
16	3,250	9,690	11,150	14,500	7,780	5,500	7,780	1,280	1,440	2,890	1,610	970
17	3,080	8,400	10,040	14,500	7,470	6,870	6,580	970	1,280	2,700	1,610	970
18	3,080	8,090	9,690	14,050	7,170	6,020	5,250	970	1,280	2,510	1,610	830
19	3,080	7,470	9,690	13,600	12,340	5,500	6,580	970	1,120	2,320	1,610	830
20	3,080	6,580	11,540	11,540	27,500	4,770	4,070	1,120	970	2,140	1,610	830
21	2,890	6,300	29,950	10,400	24,600	4,770	3,460	1,280	970	2,320	1,440	830
22	3,080	6,020	21,870	9,690	18,330	6,580	3,080	1,280	830	3,080	1,440	830
23	27,500	5,760	21,390	8,400	14,500	8,710	2,700	1,960	830	3,460	1,440	830
24	70,700	8,710	42,100	7,470	12,340	9,690	2,700	2,140	830	3,270	1,610	890
25	41,400	10,400	65,200	7,170	11,150	7,170	2,510	830	890	3,080	1,440	970
26	23,460	9,690	57,600	8,400	10,400	6,020	2,320	2,140	970	3,080	1,440	970
27	15,420	6,580	61,300	10,770	9,690	4,770	2,320	1,960	1,440	2,890	1,440	1,120
28	13,170	5,760	47,400	14,050	9,690	4,300	2,320	1,610	2,140	2,700	1,440	3,690
29	8,400	5,250	35,100	22,340	8,090	3,860	2,320	1,440	2,140	2,700	1,280	7,640
30	7,170		22,890	21,800	7,470	4,300	2,320	1,440	2,320	2,510	1,280	6,010
31	8,400		18,330		7,470		2,140	1,280		2,320		1,280

Estimated monthly discharge of West Branch of Susquehanna River at Williamsport, Pa., 1895–1904.

[Drainage area, 5,640 square miles.]

Month.	Discharge in second-feet.			Run-off.	
	Maximum.	Minimum.	Mean.	Second-feet per square mile.	Depth in inches.
1895.					
March	46,600	10,400	20,751	3.679	4.241
April	58,900	4,530	20,166	3.576	3.990
May	10,770	3,270	5,513	.978	1.128
June	19,330	830	3,480	.617	.688
July	11,540	410	2,946	.522	.602
August	3,660	600	1,898	.336	.387
September	3,660	410	1,030	.183	.204
October	970	410	746	.132	.152
November	6,870	500	1,462	.259	.289
December	23,400	830	4,523	.802	.924
The period	58,900	410	6,252	1.108	12.605
1896.					
January	22,380	2,700	5,705	1.012	1.167
February	49,000	3,080	10,861	1.926	2.077
March	76,500	4,070	13,809	2.448	2.822
April	67,900	8,710	20,118	3.567	3.980
May	8,090	2,140	3,853	.688	.787
June	23,400	2,510	7,454	1.322	1.475
July	22,380	3,270	6,276	1.118	1.288
August	22,890	1,280	6,382	1.132	1.305
September	4,070	970	1,560	.277	.309
October	49,000	1,610	13,137	2.329	2.685
November	26,300	4,300	8,770	1.554	1.734
December	18,600	2,700	6,245	1.107	1.276
The year	76,500	970	8,681	1.539	20.899

*Estimated monthly discharge of West Branch of Susquehanna River at Williams-
port, Pa., 1895-1904—Continued.*

[Drainage area, 5,640 square miles.]

Month.	Discharge in second-feet.			Run-off.	
	Maximum.	Minimum.	Mean.	Second-feet per square mile.	Depth in inches.
1897.					
January	9,360	2,890	4,955	0.878	1.012
February	34,400	3,080	9,495	1.684	1.754
March	53,000	6,870	25,589	4.537	5.231
April	34,400	5,760	13,869	2.459	2.744
May	34,400	4,070	14,294	2.534	2.921
June	5,010	2,140	3,046	.540	.602
July	12,750	1,610	3,409	.604	.696
August	8,090	1,780	3,712	.658	.759
September	5,010	600	1,706	.302	.337
October	1,960	970	1,286	.228	.263
November	16,850	1,120	6,716	1.191	1.329
December	27,500	4,070	11,475	2.034	2.345
The year	53,000	600	8,295	1.471	19.993
1898.					
January	42,100	3,460	15,799	2.801	3.230
February	31,800	4,300	12,211	2.165	2.254
March	162,600	6,020	31.357	5.560	6.410
April	33,700	5,760	12,900	2.287	2.552
May	19,830	6,800	10,536	1.868	2.154
June	8,090	2,140	4,289	.760	.848
July	4,070	1,280	2,056	.364	.420
August	22,380	1,960	4,467	.792	.914
September	2,330	1,120	1,529	.271	.302
October	35,800	1,280	7,372	1.307	1.507
November	38,600	3,660	8,513	1.509	1.684
December	31,100	2,700	7,590	1.346	1.552
The year	162,600	1,120	9,885	1.753	23.827

Estimated monthly discharge of West Branch of Susquehanna River at Williamsport, Pa., 1895–1904—Continued.

Month.	Discharge in second-feet.			Run-off.	
	Maximum.	Minimum.	Mean.	Second-feet per square mile.	Depth in inches.
1899.					
January	29,300	6,580	13,005	2.128	2.453
February	31,100	4,770	9,303	1.649	1.717
March	68,800	15,420	27,500	4.876	5.622
April	28,100	6,870	15,693	2.782	8.104
May	25,100	4,070	7,484	1.327	1.530
June	5,010	1,280	2,724	.483	.539
July	3,460	970	1,748	.310	.357
August	5,250	600	1,335	.237	.273
September	3,080	970	1,845	.327	.365
October	1,280	710	1,008	.179	.206
November	9,030	1,120	5,744	1.018	1.136
December	26,300	3,080	9,258	1.641	1.892
The year	68,800	600	7,971	1.413	19.194
1900.					
January	67,900	5,250	13,934	2.470	2.848
February	41,400	6,020	14,095	2.499	2.602
March	35,800	5,250	15,639	2.773	3.197
April	22,890	8,400	13,992	2.481	2.768
May	9,690	3,460	4,923	.873	1.006
June	8,090	1,780	4,043	.717	.800
July	3,460	1,280	2,046	.363	.418
August	2,140	710	1,811	.282	.267
September	1,780	710	931	.165	.184
October	3,660	710	1,821	.323	.372
November	110,100	1,610	9,328	1.654	1.845
December	24,600	3,660	8,562	1.518	1.750
The year	110,100	710	7,551	1.339	18.057

Estimated monthly discharge of West Branch of Susquehanna River at Williamport, Pa., 1895-1904—Continued.

Month.	Discharge in second-feet.			Run-off.	
	Maximum.	Minimum.	Mean.	Second-feet per square mile.	Depth in inches.
1901.					
January	11,540	2,140	5,182	0.919	1.060
February	5,250	1,960	3,010	.534	.556
March	52,200	2,140	20,920	3.709	4.280
April	89,900	9,690	27,538	4.882	5.447
May	77,500	8,660	15,403	2.731	3.148
June	41,400	5,500	12,311	2.183	2.436
July	6,870	1,610	2,911	.516	.595
August	28,100	1,610	7,049	1.250	1.441
September	22,380	2,510	6,296	1.116	1.245
October	4,070	1,440	2,122	.376	.433
November	21,870	1,120	4,266	.756	.844
December	150,900	8,660	20,276	3.595	4.145
The year	150,900	1,120	10,606	1.881	25.630
1902.					
January	22,130	3,270	7,090	1.257	1.449
February	49,800	3,660	8,517	1.510	1.572
March	164,100	8,710	39,585	7.019	8.092
April	105,500	4,770	20,096	3.563	3.975
May	7,780	2,890	4,711	.835	.963
June	10,770	1,440	3,371	.598	.667
July	49,000	6,870	20,095	3.563	4.108
August	13,600	1,120	4,868	.863	.995
September	6,020	830	1,723	.305	.340
October	10,040	1,960	3,546	.629	.725
November	3,860	1,780	2,461	.436	.486
December	48,200	2,140	12,508	2.217	2.556
The year	164,100	830	10,714	1.899	25.928

Estimated monthly discharge of West Branch of Susquehanna River at Williamsport, Pa., 1895–1904—Continued.

Month.	Discharge in second-feet.			Run-off.	
	Maximum.	Minimum.	Mean.	Second-feet per square mile.	Depth in inches.
1903.					
January	50,600	4,530	9,948	1.768	2.082
February	93,100	8,400	24,459	4.337	4.516
March	110,700	11,150	35,220	6.245	7.200
April	56,300	5,760	17,825	3.160	3.526
May	5,010	1,960	2,938	.521	.601
June	37,200	1,280	7,929	1.407	1.569
July	29,300	4,070	9,747	1.728	1.992
August	24,600	2,890	6,019	1.067	1.230
September	17,340	2,320	5,890	1.044	1.165
October	28,100	1,960	8,313	1.474	1.699
November	58,900	2,510	8,773	1.555	1.735
December	5,010	2,140	3,519	.624	.719
The year	110,700	1,280	11,715	2.077	27.984
1904.					
January	70,700	2,890	9,477	1.68	1.94
February	46,600	5,010	10,320	1.83	1.97
March	135,100	5,760	36,070	6.40	7.38
April	107,800	7,170	23,760	4.21	4.70
May	28,100	5,250	12,080	2.14	2.47
June	18,830	3,460	7,170	1.27	1.42
July	29,950	2,140	6,219	1.10	1.27
August	2,510	970	1,541	.273	.315
September	2,320	880	1,170	.207	.231
October	3,460	1,280	2,309	.409	.472
November	2,140	1,280	1,648	.292	.326
December	8,010	1,120	1,660	.294	.339
The year	135,100	880	9,450	1.68	22.88

WEST BRANCH OF SUSQUEHANNA RIVER AT ALLENWOOD, PA.

Observations of height of water on the West Branch have been made by the Weather Bureau at Lock Haven, Pa., 47 miles above Allenwood. The drainage area is given as 3,740 square miles, and the width of river 1,125 feet. The gage is in two sections. The lower section is painted on the side wall of the canal lock and the upper is on the highway bridge over the river. The elevation of the zero is 555.7 feet. The highest water was 18 feet, on June 1, 1889, and the danger line is at 10 feet.

A gaging station was established on the West Branch by E. G. Paul on March 25, 1899, at Allenwood, Pa., 20 miles above the junction with the main stream. Measurements are made from the public highway bridge, one-fourth of a mile east of the railroad station at Allenwood. The wire gage is 42.15 feet from zero to the end of the weight, and is referred to a pine-board scale fastened to ironwork of the bridge and divided into feet and tenths. The initial point of soundings is at the end of the iron guard rail on the right bank. The channel is straight for one-half a mile above and below the station. The current is sluggish, but unobstructed. The banks are low and subject to overflow at time of high water. The bed of the stream is rocky and permanent. The observer is Frank L. Allen, a farmer, living 200 feet from the gage. A bench mark was established on September 24, 1900. It consists of a copper bolt set in the capstone of the wing wall on the lower side of the west end of the bridge, and is 33.19 feet above datum of the gage.

This station was discontinued in April, 1902, the station at Williamsport taking its place.

Discharge measurements of West Branch of Susquehanna River at Allenwood, Pa., 1899–1902.

Date.	Hydrographer.	Gage height.	Area of section.	Mean velocity.	Discharge.
1899.		*Feet.*	*Square feet.*	*Feet per second.*	*Second-feet.*
Mar. 24	E. G. Paul	7.00	7,885	4.06	32,031
June 8	....do	3.00	3,367	1.18	3,988
July 28	....do	2.05	2,625	.52	1,360
Sept. 15	....do	1.90	2,437	.51	1,234
Oct. 17	....do	1.70	2,187	.39	842
1900.					
May 18	E. G. Paul	3.20	3,729	1.29	4,812
Sept. 24	....do	1.30	327	1.56	511
1901.					
Aug. 17	E. G. Paul	4.10	4,460	1.99	8,857
Oct. 26	....do	2.30	2,824	.81	2,308
1902.					
Apr. 21	E. G. Paul	4.40	4,736	2.09	9,896

Mean daily gage height, in feet, of West Branch of Susquehanna River at Allenwood, Pa., 1899–1902.

Day.	Jan.	Feb.	Mar.	Apr.	May.	June.	July.	Aug.	Sept.	Oct.	Nov.	Dec.
1899.												
1				6.70	3.80	3.50	2.90	2.00	2.70	2.00	2.20	2.90
2				6.30	3.80	3.50	2.70	2.00	2.70	2.00	3.60	2.80
3				5.80	3.80	3.40	2.50	2.00	2.70	1.90	4.20	2.70
4				5.35	3.90	3.40	2.50	1.80	2.50	1.90	5.20	2.70
5				5.05	3.80	3.30	2.40	1.70	2.40	1.90	4.60	2.70
6				4.90	3.60	3.20	2.40	1.70	2.30	1.90	4.00	2.60
7				4.80	3.50	3.00	2.30	1.70	2.10	1.90	3.40	2.60
8				6.45	3.80	3.00	2.40	1.70	2.00	1.90	3.20	2.60
9				7.80	3.40	2.90	2.20	1.70	2.00	1.90	3.00	2.60
10				7.40	3.50	2.90	2.00	1.70	2.00	1.80	8.00	2.60
11				6.60	3.60	2.70	2.10	1.70	1.90	1.80	8.00	2.60
12				8.20	3.70	2.60	2.30	1.70	1.90	1.80	8.20	2.50
13				6.50	3.50	2.60	2.20	1.90	1.90	1.70	3.30	8.40
14				7.00	3.40	2.60	2.20	1.90	1.90	1.70	8.40	6.50
15				6.90	3.30	2.50	2.30	1.90	1.90	1.70	8.50	6.50
16				6.80	3.20	2.50	2.20	1.90	1.90	1.70	8.60	5.80
17				6.40	3.40	2.50	2.30	1.90	1.90	1.70	8.80	5.10
18				5.60	3.80	2.40	2.40	1.90	1.90	1.70	3.90	4.80
19				5.40	7.40	2.40	2.60	1.80	1.90	1.70	4.10	4.80
20				5.00	6.50	2.40	2.80	1.70	1.90	1.70	4.80	4.70
21				4.80	5.75	2.40	3.00	1.70	1.90	1.60	4.10	4.30
22				4.70	5.15	2.30	2.70	1.70	1.90	1.60	4.00	4.20
23			7.00	4.50	4.70	2.20	2.50	1.70	1.90	1.60	3.90	4.20
24			7.00	4.40	4.85	2.20	2.30	1.60	1.90	1.60	3.80	5.15
25			6.70	4.30	4.00	2.80	2.20	1.60	1.90	1.60	3.70	7.25
26			6.30	4.80	3.80	2.50	2.20	1.60	1.90	1.60	3.40	5.00
27			6.40	4.30	3.60	2.60	2.10	1.70	2.00	1.60	3.40	5.00
28			6.20	4.20	3.50	2.70	2.00	2.00	2.00	1.60	3.80	4.50
29			6.70	4.80	3.40	2.70	1.90	3.00	2.00	1.60	3.20	4.10
30			7.80	4.10	3.40	2.80	1.80	2.60	2.00	1.60	3.10	3.60
31			7.85		3.50		2.00	2.60		1.60		8.40
1900.												
1	4.50	3.20	7.55	5.00	4.30	3.90	2.10	1.90	2.00	1.30	2.10	5.75
2	5.50	3.20	9.60	5.30	4.20	3.90	2.30	1.80	1.90	1.30	2.10	5.40
3	5.70	3.40	7.70	5.40	4.20	3.90	2.60	1.80	1.90	1.20	2.10	5.80
4	5.80	3.40	7.00	5.80	4.00	4.00	2.50	1.80	1.80	1.20	2.10	5.80
5	5.90	3.50	6.00	5.90	3.80	4.10	2.30	1.70	1.80	1.20	2.10	6.40
6	5.90	3.60	5.40	6.20	3.50	3.90	2.30	1.70	1.70	1.20	2.00	6.40
7	5.90	3.80	5.80	6.40	3.30	3.60	2.20	1.70	1.70	1.20	1.90	6.70
8	4.70	4.50	5.90	6.20	3.30	3.50	2.20	1.60	1.70	1.20	1.90	6.10
9	3.70	5.00	6.10	7.80	3.20	3.40	2.20	1.60	1.60	1.80	1.90	5.50
10	3.90	5.30	6.40	6.00	3.20	3.30	2.30	1.50	1.60	2.20	1.90	4.90
11	4.20	5.60	6.90	5.70	3.20	3.20	2.50	1.40	1.60	2.20	1.90	4.60
12	4.50	5.30	6.20	5.80	3.20	3.00	2.70	1.50	1.50	2.10	1.90	4.20
13	4.40	6.00	5.40	4.90	3.50	3.00	2.90	1.50	1.50	2.10	1.90	4.00
14	4.20	7.70	5.00	4.80	3.40	3.00	2.80	1.40	1.40	2.10	1.90	3.80
15	4.00	7.30	4.00	4.80	3.40	3.00	2.60	1.40	1.40	2.10	1.90	3.60
16	4.00	6.50	4.00	4.60	3.30	3.00	2.60	1.40	1.40	2.10	1.80	3.30
17	4.00	6.20	3.90	4.70	3.20	2.90	2.50	1.40	1.80	2.10	1.80	3.20
18	4.20	5.40	3.80	6.00	3.20	2.80	2.40	1.40	1.30	2.20	1.80	3.20
19	4.50	5.60	3.70	7.00	3.50	2.70	2.00	1.40	1.30	2.10	1.70	3.10
20	5.30	5.90	6.20	6.90	3.50	2.70	2.00	1.40	1.30	2.00	1.70	3.10
21	18.20	6.00	7.10	6.30	3.30	2.60	2.00	1.40	1.30	1.90	1.70	3.00
22	12.20	8.20	6.90	6.20	3.20	2.50	1.90	2.30	1.80	1.90	1.90	3.00
23	8.50	10.15	6.60	6.00	3.00	2.40	1.90	2.30	1.30	2.10	2.10	3.00
24	6.50	7.85	6.10	6.30	3.00	2.40	1.90	2.30	1.30	2.20	3.00	3.00
25	6.30	6.50	5.90	5.90	3.00	2.40	1.90	2.30	1.30	2.40	5.00	3.00
26	6.10	5.00	5.50	5.60	3.00	2.30	1.90	2.30	1.30	2.50	7.70	3.00
27	5.30	5.00	5.20	5.30	3.40	2.30	2.20	2.20	1.30	2.60	15.75	3.00
28	4.60	4.80	5.10	4.80	3.20	2.20	2.20	2.30	1.30	2.40	10.05	3.00
29	4.60		5.00	4.80	3.00	2.20	2.20	2.20	1.30	2.40	8.25	3.00
30	4.50		4.90	4.40	3.00	2.10	2.30	2.10	1.30	2.30	6.60	3.00
31	3.20		4.80		3.20		2.00	2.00		2.20		3.00

Mean daily gage height, in feet, of West Branch of Susquehanna River at Allenwood, Pa., 1899–1902—Continued.

Day.	Jan.	Feb.	Mar.	Apr.	May.	June.	July.	Aug.	Sept.	Oct.	Nov.	Dec.
1901.												
1	3.00	2.50	3.00	5.80	4.70	8.50	2.90	2.50	4.10	3.80	2.40	4.00
2	3.00	2.50	3.00	5.60	4.60	7.70	2.90	2.40	4.20	3.20	2.60	8.90
3	3.00	2.50	2.90	5.20	5.00	7.20	2.80	2.40	4.20	3.10	2.80	8.70
4	3.00	2.50	3.00	6.50	5.20	6.70	2.80	2.30	4.30	3.10	3.00	8.60
5	3.00	2.60	3.50	6.80	5.00	5.70	2.80	2.20	4.40	3.00	3.10	8.40
6	2.80	2.80	4.80	7.00	4.90	5.00	2.70	2.20	4.40	3.10		8.40
7	2.70	3.00	5.20	9.15	4.70	5.70	2.70	2.90	4.20	3.20	2.20	8.20
8	2.50	2.50	5.50	10.00	4.60	5.90	2.70	2.50	4.30	3.30	2.40	5.00
9	2.50	3.00	5.80	11.15	4.40	6.00	2.70	3.10	3.90	3.40	2.40	7.20
10	2.90	4.50	4.80	9.80	4.00	5.90	2.70	3.00	3.90	3.40	3.40	6.50
11	3.40	4.00	9.50	8.80	4.20	5.60	2.70	2.90	3.80	3.60	3.60	6.30
12	3.80	4.00	9.70	7.00	4.30	5.20	2.60	2.80	3.80	3.60	3.40	6.20
13	4.50	3.80	9.10	6.80	4.30	5.00	2.60	2.90	3.80	3.70	3.90	5.90
14	4.90	4.20	8.50	6.40	4.40	4.90	2.50	2.70	3.80	3.40	3.20	8.00
15	4.50	4.00	7.40	6.30	4.50	4.80	2.50	2.70	3.60	3.20	3.20	20.15
16	4.20	3.80	6.80	5.80	4.50	4.40	2.50	2.60	3.70	3.00	3.10	17.70
17	4.00	3.50	6.20	5.40	4.70	4.30	2.40	4.10	3.80	2.90	3.10	11.30
18	4.00	3.20	5.80	5.20	4.60	4.10	2.40	4.30	3.90	2.80	3.00	7.40
19	3.90	3.00	7.30	5.10	4.80	4.00	2.90	4.50	3.80	2.70	3.00	7.00
20	3.50	3.00	8.00	6.30	4.40	4.00	2.20	4.70	3.60	2.50	2.90	5.90
21	3.20	3.20	8.00	11.45	4.20	3.90	2.10	4.60	3.50	2.40	2.80	5.40
22	3.00	3.00	8.00	14.35	4.20	3.80	2.10	6.40	3.50	2.80	2.60	5.10
23	3.00	3.90	8.00	11.65	5.20	3.70	2.00	7.90	3.30	2.80	2.40	4.80
24	2.80	3.00	7.60	10.20	6.20	3.50	2.00	7.70	3.30	2.80	5.00	4.50
25	2.50	3.00	7.20	9.30	5.80	3.30	2.00	6.80	3.20	2.80	6.70	4.40
26	2.50	3.00	9.40	8.50	6.00	3.30	2.00	6.20	3.20	2.90	5.90	4.30
27	2.50	3.00	11.20	7.40	6.40	3.10	2.20	5.70	3.00	2.30	5.50	4.20
28	2.50	3.00	11.20	5.80	7.10	3.10	2.30	4.80	2.80	2.20	4.80	4.10
29	2.50		8.70	5.30	11.15	2.90	2.40	4.30	2.70	2.20	4.00	4.10
30	2.50		7.00	5.00	13.00	2.90	2.50	4.20	2.80	2.20	4.20	4.00
31	2.50		6.60		10.40		2.50	4.10		2.30		3.90
1902.												
1	3.80	5.40	21.60	6.40								
2	3.80	5.20	19.40	6.50								
3	3.60	4.90	15.50	6.50								
4	3.60	4.90	11.50	6.40								
5	3.50	4.90	8.20	5.80								
6	3.50	4.90	6.80	(a)								
7	3.50	4.80	6.40									
8	3.50	4.80	5.50									
9	3.40	4.80	4.90									
10	3.40	4.80	6.40									
11	3.40	4.80	7.80									
12	3.30	4.70	8.40									
13	3.20	4.70	10.00									
14	3.20	4.60	8.90									
15	3.20	4.50	8.60									
16	3.20	4.70	8.80									
17	3.20	4.70	12.20									
18	3.10	4.70	10.00									
19	3.10	4.70	8.60									
20	3.10	4.70	7.40									
21	3.40	4.70	6.70									
22	7.40	4.70	6.40									
23	6.80	4.70	5.70									
24	6.60	7.00	5.40									
25	6.50	7.40	5.20									
26	6.30	5.50	4.80									
27	6.20	5.90	4.70									
28	5.90	9.70	5.00									
29	5.90		5.60									
30	5.80		6.10									
31	5.40		6.20									

a Discontinued.

Rating table for West Branch of Susquehanna River at Allenwood, Pa., for 1900 to 1902.

Gage height.	Discharge.	Gage height.	Discharge.	Gage height.	Discharge.	Gage height.	Discharge.
Feet.	*Second-feet.*	*Feet.*	*Second-feet.*	*Feet.*	*Second-feet.*	*Feet.*	*Second-feet.*
1.2	430	3.5	5,970	5.8	20,500	9.2	59,800
1.3	510	3.6	6,400	5.9	21,350	9.4	62,700
1.4	600	3.7	6,830	6.0	22,200	9.6	65,700
1.5	690	3.8	7,260	6.1	23,100	9.8	68,800
1.6	790	3.9	7,700	6.2	24,000	10.0	72,000
1.7	900	4.0	8,160	6.3	24,900	10.2	75,300
1.8	1,040	4.1	8,630	6.4	25,900	10.4	78,600
1.9	1,220	4.2	9,110	6.5	26,900	10.6	82,000
2.0	1,410	4.3	9,610	6.6	27,900	10.8	85,500
2.1	1,610	4.4	10,140	6.7	28,900	11.0	89,000
2.2	1,830	4.5	10,710	6.8	29,900	11.2	92,600
2.3	2,070	4.6	11,300	6.9	31,000	11.4	96,300
2.4	2,320	4.7	11,930	7.0	32,000	11.6	100,000
2.5	2,580	4.8	12,600	7.2	34,200	11.8	103,800
2.6	2,850	4.9	13,300	7.4	36,500	12.0	107,600
2.7	3,130	5.0	14,030	7.6	38,800	12.2	111,500
2.8	3,420	5.1	14,780	7.8	41,200	12.4	115,500
2.9	3,730	5.2	15,550	8.0	43,600	12.6	119,500
3.0	4,050	5.3	16,350	8.2	46,100	12.8	123,700
3.1	4,400	5.4	17,170	8.4	48,700	13.0	128,000
3.2	4,770	5.5	17,990	8.6	51,400		
3.3	5,150	5.6	18,820	8.8	54,100		
3.4	5,550	5.7	19,650	9.0	56,900		

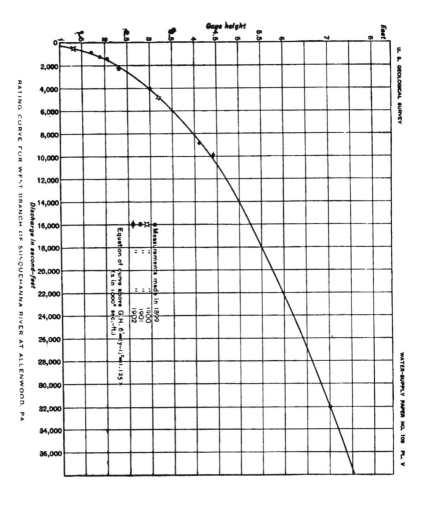

RATING CURVE FOR WEST BRANCH OF SUSQUEHANNA RIVER AT ALLENWOOD, PA

Mean daily discharge, in second-feet, of West Branch of Susquehanna River at Allenwood, Pa., 1899–1902.

Day.	Jan.	Feb.	Mar.	Apr.	May.	June.	July.	Aug.	Sept.	Oct.	Nov.	Dec.
1899.												
1				28,900	7,290	5,970	3,730	1,410	3,130	1,410	1,830	3,730
2				24,000	7,290	5,970	3,130	1,410	3,130	1,410	6,400	3,420
3				20,500	7,290	5,550	2,580	1,410	3,130	1,220	9,110	3,180
4				16,760	7,700	5,550	2,580	1,040	2,580	1,220	15,550	3,130
5				14,400	7,290	5,150	2,320	900	2,320	1,220	11,300	3,130
6				13,300	6,400	4,770	2,320	900	2,070	1,220	8,160	2,850
7				12,600	5,970	4,060	2,070	900	1,610	1,220	5,550	2,850
8				20,400	5,150	4,050	2,320	900	1,410	1,220	4,770	2,850
9				41,200	5,550	3,730	1,830	900	1,410	1,220	4,050	2,850
10				26,500	5,970	3,730	1,410	900	1,410	1,040	4,050	2,850
11				27,900	6,400	3,130	1,610	900	1,220	1,040	4,050	2,850
12				24,000	6,830	2,850	2,070	900	1,220	1,040	4,770	16,350
13				28,900	5,970	2,850	1,830	1,220	1,220	900	5,150	48,700
14				32,000	5,550	2,850	1,830	1,224	1,220	900	5,550	38,500
15				31,000	5,150	2,580	2,070	1,220	1,220	900	5,970	26,900
16				29,300	4,700	2,580	1,830	1,220	1,220	900	6,400	20,500
17				25,900	5,550	2,580	2,070	1,220	1,220	900	7,290	14,780
18				18,820	7,290	2,320	2,320	1,220	1,220	900	7,700	13,300
19				17,170	30,500	2,320	2,850	1,040	1,220	900	8,630	12,600
20				14,000	26,900	2,320	3,420	900	1,220	900	9,610	11,980
21				12,600	20,075	2,320	4,050	900	1,220	790	8,630	9,610
22				11,980	15,160	2,070	3,130	900	1,220	790	8,160	9,110
23			32,000	10,710	11,980	1,830	2,580	900	1,220	790	7,700	9,110
24			32,000	10,140	9,870	1,830	2,070	790	1,220	790	7,290	15,160
25			28,900	9,610	8,160	3,420	1,830	790	1,220	790	6,830	34,700
26			24,900	9,610	7,290	2,580	1,830	790	1,220	790	6,400	18,820
27			25,900	9,610	6,400	2,850	1,610	900	1,410	790	5,550	14,080
28			24,000	9,110	5,970	3,130	1,410	6,830	1,410	790	5,150	10,710
29			28,900	9,610	5,550	3,130	1,220	4,050	1,410	790	4,770	8,630
30			41,200	8,630	5,550	3,420	1,040	2,580	1,410	790	4,400	6,400
31			35,900		5,970		1,410	2,850		790		5,550
1900.												
1		10,710	4,770	38,200	14,080	9,610	7,700	1,610	1,220	1,410	510	1,610	20,070
2		17,990	4,770	65,700	16,350	9,110	7,700	2,070	1,040	1,220	510	1,610	17,170
3		19,650	5,550	40,000	17,170	9,110	7,700	2,580	1,040	1,220	430	1,610	14,080
4		20,500	5,550	32,000	20,500	8,160	8,160	2,580	1,040	1,040	430	1,610	20,500
5		21,350	5,970	22,200	21,350	7,290	8,630	2,070	900	1,040	430	1,610	21,350
6		21,350	6,400	17,170	24,000	5,970	7,700	2,070	900	900	430	1,410	25,900
7		21,350	7,290	20,500	25,900	5,150	6,400	1,830	900	900	430	1,220	24,900
8		11,980	10,710	21,350	24,000	5,150	5,970	1,830	790	900	430	1,220	22,200
9		6,830	14,080	23,100	35,900	4,770	6,550	1,830	790	790	1,040	1,220	17,990
10		7,700	16,350	25,900	22,200	4,770	5,150	2,070	690	790	1,830	1,220	13,300
11		9,190	18,820	31,000	19,650	4,770	2,580	2,580	690	790	1,830	1,220	11,300
12		10,710	16,350	24,000	16,350	4,770	4,050	3,130	690	690	1,610	1,220	9,110
13		10,140	22,200	17,170	13,300	5,550	4,050	3,730	600	690	1,610	1,220	8,160
14		9,110	40,000	14,080	12,600	5,550	4,050	3,420	600	600	1,610	1,220	7,290
15		8,160	35,900	8,160	12,600	5,150	4,050	2,850	600	600	1,610	1,040	6,400
16		8,160	26,900	8,160	11,300	5,150	4,050	2,850	600	600	1,610	1,040	5,150
17		8,160	24,000	7,700	11,980	4,770	3,730	2,580	600	510	1,610	1,040	4,770
18		9,110	17,170	7,290	22,200	4,770	3,420	2,320	600	510	1,830	1,040	4,770
19		10,710	18,820	6,830	22,100	5,970	3,130	1,830	600	510	1,410	900	4,400
20		16,350	21,350	24,000	31,000	6,400	3,130	1,410	600	510	1,220	900	4,400
21		32,000	32,200	33,100	24,900	5,150	2,850	1,410	600	510	1,220	900	4,400
22			20,100	21,350	22,200	4,770	2,580	1,220	2,070	510	1,220	1,220	4,050
23						4,060	2,320	1,220	2,070	510	1,410	2,320	4,050
24						4,060	2,320	1,220	2,070	510	1,830	1,610	4,050
25		900	26,900	21,350	22,200	4,060	2,070	1,830	2,070	510	2,320	11,300	4,050
26		23,100	14,080	17,990	18,820	5,550	2,070	1,220	2,070	510	2,580	40,000	4,050
27		16,350	14,080	15,550	16,350	5,550	2,070	1,830	1,830	510	2,070	26,900	4,050
28		11,300	12,600	14,780	12,600	4,770	1,830	1,830	1,830	510	1,830	22,200	4,050
29		11,300		14,080	11,300	4,060	1,830	1,830	1,610	510	2,070	40,750	4,050
30		10,710		14,300	10,140	4,770	1,610	1,830	1,610	510	2,070	25,900	4,050
31		4,770		12,600		4,770		1,410	1,410		1,830		4,050

Mean daily discharge, in second-feet, of West Branch of Susquehanna River at Allenwood, Pa., 1899-1902—Continued.

Day.	Jan.	Feb.	Mar.	Apr.	May.	June.	July.	Aug.	Sept.	Oct.	Nov.	Dec.
1901.												
1	4,050	2,580	4,050	20,500	11,930	50,000	3,730	2,580	8,650	5,150	2,320	8,180
2	4,050	2,580	4,050	18,820	11,300	40,000	3,790	2,320	9,110	4,770	2,850	7,700
3	4,050	2,580	3,730	15,550	14,030	34,200	3,430	2,320	9,110	4,400	3,420	6,880
4	4,050	2,580	4,050	26,900	15,550	28,900	3,430	2,070	9,610	4,400	4,050	6,440
5	4,050	2,850	5,970	29,900	14,030	19,650	3,430	1,830	10,140	4,050	4,400	5,550
6	3,430	3,430	12,600	32,000	13,300	14,030	3,130	1,830	10,140	4,400	4,400	5,550
7	3,130	4,050	15,550	59,000	11,930	19,650	3,130	2,070	9,110	4,770	1,830	4,770
8	2,580	2,580	17,990	72,000	11,300	21,350	3,130	2,580	8,650	5,150	2,320	14,030
9	2,580	4,050	20,500	91,700	10,140	22,200	3,130	4,400	7,700	5,550	2,320	34,200
10	3,730	5,970	12,600	61,200	8,160	21,350	3,130	4,050	7,700	5,550	5,550	26,900
11	5,550	8,160	64,200	47,400	9,110	18,820	3,130	3,730	7,260	6,400	6,400	24,900
12	7,260	8,160	67,200	32,000	9,610	15,550	2,850	3,430	7,260	6,400	5,550	24,000
13	10,710	7,260	58,300	29,900	9,610	14,030	2,850	3,430	7,260	6,830	5,150	21,350
14	13,300	9,110	50,000	25,900	10,140	13,300	2,580	3,130	7,260	5,550	4,770	43,600
15	10,710	8,160	36,500	24,900	10,710	12,600	2,580	3,130	6,400	4,770	4,770	326,000
16	9,110	7,260	29,900	20,500	10,710	10,140	2,580	2,850	6,830	4,050	4,400	247,900
17	8,160	5,970	24,000	17,170	11,930	9,610	2,320	8,650	7,260	3,730	4,400	94,400
18	8,160	4,770	20,500	15,550	11,300	8,650	1,830	9,610	7,700	3,430	4,050	36,500
19	7,700	4,050	34,200	14,780	12,600	8,160	1,830	10,710	7,260	3,130	4,050	32,000
20	5,970	4,050	43,600	24,900	10,140	8,160	1,830	11,930	6,400	2,580	3,730	21,350
21	4,770	4,770	43,600	97,200	9,110	7,700	1,610	11,300	5,970	2,320	3,420	17,170
22	4,050	4,050	43,600	158,400	9,110	7,260	1,610	25,900	5,970	2,070	2,850	14,780
23	4,050	7,700	43,600	101,000	15,550	6,830	1,410	42,400	5,150	2,070	2,320	12,600
24	3,430	4,050	38,800	75,300	24,000	5,970	1,410	40,000	5,150	2,070	14,030	10,710
25	2,580	4,050	34,200	61,200	20,500	5,150	1,410	29,900	4,770	2,070	28,900	10,140
26	2,580	4,050	62,700	50,000	22,200	5,150	1,410	24,000	4,770	2,070	21,350	9,610
27	2,580	4,050	92,600	36,500	25,900	4,400	1,830	19,650	4,050	2,070	17,990	9,110
28	2,580	4,050	92,600	20,500	33,100	4,400	2,320	12,600	3,420	1,830	12,600	8,650
29	2,580		52,700	16,350	91,700	3,730	2,320	9,610	3,130	1,830	8,160	8,630
30	2,580		32,000	14,030	128,000	3,730	2,540	9,110	3,420	1,830	9,110	8,180
31	2,580		27,100		78,600		2,580	8,650		2,070		7,700
1902.												
1	7,260	17,170	377,200	25,900								
2	7,260	15,550	380,900	26,900								
3	6,400	13,300	186,900	26,900								
4	6,400	13,300	98,100	25,900								
5	5,970	13,300	40,100	20,500								
6	5,970	13,300	29,900									
7	5,970	12,600	25,900									
8	5,970	12,600	17,990									
9	5,550	12,600	13,300									
10	5,550	12,600	25,900									
11	5,550	12,000	8,180									
12	5,150	11,930	48,700									
13	4,770	11,930	72,000									
14	4,770	11,300	55,500									
15	4,770	10,710	51,400									
16	4,770	11,930	54,100									
17	4,770	11,930	111,700									
18	4,400	11,930	72,000									
19	4,400	11,930	51,400									
20	4,400	11,930	36,500									
21	5,550	11,930	28,900									
22	96,500	11,930	25,900									
23	29,900	26,900	19,650									
24	27,900	32,000	17,170									
25	26,900	36,500	15,550									
26	24,900	17,990	12,600									
27	24,000	21,350	11,300									
28	21,350	67,200	14,030									
29	21,350		18,820									
30	20,500		25,100									
31	17,170		24,000									

Estimated monthly discharge of West Branch of Susquehanna River at Allenwood, Pa., 1899–1902.

[Drainage area, 6,538 square miles.]

Month.	Discharge in second-feet.			Run-off.	
	Maximum.	Minimum.	Mean.	Second-feet per square mile.	Depth in inches.
1899.					
March (23–31)	41,200	24,000	30,411	4.651	1.557
April	41,200	8,630	19,488	2.981	3.326
May	36,500	4,770	8,985	1.374	1.584
June	5,970	1,830	3,383	.517	.577
July	4,050	1,040	2,205	.337	.388
August	6,830	790	1,428	.218	.251
September	3,130	1,220	1,579	.242	.270
October	1,410	790	980	.150	.173
November	15,550	4,050	6,690	1.023	1.141
December	48,700	2,850	12,162	1.860	2.144
The period	48,700	790	8,731	1.335	11.411
1900.					
January	132,300	4,770	22,007	3.366	3.881
February	74,500	4,770	20,515	3.138	3.268
March	65,700	6,830	21,907	3.351	3.863
April	35,300	10,140	19,705	3.014	3.363
May	9,610	4,050	5,536	.847	.976
June	8,630	1,610	4,355	.666	.748
July	3,730	1,220	2,056	.314	.362
August	2,070	600	1,120	.171	.197
September	1,410	510	711	.109	.122
October	2,850	430	1,451	.222	.256
November	193,400	900	14,291	2.186	2.439
December	28,900	4,050	10,266	1.570	1.752
The year	193,400	430	10,327	1.578	21.222

Estimated monthly discharge of West Branch of Susquehanna River at Allenwood, Pa., 1899–1902—Continued.

Month.	Discharge in second-feet.			Run-off.	
	Maximum.	Minimum.	Mean.	Second-feet per square mile.	Depth in inches.
1901.					
January	13,300	2,580	5,054	0.773	0.891
February	9,110	2,580	4,891	.748	.779
March	92,600	3,730	35,284	5.397	6.222
April	158,400	14,030	43,702	6.684	7.457
May	128,000	8,160	22,106	3.381	3.898
June	50,000	3,730	14,822	2.267	2.529
July	3,730	1,410	2,524	.386	.445
August	42,400	1,830	10,313	1.577	1.818
September	10,140	3,130	6,886	1.053	1.175
October	6,830	1,830	3,785	.579	.668
November	28,900	1,830	6,715	1.027	1.146
December	326,000	4,770	35,785	5.473	6.310
The year	326,000	1,410	15,989	2.445	33.591
1902.					
January	36,500	4,400	11,809	1.806	2.082
February	67,200	10,710	17,151	2.623	2.731
March	377,200	11,930	61,798	9.452	10.897

JUNIATA RIVER AT NEWPORT, PA.

Juniata River rises in Center County, Pa., and flows in a general southeasterly direction into Susquehanna River 15 miles above Harrisburg. Its drainage area is mountainous and for the most part covered with forest growth.

This station was established at Newport, about 15 miles above the mouth of Juniata River, March 21, 1899, by E. G. Paul. The standard boxed chain gage was located on the covered wagon bridge which was 800 feet east of the public square at Newport, Pa. It was attached to the bridge timbers inside of the bridge near the right bank. The length of the chain from the end of the weight to the marker was 39.54 feet. The gage is read once each day by A. R. Bortel. Bench mark No. 1 is on the extreme east end of the stone doorsill, south front of Butz's store building, near end of bridge; its elevation is 28.83 feet above gage datum. Bench mark No. 2 is on shelf in southeast corner of underpinning of store of J. M. Ewing; its elevation is 27.37 feet above gage datum. This bench mark was set by the Pennsylvania Railroad, and according to their records its elevation is 390.69 feet above sea level. Discharge measurements were made from the lower side of the four-span wagon bridge to which the gage was attached. The initial point for soundings was the end of the woodwork of the bridge on the right bank downstream side. In the fall of 1904 this bridge was replaced by a steel structure. During its construction the stage of the river was obtained by means of a temporary gage staff attached to the exposed end of a sewer near the bridge. This gage was set at the same elevation as the old one. As soon as the bridge is completed a standard chain gage will be put in place. The channel is straight for one-half mile above and below the station. Both banks are high and are not subject to overflow. There is a single channel, broken by three bridge piers. The piers do not interfere with the flow of the stream and there is little eddying and boiling near them. The bed is of hard material and is probably permanent. There is a good measurable velocity at all stages.

Discharge measurements of Juniata River at Newport, Pa., 1899–1904.

Date.	Hydrographer.	Gage height.	Area of section.	Mean velocity.	Discharge.
1899.		*Feet.*	*Sq. feet.*	*Ft. per sec.*	*Sec. feet.*
Mar. 21	E. G. Paul	6.60	3,486	3.75	13,094
June 9	....do	3.20	1,158	1.64	1,903
July 31	....do	2.90	849	.80	682
Sept. 14	....do	4.55	1,755	2.64	4,625
Oct. 18	....do	2.90	661	1.25	829
1900.					
May 17	E. G. Paul	3.40	1,139	1.56	1,778
Sept. 22	....do	2.80	723	.58	418
1901.					
Aug. 14	E. G. Paul	3.40	1,080	1.77	1,915
Oct. 24	....do	3.10	881	1.46	1,288
1902.					
Apr. 19	E. G. Paul	5.00	2,093	3.24	6,779
Sept. 17	....do	2.84	702	1.05	734
1903.					
Mar. 9	E. C. Murphy	6.21	2,978	3.64	10,843
Apr. 2	....do	6.21	2,988	3.53	10,555
May 7	....do	3.96	1,409	3.10	2,963
June 3	J. C. Hoyt	3.40	1,102	1.38	1,525
Oct. 6	W. C. Sawyer	3.40	1,044	1.58	1,655
Nov. 3	Brundage and Sawyer	3.33	1,062	1.51	1,604
1904.					
July 16	N. C. Grover	4.28	1,520	2.73	4.152

Mean daily gage height, in feet, of Juniata River at Newport, Pa., 1899–1904.

Day.	Jan	Feb.	Mar.	Apr	May	June.	July.	Aug	Sept.	Oct	Nov.	Dec.
1899.												
1				7.00	3.40	3.11	2.70	3.00	3.50	3.20	2.70	3.30
2				6.10	3.40	3.60	2.70	3.00	3.50	3.10	4.00	3.80
3				5.50	3.60	3.50	2.60	3.00	3.40	3.10	4.90	3.30
4				5.10	3.70	3.40	2.50	3.00	3.40	3.10	4.60	3.30
5				4.90	3.50	3.30	3.00	3.00	3.20	3.00	4.20	3.30
6				4.50	3.40	3.40	3.00	3.00	3.30	3.00	3.90	3.20
7				4.30	3.40	3.30	3.00	3.30	3.30	3.00	3.70	3.10
8				5.60	3.40	3.20	3.20	3.30	3.50	3.30	3.50	3.10
9				6.90	3.60	3.20	3.10	3.30	3.30	2.90	3.50	3.10
10				5.50	3.60	2.80	3.30	3.30	3.30	2.90	3.40	3.10
11				5.50	4.00	2.80	3.30	3.10	3.30	2.90	3.90	3.10
12				5.10	4.10	2.80	3.10	3.10	3.40	2.90	3.30	3.70
13				4.90	4.00	2.80	3.10	3.40	4.80	2.90	3.20	4.80
14				4.80	3.80	2.70	3.10	3.10	3.80	2.90	3.20	5.10
15				4.80	3.80	2.70	3.10	3.10	3.50	2.90	3.20	4.80
16				4.70	3.60	2.70	2.70	3.00	3.30	2.90	3.20	4.30
17				5.50	3.70	2.70	2.70	3.00	3.30	2.90	3.20	4.00
18				4.40	4.10	2.60	2.70	3.00	3.10	2.90	3.10	4.00
19				4.30	8.00	2.60	2.80	3.00	3.10	2.90	3.10	3.70
20				4.10	7.30	2.60	3.00	3.00	3.10	2.90	3.10	3.70
21			6.50	4.00	7.60	2.60	3.00	3.00	3.10	2.90	3.10	5.00
22			6.00	3.90	5.10	2.60	3.00	3.00	3.10	2.80	3.10	5.00
23			5.70	3.80	4.70	2.50	3.00	2.90	3.10	2.80	3.40	5.00
24			6.00	3.80	4.40	2.50	3.00	2.90	3.10	2.80	4.00	5.80
25			5.50	3.70	4.00	2.50	3.00	2.90	3.10	2.80	4.00	5.50
26			5.20	3.60	3.70	2.50	2.90	2.90	3.10	2.80	3.80	4.50
27			5.10	3.60	3.70	2.50	2.90	2.90	3.20	2.80	3.60	4.30
28			5.10	3.60	3.70	2.50	2.90	4.40	3.20	2.80	3.50	4.10
29			8.80	3.50	3.70	2.70	2.90	4.10	3.30	2.80	3.50	4.10
30			10.30	3.40	3.70	2.70	2.90	5.00	3.30	2.70	3.40	4.10
31			8.30		3.11		2.90	4.40		2.70		4.10
1900.												
1	4.10	3.70	5.90	4.50	4.10	3.30	3.30	3.00	3.30	2.80	3.00	4.40
2	4.10	3.40	12.90	4.50	4.10	3.90	3.20	3.00	3.20	2.90	3.00	4.10
3	4.60	3.40	8.00	4.50	4.00	3.40	3.10	3.00	3.20	2.90	3.00	3.90
4	5.00	3.50	6.00	4.40	3.90	3.70	3.10	3.00	3.10	2.90	2.90	3.90
5	5.00	3.80	6.50	4.50	3.80	3.60	3.10	3.00	2.90	2.90	2.80	5.50
6	4.70	4.40	5.40	4.60	3.70	3.40	3.10	2.90	2.90	2.90	2.90	6.30
7	5.20	4.10	6.00	4.50	3.70	3.40	3.10	2.90	2.90	2.90	3.00	5.20
8	4.00	4.20	6.40	4.40	3.70	3.90	3.10	2.90	2.80	2.90	3.00	4.60
9	4.20	5.10	5.60	4.40	3.60	3.40	2.90	2.90	2.80	2.90	3.00	4.50
10	4.10	5.60	5.40	4.40	3.60	3.60	2.90	2.80	2.80	2.90	3.00	4.50
11	4.10	4.80	5.10	4.40	3.50	3.40	2.80	2.80	2.80	2.90	2.90	4.30
12	4.80	4.60	5.10	4.30	3.50	3.30	2.80	2.80	2.80	3.00	2.90	4.20
13	4.60	5.40	4.90	4.30	3.50	3.30	3.10	2.80	2.90	3.00	2.80	4.00
14	4.20	9.40	4.80	4.30	3.50	3.30	3.00	2.80	2.90	3.00	2.90	3.80
15	3.90	7.60	4.70	4.30	3.50	3.30	3.00	2.80	2.80	3.00	3.00	3.70
16	3.50	5.90	4.60	4.10	3.50	3.30	3.00	2.80	2.80	3.00	3.00	3.70
17	4.10	5.30	4.10	4.00	3.40	3.30	3.00	2.80	2.80	3.00	3.00	3.60
18	3.80	4.90	4.10	4.00	3.40	3.30	3.00	2.80	2.80	3.00	3.00	3.30
19	4.20	4.10	4.10	4.40	3.50	3.50	2.90	2.80	2.80	3.00	3.00	3.50
20	4.90	4.20	4.40	4.70	3.70	3.90	2.90	2.80	2.80	3.00	3.00	3.70
21	10.60	4.40	6.50	4.50	4.00	3.90	2.90	2.80	2.80	3.00	3.00	3.80
22	10.20	11.70	6.50	4.50	3.70	3.30	2.90	2.80	2.80	3.00	3.00	3.80
23	7.20	11.10	5.70	4.50	3.70	3.30	2.90	2.80	2.90	2.90	3.10	3.60
24	6.00	8.20	5.70	4.70	3.60	3.30	3.20	2.90	2.80	2.90	8.10	8.40
25	5.20	5.90	5.60	4.70	3.50	3.20	3.10	3.30	2.90	3.40	4.00	3.90
26	5.00	5.40	5.40	4.70	3.50	3.60	3.10	3.30	2.90	3.30	6.30	3.50
27	4.80	4.40	5.10	4.40	3.20	3.60	3.10	3.70	2.80	3.30	11.60	3.30
28	4.40	4.60	5.00	4.30	3.20	3.40	3.10	3.40	2.80	3.20	8.00	3.20
29	4.40		4.80	4.20	3.30	3.30	3.10	3.30	2.90	3.20	5.70	3.20
30	4.20		4.60	4.20	3.30	3.30	3.00	3.70	2.80	3.10	4.80	3.20
31	4.10		4.50		3.30		3.00	3.60		3.00		3.20

Mean daily gage height, in feet, of Juniata River at Newport, Pa., 1899–1904—
Continued.

Day.	Jan.	Feb.	Mar.	Apr.	May.	June.	July.	Aug.	Sept.	Oct.	Nov.	Dec.
1901.												
1	3.40	3.40	3.50	5.10	4.80	8.80	4.10	3.50	5.40	3.60	3.00	3.60
2	3.30	3.30	3.50	4.90	4.70	7.70	4.20	3.50	5.40	3.40	3.00	3.50
3	3.30	3.30	3.60	4.90	4.50	7.10	4.20	3.50	5.20	3.50	3.00	4.20
4	3.10	3.40	3.60	7.60	4.60	6.10	a4.00	3.80	5.00	3.50	3.00	4.20
5	3.30	3.80	4.40	9.00	4.50	5.20	a3.90	3.10	4.80	3.50	3.00	4.20
6	3.40	3.30	4.80	10.50	4.40	5.00	a3.80	3.10	4.20	3.40	3.00	4.20
7	3.20	3.30	4.70	11.00	4.20	4.90	a3.70	4.50	4.00	3.30	3.00	4.20
8	3.60	3.30	4.40	10.90	4.10	5.30	a3.60	6.20	3.90	3.20	3.00	3.70
9	3.30	3.30	4.20	9.50	4.00	5.10	a3.50	5.00	3.70	3.20	3.00	4.20
10	3.20	3.30	5.00	7.90	4.20	4.60	a3.40	4.10	3.60	3.10	3.00	5.00
11	3.20	4.00	15.90	7.00	4.70	4.50	3.30	4.00	3.70	3.10	3.00	7.00
12	3.50	3.80	15.40	6.20	4.80	4.50	3.30	3.70	4.10	3.20	3.00	6.20
13	3.80	3.80	10.40	5.80	4.80	4.50	3.30	3.50	4.00	3.30	3.00	5.10
14	3.80	3.30	7.80	5.40	4.70	4.50	3.40	3.40	3.80	3.40	3.00	18.00
15	3.80	3.80	7.20	5.20	4.60	4.40	3.40	3.40	3.80	3.40	3.00	18.00
16	3.80	3.80	6.50	5.60	4.40	4.40	3.50	3.40	3.80	3.30	3.00	18.00
17	3.80	3.60	5.80	5.60	4.10	4.50	4.90	3.40	3.70	3.30	3.00	10.80
18	3.80	3.50	5.50	5.40	4.20	5.00	5.00	4.10	3.80	3.30	3.00	13.65
19	3.80	3.50	5.10	5.40	4.10	4.60	5.20	4.30	8.90	3.20	3.00	6.30
20	3.80	3.50	5.00	5.40	4.10	4.40	4.80	4.30	8.70	3.20	3.00	5.30
21	3.80	3.50	5.90	10.50	4.10	4.30	4.10	4.10	8.60	3.10	3.00	12.05
22	3.90	3.50	6.90	18.80	4.50	4.60	3.80	4.10	3.50	3.10	3.00	4.10
23	4.10	3.60	6.50	11.50	13.00	5.30	3.70	4.10	8.40	3.10	3.00	4.40
24	3.70	3.70	5.80	9.00	9.50	5.60	3.50	5.50	3.20	3.10	3.80	4.40
25	3.50	3.90	5.50	7.60	9.00	5.00	3.40	5.50	8.30	3.00	4.90	4.60
26	3.40	3.40	5.80	6.80	10.60	4.60	3.70	5.10	3.30	3.00	4.80	4.80
27	3.70	3.40	5.50	6.00	8.60	4.40	3.50	4.90	3.20	3.00	4.00	4.50
28	3.70	3.50	6.60	5.60	10.30	4.20	3.50	4.30	3.20	3.00	4.00	4.50
29	3.60	......	6.60	5.30	12.60	4.00	3.40	4.20	3.50	3.00	3.90	5.20
30	3.50	......	5.90	5.00	13.30	4.00	3.40	4.30	3.50	3.00	3.70	6.40
31	3.60	......	5.40	......	11.60	......	3.40	4.30	......	3.00	......	7.70
1902.												
1	6.40	4.20	25.30	5.80	4.00	3.20	5.40	4.40	3.00	4.90	4.00	3.60
2	5.80	4.20	19.50	5.70	3.80	3.20	6.30	4.00	3.00	4.90	3.80	3.60
3	5.00	4.60	15.50	5.40	3.80	3.20	6.10	3.50	2.90	3.50	3.70	4.30
4	5.40	3.90	12.00	5.80	3.90	3.20	6.40	4.00	2.90	3.50	3.60	5.30
5	4.80	4.50	9.30	5.00	3.90	3.20	6.70	4.00	2.90	3.50	3.60	5.50
6	4.20	3.60	7.10	5.00	3.90	3.20	5.60	4.00	2.90	4.00	3.50	4.90
7	4.20	3.60	6.50	5.20	3.90	3.10	5.40	3.80	2.90	4.00	3.40	4.50
8	4.20	3.70	6.00	14.65	3.90	3.10	5.00	3.80	2.90	3.80	3.30	4.50
9	4.10	5.10	5.50	18.50	3.90	3.10	4.50	3.80	2.90	3.50	3.40	4.20
10	4.10	5.80	6.20	18.50	3.90	3.10	4.80	4.00	3.10	3.40	3.40	4.40
11	4.10	5.80	8.40	12.50	3.70	3.10	4.60	4.60	3.10	3.40	3.40	4.20
12	4.00	5.70	9.50	10.00	3.50	3.10	4.00	3.90	3.00	4.60	3.90	5.30
13	3.90	5.00	13.30	8.10	3.50	3.80	3.90	3.80	3.00	6.40	3.30	7.70
14	3.90	4.50	14.10	7.00	3.30	3.80	3.90	3.60	2.90	6.00	3.30	4.80
15	3.70	4.30	9.60	6.50	3.30	3.80	3.80	3.30	2.90	4.70	3.30	6.40
16	3.50	5.10	9.00	5.50	3.30	4.30	3.60	3.40	2.90	4.40	3.20	5.80
17	3.80	5.10	12.50	5.00	3.40	3.90	3.60	3.40	2.90	4.00	3.20	7.70
18	3.80	5.10	12.50	5.00	3.40	3.90	3.60	3.30	2.90	3.80	3.20	7.00
19	7.50	5.10	9.50	4.90	3.40	3.50	3.50	3.30	2.80	3.80	3.20	4.60
20	4.00	4.90	8.00	4.70	3.40	3.30	3.50	3.20	2.90	3.50	3.20	5.70
21	4.00	4.80	6.50	4.60	3.40	3.40	3.40	3.10	2.90	3.40	3.20	6.20
22	9.50	4.80	6.00	4.50	3.40	3.10	3.70	3.20	2.80	3.30	3.20	9.50
23	8.20	4.90	5.50	4.40	3.40	3.10	3.60	3.10	2.80	3.40	3.20	10.80
24	6.20	4.40	5.50	4.70	3.40	3.10	3.50	3.30	2.80	3.90	3.20	8.60
25	5.00	4.50	5.10	4.20	3.40	3.10	4.10	3.20	3.00	3.20	3.20	7.40
26	4.60	9.00	5.00	4.10	3.40	4.00	3.80	3.20	3.30	3.20	3.30	6.30
27	5.70	9.90	4.80	3.80	3.60	3.80	3.50	2.90	4.20	3.20	3.50	5.80
28	7.50	14.90	4.50	3.80	8.40	3.90	3.50	4.30	3.60	3.80	3.70	5.30
29	5.00	......	4.70	3.80	8.30	3.90	6.00	4.70	8.50	5.70	3.80	4.80
30	5.00	......	5.80	4.10	8.80	4.70	4.20	3.30	3.50	5.00	3.80	4.70
31	4.50	......	6.00	......	8.20	......	4.20	2.90	......	4.40	......	4.70

a Estimated.

Mean daily gage height, in feet, of Juniata River at Newport, Pa., 1899–1904—
Continued.

Day.	Jan.	Feb	Mar.	Apr.	May.	June.	July.	Aug.	Sept.	Oct.	Nov.	Dec.
1903.												
1	4.60	8.20	15.50	7.00	3.80	3.50	9.50	3.50	6.10	3.40	3.50	3.30
2	5.30	6.90	12.10	6.30	3.80	8.40	6.10	3.50	5.60	8.40	8.50	8.80
3	5.30	6.70	9.00	5.60	4.10	8.40	5.20	3.40	5.10	3.30	3.30	8.20
4	7.90	10.10	7.50	5.20	4.10	8.40	4.80	3.40	4.50	3.30	3.30	3.20
5	7.50	14.50	6.70	5.20	4.00	3.30	4.50	3.50	4.30	3.30	3.30	3.20
6	6.60	11.50	6.30	5.10	4.00	3.30	4.50	5.00	4.20	3.40	3.30	3.20
7	6.00	8.50	6.00	4.80	4.00	3.30	9.50	3.80	4.10	3.40	3.30	3.20
8	5.00	7.10	5.80	5.30	3.80	4.00	6.80	4.00	4.00	3.90	3.30	3.30
9	5.50	6.50	6.40	5.60	3.80	4.20	5.40	3.80	4.30	3.80	3.30	3.30
10	4.70	5.80	6.90	5.80	3.88	4.20	4.90	8.70	5.00	5.40	3.30	3.35
11	4.30	5.30	6.60	5.60	3.70	4.30	4.50	3.50	4.70	4.80	3.30	3.30
12	4.00	6.30	6.30	5.40	3.60	4.30	4.40	3.50	4.90	4.50	3.30	3.10
13	3.80	6.60	5.90	5.50	3.60	5.00	4.50	3.40	4.60	4.20	3.30	3.10
14	4.40	6.30	5.50	6.60	3.60	4.70	4.40	3.30	4.20	4.10	3.30	3.20
15	4.50	5.90	5.30	13.10	3.50	4.80	4.40	8.30	4.00	3.90	3.80	3.30
16	4.30	5.80	5.00	15.60	8.50	4.70	4.20	4.20	3.90	3.90	3.90	3.20
17	4.40	10.20	4.90	14.00	3.50	4.60	4.00	8.90	3.90	8.80	3.40	3.20
18	4.30	7.90	4.90	9.40	8.50	4.30	4.20	4.20	4.30	4.20	3.50	3.60
19	4.40	6.70	4.70	8.00	3.50	4.10	7.50	3.50	4.60	4.40	3.50	3.70
20	4.30	6.00	4.50	7.10	8.50	4.10	6.00	3.40	4.10	4.20	3.60	3.70
21	4.90	5.40	4.50	6.50	3.50	4.10	5.20	3.40	4.00	4.00	3.70	3.90
22	4.90	5.40	4.70	5.80	8.50	4.30	4.70	3.50	3.90	3.90	3.60	5.90
23	4.80	5.40	5.80	5.40	3.50	4.30	4.40	3.50	3.80	3.80	3.50	3.90
24	4.80	5.00	12.70	5.20	3.50	4.80	4.30	3.30	3.70	3.70	3.50	3.90
25	4.80	5.30	12.20	4.90	3.50	6.00	4.10	3.40	3.60	a3.60	3.50	3.90
26	4.60	5.10	8.50	4.80	3.40	5.60	3.90	3.80	3.50	a3.60	3.40	3.90
27	4.40	5.00	7.10	4.80	8.50	5.00	3.80	3.50	3.50	3.50	3.40	3.90
28	4.40	8.90	6.30	4.30	3.50	4.50	3.80	3.50	3.50	3.50	3.40	3.90
29	5.30		5.60	4.10	3.50	4.60	3.70	3.70	3.40	3.50	3.30	3.90
30	8.00		5.50	4.10	3.50	4.90	3.50	8.00	3.40	3.50	3.30	3.90
31	10.20		6.20		3.50		3.50	6.70		3.50		4.20
1904.												
1	4.20	4.00	7.50	6.70	6.70	4.90	8.70	3.30	3.00	2.90	2.90	2.50
2	4.20	5.00	12.00	18.40	6.10	5.60	3.70	3.30	3.00	2.90	2.90	2.80
3	4.50	5.00	7.20	9.40	5.70	6.00	3.70	3.70	3.00	2.90	2.90	2.90
4	4.60	5.00	3.50	7.70	5.30	5.40	3.70	3.60	3.00	2.90	2.90	3.20
5	4.60	8.00	8.90	6.70	5.00	5.90	8.70	3.50	3.00	2.90	2.90	2.90
6	4.60	8.50	6.00	5.70	4.80	5.90	8.70	3.50	3.00	2.90	2.80	3.10
7	4.60	11.50	5.50	5.70	4.70	5.40	5.40	4.30	2.90	2.90	2.80	3.20
8	4.50	a8.50	14.00	5.30	4.60	4.70	5.10	8.50	2.90	2.90	2.80	3.20
9	4.50	6.50	10.00	5.30	4.50	4.60	4.60	3.30	8.00	2.90	2.80	3.20
10	4.50	5.00	7.20	6.30	4.40	4.50	7.20	3.30	3.00	2.90	2.80	3.10
11	4.40	4.60	6.00	6.30	4.30	5.10	8.70	3.20	3.00	2.90	2.90	3.10
12	4.20	4.20	6.00	6.00	4.20	4.60	7.10	3.20	3.00	2.90	2.90	3.10
13	4.10	4.00	5.70	5.70	4.40	4.40	6.50	3.10	3.00	2.90	2.90	3.10
14	4.10	3.90	5.20	5.30	4.20	4.20	5.30	3.10	2.90	2.90	2.90	3.10
15	4.10	4.10	5.00	5.00	4.20	4.00	4.70	3.10	3.00	2.90	2.90	3.10
16	4.10	4.20	4.80	4.80	4.20	4.00	4.70	3.10	3.00	2.90	2.90	8.10
17	4.00	4.40	4.50	4.80	4.20	4.40	3.90	3.90	3.00	2.90	2.90	8.10
18	4.00	5.00	4.50	4.70	4.30	3.90	3.90	3.20	3.00	2.90	2.90	3.10
19	4.00	4.70	4.50	4.40	4.50	3.90	3.80	3.10	3.00	2.90	2.90	3.10
20	4.00	4.70	4.50	4.40	6.70	3.90	3.90	3.20	2.90	2.90	2.80	8.10
21	4.00	5.00	5.80	4.30	5.90	3.70	3.70	3.20	2.90	3.30	2.80	8.10
22	4.00	5.00	5.50	4.20	5.50	5.70	3.60	3.20	2.90	3.30	2.80	8.10
23	5.40	5.00	5.80	4.00	4.90	5.50	3.50	3.20	2.90	3.20	2.80	3.10
24	b11.00	5.40	8.00	4.00	4.60	5.30	8.80	3.20	2.80	3.10	2.70	3.20
25	7.00	7.20	7.50	4.00	4.50	4.40	3.80	3.10	2.90	3.10	2.70	3.20
26	5.50	7.40	6.90	4.00	4.60	4.00	3.70	3.10	2.90	3.00	2.70	5.20
27	4.50	5.90	6.20	4.60	4.70	3.80	3.70	3.00	2.90	2.90	2.70	3.50
28	4.10	4.80	6.20	4.60	4.50	3.80	3.50	8.00	2.90	2.90	2.60	3.70
29	3.80	4.50	5.60	6.50	4.40	3.70	3.40	3.10	2.90	2.90	2.60	8.80
30	3.70		5.20	7.50	4.20	3.70	3.30	3.10	2.90	2.90	2.50	3.80
31	3.80		5.00		4.60		3.30	3.00		2.90		8.80

a Interpolated.   b Ice moved out.

Rating table for Juniata River at Newport. Pa., from 1899 to 1904.

Gage height.	Discharge.	Gage height.	Discharge.	Gage height.	Discharge.	Gage height.	Discharge.
Feet.	*Second-feet.*	*Feet.*	*Second-feet.*	*Feet.*	*Second-feet.*	*Feet.*	*Second-feet.*
2.5	230	4.7	5,180	6.9	14,570	10.2	38,500
2.6	320	4.8	5,510	7.0	15,170	10.4	40,300
2.7	430	4.9	5,850	7.1	15,770	10.6	42,200
2.8	570	5.0	6,200	7.2	16,370	10.8	44,100
2.9	750	5.1	6,550	7.3	16,970	11.0	46,000
3.0	950	5.2	6,910	7.4	17,570	11.2	48,000
3.1	1,160	5.3	7,270	7.5	18,170	11.4	50,100
3.2	1,370	5.4	7,640	7.6	18,770	11.6	52,200
3.3	1,580	5.5	8,010	7.7	19,380	11.8	54,300
3.4	1,790	5.6	8,390	7.8	20,000	12.0	56,400
3.5	2,000	5.7	8,770	7.9	20,640	12.2	58,600
3.6	2,210	5.8	9,150	8.0	21,300	12.4	60,800
3.7	2,430	5.9	9,540	8.2	22,700	12.6	63,100
3.8	2,650	6.0	9,930	8.4	24,100	12.8	65,400
3.9	2,880	6.1	10,330	8.6	25,500	13.0	67,700
4.0	3,120	6.2	10,740	8.8	27,000	13.2	70,100
4.1	3,380	6.3	11,200	9.0	28,500	13.4	72,600
4.2	3,650	6.4	11,720	9.2	30,100	13.6	75,100
4.3	3,930	6.5	12,270	9.4	31,700	13.8	77,600
4.4	4,220	6.6	12,830	9.6	33,400		
4.5	4,530	6.7	13,400	9.8	35,100		
4.6	4,850	6.8	13,980	10.0	36,800		

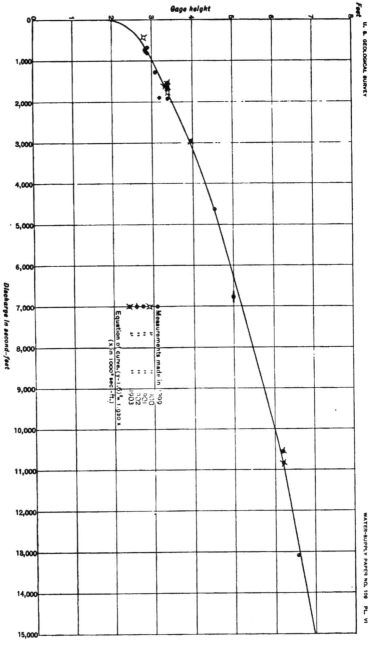

Gage height

Feet

Discharge in second-feet

RATING CURVE FOR JUNIATA RIVER AT NEWPORT, PA.

Measurements made in 1899
"   "   "   1900
"   "   "   1902
"   "   "   1903

Equation of curve, $(y-1.6)^2 = 1,920\,x$
(x in 1000's sec.-ft.)

Mean daily discharge, in second-feet, of Juniata River at Newport, Pa., 1899–1904.

Day.	Jan.	Feb.	Mar.	Apr.	May.	June.	July.	Aug.	Sept.	Oct.	Nov.	Dec.
1899.												
1				15,120	1,790	1,160	430	950	2,000	1,370	430	1,580
2				10,550	1,790	2,210	430	950	2,000	1,160	3,120	1,580
3				8,010	2,210	2,000	320	950	1,790	1,160	5,850	1,580
4				6,550	2,430	1,790	320	950	1,790	1,160	4,850	1,580
5				5,850	2,000	1,580	320	950	1,370	950	3,650	1,580
6				4,530	1,790	1,790	950	950	1,580	950	2,880	1,370
7				3,980	1,790	1,580	950	1,580	1,580	950	2,430	1,160
8				8,380	1,790	1,370	950	2,000	1,580	950	2,000	1,160
9				20,000	2,210	1,370	1,160	1,580	1,580	750	2,000	1,160
10				14,570	2,210	570	1,580	1,370	1,580	750	1,790	1,160
11				9,150	8,120	570	1,580	1,160	1,580	750	1,580	1,160
12				8,010	3,380	570	1,580	1,160	1,790	750	1,580	2,430
13				6,550	8,120	570	1,160	1,790	5,510	750	1,580	5,510
14				5,850	2,650	430	1,160	1,160	5,510	750	1,370	8,010
15				5,510	2,650	430	1,160	1,160	2,650	750	1,370	6,550
16				5,180	2,210	430	950	950	2,000	750	1,370	5,510
17				8,010	2,430	430	750	950	1,580	750	1,370	3,980
18				4,220	8,380	430	750	950	1,160	750	1,160	8,120
19				3,980	21,300	430	750	950	1,160	750	1,160	3,120
20				3,380	16,970	320	950	1,160	1,160	750	1,160	2,430
21			12,270	3,120	18,770	320	950	950	950	750	1,160	2,430
22			9,150	2,880	6,550	320	950	950	950	750	1,160	6,200
23			8,770	2,650	5,180	320	950	950	750	750	570	6,200
24			9,150	2,430	4,220	320	950	950	750	570	1,790	6,200
25			8,010	2,210	8,120	320	950	950	750	570	3,120	9,150
26			6,910	2,210	2,430	320	950	950	750	570	8,120	8,010
27			6,550	2,210	2,430	320	750	950	750	570	2,650	4,530
28			6,550	2,000	2,430	320	750	4,220	1,370	570	2,210	3,980
29			7,000	2,000	2,430	430	750	3,380	1,580	570	2,000	3,880
30			20,400	1,790	8,380	430	750	6,200	1,580	430	1,790	3,880
31			23,400		1,160		750	4,220		430		3,380
1900.												
1	8,380	2,430	9,510	4,530	3,380	1,580	1,580	950	1,580	570	950	4,220
2	8,380	1,790	66,580	4,530	3,380	1,580	1,370	950	1,370	750	950	3,380
3	4,850	1,790	21,300	4,530	3,120	1,790	1,160	950	1,370	750	950	2,880
4	6,200	2,000	9,950	4,220	2,880	2,430	1,160	950	1,160	750	950	2,880
5	6,200	2,650	8,010	4,530	2,650	2,210	1,160	950	750	750	750	8,010
6	5,180	4,220	7,640	4,530	2,430	1,790	1,160	950	750	750	750	15,170
7	6,910	3,380	9,950	4,530	2,430	1,790	1,160	950	750	750	950	11,200
8	3,120	3,650	11,720	4,220	2,430	1,580	1,160	750	570	750	950	6,910
9	3,650	6,550	8,380	4,220	2,210	1,790	1,160	750	570	750	950	4,850
10	3,380	8,380	7,640	4,220	2,210	2,000	1,160	750	570	750	950	4,530
11	3,380	5,510	6,550	4,220	2,000	1,700	1,160	570	570	750	750	3,980
12	5,510	4,850	6,550	3,650	2,000	1,580	1,160	570	570	950	750	3,650
13	4,850	7,640	5,850	3,650	2,000	1,580	1,160	570	570	950	750	8,120
14	3,650	31,700	5,510	3,650	2,000	1,580	950	570	570	950	750	2,650
15	2,880	18,770	5,180	3,650	2,000	1,580	950	570	570	950	950	2,430
16	2,000	9,540	4,850	3,650	2,000	1,580	950	570	570	950	950	2,430
17	3,380	7,270	3,380	3,120	1,790	1,580	950	570	570	950	950	2,210
18	2,650	5,850	3,380	3,120	1,790	1,580	950	570	570	950	950	2,000
19	3,650	3,380	3,380	4,220	2,000	1,580	750	570	570	950	950	2,000
20	5,850	3,650	4,220	5,180	2,430	1,580	750	570	570	950	950	2,430
21	42,200	4,220	12,270	4,530	8,120	1,580	750	570	570	950	950	2,650
22	38,500	53,200	12,270	4,530	2,430	1,580	750	570	570	950	950	2,650
23	16,370	47,000	8,770	4,530	2,430	1,580	570	570	750	1,160	2,210
24	9,950	22,700	8,770	4,530	2,210	1,370	570	570	570	1,160	1,790
25	6,910	9,540	8,380	5,180	2,000	1,370	1,160	1,580	570	1,790	3,120	2,650
26	6,200	4,530	7,640	5,180	2,000	1,370	1,160	1,580	570	1,540	11,200	2,000
27	5,510	4,220	6,550	4,530	1,370	2,210	1,160	2,430	570	1,540	52,200	1,580
28	4,220	4,850	6,200	3,980	1,580	1,580	1,160	1,790	570	1,370	21,300	1,370
29	4,220		5,510	3,650	1,580	1,580	1,160	1,580	570	1,370	8,770	1,370
30	3,650		4,850	3,650	1,580	1,580	950	2,430	570	1,160	5,510	1,370
31	3,380		4,530		1,580		950	2,210		950		1,370

Mean daily discharge, in second-feet, of Juniata River, at Newport, Pa., 1899–1904—Continued.

Day.	Jan.	Feb.	Mar.	Apr.	May.	June.	July.	Aug.	Sept.	Oct.	Nov.	Dec.
1901.												
1	1,730	1,730	2,000	6,550	5,510	27,000	3,380	2,000	7,640	2,210	950	2,210
2	1,580	1,580	2,000	5,850	5,180	19,380	3,650	2,000	7,640	1,790	950	2,000
3	1,580	1,580	2,210	5,850	4,530	15,770	3,650	2,000	6,910	2,000	950	3,650
4	1,160	1,730	2,210	18,770	4,850	10,380	3,120	1,580	6,200	2,000	950	8,650
5	1,580	2,650	4,220	28,500	4,530	6,910	2,880	1,160	4,850	2,000	950	3,650
6	1,590	3,930	5,510	41,200	4,220	6,200	2,650	1,160	3,650	1,790	950	3,650
7	1,570	3,930	5,180	46,000	3,650	5,850	2,430	4,530	3,120	1,580	950	3,650
8	2,210	3,930	4,220	45,000	3,380	7,270	2,210	10,740	2,880	1,370	950	2,430
9	1,580	3,930	3,650	22,500	3,120	6,550	2,000	6,200	2,430	1,370	950	3,650
10	1,570	3,930	6,200	20,640	3,650	4,850	1,790	3,380	2,210	1,160	950	6,200
11	1,570	3,120	106,500	15,170	5,180	4,530	1,580	3,120	2,430	1,160	950	15,170
12	2,000	2,650	99,200	10,740	5,510	4,530	1,580	2,430	3,380	1,370	950	10,740
13	2,650	2,650	40,300	9,150	5,510	4,530	1,580	2,000	3,120	1,580	950	6,550
14	2,650	3,930	30,000	7,640	5,180	4,530	1,590	1,790	2,650	1,790	950	6,910
15	2,650	2,650	16,370	6,910	4,850	4,220	1,590	1,790	2,650	1,790	950	140,100
16	2,650	2,650	12,270	8,380	4,220	4,530	2,000	1,790	2,650	1,580	950	140,100
17	2,650	2,210	9,150	8,380	3,380	4,530	5,850	1,790	2,430	1,580	950	44,100
18	2,650	2,000	8,010	7,640	3,650	6,200	6,200	3,380	2,650	1,580	950	75,000
19	2,650	2,000	6,550	7,640	3,380	4,850	6,910	3,930	2,880	1,370	950	11,200
20	2,650	2,000	6,200	7,640	3,380	4,220	5,510	7,270	2,430	1,370	950	7,270
21	2,650	2,000	9,540	41,200	3,380	3,930	3,840	3,380	2,210	1,160	950	57,000
22	2,880	2,000	14,550	77,000	4,530	8,380	2,650	3,380	2,000	1,160	950	4,220
23	3,380	2,210	12,270	51,100	67,700	7,270	2,430	3,380	1,790	1,160	950	4,220
24	2,430	2,430	9,150	28,500	32,500	8,380	2,000	8,010	1,370	1,160	2,650	4,220
25	2,000	2,880	8,010	18,770	28,500	6,200	1,790	8,010	1,580	950	5,850	4,850
26	1,790	1,790	7,270	13,980	42,200	5,510	2,430	6,550	1,580	950	5,510	4,530
27	2,430	1,790	8,010	9,150	25,500	4,220	2,000	5,850	1,370	950	3,120	4,530
28	2,430	2,000	12,550	8,380	59,100	3,650	1,650	3,930	1,370	950	3,120	6,200
29	2,210		12,270	7,270	63,100	3,120	1,730	3,650	2,000	950	2,880	6,910
30	2,000		9,540	6,200	71,300	3,120	1,580	3,930	2,000	950	2,430	11,720
31	2,210		7,640		52,200		1,730	3,980		950		19,380
1902.												
1	11,720	3,650	292,500	9,150	3,120	1,370	7,640	4,220	950	5,850	3,120	2,210
2	8,380	3,650	105,900	8,770	2,650	1,370	11,200	3,120	950	5,850	2,650	2,210
3	6,200	4,850	100,600	7,640	2,650	1,370	10,380	2,000	750	3,000	2,430	3,930
4	7,640	2,880	56,400	7,270	2,880	1,370	11,720	3,120	750	2,000	2,210	7,270
5	3,160	4,530	30,000	12,270	2,880	1,370	13,400	3,120	750	2,000	2,210	8,010
6	3,650	2,210	15,770	6,200	2,880	1,370	8,380	3,120	750	3,120	2,000	5,850
7	3,650	2,210	12,270	6,910	2,880	1,370	7,640	2,650	750	2,650	1,790	4,530
8	3,650	2,430	9,580	88,700	2,880	1,160	6,200	2,650	750	2,650	1,580	4,530
9	3,380	6,550	9,150	148,800	2,880	1,160	4,530	2,650	750	1,790	1,790	4,530
10	3,380	9,150	10,740	148,800	2,880	1,160	5,510	3,120	1,160	1,790	1,790	4,220
11	3,380	9,150	24,100	61,900	2,430	1,160	3,120	4,850	1,160	1,790	1,790	3,650
12	3,120	8,770	32,500	36,800	2,000	1,160	3,120	2,880	950	4,850	1,580	7,270
13	2,880	6,200	71,900	22,000	2,880	2,880	2,650	2,650	950	11,720	1,580	19,380
14	2,880	4,530	81,400	15,170	1,580	1,580	2,880	2,210	750	9,900	1,580	5,510
15	2,430	3,930	33,700	12,270	1,580	1,580	2,650	1,580	750	5,180	1,580	11,720
16	2,000	6,550	28,500	8,010	1,580	3,080	2,210	1,790	750	4,220	1,370	9,150
17	2,650	6,550	97,700	6,200	1,700	2,880	2,210	1,790	750	3,120	1,370	19,380
18	2,650	6,550	61,900	6,200	1,700	2,880	2,210	1,580	750	2,650	1,370	15,170
19	18,170	6,550	32,500	5,180	1,700	2,000	2,210	1,580	570	2,650	1,370	11,720
20	3,120	5,850	21,800	5,180	1,790	1,580	2,000	1,370	750	2,000	1,370	8,770
21	3,120	5,510	12,270	4,850	1,790	1,700	2,210	1,160	750	1,790	1,370	10,740
22	32,500	5,510	9,180	4,530	1,700	1,160	2,430	1,160	570	1,580	1,370	32,500
23	22,700	5,850	8,010	4,220	1,700	1,160	2,210	1,160	570	1,790	1,370	44,100
24	10,740	4,220	8,010	3,650	1,700	2,000	2,000	1,580	570	1,580	1,370	25,500
25	6,200	4,530	6,550	3,380	1,790	1,120	3,380	1,370	950	1,580	1,370	17,570
26	4,850	28,500	6,200	3,380	1,790	1,120	2,650	1,370	1,580	1,370	1,580	11,200
27	8,770	35,000	5,510	2,650	2,210	2,050	2,000	75	3,160	1,580	2,000	9,150
28	18,170	82,100	4,530	2,650	1,790	2,880	2,000	3,930	2,210	2,650	2,430	7,270
29	8,380		4,530	3,120	1,580	2,210	2,210	5,180	2,000	8,770	2,650	5,510
30	6,200		9,150	3,380	1,580	5,180	1,650	1,580	2,000	6,200	2,650	5,180
31	4,530		9,380		1,370		3,650	750		4,220		5,180

Mean daily discharge, in second-feet, of Juniata River at Newport, Pa., 1899–1904—Continued.

Day.	Jan.	Feb.	Mar.	Apr.	May.	June.	July.	Aug	Sept	Oct.	Nov.	Dec.
1903.												
1	4,850	22,700	100,600	15,170	2,650	2,000	32,500	2,000	10,330	1,780	2,000	1,580
2	7,270	14,570	57,500	11,200	2,650	1,780	10,350	2,000	8,380	1,780	2,000	1,580
3	7,270	13,400	28,500	8,380	3,380	1,780	6,910	1,780	6,550	1,580	1,580	1,370
4	20,640	37,600	18,170	6,910	3,380	1,780	5,510	1,780	4,530	1,580	1,580	1,370
5	18,170	86,700	13,400	6,910	3,120	1,580	4,530	2,000	3,380	1,580	1,580	1,370
6	12,830	51,100	11,200	6,550	3,120	1,580	6,300	2,000	3,650	1,780	1,580	1,370
7	9,930	24,800	9,930	5,510	3,120	1,580	32,500	2,650	3,380	1,780	1,580	1,370
8	6,200	15,770	9,150	7,270	2,650	3,120	18,080	3,120	3,120	2,880	1,580	1,370
9	8,010	12,270	11,720	8,380	2,650	3,650	7,640	2,650	3,180	5,510	1,580	1,580
10	5,180	9,150	14,570	9,150	2,650	3,650	5,850	2,650	6,210	7,640	1,580	1,680
11	3,930	7,270	12,830	8,380	2,650	3,930	4,530	2,000	5,180	5,510	1,580	1,580
12	3,120	10,330	11,200	7,640	2,210	3,930	4,220	2,000	5,850	4,530	1,580	1,160
13	2,650	12,830	9,540	8,010	2,210	6,200	4,530	1,780	4,850	3,650	1,580	1,160
14	4,220	11,200	8,010	12,830	2,210	5,180	4,220	1,580	3,380	3,380	1,580	1,370
15	4,530	9,540	7,270	68,900	2,000	5,510	4,220	1,580	3,120	2,880	1,580	1,580
16	3,930	13,980	6,200	102,100	2,000	5,180	3,650	3,650	2,880	2,880	1,580	1,370
17	4,220	38,500	5,850	80,100	2,000	4,850	3,120	2,880	2,880	2,650	1,780	1,370
18	3,930	20,640	5,850	31,700	2,000	3,930	3,120	3,120	3,120	3,650	2,000	2,000
19	4,220	13,400	5,180	21,300	2,000	3,380	18,170	2,000	4,850	4,220	2,000	2,430
20	3,930	9,930	4,530	15,770	2,000	3,380	9,930	1,780	3,380	3,650	2,210	2,430
21	5,850	7,640	4,530	12,270	2,000	3,380	6,910	1,780	3,120	3,120	2,430	2,880
22	5,850	7,640	5,180	9,150	2,000	3,930	3,930	2,000	2,880	2,880	2,210	2,880
23	5,510	7,640	9,150	7,640	2,000	3,930	4,220	2,000	2,650	2,650	2,000	2,880
24	5,510	6,200	64,200	6,910	2,000	5,510	3,930	1,580	2,430	2,430	2,000	2,880
25	5,510	7,270	58,600	5,850	2,000	9,930	3,380	1,780	2,210	2,210	1,780	2,880
26	4,850	6,550	24,800	5,510	1,780	8,380	2,650	2,000	2,000	2,210	1,780	2,880
27	4,220	6,210	15,770	5,510	2,000	6,200	2,650	2,000	2,000	2,000	1,780	2,880
28	4,220	27,700	11,200	3,930	2,000	4,530	2,430	2,000	2,000	2,000	1,780	2,880
29	7,270	....	8,380	3,380	2,000	4,850	2,430	2,430	1,790	2,000	1,580	2,880
30	21,300	....	8,380	3,380	2,000	5,850	2,010	21,300	1,790	2,000	1,580	2,880
31	38,500	....	10,740	....	2,000	....	2,000	13,400	....	2,000	....	3,650
1904.												
1	3,650	3,120	18,170	13,400	13,400	5,850	2,430	1,580	950	750	750	230
2	3,650	6,200	56,400	72,600	10,330	8,390	2,430	1,580	950	750	750	570
3	4,530	6,200	16,370	31,700	9,930	2,430	2,430	950	750	750	750	750
4	4,850	6,200	73,850	19,380	7,270	7,640	2,430	2,210	950	750	750	1,370
5	4,850	21,300	27,700	13,400	6,200	9,540	2,430	2,210	950	750	750	570
6	4,850	24,800	9,930	8,770	5,510	9,540	2,430	2,000	950	750	570	1,160
7	4,850	51,100	8,010	8,770	5,180	7,640	2,880	1,580	750	570	570	1,370
8	4,530	24,800	80,100	7,270	4,850	5,180	6,550	2,000	750	570	570	1,370
9	4,530	12,270	36,800	7,270	4,850	4,850	9,150	1,580	950	750	570	1,370
10	4,530	6,200	16,370	11,200	4,220	4,530	16,370	1,580	950	750	570	1,160
11	4,220	4,850	9,930	11,200	3,930	6,550	26,200	1,370	950	750	750	1,160
12	3,650	3,650	9,930	9,930	3,650	4,850	15,770	1,370	950	750	750	1,160
13	3,380	3,120	6,910	8,770	3,650	4,220	7,270	1,160	950	750	750	1,160
14	3,380	2,880	6,910	7,270	3,650	3,650	7,270	1,160	750	750	750	1,160
15	3,380	3,340	6,200	6,200	3,650	3,120	5,180	1,160	750	750	750	1,160
16	3,380	3,650	5,510	5,510	3,650	3,120	5,180	1,160	950	750	750	1,160
17	3,120	4,220	4,530	5,510	3,650	4,220	2,880	1,370	950	750	750	1,160
18	3,120	6,200	4,530	5,180	3,930	2,880	2,880	1,160	950	750	750	1,160
19	3,120	4,850	5,510	4,220	4,530	2,880	2,650	1,370	950	750	570	1,160
20	3,120	5,180	4,530	4,220	13,400	2,880	2,880	1,370	750	750	570	1,160
21	3,120	5,180	9,150	3,930	9,540	2,430	2,430	1,370	750	1,580	570	1,160
22	3,120	6,200	8,010	3,650	8,010	8,770	2,210	1,370	750	1,580	570	1,160
23	7,640	6,200	9,150	3,650	5,850	8,010	2,000	1,370	750	1,370	570	1,160
24	46,000	7,640	21,300	3,120	4,850	7,270	2,650	1,370	750	1,160	430	1,370
25	15,170	16,370	18,170	3,120	4,530	4,220	2,650	1,160	750	1,160	430	1,370
26	8,010	17,570	14,570	3,120	4,850	3,120	2,430	1,160	750	950	430	1,370
27	4,500	9,540	10,740	3,650	5,180	2,000	2,430	950	750	750	430	2,000
28	3,380	5,510	10,740	4,850	4,530	2,650	2,000	950	750	750	320	2,430
29	2,650	4,530	8,380	12,270	4,220	2,430	1,780	1,580	750	750	320	2,650
30	2,430	....	6,910	18,170	3,650	2,430	1,580	1,160	750	750	230	2,650
31	2,650	....	6,200	....	4,850	....	1,580	950	....	750	....	2,650

Estimated monthly discharge of Juniata River at Newport, Pa., 1899–1904.

[Drainage area, 3,476 square miles.]

Month.	Discharge in second-feet.			Run-off.	
	Maximum.	Minimum.	Mean	Second-feet per square mile.	Depth in inches.
1899.					
March (21–31)	39,400	6,550	14,429	4.151	1.698
April	20,000	1,790	6,042	1.738	1.939
May	21,300	1,160	4,301	1.237	1.426
June	2,210	230	760	.219	.244
July	1,580	230	904	.260	.300
August	6,200	750	1,525	.439	.506
September	5,510	1,160	1,787	.514	.573
October	1,370	430	774	.223	.257
November	5,850	430	2,095	.603	.673
December	9,150	1,160	3,628	1.044	1.204
The period	39,400	230	3,624	1.043	8.820
1900.					
January	42,200	2,000	7,263	2.089	2.408
February	53,200	1,790	10,188	2.931	3.052
March	66,500	3,380	9,523	2.740	3.159
April	5,180	3,120	4,264	1.227	1.369
May	3,380	1,370	2,226	.640	.738
June	2,430	1,370	1,692	.487	.543
July	1,580	750	1,074	.309	.356
August	2,430	570	971	.279	.322
September	1,580	570	695	.200	.223
October	2,430	570	1,016	.292	.337
November	52,200	750	4,137	1.190	1.328
December	15,170	1,370	3,596	1.035	1.193
The year	66,500	570	3,887	1.118	15.028

Estimated month y discharge of Juniata River at Newport, Pa., 1899-1904—Con.

Month	Discharge in second-feet.			Run-off.	
	Maximum.	Minimum.	Mean.	Second-feet per square mile.	Depth in inches.
1901.					
January	8,380	1,160	2,161	0.622	0.717
February	8,930	1,580	2,571	.740	.771
March	106,500	2,000	15,260	4.390	5.061
April	77,600	5,850	20,104	5.784	6.453
May	71,800	3,120	16,683	4.799	5.533
June	27,000	3,120	6,869	1.976	2.205
July	6,910	1,580	2,794	.804	.927
August	10,740	1,160	8,808	1.096	1.264
September	7,640	1,870	3,069	.883	.985
October	2,210	950	1,411	.406	.468
November	5,850	950	1,580	.455	.508
December	140,100	2,000	19,940	5.737	6.614
The year	140,100	950	8,021	2.308	31.506
1902.					
January	82,500	2,000	7,250	2.088	2.407
February	92,100	2,210	10,316	2.968	3.091
March	292,500	3,650	41,044	11.808	13.614
April	148,800	2,650	21,813	6.275	7.001
May	3,120	1,370	2,135	.614	.708
June	5,180	1,160	1,870	.538	.600
July	13,400	2,000	4,586	1.319	1.521
August	5,180	750	2,381	.671	.774
September	8,650	570	1,043	.300	.335
October	11,720	1,370	8,586	1.032	1.190
November	8,120	1,370	1,823	.524	.585
December	44,100	2,210	10,711	8.081	3.552
The year	292,500	570	9,043	2.602	35.878

Estimated monthly discharge of Juniata River at Newport, Pa., 1899-1904—Cont'd.

Month.	Discharge in second-feet.			Run-off.	
	Maximum	Minimum.	Mean.	Second-feet per square mile.	Depth in inches.
1903.					
January	38,500	2,650	7,988	2.298	2.649
February	86,700	6,200	18,304	5.266	5.484
March	100,600	4,530	18,444	5.306	6.117
April	102,100	3,380	16,857	4.850	5.411
May	3,380	1,790	2,330	.670	.772
June	9,930	1,580	4,150	1.194	1.332
July	32,500	2,000	7,322	2.106	2.428
August	21,300	1,580	3,090	.889	1.025
September	10,330	1,790	3,915	1.126	1.256
October	7,640	1,580	2,917	.839	.967
November	2,430	1,580	1,776	.511	.570
December	3,650	1,160	2,050	.590	.680
The year	102,100	1,160	7,429	2.137	28.691
1904.					
January*a*	46,000	2,430	5,722	1.65	1.90
February	51,100	2,880	9,756	2.81	3.03
March	80,100	4,530	17,150	4.93	5.68
April	72,600	3,120	10,710	3.08	3.44
May	13,400	3,650	5,742	1.65	1.90
June	9,930	2,000	5,160	1.48	1.65
July	26,200	1,580	4,968	1.43	1.65
August	2,880	950	1,460	.420	.484
September	950	750	850	.245	.273
October	1,580	750	856	.246	.284
November	750	230	607	.175	.195
December	2,650	230	1,344	.386	.445
The year	80,100	230	5,360	1.54	20.93

a Frozen January 1 to 23. Rating table assumed to apply correctly.

SUSQUEHANNA RIVER AT HARRISBURG, PA.

In 1890 regular daily observations of fluctuations of the water surface of the Susquehanna River at Harrisburg were started by E. Mather, president of the Harrisburg water board. These observations have been continued since that time and have been furnished through the courtesy of Mr. Mather.

The gage, the zero of which is the low-water mark of 1803, is located at the pump house of the waterworks in the pump well, which is connected with the river by two large mains. The original readings are taken in feet and inches, and for convenience in computations have been reduced to feet and tenths.

The first discharge measurement was made at this station in March, 1897, by Mr. E. G. Paul, who has carried on systematic measurements there since that date. The measuring section is at the lower side of the Walnut street toll bridge. The initial point for soundings is the upright at the end of the hand rail on the downstream side on the left bank.

At this point the river is divided into two channels by Fosters Island, which at the measuring section is about 1,200 feet wide. Its banks are low and sloping and during extreme floods the island is submerged.

At ordinary stages the left channel is 1,350 feet wide and is broken by six bridge piers. The right channel is 1,300 feet wide and is broken by seven piers. The main banks of the river are high. The bed is composed of a hard material and is permanent, except in the spans adjacent to the island. The velocity never becomes too sluggish to measure.

During the spring and summer of 1903 a new bridge was built across Susquehanna River at Market street, which is about 1,200 feet below the gaging section. The piers of this new bridge obstruct the channel of the river by between 10 and 15 per cent of the total cross section. The result of this obstruction, as shown by the discharge measurements taken since the erection of the piers, has been to back up the water, thus increasing the gage height at the Walnut street station. On account of this backwater the measurements taken during 1903 show that, in order to use the standard rating table after June 1, 1903, and until January 1, 1904, a deduction of 14 per cent is necessary in the daily discharges. The following table gives the data from which this deduction was made:

Date.	Gage height.	Observed discharge.	Standard rating table discharge.	Difference.	Difference.
	Feet.	Second-feet.	Second-feet.	Second-feet.	Per cent.
May 8	2.30	16,280	15,980	300	− 2
June 2	1.50	8,890	9,520	1,130	12
October 5	1.65	9,116	10,560	1,440	13
November 2	3.08	20,245	24,350	4,100	16

About January 1, 1904, the old piers which were standing at the site of the new bridge at Market street were removed, so that the river channel was left in such a condition that the stage of the river at Walnut street bridge returned to the same condition that existed before the 1903 bridge was built.

In the summer of 1904 certain changes and improvements were made at the pumping station, and a partial dam was made in the river just below the pumping station. The effect of this dam was to raise the apparent stage of the water at the gage. A correction was applied to measurements of discharge made prior to July 18, 1904, so as to eliminate the effect of the dam and alterations at the pump house upon the gage readings.

On July 18, 1904, a standard chain gage was attached to the guard rail on the upstream side of the Walnut Street Bridge in the left-hand span. The datum of this gage is the low-water mark of 1803, and it is believed that it records truly the stage of the river to that datum, and that the changes in bridges below and at the pumping station above do not affect the records obtained from it.

The length of chain is 39.38 feet; the bench mark is on the left abutment at the top upstream outer corner of the bridge seat; its elevation is 32.99 feet above low water of 1803.

Observations at the gage in the pumping station are made by the engineer, C. M. Nagle, each morning before starting the pump. Observations at the standard chain gage are made by Thomas Numbers, toll collector, once daily.

The following pages give the data which have been collected at Harrisburg gaging station since its establishment; also the results of the computation of these data.

Discharge measurements of Susquehanna River at Harrisburg, Pa., 1897-1904.

Date.	Hydrographer.	Gage height.	Area of section.	Mean velocity.	Discharge.
1897.		*Feet.*	*Square feet.*	*Feet per second.*	*Second-feet.*
Mar. 31	E. G. Paul	5.42	17,048	3.45	58,859
May 15	.....do	7.83	24,351	4.35	105,888
Aug. 30	.....do	1.50	7,444	1.29	9,568
Sept. 16	.....do	.58	3,756	1.06	3,962
Nov. 17	.....do	2.50	9,825	1.91	17,824

Discharge measurements of Susquehanna River at Harrisburg, Pa., 1897–1904—
Continued.

Date	Hydrographer.	Gage height.	Area of section.	Mean velocity.	Discharge.
1898.		*Feet.*	*Square feet.*	*Feet per second.*	*Second-feet.*
Feb. 25	E. G. Paul	6.58	19,420	3.91	76,250
Mar. 24	....do	15.75	43,715	5.73	250,485
Mar. 25	.....do	10.75	29,587	5.06	149,589
Mar. 26	.....do	14.65	39,725	5.62	223,374
July 10	....do	.83	4,400	1.22	5,466
Sept. 22	.....do	.92	4,834	1.44	6,993
Oct. 7	.....do	.72	4,459	1.31	6,121
1899.					
June 11	E. G. Paul	1.75	7,656	1.53	11,746
July 29	....do	.91	4,524	1.44	6,534
Sept. 12	.....do	.75	4,845	1.12	5,404
Oct. 25	.....do	.16	3,699	.98	3,625
1900.					
May 16	E. G. Paul	2.42	9,404	1.87	17,621
Sept. 21	....do	.08	3,313	.80	2,655
Sept. 28	.....do	.04	3,223	.72	2,357
1901.					
Aug. 12	E. G. Paul	2.70	9,775	2.05	20,023
Oct. 23	.....do	1.85	7,737	1.62	12,556
1902.					
Apr. 17	E. G. Paul	5.40	17,476	3.46	60,534
Sept. 15	.....do	1.10	5,023	1.39	6,982
1903.					
May 8	E. C. Murphy	2.30	9,810	1.65	16,280
June 2	Hoyt and Holmes	1.50	7,577	1.11	8,390
Oct. 5	Paul and Sawyer	1.65	7,290	1.25	9,116
Nov. 2	E. G. Paul and others	3.08	10,325	1.96	20,245
1904.					
Mar. 9	Sawyer and Tillinghast	15.60	....	....	a261,860
July 15	N. C. Grover	3.08	11,870	2.22	26,408
Sept. 18	J. C. Hoyt	1.10	6,646	.90	5,950
Sept. 29	.....do	1.78	8,730	1.34	11,660
Oct. 1	N. C. Grover	1.85	8,460	1.48	12,560
Nov. 4	Hoyt and Comstock	1.82	8,972	1.39	12,600

a River running full of ice. Measurement approximate.

Mean daily gage height, in feet, of Susquehanna River at Harrisburg, Pa., 1891–1904.

Day.	Jan.	Feb.	Mar.	Apr.	May.	June.	July.	Aug.	Sept.	Oct.	Nov.	Dec.
1891.												
1	2.83	10.58	11.00	8.25	3.58	2.00	2.75	3.25	4.67	1.75	2.50	4.25
2	3.00	11.50	9.00	9.00	3.50	1.92	2.50	3.17	4.00	1.67	2.50	4.00
3	3.33	11.50	7.83	8.58	3.42	2.00	2.58	3.08	3.67	1.67	2.33	3.67
4	4.50	11.17	6.67	8.75	3.42	2.00	3.17	2.92	3.33	1.58	2.25	3.50
5	5.25	10.17	5.67	8.42	3.25	2.00	4.08	3.00	3.00	1.58	2.25	4.58
6	5.00	8.92	5.67	8.00	3.08	2.00	3.50	3.08	3.00	1.58	2.25	8.75
7	5.50	7.67	5.25	7.17	3.00	2.08	3.08	3.00	3.83	1.58	2.17	9.50
8	5.42	7.50	5.00	6.42	3.00	2.17	2.67	8.33	4.67	1.75	2.17	8.33
9	4.92	7.50	4.67	6.00	2.92	2.58	2.75	3.08	4.50	2.58	2.00	7.00
10	4.50	7.42	4.67	5.67	2.75	2.75	2.67	2.83	4.08	3.00	2.00	6.00
11	4.08	7.50	6.16	5.38	2.67	3.00	2.92	2.75	3.83	2.83	2.00	5.42
12	4.25	7.42	7.08	6.08	2.67	2.75	2.83	2.58	3.50	2.67	2.67	5.00
13	6.00	7.00	8.50	7.33	2.58	2.67	2.75	2.58	8.08	2.67	3.67	4.17
14	8.75	6.42	9.67	9.00	2.50	2.67	2.50	2.58	8.00	2.58	4.00	4.33
15	7.92	5.92	10.75	8.50	2.50	2.58	2.25	2.50	3.00	2.42	4.25	4.00
16	7.50	5.58	10.00	8.00	2.42	2.50	2.17	2.50	2.67	2.33	4.08	3.83
17	6.67	5.92	8.83	7.87	2.42	2.42	2.00	2.50	2.67	2.08	3.75	3.75
18	6.00	14.25	7.75	7.42	2.33	2.33	1.83	2.42	2.58	2.00	4.00	3.67
19	5.67	19.00	6.83	6.83	2.25	2.33	1.92	2.25	2.58	1.83	4.83	4.58
20	5.08	17.83	6.17	6.75	2.25	2.33	2.08	2.42	2.50	1.92	4.75	5.00
21	4.83	13.25	5.92	6.33	2.04	3.33	2.08	2.25	2.25	2.17	4.67	4.75
22	4.50	11.75	6.33	5.92	2.00	8.58	2.08	2.08	2.17	2.50	4.25	4.17
23	7.08	11.50	6.67	5.50	2.13	5.42	2.00	2.00	2.08	8.25	4.17	3.83
24	9.17	10.25	8.08	5.17	2.25	6.17	2.00	8.08	2.08	4.67	4.08	3.92
25	9.50	9.00	10.33	5.00	2.38	5.58	4.33	6.50	2.00	4.17	5.42	4.58
26	9.42	8.25	10.83	4.75	2.29	4.58	4.00	6.58	1.92	8.67	6.42	6.33
27	8.42	11.33	10.08	4.67	2.25	4.33	3.83	5.25	1.83	8.17	6.17	8.25
28	7.50	13.08	8.92	4.25	2.21	3.75	8.33	5.67	1.75	8.00	5.42	9.33
29	7.00	......	7.83	4.08	2.17	3.50	3.00	6.00	1.75	2.83	5.00	8.58
30	7.08	......	7.50	3.83	2.08	3.50	2.75	5.83	1.75	2.67	4.67	7.83
31	9.83	......	7.67	......	2.00	......	3.92	5.17	......	2.58	......	8.50
1892.												
1	8.50	2.88	4.50	9.75	3.00	5.92	4.67	1.92	2.92	1.08	.50	1.92
2	8.25	2.92	4.00	9.00	2.83	5.50	4.33	2.00	2.50	1.25	.50	1.83
3	8.75	2.92	3.58	8.58	2.83	5.17	3.75	1.83	2.33	1.42	.50	1.75
4	9.33	3.08	3.25	11.75	2.83	7.58	3.67	2.00	2.17	1.25	.50	1.58
5	8.83	3.08	3.00	14.33	4.50	12.50	3.50	8.00	2.00	1.08	.50	1.58
6	8.00	3.00	2.67	14.67	5.83	12.00	3.58	2.83	1.83	1.08	.50	1.50
7	7.83	3.00	2.83	13.17	7.58	11.25	3.42	2.83	1.83	1.00	.50	1.50
8	6.83	2.92	2.83	11.33	7.58	9.00	8.42	3.00	1.75	1.00	.50	1.50
9	5.33	2.75	3.83	9.50	7.83	7.67	3.42	2.67	1.67	1.00	.75	1.58
10	5.67	2.50	5.25	7.83	6.67	7.00	3.00	2.42	1.50	1.00	.92	1.67
11	4.17	2.58	6.17	7.00	5.58	7.42	2.83	2.17	1.50	1.00	1.00	2.42
12	3.67	2.50	5.92	6.42	5.00	7.00	2.50	2.08	1.42	.92	1.17	4.25
13	3.75	2.00	5.67	5.67	4.75	6.42	2.17	2.42	1.42	.92	1.17	8.50
14	5.50	1.80	5.00	5.33	4.25	5.42	2.17	2.50	1.50	.83	1.17	8.50
15	11.83	1.75	4.42	4.75	4.17	4.67	2.33	3.50	2.33	.83	1.25	3.08
16	13.17	1.83	4.00	4.75	4.17	4.17	2.42	4.17	2.33	.83	1.25	2.83
17	10.83	1.67	3.50	4.33	4.42	3.75	2.42	4.00	2.08	.83	1.25	2.67
18	9.08	1.75	3.33	4.33	4.83	3.58	2.25	3.50	1.83	.83	1.25	2.67
19	7.75	2.00	3.08	4.00	4.92	3.50	2.25	2.83	1.67	.83	1.92	2.54
20	7.67	2.38	3.00	3.83	5.67	3.50	2.08	2.67	1.50	.83	2.50	2.50
21	7.00	2.17	2.92	3.67	7.25	3.67	2.00	2.33	1.50	.83	2.50	2.42
22	6.17	2.50	2.67	3.50	8.25	4.00	1.75	2.17	1.50	.83	2.92	2.08
23	5.33	2.67	2.50	3.42	8.83	3.67	1.67	1.90	1.83	.83	3.58	1.50
24	4.75	3.17	2.50	3.50	8.75	3.50	1.67	1.83	1.17	.83	3.33	.92
25	4.50	3.50	2.67	3.50	8.58	3.58	1.67	1.92	1.17	.75	2.92	1.08
26	4.33	4.33	3.50	3.58	7.33	4.17	1.58	2.17	1.25	.58	2.50	2.58
27	3.58	4.50	4.50	3.58	6.67	3.58	1.50	2.00	1.25	.58	2.08	2.00
28	2.50	4.83	10.83	3.50	6.50	3.25	1.50	2.00	1.25	.58	2.00	2.25
29	2.08	4.67	10.00	3.33	6.83	3.50	1.50	2.00	1.08	.58	2.00	2.25
30	2.83	......	12.00	8.17	7.08	4.88	1.42	2.25	1.08	.58	1.92	2.25
31	2.83	......	10.58	......	6.42	......	1.67	8.00	......	.50	......	2.17

Mean daily gage height, in feet, of Susquehanna River at Harrisburg, Pa., 1891–1904—Continued.

Day.	Jan.	Feb.	Mar.	Apr.	May.	June.	July.	Aug.	Sept.	Oct.	Nov.	Dec.
1893.												
1	2.00	2.67	2.58	6.08	4.92	3.67	2.33	.92	3.58	2.00	2.17	4.00
2	2.50	3.00	2.58	6.00	4.83	3.67	2.17	.83	4.17	2.00	2.17	3.83
3	2.83	4.00	2.75	6.42	5.50	3.50	2.08	.83	3.92	1.83	2.17	3.67
4	2.83	4.17	2.75	7.50	6.83	3.58	1.92	.83	3.50	1.67	2.17	3.67
5	2.75	5.00	2.75	7.92	16.17	3.58	1.92	.75	2.67	1.50	2.33	3.67
6	2.67	5.08	2.50	8.92	16.50	3.17	1.67	.75	2.25	1.50	3.00	3.50
7	2.50	5.00	2.50	9.50	14.58	3.00	1.67	.67	2.00	1.42	3.25	3.17
8	2.50	5.33	2.67	8.83	12.00	3.00	1.58	.75	1.75	1.42	2.83	8.00
9	2.50	5.42	3.08	8.00	9.92	3.00	1.50	.58	1.67	1.42	2.75	3.00
10	2.50	6.42	6.50	8.42	8.25	2.83	1.50	.58	1.60	1.33	2.50	2.92
11	2.25	7.75	12.50	10.00	7.00	2.67	1.50	.60	1.50	1.33	2.50	2.83
12	2.25	11.58	13.83	9.42	6.17	2.58	1.50	.60	1.67	1.33	2.42	2.83
13	2.08	7.50	14.50	8.42	5.50	2.50	1.50	.42	2.00	1.25	2.33	2.83
14	2.08	6.50	14.58	7.75	5.00	2.33	1.50	.42	2.00	1.67	2.17	2.50
15	2.08	5.58	13.00	7.42	4.75	2.08	1.75	.42	1.83	4.67	2.08	2.00
16	2.00	5.25	12.25	8.08	4.58	2.00	1.83	.38	2.00	5.33	2.00	2.25
17	2.00	7.75	10.50	8.83	5.92	1.92	1.83	.33	2.50	5.25	1.92	-2.42
18	2.00	6.75	8.83	8.92	8.50	1.83	1.67	.33	2.67	4.25	1.83	5.75
19	2.00	5.83	7.33	7.75	9.75	1.75	1.67	.33	4.42	3.83	1.75	8.83
20	2.00	5.33	6.67	6.92	9.00	1.75	1.67	.67	3.67	3.42	1.75	7.00
21	2.00	4.67	5.92	7.00	7.58	1.75	1.67	.58	3.25	8.00	1.67	6.00
22	2.00	4.25	5.58	10.00	7.00	1.58	1.50	.60	2.83	2.50	1.58	5.92
23	2.00	3.50	5.67	10.92	6.25	1.58	1.42	.42	2.50	2.50	1.58	4.42
24	2.00	3.00	6.83	10.50	5.58	1.75	1.83	.42	2.33	2.33	1.67	3.92
25	2.00	3.00	7.25	8.92	5.42	1.75	1.25	.83	3.33	2.25	1.67	3.83
26	2.00	3.00	7.75	7.67	4.92	2.00	1.17	.83	2.17	2.25	1.58	3.83
27	2.00	2.92	9.42	6.83	4.50	2.25	1.08	.50	2.00	2.25	1.58	4.83
28	2.00	2.75	8.67	6.17	4.33	2.50	1.08	.50	2.00	2.00	1.75	5.92
29	2.00		7.83	5.67	4.17	2.75	1.83	1.00	2.00	2.00	2.83	5.83
30	2.33		7.83	5.17	3.92	2.50	.92	.92	2.00	2.00	3.67	5.17
31	2.50		6.50		3.67		.92	.92		3.08		4.67
1894.												
1	4.50	2.41	3.16	3.83	4.58	9.50	2.58	1.08	.33	1.91	5.08	2.41
2	4.50	2.33	3.33	3.66	4.50	9.66	2.41	1.08	.33	1.83	5.25	2.33
3	4.00	2.25	8.50	3.50	4.16	9.16	2.33	1.83	.33	1.58	5.41	2.50
4	3.66	2.16	3.75	3.25	3.83	8.58	2.25	1.50	.33	1.58	7.50	2.91
5	3.50	2.08	4.08	3.16	3.50	8.41	2.00	1.66	.25	1.41	7.66	3.50
6	3.33	2.00	7.66	3.00	3.16	7.91	2.00	1.58	.25	1.41	7.58	3.58
7	3.41	2.00	7.66	2.91	2.91	6.75	1.83	1.50	.33	1.33	7.16	3.58
8	5.16	2.00	11.33	2.83	3.33	6.00	1.83	1.50	.33	1.83	7.00	3.33
9	5.25	2.08	12.16	2.75	3.50	5.50	1.75	1.08	.41	1.25	6.50	3.00
10	4.58	3.50	10.83	2.75	3.50	5.00	1.66	1.08	1.00	1.83	6.00	3.00
11	3.75	5.00	8.50	2.83	8.50	4.66	1.58	1.08	1.91	2.08	5.50	3.33
12	3.83	6.00	9.83	3.00	3.08	4.00	1.50	1.00	1.50	4.91	5.33	4.00
13	2.50	5.66	7.16	3.25	2.91	3.75	1.41	1.00	1.33	5.58	4.66	4.33
14	3.16	4.58	7.00	3.66	2.75	3.66	1.41	1.00	1.25	5.08	4.50	5.75
15	3.16	4.33	6.41	6.33	2.50	3.66	1.33	1.00	1.25	4.66	4.00	6.33
16	2.83	3.66	5.83	7.58	2.50	3.58	1.33	1.00	1.16	4.16	3.91	5.75
17	2.66	3.33	5.50	9.08	2.33	3.41	1.33	1.00	1.08	3.83	3.66	5.75
18	2.83	3.33	5.08	9.08	2.33	3.16	1.16	1.00	1.08	3.66	3.50	5.16
19	2.83	3.33	4.83	8.50	2.33	3.00	1.08	.91	2.16	3.41	3.25	4.66
20	3.00	4.16	4.58	7.50	5.33	3.50	1.08	.91	4.08	3.00	3.16	4.33
21	2.83	6.66	4.50	6.75	16.33	3.41	1.08	.83	5.00	2.75	3.08	4.08
22	2.83	5.33	4.33	8.50	25.58	3.08	1.00	.83	5.50	2.50	3.25	3.83
23	2.58	5.16	4.50	9.41	21.41	2.83	1.00	.66	5.66	2.33	3.16	3.50
24	2.41	4.33	4.66	9.58	15.25	2.50	1.08	.75	4.83	2.16	3.00	3.50
25	2.41	3.33	5.50	9.91	11.83	2.50	1.25	.75	4.00	2.33	3.00	3.33
26	2.41	2.91	7.00	9.00	11.33	2.66	1.41	.75	3.41	3.58	2.83	3.16
27	2.41	2.33	6.33	7.25	11.66	2.58	1.50	.66	3.00	4.75	2.66	3.00
28	2.50	2.50	5.50	6.00	9.50	2.66	1.50	.66	2.58	4.83	2.58	3.00
29	2.58		4.91	5.41	7.91	2.41	1.41	.58	2.25	4.33	2.58	4.00
30	2.58		4.33	5.00	7.00	2.75	1.16	.50	2.08	4.00	2.50	3.66
31	2.50		4.00		7.50		1.08	.41		3.75		3.66

Mean daily gage height, in feet, of Susquehanna River at Harrisburg, Pa., 1891–1904—Continued.

Day.	Jan.	Feb.	Mar.	Apr.	May.	June.	July.	Aug.	Sept.	Oct.	Nov.	Dec.
1895.												
1	3.92	2.92	6.00	5.75	3.42	2.67	2.83	.58	.75	.42	.21	3.08
2	4.00	2.83	8.58	5.67	3.33	2.58	2.67	.67	.75	.42	.21	3.08
3	4.25	3.00	8.08	6.17	3.25	2.50	2.92	.67	.67	.33	.25	2.75
4	4.33	3.00	10.50	6.83	3.00	2.25	2.50	.67	.67	.33	.25	2.50
5	4.33	7.00	7.88	6.67	2.75	2.08	2.25	.58	.58	.33	.33	2.25
6	4.33	5.67	7.67	6.17	2.67	1.92	2.00	.50	.58	.33	.38	2.00
7	4.33	5.75	6.67	6.00	2.50	1.83	1.92	.50	.75	.33	.38	1.92
8	4.50	5.67	6.25	5.75	2.42	1.75	1.75	.83	.75	.25	.42	1.92
9	4.75	5.50	5.88	8.08	2.25	1.75	1.58	.75	.67	.25	.42	1.92
10	6.17	5.50	6.17	12.00	2.75	1.58	1.50	1.00	.50	.21	.42	1.83
11	7.42	5.58	6.17	13.67	3.00	1.33	1.50	1.08	1.00	.21	.42	1.50
12	7.83	5.92	6.33	12.50	3.33	1.42	1.42	1.08	1.50	.21	.46	1.50
13	8.50	5.83	6.17	10.92	3.67	1.33	1.33	1.08	1.58	.33	.50	.95
14	7.83	5.83	6.00	9.50	4.33	1.25	1.33	.92	1.42	.29	.58	.75
15	6.75	5.67	6.50	10.00	4.33	1.25	1.25	1.33	1.00	.29	.58	1.00
16	6.25	5.58	6.50	9.75	4.17	1.25	1.25	1.33	.83	.25	.58	1.00
17	5.75	5.50	6.67	8.75	4.08	1.25	1.08	1.08	.67	.25	.67	1.33
18	5.42	5.50	6.33	7.58	3.67	1.25	1.00	1.00	.58	.42	.83	1.33
19	5.00	5.33	5.67	6.67	3.50	1.25	.92	1.00	.67	.58	1.00	1.33
20	4.50	5.25	5.50	6.00	3.38	1.25	.92	.82	.67	.50	1.00	1.33
21	4.42	5.17	5.33	5.50	3.17	1.17	.83	.83	.67	.42	.92	1.50
22	4.33	5.08	5.17	5.00	3.08	1.00	.83	.58	.58	.42	.79	1.83
23	4.00	5.00	5.00	4.58	2.92	.75	.83	.50	.58	.33	.67	2.00
24	4.00	4.92	5.00	4.33	2.75	.75	.83	.50	.58	.25	.75	2.67
25	3.33	4.75	5.00	4.00	2.58	.75	.83	.42	.58	.25	.75	2.75
26	3.25	4.58	5.83	3.75	2.50	1.50	.83	.33	.50	.21	.75	2.83
27	3.08	4.50	8.00	3.58	2.50	1.50	.83	.33	.50	.13	.75	3.33
28	3.08	4.75	9.00	3.75	2.42	1.50	.83	.33	.42	.08	2.67	3.50
29	3.08	......	8.00	3.75	2.42	2.00	.75	.33	.42	.06	2.83	5.08
30	3.25	......	7.17	3.50	3.08	3.50	.58	.33	.42	.04	2.83	5.67
31	3.00	......	6.33	......	3.00	......	.42	.50	......	.04	......	5.67
1896.												
1	9.92	4.50	7.17	14.58	3.00	1.50	2.67	4.67	.33	5.42	2.08	3.92
2	9.17	3.75	9.17	14.58	3.00	1.50	2.42	4.33	.33	4.25	1.92	3.92
3	8.42	3.58	9.75	13.75	2.83	1.75	2.08	4.33	.33	4.00	1.83	3.83
4	6.50	3.58	8.42	12.33	2.83	1.83	1.83	3.75	.83	3.17	1.83	3.33
5	5.08	3.50	7.17	10.50	2.67	1.67	1.75	8.67	.25	2.67	1.83	3.00
6	4.00	4.00	5.50	8.83	2.50	1.67	1.67	3.58	.25	2.08	7.25	2.75
7	3.88	11.50	5.00	7.25	2.42	1.67	1.67	2.50	.25	1.83	10.08	2.67
8	3.00	12.50	4.75	6.50	2.17	1.58	2.00	2.33	.25	1.67	7.75	2.50
9	4.67	10.33	4.50	6.17	2.08	1.42	1.92	2.33	.25	1.50	6.50	2.50
10	4.33	8.50	4.83	5.88	2.00	1.75	2.33	2.25	.25	1.50	5.67	2.67
11	4.08	6.83	5.08	5.50	2.00	2.50	2.75	2.25	.26	1.50	4.75	3.42
12	4.00	5.33	4.67	5.50	1.92	2.58	2.75	2.00	.25	1.92	4.42	3.75
13	3.92	4.92	4.00	6.00	1.75	3.42	2.50	1.83	.25	1.92	4.17	4.00
14	4.00	4.25	3.50	6.42	1.67	3.25	2.17	1.67	.83	7.33	4.00	4.25
15	3.83	3.75	2.67	8.00	1.67	2.92	2.00	1.67	.83	7.00	3.83	3.83
16	3.83	3.75	2.67	8.42	1.75	2.58	1.83	1.58	.83	9.50	3.67	3.67
17	3.75	3.83	2.33	8.17	1.58	2.58	1.67	1.58	.50	7.67	3.50	3.42
18	3.58	3.58	2.50	7.33	1.50	2.83	1.58	1.58	.50	5.58	3.33	3.08
19	3.67	2.92	3.17	6.83	1.50	2.67	1.67	1.33	.58	4.83	3.17	2.92
20	4.00	3.00	4.00	6.33	1.50	8.00	1.67	1.25	.58	4.08	3.00	2.58
21	3.67	2.38	6.00	5.75	1.50	3.17	1.92	1.00	.67	3.58	2.83	2.33
22	3.50	3.67	5.75	5.25	1.42	3.00	1.67	.83	.83	3.42	2.67	2.00
23	3.50	5.42	5.75	4.83	1.42	2.42	1.58	.83	1.17	3.25	2.58	2.00
24	3.50	5.42	6.25	4.58	1.42	2.33	1.67	.83	1.17	3.00	2.50	1.60
25	4.00	3.42	5.58	4.83	1.38	2.25	1.67	.83	.92	3.00	2.50	1.50
26	7.25	3.50	5.00	4.08	1.25	2.67	1.75	.75	.75	3.00	2.33	1.50
27	7.33	3.67	5.25	4.00	1.17	4.75	1.92	.75	.58	2.75	2.33	1.50
28	6.17	3.17	6.08	3.58	1.25	4.00	2.50	.67	.50	2.67	2.42	1.50
29	6.00	3.17	6.50	3.58	1.50	3.50	3.50	.58	.42	2.50	2.67	1.33
30	5.75	......	9.25	3.25	1.50	8.08	3.75	.50	.83	2.42	3.50	1.58
31	5.42	......	12.50	......	1.50	......	4.88	.38	......	2.25	......	1.75

Mean daily gage height, in feet, of Susquehanna River at Harrisburg, Pa., 1891–1904—Continued.

Day.	Jan.	Feb.	Mar.	Apr.	May.	June.	July.	Aug.	Sept.	Oct.	Nov.	Dec.
1897.												
1	1.83	3.88	4.25	5.00	3.08	2.92	1.42	4.00	1.25	1.75	.67	5.00
2	2.00	3.17	3.67	4.67	3.08	2.83	1.83	4.38	1.08	1.50	1.17	4.50
3	2.00	3.17	3.25	4.83	5.50	2.67	1.25	3.83	1.00	1.33	8.08	4.00
4	2.08	3.17	3.83	4.17	6.50	2.58	1.25	3.25	1.00	1.17	4.08	3.75
5	2.50	3.08	4.92	4.00	7.50	2.67	1.25	2.83	1.00	1.08	3.50	3.33
6	3.00	3.00	5.92	3.83	7.08	3.00	1.25	2.67	.92	1.00	3.08	4.75
7	3.67	4.25	7.67	3.75	7.00	2.67	1.42	2.42	.83	1.00	3.00	5.17
8	3.67	7.50	8.58	3.75	6.33	2.50	1.42	2.67	.83	.92	2.75	5.08
9	3.67	6.58	8.00	3.75	5.50	2.67	1.25	2.50	.83	.83	2.50	5.42
10	3.33	5.42	6.92	5.92	4.83	2.67	1.25	2.08	.66	.67	2.41	4.92
11	3.08	4.98	6.50	9.00	4.50	2.67	1.17	2.08	.58	.67	2.67	4.38
12	2.83	4.50	7.25	9.50	4.00	2.67	1.08	2.00	.67	.58	2.67	4.17
13	2.42	3.92	8.67	8.00	4.00	3.08	1.00	1.83	.67	.75	2.50	4.17
14	2.00	3.88	8.42	6.83	6.00	3.50	1.08	1.75	.67	.75	2.50	4.33
15	2.00	3.83	7.75	6.00	7.75	3.25	1.00	1.58	.50	.75	2.50	4.58
16	2.00	3.50	7.00	6.00	7.92	2.92	1.00	1.58	.58	.75	2.50	6.58
17	2.00	3.50	6.92	6.58	7.33	2.67	1.17	1.50	.67	.67	2.50	7.67
18	2.17	3.33	5.50	7.00	6.50	2.50	1.17	1.50	.75	.67	2.67	8.17
19	2.33	3.58	5.00	6.58	5.75	2.25	1.08	1.42	.75	.58	2.92	7.33
20	2.00	4.08	5.83	6.00	5.00	2.17	1.08	1.42	.67	.58	3.42	6.33
21	1.83	4.00	7.42	5.50	4.25	2.17	1.50	1.33	.58	.50	8.25	5.58
22	1.83	4.25	8.25	4.92	4.25	2.17	1.50	1.17	.58	.58	8.17	5.00
23	1.92	5.92	9.75	4.50	3.58	2.00	1.33	1.17	.58	.75	2.83	4.08
24	1.67	7.92	9.50	4.17	3.50	1.83	1.42	1.25	1.00	.75	2.50	3.83
25	1.67	7.50	10.17	3.83	3.75	1.75	1.58	1.67	1.50	1.00	2.50	3.42
26	1.50	6.50	11.50	3.67	3.75	1.75	1.75	2.67	1.50	1.00	2.50	2.83
27	3.33	5.50	10.67	3.58	3.50	1.67	1.75	2.08	1.83	1.00	2.33	2.75
28	3.33	4.50	8.00	3.60	3.58	1.58	2.17	1.75	1.92	.92	2.50	2.67
29	3.00		7.42	3.33	3.92	1.58	3.83	1.58	2.25	.83	3.50	2.67
30	3.25		6.33	3.17	3.50	1.50	4.50	1.50	2.00	.75	4.92	2.58
31	3.33		5.58		3.25		4.08	1.33		.75		2.50
1898.												
1	2.66	3.91	4.66	8.66	6.00	4.33	2.00	1.41	2.66	.75	4.66	3.08
2	2.33	3.41	4.33	7.41	5.41	4.16	2.16	1.50	2.33	.75	4.00	3.16
3	2.16	3.00	4.16	6.41	4.83	3.91	2.00	1.41	3.00	.66	3.66	3.08
4	2.66	2.66	3.91	5.75	4.66	3.58	1.75	2.33	2.50	.66	3.50	3.00
5	1.91	2.66	3.66	5.41	4.41	3.33	1.66	4.58	2.08	.66	3.16	3.66
6	1.91	2.66	3.58	4.91	4.43	3.00	1.58	5.33	1.91	.66	3.00	5.00
7	2.25	2.66	3.50	4.50	4.66	2.83	1.50	4.00	1.66	.66	2.91	4.50
8	2.50	3.08	3.50	4.41	5.50	2.66	1.41	3.50	1.66	1.00	2.50	4.08
9	2.66	3.41	3.33	4.16	6.25	2.50	1.33	3.08	1.66	1.33	2.50	3.83
10	2.75	3.50	3.83	3.83	5.58	2.50	1.25	3.66	2.00	1.41	2.50	3.58
11	3.00	3.41	3.83	3.66	5.16	2.83	1.16	4.25	2.83	2.25	2.58	3.08
12	8.00	3.75	4.91	3.50	4.75	2.83	1.08	3.75	2.75	2.40	4.00	2.50
13	8.33	4.41	8.66	3.33	4.50	2.25	1.00	3.33	2.33	2.33	8.75	2.25
14	4.00	7.66	8.66	3.25	4.00	2.25	.91	2.66	2.08	2.00	8.00	2.25
15	6.95	8.16	9.83	3.16	4.00	2.41	.83	2.50	1.91	2.00	6.54	2.08
16	8.08	7.50	9.33	3.66	4.25	2.75	.83	2.25	1.75	2.08	5.50	2.00
17	7.83	6.50	8.08	4.08	5.16	3.25	.75	2.00	1.41	2.16	4.83	2.00
18	7.58	5.83	7.16	3.91	6.08	3.00	.66	1.91	1.33	3.25	4.33	1.91
19	6.58	5.00	6.33	3.66	5.33	2.66	.66	2.33	1.16	3.75	4.16	2.00
20	5.83	4.33	5.83	3.50	5.50	2.41	.75	3.00	1.00	4.00	4.16	2.50
21	5.75	4.66	7.83	3.41	6.66	2.33	.91	4.41	.91	4.33	4.25	2.91
22	6.16	6.89	9.25	3.33	6.66	2.33	.75	4.33	.91	4.25	4.58	3.08
23	7.41	6.91	10.91	3.16	6.50	2.08	.91	3.75	.91	7.33	4.83	3.50
24	9.25	7.75	15.83	3.00	6.00	2.00	.83	3.41	.83	7.41	4.33	7.83
25	10.50	6.66	15.25	3.50	7.00	2.16	.83	3.00	.83	7.16	4.00	7.66
26	9.50	6.25	11.66	6.66	6.66	2.08	.83	2.66	.75	6.16	3.91	6.33
27	8.00	5.66	9.25	10.33	6.50	2.00	1.83	2.50	.91	5.66	3.66	5.33
28	7.00	5.00	9.50	9.50	6.16	1.91	1.16	2.41	.91	5.58	3.50	4.83
29	6.08		6.66	8.16	5.75	1.83	1.83	4.16	.75	5.66	3.50	4.83
30	5.50		7.00	6.66	5.33	1.66	1.58	4.16	.75	6.08	3.83	4.83
31	4.83		9.00		4.91		1.33	8.00		5.83		3.83

Mean daily gage height, in feet, of Susquehanna River at Harrisburg, Pa., 1891–1904—Continued.

Day.	Jan.	Feb	Mar.	Apr.	May.	June.	July.	Aug.	Sept.	Oct.	Nov.	Dec.
1899.												
1	3.25	2.50	8.41	7.25	8 41	2.50	1.75	.75	1.83	1.08	.50	1.75
2	3.16	2.00	8.16	6.41	3.08	2.58	1.66	.75	1.50	.83	1.66	1.58
3	2.75	1.91	7.83	5.83	3.08	2.50	1.66	.75	1.25	.88	2.50	1.50
4	3.25	2.25	7.41	5.33	3.41	2.50	1.50	.75	1.08	.75	3.25	1.50
5	3.50	2.58	8 00	4.91	3.16	2.50	1.33	.75	1.08	.66	4.50	1.50
6	5.00	2.66	12.50	4.41	3.16	2.38	1.25	.91	1.00	.66	3.91	1.50
7	8.00	2.83	13.00	4.25	3.00	2.08	1.25	.75	.91	.58	3.75	1.50
8	6.83	2.41	11.41	4.75	2.75	1.91	1.16	.75	.91	.58	3.16	1.50
9	6.08	2.50	9.25	6.83	2.83	1.91	1.16	.83	.83	.58	2.83	1.50
10	5.41	2.41	7.66	8.75	2.66	1.91	1.16	.75	1.00	.66	2.50	1.50
11	4.58	2.41	6.50	8.41	2.75	1.75	1.41	.66	1.00	.58	2.25	1.50
12	4.00	4.41	5.75	7.75	2.75	1.66	1.25	.66	.75	.58	2.16	1.50
13	3.33	4.41	5.75	6.75	2.91	1.66	1.16	1.08	.83	.50	2.08	2.75
14	3.16	4.58	7.50	6.75	2.83	1.58	1.16	.91	1.41	.50	2.00	5.50
15	3.33	4.58	8.41	8.00	2.58	1.50	1.16	1.25	1.25	.51	2.25	6.33
16	3.66	4.66	8.00	8.00	2.50	1.50	1.08	.91	.83	.41	2.41	6.00
17	4.83	4.83	7.41	7.83	2.50	1.41	1.00	.66	.75	.41	2.41	5.33
18	7.00	4.83	6.41	7.33	2.58	1.25	1.25	.66	.75	.41	2.41	4.58
19	6.33	4.91	4.33	6.83	3.75	1.25	1.25	.50	.58	.41	2.83	4.08
20	5.66	4.75	7.16	6.00	4.75	1.25	1.25	.50	.66	.83	3.00	3.75
21	4.91	4.91	8.50	5.41	5.16	1.25	1.25	.50	.75	.83	2.91	3.75
22	4.38	5.33	8.16	5.08	4.25	1.16	1.25	.50	.66	.83	2.58	3.83
23	4.25	7.50	7.50	4.91	3.91	1.08	1.33	.50	.66	.33	2.50	4.50
24	4.08	7.50	7.16	4.50	3.58	1.00	1.33	.50	.66	.16	2.25	4.25
25	4.16	7.16	7.41	4.41	3.16	1.41	1.16	.50	.66	.16	2.25	5.83
26	5.25	6.83	7.41	4.00	3.00	2.00	1.00	.41	.66	.25	2.25	6.75
27	4.50	7.33	6.83	3.91	2.91	1.66	1.00	.66	1.00	.83	2.16	5.25
28	3.83	9.00	6.33	3.75	2.66	1.50	1.00	4.00	1.33	.88	2.00	4.50
29	3.25		6.83	3.66	2.50	1.50	.91	2.66	1.16	.41	2.00	3.83
30	3.00		7.83	3.50	2.50	1.75	.83	2.50	1.08	.88	1.83	3.00
31	3.00		8.08		2.50		.75	2.16		.33		2.25
1900.												
1	1.83	2.91	4.00	4.16	4.00	2.58	1.17	1.25	1.00	.04	.88	7.00
2	1.66	1.88	13.12	4.00	3.75	2.50	1.08	1.00	1.00	.04	.88	5.83
3	1.50	3.91	12.33	4.16	3.50	2.33	1.00	1.00	.83	.04	.75	5.25
4	4.91	4.00	9.50	4.41	3.33	2.17	1.08	.92	1.17	.06	.75	4.50
5	4.83	4.66	7.91	5.33	3.08	2.50	1.33	.75	.92	.04	.75	5.00
6	5.25	4.33	6.91	6.00	2.83	2.67	1.17	.67	.83	.04	.66	7.25
7	5.50	5.50	6.00	5.41	2.83	2.50	1.17	.67	.58	.04	.66	7.41
8	5.33	5.00	6.16	5.08	2.75	2.17	1.17	.58	.58	.08	.66	7.08
9	4.91	4.00	6.50	6.16	2.50	2.17	1.42	.58	.58	.04	.75	6.00
10	4.58	4.83	5.83	6.75	2.50	2.08	1.42	.58	.50	.04	.58	5.25
11	4.50	5.75	5.66	6.50	2.42	2.00	1.33	.50	.42	.04	.66	4.75
12	5.50	5.50	6.25	5.58	2.33	2.00	1.17	.33	.33	.04	.50	4.08
13	4.91	5.66	5.75	5.00	2.42	1.92	1.00	.33	.17	.25	.58	3.83
14	5.25	7.66	4.66	4.50	2.42	1.92	1.04	.25	.25	.83	.75	3.60
15	5.25	8.00	4.50	4.33	2.50	2.00	1.00	.17	.25	.83	.66	2.91
16	5.25	8.25	4.00	4.50	2.40	2.17	1.00	.17	.25	.75	.66	2 85
17	4.66	7.41	3.66	4.41	2.33	2.17	1.00	.25	.25	.58	.83	2.25
18	5.00	6.00	3.16	4.33	2.33	2.00	1.08	.17	.17	.66	.91	2.08
19	4.83	4.75	3.00	5.08	2.25	1.83	.92	.17	.08	.66	.75	2.08
20	4.00	3.91	3.00	7.08	2.50	1.83	.92	.17	.12	.58	.91	2.08
21	4.25	2.16	3.91	7.83	2.92	1.82	.83	.83	.08	.50	.91	2.00
22	10.66	3.58	6.87	6.83	2.17	1.75	.75	.42	.07	.50	.91	2.16
23	12.00	9.50	6.83	6.08	2.83	1.75	.83	.06	.50	.83		2.41
24	9.16	11.16	6.00	5.83	2.58	1.58	.75	.50	.04	.50	1.00	2.16
25	7.25	9.75	5.75	6.08	2.42	1.42	.75	1.25	.04	1.00	1.08	2.33
26	6.08	6.83	5.83	6.25	2.25	1.33	.83	1.00	.02	1.08	1.66	2.41
27	5.00	5.50	5.50	5.08	2.17	1.33	1.50	1.17	1.00	1.00	5.91	2.08
28	4.50	4.50	5.25	5.08	2.00	1.33	1.25	1.50	-.04	1.25	13.04	2.66
29	4.08		4.83	4.58	2.00	1.33	1.25	1.33	-.04	1.16	12.33	2.91
30	3.33		4.50	4.17	2.00	1.17	1.42	1.00	+.04	1.00	8.91	2.58
31	2.50		4.41		1.92		1.25	1.08		.91		2.50

Mean daily gage height, in feet, of Susquehanna River at Harrisburg, Pa., 1891-1904—Continued.

Day.	Jan.	Feb.	Mar.	Apr.	May.	June.	July.	Aug.	Sept.	Oct.	Nov.	Dec.
1901.												
1	2.25	2.58	1.75	7.16	5.16	12.58	3.08	1.66	3.50	2.08	1.41	3.08
2	2.08	4.00	1.66	6.00	4.58	10.41	2.83	1.83	3.75	2.41	1.41	8.00
3	1.66	3.83	1.75	5.66	4.50	8.91	2.58	1.75	4.75	2.33	1.33	2.75
4	1.66	3.25	1.83	6.25	4.41	7.83	2.33	1.58	5.16	2.33	1.33	2.75
5	1.75	3.25	2.33	7.50	5.16	7.16	2.25	1.50	4.83	2.33	1.25	3.08
6	1.66	3.08	2.50	7.83	5.00	6.33	2.16	1.25	4.16	2.41	1.25	2.66
7	1.41	3.16	2.58	8.66	4.58	6.50	2.33	1.66	3.58	2.16	1.25	2.75
8	1.16	3.16	2.50	11.41	4.08	5.50	2.16	2.58	8.16	1.83	1.16	2.25
9	1.50	3.16	3.00	12.75	3.75	6.00	2.08	2.75	2.83	1.75	1.16	2.16
10	1.50	3.00	3.25	11.50	3.66	5.75	2.08	2.50	2.50	1.75	1.16	2.58
11	1.66	2.83	6.41	10.00	3.41	5.50	2.00	2.33	2.50	1.66	1.16	4.50
12	2.00	2.91	11.75	8.66	3.83	5.00	1.91	2.75	2.33	1.66	1.00	7.00
13	2.00	2.83	11.83	7.50	4.16	4.66	1.91	2.41	2.41	1.66	1.08	7.00
14	2.50	2.75	9.33	6.91	4.50	4.25	1.83	2.00	2.33	1.83	1.25	6.16
15	3.50	2.75	7.50	6.16	5.16	3.91	1.91	1.75	2.33	2.41	1.33	9.25
16	3.33	2.58	6.66	5.91	5.08	3.50	1.75	1.66	2.25	2.66	1.58	21.41
17	3.41	2.75	6.25	5.75	4.66	3.75	1.66	1.66	2.41	2.50	1.66	18.58
18	2.91	2.58	5.75	5.33	4.16	3.58	2.08	1.75	2.41	2.08	1.91	14.16
19	2.58	2.50	5.25	5.00	4.00	3.50	2.41	5.50	2.50	2.08	1.91	9.83
20	1.75	2.50	5.00	4.75	4.25	3.25	2.25	5.83	2.66	2.00	1.91	7.41
21	1.75	3.08	5.91	5.50	4.08	8.08	2.00	5.00	2.58	2.00	1.75	6.16
22	1.83	2.00	8.50	11.00	4.00	8.25	1.83	4.08	2.58	1.91	1.75	4.83
23	2.00	2.00	9.50	13.58	5.50	3.75	1.75	4.16	2.41	1.91	1.58	8.83
24	1.75	1.91	9.08	12.16	8.41	3.83	1.66	4.75	2.33	1.83	1.83	3.58
25	2.00	1.91	8.00	10.16	7.50	4.00	1.58	7.75	2.08	1.83	2.50	3.75
26	1.75	1.91	7.66	9.16	8.00	8.91	1.50	9.00	1.66	3.08	3.75	
27	2.00	1.75	8.83	8.50	7.50	3.75	1.58	7.25	1.83	1.58	5.41	8.91
28	2.00	1.75	11.75	7.25	7.00	3.50	1.66	5.75	1.75	1.66	5.25	8.91
29	2.00		12.91	6.50	8.75	3.25	1.50	4.75	1.66	1.50	4.00	3.91
30	1.75		11.16	5.75	12.25	8.16	1.50	4.00	1.66	1.41	3.58	5.58
31	1.66		9.00		13.91		1.50	3.50		1.41		6.25
1902.												
1	5.25	3.58	20.83	6.25	2.75	1.75	3.58	5.83	1.25	4.83	5.50	2.41
2	4.75	3.66	23.91	5.58	2.83	1.75	6.16	5.33	1.25	6.00	4.75	2.41
3	4.25	3.50	23.83	5.33	2.83	1.66	7.33	5.50	1.25	5.91	4.50	2.58
4	3.83	3.25	21.41	5.00	2.66	1.66	6.66	6.25	1.25	5.66	4.00	3.33
5	3.00	2.41	16.83	4.75	2.66	1.66	7.33	5.50	1.16	4.66	3.50	3.75
6	3.00	2.00	12.25	4.50	2.83	1.66	7.50	4.83	1.08	4.66	3.25	3.25
7	8.00	5.08	9.50	4.50	2.75	1.50	6.83	4.50	1.00	4.66	3.25	3.50
8	2.83	5.25	7.00	4.50	2.66	1.25	7.33	4.00	.91	4.41	3.08	3.41
9	2.75	5.25	5.25	9.00	2.66	1.50	8.50	3.58	.91	3.83	2.91	3.41
10	8.00	5.08	5.00	14.66	2.66	1.58	7.16	3.25	.91	3.83	2.75	8.16
11	2.91	5.33	6.66	14.16	2.66	1.50	6.16	3.58	.91	3.50	2.66	3.00
12	2.66	5.16	8.33	11.58	2.50	1.50	6.25	3.58	1.25	4.75	2.41	8.83
13	2.58	4.83	10.91	13.41	2.41	1.50	5.50	3.08	1.08	4.83	2.33	8.66
14	2.25	4.41	13.41	7.00	2.33	1.75	5.50	2.83	1.16	3.75	2.33	4.00
15	2.25	4.41	13.58	6.41	2.25	1.75	4.00	2.75	1.00	3.91	2.25	4.00
16	2.25	4.25	12.00	5.66	2.16	2.25	3.50	2.50	1.00	3.75	2.16	5.33
17	2.16	4.08	12.16	5.66	2.16	2.25	3.50	2.50	1.00	3.16	2.16	8.58
18	2.00	3.83	15.00	5.00	2.00	2.41	3.25	2.16	1.00	3.33	2.16	8.33
19	2.00	3.75	13.66	4.75	1.83	2.41	3.25	2.00	1.00	3.00	1.91	7.66
20	2.16	3.75	11.33	4.41	1.83	2.33	8.16	2.00	1.00	2.91	1.83	7.16
21	2.16	3.75	9.50	4.08	1.75	2.16	3.33	2.00	.91	2.66	1.75	8.50
22	5.16	4.00	6.00	3.83	1.83	2.16	4.33	1.91	.91	2.58	1.75	12.50
23	10.00	4.00	5.50	3.50	1.83	2.16	8.00	1.91	.83	2.41	1.66	12.66
24	6.75	4.08	5.33	3.41	1.66	2.00	7.25	1.75	.83	2.25	1.66	11.50
25	6.50	4.16	5.33	3.25	1.66	2.00	7.75	1.58	1.66	2.41	1.91	8.25
26	5.41	6.41	4.66	3.00	1.66	2.00	8.00	1.58	1.58	2.83	2.00	7.25
27	5.08	9.41	3.86	2.91	1.66	2.16	8.08	1.83	5.16	2.33	2.00	6.16
28	5.83	9.66	3.66	2.75	1.66	2.41	6.83	1.50	5.50	2.33	2.25	5.58
29	5.33		4.41	2.75	1.66	2.41	5.83	1.41	4.83	3.66	2.33	5.58
30	4.33		4.41	2.75	1.66	3.00	6.16	1.25	4.33	5.66	2.41	4.83
31	8.91		5.33		1.75		6.16	1.25		6.00		4.58

Mean daily gage height, in feet, of Susquehanna River at Harrisburg, Pa., 1891–1904—Continued.

Day.	Jan	Feb.	Mar.	Apr.	May.	June.	July.	Aug.	Sept.	Oct.	Nov.	Dec.
1903.												
1	4.16	11.50	13.41	6.50	8.41	1.50	7.88	8.50	10.29	1.75	3.16	2.50
2	3.66	10.50	16.83	7.25	3.25	1.50	6.00	3.38	8.33	1.75	3.08	2.41
3	3.83	8.75	14.50	7.50	3.00	1.50	5.16	2.91	6.83	1.75	3.00	2.33
4	4.83	8.91	11.00	6.75	2.58	1.50	4.66	2.66	5.66	1.58	2.83	2.16
5	5.58	13.83	9.00	6.50	2.50	1.50	4.08	2.41	5.16	1.58	2.83	2.08
6	5.91	14.58	8.75	5.75	2.83	1.50	4.41	2.50	4.58	1.50	2.66	2.00
7	6.33	12.25	7.66	5.75	2.33	1.50	4.66	2.91	4.00	1.41	2.58	2.00
8	5.83	9.33	8.16	5.75	2.25	1.50	5.33	4.75	3.33	1.83	2.50	2.00
9	5.00	8.25	8.00	5.75	2.16	1.66	5.33	4.66	3.66	2.66	2.50	2.33
10	4.33	7.00	10.58	6.83	2.16	2.25	4.33	4.08	3.83	5.00	2.41	2.16
11	3.41	6.00	12.50	7.00	2.16	2.25	3.83	3.06	3.50	10.66	2.41	1.91
12	2.91	6.16	11.41	6.83	2.16	2.25	3.16	3.50	3.50	11.25	2.41	1.91
13	2.66	6.50	11.91	6.50	2.08	3.16	3.25	3.50	3.50	11.08	2.41	1.91
14	2.25	6.66	10.83	6.50	2.08	3.66	3.25	3.50	3.33	9.25	2.33	2.00
15	2.25	7.50	9.75	8.83	2.08	4.08	2.75	3.00	3.50	7.33	2.33	1.00
16	2.66	7.66	8.33	12.66	2.08	4.33	2.75	3.16	3.16	5.91	2.33	1.00
17	3.00	7.66	7.83	12.75	2.08	4.41	2.58	3.50	2.83	5.16	2.33	1.00
18	3.16	7.00	7.16	10.66	1.83	4.25	2.33	3.83	2.88	4.83	2.50	1.33
19	3.16	6.00	6.50	9.33	1.83	3.83	3.08	3.16	3.16	5.33	8.66	3.16
20	3.16	5.25	6.50	8.00	1.75	3.41	4.50	2.83	3.38	6.50	8.25	4.00
21	3.16	4.08	5.50	6.50	1.75	8.33	5.66	2.58	3.00	6.58	6.50	5.66
22	3.25	4.50	5.66	6.83	1.66	3.33	5.41	2.50	2.83	6.16	6.16	5.58
23	4.16	4.50	6.00	5.83	1.66	3.66	4.33	2.33	2.66	5.50	4.66	5.58
24	4.00	4.33	9.41	5.08	1.66	4.33	3.91	2.41	2.50	4.83	4.33	4.58
25	3.91	4.16	15.16	5.25	1.66	5.58	3.58	2.33	2.41	4.41	4.00	4.41
26	3.50	4.08	14.16	4.58	1.66	6.50	3.16	2.16	2.33	3.66	3.75	4.00
27	3.50	4.58	11.00	4.50	1.66	7.16	3.00	2.16	2.16	3.75	3.83	3.50
28	3.58	5.50	9.58	4.00	1.66	6.50	3.00	2.25	2.08	3.66	2.50	3.08
29	3.75	......	8.16	3.50	1.58	6.00	2.83	4.16	1.83	5.30	2.50	2.91
30	4.66	......	6.83	3.50	1.58	5.50	3.00	5.91	1.83	3.33	2.50	2.66
31	3.08	......	6.83	......	1.50	......	3.33	9.25	......	3.16	......	2.08
1904. a												
1	2.16	4.41	9.41	6.40	7.65	3.65	1.90	1.58	1.48	1.78	2.08	1.79
2	2.16	4.16	11.50	10.15	6.65	3.90	1.73	1.68	1.28	1.68	1.96	1.54
3	4.00	4.00	11.91	13.06	6.40	4.23	1.98	1.93	1.23	1.53	1.88	1.44
4	8.16	4.75	13.50	11.15	5.65	4.23	1.90	1.93	1.23	1.78	1.78	1.24
5	8.16	3.41	22.00	9.40	4.90	3.98	1.65	1.88	1.18	1.93	1.68	1.29
6	2.91	4.41	19.41	7.73	4.08	4.90	1.73	1.78	1.13	1.73	1.64	.94
7	2.91	8.75	16.33	6.73	3.98	5.23	1.73	2.08	1.08	1.58	1.60	1.29
8	2.83	3.83	21.16	6.15	3.81	4.73	2.23	2.03	.93	1.48	1.54	1.00
9	2.83	5.50	15.91	6.06	3.48	3.98	2.56	1.78	.88	1.38	1.54	1.24
10	b2.83	9.08	15.00	6.40	3.40	3.56	2.56	1.68	1.18	1.23	1.49	1.19
11	3.00	9.33	12.00	8.48	3.15	4.31	4.48	1.88	1.18	1.18	1.59	.84
12	3.58	8.41	9.16	9.15	2.98	5.40	5.06	1.63	1.18	1.23	1.54	.94
13	3.83	9.91	7.91	7.98	2.90	4.65	4.40	1.58	1.13	1.23	1.59	1.69
14	4.91	13.50	6.58	7.15	2.56	3.90	3.73	1.48	1.08	1.23	1.69	1.44
15	4.66	12.50	6.08	6.31	2.81	3.23	3.23	1.33	1.38	1.38	1.64	1.49
16	4.50	11.58	5.58	5.25	3.15	2.90	2.90	1.33	1.58	2.93	1.59	1.39
17	5.00	10.16	5.25	5.15	3.40	2.65	2.56	1.28	1.98	2.73	1.54	1.30
18	5.00	9.91	4.83	5.06	3.65	2.81	2.24	1.23	2.18	2.38	1.49	1.50
19	4.25	9.16	4.66	4.56	3.98	2.56	2.08	1.13	1.78	2.13	1.58	1.50
20	4.08	9.16	4.66	4.48	4.98	2.56	1.98	1.18	1.78	1.88	1.59	1.50
21	4.16	8.66	5.00	3.90	6.06	2.56	2.03	1.28	1.63	1.73	1.49	1.40
22	4.66	9.16	5.58	3.31	6.56	2.65	1.88	1.18	1.43	1.88	1.54	1.40
23	5.50	10.16	6.66	3.73	5.31	2.56	1.93	1.28	1.33	2.93	1.50	1.40
24	5.50	10.16	7.08	3.56	4.56	2.56	2.98	1.28	1.18	3.76	1.69	1.63
25	11.50	10.75	10.41	3.48	4.23	2.73	2.13	1.28	1.18	4.06	1.69	1.63
26	10.16	10.41	11.00	3.48	3.81	2.48	1.83	1.68	1.08	3.58	1.79	1.40
27	7.66	10.58	15.25	3.48	3.98	2.31	1.73	2.33	1.03	3.08	1.89	1.40
28	6.83	9.50	13.83	3.73	3.90	2.06	1.68	2.08	1.13	2.68	1.84	1.90
29	5.83	9.08	12.66	3.65	3.65	1.98	1.78	1.68	1.68	2.53	1.74	2.10
30	4.75	......	10.16	6.98	3.31	1.81	1.68	1.63	1.73	2.48	1.84	9.40
31	4.50	......	8.41	......	3.40	......	1.63	1.58	......	2.28	......	8.40

a From January 1 to July 17, inclusive, gage readings were taken at the pump house. From July 18 to the end of the year the readings were taken at the Walnut Street Bridge. Beginning with April 1 the readings at the pump house were too high by 0.6 foot, owing to the fact that a cofferdam was built just below the intake. This correction has been applied; therefore the gage readings for the complete year are referred to the low-water datum of 1903.

b River frozen over at 5 a. m

c Several ice gorges existed both above and below Harrisburg from January 24 to March 18. These caused the backing up of the water, thus increasing the gage height.

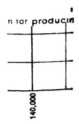

n for producin

140,000

Γ HARRISB

Rating table for Susquehanna River at Harrisburg, Pa., from 1891 to 1904.

Gage height.	Discharge	Gage height.	Discharge	Gage height	Discharge.	Gage height.	Discharge.
Feet.	Second-feet	Feet.	Second-feet.	Feet.	Second-feet.	Feet.	Second-feet.
−0.05	2,330	2.4	16,950	5.8	65,000	12.0	174,500
+0.0	2,440	2.5	17,960	6.0	68,400	12.5	183,600
.1	2,710	2.6	19,010	6.2	71,900	13.0	198,000
.2	3,000	2.7	20,100	6.4	75,500	13.5	202,500
.8	3,330	2.8	21,210	6.6	79,200	14.0	212,000
.4	3,680	2.9	22,340	6.8	82,900	14.5	221,300
.5	4,070	3.0	23,480	7.0	86,500	15.0	231,000
.6	4,500	3.1	24,620	7.2	90,000	15.5	242,800
.7	4,980	3.2	25,760	7.4	93,400	16.0	254,500
.8	5,500	3.3	26,910	7.6	96,700	16.5	267,400
.9	6,020	3.4	28,130	7.8	100,100	17.0	280,400
1.0	6,550	3.5	29,430	8.0	103,500	17.5	293,600
1.1	7,090	3.6	30,800	8.2	106,900	18.0	306,700
1.2	7,650	3.7	32,200	8.4	110,300	19.0	334,500
1.3	8,240	3.8	33,600	8.6	113,800	20.0	363,100
1.4	8,850	3.9	35,000	8.8	117,300	21.0	392,600
1.5	9,520	4.0	36,400	9.0	120,800	22.0	423,100
1.6	10,200	4.2	39,200	9.2	124,300	23.0	454,600
1.7	10,930	4.4	42,200	9.4	127,800	24.0	487,000
1.8	11,700	4.6	45,400	9.6	131,400	25.0	520,200
1.9	12,500	4.8	48,600	9.8	134,900	26.0	554,400
2.0	13,300	5.0	51,900	10.0	138,400	27.0	589,400
2.1	14,160	5.2	55,100	10.5	147,200		
2.2	15,050	5.4	58,400	11.0	156,300		
2.3	15,980	5.6	61,700	11.5	165,300		

*Mean daily discharge, in second-feet, of Susquehanna River at Harrisburg, Pa.,
1891–1904.*

Day.	Jan	Feb.	Mar.	Apr.	May.	June.	July.	Aug.	Sept.	Oct	Nov.	Dec
1891.												
1	21,770	149,000	156,300	107,800	30,800	13,300	20,650					
2	23,440	165,300	120,800	120,800	29,430	12,500	17,960	25,190	36,400	10,580	17,960	36,400
3	27,510	165,300	92,400	113,800	28,180	13,300	19,010	24,020	31,500	10,580	16,460	31,500
4	43,800	159,000	80,100	116,400	28,180	13,300	25,190	22,340	27,510	10,200	15,510	29,430
5	55,100	141,	62,500	110,300	26,330	13,300	37,800	23,480	21,480	10,200	15,510	45,400
6	51,	119,			24,620	13,300	29,430	24,620	21,480	10,200	15,510	116,400
7	60,	97,	55,900	80,	23,480	14,180	24,620	23,480	31,300	10,200	14,600	129,000
8	58,	95,	51,900	75,	23,480	14,600	19,550	27,510	46,200	11,310	14,600	109,400
9	50,2	95,100	46,200	68,	22,340	19,010	20,650	24,620	43,800	19,010	13,300	86,500
10	43,8	93,400	46,200	62,	20,650	20,650	19,550	21,770	37,800	23,480	13,300	68,400
11	37,8	95,100	71,630	57,	19,550	23,480	22,340	20,650	31,300	21,770	13,300	58,400
12	30,90	93,400	88,300	70,	19,550	20,650	21,770	19,010	29,430	19,550	15,510	51,900
13	63,40	86,500	112,000	92,	19,010	19,550	20,650	19,010	21,620	19,550	31,500	38,500
14	116,	75,500	132,300	120,800	17,960	19,550	17,960	19,010	21,480	19,010	36,400	41,400
15	101,	93,400	151,700	112,	17,960	19,010	15,510	17,960	23,480	16,950	39,900	36,400
16	95,	61,700	138,400	103,	16,950	17,960	14,600	17,400	19,550	16,460	37,800	34,300
17	84,	66,000	118,300	97,	16,950	16,950	13,300	17,960	19,750	14,180	32,900	32,900
18	68,400	216,000	99,300	93,400	16,460	16,460	12,100	16,950	19,010	13,300	36,400	31,500
19	62,500	284,500	83,800	83,800	15,510	16,460	12,500	15,510	19,010	12,100	49,400	45,400
20	53,500	302,800	71,000	82,400	15,510	16,460	14,600	16,950	17,960	12,500	47,800	51,900
21	49,400	197,800	66,600	74,000	13,720	27,510	14,160	15,510	15,510	14,600	46,200	47,800
22	43,800	169,800	74,600	68,000	13,300	30,800	14,600	14,160	14,000	17,960	39,900	38,500
23	88,300	165,300	80,100	60,000	14,600	68,400	13,300	13,300	14,160	28,330	38,500	34,300
24	123,400	142,800	105,200	51,300	15,510	71,000	13,300	24,620	16,160	46,200	37,800	35,000
25	120,800	120,800	144,500	51,900	16,460	61,700	14,400	77,300	13,300	38,500	58,400	45,400
						400	36,400	79,200	12,700	31,500	75,500	74,600
27	110,300	162,000	140,100	46,200	15,510	41,400	34,300	55,100	12,100	25,190	71,000	107,800
28	95,100	114,000	119,000	39,900	15,050	32,900	27,510	62,500	11,310	23,480	58,400	126,900
29	89,500		101,000	37,800	14,600	29,430	23,480	68,400	11,310	21,770	51,900	113,800
30	88,300		95,100	34,300	14,180	29,430	20,650	57,700	11,310	19,550	46,200	101,000
31	135,800		97,600		13,300		35,000	54,300		19,010		112,000
1892.												
1	112,000	21,770	43,800	134,000	23,480	68,400	46,200	12,500	22,340	7,090	4,070	12,500
2	107,800	22,340	36,400	120,800	21,770	60,000	41,400	13,300	7,940	7,090	4,070	12,100
3	116,400	22,340	30,800	112,000	21,770	54,300	32,900	12,100	16,460	7,090	4,070	11,310
4	120,800	24,620	26,330	109,800	21,770	46,200	13,300	14,600	8,850	7,090	4,070	10,200
5	118,300	24,620	23,480	218,400	43,800	183,000	29,430	23,480	13,300	7,090	4,070	10,200
6	108,500	23,480	19,550	224,200	65,800	171,500	30,800	21,770	12,100	7,090	4,070	9,520
7	101,000	23,480	21,770	116,800	96,700	101,000	28,130	21,770	12,100	6,550	4,070	9,520
8	88,300	22,340	21,770	102,000	96,700	120,800	28,130	23,480	11,310	6,550	4,070	9,520
9	57,500	30,050	34,300	120,600	101,000	97,600	28,130	19,550	10,500	6,550	5,210	10,200
10	69,500	17,960	55,500	101,000	80,100	86,500	23,480	29,430	9,520	6,550	6,020	10,500
11	36,500	19,010	71,000	80,500	61,700	98,400	21,770	14,600	9,520	6,020	7,350	16,950
12	31,500	17,960	66,000	75,500	51,900	86,500	17,960	16,950	8,850	6,020	7,350	39,900
13	32,900	13,300	62,500	62,500	47,800	75,500	14,600	16,950	8,850	5,700	7,350	36,400
14	60,000	11,700	51,900	57,500	39,900	58,400	14,600	16,950	8,850	5,700	7,350	29,430
15	171,700	11,310	42,500	47,800	38,500	46,200	16,460	29,430	16,460	5,700	7,940	24,620
16	195,800	12,100	36,400	47,800	38,500	38,500	16,950	38,500	11,160	5,700	21,770	21,770
17	153,500	10,580	29,430	41,400	42,200	32,900	16,950	36,100	11,160	5,700	7,940	22,340
18	122,500	11,310	27,510	41,400	49,400	90,800	15,510	29,430	12,500	5,700	7,940	19,550
19	99,200	13,300	24,620	36,400	50,200	29,430	15,510	21,770	10,500	5,780	12,500	19,010
20	97,600	16,460	23,480	34,300	62,500	20,130	14,160	29,430	12,100	5,760	17,960	17,960
21	86,500	14,000	22,340	31,500	90,800	31,500	13,300	16,460	9,520	5,760	22,340	16,950
22	71,000	17,960	19,550	29,430	107,800	26,330	11,310	14,160	9,520	5,780	22,340	14,160
23	57,500	19,550	17,960	28,130	118,300	31,500	10,580	12,500	8,510	5,760	30,800	9,520
24	47,800	25,190	17,960	24,620	116,400	23,480	10,580	12,100	7,350	5,240	27,510	6,020
25	43,800	29,430	19,550	29,430	107,800	31,500	10,580	12,500	7,350	5,240	22,340	7,090
26	41,400	41,400	29,430	92,600	38,500		7,940		4,500	17,960	19,010	
27	30,800	43,800	43,800	30,800	80,100	30,800	9,520	13,300	7,090	4,500	14,160	13,300
28	17,960	49,400	153,500	30,800	77,300	26,330	9,520	13,300	7,090	4,500	13,300	15,510
29	14,160	46,200	186,000	27,510	74,600	29,430	9,520	13,300	7,090	4,500	13,300	15,510
30	21,770		174,500	25,190	88,300	49,400	8,850	15,510	7,090	4,700	12,500	15,510
31	21,770		149,000		75,500		10,580	23,180		4,070		14,000

Mean daily discharge, in second-feet, of Susquehanna River at Harrisburg, Pa., 1891-1904—Continued.

Day.	Jan.	Feb.	Mar.	Apr.	May.	June	July	Aug	Sept	Oct	Nov	Dec
1893.												
1	13,300	19,550	19,010	70,100	50,200	31,500	16,460	6,020	30,800	13,300	14,600	36,400
2	17,980	23,480	19,010	68,400	49,400	31,500	14,600	5,760	38,540	13,300	14,600	34,300
3	21,770	36,400	20,650	75,500	60,000	29,430	14,160	5,760	35,000	12,100	14,600	31,500
4	21,770	38,540	20,650	95,100	83,800	30,800	12,500	5,760	29,430	10,560	14,600	31,500
5	20,650	51,900	20,650	101,800	258,400	30,800	12,500	5,240	19,550	9,520	16,460	31,500
6	19,550	53,500	17,980	119,000	267,400	25,190	10,560	5,240	15,510	9,520	23,480	29,430
7	17,980	51,900	17,980	129,600	223,200	23,480	10,560	4,740	13,300	8,850	26,330	25,190
8	17,960	57,500	19,550	118,200	174,500	23,480	10,200	4,740	11,310	8,850	21,770	23,480
9	17,960	58,400	24,620	108,500	136,600	23,480	9,520	4,500	10,560	8,850	20,650	23,480
10	17,960	75,500	77,300	110,300	107,800	21,770	9,520	4,500	9,520	8,540	17,980	22,340
11	15,510	99,200	183,000	138,400	86,500	19,550	9,520	4,070	9,520	8,540	17,980	21,770
12	15,510	167,100	290,200	127,800	71,000	19,010	9,520	4,070	10,560	8,540	16,950	21,770
13	14,160	95,100	221,300	110,300	60,000	17,980	9,520	3,680	13,300	7,940	16,460	21,770
14	14,160	77,300	223,300	99,200	51,500	16,460	9,520	3,680	13,300	10,560	14,600	17,980
15	14,160	61,700	193,000	93,400	47,800	14,160	11,310	3,680	12,100	46,200	14,160	13,300
16	13,300	55,900	178,900	105,200	45,400	13,300	12,100	3,500	13,300	57,500	13,300	15,510
17	13,300	99,200	147,200	118,200	66,600	12,500	12,100	3,500	17,980	55,900	12,500	16,950
18	13,300	82,000	118,200	119,000	112,600	12,100	10,560	3,500	19,550	39,900	12,100	64,100
19	13,300	65,800	92,600	99,200	134,000	11,310	10,560	3,500	42,200	34,300	11,310	118,200
20	13,300	57,500	80,100	84,700	120,800	11,310	10,560	4,740	31,500	28,130	11,310	88,300
21	13,300	46,200	66,600	86,500	96,700	11,310	10,560	4,500	26,330	23,480	10,560	68,400
22	13,300	39,900	61,700	138,400	86,500	10,200	9,520	4,070	21,770	17,980	10,200	66,600
23	13,300	29,430	62,500	154,400	72,800	10,200	8,850	3,680	17,980	17,080	10,200	42,200
24	13,300	23,480	83,800	147,200	61,700	11,310	8,540	3,680	16,460	16,460	10,560	35,000
25	13,300	23,480	90,800	119,000	58,400	11,310	7,940	3,500	16,460	15,510	10,560	34,300
26	13,300	23,480	99,200	97,600	50,200	13,300	7,370	3,680	14,600	15,510	10,200	34,300
27	13,300	22,340	127,800	83,800	43,800	15,510	7,080	4,070	18,300	15,510	10,200	49,400
28	13,300	20,650	114,600	71,000	41,400	17,980	7,080	4,070	13,300	18,300	11,310	66,600
29	18,300		101,000	62,500	38,500	20,650	12,100	6,550	13,300	13,300	21,770	65,800
30	16,460		101,000	54,300	35,000	17,980		6,020	23,480	13,300	31,500	54,300
31	17,980		77,300		31,500		6,020	24,620		14,600		46,200
1894.												
1	43,800	16,950	25,190	34,300	45,400	129,600	19,010	7,090	3,500	12,500	53,500	16,950
2	43,800	16,460	27,510	31,500	43,800	132,300	16,950	7,090	3,500	12,100	55,900	16,460
3	36,400	15,510	29,430	29,430	34,500	123,400	16,460	8,540	3,500	10,200	58,400	17,980
4	31,500	14,600	32,900	26,330	34,300	113,800	15,510	9,520	3,500	10,200	95,100	22,340
5	29,430	14,160	37,800	25,190	29,430	101,200	13,300	10,200	3,160	8,850	97,600	30,800
6	27,510	13,300	62,500	23,480	25,190	101,800	13,300	10,200	3,500	8,850	96,700	30,800
7	23,130	13,300	97,600	22,340	26,330	82,000	12,100	9,520	3,500	8,540	86,500	30,800
8	54,300	13,000	102,600	21,770	27,510	68,400	12,100	9,520	3,500	8,540	86,500	27,510
9	55,900	14,160	177,100	20,650	29,430	60,000	12,100	7,080	3,680	7,940	77,300	23,480
10	45,400	14,160	154,500	20,650	29,430	51,900	10,560	7,080	6,550	8,540	68,400	23,480
11	32,900	51,900	112,600	21,770	29,430	46,200	10,200	7,090	12,500	14,160	60,000	37,800
12	27,510	68,400	135,800	23,480	24,620	36,400	9,520	9,520	9,520	50,200	57,500	36,400
13	17,000	62,500	89,200	26,330	22,340	32,900	8,850	6,550	8,540	61,700	61,200	64,100
14	25,190	45,400	86,500	31,500	20,650	31,500	8,850	6,550	7,940	53,500	43,800	64,100
15	25,190	41,400	75,500	74,600	17,180	31,500	8,540	6,550	7,370	46,200	36,400	77,300
16	21,770	31,500	65,800	86,700	17,180	30,800	8,540	6,550	7,090	38,500	35,000	74,600
17	19,550	27,510	60,000	122,500	16,460	29,430	7,940	6,550	7,090	34,300	31,500	64,100
18	21,770	27,510	58,500	122,500	16,460	25,190	7,370	6,550	7,090	31,500	29,430	54,300
19	21,770	27,510	49,400	112,600	16,460	23,180	7,090	6,020	14,600	28,200	26,330	46,200
20	23,480	38,500	45,400	95,100	57,500	20,430	7,090	6,020	37,800	24,400	25,190	41,400
21	21,770	62,500	43,800	82,000	263,770	25,130	7,090	5,760	60,000	17,780	26,330	34,300
22	21,770	57,500	41,400	112,000	543,500	24,620	7,090	5,760	51,900	20,610	24,620	37,800
23	19,010	54,300	43,800	127,800	405,100	21,770	6,550	5,240	62,500	16,460	25,190	30,800
24	16,950	41,400	46,200	131,400	236,400	21,770	7,090	5,240	49,400	14,600	23,480	29,430
25	16,950	27,510	60,000	136,600	171,740	17,980	7,940	5,240	38,400	16,460	23,480	27,510
26	16,950	22,340	86,500	129,600	102,600	19,010	8,850	5,240	28,130	30,800	21,770	24,620
27	16,950	16,460	74,600	90,800	168,000	19,010	9,520	4,740	23,480	47,800	19,550	23,480
28	17,080	17,980	60,000	68,400	129,600	19,550	9,520	4,740	19,010	49,400	19,010	23,420
29	19,010		50,200	58,400	101,800	16,950	8,850	4,500	15,510	41,400	19,010	36,400
30	19,010		41,400	51,900	86,500	20,650	7,370	4,070	14,160	36,400	17,980	31,500
31	17,980		36,400		95,100		7,090	3,680		32,900		31,500

Mean daily discharge, in second-feet, of Susquehanna River at Harrisburg, Pa., 1891–1904—Continued.

Day.	Jan.	Feb.	Mar.	Apr.	May.	June.	July.	Aug.	Sept.	Oct.	Nov.	Dec.
1895.												
1	35,000	22,340	68,400	64,100	28,130	19,550	21,770	4,500	5,240	3,680	3,000	24,620
2	36,400	21,770	113,800	62,500	27,510	19,010	19,550	4,740	5,240	3,680	3,000	24,620
3	39,900	23,480	105,200	71,000	26,330	17,980	22,340	4,740	4,740	3,500	3,160	20,650
4	41,400	23,480	147,200	83,800	23,480	15,510	17,980	4,740	4,740	3,500	3,160	17,980
5	41,400	85,500	101,000	80,100	20,650	14,160	15,510	4,500	4,500	3,500	3,500	15,510
6	41,400	62,500	97,600	71,000	19,550	12,500	13,300	4,070	4,500	3,500	3,680	18,300
7	41,400	84,100	80,100	68,100	17,980	12,100	12,500	4,070	5,240	3,500	3,680	12,500
8	43,800	82,500	72,800	64,160	16,950	11,310	11,310	5,760	5,240	3,160	3,680	12,500
9	47,800	60,000	65,800	105,200	15,510	11,310	10,390	5,240	4,740	3,160	3,680	12,500
10	71,000	60,000	71,000	174,500	20,650	10,390	9,520	6,550	4,070	3,000	3,680	12,100
11	83,400	61,700	71,000	205,400	23,480	8,540	9,520	7,090	6,550	3,000	3,680	9,520
12	101,000	96,600	74,690	183,000	27,510	8,850	8,850	7,090	9,520	3,000	3,870	9,520
13	112,000	65,800	71,000	154,400	31,500	8,540	8,540	7,090	10,200	3,500	4,070	6,280
14	101,000	85,800	68,400	129,600	41,400	7,940	8,540	6,020	8,850	3,330	4,500	5,240
15	82,000	62,500	77,300	138,400	41,400	7,940	7,940	8,540	6,550	3,330	4,500	6,550
16	72,800	61,700	82,000	131,000	38,500	7,940	7,940	8,540	5,760	3,160	4,500	6,550
17	64,100	80,000	80,100	116,400	37,800	7,940	7,090	7,000	4,740	3,160	4,740	8,540
18	58,400	60,000	74,690	96,700	31,500	7,940	6,550	6,550	4,500	3,680	5,760	8,540
19	51,900	57,500	62,500	80,100	29,430	7,940	6,020	6,550	4,740	4,500	6,550	8,540
20	48,800	55,900	60,000	68,400	27,510	7,940	6,020	6,550	4,740	4,070	6,550	8,540
21	42,200	54,300	57,500	60,000	25,190	7,370	5,760	5,760	4,740	3,680	6,020	9,520
22	41,400	53,500	54,300	51,900	24,620	6,550	5,760	4,500	3,680	5,500	12,100
23	36,400	51,900	51,900	45,400	22,340	5,240	5,760	4,070	4,500	3,500	4,740	13,300
24	36,400	50,200	51,900	41,400	20,650	5,240	5,760	4,070	4,500	3,160	5,240	19,550
25	27,510	47,800	51,900	36,400	19,010	5,240	5,760	3,680	4,500	3,160	5,240	20,650
26	26,330	45,400	65,800	32,900	18,500	5,760	5,760	3,500	4,070	3,000	5,240	21,770
27	21,620	43,800	103,500	32,900	17,980	9,520	5,760	3,500	4,070	2,850	5,240	27,510
28	24,620	47,800	129,600	32,900	16,950	8,850	5,760	3,500	3,680	2,710	19,550	29,430
29	24,620		103,500	32,900	16,950	13,300	5,240	3,500	3,680	2,710	21,770	
30	26,330		89,300	20,430	24,620	20,430	4,500	3,500	3,680	2,570	21,770	62,500
31	23,480		74,000		23,480		3,680	4,070		2,570		62,500
1896.												
1	136,600	41,800	80,300	223,200	23,480	19,520	19,550	46,200	3,500	58,400	14,180	35,000
2	124,400	32,900	123,100	223,200	23,480	19,520	16,160	41,400	3,500	39,900	12,500	35,000
3	110,300	30,800	131,000	297,200	21,770	17,160	14,160	34,300	3,500	36,400	12,100	34,300
4	77,300	30,800	110,300	180,800	21,770	12,100	12,100	32,900	3,500	25,190	12,100	27,510
5	53,500	29,430	80,300	147,200	19,550	10,560	11,310	31,500	3,160	19,550	12,100	23,480
6	36,400	36,100	72,000	118,200	17,980	10,560	10,560	30,800	3,160	14,160	90,800	20,650
7	31,500	105,300	51,900	80,300	16,950	10,560	14,160	17,980	3,160	12,100	140,100	19,550
8	23,480	183,600	47,800	77,300	14,160	10,200	13,300	16,400	3,160	10,560	99,200	19,550
9	46,200	114,500	43,800	71,000	14,160	8,550	12,500	16,460	3,160	9,520	77,300	17,980
10	41,400	112,000	49,400	65,800	13,300	11,310	16,460	15,510	3,160	9,520	62,500	17,980
11	37,800	83,800	53,500	60,000	13,300	17,980	20,650	15,510	3,160	9,520	47,800	28,130
12	36,400	57,500	46,200	60,000	12,500	19,010	20,650	13,300	3,160	9,520	42,200	32,900
13	35,000	50,200	36,100	60,400	11,310	28,130	17,980	12,100	3,160	12,500	38,500	36,400
14	36,400	36,100	29,430	75,500	10,560	26,330	14,160	10,560	3,500	92,600	36,400	36,400
15	34,300	32,900	19,550	161,500	10,560	23,480	18,300	10,560	3,500	86,500	34,300	34,300
16	34,300	32,900	19,550	110,300	11,310	19,010	12,100	10,200	3,500	129,600	31,500	31,500
17	32,900	34,300	16,460	106,600	11,310	19,010	10,560	10,200	4,070	97,600	29,480	28,130
18	30,800	30,800	17,980	92,600	9,520	21,770	10,200	10,200	4,070	61,700	27,510	24,620
19	31,500	22,340	25,190	83,800	9,520	19,550	10,560	10,560	4,500	49,400	25,190	22,340
20	36,100	23,480	86,400	71,000	9,520	23,480	10,560	8,540	4,500	37,800	23,480	19,010
21	31,500	16,460	68,400	64,100	9,520	25,190	12,500	6,550	4,740	30,800	21,770	16,460
22	29,130	31,500	64,100	55,900	8,850	23,480	10,560	6,550	5,760	28,130	19,550	13,300
23	29,430	58,400	64,100	49,400	8,850	16,950	10,200	5,760	7,370	26,330	19,010	13,300
24	29,430	58,400	72,800	45,400	8,850	16,460	10,200	5,760	7,370	26,330	17,980	9,520
25	36,400	28,130	41,700	41,400	8,540	15,510	10,560	5,760	6,020	23,480	17,980	9,520
26	30,800	29,430	51,900	37,800	19,550	11,310	5,240	23,480	16,460	9,520
27	92,600	31,500	55,900	36,400	47,800	12,500	5,240	4,500	20,650	16,460	9,520
28	71,000	25,190	55,900	36,400	46,100	12,500	5,240	4,500	20,650	16,460	8,540
29	68,400	25,190	77,300	28,130	26,330	17,980	4,500	3,680	17,980	19,550	10,200
30	64,100		125,200	26,330	24,620	32,900	4,070	6,760	16,950	29,430	10,200
31	58,400		183,600		41,400	3,500		15,510		11,310

Mean daily discharge, in second-feet, of Susquehanna River at Harrisburg, Pa., 1891–1904—Continued.

Day.	Jan.	Feb.	Mar.	Apr.	May.	June.	July.	Aug.	Sept.	Oct.	Nov.	Dec.
1897.												
1	12,100	27,510	89,900	51,900	24,620	22,340	8,850	36,400	7,940	11,310	4,740	51,900
2	13,800	25,190	31,500	46,200	24,620	21,770	8,540	41,400	7,080	9,520	4,870	43,800
3	13,800	25,190	26,390	41,400	60,000	19,550	7,940	34,300	6,550	8,540	24,620	36,400
4	14,160	25,190	34,300	38,500	77,300	19,010	7,940	26,330	6,550	7,370	37,800	32,900
5	17,980	24,620	50,200	36,400	95,100	19,550	7,940	21,770	6,550	7,080	29,430	27,510
6	23,480	23,480	66,600	34,300	88,300	23,480	7,940	19,550	6,020	6,550	24,620	47,800
7	31,500	39,900	97,600	32,900	86,500	19,550	8,850	16,950	5,780	6,550	23,480	54,300
8	31,500	95,100	118,800	32,900	74,600	17,980	8,850	19,550	5,780	6,020	20,650	53,500
9	31,500	79,200	108,500	32,900	60,000	19,550	7,940	17,980	5,780	5,760	17,980	58,400
10	27,510	58,400	84,700	66,800	49,400	19,550	7,940	14,160	4,740	4,740	16,950	50,200
11	24,620	49,400	77,300	120,800	43,800	19,550	7,370	14,160	4,500	4,740	19,550	41,400
12	21,770	43,800	90,800	129,600	36,400	19,550	7,090	13,300	4,740	4,740	19,550	84,500
13	16,950	35,000	114,600	108,500	36,400	24,620	6,550	12,100	4,740	5,240	17,980	88,500
14	13,300	34,300	110,300	83,800	68,400	29,430	7,090	11,310	4,740	5,240	17,980	41,400
15	13,300	34,300	99,200	68,400	99,200	26,330	6,550	10,200	4,070	5,240	17,980	45,400
16	13,300	29,430	86,500	68,400	101,800	22,340	6,550	10,200	4,500	5,240	17,980	79,200
17	13,300	29,430	84,700	79,200	92,600	19,550	7,370	9,520	4,740	4,740	17,980	97,600
18	14,600	27,510	60,000	86,500	77,300	17,980	7,090	9,520	5,240	4,740	19,550	108,000
19	16,480	30,800	51,900	79,200	64,100	15,510	7,090	8,850	5,240	4,500	22,340	92,500
20	14,900	37,800	57,500	68,400	51,900	14,600	7,090	8,850	4,740	4,500	28,130	74,600
21	12,100	36,400	93,400	60,000	39,900	14,600	9,520	8,540	4,500	4,070	26,330	61,700
22	12,100	39,900	107,800	50,200	39,900	14,600	9,520	7,370	4,500	4,500	25,190	51,900
23	12,500	66,600	134,000	43,800	30,800	13,300	8,540	7,370	4,500	5,240	21,770	37,800
24	10,580	101,800	129,600	38,500	29,430	12,100	8,850	7,940	6,550	5,240	17,980	34,300
25	10,580	95,100	141,000	34,300	32,900	11,310	10,200	10,580	9,520	6,550	17,980	28,130
26	9,520	77,300	165,800	31,500	32,900	11,310	11,310	19,550	9,520	6,550	17,980	21,770
27	27,510	60,000	149,900	30,800	29,430	10,580	11,310	14,160	12,100	6,550	16,460	20,650
28	27,510	43,800	108,500	29,430	30,800	10,200	14,600	11,310	12,500	6,020	17,980	19,550
29	23,480	......	93,400	27,510	35,000	10,200	34,300	10,200	15,510	5,780	29,430	19,560
30	25,330	......	74,600	25,190	29,430	9,520	43,800	9,520	13,300	5,240	50,200	19,010
31	27,510	......	61,700	......	26,330	......	37,800	8,540	......	5,240	......	17,980
1898.												
1	19,550	35,000	84,200	114,600	68,400	41,400	13,300	8,850	19,550	5,240	46,200	24,620
2	16,460	28,130	41,400	83,400	58,400	38,500	14,600	9,520	16,460	5,240	36,400	25,190
3	14,620	23,480	36,500	75,500	49,400	35,000	13,300	8,850	22,480	4,740	31,500	24,620
4	19,550	19,550	35,000	64,100	46,200	30,800	11,310	16,480	17,980	4,740	29,430	23,480
5	12,500	19,550	31,500	58,400	42,300	27,510	10,580	45,400	14,160	4,740	25,190	31,500
6	12,500	19,550	30,800	50,300	43,000	23,480	10,200	57,500	12,500	4,740	23,480	51,900
7	15,510	19,550	29,430	43,800	46,200	21,770	9,520	36,400	10,580	4,740	22,680	47,800
8	17,980	24,620	29,430	42,300	60,000	19,550	8,850	29,430	10,580	6,550	17,980	37,800
9	19,550	28,130	27,510	34,300	72,800	17,980	8,540	24,620	13,300	8,540	17,980	34,300
10	20,650	41,400	27,510	34,300	61,700	17,980	7,940	31,500	13,300	8,850	19,980	30,800
11	23,480	28,130	34,300	31,500	54,300	16,460	7,370	39,900	21,770	15,510	19,010	24,620
12	23,480	32,900	50,200	29,430	47,800	16,460	7,090	32,900	20,650	16,950	36,400	17,980
13	27,510	42,300	77,300	27,510	43,800	15,510	6,020	19,550	14,160	18,300	103,500	15,510
14	36,400	97,600	114,600	26,330	36,400	15,510	5,720	17,980	12,500	18,300	79,200	14,160
15	85,600	101,800	155,800	25,190	36,400	16,950	5,780	15,510	11,310	14,160	60,000	13,300
16	105,200	95,100	129,600	31,500	39,900	20,650	5,760	15,510	13,300	8,850	49,400	13,300
17	101,000	77,300	106,200	37,800	54,300	23,480	4,740	12,500	8,540	26,330	41,400	12,500
18	96,700	65,800	89,200	35,000	70,100	23,480	4,740	16,480	7,370	32,900	38,500	13,300
19	79,200	51,900	74,600	31,500	57,500	19,550	5,240	23,480	6,550	36,400	38,500	17,980
20	65,800	41,400	65,800	29,430	60,000	16,950	6,020	42,300	6,020	41,400	39,900	22,340
21	64,100	46,200	62,900	28,130	80,100	16,460	5,240	41,400	6,020	92,600	45,400	24,620
22	71,000	83,800	125,200	27,510	80,100	16,460	6,020	32,900	6,020	92,600	41,400	29,430
23	93,400	84,700	154,400	25,190	77,300	14,160	5,780	28,130	5,780	109,400	46,200	58,400
24	125,200	99,200	245,900	23,480	68,400	13,300	5,780	23,480	5,780	93,400	41,400	101,000
25	147,300	80,100	236,600	29,430	86,500	14,600	5,780	19,550	5,240	71,000	36,400	97,600
26	129,600	72,800	168,400	30,100	77,300	14,160	8,540	17,980	6,020	62,500	35,000	57,500
27	103,500	62,500	125,200	144,500	77,300	13,300	8,540	16,950	6,020	61,700	31,500	57,500
28	86,500	51,900	90,200	129,000	71,000	12,500	7,370	38,500	5,240	62,500	29,430	49,400
29	70,100	......	80,100	106,000	64,100	12,100	12,100	38,500	5,240	70,100	27,510	41,400
30	60,000	......	80,500	80,100	57,500	10,580	10,200	34,300	......	57,500	......	84,300
31	49,400	......	120,800	......	50,200	......	8,540	23,480	......	57,500	......	

*Mean daily discharge, in second-feet, of Susquehanna River at Harrisburg, Pa.,
1891–1904—Continued.*

Day.	Jan.	Feb.	Mar.	Apr.	May	June	July.	Aug.	Sept.	Oct.	Nov	Dec.
1899.												
1	26,350	17,980	110,300	90,800	28,130	17,980	11,310	5,240	12,160	7,000	4,070	11,310
2	25,190	13,310	106,000	75,500	24,620	19,010	10,560	5,240	9,520	5,760	10,560	10,230
3	20,650	12,560	101,000	65,800	24,620	17,980	10,560	5,240	7,910	5,760	17,980	9,520
4	26,330	15,510	93,400	57,500	28,130	17,980	9,520	5,240	7,090	5,240	26,330	9,520
5	29,430	19,010	103,500	50,200	25,190	17,980	8,540	5,240	7,090	4,740	43,800	9,520
6	51,900	19,550	183,600	42,300	25,190	16,460	7,910	6,020	6,550	4,740	35,000	9,520
7	103,500	21,770	193,000	39,900	23,480	14,160	7,910	5,240	6,020	4,500	32,900	9,520
8	83,800	16,950	163,500	47,400	20,650	12,500	7,370	5,240	6,020	4,500	25,190	9,520
9	70,100	17,980	125,200	83,800	21,770	12,500	7,370	5,760	5,760	4,500	21,770	9,520
10	58,400	16,950	97,400	116,100	19,550	12,500	7,370	5,240	6,550	4,740	17,980	9,520
11	45,400	16,950	77,300	110,300	20,650	11,310	8,850	4,740	6,550	4,500	15,510	9,520
12	36,400	42,300	64,100	99,200	20,650	10,560	7,910	4,740	5,240	4,500	14,160	9,520
13	27,510	42,300	64,100	82,000	22,340	10,560	7,370	7,090	5,760	4,070	14,160	20,650
14	25,190	45,400	64,100	82,000	21,770	10,230	7,370	7,090	8,850	4,070	13,300	60,000
15	27,510	45,400	110,300	103,500	19,010	9,520	7,370	7,940	7,910	4,070	15,510	74,400
16	31,500	46,200	103,500	103,500	17,980	9,520	7,180	6,020	5,760	3,680	16,950	68,400
17	49,400	49,400	93,400	101,000	17,980	8,850	6,550	4,740	5,240	3,680	16,950	57,500
18	80,500	49,400	75,500	92,600	19,010	7,910	7,180	4,740	5,240	3,680	16,950	45,400
19	74,600	50,200	41,400	83,800	32,900	7,910	7,940	4,070	4,500	3,680	21,770	37,800
20	62,500	47,800	80,200	68,400	47,800	7,910	7,910	4,070	4,740	3,500	23,480	32,900
21	50,200	50,200	112,000	58,400	54,300	7,370	7,340	4,070	5,240	3,790	22,340	32,900
22	41,400	57,500	106,000	53,500	39,900	7,180	8,540	4,070	4,740	3,500	19,010	34,300
23	39,900	55,100	95,100	50,200	35,000	6,550	8,540	4,070	4,740	3,500	17,980	43,800
24	37,800	55,100	89,200	43,800	30,800	8,850	8,540	4,070	4,740	2,850	15,510	39,900
25	38,500	89,200	93,400	42,300	25,190	13,300	7,370	4,070	4,740	2,850	15,510	65,800
26	55,900	83,800	93,400	39,100	23,480	10,560	6,550	3,680	4,740	3,160	15,510	82,000
27	43,800	92,600	83,800	35,000	22,340	9,520	6,550	4,740	6,550	3,500	14,600	55,900
28	34,300	120,800	82,000	32,900	19,550	9,520	6,550	6,400	8,540	3,790	13,300	45,400
29	26,330		120,800	31,500	17,980	11,310	6,020	19,550	7,370	3,680	13,300	34,300
30	23,480			29,430	17,980		5,760	17,980	7,090	3,500	12,100	33,400
31	23,480				17,980		5,240	14,600		3,500		15,510
1900.												
1	12,100	22,340	36,400	38,500	36,400	19,010	7,370	7,910	6,550	2,570	5,760	88,500
2	10,560	12,100	134,000	36,400	32,900	17,980	7,090	6,550	6,550	2,570	5,760	65,800
3	43,800	35,000	180,800	38,500	29,430	16,460	6,550	6,550	5,760	2,570	5,240	55,900
4	50,200	36,400	129,000	42,300	27,510	14,600	7,090	6,020	7,370	2,570	5,240	43,800
5	49,400	46,200	101,800	57,500	24,620	17,980	8,540	5,240	6,020	2,570	5,240	61,800
6	55,900	41,400	84,700	68,400	21,770	19,550	7,370	4,740	5,760	2,570	4,740	90,800
7	60,000	70,100	68,400	58,400	21,770	17,980	8,540	4,740	4,500	2,570	4,740	93,400
8	57,500	51,900	71,000	53,500	20,650	14,600	7,370	4,500	4,500	2,710	4,740	88,300
9	50,200	36,400	77,300	71,000	17,980	14,600	8,850	4,070	4,500	2,570	5,240	68,400
10	45,400	49,400	65,800	82,000	17,980	14,160	8,850	4,500	4,070	2,570	4,500	55,900
11	43,800	64,100	62,500	77,300	16,950	13,300	8,540	4,070	3,160	2,570	4,740	47,800
12	60,000	60,000	72,800	61,700	16,460	13,300	7,370	3,790	3,500	2,570	4,070	37,800
13	50,200	62,500	64,100	51,900	16,950	12,500	7,090	3,500	2,850	5,760	4,740	34,300
14	55,900	97,600	46,200	43,800	16,950	12,500	6,550	2,850	3,160	5,760	5,240	30,800
15	55,900	103,500	43,800	41,400	17,980	13,300	6,550	2,850	3,160	5,760	4,740	22,340
16	55,900	107,800	36,400	43,800	16,950	14,600	6,550	2,850	3,160	5,240	4,740	21,770
17	46,200	93,400	31,500	37,200	16,460	14,600	6,550	3,160	3,160	4,500	5,760	15,510
18	51,900	68,400	25,190	41,100	16,460	13,300	7,090	2,850		4,740	6,020	14,160
19	49,400	47,800	21,480	70,100	15,510	12,100	6,020	2,850	2,710	4,740	5,240	14,160
20	36,400	35,000	21,480	88,300	17,980	12,100	6,020	2,850	2,710	4,500	6,020	14,160
21	39,900	11,400	35,000	92,600	22,340	11,710	5,760	3,500	2,710	4,070	6,020	13,300
22	149,900	99,800	83,800	81,800	14,600	11,310	5,240	3,680	2,570	4,070	6,020	14,600
23	174,500	129,600	89,200	70,100	21,770	11,310	5,240	5,760	2,570	4,070	5,760	16,950
24	123,400	150,000	68,400	65,800	19,010	10,230	5,240	4,070	2,570	4,070	6,550	14,600
25	90,800	134,000	64,100	51,900	16,950	8,850	5,240	7,910	2,570	6,550	7,090	16,460
26	70,100	81,800	65,800	72,800	15,510	8,540	5,760	6,550	2,410	7,090	10,560	16,950
27	51,900	60,000	64,100	51,900	14,600	8,540	9,520	7,370	2,410	6,550	61,400	13,300
28	43,800	43,800	55,900	53,500	13,300	8,540	7,910	6,020	2,570	5,210	194,000	19,550
29	37,800		49,400	45,400	13,300	8,540	7,910	8,540	2,850	7,370	198,800	22,340
30	27,510		43,800	38,500	13,300	7,370	8,850	6,550	2,570	6,550	119,000	19,010
31	17,980		42,300		12,500		7,910	7,090		6,020		17,980

Mean daily discharge, in second-feet, of Susquehanna River at Harrisburg, Pa., 1891–1904—Continued.

Day.	Jan.	Feb.	Mar.	Apr.	May.	June	July.	Aug.	Sept.	Oct.	Nov.	Dec.
1901.												
1	15,510	19,000	11,310	80,200	54,800	185,500	24,620	10,560	29,430	14,160	8,850	24,620
2	14,160	36,400	10,560	68,400	45,400	145,400	21,770	12,100	32,900	16,950	8,850	23,480
3	10,560	27,580	11,310	62,500	43,800	119,000	19,010	11,310	47,800	16,460	8,540	20,650
4	10,560	26,380	12,100	72,800	42,200	101,000	14,480	10,200	54,300	16,460	8,540	20,650
5	11,310	26,380	16,460	95,100	54,300	89,200	15,510	9,520	49,400	16,460	7,940	24,620
6	10,560	24,570	17,980	101,000	51,900	74,600	14,600	7,940	38,500	16,950	7,940	19,550
7	8,850	25,190	19,010	114,000	45,400	60,000	16,460	10,560	30,800	14,600	7,940	20,650
8	7,370	25,190	17,980	163,400	37,800	60,000	14,600	19,010	25,190	12,100	7,370	15,510
9	9,520	25,190	23,480	188,400	32,900	68,400	14,180	20,650	21,770	11,310	7,370	14,600
10	9,520	23,400	26,330	165,300	31,500	64,100	14,160	17,980	17,980	11,310	7,370	19,010
11	10,560	21,700	75,500	138,900	28,130	60,000	13,300	16,460	17,980	10,560	7,370	43,800
12	13,300	22,250	169,800	114,000	34,300	51,900	12,500	20,650	16,460	10,560	6,550	86,500
13	13,300	21,700	171,700	95,100	38,510	46,200	12,500	16,950	16,950	10,560	7,090	86,500
14	17,980	20,610	126,900	84,700	43,800	39,900	12,100	13,300	16,460	12,100	7,940	71,000
15	29,430	20,610	95,100	71,000	54,300	35,000	12,500	11,310	16,460	16,950	8,540	125,200
16	27,510	19,000	80,100	60,000	53,500	29,430	11,310	10,560	15,510	19,550	10,200	405,100
17	24,130	20,610	72,800	64,100	46,200	32,900	10,560	10,560	16,950	17,980	10,580	322,700
18	22,340	19,010	64,100	57,500	38,500	30,800	14,160	11,310	16,950	14,160	12,500	214,800
19	19,010	17,780	55,900	51,900	36,400	29,430	16,950	60,000	17,980	14,160	12,500	135,800
20	11,310	17,780	50,300	47,800	39,900	26,330	15,510	65,800	19,550	13,300	12,500	93,400
21	11,310	14,160	66,600	42,200	37,800	24,620	18,300	51,900	19,010	13,300	11,310	71,000
22	12,100	13,300	112,000	156,300	36,400	26,330	12,100	37,800	19,010	12,500	11,310	49,400
23	13,300	13,300	129,000	204,400	60,000	32,900	11,310	38,500	16,950	12,500	10,200	34,300
24	11,310	12,500	122,500	177,100	110,300	34,300	10,560	47,800	16,460	12,100	12,100	30,800
25	13,300	12,500	163,500	141,000	95,100	36,400	10,200	99,200	14,160	12,100	17,980	82,900
26	13,300	12,500	97,400	123,400	103,500	35,000	9,520	120,800	13,300	10,560	24,620	82,900
27	13,300	11,310	109,400	112,000	95,100	32,900	10,200	90,800	12,100	10,200	58,400	85,000
28	13,300	11,310	102,600	80,800	86,500	29,430	10,560	64,100	11,310	10,560	55,900	35,000
29	18,300		191,100	77,300	116,400	26,330	9,520	47,800	10,560	9,520	86,400	35,000
30	11,310		159,400	64,100	178,900	25,190	9,520	36,400	10,560		30,800	61,700
31	10,500		120,800		210,100		9,520	29,430		8,850		72,800
1902.												
1	55,900	30,800	372,800	72,800	20,650	11,310	30,800	65,800	7,940	49,400	60,000	16,950
2	47,800	31,500	484,100	61,700	21,770	11,310	71,000	57,500	7,940	68,400	47,800	16,950
3	30,800	29,430	465,300	57,500	21,770	10,560	92,000	60,000	7,940	66,600	43,800	19,010
4	34,300	26,330	465,100	51,100	19,550	10,560	80,100	72,800	7,940	62,500	36,400	27,510
5	23,480	16,950	253,600	47,400	19,550	101,000	69,000	7,370	46,200	29,430	26,330
6	23,480	13,300	178,900	43,400	21,770	10,560	95,100	49,400	7,090	46,200	29,430	26,330
7	23,480	70,100	129,600	43,400	20,650	9,520	83,800	43,800	6,550	46,200	26,330	29,430
8	21,770	55,900	86,500	43,400	19,550	7,940	92,600	36,400	6,020	42,200	24,620	28,130
9	29,650	51,900	53,900	120,500	19,550	9,520	112,000	30,800	6,020	34,300	22,340	28,130
10	23,480	53,500	51,100	224,300	19,550	10,200	89,200	26,330	6,020	34,300	20,650	25,190
11	22,340	57,500	80,100	214,300	17,980	9,520	71,000	30,800	6,020	29,430	19,550	23,480
12	19,550	54,300	100,400	167,100	17,980	9,520	71,000	30,800	7,940	30,800	16,950	23,480
13	19,010	49,400	154,400	154,000	16,950	9,520	60,000	24,620	7,940	47,800	16,950	34,300
14	15,510	42,200	210,600	105,100	16,460	9,520	60,000	24,620	7,090	49,400	16,460	31,500
15	15,510	12,290	234,400	88,100	15,510	11,310	45,400	21,770	7,370	32,900	16,460	36,400
16	15,510	39,900	174,500	75,500	14,600	11,310	36,400	20,650	7,090	35,000	15,510	36,400
17	14,600	37,800	177,100	62,400	14,600	11,310	36,400	20,650	7,090	32,900	14,600	57,500
18	13,300	34,300	231,000	53,500	13,300	16,950	36,330	17,980	6,550	25,190	14,600	113,800
19	13,300	32,900	205,400	47,400	12,100	16,950	34,300	14,600	6,550	27,510	14,600	109,400
20	14,600	32,900	162,600	42,200	12,100	16,460	25,190	13,300	6,550	23,480	12,500	97,600
21	14,600	32,900	129,000	37,800	11,310	14,600	27,510	13,300	6,550	22,340	12,100	89,200
22	54,300	36,400	68,400	34,300	11,310	14,600	41,400	12,500	6,020	19,550	11,310	112,000
23	188,400	39,400	69,000	28,130	12,100	14,600	105,200	12,500	5,780	19,010	11,310	183,600
24	82,000	37,800	57,500	28,130	10,560	13,300	163,500	11,310	5,780	16,950	10,560	186,400
25	77,300	34,300	57,500	28,130	10,560	13,300	90,800	11,310	5,780	15,510	10,560	185,300
26	58,400	75,500	48,200	25,190	10,560	13,300	99,200	10,200	10,560	16,950	10,560	107,800
27	53,500	127,700	31,500	21,770	10,560	16,950	83,800	29,430	32,900	16,460	13,300	71,000
28	57,500	132,300	31,500	20,650	10,560	16,950	83,800	9,520	54,300	16,460	13,300	71,000
29	57,500			20,650	10,560	23,480	83,800	8,850	41,400	31,500	15,510	61,700
30	41,400			20,650	10,560	23,480	71,000	7,940	41,400	31,500	16,460	49,400
31	85,000		57,500		11,310		71,000	7,940		68,400	16,950	45,400

*Mean daily discharge, in second-feet, of Susquehanna River at Harrisburg, Pa.,
1891–1904—Continued.*

Day.	Jan.	Feb.	Mar.	Apr.	May	June	July	Aug.	Sept.	Oct.	Nov.	Dec.
1903.												
1	38,500	165,300	290,600	77,300	28,130	8,190	79,640	25,310	123,500	9,730	21,690	15,450
2	31,500	147,200	276,500	90,800	26,330	8,190	58,820	23,660	94,080	9,730	21,170	14,580
3	34,800	116,400	221,300	95,100	23,480	8,190	46,700	19,210	72,070	9,730	20,190	14,160
4	49,400	119,000	156,300	82,000	19,010	8,190	39,730	16,810	53,750	8,770	18,720	12,560
5	61,700	209,200	120,800	77,300	17,960	8,190	32,510	14,580	46,700	8,770	18,720	12,180
6	66,600	223,200	116,400	64,100	16,460	8,190	36,290	15,450	39,040	8,190	16,810	11,440
7	74,600	178,900	97,600	64,100	16,460	8,190	39,730	19,210	31,300	7,610	16,350	11,440
8	65,800	128,900	105,000	64,100	15,510	8,190	49,450	41,110	23,660	10,410	15,450	11,440
9	51,900	107,800	103,500	64,100	14,600	9,080	49,450	39,730	27,000	16,810	15,450	14,160
10	41,400	86,500	149,000	83,800	14,600	13,340	35,000	32,510	24,500	44,630	14,580	12,560
11	28,130	68,400	183,600	86,500	14,600	13,310	29,500	27,060	25,310	128,900	14,580	10,750
12	22,340	71,000	183,500	83,800	14,600	13,340	21,690	25,310	25,310	138,300	14,580	10,750
13	19,650	77,300	172,600	77,300	14,160	21,660	22,640	25,310	25,310	136,000	14,580	10,750
14	15,510	80,100	153,500	77,300	14,160	27,000	22,640	25,310	23,660	107,700	14,160	11,440
15	15,510	95,100	131,000	118,200	14,160	32,510	17,760	20,190	25,310	79,640	14,160	5,630
16	19,550	97,600	109,400	186,400	14,160	35,600	17,760	21,660	57,290	14,160	5,630
17	23,480	97,600	101,000	188,400	14,160	36,290	16,350	25,310	18,720	46,700	14,160	5,630
18	25,190	86,500	89,200	149,900	12,310	34,310	14,160	23,660	18,720	42,480	15,450	7,340
19	25,190	68,400	77,300	126,900	12,100	29,500	21,170	21,660	21,660	49,450	18,560	21,690
20	25,190	55,900	77,300	103,500	11,310	24,190	37,660	18,720	23,660	66,480	32,710	53,750
21	25,190	37,800	60,000	77,300	11,310	23,660	53,750	16,350	20,190	68,110	66,480	53,750
22	26,330	43,800	62,500	71,600	10,560	23,660	59,220	15,450	18,720	61,060	61,060	53,060
23	38,500	43,800	68,400	65,800	10,560	27,000	35,000	14,160	16,810	51,800	59,730	53,060
24	36,400	41,400	127,800	62,500	10,560	30,100	30,100	14,580	15,450	42,480	35,600	39,040
25	35,000	38,500	234,300	55,900	10,560	53,000	26,490	14,160	14,580	36,290	31,300	36,290
26	29,430	37,800	214,800	45,400	10,560	66,480	21,690	14,160	27,000	28,290	31,300
27	29,430	45,400	156,300	43,800	10,560	76,710	20,190	12,560	28,290	23,660	25,310
28	31,800	43,800	131,400	36,400	10,560	66,480	20,190	13,340	12,180	27,000	15,450	21,170
29	32,900	106,000	20,430	10,290	58,820	18,720	33,110	10,410	25,310	15,450	19,210
30	46,200	83,800	20,430	10,290	51,000	20,190	57,290	10,410	23,660	15,450	16,810
31	105,200	83,800	9,520	23,660	107,670	21,660	12,180
1904.												
1	(a)	(a)	(c)	75,500	97,600	31,500	12,500	10,090	9,048	11,540	13,980	11,620
2	(a)	(a)	(c)	141,000	80,100	35,000	11,160	10,780	8,120	10,780	13,140	9,782
3	(a)	(a)	(c)	194,300	75,500	39,000	13,140	12,740	7,824	9,724	12,340	9,114
4	(a)	(a)	(c)	170,000	62,500	30,600	12,500	12,740	7,824	11,540	11,540	7,882
5	(a)	(a)	(c)	127,800	50,200	36,120	10,560	12,340	7,558	12,740	10,780	8,180
6	(a)	(a)	(c)	98,300	37,290	50,200	11,160	11,540	7,258	11,160	10,490	6,228
7	(a)	(a)	(c)	81,400	36,120	55,900	11,160	13,980	6,982	10,060	10,390	8,180
8	(a)	(a)	(c)	71,000	33,740	47,500	15,320	13,550	6,442	9,384	9,702	7,086
9	(a)	(a)	(c)	69,100	29,170	36,120	18,500	11,540	6,442	8,728	9,702	7,882
10	(a)	(a)	(c)	75,500	28,190	30,250	18,500	10,780	7,558	7,824	9,452	7,594
11	(a)	(a)	(c)	111,160	25,190	40,800	43,480	12,340	7,558	7,558	10,130	5,708
12	(a)	(a)	(c)	123,400	23,250	58,400	52,000	10,430	7,558	7,824	9,702	6,228
13	(a)	(a)	(c)	103,500	22,340	46,200	42,200	10,490	7,258	7,824	10,430	10,860
14	(a)	(a)	(c)	89,200	18,500	35,000	32,620	9,384	6,982	7,824	8,728	9,114
15	(a)	(a)	(c)	73,000	21,320	36,100	26,100	8,420	8,796	20,440	9,702	9,452
16	(a)	(a)	(c)	55,900	25,190	22,340	22,340	8,420	10,060	22,680	10,130	8,788
17	(a)	(a)	(c)	51,300	28,190	19,500	18,500	8,120	13,140	20,440	9,702	8,240
18	(a)	(a)	(c)	52,900	31,500	21,320	15,790	7,824	14,870	16,750	9,452	9,520
19	(a)	(a)	(c)	44,800	26,120	21,320	13,980	7,258	11,540	14,420	10,130	9,520
20	(a)	(a)	(c)	43,500	51,510	18,500	13,140	7,558	11,540	12,340	10,130	9,520
21	(a)	(a)	(c)	35,000	49,400	18,500	13,550	8,120	10,420	11,160	9,452	8,850
22	(a)	(a)	(c)	27,000	78,400	19,550	12,740	7,558	9,048	12,340	9,702	8,850
23	(a)	(a)	(c)	32,620	50,100	18,500	12,740	8,120	22,680	10,130	9,520
24	(a)	(a)	(c)	30,250	44,800	18,500	23,250	8,120	7,558	33,040	10,860	10,200
25	(a)	(a)	(c)	28,130	40,800	14,420	14,420	8,120	7,558	37,240	10,800	10,200
26	(a)	(a)	(c)	29,170	33,740	17,760	11,340	10,780	6,982	30,520	11,620	10,200
27	(a)	(a)	(c)	29,170	36,120	16,100	11,540	10,780	6,712	23,820	12,020	11,700
28	(a)	(a)	(c)	32,620	35,000	13,980	10,780	13,980	7,258	19,880	12,020	12,500
29	(a)	(a)	(c)	50,200	31,500	13,140	11,540	11,940	10,420	18,270	11,220	14,160
30	(a)	(a)	(c)	86,100	27,000	11,780	10,780	10,420	11,160	12,020	b51,120
31	(a)	(a)	(c)	28,130	10,430	9,724	15,730	b44,120

a The ice gorges during January, February, and March make it impossible to estimate daily
flow.
b Discharge for December 30 and 31 reduced to 40 per cent on account of ice gorge.

Estimated monthly discharge of Susquehanna River at Harrisburg, Pa., 1891-1904.

[Drainage area, 24,080 square miles.]

Month.	Discharge in second-feet.			Run-off.	
	Maximum.	Minimum.	Mean.	Second-feet per square mile.	Depth in inches.
1891.					
January	135,800	21,770	72,224	3.006	3.466
February	334,500	61,700	140,746	5.857	6.099
March	156,300	46,200	97,861	4.052	4.672
April	120,800	34,300	79,830	3.322	3.706
May	30,800	13,300	19,193	.799	.921
June	71,000	12,500	25,397	1.057	1.179
July	41,400	12,100	21,708	.908	1.041
August	79,200	13,300	30,568	1.272	1.467
September	46,200	11,310	23,711	.987	1.101
October	46,200	10,200	18,596	.774	.892
November	75,500	13,300	34,115	1.419	1.583
December	129,600	29,430	62,988	2.621	3.022
The year	334,500	10,200	52,201	2.172	29.149
1892.					
January	195,800	14,160	78,944	3.285	3.787
February	49,400	10,560	22,350	.930	1.003
March	193,000	17,960	51,301	2.135	2.461
April	224,200	25,190	79,705	3.317	3.701
May	118,200	21,770	67,255	2.799	3.227
June	183,600	26,330	65,242	2.715	3.029
July	46,200	8,850	19,324	.804	.927
August	38,500	12,100	18,664	.777	.896
September	22,340	7,090	11,219	.467	.521
October	8,850	4,070	5,999	.250	.288
November	30,800	4,070	10,896	.453	.505
December	39,900	6,020	16,153	.672	.775
The year	224,200	4,070	37,254	1.550	21.120

Estimated monthly discharge of Susquehanna River at Harrisburg, Pa., 1891-1904—Continued.

Month.	Discharge in second-feet.			Run-off.	
	Maximum.	Minimum.	Mean.	Second-feet per square mile.	Depth in inches.
1893.					
January	21,770	18,300	15,515	0.646	0.745
February	167,100	19,550	55,585	2.313	2.409
March	223,200	17,960	93,257	3.881	4.474
April	154,400	54,300	103,387	4.302	4.800
May	267,400	31,500	91,090	3.791	4.371
June	31,500	10,200	18,627	.775	.865
July	16,460	6,020	10,224	.425	.490
August	24,620	3,500	5,680	.236	.272
September	42,200	9,520	18,785	.782	.872
October	57,500	7,940	18,638	.776	.895
November	31,500	10,200	15,425	.642	.716
December	118,200	13,300	40,382	1.681	1.938
The year	267,400	3,500	40,549	1.688	22.847
1894.					
January	55,900	16,950	27,018	1.124	1.296
February	68,400	13,300	31,545	1.313	1.367
March	177,100	25,190	69,791	2.904	3.348
April	136,600	20,650	65,407	2.722	3.037
May	543,500	16,460	94,621	3.938	4.540
June	132,300	16,950	49,839	2.074	2.314
July	19,010	6,550	10,050	.418	.482
August	10,560	3,680	6,626	.276	.318
September	62,500	3,500	17,281	.719	.802
October	61,700	7,940	25,888	1.077	1.242
November	97,600	17,960	46,345	1.929	2.152
December	74,600	16,460	35,195	1.465	1.689
The year	543,500	3,500	39,967	1.663	22.587

Estimated monthly discharge of Susquehanna River at Harrisburg, Pa., 1891–1904—Continued.

Month.	Discharge in second-feet.			Run-off.	
	Maximum.	Minimum.	Mean.	Second-feet per square mile.	Depth in inches.
1895.					
January	112,000	23,480	50,123	2.086	2.405
February	86,500	21,770	53,531	2.228	2.320
March	147,200	51,900	79,655	3.315	3.822
April	205,400	29,430	84,858	3.531	3.940
May	41,400	15,510	25,048	1.042	1.201
June	29,430	5,240	10,868	.452	.504
July	22,340	3,680	9,370	.390	.450
August	8,540	3,500	5,263	.219	.252
September	10,200	3,680	5,211	.217	.242
October	4,500	2,570	3,806	.158	.159
November	21,770	3,000	6,108	.254	.283
December	62,500	5,240	18,594	.774	.892
The year	205,400	2,570	29,328	1.220	16.470
1896.					
January	136,600	23,480	52,586	2.188	2.523
February	183,600	16,460	52,478	2.184	2.355
March	183,600	16,460	64,346	2.678	3.087
April	223,200	26,330	88,502	3.683	4.109
May	23,480	7,370	12,637	.526	.606
June	47,800	8,850	19,216	.800	.893
July	41,400	10,200	15,195	.632	.729
August	46,200	3,500	14,499	.603	.695
September	7,870	3,160	4,153	.173	.193
October	129,600	9,520	34,463	1.434	1.653
November	140,100	12,100	35,476	1.476	1.647
December	89,900	8,540	21,577	.898	1.035
The year	223,200	3,160	34,594	1.439	19.525

Estimated monthly discharge of Susquehanna River at Harrisburg, Pa., 1891–1904—Continued.

Month.	Discharge in second-feet.			Run-off.	
	Maximum.	Minimum.	Mean.	Second-feet per square mile.	Depth in inches.
1897.					
January	31,500	9,520	18,609	0.774	0.892
February	101,800	23,480	46,302	1.927	2.007
March	165,300	26,330	88,240	3.672	4.233
April	129,600	25,190	55,768	2.321	2.590
May	101,800	24,620	53,844	2.241	2.584
June	29,430	9,520	17,648	.734	.819
July	43,800	6,550	11,374	.473	.545
August	41,400	7,370	15,208	.633	.730
September	15,510	4,070	6,749	.281	.314
October	11,310	4,070	5,906	.246	.284
November	50,200	4,740	21,592	.899	1.003
December	106,000	17,960	46,585	1.939	2.235
The year	165,300	4,070	32,319	1.345	18.246
1898.					
January	147,200	12,500	58,490	2.434	2.806
February	106,000	19,550	52,376	2.199	2.290
March	245,900	27,510	88,570	3.686	4.250
April	144,500	23,480	53,141	2.211	2.467
May	86,500	36,400	59,310	2.468	2.845
June	41,400	10,560	19,979	.831	.927
July	14,600	4,740	7,998	.333	.384
August	57,500	8,850	26,014	1.083	1.249
September	23,480	5,240	11,288	.468	.522
October	109,400	4,740	32,904	1.369	1.578
November	116,400	17,960	41,096	1.710	1.908
December	101,000	12,500	34,733	1.445	1.666
The year	245,900	4,740	40,487	1.686	22.892

Estimated monthly discharge of Susquehanna River at Harrisburg, Pa., 1891-1904—Continued.

Month.	Discharge in second-feet.			Run-off.	
	Maximum.	Minimum.	Mean.	Second-feet per square mile.	Depth in inches.
1899.					
January	103,500	20,650	44,427	1.849	2.132
February	120,800	12,500	46,106	1.919	1.998
March	193,000	41,400	100,920	4.200	4.842
April	116,400	29,480	66,984	2.788	3.111
May	54,300	17,960	25,349	1.055	1.216
June	19,010	6,550	11,511	.479	.534
July	11,310	5,240	7,820	.325	.375
August	36,400	8,680	7,297	.304	.350
September	12,100	4,500	6,432	.268	.299
October	7,090	2,850	4,130	.172	.198
November	48,800	4,070	18,795	.782	.872
December	82,000	9,520	32,169	1.340	1.545
The year	193,000	2,850	30,995	1.290	17.472
1900.					
January	174,500	10,560	57,040	2.374	2.737
February	159,000	12,100	63,816	2.656	2.766
March	194,900	23,480	67,494	2.809	3.238
April	92,600	36,400	58,223	2.423	2.703
May	36,400	12,500	19,250	.801	.923
June	19,550	7,370	13,112	.546	.609
July	9,520	5,240	7,134	.297	.342
August	9,520	2,850	5,066	.211	.243
September	7,370	2,330	3,721	.155	.173
October	7,940	2,570	4,314	.180	.208
November	194,000	4,070	23,489	.977	1.091
December	93,400	13,300	36,726	1.528	1.762
The year	194,900	2,330	29,949	1.246	16.595

Estimated monthly discharge of Susquehanna River at Harrisburg, Pa., 1891–1904—Continued.

Month.	Discharge in second-feet.			Run-off.	
	Maximum.	Minimum.	Mean.	Second-feet per square mile.	Depth in inches.
1901.					
January	29,430	7,370	14,038	0.584	0.673
February	86,400	11,810	20,038	.834	.868
March	191,100	10,560	81,085	3.372	3.888
April	204,400	47,800	103,968	4.326	4.827
May	210,100	28,130	63,972	2.662	3.069
June	185,500	25,190	55,088	2.292	2.557
July	24,620	9,520	13,518	.563	.649
August	120,800	7,940	33,266	1.384	1.596
September	54,300	10,560	22,089	.919	1.025
October	19,550	8,850	13,150	.547	.631
November	58,400	6,550	14,849	.618	.689
December	405,100	14,600	73,514	3.059	3.527
The year	405,100	6,550	42,376	1.738	23.999
1902.					
January	138,400	18,300	37,012	1.540	1.775
February	132,300	13,300	47,168	1.963	2.044
March	484,100	31,500	155,396	6.467	7.456
April	224,200	20,650	68,132	2.835	3.163
May	21,770	10,560	15,401	.641	.739
June	23,480	7,940	12,810	.533	.595
July	112,000	25,190	70,209	2.922	3.369
August	72,800	7,940	26,962	1.122	1.294
September	54,300	5,760	11,714	.488	.544
October	68,400	15,510	35,656	1.484	1.711
November	60,000	10,560	20,985	.873	.974
December	186,400	16,950	63,774	2.654	3.060
The year	484,100	5,760	47,102	1.960	26.724

Estimated monthly discharge of Susquehanna River at Harrisburg, Pa., 1891–1904—Continued.

Month.	Discharge in second-feet.			Run-off.	
	Maximum.	Minimum.	Mean.	Second-feet per square mile.	Depth in inches.
1903.					
January	105,200	15,510	37,765	1.572	1.812
February	223,200	37,800	93,236	3.880	4.040
March	276,500	60,000	133,500	5.556	6.405
April	188,400	29,430	82,715	3.442	3.840
May	28,130	9,520	14,297	.595	.686
June	76,710	8,190	27,964	1.163	1.298
July	79,640	14,160	32,581	1.355	1.560
August	107,670	12,560	25,581	1.064	1.227
September	123,500	10,410	30,511	1.270	1.417
October	138,300	7,610	45,160	1.880	2.167
November	98,560	14,160	27,289	1.135	1.266
December	53,750	5,630	19,743	.822	.948
The year	276,500	5,630	47,528	1.978	26.666
1904.					
January a			30,410	1.27	1.47
February a			38,590	1.61	1.74
March a			102,000	4.24	4.89
April	194,200	27,030	74,280	3.09	3.45
May	97,600	18,590	41,740	1.74	2.01
June	58,400	11,780	29,320	1.22	1.36
July	52,900	10,420	18,020	.750	.865
August	16,270	7,258	10,420	.434	.500
September	14,870	6,442	8,657	.360	.402
October	37,240	7,538	15,240	.634	.731
November	13,980	9,452	10,760	.448	.500
December	51,120	5,708	8,448	.352	.405
The year			32,320	1.35	18.32

a Owing to an ice gorge below Harrisburg the monthly mean for January, February, and March has been estimated by taking 89 per cent of means for McCalls Ferry. Practically open conditions existed at the latter station (see p. 188).

SUSQUEHANNA RIVER AT McCALLS FERRY, PA.

The McCalls Ferry gaging station is located, as shown in Pl. VIII, at a narrow and rocky part of Susquehanna River, about 20 miles above its mouth and 1 mile above the village of that name. It was established on May 17, 1902, by Boyd Ehle while investigating a power development there. For a considerable distance along this portion of the river the bank on the York County shore is the retaining wall of an abandoned canal which can be overtopped only in the greatest floods. The Lancaster shore, on the opposite side, is made up of almost vertical rock, and the railroad which skirts it has never yet been flooded at this point.

The gaging section first selected for the station is located at Duncans Run (A-A, Pl. VIII), where two islands, Hartman and Streepers, divide the river into three channels, ranging in width from 100 to 500 feet. At ordinary low water, however, two of these run dry, thus confining the discharge to the main or westernmost channel. The river bed at the section is composed of schistose rock, with some projecting bowlders and large irregularities. The flow, however, is comparatively free from the boils so common in a river of this character.

The discharge measurements are made from a boat held in place by a rope stretched between the towpath and Streepers Island, the gaging points, 10 feet apart, being indicated by a tagged wire, which is also used for keeping the boat parallel to the current.

In order to provide for measuring the large floods which occur in the winter and spring months a cable station was established by Mr. Ehle in the fall of 1902, about 1,000 feet downstream from the Duncans Run section (B-B, Pl. VIII). The banks of the river and the condition of the river bed are very similar to those at the upper section, though the latter is somewhat more irregular, as shown by Pl. I, B. During the low-water period of the fall of 1902 a careful survey was made of the section at the cable station, and a contour map with 1-foot intervals was prepared from which the effective areas could be accurately determined, thus eliminating the error in discharge due to possible inaccuracies in soundings made at the time of the measurements. The width of the stream at this point is about 1,300 feet, and the maximum depth during a gaging was 46 feet.

The car cable, a ¾-inch 37-wire strand, with a span of 1,450 feet, is anchored to 3-inch eyebolts set in cement in the solid rock on either side of the river. A 2-inch turn-buckle is provided at the York County end to regulate its height above the water. A high cliff on one shore and a large red oak on the other give the cable a 10-foot clearance over the highest floods on record. The car which runs on the cable, as shown in Pl. IX, B, accommodates two people, and is propelled by a crank turning one of the sheaves.

VIEW OF SUSQUEHANNA RIVER ABOVE McCALLS FERRY.

A A, Duncans Run gaging station, B B, cable gaging station.

Eighty feet upstream from the main cable is suspended a ⅜-inch secondary cable, along which runs a trolley carrying a guy rope to hold the meter against the current (Pl. IX, *A*). Measuring points for this section are 50 feet apart and are indicated by red and white bands painted on the main cable, the intermediate distances being readily estimated by counting the revolutions of the sheave.

The measurements at both of the above stations are referred to two permanent gages, designated Nos. 2 and 5. These are painted on the rock and give elevations directly above sea level. Gage No. 2 is located about three-fourths of a mile below the village of McCalls Ferry in the tailrace of the proposed power house and has been read daily since June, 1902. The records in the following tables have been referred to this gage. Gage No. 5 is placed about 2 miles below McCalls Ferry, at the foot of Cullys Falls, and was thus located in order to be entirely out of the influence of the proposed dam. One of the purposes of the extensive investigations carried on at McCalls Ferry was to obtain data for determining the coefficient of discharge over ogee-faced weirs under high heads, and it is for use in these investigations that gage No. 5 was established.

The methods used in carrying on the work at the McCalls Ferry station were practically the same as those employed by the United States Geological Survey. Every effort was made to eliminate any source of error, and vertical velocity determinations were taken whenever possible. At Duncans Run, in order to get satisfactory vertical velocity curves, an 80-pound weight, with pulley and rope attached, was dropped to the bottom, so that the meter could be pulled down without being washed too far from the section. When the surface velocity or 0.6 method was used the results were reduced by coefficients determined from these vertical velocity curves. At the cable station the secondary cable with the aid of the guy rope made it possible to get vertical velocity measurements at exceptionally great velocities and depths. A No. 12 telegraph wire was found to be more satisfactory at such times for holding the meter than the insulated cable ordinarily used, as it offered less resistance to the current, would allow the meter to sink deeper, and being less bowed by the water would show more accurately its depth below the surface. In this way curves were obtained to depths of 20 feet and in currents of 10 feet per second.

During the highest stages, when the velocity sometimes reaches 17 feet per second, readings could only be taken at the surface. These results were, however, reduced by coefficients determined from the vertical velocity curves for each measuring point.

Discharge measurements of Susquehanna River at Duncans Run station above McCalls Ferry, Pa., 1902-1904.

Date.	Hydrographer.	Gage height.a	Area of section.	Mean velocity.	Discharge.
1902.		*Feet.*	*Square feet.*	*Feet per second.*	*Second-feet.*
May 17	Boyd Ehle	116.62	4,570	3.70	16,880
24	...do	115.83	4,840	2.93	12,710
June 9	...do	115.30	3,990	2.59	10,330
28	...do	116.82	4,564	3.17	14,440
July 14	...do	121.90	9,180	6.00	55,100
16	...do	120.12	7,400	5.15	38,100
21	...do	117.90	6,020	4.02	24,200
24	...do	125.10	11,900	8.01	95,300
26	...do	123.82	11,000	7.41	81,500
Sept. 3	...do	114.82	3,800	2.14	8,130
25	...do	114.84	3,500	1.82	6,370
1903.					
June 5	R. H. Anderson	115.17	3,850	2.60	10,000
1904.					
Sept. 29	W. G. Steward	114.75	3,717	216	7,940

a At gage No. 2.

A

B

GAGING CAR AT McCALLS FERRY CABLE STATION.

A, Gaging car in operation, *B*, gaging car

Discharge measurements of Susquehanna River at cable station above McCalls Ferry, Pa., 1903-1904.

Date.	Hydrographer.	Gage height.a	Area of section.	Mean velocity.	Discharge.
1903.		*Feet.*	*Square feet.*	*Feet per second.*	*Second-feet.*
Feb. 10	R. H. Anderson	123.90	14,300	5.97	b 85,400
Mar. 2	.....do	135.90	33,800	8.50	b290,550
3	.....do	133.60	30,365	8.23	b250,000
4	.....do	130.00	28,050	7.55	b174,060
5	.....do	127.20	19,000	6.80	b129,300
6	.....do	125.20	16,175	6.41	c104,600
7	.....do	124.20	14,780	5.77	c 85,300
12	.....do	129.40	22,460	7.16	c160,600
18	.....do	123.40	18,220	5.84	c 77,240
25	.....do	134.30	31,220	8.75	b273,300
27	.....do	130.10	28,720	7.38	b175,210
28	.....do	127.60	19,780	6.90	b136,400
Apr. 3	.....do	123.80	14,060	5.72	b 80,400
9	.....do	123.80	13,310	5.75	c 76,600
16	.....do	131.50	26,445	7.91	b209,200
18	.... do	128.80	21,350	7.15	b152,500
22	.....do	122.60	11,840	5.62	b 66,600
25	... .do	120.70	9,400	4.96	c 46,660
May 4	.....do	117.85	5,870	4.16	c 24,400
14	.....do	116.50	4,410	3.63	c 16,000
28	.....do	115.72	4,120	3.19	c 13,140
June 5	.....do	115.17	2,885	3.40	c 9,810
17	.....do	120.00	8,180	4.67	c 38,200
1904.					
Mar. 8	R. H. Anderson	146.6	54,500	11.6	d631,000
May 11	.....do	119.00	7,035	4.7	b 84,400

a At gage No. 2.
b Surface velocities.

c Multiple points.
d See page 177.

Mean daily gage height, in feet, of Susquehanna River at McCalls Ferry, Pa., for 1902–1904.

Day.	Jan.	Feb.	Mar.	Apr.	May.	June.	July.	Aug.	Sept.	Oct.	Nov.	Dec.
1902.												
1					116.15	117.50	122.10	114.90	120.50	122.10	117.15
2					116.15		121.70	114.90	122.60	121.30	117.40
3					115.80	123.70	121.50	114.80	122.70	120.10	118.45
4					115.80	123.10	122.30	114.85	122.10	119.60	119.25
5					115.80	123.15	122.00	114.80	121.50	119.00	119.60
6					115.35	124.30	121.20	114.60	121.40	118.50	119.40
7					115.25	123.55	120.60	114.55	121.30	118.20	119.40
8					115.20	123.55	119.40	114.50	120.90	118.00	119.10
9					115.20	125.50	118.85	114.60	120.00	117.80	119.10
10					115.50	124.60	118.50	114.65	119.50	117.55	
11					115.65	122.90	118.90	114.55	118.80	117.40	
12					115.60	122.10	119.00	114.65	119.20	117.10	118.10
13					115.60	122.50	118.70	114.80	121.40	117.00	119.50
14					115.70	121.85	118.10	114.75	121.00	116.90	120.10
15					116.20	120.80	117.75	114.75	120.50	116.70	119.30
16					116.20	120.20	117.50	114.70	119.60	116.60	119.40
17					116.35	119.30	117.20	114.65	119.00	116.50	123.00
18					116.80	118.65	116.95	114.65		116.40	126.35
19					116.45	118.20	116.70	114.55	118.70	116.35	125.85
20					116.65	117.80	116.80	114.50	118.20	116.30	125.00
21					116.60	117.90	116.20	114.40	117.80	116.20	124.50
22					116.35	117.30	116.00	114.50	117.50	116.10	127.65
23					116.90	122.85	115.80	114.40	117.20	116.00	131.50
24					116.15		115.75	114.30	116.90	115.95	131.50
25					115.95	124.05	115.75	114.35	117.00	116.00	129.95
26					116.15	123.85	115.70	114.60	117.00	116.20	126.55
27					116.40	124.70	115.55	118.55	116.90	116.85	124.30
28					116.65	123.85	115.40	121.00	117.10		122.90
29					116.75	122.20	115.30	120.00	118.70	117.15	122.30
30					116.95	121.90	115.30	119.85	122.00	117.15	121.00
31						122.60	115.00				120.60
1903.												
1		120.10	131.00	132.80	123.10	118.60	115.55	123.00	118.00	127.00	115.75	117.80	117.00
2		119.50	129.20	136.00	123.40	118.20	115.50	122.30	117.70	124.80	115.75	117.75	116.80
3		121.30	126.80	133.60	123.80	118.00	115.40	119.90	117.50	123.20	115.50	117.60	116.60
4		122.10	126.50	129.90	123.40	117.80	115.30	119.40	117.20	122.20	115.40	117.50	116.30
5		122.70	131.50	127.00	122.60	117.75	115.20	120.10	117.10	121.00	115.40	117.40	116.50
6		122.90	133.10	125.20	122.10	117.60	115.10	120.00	117.10	120.00	115.40	117.25	116.80
7		123.10	131.20	124.20	122.10	117.50	115.10	118.00	119.60	115.50	117.10	116.80
8		122.30	124.70	124.90	122.30	117.30	115.50	121.60	119.70	119.00	116.10	116.95	116.60
9		121.10	125.60	124.70	123.10	117.10	115.10	121.60	119.90	119.00	116.70	117.00	116.50
10		(a)	124.00	127.10	123.80	117.00	116.20	120.00	119.60	118.75	120.80	117.00	116.60
11			122.90	131.00	124.00	116.90	116.90	119.00	118.75	127.80	117.00	116.10
12			122.80	129.70	123.50	116.60	117.45	118.70	118.60	118.50	129.20	117.00
13			123.00		123.00	116.55	117.50	118.85	118.10	118.75	128.50	116.85	116.00
14		(b)	123.30	129.40	123.00	116.50	118.30	118.50	118.00	118.60	126.40	116.60	116.00
15			123.60	127.50		116.25	119.60	117.65	117.90	118.55	123.80	116.50	115.50
16			124.50	125.90		116.20	119.50	117.50	117.90	118.00	122.00	116.40	115.30
17		118.40	124.90	124.20	131.70	116.15	120.10	117.20	117.90	120.90	116.70	115.00
18		118.70	124.50	123.30	129.50	116.05	119.70	117.15	118.00	118.00	120.80	116.80	114.70
19		119.00		122.70	126.60	115.95	119.50	119.50	117.70	121.10	125.60	115.00
20			120.30	122.00	124.80	115.95	118.75	120.60	117.40	118.50	122.50	125.00	116.00
21		119.50	119.30	121.70	123.60	115.95	121.80	117.10	118.00	123.10	123.10	115.50
22		119.80	119.10	121.80	122.60	115.85	118.40	120.60	116.80	117.60	122.30	121.60	118.60
23		120.00	118.70	120.60	121.10	115.73	118.10	119.50	117.00	121.30	120.40	119.70
24		120.10	119.50	126.80	121.10	115.85	118.50	119.50	117.00	117.10	120.50	119.70	120.50
25		119.60	120.60	134.10	120.50	115.85		118.10	116.95	116.60	119.60	118.30	120.20
26		119.30	120.40	132.80	120.10	115.45	121.80	118.10	116.95	116.60	119.60	118.80	119.40
27		119.20	120.50	120.80	119.50	115.80	123.00	117.85	116.70	116.30	118.70	119.20	117.50
28		119.50	122.30	127.00	119.50	115.80	123.00	118.20	117.80	116.20	118.70	118.20	117.50
29		120.40		125.20	131.10		122.30	117.80	121.80	116.00	118.40	117.70	117.50
30		121.10		123.90		115.70	122.40	117.50	122.40	115.90	118.10	117.30	117.40
31		122.70		123.50		115.60		118.00	124.20		118.00		116.90

a Slush ice filled in above gage.
b River frozen over at neck and foot of Gullys Falls.

Mean daily gage height, in feet, of Susquehanna River at McCalls Ferry, Pa., for 1902-1904—Continued.

Day.	Jan.	Feb.	Mar	Apr.	May,	June	July.	Aug	Sept	Oct.	Nov.	Dec.
1904.												
1	116.6	120.0	120.0	123.9	125.2	119.4	116.8	115.8	115.5	116.2	117.1	116.0
2	116.3	119.0	121.0	129.2	124.4	119.3	116.5	115.7	115.3	116.1	116.9	115.9
3	115.9	118.5	122.0	132.6	123.3	119.9	116.3	116.0	115.1	116.0	116.8	115.8
4	a115.8	117.9	122.9	130.0	122.5	120.4	116.0	116.6	115.5	115.8	116.5	115.5
5	116.0	117.3	124.0	127.0	121.5	120.2	116.0	116.6	115.3	115.9	116.3	115.3
6	116.5	117.0	124.0	125.0	120.9	120.8	116.0	116.4	115.1	115.6	116.2	115.3
7	116.9	118.5	126.4	123.9	120.0	122.3	116.2	116.6	115.0	115.8	115.9	115.1
8	115.8	119.4	b146.6	123.1	119.8	121.4	116.5	116.7	114.9	115.7	115.7	115.0
9	115.5	121.5	130.2	123.2	119.5	120.1	117.0	116.7	114.8	115.6	115.5	114.8
10	115.5	125.0	130.4	123.4	119.3	119.9	117.5	117.0	114.7	115.4	115.7	114.7
11	116.0	125.7	130.9	124.6	119.0	119.6	119.9	117.5	114.7	115.8	115.5	114.5
12	116.8	124.3	126.6	127.3	118.6	121.7	121.0	117.0	114.8	115.4	115.5	114.4
13	117.1	122.7	124.9	125.9	118.3	121.0	121.1	116.4	115.0	115.4	115.6	114.2
14	117.3	121.9	123.6	124.4	118.3	119.9	119.9	116.0	115.3	115.4	115.9	114.2
15	117.3	121.0	122.3	123.6	118.2	119.3	119.0	115.7	115.8	115.4	116.0	114.4
16	117.4	120.4	121.5	122.6	119.0	118.5	118.5	115.5	116.1	115.4	116.0	115.3
17	117.0	119.5	121.1	121.9	119.5	118.3	118.7	115.3	116.4	118.2	115.9	114.6
18	116.6	118.6	120.7	121.6	119.7	117.8	117.4	115.2	117.0	118.0	115.8	114.6
19	116.4	118.0	120.9	121.0	120.8	118.0	117.0	115.2	116.8	117.5	115.7	114.6
20	116.0	117.8	121.0	120.6	121.3	117.9	116.8	115.3	116.5	116.8	115.7	114.5
21	116.0	118.0	121.6	120.2	122.7	117.8	116.6	115.7	116.3	117.0	115.7	114.6
22	117.4	120.0	122.6	120.1	123.8	117.2	116.5	115.6	116.0	117.3	115.6	114.5
23	122.3	120.9	123.0	119.9	122.8	118.0	116.4	115.5	115.8	117.5	115.5	114.5
24	c120.7	120.1	123.9	119.5	121.0	117.9	.16.4	115.6		118.7	115.7	114.8
25	129.3	120.7	124.3	119.3	120.6	118.0	117.8	115.3	115.2	119.7	115.7	115.0
26	126.8	120.7	130.0	119.2	119.9	117.8	117.4	115.4	114.9	120.0	115.8	114.9
27	124.0	120.3	131.6	119.3	120.2	117.3	116.5	115.7	114.8	119.3	116.0	115.0
28	123.0	119.8	132.9	119.7	119.9	116.9	116.3	116.9	114.8	118.5	116.3	115.1
29	122.3	119.0	130.7	121.0	119.6	116.8	116.0	116.6	114.8	117.9	115.7	115.5
30	121.4	......	128.9	122.1	119.6	116.7	116.0	116.1	115.8	117.8	116.1	116.2
31	120.5	......	125.3	......	119.6	......	115.9	115.8	......	117.5	......	123.0

a Entire river covered with 14 to 18 inch ice
b Ice moved 2 p.m.
c Ice broke and went out of deeps at 5.80 p. m.; 133.8 maximum reading during night, 24th and 25th.

Rating table for Susquehanna River at McCalls Ferry, Pa., for 1902 to 1904.

Gage height.	Discharge.	Gage height.	Discharge.	Gage height.	Discharge	Gage height.	Discharge.
Feet.	*Second-feet.*	*Feet.*	*Second-feet.*	*Feet.*	*Second-feet.*	*Feet.*	*Second-feet.*
114.0	5,160	116.4	15.610	120.6	44,200	126.0	112,900
114.1	5,500	116.5	16.150	120.8	46.100	126.5	119,900
114.2	5,840	116.6	16.690	121.0	48,000	127.0	127,000
114.3	6,200	116.7	17,240	121.2	50,000	127.5	134,100
114.4	6,560	116.8	17,800	121.4	52,100	128.0	141,100
114.5	6,930	116.9	18,360	121.6	54,300	128.5	148,300
114.6	7.310	117.0	18,930	121.8	56,600	129.0	155,300
114.7	7,700	117.2	20,120	122.0	59,000	129.5	163,400
114.8	8,100	117.4	21,320	122.2	61,500	130.0	172,500
114.9	8,500	117.6	22,560	122.4	64,000	130.5	182,800
115.0	8,920	117.8	23,820	122.6	66,500	131.0	194,100
115.1	9,340	118.0	25,110	122.8	69,000	131.5	205,800
115.2	9,770	118.2	26,430	123.0	71,500	132.0	217,300
115.3	10,210	118.4	27,780	123.2	74,000	132.5	228,600
115.4	10,660	118.6	29,140	123.4	76,400	133.0	240,000
115.5	11,120	118.8	30,500	123.6	78,900	133.5	251,200
115.6	11,580	119.0	31,900	123.8	81,500	134.0	262,000
115.7	12,060	119.2	33,300	124.0	84,200	134.5	273,600
115.8	12,540	119.4	34,700	124.2	87,000	135.0	285,300
115.9	13,040	119.6	36,100	124.4	89,900	135.5	297,200
116.0	13,540	119.8	37,500	124.6	92,800	136.0	309,300
116.1	14,040	120.0	39,100	124.8	95,700		
116.2	14,560	120.2	40,700	125.0	98,600		
116.3	15,080	120.4	42,400	125.5	105,900		

130

240,000

29

128

127

$-111)^2=2.020 x$
x in 1000' sec.-ft.

126

290,000

300,000

310,000

r< Furnace
ncan Run

cable
all Price meter
ge   "   "

G.H. 120 (y-111)=(1.032 ×
whole curve (y-111)²=2.020
ec.-ft.

140,000

180,000

/ER AT McCALLS

Mean daily discharge, in second-feet, of Susquehanna River at McCalls Ferry, Pa., for 1902–1904.

Day.	Jan.	Feb.	Mar.	Apr.	May.	June.	July.	Aug.	Sept.	Oct.	Nov.	Dec.
1902.												
1						14,300	21,940	60,300	8,510	43,300	60,200	19,880
2						14,300	60,000	56,400	8,510	66,500	51,100	21,320
3						12,550	60,200	53,300	8,100	67,700	39,400	28,120
4						12,550	72,700	61,500	8,300	60,300	38,000	33,500
5						12,550	73,350	50,000	8,100	53,200	31,400	36,000
6						10,430	88,500	50,000	7,300	52,100	28,400	34,600
7						9,300	78,250	44,300	7,120	51,100	26,400	34,600
8						9,770	78,250	34,600	6,800	47,050	25,110	32,500
9						9,770	105,400	30,870	7,300	30,100	23,820	32,500
10						11,120	91,300	28,400	7,500	35,300	22,250	31,000
11						11,830	70,300	31,210	7,120	30,530	21,320	28,300
12						11,580	60,300	31,900	7,500	39,300	19,550	25,770
13						11,580	65,300	29,840	8,100	52,100	18,940	35,300
14						12,000	57,300	25,770	7,000	48,000	18,300	39,400
15						14,580	46,100	23,500	7,900	43,300	17,250	33,900
16						14,580	40,700	21,940	7,700	36,000	16,680	34,600
17						15,340	33,900	20,130	7,500	31,900	16,150	71,300
18						17,810	29,500	18,640	7,500	31,000	15,610	117,800
19						15,880	26,430	17,250	7,120	20,840	15,340	110,800
20						16,970	23,820	15,080	6,800	26,480	15,080	98,600
21						16,600	24,460	14,580	6,500	23,820	14,580	91,300
22						15,340	20,720	13,540	6,800	21,940	14,050	136,300
23						15,080	69,400	12,550	6,600	20,120	13,540	205,800
24						14,300	77,400	12,300	6,300	18,300	13,290	205,800
25						13,290	84,500	12,300	6,340	18,910	13,540	170,600
26						14,300	82,150	12,000	7,300	18,240	14,580	119,300
27						15,610	94,300	11,350	28,800	18,300	18,060	88,500
28						16,970	82,150	10,620	48,000	19,530	19,000	70,300
29						17,530	61,500	10,210	39,100	20,840	19,830	62,800
30						18,640	57,300	9,770	37,300	50,000	19,830	48,000
31						66,500		8,930		50,000		44,200
1903.												
1	39,900	194,100	235,400	72,700	29,150	11,350	71,500	25,110	127,000	12,300	23,820	18,940
2	35,300	158,400	309,300	76,400	26,430	11,120	62,800	25,340	95,700	12,300	23,500	17,800
3	51,000	124,100	253,400	81,500	25,110	10,620	38,300	21,940	74,000	11,120	22,560	16,380
4	60,300	119,900	170,600	76,400	23,820	10,210	34,000	20,120	59,000	10,680	21,940	15,080
5	67,700	205,800	127,000	66,500	23,500	9,770	30,600	19,530	48,000	10,680	21,320	17,800
6	70,300	242,300	101,500	60,300	22,560	9,350	30,100	19,530	39,100	10,600	20,420	17,800
7	72,700	198,800	87,000	60,300	21,940	9,350	37,500	25,110	35,300	11,120	19,530	17,800
8	62,800	151,100	88,500	62,800	20,720	11,120	54,300	30,750	31,900	14,050	18,640	16,680
9	49,000	107,300	94,300	72,700	19,530	11,350	46,100	38,300	29,840	17,250	18,940	16,150
10	46,200	84,200	126,500	81,500	18,940	14,580	30,100	36,000	30,100	46,100	18,940	16,680
11	43,600	70,300	194,100	84,200	18,300	31,900	31,900	30,100	138,300	19,230	16,150	13,540
12	41,000	69,000	167,000	77,600	16,680	21,630	20,840	29,150	24,460	158,400	18,940	13,540
13	38,400	71,500	164,000	71,500	16,430	21,940	30,870	25,770	30,100	118,300	18,060	13,540
14	85,800	75,300	161,700	71,500	16,150	27,100	28,400	25,110	29,150	118,500	16,680	13,540
15	83,300	78,900	134,100	120,000	14,820	36,000	25,110	22,870	28,800	81,500	16,150	11,120
16	90,600	91,300	103,000	230,000	14,580	35,300	21,940	24,460	25,110	59,000	15,610	10,210
17	27,780	97,100	87,000	210,400	14,300	39,100	30,120	24,460	24,780	47,050	17,250	9,350
18	29,840	91,300	75,300	163,400	13,800	36,750	19,830	25,110	25,110	36,100	17,800	7,700
19	31,900	66,000	72,700	121,300	13,300	32,800	35,300	25,110	25,440	49,000	107,300	8,930
20	53,000	41,550	59,000	96,700	13,290	30,100	44,300	21,320	26,400	65,300	98,600	13,540
21	35,300	33,900	55,400	78,900	13,290	39,100	19,530	25,110	72,700	72,700	28,400	
22	37,500	32,500	58,000	66,500	12,800	25,780	44,300	17,800	22,560	62,800	54,300	29,150
23	39,100	29,840	66,500	66,500	12,300	25,770	30,100	16,680	21,320	51,100	42,400	36,750
24	39,900	35,300	124,100	49,000	12,800	28,400	35,300	18,590	19,530	43,300	36,750	43,300
25	36,000	44,300	264,300	43,300	12,800	42,100	31,550	30,720	18,300	37,500	33,900	36,750
26	33,900	42,400	235,400	30,100	12,800	56,000	25,770	18,640	16,620	36,000	30,530	34,000
27	38,200	43,300	168,800	37,500	12,550	78,900	24,140	17,250	15,080	31,900	28,400	25,780
28	85,300	62,800	127,000	35,300	12,550	71,500	26,430	23,820	14,580	20,840	26,430	23,180
29	42,400		101,500	32,500	12,300	62,800	23,820	51,100	13,540	27,780	23,180	21,940
30	49,000		82,800	31,000	12,000	61,000	21,940	64,000	13,040	25,770	20,720	21,320
31	67,700		77,600		11,580		25,110	87,000		25,110		18,300

a Estimated.

*Mean daily discharge, in second-feet, of Susquehanna River at McCalls Ferry.
Pa., for 1902–1904—Continued.*

Day.	Jan.	Feb.	Mar.	Apr.	May.	June.	July.	Aug.	Sept.	Oct.	Nov.	Dec.
1904.												
1	16,630	59,100	39,100	82,800	101,500	34,700	17,800	12,540	11,120	14,580	19,520	13,540
2	15,080	31,900	48,000	158,400	89,300	34,000	16,150	12,060	10,210	14,040	18,360	13,040
3	13,040	28,460	59,000	230,900	75,200	38,300	15,080	13,540	9,340	13,540	17,800	12,540
4	12,540	24,400	70,200	172,500	65,700	42,400	13,540	16,690	11,120	12,540	16,150	11,120
5	13,540	20,720	141,100	127,000	53,200	40,700	13,540	16,690	10,210	13,040	15,080	10,210
6	16,150	18,930	141,100	98,800	47,050	46,100	13,540	15,610	9,340	11,120	14,560	10,210
7	18,360	28,460	118,500	82,800	59,100	62,800	14,560	16,690	8,920	12,540	13,040	9,340
8	12,540	34,710	300,000	72,700	37,500	52,100	16,150	17,240	8,590	12,060	12,060	8,920
9	11,120	53,200	176,500	74,000	35,400	39,900	18,930	17,240	8,100	11,120	11,120	8,100
10	11,120	98,400	180,700	76,400	34,000	38,300	21,340	18,930	7,700	10,690	12,060	7,700
11	13,540	108,700	182,000	92,800	31,900	36,100	38,300	21,940	7,700	10,210	11,120	6,930
12	17,800	88,500	121,300	131,300	29,140	55,400	48,000	18,930	8,100	10,690	11,120	6,560
13	19,720	67,700	97,100	111,500	27,100	48,000	49,000	15,610	8,120	10,690	11,580	5,840
14	20,720	57,800	78,900	89,800	27,100	38,300	38,300	13,540	10,210	10,690	13,040	5,840
15	20,720	48,000	62,800	78,900	26,430	34,000	31,900	12,060	12,540	10,210	13,540	5,560
16	21,320	42,400	53,200	66,500	31,900	28,480	28,460	11,120	14,040	10,690	18,540	10,210
17	18,930	35,400	49,000	57,800	35,400	27,100	28,820	10,210	15,610	26,430	13,040	7,310
18	16,690	21,140	45,100	54,300	36,800	25,110	21,320	9,770	18,930	29,820	12,540	7,310
19	15,610	25,110	47,050	48,000	41,550	25,110	18,930	9,770	17,800	21,940	12,600	7,310
20	13,540	23,820	48,000	44,200	51,100	24,480	17,800	10,210	16,150	17,800	12,060	6,930
21	13,540	25,110	54,300	40,700	47,700	23,820	16,690	12,060	15,080	18,930	12,060	7,310
22	21,320	20,100	65,700	39,900	81,500	20,120	16,150	11,580	13,540	20,720	11,580	6,930
23	62,800	47,050	71,500	38,300	69,000	25,110	15,610	11,120	12,540	21,940	11,120	6,560
24	45,100	39,900	82,800	45,000	48,000	24,480	15,610	10,660	11,580	29,820	12,060	8,100
25	160,000	45,100	145,500	34,000	44,200	25,110	28,820	10,210	9,770	36,800	12,060	8,920
26	124,100	45,100	172,700	37,300	38,300	25,820	21,320	10,690	8,500	39,100	12,540	8,540
27	84,300	41,550	208,100	34,000	40,700	20,720	16,150	12,060	8,100	34,000	13,540	8,920
28	71,500	57,500	237,700	30,900	40,100	18,080	15,080	18,360	7,310	28,460	15,080	9,340
29			187,200	48,000	39,100	17,800	13,540	16,690	8,100	24,460	12,060	11,120
30			153,800	42,200	31,900	17,240	13,540	14,040	12,540	23,820	14,040	14,560
31	43,300	...	100,000		36,100		13,040	12,540	..	21,940		71,500

a Maximum discharge, 631,000.　Mean daily discharge estimated

*Estimated monthly discharge of Susquehanna River at McCalls Ferry, Pa.,
1902–1904.*

[Drainage area 26,766 square miles]

Month.	Discharge in second-feet.			Run-off.	
	Maximum	Minimum.	Mean.	Second-feet per square mile.	Depth in inches.
1902.					
June	18,640	9,770	13,908	0.519	0.580
July	105,900	20,720	61,768	2.307	2.658
August	61,500	8,920	27,126	1.013	1.168
September	48,000	6,200	11,556	.431	.481
October	67,700	18,360	38,248	1.429	1.649
November	60,200	13,290	22,657	.846	.944
December	205,800	19,830	69,111	2.582	2.977

Estimated monthly discharge of Susquehanna River at McCalls Ferry, Pa., 1902-1904—Continued.

Month.	Discharge in second-feet.			Run-off.	
	Maximum.	Minimum.	Mean.	Second-feet per square mile.	Depth in inches.
1903.					
January	72,700	27,780	43,533	1.626	1.877
February	242,300	29,840	95,082	3.552	3.698
March	309,300	55,400	134,461	5.023	5.791
April	210,400	31,000	79,900	2.910	3.247
May	29,150	11,580	16,826	.628	.724
June	78,900	9,350	29,859	1.115	1.244
July	71,500	19,830	35,636	1.331	1.535
August	87,000	16,690	28,206	1.053	1.214
September	127,000	13,040	34,183	1.277	1.426
October	158,400	10,660	48,757	1.822	2.102
November	107,300	15,610	30,797	1.151	1.284
December	43,300	7,700	19,751	.737	.848
The year	309,300	7,700	49,638	1.854	25.019
1904.					
January	160,000	11,120	34,170	1.280	1.480
February	108,700	18,930	43,360	1.620	1.750
March	300,000	39,100	114,600	4.280	4.980
April	230,900	33,300	78,400	2.930	3.270
May	101,500	26,430	46,720	1.750	2.020
June	62,800	17,240	34,580	1.290	1.440
July	49,000	13,040	21,410	.800	.922
August	21,940	9,770	13,880	.519	.598
September	18,930	7,310	11,050	.413	.461
October	39,100	10,210	18,700	.698	.805
November	19,520	11,120	13,320	.498	.556
December	71,500	5,840	10,890	.407	.469
The year	300,000	5,840	36,760	1.370	18.700

CHEMUNG RIVER AT CHEMUNG, N. Y.[a]

A gaging station was established at the suspension bridge across Chemung River near Chemung station, September 7, 1903. Gage heights are taken each morning and night, by Daniel L. Orcutt, by a chain gage attached to the bridge. Current-meter measurements which have been made, and the mean daily stage of the stream, are shown in the accompanying tables. The gaging station is located 1 mile upstream from the New York-Pennsylvania line, and is shown on the Waverly sheet of the United States Geological Survey's topographic map of the country.

Chemung River is formed at Painted Post, N. Y., by the union of Tioga and Cohocton rivers. The Cohocton branch lies entirely in the State of New York. Tioga River receives, just above its mouth, Canisteo River, a large tributary, which also has its drainage basin in New York to the south of the Cohocton. The drainage of Tioga River above the Canisteo is mainly in Pennsylvania. The concentration, just above Corning, of the storm waters of these three main branches favors the formation of excessive floods.

Chemung River flows southeasterly through Corning, Elmira, and Chemung, crosses the State line, flows for a short distance in Pennsylvania, then returns to New York and again crosses to Pennsylvania near Waverly, finally emptying into Susquehanna River near Athens, Bradford County, Pa. The total length of the stream is about 40 miles, about 30 miles of which is in New York State. Chemung River is a sluggish stream with low banks and a broad valley or flood plain, which is often overflowed. It was formerly paralleled by a canal taking its supply from dams across the stream. This has been abandoned and at present the largest water-power development on the main river is at Elmira.

The topographic features of the drainage basin are, as a rule, bold and broad. The hills rise within a short distance of the stream several hundred feet on either side, and the upland plateau is to a large extent wooded, with impervious soil, no lake storage, and few marsh areas. Tributaries are ramifying and uniformly distributed, though not numerous, and dry gulleys or flood channels are common. Dikes have been erected in the cities of Elmira and Corning for protection against floods. One of the highest recorded freshets in the stream occurred June 1, 1889. It was preceded by phenomenal rainfall, on the night of May 31 and June 1, aggregating several inches in the course of a few hours. The discharge has been estimated at 67 second-feet per square mile from 2,055 square miles, or 138,000 cubic feet per second.[b]

---

[a] Data on pages 140-168, inclusive, from Supplement of 1903 Report of New York State Engineer.
[b] Report of Francis Collingwood, C. E., on The Protection of the City of Elmira, N. Y., against Floods.

Discharge measurements of Chemung River at Chemung, N. Y.

Date.	Hydrographer	Gage height.	Discharge.
1903.		*Feet.*	*Second-feet.*
August 27	C. C. Covert	2.89	809
September 7	R. E. Horton	3.29	1,354
October 2	H. H. Halsey	2.47	611
October 12	C. C. Covert	6.72	8,766
1904.			
March 11	C. C. Covert	5.75	6,170
April 9	R. E. Horton	5.64	5,717
July 15	C. C. Covert	3.05	1,042
September 9	do	1.90	220

142   HYDROGRAPHY OF SUSQUEHANNA BASIN.   [NO. 109.

Mean daily gage height, in feet, of Chemung River at Chemung, N. Y.

Day.	Jan.	Feb	Mar.	Apr	May.	June.	July.	Aug.	Sept.	Oct	Nov.	Dec.
1903.												
1										2.24	2.98	2.90
2										2.40	2.88	2.88
3										2.52	3.88	2.88
4										2.57	2.83	2.82
5										2.74	2.86	2.82
6										3.30	2.90	2.59
7									3.29	3.37	3.08	3.09
8									2.24	4.62	2.98	2.79
9									3.19	9.97	4.93	2.69
10									3.16	7.78	2.90	2.64
11									4.84	8.80	2.88	2.49
12									4.56	6.74	2.86	2.49
13									3.84	6.12	2.80	2.69
14									3.46	4.97	2.73	2.69
15									3.22	4.47	2.68	2.69
16									3.06	4.20	2.76	2.74
17									2.96	3.92	7.06	2.74
18									3.44	7.04	8.13	2.64
19									3.46	6.24	5.88	2.64
20									3.29	4.90	4.88	2.64
21									2.99	4.42	4.26	2.69
22									2.84	4.12	3.98	2.74
23									2.54	3.87	3.88	2.79
24									2.34	4.72	3.83	2.79
25									2.34	3.54	3.78	2.79
26									2.29	3.44	3.38	2.74
27									2.24	3.32	3.23	2.69
28									2.24	3.30	3.10	2.54
29									2.22	3.24	3.10	2.44
30									2.26	3.22	3.10	2.54
31										3.18		2.64
1904.												
1	3.00	a 3.85	3.57	6.50	7.20	7.05	2.60	2.50	2.00	2.85	2.40	2.05
2	2.95	3.50	3.37	9.00	6.25	5.85	2.88	2.42	2.02	2.42	2.30	2.00
3	2.90	3.45	3.67	7.05	5.45	5.35	2.70	2.96	2.00	2.22	2.22	1.90
4	2.90	3.35	8.57	5.75	5.02	4.85	2.62	2.82	1.95	2.10	2.20	1.90
5	2.90	4.00	5.72	5.88	4.62	7.70	2.60	2.70	2.00	2.15	2.25	1.90
6	2.90	4.20	4.72	5.15	4.40	5.95	2.58	2.60	1.98	2.18	2.20	1.90
7	2.85	5.90	7.69	5.20	4.18	5.10	2.95	2.45	1.92	1.88	2.20	1.85
8	2.90	a16.70	b15.97	5.25	4.00	4.62	2.85	2.35	1.95	1.95	2.22	1.65
9	3.00	8.70	9.68	5.75	8.80	5.85	2.72	2.20	1.90	1.90	2.12	2.25
10	3.00	6.85	6.48	9.55	8.70	6.15	2.75	2.15	1.90	1.95	2.18	2.10
11	3.00	5.85	5.02	7.40	3.58	4.90	8.90	2.20	1.95	1.95	2.20	2.10
12	3.00	5.40	4.90	6.55	3.40	4.42	3.68	2.18	1.95	2.10	2.20	2.00
13	3.00	4.75	4.50	5.75	3.38	4.00	3.45	2.10	1.95	2.62	2.12	2.00
14	3.00	4.22	4.30	5.15	8.90	3.70	3.45	2.08	1.90	8.65	2.08	2.00
15	3.00	3.95	4.05	4.80	5.15	8.48	3.02	2.00	1.90	8.15	2.05	2.00
16	a 3.15	3.65	8.88	4.80	6.75	4.05	2.82	2.00	1.90	2.82	2.25	2.00
17	3.20	d4.85	3.62	4.80	5.65	8.80	2.70	2.00	1.90	2.70	2.15	1.90
18	3.20	4.55	3.78	5.10	5.00	3.42	2.62	1.95	1.90	2.60	2.20	1.90
19	3.20	e 4.30	3.92	5.10	9.45	3.22	2.50	1.95	1.90	2.50	2.05	1.95
20	3.20	4.15	5.98	4.85	8.40	3.12	2.40	2.05	1.88	2.45	2.00	2.00
21	3.35	4.00	6.78	4.42	6.60	3.02	2.30	2.05	1.80	2.52	2.00	2.10
22	3.50	f4.12	5.20	4.55	5.40	3.10	2.85	2.80	1.75	3.40	2.00	2.05
23	g11.35	4.05	h10.90	4.60	4.95	8.05	2.25	2.75	1.80	3.40	2.00	2.18
24	a 9.55	4.32	11.40	4.50	5.35	3.05	2.72	2.88	1.82	8.18	2.02	2.10
25	6.65	4.12	10.25	4.55	5.25	2.88	2.78	2.70	2.00	8.05	2.20	2.10
26	5.30	4.05	h13.20	4.82	4.75	2.80	2.55	2.45	2.15	2.85	2.15	2.15
27	4.90	3.90	11.05	4.65	4.82	2.70	2.60	2.30	2.38	2.75	2.00	2.60
28	4.20	3.37	7.28	9.10	5.40	2.65	2.50	2.12	2.85	2.65	1.95	6.40
29	4.22	8.57	5.95	8.50	4.25	2.60	2.70	2.10	2.35	2.60	2.10	5.15
30	4.25		5.60	7.42	4.00	2.60	2.80	2.08	2.35	2.45	1.96	3.90
31	4.05		5.70		5.85		2.62	2.00		2.30		3.80

a No ice.
b Water over flats highest point 17 feet.
c River freezing over below gage.
d River frozen over.

e Thickness of ice 5 inches.
f Thickness of ice 12 inches.
g Ice running.
h River over the flats.

Rating table for Chemung River at Chemung, N. Y., from August 27, 1903, to December 31, 1904.

Gage height.	Discharge.	Gage height.	Discharge.	Gage height.	Discharge.	Gage height.	Discharge.
Feet.	*Second-feet.*	*Feet.*	*Second-feet.*	*Feet.*	*Second-feet.*	*Feet.*	*Second-feet.*
1.75	146	4.00	2,255	6.30	7,575	8.60	14,260
1.80	170	4.10	2,420	6.40	7,855	8.70	14,560
1.90	220	4.20	2,590	6.50	8,135	8.80	14,860
2.00	273	4.30	2,765	6.60	8,415	8.90	15,160
2.10	328	4.40	2,950	6.70	8,700	9.00	15,460
2.20	385	4.50	3,140	6.80	8,985	9.10	15,760
2.30	445	4.60	3,340	6.90	9,270	9.20	16,060
2.40	510	4.70	3,550	7.00	9,560	9.30	16,360
2.50	575	4.80	3,765	7.10	9,850	9.40	16,660
2.60	645	4.90	3,990	7.20	10,140	9.50	16,960
2.70	720	5.00	4,220	7.30	10,430	9.60	17,260
2.80	800	5.10	4,455	7.40	10,720	9.70	17,560
2.90	890	5.20	4,695	7.50	11,010	9.80	17,860
3.00	985	5.30	4,940	7.60	11,300	9.90	18,160
3.10	1,085	5.40	5,190	7.70	11,590	10.00	18,460
3.20	1,190	5.50	5,445	7.80	11,880	11.00	2,146
3.30	1,300	5.60	5,700	7.90	12,170	12.00	24,460
3.40	1,415	5.70	5,960	8.00	12,460	13.00	27,460
3.50	1,540	5.80	6,220	8.10	12,760	14.00	30,460
3.60	1,670	5.90	6,485	8.20	13,060	15.00	33,460
3.70	1,805	6.00	6,750	8.30	13,360	16.00	36,460
3.80	1,945	6.10	7,020	8.40	13,660		
3.90	2,095	6.20	7,295	8.50	13,960		

The above table is applicable only for open-channel conditions. It is based upon 8 discharge measurements made during 1903 and 1904. It is fairly well defined between gage heights 1.90 and 3.30 feet. The table has been extended above gage height 6.70 feet. Above gage height 8.0 feet the rating curve is a tangent, the difference being 300 per tenth. The rating table has been applied to the nearest hundredth of a foot to gage height 6.00, to the nearest half-tenth of a foot to gage height 9.00, to the nearest tenth of a foot above gage height 9.00 feet.

Mean daily discharge, in second-feet, of Chemung River at Chemung, N. Y.

Day.	Jan.	Feb.	Mar.	Apr.	May.	June.	July.	Aug.	Sept.	Oct.	Nov.	Dec.
1903.												
1										409	986	890
2										510	872	872
3										589	2,085	872
4										624	827	872
5										752	864	818
6										1,500	890	638
7									1,280	1,180	1,085	1,791
8									409	3,382	986	792
9									1,180	18,460	986	712
10									1,148	11,880	4,059	712
11									8,855	14,860	890	675
12									8,280	8,340	872	562
13									2,005	7,080	854	569
14									1,490	4,157	800	712
15									1,212	3,063	748	712
16									1,045	2,590	705	712
17									947	2,127	768	752
18									1,465	9,705	9,705	752
19									1,490	7,485	12,910	675
20									1,280	3,990	6,432	675
21									975	2,968	3,945	675
22									836	2,454	2,695	712
23									603	2,050	2,223	752
24									471	3,593	2,085	792
25									471	1,582	1,990	792
26									439	1,465	1,917	792
27									409	1,383	1,392	752
28									409	1,300	1,223	712
29									397	1,284	1,085	803
30									421	1,212	1,085	536
31										1,116		675
1904.												
1				8,185	10,140	9,705	645	575	273	478	510	300
2				15,460	7,485	35,860	872	523	284	523	445	273
3				9,705	5,318	5,065	720	966	273	397	397	230
4				8,090	4,267	3,877	660	818	246	328	385	230
5				5,140	3,382	11,590	645	720	273	156	415	230
6				4,575	2,950	6,617	631	645	262	374	385	195
7				4,695	2,556	4,455	938	542	231	210	385	100
8			30,460	4,817	2,255	3,382	845	477	246	246	397	100
9			17,560	6,090	1,945	5,065	736	385	230	230	339	415
10			8,135	17,280	1,805	7,158	760	356	220	246	374	328
11			4,267	10,720	1,644	3,990	2,095	385	246	246	385	328
12			3,990	3,275	1,415	2,988	1,778	374	246	328	385	273
13			3,140	6,090	1,392	2,255	1,477	328	246	660	339	273
14			2,705	4,575	1,300	1,805	1,477	317	220	1,782	317	273
15			2,337	3,785	4,575	1,515	1,005	273	220	1,138	300	273
16			2,065	3,842	8,842	2,337	818	273	220	818	415	273
17			1,697	3,785	5,830	1,945	720	273	220	730	356	230
18			1,917	4,455	4,220	1,440	660	246	220	645	385	220
19			2,127	4,455	16,660	1,212	575	246	220	575	300	246
20			6,697	3,877	13,660	1,106	510	300	210	542	273	328
21			8,985	2,968	8,415	1,005	445	300	170	589	273	328
22			4,695	3,240	5,190	1,065	477	445	146	1,415	273	300
23			21,160	3,840	4,105	1,085	415	760	170	1,415	273	374
24			22,000	3,140	5,065	1,085	736	872	180	1,169	284	328
25			19,060	3,240	4,817	872	784	720	273	1,085	385	328
26			28,060	3,658	800	800	610	542	356	845	356	356
27			21,460	3,445	3,810	720	645	445	497	780	273	645
28			10,430	15,760	5,190	683	575	339	477	682	246	7,855
29			6,617	13,960	2,678	645	720	328	477	645	328	4,575
30			5,700	10,720	2,255	645	800	317	477	542	246	2,095
31			5,960		6,852		660	273		445		1,945

Estimated monthly discharge of Chemung River near Chemung, N. Y., for 1903-4.

[Drainage area, 2,440 square miles.]

Month.	Discharge in second-feet.			Run-off.	
	Maximum.	Minimum.	Mean.	Second-feet per square mile	Depth in inches.
1903.					
September 7-30	8,855	397	1,146	0.47	0.42
October	18,460	409	3,981	1.63	1.88
November	12,910	705	2,265	.93	1.04
December	1,791	536	757	.31	.36
1904.					
March 8-31	36,460	1,697	10,331	4.23	3.90
April	17,260	2,988	6,645	2.72	3.03
May	16,660	1,300	4,940	2.02	2.33
June	35,860	645	4,063	1.67	1.86
July	2,095	415	820	.336	.387
August	966	246	463	.190	.219
September	497	146	267	.109	.122
October	1,732	210	656	.269	.310
November	510	246	347	.142	.158
December	7,855	100	785	.322	.371
The period	36,460	100	2,932	1.20	12.69

TIOUGHNIOGA RIVER AT CHENANGO FORKS, N. Y.

During the fall of 1903 the gaging station was established at this point in order to determine the low-water flow. Owing to the heavy rains which occurred that fall, as shown by the following table, the stage of the river did not fall as low as was expected.

Rainfall at Deruyter, N. Y., 1903.

	Inches.		Inches.
September 1 to 10	0.00	October 8 to 11	8.00
September 11	.96	October 16 to 19	1.38
September 17 and 18	.71	October 23 to 28	.39
September 27	.40	November 5	.34
October 1 and 2	.71	November 6 to 15	.12
October 5	.99		

The measurements were made at the highway bridge across the river at Chenango Forks. This bridge is located straight across the section of the channel and affords an excellent opportunity for gagings, except at extreme high waters. Gage readings were taken during October and part of November from a staff gage fastened to the right-hand face of the center pier of the bridge. The drainage area of Tioughnioga River above the mouth at Chenango Forks, including the areas naturally tributary to the Tioughnioga, but now diverted to supply Erie Canal through the Erieville and Deruyter reservoirs is 735 square miles.

The following measurements were made at the station:

Date.	Hydrographer.	Gage height.	Discharge.
1903.			
September 11	C. C. Covert	2.0	992
September 30	H. H. Halsey	1.2	858

Mean daily gage height, in feet, of Tioughnioga River at Chenango Forks, N. Y.

Day.	Oct.	Nov.	Day.	Oct	Nov.	Day.	Oct.	Nov	Day.	Oct.	Nov.
1903.			1903.			1903.			1903.		
1	1.12	2.15	9	4.00	1.90	17	3.40		25	2.32	
2	1.20	1.95	10	(a)	1.90	18	4.50		26	2.30	
3	1.45	2.00	11	(a)		19	3.65		27	2.20	
4	1.22	1.96	12	4.30		20	3.10		28	2.15	
5	1.50	1.90	13	3.15		21	2.70		29	2.25	
6	2.45	2.06	14	2.80		22	2.45		30	2.25	
7	1.90	2.00	15	3.88		23	2.45		31	2.20	
8	2.10	1.95	16	3.35		24	2.45				

a Above gage.

CAYUTA CREEK AT WAVERLY, N. Y.

A record of the daily stage of Cayuta Creek at the Ithaca Street Bridge, a short distance below the milldam in Waverly, was kept by T. P. Yates, covering the period March 1, 1898, to March 31, 1902. The accompanying tables show the observed distance from the reference point on bridge to water surface, the mean of the several readings being used where more than one daily observation was taken.[a] Discharge measurements by means of floats were also made by Mr. Yates.

Cayuta Creek drains a long, narrow valley extending from eastern Schuyler County in a direction somewhat east of southerly a distance of 30 miles, the stream crossing the New York State line at Waverly and emptying into Susquehanna River at Sayre, Pa. In cross section the valley consists of a plain about one-half mile wide, through which the stream flows, bordered on both sides by abrupt slopes rising 500 feet within a distance of 1 or 2 miles from the foot on each side,

---

[a] Reference point is top iron hand rail at left-hand side second iron post from left-hand end of bridge on upstream side

beyond which lies a plateau, cut by the numerous short lateral tributaries and their branches.

Cayuta Lake drains an area of 16.5 square miles at the head of the stream. The area of the lake is 0.78 square mile, and this constitutes the only storage in the drainage basin. The average width of the valley is about 6 miles. The conditions favor rapid concentration of the run-off in the main stream, there being no large branches. Maximum floods result, however, only from rapid inflow of sufficient duration to enable the waters from the whole length of the valley to reach the lower stretches of the stream at the same time. Cayuta Lake is at elevation 1,272 feet. The stream descends to elevation 800 feet at Waverly in a distance of 18 miles from Cayuta Lake, following the general trend of the valley, a limited amount of water power being developed at small dams.

Drainage areas of Cayuta Creek.[a]

	Area	Total.
	Sq. miles.	*Sq. miles.*
Above outlet, Cayuta Lake...	16	16
Above Van Etten	92	108
Above Ithaca Street Bridge. Waverly	41	149

[a] From Watkins, Ithaca, and Waverly sheets, U. S. G. S. topographic map

Discharge measurements of Cayuta Creek at Waverly, N. Y.

Date.	Hydrographer.	Gage height.[a]	Discharge.
1903.		*Feet.*	*Second-feet.*
June 13	R. E. Horton	17.11	24.9
August 27	C. C. Covert.	17.25	46.3
October 2	H. H. Halsey	17.00	25.4
October 12	H. H. Halsey	14.45	698

[a] Gage inverted.

Mean daily gage height, in feet, of Cayuta Creek at Waverly, N. Y., 1898–1902.

Day.	Jan.	Feb.	Mar.	Apr.	May	June	July	Aug.	Sept.	Oct.	Nov.	Dec.
1898.												
1			16.90	16.00	16.80	16.80	17.50	17.70	17.20	17.80	16.70	16.70
2				16.20	16.40		17.60		17.30		16.80	16.80
3			16.80	16.40		16.90		17.80	17.40		16.90	
4				16.50	16.30			17.30	17.50			
5			16.90	16.60	16.50		17.00	17.05			17.00	16.70
6					15.60		17.10	17.10	17.60			
7			16.80	16.70	16.20		17.30	17.40	16.87		17.10	
8			16.50	16.80	16.40		17.00	17.70	16.70	17.30		16.80
9				16.90	16.50		17.10		17.00	17.40	17.20	16.90
10			16.30		16.60		17.20		17.30	17.60	14.87	17.00
11			14.60		16.70	16.63			17.40	17.70	13.30	17.10
12			13.00		16.60	16.90				17.80	15.25	
13			15.00		16.40		17.00	17.60	17.50	17.70	15.70	17.20
14			15.30	17.00	16.60		17.00	17.70	17.50	16.54	16.00	
15			15.50		16.70		17.10	17.60		16.30	16.30	
16			15.80		16.23		17.20			16.85	16.60	
17			16.00	17.10	16.20		17.20	17.50		17.00	16.70	
18			16.20	17.20	16.40		17.40			17.10	16.70	
19			16.30		15.37			17.25		17.00	16.40	
20			16.00	17.30	14.50	17.30		17.40		16.68	15.73	17.00
21			16.40		15.20			17.50		16.60	16.10	16.35
22			15.40		15.70	17.40	17.50			14.52	16.30	16.15
23			14.30	17.30	16.20		17.60	17.60		15.05	16.40	13.80
24			15.00	12.05	15.33		17.70	17.70		16.00	16.50	14.70
25			15.40	12.25		17.50	17.80	17.35		16.30	16.60	15.80
26			15.80	13.40	15.95		17.50	17.30	17.80	16.36		16.00
27			16.00	14.90	16.00	17.40	17.60			16.20	16.70	16.50
28			16.10	15.50	16.30		17.70	17.50		16.10	16.60	16.60
29			15.40	15.80	16.50		17.60	17.60		15.40		16.70
30				16.00	16.60		17.70	16.57		16.60	16.70	16.80
31			15.80		16.70			17.00				16.90
1899.												
1	15.80	17.80	15.50	15.80	17.00	17.30	17.80	17.90	17.90	17.90	15.08	17.70
2	16.20		16.00	15.90	17.10	16.45					15.40	
3	16.30	17.60	16.20	16.00		17.00					15.90	
4			14.40	16.30		17.20						
5	14.03	17.20	13.20	16.50		17.30		18.00			16.50	17.80
6	15.55		14.00	16.60		17.40					16.60	
7	15.80		14.60	16.70	17.30						16.70	
8	16.00		16.00	14.80		17.80	17.90				16.80	
9	16.10		16.30	15.00		17.40	15.80				16.90	
10	16.40			15.80		17.50	17.20	18.10	18.00		17.00	
11			16.40	16.20			17.30				17.10	
12	16.50	17.10	15.20	15.35	17.20						16.50	16.02
13	16.70		14.48	14.40	17.30	17.60			18.00		16.90	16.40
14			15.70	14.90	17.40							
15	14.90	17.20	16.00	15.40		17.80						66.50
16	15.80		16.00				17.70	18.00		18.10	17.00	
17	15.80	17.30	16.60	15.80	17.30		17.30					16.70
18	15.90		16.60	16.30	17.26		16.80					16.70
19			16.90	16.40	17.20							
20	16.00	17.20	16.10				17.40					
21	16.80	16.70	16.60								17.20	16.80
22	16.50	16.00	16.30		17.30		17.60					
23	16.80	14.80	15.40	16.60								
24	17.20	15.40	15.90	16.60								16.90
25	17.40		16.70	16.70	17.55		18.10				17.40	15.00
26	17.60	16.30	16.50	16.60	17.40							15.70
27	17.70	15.60	16.00	16.70		17.70		17.77	17.90			15.90
28	17.80		15.70	16.80								16.00
29			15.50		17.50						17.60	16.30
30			15.90	16.80	17.20	17.80	17.90	17.90			17.70	16.60
31			15.80		17.20							16.90

Mean daily gage height, in feet, of Cayuta Creek at Waverly, N. Y., 1898-1902—
Continued.

Day.	Jan.	Feb.	Mar.	Apr.	May.	June.	July.	Aug.	Sept.	Oct.	Nov.	Dec.
1900.												
1	16.90	16.90	14.08	16.10	14.30	17.30	17.90	18.10	18.20	18.30	18.80	15.00
2			14.00	15.20								15.50
3		17.00	15.40	14.30	16.40	17.10						15.70
4		17.10	15.80	14.70		17.20						15.50
5		16.50	16.10	15.60	16.50							14.08
6		16.40	15.05			17.30	17.60					14.60
7		16.80	16.70	13.76	16.10	17.50	17.50					15.00
8		16.60	13.80			17.50	17.70					15.20
9	17.00	16.50	16.40	15.50	16.70	17.40						15.60
10		14.00	15.40	16.00		17.50						15.80
11		15.40	15.90				17.80					16.00
12		16.40	16.00	16.10								16.20
13		15.50	16.20		16.80							16.30
14		15.60	16.40			17.60						16.40
15		16.00	16.70	15.90							18.20	16.50
16		16.40	16.90									
17		16.60		15.15	16.90	17.70						16.60
18		16.80		14.45			17.90					
19	17.00	17.00	16.60	14.90		17.80						16.70
20	13.50		13.70	15.50		17.60						
21	12.55	17.20	15.40	16.20	12.00							16.80
22	15.10	12.18	16.40	16.00								
23	15.70	12.20	15.20	15.50			18.00				18.10	16.80
24	16.10	15.40	15.05	16.00		17.70						16.80
25	16.70	15.00			17.10				18.20	18.00		15.60
26	14.70	15.80		16.10		17.80			18.25	10.30		15.90
27	15.90	16.20	16.20				18.10		18.30	11.75		16.20
28	16.40	16.70	16.10	16.20	17.20					14.40		16.40
29	16.80		16.20			18.30				14.50		16.5u
30				16.30		17.90				14.70		
31	16.90		16.30		17.80							
1901.												
1	16.5	17.8	17.5	15.7	16.1	14.8	17.8	17.6	17.1	17.4	17.8	17.3
2	16.7			15.95	16.2	15.08		17.6	16.9	17.5		
3	16.8	17.4		15.45	15.85	15.2			17.1	17.4		16.8
4	16.9		17.45	15.1	16.2	15.5	17.4		17.2			
5			17.4	15.2	16.3	15.9	17.4	17.1	17.3			
6			17.5	14.05	16.5	16.1	17.0	17.8	17.4	17.5		16.9
7	17.0			12.35	16.6	15.86	17.1	17.7	17.6			
8			17.3	12.90	16.7	15.3	17.3	17.4	17.6			17.0
9			18.90		16.0			17.7				17.05
10	16.95	17.5	16.9	14.1		16.3		17.6				14.86
11	16.4		18.25	14.7	16.6	16.5						15.0
12	15.63		14.0	14.9	16.5	16.8	17.5	17.7				16.2
13	16.1		15.2	15.0	16.4							16.3
14	16.5		15.65	15.3	16.5				17.5	17.5	17.7	13.48
15	16.5		15.15	15.6	16.7			17.7	17.4			9.80
16	16.0		15.35	15.8	16.8		17.0		17.3	17.3		13.35
17	16.1	17.5	15.6	15.9	16.9	17.0		17.3				14.4
18	16.5		15.36	16.0	16.9							15.0
19			14.2	16.1	16.7		17.4	17.4	17.4			15.1
20	16.6		15.4	14.75	16.9	16.9	17.5	17.4			17.8	15.6
21	16.7		12.52	11.75				17.0				15.7
22	16.8		14.26	11.83	16.85	17.0		17.3	17.5			15.8
23	16.9		15.1	13.4	16.80			17.4		17.6		16.0
24	17.1	17.5	14.2	13.75	16.50		17.6		17.6		15.16	16.2
25			13.3	14.1	16.40	17.1	17.7		17.6		14.7	
26			12.26	14.6	16.5	17.2		17.7		15.0	16.3	
27			11.73	15.1	16.5		17.5		17.4	17.7	16.6	
28			13.5	15.5	15.58	17.3		17.6			16.6	
29			14.8	15.8	13.3		17.6			17.1	16.4	
30	17.2		15.2	16.0	12.85		17.4	17.5		17.8	17.2	16.0
31			15.5		14.6		17.5					16.1

Mean daily gage height, in feet, of Cayuta Creek at Waverly, N. Y., 1898–1902—
Continued.

Day.	Jan	Feb.	Mar.	Apr.	May.	June.	July.	Aug.	Sept.	Oct.	Nov.	Dec.
1902.												
1	16.2	15.6	9.5									
2	16.4		11.2									
3	16.3		11.5									
4	16.4		13.5									
5			14.8									
6			15.4									
7	16.5	16.0	15.0									
8			15.0									
9			14.8									
10	16.7		14.3									
11			14.0									
12	16.9		13.9									
13			12.7									
14	17.0		13.5									
15		16.4	14.3									
16	17.1		14.2									
17	17.2		11.5									
18	17.3		14.0									
19	17.4		14.8									
20	17.5	16.5	15.0									
21			15.2									
22	15.5		15.4									
23	13.15		15.4									
24	14.0		15.6									
25	15.0	16.8	15.8									
26	15.4		16.0									
27	15.4	16.65	16.2									
28	15.0	12.4	16.4									
29	15.4		16.4									
30	15.8											
31			16.4									

CHENANGO RIVER AT OXFORD, N. Y.

A temporary board gage was attached to the upstream side of the left-hand abutment of the highway bridge across Chenango River at South Oxford, N. Y., September 29, 1903, and observations of the stream stage were taken twice daily from that date until November 7, 1903. The desired data relative to low-water flow could not be obtained on account of heavy rains. The precipitation during the period of observation, as recorded at Oxford, is given below:

Precipitation at Oxford, N. Y.

1903.	Depth *Inches.*
September 1–10	T.
September 11	0.64
September 17	.72
September 27–28	.16
October 5	1.14
October 8–12	3.71
October 16–19	1.72
October 23–27	.49
November 5	.34
November 6–15	.12

South Oxford is located on Chenango River 18 miles above the inflow of Tioughnioga River. The drainage area is 453 square miles gross, or 423 square miles net, excluding 30 square miles tributary to the reservoirs which supply Erie Canal summit level during the navigation period.

Mean daily gage height, in feet, of Chenango River at South Oxford, N. Y.

Day.	Sept.	Oct.	Nov.	Day.	Sept.	Oct.	Nov.	Day.	Sept.	Oct.	Nov.
1906.				**1906.**				**1906.**			
1		0.85	1.80	12		4.55		22		2.55	
2		.85	1.70	13		3.65		23		2.40	
3		1.00	1.70	14		2.90		24		2.20	
4		.90	1.60	15		2.55		25		2.15	
5		1.35	1.65	16		2.30		26		2.00	
6		1.80	1.75	17		2.35		27		1.90	
7		1.45	1.65	18		4.90		28		1.95	
8		1.65		19		4.30		29		0.90	1.95
9		4.35		20		3.40		30		.85	1.90
10		7.40		21		2.90		31			1.85
11		6.50									

EATON AND MADISON BROOKS, MADISON COUNTY, N. Y.

Records of the flow of Eaton and Madison brooks, two small streams near the headwaters of Chenango River, are among the earliest, if not the first, systematic stream gagings in the United States. The flow of these streams was determined by John B. Jervis in 1835 in an investigation of water supply for the summit level of Chenango Canal, extending from Utica to Binghamton, and now abandoned.

The headwaters of Chenango River, including Eaton and Madison brooks and the storage reservoirs which have been constructed to supply the summit level of Erie Canal through Oriskany Creek, are shown on the Morrisville, Cazenovia, Norwich, and Pitcher sheets of the United States Geological Survey topographic map.

Eaton Brook drainage basin is from 1½ to 3 miles in width and 7 miles in length. It contains near its head Eaton reservoir, at an elevation of about 1,430 feet. The slopes are steep; the soil is close textured, with shale near the surface. Tributaries are few, and the fall is rapid.

The soil and topography of Madison Brook are similar, the area consisting of rounded hill slopes with a somewhat more porous soil, greater breadth, and more tributaries than in the Eaton Brook area.

It is stated that the Eaton Brook and Madison Brook gagings show only the volume of water passed downstream from the reservoirs.

Estimated monthly discharge of Eaton Brook, Madison County, N. Y.

[Drainage area, 10.68 square miles.]

Month.	Mean di - charge in second- feet.	Run-off.		Rainfall, inches.
		Second-feet per square mile.	Depth in inches.	
1835.				
January				
February				
March				
April				
May				
June	22.15	2.06	2.32	6.72
July	10.46	.98	1.13	2.74
August	5.06	.48	.55	2.86
September	3.70	.35	.39	1.34
October	7.78	.78	.84	3.00
November	9.17	.86	.96	2.20
December	12.89	1.21	1.39	.96
The period			7.58	19.82
Per cent run-off				38

Estimated monthly discharge of Madison Bro)k, Madison County, N. Y.

[Drainage area, 9.¬7 square miles.]

Month.	Mean discharge in second-feet.	Run-off.		Rainfall, inches.a
		Second-feet per square mile.	Depth in inches.	
1885.				
January	8.66	0.93	1.07	2.17
February	10.49	1.12	1.16	2.50
March	16.16	1.78	1.99	1.03
April	31.16	3.33	3.71	5.00
May	21.66	2.32	2.67	1.98
June	7.77	.83	.93	8 05
July	8.64	.92	1.06	3.87
August	8.86	.95	1.10	3.06
September	7.39	.79	.88	.88
October	7.30	.78	.90	3.86
November	7.03	.75	.84	2.10
December	7.24	.77	.89	.76
The year			17.20	39.26
Per cent run-off				44

Estimated monthly discharge of Eaton Brook, Madison County, N. Y.

[Drainage area, 10.62 square miles.]

Month.	Mean discharge in second-feet.	Run-off.		Rainfall, inches.
		Second-feet per square mile.	Depth in inches.	
1835.				
January				
February				
March				
April				
May				
June	22.15	2.08	2.32	6.72
July	10.46	.98	1.13	2.74
August	5.06	.48	.55	2.86
September	3.70	.35	.39	1.34
October	7.73	.73	.84	3.00
November	9.17	.86	.96	2.20
December	12.89	1.21	1.39	.96
The period			7.58	19.82
Per cent run-off				38

Estimated monthly discharge of Madison Brook, Madison County, N. Y.

[Drainage area, 9.57 square miles.]

Month.	Mean discharge in second-feet.	Run-off.		Rainfall, inches.[a]
		Second-feet per square mile.	Depth in inches.	
1835.				
January	8.66	0.93	1.07	2.17
February	10.49	1.12	1.16	2.50
March	16.16	1.73	1.99	1.03
April	31.16	3.33	3.71	5.00
May	21.66	2.32	2.67	1.98
June	7.77	.83	.93	8.05
July	8.64	.92	1.06	3.87
August	8.86	.95	1.10	3.06
September	7.39	.79	.88	.88
October	7.30	.78	.90	3.86
November	7.03	.75	.84	2.10
December	7.24	.77	.89	.76
The year			17.20	39.26
Per cent run-off				44

Rainfall and run-off in the portion of the Susquehanna River drainage basin above Harrisburg, Pa.

Month.	1891.			1892.			1893.		
	Rainfall, inches	Run-off		Rainfall, inches	Run-off		Rainfall, inches	Run-off	
		Inches.	Per cent of rainfall.		Inches.	Per cent of rainfall.		Inches.	Per cent of rainfall
January	3.96	3.466	87	4.40	3.787	86	2.30	0.745	32
February	3.77	6.009	162	1.72	1.008	58	4.55	2.409	53
March	3.89	4.672	120	4.11	2.461	60	2.68	4.474	167
April	1.97	3.706	188	1.49	3.701	25	4.06	4.800	118
May	1.56	.921	59	5.97	3.227	54	6.05	4.371	72
June	3.93	1.178	30	5.71	3.029	53	3.15	.865	27
July	5.07	1.041	21	4.62	.777	17	3.26	.490	15
August	4.84	1.467	30	4.60	.896	19	4.84	.272	6
September	1.91	1.101	58	2.30	.521	23	3.00	.872	29
October	3.49	.892	26	.95	.288	30	2.76	.895	32
November	2.63	1.583	60	3.45	.505	15	2.08	.716	35
December	4.13	3.022	73	1.28	.775	61	2.69	1.989	72
The year	41,17	29.148	71	40.60	20.970	52	41.37	22.848	55

Month.	1894.			1895.			1896.		
January	2.25	1.296	58	3.32	2.405	72	1.90	2.523	133
February	2.93	1.387	47	1.11	2.380	209	4.49	2.955	52
March	1.21	3.348	277	1.78	3.822	214	3.98	3.087	78
April	4.41	3.037	69	2.50	3.940	158	1.27	4.109	324
May	7.70	4.540	59	2.84	1.201	42	2.89	.606	21
June	2.81	2.314	82	3.47	.504	14	4.84	.898	21
July	2.42	.482	20	2.66	.450	17	5.14	.729	14
August	2.19	.318	15	3.93	.252	6	1.92	.695	36
September	5.61	.802	14	2.17	.242	11	4.01	.193	5
October	4.64	1.242	27	1.46	.159	11	3.88	1.653	43
November	2.04	2.152	105	2.52	.283	11	2.89	1.647	57
December	3.28	1.689	51	3.65	.892	24	1.04	1.065	100
The year	41.49	22.587	54	51.41	16.470	52	37.75	19.525	52

Rainfall and run-off in the portion of the Susquehanna River drainage basin above Harrisburg, Pa.—Continued.

Month.	1897.			1898.			1899.			1900.		
	Rain-fall, inches	Run-off Inches	Per cent of rain-fall.	Rain-fall, inches	Run-off Inches	Per cent of rain-fall.	Rain-fall, inches	Run-off Inches	Per cent of rain-fall.	Rain-fall, inches	Run-off Inches	Per cent of rain-fall.
January	1.77	0.892	50	3.65	2.808	77	2.29	2.133	96	2.28	2.737	120
February	2.38	2.007	86	1.79	2.290	128	3.22	1.996	62	3.69	2.766	75
March	3.22	4.238	131	3.46	4.250	123	3.94	4.842	123	3.52	3.238	92
April	3.06	2.590	85	2.97	2.467	83	1.63	3.111	191	1.52	2.705	178
May	4.72	2.584	55	4.74	2.845	60	3.48	1.216	35	2.20	.923	42
June	3.24	.819	25	2.77	.927	33	3.25	.584	16	2.95	.609	21
July	4.53	.545	12	3.12	.384	12	2.76	.375	14	3.68	.342	9
August	3.11	.730	23	6.35	1.249	20	4.08	.350	9	3.04	.243	8
September	2.90	.814	11	2.04	.522	26	3.70	.299	8	1.41	.173	12
October	1.19	.284	24	5.74	1.573	28	1.68	.198	12	3.35	.206	6
November	4.42	1.008	23	3.23	1.908	59	2.70	.872	32	4.43	1.091	25
December	3.27	2.235	68	2.43	1.666	69	2.95	1.545	52	2.12	1.762	83
The year	37.73	18.246	48	42.29	22.898	54	35.68	17.472	49	34.19	16.595	49

Month.	1901.			1902.			1903.			1904.		
January	1.81	0.673	37	2.31	1.775	77	3.29	1.812	56	3.31	1.470	44
February	.93	.868	96	3.41	2.044	60	3.71	4.040	109	2.16	1.740	81
March	3.52	3.888	110	3.88	7.456	192	4.58	6.405	140	3.43	4.890	142
April	4.46	4.827	108	2.87	3.163	110	2.76	3.840	139	3.28	3.450	105
May	5.68	3.069	54	1.68	.789	45	1.27	.686	54	3.82	2.010	53
June	2.96	2.557	86	6.17	.595	10	6.44	1.298	20	3.37	1.380	40
July	3.96	.649	16	7.24	3.252	47	4.52	1.560	35	4.95	.865	17
August	6.24	1.596	26	2.76	1.294	47	6.48	1.227	19	3.94	.500	13
September	3.01	1.085	34	4.12	.544	13	1.95	1.417	73	3.20	.402	13
October	1.43	.631	44	4.13	1.711	41	4.94	2.167	44	2.71	.731	27
November	2.30	.689	30	1.24	.974	79	2.02	1.266	63	.92	.500	54
December	5.68	3.527	68	4.56	3.080	67	2.42	.948	39	2.13	.405	19
The year	41.98	23.999	57	44.32	26.724	60	44.32	26.666	60	37.22	18.320	49

Rainfall stations in the portion of the Susquehanna River drainage basin above Wilkesbarre.

NEW YORK.

1. Richmondville.
2. Cooperstown.
3. Bouckville.
4. New Lisbon.
5. Oneonta.
6. South Kortright.
7. Oxford.
8. Cortland.
9. Binghamton.
10. Perry City.
11. Wedgwood.
12. Atlanta.
13. Angelica.
14. South Canisteo.
15. Addison.
16. Elmira.
17. Waverly.

PENNSYLVANIA.

18. Athens.
19. Lawrenceville.
20. Wellsboro.
21. Leroy.
22. Towanda.
23. Dushore.
24. South Eaton.
25. Scranton.
26. Wilkesbarre.
34. Girardville.

In the following table are shown the rainfall and run-off in the portion of the Susquehanna basin above Wilkesbarre. The computations are based on the flow at the Wilkesbarre gaging station and the rainfall at the 27 stations listed above.

Rainfall and run-off in the portion of the Susquehanna River drainage basin above Wilkesbarre, Pa.

Month.	1899.			1900.			1901.		
	Rainfall, inches.	Run-off. Inches.	Per cent of rainfall.	Rainfall, inches.	Run-off. Inches.	Per cent of rainfall.	Rainfall, inches.	Run-off. Inches.	Per cent of rainfall.
January	2.14	......	......	2.43	2.078	85	1.69	3.402	201
February	2.67			3.46	2.987	86	1.17	1.696	145
March	3.60			3.59	2.773	77	3.36	4.044	120
April	1.63	3.262	200	1.50	2.988	199	4.67	4.465	96
May	2.78	.876	32	1.97	.680	33	5.30	2.490	48
June	3.11	.854	11	2.94	.364	12	3.11	1.712	55
July	3.13	.235	8	4.13	.269	7	4.03	.337	8
August	3.76	.197	5	2.73	.201	7	5.96	.831	14
September	3.14	.138	4	1.40	.148	11	2.94	.434	15
October	1.85	.136	7	3.58	.141	4	1.69	.382	23
November	2.58	.724	28	4.70	1.226	26	2.68	.563	21
December	3.19	1.470	46	2.29	3.206	140	5.58	4.902	88
The year	33.53	7.571	......	34.73	16.977	49	42.27	25.258	60

Month.	1902.			1903.			1904.		
January	2.00	3.144	157	2.64	3.441	130	3.40	2.570	76
February	3.03	2.432	80	2.98	3.715	127	1.99	3.920	197
March	3.51	7.838	223	4.77	6.289	132	3.17	6.160	195
April	2.54	2.441	96	2.30	2.654	115	2.79	3.560	128
May	2.17	.495	23	1.11	.366	33	3.69	1.860	50
June	5.87	.489	8	6.38	1.134	18	3.27	1.270	39
July	7.86	3.401	43	4.39	.842	19	4.95	.428	9
August	2.88	1.115	39	6.51	1.446	22	4.26	.529	14
September	4.32	.543	13	1.67	1.157	69	3.69	.469	13
October	3.83	1.674	44	6.04	3.183	53	3.00	1.330	44
November	1.13	.861	76	2.21	1.382	62	1.18	.679	58
December	4.04	2.999	74	2.44	1.543	63	2.24	.900	40
The year	43.18	27.817	63	43.82	27.153	63	37.64	23.760	63

Rainfall stations in the portion of the West Branch of the Susquehanna River drainage basin above Williamsport.

20. Wellsboro.	31. Lock Haven.
21. Leroy.	36. Center Hall.
27. Williamsport.	38. State College.
29. Emporium.	39. Grampian.

In the following table are given the rainfall and run-off in the portion of the West Branch of Susquehanna River drainage basin above Williamsport. The computations are based on the flow at the Williamsport gaging station and the rainfall at the eight stations listed above.

Rainfall and run-off in the portion of the West Branch of the Susquehanna River drainage basin above Williamsport.

Month.	1895. Rainfall, inches.	1895. Run-off. Inches.	1895. Run-off. Per cent of rainfall.	1896. Rainfall, inches	1896. Run-off. Inches.	1896. Run-off. Per cent of rainfall.	1897. Rainfall, inches.	1897. Run-off. Inches.	1897. Run-off. Per cent of rainfall.
January	3.74			1.51	1.167	77	2.04	1.012	50
February	1.04			4.00	2.077	52	2.95	1.754	59
March	2.02	4.241	210	3.84	2.822	74	3.77	5.231	139
April	2.33	3.990	171	1.44	3.980	276	3.21	2.744	85
May	3.33	1.128	34	2.05	.787	38	4.47	2.921	65
June	4.66	.688	15	4.48	1.475	33	3.18	.608	19
July	3.00	.602	20	5.75	1.283	22	5.28	.696	13
August	3.57	.987	11	2.26	1.305	58	3.30	.759	23
September	2.31	.204	9	4.70	.309	7	3.37	.337	10
October	1.26	.152	12	4.22	2.685	64	1.16	.263	23
November	2.42	.289	12	2.75	1.784	63	4.91	1.329	27
December	3.74	.924	25	1.25	1.276	102	3.54	2.345	66
The year	38.43			38.26	20.899	55	41.18	19.908	49

Month.	1898. Rainfall	1898. Run-off Inches	1898. Per cent	1899. Rainfall	1899. Run-off Inches	1899. Per cent	1900. Rainfall	1900. Run-off Inches	1900. Per cent
January	3.69	3.230	87	2.49	2.453	99	2.46	2.848	116
February	1.54	2.254	146	3.46	1.717	50	3.71	2.602	70
March	5.20	6.410	123	3.89	5.622	144	3.87	3.197	83
April	2.96	2.552	86	1.86	3.104	166	1.33	2.768	208
May	4.26	2.154	50	3.70	1.530	41	2.22	1.006	45
June	3.37	.848	25	3.60	.539	15	2.94	.800	27
July	2.92	.420	14	2.77	.357	13	3.63	.418	12
August	5.47	.914	17	4.18	.273	7	3.24	.267	82
September	1.23	.302	25	3.50	.365	10	1.05	.184	17
October	6.22	1.507	24	1.87	.206	11	3.71	.872	10
November	2.68	1.684	63	2.77	1.136	41	4.43	1.845	42
December	2.81	1.552	55	3.95	1.898	48	2.05	1.750	85
The year	42.38	23.827	56	38.02	19.194	50	34.64	18.057	52

Month.	1901. Rainfall	1901. Run-off Inches	1901. Per cent	1902. Rainfall	1902. Run-off Inches	1902. Per cent	1903. Rainfall	1903. Run-off Inches	1903. Per cent
January	1.83	1.080	58	2.46	1.449	59	3.09	2.082	66
February	1.28	.556	43	3.19	1.572	49	3.68	4.516	123
March	3.42	4.280	126	4.04	8.092	200	4.41	7.200	163
April	4.69	5.447	116	3.24	3.975	123	3.23	3.526	109
May	5.41	3.148	58	1.90	.963	51	1.74	.601	34
June	3.69	2.436	66	5.72	.667	12	6.03	1.569	26
July	3.79	.595	16	7.56	4.108	54	5.30	1.992	38
August	6.62	1.441	22	2.72	.995	37	5.44	1.230	23
September	3.19	1.245	39	3.68	.340	9	2.06	1.165	56
October	.89	.433	49	3.18	.725	23	4.32	1.699	39
November	2.89	.844	29	1.43	.486	34	2.55	1.735	68
December	5.48	4.145	76	4.12	2.556	62	2.36	.719	30
The year	43.18	25.630	59	43.26	25.928	60	44.23	27.984	63

Month.	1904. Rainfall	1904. Run-off Inches	1904. Per cent
January	3.44	1.940	56
February	2.30	1.970	86
March	5.08	7.880	147
April	4.44	4.700	106
May	3.69	2.470	69
June	3.73	1.420	38
July	4.70	1.270	27
August	3.32	.315	9
September	2.63	.281	9
October	2.20	.472	21
November	.54	.326	60
December	2.18	.334	15
The year	38.20	22.830	60

Rainfall stations in Susquehanna drainage basin.

No.a	Station.	County.	Elevation above sea level.
	NEW YORK.		*Feet.*
1	Richmondville	Schoharie	500
2	Cooperstown	Otsego	1,250
3	Bouckville	Madison	1,350
4	New Lisbon	Otsego	1,234
5	Oneonta	do	1,100
6	South Kortright	Delaware	1,700
7	Oxford	Chenango	550
8	Cortland	Cortland	1,130
9	Binghamton	Broome	854
10	Perry City	Schuyler	1,038
11	Wedgwood	do	1,350
12	Atlanta	Steuben	1,200
13	Angelica	Allegany	1,340
14	South Canisteo	Steuben	1,480
15	Addison	do	993
16	Elmira	Chemung	856
17	Waverly	Tioga	824
	PENNSYLVANIA.		
18	Athens	Bradford	768
19	Lawrenceville	Tioga	1,006
20	Wellsboro	do	1,827
21	Leroy	Bradford	1,400
22	Towanda	do	754
23	Dushore	Sullivan	1,590
24	South Eaton	Wyoming	660
25	Scranton	Lackawanna	805
26	Wilkesbarre	Luzerne	541
27	Williamsport	Lycoming	530
28b	Renovo	Clinton	672
29	Emporium	Cameron	1,029
30b	St. Marys	Elk	1,740
31	Lock Haven	Clinton	560
32	Lewisburg	Union	450
33b	Drifton	Luzerne	1,633
34	Girardville	Schuylkill	1,018
35	Selinsgrove	Snyder	455
36	Center Hall	Center	1,272
37b	Bellefonte	do	744

a The numbers indicate locations on map, fig. 1, p. 11.   b Data incomplete, not used.

Rainfall stations in Susquehanna drainage basin—Continued.

No.	Station.	County.	Elevation above sea level.
	PENNSYLVANIA—continued.		Feet.
38	State College	Center	1,191
39	Grampion	Clearfield	1,570
40	Altoona	Blair	1,179
41	Huntingdon	Huntingdon	650
42	Harrisburg	Dauphin	317
43	Lebanon	Lebanon	458
44a	Ephrata	Lancaster	381
45a	Lancaster	do	413
46	York	York	381
47a	Everett	Bedford	1,060

a Data incomplete, not used.

Monthly and annual precipitation at stations in Susquehanna drainage basin.

1.a RICHMONDVILLE, N. Y.

Year.	Jan.	Feb.	Mar.	Apr.	May.	June	July.	Aug.	Sept.	Oct.	Nov.	Dec.	Annual.
1899	[2.02]	[2.48]	6.24	1.02	2.75	2.32	[5.74]	1.20	3.22	1.15	1.58	2.85	33.17
1900	3.21	3.61	4.06	2.35	2.23	2.37	5.63	3.39	1.34	2.61	3.74	1.96	36.50
1901	1.69	.66	2.09	6.82	5.22	2.54	7.24	5.38	3.24	2.19	1.62	3.83	42.52
1902	1.38	3.11	3.54	3.99	2.39	4.81	6.95	4.49	3.81	1.05	4.45	43.50
1903	1.78	2.54	5.16	1.06	.22	8.84	3.12	5.66	1.23	6.78	1.68	2.42	40.46
1904	3.21	2.18	3.27	2.47	1.10	3.61	3.27	4.20	3.86	4.16	1.26	2.62	35.21
Mean	2.22	2.43	4.06	3.05	2.82	4.08	5.32	3.81	2.90	3.45	1.82	3.02	38.48

2. COOPERSTOWN, N. Y.

Year.	Jan.	Feb.	Mar.	Apr.	May.	June	July.	Aug.	Sept.	Oct.	Nov.	Dec.	Annual.
1891	5.54	4.76	2.60	2.22	2.16	1.96	5.02	4.26	1.41	3.01	8.15	4.96	41.07
1892	4.90	2.23	3.43	1.38	7.82	4.86	7.80	7.96	3.57	1.79	3.19	1.53	50.55
1893	1.80	4.90	2.13	2.96	6.74	2.20	4.85	7.59	4.08	1.27	2.20	4.02	44.87
1894	2.84	2.09	1.92	3.54	5.29	2.62	3.41	1.88	5.55	4.73	2.72	2.38	37.92
1895	2.34	1.43	1.93	2.89	2.44	2.18	3.80	7.15	2.86	2.17	3.65	3.89	36.73
1896	1.48	5.36	4.74	1.25	2.33	4.70	4.60	3.49	4.33	2.23	3.56	1.21	39.28
1897	1.72	2.06	3.31	3.65	5.21	5.22	4.86	6.60	3.40	.64	5.21	4.64	46.52
1898	4.90	2.93	2.14	4.00	4.70	3.80	3.02	9.75	4.20	5.36	4.64	2.44	51.88
1899	2.22	2.81	6.04	1.87	4.52	2.85	3.92	2.72	3.17	2.25	1.93	4.10	37.90
1900	3.08	5.59	2.91	1.94	1.98	3.08	6.61	4.62	1.92	2.57	4.62	2.59	41.46
1901	2.47	1.12	3.00	4.73	4.94	3.65	6.79	5.96	3.08	2.48	2.74	4.85	45.81
1902	1.04	2.89	3.70	3.10	2.76	5.43	9.17	3.05	4.89	4.00	1.48	4.30	45.31
1903	3.30	3.61	5.84	1.57	.17	7.35	5.52	7.26	1.64	8.32	2.21	2.66	49.45
1904	4.29	3.00	3.06	2.84	2.40	4.00	4.74	4.55	4.08	3.49	1.18	2.49	40.12
Mean	3.01	3.17	3.34	2.64	3.82	3.85	5.29	5.49	3.40	3.16	3.06	3.29	43.49

3. BOUCKVILLE, N. Y.

Year.	Jan.	Feb.	Mar.	Apr.	May.	June	July.	Aug.	Sept.	Oct.	Nov.	Dec.	Annual.
1899	2.43	2.19	4.80	2.90	3.35	3.08	2.86	1.97	2.28	2.53	2.85	3.25	33.79
1900	3.82	2.60	6.73	1.21	1.93	2.21	5.09	3.82	1.21	3.60	6.03	3.72	41.47
1901	3.85	3.30	3.18	3.87	5.79	4.14	3.54	3.44	2.30	2.38	3.74	4.50	44.03
1902	1.88	[4.61]	[3.70]	[1.56]	[3.53]	[6.25]	[7.25]	[3.13]	[2.99]	[5.59]	[1.53]	[5.37]	[47.39]
1903	3.60	3.60	4.70	1.80	.00	10.25	2.49	5.91	1.66	8.09	2.32	4.72	48.57
1904	5.39	3.24	2.68	3.80	2.49	2.35	8.85	4.79	3.28	3.06	1.11	3.88	44.92
Mean	3.50	3.16	4.30	2.41	2.85	4.71	5.01	3.76	2.29	4.21	2.93	4.24	43.37

a The numbers indicate locations on map, fig. 1, p. 11.
[] Interpolated.

Monthly and annual precipitation at stations in Susquehanna drainage basin—
Continued.

4. NEW LISBON, N. Y.

Year.	Jan.	Feb.	Mar.	Apr.	May.	June.	July.	Aug.	Sept.	Oct.	Nov.	Dec.	Annual.
1891	4.11	3.56	2.09	1.89	2.50	3.72	4.68	5.59	1.39	3.26	2.25	4.78	39.77
1892	4.40	1.52	8.44	1.25	7.27	3.86	6.23	8.70	2.76	1.61	3.63	1.00	45.67
1893	1.65	4.86	2.12	3.30	4.90	1.97	5.13	8.38	4.05	1.25	.95	2.38	40.94
1894	2.13	1.75	1.40	1.50	4.82	3.88	2.13	2.04	5.74	4.67	2.00	1.92	33.98
1895	2.08	1.98	1.41	8.21	2.50	2.00	2.53	5.76	2.16	1.45	2.98	4.04	32.05
1896	.86	4.31	8.96	.80	2.42	3.77	5.12	2.45	5.07	2.09	2.96	.95	34.76
1897	1.14	1.53	2.90	2.63	4.40	4.10	5.56	3.17	3.19	.73	4.04	4.20	37.61
1898	4.37	2.13	1.68	2.77	3.92	3.04	6.50	7.88	4.95	7.19	8.64	1.48	49.05
1899	1.46	1.96	4.49	2.04	3.44	3.67	3.19	3.49	3.25	1.70	1.93	8.17	33.79
1900	2.04	3.29	8.82	1.30	1.63	2.98	7.27	8.50	2.33	2.87	3.89	2.54	37.46
1901	1.27	.83	2.78	3.88	5.51	4.21	8.68	5.60	3.60	1.54	2.08	4.58	39.01
1902	1.00	2.81	4.13	1.72	2.94	4.61	10.08	3.93	3.05	4.11	1.12	5.85	45.05
1903	2.88	3.19	5.77	1.26	.25	7.04	5.24	6.54	1.57	7.36	2.04	8.35	46.49
1904	3.73	1.75	2.98	2.59	2.62	4.60	5.92	4.41	4.51	3.09	1.86	2.08	40.14
Mean	2.36	2.53	8.07	2.12	3.51	3.82	5.23	5.07	3.40	3.07	2.58	3.00	39.71

5. ONEONTA, N. Y.

Year.	Jan.	Feb.	Mar.	Apr.	May.	June.	July.	Aug.	Sept.	Oct.	Nov.	Dec.	Annual.
1899	2.33	2.60	5.51	0.81	2.79	4.82	4.05	2.72	4.96	1.77	1.70	3.58	37.59
1900	2.63	[2.44]	2.23	1.35	1.26	8.41	5.14	6.24	2.44	3.07	2.65	2.06	34.92
1901	1.80	.92	2.41	3.93	4.54	[5.00]	3.85	4.45	3.84	2.64	2.15	4.96	39.39
1902	1.09	2.97	3.45	1.30	2.82	4.96	7.71	2.54	2.59	4.91	1.11	4.61	40.08
1903	2.46	3.29	5.90	1.05	.36	6.83	4.81	7.70	1.44	7.9;	2.31	2.35	46.48
1904	3.57	2.80	5.28	3.59	2.82	2.71	5.20	7.13	4.66	4.45	2.07	2.64	46.92
Mean	2.31	2.50	4.13	2.00	2.43	4.62	5.13	5.13	3.24	4.14	2.00	3.26	40.89

6. SOUTH KORTRIGHT, N. Y.

Year.	Jan.	Feb.	Mar.	Apr.	May.	June.	July.	Aug.	Sept.	Oct.	Nov.	Dec.	Annual.
1891	4.67	8.31	2.37	1.65	3.57	3.04	3.67	4.21	1.45	[2.70]	2.63	4.57	[37.84]
1892	3.30	1.20	2.32	.77	6.35	2.80	5.14	6.55	2.98	1.13	2.61	1.11	36.26
1893	1.27	4.22	2.82	8.35	5.81	5.76	8.50	7.26	3.76	2.05	1.10	1.99	42.80
1894	2.28	1.19	1.25	2.25	6.67	4.16	4.10	.84	3.08	4.04	2.30	3.08	35.24
1895	1.76	1.40	1.69	3.31	2.10	1.53	3.11	4.68	2.69	2.71	3.70	3.23	31.91
1896	[2.19]	4.81	8.76	1.48	2.94	2.75	5.50	2.12	8.68	2.35	2.83	1.37	35.78
1897	.94	1.53	2.59	2.91	5.33	5.00	5.56	6.08	4.67	.98	4.35	4.02	43.91
1898	2.84	2.38	1.82	2.54	4.06	3.70	2.56	8.21	2.98	5.23	3.88	1.87	43.07
1899	1.35	2.35	3.53	1.79	2.81	4.24	4.31	4.89	.90	1.48	2.44	3.80	32.23
1900	1.91	3.55	2.31	1.71	1.66	4.74	2.84	3.18	2.50	2.09	2.37	[8.07]	31.93
1901	1.84	1.23	8.64	3.06	4.97	[4.37]	[4.17]	3.87	4.26	8.57	2.57	5.75	43.59
1902	1.61	3.56	8.28	3.30	2.48	8.41	6.99	8.55	5.24	5.11	.81	4.11	47.85
1903	2.55	8.31	4.74	1.71	.25	6.21	3.89	5.44	1.64	8.30	2.23	8.25	45.02
1904	2.37	1.67	2.75	1.99	2.19	1.73	4.54	6.33	4.34	4.61	1.98	1.87	36.37
Mean	2.21	2.55	2.78	2.27	3.66	4.17	4.20	4.60	3.44	3.29	2.48	2.98	38.63

7. OXFORD, N. Y.

Year.	Jan.	Feb.	Mar.	Apr.	May.	June.	July.	Aug.	Sept.	Oct.	Nov.	Dec.	Annual.
1891	4.83	4.15	2.78	2.44	1.80	5.44	4.27	6.02	2.73	4.42	2.65	5.88	46.49
1892	6.47	1.66	4.87	1.74	9.37	4.12	5.62	7.90	2.50	1.62	8.44	1.27	50.58
1893	2.57	4.47	2.58	4.89	6.28	3.70	6.01	7.87	3.94	1.46	1.72	8.28	48.22
1894	2.85	2.46	1.86	2.79	5.08	4.02	2.73	2.36	6.11	5.97	2.58	2.00	41.36
1895	8.48	2.00	2.13	2.76	2.78	1.74	2.48	4.59	2.64	1.08	8.95	4.23	53.82
1896	1.99	4.97	5.56	.77	3.53	2.96	5.37	2.71	2.17	2.69	2.66	1.72	37.10
1897	1.76	2.09	4.08	8.76	5.47	4.80	8.04	2.68	3.18	.80	4.85	4.01	45.47
1898	4.76	3.11	2.75	4.90	3.90	3.58	3.41	9.82	4.99	7.08	4.58	8.35	56.23
1899	2.22	3.29	5.44	1.70	3.43	4.30	5.22	3.20	3.05	2.52	2.03	3.54	39.94
1900	3.19	4.76	5.31	1.70	2.00	3.77	3.72	2.89	2.53	3.62	5.31	3.43	42.23
1901	2.89	2.05	8.70	3.88	7.60	2.96	8.93	4.88	3.61	3.04	8.12	6.21	46.86
1902	1.82	4.02	4.32	1.78	2.78	6.46	8.65	2.62	3.97	4.80	1.25	6.11	48.53
1903	8.92	2.94	5.64	1.69	.42	7.56	8.99	7.89	1.52	7.06	1.88	5.58	50.08
1904	4.63	2.85	8.72	8.09	3.06	1.22	5.98	4.49	5.25	3.06	1.50	8.75	42.60
Mean	8.88	8.20	8.91	2.67	4.07	4.04	4.96	4.92	8.44	8.51	2.97	8.89	44.96

Monthly and annual precipitation at stations in Susquehanna drainage basin—
Continued.

8. CORTLAND, N. Y.

Year.	Jan.	Feb.	Mar.	Apr.	May.	June.	July.	Aug.	Sept.	Oct.	Nov	Dec.	Annual.
1899	1.88	0.69	1.83	0.56	2.50	2.25	4.69	2 64	2.40	2.99	2.99	3.98	29.40
1900	3.28	1.84	1.49	1.56	1.17	2 40	4.78	1.92	2 00	4.59	7.17	2.58	34.78
1901	1.22	1.44	2.76	3.81	3.25	2.96	3.49	3.83	2 90	1.02	3.47	6 41	35.06
1902	1.95	1.35	3.20	1.21	2.79	5 03	10 12	3.68	2.51	3.59	1 07	4.78	40.58
1903	1.70	1.71	5.13	1.12	[2.43]	6.12	3.99	8.21	2 07	11.47	2.24	1.62	47.81
1904	3.02	2.10	2.85	[1.55]	4.08	2.57	7.55	4.50	5 02	3.29	.84	2.68	40.60
Mean	2.16	1.52	2.88	1.55	2.70	3.55	5.77	4 13	2.82	4.49	2 96	3.68	38.21

9. BINGHAMTON, N. Y.

Year.	Jan.	Feb.	Mar.	Apr.	May.	June.	July.	Aug.	Sept.	Oct.	Nov	Dec.	Annual.
1891	3.90	3.27	4.46	2.16	1.16	3.55	3.30	6.59	1.54	4.24	2.65	3.24	39.46
1892	4.21	1.90	3.98	1.13	6.08	5.43	2.92	6.04	1.33	1.54	2.65	1.27	38.48
1893	2.42	4.16	2.80	3.36	5.16	2.58	4.10	4.88	4.50	1.68	1.98	2.91	39.93
1894	2.18	2.98	1.51	3.53	5.34	1.97	2.88	1.47	4.98	5.62	1.98	3.31	37.75
1895	3.18	1.60	1.58	2.29	2.92	2.05	4.06	3.39	2.11	.82	2.94	3.68	30.57
1896	2.25	4.28	4.68	.63	3.11	2.64	3.85	1.42	4.62	3.68	2.66	1.20	35.02
1897	1.12	1.37	2.65	1.98	4.01	2.98	2.30	1.87	3.03	.66	2.43	3.23	27.14
1898	2.86	2.51	2.31	2.79	4.02	2.16	2.05	6.48	2.70	5.79	3.15	1.45	38.27
1899	1.79	2.63	2.84	.96	2.43	2.15	1.84	2.44	1.45	1.12	1.83	2.02	23.50
1900	1.59	2.65	3.17	1.35	.53	1.54	2.29	.67	2.10	2.05	3.08	1.40	22.42
1901	.76	1.09	2.95	4.20	5.49	1.77	3.47	3.76	8.10	1.48	2.31	5.41	35.77
1902	1.13	2.31	3.54	1.49	1.93	6.84	5.51	2.13	4.75	3.08	1.07	2.92	36.70
1903	2.41	2.24	3.84	1.57	.42	5.79	2.67	6.85	1.21	5.74	2.26	2.12	37.12
1904	2.11	1.16	2.11	2.51	2.66	2.76	4.73	3.12	[2.88]	3.81	.49	1.12	28.96
Mean	2.24	2.44	3.08	2.14	3.23	3.16	3.28	3.62	2.88	2.91	2.21	2.52	33.66

10. PERRY CITY, N. Y

Year.	Jan.	Feb.	Mar.	Apr.	May.	June.	July.	Aug.	Sept.	Oct.	Nov	Dec.	Annual.
1891	3.34	4.23	3.45	2.16	0.74	4.13	3.54	3.90	0.98	5.46	2.19	4.48	38.60
1892	4.56	1.54	3.95	1.65	6.08	6.65	6.86	4.12	.84	1.64	4.63	.78	43.30
1893	1.75	2.80	2.43	3.58	5.37	2.18	4.99	5.21	4.12	2.74	.91	1.87	37.90
1894	3.13	2.54	.99	6.10	6.55	4.05	2.86	1.38	5.46	4.33	2.10	3.06	42.55
1895	2.82	1.40	2.06	1.87	2.49	3.54	2.72	4.67	2.00	.91	4.16	3.08	31.22
1896	1.68	3.58	3.70	1.58	3.81	3.67	4.18	2.54	3.97	4.07	2.44	1.40	36.62
1897	1.81	1.33	2.66	2.56	3.69	4.18	3.55	2.58	.86	3.74	2.86	32.12
1898	2.47	1.68	1.85	3.64	3.96	3.47	1.82	4.68	2.12	6.26	3.90	2.35	37.60
1899	2.03	1.42	2.93	1.46	2.73	2.38	4.30	.96	2.42	3.22	3.34	3.02	30.21
1900	2.52	3.84	3.64	2.00	2.29	1.51	2.66	2.48	1.07	4.76	6.58	2.42	35.77
1901	2.10	1.42	3.12	4.85	4.80	2.85	5.39	7.37	2.22	.86	3.96	5.28	43.62
1902	2.18	1.46	2.28	1.67	2.14	5.52	9.46	4.82	2.40	4.03	1.20	3.69	40.85
1903	2.28	2.03	5.34	1.86	.72	7.04	4.94	8.60	.99	5.79	2.56	1.52	43.67
1904	2.70	1.83	2.92	3.54	5.61	2.01	5.48	3.10	2.80	3.82	1.07	1.80	36.68
Mean	2.58	2.22	2.95	2.72	3.60	3.80	4.48	4.01	2.43	3.48	3.01	2.69	37.92

11. WEDGWOOD, N. Y.

Year.	Jan.	Feb.	Mar.	Apr.	May.	June.	July.	Aug.	Sept.	Oct.	Nov	Dec.	Annual.
1891	2.48	3.88	3.11	2.46	0.89	2.43	2.45	4.58	0.66	4.19	1.77	3.85	32.75
1892	3.50	2.50	3.81	1.08	5.17	4.35	7.24	4.02	.75	2.20	3.25	.71	38.58
1893	2.23	2.49	2.93	3.55	5.37	5.51	3.55	5.61	2.83	2.57	1.60	1.71	39.95
1894	3.10	3.09	1.00	6.67	8.01	2.59	2.49	1.41	5.91	4.22	1.86	3.15	43.50
1895	2.30	.85	1.00	1.55	2.71	4.03	2 31	8.27	1.32	1.02	3.37	3.51	32.24
1896	1.72	5.02	3.48	2.52	2.98	6.23	5.02	1.54	5.02	4.42	2.03	1.42	41.35
1897	1.85	.87	2.54	2.72	3.72	2.74	3.43	3.04	2.66	.74	3.20	1.98	29.44
1898	2.73	1.86	2.62	2.91	3.40	2.72	3.48	4.73	1.86	5.95	2.73	1.98	36.99
1899	1.72	2.07	2.80	1.03	2.04	2.11	3.77	2.55	2.48	2.62	3.50	2.90	29.59
1900	2.56	2.57	3.74	1.80	2.72	1.91	3.19	1.71	.90	5.33	6.79	2.53	35.75
1901	2.05	1.37	3.32	5.44	4.82	4.09	2.84	4.92	2.46	.81	2.90	5.29	44.81
1902	2.04	2.02	2.87	2.96	2.33	6.25	9.23	3.70	2.73	3.41	1.24	3.25	42.03
1903	3.29	2.25	5.42	2.08	.87	5.53	3.26	10.34	1.51	5.05	1.81	1.93	43.32
1904	3.68	1.77	3.12	3.87	5.31	3.39	4.79	4.85	2.13	2.02	.62	1.87	37.42
Mean	2.52	2.33	2.98	2.90	3 60	3.85	4.08	4.70	2.37	3.18	2.62	2.57	37.70

Monthly and annual precipitation at stations in Susquehanna drainage basin—
Continued.

12. ATLANTA, N. Y.

Year.	Jan.	Feb.	Mar.	Apr.	May.	June.	July	Aug.	Sept.	Oct.	Nov.	Dec.	Annual.
1899	1.31	1.54	2.45	1.51	2.85	1.18	2.10	3.14	2.54	1.91	2.38	3.57	26.48
1900	2 64	3.00	4.04	2.08	1.77	2.17	3.06	2.41	3.66	3.79	5.80	1.87	33.81
1901	2.18	2.04	2.60	5.97	5.97	2.10	7.59	9.08		1.31	2.99	4.82	49.73
1902	2.83	1.94	2 27	3.60	2.97	5.19	10.21	1.93		3.25	1.39	2.59	41.00
1903	2 41	2.46	5.02	2.92	1.16	4.66	4.27	5.58		3.86	1.84	1.67	37.91
1904	4.56	2.39	3.59	2.99	4.39	4.81	6.35	3.08		2.79	.98	2.05	41.17
Mean	2 66	2.31	3.33	3.18	3.18	3.27	5.60	4.20	2.46	2.82	2.58	2.76	38.85

13. ANGELICA, N. Y.

Year.	Jan.	Feb.	Mar.	Apr.	May.	June.	July	Aug.	Sept.	Oct.	Nov.	Dec.	Annual.
1899	2.04	1.64	2.72	0.90	2.39	1.81	2.56	2.05	2.86	2.99	2.09	3.97	26.02
1900	2.61	2.33	3.76	8.44	2.62	2.56	4.04	2.59	1.47	4.52	5.40	2.15	35.49
1901	2 62	2.04	2.95	.29	5.23	3.69	3.34	4.87	3.11	1.15	2.88	4.77	41.94
1902	2.80	1.80	2.58	.76	3.97	5.79	12.46	3.85	4.46	2.08	.79	1.95	45.72
1903	1.78	1.45	4.60	.65	1.16	4.54	4.11	7.51	1.80	[2.68]	2.57	.77	36.62
1904	2.69	1.48	2.47	.97	4.00	[3.68]	6.54	[4.07]	[2.74]	[2.68]	[2.75]	[2.72]	[37.79]
Mean	2.42	1.79	3.17	2.67	3.23	3.68	5.51	4.07	2.74	2.68	2.75	2.72	37.43

14. SOUTH CANISTEO, N. Y.

Year.	Jan.	Feb.	Mar.	Apr.	May.	June.	July	Aug.	Sept.	Oct.	Nov.	Dec.	Annual.
1891	2.53	4.72	3.43	2.22	1.41	2.68	4.62	5.80	1.20	3.48	2.74	3.30	38.13
1892	3.50	3.40	3.42	1.57	6.74	3.99	4.56	4.83	1.40	2.44	3.60	1.01	40.46
1893	2.96	3.58	3.51	5.84	5.25	4.78	2.70	4.13	2.76	4.05	2.08	2.91	44.50
1894	3.41	3.21	1.64	7.80	11.46	3.51	3.34	2.71	7.12	4.40	2.13	8.41	54.14
1895	3.32	.97	1.63	1.49	2.79	4.75	2.77	3.88	1.15	1.17	3.39	4.84	31.65
1896	2.76	5.62	3.62	1.25	4.03	6.22	5.01	1.62	5.10	6.49	1.82	1.14	44.68
1897	2.34	1.60	3.01	3.13	3.18	3.48	5.62	2.69	3.47	1.04	3.56	2.71	35.83
1898	3.90	2.09	4.53	3.35	3.87	2.90	1.75	4.45	2.28	4.80	3.33	2.62	39.87
1899	1.99	1.95	2.60	1.51	3.29	2.48	2.90	1.99	3.15	3.21	1.80	4.27	31.23
1900	2.40	5.62	2.62	1.60	3.05	5.11	4.10	3.37	1.43	5.81	6.08	1.60	43.74
1901	1.95	1.32	3.13	7.07	5.15	3.53	3.97	5.93	3.24	.62	2.64	4.66	43.21
1902	2.90	2.37	2.73	2.86	1.77	6.24	8.40	2.56	3.32	1.49	1.41	3.05	39.10
1903	3.25	2.15	4.64	3.24	1.94	5.49	4.59	7.18	1.96	4.47	2.48	1.38	42.74
1904	3.45	3.85	3.15	2.81	5.06	2.03	4.20	3.80	3.01	2.46	1.05	2.10	36.97
Mean	2.90	3.03	3.12	3.27	4.21	4.09	4.18	3.92	2.90	3.28	2.72	2.75	40.37

15. ADDISON, N. Y.

Year.	Jan.	Feb.	Mar.	Apr.	May.	June.	July	Aug.	Sept.	Oct.	Nov.	Dec.	Annual.
1891	1.84	2.89	2.12	1.44	0.32	2.05	2.91	4.24	0.49	2.94	1.64	2.96	25.84
1892	2.97	1.58	3.68	.94	5.85	3.18	4.94	3.62	.91	1.50	3.46	.48	33.11
1893	1.64	2.27	2.62	3.50	7.87	3.04	2.37	3.69	2.34	2.89	1.22	1.98	35.33
1894	1.94	1.89	1.06	6.60	9.70	1.82	2.06	1.44	5.62	4.08	1.42	2.93	40.51
1895	3.11	1.12	.88	1.31	2.11	4.15	2.02	3.82	1.22	.80	2.44	2.92	25.90
1896	1.47	3.18	3.05	1.07	4.50	5.78	4.45	.77	3.67	5.73	.83	.88	35.38
1897	1.54	.76	2.29	2.41	4.26	2.56	4.52	2.05	2.90	.94	8.10	1.91	29.24
1898	3.91	1.80	2.30	2.51	4.12	3.67	2.16	2.92	1.31	5.99	2.13	2.15	34.97
1899	1.87	1.49	2.24	1.17	2.88	2.93	3.31	2.90	4.25	1.93	3.58	3.04	31.62
1900	1.92	2.15	2.86	1.49	2.92	2.86	1.93	2.39	1.01	4.80	6.00	1.66	31.90
1901	1.23	.71	3.05	5.82	4.94	2.14	2.01	6.22	2.55	.93	2.00	4.86	36.47
1902	2.30	1.42	2.57	2.41	2.26	5.37	6.85	2.91	3.55	2.84	.89	2.50	35.47
1903	1.87	1.81	4.56	2.67	1.90	5.90	5.51	7.25	1.81	4.42	1.84	.79	40.33
1904	2.47	1.56	2.79	2.27	4.44	1.94	4.53	3.76	2.63	1.57	.56	1.13	29.65
Mean	2.15	1 76	2.58	2.54	4.15	3.89	3.54	3.43	2.45	2.95	2.22	2.15	33.31

Monthly and annual precipitation at stations in Susquehanna drainage basin—
Continued.

16. ELMIRA, N. Y.

Year.	Jan.	Feb.	Mar.	Apr.	May.	June.	July.	Aug.	Sept.	Oct.	Nov.	Dec.	Annual.
1891	2.38	2.19	1.98	1.73	0.50	4.57	2.13	3.72	3.25	[4.30]	[1.80]	[3.80]	32.30
1892	3.01	[1.76]	2.96	1.01	5.30	4.11	3.80	3.28	1.18	1.30	[3.10]	[2.31]	31.71
1893	.68	1.61	2.05	3.55	6.84	3.62	3.80	5.54	3.72	2.66	[2.10]	[2.31]	38.51
1894	2.73	1.89	1.05	4.42	7.65	1.94	1.62	1.23	5.16	4.21	1.28	2.89	36.07
1895	2.70	1.20	1.87	1.56	3.03	3.51	2.34	4.04	1.89	.78	1.25	2.70	26.37
1896	1.56	3.40	3.22	.77	3.14	3.31	5.55	.94	2.73	4.86	1.40	.61	31.49
1897	1.40	.98	2.41	2.80	5.56	1.76	3.23	3.70	3.70	.65	2.89	1.60	30.13
1898	2.45	1.45	2.58	2.84	4.29	3.43	2.24	4.70	1.78	4.49	2.24	2.25	34.69
1899	1.51	1.65	2.94	1.52	2.52	2.84	2.69	3.16	3.23	3.07	1.68	1.82	28.63
1900	[1.95]	2.26	3.85	1.58	1.43	1.82	3.48	1.25	1.16	4.19	5.09	1.72	29.28
1901	1.09	.59	2.84	5.56	4.83	1.84	4.23	4.07	2.86	.98	2.75	5.22	36.80
1902	1.98	1.46	2.68	1.71	2.02	4.12	7.84	2.91	3.53	3.90	.88	1.96	34.29
1903	2.08	2.50	4.25	2.24	1.52	7.18	4.78	6.28	1.47	5.10	1.87	.81	40.08
1904	3.18	2.21	2.59	2.77	5.00	4.56	3.80	3.61	3.52	2.01	.57	1.15	34.90
Mean	2.04	1.79	2.58	2.40	3.88	3.47	3.66	3.46	2.80	2.99	1.99	2.22	33.23

17. WAVERLY, N. Y.

Year.	Jan.	Feb.	Mar.	Apr.	May.	June.	July.	Aug.	Sept.	Oct.	Nov.	Dec.	Annual.
1899	1.77	2.26	2.86	1.23	3.26	2.77	4.08	5.23	2.40	1.53	3.37	2.48	33.26
1900	2.00	3.35	4.08	1.58	1.11	2.75	3.07	1.64	1.12	3.72	5.20	2.76	32.38
1901	1.22	.86	4.42	5.87	5.96	2.59	3.35	5.08	2.59	1.42	3.47	6.61	44.19
1902	2.48	2.20	4.56	2.76	1.97	5.50	7.29	2.36	3.98	3.46	1.05	3.19	40.80
1903	2.52	2.23	4.27	2.25	.76	6.67	3.87	6.52	1.85	5.60	2.30	1.49	40.33
1904	3.47	1.53	3.67	2.57	4.02	3.33	2.70	3.31	3.38	2.08	.69	1.81	32.56
Mean	2.24	2.07	3.98	2.71	2.85	3.94	4.06	4.15	2.55	2.97	2.68	3.06	37.26

18. ATHENS, PA.

Year.	Jan.	Feb.	Mar.	Apr.	May.	June.	July.	Aug.	Sept.	Oct.	Nov.	Dec.	Annual.
1899	2.53	2.84	2.75	1.41	3.15	1.98	3.90	4.32	2.49	1.38	3.26	2.57	32.53
1900	1.59	3.84	3.89	1.73	1.26	2.16	2.70	1.48	1.15	3.10	4.60	2.14	28.14
1901	.74	.90	3.82	5.40	5.14	4.11	2.99	5.08	2.33	1.48	3.10	4.47	39.15
1902	2.05	1.80	3.41	2.71	1.65	5.18	5.66	2.17	4.01	3.08	1.11	2.93	35.87
1903	2.60	2.54	4.33	[2.81]	2.00	5.42	3.57	5.79	1.71	5.91	2.40	1.42	40.50
1904	3.02	1.15	(a)										
Mean	2.09	1.95	3.54	2.81	2.64	3.76	3.83	3.71	2.34	2.99	2.89	2.71	35.24

19. LAWRENCEVILLE, PA.

Year.	Jan.	Feb.	Mar.	Apr.	May.	June.	July.	Aug.	Sept.	Oct.	Nov.	Dec.	Annual.
1899	1.85	3.22	2.26	2.10	2.81	3.78	3.15	6.06	3.05	0.41	3.46	2.60	33.75
1900	3.48	5.10	[3.18]	1.11	2.47	2.02	3.50	2.05	.95	4.85	6.36	1.60	36.67
1901	1.60	.90	3.45	5.64	3.90	1.61	2.99	5.08	2.05	1.54	2.78	6.22	37.76
1902	1.75	1.95	2.30	2.70	2.16	5.54	7.37	2.14	4.30	2.22	1.19	3.21	36.83
1903	2.62	2.33	4.67	2.67	1.65	8.60	5.60	5.31	1.99	5.10	2.85	1.92	45.31
1904	3.08	3.05	2.60	2.95	4.32	3.04	3.78	2.68	2.30	2.24	.40	1.60	32.05
Mean	2.40	2.59	3.08	2.86	2.88	4.10	4.40	3.89	2.44	2.73	2.84	2.86	37.07

a No record.

Monthly and annual precipitation at stations in Susquehanna drainage basin—
Continued.

20. WELLSBORO, PA.

Year.	Jan.	Feb.	Mar.	Apr.	May.	June.	July.	Aug.	Sept.	Oct.	Nov.	Dec.	Annual.
1891	6.53	8.46	2.72	1.07	1.30	4.07	3.43	3.57	2.30	2.44	4.11	4.01	39.01
1892	3.67	2.21	4.56	.61	6.69	8.84	2.15	4.73	1.18	.33	2.55	.40	37.92
1893	4.92	6.55	5.09	5.88	6.58	1.42	2.50	4.50	2.08	2.88	3.00	4.21	49.15
1894	2.25	2.25	.24	8.69	10.23	1.89	3.86	2.05	5.85	3.81	3.06	4.07	48.27
1895	3.00	.85	2.90	2.21	6.44	3.50	3.22	4.65	1.12	1.62	2.67	6.55	38.73
1896	1.50	4.34	3.00	.91	1.87	3.92	5.67	.88	3.08	5.40	.82	.95	32.29
1897	2.23	2.30	3.55	2.55	5.53	2.85	5.46	1.84	3.40	.07	5.21	3.09	38.96
1898	1.72	1.33	4.78	4.43	4.70	2.70	2.04	5.13	2.24	3.62	2.83	2.68	43.20
1899	3.42	2.54	2.75	3.07	2.15	4.09	3.87	3.49	2.97	2.63	2.90	3.78	37.16
1900	3.04	4.90	2.90	1.22	2.50	2.90	2.90	3.67	.55	5.01	6.11	.97	36.67
1901	1.27	.80	2.58	4.46	4.28	4.17	2.27	5.04	2.14	.39	3.59	5.66	36.55
1902	1.54	2.70	2.67	2.86	2.05	6.17	9.48	1.29	3.39	2.14	.50	5.18	39.90
1903	1.86	3.55	5.19	2.76	2.12	4.87	5.27	3.37	1.10	5.68	2.42	1.35	39.54
1904	2.95	(a)											
Mean	2.85	2.91	3.30	3.09	4.34	3.95	3.97	3.41	2.40	3.20	3.06	3.30	39.77

21. LEROY, PA.

Year	Jan.	Feb.	Mar.	Apr.	May.	June.	July.	Aug.	Sept.	Oct.	Nov.	Dec.	Annual.
1891	4.63	3.13	3.15	2.01	1.18	4.75	3.05	4.33	2.00	4.25	3.24	4.34	40.06
1892	4.60	1.09	4.25	.96	5.14	7.97	2.39	4.04	2.04	.91	3.22	.98	37.54
1893	2.59	3.86	3.10	4.19	7.76	1.96	2.18	5.92	2.70	3.91	2.07	2.71	42.95
1894	2.43	3.04	1.00	6.12	8.35	1.64	2.98	1.23	5.44	5.29	2.47	3.39	43.38
1895	3.27	.80	1.55	2.65	3.24	3.09	3.42	3.81	3.11	.65	3.06	4.05	33.30
1896	2.00	4.66	4.58	1.44	2.46	2.66	5.84	2.22	3.87	5.04	2.92	.79	36.48
1897	2.18	2.28	2.55	2.70	4.84	3.77	3.95	4.40	3.08	1.30	3.31	2.89	37.70
1898	3.30	2.05	3.39	4.61	3.65	2.75	3.06	6.95	.81	5.37	2.62	1.58	40.14
1899	2.19	3.05	3.02	2.15	2.07	4.90	1.93	6.84	2.85	1.34	3.64	4.47	38.45
1900	1.94	3.07	5.45	1.34	1.50	3.40	4.06	2.14	.54	3.88	4.71	2.12	34.15
1901	.99	.75	4.21	4.68	5.34	3.44	3.22	5.40	6.70	1.16	2.83	3.26	43.98
1902	2.50	3.02	4.76	3.16	1.47	5.40	9.46	4.31	4.67	3.29	.90	3.46	46.49
1903	2.95	3.00	4.37	2.97	2.00	5.13	4.17	4.40	1.57	5.08	2.76	2.60	41.00
1904	3.88	1.13	3.94	3.15	5.45	3.50	2.21	4.80	3.53	2.58	.65	1.65	36.42
Mean	2.75	2 50	3.52	3.01	3.89	3.93	3.71	4.34	2.85	3.15	2.78	3.20	39.52

22. TOWANDA, PA.

Year	Jan.	Feb.	Mar.	Apr.	May.	June.	July.	Aug.	Sept.	Oct.	Nov.	Dec.	Annual.
1899	1.80	2.52	2.55	1.84	2.10	3.52	2.47	5.43	2.08	1.21	3.39	2.82	32.68
1900	1.36	2.90	5.48	1.31	1.38	.49	3.49	3.44	.69	2.83	5.53	1.99	29.89
1901	.91	.45	.92	4.65	7.58	.26	3.51	4.79	3.95	1.31	.43	6.00	43.76
1902	1.72	3.35	.07	2.36	1.06	.86	7.77	2.02	4.58	3.35	.11	2.95	39.29
1903	2.62	2.73	.83	2.37	.89	.05	4.85	4.63	1.24	4.98	.66	2.42	38.27
1904	2.72	1.06	.73	2.48	4.89	.08	3.96	4.32	4.70	2.18	.69	1.59	36.35
Mean	1.86	2.17	3.43	2.50	2.98	4.54	4.84	4.10	2.86	2.64	2.30	2.96	35.68

23. DUSHORE, PA.

Year	Jan.	Feb.	Mar.	Apr.	May.	June.	July.	Aug.	Sept.	Oct.	Nov.	Dec.	Annual.
1899	1.94	3.48	3.79	1.82	2.30	3.13	2.03	3.79	2.80	1.36	2.84	5.09	34.27
1900	1.97	4.01	3.19	1.05	2.31	4.10	4.68	2.25	1.13	2.35	3.38	2.09	32.51
1901	1.10	.78	4.37	5.50	6.90	3.34	5.34	10.59	3.33	2.71	2.87	7.13	53.96
1902	2.58	4.45	5.66	3.91	1.16	7.39	8.95	3.28	5.29	3.37	1.20	4.35	51.89
1903	2.61	4.02	3.36	2.66	1.25	5.34	5.05	5.29	1.52	4.98	2.38	3.48	41.94
1904	3.34	.99	3.26	2.68	4.94	[4.66]	2.98	3.95	3.18	2.15	.97	2.19	35.29
Mean	2.26	2.96	3.94	2.94	3.13	4.66	4.84	4.86	2.88	2.82	2.27	4.10	41.66

a No record.

Monthly and annual precipitation at stations in Susquehanna drainage basin—
Continued.

24. SOUTH EATON, PA.

Year.	Jan.	Feb.	Mar.	Apr.	May.	June.	July	Aug.	Sept.	Oct.	Nov.	Dec.	Annual.
1891	5.47	3.48	4.54	2.85	1.06	2.17	4.88	4.15	1.35	3.71	2.84	3.88	40.38
1892	5.98	.91	4.53	1.20	5.49	4.50	3.14	2.85	2.97	.77	2.88	.86	35.48
1893	2.69	5.49	3.06	3.53	5.12	2.94	3.83	5.41	2.21	1.88	1.94	2.48	40.57
1894	1.65	2.79	.90	2.76	7.28	1.09	1.98	2.22	3.69	6.50	2.27	3.41	36.42
1895	2.35	1.83	1.62	3.60	3.40	4.50	2.81	2.07	1.68	2.96	2.44	4.98	32.32
1896	10.52	4.11	4.45	1.18	2.86	2.62	4.66	3.06	2.45	4.94	4.16	1.11	46.07
1897	1.80	2.49	2.40	3.11	5.29	3.92	3.39	3.23	2.24	1.12	3.96	4.13	37.16
1898	3.93	1.43	3.16	2.78	3.67	1.63	1.04	6.30	1.90	4.49	3.27	2.02	36.17
1899	1.98	3.58	3.96	2.30	2.24	2.58	2.59	3.38	2.16	1.16	2.71	2.81	31.25
1900	2.10	3.47	3.75	.97	1.97	3.52	4.09	1.93	1.84	1.98	3.21	2.17	31.00
1901	.92	.81	3.73	4.21	6.70	3.01	5.32	5.78	2.66	1.94	1.69	6.16	42.91
1902	1.42	5.96	4.06	2.22	1.34	6.61	5.41	2.27	8.15	7.05	1.00	6.09	50.88
1903	2.78	4.53	4.83	3.29	1.31	6.74	3.46	6.19	1.98	5.23	2.09	3.85	46.63
1904	2.97	1.67	2.56	3.21	3.00	8.74	5.94	3.40	3.71	3.54	1.08	1.90	36.70
Mean	3.29	2.96	3.39	2.65	3.62	3.54	3.81	3.73	2.78	3.33	2.54	3.22	34.85

25. SCRANTON, PA.

Year.	Jan.	Feb.	Mar.	Apr.	May.	June.	July	Aug.	Sept.	Oct.	Nov.	Dec.	Annual.
1899	3.03	6.30	4.46	1.96	2.73	2.66	4.73	3.62	3 47	0.63	2.11	2.10	37.80
1900	2.13	2.75	2.98	1.81	2.81	3.54	4.63	1.27	1.72	2.66	2.87	2 61	31.24
1901	1.17	1.84	3.23	3.44	5.58	1.82	4.12	6.88	2.35	1.11	2.58	5.84	39.26
1902	2.14	4.73	3.14	2.27	1.61	6.69	4.60	3.24	6.23	4.94	1.05	4.95	45.05
1903	2.73	3.54	4.40	2.55	.96	7.73	4.99	6.03	1.27	6.42	1.86	2.59	44.97
1904	3.23	.92	2.10	2.82	2.17	3.46	5.94	4.69	3.83	3.90	1.51	3.71	37.18
Mean	2.40	3.26	3.38	2.39	2.64	4.32	4.82	4.30	3.06	3.26	1.92	3.50	39.25

26. WILKESBARRE, PA.

Year.	Jan.	Feb.	Mar.	Apr.	May.	June.	July	Aug.	Sept.	Oct.	Nov.	Dec.	Annual.
1891	4.59	4.00	3.67	2.28	1.53	2.84	4.48	3.46	1.80	1.63	2.54	4.38	37.24
1892	7.02	1.11	6.41	1.55	5.69	10.55	4.71	5.56	2.51	.72	4.37	1.53	51.93
1893	3.34	7.23	3.83	3.27	4.15	1.43	3.00	3.74	3.74	2.97	4.07	4.42	42.49
1894	1.63	4.50	1.68	3.41	8.56	1.78	.74	5.05	5.53	2.29	3.65	3.65	39.97
1895	3.43	2.32	2.94	2.71	4.16	2.89	2.59	4.97	1.59	2.51	1.37	4.13	35.61
1896	1.14	6.17	6.31	1.06	3.17	2.40	6.20	2.99	2.26	2.74	3.44	1.08	38.96
1897	1.40	2.06	3.78	3.34	5.81	3.72	3.76	2.57	1.49	1.47	4.35	3.80	37.55
1898	2.90	.96	2.76	2.46	6.04	3.29	2.33	5.16	3.44	2.36	3.90	1.95	37.55
1899	3.21	4.48	4.49	1.37	2.07	2.82	3.91	2.67	4.29	1.29	2.70	1.72	35.02
1900	1.98	3.21	2.91	1.01	3.81	3.89	5.74	3.16	.52	2.59	3.05	3.02	34.39
1901	2.10	.75	3.81	3.11	5.36	2.48	4.74	7.23	1.64	2.55	1.23	5.98	34.90
1902	2.23	5.60	3.19	1.58	.98	6.10	5.01	1.80	6.82	4.29	1.14	4.95	43.78
1903	2.09	4.13	4.33	3.07	1.12	8.38	4.42	7.13	2.16	4.88	1.98	3.05	46.75
1904	2.86	1.69	3.62	2.34	2.15	2.95	5.83	5.58	3.34	3.68	1.18	3.38	38.50
Mean	2.85	3.44	3.84	2.33	3.91	3.93	3 96	4.09	2.90	2.71	2.61	3.84	39.91

27 WILLIAMSPORT, PA.

Year.	Jan.	Feb.	Mar.	Apr.	May.	June.	July	Aug.	Sept.	Oct.	Nov.	Dec.	Annual.
1899	1.46	3.71	4.36	1.71	2.36	4.25	2.00	4.15	2.94	3.26	2.13	4.63	36.96
1900	2.31	3.72	3.63	.81	2.35	2.89	2.57	2.89	1.01	2.35	3.26	2.15	29.94
1901	1.40	.86	3.63	5.57	6 34	2.99	5.99	3 21	1.59	2.59	5 46	..	42.31
1902	3.61	4.81	4.05	2.43	1.45	5.61	6.02	1.69	5.85	2 10	1.31	3 74	42.47
1903	3.44	3.24	3.96	3.67	1.89	5.49	6.08	5.05	1.43	4.22	2.33	2.95	43.64
1904	3.64	1.10	5.11	3.68	5.28	3 07	5.59	2.13	2 60	2.24	.51	2.63	37.53
Mean	2.64	2.87	4.12	2.97	3.28	4.05	4.26	3 52	2.81	2.63	2.02	3.64	38.81

Monthly and annual precipitation at stations in Susquehanna drainage basin—
Continued.

29. EMPORIUM, PA.

Year.	Jan.	Feb.	Mar.	Apr.	May.	June.	July.	Aug.	Sept.	Oct.	Nov.	Dec.	Annual.
1891	3.47	4.56	5.12	2.38	1.06	4.45	8.46	5.40	1.17	3.48	4.01	4.96	48.47
1892	3.29	3.77	3.87	1.64	7.38	6.13	2.67	3.02	2.78	1.35	3.24	.94	40.08
1893	3.11	5.91	2.92	4.21	4.99	4.88	2.37	3.00	2.10	3.36	2.05	4.07	42.12
1894	3.85	3.08	1.24	3.89	9.45	3.05	2.09	1.37	5.26	3.94	1.81	2.93	41.97
1895	4.79	.50	1.60	2.53	3.08	4.95	3.06	2.98	2.89	1.82	2.59	3.37	34.16
1896	1.17	3.68	4.36	1.88	3.36	6.75	5.11	1.62	5.69	3.31	3.60	1.82	42.85
1897	2.80	3.20	4.08	3.49	3.42	2.04	5.26	2.13	2.73	.94	5.13	4.20	38.49
1898	4.54	1.47	5.80	2.59	4.21	3.90	4.13	5.87	1.89	6.24	3.37	2.66	46.67
1899	2.91	3.66	4.69	2.57	3.98	3.82	4.32	3.73	4.89	3.21	2.86	4.80	43.93
1900	3.16	2.85	5.50	1.29	3.46	2.43	4.48	3.50	1.36	3.84	5.05	2.08	38.00
1901	2.55	1.08	3.01	5.03	6.74	4.39	4.07	6.29	4.05	1.23	2.94	5.22	46.60
1902	2.27	3.23	3.78	3.32	2.29	7.15	12.86	2.49	2.93	2.05	1.72	5.00	48.59
1903	4.07	5.21	4.84	2.76	1.37	5.44	8.42	5.22	1.56	4.03	8.67	2.88	50 17
1904	3.04	3.09	6.18	4.74	3.28	5.11	5.46	4.13	4.59	2.08	.64	2.89	45.23
Mean	3.18	3.24	4.00	3.02	4.14	4.57	5.16	3.68	3.14	2.85	3.05	3.42	43.45

31. LOCK HAVEN, PA.

Year.	Jan.	Feb.	Mar.	Apr.	May.	June.	July.	Aug.	Sept.	Oct.	Nov.	Dec.	Annual.
1891	4.21	4.21	4.06	1.48	1.85	5.14	6.95	4.40	3.41	2.81	2.82	4.44	45.78
1892	4.86	1.37	4.73	1.21	4.91	9.66	3.92	3.72	1.84	.38	3.34	1.85	40.79
1893	2.71	5.28	2.26	4.72	4.89	2.51	3.34	2.82	3.70	2.67	1.09	2.14	38.13
1894	1.77	3.67	.84	5.81	[3.19]	3.52	2.96	5.51	6.46	6.73	1.99	3.73	45.18
1895	4.73	1.00	1.69	.79	2.35	4.84	2.83	3.27	3.18	1.35	2.48	3.46	31.97
1896	.85	4.44	4.05	1.02	1.49	3.67	5.16	3.59	5.46	4.44	2.64	1.02	37.83
1897	1.67	2.67	3.17	2.90	4.65	2.72	5.14	3.94	2.98	.77	4.93	2.59	39.00
1898	4.11	1.51	5.02	2.34	4.10	3.45	3.76	4.90	.36	5.19	2.24	2.14	39.02
1899	2.16	3.72	3.27	1.06	3.80	3.80	2.16	5.05	8.57	.43	8.26	3.56	35.34
1900	2.40	4.04	3.42	1.20	.94	1.59	3.03	4.45	.65	4.92	4.95	1.70	33.23
1901	2.32	.80	4.11	5.67	7.42	3.53	3.21	6.54	4.38	1.37	2.90	5.52	47.97
1902	2.70	3.50	4.93	5.01	.70	6.12	8.34	1.86	4.52	3.98	1.06	4.27	47.08
1903	3.73	2.99	3.97	2.81	1.69	7.44	5.34	6.37	8.20	8.76	1.67	2.37	45.34
1904	3.66	2.33	4.99	4.52	8.66	2.78	2.92	4.09	1.95	1.92	.48	2.83	36.08
Mean	2.99	2.97	3.61	2.89	3.22	4.38	4.22	4.32	3.29	2.88	2.56	2.95	40.18

32. LEWISBURG, PA.

Year.	Jan.	Feb.	Mar.	Apr.	May.	June.	July.	Aug.	Sept.	Oct.	Nov.	Dec.	Annual.
1891	3.38	3.75	6.40	2.39	0.67	5.21	5.09	9.42	2.90	3.75	2.40	4.40	49.71
1892	[2.88]	[3.34]	5.53	2.34	4.96	5.21	3.40	4.55	4.18	.22	3.94	.70	41.25
1893	2.40	4.57	3.07	4.62	6.42	4.96	2.35	[5.11]	1.74	3.20	1.61	[3.43]	42.88
1894	2.84	2.46	1.13	5.33	9.40	2.99	1.96	2.06	5.09	6.02	1.86	4.06	44.00
1895	3.10	1.35	1.88	2.41	3.66	4.18	2.54	4.22	4.11	1.29	2.96	4.09	35.24
1896	1.98	4.46	3.74	1.11	2.16	4.70	5.62	1.39	3.66	5.58	5.85	1.29	41.04
1897	3.26	2.54	4.74	3.21	4.30	2.31	4.72	2.52	2.01	2.08	4.76	3.94	40.39
1898	3.62	2.27	4.23	2.88	6.04	2.79	4.21	9.68	.98	5.76	2.33	2.44	47.13
1899	2.55	4.57	4.36	1.89	4.82	3.83	1.53	5.49	4.36	1.36	2.88	3.98	41.12
1900	2.33	3.92	5.60	1.07	3.16	3.21	3.26	4.08	.65	8.05	4.24	2.38	36.95
1901	1.67	.74	4.49	4.39	7.95	2.09	5.02	10.60	3.85	1.15	1.75	6.90	50.61
1902	3.53	4.41	5.84	2.76	.62	8.28	6.86	2.12	6.40	4.86	1.80	4.96	52.44
1903	3.95	4.86	3.53	4.34	2.40	8.02	5.73	5.21	2.21	3.47	1.09	2.00	47.19
1904	4.52	1.62	3.75	3.78	5.40	1.94	3.61	3.76	3.41	2.69	.72	1.79	36.99
Mean	3.00	3.20	4.11	3.03	4.39	4.18	3.95	5.02	3.25	3.18	2.74	3.31	43.86

34. GIRARDVILLE, PA.

Year.	Jan.	Feb.	Mar.	Apr.	May.	June.	July.	Aug.	Sept.	Oct.	Nov.	Dec.	Annual.
1899	2.76	6.69	4.85	2.02	3.53	5.40	4.99	7.40	6.65	1.02	2.63	4.19	52.13
1900	2.65	5.63	5.50	.94	1.29	3.70	6.96	4.77	1.22	3.82	3.77	3.03	42.78
1901	2.48	11.06	5.68	2.52	5.59	1.39	3.21	12.05	4.92	2.81	2.51	7.87	51.34
1902	4.22	6.45	6.39	3.57	1.31	7.70	5.02	2.83	8.44	6.92	1.90	7.04	61.79
1903	4.28	5.86	4.72	4.23	2.28	7.95	6.19	15.15	3.05	6.75	1.87	4.83	57.16
1904	5.78	2.91	5.89	3.42	4.01	5.95	4.26	4.04	6.50	[4.16]	2.55	[5.89]	54.36
Mean	3.70	4.76	5.42	2.78	3.00	5.35	5.10	6.04	0.01	4.16	2.54	5.39	53.25

Monthly and annual precipitation at stations in Susquehanna drainage basin—
Continued.

35. SELINSGROVE, PA.

Year.	Jan.	Feb.	Mar	Apr.	May	June	July.	Aug.	Sept.	Oct.	Nov.	Dec.	Annual.
1891	4.70	3.09	8.39	1.82	1.36	4.74	6.69	7.18	4.12	4.46	3.85	3.97	54.37
1892	5.13	.88	3.92	1.80	6.25	8.18	4.77	3.17	3.29	.37	4.30	1.99	43.76
1893	2.78	5.63	3.57	4.64	6.85	4.44	2.82	4.07	3.12	4.21	2.40	2.75	46.78
1894	1.22	3.87	1.09	5.45	10.08	2.40	1.20	2.47	4.25	5.58	2.08	3.76	43.40
1895	[2.88]	1.26	2.92	2.55	3.26	3.39	2.54	4.58	1.53	1.80	1.50	3.06	31.27
1896	.90	5.71	4.04	1.16	2.40	2.49	3.36	2.18	3.81	4.36	3.47	.73	37.61
1897	1.85	3.26	3.74	3.25	4.74	2.62	5.08	1.88	2.56	1.89	6.35	3.56	40.78
1898	4.08	2.06	3.87	2.98	5.28	1.61	5.63	6.86	.91	6.22	2.90	2.72	45.12
1899	1.76	4.87	4.58	1.37	4.45	4.04	2.42	4.63	4.72	1.53	3.26	2.61	40.24
1900	2.60	3.50	3.69	1.16	.72	2.09	3.74	2.38	1.59	3.65	3.89	2.18	31.28
1901	2.03	.80	4.11	3.73	7.73	2.50	5.59	8.50	3.52	1.34	1.66	4.84	46.35
1902	3.28	3.23	5.08	3.23	.94	8.11	4.79	1.69	5.16	4.90	1.54	4.26	46.21
1903	4.20	4.84	3.29	4.39	1.78	1.57	4.39	4.91	3.01	3.72	1.53	3.98	47.61
1904	3.99	3.76	3.36	3.70	6.27	3.02	5.04	2.53	4.68	2.40	.70	2.45	41.90
Mean	2.96	3.85	3.98	2.98	4.43	4.09	4.33	4.07	3.30	3.32	2.82	3.06	42.64

36. CENTERHALL, PA.

Year.	Jan.	Feb.	Mar	Apr.	May	June	July.	Aug.	Sept.	Oct.	Nov.	Dec.	Annual.
1895	[2.30]	[3.48]	[4.32]	[2.27]	[3 56]	5.70	3.60	4.70	2.10	1.20	2.33	3 94	39.45
1896	2.18	[3.43]	3.77	1.41	2 00	4.06	5.66	1.26	6.23	3.92	3 11	1.63	38.66
1897	2.20	4.17	5.08	3.84	5.79	4.03	4.96	2.43	4.06	1 78	5 43	4 19	47.96
1898	3.89	1 16	5.16	2.60	4.87	2.89	2.86	7.37	1.26	6.70	2 60	3 90	45 26
1899	2 07	4.54	4.42	.88	5.66	3.05	2.36	3.79	3.90	2 12	1 96	3.87	38 62
1900	1.95	4.09	3.58	1.52	1.92	3.70	3 48	2 56	.88	[3 17]	[2.57]	[3.23]	32.65
1901	[2.30]	[3.43]	[4.32]	[2.27]	[3.56]	[4.43]	5.45	11.30	2.73	.71	2.46	[3 23]	[46.19]
1902	1.50	[3 43]	[4.32]	[2.27]	[3.56]	[4.43]	[4.04]	[5 00]	[3 04]	5 20	.80	[3.23]	[40.82]
1903	[2.80]	3.21	3.90	3.35	1.10	7.59	3.91	6 61	3.19	3 75	1.89	1.84	42.64
1904	2.90	2.07	4.91	5.18	2.84	8.79	5.72	3.01	1.26	[3.17]	[2.57]	1.82	38.28
Mean	2.36	3.30	4.38	2.56	3.44	4.37	4 20	4.80	2.86	3.17	2 57	3.04	41.05

38. STATE COLLEGE, PA.

Year.	Jan.	Feb.	Mar	Apr.	May	June	July.	Aug.	Sept.	Oct.	Nov.	Dec.	Annual.
1891	4.11	5.29	4.07	1.47	1.94	4.24	5.65	5.40	2.20	4.38	2.98	4.08	45.81
1892	3.98	1.73	3.78	2.09	5.79	7.36	3.26	5.78	2.24	.24	3.62	1.07	40.98
1893	1.94	6.71	1.88	5.13	6.46	3.94	4.10	3.14	2.22	3.23	3.04	2.26	43.05
1894	1.75	3.39	1.14	3.85	9.45	4.80	2.10	2.13	5.78	3.13	1.59	3.14	42.05
1895	4.18	.22	1.03	2.23	2.21	6.74	3.11	3.70	1.75	1 03	1.74	2.75	30.69
1896	1.40	4.10	2.82	1.47	1.37	5.02	5.56	1.56	5.02	3.29	3.11	1.04	35.76
1897	2.21	3.19	4.53	3.78	4.13	3.03	5 69	3.39	3.60	1.45	5.26	3.18	43.44
1898	4.40	1.14	5.63	2.29	4.28	3.53	2.95	4.70	.93	6.51	2.28	3.07	41.71
1899	2.60	3.42	4.23	1.71	4.77	2.41	2.14	2.76	3.84	1.40	3.06	2.53	34.87
1900	1.65	3.39	3.81	1.93	2.30	2.54	3.36	2.95	.63	3.22	4.10	1.77	31.65
1901	1.82	.73	3.71	6.14	2.46	3.80	6.30	8.97	2.35	.40	2.06	6.59	45.43
1902	3.02	2.92	4.91	3.13	.92	6.71	5.76	1.37	2.59	4.25	1.44	4.82	41.84
1903	3.50	3.61	4.18	3.81	1.24	7.28	4.04	6.85	2.61	3.51	1.89	1.67	44.19
1904	2.72	3.28	4.04	5.42	2.10	4.19	6.30	1.74	1.86	2.18	.42	1.78	36.03
Mean	2.81	3.01	3.55	3.07	3.79	4.58	4 12	3.89	2.69	2.73	2.61	2 84	39.69

39. GRAMPIAN, PA.

Year.	Jan.	Feb.	Mar	Apr.	May	June	July.	Aug.	Sept.	Oct.	Nov.	Dec.	Annual.
1895	5.19	0 96	1.90	3.81	2 38	2.87	2.85	3 08	2.20	1 26	2.57	3.48	32.55
1896	1.22	3.57	4 02	2.40	2 20	5.76	8.83	3.98	4.45	2.62	3.26	1.82	44 13
1897	2.15	2.78	4.25	4.14	4.55	3.14	7 02	2.46	3.16	.64	6.04	4 57	44 94
1898	3.81	2.06	8 40	2 30	3.30	5.03	3 41	4 12	1.54	5 21	3 55	3 56	46.29
1899	3.12	3.03	4 42	1.67	5.34	3 00	3 84	3 54	3.00	1.56	2.31	3 98	38.79
1900	3.21	3.63	3.64	1 36	2.77	[4.13]	[5 18]	[3.76]	[2.75]	3.82	4.71	2 40	40 86
1901	2.03	1.96	1 88	3 22	3.51	[4 13]	[5.18]	[3.76]	4 22	2.95	.26	[3.74]	[3.30]	38.40
1902	2.42	1.84	2.87	3.71	2.81	[4.13]	[5.18]	[3.76]	[2.75]	[2.43]	[3 74]	[3 30]	[38.94]
1903	[2.80]	4.64	4.89	3.72	2.51	4.98	5.15	4.94	1.98	4.55	[3.74]	3.30	47.29
1904	5.75	3.09	6.08	(a)									
Mean	3.18	2.76	4.23	3.15	3.26	4.13	5.18	3.76	2.75	2.43	3.74	3.30	41.86

a No record.

Monthly and annual precipitation at stations in Susquehanna drainage basin—
Continued.

40. ALTOONA, PA.

Year.	Jan.	Feb.	Mar.	Apr.	May.	June.	July.	Aug.	Sept.	Oct.	Nov.	Dec.	Annual.
1891	2.85	4.50	2.64	1.39	1.97	7.73	3.99	3.13	2.71	2.54	1.89	2.96	37.89
1892	2.06	1.57	2.87	1.66	5.35	5.38	2.50	2.96	1.94	.10	2.69	(3.64)	31.19
1893	1.65	3.21	1.06	3.48	4.67	2.94	2.50	2.92	1.85	2.71	1.48	2.15	31.62
1894	.90	1.82	.80	1.69	9.32	2.66	1.01	3.18	5.25	1.77	.74	3.30	31.53
1895	3.22	.17	1.05	2.16	.80	3.75	1.75	1.64	2.28	.55	1.30	2.50	21.17
1896	.87	1.94	1.77	1.38	2.70	7.69	4.22	1.70	6.03	1.66	2.59	.89	33.44
1897	.95	2.09	3.44	2.91	2.52	2.44	3.22	2.08	2.89	.71	4.31	3.17	29.73
1898	4.05	1.23	5.81	2.22	6.55	1.99	1.91	3.75	.76	7.44	2.14	3.67	40.52
1899	2.41	3.83	4.79	1.64	5.62	1.79	3.67	4.46	3.82	1.28	2.89	2.70	38.35
1900	2.21	3.55	3.12	1.22	3.91	2.53	3.25	3.90	1.48	3.69	4.54	1.50	34.84
1901	1.89	.78	4.07	6.22	5.85	4.04	5.83	5.34	2.29	.59	3.08	4.98	43.86
1902	2.85	2.60	3.96	5.30	1.30	4.95	6.88	1.12	1.58	4.36	1.05	5.37	41.32
1903	3.84	4.59	4.38	2.99	2.63	4.34	4.51	5.08	1.98	3.36	1.82	1.50	40.97
1904	3.08	2.39	4.12	4.40	2.93	3.09	4.68	1.69	1.73	1.43	.63	1.98	32.10
Mean	2.31	2.42	3.10	2.76	4.01	3.95	3.57	3.07	2.61	2.29	2.15	2.89	34.63

41. HUNTINGDON, PA.

Year.	Jan.	Feb.	Mar.	Apr.	May.	June.	July.	Aug.	Sept.	Oct.	Nov.	Dec.	Annual.
1891	3.58	3.84	4.48	1.92	1.84	4.24	4.49	3.80	2.07	3.13	2.39	4.18	39.95
1892	4.22	1.86	5.11	2.29	6.24	6.44	3.48	4.08	2.81	.12	3.04	1.55	41.19
1893	2.10	5.27	2.07	4.61	7.79	2.37	2.39	3.49	8.50	3.70	2.46	2.46	42.21
1894	1.82	3.44	1.07	3.19	9.20	3.56	1.57	1.26	7.56	2.98	1.81	4.21	41.62
1895	5.16	.46	1.42	1.97	3.01	4.78	3.15	1.46	1.26	1.09	1.07	2.99	27.82
1896	2.13	2.90	3.82	1.85	2.56	7.93	8.60	2.29	7.43	3.24	3.04	.76	40.13
1897	1.65	4.69	3.95	3.86	4.69	4.27	3.13	3.81	1.74	5.16	3.19	1.70	43.02
1898	4.60	1.12	4.79	1.73	4.60	2.07	2.08	4.68	.67	6.54	2.02	2.41	37.26
1899	2.10	3.49	4.55	1.07	3.83	2.43	3.68	4.96	8.57	.49	8.25	2.60	36.02
1900	1.07	2.68	2.61	2.64	3.11	2.77	1.38	1.78	.64	2.51	4.33	1.38	26.85
1901	1.32	.67	3.30	4.18	5.19	1.59	5.20	5.68	2.49	1.50	.94	5.61	36.60
1902	2.44	2.98	5.24	3.79	1.30	7.18	4.80	1.72	3.21	5.67	.96	5.50	44.29
1903	3.80	5.38	4.18	3.04	1.76	6.32	4.84	6.43	3.02	3.64	1.83	1.40	45.50
1904	3.07	2.39	4.00	4.05	2.41	6.42	7.61	4.38	.84	1.91	.61	1.78	39.47
Mean	2.79	2.95	3.57	2.87	4.11	4.46	3.63	3.52	3.03	2.66	2.35	2.86	38.80

42. HARRISBURG, PA.

Year.	Jan.	Feb.	Mar.	Apr.	May.	June.	July.	Aug.	Sept.	Oct.	Nov.	Dec.	Annual.
1891	4.73	3.31	4.25	1.70	1.77	3.76	8.40	5.20	1.75	2.87	1.95	3.71	43.40
1892	5.14	1.02	4.81	2.15	3.95	4.93	6.48	2.80	3.31	.15	4.15	1.17	39.65
1893	2.05	4.66	1.97	3.67	5.32	2.46	1.92	1.74	3.25	2.54	1.91		35.13
1894	1.77	4.56	1.30	2.27	6.07	3.25	1.89	4.08	5.53	4.60	1.90	3.34	40.56
1895	3.80	.54	1.94	3.67	1.98	1.66	1.16	2.36	2.18	1.63	1.72	3.86	26.02
1896	1.00	5.48	3.85	1.19	2.99	3.82	6.32	1.45	1.81	3.45	3.80	.40	35.05
1897	1.60	2.77	2.87	2.53	5.30	1.83	3.68	3.13	1.30	1.85	4.09	3.21	33.46
1898	3.23	1.60	3.04	1.95	6.13	1.98	5.07	8.44	2.08	5.26	3.15	3.16	45.09
1899	2.27	3.71	8.69	1.15	4.49	2.93	1.90	4.85	4.25	.78	2.13	1.83	38.94
1900	2.07	3.40	3.00	1.43	1.83	2.88	3.14	4.72	1.41	1.25	2.69	1.62	28.94
1901	1.83	.53	3.60	2.88	5.98	1.13	1.52	2.99	2.16	1.15	1.99	4.75	29.81
1902	3.28	5.49	2.98	2.73	.29	4.76	3.68	2.26	4.01	5.81	1.49	4.57	39.35
1903	3.67	4.19	3.76	3.24	.46	5.63	1.76	5.82	1.95	2.62	.88	1.92	35.90
1904	3.11	1.54	2.72	2.07	3.45	8.99	4.76	2.95	1.69	2.78	.54	2.59	31.99
Mean	2.82	3.06	3.13	2.38	3.54	3.22	3.69	3.88	2.51	2.50	2.27	2.67	35.62

Monthly and annual precipitation at stations in Susquehanna drainage basin—
Continued.

43. LEBANON, PA.

Year.	Jan.	Feb.	Mar.	Apr.	May.	June.	July.	Aug.	Sept.	Oct.	Nov.	Dec.	Annual.
1891	[5.30]	3.33	5.30	2.19	3.00	3.40	8.70	5.06	1.07	3.14	2.44	4.34	47.27
1892	6.27	.95	4.91	2.22	5.14	4.75	4.75	3.80	3.63	.29	4.55	1.96	43.22
1893	2.10	5.67	2.63	3.67	8.05	2.21	2.67	5.30	3.79	3.95	3.42	2.35	45.81
1894	2.17	4.23	1.48	4.77	9.45	1.91	4.42	4.17	5.47	6.14	2.57	4.17	50.95
1895	4.70	.87	2.49	5.10	1.85	1.88	2.10	1.97	1.32	2.31	1.95	4.14	30.68
1896	1.11	6.31	5.29	1.29	4.54	4.51	6.88	.56	2.92	4.70	4.76	.68	48.05
1897	2.96	3.75	3.46	3.51	6.52	3.00	5.89	2.51	1.57	2.95	5.76	4.05	44.84
1898	4.27	1.59	3.20	3.18	7.90	1.90	3.58	10.43	.99	5.38	5.54	3.41	50.77
1899	3.67	5.16	5.21	1.51	4.53	5.54	1.91	3.18	6.20	.95	2.50	1.75	42.20
1900	2.81	5.50	2.94	2.08	2.13	3.64	5.43	4.26	1.84	1.85	2.85	2.39	37.22
1901	2.46	.84	4.36	4.02	6.05	3.24	3.61	3.65	3.65	1.40	1.39	6.35	46.03
1902	3.62	5.67	4.79	3.38	.43	6.18	4.21	5.49	4.43	5.93	1.45	7.46	53.04
1903	4.68	5.95	4.65	3.67	.94	6.08	3.94	7.23	2.55	4.48	1.28	3.15	48.65
1904	3.58	2.22	3.50	2.48	5.60	5.22	5.89	5.56	3.81	3.05	1.68	2.71	45.26
Mean	3.50	3.72	3.87	3.08	4.72	3.78	4.53	4.87	3.09	3.25	3.01	3.49	44.91

46. YORK, PA.

Year.	Jan.	Feb.	Mar.	Apr.	May.	June.	July.	Aug.	Sept.	Oct.	Nov.	Dec.	Annual.
1891	3.65	3.37	6.07	2.01	2.39	3.98	10.77	3.29	1.88	3.20	2.13	4.20	46.94
1892	6.08	.10	3.94	1.70	4.10	3.81	8.59	2.81	2.66	.14	4.44	2.13	40.50
1893	1.76	4.76	1.76	4.37	6.53	2.50	1.58	3.40	1.57	3.03	8.55	2.22	37.03
1894	1.34	4.20	1.58	4.48	4.40	3.08	2.22	2.93	9.16	4.24	2.09	3.90	43.60
1895	4.08	.98	2.50	3.74	2.73	3.10	1.41	2.41	4.01	2.36	1.80	3.38	32.40
1896	.94	4.88	4.20	1.45	2.58	3.92	4.00	1.05	2.54	3.44	3.00	.45	32.40
1897	1.55	4.59	2.51	3.42	6.61	2.42	3.69	4.04	2.73	2.60	5.69	3.37	43.22
1898	3.67	1.15	3.00	2.71	6.86	1.08	3.47	6.44	1.82	4.81	4.75	3.58	42.84
1899	3.61	6.64	5.16	1.28	5.71	3.54	5.32	6.76	6.07	.92	3.59	1.18	49.78
1900	2.12	4.62	3.08	1.35	1.85	4.81	2.36	4.09	3.18	1.51	2.81	2.52	34.30
1901	2.72	3.53	3.94	2.51	2.55	1.55	3.33	6.27	2.36	1.59	2.50	6.17	36.02
1902	2.73	6.74	4.80	3.41	1.24	5.15	5.74	4.22	4.12	6.40	2.39	6.15	53.09
1903	4.67	6.13	4.73	3.21	1.18	6.21	4.01	6.96	2.72	3.51	1.89	2.90	48.11
1904	4.39	1.03	2.98	(a)									
Mean	3.09	3.56	3.58	2.74	3.74	3.47	4.36	4.21	3.45	2.87	3.13	3.24	41.56

a No record.

FLOODS.

During the last century there have been several great floods on Susquehanna River, the most notable of which are those of **March, 1865**; June, 1889 (the Johnstown flood); May, 1894, and March, 1904.

The flood of 1865 was the result of the rapid melting and passing away of a large quantity of ice and snow which had accumulated during an exceptionally severe winter. The amplitude of this flood was probably increased by ice gorges. No information in regard to the height of this flood has been obtained except that at the junction with the West Branch the river was 2 feet higher than during the June flood of 1889; and the old residents along other portions of the main river state that this flood was approximately the same as the June flood of 1889.

The flood of June, 1889, caused by the heavy rainfall of May 30 to June 1, probably exceeded any flood which has ever occurred on this stream. Being in the summer months, it was not augmented by ice gorges, and therefore illustrates the normal effect of high-water conditions. The table below, taken from the report of the Chief of Engineers, U. S. Army, shows the extent and duration of rainfall within the limits of the West Branch; it was upon the high table-lands of this portion of the basin that the heaviest precipitation took place.

Rainfall over drainage area of West Branch, May 30 to June 1, 1889.

Station.	County.	Storm began—	Storm ended—	Duration.	Rainfall.
				Hrs.	*Ins.*
Siglerville	Mifflin	3 p. m. May 30	1 a. m. June 1	34	
Hollidaysburg	Blair	do	3 a. m. June 1	36	6.10
State College	Center	3.30 p.m. May 30	do	37	5.04
Lewistown	Mifflin	4 p. m. May 30	2 a. m. June 1	34	
Huntingdon	Huntingdon	do	do	34	7.50
Philipsburg	Center	do	3 a. m. June 1	35	6.09
Grampian	Clearfield	4.30 p.m. May 30	11.30 p.m. May 31	32	8.60
Emporium	Cameron	5 p. m. May 30	11 p. m. May 31	32	5.97
Coudersport	Potter	6 p. m. May 30	12 p. m. May 31	30	5.40
Selinsgrove	Snyder	do	3 a. m. June 1	33	7.53
Charlesville	Bedford	8 p. m. May 30	3 p. m. May 31	36	7.60
Williamsport	Lycoming	9 p. m. May 30	5 a. m. June 1	32	
Ralston	do	1 a. m. May 31	12 m. June 1	32	
Muncy	do	3 a. m. May 31	1 p. m. June 1	34	

From this table it is seen that the average duration of the rainfall was about thirty-four hours and that the average depth was about 6.6 inches. Under ordinary conditions about 50 per cent of the rainfall

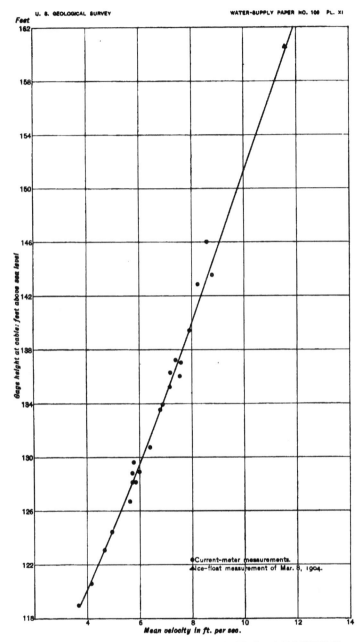

CURVE OF MEAN VELOCITY FOR SUSQUEHANNA RIVER AT McCALLS FERRY, PA.,
CABLE STATION.

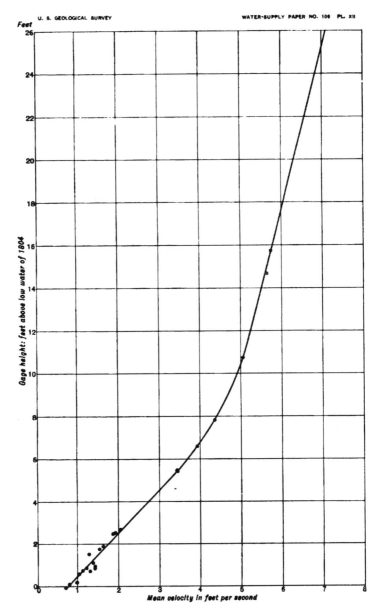

CURVE OF MEAN VELOCITY FOR SUSQUEHANNA RIVER AT HARRISBURG, PA.

in the Susquehanna drainage area reaches the outlet of the river. It is probable, however, that under extraordinary conditions, such as mentioned above, there was a run-off of at least 75 per cent of the rainfall.

Various methods of estimating the maximum discharge of the 1889 flood have been used, perhaps the most reliable indicating that about 593,000 second-feet flowed past Harrisburg, and 671,000 second-feet past McCalls Ferry. The basis of these estimates is shown in Pls. XI and XII, the other methods and results being given on pages 177 to 180.

Pls. XI and XII were prepared as follows: The mean velocities for the various discharge measurements taken at the respective stations were plotted with gage heights as ordinates and mean velocity in feet per second as abscissæ. Through these points a mean velocity curve was drawn and extended to reach the highest gage height of the flood. This curve shows the mean velocity for any stage of the river. The crest of the 1889 flood at Harrisburg was 27.1 feet above the low water of 1803 and at McCalls Ferry cable station about 162 feet above mean sea level. The curves show that the mean velocities for these heights are 7.24 feet per second and 11.90 feet per second, respectively. At each of these stations an accurate cross section was determined, and the product of the area below the flood line and the mean velocity for that gage height, as taken from the extended mean velocity curve, gives the flow of the river. In this method of estimating flood discharges the uncertainty due to the area of the cross section, as when the discharge curve is produced, is eliminated. A study of other mean velocity curves made in this manner shows that the liability to error in the mean velocity is comparatively small, and it is probable that this method gives a better estimate than either Kutter's formula or the discharge curve.

The result is a maximum flow at McCall Ferry about 13 per cent greater than at Harrisburg, which accords with the assumption that the discharge between two points on the same river where the drainage area is similar should increase in proportion to the drainage area. At McCalls Ferry the drainage area is 11.4 per cent greater than at Harrisburg.

The loss of life caused by the flood within the drainage area of the West Branch was 78, and the flood relief commission disbursed nearly $300,000 to the sufferers within this district, but no attempt was made to secure even an approximate estimate of the damage. The flood of May, 1894, near McCalls Ferry was 2 or 3 feet lower than the 1889 flood.

The primary cause of the flood of March, 1904, was the breaking up of the ice in January without enough water behind it to force it down the river. Gorges were formed at various points along the river and its branches, which were greatly solidified by the exceptionally cold weather in the following month. When the final break came these gorges were still further augmented and acted as dams, impounding the large quantity of water which was so destructive to property along the shores.

On March 6 and 7 there were heavy rains all over the drainage area, and on the morning of March 8 the floods so caused began to break through the various barriers. It finally forced the big gorges at Highspire and Bainbridge, wiping out islands and doing much damage in its course.

After the flood had subsided at York Haven, the gorge moved to Turkey Hill, where it stood for several hours and backed the water to within a few feet of the Columbia Bridge. Between 1 and 2 p. m. this gorge in turn gave way and moved to Shanks Ferry, where it gorged for the last time. Although it held here for only a few moments, it raised the water and ice 6 feet above the railroad track at Safe Harbor, completely destroying the stone-arch bridge there and leaving ice throughout the village to the height of the second-story windows.

The elevation of the crest of the flood, as shown for a portion of the river by the table on page 175, varied in height at various places along its course, as compared with the June flood of 1889. At York Furnace the height was about 3 feet greater; about a mile above McCalls Ferry it was practically the same; at McCalls Ferry station it was 3 feet lower, and at the head of Cullys Falls it was again about the same height.

There came down with the flood wave a large amount of ice, which varied from 3 to 10 feet in thickness, as shown by the blocks left on the shores. Owing to the cross currents in the river, the greater portion of the ice went down on the York County side, and it was on this side that most of it was left piled up on the shores. The channel on the Lancaster County shore soon cleared itself, and but little ice accumulated upon that bank.

The gorge at Turkey Hill broke about 2 o'clock in the afternoon, and at 3.30 p. m. the water reached a maximum height at McCalls Ferry. At the cable station it was 161.3 feet above sea level on the Lancaster County side and 159.8 feet on the York County side. Within half an hour from the time the maximum height was reached the water had fallen from 2 to 3 feet, and on the morning of March 9 it had fallen 15 feet.

Between Shanks Ferry and Port Deposit no more ice jams were formed, and the ice passed through the channel of the river very rapidly and caused but little damage. The history of nearly all floods has been that between "The Neck" and Port Deposit but little gorging takes place and that the river rapidly clears itself from any

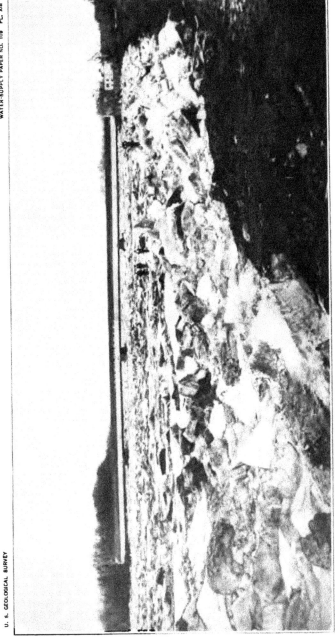

ICE FLOOD OF 1875 AT WILKESBARRE, PA.

FLOOD OF MARCH 8, 1904, AT ITS HEIGHT AT YORK HAVEN, PA.

ice and seldom rises to such a height as to cause particular damage
along the shores. At Port Deposit there is frequent trouble, for the
shallow sand bars and tidal backwater often cause gorges which flood
the tracks and lower part of the town.

Elevations of flood on lower portion of Susquehanna River, March 8, 1904.

Locality.	Elevation.	Remarks.
	Feet.	
Fort Cullys Falls, gage No. 5	139.5	Approximate.
Lock 13 (behind ice)	136.3	Ice gorged in channel above.
600 feet above Lock 13...............	140.1	Made of drift.
500 feet above Lock 12...............	143.0	Observed during flood.
Power house, gage 2.................	146.6	Do.
Dam line, York side	146.7	Do.
High-water gage 10..................	147.7	Do.
McCalls Ferry, York County........	150.7	Observed during flood; in backwater behind ice.
McCalls Ferry, Lancaster County...	151.8	Observed during flood.
At telegraph line on T. P	156.3	Do.
Station 71+80 on T. P	158.8	Do.
At cable, York County...............	159.8	Observed during flood; behind ice.
At cable, Lancaster County.	161.8	Drift marks.
Tucquan culvert	167.5	Do.
Milepost 29.........................	175.5	Watermark on post.
York Furnace station..........	179.5	Watermark on station.
York Furnace Hotel.................	178.6	Observed during flood.
Pequea Bridge	182.6	Watermarks on house and post.
Milepost 31.........................	182.7	Watermarks on post.
Shanks Ferry Hotel	185.7	Observed during flood.
Milepost 32.........................	186.3	Watermarks on posts.
Safe Harbor........................	204.0	Watermarks on station.

Above Shanks Ferry much damage was done, and the loss of property was great at many points. The facts are interesting to those
who contemplate power development in the lower portion of Susquehanna River, as the possible damage from ice has been one of the
great objections to such development.

The full effect of the flood on the main stream was not felt below
Sunbury, being restrained by the big gorges at Kipps Run, Catawissa,
and Nanticoke, which held several days longer. It was at its worst
in Wyoming Valley on the 9th, doing much damage to Plymouth,
Wilkesbarre, and Pittston, and then quietly passed away without
noticeable effect on the lower river.

A rough estimate of damage due to flood, as given by press reports, is as follows:

Damage due to flood of March, 1904.

Pittston to Sunbury *a*	$6,500.000
York County *b*	200.000
Lancaster County	275.000
Dauphin County *c*	275.000
Cumberland County	200.000
Perry County	200.000
Snyder County	125.000
Juniata County	100.000
Maryland	100,000
Total	7,975,000

The loss and damage to State bridges was reported as $800,000.

The table below gives a comparison of the heights during the flood period at various points along the river.

1904 flood heights, in feet, above low water of September, 1900.

Date.	Main river at McCalls Ferry (4 p. m.).	Main river at Harrisburg (7 a. m.).	Main river at Wilkesbarre (8 a. m.).	West. Branch at Williamsport (7.30 a. m.)	Juniata at Newport (12 m.).
1904.					
March 3	9.0	11.9	9.0	7.4	4.4
March 4	9.9	13.5	11.2	18.9	10.7
March 5	15.0	22.0	16.0	16.4	6.1
March 6	15.0	19.4	14.9	9.1	3.3
March 7	13.4	16.3	15.4	7.8	2.7
March 8	33.6	21.2	26.8	17.6	11.2
March 9	17.2	15.9	28.5	18.4	7.3
March 10	17.4	15.0	24.0	9.7	4.4
March 11	17.9	12.0	21.9	7.5	3.3
March 12	13.6	9.2	19.9	6.4	3.2
Maximum height attained.	*a* 33.6	*b* 23.3	*c* 28.5	*d* 18.9	....

a March 8, 4 p. m.   *b* March 4, 3 p. m.   *c* March 9, 8 a. m.   *d* March 4, 7 a. m.

NOTE.—Maximum heights other than at McCalls Ferry were caused by backwater from gorges.

a Of which one to two millions were in Wyoming Valley.

b Most damage at York Haven and vicinity.

c Of which Middletown losses amounted to about $100,000.

A

B

McCALLS FERRY IN FLOOD OF MARCH 8, 1904

A. At beginning of flood. *B*. after flood

A

B

ICE LEFT BY FLOOD OF MARCH 8, 1904

A, At York Haven, Pa , *B*, below McCalls Ferry, Pa.

The cable gaging station about three-fourths mile above McCalls Ferry offered a good opportunity for determining the amount of water flowing at the maximum stage. At this point two cables are stretched across the river 80 feet apart, and at the time of the flood the sun was shining in line with these and bright enough to cast their shadows on the white ice, thus enabling the determination of the velocity at this point with considerable degree of accuracy. The velocity was determined in four different portions of the river, and several individual determinations were made in each portion. The result of this measurement is shown in the table below.

Flood discharge at cable station, McCalls Ferry, Pa., March 8, 1904, 4 p. m.

[Elevation water surface, Lancaster County side, 161.3 feet; York County side, 159.8 feet; mean 160.6 feet.a]

Stations.	Surface velocities.	Mean velocity 90 per cent of surface.	Area.	Discharge.	Remarks.
	Ft. per sec.	*Ft. per sec.*	*Sq. feet.*	*Sec.-feet.*	
50 to 125	0	--------	4,710	0	Ice piled along towpath. No apparent velocity.
125 to 625	20	18	23,560	424,000	Velocity obtained by timing ice cakes between cables 80 feet apart.
625 to 725	13.3	12	4,600	55,200	Do.
725 to 825	0	--------	4,370	0	Backwater behind Streepers Island.
825 to 975	13.3	12	6,960	83,500	Velocity obtained by timing ice cakes between cables 80 feet apart.
975 to 1180	11.4	10.2	6,700	68,300	Do.
1180 to 1320	0	--------	3,600	0	Ice and backwater.
Total	--------	--------	54,500	631,000	Mean velocity 11.6 feet per second.

a Corresponding gage height for 1889 flood was about 162 feet, with discharge of 671,000 second-feet.

The table on page 178 gives the estimated maximum, minimum, and mean discharge of Susquehanna River at Harrisburg for 1891 to 1904, inclusive.

Minimum, maximum, and mean discharge of Susquehanna River at Harrisburg, Pa., for 1891 to 1904, inclusive.

Year.	Minimum.			Maximum.			Mean discharge.
	Date.	Gage height.	Discharge.	Date.	Gage height.	Discharge.	
		Feet.	*Sec.-ft.*		*Feet.*	*Sec.-ft.*	*Sec.-ft.*
1891	Oct. 4–7, inclusive	1.60	10,200	Feb. 19	19.00	334,500	52,200
1892	Oct. 31–Nov. 8, inclusive	.50	4,070	Apr. 6	14.65	224,200	37,250
1893	Aug. 16–19, inclusive, 25	.35	3,500	May 6	16.50	267,400	40,550
1894	Sept. 5–6	.25	3,160	May 22	25.60	543,500	39,970
1895	Oct. 30–31	.05	2,570	Apr. 11	18.65	205,400	29,330
1896	Sept. 5–13	.25	3,160	Apr. 1–2	14.60	223,200	34,600
1897	Sept. 15, Oct. 21	.50	4,070	Mar. 26	11.50	165,306	32,320
1898	Oct. 3–7	.65	4,740	Mar. 24	15.65	245,900	40,490
1899	Oct. 24 and 25	.15	2,850	Mar. 7	18.00	193,000	31,000
1900	Sept. 28 and 29	—.04	2,360	Mar. 2	18.10	194,900	29,950
1901	Nov. 12	1.00	6,550	Dec. 16	21.40	405,100	42,380
1902	Sept. 23, 24, 25	.85	5,760	Mar. 2	23.90	484,100	47,100
1903	Oct. 7	1.40	8,850	...do ...	16.85	276,500	54,510
1849	Dec. 11	0.84	5,708				32,318
For the 14 years	Sept. 28–29, 1900.	—.04	2,360	1894. May 22	25.60	543,500	38,855

FLOOD DISCHARGES AND VALUES OF "N" BY KUTTER'S FORMULA.

Owing to the lack of high-water gagings on Susquehanna River, it became necessary to estimate the flood discharges by means of the slope formula, $v = c\sqrt{Rs}$, using Kutter's formula to fix the value of c. The 1889 flood is the highest on record, and as there remain many of its high-water marks made by eyewitnesses along the railroad and canal above McCalls Ferry, Pa., the mean slope along this part of the river could be closely approximated. These marks consist of notches on posts, rocks, hotels, bridge piers, and locks, and their elevations were accurately determined, as shown on the profile.

Ten sections, located as shown on Pl. XVIII, were then chosen from the contour map. These were selected so as to show as far as possible the average for the portions of the river represented, so that the mean slope between the nearest reliable high-water marks could be used in connection with them. The sections were carefully surveyed and sounded to determine their area and wetted perimeter.

In order to get a value for n in Kutter's formula the slopes were measured on the west channel of the Duncans Run section during

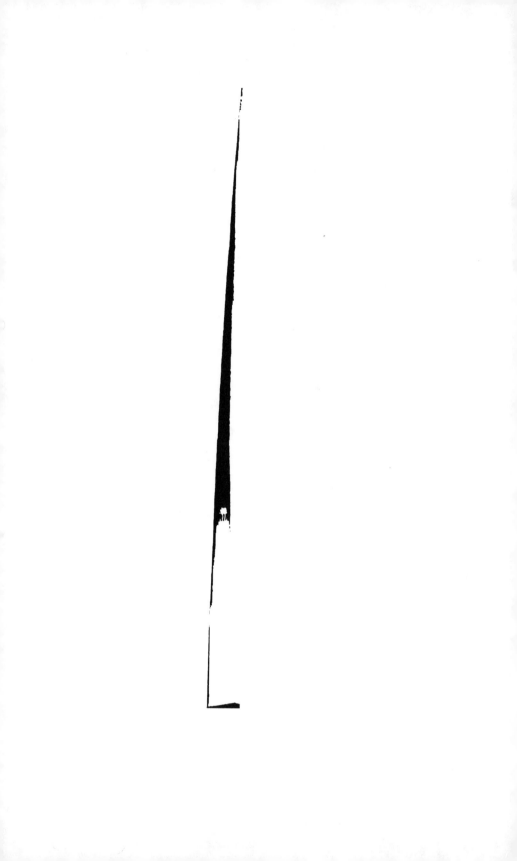

ιgs
ιhe

—

—

0.72′
0.30′
———
0.42′

Calls

of
for
.ses
the

ιec-

.um
ιter
See

ιal-
10,
the
,000
by

X is
οr a
ond-

dis-
'eet.

several gagings. With these slopes and the data from the gagings made on July 24 and 26, 1902, June 5, 1903, and March 8, 1904, the coefficients c and n have been computed by the formulas—

$$Q = A v; \quad v = c\sqrt{Rs}; \quad c = \frac{41.6 + \dfrac{.00281}{s} + \dfrac{1.811}{n}}{1 + \dfrac{\left(41.6 + \dfrac{.00281}{s}\right)n}{\sqrt{R}}},$$

as shown in the table below.

Values of c and n, with data used in their determination.

Date.	Discharge.	Area.	Wetted perimeter	(R) Hydraulic radius.	(V) Mean velocity	Coefficient (c).	Observed slope (s).	Computed coefficient (n).	Remarks.
	Sec. ft.	Sq ft.	Feet.	Feet.	Feet per sec.				
July 24,1902a.	78,300	9,340	560	16.68	8 38	54.9	0.0014	0.0468	El. W. S 150' above line = 130.72' El. W. S. 150' below line = 130.30' Fall in 300' 0.42'
July 26,1902a.	68,000	8,650	557	15 51	7 86	54.8	.00138	0462	Fall in 300' as above = 0.40'
June 5, 1903a.	10,000	3,846	380	10 12	2. 60	52.3	.000244	.0460	Fall in 900' = 0.22'
Mar. 8. 1904b.	631,000	63,400	2,420	26.20	9 96	52.45	.00138	.0645	Slope taken between McCalls Ferry and Gage No. 2.

a At Duncans Run.   b At section No 10.

The three measurements at Duncans Run give a coefficient of about 0.046. The conditions there are exceptionally favorable for this part of the river, so that as the flood sections in many cases included brushy and wooded islands, the value of n as used in the computations was increased to 0.05.

The data and results showing the discharge at the respective sections during the 1889 flood are shown in table on page 180.

The mean of the discharges of these 10 sections gives a maximum for the 1889 flood of about 730,000 second-feet, or 9 per cent greater than the mean velocity curve estimate of 671,100 second-feet. (See pages 177 and 180.)

In this connection it is of interest to note that if a coefficient equaling 0.055, as determined by the single measurement at section 10, based upon the flood gaging of March 8, 1904, had been used, the mean discharge for the 1889 flood would have been about 685,000 second-feet, or only 2 per cent greater than the results obtained by using the mean velocity curve.

The general equation of the discharge curve shown on Pl. X is approximately that of the parabola $(y-111)^2 = .00202\ x$, which for a gage height of 149.5 gives the 1889 flood discharge as 733,800 second-feet.

From these estimates it may be assumed that the maximum discharge of the 1889 flood was between 670,000 and 735,000 second-feet.

In determining n at section 10 by means of the flood measurement of March 8, 1904, the slope used was between McCalls Ferry and gage No. 2, the same points as were taken for the 1889 flood slope, thus making the two comparable and indicating that the assumed value of $n=.05$ is on the safe side.

Discharge of Susquehanna River during 1889 flood as computed by Kutter's formula.

$$Q = Ac \sqrt{Rs}; \quad c = \cfrac{77.82 + \cfrac{.00281}{s}}{1 + \cfrac{2.08 + \cfrac{.00014}{s}}{\sqrt{R}}}$$

No. of section.	Area.	Wetted perimeter.	Hydraulic radius.	Mean slope.	Coefficient (N).	Mean velocity.	Discharge.	Remarks.
	Sq. feet.	Feet.	Feet.			Ft. per sec.	Sec. ft.	
1.....	89,300	4,750	18.80	0.0012	0.05	7.98	713,000	One-fourth of section is brushy island.
2.....	105,500	4,210	25.06	.00060	.05	6.91	730,000	One-third of section is low, brushy, rocky island.
3.....	110,400	4,300	25.66	.00060	.05	7.02	775,000	Do.
4.....	113,600	5,020	22.63	.00064	.05	6.67	758,000	One-fourth of section covered with trees or brush.
5.....	110,500	3,220	34.32	.00035	.05	6.61	730,000	One-sixth of section covered with brush.
6.....	63,700	2,800	22.75	.00130	.05	9.43	602,000	One-fourth of section is covered with trees.
7.....							739,000	
8.....	89,500	2,800	31.96	.00070	.05	8.72	780,000	One-fourth of section is rocky island.
9.....							720,000	
10.....	72,800	2,430	29.95	.00110	.05	10.38	756,000	One-fourth of section covered with brush or trees.
Mean							730,300	

LOW-WATER CONDITIONS.

At the time of the establishment of the gage at Harrisburg, in 1891, the lowest-known water on Susquehanna River was in 1803, and the zero of the gage was placed at the elevation of this low water.

The months of August and September, 1900, were periods of extreme drought, and beginning with the 1st of September the observations at Harrisburg showed a gradual falling of the river until September

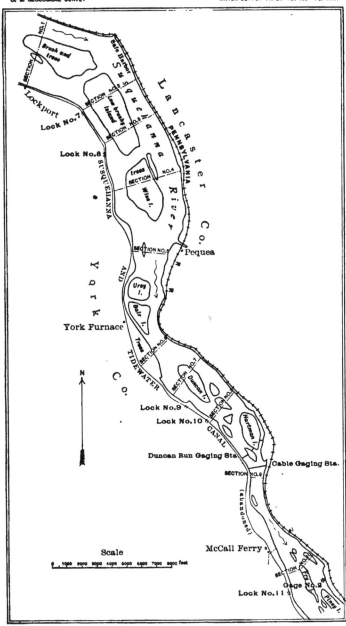

MAP SHOWING SECTIONS USED IN KUTTER'S FORMULA DETERMINATIONS NEAR McCALLS FERRY, PA.

28–29, when the gage read 0.04 of a foot below the low-water mark of 1803. During this period of low water Mr. E. G. Paul, hydrographer, United States Geological Survey, spent considerable time in measuring the flow at the various stations in the Susquehanna drainage basin. On September 21 a measurement was made at Harrisburg at a gage height of 0.08 of a foot and a discharge of 2,655 second-feet. Mr. Paul returned to Harrisburg on September 28, at which date the river reached its extreme low point of −0.04 of a foot, and made a measurement giving a discharge of 2,357 second-feet.

The measurements made by Mr. Paul during the week of September 28, 1900, at Allenwood, Danville, and Newport, Pa., as shown by the table below, gave a very close check upon the Harrisburg work, and show that the measurements as made at the various points along the river are consistent among themselves and that no errors greater than would be expected in work of this kind exist.

Comparison of minimum discharges of Susquehanna River and its branches.

Date.	Stream.	Station.	Dis-charge.	Remarks.
			Sec.-feet.	
Sept. 24, 1900	West Branch	Allenwood, Pa.	511	Gage same height as on Sept. 28.
Sept. 25, 1900	Susquehanna	Danville, Pa.	822	Gage 0.1 of a foot lower than Sept. 26–28.
Sept. 22, 1900	Juniata	Newport, Pa.	418	Gage same as Sept. 28.
Total discharge from gagings above Harrisburg			1,751	
Add 14 per cent for increase in drainage area			258	
Add for 0.1 lower gage height at Danville			140	
Total estimated discharge above Harrisburg			2,149	
Gaging at Harrisburg Sept. 28			2,357	
Difference			208	

From the best available authorities the elevation of lowest water, in September, 1900, at McCalls Ferry, gage No. 2, was about 112.6 feet. The measured minimum discharge at Harrisburg for that month was 2,357 second-feet, and by increasing this figure 11.4 per cent, to allow for the increase in drainage area, we find the corresponding maximum discharge at McCalls Ferry to be about 2,620 second-feet. In order to check this result, the mean velocities of the various discharge measurements made at Duncans Run have been plotted as abscissæ and their respective gage heights as ordinates, as shown in Pl. XIX. These points, it will be seen, seem to follow a general law, and a curve has been drawn through them

which has been extended through the gage height of the lowest
water, which at Duncans Run was about 114.2 feet. The velocity
from the curve for that gage height is 1.0 foot per second, and the area
of the section is 2,940 square feet, the product of these two giving
a discharge of 2,940 second-feet as a rough check on the above.
The lowest water actually measured at McCalls Ferry was on September 25, 1902, at a gage height on gage No. 2 of 114.34 feet, giving a
discharge of 6,370 second-feet. The mean discharge from the rating
table at Harrisburg on that date was 5,760 second-feet, corresponding
to a difference in drainage area of 10.6 per cent. The table on page
178 gives the minimum estimated discharge at Harrisburg for the
years 1891 to 1904, inclusive.

ACCURACY OF STREAM MEASUREMENTS.

Considerable comment has been made upon the hydrographic
work of the United States Geological Survey on Susquehanna River
by engineers and others who are promoting power schemes in the
lower portion of the river, and it was to obtain varying data that
the late George S. Morison, engineer for the McCalls Ferry project,
established a gaging station at that point.

As stated on page 130, the McCalls Ferry station was established
in May, 1902, and during the following year 35 discharge measurements were made at stages which ranged between the highest and
lowest gage heights during this period. These measurements were
taken with great care, vertical velocity curves being used in most
cases. From the measurements a rating curve and table was prepared, by which, in connection with the daily gage heights, both the
daily and the monthly discharges of the river were computed, as
shown on pages 137–139.

On comparing the monthly discharges at McCalls Ferry from June
1, 1902, to December 31, 1904, as obtained by Mr. Morison's engineers,
with those obtained by the United States Geological Survey at Harrisburg, as shown in the table on page 183, it is found that the mean
monthly discharge is approximately between 7 and 25 per cent greater
at McCalls than at Harrisburg. This difference is what would be
expected, as the drainage area at McCalls Ferry is 11.4 per cent greater
than that at Harrisburg.

It is thus seen that the methods of stream measurement used by
the Geological Survey give results which agree with those obtained by
private engineers, whose work is generally carried on in greater detail
and at much greater cost.

An inspection of the discharge curves shows that almost all of the
individual measurements plot nearly on the curve, very few of them
varying from it by more than 3 per cent. This fact, while it does not
prove their accuracy, indicates that the measurements were carefully
made and that the results are consistent.

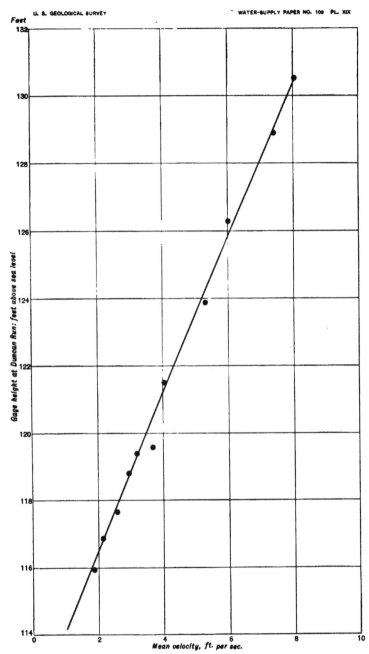

CURVE OF MEAN VELOCITIES FOR SUSQUEHANNA RIVER AT DUNCANS RUN,
NEAR McCALLS FERRY, PA

Comparison of the estimated monthly discharge of Susquehanna River at Harrisburg and McCalls Ferry, Pa.

Month.	Mean discharge in second-feet.			
	Harrisburg.	McCalls Ferry.	Difference.	
			Second-feet.	Per cent.
1902.				
June	12,810	13,908	1,098	+ 7.9
July	70,209	61,768	−8,441	−13.7
August	26,962	27,126	164	+ .6
September	11,714	11,556	− 158	− 1.4
October	35,656	38,248	2,592	+ 6.8
November	20,985	22,657	1,672	+ 7.4
December	63,774	69,111	5,337	+ 7.7
The period	34,587	34,911	324	+ .9
1903.				
January	37,765	43,533	5,768	+13.2
February	93,236	95,082	1,846	+ 1.9
March	133,500	134,461	961	+ .7
April	82,715	79,900	−2,815	− 3.4
May	14,297	16,826	2,529	+15.0
June	27,964	29,859	1,895	+ 6.4
July	32,581	35,636	3,055	+ 8.6
August	25,581	28,206	2,625	+ 9.3
September	30,511	34,188	3,672	+10.7
October	45,160	48,757	3,597	+ 7.4
November	27,289	30,797	3,508	+11.4
December	19,743	19,751	− 8	0
The year	47,528	49,638	2,110	+4.3
1904.				
April	74,230	78,400	4,170	+ 5.3
May	41,740	46,720	4,980	+10.7
June	29,320	34,580	5,260	+15.2
July	18,020	21,410	3,890	+15.8
August	10,420	13,880	3,460	+24.9
September	8,657	11,050	2,893	+21.7
October	15,240	18,700	3,460	+18.5
November	10,760	13,320	2,560	+19.3
December	8,448	10,890	2,442	+22.4
The period	24,090	27,660	3,570	+12.9

NOTE.—Owing to an ice gorge below Harrisburg the monthly means for January, February, and March have been estimated by taking 89 per cent of means for McCalls Ferry.

VERTICAL VELOCITY MEASUREMENTS.

The standard with which all velocity determinations in stream-measurement work are compared is the mean velocity obtained by the vertical velocity method. This method consists in taking, in a vertical line, a series of velocity determinations, which when plotted with depths as ordinates and velocities as abscissæ give the basis for the construction of a velocity curve along the vertical in question. This curve shows the variation in velocity from the surface to the bottom of the stream, and from it the mean velocity for the vertical can be determined by dividing the area included within the curve by the depth. From these curves not only the depth at which the mean velocity occurs can be found, but also coefficients for reducing to the mean the velocities found at the top, bottom, or at other points.

In the work in the Susquehanna drainage area three series of vertical velocity measurements have been made, as follows: At McCalls Ferry, Pa.; at Binghamton, N. Y., and at Harrisburg, Pa.

The series at McCalls Ferry, Pa., was made during the years 1902 and 1903 by Messrs. Boyd Ehle and R. H. Anderson and consisted of 73 determinations at the Duncans Run section and 104 measurements at the cable section. The depths at the first section varied from 3 to 30 feet and the mean velocities from 1.2 to 5.8 feet per second. At the second section the depths ranged from 3 to 36 feet and the mean velocities from 1.2 to 9.7 feet per second. These great depths and the high velocities at which these measurements were made make them by far the most interesting series of the kind that have been made.

The bed of the stream at both of these points is very irregular and is made up mostly of solid rock, strewn with large bowlders, as shown in Pl. I, B, thus making the velocities near the bottom hard to determine.

The secondary guy cable with which the station is equipped, as noted on page 131 and shown on Pl. IX, A, enabled the observer to hold the meter at a depth which it is very difficult to reach under ordinary conditions.

The results of the measurements have been tabulated and are given in the tables on pages 185-187, and the platted curves are shown in Pls. XX to XXVI, inclusive.

A study of these tables shows that in order to draw any conclusions from the results the individual determinations must be grouped, in order to bring together those which were taken under the same conditions. The grouping for the Duncans Run series was made according to depth as follows: Group 1, 4 to 10 feet; group 2, 10 to 20 feet; group 3, 20 to 30 feet, and those for the cable station according to the distance from the initial point.

Rejecting disturbed and discordant observations, the averages from these groups give the results shown in the table on page 188.

Vertical velocity measurements at Duncans Run, above McCalls Ferry, Pa.

Distance from initial point, in feet.	Depth, in feet.	Velocity, in feet, per second by following methods:				Coefficient for reducing to mean velocity			Depth of thread of mean velocity.*	
		Vertical velocity.	0.6 depth *	Top and bottom.	Top.	0.6 depth.	Top and bottom.	Top.	In feet.	In per cent of depth.
9	15.5	2.52	2.73	1.77	3.00	0.92	1.42	0.84	10.5	68
10 a	18.0	2.36	2.50	1.80	2.40	.90	1.26	.94	13.6	76
10	22.5	3.12	3.40	2.63	3.32	.92	1.19	.94	15.2	68
10	16.0	2.02	2.20	1.69	2.42	.92	1.20	.84	11.7	73
15	19.0	2.74	2.92	2.25	3.35	.94	1.22	.82	12.7	67
15 a	18.0	2.58	2.63	2.40	2.73	.98	1.08	.94	12.7	71
20	16.0	1.79	2.10	1.38	1.20	.86	1.30	1.49	13.7	86
20 a	17.8	2.43	2.68	1.72	2.62	.90	1.41	.98	12.8	72
20	22.0	2.96	3.30	2.16	3.32	.90	1.37	.89	15.5	71
25 a	20.5	2.62	2.72	2.14	2.85	.96	1.22	.92	15.4	75
28 a	19.0	1.83	2.32	1.18	1.10	.79	1.55	1.66	15.5	82
30 a	18.0	1.68	2.10	1.28	1.18	.80	1.31	1.48	16.2	90
30 a	23.0	2.64	2.82	2.31	2.90	.94	1.14	.91	17.5	76
40 b	4.0	2.68	2.88	2.58	3.05	.93	1.04	.88	2.5	63
50 b	4.3	3.30	3.55	3.44	3.46	.93	.96	.96	3.1	72
60 b	3.3	3.10	3.64	2.52	4.36	.85	1.23	.71	2.2	67
70 b	5.0	3.60	3.62	3.60	3.83	1.00	1.00	.94	3.2	64
80 b	9.0	3.55	3.45	3.51	4.50	1.03	1.01	.79	4.8	53
90 b	5.0	4.66	4.65	4.48	4.73	1.00	1.04	.98	3.0	60
100 b	4.5	5.80	6.05	4.43	5.30	.96	1.31	1.10	3.0	67
110 b	6.0	3.86	4.13	3.70	4.22	.94	1.04	.91	4.0	67
120 b	7.5	2.42	2.48	2.53	2.72	.98	.96	.89	5.3	71
122 b	14.0	3.04	3.28	2.28	3.70	.93	1.33	.82	9.1	65
130 a	12.0	2.12	2.30	1.95	2.06	.92	1.09	1.03	9.5	79
130	14.0	2.38	2.42	2.15	3.10	.98	1.11	.77	8.8	63
132 a	13.5	3.20	3.30	3.14	3.05	.97	1.02	1.05	12.6	98
130	20.0	3.41	3.50	2.96	3.83	.98	1.15	.89	12.7	64
140	20.5	2.24	2.30	1.97	2.58	.97	1.14	.87	14.4	70
140	22.0	2.46	2.62	2.20	2.58	.94	1.12	.95	15.1	69
140	25.0	3.48	3.71	2.70	4.03	.94	1.29	.86	16.7	67
140	25.0	2.63	2.80	2.01	3.08	.94	1.31	.86	16.5	66
150 a	20.0	2.20	2.27	2.08	2.34	.97	1.07	.94	14.7	73
150	21.5	2.98	3.05	2.83	2.96	.98	1.04	.99	15.7	73
150	22.5	2.65	2.75	2.59	2.76	.96	1.02	.96	16.4	73
150	27.5	3.38	3.58	2.55	3.83	.94	1.32	.88	20.3	74
160 a	24.0	1.97	2.02	1.66	2.13	.98	1.19	.92	15.8	64
160	26.5	2.54	2.67	2.25	2.62	.96	1.13	.97	18.0	68
160	31.0	3.06	3.06	2.62	3.83	.99	1.16	.79	19.3	62
160	27.0	2.72	2.98	2.30	3.05	.92	1.18	.89	19.3	72
170 b	24.5	2.02	2.22	1.73	2.10	.91	1.17	.96	18.0	73
170	25.5	2.35	2.54	2.06	2.48	.92	1.14	.95	17.8	70
170	28.0	3.22	3.18	2.75	3.79	1.01	1.17	.85	16.0	57
170	24.0	2.80	3.00	2.37	2.84	.94	1.18	.99	19.7	82
180 b	17.0	2.10	2.14	2.12	2.10	.98	.99	1.00	16.0	94
180	25.0	1.87	2.20	1.42	2.25	.85	1.32	.83	17.0	68
180	29.0	2.77	3.00	2.20	3.79	.92	1.26	.73	18.1	62
180	16.0	2.82	2.94	2.70	2.60	.96	1.04	1.08	15.0	94
190 b	25.0	1.84	1.92	1.67	1.92	.96	1.10	.96	16.7	67
190	27.0	2.16	2.33	1.88	2.34	.93	1.15	.92	20.4	75
190	30.0	2.96	2.87	2.46	3.70	1.04	1.21	.80	17.4	58
190	25.0	2.75	2.75	2.69	2.81	1.00	1.02	.98	15.0	60
200 b	25.0	1.70	1.83	1.28	1.92	.92	1.33	.88	17.0	68
200	26.0	2.20	2.38	1.72	2.25	.92	1.28	.98	18.7	72
200	26.5	2.30	2.46	1.93	2.72	.97	1.24	.88	17.4	66
210 a	21.0	1.69	1.78	1.50	1.79	.95	1.13	.94	16.5	78
210	22.5	2.17	2.30	1.88	2.20	.94	1.15	.98	17.7	79
210	21.5	2.78	2.78	2.77	3.11	1.00	1.00	.90	12.9	60
220 a	21.5	1.58	1.66	1.37	1.73	.95	1.15	.92	15.5	84
220	19.5	2.06	2.09	2.08	2.10	.98	.99	.98	14.5	74
220	20.0	2.52	2.58	2.40	2.63	.98	1.05	.96	18.8	69
230 a	16.3	1.45	1.57	.94	1.56	.92	1.54	.93	11.0	68
230	16.0	2.40	2.50	2.25	2.54	.96	1.07	.94	11.0	69
230	17.0	1.75	1.85	1.62	2.02	.95	1.08	.87	11.5	68
240 a	13.0	1.31	1.36	1.30	1.28	.96	1.01	1.02	11.0	85
240	15.0	1.67	1.80	1.60	1.83	.93	1.04	.91	10.8	72
240	14.5	2.37	2.41	2.27	2.47	.98	1.04	.96	9.5	66
250 a	10.0	1.21	1.35	1.11	1.09	.90	1.09	1.11	8.6	86
250	12.5	1.55	1.73	1.44	1.56	.90	1.08	1.00	9.7	78
250	18.5	1.90	2.00	1.62	2.10	.95	1.17	.90	9.0	67
260 b	8.0	1.34	1.40	1.24	1.25	.88	1.09	.99	6.0	75
260	8.0	1.25	1.35	1.21	1.56	.93	1.03	.80	5.5	69
260	9.5	1.70	1.65	1.66	2.04	1.03	1.02	.84	5.0	53

* From vertical velocity curve.   a Even rock bottom.   b Uneven rock bottom.

Vertical velocity measurements at cable station above McCalls Ferry, Pa.

Distance from initial point, in feet	Depth, in feet.	Velocity, in feet per second, by following methods—			Coefficient for reducing to mean velocity.		Depth of thread of mean velocity. *a*	
		Vertical velocity	0.6 depth.*a*	Top.	0.6 depth.	Top.	In feet.	In per cent of depth.
150 *b*	8.0	3.26	3.22	3.70	1.01	0.88	4.6	5*x*
	10.0	4.30	4.40	4.82	.98	.89	6.5	65
	10.0	4.06	4.24	4.48	.96	.91	7.3	73
	12.0	4.15	4.68	4.45	.89	.93	9.3	77
	13.0	4.80	5.20	5.27	.92	.91	9.6	74
	19.0	5.76	6.40	5.75	.90	1.00	15.0	79
200 *b*	8.7	4.00	4.08	4.38	.98	.91	6.7	77
	10.0	5.20	5.45	5.75	.95	.90	7.3	73
	11.0	5.00	5.30	5.33	.94	.94	8.2	75
	14.0	6.75	7.06	7.07	.96	.96	11.0	78
250 *b*	7.0	3.42	3.68	3.67	.93	.93	5.6	80
	9.0	4.90	5.00	5.43	.98	.90	6.3	70
	16.5	7.50	7.45	7.77	1.01	.96	10.6	64
300 *c*	7.0	4.64	5.05	5.30	.92	.88	5.3	76
	8.0	4.86	5.15	5.45	.94	.89	6.0	75
	16.5	7.60	6.63	9.60	1.14	.79	12.6	76
350 *b*	6.0	4.20	4.27	4.85	.98	.96	5.0	83
	8.0	4.76	4.88	5.27	.98	.90	6.5	81
	9.0	5.40	5.65	5.75	.96	.94	7.0	78
	16.0	8.12	8.70	9.60	.93	.85	12.7	79
385 *c*	13.0	2.47	2.57	2.70	.96	.92	9.0	69
400 *c*	10.0	1.32	1.01	1.73	1.21	.71	3.5	35
	14.0	3.28	3.28	3.70	1.00	.89	8.4	60
	15.0	2.96	3.00	3.63	.99	.82	9.2	61
	15.0	3.74	3.55	4.78	1.05	.78	7.7	51
	15.0	5.20	5.72	5.30	.91	.98	11.6	77
	16.0	4.13	4.28	5.58	.97	.74	11.0	69
	18.0	5.13	4.93	6.83	1.04	.75	8.2	46
	22.5	7.62	8.12	8.90	.94	.86	16.2	72
450 *c*	8.0	3.18	3.30	3.38	.96	.94	6.0	75
	10.0	5.09	6.13	5.87	.93	.97	7.7	77
	15.5	5.75	6.10	6.20	.94	.93	10.7	69
	14.0	8.15	8.47	9.35	.96	.87	9.8	70
	16.0	9.16	9.60	10.90	.95	.84	11.3	70
500 *b*	16.0	3.80	4.12	3.90	.92	.98	13.1	82
	16.5	3.74	3.83	3.93	.98	.95	15.3	93
	21.5	5.03	5.17	5.17	.97	.97	19.0	88
	24.5	6.02	6.00	6.88	1.00	.88	14.4	59
	27.0	7.77	7.70	9.10	1.01	.85	15.8	59
	28.0	7.50	7.80	8.75	.96	.86	18.7	67
	36.0	9.00	9.22	10.00	.98	.90	23.8	66
550 *b*	16.0	4.30	4.30	5.17	1.00	.83	9.6	60
	19.0	4.24	4.41	4.85	.96	.87	12.6	66
	21.0	4.33	4.42	5.00	.98	.87	13.1	62
	24.5	6.38	6.38	7.50	1.00	.85	14.7	60
	28.0	7.20	7.22	8.15	1.00	.88	17.0	61
	28.0	7.47	7.62	7.97	.98	.94	20.2	72
600 *b*	35.0	9.70	9.80	10.66	.99	.91	22.2	63
	17.0	3.95	4.10	4.55	.96	.87	11.3	66
	20.0	4.30	4.50	4.90	.96	.88	13.3	66
	21.0	4.97	5.02	5.40	.99	.92	14.1	67
	25.0	6.30	6.43	6.63	.98	.95	17.8	71
	28.5	7.40	7.42	7.47	1.00	.99	17.5	61
	29.0	7.54	7.64	8.05	.99	.94	22.0	76
625 *c*	35.0	8.23	8.62	9.25	.96	.89	25.2	72
650 *c*	15.0	3.27	3.00	4.20	1.09	.78	7.9	58
	5.5	5.15	5.57	6.05	.92	.85	3.9	71
	11.0	5.80	5.65	6.53	1.03	.89	6.0	55
	15.0	6.84	6.45	7.73	1.06	.88	6.9	46
650	17.0	6.83	6.50	7.73	1.05	.88	8.5	50
	18.0	6.70	6.60	8.17	1.01	.82	10.5	58
	21.0	7.64	8.07	8.51	.95	.90	16.6	79
	26.0	7.44	7.70	8.92	.97	.83	17.6	64
700 *b*	4.5	4.70	4.97	5.35	.95	.88	3.1	69
	8.0	5.28	5.60	6.08	.94	.87	5.8	73
	8.0	4.97	5.20	5.20	.96	.96	6.2	78
	13.7	6.24	6.45	7.25	.97	.86	9.2	67
	15.0	6.12	6.30	6.75	.97	.91	10.1	67
	15.5	6.00	6.12	6.85	.98	.88	10.4	67
	20.0	6.67	7.00	7.42	.95	.90	16.7	84
	24.5	7.00	7.87	7.87	.95	.89	19.3	79
750 *c*	5.5	5.00	5.60	6.10	.89	.82	4.0	73
	12.0	5.56	5.70	6.20	.98	.90	7.9	66
	12.0	5.22	5.25	6.40	.99	.82	7.3	61
	13.5	5.30	5.47	6.33	.97	.84	8.8	65
	15.0	6.33	6.85	7.07	.93	.90	12.3	82
	20.0	5.50	5.50	6.65	1.00	.83	12.0	60

a From vertical velocity curve   *b* Regular bottom.   *c* Rough and irregular bottom.

Vertical velocity measurements at cable station above McCalls Ferry, Pa.—Continued.

Distance from initial point, in feet.	Depth, in feet.	Velocity, in feet per second, by following methods			Coefficient for reducing to mean velocity		Depth of thread of mean velocity.	
		Vertical velocity.	0.6 depth.	Top.	0.6 depth.	Top.	In feet.	In per cent of depth.
380 a	6.0	5.60	5.73	6.33	0.98	0.89	3.8	63
	11.0	5.80	6.20	6.80	.94	.85	7.9	72
	11.5	6.17	6.20	7.00	1.01	.88	7.2	63
	15.0	5.78	6.12	6.20	.94	.93	12.1	81
	16.0	6.12	6.40	7.00	.96	.87	12.3	77
	21.5	5.86	5.55	5.80	.97	.96	16.6	77
450 a	6.0	3.83	3.95	4.13	.97	.93	4.2	70
	11.0	4.97	5.15	5.63	.96	.88	7.5	68
	13.0	4.87	5.15	5.05	.95	.96	9.7	75
	15.0	4.80	4.95	5.45	.97	.88	10.6	71
	15.0	4.66	4.82	5.63	.97	.83	10.6	71
	16.0	5.54	5.85	5.72	.95	.97	13.0	81
	21.0	6.82	7.17	7.23	.95	.94	16.5	79
900 a	7.0	1.34	1.45	1.62	.95	.85	4.8	69
	9.0	3.14	3.35	4.00	.94	.79	6.7	74
	13.0	3.38	3.56	3.77	.95	.90	9.7	75
	16.0	5.00	5.43	5.38	.92	.93	12.3	77
	16.0	4.94	5.20	5.32	.95	.93	11.2	70
	18.0	5.30	5.35	5.87	.99	.90	12.0	67
	19.0	6.08	6.23	6.32	.97	.96	16.0	84
	25.0	7.20	7.35	8.05	.98	.90	19.7	79
950 a	7.7	1.85	1.98	2.02	.98	.92	5.5	71
	10.0	2.67	2.75	3.14	.97	.85	6.3	63
	12.7	3.32	3.43	4.00	.97	.83	8.6	68
	16.0	4.90	5.07	5.50	.97	.89	11.3	71
	16.5	5.07	5.10	5.80	.99	.87	10.2	62
	17.7	6.40	6.66	7.07	.96	.91	14.0	79
	2.4	7.70	7.80	8.28	.99	.93	17.4	73

a Regular bottom.

Recapitulation and deductions from vertical velocity measurements at Duncans Run.

Group.	No of observations.	Depth	Coefficients for reducing to mean velocity.			Depth of thread of mean velocity in per cent of total depth.
			Six-tenths depth.	Top and bottom.	Top.	
		Feet.	Per cent.	Per cent.	Per cent.	
1	12	4 to 10	94.3	106.7	92.2	67.8
2	23	10 to 20	94.8	115.5	92.2	71.7
3	25	20 +	94.8	118.4	91.7	70.1

From the above table we find, first, that the depth of the thread of mean velocity ranges from about 68 to 72 per cent of the total depth, and that holding the meter at 0.6 depth gives a result about 5 per cent too large; second, that the coefficient for reducing top velocity to mean velocity is practically 92 per cent; third, that the coefficient for reducing the mean of the top and bottom velocities to mean velocity ranges from 106 to 118 per cent. The discordance here is due to the roughness of bed, which reduces the bottom velocity to a minimum.

Recapitulation and deductions from vertical velocity measurements at cable station, McCalls Ferry, Pa.

Distance from initial point, in feet.	Depths, in feet.	Velocities, in feet per second.	Number of observations.	Coefficients for reducing to mean velocity.		Depth of thread of mean velocity in per cent of total depth.
				Six-tenths depth.	Top.	
150	8 to 19	3.3 to 5.8	6	0.94	0.92	71
200	9 to 14	4.0 to 6.8	4	.95	.93	76
300	7 to 16	5.0 to 6.6	8	1.00	.85	76
850	6 to 16	4.2 to 8.1	4	.96	.91	80
500	16 to 36	3.8 to 9.2	7	.97	.91	73
550	16 to 35	4.3 to 9.7	7	.99	.88	63
600	17 to 29	4.0 to 7.5	7	.98	.92	68
700	4 to 24	4.7 to 7.0	8	.96	.89	73
850	6 to 21	3.8 to 6.8	7	.96	.91	74
900	7 to 25	1.4 to 7.3	8	.96	.90	74
950	8 to 24	1.9 to 7.7	7	.97	.89	70
Mean	5 to 36	1.4 to 9.7	68	.97	.90	72

An examination of the above table shows, first, that the thread of mean velocity varies between about 63 and 80 per cent of the total depth, and that holding the meter at 0.6 depth gives a result between 0 and 6 per cent too large, with an average of about 3 per cent. Second, that the coefficient for reducing top to mean velocity ranges from about 85 to 93 per cent, with a mean of 90 per cent.

From July 1, 1901, to August 15, 1902, Mr. E. C. Murphy made a special study of the accuracy of current-meter work and the laws of flowing water, on Chenango and Susquehanna rivers, at Binghamton, N. Y. A detailed account of these studies can be found in Water-Supply and Irrigation Paper No. 95, from which paper the data used in the following are taken.

Figs. 4 and 5 show contours of the bed and position of the piers and abutments at the two measuring stations. The Chenango River

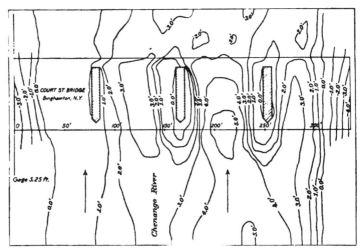

FIG. 4.—Contour of bottom of Chenango River at Court Street Bridge, Binghamton, N. Y.

station is at Court Street Bridge, Binghamton, where the observations were taken. The channel there is straight for about 1,000 feet on each side of the station, has a width of about 300 feet at low water and 340 feet at high water, and is broken by three piers. The bed is gravel and cobbles, with large rough stones around the piers. The bed is seen to be irregular in shape, as well as rough, but is permanent. The station is about 2,500 feet from Susquehanna River, and is subject to backwater at certain stages. Although the channel is broken by three piers, the bridge projects over the piers on each side, so that the section of measurement is continuous.

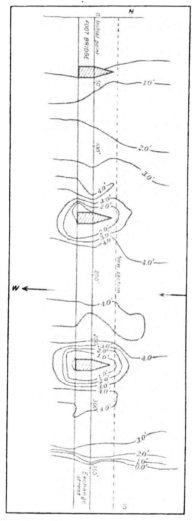

FIG. 5.—Contours of Susquehanna River bed at Exchange Street Bridge, Binghamton, N Y.

At the Exchange Street Bridge, where the observations on Susquehanna River were made, the channel is straight for about 500 feet above and below the station, has a width of about 300 feet at low water and about 450 feet at high water, broken by 3 piers. The bed is of gravel and cobbles, with large irregular-shaped-rock filling around the piers. The velocity is rather high, especially at the higher stages. About 900 feet above the station is a dam whose height is about 6 feet.

The methods of work and computations at each station were as follows: The vertical velocity curve observations consisted in measuring velocity at from three to five points in each of the verticals, the lowest point being one-half foot above the bed, and the highest 1 foot below the surface. Each observation covered four periods of 25 seconds each. The velocities computed from these observations were plotted on section paper, and a smooth curve was drawn among these called the velocity curve. These points gave, as a rule, a well-defined curve, except near the bottom, where the bed was rough.

The curves for each vertical were grouped according to gage height, so that the range for each group was not greater than 1 foot. A mean vertical velocity curve was then drawn for each group. In making these mean curves the means of the velocity at the surface and at each two-tenths depth of the original curves were used. The resulting mean curves are shown in figs. 6, 7, 8, and 9, and the deductions from these are given in the tables headed " Vertical Velocity Measurements on Susquehanna River at Binghamton, N. Y.," and " Vertical Velocity Curves on Chenango River at Binghamton, N, Y."

In the tables, top velocity means velocity one-half foot below the surface, and bottom velocity means velocity one-half foot above the bed. Columns 9, 10, and 11 give the mean velocities in each vertical, as obtained by three methods, and columns 12, 13, and 14 the coefficients for reducing velocities obtained by either of these methods to mean velocity as obtained from the vertical velocity curves.

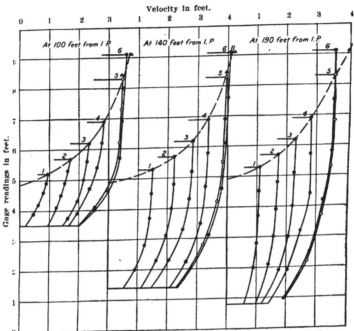

FIG. 6.—Mean vertical velocity curves, Chenango River, Binghamton, N. Y.

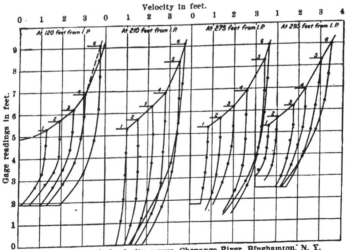

FIG. 7.—Mean vertical velocity curves, Chenango River, Binghamton, N. Y.

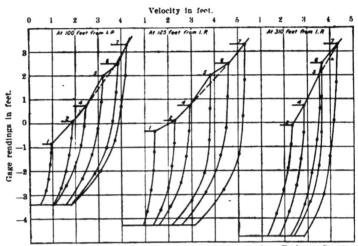

FIG. 8.—Mean vertical velocity curves, Susquehanna River, Exchange Street
Bridge, Binghamton, N. Y.

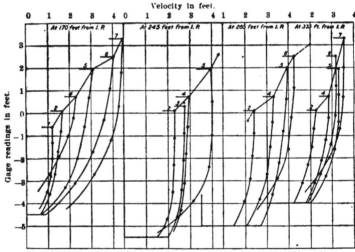

FIG. 9.—Mean vertical velocity curves, Susquehanna River, upper side of Exchange
Street Bridge, Binghamton, N. Y.

Vertical velocity measurements on Chenango River, Binghamton, N. Y.

No. of curve.	Gage height.	Distance from initial point.	Depth.	Velocity in feet per second from the mean curves by following method:							Coefficient for reduction to mean velocity.			Position of thread of mean velocity. In per cent of depth.	Character of bed.
				Top.	Middle depth.	0.6 depth.	Bottom.	V. V. curve.	$T+\frac{B}{2}$	$\frac{T+2M+B}{4}$	0.6 depth.	$T+\frac{B}{2}$	$\frac{T+2M+B}{4}$		
1	5.2	100	1.7	0.83	0.68	0.62	0.50	0.64	0.67	0.67	1.08	0.96	0.96	56	G
2	5.7	100	2.2	1.59	1.42	1.38	1.18	1.38	1.39	1.40	1.00	.99	.99	60	G
3	6.2	100	2.9	2.26	2.15	2.02	1.77	2.05	2.02	2.08	1.01	1.02	.98	60	G
4	6.9	100	3.4	2.77	2.55	2.42	1.97	2.43	2.37	2.46	1.00	1.03	.99	60	G
5	8.3	100	4.8	3.41	3.31	3.22	2.43	3.11	2.94	3.11	.97	1.06	1.00	67	G
1	5.3	140	3.9	1.45	1.35	1.27	.83	1.22	1.14	1.25	.96	1.07	.98	66	G
2	5.7	140	4.3	2.20	2.05	1.95	1.42	1.92	1.81	1.93	.98	1.06	.99	61	G
3	6.2	140	5.0	2.80	2.65	2.54	1.80	2.46	2.30	2.47	.97	1.07	1.00	61	G
4	6.9	140	5.5	3.35	3.20	3.10	2.43	3.06	2.89	3.05	.98	1.07	.99	65	G
5	8.3	140	6.9	3.90	3.75	3.73	2.67	3.53	3.29	3.52	.95	1.07	1.00	68	G
1	5.3	190	4.5	1.08	1.00	.97	.74	.94	.91	.95	.97	1.03	.99	61	G
2	5.7	190	4.9	1.70	1.60	1.50	1.12	1.50	1.46	1.53	1.00	1.03	.98	60	G
3	6.2	190	5.4	2.26	2.10	2.04	1.40	2.00	1.88	1.96	.98	1.09	1.02	68	G
4	6.9	190	6.1	2.80	2.67	2.45	1.63	2.44	2.22	2.44	1.00	1.10	1.00	60	G
5	8.3	190	7.4	3.54	3.35	3.20	2.02	3.14	2.78	3.06	.98	1.13	1.08	65	G
1	5.3	210	5.3	1.00	.95	.90	.66	.89	.83	.89	.99	1.07	1.00	60	G
2	5.7	210	5.7	1.59	1.52	1.49	1.05	1.43	1.32	1.42	.96	1.08	1.00	68	G
3	6.2	210	6.2	2.26	2.15	2.05	1.53	2.02	1.90	2.02	.99	1.06	1.00	63	G
4	6.9	210	6.9	2.80	2.70	2.65	1.80	2.53	2.30	2.50	.96	1.10	1.01	68	G
5	8.3	210	8.3	3.44	3.28	3.15	2.05	3.06	2.75	3.01	.97	1.11	1.02	64	G
1	5.3	120	3.4	1.26	1.13	1.05	.50	.98	.88	1.01	.98	1.11	.96	66	G
2	5.7	120	3.8	1.90	1.80	1.62	.93	1.65	1.22	1.61	1.02	1.08	1.08	59	G
3	6.2	120	4.3	2.53	2.30	2.09	1.87	2.10	1.95	2.12	1.00	1.08	.99	60	G
4	6.9	120	5.0	3.06	2.77	2.62	1.70	2.60	2.38	2.58	.99	1.09	1.01	62	G
1	5.3	275	3.5	.90	.73	.70	.58	.70	.69	.71	1.00	1.01	.99	60	B
2	5.7	275	4.2	1.41	1.29	1.23	.89	1.19	1.15	1.22	.97	1.04	.99	65	B
3	6.2	275	4.4	1.97	1.89	1.76	1.29	1.76	1.63	1.78	1.00	1.08	.99	60	B
4	6.9	275	5.4	2.58	2.52	2.44	1.89	2.38	2.24	2.38	.97	1.06	1.00	67	B
5	8.3	275	6.9	3.40	3.22	3.13	2.22	3.06	2.81	3.02	.98	1.09	1.01	65	B
1	5.3	295	2.3	.58	.53	.49	.34	.46	.46	.48	.94	1.00	.96	65	G
2	5.7	295	2.9	.98	.92	.86	.60	.87	.79	.88	1.01	1.10	1.01	55	G
3	6.2	295	3.7	1.75	1.59	1.52	1.20	1.52	1.48	1.53	1.00	1.03	.99	60	G
4	6.9	295	4.4	2.30	2.06	1.95	1.47	1.97	1.88	1.97	1.01	1.05	1.00	59	G
5	8.3	295	5.8	2.98	2.60	2.47	1.75	2.50	2.37	2.48	1.01	1.06	1.01	59	G
Mean											.984	1.041	.996	65.6	

NOTE.—"No. of curve" refers to figs. 6 and 7.

Vertical velocity measurements on Susquehanna River, Binghamton, N. Y.

No. of curve.	Gage height.	Distance from initial point.	Depth.	Velocity in feet per second from the mean curves by following method							Coefficient for reduction to mean velocity			Position of thread of mean velocity	Character of bed.
				Top.	Middle depth.	0.6 depth.	Bottom.	V.V. curve.	$\frac{T+B}{2}$	$\frac{T+2M+B}{4}$	0.6 depth.	$\frac{T+B}{2}$	$\frac{T+2M+B}{4}$	In per cent of depth.	
1	−0.85	100	2.5	0.94	0.82	0 81	0.66	0.80	0.80	0.81	0.99	1.01	0.99	61	G
2	+.10	100	3.5	1.85	1.65	1.52	1.27	1.58	1.56	1.60	1 04	1.01	1 03	57	G
4	+.73	100	4.1	2.42	2.20	1.99	1.42	2.03	1.92	2.05	1 02	1.00	99	56	G
5	+2.00	100	5.4	3.20	2.87	2.74	1.90	2.71	2.55	2.71	99	1.01	1 00	61	G
6	+2.50	100	5.9	3.84	3.50	3.35	2.23	3.26	3.02	3.27	.97	1.08	1 00	61	G
7	+3.30	100	6.7	4.22	3.87	3.72	2.35	3.58	3.29	3.58	.96	1.00	1.00	64	G
1	.30	125	4.0	1 42	1.32	1.26	1.05	1.27	1.24	1.28	1.01	1.03	.99	56	G
2	+.10	125	4.4	2.29	2.10	2.03	1.57	2.03	1.93	2.02	1 00	1 05	1.00	60	G
4	0.73	125	5.0	2.92	2.70	2.59	1.90	2.57	1.41	2.56	.99	1 11	1.00	61	G
5	2.00	125	6.3	3.82	3.74	2.63	2.55	3.48	3.19	8.46	.96	1.00	1.01	68	G
6	2.50	125	6.8	4.58	4.40	4.20	2.95	4.10	3.77	4.08	.98	1.00	1.00	65	G
7	3.30	125	7.6	5.29	5.15	5.04	3.60	4.86	4.45	4.80	.96	1.00	1.01	65	G
2	−.10	310	4.7	2.30	2.06	1.95	1.48	1.98	1.89	1.97	1 01	1.05	1 00	60	G
4	.73	310	5.5	2.85	2.62	2.53	2.00	2.53	2.42	2.52	1 00	1.05	1 00	60	G
5	+2.00	310	6.8	3.52	3.32	3.15	2.51	3.18	3.02	3 17	1.01	1.05	1 00	60	G
6	2.50	310	7.3	3.63	3.57	3.43	2.65	3.37	3.14	3.35	.98	1.07	1.01	68	G
7	3.30	310	8.1	4.30	4.05	3 97	3.13	3.93	3.72	3.88	.99	1.06	1.02	63	G
1	−.60	170	8.6	1.20	1.15	1.10	.90	1.08	1.02	1.09	.98	1.01	.99	61	B
2	+.10	170	4.6	1.65	1.40	1.30	.90	1.34	1.27	1.34	1.03	1.06	1 00	56	B
4	+.73	170	4.2	2.24	1.75	1.57	.85	1.67	1.55	1.65	1.06	1.08	1.01	55	B
5	2.00	170	6.5	3.02	2.53	2.40	1.20	2.36	2.11	2.34	.98	1.12	1.01	62	B
6	2.50	170	7.0	3.98	3.32	3.07	1.47	3.08	2.73	3.02	.00	1.13	1.02	60	B
7	3.30	170	7.5	4.45	4.12	3.90	2.35	3.79	3.40	3.76	.97	1 11	1.01	64	B
2	−.10	245	5.6	2.30	2.23	2.20	1.80	2.16	2.05	2.14	.98	1.05	1.01	66	G
3	+.30	245	5.8	2.76	2.65	2.62	2.25	2.60	2.50	2.58	1.00	1.04	1.01	60	G
4	+.70	245	6.0	2.94	2.80	2.75	2.45	2.78	2.75	2.75	1.01	1.01	1.01	58	G
5	2.00	245	7.5	3.96	3.95	3.78	2.20	3.59	3 08	3 51	.95	1.17	1.02	68	G
2	+0.10	285	5.1	2.45	2.36	2.19	1.69	2.16	2.07	2.16	.99	1.01	1.00	61	R
4	.73	285	5.7	3.30	3.05	3.00	2.32	2.95	2.81	2.93	.98	1.05	1.01	63	R
5	2.00	285	7.0	3.98	3.65	3.55	2.88	3.55	3.40	3.53	1.00	1.04	1.01	60	R
6	2.50	285	6.8	4.25	4.05	3 95	2.92	3.86	3.59	3.82	.98	1.08	1.01	63	R
2	+.10	335	4.1	2.20	1.96	1.85	1.45	1.88	1.83	1.89	1.01	1.04	1.00	59	G
4	.78	335	4.7	2.85	2.61	2.50	2.01	2.52	2.43	2.52	1.01	1.04	1.00	58	G
5	2.00	335	6.0	3.30	3.10	3.08	2.33	3.08	2.82	3.01	.98	1.01	1.01	63	G
6	2.50	335	6.5	3.18	3.05	2.87	2.12	2.86	2.65	2.85	1.00	1.08	1.00	60	G
7	8.80	335	6.8	3.55	3.44	3.35	2.15	3.18	2.85	3.15	.95	1 11	1.01	66	G
Mean											.992	1.068	1.005	61 2	

NOTE.—" No. of curve " in column 1 refers to figs. 8 and 9.

From the curves and table for Chenango River it is seen that the value of the coefficient for reducing velocity obtained by the six-tenths-depth method varies from 0.93 to 1.03, the mean being 0.984. The coefficient for reducing velocity obtained by the top and bottom method to that obtained from the vertical velocity curve varies from 0.96 to 1.13, the mean being 1.041, the error of this method increasing as the depth increases. The coefficient for reducing velocity obtained by the third method to mean velocity obtained from the vertical velocity curve varies from 0.96 to 1.03, the mean being 0.996.

From the curves and table for Susquehanna River it is seen that the coefficient for reducing velocity at six-tenths depth to mean velocity obtained from vertical velocity curves varies from 0.95 to 1.06, the mean being 0.992. The coefficient for reducing velocity by the top and bottom method varies from 1 to 1.17, the mean being 1.068. The coefficient for reducing velocity obtained by the third method to mean velocity varies from 0.99 to 1.03, the mean being 1.005.

It is seen from the result in these tables: (1) That the third method of obtaining mean velocity by observing velocity one-half foot above the bed and one-half foot beneath the surface and at mid depth gives results agreeing very closely with that obtained from vertical velocity curves if the bed is smooth; (2) that results obtained by the top and bottom method agree quite closely with those obtained from vertical velocity curves if the depth is small and bed smooth, and that the error by this method increases as the depth increases; (3) that velocities obtained by the six-tenths-depth method are somewhat larger than those obtained from vertical velocity curves if the average depth is greater than about 4 feet.

The series of vertical velocity measurements made at Harrisburg were taken on November 2, 1903. They consisted of 20 measurements at depths ranging from 3 to 8 feet and mean velocity varying from 1.5 to 2.6 feet per second. The results of these measurements are shown in the following table and by the curves on Pl. XXVI.

Vertical velocity measurements made on Susquehanna River at Harrisburg. Pa., November 2, 1903.

Distance from initial point, in feet.	Depth at measuring point, in feet.	Velocity in feet per second by following methods.					Coefficients for reducing to mean velocity.				Depth of thread of mean velocity.	
		Vertical veloc- ity.	Six-tenths.	Top and bot- tom.	Integration.	Top.	Six-tenths.	Top and bot- tom.	Integration.	Top.	In feet.	In per cent of depth.
140	3.2	2.00	1.96	-----	1.92	-----	1.02	-----	1.04	-----	2.0	62
120	4.3	1.52	1.79	1.83	1.74	1.96	.85	0.83	.87	0.78	2.8	65
220	4.3	1.95	1.98	-----	2.08	-----	.99	-----	.94	-----	2.6	60
200	4.7	1.85	1.67	-----	1.93	-----	1.11	-----	.96	-----	2.6	55
160	4.8	1.82	1.87	-----	1.74	-----	.97	-----	1.05	-----	3.3	69
180	5.0	1.67	1.70	-----	1.74	-----	.98	-----	.96	-----	2.9	58
260	5.2	2.02	2.05	1.68	2.01	2.37	.99	1.21	1.00	.85	3.6	69
320	5.4	2.55	2.88	2.34	2.64	2.92	.89	1.09	.97	..87	3.9	72
280	5.8	2.15	1.73	2.00	2.06	2.67	1.24	1.07	1.04	.81	3.6	62
340	5.9	2.57	2.62	2.72	2.80	2.83	.98	.95	.92	.91	3.5	59
380	6.0	2.63	2.35	2.81	2.62	3.02	1.12	.94	1.00	.87	3.9	65
300	6.0	2.44	2.48	2.57	2.37	2.79	.98	.95	1.03	.87	3.7	62
360	6.1	2.71	2.85	2.75	2.72	2.99	.95	.99	1.00	.91	3.7	61
560	7.6	2.16	2.28	2.14	2.31	2.63	.95	1.01	.94	.82	4.6	61
590	7.7	2.40	2.40	2.34	2.41	2.92	1.00	1.02	1.00	.82	4.3	56
540	7.9	2.18	2.09	2.23	2.29	2.87	1.04	.98	.95	.76	4.4	56
520	8.0	2.57	2.73	2.66	2.52	3.08	.94	.97	1.02	.83	5.2	65
585	8.0	2.48	2.28	2.42	2.62	2.85	1.09	1.02	.95	.87	4.6	58
580	8.0	2.48	2.33	2.32	2.46	2.80	1.06	1.07	1.01	.89	4.1	51
580	8.0	2.49	2.49	-----	2.48	-----	1.00	-----	1.00	-----	5.5	60
Mean							1.01	1.08	.98	.85	-----	61

From these observations at Harrisburg we find, first, that the depth of the thread of mean velocity ranges from 51 to 72 per cent of the total depth and that the mean is 61 per cent. The error, therefore, introduced by holding the meter at 0.6 depth is only about 1 per cent. Second, the mean coefficient found for reducing top and bottom velocities to mean velocities is 1.08. Third, the coefficient for reducing velocities by the integration method to mean velocity is 0.98. Fourth, the coefficient for reducing top velocity to mean velocity is 0.85.

An interstudy of these various series of vertical velocity measurements shows that at these stations for depths up to about 10 feet and velocities not over 5 feet per second the depth of the thread of mean velocity is practically 60 per cent of the total depth, while for depths over 10 feet and velocities over 5 feet per second the depth of the thread of mean velocity becomes greater, averaging about 70 per cent of the total depth.

The coefficient for reducing top velocities to mean velocity for depths under 10 feet and velocities under 5 feet is about 0.85, while for greater depths and velocities it increases to a maximum of about 0.92.

The top and bottom velocities invariably give too small results, depending upon the roughness of the bed.

Furthermore, it is found that although the depth of the thread of mean velocity may vary between 50 and 80 per cent of the total depth, the error caused by holding the meter at 60 per cent of the depth does not exceed 5 or 6 per cent, which is within the limits of the accuracy one can expect in stream-measurement work.

The following table gives a summary of the results of the various series of vertical velocity measurements in the Susqehanna drainage:

Summary of results of vertical velocity measurements.

Place.	Number of curves.	Range of depths.	Range of velocities.	Depth of thread of mean velocity in per cent of depth.	Coefficient for reducing to mean velocity.				
					Six-tenths.	Top and bottom.	Top.	$\frac{T+2M+B}{4}$	Integration.
		Feet	*Ft per sec*						
McCalls Ferry, Duncan Run	73	3.3–30.0	1.21–5.80	68	0.94	1.07	0.92	....	
McCalls Ferry, cable station....	68	5.0–36.0	1.40–9.70	72	.97	......	.90	....	
Binghamton (Susquehanna River)	36	2.5– 8.1	.80–4.86	61	.99	1.07	......	1.00	......
Binghamton (Chenango River)	34	1 7– 8 3	.46–3.38	66	.98	1.04	......	1.00	......
Harrisburg (Susquehanna River)	20	3.2– 8.0	1.52–2 71	61	1 01	1.08	.85	......	0.98

NOTE In the above table erratic observations were not used.

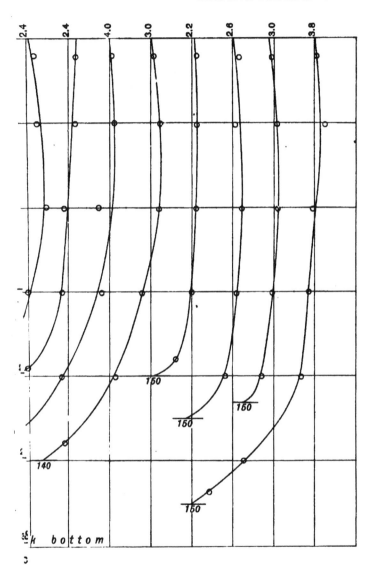

s_3, 1 inch=5 feet depth.

NEA

m.

7.2

00

NEAR M∝

). The cι

5.4

NEAR McC*

n. The curv

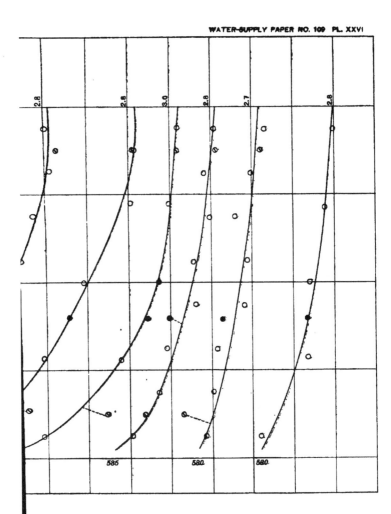

585   580   580

ttom. The curves terminate at their tops with the surface of the water.

In marked contrast to the New England streams, the power resources of the Susquehanna River basin, one of the largest draining into the Atlantic Ocean, are little developed.

As shown by the tables on pages 204, 205, taken from schedules furnished by the manufacturers' division of the Twelfth Census, 1900, a maximum of 10,375 horsepower is utilized in the portion of the drainage area in New York and 38,812 horsepower in Pennsylvania. This makes a total of less than 50,000 horsepower—an amount which, according to the estimates of various engineers, can be developed at any of several points on the lower river. By far the greater part of this is developed intermittently upon the smaller tributary streams by mills of from 20 to 50 horsepower. Pls. XXVIII and XXIX show the profile of Susquehanna River and its principal tributaries. These profiles are made up from data obtained from the army engineers, the report of the Tenth Census, Vol. XVI, and from levels furnished by private engineers, as shown in the tables on pages 207-210.

Over the greater portion of the river above Harrisburg the fall per mile is from 1 to 2 feet, while below Harrisburg the fall increases to between 5 and 8 feet, and it is here that the greatest opportunities for large power developments exist. The only point on the entire river at which this fall is now being utilized to any great extent is at York Haven, where a paper mill uses 2,000 horsepower, and a large electric-power plant in course of construction will soon use 10,000 or 20,000 more.

Mr. W. F. Bay Stewart, of York, Pa., describes the York Haven Power Plant, as follows:

The York Haven Water and Power Company's plant is located at the foot of the Conewago Falls on the Susquehanna River, ten miles from York and sixteen miles below Harrisburg. The natural fall at this point is about 23 feet in about three-quarters of a mile. The method of utilizing this fall is by building a wing dam out into the river above the falls and turning the greater portion of the flow by means of this wing dam within a retaining wall 3,500 feet long, constructed of masonry. This wall is built along the river shore just above low water. The wall is 16 feet high at the upper end and 32 feet high at the lower end, it is 6 feet wide on top all the way, and is built vertical on the inside and with a batter on the outside toward the river. The width of the foundation increases with the height of the wall, so that at the lower end it is about 22 to 24 feet in width. It is built of rubble masonry laid in cement.

The power house begins at the lower end of this wall, and is about 50 feet wide and 480 feet long. It contains twenty full-sized chambers and one smaller chamber. The design is to install in each of these chambers two 600-horsepower water wheels, and to connect the shafts of these water wheels by means of beveled gears at their top with the shaft of a 750-kilowatt generator, which runs horizontally and which is intended to develop at least 1,000 horsepower. To

equip the plant will require forty 600-horsepower water wheels and twenty
generators. In addition to this, in the smaller chamber there will be installed
two 300-horsepower water wheels which drive two exciters, duplicates, either
one of which is capable of exciting the whole plant. This building up to a
height of 34 feet is of the same class of masonry as the retaining wall, and these
chambers for water wheels are practically openings in an otherwise solid mass
of masonry 480 feet long by 50 feet wide and 34 feet high. On top of this
foundation is a brick building, one portion of which is two story and the remaining, one story. In the two-story part the switch boards and controlling devices
are located. At the lower end of this building and at right angles to it another
wall is constructed the same height as the high part of the retaining wall and
about 170 feet long. This wall then extends in an irregular form around the
buildings of the York Haven Paper Company's plant to the main land. On the
angle of this wall is constructed a tranformer house sufficient to receive the machinery for transforming all the current generated in the generating plant. The
current is developed at 2,400 volts and stepped up to 24,000 volts in this transformer house and is transmitted at this voltage to points of consumption. The
company has built a transmission line capable of transmitting 6,000 horsepower
from York Haven to York, where another transformer house has been built capable of transforming 24,000 volt current down to 2,200 volts, at which voltage it
will be delivered to customers. It is the purpose of the company to build a like
transmission line to Harrisburg, with a like transformer house at that city, and.
possibly, also to Lancaster, Pa., which is about 20 miles from the plant. The
machinery installed and to be installed in this plant is capable of an overload
of 25 per cent, thus increasing the capacity to 25,000 horsepower, and of course
it could be more largely increased by raising the head.

Between York Haven and the mouth of the river there is a fall of
about 270 feet. The mean annual discharge at York Haven from
1891 to 1904, inclusive, is about 40,000 second-feet. By applying the
rule that 11 second-feet of water falling 1 foot equals a horsepower
with 80 per cent efficiency it is seen that between York Haven and
the outlet of the river there about one million horsepower running
to waste, though several neighboring cities would afford an eager
market for all that could be developed. There are, of course, several
obstacles in the way of development, perhaps the most serious of
which would be the occasional ice freshets and gorges, making substantial protective works necessary and reducing or obliterating the
available head. Between the narrows above McCalls Ferry and Port
Deposit, however, the ice passes down through either a deep or a
broad channel, with no tendency to gorge and seldom doing damage.
At present there are several individuals and companies who are promoting power schemes on the lower river, and a large plant at York
Haven has recently been completed.

Mr. H. F. Labelle, who spent several years in the study of the
power possibilities of the lower Susquehanna, states the following in
regard to the power developments on the lower Susquehanna River:

The bed of the stream from Columbia to Port Deposit is for the most part
very wide, varying from 3,500 feet to about 2½ miles opposite Washingtonboro
There are, however, a few "narrows," as at Conowingo and McCalls Ferry. The
stream being wide and rapid, it naturally follows that at low water it is very
shallow and can be forded in many places. The water in the narrows is, how-

ver. very deep. At Conowingo Bridge, on the west side, there is a narrow channel over one-half mile long in which depths of 75 feet have been found. At
McCalls Ferry, where the river narrows to about 300 feet, the depth is also considerable. These deep channels are also met here and there on the wider parts of
the river—namely, between Turkey Hill and Star Rock station, on the east side,
where depths of over 90 feet have been found.

The Susquehanna and Tide-water canal skirts the west side of the river from
Wrightsville to Havre de Grace. Before the building of the Philadelphia, Baltimore and Washington Railroad and the Frederick Branch of the Pennsylvania
Railroad this canal had a brisk carrying trade, chiefly in coal from the anthracite regions. The flood of June, 1889, wrecked the canal in many places. The
cost of repairs was very high, and the canal continued in operation until May,
1894, when another flood caused considerable damage to the property. Since that
time it has been practically out of operation. After changing hands several
times, it was finally bought by the Susquehanna Electric Power Company, of
Baltimore. This company is about to begin the construction of their first plant,
below Peach Bottom. The Frederick Branch of the Pennsylvania Railroad runs
on the west side of the river from Columbia to Perryville, where it connects with
the main line of the Philadelphia, Baltimore and Washington Railroad.

The minimum discharge of the river at Shures Landing can be taken safely at
6,000 second-feet. This would give a minimum gross power to be developed from
Columbia to tide water of 153,000 horsepower. The proposed plants, however,
have been designed for a supply of 10,000 second-feet, which is available most of
the time.

This would give a possible power of about 255,000 horsepower. This available
power can almost be totally utilized, and the writer knows of projects on the
river aggregating over 185,000 horsepower.

The power available on the Susquehanna has at its disposal a much better
market than any other in the United States, not barring Niagara Falls. Baltimore is a little more than 40 miles from the half of the minimum power and Philadelphia is within 65 miles of the two lower plants, taking on the way Wilmington, with its heavy power consumption.

The upper plants are within easy reach of Lancaster, York, Harrisburg, Reading, and other manufacturing centers. Eastern Pennsylvania, with its great
manufacturing activity, will surely avail itself of whatever amount of power can
be developed on the river, and towns like Havre de Grace (10 miles below Shures
Landing), located on two of the large trunk lines between the North and the
South and also at the head of Chesapeake Bay, can be transformed by cheap
power into manufacturing centers of no mean importance.

There is no doubt that with the help of steam plants—and there are many
already established in the larger cities of the district—400,000 horsepower could be
developed on the river below Columbia and find a ready and remunerative market.

Starting from tide water the principal plants projected are as follows: (1) Conowingo plant, 25,000 to 35,000 horsepower; (2) the Peach Bottom plant, 40,000 horsepower; (3) the Fites-Eddy plant, 40,000 horsepower; (4) the York Furnace, Mc-
Calls Ferry plant, 45,000 horsepower; (5) the Turkey Hill plant, 30,000 horsepower.

There is about 9 feet fall available below the Conowingo works, but it is
believed that the conditions would not make it advisable to develop any power at
that point.

At Conowingo the power house is located a short distance above Shures Landing.
The building extends for a distance of about 500 feet, square across the stream
from the west shore. The original development is to be of 25,000 horsepower, but
provision is made in the power house for the development of 10,000 additional
horsepower. From the river end of the power house the dam extends upstream

a distance of 1,200 feet, the crest being at an elevation of 50.5 feet. The dam then turns toward the foot of McDowells Island, 800 feet away; thence it follows the center of McDowells Island for 3,600 feet to its head, and thence it goes diagonally to the east shore, a distance of 2,600 feet. The last 7,000 feet have their crest at an elevation of 43 feet, except 200 feet close to the high part of the dam, where a spillway for ice has been located, its crest being at an elevation of 41 feet. A needle dam will close this spillway at ordinary stages. The river above McDowells Island is over 3,000 feet wide and the dam forms a pool over 4 miles long. It has a sufficient rollway to pass the highest known floods without endangering the riparian property above it. The high part of the dam and the McDowells Island section are 8 feet wide on the crest. The remainder of the dam has a crest 12 feet wide. The whole dam will be of rubble, with ashlar facing on the downstream side. Borings have shown that a continuous rock bottom will be obtained on McDowells Island at an average depth of 11 feet. The generating plant will probably be divided into 1,250 kilowatt units. The turbines will be vertical, with draft tube. One pair of turbines will serve each dynamo, the connection between turbines and horizontal shaft of dynamo being made by two crown wheels engaging bevel gears on this shaft.

The working head will be 34 feet at low water and 30 feet at ordinary stages.

The Turkey Hill plant is located between Turkey Hill and Safe Harbor, on the east side of the river. At Turkey Hill the river is about 1 mile wide, and a low diverting dam about 5 feet high will form a large pond above it. This pond extends to Columbia, a distance of 5 miles, and its width varies between 1 and . miles. The head and tail race canals are formed by an embankment paralleling the railroad track and forming a canal varying from 190 to 250 feet in width at the bottom. This embankment is about 3 miles long. It is composed of a river wall in cement battering 1½ inch per foot on the river side and 2½ inches on the back. Next to this is the loose rock embankment proper, 40 wide on top and sloping 1 to 1 on the power-canal side. This mode of construction will meet the impact of the ice and prevent it from overtopping the embankment. At the main dam, and close to the head works, there will be a raft chute and a raft channel leading from it and close to the embankment on the river side. The average working head will be 30 feet, and the power house will be located at Star Rock.

DURATION OF THE STAGES OF THE LOWER SUSQUEHANNA.

In order to show the mean conditions and the duration of flow which have existed on the lower Susquehanna River during the last twelve years—1891 to 1902, inclusive—the curves in Pl. XXVII have been constructed. The dotted-line curve is plotted with gage heights as ordinates, and with the number of days during the mean year on which the stage of the river was less than the given gage height as abscissas. The full-line curve shows the number of days during the mean year when the discharge was below any given amount. In the preparation of these curves the Harrisburg gage heights for each year, shown on pages 108 to 114, were tabulated according to magnitude. The number of days during the year when the water stood at each gage height were then tabulated, and from these the number of days during the year when the river was lower than the various gage heights was determined. The curves were constructed from the mean of these yearly tables, and in the case of the full-line curve the discharges as given in the rating table on page 115 were substituted for the gage heights.

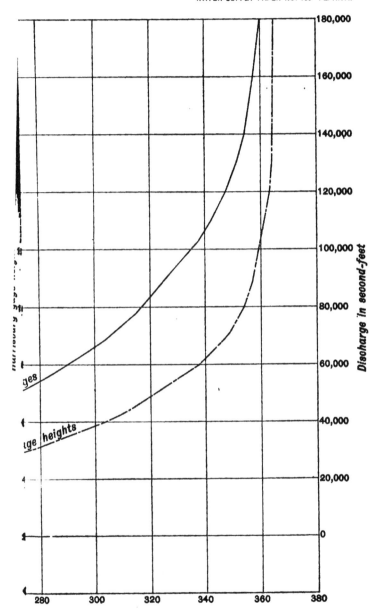

SBURG FOR 1891-1902, INCLUSIVE.

To use the two curves in conjunction with each other, enter the iagram with a certain gage height, find where it intersects the gage-height curve, then follow the ordinate of this intersection until it cuts he discharge curve, and the discharge for that particular gage height s found on the right side of the diagram.

Assuming that the discharges at the various points in this portion of he river vary in proportion to the drainage area above, one can readily letermine by the use of the curves the conditions which may reason-ibly be expected at any point below Harrisburg. For example, sup-)ose one wishes to know how many days during the mean year the lischarge will be less than 5,500 second-feet at the Pennsylvania-Maryland line, where the drainage area is 27,150 square miles, or 13)er cent more than at Harrisburg. As the drainage area at Harris->urg is 88.6 per cent of that at the State line, 5,500 second-feet would correspond to a discharge of 4,870 second-feet at Harrisburg. From .he full-line curve on Pl. XXVII we find that for twenty days during the mean year the discharge is less than 4,870 second-feet at Harrisburg, or 5,500 second-feet at the Maryland-Pennsylvania line.

By applying the following simple rule for horsepower it is possible to determine the probable power which could be developed during a mean year at any point in the lower Susquehanna:

Rule: Horsepower on the turbine shaft equals the discharge in sec-ond-feet multiplied by the fall divided by 11. This is based upon an assumption of 80 per cent efficiency for the turbines.

Applying this to the above example, we find that for three hundred and forty-five days during the mean year 500 horsepower for 80 per cent efficiency can be developed for each foot fall at the Maryland-Pennsylvania line.

RULES FOR ESTIMATING DISCHARGE.

The approximate mean monthly discharge in second-feet for any stream in the Susquehanna drainage basin, may be determined in either of two ways—

First. Its drainage area in square miles can be taken from the table on page 15, or measured on a map, and multiplied by the monthly run-off in second-feet per square mile given in the tables of the nearest gaging station.

Second. The monthly rainfall in inches for the district, as deter-mined from the tables on pages 161 to 171, can be multiplied by the per cent of run-off for that month at the nearest of the three gaging stations—Wilkesbarre, Williamsport, or Harrisburg—giving the total monthly run-off in inches. This result multiplied by one of the fol-lowing coefficients gives the mean monthly run-off in second-feet per square mile:

For month of 28 days 0.9603
For month of 30 days .... .8963
For month of 31 days .... .8674

The drainage area in square miles may be found as before, and if multiplied by the above product will give the mean discharge of the stream for that month in second-feet.

The horsepower may then be computed by the rule on page 203.

TABLES SHOWING DEVELOPED HORSEPOWER AND ELEVATIONS.

Horsepower developed in New York on Susquehanna River and tributaries.[a]

County.	Grist and flour mills.		Sawmills.		Miscellaneous.[b]		Total horsepower in county
	Number of mills.	Total horsepower.	Number of mills.	Total horsepower.	Number of mills	Total horsepower.	
Broome	13	840	9	291	3	33	1,164
Chemung	9	426	0	0	0	0	426
Chenango	20	963	23	759	6	163	1,885
Cortland	12	668	11	463	4	77	1,208
Delaware	9	314	10	276	0	....	590
Madison	9	367	8	359	2	175	901
Otsego	23	748	35	1,453	2	155	2,356
Schoharie	0	....	2	45	0	....	45
Steuben	23	1,155	3	121	6	27	1,303
Tioga	12	402	1	55	1	40	497
Total in State	130	5,883	102	3,822	24	670	10,375

[a] From manuscript schedules of the Twelfth Census.
[b] Includes woolen mills, tanneries, printing, cordage, and carriage works.

Horsepower developed in Pennsylvania on Susquehanna River and tributaries.a

County	Flour and grist mills.		Sawmills.		Creameries and paper mills.		Electric power plants.		Total horse-power in county.
	Number of mills.	Total horse-power.	Number of mills.	Total horse-power.	Number of mills.	Total horse-power.	Number of mills	Total horse-power.	
Adams	24	734	5	90					824
Bedford	34	699	5	100					799
Blair	26	597	2	40	1	25			662
Bradford	29	1,175	5	186					1,361
Cambria	4	111	8	218					329
Center	26	1,022	7	125	1	10			1,157
Clearfield	11	350	7	210					560
Clinton	11	451	6	213	1	120			784
Columbia	35	1,217	9	166	2	270			1,653
Cumberland	40	1,179	1	20	2	355	1	121	1,675
Dauphin	39	1,004	4	63			2	360	1,427
Elk	1	13							13
Franklin	9	169	1	10					179
Fulton	2	51	2	27					78
Huntingdon	30	979	2	40					1,019
Juniata	20	487	2	50					537
Lackawanna	7	324	3	90					414
Lancaster	176	5,451	11	667	9	225	4	1,262	7,605
Lebanon	22	615	2	30					645
Luzerne	24	712	8	205	1	125	1	208	1,250
Lycoming	31	1,530	6	140					1,670
Mifflin	16	605							605
Montour	6	135							135
Northumberland	22	445							445
Perry	31	697	7	154					851
Potter	1	20							20
Snyder	21	488	6	176					664
Schuylkill	17	277	2	45					322
Sullivan	7	224	5	129			1	250	603
Susquehanna	29	965	17	619			1	275	1,859
Tioga	15	554	1	55					609
Union	18	632	2	32					664
Wyoming	23	835	5	194					1,029
York	145	3,596	8	94	3	2,175	1	500	6,365
Total in State	952	28,343	149	4,188	20	3,305	11	2,976	38,812

a From manuscript schedules of the Twelfth Census.

Water power used for electric light and power development in Susquehanna drainage.[a]

Name of establishment.	County.	Post-office.	Water wheels.		Steam.		Electric	
			Number.	Power.	Number.	Power.	Number.	Power.
West Earl Electric Light and Power Co.	Lancaster	Brownstown	1	50			2	50
Eagles Mere Light Co	Sullivan	Eagles Mere	1	250			1	70
Harrisburg Light, Heat and Power Co.	Dauphin	Harrisburg	4	300	10	2,980	38	3,386
Lancaster Electric Light, Heat and Power Co.	Lancaster	Lancaster	8	1,050	1	325	12	1,762
Manheim Electric Light, Heat and Power Co	do	Manheim	2	100	1	150	1	130
Millersburg Electric Light, Heat and Power Co.	Dauphin	Millersburg	2	60	2	175	2	250
Delta Electric Power Co	York	Peach Bottom	2	500			1	470
John Hoafeld Co	Cumberland	Shippensburg	4	121	1	40	4	30
Strasburg Electric Light Plant	Lancaster	Strasburg	2	62			1	65
Susquehanna Electric Light, Heat and Power Co.	Susquehanna	Susquehanna	1	275	2	320	4	284
White Haven Electric Illuminating Plant	Luzerne	Whitehaven	2	208			4	270
Total			29	2,976	17	3,990	70	7,497

[a] From manuscript schedules of the Twelfth Census.

Approximate elevations and slope of Susquehanna River and North Branch.

Locality.	Distance from mouth.	Elevation above tide.	Distance between points.	Fall between points.
	Miles.	*Feet.*	*Miles.*	*Feet.*	*Ft per mile.*
Mouth	0	0			
Port Deposit	5	2	5	2	0.4
Stateline	15	69	10	67	6.7
Peach Bottom	18	85	3	16	5.3
Muddy Creek	21	98	3	13	4.3
McCalls Ferry	26	115	5	17	5.4
York Furnace	30	140	4	25	6.2
Safe Harbor	34	168	4	28	7.0
Turkey Hill	39	210	5	42	8.4
Columbia	45	225	6	15	2.5
Head Conewago Falls	58	273	13	48	3.7
Harrisburg	73	290	15	17	1.1
Mouth Juniata River	88	336	15	46	3.1
Liverpool	107	379	19	43	2.3
Selinsgrove	126	422	19	43	2.3
Below Sunbury dam	131	423	5	1	.2
Below Nanticoke dam	189	509	58	86	1.5
Wilkesbarre	197	525	8	16	2.0
Pittston	204	539	7	14	2.0
Gardners Creek	210	551	6	12	2.0
Tunkhannock	228	587	18	36	2.0
Mehoopany Creek	239	615	11	28	2.5
Tuscarora Creek	249	630	10	15	1.5
Wyalusing	261	656	12	26	2.2
Rummerfield Creek	270	678	9	22	2.4
Big Wysox Creek	276	694	6	16	2.7
Towanda	281	706	5	12	2.4
Ulster Ferry	289	727	8	21	2.6
Mouth Chemung River	294	742	5	15	3.0
Athens	297	752	3	10	3.3

Approximate elevations and slope of Juniata River.

Locality.	Distance from mouth.	Elevation above tide.	Distance between points.	Fall between points.	
	Miles.	*Feet.*	*Miles.*	*Feet.*	*Ft.per mil.*
Mouth	0	336	....	....	....
Millerstown dam, water below.	16	380	16	44	2.7
Millerstown dam, crest	16	388	0	8	....
Mifflin	34	417	18	29	1.6
Lewistown dam, water below.	44	442	10	25	2.5
Lewistown dam, crest	44	450	0	8	....
McVeytown	61	476	17	26	1.5
Newton Hamilton dam, water below	68	512	7	36	5.1
Newton Hamilton dam, crest..	68	520	0	8	....
Huntingdon dam, water below.	90	±610	22	90	4.1
Huntingdon dam, crest	90	±622	0	12	....

Approximate elevations and slope of Raystown Branch of Juniata River.

Locality.	Distance from mouth.	Elevation above tide.	Distance between points.	Fall between points.	
	Miles.	*Feet.*	*Miles.*	*Feet.*	*Ft.per mile*
Mouth	0	595	....	....	....
Near Saxton	40	837	40	242	6.0
Pipers Run..................	53	891	13	54	4.2
Mount Dallas................	79	1,016	26	125	4.8

Approximate elevations and slope of Frankstown Branch of Juniata River.

Locality.	Distance from Huntingdon.	Elevation above tide.	Distance between points.	Fall between points.	
	Miles.	*Feet.*	*Miles.*	*Feet.*	*Ft. per mile.*
Huntingdon dam, crest........	0.0	622	....	....	....
Piper's dam, water below.....	2.5	628	2.5	6.0	2.4
Piper's dam, crest............	2.5	636	0	8.0	....
Petersburg dam, water below.	4.1	641	1.6	5.0	2.1
Petersburg dam, crest........	4.1	648	0	6.5	....
Big.Water Street dam, water below.....................	10.0	698	5.9	45.0	7.6
Big Water Street dam, crest..	10.0	712	0	19.3	....
Little Water Street dam,water below.....................	12.4	714	2.4	2.0	.8
Little Water Street dam, crest.	12.4	726	0	12.0	....
Willow dam, water below	14.4	728	2.0	2.0	1.0
Willow dam, crest	14.4	741	0	13.0	....
Donnelly's dam, water below .	17.0	770	2.6	29.0	11.2
Donnelly's dam, crest	17.0	784	0	14.0	....
Smoker's dam, water below...	18.7	787	1.7	3.0	1.7
Smoker's dam, crest..........	18.7	799	0	12.0	....
Mud dam, water below.......	20.1	800	1.4	1.0	.7
Mud dam, crest	20.1	808	0	7.5	....
Williamsburg dam, water below	23.0	831	2.9	23.0	7.9
Williamsburg dam, crest.....	23.0	839	0	10.0	....
Threemile dam, water below..	24.1	839	1.1	0	0
Threemile dam, crest.........	24.1	856	0	17.5	....
Crooked dam, water below ...	27.2	856	3.1	0	0
Crooked dam, crest...........	27.2	866	0	10.0	....
Frankstown dam, water below	33.5	895	6.3	29.0	4.6
Frankstown dam, crest........	33.5	899	0	3.5	....
Hollidaysburg dam, water below	36.4	923	2.9	24.0	8.3
Hollidaysburg dam, crest.....	36.4	927	0	4.5	....

Elevation and slope of West Branch of Susquehanna River.

Locality.	Distance from mouth.	Elevation above tide.	Distance between points.	Fall between points.	
	Miles.	*Feet.*	*Miles.*	*Feet.*	*Ft.per mile.*
Mouth	0	429			
Lewisburg dam, water below	7	431	7	2	0.3
Lewisburg dam, crest	7	434	0	3	
Muncy dam, water below	23	462	16	28	1.8
Muncy dam, crest	23	469	0	7	
Williamsport dam, water below	39	498	16	29	1.8
Williamsport dam, crest	39	508	0	10	
Lock Haven dam, water below	65	539	26	31	1.2
Lock Haven dam, crest	65	550	0	11	
Queen's Run dam, water below	69	551	4	1	0.2
Queen's Run dam, crest	69	557	0	6	
Keating	105	695	36	138	3.8
Curwinsville	160	1,117	55	422	7.7

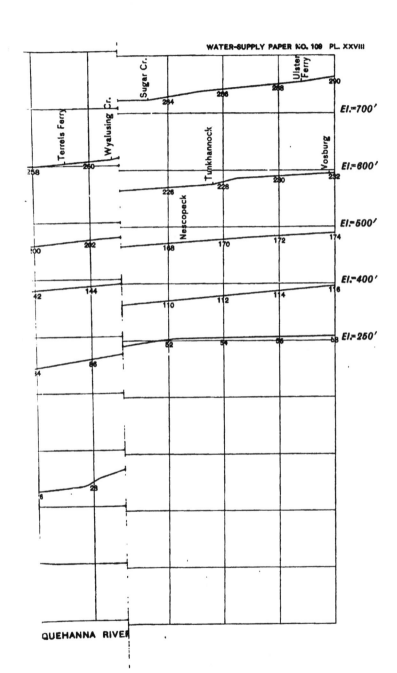

QUEHANNA RIVER

rg dam
ile dam

id dam

noh

et dam

rstreet dam

am

nnellys dam
mokers dam

Mouth Ra stown Branch Juniata River

El.=800'

Mouth Juniata

El.=800'

eating

Mouth West Branch

dam

El.=400' 0 Miles   10   100   110   120   130

INDEX.

211

O

Author.

Hoyt, John C[layton], 1874—

. . . Hydrography of the Susquehanna River drainage basin, by John C. Hoyt and Robert H. Anderson. Washington, Gov't print. off., 1905.

215 p., 1 l. illus., XXIX pl. (incl. map) diagrs. 23ᶜᵐ. (U. S. Geological survey. Water-supply and irrigation paper no. 109)

Subject series: M, General Hydrographic investigations, 13; N, Water power, 9.

1. Water-supply—Pennsylvania. 2. Water-supply—New York (State) 3. Susquehanna River. I. Anderson, Robert H. joint author.

Subject.

Hoyt, John C[layton], 1874—

. . . Hydrography of the Susquehanna River drainage basin, by John C. Hoyt and Robert H. Anderson. Washington, Gov't print. off., 1905.

215 p., 1 l. illus., XXIX pl. (incl. map) diagrs. 23ᶜᵐ. (U. S. Geological survey. Water-supply and irrigation paper no. 109)

Subject series: M, General hydrographic investigations, 13; N. Water power, 9.

1. Water-supply—Pennsylvania. 2. Water-supply—New York (State) 3. Susquehanna River. I. Anderson, Robert H. joint author.

Series.

U. S. Geological survey.

Water-supply and irrigation papers.

no. 109. Hoyt, J. C. Hydrography of the Susquehanna River drainage basin, by J. C. Hoyt and R. H. Anderson. 1905.

Reference.

U. S. Dept. of the Interior.

see also

U. S. Geological survey.

Water-Supply and Irrigation Paper No. 110   Series { B, Descriptive Geology, 46
                                                        { 0, Underground Waters, 27

DEPARTMENT OF THE INTERIOR
UNITED STATES GEOLOGICAL SURVEY
CHARLES D. WALCOTT, DIRECTOR

CONTRIBUTIONS

TO THE

HYDROLOGY OF EASTERN UNITED STATES

1904

MYRON L. FULLER
GEOLOGIST IN CHARGE

WASHINGTON
GOVERNMENT PRINTING OFFICE
1905

CONTENTS.

3

ILLUSTRATIONS.

5

LETTER OF TRANSMITTAL.

DEPARTMENT OF THE INTERIOR,
UNITED STATES GEOLOGICAL SURVEY,
Washington, D. C., March 31, 1904.

SIR: I have the honor to transmit herewith a manuscript, entitled "Contributions to the Hydrology of Eastern United States, 1904." This paper is the second of a series of contributions relating largely to the hydro-geology of the eastern portion of the country, the first having been published as Water-Supply and Irrigation Paper No. 102, and has been prepared as the result of investigations of underground currents and artesian wells, as authorized by law. It includes 23 short papers by 19 geologists, physicists, and others connected with the eastern section of the division of hydrology. The aim is to present the results of subordinate lines of investigation which, because of their limited length, do not warrant separate publication.

A number of points of unusual interest are considered in the report. A full description of the electrical apparatus devised by Prof. C. S. Slichter for measuring underflow is presented for the first time. The description of the experiment at Quitman, Ga., for determining the liability to pollution of deep wells is an account of a practical investigation which proved to be of great importance in preventing steps that might have led to a serious contamination of the underground waters and a possible epidemic.

Very respectfully,

F. H. NEWELL,
Chief Engineer.

Hon. CHARLES D. WALCOTT,
Director United States Geological Survey.

CONTRIBUTIONS TO THE HYDROLOGY OF EASTERN UNITED STATES, 1904.

M. L. FULLER,

Geologist in Charge.

INTRODUCTION.

By M. L. FULLER.

OBJECT OF REPORT.

The present paper, which is the second of the series of "Contributions to the Hydrology of Eastern United States," includes 23 short reports by 19 geologists and others. Of these the longer papers have been contributed by those connected with the eastern section of the division of hydrology, but several that embody summaries of the water resources of regions covered by geologic investigations have been prepared by members of the geologic branch. The aim in preparing the report is to present the results of subordinate investigations, the length or scope of the reports of which do not warrant their publication as separate papers. In this way there is presented a considerable amount of material of local interest, especially in regions where complete investigations are not practicable. The paper is also intended to afford opportunity for publication of the results of laboratory or other physical, mechanical, or mathematical investigations bearing on underground water supplies.

SUMMARY.

The papers included may be briefly summarized as follows:

Description of the Underflow Meter used in Measuring the Velocity and Direction of Underground Water, by Charles S. Slichter.

The electrical apparatus described in this paper is intended to replace the inaccurate estimation of underflow based on size of material and head, and the troublesome chemical method. Commonly the test wells required by the present method consist of a group of four

9

2-inch driven wells, one located on the side from which the flow is expected and the other three in the arc of a circle 4 or 6 feet in radius on the downflow side. These are from 24 to 30 inches apart. Electrical connection is made with each well and the upstream well is charged with sal ammoniac or other electrolyte. The appearance of the salt at or near any one of the lower wells is recorded by an electric device, of which there are two types, nonrecording and recording. A full description of the construction and use of the apparatus is given.

The California or "Stove-Pipe" Method of Well Construction, by Charles S. Slichter.

In the larger number of the wells in the gravels, sands, etc., of the Coastal Plain regions of the Atlantic States and elsewhere, the hole is tightly cased throughout, the only point of entrance for the water being at the bottom. Only one water bed can be drawn upon in such wells. In California, and to some extent elsewhere, however, a method is in use whereby, by means of casings perforated at intervals, water can enter the well at a number of different levels. The casing consists of a steel shod " starter," 15 to 25 feet long, and sections of steel pipe 2 feet in length overlapping with flush joints. The casing is forced downward, length by length, by hydraulic jacks to the desired depth. In one well a depth of over 1,300 feet has been reached. After the well has been forced to the required depth a cutting knife, two types of which are figured in the paper, is lowered and slits or holes are cut through the casing at the points where, according to the record, water is known to occur. As much as 400 feet of a 500-foot well may be perforated if the conditions demand it. The type of well described has many advantages in addition to that of rendering several sources of water simultaneously available, and will doubtless be of great value in many localities in the east when the waters at the different levels are under similar heads.

Approximate Methods of Measuring the Yield of Flowing Wells, by Charles S. Slichter.

In this paper descriptions of simple methods and tables for the approximate field determinations of the yield of artesian wells are given. The tables relating to completely filled pipes, both horizontal and vertical, are reprinted from a private report by J. E. Todd, while those relating to measurement of flows from partially filled horizontal and inclined pipes are new. The only instrument required is a foot rule, and the measurement, which consists in measuring the height of the jet in the case of a vertical pipe or its lateral projection in the case of a horizontal pipe, can be made in a few moments. The results are within about 10 per cent of the actual flow, which error is no greater than the average daily variation of flow due to changes of barometric pressure.

Corrections Necessary in Accurate Determinations of Flow From Vertical Well Casings, from Notes furnished by A. N. Talbot.

This paper deals with certain corrections which it is necessary to apply to the figures of the field tables compiled by J. E. Todd and Charles S. Slichter in those cases where refined measurements of flows from vertical well casings are desired. A simple apparatus for measuring the height of the water jet is illustrated, and a diagram showing by curves the coefficients of discharge is given. It was found that with jets less than a foot in height the actual discharge is from 7½ to 12½ per cent lower than the discharge computed from the tables for 2-inch to 6-inch pipes.

Experiment Relating to Problems of Well Contamination at Quitman, Ga., by S. W. McCallie.

Many of the disastrous epidemics which have visited the towns and cities of this country have been traced to polluted drinking water. The present paper deals with a successful effort made to determine the possibility of pollution of the deep wells at Quitman, Ga., the result of which was to prevent a step that might have had fatal results. In a well drilled in 1903 at Quitman a cavity containing what was regarded as a subterranean stream was encountered, and it was thought that such a stream would afford an admirable method of disposing of the city's sewage. Objections were at once raised, however, on every hand, because of the liability of pollution of wells or springs of the region. To test this possibility 2 tons of salt were put into one of the wells, while samples of water from all other wells and springs in the vicinity were taken at short intervals and analyzed. The results showed that the salt had penetrated the deep wells and demonstrated that the emptying of sewage into the underground stream would have resulted in the pollution of the waters of all the deep wells in town and would possibly have led to a serious epidemic.

The New Artesian Water Supply at Ithaca, N. Y., by Francis L. Whitney.

In this paper Mr. Whitney presents an account of the outbreak of the typhoid epidemic of 1903 and the steps taken by various local bodies to obtain a pure supply. The deep wells sunk in the gravels, sands, and clays in the valley of Cayuga Inlet just above Ithaca are described, and the source and geologic occurrence of the supply are discussed. Records of all of the wells are given, the more important being shown by diagrams. The success of the wells, both as to the quantity and the quality of the water, is of special interest, as like supplies could doubtless be obtained at many points in New York and New England which are similarly situated.

Drilled Wells of the Triassic Area of the Connecticut Valley, by W. H. C. Pynchon.

The paper gives a sketch map and section showing the principal geologic features of the area and describes the character and succession of the eastward-dipping series of sandstones, shales, and interbedded or intruded traps throughout the Triassic area of Massachusetts and Connecticut. The geologic discussion is followed by descriptions of a considerable number of wells, in which several points of interest are emphasized, including the nearly uniform water-bearing character of the sandstones, the high percentage of mineral matter present in all the water, and the general absence of flowing wells.

Triassic Rocks of the Connecticut Valley as a Source of Water Supply, by M. L. Fuller.

In this paper a review is given of the occurrence of waters in Triassic rocks of various types, including conglomerates, sandstones, shales, and traps, and the structure, jointing, and faulting of the rocks and their influence on the underground waters are described. Summaries of the conditions favorable and unfavorable to flowing water are given and a number of important conclusions presented. While all of the Triassic rocks except the traps will usually be found to be water bearing, the conditions, because of the interruption of the beds due to faulting or jointing, will rarely be favorable to flowing wells, and high heads can never be expected. The water will in most instances be found to be highly mineralized, but, except possibly in the shallower wells in crowded cities, will rarely be subject to pollution. Attention is called to the need of keeping accurate records and of thoroughly testing each well; and the question of the proper depth of wells is discussed.

Spring System of the Decaturville Dome, Camden County, Mo., by E. M. Shepard.

In the center of this area is a mass of granite (pegmatite) that has apparently been thrust upward through the surrounding Paleozoic limestones and other rocks, which are thereby tilted away from it in all directions. The dome thus formed is surrounded at a distance of several miles by a line of springs, the channels of which seem to radiate from the center of the dome, from which direction the waters appear to come. Several deep flowing wells in similar situations also derive water from the outward-sloping rocks. Descriptions are given of many of the springs, some of which are of immense size and present many points of interest.

Water Resources of the Fort Ticonderoga Quadrangle, Vermont and New York, by T. Nelson Dale.

In this area, which lies on the line between Vermont and New York, there are several important towns, including Ticonderoga in New York, and Proctor, Brandon, Poultney, and West Rutland in Vermont. The land varies considerably in altitude, ranging from low plains near Lake Champlain to ridges 2,700 feet high, such as the one southwest of West Rutland. In general the region is well watered, having numerous springs and streams. The larger streams have, however, become subject to pollution as the industries and towns along their banks have grown up. The limestones or dolomites of the area would doubtless yield water if penetrated by deep wells, but it would probably be hard and in some cases might be liable to pollution. The sandy portion of the drift yields considerable quantities of water, but the clays near the lake give unsatisfactory supplies.

Water Resources of the Taconic Quadrangle, New York, Massachusetts, and Vermont, by F. B. Taylor.

The center of this area falls almost exactly at the point at which the boundaries of the three States mentioned come together. The area is mountainous except in the western third, although it contains valleys with bottoms as low as 350 feet above sea. The drainage is by the Hoosac River, along which, as well as along its tributaries, there are water powers that are either utilized or available. Practically all the cities and towns in the quadrangle obtain their water supplies from mountain streams, but a large proportion of the rural inhabitants procure their supplies from springs. One important mineral spring, developed at a sanitarium, occurs along a probable fault line.

Water Resources of the Watkins Quadrangle, New York, by Ralph S. Tarr.

The paper gives a general description of the water resources of this region, which includes the cities of Elmira and Ithaca, in the southern portion of the State. Special attention is given to the subject of obtaining artesian supplies from the deep gravel-filled valleys, and the steps taken to obtain pure supplies from such a source at Ithaca after the typhoid epidemic of 1903, are described. A number of analyses are given.

Water Resources of the Central and Southwestern Highlands of New Jersey, by Laurence La Forge.

The region treated in this paper is that part of the Highlands which lies south of Andover and Pompton, including about two-thirds of the Highland area of the State. The population is mainly located in villages and is dependent largely upon manufacturing industries for support. The rocks are principally of granitic types, but some conglomerate and quartzite occurs. Lakes and ponds are numerous in a part of the area and streams are abundant and afford numerous water powers as well as the supply for the Morris Canal, while springs are common and furnish water to a considerable number of towns. The surplus water, of which there is considerable, may in the future become of great importance as a source of supply for the large and rapidly growing urban district near New York. The paper describes the needs of this district and the amount and character of the water and its availability as a source of supply.

14 HYDROLOGY OF EASTERN UNITED STATES, 1904. [no. 110.

Water Resources of the Chambersburg and Mercersburg Quadrangles, Pennsylvania, by George W. Stose.

The Chambersburg and Mercersburg quadrangles are located in the Cumberland Valley, in southern Pennsylvania, and include the two important towns from which they are named. They are crossed by several mountain ridges and by many streams that afford water powers, some of which have already been utilized. Many springs are found in the various rocks, especially in the limestone, some of which have been developed as attractive resorts. The public water supplies are obtained largely from spring-fed mountain streams, and in several instances directly from springs.

Water Resources of the Curwensville, Patton, Ebensburg, and Barnesboro Quadrangles, Pennsylvania, by F. G. Clapp.

These quadrangles are situated near the eastern edge of the bituminous coal field, and lie mostly within the limits of Clearfield and Cambria counties. The region is one of high ridges alternating with valleys. Springs are abundant and, except in towns, constitute the main source of water supply. The once noted Cresson Springs are in the area. Many wells obtain abundant supplies, both from the stream gravel and from the rocks. In some places flowing wells are obtained. More than ten towns are equipped with water systems. The majority obtain their supplies from streams, but several procure water from springs or deep wells.

Water Resources of the Elders Ridge Quadrangle, Pennsylvania, by Ralph W. Stone.

This quadrangle lies in Armstrong and Indiana counties, in the west-central part of the State. In general the surface is moderately hilly and is drained by Kiskiminitas River and other streams that flow westward into Allegheny River. Some present available water powers. The region is distinctly a rural district, without large towns, and springs and shallow wells afford the only source of water supply. The Mahoning and Pittsburg sandstones, which overlie the coals of the same name, are the best water bearers.

Water Resources of the Waynesburg Quadrangle, Pennsylvania, by Ralph W. Stone.

The Waynesburg quadrangle is located in Greene County, in the southwest corner of the State. The topography is uniformly hilly, the crests standing generally not over 500 feet above the valleys. The area is drained eastward into the Monongahela by streams of low grade and small volume, with few if any available water powers. The city of Waynesburg obtains its public water supply from a near-by stream, but the water is frequently highly charged with silt and is generally unsatisfactory. The smaller towns depend on shallow private wells. Some of the rock wells yield good supplies, especially those

drawing water from the Waynesburg sandstone. In drilling wells to
the rock, however, care must be taken not to penetrate the coal.
Springs are numerous but small.

Water Resources of the Accident and Grantsville Quadrangles, Maryland, by
G. C. Martin.

The area covered by this paper is located in the "handle," in the
extreme western portion of Maryland. The topography is essentially
that of a plateau, standing between 2,500 and 3,000 feet, above which
rise a number of ridges. It is drained by Youghiogheny, Castleman,
and Savage rivers, all of which would afford good water supplies.
Springs are numerous, especially along the outcrop of the Greenbrier
limestone. The possibilities of artesian waters have not been tested,
although bore holes sunk for other purposes have given flowing water.
It is probable that such flows would be afforded by each of the three
synclines which cross the area.

Water Resources of the Frostburg and Flintstone Quadrangles, Maryland and
West Virginia, by G. C. Martin.

These quadrangles lie just east of the Grantsville, mainly in western Maryland. They are crossed by a number of ridges that rise as
high as 3,000 feet, between which are valleys of considerably lower
level. Through these valleys pass the north and south branches of
Potomac River and a number of smaller streams. The smaller
streams are unpolluted and generally afford good water. Springs
abound in the limestone regions. Artesian water is found in sandstones of Carboniferous age, and it is thought that the Oriskany and
Tuscarora sandstones probably carry artesian water in the synclines.
A large part of the water supply of Frostburg is obtained from an
artesian well, but except at Cumberland, there is otherwise little
demand for artesian water.

Water Resources of the Cowee and Pisgah Quadrangles, North Carolina, by
Hoyt S. Gale.

The Cowee and Pisgah quadrangles are located in the heart of the
Southern Appalachians, in the extreme western part of North Carolina. They are traversed by French Broad River and other streams
which present available water power at many points. The whole
region abounds with springs, generally of pure water, but a few mineral springs, especially chalybeate, occur. Of these the carbonate
springs are usually associated with hornblendic gneiss, while the chalybeate waters are associated with pyrite deposits along faults. Very
few wells have been sunk in the area.

Water Resources of the Middlesboro-Harlan Region of Southeastern Kentucky,
by George H. Ashley.

This region is in a general way a broad basin lying between two high
mountain ridges. The surface is cut by deep ravines separated by

sharp crests. The rocks are of the Coal Measure series and yield abundant springs, which form the main source of supply of the scattered inhabitants. Shallow wells are frequently relied upon in the bottom lands. No deep wells have been drilled for water, but flowing water has sometimes been obtained from wells drilled for oil. The public supplies of Middlesboro and Pineville, the two principal towns, are obtained from spring-fed mountain streams. Some available water powers exist.

Summary of the Water Supply of the Ozark Region in Northern Arkansas, I c George I. Adams.

This area includes portions of the Boston Mountain belt and the Springfield and Salem uplands, the limits of which are shown in fig. 32. The rocks consist largely of Ordovician dolomites and Carboniferous limestones, the former being confined mainly to the Salem upland. Springs are especially numerous in the Boone limestone and chert and the Ordovician dolomites, but some are also found in the Key sandstone. Some of the springs are of immense size, and many have been developed as resorts.

Notes on the Hydrology of Cuba, by M. L. Fuller.

During the American occupation of Cuba much interest was aroused in the water resources of the island, and special attention was given to the problem of water supplies for the various cities. This paper gives, after a résumé of geology, topography, and drainage, a summary of the natural water resources, including underground streams and springs, accounts of the public water supplies, and descriptions of the wells sunk by the War Department for the various military posts. Among the principal water supplies are those of Habana, Matanzas, Cardenas, Cienfuegos, Guantanamo, and Santiago. Underground water courses in the soft limestone, some of them of considerable size, are everywhere present, and springs are very common in many regions. The waters are generally pure, except for the lime, but mineral waters have been exploited for drinking or bathing purposes at a number of localities. The wells drilled by the War Department were located at a considerable number of scattered points. In general they were successful in obtaining supplies at a moderate depth.

DESCRIPTION OF UNDERFLOW METER USED IN MEASURING THE VELOCITY AND DIRECTION OF MOVEMENT OF UNDERGROUND WATER.

By CHARLES S. SLICHTER.

A brief description of the writer's electrical method of measuring the velocities of underground water was printed in the Engineering News of February 20, 1902, and in Water-Supply and Irrigation Paper No. 67 of the United States Geological Survey. The present account will give a more detailed description of the form of the apparatus.

APPARATUS.

The apparatus used is of two types: (1) direct-reading, or hand apparatus, which requires the personal presence of the operator every hour for reading of instruments, and (2) recording apparatus, which requires attention but once in a day. The arrangement of the test wells and manner of wiring the wells is essentially the same in both forms and will now be described.

TEST WELLS.

The test wells may be common 1¼ or 2 inch drive wells if the soil and water-bearing material are easily penetrated, and if the depths to be reached do not exceed 40 feet. For greater depths and harder materials wells of heavier construction should be used. The test wells put down by the Commission on Additional Water Supply for Greater New York in 1903 are suitable for ordinary conditions in the eastern part of the United States, or in any place where the gravels are not too coarse or too compact. These test wells were made of full-weight standard wrought-iron 2-inch pipe, in lengths of 6 or 7 feet, with long threads (1¼-inch) and heavy wrought nipples. The well points were 4-foot standard brass jacket points, No. 60 wire gauze. For wells no deeper than 30 feet, closed-end points were driven, but for deeper work open-end points were used. The test wells were driven in place by use of a ram from 150 to 250 pounds in weight, simultaneously hydraulicking a passage for the pipe with water jet in three-fourth inch standard wash pipe. In fine material there was coupled ahead of the open end well point 3 or 4 feet of pipe carrying a shoe coupling, so that the sand, in running in through the open end of the well, would not rise above the bottom of the screen inside of the finished well.

Mr. Homer Hamlin, of Los Angeles, Cal., has devised a powerful drilling rig run by a gasoline engine, which enables him to sink test wells in the bowlder gravel of that locality. He uses a special double-strength casing with flush joints, which he has been able to sink to great depths with his remarkable drilling machine. He has used the electrical method for determining underflow velocities with great success.

The test wells are grouped as shown in fig. 1.

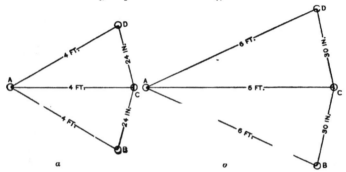

FIG. 1.—Plan of arrangement of test wells used in determining the velocity and direction of motion of ground waters. A, B, C, D are the test wells. The direction A-C is the direction of probable motion of the ground waters. The dimensions given in plan a are suitable for depths up to about 25 or 30 feet; those in plan b for depths up to about 75 feet. For greater depths the distances A—B, A—C, A—D should be increased to 9 or 10 feet and the distances B—C and C—D to 4 feet. The well A is the "salt well" or well into which the electrolyte is placed.

In case the wells are not driven deeper than 25 feet, an "upstream" or "salt well," A, is located, and three other wells, B, C, and D, are driven at a distance of 4 feet from A, the distance between B and C and C and D being about 2 feet. The well C is located so that the line from A to C will coincide with the probable direction of the ground-water movement. This direction should coincide, of course, with the local slope of the water plane. For deeper work the wells should be placed farther apart, as shown in the right portion of fig. 1. For depths exceeding 75 feet, a radius of 8 or 9 feet and chords of 4 feet should be used, the general requirement being that the wells should be as close together as possible, so as to cut down to a minimum the time required for a single measurement, but not so close that important errors are liable to be introduced by the inability to drive the wells perfectly straight and plumb. On this account the deeper the wells the farther apart they should be placed. The angles B A C and C A D should not exceed 30 degrees.

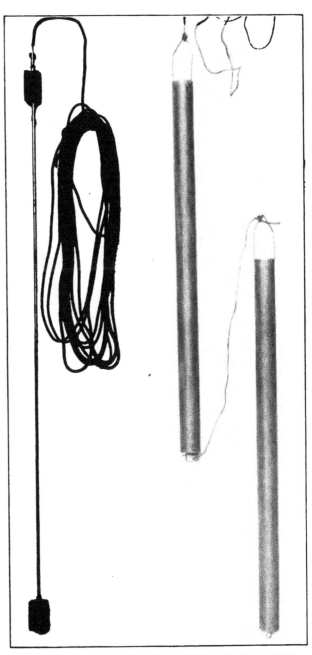

ELECTRODE AND PERFORATED BRASS BUCKETS USED IN CHARGING
WELLS.

METHOD OF WIRING.

Electrical connection is made with the casing of each test well by means of drilled coupling carrying a binding post. Each of the downstream wells, B, C, D, contains within the well point or screen section an electrode consisting of a nickeled brass rod three-eighths inch by 4 feet, insulated from the casing by wooden spools. This electrode communicates with the surface by means of No. 14 rubber-covered

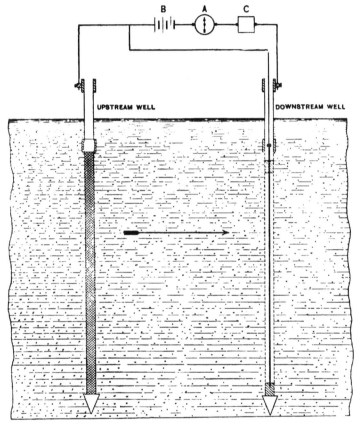

FIG. 2.—Diagram illustrating electrical method of determining the velocity of flow of ground water. The ground water is supposed to be moving in the direction of the arrow. The upstream well is charged with an electrolyte. The gradual motion of the ground water toward the lower well and its final arrival at that well are registered by the ammeter A. B is the battery and C a commutator clock which is used if A is a recording ammeter.

copper wire. Fig. 2 illustrates the arrangement of electric circuits between the upstream well and one of the downstream wells. An electrode is shown in Pl. I. At left of cut there is shown an electrode, such as is used in a downstream well. The electrode is 4 feet

long, made of three-eighths inch nickeled brass rod. Insulators are
of wood. The end of rod receives a No. 14 rubber-covered wire, to
which good contact is made by a simple chuck clutch. At right of cut
are shown two buckets of perforated brass used in charging wells with
granulated sal ammoniac. Size of each is 1¾ by 30 inches. Each of
the downstream wells is connected to the upstream well in the manner
shown in this figure.

DIRECT-READING METER.

A view of the direct-reading underflow meter is shown in Pl. II,
A1. Six standard dry cells are contained in the bottom of the box,
their poles being connected to the six switches shown at the rear of
the case. By means of these switches any number of the six cells
may be thrown into the circuit in series. One side of the circuit ter-
minates in eight press keys, shown at the left end of the box. The
other side of the circuit passes through an ammeter, shown in the
center of the box, to two three-way switches at right end of the box.
Four of the binding posts at the left end of the box are connected
to the casing of well A and to the three electrodes of wells B, C,
and D, in order. The binding posts at the right end of the box are
connected to the casings of wells B, C, and D. There are enough
binding posts to permit two different groups of wells to be connected
with the same instrument. When the three-way switch occupies the
position shown in the plate, to press the first key at the left end of
box will cause the ammeter to show the amount of current between
casing of well A and casing of well B. When the next key is pressed
the ammeter will indicate the current between the casing of well B
and the electrode contained within it. In the first case the current
is conducted between the two well casings by means of the ground
water in the soil; in the second case by means of the water within well
B. By putting the three-way switch in second position and pressing
the first and third keys in turn, similar readings can be had for current between casings A and C, and between casing C and its internal
electrode. Similarly, with switch in third position, readings are taken
by pressing first and fourth keys. The results may be entered in
notebook, as shown in Table I.

The principles involved in the working of the apparatus are very
simple. The upstream well A is charged with a strong electrolyte,
such as sal ammoniac, which passes downstream with the moving
ground water, making the ground water a good conductor of electricity.
If the ground water moves in the direction of one of the lower wells,
B, C, D, etc., the electric current between A and B, A and C, or A
and D will gradually rise, mounting rapidly when the electrolyte
begins to touch one of the lower wells. When the electrolyte finally

A UNDERFLOW METER, SHOWING CONNECTIONS WHEN USED AS
DIRECT READING APPARATUS

The switches at back of case throw any of the dry cells in bottom of box in or out of
circu t. When used with recording ammeter, only two connections are made,
one to each side of battery circuit, but the ammeter is left in circuit with the
recording instrument, to indicate whether the latter is working properly

B. COMMUTATOR CLOCK, FOR USE WITH RECORDING AMMETER.

The clock makes e'ectrical contact at any five-minute interval

reaches and enters inside of one of the wells B, C, D, it forms a short circuit between the casing of the well and the internal electrode, causing an abrupt rise in the electric current. The result can be easily understood by consulting Table I and fig. 3, in which the current is depicted graphically.

TABLE I.—*Field record of electric current during underflow measurements at station 5, Rio Hondo and San Gabriel River, California, August 5 and 6, 1902.*

[Readings in amperes and decimals of an ampere.]

Time.	Well B.		Well C.		Well D.		Remarks.a	
	Casing.	Electrode.	Casing.	Electrode.	Casing.	Electrode.		
8 a. m	0.140	0.360	0.142	0.332	0.150	0.390	....	....
8.15 a. m	Salt.	....	Salt.	....	Salt.	....	1 NaCl	2 NH₄Cl
8.30 a. m	.160	....	.163	....	.170	....	....	1 NH₄Cl
9 a. m	.168	....	.170	....	.180	....	1 NaCl	....
10 a. m	.180	.360	.182	.330	.192	.390	....	1 NH₄Cl
11.40 a. m	.192	.345	.195	.325	.202	.380	....	....
1 p. m	.202	.340	.202	.320	.210	.370	....	1 NH₄Cl
2 p. m	.205	.345	.204	.340	.210	.370	....	....
3 p. m	.208	.342	.205	.320	.210	.360	1 NaCl	....
4 p. m	.210	.350	.205	.320	.210	.370	....	....
5 p. m	.218	.380	.210	.310	.212	.360	....	1 NH₄Cl
6 p. m	.225	.380	.210	.310	.218	.360	....	....
7 p. m	.230	.380	.218	.310	.220	.360	....	....
8 p. m	.240	.380	.222	.310	.223	.350	1 NaCl	....
9 p. m	.250	.380	.222	.320	.225	.352	....	....
10.30 p. m	.275	.340	.225	.315	.225	.360	....	....
12 p. m	.350	.600	.230	.310	.230	.340	....	1 NH₄Cl
1 a. m.b	.420	.850	.240	.310	.230	.340	....	....
2.30 a. m	.510	1.550	.240	.310	.230	.340	....	....
4.15 a. m	.560	2.000	.240	.310	.230	.340	....	....
5.30 a. m	.550	2.200	.230	.310	.230	.330	....	....
7.45 a. m	.520	2.250	.230	.310	.225	.330	....	....
8.15 a. m	....	2.250	....	....	....	....	....	....
9 a. m	....	2.200	....	....	....	....	....	....

a The electrolyte was lowered into well A by means of a perforated brass bucket, 11 by 30 inches in size. The formula "2 NH₄Cl" means that two of these buckets, full of ammonium chloride, were introduced into well A at the time indicated. Each of these buckets held 2 pounds of the salt.
b August 6.

The time which elapses from the charging of the well A to the arrival of the electrolyte at the lower well gives the time necessary for the ground water to cover the distance between these two wells.   Hence, if the distance between the wells be divided by this elapsed time, the result will be the velocity of the ground water.   The electrolyte does not appear at one of the downstream wells abruptly; its appearance there

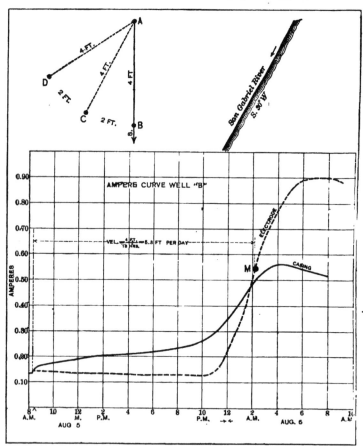

Fig. 3.—Curves showing electric current between casing of well A and casing of well B (heavy curve), and between casing of well B and its internal electrode (dotted curve), at station 5, San Gabriel River, California. These curves illustrate results obtained with the direct-reading form of apparatus.

is somewhat gradual, as is shown by the curves in figs. 3 and 4.   The time required for the electrolyte to reach its maximum strength in one of the downstream wells after its arrival at that well (and, hence, the time required for the current to reach its maximum value) may vary

from a few minutes in a case of high ground water velocity to several
hours in a case of low velocity. The writer formerly supposed that
the gradual appearance of the electrolyte at the downstream well was
largely due to the diffusion of the dissolved salt, but it is now known
that diffusion plays but a small part in the result. The principal cause
of the phenomenon is the fact that the central thread of water in each
capillary pore of the soil moves faster than the water at the walls of the
capillary pore, just as the water near the central line of a river channel usually flows faster than the water near the banks. For this
reason, if the water of a river be made suddenly muddy at a certain

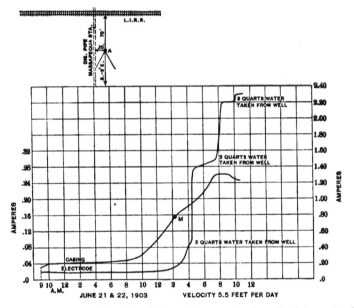

FIG. 4.—Curves showing possibility of use of direct-reading apparatus when well points are not used.
The casing in this instance consisted of common black 2-inch pipe, with a few small holes in bottom
section. The "casing" curve must be relied upon for determining velocity. The "electrode"
curve was obtained by drawing water from well C, as shown on diagram, the charged water being
drawn into the well through the small holes and the open end of well.

upstream point, the muddiness will appear somewhat gradually at a
downstream point, being first brought down by the rapidly moving
water in the center of the channel and later by the more slowly moving
water near the banks. The effect of the analagous gradual rise in the
electrolyte in the downstream well requires us to select the "point of
inflection" of the curve of electric current as the proper point to
determine the true time at which the arrival of the electrolyte should
be counted. This point is designated by the letter M in figs. 3 and 4.

Owing to the repeated branching and subdivision of the capillary pores around grains of sand or gravel the stream of electrolyte issuing from the well will gradually broaden as it passes downstream. The actual width of this charged water varies somewhat with the velocity of the ground water, but in no case is the rate of the divergence very

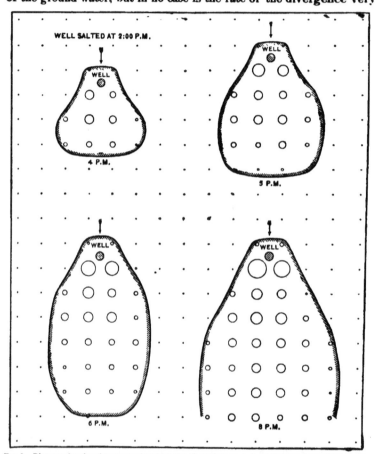

Fig. 5.—Diagram showing the manner in which the electrolyte spreads in passing downstream with the ground water. The shaded circle shows the location of the salted well, and samples were taken from small test wells placed in the sand in rows and columns at intervals of 6 inches, shown by dots in the diagram. The areas of the circles are proportional to the strength of the electrolyte found at their centers. The rough outline indicates the area covered by the charged water at the times specified. The velocity of the ground water (in the direction of the arrows) was 12 feet a day. It can be seen that the electrolyte barely reached a distance of 3 inches against the direction of flow.

great. Figs. 5 and 6 show some actual determination of the spread of the electrolyte around a well in a coarse sand, in one case the ground water moving 12 feet a day and in the other case moving 23 feet a day.

Samples of ground water were taken from small test wells placed only 6 inches apart, and the amount of salt or electrolyte was determined chemically. The amount at any point is indicated by the area of the circles shown in the diagrams. It will be seen that the salt barely showed itself at a distance of 3 inches upstream from the well. Three feet downstream from the well the width of the salt stream was about 3 feet in the first case and about 2 feet in the other.

It is possible to dispense with the circuit between the casing of well A to the casing of each of the other wells, as the short circuit between the well and electrode forms the best possible indication of the arrival of the electrolyte at the downstream well. For cases in which the velocity of ground water is high the circuit to well A is practically of no value, but for slow motions this circuit shows a rising current before the arrival of the electrolyte at the lower well, often giving indications that are of much value to the observer.

The method can be used successfully even if only common pipe be used for the wells. In this case, however, the absence of screen or perforations in the wells renders the internal electrodes useless, and one must depend upon the circuit from the well casing of the upstream well to the well casing of the downstream well. The results in Table II and fig. 4 present such a case.

TABLE II.—*Field record of electric current obtained at station 1, Massapequa, Long Island, June 21, 1903, with direct-reading underflow meter.*

[Readings in amperes and decimals of an ampere.]

Time.	Well B.		Well C.		Well D.	
	Casing.	Electrode.	Casing.	Electrode.	Casing.	Electrode.
8.45 a. m	0.03	0.08	0.03	0.10	0.03	0.09
9 a. m. a						
9.30 a. m. b	.04	.08	.04	.095	.036	.088
10 a. m	.04	.079	.039	.092	.036	.088
10.30 a. m	.04	.079	.04	.097	.039	.087
11 a. m	.04	.079	.04	.095	.059	.087
11.30 a. m	.04	.079	.04	.091	.039	.087
12 m	.041	.079	.04	.092	.040	.087
1 p. m	.042	.079	.04	.090	.040	.088
1.30 p. m	.042	.079	.04	.092	.040	.088
2 p. m	.043	.079	.04	.092	.040	.089
2.30 p. m	.043	.078	.041	.094	.040	.088
3 p. m. b	.043	.078	.041	.094	.040	.090
3.30 p. m	.043	.078	.040	.094	.041	.090
4 p. m	.043	.078	.042	.094	.041	.090

a 10 pounds of sal ammoniac placed in well A.
b 2 pounds of sal ammoniac placed in well A.

TABLE II.—*Field record of electric current obtained at station 1, Massapequa, Long Island, June 21, 1903, with direct-reading underflow meter*—Continued.

Time.	Well B.		Well C.		Well D.	
	Casing.	Electrode	Casing.	Electrode.	Casing.	Electrode.
4.30 p. m	0.043	0.078	0.042	0.095	0.041	0.098
5 p. m	.043	.078	.042	.096	.041	.098
5.30 p. m	.045	.078	.043	.096	.041	.090
6.30 p. m	.045	.078	.043	.097	.042	.091
7 p. m	.045	.078	.046	.099	.041	.091
7.30 p. m. a	.045	.078	.046	.099	.041	.090
8 p. m	.045	.080	.048	.099	.042	.093
8.30 p. m	.049	.080	.049	.100	.043	.094
9 p. m	.048	.079	.050	.100	.043	.094
10.30 p. m	.050	.079	.070	.101	.045	.095
12 p. m	.050	.079	.095	.106	.047	.095
1 a. m. b	.051	.079	.120	.122	.049	.099
2 a. m	.051	.079	.147	.152	.050	.100
3 a. m. a	.050	.079	.168	.195	.050	.100
4 a. m	.053	.079	.178	.430	.050	.100
4.30 a. m	.053	.079	.188	.470	.050	.100
4.40 a. m				c 1.3		
5 a. m	.053	.075	.200	1.4	.050	.100
6 a. m			.200	1.4		
7.45 a. m			.260	1.5		
8 a. m	.052	.075	.260	d 1.9	.050	d .100
8.15 a. m				d 2.20		
8.30 a. m				d 2.20		
9.15 a. m			.26	2.20	.049	.099
10 a. m	.050	.072		2.20	.049	.099
10 a. m			.25	e 2.30		
11 a. m			.245	2.30		
11 a. m				e 2.30		

a 2 pounds of sal ammoniac placed in well A.
b June 22, 1903
c Before this reading some water was taken from well C.
d About 2 quarts of water was taken from well C before this reading.
e After 6 quarts of water was taken from well C.

In this case the wells were not provided with well points, but did possess a 4-foot length of pipe, provided with four or five holes on opposite sides of the pipe, containing small one-half inch washer screens. These few openings are not sufficient to permit the electrolyte to freely enter the well, so that readings between casings must be relied upon for results. As a matter of fact, enough of the electrolyte did get into the well to give small increased readings, but in order to obtain the electrode readings given in the table, water was removed from the downstream wells by a small bucket holding about 6 ounces, so as to force a quantity of the water surrounding the well into the perforated sections. The notes appended to Table II show times of "bucketing" well C, and comparison with the column headed " well C, electrode" shows the effect on the strength of solution in the well. In cases where good well points are used the ground water charged with the electrolyte finds its way gradually and naturally into the well. The well point should be clean enough to allow as free passage into the well as through the soil itself. Second-hand points used for this purpose may show a marked lag in the entry of the electrolyte.

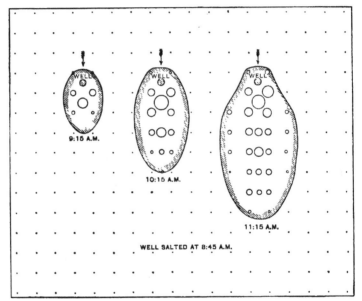

FIG. 6.—Figure showing conditions similar to those shown in fig. 5, but with ground water moving about twice as fast, or 22.9 feet a day. The electrolyte spreads less rapidly for the higher velocities as is seen by comparing this diagram with fig. 5.

Granulated sal ammoniac is used to dose well A. A single charge may vary from 4 to 10 pounds. If common pipe without points or screen is used for the wells, so that internal electrodes must be dispensed with, doses of about 2 pounds each should be repeated about every hour. The dry salt should not be poured directly into the well, but should be lowered in perforated buckets, as shown in Pl. I. These buckets are 1¾ by 30 inches and hold about 2 pounds of the salt. Two of these buckets may be tied one above the other for the initial charge and followed by two more in ten or twenty minutes.

If the wells are not too deep the sal ammoniac may be introduced into the well in the form of a solution. A common bucket full of saturated solution is sufficient. There is an uncertainty in introducing the sal ammoniac in solution in deep wells, as the time required for the solution to sink to the bottom of the well may be considerable.

The ammeter used in the work has two scales, one reading from 0 to 1.5 amperes and the other from 0 to 5 amperes. With a given

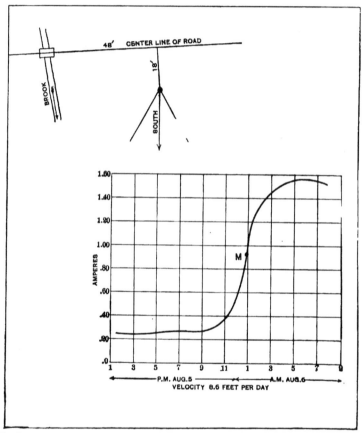

Fig. 7.—Electrical current for well C, station 14. Long Island, plotted on coordinate paper. The record was taken from the chart shown in upper part of fig. 13.

number of cells the amount of current between the upstream and a downstream well will depend, of course, upon several factors, such as the depth of the wells and their distance apart, but more especially upon the amount of dissolved mineral matter in the ground water. The initial strength of the current can be readily adjusted, however, after

A   COMMUTATOR CLOCK, FOR USE WITH RECORDING AMMETER

The clock works can be removed for cleaning or oiling without disturbing the rest
of the mechanism

B.   RECORDING AMMETER, COMMUTATOR CLOCK, AND BATTERY BOX, IN USE
IN THE FIELD

The instruments are shown as arranged in a rough box 16 by 22 by 36 inches, covered with tar paper

the wells have been connected with the instruments, by turning on or off some of the battery cells by means of the switches at the rear of the box. It is a good plan to use cells enough to make the initial current between one-tenth and two-tenths of an ampere.

SELF-RECORDING METER.

In the second form of underflow meter self-recording instruments are used, so as to do away with the tedious work of taking the frequent observations, day and night, required when direct-reading instru-

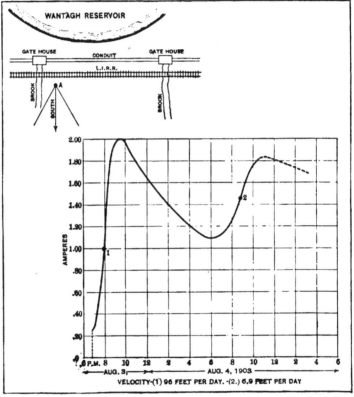

FIG. 8 —Electrical current for well C, station 13, Long Island, plotted on coordinate paper. The record was taken from the chart shown in the lower part of fig. 7. The curve indicates two different velocities in different beds of gravel penetrated by the test wells. One velocity is 96 feet a day; the other 6.9 feet a day.

ments are used. The arrangement of the apparatus is not materially different from that described above. In the place of the direct reading ammeter a special recording ammeter is used, having a range of 0 to 2 amperes. It has been found practicable, although it is a

matter of no small difficulty, to construct an instrument of this low
range sufficiently portable for field use and not too delicate for the
purpose for which it is intended.   The ammeter has a resistance of
about 1.6 ohms and is provided with oil dash pot to dampen swing of
arm carrying the recording pen.   The instruments were manufactured
by the Bristol Company.   The ammeters have gone through hard usage
in the field without breakage or mishap.   The portability of the instruments will be materially increased by changes in design which are now
being made.

The methods of wiring the wells when the recording instruments
are used is slightly changed.   In this case one side of the battery circuit is connected to casing of well A and to all of the electrodes of
wells B, C, and D.   The other side of the battery is run through the
recording ammeter to a commutator clock, which once every hour
makes a contact and completes the circuit, one after the other, to a
series of binding posts.   One of these binding posts is connected to
the casing of well B, one to the casing of well C, and one to the casing
of well D.   The time of contact is ten seconds, which gives an abundance of time for the pen to reach its proper position and to properly
ink its record.

Pl. II, B, and Pl. III, A, show two commutator clocks made for this
purpose by the instrument maker of the College of Engineering,
University of Wisconsin.   The clock movement is a standard movement
of fair grade, costing about $5.   It can readily be taken from the case
for cleaning or oiling and replaced.   A seven-day marine movement,
with powerful springs, is best for this purpose.

It will be seen from the method of wiring the wells that the record
will show the sum of the current between well A and well B added
to the current between the casing of well B and its electrode.   The
removal of the connection to well A would permit the record to show
the current between the casing of a downstream well and its electrode,
but the connection to the upstream well involves no additional trouble,
and occasionally its indications are of much service, especially if the
velocities are low.

All of the instruments above mentioned can be placed in a common
box 16 by 22 by 36 inches, covered with tar paper and locked up.
Pl. III, B, shows a photograph of the instruments thus arranged.
The shelf contains the recording ammeter (shown at left of cut) and
the commutator clock (shown at right of cut).

The contacts on the commutator clock are arranged about five minutes apart, so that the record made for the wells will appear on the
chart as a group of lines, one for each downstream well, of length
corresponding to the strength of the current.   The increasing current
corresponding to one of the wells will finally be indicated by the
lengthening of the record lines for that well.   This can be seen by

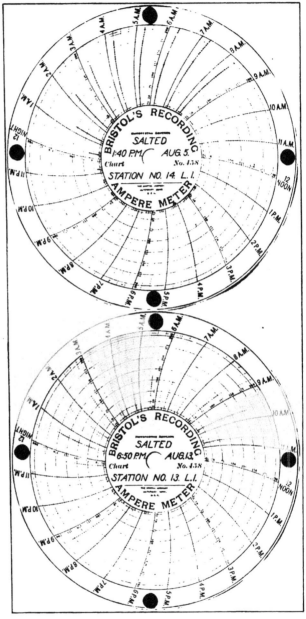

CHARTS MADE BY RECORDING AMMETER.

consulting the photographs of records shown in Pl. IV. In the upper chart the electrical current for wells B, C, and D at station 14, Long Island, is recorded, in the order named, at 2.10, 2.15, and 2.20 p. m., and hourly thereafter, the current remaining nearly constant at twenty-two to twenty-four hundredths ampere until 10.15 p. m., when the current for well C rises as indicated in the chart. In the lower chart the electrical current for wells B, C, and D is recorded, in the order named, at 6.30, 6.35, and 6.40 p. m., and hourly thereafter. The current for wells B and D remains constant at twenty-five hundredths ampere, but the current for well C is seen to rise as shown in the chart. The record charts are printed in light-green ink, and red ink is used in the recording pen, so that record lines can be readily distinguished when superimposed upon the lines of the chart. A special chart has been designed for this work and is furnished by the Bristol Company as chart 458.

The recording instruments in use have given perfect satisfaction, and the method is a great improvement in accuracy and convenience over the direct-reading method. The highest, as well as the lowest, ground-water velocities yet found have been successfully measured by the recording instruments. By using one or two additional dry cells the instrument may be made quite as sensitive as the direct-reading type.

. In using the recording instruments, but a single dose of salt need be placed in the upstream well. If the wells are deep it is important to use enough salt solution to make sure that the salt reaches as far down as the screen of the well point immediately after the solution is poured into the well. A gallon of solution will fill about 6 feet of full weight wrought-iron pipe, so that 10 gallons of solution should be used if the well is 60 feet deep. If the proper amount of solution be not used it will take an appreciable time for the solution to reach the bottom of the well by convection currents, and the results will be vitiated to that extent. As before stated, it is preferable to introduce granulated sal ammoniac into the well in a suitable bucket, in case the depth of the well renders the use of a solution uncertain.

If the ground water runs high in the quantity of total dissolved solids, say more than 100 parts per 100,000 total solids, it is important to run the electric current through the apparatus in a particular direction so as to reduce polarization and lessen the deposit of material upon the electrode. This is accomplished by arranging the batteries so that the positive or carbon pole is connected to the internal electrodes in the downstream wells. The well casings form in this case the negative or other end of the battery circuit, and the available surface is so much greater than it would be if the internal electrodes formed this end of the circuit that little trouble from polarization or falling off of current need be experienced.

THE CALIFORNIA OR "STOVEPIPE" METHOD OF WELL CONSTRUCTION.

By CHARLES S. SLICHTER.

INTRODUCTION.

The peculiar conditions of water supply existing in southern California have led to the development of a special type of well which the writer believes to be admirably adapted to conditions found in many places in the eastern part of the United States, especially on Long Island and in the Coastal Plain region of the Atlantic States. It is with the object of calling attention to the importance of this method, by which a large number of water-bearing beds can be drawn upon simultaneously, that the many points of excellence of the California type of wells and methods of well construction are pointed out. In addition to the illustrations accompanying this paper, a number of others showing the rigs, casing, and other appliances used in connection with wells in southern California may be found in the issue of the Engineering News for November 19, 1903.

CONDITIONS IN CALIFORNIA.

The valleys in southern California are filled with deposits of mountain débris, gravels, sands, bowlders, clays, etc., to a depth of several hundred feet, into which a considerable part of the run-off of the mountains sinks. The development of irrigation upon these valleys soon became so extensive that it was necessary to supplement more and more the perennial flow of the canyon streams by ground water drawn from wells in the gravels. This necessity was greatly accentuated by a series of dry years, so that ground waters became a most valuable source of auxiliary supply for irrigation in the important citrus areas in southern California. The type of well that came to the front and developed under these circumstances is locally known as the "stovepipe" well. It seems to suit admirably the conditions prevailing in southern California. In procuring water for irrigation the item of cost is, of course, much more strongly emphasized than in obtaining water for municipal use. The drillers of wells in California were not only confronted with a material which is almost everywhere

32

full of bowlders and similar mountain débris, but also by a high cost of labor and of well casings. It was, undoubtedly, these difficulties that led to the very general adoption in California of the "stovepipe" well.

DESCRIPTION OF APPARATUS AND METHODS.

The wells are put down in the gravel and bowlder mountain outwash or other unconsolidated material to any of the depths common in other localities. One string of casing in favorable location has been put down over 1,300 feet. The usual sizes of casings are 7, 10, 12, and 14 inches, or even larger. A common size is 12 inches. The well casing consists of, first, a riveted sheet-steel "starter," from 15 to 25 feet long, made of two or three thicknesses of No. 10 sheet steel, with a forged steel shoe at lower end. In ground where large bowlders are encountered these starters are made heavier, the shoe 1 inch thick and 12 inches deep, and three-ply instead of two-ply No. 10 sheet-steel body.

The rest of the well casing, above the starter, consists of two thicknesses of No. 12 sheet steel made into riveted lengths, each 2 feet long. One set of sections is made just enough smaller than the other to permit them to telescope together. Each outside section overlaps the inside section 1 foot, so that a smooth surface results both outside and inside of the well when the casing is in place, and so that the break in the joint is always opposite the middle of a 2-foot length. It is these short overlapping sections which are popularly known as "stovepiping."

The casing is sunk by large steam machinery of the usual oil-well type, but with certain very important modifications. In ordinary material the "sand pump" or "sand bucket" is relied upon to loosen and remove the material from the inside of the casing. The casing itself is forced down, length by length, by hydraulic jacks, buried in the ground and anchored to two timbers 14 by 14 inches and 16 feet long, which are planked over and buried in 9 or 10 feet of soil. These jacks press upon the upper sections of the stovepiping by means of a suitable head. The driller, who stands at the front of the rig, has complete control of the engine, the hydraulic pump, and the valves by which pistons are moved up or down, and also of the lever that controls the two clutches which cause tools to work up and down or to be hoisted.

The sand pumps used are usually large and heavy. For 12-inch work they vary in length from 12 to 16 feet, are 10⅜ inches in diameter, and weigh, with lower half of jars, from 1,100 to 1,400 pounds.

After the well has been forced to the required depth, a cutting knife is lowered into the well and vertical slits are cut in the casing where desired. A record of material encountered in digging the well is kept and the perforations are made opposite such water-bearing materials as may be most advantageously drawn upon. A well 500 feet deep may have 400 feet of screen, if circumstances justify it.

The perforator (see fig. 9) for slitting stovepipe casing is handled with 3-inch standard pipe with three-fourths-inch standard pipe on the inside. In going down or in coming out of the well the weight of the three-fourths-inch line holds the point of the knife up. When ready to "stick" the three-fourths-inch line is raised. By raising slowly on the 3-inch line with hydraulic jacks, cuts are made from three-eighths to three-fourths inch wide and from 6 to 12 inches long. according to the material at that particular depth. In another type of perforating apparatus (fig. 10) a revolving cutter punches fine holes at each revolution of the wheel. This style of perforator is called a " rolling knife." Besides these many other different kinds of perforators are in use in California. In fact, the perforator is a favorite hobby of the local inventors. They all seem to work well.

ADVANTAGES OF CALIFORNIA METHOD.

FIG. 9.—Perforator for slitting stovepipe casing.

The advantages of this method of well construction are quite obvious. For wells in unconsolidated material. the California type is undoubtedly the best yet devised. I believe that wells of this type would be highly successful in the unconsolidated coastal deposits on Long Island, in New Jersey, and at similar localities. The absence of bowlders and very coarse gravels in those deposits might possibly make it more advantageous to use the hydraulic jet instead of the ponderous sand bucket in soft material, but this is the only modification that Eastern conditions seem to suggest.

Among the special advantages in the stovepipe construction we may enumerate the following:

1. The absence of screw joints liable to break and give out.

2. The flush outer surface of the casing, without couplings to catch on bowlders or to hang in clay.

3. The elastic character of the casing, permitting it to adjust itself in direction and otherwise to dangerous stresses, to obstacles, etc.

4. The absence of screen or perforation in any part of the casing when first put down, permitting the easy use of sand pump and the penetration of quicksand, etc., without loss of well.

5. The cheapness of large-size casings, because made of riveted sheet steel.

6. The advantage of short sections, permitting use of hydraulic jacks in forcing casing through the ground.

7. The ability to perforate the casing at any level at pleasure is a decided advantage over other construction. Deep wells with much screen may thus be heavily drawn upon with little loss of suction head.

8. The character of the perforations made by the cutting knife are the best possible for the delivery of water and avoidance of clogging. The large side of the perforation is inward, so that the casing is not likely to clog with silt and débris.

9. The large size of casing possible in this system permits a well to be put down in bowlder wash where a common well could not possibly be driven.

10. The uniform pressure produced by the hydraulic jacks is a great advantage in safety and in convenience and speed over any system relying upon driving the casing by a weight or ram.

11. The cost of construction is kept at a minimum by the limited amount of labor required to man the rig, as well as by the good rate of progress possible in what would be considered in many places impossible material to drive in, and by the cheap form of casing.

COST OF THE WELLS.

An idea of the cost of constructing these wells can best be given by quoting actual prices on some recent construction in California. According to contracts recently let near Los Angeles, the cost of 12-inch wells was:

Fifty cents per foot for the first 100 feet, and 25 cents additional per foot for each succeeding 50 feet, casing to be furnished by the well owner. This makes the cost of a 500-foot well $700 in addition to casing. The usual type of No. 12 gauge, double stovepipe casing, is about $1.05 a foot, with $40 for 12-foot starter with three-quarter inch by 8-inch steel ring. A good driller gets $5 a day; helpers, $2.50 a day. The cost of drilling runs higher than that given above in localities where large and numerous bowlders are encountered.

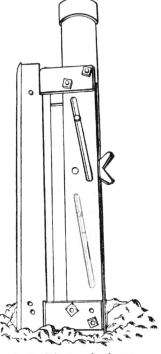

Fig. 10.—Roller type of perforator.

The drillers build their own rigs according to their own ideas, so that no two rigs are exactly alike; that is, the drillers pick out the castings and working parts and mount them according to ideas that experience has taught them are the best for the wash formations in which they must work.   Fig. 11 shows a common form of rig.

It is not very profitable to name individual wells of this type and give their flow or yield, since conditions vary so much from place to place.   From the method of construction it must be evident that this type of well is designed to give the very maximum yield, as every water-bearing stratum may be drawn upon.   The yield from a number of wells in California of average depth of about 250 feet, pumped by centrifugal pumps, varied from about 25 to 150 miners' inches, or

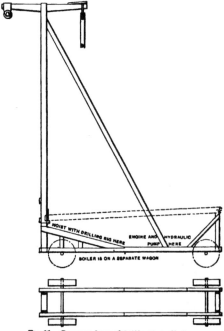

FIG. 11.—Common form of California well rig.

from 300,000 to 2,000,000 gallons a day.   These are actual measured yields of water supplied for irrigation.

Among the very best flowing wells in southern California are those near Long Beach.   The Boughton well, the Bixby wells, and the wells of the Sea Side Water Company are 12-inch wells, varying in depth from 500 to 700 feet and flowing about 250 miners' inches each, or over 3,000,000 gallons per twenty-four hours.   The flow of one of these wells is the greatest I have seen reported.   Among the records for depth are those of 1,360 feet for a 10-inch well, and 915 feet for a 12-inch well.   A new 14-inch well has already reached a depth of 704 feet.

APPROXIMATE METHODS OF MEASURING THE YIELD OF FLOWING WELLS.

By CHARLES S. SLICHTER.

Tables for determining the discharge of water from completely filled vertical and horizontal pipes were prepared a number of years ago by Prof. J. E. Todd, State geologist of South Dakota, who issued a private bulletin describing simple methods of determining quickly, with fair accuracy and with little trouble, the yield of artesian wells. In the following notes the tables and explanations relating to vertical and horizontal pipes are taken from this bulletin. The explanations and tables relating to the measurement in the partially filled horizontal and inclined pipes have been appended by the present writer.

MEASUREMENT OF FLOWS FROM FILLED PIPES.

In determining the flow of water discharged through a pipe of uniform diameter all that is necessary is a foot rule, still air, and care in taking measurements. Two methods are proposed, one for pipes discharging vertically, which is particularly applicable before the well is permanently finished, and one for horizontal discharge, which is the most usual way of finishing a well.

The table on page 39 is adapted to wells of moderate size as well as to large wells. In case the well is of other diameter than given in the table its discharge can without much difficulty be obtained from the table by remembering that, other things being equal, the discharge varies as the square of the diameter of the pipe. If, for example, the pipe is one-half inch in diameter its discharge will be one-fourth of that of a pipe 1 inch in diameter for a stream of the same height. In a similar manner the discharge of a pipe 8 inches in diameter can be obtained by multiplying the discharge of the 4-inch pipe by 4.

In the first method the inside diameter of the pipe should first be measured, then the distance from the end of the pipe to the highest point of the dome of the water above, in a strictly vertical direction— a to b in the diagram, fig. 12. Find these distances in Table I (A), and the corresponding figure will give the number of gallons discharged each minute. Wind would not interfere in this case, so long as the measurements are taken vertically.

37

The method for determining the discharge of horizontal pipe requires a little more care. First, measure the diameter of the pipe. as before, then the vertical distance from the center of the opening of the pipe, or some convenient point corresponding to it on the side of the pipe, vertically downward 6 inches, *a* to *b* of the diagram, then from this point strictly horizontally to the center of the stream, *b* to *c*. With these data the flow in gallons per minute can be obtained from Table I (B). It will readily be seen that a slight error may make much difference in the discharge. Care must be taken to measure horizontally and also to the center of the stream. Because of this difficulty it is desirable to check the first determination by a second. For this purpose columns are given in the tables for corresponding measurements 12 inches below the center of the pipe. Of course the discharge from the same pipe should be the same in the two measurements

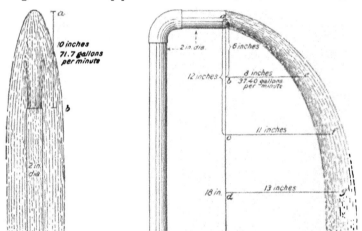

Fig. 12.—Diagram illustrating flow from vertical and horizontal pipes.

of the same stream. Wind blowing either with or against the water may vitiate results to an indefinite amount; therefore measurements should be taken while the air is still.

Whenever fractions occur in the height or horizontal distance of the stream, the number of gallons can be obtained by apportioning the difference between the readings in the table for the nearest whole numbers, according to the size of the fraction. For example, if the distance from the top of the pipe to the top of the stream in the first case is 9¼ inches, one-third of the difference between the reading in the table for 9 and 10 inches must be added to the former to give the correct result.

In case one measures the flow of a well by both methods he may think that the results should agree, but such is not the case. In the vertical discharge, there being less friction, the flow will be larger; so also in the second method differences will be found according to the length of the horizontal pipe used.

As pipes are occasionally at an angle, it is well to know that the second method can be applied to them if the first measurement is taken strictly vertically from the center of the opening, and the second measurement from that point parallel with the axis of the pipe to the center of the stream, as before. The measurements can then be read from the table.

TABLE I.—*For determining yield of artesian wells.*

[In gallons per minute.]

Height of Jet.	A. Flow from vertical pipes.					Horizontal length of Jet.	B. Flow from horizontal pipes.			
	Diameter of pipe in inches.						1-inch pipe.		2-inch pipe.	
	1	1¼	1½	2	3		6-inch level.	12-inch level.	6-inch level.	12-inch level.
Ins.						*Ins.*				
½	3.96	6.2	8.91	15.8	35.6	6	7.01	4.95	27.71	19.63
1	5.60	8.7	12.6	22.4	51.4	7	8.18	5.77	32.33	22.90
2	7.99	12.5	18.0	32.0	71.9	8	9.35	6.60	36.94	26.18
3	9.81	15.3	22.1	39.2	88.3	9	10.51	7.42	41.56	29.45
4	11.33	17.7	25.5	45.3	102.0	10	11.68	8.25	46.18	32.72
5	12.68	19.8	28.5	50.7	113.8	11	12.85	9.08	50.80	35.99
6	13.88	21.7	31.2	55.5	124.9	12	14.02	9.91	55.42	39.26
7	14.96	23.6	33.7	59.8	134.9	13	15.19	10.73	60.03	42.54
8	16.00	25.1	36.0	64.0	144.1	14	16.36	11.56	64.65	45.81
9	17.01	26.6	38.3	68.0	153.1	15	17.53	12.38	69.27	49.08
10	17.93	28.1	40.3	71.6	161.3	16	18.70	13.21	73.89	52.35
11	18.80	29.5	42.3	75.2	169.3	17	19.87	14.04	78.51	55.62
12	19.65	30.7	44.2	78.6	176.9	18	21.04	14.86	83.12	58.90
13	20.46	31.8	45.9	81.8	184.1	19	22.21	15.69	87.74	62.17
14	21.22	33.0	47.6	84.9	190.9	20	23.37	16.51	92.36	65.44
15	21.95	34.2	49.3	87.8	197.5	21	24.54	17.34	96.98	68.71
16	22.67	35.2	50.9	90.7	203.9	22	25.71	18.17	101.60	71.98
17	23.37	36.3	52.5	93.5	210.3	23	26.88	18.99	106.21	75.26
18	24.06	37.5	54.1	96.2	216.5	24	28.04	19.82	110.83	78.53
19	24.72	38.6	55.6	98.9	222.5	25	29.11	20.64	115.45	81.80
20	25.37	39.6	57.0	101.6	228.5	26	30.38	21.47	120.07	85.07
21	26.02	40.6	58.4	104.2	234.3	27	31.55	22.29	124.69	88.34
22	26.66	41.6	59.9	106.7	240.0	28	32.72	23.12	129.30	91.62
23	27.28	42.6	61.4	109.2	245.6	29	33.89	23.95	133.92	94.89
24	27.90	43.5	62.8	111.6	251.1	30	35.06	24.77	138.54	98.16
25	28.49	44.4	64.1	114.0	256.4	31	36.23	25.59	143.16	101.43
26	29.05	45.3	65.3	116.2	261.4	32	37.40	26.42	147.78	104.70

TABLE I.—*For determining yield of artesian wells*—Continued.

	A. Flow from vertical pipes.						B. Flow from horizontal pipes.			
Height of jet.	Diameter of pipe in inches.					Horizontal length of jet.	1-inch pipe.		2-inch pipe.	
	1	1¼	1½	2	3		6-inch level.	12-inch level.	6-inch level.	12-inch level.
Ins.						*Ins.*				
27	29.59	46.1	66.4	118.2	266.1	33	38.57	27.25	152.39	107.98
28	30.08	46.9	67.5	120.3	270.4	34	39.64	28.08	157.01	111.25
29	30.55	47.5	68.5	121.9	274.1	35	40.45	28.64	161.63	114.52
30	30.94	48.2	69.4	123.4	277.6	36	41.60	29.46	166.25	117.79
36	34.1	53.2	76.7	136.3	306.6					
48	39.1	61.0	88.0	156.5	352.1					
60	43.8	68.4	98.6	175.2	394.3					
72	48.2	75.2	108.0	192.9	434.0					
84	51.9	81.0	116.8	207.6	467.0	Continue by adding for each inch—				
96	55.6	86.7	125.0	222.2	500.0					
108	58.9	92.0	132.6	235.9	530.8					
120	62.2	98.0	139.9	248.7	559.5					
132	65.1	102.6	146.5	260.4	585.9		1.15	0.82	4.62	3.27
144	68.0	106.4	153.1	272.2	612.5					

NOTE.—To convert results into cubic feet, divide the number of gallons by 7.5, or, more accurately, by 7.48.

The flow in pipes of diameters not given in the table can easily be obtained in the following manner:

For ¼-inch pipe, multiply discharge of 1-inch pipe by............... 0.25
For ½-inch pipe, multiply discharge of 1-inch pipe by............... 0.56
For 1¼-inch pipe, multiply discharge of 1-inch pipe by.............. 1.56
For 1½-inch pipe, multiply discharge of 1-inch pipe by.............. 2.25
For 3-inch pipe, multiply discharge of 2-inch pipe by............... 2.25
For 4-inch pipe, multiply discharge of 2-inch pipe by............... 4.00
For 4½-inch pipe, multiply discharge of 2-inch pipe by.............. 5.06
For 5-inch pipe, multiply discharge of 2-inch pipe by............... 6.25
For 6-inch pipe, multiply discharge of 2-inch pipe by............... 9.00
For 8-inch pipe, multiply discharge of 2-inch pipe by............... 16.00

MEASUREMENT OF FLOWS FROM PARTLY FILLED PIPES.

From Table II, given below, it should be possible, if the wind is not blowing too hard, to determine the flow with an error probably not exceeding 10 per cent. This error is no greater than the fluctuation of the flow due to the daily variations of the barometric pressure in the case of most wells. The results in extreme cases, such as a one-fourth inch stream in a 6-inch pipe, are, of course, still less accurate, and actual measurement by collecting the water in a vessel of definite capacity should be resorted to if possible.

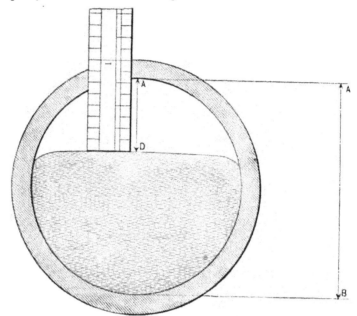

FIG. 13.—Method of measuring partly filled pipes.

TABLE II.— *For estimating the discharge from partly filled horizontal or sloping pipes.*

Fractional part of diameter of pipe not occupied by water. [Obtained by dividing A D by A B in fig. 1.]	Discharge expressed as percentage of discharge from full pipe, same size.	Fractional part of diameter of pipe not occupied by water. [Obtained by dividing A D by A B in fig. 1.]	Discharge expressed as percentage of discharge from full pipe, same size.
0.05	0.98	0.55	0.44
.10	.95	.60	.37
.15	.91	.65	.31
.20	.86	.70	.25
.25	.80	.75	.20
.30	.75	.80	.14
.35	.69	.85	.092
.40	.63	.90	.054
.45	.55	.95	.015
.50	.50	1.00	.000

To estimate discharge from partly filled horizontal or sloping pipe, first estimate discharge from full pipe of same size by means of Table I.

Next measure with a foot rule the dimension A D (see fig. 13) of the empty portion of cross section of the pipe. Divide this by the inside diameter of the pipe, which will give the fractional part of the diameter that is occupied by the empty part of pipe. In Table II find in the first column the number nearest to the above quotient. Opposite this number will be found the per cent of discharge of full pipe which the partially filled pipe is yielding. It is sufficient to measure the distance A D to the nearest eighth of an inch, or at most to the nearest sixteenth of an inch.

EXAMPLE.

Suppose that a 2-inch horizontal pipe has a length of jet of 13 inches at 6-inch level. From Table I this would represent a discharge of 60 gallons per minute from full pipe. Suppose that the distance A D is five-eighths of an inch and A B is 2 inches, or sixteen-eighths of an inch. Dividing 5 by 16 gives 0.31. In the first column of Table II we find 0.30, the nearest to 0.31, and opposite 0.30 in the second column of the table appears 0.75. The discharge from the partly filled pipe is therefore $60 \times 0.75 = 45$ gallons per minute.

CORRECTIONS NECESSARY IN ACCURATE DETERMINATIONS OF FLOW FROM VERTICAL WELL CASINGS.

From notes furnished by A. N. TALBOT.

The method described by Professor Todd is based on certain assumed theoretical conditions. While the error resulting from the method of measurement is usually less than 10 per cent, and while it is of great value in approximate field determinations, certain corrections must be made in order to exactly determine the amount of flow.

The considerations to be taken into account are that in vertical pipes the water will not rise to the height of its full velocity head and that the velocity is not uniformly distributed over the cross section. For velocity heads equal to the diameter of the pipe up to a height of 10 feet this discrepancy is not very great, but for lower values it is considerable.

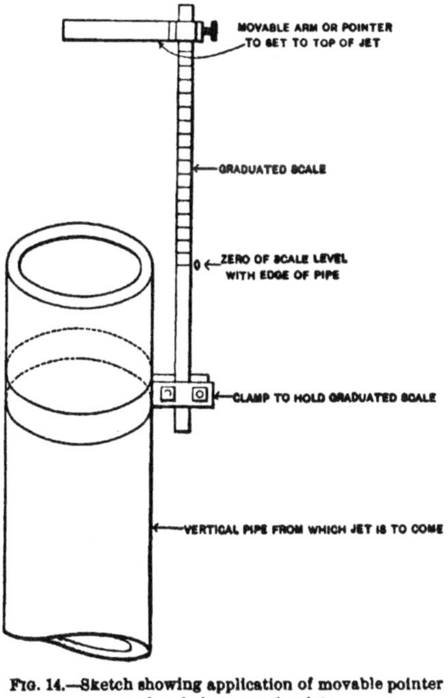

Fig. 15 shows the coefficients for 2-inch, 4-inch, and 6-inch pipes from low heads up to heads of 2 to 3 feet. These range from 0.80 to

FIG. 14.—Sketch showing application of movable pointer and scale in measuring jets.

0.96. The values above this, until atmospheric resistance becomes great, will not be much different. The quantities found by Professor Todd's table should be multiplied by these coefficients. These values

43

are fixed on the assumption that the height of water may be determined by finding the distance from the end of the pipe to the highest point

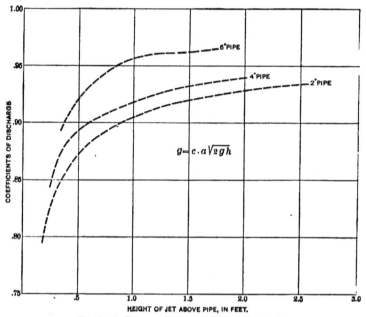

FIG. 15.—Coefficients of discharge for high and low heads.

of the jet by means of a movable pointer and scale, as shown in fig. 14, or by some similar device.

EXPERIMENT RELATING TO PROBLEMS OF WELL CONTAMINATION AT QUITMAN, GA.

By S. W. McCallie.

INTRODUCTION.

During the summer of 1903, a boring for a deep well, constructed by the town of Quitman, Ga., to improve its water supply, penetrated, at a depth of 123 feet from the surface, in limestone, what appeared to be a cavity 6¼ feet deep. Immediately after the cavity had been penetrated by the drill the water rose to within 77 feet of the surface, at which point it remained. In extending the bore hole beyond this depth it was found that all of the water forced into the well to carry out the drillings, and also the drillings themselves, appeared to pass off by the cavity. It was further discovered that any quantity of water, however great, that was forced into the bore hole, did not raise the level of the water in the well above 77 feet, and, on the other hand, that continuous pumping was equally ineffective in lowering the level.

After the sinking of the well to a second water-bearing stratum, at a depth of 321 feet, another well was constructed with a view of testing more fully the water-carrying capacity of the underground cavity, which was supposed to be a channel of a large subterranean stream. The second well, 6 inches in diameter, was put down a few hundred yards southwest of the first well, and only a few feet from the margin of Russell Pond, a small body of stagnant water occupying a nearly circular depression having the appearance of a partially filled lime sink. This well having been extended to the cavernous limestone, a canal was dug to connect it with the pond, and the water was allowed to flow from the pond into the well. The pond, which contained about one-half million gallons of water, was drained by the well in a few hours, without apparently affecting the level of the water in the bore hole, which remained constant at 77 feet from the surface. This test was conclusive to the town authorities that the underground cavity had a capacity to carry off an illimitable amount of water, and it was at once suggested that the town might be able to make use of this so-called underground stream for sewage disposal. This suggestion was soon taken up by the press of south Georgia, and within a short time Governor Terrell received a number of letters importuning

45

him to interfere to prevent the Quitman authorities from using deep
wells for sewerage purposes.

At the request of the governor, the writer, who at that time was
engaged in the study of the underground waters of Georgia for the
State and the United States Geological Survey, made a trip to Quitman to investigate the reports and, if they should be found true.
to point out to the people of Quitman the possibility that such a sewerage system might contaminate the wells and springs in that region.
When the writer arrived at Quitman he found that the town authorities were seriously considering the question of disposing of the sewage
as reported. They were willing, however, to give up the idea of using
the supposed underground stream for sewerage purposes if it could be
shown that such use would prove injurious in any way whatever.
Furthermore, they were willing to cooperate with the State and the
United States surveys by paying the greater part of the expense of
any experiments that might be necessary to establish this fact.

After some delay arrangements were finally made with the United
States Geological Survey for conducting an experiment to determine
the possibility of contamination of adjacent wells and springs, the
plan adopted for this purpose being what is known as the chlorine
method of tracing underground watercourses.

The well into which the chlorine (sodium chloride) was introduced
in carrying out this experiment was the Russell Pond well, referred to
above. The well has a depth of 120 feet, and shows the following section:

Section of well at Russell Pond, near Quitman, Ga.

	Feet.
1. Surface sand	2
2. Varicolored clay	60
3. Yellow sand	15
4. Gray sandy clay	43
5. Limestone (water bearing).	

The water, which rises to about 77 feet of the surface, comes from
the cavernous limestone found in all of the deep wells of Quitman and
vicinity.

GEOGRAPHY AND GEOLOGY.

Before the experiment is described in detail a few general notes will
be given on the geography and geology of the region, both of which
have a certain bearing on the question under consideration.

That part of south Georgia covered in conducting the Quitman
experiment lies along the Georgia-Florida State line, and includes the
southern part of Brooks and the adjacent portions of Thomas and
Lowndes counties. This part of the State is comparatively level, but
in parts of Brooks County the surface is more or less rolling, and
depressions caused by lime sinks are occasionally seen. The streams.
which all flow southward, are usually sluggish, and at points in their
course frequently traverse cypress swamps of considerable extent.

The springs of the region are few in number, but are usually large. They are generally found in or near the larger streams, and are often submerged during the wet season.

The geology of the region is typical of the portion of south Georgia that lies along the State line west of Thomasville, and that has been described by Spencer and others. Nearly everywhere throughout the piny woods or the cultivated fields of this section is to be seen a superficial covering or veneer of fine sand, which at some points attains a thickness of 2 feet or more. This sand corresponds probably to McGee's Columbia formation, as it lies directly upon the orange and reddish Lafayette clays.

The Lafayette clays, which are well exposed in numerous cuts along the Coast Line Railroad both east and west of Quitman, are usually stratified below and massive above. On the more elevated lands they attain a thickness of many feet, but along the larger streams they have been partially or wholly removed by erosion so as to expose underlying clays which at certain points along Withlacoochee River, notably at McIntyre Spring, contain large masses of coral. These lower clays are probably Miocene, and belong, no doubt, to Langdon's Chattahoochee group.

Beneath these Miocene clays there is a thick limestone, which is the source of the water supply of all the deep wells at Quitman, Boston, and Valdosta. This limestone seems to belong to Conrad's Vicksburg group, which, according to Dall, forms the lowest member of the Oligocene beds in the southern Tertiary series. As is shown by samples of borings from deep wells, this rock is somewhat variable in character, but seems to consist largely of thick beds of comminuted shells and corals interlaminated with layers of hard, compact limestone. The upper beds of this limestone appear to be cavernous, for cavities, shown by the dropping of the drill, are frequently found in it in the deep wells throughout the region. These cavities are said to vary in depth from 3 to 9 feet, and are always filled with water having a static head that is 40 feet or more higher than the point at which the cavities are struck.

The order of occurrence and the approximate thickness of these several formations are shown in the section of the Russell Pond well, above given. The Boston and the Quitman wells also show similar sections. The dip of the formations is supposed to be toward the south at a few feet per mile.

DESCRIPTION OF EXPERIMENT.

The chlorine method of tracing underground streams, as carried out in the Quitman experiment, was adopted at the suggestion of Prof. C. S. Slichter, consulting engineer of the United States Geological Survey.

A reconnaissance topographic survey of the region surrounding Quitman was made, with the view of finding all springs and wells whose waters stood at a lower level than the water in the Russell Pond well. where the salt was to be introduced, altitudes being determined by aneroid barometer. From all such springs samples of water were collected for determination of the amount of chlorine present. Each of the springs and wells thus sampled was made a station from which. at regular intervals, samples of water were collected by observers who were advised how and when to collect samples. The various stations having been established and samples collected for the determination of the normal amount of chlorine contained in the waters, 2 tons of salt (sodium chloride) were introduced into the Russell Pond well. The salt was put into the well in the form of solution, its introduction beginning at 8 a. m. October 15, and continuing until 8 p. m. October 20. The first ton of salt put into the well was introduced continuously for twelve hours at the rate of 166⅔ pounds an hour, while the second ton was put into the well continuously for one hundred and twenty hours, or five days, at the rate of 16⅔ pounds an hour.

In order to introduce the salt into the well at the above rate and also to insure as complete saturation of the water as possible, five 50-gallon barrels were set about 20 inches apart upon a level platform at the well and connected with one another by pieces of 2-inch iron pipe, firmly screwed into the barrels by right and left screws cut on their opposite ends. The iron pipe was inserted into the barrel nearest the well about 2 inches from the bottom, while the pipe between each succeeding barrel was elevated about 4 inches, so that the one uniting the last two barrels entered them about halfway up. When everything was in readiness for the experiment to begin. each barrel was filled with salt up to the point at which the outflow pipe entered it. The water was then turned into the barrel farthest from the well through a hydrant connected with the public water supply. As the water rose in this barrel to the point of entrance of the pipe it flowed into the next barrel, and so on until the entire chain of barrels was filled. The water now being more or less completely saturated. having flowed over the salt in the bottom of each barrel, was turned into the well through a stopcock so adjusted as to deliver the desired amount of water per minute. By a nice adjustment of the stopcock at the well and of that at the hydrant it was found that this plan of delivering a given amount of brine to the well in a stated time was practically automatic, the attendant having nothing to do except to keep the bottoms of the barrels supplied with salt.

DESCRIPTION OF STATIONS.

The stations from which water was collected were seven in number and are described below. Their locations are shown by the sketch map (fig. 16).

Station No. 1.—The station is the well at the public waterworks, 1,000 feet northeast of the Russell Pond well. The well is 321 feet deep and supplies the town with water. It is cased with 8-inch casing to 122 feet, only a short distance above the first water-bearing stratum. Within the 8-inch casing is inserted a 4½-inch casing which extends from the surface to a depth of 309 feet, at which point it was driven securely into a hard rock so as to form what was supposed to be a water-tight joint, cutting off all the water from the 123-foot water-bearing stratum above. The town water supply is pumped through this 4½-inch casing from the second water-bearing stratum, 311 feet below the surface. The second water-bearing stratum has a static head 2¾ feet greater than that of the first stratum. Nevertheless, it was thought that in case the 4½-inch casing was not absolutely water-tight at the point where it was driven into the rock, continuous pump-

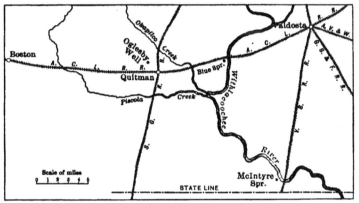

Fig. 16.—Sketch map showing location of Quitman experiment.

ing might lower the static head of the water from the second stratum sufficiently to set up a current from the stratum above, in which event sewage poured into the upper stratum would contaminate the lower.

Station No. 2.—This is the "Old City well," 1,500 feet northeast of the Russell Pond well. The well used for this station is 500 feet deep, and was completed in 1884. It was originally cased to a point some distance below the first water-bearing stratum, but its lower part becoming obstructed by sand or some other material, the 6-inch casing was burst by an explosive at the first water-bearing stratum, in order to obtain water for public use. While the experiment here reported was being carried on, this well, which had not been in use for some months, was connected with the pumps at the waterworks station, so that the water might be procured at regular intervals.

Station No. 3.—This is the Oglesby well, located at the Oglesby Mills, three-fourths of a mile northwest of Quitman. This well was completed in May, 1903, and is only 92 feet deep. It is 6 inches in diameter and is cased to 78 feet. The water-bearing stratum was struck at 87 feet, at which point, it is said, the drill dropped into a cavity 4 feet in depth and immediately thereafter the water rose to within 48 feet of the surface. It will here be noted that the water-bearing stratum in the Oglesby well is much nearer the surface than the first water-bearing stratum encountered in the wells within the corporate limits of Quitman, and also that the static head of the water is nearer the surface. This difference, however, is due entirely to the fact that the ground at this point is lower than that at neighboring wells, and not, as might be supposed, to the occurrence of a different water-bearing stratum.

Station No. 4.—This station was established at Blue Spring, near the right bank of the Withlacoochee River, 6 miles east of Quitman. By aneroid measurement the spring was found to be about on a level with the static head of the water in the Russell Pond well, so that it seemed to be a possible outlet for the supposed underground stream into which the salt was introduced. At the time the samples were collected Blue Spring was flowing about 15,000,000 gallons in twenty-four hours. It was learned, however, that at one time during an extremely dry season, several years ago, the spring entirely ceased flowing.

Station No. 5.—This is McIntyre Spring, about 15 miles southeast of Quitman, only a short distance from the Georgia-Florida State line. The spring is located partially in the Withlacoochee River and appears to furnish a much greater volume of water than Blue Spring. Owing to the great distance of this spring from Quitman it was found impracticable to determine its elevation by aneroid barometer, but judging from the fall of the river it must be several feet below Blue Spring. A flood in the river during the time the samples were being collected submerged McIntyre Spring for some time, so that only a limited number of samples were procured from this station.

Station No. 6.—This is a deep well at Boston, Thomas County, 14 miles west of Quitman. The Boston well has a depth of 290 feet and was completed in June, 1900. It is reported that three water-bearing strata were penetrated in this well, at 120, 160, and 286 feet. The static head of the water in the third water-bearing stratum is said to be 69 feet above sea level, which is 70 feet below the static head of the water in the Russell Pond well.

Station No. 7.—This station is one of the deep wells at Valdosta, 17 miles east of Quitman. The well from which the samples were taken at Valdosta was the 8-inch well, now used to supply the city water-works. This well is 500 feet deep, but its main water supply is said to come from a stratum at a depth of 260 feet, where the drill is reported to have entered a cavity 4 feet in depth. The static head of the water now in this well is 105 feet above sea level, which is 24 feet below the static head of the water in the Russell Pond well.

Additional station.—In addition to the stations here described there was also one kept up for a short time at the Quitman cotton factory; but, as the well was afterwards found not to be of sufficient depth to reach the water-bearing stratum into which the salt was introduced, it has been omitted in the tables and on the diagrams showing the variation of chlorine in the samples of water collected at the various stations.

SAMPLES OF WATER TAKEN.

The time at which the samples of water were taken at each station, the intervals between samples, and the number of samples themselves are shown in the following table:

Place, time, and number of samples of water taken.

Stations.	Date of first sample.	Interval between samples.	No.
		Hours.	
No. 1	Oct. 15, 12 noon	4	36
No. 2	....do	4	31
No. 3	Oct. 15, 4 p. m.	8	30
No. 4	Oct. 16, 6 a. m.	12	40
No. 5	Oct. 19, 6 a. m.	24	a 11
No. 6	Oct. 19, 8 a. m.	24	b 16
No. 7	....do	24	31

a Full number of samples not collected on account of high water.
b Fourteen samples lost by breakage in shipment.

CORRELATION OF WATER-BEARING STRATA.

In regard to the different water-bearing strata above noted in the Valdosta and Boston wells it might be stated that it was found impossible, with the meager data at hand, to correlate any of them with the water-bearing stratum in the Russell Pond well. In the absence of other data, an attempt was made to correlate the strata by means of chemical analyses of the water, but the results were unsatisfactory.

While it does not seem possible at present to correlate any of the strata with the Quitman stratum, into which the salt was introduced, there is but little doubt that they all occur in the Vicksburg limestone. Furthermore, as previously noted, the static head of the water in the Valdosta and Boston wells is greater than the static head of the water in the Quitman well by 24 and 69 feet, respectively. This would seem to indicate that the water-bearing strata are not continuous throughout the entire region, or that there is a flow converging toward Quitman, a condition not probable.

Station No. 1.—The normal chlorine in the waterworks well was determined as 5.44 parts per 1,000,000. Four hours after the introduction of the salt in the Russell Pond well the amount of chlorine in the water, as shown by the samples, began to increase, reaching a maximum of 6.80 at 8 p. m. October 15, or twelve hours after the introduction of the salt. From this time on the water continued to show excess of chlorine in varying quantities for about five days, or as long as the salt water was poured in at the other well, finally subsiding to its normal amount on October 21.

Station No. 2.—The normal chlorine in the old Waterworks well appears to be 5.44, although a sample taken just before the experiment showed 6.12. The chlorine content of the water of this well began to rise within four hours after the insertion of the salt, reaching a maximum of 6.97 at 8 p. m. on October 15, from which time it gradually declined, with several fluctuations, until 8 p. m. October 19, after which it remained constant until October 21, at its normal amount.

Station No. 3.—The normal chlorine in the Oglesby well is 5.44. The water of this well was examined at intervals from October 15 to October 29. From October 15 to October 21 there were some fluctuations in the amount of chlorine, which, however, appear to have no relation to the introduction of the salt in the Russell Pond well. At 12 p. m. on October 21 a decided rise was noticed, which continued until 4 p. m. October 22, when it reached a maximum of 6.46. It remained at this point until 12 p. m. of the same day, after which it declined gradually, though with some fluctuations, until 12 p. m. October 26, when the normal was again reached.

Station No. 4.—The normal chlorine at Blue Spring appears to be 5.78 parts per 1,000,000. Tests were conducted from October 16 to November 4, but no variations in the amount of chlorine which could be attributed to the salt inserted at Quitman were observed.

Station No. 5.—The normal chlorine at McIntyre Spring is 5.78, the same as at Blue Spring. Samples were taken from October 19 to November 6, but no variations in the amount of chlorine referable to the introduction of salt at Quitman were noted.

Station No. 6.—The normal chlorine in the Boston well appears to be 6.80, and no persistent variations were observed during the interval from October 19 to November 24.

Station No. 7.—The normal chlorine in the Valdosta well is 5.44. No variations due to the insertion of salt at Quitman were noted.

From the preceding it will be noted that only three stations—namely, the Waterworks well, the Old City well, and the Oglesby well—show variations which can with any degree of certainty be attributed to the effect of salt introduced in the Russell Pond well.   It will be observed that the maximum amount of chlorine in the two wells first named occurred at 8 p. m. October 15, just twelve hours after the first introduction of salt into the well.   As both of these wells are within a short distance of the Russell Pond well, this result would naturally be expected; yet at the same time it is difficult to explain how the salt made its appearance in the Waterworks well, as it obtains its supply from a water-bearing stratum that lies 200 feet below the one into which the salt was introduced.   It was surmised that the casing cutting off the upper water-bearing stratum from the one below did not form a water-tight joint at the point where it was driven into the rock, but allowed the water to flow downward when the pump was in action.   In order to determine whether this supposition was true or not, the following test was made: The pump was started and run for a few minutes when a sample of the water was taken to determine the normal amount of chlorin present.   There was then introduced through the 8-inch casing into the upper water-bearing stratum 15 pounds of salt in solution, and the pump was again started and run continuously for a half hour.   During the time the pump was in operation samples were taken every five minutes.   Analyses of these samples showed no increase of chlorine, demonstrating that the salt had not reached the second water-bearing stratum by the way of the suspected joint at the lower end of the 4½-inch casing elsewhere described.   This test seemed to show that the chlorine in the original experiment did not reach the water in the second stratum by the way of any defect of the joint at the end of the 4½-inch casing.   The test, however, can hardly be considered conclusive, owing to the small amount of salt used and the limited time the pump was operated after the salt was introduced.   It is very probable that if the amount of salt had been larger and the samples had been taken at longer intervals, the presence of the salt would have been detected.

In regard to the Oglesby well it will be noticed that the salt was transmitted in a northwesterly direction, notwithstanding the fact that the general flow of underground waters of the region is supposed to be southward rather than northwestward.

In addition to the variation in the amount of chlorine here explained as probably due to the presence of the salt introduced into the Russell Pond well, there are also other variations which, with one exception, are unexplained.   The exception referred to occurred at the old City Waterworks well.   The diagram for this well shows that the amount of chlorine in the sample taken before the experiment was much higher than the normal.   This was probably due to the presence of surface waters which had reached the water-bearing stratum below through the defective casing.

CONCLUSIONS.

From the notes above given on the Quitman experiment the following conclusions may be drawn:

1. The so-called underground stream, in the ordinary meaning of that term, does not exist in the wells investigated.

2. The water, which has a motion of probably not over 200 feet per hour, occurs in a porous, cavernous limestone, several feet in thickness.

3. Sewage introduced into the first water-bearing stratum will contaminate all of the wells in the vicinity that attain a depth of 120 feet or more.

4. The upper water-bearing stratum in the Waterworks well is not completely cut off from the water-bearing stratum below, so that the water from the lower stratum is likely to be contaminated from the stratum above.

THE NEW ARTESIAN WATER SUPPLY AT ITHACA, N. Y.

By Francis L. Whitney.

The object of this paper is to call attention to steps taken at Ithaca to obtain a pure water supply from underground sources after the pollution of the surface sources had given rise to a severe typhoid epidemic. The conditions existing at Ithaca, as regards underground waters, are probably duplicated at many points in New York, New England, and elsewhere, and attention is called to the possibility, under favorable conditions, of procuring adequate supplies, even for towns or cities of some size, from the glacial gravels of the deeply filled valleys of these regions. Such supplies may not always be available, but the possibility that they may be procured can usually be determined by a few inexpensive test wells.

FACTORS LEADING TO CHANGE OF SUPPLY.

Old water supply.—Up to the early part of 1903 the city of Ithaca obtained the larger part of its water supply from Sixmile Creek, a stream entering Cayuga Lake from the southeast at Ithaca. The pumping station was located in the outskirts of the city, on the southeast side. The creek is between 15 and 20 miles long. In general its drainage area is thinly populated, but two towns having an aggregate population of several hundred inhabitants, are located directly on the stream, which, in fact, receives all the drainage from them.

An additional water supply was obtained at Buttermilk Falls, on a small creek about 1½ miles south of Ithaca. There are very few houses in its drainage area.

Cornell University derives its supply from Fall Creek, a stream that empties into Cayuga Lake at its south end.

Typhoid epidemic.—It had long been recognized that these waters, especially that of Sixmile Creek, were liable to pollution, and precautionary systematic examinations of the water were made from time to time. These analyses were made by Prof. E. M. Chamot, of Cornell University. Some difficulty existed in the interpretation of the analyses, because of the fact that a number of flowing salt wells were tributary to the stream, so that it was not possible to determine what part, if any, of the chlorine that the water contained was due to sewage or similar pollution.

The water gave general satisfaction until about the beginning of 1903, when an exceptionally severe epidemic of typhoid fever broke out in the city. Hundreds of residents, including many students of Cornell University, were stricken, and the percentage of deaths was very high. The public was thoroughly aroused to the necessity of taking immediate and radical steps for determining the cause of the epidemic and for furnishing a new water supply, provided the source of danger was found to lie in the old supply.

It was found that the city supply from Buttermilk Falls and the college supply from Fall Creek were unpolluted. Sixmile Creek, on the contrary, was proved to be highly polluted, and it was seen that this was the probable cause of the epidemic.

Proposed dam on Sixmile Creek.—During 1902 the Waterworks Company began the erection of a high dam on Sixmile Creek, in order to provide a supplementary source of supply. The dam was designed to be 90 feet high and 100 feet wide. Many of the citizens of the town and some experts considered the dam unsafe and vigorous protests were made. The protests were, however, without result, and the company continued the work of construction. The feeling against the company became very strong and was doubtless a factor in the controversy which arose over the question of securing artesian water.

Action of water commissioners.—On February 16, 1903, the water commissioners met and discussed the feasibility of installing a filtration plant. The plan was proposed by President Schurman, of the university. Prof. E. M. Chamot testified that such a plant would not act as a perfect sterilizer, although if properly constructed and controlled it would remove 98 to 99 per cent of the bacteria. The plan received little support.

Commissioner Horton called attention to the expense of filtering and the ease with which a company can furnish, without detection, unfiltered water, and advocated artesian wells as a source of supply. It was thought that such wells could be obtained within a radius of 10 miles of the city.

It was urged by Professor Chamot that most of the artesian-well water of the locality was too hard for household purposes, but that it could be used in boilers, as attested by the fact that it had for eight years been used in those at the salt plant without bad effects. A further objection to the use of artesian wells lay in the fact that the city might be liable for damages resulting from the loss of water in adjacent wells. The drilling of test wells, the effect of which it was proposed to observe carefully, was advised as a precautionary measure.

Action of common council. —At a meeting of the common council early in the year President Schurman urged immediate efforts to obtain uncontaminated supplies, stating that it would become necessary to close the university unless a satisfactory supply was obtained by September 1, 1903. He reported that the executive committee and a number of trustees of the university had considered the advisability of sinking artesian wells, and had reported that it would probably not be possible to obtain from such wells the 2,000,000 gallons of water consumed daily by the city. In consideration of the necessity of obtaining pure supplies by September 1, it was recommended that the filtration plant be installed. This proposition was adopted by the council, but it was decided, before proceeding further, to submit the subject to the referendum of the people. A special election was ordered, to take place on March 2, to determine whether the city should undertake the ownership of its water supply. This election resulted in favor of municipal ownership.

Action of committee of one hundred.—Following the election a committee of one hundred was appointed from among the representative men of the city which agreed to subscribe to the amount of $10,000 or more toward procuring an immediate pure supply and for conducting experiments to determine the probability of a permanent supply of artesian water from the Freeville basin. The Fourth street test well, described on a subsequent page, was sunk by this committee.

FREEVILLE ARTESIAN BASIN.

The Freeville basin is a portion of the deep, gravel-filled valley of Fall Creek lying in the vicinity of Freeville, about 10 miles northeast of Ithaca. The watershed contributing to the basin covers 75 square miles, while the basin proper has an area of from 15 to 25 square miles. It was estimated that 10,000,000 gallons of water daily could be obtained from this basin, although the nature of the material and the slope of the hydraulic grade, or underground water surface, was practically unknown. Because of the lack of definite information in regard to the basin and its distance from the city the project of obtaining water supply from it was finally abandoned.

ITHACA ARTESIAN BASIN.

Attention was then directed to the valley in which the city of Ithaca lies. An examination of the geology of the region showed that because of the slope of the rocks to the south, while their outcrop was at a lower level to the north, they could not be expected to yield flowing water. The fineness of the texture of the rocks is also such that their water-holding capacity is small.

The conditions for a supply from the gravels filling the valley. however, are much more favorable.   Such wells as had been drilled obtained considerable supplies, and it appeared probable that with additional wells a supply sufficient to meet the needs of the city could be obtained.

The underground water of the region is probably derived by absorption from the waters brought in from the south by Coy Glen, Buttermilk Creek, Enfield Creek, and other streams tributary to the main valley occupied by Cayuga inlet.   Cayuga Lake will apparently have little bearing on the water supplies of the gravels, its surface being considerably lower than the mouth of the wells.   Prof. G. D. Harris suggested the use of screens adjusted in the casing of the wells at each water bed in order that all producing horizons might be drawn upon.

It was not regarded as essential to the success of the artesian supply that the wells should actually flow, it making very little difference in the cost of distribution whether the water stood a few feet above or a few feet below the surface.   Jamestown, in western New York, is supplied with water from a deeply gravel-filled channel.   At the start the wells rose perhaps 25 feet above the surface, but as more wells have been sunk into the water-bearing gravels and a considerable supply drawn from them the old wells have ceased to flow, all the larger wells being connected with powerful steam pumps.

It was conceded that the water supply of Cayuga inlet valley would easily supply Ithaca proper, while if necessary the university and portions of the city in its vicinity could be supplied by Fall Creek.

WELLS IN THE ITHACA BASIN.

The Illston well.—The Illston well, which is about 287 feet deep. was sunk by Mr. Illston for the purpose of obtaining pure water for the manufacture of ice.   After much labor and expense he succeeded in obtaining a well which gave a pressure of about 13 pounds per square inch and a temperature of 54° F.   Tests made by Mr. E. B. Kay on March 6, 1903, showed that, with the pipe 8 feet above ground, 403,142 gallons of water ran from the well daily, the test being made with a weir after the well had run freely for one hour.   Mr. Kay estimated that had the test been made at the surface of the ground the result would have been more than 425,000 gallons.   By the use of an air lift or other pump the capacity would probably be increased at least 50 per cent, and it is not impossible that 1,000,000 gallons daily might be procured from this well alone.   A second test, made March 8, after the well had flowed twenty-four hours, showed an output of 405,000 gallons a day.   The flow was even stronger than in the former test. Records at the Cornell chemical laboratories show that the analysis of water from this well is as follows:

Free ammonia	0.480
Albuminoid ammonia	0.005
Nitrogen as nitrites	None.
Nitrogen as nitrates	Trace.
Oxygen consumed	0.644
Chlorine as chlorides	61.640
Total solid residue	304.000
Loss of solids on ignition	78.000

Comparative analyses made by Professor Chamot on the Illston and other waters of this vicinity showed that in 1,000,000 parts the Illston water contained 309 parts solid matter; Fall Creek water, 309 parts; Buttermilk Creek water, 200 parts; springs in vicinity of the city, 165 to 200 parts; city well water, 500 parts, and Forest Home well water, 670 parts per million.

From the analysis it appears that the Illston well water is softer than that of most of the wells and springs in the city or at Forest Home, as soft as the Fall Creek water, and almost as soft as Buttermilk Creek water.

Salt well.—About eight years ago the National Salt Company began to sink its first well to the salt bed. A 14-inch hole was drilled. For the first hundred feet the drill went through a stratum of clay, in which no water was found. At the depth of 100 feet the drill broke through the clay into a gravel bed, upon which the hole was instantly filled with pure, clear water.

After going through the gravel, 100 feet beyond the water vein, the drill was brought to the surface, and in the corner of the derrick house another well, this one 6¼ inches in diameter, was sunk to a depth of about 120 feet, the lower 20 feet being in the gravel. The water from this well was clear, despite the fact that the drill had ground up the gravel and had created a condition which, naturally, would have caused the water to be turbid. A third well, of the same size as the last, was sunk, and then the drilling of the first and deeper well was resumed. After the completion of the plant a pump was attached to one of the wells. The flow was so copious that the engine was shut off to half its power, and for eight years only a small part of the water which might be drawn from the well has been used, although thousands of gallons are consumed daily at the plant.

Fourth street test well.—With the belief that other wells sunk in this vicinity would prove successful a test well was bored near Fourth street. This well was readily drilled to the water-bearing stratum, which was struck at the depth of 91 feet. When the gravel was reached the 6-inch pipe was quickly filled with water. Another length of pipe was then added, and the well was completed at a depth of 123 feet, with 32 feet of pipe in the gravel.

This well was then tested by pumping for several hours with a steam fire engine. Although the engine was worked to its full capacity there was no diminution of the water supply. The well, however, has since become choked with fine sand, which has gradually filled the screen. It is believed, however, that if properly developed it may again be made to yield abundantly.

Wells drilled for temporary supply.—Meanwhile workmen, under the direction of the water board, were drilling several wells for a

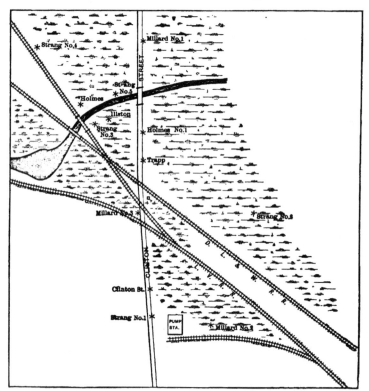

Fig. 17.—Sketch map showing location of temporary supply wells near Ithaca.

temporary supply. These wells are located on the lowlands in the valley of Cayuga Lake, along the lines of the Delaware, Lackawanna, and Western Railroad Company and the Lehigh Valley Railroad Company, near the intersection of Clinton street with the railroads. The lowlands are level and swampy, being only about 3 feet above Cayuga Lake. As shown by the accompanying map (fig. 17) they are within a few feet of the Illston well, south of the city.

The wells were all sunk by a walking-beam drill, assisted by hydraulic jet to remove the fine sand when the water-bearing strata was reached.

Of the 13 wells put down by the board only 3 are dry, 2 ending in black shale, the others in fine sand and gravel. Many have an excellent head, those nearest the Illston well, already described, having the greatest. One, known as Strang No. 3, had, at the time of boring, a static head of 17½ pounds. Some of the wells nevertheless required developing, and were opened by exploding dynamite at their bottoms, the loosened material being then removed by washing. In the Trapp well over 30 pounds of dynamite were used for this purpose.

As already stated, the Illston well yields with an air lift 425,000 gallons daily. The wells sunk for the temporary supply draw their water from the same source. What the wells will yield, however, is not yet known, although it had been found that the water in the various wells is affected by the pumping of others in the series. The water board estimates that the wells when pumped will have a capacity of 3,000,000 gallons daily.

GEOLOGY OF THE SUPPLY.

The diagrammatic sections in Pl. V, together with the written records on pages 64 to 66 indicate the manner in which the materials are distributed in the different wells. It will be seen that there is considerable variation in the upper and lower portions of the wells, although a certain system in the character of the deposits prevails. At the top there is a bed of mucky clay about 6 feet thick. Below this is a pure clay, or a clay containing in some instances considerable amounts of sand or gravel, which extends to a depth from about 50 or more feet, beneath which is from 20 to 50 feet or more of prevailingly sandy or gravelly material, in which some clay may also occur. These strata carry water, which will rise to within about 10 feet of the surface. None of this water is admitted to the wells, but is cased off so as to avoid any possibility of contamination from surface water. The stratum of sand and gravel last mentioned is frequently overlain by logs in a good state of preservation, which the drills penetrate with difficulty. Underlying the gravel and sand is a stratum of clay that extends to depths ranging from 210 to 250 feet, where a mixture of gravel and clay is encountered that extends downward to depths varying from 280 to 290 feet below the surface. The lower part of this stratum was in many cases a very hard and compact hardpan, directly below which is the stratum of gravel and sand from which the supply of water is drawn. As the water is taken only from this stratum, which lies 285 feet below the surface and is protected by a great thickness of clay, surface pollution becomes practically impossible.

The abundant supply can be attributed to the general conditions of the valley. On either side are hills attaining a height of about

1,000 feet and forming an immense catchment area. Underneath the drift are the Portage, Genesee, and Hamilton shales, and other rocks. dipping southward. The slope of the hills being nearly at right angles to the dip of the beds, the water flows downward into the valley instead of percolating through the underlying rocks. At a distance of 8 miles south of the city the floor of the valley·is about 472 feet higher than at the wells. The water from the hills of this region finds its way into the gravel and sand deposits of the valley. and by these is conveyed toward the lake.

Wells drilled outside a rather definitely limited area either fail to obtain water or procure only small supplies with little head. Wells drilled along a certain line, however, procure water under considerable pressure. The facts indicate the probable presence of a rather definite and limited water course through sand and gravel overlain by impervious clays.

WELL RECORDS.

The sections of six of the artesian wells in the vicinity of Ithaca have been given in Pl. V. Below are given records of a number of additional wells.

SOUTH WELL.

Material.	Thickness.	Depth to base.
	Feet.	*Feet.*
Soil	5	5
Blue clay	7	12
Clay	8	20
Sandy clay and wood	10	30
Sandy clay	10	40
Gravel and clay	10	50
Clay	10	60
Clay and gravel	10	70
Clay	65	135
Sandy clay	65	200
Sand	10	210
Coarse sand	10	220
Coarse to fine sand	10	230
Coarse sand; struck water	2	232

Located 1,500 feet up the valley from Clinton street, Ithaca, and about 70 feet east from the Delaware, Lackawanna and Western Railroad. Authority, Mr. Partridge.

SECTIONS OF TEMPORARY SUPPLY WELLS.

MILLARD WELL NO. 2.

Material.	Thick-ness.	Depth to base.
	Feet.	*Feet.*
Clay	44	44
Sand and gravel	31	75
Quicksand	3	78
Clay	123	201
Clay and gravel	21	222
Quicksand	18	240
Cemented gravel	4	244
Fine sand and clay mixed	7	251
Hardpan	6	257
1 bucket coarse gravel; black sand	1	258
Hardpan	1	259
Struck flowing water	--------	259

STRANG WELL NO. 2.

Driven with plugged pipe	87	87
Fine sand	10	97
Clay	116	213
Clay with large stones	22	235
Thin clay	33	268
Thin clay and fine sand	11	279
Fine sand	8.	287
3 or 4 buckets of sand and coarse gravel.		
Fine sand	13	300
Clay and gravel, hardpan	7	307
Clay and gravel with large stone	6	313
Clay and gravel with streaks of fine sand	3	316
Clay and sharp black sand	11	327
Clay, fine sand, and gravel	1	328
Clay and black sand	2	330
Soft black shale rock and sand	12	342
Black shale	10	352
Dry hole.		

STRANG WELL NO. 3.

Pipe driven with end plugged	110	110
Log	2	112
Clay and sand	11	123
Clay	47	170
Clay and sand	70	240
Quicksand	19	259
Black sand and some clay and fine sand mixed	16	275
Do	5	280
Struck water	--------	280

STRANG WELL NO. 4.

Material.	Thick-ness.	Depth to base.
	Feet.	*Feet.*
Driven with plugged pipe	110	110
Clay and large gravel	10	120
Fine sand	10	130
Clay	112	242
Hard clay and stone	6	248
Fine sand	32	280
Struck water		280

STRANG WELL NO. 5.

Driven with plugged pipe	108	108
Clay and stone	7	115
Fine sand	8	123
Clay	121	244
Clay and stone	12	256
Clay	14	270
Gravel; struck water	6	276

PRESENT CONDITIONS.

The total number of wells sunk is 14, of which two, put down in the first prospecting, are at such a distance from the station and collecting well as to make it impracticable to connect with them on account of cost, and three are dry or nearly dry. The remaining nine productive wells vary in flow from 50,000 gallons to 430,000 gallons each per day of twenty-four hours, and the static heads vary from 16 to 18 pounds.

The flow is collected by a system of pipes and laterals and delivered to a circular brick cistern or well, 16 feet in diameter and 18 feet deep, where any gravel or sand is deposited, so that it may not reach the pumps. The walls of the collecting well are 17 inches thick, and are made water-tight by a thick wall of puddled blue clay that entirely surrounds the well and extends from the surface downward to a depth of 6 feet, reaching the stratum of clay underlying the soil.

In the meantime the waterworks company has constructed a filtration plant of the mechanical type, from which those who are willing to take real or fancied filtered water obtain their supplies.

For a permanent supply deep wells in the valley of Sixmile Creek are contemplated, although not yet decided upon. These can be so located that the water may be distributed by a gravity system.

DRILLED WELLS OF THE TRIASSIC AREA OF THE CONNECTICUT VALLEY.

By W. H. C. PYNCHON.

INTRODUCTION.

Location and area.— The Triassic area, so called from the fact that it is underlain by rocks of Triassic age, is the lowland belt extending from N·w Haven in a direction slightly east of north to Northfield, Mass., not far from the New Hampshire line—a distance of about 110 miles (fig. 18). Its width at New Haven is about 3 miles. To the north it expands until at the Massachusetts line it has a width of 20 miles, from which it gradually declines to a width of about a mile at its northern end.

It is an area of soft sandstones and shales, with occasional trap ridges, lying in a broad depression between harder, more or less crystalline rocks. It is traversed from its northern end southward to beyond Middletown by the Connecticut River and is frequently spoken of as the Connecticut Valley lowland.

General conditions.—The sedimentary rocks of the Triassic area vary from coarse conglomerates through sandstone to fine clay shales, interbedded with which are thick sheets of trap. The dip is a little south of east and averages about 14 degrees. The rocks, except the traps, are porous, holding up to 30 per cent of their own volume of water, and the dips are such that artesian flows would ordinarily be expected in the lower valleys. Such flows are sometimes obtained, though, because of disturbance by faulting and the presence of numerous joints, the waters of the area are uncertain as to quantity and as to the height to which they will rise. The Triassic rocks are, nevertheless, a very important source of water, and it is with a view of describing the conditions to drillers and well owners that the following account has been prepared. The region involved in the problem is not limited by geographical but by geological boundaries, being confined to the area of outcrop of the Triassic rocks.

Nature of investigation.—In the Triassic area the occurrence of water is dependent directly upon the texture and structure of the rocks, and is therefore strictly geological. To fully understand its occurrence a knowledge of the geological conditions is necessary. This knowledge

65

may be obtained in part by an examination of the rocks where they are exposed at the surface, but a full insight into the question requires an intimate understanding of the conditions as they are disclosed by borings made for wells. No one is in a better position to know the underground conditions in his own wells than the driller, but his detailed knowledge rarely extends to the wells of other drillers and almost never covers the whole field. It is by combining information obtained from local authorities throughout the entire field and from his own personal observation that the geologist is able to present a comprehensive view of the entire region.

The present paper is a result of the method of investigation outlined above. Published descriptions of the region were consulted, additional field work was done in places concerning which further information was desired, and well records and other information furnished by drillers was collected from all parts of the area.

Collection of well data.—At the beginning of the investigation much information concerning well records was obtained by correspondence. Printed or typewritten blanks, containing a number of questions relating to the owner, location, diameter, and depth of well; quantity, quality, and height of water; materials encountered, additional well owners, drillers, etc., were sent out. The typewritten blanks were returned in almost every case, and a large number of the printed forms were also received. These furnished a good basis for planning field work, which was then undertaken. In the field, personal interviews were had with drillers and well owners and facts relating to the wells were obtained at first hand. These facts were afterwards studied and compared with one another and the resulting information was compiled for publication in the present report.

Special thanks are due to Mr. C. L. Grant and Mr. H. B. King, well drillers, of Hartford, Conn., for records and valuable information relating to wells of the area.

TOPOGRAPHY.

The topography[a] of the region is in general such as might be expected from the general geological structure. The hard crystalline rocks of the eastern and western uplands stand well above the region of soft Triassic sediments, surrounding them like a barrier. At its northern limit the lowland has an elevation of from 400 to 500 feet above sea level, but the land slopes steadily downward toward the south until, in the region of New Haven, there are portions which are only slightly above sea level. As would be expected, the harder sedi-

---

[a] The entire areas of the States of Connecticut and Massachusetts have been mapped by the United States Geological Survey. The complete series of maps includes about 70 sheets, of which about 15 lie within the area under discussion. The separate sheets are sold by the Survey for 5 cents each—the approximate cost of printing. Lists can be had on application to the Survey.

mentary beds find topographic expression in low ridges following the strike of the rocks in a generally north-to-south direction, their steeper faces being toward the west, as the beds dip eastward. But it is to the highly resistant trap sheets that the valley owes its most striking topographic features. Throughout the whole Triassic area the trap ranges dominate the lowlands in strong ridges that run from north to south and present bold cliffs on the west and gentle slopes on the east. The western cliffs of these ridges are buried in some cases almost to their summits by a heavy talus of fragments which the frost has split from their faces. These slopes of talus, topped by perpendicular cliffs, have a marked individuality, and though of no great elevation have long been spoken of as "mountains" throughout the region. The highest of these ridges in Massachusetts is Mount Tom, about 10 miles northwest of Springfield, which has an elevation of 1,200 feet above sea level, while the most conspicuous in Connecticut is West Peak, in the Hanging Hills of Meriden, with an elevation of 1,007 feet.

GEOLOGY.

PREVIOUS INVESTIGATIONS.

Soon after the beginning of the nineteenth century the geology and the mineralogy of the Triassic area of Connecticut and Massachusetts began to attract attention, papers on the subject having been published by the elder Professor Silliman, of Yale, as early as 1814. The first geologist to make a thorough study of the Connecticut section was Dr. James Gates Percival, who published in 1842, under State auspices, a "Report of the Geology of Connecticut." Dr. Charles Upham Shepard had published in 1837 his "Report on the Geological Survey of Connecticut," but this work was confined almost wholly to the mineral resources of the State. At about the same time the portion lying in Massachusetts received attention from Prof. Edward Hitchcock, of Amherst College. Among the geologists who have since devoted special attention to the region are Prof. J. D. Dana, of Yale, Prof. B. K. Emerson, of Amherst, and Prof. W. M. Davis, of Harvard. The latest and most exhaustive treatise on the Triassic of Massachusetts is Prof. B. K. Emerson's "Geology of Old Hampshire County, Massachusetts."[a] The latest and fullest treatise on the Connecticut section is Prof. W. M. Davis's "Triassic Formation of Connecticut."[b]

The geology of the Connecticut Valley lowland is interesting from many standpoints. The manner in which quiet processes of deposition have alternated with volcanic or other igneous action presents in itself a broad field of study, but the elaborate manifestations of faulting and displacement and the complicated structures and topographic forms

---

[a] Mon. U. S. Geol Survey, vol 29, 1898.
[b] Eighteenth Ann. Rept. U. S. Geol. Survey, pt. 2, 1898, pp. 1-192.

which have been developed in the region are the chief sources of interest to the geologist. It is not within the province of this paper, however, to take up the geological history of the area, attention being

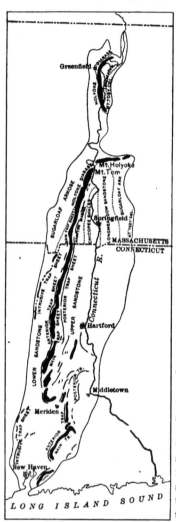

necessarily confined to the existing structure without regard to the processes by which this structure has been produced, except in so far as they influence the underground water supply. It is from this standpoint alone that the following description of the geology and topography of the region is given, the reader being referred for further information to the publications mentioned above.

GENERAL RELATIONS.

Both the eastern and the western portions of Connecticut are occupied by extensive highlands. These consist of ancient metamorphic rocks, deeply eroded by streams and overlain by glacial drift. The western upland is the more elevated, attaining at one point, Bear Mountain, a height of 2,355 feet above the sea. In Massachusetts this western upland is even more rugged than in Connecticut, and contains within its borders the famous Berkshire Hills. The eastern upland of Massachusetts is, like that of Connecticut, much less rugged than the western. The rocks are highly metamorphosed and deeply dissected by the streams which traverse them. Professor Emerson assigns the chief part of the metamorphosed rocks in Franklin, Hampshire, and Hampden counties, Mass., to the Silurian, a considerable portion to the Cambrian, two much smaller portions to the Upper Devonian and the Carboniferous, and

Fig. 18.—Geologic sketch map of the Triassic area of the Connecticut Valley.

a very small portion to the pre-Cambrian. It is probable that a detailed study of the uplands of Connecticut would show that the rocks comprising them would fall under a similar classification.

CHARACTER OF DEPOSITS.

Nature of the rocks.—Between the eastern and the western uplands, occupying a depression in the older rocks, lies the Connecticut Valley lowland, the limits of which have already been stated (p. 67). Its rocks all belong to the Triassic system. The greater portion of these rocks consists of comparatively soft sedimentary beds which are now so tilted that they have a rather uniform dip eastward of about 15 degrees. Interspersed with these rocks are to be found, however, certain sheets of volcanic lava (trap), generally interbedded between the strata of the sedimentary rocks, in nearly all cases in perfect conformity with the beds they overlie.

General structure.—A section across the Triassic area, as constructed by Professor Davis, is shown in fig. 19. The succession of the sandstones and traps, the uniform eastward dip of the beds of about 15 degrees and their displacement by faulting are strikingly brought out.

Succession and distribution of beds.—The portion of the Triassic area in Massachusetts and that in Connecticut were studied, as stated, by two geologists working independently, and in general the same conclusions were reached by both. The uniformly eastward dip, the

FIG. 19.—Generalized section across the Triassic area of the Connecticut Valley.

succession from lower and older rocks on the west to higher and younger at the center of the area, the presence of conglomerates along the border, the succession of trap sheets, and the presence of great numbers of faults are recognized by both. In the details of mapping, however, different subdivisions of the rocks were made in the two States, and there is some difference of opinion as to the succession of beds in the eastern portion of the valley and the character of the contact along that margin. Fig. 18 shows the distribution of the rocks in both the Massachusetts and Connecticut areas as mapped by Professor Emerson[a] and Professor Davis[b] respectively. In the following table are shown the equivalency of the divisions recognized in the two States. In addition to the formations given for the Massachusetts area there are the minor Black Rock diabase and the Granby tuff or ash beds. In Connecticut there is in addition what has been known as the intrusive trap sheet. These minor beds will be considered on a subsequent page.

a Emerson, B. K., Geology of old Hampshire County, Massachusetts: Mon. U. S. Geol. Survey, vol. 29, 1898.

b Davis, W. M., The Triassic formation in Connecticut: Eighteenth Ann. Rept. U. S. Geol. Survey, pt. 2, 1898.

Table showing correlation of Triassic rocks in Massachusetts and Connecticut.

Subdivisions of B. K. Emerson in Massachusetts.		Subdivisions of J. G. Percival and W. M. Davis in Connecticut.
Chicopee shales............................		Upper sandstones, including also Mount Toby conglomerate and part of Sugarloaf arkose.
Longmeadow sandstone and included traps.	Sandstone, shale, etc	
	Hampton diabase (trap).	Posterior trap sheet.
	Sandstone, shale, etc	Posterior shale.
	Holyoke diabase (trap)..	Main trap sheet.
	Sandstone, shale, etc	Anterior shales.
	Talcott diabase (trap) ...	Anterior trap sheet.
	Sandstone, shale, etc	Lower sandstone.
Sugarloaf arkose........................		
Mount Toby conglomerate...............		Not differentiated.

DESCRIPTION OF ROCKS.

Mount Toby conglomerate.—This is a coarse conglomerate, composed of fragments of slaty rocks, schist, and quartz, ranging from 2 inches to 4 feet in diameter, which forms a narrow band along the eastern border of the Triassic area in Massachusetts, extending from the Boston and Albany Railroad, near Ellis, southward to the State line, and on the same side from a point just north of Amherst to the extreme northern end of the area. It has not been differentiated in Connecticut, but undoubtedly extends along the eastern border in at least the northern portion of the State.

Sugarloaf arkose ("Lower sandstone" of Davis in part).—This is a coarse buff to pale red sandstone or conglomerate, made up largely of granite fragments that were derived from rocks lying farther west. It forms the western border of the Triassic area in Massachusetts, and bends around and forms the eastern border, except where it is separated from the crystalline rocks by the Mount Toby conglomerate. It is the equivalent of the conglomeratic portions of the "Lower sandstone" of Connecticut, where the latter rests in contact with the western border of the area.

Longmeadow sandstone and its Connecticut equivalents.—The sandstone occurs as a broad band inside of the Sugarloaf arkose belt. It is largely of the red or brown variety of sandstone, known as brownstone, which is so largely quarried, but includes many thin beds of shale. Interbedded with the sandstone are also thick beds of traps, which were laid down as lava sheets at various times during the period when the sandstones were deposited. The Longmeadow sandstone includes the nonconglomeratic portion of the lower sandstones of Connecticut, and the Anterior, Posterior, and a portion of the Upper shales or sandstones of Connecticut.

Talcott diabase (*Anterior trap sheet*).—This is a massive dark-colored trap that forms an important 'feature in the Connecticut area, where it is known as the Western trap range, but does not enter Massachusetts. It is a bedded sheet, originally formed as a volcanic flow on an old surface during the period of deposition of the sandstone.

Holyoke diabase (*Main trap sheet*).—This is a dark, dense trap bed, several hundred feet thick, that extends with a few breaks from the northern to the southern portion of the area. Because of its resistance to erosion it forms a conspicuous ridge, frequently called a mountain, as at Mount Tom and Mount Holyoke. It is sometimes known as the Eastern trap range. Like the Talcott diabase it was laid down as a surface lava flow.

Hampton diabase (*Posterior trap sheet*).—This is a relatively thin but at the same time persistent sheet that extends with some interruptions from near Mount Tom on the north to Long Island Sound on the south. It does not commonly give rise to a conspicuous ridge. Like the preceding sheets it was a surface flow.

Black rock diabase.—This is a diabase which, instead of flowing out over the old sandstone surfaces, was forced or intruded through the sandstones and the older Holyoke and Hampton traps as dikes or other igneous masses. It is found only in the region southeast of the Mount Tom and Mount Holyoke ridges in Massachusetts.

Granby tuff.—This is a sandstone or conglomerate, made up of volcanic débris, possibly derived from the volcanic eruptions accompanying the intrusion of the Black Rock diabase. Its distribution is much the same as the latter.

Intrusive traps in Connecticut.—The intrusive traps are of two types—intrusive sheets and dikes. The former is represented by the important trap sheet that was injected between certain of the lower sandstone beds near the western border of the Triassic area, and that extends from the Massachusetts line southward for nearly 20 miles, and again from near Southington to New Haven. The dikes are represented by smaller and more isolated masses, having approximately vertical attitudes cutting across the bedding of the sandstones.

Chicopee shale.—This is a band of dark-gray shaly sandstone, with some shales, that extends along the Connecticut River from Holyoke southward into Connecticut. In that State it has not been differentiated.

FAULTING.

When first formed both the sedimentary beds and the trap sheets were continuous and unbroken. If simply tilted and subjected to erosion such beds would give rise to continuous ridges, but in reality the latter are broken and their component parts shifted in relation to

one another. This shifting is due to movements which have taken
place along certain lines of fracture, and which in some regions have
broken the rocks into a series of blocks bounded by faults and known
as fault blocks. The faults are abundant in the Triassic area.
both in Connecticut and in Massachusetts, reaching a high development in the region of the Hanging Hills of Meriden. In direction they commonly run from northeast to southwest, transversely
to the strike of the beds, but some northwest-southeast faults occur.
In a given district they are usually parallel to one another. In many
places they are separated by intervals of only a fraction of a mile.
the result being the subdivision of the rocks into long narrow blocks
slightly offset as regards one another. In most cases there has also
been much movement in a vertical direction.

In these dislocations there is to be found the greatest variation in
length of fault line and amount of displacement. At one end of the
series stand the small local faults, which can be traced for only a few
feet and which show only a few inches of slipping; at the other end
may be placed the great fault which separates the Hanging Hills from
Lamentation Mountain. This fault crosses the whole lowland from
northeast to southwest, penetrating both the eastern and the western
uplands and having a total length of about 40 miles.[a] The upthrow is
on the east of the fault line and is so great that the lower sandstones
under Lamentation Mountain abut against the Posterior shales of Cathole Peaks, the most eastern member of the Hanging Hills. The fault
gap between these two portions of the main trap sheet is a mile across.
Faults of considerable movement everywhere leave distinct notches in
the crests of the various trap ridges. At these points erosion, acting
along the fault line, has cut deep gorges, some of which reach the
dignity of passes, flanked by long slopes of rocky talus topped by perpendicular cliffs. Excellent examples of these fault gaps are Cathole
Pass, between Cathole Peaks and Notch Mountain, and Reservoir
Notch, between Notch Mountain and West Peak, both in the Hanging
Hills.

Throughout the southern half of the Connecticut area, from about
the latitude of Hartford to Long Island Sound, the faults are closely
parallel, and run from northeast to southwest, the upthrow always
being on the eastern side. As a combined result of this displacement
and subsequent erosion there is, as we proceed southward, a constantly
recurring recession of the outcrop of the various members of the
series, this overlapping arrangement being specially well developed in
the vicinity of Meriden. From about the latitude of Hartford northward to the Massachusetts State line the faults, though still oblique in
character, run from northwest to southeast. The upthrow is still on

---

a Davis, W. M., The Triassic formation of Connecticut: Eighteenth Ann. Rept. U. S. Geol. Survey, pt. 2
1898, p. 100.

the eastern side of the fault lines, so that there is still a tendency to an eastward recession of the outcrops, but in a manner the reverse of that which is seen in the southern portion of the valley.

In the Massachusetts area the faults in the Triassic are fewer and of less importance. Faults of both the varieties mentioned occur, as will be seen from Professor Emerson's map,[a] but they produce a less striking effect on the arrangement of the chief ridges of the region.

Certain faults are supposed to occur in connection with the boundaries of the Triassic both in Connecticut and Massachusetts, principally in connection with the eastern boundary, and they have been indicated on the map[b] accompanying the report of Professor Davis.

JOINTING.

Jointing in traps.—Besides the faults there are certain minor lines of regular separation which demand attention. These are the joints which exist in great numbers throughout the mass. Every trap ridge in the lowland shows them in a remarkable degree, but they are especially prominent in the mountains of the main range. The trap sheets are divided by a vast number of these joints, the planes of which run in a direction generally perpendicular to the cooling surface, and along these planes the rock splits up into approximately rectangular blocks of all imaginable sizes. It is these joint planes that form the nearly vertical western faces of the mountains, and in cases where erosion, working on some local fault, has produced an escarpment facing eastward, the cliffs actually overhang. Every trap eminence of any considerable height is fronted by a long talus of broken trap, formed of joint blocks of all sizes which have fallen down from the cliffs above. A little gorge marking the line of a small fault near the crest of West Peak is fairly choked with slabs, some of them 12 feet square and several feet thick, which have been derived from the walls of the gorge. On the top of Rattlesnake Mountain are large blocks of trap that have been detached along joint planes, while at the foot of Talcott Mountain, on the line of the old Albany turnpike, lies a joint block as large as a small house. The blocks range from this size down to an inch square. In some localities the joints divide the trap into sheaves of blade-shaped fragments.

Jointing in sedimentary rocks.—Jointing in the sedimentary beds does not reach the highly developed condition seen in the trap sheets. In the heavy sandstones at the great brownstone quarries at Portland the joints are on a large scale and are indispensable aids in separating the great blocks of building stone. Yet here these lines of fracture

a Emerson, B. K., Geology of old Hampshire County, Massachusetts· Mon. U. S. Geol. Survey, vol. 39, Pl. XXXIV.
b Davis, W. M., The Triassic formation of Connecticut: Eighteenth Ann. Rept. U. S. Geol. Survey, Pl. XIX.

are potential rather than actual.   In the shales the joints become much
more numerous and may be regarded as universal.   In the city stone
pits near Trinity College at Hartford the Posterior trap has been
removed for road metal, leaving broad sheets of Posterior shale,
marked with ripple marks and mud cracks, exposed to view.   One
looking down upon this floor from the top of the cliff may see two sets
of joint cracks that cross each other at an angle of about 60 degrees
and divide the mass into lozenge-shaped sections.   Working along these
two sets of joint planes and the plane of the bedding, the quarrymen
remove blocks of shale averaging about a foot square and 6 to 8 inches
thick, which are used as the foundation rock in the construction of the
streets.   Joint blocks in this shale may be extremely variable in size,
and are often not over an inch in any dimension.   An examination of
almost any mass of shale throughout the lowland will show that the
jointing seen in this quarry is typical.

Although jointing in the sediments does not produce such striking
results as in the trap, yet the joints themselves extend, the writer
believes, to much greater distances, both vertically and horizontally,
in the sedimentary than in the igneous rocks.   Apparently they frequently become actual fissures in the deep rocks and have a profound
effect on the water supply, as will be shown later.

GLACIAL AND RECENT DEPOSITS.

The entire Triassic area has been subjected to the action of the ice
sheets that passed over the region in Pleistocene times.   The outlines
of the harder ridges have been softened and the softer rocks have been
worn away in many places by the scouring action of the ice.   Deposits
of till were formed beneath the sheet and now remain in the shape of
the well-known hardpan, while gravel and sand deposits were formed
as terraces or valley fillings by the glacial streams throughout the
region.   Since these deposits were laid down much material has been
carried from them by streams and redeposited to form the present
flood plains.

The presence of drift over the region is of great importance to the
water supply.   If the bare surfaces of the rocks were exposed the water
would run off rapidly instead of being absorbed, but the glacial deposits, especially the sands and gravels, hold in storage large quantities
of water, which is thus kept in contact with the underlying sandstones, by which the water is continually being absorbed.

WELLS OF THE CONNECTICUT VALLEY LOWLAND.

Although there are few flowing wells in the Triassic area there are
many wells from which water is pumped, and some of these would
yield a supply much in excess of the capacity of the pumps that are
used in them.

In the description of wells which follows an attempt is made to present data bearing upon the occurrence, volume, and quality of the water. Most of the records were gathered several years ago, but the area is fairly well covered and the information is comprehensive. In interpreting the figures due weight should be given to the source of information. Depths reported by drillers are likely to be more accurate than those furnished by owners, but on the other hand the quantity is more likely to be accurately stated by the owner. The reports of both owner and driller as to quality are likely to be more optimistic than the facts warrant, but the use to which the water is put—or, better, the uses to which it is not put—usually gives a clue to its actual character.

The records are far from uniform in the character of the points presented, certain features that are of special interest and consequence in one well being of little interest in another, which may show an entirely different set. The greater abundance of the records in the vicinity of the towns along the center of the area makes the data in such situations more complete than for the outlying portions, but the value of the data presented for wells in the outlying regions is probably greater. Data relating to wells that were failures are especially difficult to obtain from either owner or driller. The driller especially is loath to give details which might tend to discourage further drilling. This is unfortunate, as perhaps fully as much information as to the actual underground conditions is afforded by the dry holes as by the successful wells.

WELLS ENTERING THE ROCK.

The typical wells of this area are those which are sunk for the greater part of their depth in the rock, but a short list of wells of the drift, including those which lie in unconsolidated deposits, generally of glacial origin, overlying the Triassic rocks is also given. Certain of these wells are distinctly interesting, as their water supply seems to depend not on the arrangement of the deposits in which they are located, but on the conformation of the surface of the rocks which lie below.

WELLS IN MASSACHUSETTS.

NORTHAMPTON.

No. 1. Belding Brothers' Silk Company.—This well is in its way the most remarkable in the valley, as it was carried to the great depth of 3,700 feet without striking water. Mr. E. F. Crooks, writing for the company, states that at the " depth of 150 feet the drill entered sandstone or conglomerate rock which did not change at any time down to 3,700 feet." Professor Emerson classes the rocks which underlie Northampton as Sugarloaf arkose and gives a list of the beds traversed in drilling this well.[a] It seems probable that this well was driven through a water-bearing stratum at the point where it enters the rock, since a shallower well sunk by the same company to the rock, within a few feet of the former well, draws good water from that depth. This second well will be referred to again among the wells of the drift. The dry well is 8 inches in diameter.

SOUTH HADLEY FALLS.

No. 2. Sans Souci Club.—At the clubhouse of this organization is a private well 74 feet deep and 6 inches in diameter, entering the Long-meadow sandstone at a point 10 or 15 feet below the surface of the ground. The water, which is of medium hardness, supplies all needs of the clubhouse. The well is pumped.

HOLYOKE.

Nos. 3 and 4. American Writing Paper Company (Albion Paper Company Division).—These two wells, which are identical in detail, are 720 feet deep and have a diameter of 8 inches. The water rises within 14 feet of the surface, from which point it is pumped. Each well yields 450 gallons a minute throughout the twenty-four hours of the day, the water being slightly tinged with iron and not used in boilers. The wells enter the Chicopee shale about 50 feet below the surface.

No. 5. Holyoke Cold-Storage and Provision Company.—This well is over 500 feet deep, has a diameter of 3½ inches, and has its bottom in the Chicopee shale. The water, which has been analyzed, can be used in boilers, but is employed mostly for refrigerating purposes. The water is pumped at the rate of 20 to 25 gallons a minute.

No. 6. Riverside Paper Company.—A well, concerning which no data could be obtained, was drilled by this company, but was aban-doned because of the mineral matter contained in the water.

WILLIMANSETT.

No. 7. Hamden Brewing Company.—This well is 112 feet deep and has a diameter of 4 inches. It is sunk partly in rock, which is appar-ently Chicopee shale. The well is pumped at the rate of 60 barrels an hour—that is, about 40 gallons a minute—without lowering the level at which the water stands. The water has been analyzed and is too hard to be used in boilers.

LUDLOW.

No. 8. Ludlow Manufacturing Company.—This well is 150 feet deep and has a diameter of 8 inches for the first 50 feet, below which it is reduced to 6 inches. After passing through 6 feet of made land the

---

[a] Emerson, B. K., Geology of old Hampshire County, Massachusetts: Mon. U. S. Geol. Survey, vol. 29, 1898, pp. 385–388.

well lies for the rest of its depth in rock, which is mapped as Sugarloaf arkose. The water level is about 21 feet below the surface, but, when the well is pumped at the rate of 20 gallons a minute, the level drops to a point 20 feet lower. The water is called medium hard, but can not be used in boilers.

CHICOPEE FALLS.

No. 9. Overman Wheel Company.—This well is 475 feet deep and 6 inches in diameter, the lower 175 feet lying in the Chicopee shales. The well was drilled to obtain water for drinking purposes only and it supplied all the employees of the big establishment abundantly. It was found, however, that the water was so full of mineral matter that it had a purgative effect on those who drank it and now its use is wholly discontinued. The water has been analyzed.

BRIGHTWOOD.

No. 10. Springfield Rendering Company.—This well was unfinished when the locality was visited in July, 1900. It was drilled by Messrs. King & Mather, of Hartford, who very kindly furnished the data concerning it on the completion of the work, a few weeks later. The well is 200 feet deep and has a diameter of 6 inches. It pumps 150 gallons a minute and meets the demands made upon it—90,000 gallons for a working day of ten hours. The water when analyzed proved to be sufficiently pure for use in boilers. It is sunk 59 feet through earth, the remainder being in the Chicopee shale.

Nos. 11 and 12. Springfield Provision Company.—This company has two wells, concerning which there is some uncertainty as to depth and diameter. They are only a few hundred feet south of the well just mentioned, and it is a matter of no small interest to note the great difference in the character of the water. Well No. 11 has a depth of 160(?) feet and a diameter of 6(?) inches, and pumps 150 gallons a minute without being pumped dry. No. 12 has a depth of 300(?) feet and a diameter of 6(?) inches, pumps 300 gallons a minute, and has never failed. Both wells enter the Chicopee shale at a depth of 60 feet. The water of these wells has been analyzed. It is too hard for boilers, and can not even be used for drinking, because of the mineral matter it contains.

SPRINGFIELD.

No. 13. Springfield Cold Storage Company.—This well is 325 feet deep and 8 inches in diameter, 290 feet of its depth lying in the Chicopee shale. The company has also a shallower well, 35 feet deep, which would bring it down to the surface of the rock. The two wells together yield 30 gallons a minute to the pump. The water is not fit for boilers, and is used only for condensing purposes.

Nos. 14, 15, and 16. Springfield Brewing Company.—No. 14 is 100 feet deep and 4 inches in diameter. The pump renders 150 gallons a minute.   The analysis shows that the water carries considerable mineral matter.

No. 15 is identical in every respect with No. 14.

No. 16 is 300 feet deep, 4 inches in diameter, and yields only 60 gallons a minute.   The water is poor in quality, and is used only for washing purposes.

All of these wells are pumped.   The water is not used for boilers as the city water, which is very satisfactory, is already in use.   Wells Nos. 14 and 15 go only to the rock, and are therefore to be reckoned among the wells of the drift, under which head they will be mentioned again.

No. 17. Morgan Envelope Company.—This well was originally 212 feet deep and flowed at the surface at the rate of 7 gallons a minute. It was afterwards sunk to 325 feet, but there was no gain in the supply.   A deep-well pump was then put in at a depth of nearly 200 feet, and a supply of 25 gallons a minute is now obtained.   The well runs for "60 feet through dirt, clay, hardpan with gravel," and then for "265 feet in sandstone rock."   This rock is apparently the Chicopee shale.   The water has been analyzed and proves to be a very good drinking water, though hard.   In a working day of 10 hours from 5,000 to 6,000 gallons are commonly used.   The diameter of the well is 6 inches.

No. 18. Kibbe Brothers Company, candy manufacturers.—This well is 252½ feet deep, 6 inches in diameter, and pumps 25 gallons a minute— the capacity of the pump.   An analysis of the water made at Amherst College showed that it contains only a small amount of solid matter. It is not used for boilers, but is employed in the making of candy and for drinking purposes.   The well runs first through about 40 feet of clay, the balance being in "sandstone"—probably Chicopee shale.

No. 19. Fiske Manufacturing Company.—The well of this company is 365 feet deep and 4 inches in diameter.   It is pumped, but the amount of water obtained is unknown.   The water has been analyzed and is not used in boilers.   About two-thirds of the well lies in rock, probably the Chicopee shale.

Though the map indicates that many of the foregoing wells are sunk in the Chicopee shale, they nevertheless lie very close to its contact with the eastern area of Longmeadow sandstone.[a]   This will perhaps account for the "sandstones" noted in the original records furnished.

Nos. 20 and 21. Highland Brewing Company.—These wells are used by the Highland branch of the Springfield Breweries Company.   No. 20 is 424 feet deep, of which about 365 feet are in rock.   Its diameter is 10 inches.   It pumps 80 gallons a minute, the water rising to within 30 feet of the surface.

---

a See map accompanying Professor Emerson's monograph.

No. 21 is 200 feet deep, of which about 160 feet are in rock. Its diameter is 8 inches. It pumps 60 gallons a minute, the water rising to within 15 feet of the surface.

Both of these wells yield water of excellent quality, which can be used in boilers. The brewery stands upon the terraces lying east of Connecticut River and in consequence is about 100 feet above the level of the Springfield wells previously mentioned, which are pretty close to the east bank of the river. The rock underlying this portion of the town is mapped by Professor Emerson as Longmeadow sandstone.

No. 22. Joseph II. Wesson, 13 Federal street.—This well is 276 feet deep and 6 inches in diameter. No estimate could be obtained of the quantity yielded, except that the pump would "furnish a 1-inch stream constantly." "The water is used only for drinking, but is pronounced good for boilers by Professor Wood, of Harvard." One hundred and fifty-one feet of the well lie in rock which is presumably Longmeadow sandstone.

Professor Emerson's monograph[a] contains a number of valuable items concerning the water supply of the Triassic area. Notes relating to several of the wells mentioned by him are subjoined.

OTHER WELLS IN TRIASSIC AREA IN MASSACHUSETTS.

No. 23. Daniel Brothers' paper mill.—This mill is on Westfield Little River, south of Westfield. The well was carried to a depth of 1,100 feet, but proved unsuccessful and is now closed up.[b] The rock at this locality is Sugarloaf arkose. It is an interesting and suggestive fact, as has been remarked before, that the only wells in the lowland which have been carried to exceptional depths, merely to meet with failure, have been sunk in this rock.

No. 24. Mount Holyoke College, South Hadley.—This well was carried to a depth of 450 feet in Longmeadow sandstone, the full record of the borings being given by Professor Emerson.[c] The water, when analyzed, was found to contain common salt in large amount.

No. 25. Montague Paper Company at Turners Falls.—This well was carried to a depth of 875 feet in Longmeadow sandstone. The record of the borings is given by Professor Emerson.[d] The appended analysis of a sample from the well shows that the water has nearly the composition of bittern.[e]

a Mon. U. S. Geol. Survey, vol. 29.
b Ibid., p. 389.
c Ibid., p. 382.
d Ibid., pp. 380–381.
e Ibid., p. 750.

Analysis of deep well water at Turners Falls, Mass.

[C. A. Goessman, analyst, 1874.]

	Parts per million.
Potassa	6
Soda	51
Magnesia	63
Lime	633
Chlorine	6
Sulphuric acid	996
Silicon	Trace.
Total	1,755

Other wells.—Other wells mentioned are two of the Parsons Paper Company, of Holyoke, which reached depths of 510 feet and 685 feet, respectively.[a]

WELLS IN CONNECTICUT.

THOMPSONVILLE.

No. 26. Connecticut Valley Brewing Company.—This well is 52 feet deep and 6 inches in diameter, having 40 feet of its depth in the rock. The water rises within 7 feet of the top, a good many feet above the level of the Connecticut River, which is only a stone's throw away. The pump raises 31 gallons a minute, which lowers the well 8 feet, but no farther. The water is moderately hard, has been used in boilers, and keeps a uniform temperature of 51° F. the year round.

No. 27. Isaac Allen, Enfield street.—This is a private well, 67 feet deep and 6 inches in diamater. It is sunk 50 feet in rock, the last 5 feet, it is said, being in granite—a manifest impossibility. Mr. Allen states that a spring of soft water was struck at a depth of about 45 feet and a spring of hard water at the bottom, the mixture making an excellent drinking water. The well furnishes more than double the quantity of water that can be used in house and barn. The water is 45 feet deep in wet weather and 30 feet in dry. The temperature is 52° F.

SUFFIELD.

Nos. 28 and 29. Town water supply of Suffield.—These wells are owned by Paulus Fuller. No. 28 is 230 feet deep and has a diameter of 6 inches, and No. 29 is 240 feet deep and has a diameter of 8 inches. These two wells together pump at the rate of 300 gallons a minute into a standpipe containing 293,000 gallons. The water rises within about 60 feet of the surface and is rather highly mineralized. It can be used in boilers, however, but gives some scale. The wells enter rock about 10 feet below the surface of the ground.

No. 30. Public well, Suffield Village.—The depth of this well is 140 feet and its diameter is 6 inches. The water stands about 90 feet below

---

[a] See Mon. U. S. Geol. Survey, vol. 29, pp. 383–385, where a full record of the borings is given.

the surface and is raised by a common force-pump. It is used for drinking purposes only.

Additional Suffield wells.—The wells tabulated below are 6 inches in diameter and are in greater part in rock. The data were furnished by Mr. C. L. Grant, well driller.

Serial No.	Owner.	Depth of well.	Depth of water.	Yield per minute.
		Feet.	*Feet.*	*Gallons.*
31	F. A Fuller...................................	40	34	8
32	A. C. Harmon................................	45	39	3
33	Second Baptist Church......................	60	47	4
34	M. T. Newton...............................	63	49	10

WEST SUFFIELD.

No. 35. West Suffield Hotel.—This well is 78 feet deep and 4 inches in diameter, and pumps 50 gallons of moderately hard water a minute. The water is said to have grown harder since the well was sunk. About 50 feet of the well is in rock.

No. 36. Mrs. John D. Loomis.—Depth, 100 feet; diameter, 4 inches; yield, 4 to 5 gallons a minute. Almost no water was found until a depth of nearly 95 feet was reached.

TARIFFVILLE.

No. 37. A. B. Hendryx.—Depth of well, 284 feet; depth of water, 278 feet; yield, 48 gallons a minute. Greater part of well is in rock. Data furnished by C. L. Grant, well driller, of Hartford.

BLOOMFIELD.

No. 38. Douglas & Cowles.—Depth, 100 feet; diameter, 6 inches. Well flows at the surface. Greater part of well is in rock. No data of yield given. Data furnished by Grant.

WINDSOR.

No. 39. Windsor Water Company.—Depth of well, 386 feet; depth of water, 326 feet; diameter, 6 inches. The pump yields 30 gallons a minute, which lowers the water 5 feet. The series through which the well was sunk is as follows: Sand, 17 feet; clay, 56 feet; hard red gravel, 50 feet; the remainder in sandstone with the exception of two layers of slate. Four analyses of the water have been made, according to which it ranges from moderately hard to excessively hard. It would seem that the water is extremely free from organic impurities, but shows sulphate of lime to the extent of 590 parts to the million.

It is extremely hard to the soap test.   When first drawn the water is said to give off a strong odor of sulphur.

No. 40. Christianson Brothers, Wilson Station.—Depth, 113 feet: diameter, 6 inches; yield, 50 gallons a minute.   The well was drilled in the bottom of an old open well and lies in the rock.   The water is good only for drinking and garden use.   It is said that when the well was first drilled the water had a very strong odor, which disappeared after the well had been used a while.   Data furnished by King and Mather.

No. 41. Misses Crompton, Windsor Heights.—One hundred and forty feet deep, but of small diameter.   The water, which is raised by a pump, is of medium hardness.   About 500 gallons per week is the ordinary consumption.   The bottom of the well is in rock.

No. 42. Dr. H. J. Fisk, Windsor Heights.—About 135 feet deep and of small diameter.   The water is pumped and hardly supplies the needs of the house and barn.   It is of medium hardness, much softer than that of the neighboring surface well.   The well was originally 85 feet deep, but was always more or less turbid.   It was then deepened and the trouble disappeared.   It enters the rock.   For analysis see No. 7, table on pages 108–109.

No· 43. C. D. Reed.—Depth of well, 101 feet; depth of water, 89 feet; diameter, 6 inches; yield, 25 gallons a minute.   Greater portion lies in rock.   Data furnished by Grant.

HARTFORD.

No. 44. Hubert Fischer Brewing Company.—This well is 500 feet deep and has a diameter of 8 inches.   The water rises to the surface of the engine-room floor, and is pumped at the rate of 75 gallons a minute. There is a difference of opinion as to whether the water is suitable for boilers.   The well is drilled in the Upper sandstones, but unquestionably pierces the Posterior trap sheet, which outcrops at no great distance to the west.

Nos. 45, 46, 47, 48, and 49. Hartford Light and Power Company.— The wells of this company all lie in the rock.   The data were furnished by Mr. Grant, who drilled them.

Data concerning wells of Hartford Light and Power Company.

Serial No.	Depth	Yield per minute.
	Feet.	*Gallons.*
45	200	120
46	228	150
47	201	120
48	200	150

The above wells are all 6 inches in diameter. The deeper well described below has greatly diminished the supply of the earlier wells enumerated above. The data concerning this fifth well (No. 49) are as follows:

Depth, 620 feet; diameter, 12 inches. The meter has shown a capacity of 125 gallons a minute. The well is pumped and gives a more copious supply of water in rainy weather than in dry. The water is very hard from sulphate of lime and can not be used in boilers, but is employed for condensing. The well lies in rock. These wells are said to have seriously depleted the supply of water in the well of the neighboring Plimpton Company.

No. 50. Capewell Horsenail Company.—This well was drilled by Mr. C. L. Grant, who furnished the following data: Depth, 250 feet; diameter, 8 inches. The well is in rock and flows. Elisha Gregory, a well driller of New York City, states in his "Torpedo Circular" that the well was torpedoed by him at a latter date and that as a result the yield of the well was increased from 15 gallons a minute to 35 gallons. It is reported that the quality of the water was injured by the process. Inquiry at the office of the company shows that at last accounts the water was not used for anything, so heavily is it charged with mineral matter. For analysis see No. 3, table on pages 108–109.

No. 51. Patten's dyeing and carpet-cleaning establishment.—Depth, 110 feet; diameter, 6 inches; yield, 150 gallons a minute (data furnished by Grant). The amount which is ordinarily pumped is about 70 gallons a minute. The well lies in the rock. The water is unfit for boilers.

No. 52. Ropkins & Company, brewery.—Depth, 200 feet; diameter, 6 inches; yield, 60 gallons a minute (data by Grant). The ordinary yield of the well is 25 gallons a minute and it flows if left standing. The water is too hard for boilers.

No. 53. New England Brewing Company.—Depth, 462 feet; diameter, 10 inches. The ordinary yield to the pump is 350 gallons a minute, but 400 gallons a minute have been pumped without making any apparent impression on the well. The water is too hard for boilers.

No. 54. Columbia Brewing Company—This was formerly the Herold Capitol Brewing Company. Depth, 300 feet; diameter, 6 inches; yield, 250 gallons a minute (data by Grant). The amount of water pumped for common use is about 80 to 90 gallons a minute. The water is too hard for boilers.

No. 55. Armour & Company.—Depth below grade, 436 feet; actual depth below floor of basement, 420 feet; diameter, 6 inches. The water flows about 1 inch over the top of the pipe when allowed to stand, which brings the water level above the surface of a large part of the neighboring land. The ordinary yield is about 150 gallons a minute throughout the twenty-four hours, but a much larger yield could be obtained if desired.   At the time these data were obtained no report had been received on the sample of water sent for analysis.

No. 56. Long Brothers hotel.—Depth, 200 feet, of which 186 feet are in the rock.   Diameter of the portion which is in the rock, 6 inches. The well is pumped and yields a maximum supply of 35 gallons a minute.   The water is too hard for boilers or for laundry work, but it is claimed that its quality is steadily improving.

No. 57. Brady Brothers, bottlers.—Depth, 277 feet, of which 244 feet lie in the rock.   Diameter, 6 (?) inches.   The well will yield 25 gallons a minute, but the amount usually pumped is 9 gallons a minute.   The water has never been tried in boilers, but is used for making all kinds of soft drinks.   At times it becomes clouded, but in a few days it clears again.

The above data were obtained from Brady Brothers.

No. 58. Brady Brothers.—Mr. Grant reports the following data concerning a well drilled for Brady Brothers, which is probably another than the one just described: Depth, 159 feet; diameter, 6 inches; yield, 29 gallons a minute.

No 59. "Allyn House."—Depth, 318 feet, of which 288 feet are in rock.   Diameter, 4½ inches; yield, 60 gallons a minute.   The ordinary consumption for eighteen hours is 30,000 gallons.   The water, which has been analyzed, is too hard for boilers, but has improved in quality with time

No. 60. W. C. Wade, corner State and Front streets.—Depth, 125 feet, of which 113 feet are in rock.   The maximum yield is 200 gallons a minute, though only 50 gallons a minute are in common use, chiefly for refrigerating purposes.   The water, which rises to within 13 feet of the surface, is a little too hard for boilers.   Its summer temperature is 54° F.

No. 61. W. C. Wade, "Public Market."—Mr. Grant sends data of a well drilled for Mr. Wade at the "Public Market," about one-third of a mile from the well above mentioned, as follows: Depth, 225 feet; depth of water, 183 feet; yield, 28 gallons a minute.

No. 62. Keney Park.—Depth, 200 feet; diameter, 6 inches.   The well flows, but the yield was not given.   Data by Grant.

No. 63. Keney Park.—Depth of well, 170 feet; depth of water, 138 feet; yield, 5 gallons a minute.   Data by Grant.

Other wells in Hartford.—The following list of wells in Hartford was obtained from Mr. Grant.   The wells are all 6 inches in diameter and have the greater part of their depth in rock.

Additional wells in Hartford.

Serial No.	Owner.	Depth of well.	Depth to water.	Yield per minut
		Feet.	*Feet.*	*Gallons.*
64	F. C. Rockwell	115	111	60
65	E. C. McCune	110	87	18
66	Frank S. Tarbox	57	48	5
67	....do	67	56	3
68	T. C. Moore	60	40	3
69	G. E. Hubbard	37	37	8
70	....do	62	40	12
71	St. Mary's Home	100	97	15
72	Walter S. Mather	37	37	3
73	J. Dart & Son	24	12	3
74	....do	74	Flows.	22
75	Hotopp & Carlsson	155	130	10
76	Wm. O'Brien	50	30	11
77	Addison & Impey	63	36	5
78	Johnson & Weeks	49	29	10
79	Wm. Rogers	110	35	5
80	Peter Peterson	50	40	3
81	A. Hepburn	50	39	5
82	C. L. Bailey	48	27	12
83	H. J. Abbey	35	18	3
84	Andrew Nason	28	16	4
85	B. L. Chappell	50	34	4
86	G. E. Hurd	50	39	4
87	M. H. Ericksen	37	24	10
88	O. W. Crane	60	30	2
89	C. A. Green	50	38	3
90	J. C. Parsons	136	86	1
91	O. Bengston	50	25	10
92	F. H. Seymour	70	40	8
93	G. J. Maher	65	47	11
94	A. M. Weber	75	50	5
95	D. F. Keenan	140	111	10
96	The "Linden"	240	215	45
97	Hartford Woven Wire Mattress Co	246	230	10

WEST HARTFORD.

No. 98. H. O. Griswold.—Depth, 152 feet; diameter, 6 inches. The water has been analyzed, and contains three times as much lime as is desirable in drinking water. In consequence it is very hard and is used very little. The ordinary depth of the water is 137 feet, but it has been lowered by pumping to within 40 feet of the bottom. The well runs through loam and hardpan for 10 feet, the balance being in rock.

Other wells in West Hartford.—The following supplementary list is furnished by Mr. Grant. All the wells are 6 inches in diameter, except No. 101, which is 4 inches.

Additional wells in West Hartford.

Serial No.	Owner.	Depth of well.	Depth to water.	Yield per minute.
		Feet.	*Feet.*	*Gallons.*
99	Paul Thomson	40	18	12
100	James Thomson	110	95	35
101	E. C. Wheaton	135	115	34
102	Mrs. K. Gallagher	53	44	2½
103	G. V. Brickley	30	15	12
104	W. E. Howe	62	18	2
105	H. C. Long	41	25	12
106	L. N. Burt	52	39	4
107	P. H. Reilly	206	160	42
108	D. F. Crozers	101	87	32

Well No. 107, which is in the trap, lies on the ridge south of the village of West Hartford. The trap belongs to the Posterior sheet, and is comparatively thin. The thickness of the trap at this point has not been determined, but at least two-thirds, and possibly three-fourths, of the depth of the well must be in the Posterior shales which underlie the sheet. On comparing this depth with the depth of water in the well, it will be readily seen that it is very probable that the water does not come from the trap at all, but from the underlying shale. In consequence the well can not with propriety be classed as a well of the trap.

BURNSIDE.

No. 109. East Hartford Manufacturing Company.—Depth of well, 398 feet; diameter, 6 inches. No estimate of the yield in gallons could be obtained, but the well is said to flow constantly 1 inch deep over the top of a 10-inch pipe. If this statement is correct, this is the largest flowing well in the lowland. The water has been analyzed by Springfield chemists and found to carry considerable mineral matter, and is too hard for boilers. The well is entirely in rock.

No. 110. J. H. Walker, paper mill.—The water of the well is excellent for drinking, but hard. Measurement showed the well to be pumping 8 to 10 gallons a minute. As it is separated from the last-mentioned well only by the breadth of the Hockanum River, the rock conditions are probably the same as in that well.

SOUTH MANCHESTER.

M. S. Chapman.—These are four private wells, each 6 inches in diameter, used, as I understand, to supply fish ponds. Their depths and the amounts of water they yield are shown below:

Wells of M. S. Chapman.

Serial No.	Depth.	Yield per minute.
	Feet.	*Gallons.*
111	225	30 to 40
112	125	10 to 12
113	75	2½
114	75	None.

The wells enter the rock at a depth of 20 feet, and the first three flow at the surface of the ground. The water is of good quality and, it is said, can be used for any purpose, although it is distinctly hard by the soap test.

No. 115. Cheney Brothers, silk manufacturers.—This is one of the most remarkable wells in the lowland, and the data concerning it have been carefully preserved. The well is 457 feet deep, and its diameter is 8 inches for the first 383 feet, below which it is only 6 inches. The well flows at the surface at the rate of 250 gallons a minute, while the experiment of pumping 650 gallons a minute for ten consecutive minutes lowered the water level only 10 or 12 feet below the surface. The water is unfortunately very hard, but the well is used to its full capacity in washing colored silks, the water of the well remaining at the proper temperature summer and winter. Below the first 12 feet the entire well is in red sandstone. For analysis see No. 6, table on pages 108–109.

NORTH MANCHESTER.

No. 116. F. J. Sharp.—Depth of well, 100 feet; depth of water, 99 feet; diameter, 6 inches; yield, 10 gallons a minute. The water has been analyzed and proves to be very hard. Eighty-one feet of the well lie in red sandstone.

No. 117. American Writing Paper Company (Oakland Paper Company division).—Depth, 300 feet; diameter, 4 (?) inches. The water flows over the top of the pipe at a rate of probably 10 to 12 gallons a minute. The water is hard and is used for drinking purposes only. The well lies entirely in the rock, which, in the stream near by, is seen to be rather loose in texture and to be filled with fragments of the crystalline rocks of the neighboring eastern upland.

The wells east of Hartford along the Valley of Hockanum River, comprising Nos. 109 to 117, inclusive, all flow at the surface of the ground, except No. 114, which is dry, and No. 116, which rises to within 1 foot of the surface. In general the yield of these wells is exceptionally large, but the water is uniformly hard.

WETHERSFIELD.

The following data were kindly furnished by Mr. Grant concerning wells at this place.

Wells in Wethersfield.

Serial No.	Owner.	Depth of well.	Depth to water.	Yield per minute.
		Feet.	*Feet.*	*Gallons.*
118	J. H. Rabbett	30	28	8
119	Rev. Lynch	70	40	35
120	C. I. Allen	192	85	15

No. 120 is in trap rock. All are 6 inches in diameter.

ROCKY HILL.

No. 122. W. E. Pratt.—This well is located in the thin posterior sheet of trap. It was drilled in the bottom of an old open well, 20 feet deep, which entered the rock for a distance of 6 feet. From this point the well was drilled 30 feet through trap when it broke into the underlying sedimentaries, which it pierced to the depth of 1½ feet. This well therefore gives a section of 14 feet of soil, 36 feet of trap, and 1½ feet of sedimentary rock—a total depth of 51½ feet. This brings the bottom of the well about 50 feet above the surface of Connecticut River, which flows by it only a few hundred feet eastward. The diameter of the well is 6 inches and the maximum amount of water obtainable is a little less than 1 gallon a minute. The well pumps dry in thirty minutes. The water is fair for drinking, but is excessively hard.

No. 123. J. K. Green.—Depth of well, 26 feet; depth of water, 25 feet; diameter, 6 inches; yield, not given. The well is in trap rock. Data by Grant.

KENSINGTON, BERLIN JUNCTION.

No. 124. New York, New Haven and Hartford Railroad power house.—Depth, 300 feet; diameter, 6 inches; yield, 120 gallons a minute. Data by Grant. Water is raised by a common pump. It is permantly hard and unfit for boilers.

Wells at Berlin and Berlin Junction.—The following list of wells at Berlin and Berlin Junction was furnished by Mr. Grant. All are 6 inches in diameter.

Additional wells at Berlin and Berlin Junction.

Serial No.	Owner.	Depth of well.	Depth to water.	Yield per minute.
		Feet.	*Feet.*	*Gallons.*
125	Berlin Brick Co	60	49	....
126	.....do	70	62	....
127	Yale Brick Co	100	81	10
128	F. L. Wilcox	71	40	....
129	J. B. Smith	200	130	6

NEW BRITAIN.

No. 130. Hotel Russwin.—The depth of the well is 152 feet, and the depth of the water at the lowest 130 feet, but if the well is allowed to stand the water flows at the level of the engine-room floor, which is 10 feet below grade. The ordinary consumption is fully 10,000 gallons a day. The water is very pure and can be used for all purposes. Certain features of this well which have a distinct bearing on the theory of underground flow will be considered in connection with the discussion of that subject.

Other wells.—The following list is from Mr. Grant:

Additional wells at New Britain.

Serial. No.	Owner.	Depth of well.	Depth to water.	Yield per minute.
		Feet.	*Feet.*	*Gallons.*
131	Dennis & Co	195	....	40
132	A. J. Sloper	70	....	45
133	A. B. Johnson	60	....	12
134	W. E. Bradley	40	22	4
135	J. P. Curtis	250	237	12
136	William Derby	50	.31	6
137	F. B. Wischek	50	38	6

Nos. 132 and 134 are sunk in the trap of the Posterior sheet and may lie entirely in volcanic rock. All the wells are 6 inches in diameter.

CROMWELL.

Nos. 138 and 139. New England Brownstone quarry.—The wells are 63 feet and 132 feet deep, respectively. The water, which was raised with a pump, was thoroughly tried in the boilers and was found totally unfit for the purpose.

No. 140. Goodyear Rubber Company.—The well, which is 384 feet deep and 6 inches in diameter, pumps about 40,000 gallons in twenty-four hours, or nearly 30 gallons a minute. The water is called good, but has not been tried in the boilers on the ground that it would probably prove too hard. It is reported that the well flows at the surface. For analysis of water from a Middletown well, see No. 9, table on pages 108-109.

Nos. 141-145. Edward Miller & Co.—There are on the premises of this company five artesian wells, all 8 inches in diameter, ranging from 250 to 300 feet deep, and bored into red sandstone rock, which at this place lies from 6 to 10 feet below the surface of the ground. The output of these wells was measured when they were first bored and varied. in the different wells from 10 to 80 gallons per minute. The supply thus measured was obtained by means of an ordinary suction pump, which, operated at the rate named, lowered the water about 25 feet below the surface—as low as it could be pumped with that form of apparatus. A few years ago the Pohle air system was installed for raising the water from the wells. This system works by compressed air and forces the water from depths of 70 to 90 feet. With this apparatus a very greatly increased output was obtained, which now supplies all the needs of the company. The water from these wells is all discharged into one large system, from which it is circulated through the factory. It is not easy to determine exactly the amount of water used, but by estimate it is between 75,000 and 100,000 gallons a day of ten hours. There is no doubt a much larger quantity than this could be obtained if needed. The water is satisfactory for all manufacturing purposes except for making steam, for it holds considerable mineral matter in solution and is rather hard for use in boilers.

No. 146. Foster, Merriam & Co.—The well is 300 feet deep, 200 feet of which are in rock. The well flows at the surface if allowed to stand, and pumps, at the maximum, 10,000 gallons an hour—about 170 gallons a minute. The water is too hard for boilers and contains a certain amount of sand. The outer pipe is 8 inches in diameter and the inner 6 inches.

No. 147. International Silver Company, Factory " E" (Meriden Britannia Company).—Depth of well, 560 feet; depth of water when the well is not pumped, 555 feet; diameter, 6 inches. No data of yield could be obtained. The water is very hard and wholly unfit for boilers. The section traversed in drilling is as follows: Soil, 9 feet; gravel, 18 inches; quicksand, 90 feet; balance mostly rock.

No. 148. Meriden Curtain Fixture Company.—Depth, 305 feet; diameter, 6 inches. The amount of water in common use for a working day of 10 hours is 15,000 gallons, or about 25 gallons a minute. The water is suitable for use in boilers. About 255 feet of the well are in rock.

WELLS AT OTHER PLACES IN NEW HAVEN COUNTY, CONN.

The following data, which have been put in tabular form, were received from Mr. Grant:

Additional wells of Connecticut Valley.

Serial No.	Town.	Owner.	Depth of well.	Depth to water.	Yield per minute.
			Feet.	*Feet.*	*Gallons.*
149	Yalesville	G. I. Mix & Co	100	....	40
150		J. H. Yale	67	22	3
151		C. W. Michaels	90	72	....
152	West Cheshire	Cheshire Manufacturing Co.	150	....	100
153	Mount Carmel	Chas. Wheeler	50	17	6
154		Sylvester Peck	36	21	5
155		H. D. Clark	60	Flows.	22
156	Quinnipiac	C. T. Stevens	40	Flows.	4
157	North Haven	F. L. Stiles	83	....	35
158		I. L. Stiles & Son	102	....	30
159		H. P. Smith	116	100	18
160		N. W. Hine	125	115	16
161	Whitneyville	H. Stadtmiller	85	58	5
162		J. H. Burton	50	40	22
163		W. F. Downer	67	40	5
164		W. F. Downer	50	42	7
165		Edward Davis	56	47	11
166		Edward Davis	13	11	20
167		G. W. Ives	65	52	8
168		Mr. Brock	50	35	24
169		Mr. Johnson	58	33	25
170		W. Mansfield	48	37	....

These wells are all 6 inches in diameter, and probably all of them enter rock, except, possibly, No. 166.

NEW HAVEN.

No. 171. Hoyt Beef and Produce Company.—Depth, 600 feet, of which 200 feet only are in rock; diameter, 6 inches. The well flows at the rate of 3 gallons a minute, yielding water of very good quality which has not been analyzed.

No. 172. Lavigne Automatic Machine Company.—Depth, 95 feet: diameter, 6 inches; yield, 12 to 15 gallons a minute. The data concerning the quality of the water seem to be favorable in a general way. The well passes through 53 feet of quicksand and 42 feet of red sandstone.

No. 173. New England Dairy Company.—Depth about 160 feet: diameter of outside pipe 6 inches, and of inside pipe 4 inches; yield, 20 gallons a minute. About 95 feet of the well are in rock. The water can be used in boilers. For analysis see No. 12, table on pages 108–109.

OTHER WELLS NEAR NEW HAVEN.

The following data from New Haven and other towns of the lower valley are from Mr. Grant:

Additional wells at New Haven, Northford, and Montowese.

Serial No.	Town.	Owner	Depth of well.	Depth to water.	Yield per minute.
			Feet.	*Feet.*	*Gallons.*
174	New Haven........	G. W. Ives & Son..........	50	Flows.	....
175		.....do	200	....	10
176		Seamless Rubber Co.......	160	....	10
177		H. H. Olds & Co...........	75	40	5
178		N. W. Capin...............	245	134	40
179	Northford	L. A. Smith	50	Flows.	....
180	Montowese	J. F. Barnes & Co..........	36	32	7

All are 6 inches in diameter.

FAIR HAVEN.

No. 181. J. R. King.—Depth, 203 feet; diameter, 6 inches. When pumped at the rate of 40 gallons a minute the water level did not fall at all. The well is entirely in rock, and no water was found until after a depth of 100 feet was reached. The water gave no scale when used in boilers. Data are from R. A. Mather, of King & Mather, well drillers.

WELLS STOPPING AT SURFACE OF ROCK.

Another group of wells of an entirely different nature from those of the previous list comprises the deep wells of the drift, which were sunk to the top of the rock and stopped at that point. Data concerning a few of these are given on the following pages.

WELLS IN MASSACHUSETTS.

NORTHAMPTON.

No. 182. Belding Brothers' Silk Company.—This well is a close neighbor of the famous dry well, No. 1, of this paper. It passes at a considerable depth through a dense layer of some sort and then continues through gravel to the rock, which is here 150 feet below the surface. The yield of 20 gallons a minute does not materially affect the water level, which is about 20 feet below the surface of the ground. The water, which has not been analyzed, can be used in boilers, but is principally employed in dyeing. The diameter of the well is 12 inches.

SPRINGFIELD.

Nos. 14 and 15. Springfield Breweries Company.—These wells and the rock well belonging to the same company have been mentioned on an earlier page of this paper, under the numbers here given, and do not need further description. It is interesting, however, to note that the quality and quantity of the water changes as soon as the rock is entered.

No. 183. Phelps Publishing Company.—Depth, 78 feet; diameter, 2¼ inches; yield, 10 gallons a minute. The water is too hard for boilers but is excellent as drinking water. This well, which stops at the rock, is pumped.

No. 184. Cooley's Hotel.—This is a driven well, 86 feet deep and 2¼ inches in diameter, that yields over 20 gallons of water a minute. The water is soft and could probably be used for boilers. The well runs to the surface of the rock, at which point the water is obtained.

WELLS IN CONNECTICUT.

MERIDEN.

Nos. 185, 186, 187, 188, and 189. Bradley & Hubbard Manufacturing Company.—In 1890, five wells, each 8 inches in diameter, were driven to the rock by this company, reaching the following depths: No. 1, 203 feet; No. 2, 208 feet; No. 3, 256 feet; No. 4, 327 feet; No. 5, 281 feet. Only Nos. 1, 2, and 3 are used, as Nos. 4 and 5 did not produce water enough to pay for connecting. No estimate of the amount of water obtained from Nos. 1, 2, and 3, separately, can be given; but the three wells together furnish about 6,000 gallons an hour. None of the wells flow. The water is of fair quality, but is rather hard for use in boilers.

These five wells, all over 200 feet deep, run to the rock only. Besides their interest from the standpoint of the water supply, they give some very interesting suggestions concerning the depth of the unconsolidated deposits at this point, and they hint at the presence of a very deep depression in the underlying sedimentary rocks along the general line of the great fault which separates the Hanging Hills on the west from Lamentation Mountain on the east.

It should be noted that in this paper no account is given of those wells which lie wholly in the unconsolidated deposits of the area and which do not come into relation with the rock at all. There are many of these wells, and they seem to be especially in favor in the region lying between Springfield and Holyoke and along the Chicopee River. They present innumerable local problems of a diverse and often complicated nature, and any results which might be obtained from their study would be of relatively little economic importance.

TRIASSIC ROCKS OF THE CONNECTICUT VALLEY AS A SOURCE OF WATER SUPPLY.

By M. L. FULLER. [a]

INTRODUCTION.

In the foregoing paper of Mr. W. H. C. Pynchon detailed descriptions have been given of a large number of wells of the Triassic area of Massachusetts and Connecticut, together with a considerable amount of data relating to the quantity and quality of the water supplies which are obtained. Many facts bearing upon the occurrence of the waters have been presented and certain deductions that are of local application have been made.

The purpose of this paper is to supplement the foregoing by a general summary of the water conditions in their broader relations and to point out the significant features of the water supply of the region. The effort has been made to describe these features as concisely as possible rather than to present elaborate discussions.

In addition to the conclusions presented, a few notes on points of special interest to the driller have been appended.

UNDERGROUND WATER CONDITIONS IN THE AREA.

SOURCES OF WATER.

CONGLOMERATES.

Near the edge of the area, where the Triassic beds are in contact with the adjacent crystalline rocks, both on the eastern and western sides, there are somewhat marked bands of conglomerates which were deposited along the shores of the Triassic sea that occupied the basin.

---

[a] The descriptive portions relating to wells are based on the field observations of Mr. W. H. C. Pynchon and are compiled from notes furnished by him. The present writer alone, however, is responsible for the interpretations of the various features and for the views expressed in regard to the nature, condition, and prospects of the water supplies in the Triassic area.

The pebbles of the conglomerate were derived from the adjacent crystalline rocks, and some of them are of large size. The matrix is of various degrees of fineness, ranging from coarse sand to clay. In Massachusetts the conglomerate on the west side of the Triassic area is known as the Sugarloaf arkose and on the east side in part as the Sugarloaf arkose and in part as the Mount Toby conglomerate, the latter occurring in a narrow belt between the Sugarloaf arkose and the crystalline border in the southern part of the State. The same bands extend into Connecticut and possibly continue throughout the entire southern part of the area, but they have not yet been differentiated in that region.

The conglomerates of the western border would at first sight seem to offer greater possibilities for the storage of artesian waters than any other rocks in the valley. They present their upturned edges to the waters that pour down from the western upland and are comparatively undisturbed by fracture and displacement, but because of their situation, outside of the immediate Connecticut Valley, where the principal towns are located, very few deep wells have been sunk in them, and it is therefore impossible to say whether actual results would agree with the assumed conditions. The few deep wells near the western border of which records are at hand do not reach the average depth of those of the central portion of the valley, and the fact that these few wells obtain only very moderate supplies is therefore inconclusive. Farther away from the border, as at Northampton, a well was sunk to a depth of 3,700 feet through rocks classed by Professor Emerson as Sugarloaf arkose without encountering water. Again, near Westfield, a well was sunk at Daniel Brothers' paper mill to a depth of 1,100 feet in the Sugarloaf arkose, likewise without success. In fact, so far as is known, nearly all wells that have reached great depths without procuring water have been sunk in the conglomerate or arkose.

On the eastern side of the valley the conditions are less favorable than on the west. The eastward sloping beds of the narrow belt of Mount Toby conglomerate that lies at its contact with the crystalline rocks along the eastern edge of the area abut against a system of faults that presumably constitutes an extended line of leakage, drawing water from the adjacent conglomerates.

SANDSTONES.

Distribution and character.—The sandstones, because of their great area and their water-holding capacity, are the principal source of water in the Triassic area. Except for the conglomerate belts along the eastern and western boundaries, the shale band near Connecticut

River at the center of the valley, and the trap ridges, the main portion of the area of Triassic rocks is underlain by sandstone. Many important towns are located in the sandstone area and a large number of wells obtain good supplies of water from this rock.

In character the sandstones vary from fine to coarse grained, and are usually some shade of red or brown in color, though buff and other varieties are known. They are often of very even texture and frequently occur in thick beds, so that from many standpoints they constitute an admirable building stone, for which purposes they are extensively quarried, especially at Longmeadow in Massachusetts and Portland, Cromwell, and Tolland in Connecticut. They are interspersed with beds of shale ranging in thickness from mere films up to beds many feet thick, which frequently serve as confining strata for the water-bearing beds.

A test of the brownstone from Portland, Conn.,[a] shows that this sandstone when previously dried and then immersed in water for three months will increase in weight from 150 to 154 pounds, showing an absorptive ratio of 4 to 150 or about one-fortieth of the weight of the block. This is equivalent to approximately 2 quarts of water to every cubic foot, and shows that, under favorable conditions, the rock has immense capacity as a reservoir.

Water supply.—The sandstones may be studied to good advantage in the quarries which lie within their borders at Longmeadow, Cromwell, or perhaps best of all in the great brownstone quarries at Portland, directly across the Connecticut River from Middletown, Conn. Here great pits have been opened in the sandstone to a depth of 200 feet, and the water-bearing character of the rocks may be studied. Although the sandstone when quarried contains a certain amount of "quarry water" when first removed from the pit, no water appears to percolate from the face of the stone, but water everywhere emerges from along the bedding planes, especially along the shaly partings, where, dripping down, it darkens the quarry walls over large surfaces.[b] The water flows chiefly down the slope of the dip and enters the rock in greatest amount during the spring months, at which period slight amounts may also enter from the opposite side. The beds slope from the Connecticut River toward the quarry, but the fluctuation in the amount of water entering the quarry with the season indicates its derivation directly from rainfall or melting snows. Throughout the busy season pumps keep the quarries sufficiently clear of water to permit work to be done.

A well showing features somewhat analogous to those seen at the

---

a Stone, vol. 9, June, 1894, p. 20.   b Stone, vol. 9, pp. 42–43, 1896.

quarries is located at the bottling establishment of M. M. Bacon, on Morris street, Hartford. It is an open well, 34 feet deep and about 2½ feet in diameter, about 14 feet of its depth being in shales. In one corner of the bottom of this well is a fissure large enough to allow the insertion of the fingers, from which the water flows in such abundance that it is impossible to keep the well dry enough for convenient cleaning. The draft on this well is a thousand gallons a day the year round, and in summer it will amount to half as much again. Though this well is in a city of 90,000 inhabitants, and is situated at the foot of a considerable drainage slope, yet the analysis which the city made of the water shows that it is very pure.

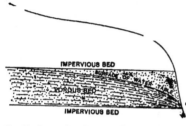

FIG. 20.—Depression of water surface at point of leakage in sandstone.

From the fact that the entrance of water into the quarries and in the well mentioned is from the bedding planes, it might appear that but little was contained in the body of the sandstones themselves. This appearance is, however, probably due to the artificial conditions of the exposures in the quarries. The sandstone, because of its open and porous texture, probably behaves much like a homogeneous mass of sand, from which the flow or leakage takes place along its contact with an underlying impervious bed or at the level of saturation. In such a mass the water table, as the upper surface of the water body which it contains is called, is depressed at the point of leakage, but rises gradually as the distance from the point of outflow increases (see fig. 20). Thus, in a sandstone bed it is but natural that the outflow should be only at its base, although perhaps a few feet back from the exposed face the whole bed may be saturated. This is true of practically all natural outcrops, even of the famous water-bearing Dakota and similar sandstones of the central West.

SHALES.

Character.—The largest body of shales in the Triassic area occurs near the central part of the valley, in the vicinity of Connecticut River. They are known in Massachusetts as the Chicopee shale, are grayish in color, and are interbedded with many sandstone layers. At the bottom they grade into the Longmeadow sandstone. In addition to the Chicopee shales, occasional thin shaly beds occur in the sandstones or conglomerates throughout the entire area.

Theoretically shales are not at all adapted to hold ground water, but in the shaly rocks under discussion there is considerable lime, and possibly, in the deeper portions, salt and gypsum, which by solution may readily give rise to channels for underground waters. The interbedded sandy layers and the bedding planes themselves would also furnish numerous channels for the passage of such waters.

Water supply.—That the shales bear considerable quantities of water is apparently shown by the success of the wells at Holyoke, Springfield, and other points in the belt of the Chicopee shales, although it is not impossible that the deeper wells enter the underlying Longmeadow sandstone and obtain a part of their supply from that formation. Among such wells may be mentioned Nos. 3, 4, 5, 7, 9, 10, 11, and 12, mentioned by Mr. Pynchon in the preceding paper. Wells 16 to 22 generally start in the Chicopee shales, although several of them may penetrate sandy beds, probably of the underlying Longmeadow sandstones.

Depth of wells.—Of the wells noted, two are between 100 and 200 feet, five between 200 and 300 feet, four between 300 and 400 feet, two between 400 and 500 feet, and three over 500 feet. This indicates that in general the wells must be fairly deep in order to obtain satisfactory supplies in the shale area.

Character of water.—Because of the presence of lime and other soluble mineral substances the waters from the shales are generally highly mineralized and can seldom be used in boilers.

TRAPS.

Catchment conditions.—Of the rocks of the Triassic area the traps present the most unfavorable conditions for water storage. This is because of the lack of porosity of the rock itself and of the absence of bedding planes along which the water might flow. Superficially they are extensively jointed and would absorb and hold much water if the open joints reached any considerable depth. It is probable, however, that below a relatively few feet from the surface the joints are in many places not sufficiently open to admit the passage of water. Moreover, the form of their outcrops is most unfavorable for the reception of the surface waters, as their exposed edges stand high above the surrounding plain, and usually rest upon an elevated pedestal of sediments. The western faces of the trap ridges, on which side they are most eroded, are precipitous cliffs, while their unbroken eastern slopes, plunging downward toward the plain at an angle of about 15°, send the waters collected from the rains in rushing brooks to the lowland. Indeed, throughout the valley the foot of the eastern slope of the Main sheet of trap is the favorite location for reservoirs. Hartford receives its water supply from a chain of such lakes that stretches for several miles along the eastern foot of the Talcott Range, while there are at least five reservoirs that use as their watershed the back slopes of the Hanging Hills. Farther down the valley Beseck Lake, Paug Pond, and Lake Saltonstall derive their supply from similar sources.

Wells in the trap.—Among the more interesting of the wells deriving their supply from the trap are those at Cedar Mountain, Wethersfield, just southwest of Hartford (No. 121, B. and C., of Mr. Pynchon). These wells are on the property of Dr. Gordon W. Russell, who sunk the wells largely as an experiment, and who has shown much interest in scientific matters. The mountain is a part of the ridge of the Main trap sheet, and has a maximum elevation of about 360 feet above the sea. Its western face is very steep, dropping 250 feet to the plain within a distance of two-fifths of a mile, and is actually precipitous near the summit. The eastern face slopes more gradually, dropping about 120 feet to the valley, occupied by shales, about three-fifths of a mile distant. It is in all respects a typical trap ridge.

Well A is an ordinary open well and was dug to supply the needs of the farm house. It was opened 9 or 10 feet to the rock, but the water became shallow in summer. It was then sunk 1 or 2 feet into the loose, greatly jointed surface trap, and has since given an abundance of water for domestic uses. The supply, however, fluctuates regularly with the wetness or dryness of the season. From the well mouth the mountain side with its drift covering rises steadily for two-fifths of a mile to the west till it reaches the crest, which is about 100 feet above the well. The supply is clearly the surface water contained in the soil and in the heavily jointed upper surface of the trap, the source also of a little stream which lies a little farther up the ridge.

Well B is located about 200 feet south of well A. It passes through 9 feet of soil and then through about 290 feet of trap rock, at which point the string of drilling tools wedged fast, possibly along a joint plane. The well is 6 inches in diameter. On illuminating it brightly for a considerable depth, by light reflected from a mirror, it appeared that no water came into it except from the shattered upper surface of the trap sheet, as in well A. The drill had not entered the underlying sediments when the well was visited in 1902, notwithstanding the fact that the bottom of the well was much below the level of the western plain.

Well C is located about three-fourths of a mile farther south and a little farther east than the other two wells. It has a depth of 103 feet. It was thought from the residue brought up by the sand bucket that the well entered the sedimentary beds below, but in view of the record of well B and the thickness of the Main sheet, this is extremely doubtful. Water was struck at a depth of 38 feet from the surface, in a joint, the yield being 32 gallons an hour. At the present depth the well is capable of giving 90 gallons an hour, the water probably coming through joints from a level below the well bottom, and rising to within 25 feet of the surface. During the drilling light was reflected down the bore and water was seen coming in through the trap, and, in the opinion of the driller, Mr. H. B. King, of Hartford, from the down-hill side. However this may be, we have here a well near the summit of the mountain whose bottom is above the level of the lowlands on either side, with water coming in through the trap and rising to a point about 175 feet above the plain three-fourths of a mile to the west and about 70 feet above the valley one-fourth of a mile to the east. The top of the well is about 280 feet above sea level.

GEOLOGY.

STRUCTURE.

Attitude of beds.—The dip of the beds of the Triassic area averages about 15°, or nearly 2,000 feet to the mile. The catchment area of an individual bed having a dip as steep as this is very narrow as compared

FIG. 21.—Outcrop of bed dipping 15°.   FIG. 22.—Outcrop of bed dipping 5°.

with that of a more gently inclined bed. Thus, in figs. 21 and 22 beds having the same thickness and dipping 15° and 5°, respectively, are represented as outcropping on a level surface. It will be seen that the catchment area is smaller at the outcrop of the more steeply dipping bed.

FIG. 23.—Conditions and beds encountered by wells.

Because of the steep dips and the rapidity with which the beds are carried to great depths a single water horizon is available only within very narrow limits. In fact, in going from west to east practically every well, unless the wells are very close to one another or penetrate to a great depth, draws from a separate horizon (see fig. 23). This is the explanation of the great variability in the quantity and quality of the water drawn from wells in the Triassic area.

Along the western rim of the area the Triassic rocks lie about 200 feet higher than in the lowlands of the Connecticut Valley. If the beds were continuous and were near enough to the surface in the center of the valley to be reached by the drill, artesian flows would probably be uniformly obtained.   In reality, however, the beds are not continuous, but are broken by faults along which the waters can escape, and, moreover, they lie at such great depths that they can not be economically reached.   The bed from which a 500-foot well draws its supply will outcrop within about a quarter of a mile of the well.

Direction of movement of water.—Although extensive artesian fields are unknown, and are in fact not to be expected, isolated flowing wells sometimes occur.   A number have been mentioned in the description of the wells by Mr. Pynchon, pages 76 to 96 of this report.   In most instances the water moves down the dip, but in a few wells the reverse seems to be true.

A well of the Connecticut Valley Brewing Company, at Thompsonville (No. 26 of Mr. Pynchon's list), is located on a neck of land that projects westward into the Connecticut River, to which a steeply sloping bank is presented.   The water is obtained from the rock about 30 feet from the surface, the water-bearing bed rising toward the

Fig. 24.—Underground conditions in Thompsonville well.

river and outcropping in the steep bank mentioned.   The well yields 8 gallons a minute, a quantity many times greater than any amount that could be absorbed from the steep and narrow catchment area. The water can not possibly come from the river, as its level is about 25 feet higher than the level of the water in the river, but must come up the dip from the east.   As the beds never bend upward and appear at the surface, but maintain their uniform dip to the east, the water can not be derived from any outcrop lying farther east.   Joints or faults apparently present the only means by which the water can be supplied (fig. 24).

A well drilled a few years ago on Talcott Mountain, near the site of the "Old Talcott tower," of local fame, was situated on the crest of the ridge at a point a little over 900 feet above the sea.   The face of the mountain falls away steeply to the west 700 feet in two-fifths of a mile.   At a depth of 200 feet the drill passed through the trap into the underlying sediments, and at this point the only water was found. Presumably it was in small quantity, as the drilling was carried 700 feet farther, but without results.   As the crest of the mountain is 700 feet above the western plain, and as the trap was left at a point only 200 feet down, the contact upon the face of the mountain must be over 500 feet above the plain.   This would bring the outcrop of the contact only a little over one-tenth of a mile west of the well and on a precipitous portion of the mountain face.   It is thought probable by Mr. Pynchon, who reported the well, that the water found, however small the quantity, came up rather than down the slope of the beds. Well No. 52, Ropkins & Co., Hartford, is probably of the same general type, as when allowed to stand the water rises in the well far above the water levels of the Park and Connecticut rivers, which are close by.

JOINTS.

Occurrence.—Joints are especially numerous throughout the Triassic area, the traps, conglomerates, sandstones, and shales all showing many of them.   In the traps, as has been pointed out, the rock is sometimes superficially broken up into a mass of small blocks, among which the water freely circulates.

Joints are also well developed in the shales.   This may be seen readily in the city quarry at Hartford, previously mentioned, where the shales forming the floor of the quarry are traversed for long distances by joint cracks that divide the rock into the diamond-shaped blocks already described.   In this quarry a number of feet of the indurated shale have been removed, leaving an escarpment of sedimentary rocks, averaging fully 10 feet high, immediately below the trap.   The joint planes bounding this face cut through many layers, some of them probably traversing the whole vertical distance.   An examination of any fairly extensive outcrop of shale within the lowland will show that jointing is developed everywhere to the same extent.

The sandstone beds are likewise commonly jointed when exposed at the surface, but to what depth pronounced jointing extends is unknown, although the jamming of drills in certain wells may indicate that fissures extend to considerable depth in the sandstone.   Thus in the well of the Hubert Fischer Brewing Company, of Hartford (No. 44), the drill, after penetrating "blue stone," entered what seems to have been a conglomerate, soon after which it became jammed, presumably by encountering a joint plane.   On its liberation water began to flow. Again, at Hotel Russwin, New Britain, a well penetrated 140 feet of sandstone, etc., without finding enough water to keep the drill wet, but at that depth the drill became jammed in what appeared to be a crack about 2 inches wide.   As in the preceding well, an ingress of water followed the loosening of the drill.   The apparent breadth of the crack may in reality be due rather to the presence of soft, decomposed rock along the joint plane than to an actual opening of the size indicated.

Effect.—Several different, and in two instances entirely opposite, effects are produced by the joints of the Triassic rocks. Superficially they greatly favor absorption of water, and it is probable that the shallow supplies, especially those reached by wells 50 feet or less in depth, are greatly increased by their presence. Doubtless the most common result of the jointing, however, is to break the continuity of the shale or other confining beds associated with the water horizons, and thus to permit sufficient leakage in most instances to destroy the prospects of artesian flows. In a few cases, on the other hand, the supplies of deep beds seem to be derived from joints rather than from outcrops, as was notably the case of the Thompsonville well (No. 26), described on page 82, where the water, as indicated in fig. 24, appears to have passed down the joint and up the dip of the sandstone bed nearly to its outcrop, giving, in fact, true artesian conditions.

FAULTING.

Occurrence.—Although jointing disturbs the continuity of the beds almost universally, it does not do this so profoundly as do the manifold faults that cross the area in several directions, especially from north to south. Reference to the map accompanying Professor Davis's paper[a] will show that important faults exist in great numbers, but numerous as they are, as shown on the map, a many times greater number of faults, too small to be recorded, yet of sufficient size to entirely destroy the continuity of the beds in their immediate vicinity, occur in the area. When the extensive brecciation and local bending of the strata that must take place along the greater faults—faults having a length to be reckoned in miles and an upthrow to be calculated in hundreds and in some cases thousands of feet—are taken into consideration, it may be readily seen how thoroughly and extensively the continuity of the strata is destroyed.

Effect.—The effect of faults is much the same as that of joints, but, because of the relatively greater influence of faults at considerable depths, it is even more important. As the jointed upper portions of traps and other rocks are reservoirs of water, so the fault breccia, or the crushed material along faults, forms an important source of water. Among successful wells located almost upon the breccia of faults may be mentioned Nos. 44, 45, 46, 47, 48, 49, and 55 of Mr. Pynchon, lying near the fault cutting off the north end of Cedar Mountain, near Hartford. Wells Nos. 124, 125, 126, and 127 at Berlin and No. 130 at New Britain are almost on fault lines, while the town of Meriden

---

[a] Davis, W. M., The Triassic formation of Connecticut: Eighteenth Ann. Rept. U. S. Geol. Survey. Pl. XIX.

with its wells is bounded on the northwest and the southeast by the two greatest faults in the whole lowland, the distance between the faults at this point being only 1¼ miles.

COMPOSITION OF TRIASSIC WATERS.

Many facts bearing on the composition of the Triassic waters are given by Mr. Pynchon in the preceding paper, especially those bearing on the practical question of their adaptability for boiler use. The analyses of the waters of most of the wells described by Mr. Pynchon are not now available, but the results of the chemical examinations of a number of similar waters from Triassic rocks are given in the following table.

Analyses of waters from Triassic rocks of Connecticut.

[Compiled from H. E. Gregory's report, Water-Supply and Irrigation Paper No. 102, pp. 127-159.]

Number	County	Town	Owner	Source of water	Depth of well	Character of material	Iron	Calcium	Magnesium	Iron oxide	Iron and aluminum oxides	Lime	Lithia	Iron carbonate	Calcium carbonate	Magnesium carbonate	Sodium carbonate	Calcium sulphate	Magnesium sulphate	Calcium nitrate
1	Hartford	Bloomfield	H. C. Douglas et al	Well	101	Rock	15						1		54			1,106	525	
2	Hartford	Crescent Beach	Atwood Collins	do	94	do	Tr.			1				1	158			272	155	158
3	Hartford	Hartford	Capewell Horse Nail Co	do	250	Red rock														
4	Hartford	do	City	do	250	Blue slate														
5	Hartford	Simsbury	W. L. Cushing	do	218	Sandstone					1	241				41	25	1,503	241	
6	Hartford	South Manchester	Cheney Bros	do	457	Red sandstone														
7	Hartford	Windsor	H. J. Fisk	Spring	185															
8	Hartford	Windsor Locks	Windsor Locks Water Co	Well	292	Trap														
9	Middlesex	Middletown	L. D. Brown & Co.			Brown sandstone									24	22		2		
10	Middlesex	do	W. E. Wilcox	Spring																
11	New Haven	Highwood	Marcus Wooding	do	157	Sandstone														
12	New Haven	New Haven	New England Dairy Co	Well	72															
13	New Haven	do	F. B. Shuster Co	do	75			74	11											
14	New Haven	do	G. D. Watrous	do		Rock														
15	New Haven	do	W. S. Swayne	Spring		do														
16	New Haven	North Haven	B. F. Judd	do		Trap														

Analyses of waters from Triassic rocks of Connecticut—Continued.

[Compiled from H. E. Gregory's report, Water-Supply and Irrigation Paper No. 102, pp. 127-159.]

Number	County	Town	Owner	Source of water	Depth of well	Character of material	Sodium chloride	Potassium chloride	Chloride	Carbon dioxide	Sulphuric acid	Phosphoric acid	Organic and volatile matter	Total mineral matter	Hardness	Silica	Sodium sulphate	Potassium sulphate	Analyst
1	Hartford	Bloomfield	H. C. Douglas et al	Well	101	Rock	887		22			60	410	2,528	28	20			Unknown
2		Crescent Beach	Atwood Collins	do	94	...do	87							182	28	18	222	6	Do.
3		Hartford	Capewell Horse Nail Co	do	250	Red rock								940	240				Do.
4		do	City	do	250	Blue slate			12		358			628	24	21	32	8	H. E. Smith
5		Simsbury	W. L. Cushing	do	218	Sandstone			2					69					Do.
6		South Manchester	Cheney Bros	do	457	Red sandstone	4							1,636					Unknown
7		Windsor	H. J. Fisk	do	185				10					56	29	10			Do.
8		Windsor Locks	Windsor Locks Water Co	Spring		Trap			10				20	110	52				S. P. Wheeler
9		Middletown	L. D. Brown & Co	Well	292	Brown sandstone								61	23				Unknown
10	New Haven	do	W. E. Wilcox	Spring		Sandstone	2		8					60	52				H. E. Smith
11		Highwood	Marcus Wooding	do		...do			10					110	35				Do.
12		New Haven	New England Dairy Co	Well	157				44		25			404	28	10	32		Unknown
13		do	F. B. Shuster Co	do	72	Rock			8					48					H. E. Smith
14		do	G. D. Watrous	do	75	...do			8					93					Do.
15	Middlesex	do	W. S. Swayne	Spring					2					32					Do.
16		North Haven	B. F. Judd	do		Trap									6				

One of the striking features of these wells and springs is the great variability of the waters. Those of the traps are comparatively low in mineral matter, carrying in one case, at North Haven (No. 16), as low as 32 parts per million of dissolved solids. From 400 to 2 500 parts per million are common in waters from the sandstones and shales (Nos. 1, 3, 6, and 13). Such amounts make the water unfit for boiler use, even where the mineral matter is mainly present as carbonate (lime). The analyses show that not only is lime present in large amounts, but that there are considerable quantities of the much more objectionable sulphates. The three springs mentioned are very low in mineral, the water in two of them, if not in all three, coming from the trap.

SUMMARY OF WATER CONDITIONS.

ESSENTIALS OF ARTESIAN FLOW.

In 1885 Professor Chamberlin published a paper[a] that explained the principles of artesian flow and called attention to many of the special conditions that tend to determine the success or failure of

FIG. 25.—Ideal section showing the requisite conditions of artesian wells.

artesian wells. The conditions favorable to true artesian flow are as follows:[b]

1. A pervious stratum to permit the entrance and the passage of water.

2. A water-tight bed below to prevent the escape of water downward.

3. A like impervious bed above to prevent the escape upward, for the water, being under pressure from the fountain head, would otherwise find relief in that direction.

4. An inclination of these beds, so that the edge at which the waters enter will be higher than the surface at the well.

5. A suitable exposure of the edge of the porous stratum, so that it may take in a sufficient supply of water.

6. An adequate rainfall to furnish this supply.

7. An absence of any escape for the water at a lower level than the surface at the well.

The requisites as outlined above apply only to sedimentary rocks, but the principle on which artesian flow depends is equally applicable

FIG. 26.—Section illustrating thinning out of porous water-bearing bed.

to other rocks, the essential feature being simply that the water be confined and that the outlet through the well be considerably lower

[a] Chamberlin, T. C., Requisite and qualifying conditions of artesian wells: Fifth Ann. Rep. U. S. Geol. Survey, pp. 125–173.

[b] Ibid., p. 13.

than the catchment area at which the water enters. A bedding, joint, or fault plane or other fissure in the insoluble rocks, or a solution passage in limestones sometimes affords favorable conditions for

FIG. 27.—Section showing transition from porous to impervious bed.

artesian flow, but such features are usually of limited and local development and are therefore a far less common source of water than the widely distributed beds of sedimentary origin.

CONDITIONS IN THE TRIASSIC AREA FAVORABLE TO FLOWS.

The conditions in the area that are favorable to artesian flow may be summarized as follows:

1. There are many porous beds which water may readily enter and through which it may find ready passage.
2. Water-tight beds, serving to confine the waters, frequently occur both above and below the water-bearing sandstones.
3. The beds are so inclined that the edges at which the waters enter are often higher than mouths of the wells.
4. The edges of the beds are so exposed that water can be readily absorbed.
5. The rainfall is adequate for large supplies.
6. In some places joints and faults appear to conduct water downward to the porous sandstones, which thus obtain their supplies.

CONDITIONS IN THE TRIASSIC AREA UNFAVORABLE TO FLOWS.

The conditions unfavorable for 'flowing wells in the area may be stated as follows:

1. The beds are so highly inclined that the areas exposed at the surface for the catchment of waters are small.
2. The inclination of the beds is such that, except in very deep borings, the outcrop is near the well and is at only a slightly greater altitude.
3. The rocks are so cut by joints and faults that ready escape is frequently afforded for the rock waters, thus neutralizing the several favorable conditions noted.

CONCLUSIONS.

The conclusions reached as to the underground water conditions on the area may be summarized as follows:

1. Rock waters are generally present in almost all of the rocks of the Triassic area.
2. In the sedimentary beds the rock waters appear to be most abundant in the sandstones, less abundant in the shales, and least abundant in the arkose and conglomerates occurring around the rim of the area. Little water is ordinarily afforded by the traps.
3. Many of the conditions favorable to artesian flows are locally present, and flowing wells will occasionally be found.

4. Because of the frequent interruption of the continuity of the water-bearing beds by joints and faults, artesian flows will be relatively rare and confined to small areas.

5. The dips are so steep that the water-bearing beds usually outcrop within a short distance from the wells and at only a slightly greater elevation; hence the head will probably never be great, and for ordinary purposes pumping will always be necessary.

6. Because of the steepness of the dip different wells, except where close together, will rarely pass through the same series of beds, but will generally show wide differences in the materials encountered and the quantity and quality of water procured.

7. Supplies sufficient for many ordinary manufacturing processes will be obtained from wells in the Triassic sandstones, but such enormous supplies, amounting to several thousand gallons a minute, as are obtained from some of the wells in the loose, porous beds of the Coastal Plain will not be found. The amounts given in the description of the wells in the paper by Mr. Pynchon indicate the character and limitations of the supplies that may reasonably be expected.

8. The waters will generally be highly mineralized, and it will usually be impossible to use them in boilers. The waters of the traps will on the whole probably be found freest from mineral matter and those of the shales will show the highest mineral content. The contrast between these waters and the pure soft waters of the crystalline rocks of the surrounding area is very striking.

9. The water will commonly be free from pollution by sewage or other extraneous matter. High chlorine, usually regarded as an indication of sewage contamination, may result from the solution of matter from the rocks, especially from the shales. In cities, however, where the rocks outcrop, as they frequently do, within the corporate limits, the water should be used for drinking only after it has been thoroughly tested, as surface waters may in some instances pass readily into the rocks. The wells should be carefully cased and packed to prevent water from passing downward along the pipe.

10. The source of the water supply is mainly the rainfall in adjacent regions. When the rocks outcrop at the surface, the water may be absorbed directly, but probably the greater amount is absorbed from the ground water in the glacial drift which overlies the rocks over large areas.

11. The waters pass through the rocks in part by general seepage and in part by following bedding, joint, or fault planes. The former method is more common in the porous sandstones, the latter in the shales and traps.

NOTES ON WELL DRILLING.

TESTING OF WELLS.

Attention has been called by Mr. Pynchon to the failure of a number of very deep wells, notably the 3,700-foot well at Northampton and the 1,100-foot well near Westfield. Although such wells have failed to obtain supplies, practically all of them encounter water in small amounts, which is generally cased off as the well proceeds. No such well should ever be abandoned without a thorough test of every water horizon, however small. The casing, if possible, should be raised above the level of the water-bearing bed, a pump inserted, and the supply measured. When it is impossible to remove the casing, it can be destroyed at the water horizon by a shot of nitroglycerin, which will also at the same time tend to loosen up the surrounding rock and increase the flow. Supplies have frequently been developed at horizons which were not at first thought worthy of testing and which were originally drilled through without stopping and cased off.

Anything that tends to increase the chances of success of wells is to the ultimate advantage of the driller. A complete failure, such as the well at Northampton, while it may cause the owner to sink the particular well to a great depth and thus give temporary employment to the driller, will tend to deter others in the vicinity from drilling, and will in the end result in a decrease in the total amount of drilling done. If such failures can be partially or wholly prevented, the drilling business will be proportionately benefited. The drillers themselves should use all possible effort to make every well successful, for each successful well is likely to lead, sooner or later, to the sinking of others.

KEEPING OF RECORDS.

One of the chief factors in the development of a well, as has been pointed out, is the thorough testing of each water horizon. Detailed notes should therefore be made of the character of material penetrated by the drill and the amount and quality of the water from each horizon, wherever encountered, and the depth of each horizon should be determined by actual measurement. The memory can not be relied upon and should not be trusted to recall the depths at which water may be struck, for although the facts are firmly impressed upon one at the time, they are gradually lost sight of during later development, so that if, at the completion of the well, it becomes necessary to test any particular horizon only a general guess as to its depth and as to the character of the material at that point can, in many instances, be made. Many wells have resulted in failure because the water horizons passed through in the earlier stages of the drilling and cased off could not be relocated.

PROPER DEPTH OF WELLS.

Nothing can be farther from the truth than the popular fallacy that water can always be found in great quantities if the well is only drilled deep enough. On the contrary, it is almost universally the case in this region that wells of moderate depth yield the most abundant supplies. The reason for this lies in the fact that all of the rock waters originally come from the surface and decrease in amount as the depth becomes greater. In the upper rocks joints and other crevices are both more numerous and more open than in the deeply buried rocks, and the absorptive and water-holding capacities of the upper rocks are correspondingly greater. Being subject to weathering and solution by circulating waters, the rocks near the surface are also much more porous than those lying deeper.

The depths to which wells should be sunk will differ greatly in different areas. What would be a shallow well in many regions would be a deep well in the Triassic region. In general, wells in this area that are under 300 feet in depth give the best supplies, while very few find their main supplies below 500 feet. There would probably be few exceptions to the principle that two 250-foot wells will give more water than one 500-foot well. This is even more true of the deeper wells. For the cost of the 3,700-foot well at Northampton fifteen or more 200-foot wells could have been sunk, and a good supply of water would almost certainly have been obtained.

It can therefore be said that success in procuring large supplies in the Triassic area is generally to be obtained rather by drilling a number of shallow wells than a single deep one.

SPRING SYSTEM OF THE DECATURVILLE DOME, CAMDEN COUNTY, MISSOURI.

By E. M. SHEPARD.

DESCRIPTION OF DOME.

The magnitude of the springs of the Ozarks, especially of the group gushing to the surface in Camden and adjacent counties in Missouri, has long been noticed, and the remarkable rounded mass or boss of

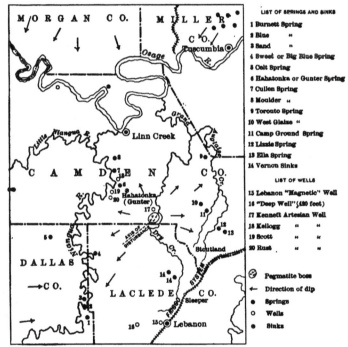

LIST OF SPRINGS AND SINKS

1 Burnett Spring
2 Blue "
3 Sand "
4 Sweet or Big Blue Spring
5 Colt Spring
6 Hahatonka or Gunter Spring
7 Cullen Spring
8 Moulder "
9 Toronto Spring
10 West Glaize "
11 Camp Ground Spring
12 Lizzie Spring
13 Ella Spring
14 Vernon Sinks

LIST OF WELLS

15 Lebanon "Magnetic" Well
16 "Deep Well" (490 feet)
17 Kennett Artesian Well
18 Kellogg " "
19 Scott " "
20 Rust " "

Pegmatite boss
Direction of dip
Springs
Wells
Sinks

FIG. 28.—Sketch map of region of Decaturville dome, Missouri.

coarsely crystalline granite, known as pegmatite, in the neighborhood of Decaturville, in southern Camden County (see fig. 28), has also been for many years an object of interest. During a trip made

113

through this region early in the summer of 1903 to investigate the
underground waters of the State, the idea occurred to the writer that
some relation existed between the granite mass and the system of
great springs and sinks that surrounds it for a distance of from 8 to
12 miles, or possibly farther. Subsequent study has confirmed this
opinion and has revealed a unique condition of underground drainage.

In 1869 some miners who were prospecting for lead laid bare, close
to the surface, a small area of granite or pegmatite, frequently called
graphic granite because of a peculiar crystalline structure resembling
the letters of an ancient alphabet. This is situated at the crest of a
low, dome-shaped hill covering about 6 acres and elevated about 40
feet above the level of the land in the immediate neighborhood. The
miners were led to prospect here by finding some small rounded
bowlders of quartzite and graphic granite at the surface. This small
exposure of granite is located near the town of Decaturville, in the
southeast corner of the SW. ¼ of sec. 32, T. 37 N., R. 16 W., near the

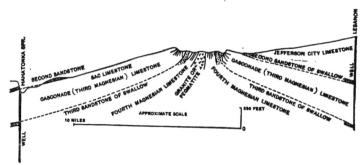

FIG. 29.—Ideal cross section from Hahatonka Spring to Lebanon.

line between Camden and Laclede counties, on land now owned by
Mr. Samuel Wheeler. The excavation was only 20 feet long, 10 feet
wide, and probably 30 feet deep, though it is now partially filled up.
When first opened the material immediately beneath the soil consisted
of an incoherent layer, several inches in thickness, of kaolinized feldspar and mica, which gradually merged into a true pegmatite, made
up mainly of the variety of feldspar known as microcline, with some
albite and oligoclase feldspars and numberless small masses of quartz.
On the south side an irregular, sheet-like mass of fine-scaled to somewhat granular, ferruginous, micaceous schist (evidently metamorphosed gouge) was found. The mica was all common muscovite.
Scattered over the surface of the hill were lenticular or irregular
masses of polished, compact quartzite, somewhat resembling glacial
bowlders of the same material. If, as the writer believes, these are
the metamorphosed remnants of a fractured intercalated layer or bed

of sandstone in the original dolomite, they form, with the mica-schist, the only evidence of metamorphism found. Most of these bowlders have been carried away, many of them having been thrown into the old pit and many used in the construction of an adjacent stone wall.

The dike or boss is surrounded by siliceous magnesian limestone, corresponding to Swallow's Fourth Magnesian limestone. The greatly tilted and in many places nearly vertical ledges indicate great disturbance. The limestone here, while not metamorphosed, is decidedly altered, being highly silicified as well as pitted and honeycomed to a remarkable degree. A quarter of a mile away to the north, east, south, and west, these ledges outcrop in a way that reminds one strongly of coral reefs exposed at low tide. Great parallel ridges, tilted at every angle and some nearly vertical, stand up in irregular segments about the granite hill. To the southwest for over 2 miles, along the narrow valley of Spencer Branch, the disturbance is very great. The axis of this lateral disturbance extends S. 52° W., and the rocks on both sides of the valley dip strongly toward this axis.

About 500 feet southwest of the hill three small shafts were sunk several years ago. The surface was covered with soil mixed with an earthy hydrated iron sesquioxide. In the first shaft, 27 feet deep, a mass of very soft earthy galena, 2 feet wide and 3 feet thick, was found. This seemed to extend about 8 feet to the northeast. The second shaft was 12 feet deep, contents unknown. The third was 53 feet deep, and Mr. Wheeler, the owner of the land, states that granite was found at the bottom. A very impure earthy galena, mixed with zinc silicate and iron pyrites, was found in three V-shaped pockets or flat openings, one above the other, which pitched strongly to the west, toward the prolonged axis of disturbance. About a quarter of a mile to the southwest, along this axis, Mr. Wheeler sunk a fourth shaft, depth unknown, at which, on the dump, there were exposed masses of a highly silicified rock mixed with a curious iron breccia and conglomerate, with scattered masses of earthy limonite.

Surrounding the pegmatite outcrop, except at the southwest, is an irregularly broken circle of low hills that stand about 60 feet above the general level of the basin, making the whole area resemble somewhat the contour of an old crater. The Third sandstone of Swallow outcrops in vertical ledges that form an irregular circle about three-fourths of a mile in diameter around the pegmatite outcrop. This sandstone was first observed just south of the hotel at Decaturville, where a large exposure may be seen. The vertical beds here are about 25 feet thick. The false bedding and the more siliceous, irregular layers, standing out as ridges, are characteristic of this formation. The sandstone is somewhat hardened at the surface. Farther west, other outcrops of vertical sandstone ledges are found in juxtaposition with vertical

ledges of the Fourth Magnesian limestone on the inner side, and vertical ledges of Gasconade limestone, or the Third Magnesian limestone
of Swallow, on the outer side.   Two of these outcrops are particularly
striking, one near the center of the south line of section 29 and the
other near the point where the road crosses the dry branch of Banks
Creek, which heads to the southwest, close to the granite outcrop near
the southwest corner of the same section.   Farther west and south
another striking vertical outcrop of this sandstone forms a "back-
bone."   An interesting feature of these sandstone outcrops is the fact
that, on either side of them, the dips of the Gasconade limestone and
Fourth Magnesian limestone range from vertical to a strong inclination
inward toward the sandstone, a condition very difficult to explain, but
which may possibly indicate a second and more local uplift.   A few
hundred feet southwest of the pegmatite outcrop several shafts already
described have been opened and small amounts of earth, impure galena,
and zinc ore have been found.

Within a radius of a mile or two of this pegmatite outcrop Mr. W. S.
Rush, of St. Joseph, has sunk a number of diamond-drill holes to a
depth of several hundred feet, and in no case did he strike the pegmatite.   These holes were all sunk in search of mineral.   A third of a
mile north of Decaturville, in the southwest quarter of section 28, is
a small outcrop of Hannibal shales and Sac limestones (Devonian).
The limestone contains cavities filled with calcite and is somewhat
altered structurally, though its lithological characters serve to identify
it without much doubt.   The shales, presumably the Hannibal, are
greatly altered and are of a bright-yellow color and of a soft, clay-like
texture.   The blue color which everywhere else characterizes these
shales has evidently been destroyed by the heat attending the uplift.
In the valley a few hundred feet to the south is a tilted outcrop of
greatly altered dolomitic limestone, which is doubtfully referred to
the Jefferson City limestone (Second Magnesian of Swallow), or possibly to the First Magnesian of Swallow.[a]   As far as the writer knows,
these are the highest rocks exposed anywhere in this county or the
adjoining counties to the south.   It is significant to note that these
upper beds retain to a considerable degree their dolomitic character.
In other words, these remnants of superincumbent beds have not
undergone replacement by silicification, as has been the case in lower
beds, which will be described later.

There is evidence of two distinct periods of uplift, both of which
must be post-Devonian.   From data obtained in other parts of the
Southwest it is presumed that the first uplift was post-Carboniferous,
when the great dome of the Ozarks was first elevated.   From the

---

[a]The above locality was recently visited by Dr. E. O. Ulrich and the writer. From fossils there
obtained Doctor Ulrich determined that these beds probably belong to the Richmond formation
of the Ordovician.

obscure evidence furnished by the lateral anticlines that merge into the main axis of the Ozarks at divergent angles—anticlines which, the writer believes, were formed in post-Tertiary time, when the lead and zinc ores of the Southwest were probably deposited—he would date the second uplift at this point in geological time.

In passing in any direction outward from the center of maximum disturbance one may readily note that the rocks tilt less and less strongly away from the pegmatite hill (see figs. 28, 29).[a] This is strikingly shown along the bluffs of Banks and Spencer creeks.

SPRINGS.

From what has been said it is evident that the area of upturned strata surrounding the pegmatite hill—an area 3 or 4 miles in diameter—forms a splendid catchment basin, which contributes to the supply of the great fountains that lie in the surrounding region. Studies of the counties adjoining Camden, particularly on the west, northwest, north, and northeast, indicate a general dip of the strata toward the dome. The water from the catchment basin around the dome therefore meets in a trough, at considerable depth, the circulating water derived from the outside area, and the body of water thus gathered from the drainage on both sides, being under pressure, is forced up to the surface. Where erosion has cut sufficiently deep into the valleys of the Big Niangua, Osage, and Auglaize rivers, which surround the pegmatite hill on the west, north, and east, great springs issue at a distance of from 8 to 15 miles from the granite outcrop. Along the course of Spencer and Banks creeks, which drain this same territory to the northwest, numerous small springs issue, the valleys not being cut deep enough to tap the larger supplies below.

Bennett or Bryce Spring.—The springs of the Decaturville system will be taken up in order from the southwest. The first and largest of the group is the Bennett or Bryce Spring, situated on the west side of Dallas County near the Laclede County line, in sec. 1, T. 34, R. 18, at an altitude[b] of 940 feet, 150 feet below the pegmatite hill and about 15 miles southwest of it. It issues from the lower beds of the Gasconade limestone, the Third Magnesian of Swallow. The temperature of the spring was 57.5° F. when the air temperature was 86° F. Swallow, who visited it in 1853, gives the temperature of the water at 58° F. and that of the air at 60° F. The calculated flow was 172,388,800 gallons a day, but Swallow's figures give only 81,084,810 gallons a day, a remarkable variation in volume. This wonderful

---

[a] Except at a point 10 miles northeast of Decaturville, in the SW. ¼ of T. 38, R. 14, where the rocks are greatly disturbed and tilted along a northeast and southwest axis. Whether this is a continuation of the Decaturville axis or not can not be determined at this time.

[b] All the altitudes given in this paper, except that of Lebanon, were determined by a series of barometric observations and are believed to be approximately correct.

spring lies in a narrow valley about 600 feet wide. It boils up with great force from a vertical, cave-like opening into a large oval basin (see fig. 30). The spring was sounded to a depth of 22 feet. Mr. Trueman Atchley, who has lived near it for many years, states that he has seen it with one-tenth less water than normal and once 3 feet higher than normal, a condition which would give it twice the flow above calculated. The channel just below the outlet of the spring has become somewhat filled up with gravel, so that the water surface has been raised above the height of the place of outlet. Mr. Atchley states that before the outlet was thus choked he has seen the spring so low that he could walk around the upper portion of the cave-like opening. He states, also, that this opening descends 20 feet vertically

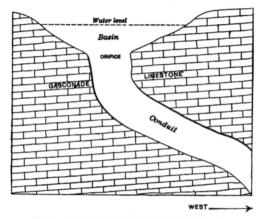

FIG. 30.—Diagrammatic cross section of Bennett Spring.

and then slants southwestward to an unknown depth at an angle of 40° to 50°. When visited by the writer the water of this spring boiled up strongly over an area about 20 feet in diameter and to a height of 2 or 3 inches above the general level of the pool. The spring forms a river that is 100 feet wide and averages 1½ feet in depth, with a rapid current.

Farther west, in Dallas County, the rocks have a general eastward dip. To the southeast, toward the town of Lebanon, the strata are nearly horizontal. There is apparently a tendency to form a synclinal trough at this point.[a] The water from the catchment basin passing along the line of disturbance extending from the pegmatite hill to the southwest, through the lower beds of the Gasconade limestone (the

[a] There seems to be here, as elsewhere in the Ozarks, a tendency to torsional strain, or twisting, in the flexures, which would operate to elevate the area to the south and depress that to the north. This would account for the absence of springs and the great depth of water-bearing strata to the south, as well as for the "extinct spring" at the Vernon Sink, and the immense size of the springs to the northwest.

Third Magnesian of Swallow) and the Third sandstone of Swallow below, meet the drainage from the west in this syncline, and the body of water thus gathered is forced to the surface through a crevice that has been gradually enlarged by continued erosion. A year ago a field was plowed just above this spring, and a heavy freshet washed down a large quantity of gravel which partially stopped its outlet. For several hours the water pushed itself through this obstruction in great intermittent jets, 3 or 4 feet high, with a noise that could be heard a quarter of a mile away.

Blue Spring.—About 1⅓ miles north of this point is Blue Spring, situated on the west side of the Niangua River. It showed a temperature of 59° F. when the air temperature was 70° F., and is at an altitude of about 900 feet. It issues from a whitish, clayey sand, and is badly choked with water cress, logs, and rubbish. The outcropping rocks indicate the middle or lower beds of the Gasconade limestone. It has a flow of 5,170,000 gallons a day.

Sand Spring.—About a quarter of a mile north of the last spring, on the same side of the river and in the same geological horizon, is Sand Spring, at about the same altitude, also badly choked with water cress, and with a flow estimated to be about the same as that of Blue Spring.

Sweet or Big Blue Spring.—This spring is in the northwest corner of Laclede County, in the NE. ¼ of the NE. ¼ of sec. 30, T. 36, R. 17, at an altitude of about 850 feet. The temperature of the water was about 58° F. when the air was 94° F. This spring issues from a vertical cave opening at the base of a bluff 40 feet high; probably the lower or middle beds of the Gasconade limestone. It has been sounded to a depth of 150 feet. The water is clear, cold, and free from weeds. It flows a distance of 150 feet westward into Niangua River. About 50 feet farther north Sweet Branch, a small, sluggish stream, empties into the Niangua. Sweet Spring is but about 5 feet above the level of the river and is subject to overflow at every rise of the stream. Situated as it is, at the junction of Sweet Branch and the Niangua, it is frequently choked with sediment from these two streams. Between the spring and the river a large area covered with treacherous mud and sand indicates that the basin of the spring was once much larger than it now is. Farmers in the vicinity state that the flow has decreased fully one-half within the last few years. The flow was calculated at 14,037,000 gallons a day. This spring lies just north of the extension of line of disturbance which runs southwestward from the Decaturville uplift.

Celt Springs.—Situated in the northeast corner of Dallas County, 7 miles northwest of Sweet Spring, are found three large springs whose waters unite to form Mill Creek. These were not visited, but they undoubtedly belong to the drainage system of the pegmatite hill.

Hahatonka or Gunter Spring.—This spring is situated 9 miles
northwest of the pegmatite hill in Camden County, sec. 2, T. 37,
R. 17, at an altitude of about 745 feet, or 345 feet below the height of
the hill. The temperature of the water was 57° F. when the air was
88° F. The spring issues from the foot of a bluff 250 feet high, and
after running for about 800 feet through a deep, narrow canyon
whose walls rise at one point to a height of a little over 300 feet, it
empties its waters into an artificial lake, covering 90 acres, situated in
the valley of the Niangua. A small island about 100 feet high divides
the stream near its outlet into the lake. The scenery in the vicinity
of this spring is extremely picturesque. The spring has been sounded
to a depth of about 40 feet, at which point the conduit bends to the
south. The spring has its source in the Gasconade limestone, 150 feet
below the base of the Moreau sandstone (Second sandstone of Swallow).
It is very difficult to measure the flow of this spring, owing to the
irregularities of the stream. The writer calculated 14,760 cubic feet
per minute, or 158,982,912 gallons a day. A Philadelphia engineer
estimated 13,800 cubic feet per minute. Professor Shaw, of the Engineering School of Missouri State University, calculated 15,000 cubic
feet per minute, while Colonel Scott, an electrical engineer residing
at Hahatonka, obtained the same result as Professor Shaw. The district in the vicinity of this spring possesses great interest both from a
scenic and a geological standpoint. Evidences of great erosion from
underground waters, showing former outlets of this great spring, are
everywhere seen. At one time it doubtless had its outlet at the Big
Red Sink, one-half mile to the southwest. A few hundred feet on the
other side of the bluff from which the spring now issues is a sink, 150
feet deep and 600 feet long, which is open at one end and spanned by
a great natural bridge. This undoubtedly marks an old cavernous
channel of the great spring. In fact, most of these sinks and canyons
mark the site of ancient cavernous outlets.

Cullen Spring.—This spring flows out of a bluff of Gasconade limestone, about 1 mile north of Hahatonka. Its water is clear and cold,
and its flow is estimated at about 1,000,000 gallons a day.

Moulder Spring.[a]—In the northeast quarter of sec. 18, T. 38, R.
17, about 4 miles northwest of Hahatonka and about 12 miles northwest of Decaturville, is Moulder Spring, heading in a ravine about 1½
miles long that extends northward from Niangua River. The spring
boils up strongly from a bed of coarse gravel 25 feet in diameter, and
is estimated by Colonel Scott to be about one-sixth the size of Hahatonka Spring. It issues from the Gasconade limestone, and has its
head in the tilted beds which surround the pegmatite hill.

Toronto Springs.—At Toronto, 6 miles northeast of Montreal, in

_____
a The writer was unable to visit either Moulder, Cullen, or Toronto springs, all probably belonging
to this system, and is indebted to Colonel Scott, of Hahatonka, for data concerning them.

the SW. ¼ of the NE. ¼ of sec. 25, T. 38, R. 15, are several large springs that boil up with great force out of gravelly beds in the Gasconade limestone, flow eastward a short distance, and empty into Auglaize River.

Wet Glaize Springs.—These are two fine springs situated in the NE. ¼ of sec. 24, T. 37, R. 15, about 8 miles a little north of east of Decaturville. One is located near the foot of a bluff of Gasconade limestone 40 feet high, and the other, which is the larger of the two, is about 300 feet farther east, rising from a gravel bed at about the same level. A dam 16 feet high forms a pond of about 10 acres, which covers these two springs. As the pond had been lowered and the gate closed just before the writer arrived at these springs it was impossible to measure their flow, but from what could be learned in regard to them it is probable that their capacity is about 40,000,000 gallons a day. As described by their owner, they both boil up with considerable force out of gravel beds, rolling and tossing up large pebbles. Their location can be marked on the surface of the pond from the disturbance of the waters. They also receive their supply from the catchment basin around the pegmatite hill and the cavernous Gasconade limestone, which, with the Third sandstone of Swallow, forms a great reservoir for all this spring system. Their altitude is about 825 feet, or 265 feet below the outcrop of pegmatite.

Camp Ground Spring.—This spring is located about one-half mile south of the Wet Glaize Springs. A large basin has been hollowed out from the side of the hill by this spring, which boils up from a bed of gravel in a remarkable manner. It issues from a number of holes about 5 inches in diameter, raising and tossing the gravel with great force and making the surface of the spring somewhat resemble a cauldron of boiling water. The water temperature was 55° F. when the air was 84° F. The flow could not be determined, owing to the masses of water cress that choked the outlet. It is stated that it takes thirty-six hours after a heavy rain for this spring to show an increase in its flow.

Armstrong Springs.—These are situated about 9 miles a little south of east of Decaturville, and are at an altitude of about 895 feet.

Ella or West Spring.—This is located in the NW. ¼ of the NE. ¼ of sec. 6, T. 36, R. 14. The temperature of the water was 56° F. when the air was 86° F. The water boils up powerfully from holes in the rock, one of which, 3 feet in diameter, has been sounded to a depth of 30 feet. A bluff of Gasconade limestone, 50 feet high, outcrops just above this spring. The flow was calculated to be 19,112,896 gallons a day. About 12 hours after a heavy rain the volume increases, and the waters become slightly turbid.

Lizzie or East Spring.—This rises from a deep, narrow gorge, on the side of the valley opposite the Ella Spring. Its flow was estimated to be about 3,000,000 gallons a day. The temperature of the water was 56° F. when that of the air was 90° F.

SINKS.

Vernon Sinks.—About 8 miles a little east of south of Decaturville are several sink holes of considerable interest. These lie in a line away from the pegmatite outcrop, in sec. 18, T. 35, R. 15. The main, or Vernon Sink, is about 120 feet in diameter, and about 50 feet deep to the water level. In very dry times the water level sinks 40 feet and exposes a cave-like opening that extends 300 feet or more on an incline due west. A drill well was sunk about 200 feet west of this sink to a depth of 60 feet, and reached the cave opening. A pump was inserted,

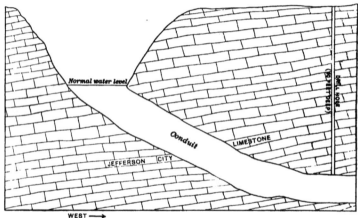

FIG. 31.—Diagrammatic cross section of Vernon Sink.

and the water obtained is constantly used, being clear and cold and never muddy or turbid. It is possible that this may be the mouth of a former outlet of the Decaturville system of springs, the water level standing at an altitude of about 1,080 feet, 10 feet below the catchment basin. Some of the enormous springs in southern Morgan County, and in western Miller and Pulaski counties, possibly belong to this system, and will furnish interesting material for future study.

ARTESIAN WELLS.

In addition to the springs that have been described, artesian wells may, be expected in this district wherever the general surface of the country is sufficiently below the level of the catchment basins. Mr. J. N. Kennett, of Decaturville, sunk a drill hole 300 feet for ore one-third of a mile northeast of the pegmatite hill and 40 feet lower than that elevation. A weak flow of water was struck at 200 feet. On the south edge of Hahatonka Lake, Colonel Kellogg sunk a drill hole for mineral to a depth of 864 feet. Water rose in this pipe 30 feet above the ground and threw a jet 6 feet higher. The strong flow was struck at 840 feet, but the force has gradually decreased until the water only flows gently over the pipe at the surface of the ground. This water is slightly charged with sulphuretted hydrogen. It is to be noted that elsewhere artesian wells from this formation are strongly saline, and highly charged with sulphuretted hydrogen, the absence of which in this region is significant of one of the great changes wrought by powerful underground drainage. The Rush artesian well, on the west side of Banks Creek, about 2 miles southwest of Hahatonka and at about the same altitude as the Kellogg well, was sunk to a depth of 780 feet. Water rises 30 feet above the ground, the first flow having been struck at 630 feet. The water temperature is 59° F. when the air is 84° F. There is a slight odor of sulphur. The flow of this well has never decreased. The Scott well, one-eighth of a mile north of the Rush well, on Banks Creek, was sunk to a depth of 740 feet. The first water was struck at 680 feet, and it flows only slowly from the pipe at the surface of the ground. The last three wells all passed through beds of siliceous limestone, only the Kellogg well entering sandstone (Third sandstone of swallow) to a depth of 3 feet. These wells show the enormous thickness of the Gasconade limestone at this point. Another artesian well, situated in the big bend of the Osage, in the NW. ¼ of sec. 22, T. 40, R. 18, was sunk to a depth of 800 feet, and obtained a strong flow of water. This well was not visited, but it probably belongs to the Decaturville drainage system.

SURFACE WATERS.

The region around Decaturville, bounded by Bennett, Sweet, Hahatonka, Wet Glaize, and Armstrong springs, is almost destitute of surface water, the inhabitants being dependent for their supply upon ponds, cisterns, and a few drilled wells. For example, Mr. Haun, who lives between Sweet Spring and Decaturville, drilled 130 feet, and has only from 3 to 6 feet of water, which is quickly pumped dry. The little town of Eldridge, nearby, is wholly dependent on cisterns and ponds for water. Mr. F. M. Cossairt, at Chauncey, Camden County, 6 miles due north from Decaturville, sunk a well 144 feet deep, and struck water at 119 feet, at which level it remains. He states that wells in this vicinity reach water at depths ranging from 100 to 125 feet, and generally have to go to sandstone. At Lebanon, 15 miles due south of Decaturville, in a well sunk to a depth of 1,000 feet, the water stands 360 feet below the surface. About 5 miles west of Lebanon, at nearly the same altitude, a farmer sunk a well 420 feet deep, in which water stands 400 feet below the surface. A pump can not lower it.

COMPOSITION OF SPRING WATERS.

: Partial quantitative analyses of a number of the springs of the Decaturville system show a nearly uniform composition quite similar to that of the Lebanon "Magnetic" well, except in a slight increase in chlorides.[a]

The following is the analysis of the Lebanon well, made by Mr. L. G. Eakins, of the U. S. Geological Survey:

Analysis of water from Lebanon well.

	Parts per million.
Silica	11.2
Alumina	32.0
Sodium chloride	Trace.
Magnesium sulphate	6.0
Ferrous carbonate	Trace.
Calcium carbonate	81.1
Magnesium carbonate	52.8
Sodium carbonate	29.1
Total solids	212.2
Free carbonic acid	69.8

CHERT BEDS.

One of the most interesting deductions drawn from the study of the underground waters of this region is their probable connection with the origin of the enormous beds of chert that occur here. Though this point can be but briefly touched upon at this stage of the investigation, it should not be left without notice. All who have visited this region have observed its wonderful deposits of chert, particularly in the Jefferson City and Gasconade limestones. These formations are usually largely dolomitic, particularly the Jefferson City limestone, which contains large masses of dolomite ("cotton rock"). In the district surrounding Decaturville, however, the dolomite is almost entirely absent, and beds of chert seem to have taken its place, especially above and below the sandstone horizons. The chert beds are of great thickness, and are composed of knotted, irregular, frequently lenticular, sometimes agatized, shattered masses. Frequently these shattered masses are recemented into chert breccias. The shattering and cementing by resilicification of the chert beds in many places is another evidence of the second uplift, before referred to, in this region.

The writer has long believed that many of our chert deposits have been formed by replacement, and the evidence in this district points strongly to this conclusion. Briefly, it may be stated that during the period of the uplift of the pegmatite boss, the great heat generated

---

[a] The presence of sodium carbonate in the waters of this district—an unusual occurrence elsewhere in these formations—is significant of its probable source from the albite (soda feldspar) of the pegmatite; and the same fact also calls attention to the probable magnitude of the pegmatite boss.

was retained by the superincumbent beds (since eroded), and that percolating water, intensely heated, aided by the alkalies derived from the decomposed feldspar and lime, dissolved from the rocks their silica, which gradually replaced the lime and magnesia in the dolomites and converted them into cherts. This replacement seems to have been especially marked above and below the sandstone horizons—so much so as to make it very difficult to separate the Moreau sandstone from the Gasconade limestone below, and the Gasconade limestone from the Gunter sandstone, by which it is underlain, for the reason that the Gasconade limestone merges, through massive chert beds, into the sandstone horizons, both above and below. This is so strikingly the condition in Miller County that State Geologist Buckley, in his report on Miller County, now in press, includes the Moreau sandstone (Second sandstone of Swallow), a part of the Jefferson City limestone (Second Magnesian of Swallow), and part of the Gasconade limestone (Third Magnesian of Swallow) in one formation, which he has named the St. Elizabeth formation, or complex.[a] It will be noted that this special chert replacement predominates along the main line of water flow on both sides of the sandstone horizons.

The study of the Decaturville dome has led the writer to believe that the association of massive chert beds with several other great spring districts in the Ozarks has a deeper significance than that which is usually recognized in considering these beds as merely the reservoirs or channels for the distribution of underground waters, and it is hoped that this brief sketch may throw some light on the origin of the chert beds, as well as emphasize the importance of underground drainage as a geologic agent.

---

[a] See also Biennial Report of the State Geologist of Missouri to the Forty-second General Assembly, Jefferson City, 1903, p. 14.

WATER RESOURCES OF FORT TICONDEROGA QUADRANGLE, VERMONT AND NEW YORK.

By T. Nelson Dale.

INTRODUCTION.

Geography.—The Fort Ticonderoga quadrangle is located on the Vermont-New York boundary line, about midway of the length of Vermont. Three-fourths of its area is in Vermont and the remainder in New York. It comprises four small quadrangles—the Ticonderoga, Brandon, Whitehall, and Castleton—each of which contains about 225 square miles. The area includes considerable portions of Rutland and Addison counties, Vermont, and of Essex, Warren, and Washington counties, New York. The principal towns are Ticonderoga and Whitehall in New York, and Proctor, Brandon, Poultney, Fair Haven, and West Rutland in Vermont.

Relief.—The relief of the region varies from an altitude of 100 feet near the shores of Lake Champlain to over 2,700 feet a few miles southwest of West Rutland. Roughly it may be divided into three belts—an eastern belt lying mainly east of a line extending from East Middlebury, through Bomoseen Lake, to Poultney; a middle belt lying between the eastern belt and Lake Champlain, and a western belt, west of Lake Champlain. The first includes the northern end of the Taconic Range, which consists of high, irregular ridges and isolated hills, several points of which reach 2,000 feet or more, and one, Herrick Mountain, rises to over 2,700 feet. It includes also in its northern half a part of the west side of the Green Mountain Range, consisting of ridges, with Mount Moosalamoo rising to over 2,600 feet. In the second belt the land generally varies from 100 to 1,000 feet in altitude. Many extensive terraces are developed at altitudes between 300 and 550 feet. Above these many hills rise to a level of 1,000 feet or more Much of the area is covered by a considerable thickness of drift which has much modified the topography of the region. The third belt is mountainous, with maximum elevations varying from 1,400 feet in the northern part of the area to 1,900 feet in the southern part. The major features of the topography are due to rock structure or compo-

126

sition, but some are due to ice action or to Glacial or post-Glacial deposits.

Drainage.—The drainage of the southern half of the area is westward to Lake Champlain and then northward to the St. Lawrence. In the northern half it is northward by Lake Champlain and Otter Creek. Two other considerable streams—the Middlebury and Leicester rivers—besides Mill Brook, a smaller stream farther south, issue from the mountains and empty into Otter Creek. These streams drain a considerable region lying east of the quadrangle, and comprise the output of innumerable forest springs and rivulets.

WATER SUPPLIES.

General conditions.—The southern half of western Vermont is generally well watered, both because of a copious rainfall and because of the presence of two mountain ranges—the Taconic on the west and the Green Mountain on the east—which are both well wooded and drained by numerous brooks. As the Taconic Range ends in about the center of the quadrangle, one of the chief sources of water supply there ceases, the conditions becoming very different farther northward. The western border of the Green Mountain Range, with the exception of certain deforested tracts in Goshen and minor clearings in Ripton, is densely forested, and is therefore rich in springs and brooks, so that the question of water supply in the eastern third of the quadrangle is rarely a serious one. In fact, no one familiar with this area can fail to be impressed with the abundance and purity of the water.

Cold springs and brooks.—Some of the water is unusually cold. About three-fourths of a mile southwest of Brandon village is a well dug in a mass of glacial gravel, the pebbles of which are cemented together by calcium carbonate, forming a quasi conglomerate. This well has long attracted attention on account of the very low temperature of its water throughout the summer months, and has also been investigated by geologists.[a] It is known locally as "the frozen well."

As pertinent to this subject, although belonging to the Mettawee quadrangle, the following data are added: Three-fourths of a mile northwest of the top of Dorset Mountain and 1,700 feet below it, in the township of Danby, is a brook of crystal purity, in which the temperature on September 22, 1891, was but little above that of ice water. As the top of Dorset Mountain is only 3,804 feet high the following explanation was adopted: The summits of the Taconic range usually consist of shattered schist ledges covered with thick moss and forest. Ice formed between the blocks in winter may be protected by the dense vegetation and may be the source of such ice-water brooks. Ice-water springs also occur occasionally at the foot of exposed, sun-heated slate-quarry dumps in Washington County, N. Y., in places where no

---

a See Hitchcock and Hager, Rept. Geol. Vermont, vol. 1, 1861, pp. 192-208.

mountain drainage could possibly come. These springs also maintain a low temperature throughout the hot summers. Another brook of low temperature, known as Cold Brook, occurs 3½ miles southeast of Dorset Mountain, in Dorset Hollow. On July 20, 1900, a very hot day, it was cold enough to form a mist about itself. It issues from a limestone cave, through which its course lies for one-half to one-fourth of a mile, and which is at times presumably more or less filled with snow or ice.

Town supplies.—Because of the vitiation of the water, owing to the building of towns, as in the case of the portion of Otter Creek below Rutland and Proctor, streams are no longer safe sources of drinking water. This has been amply demonstrated by the prevalence of typhoid at Middlebury, just north of this quadrangle, as long as Otter Creek water was used, and by its cessation as soon as that use ceased. Where springs or brooks are not available, shallow wells, especially in the lowlands, usually obtain sufficient supplies. The practical utilization of springs as a source of water supply has been demonstrated within a few years at two points north and south of this quadrangle. Water has been brought from a spring on Niquaket, a mountain a mile north-northeast of the southeast corner of the quadrangle, through pipes for a distance of over 4 miles to the village of Pittsford, which is located on a sandy terrace 130 to 150 feet above Otter Creek. A similar system now supplies the village of Middlebury with water from a spring on the Green Mountain range. But for the use of Silver Lake (altitude 1,241 feet) as a summer resort, it would be admirably adapted for a similar purpose, as it is a natural reservoir fed by forest mountain springs.

Geology of the supplies.—The geologic structure of Otter Creek valley seems unfavorable for artesian wells because the marble and dolomite beds, which underlie its gravel and clay, generally lie in small and overturned folds. It is possible, however, that borings which should penetrate that formation (1,000 feet or more thick) and the underlying quartzite and schist (1,600 feet or over) to the basal gneiss of the Green Mountain range (pre-Cambrian) might be productive.

There is a longitudinal belt about 7 miles wide, measured westward from the foot of the Green Mountain range, underlain by dolomite and marble, in which any wells reaching those beds would inevitably furnish water more or less calcareous. There are also on the Green Mountain range areas of dolomite which would have a like effect. Such an area covers about 6 square miles at the southeast corner of the quadrangle. A narrow strip of dolomite extends also 5 miles from Sucker Brook south, inclosing Silver Lake, and must needs add a small percentage of lime to the tributaries of Mill and Sucker brooks. The rocks of the Green Mountain range, with rare intercalations of dolomite, are in other places quartzose, micaceous, and feldspathic, and are thus adapted to furnish the very best of water, except where springs or brooks drain extensive swamps and add organic matter. The occasional disappearance of brooks into caves in the limestone areas is a factor which should be taken into consideration in some of the hydrographic problems.

In the lower levels of the towns of Orwell, Whiting, and Cornwall, along the western edge of the quadrangle, plastic clays, which probably formed the sea bottom of the Lake Champlain region during the "Champlain submergence," introduce conditions very different from those prevailing along the foot of the Green Mountain range, where the superficial deposits are generally more sandy. The effect of these impervious clays, together with that of deforestation in the last generation or two, has been to bring about great scarcity of water. This is very apparent in the northern part of the middle belt, where driven wells, reaching to or into the shales underlying the clays and operated by windmills, are a necessity on many farms, and where the rarity of public watering troughs on the highways is as marked as the poverty of the brooks.

Water power.—East Middlebury River, Mill Brook, and Leicester River have long been utilized in a small way for driving sawmills and other wood-working machinery, etc. The fall of all these streams is considerable. Middlebury River falls 517 feet between Ripton and East Middlebury, a distance of not quite 3 miles. Leicester River, the outlet of Lake Dunmore, falls 171 feet in its course of 1¼ miles from the lake to Salisbury village. Mill Brook falls 636 feet between Goshen and Forestdale, a distance of 3 miles. Besides the above, and as yet not utilized for power, there is Sucker Brook, which has a fall of 529 feet in a distance of 1¼ miles above Lake Dunmore. Silver Lake, the outlet of which flows into Sucker Brook, is 670 feet above Lake Dunmore. The narrow gorges through which these streams flow and their greatly swollen condition in spring offer difficulties, however, which must be considered in any scheme for their complete utilization.

WATER RESOURCES OF THE TACONIC QUADRANGLE, NEW YORK, MASSACHUSETTS, VERMONT.[a]

By F. B. TAYLOR.

INTRODUCTION.

Geography.—The center of the Taconic quadrangle falls almost exactly on the point at which the States of New York, Massachusetts, and Vermont come together. The area included is about 35 miles long from north to south, and 20 miles wide from east to west, containing a little over 900 square miles. The region lies mainly in Rensselaer County, N. Y., Berkshire County, Mass., and Bennington County, Vt. North Adams, Mass., Bennington, Vt., and Hoosick Falls, N. Y., are the leading towns.

Relief. —All of the land of the quadrangle is moderately high, except along the larger streams, where it is sometimes as low as 350 feet above sea level. The area is mountainous, except in the western third, the ranges having a trend a few degrees east of north. The principal mountain masses are the Hoosac on the east, rising to a maximum height of about 2,800 feet, Greylock, and the Green Mountains in western Massachusetts and Vermont, rising to heights of from 2,500 to 3,700 feet, and the Taconic Mountains, along the New York boundary line, rising from 2,200 to 2,800 feet. The western third of the quadrangle partakes of the nature of a compound plateau, the northern portion standing at an altitude of from 800 to 1,200 feet, and the southern at from 1,500 to 1,800 feet. The topographic expression of the region depends to a considerable degree upon the structure and composition of the rocks.

Drainage.—The drainage of nearly the entire area of the quadrangle is by the Hoosic River northwestward to the Hudson River above Troy. A small area in the southwest portion, however, drains southwestward to the Hudson, 25 miles below Albany.

WATER RESOURCES.

SURFICIAL WATERS.

The surface waters include all the running and ponded surface waters which result from rains and melting snows or which are fed by springs.

---

[a] The field work on the surficial geology of this quadrangle is only about half completed, the unfinished part being confined mainly to the western half. The conditions as regards water supply are, however, much the same throughout the area.

130

Rainfall.—The mean annual rainfall of this area is about 42 inches, but the rainfall varies by 35 to 40 per cent of this amount between extremes in different years.

Water power.—The relief of this area is considerable, the altitudes ranging from 3,764 feet above tide on Glastenbury Mountain, and 3,505 feet on Greylock Mountain, to about 300 feet above tide on Hoosic River, where it passes out of the quadrangle. Many of the smaller streams descend 1,500 to 2,000 feet in a few miles and afford excellent opportunities for the development of small water powers. The mountains are mostly forest clad, so that the run-off of rain water is somewhat checked and opportunity is afforded for its absorption into the glacial drift, and to some extent into the rocks. The drift, though patchy and irregular in thickness and distribution, covers most of the surface, except the steep slopes and the highest knobs and ridges. In a general way the drift is slightly coarser and more porous on the mountains than in the valleys. The rocks—mainly gneisses, quartzites, and schists in the mountains—are not of themselves porous enough to absorb much water, but they are cut by numerous joint cracks, and some of the rainfall finds its way into these, from which it issues in springs. The great number of perennial mountain brooks shows that the drift and the rocks are effective reservoirs for water storage.

The Hoosic, the largest river, crosses the quadrangle diagonally from southeast to northwest. It descends within the area from an altitude of about 1,100 feet above tide to 300 feet. The principal water powers on this river are at Adams, North Adams, and Hoosick Falls, though there are several others at smaller villages, such as Cheshire, Brayton-ville, Blackinton, Williamstown Station, and North Pownal. The other principal water powers in this area are on the Walloomsac at Bennington, Papermill Village, and North Hoosick; on Paran Creek at North Bennington; on Green River at Williamstown, and on the north branch of the Hoosic at Briggsville and in North Adams. Small powers are also used at Hancock and Garfield on Kinderhook Creek, at Berlin and Petersburg on the Little Hoosic, and at South Shaftsbury on Paran Creek. Besides these there are a great many smaller developments of power for private use in all parts of the area. Many farmers utilize brooks for power to run dairy and grinding machinery. The demand for power on the rivers is greater than the supply, so that even when the streams are at more than their average volume the supply is insufficient for the mills of the larger towns. In dry seasons the supply is much reduced, but does not fail altogether.

Lakes and ponds.—Natural ponds are not so numerous in this area as in some other parts of New England. There are about 30, large and small, shown on the map, and some artificial ones also. Only two or three are as long as 1 mile. The largest is Pontoosuc Lake, on the southern boundary, but this lies mostly outside of the quadrangle and its outlet is southward. The level of many ponds has been raised by dams so as to increase their capacity as reservoirs, and other reservoirs have been made by damming narrow valleys. The largest reservoir, which is just south of Cheshire, is 3 miles long and one-fourth to one-third of a mile wide.

City and domestic water supplies.—So far as known all the cities and most of the villages get their water supply from mountain brooks. Some farmers get their water from brooks, but most of them derive their supply from springs.

UNDERGROUND WATERS.

Except so far as revealed by springs, nothing is known of the distribution or quality of the underground waters of this area. The water of the mountain springs, whether issuing from drift or from the rock, is generally of excellent quality. No bored wells have yet been noted within the area and dug wells are few. Those seen were all sunk into the drift, generally to depths less than 20 feet. In the relatively low area north and west of Bennington and Hoosick Falls wells are probably more common, but most of this area remains to be investigated. In a general way the character of the rocks and the geologic structure seem to afford no certain promise of success in boring artesian or flowing wells. The artesian wells at Dalton, a few miles south of this quadrangle, appear to flow from a great fault crack which was struck accidentally by an experimental boring.

Mineral springs.—The only important mineral spring thus far noted is Sand Spring, on the north side of Broad Brook, about 1½ miles north-northwest of Williamstown station. The following information relating to this spring was obtained from Dr. S. Louis Lloyd, who owns the spring and runs a sanitarium and baths at that place. The spring boils up through white sand and flows 400 gallons a minute without noticeable variation, even during periods of extreme drought. The temperature of the water is 76° F. winter and summer. Its purity and mineral contents give it valuable medicinal properties for both drinking and bathing. A circular issued by the proprietor gives the following analyses, made by Leverett H. Mears, professor of chemistry in Williams College.

Chemical analysis of water of Sand Springs.

Parts per 100,000.

Lithium chloride	0.0353
Sodium chloride	.0768
Acid calcium carbonate	3.2249
Acid magnesium carbonate	2.6479
Calcium sulphate	.7262
Aluminum sesquioxide	.0325
Iron sesquioxide	.0075
Silica	.7026
Sodium carbonate	.4641
	7.9178

Sanitary analysis.

	Parts per 100,000.
Color, odor and sediment	None
Solids — Total	12.6200
Solids — Volatile	2.4600
Solids — Fixed	10.1600
Chlorine	.1300
Ammonia, free	.0000
Ammonia, albuminoid	.0018
Nitrogen as nitrites	.0000
Nitrogen as nitrates	.0152

Gas analysis.

Carbon dioxide	.30
Oxygen	11.34
Nitrogen	88.36

This spring issues from the drift-covered slope north of Hoosic River and stands somewhat more than 100 feet above it. It appears to lie almost directly in line with a prominent fault which runs north and south along the east side of Mason Hill in Vermont, emerging a little more than a mile south of the point where the fault passes from view under the drift. Considering its relatively high temperature, its constancy of volume and temperature, and its mineral properties, it seems probable that the water of this spring comes from the fault and hence from a deep-seated source.

There may be other mineral springs in the quadrangle, but none are known at present.

INFLUENCE OF WATER SUPPLY ON SETTLEMENT.

The presence of water power has been the main influence in locating the cities and villages of this area. Four cities—Adams, North Adams, Hoosick Falls, and Bennington—all owe their location to this cause, and the positions of most of the villages have been determined by the same influence. Farmhouses in this area are usually located with reference to convenience of access to some spring or brook. The area is comparatively poor in agriculture. Except in the northwestern part, the valleys are narrow and the hill farms are generally stony and barren. With the decline of agriculture, farmers had to find other ways of making a living. Many of them—the hill farmers especially—have met the difficulty by keeping "city boarders," and most of them have found it profitable, as is abundantly proved by the better appearance of their places. The region is justly famed as a summer and autumn resort for health and pleasure, and the prosperity of a considerable part of its population depends much upon the annual coming of visitors. Pure drinking water and the scenic charms of clear, rapid, mountain brooks constitute important factors among the many attractions that draw summer visitors to this region.

WATER RESOURCES OF THE WATKINS GLEN QUADRANGLE NEW YORK.

By Ralph S. Tarr.

LOCATION AND DRAINAGE.

The Watkins Glen quadrangle is located in south-central New York. and includes the areas mapped on the Elmira, Waverly, Watkins, and Ithaca topographic sheets of the United States Geological Survey. In it are located Elmira, Ithaca, and a number of smaller cities and towns. It is crossed by a portion of the divide that rises between the Oswego and Susquehanna drainage systems, and includes some of the headwaters of each system. Chemung River crosses the Elmira and Waverly quadrangles in the southern part of the area, carrying through Elmira the drainage of 2,055 square miles of country. Next in size to the Chemung Valley are the two northward sloping valleys occupied by Cayuga and Seneca lakes, with drainage areas of 1,593 and 707 square miles, respectively. Throughout the quadrangle one effect of glaciation has been to cause changes in divides, which have turned streams from one drainage area into another. As a result, many of the streams flow in valleys that are out of proportion to the volume of water they carry; some large valleys contain small streams, some large streams flow in small valleys. This condition has an important bearing on the distribution of underground water, whose direction of movement may consequently often be directly opposite to that of the surface water.

None of the streams of this area are of value in navigation; but the Chemung Valley was the site of a canal, now abandoned after having been operated at a loss. Both Cayuga and Seneca lakes are seats of navigation, but this shared the decline of the Erie Canal, with which the lakes are connected by short branch canals. Many hydrographic problems connected with these lakes have not yet been investigated.

The overdeepening of the two lake valleys, possibly by ice erosion, has left the tributaries hanging far above the level of the lakes. In descending to lake level the tributaries therefore have high grades, and in the short time since the Glacial epoch have been unable to do more than form steep-walled gorges through which the water flows in a series of rapids and falls. These gorges, falls, and rapids are exceedingly picturesque, and in the case of Taghanic Falls, Watkins Glen, and Havana Glen, have won wide reputations. The water power of

134

some of these streams is used for running grist and other mills; that of Fall Creek furnishes power for generating the electricity used at Cornell University and gave opportunity for the construction of the hydraulic laboratory. Through general deforesting the value of the streams for purposes of power is decreasing.

WATER SUPPLY.[a]

The surface streams are used by the towns in the area as a source of municipal water supply, in some cases, as at Ithaca and Elmira, after filtration. The need of such filtration was forcibly impressed upon the people of Ithaca by an epidemic of typhoid fever in 1903, which resulted from the pollution of creek water, then supplied unfiltered. All the surface waters are notably hard, as the following analyses show. The hardness results from the solution of lime and other salts from the shales and from the glacial deposits, which consist in large measure of ground-up shale and limestone. So much lime is at times carried in solution that springs, on emerging, precipitate it, forming deposits of calcareous tufa.

The condition of this surface water is clearly illustrated by the following analyses of two creeks whose mouths only are within the Watkins quadrangle, but whose waters have been studied with especial care, because of their relation to the Ithacan water supply.

Results of examination of water from Fall Creek at Ithaca.

[Parts per million. Prof. E. M. Chamot, analyst.]

Date.	Turbidity.	Color.	Nitrogen as—				Chlorine.	Hardness (alkalinity).	Carbon dioxide.
			Albuminoid ammonia.	Free ammonia.	Nitrites.	Nitrates.			
1903.									
February 13	35.00	....	0.071	0.081	Tr.	2.08	......	....	......
February 14	55.00	....	.071	.081	Tr.	1.82	......	....	......
February 16	33.00	....	.078	.023	Tr.	1.54	2.25	65.8	1.24
February 17	18.00	....	.094	.028	0.002	1.35	1.25	80.1	2.24
February 18	24.00	....	.082	.016	.002	1.818	1.50	80.1	1.714
February 19	22.00	....	.082	.023	.002	1.667	1.50	84.7	2.19
October:[b]									
Maximum	42.00	150.00	......	......	......	......	......	108.0	......
Minimum	.25	13.00	......	......	......	......	......	55.0	......
November:[b]									
Maximum	10.00	104.00	......	......	......	......	......	108.8	......
Minimum	.25	13.00	......	......	......	......	......	63.8	......
December:[b]									
Maximum	1.00	25.00	......	......	......	......	......	115.0	......
Minimum	.20	6.00	......	......	......	......	......	87.5	......

[a] I take pleasure in acknowledging the assistance of Mr. Lawrence Martin in obtaining data concerning the water resources; of Mr. C. C. Vermeule, engineer for the Ithaca water board, and his assistant, Mr. Getman, who supplied me with carefully kept records and samples; and of Prof. E. M. Chamot, of Cornell University, for analyses and other information about the chemical features of the problem.

[b] Daily observations.

Results of examination of water from Sixmile Creek at Ithaca.

[Prof. E. M. Chamot, analyst.]

	Raw water before filtration.										
	Bacteria per c. c.			Turbidity.			Color.			Alkalinity.		
	Aver-age.	Maxi-mum.	Mini-mum.	Aver-age.	Maxi-mum.	Mini-mum.	Aver-age.	Maxi-mum.	Mini-mum.	Aver-age.	Maxi-mum.	Mini-mum.
1903–4.												
October	7,220	36,000	850	(a)	1,800	0.25	45	128	2.5	81	134.0	47.0
November .	5,460	23,000	800	31	160	.8	63	160	18.0	82	92.5	44.0
December..	2,900	8,600	900	8	35	.75	33	85	14.0	89	97.5	40.0
January ...	6,200	520,000	1,000	13	2,500	5.0	14	40	10.0	76	100.0	42.0
February ..	13,400	97,000	1,500	46	400	6.0	11	24	4.0	72	85.0	35.0
March	28,600	115,000	2,500	220	3,000	5.0	13	24	5.0	57	87.5	41.0
April.......	7,330	50,500	1,500	104	400	40.0	11	20	5.0	54	67.5	37.5
May........	7,300	85,000	450	40	500	10.0	18	24	5.0	72	95.0	55.0
June	7,500	62,000	450	47	250	7.0	15	34	6.0	94	125.0	64.0
July........	1,050	6,500	350	57	95	28.0	9	27	3.0	127	130.0	120.0
August.....	2,400	18,000	400	76	200	45.0	20	45	1.0	127	140.0	110.0
September .	1,200	2,800	500	57	80	18.0	9	20	3.0	131	140.0	125.0
Average ...	7,550	....		58	....		21	....		91	....	

a Too variable to average.

It goes almost without saying, that springs and wells are common in this area, and that their waters resemble in hardness the creek water at low stage. Many of the wells, especially on the hill farms, are in rock; others are in glacial drift; and some are in post-Glacial deposits. Most of the shallow wells show decided fluctuation with season. The most copious springs occur in the regions of outwash gravels and near the margins of alluvial fans, into which the stream water often sinks, reappearing as springs.

In a number of places in the area there are deep wells in which the water rises nearly or quite to the surface. This is true, for example, just south of Watkins, at Montour Falls, near Breesport, and in the outwash gravels in and near Elmira. In only one place, however, have the water resources of the glacial deposits been extensively exploited, namely, at Ithaca. Prior to the introduction of a city water supply the town of Ithaca depended upon wells, and even at the present time many of these are in use. Some of these are shallow dug wells that receive the drainage of a densely populated hill slope. Others are in gravels of alluvial fans, built where Sixmile and Cascadilla creeks emerge from their gorges upon the Ithaca delta plain. Many of the Ithaca wells are, however, driven, and furnish artesian water. The depth which these wells reach varies, being in most cases from 80 to 100 feet, and in some of the wells the water has head enough to permit piping to the second floor. They are both overlain and underlain by an extensive stratum of clay. In the table following the first seven analyses show the general character of this water.

Sanitary analyses of well waters at Ithaca.   Prof. E. M. Chamot, analyst.

[In parts per million.]

No.	Location of well	Date	Depth of well.	Turbidity.	Color.	Odor.	Total solids.	Loss on ignition.	Total hardness.	Alkalinity.	Chlorine.	Free ammonia.	Albuminoid ammonia.
		1903.	Feet.										
1	Tompkins street	June 13	100							280.0	27.0		
2	Utica street	Mar. 19	80							197.5	32.6		
3	Wood street	July 7	120							140.0	59.0		
4	Madison street	Apr. 3	85							135.0	14.8		
5	Plain (South) street	July 7	96							100.5	16.0		
6	Adams street	June 15	100							165.0	39.5		
7	Albany (South) street	July 7	104							155.0	60.0		
8	Salt works well	Feb. 17	120	None.	None.	H_2S	314.5	109.5	176.5	145.8	38.5	0.083	0.013
9	Illston well	Feb. 7	280	None.	Slight.	H_2S			200.0	197.5	45.0	.565	None.
		1899.											
10	Do	Apr. 10	280	None.	Slight.	H_2S	304.0	78.0			61.6	.480	.005

Sanitary analyses of well waters at Ithaca—Continued.

No.	Location of well.	Date.	Nitrogen as nitrites.	Nitrogen as nitrates.	Oxygen consumed.	Bacterial colonies.	Growth in phenol-peptone-glucose.	Behavior in Dunham solution.	Remarks.
		1903.							
1	Tompkins street	June 13	0.0004	None.	1.375	1,800	Coli-like; gas	Uniformly turbid; strong indol.	
2	Utica street	Mar. 19	.005	0.45	1.40	1,440	Slight; gas	Coli-like; very foul; strong indol.	Sunk in 1886. Flows in cellar.
3	Wood street	July 7	None.	None.	.425	12	No growth	No growth	CO_2=5.0. Flows.
4	Madison street	Apr. 3	None.	Trace.	1.30	70	Weak; no gas	Slight turbidity; no indol.	CO_2=5.0. Flows. Sunk about 1878.
5	Plain (South) street	July 7	None.	None.	.675	430	Vigorous; gas	Turbid; indol	Driven.
6	Adams street	June 15	None.	Trace.	.50	10	No growth	No growth	Flows.
7	Albany (South) street	July 7	None.	None.	.525	46	None	...do	CO_2=4.5. Flows in cellar.
8	Salt works well	Feb. 17	None.	None.	.450	25			CO_2=6.
9	Illston well	Feb. 7	None.	Trace.	.450				CO_2=3.309.
		1899.							
10	Do	Apr. 10	None.	Trace.	.644	10	No growth	Very slight turbidity.	

Toward the western side of the valley a well, 120 feet deep, known as the "Fourth street well," was drilled by a committee of citizens during the typhoid epidemic. The flow of water from it was estimated to be about 100,000 gallons a day. Near it is a well to the same water-bearing bed, which is pumped by a salt company for use in dissolving salt to form brine. It yields from 300,000 to 400,000 gallons a day by pumping. Its composition is shown by analysis No. 8, in the accompanying table. Just south of the city, near the western side of the delta plain, 2 deep wells have been put down by a committee of citizens, and 11 by the Ithaca water board. These 13 wells are not far apart, and most of them are very near an artesian well, known as the "Illston well," that was bored several years ago. This well showed a flow of over 300,000 gallons a day, by weir measurement, in the spring of 1903. Two of the wells reached rock, one at a depth of 260 feet, the other at a depth of 342 feet. Of the 13 wells, 9 are flowing, and the engineer of the water board estimates that they have a total capacity, when pumped, of between 2,000,000 and 3,000,000 gallons daily, which is more than ample for the city of Ithaca. A number of wells are clustered around the Illston well, and reach the same water-bearing series at a depth of about 280 feet. That the various wells tap the same water-bearing layers is evident from their influence upon one another. The pressure in the Illston well, for example, is reduced from 18 to 12 pounds when all the adjacent wells are flowing. Whether the estimated capacity of the wells will be actually reached when all the wells have been pumped for a while remains to be seen.

These 14 wells reveal a set of conditions in the Ithaca delta as follows:[a] Below an upper series of beds of soil and muck about 10 feet thick is a bed of clay varying in thickness from place to place and underlain at depths ranging from 60 to 90 feet by water-bearing sands and gravels that vary in thickness from 30 to 60 feet. Beneath the sand and gravel is a great depth of uniform clay, attaining a thickness of 120 to 150 feet. Then sandy clays, quicksands, sands, and gravels are encountered. Even in wells drilled but a few feet apart the sections through the water-bearing layers differ, as one would expect in glacial sands and gravels. This variation makes well boring very uncertain. Several of the wells reached a quicksand so filled with water and under such pressure that it rose in the pipe faster than it could be removed. The more successful wells reached either gravel, or sand with gravel, from which the sand could be jettied out, leaving a protecting mass of pebbles around the pipe mouth. Amid such varying conditions the sinking of a well in this region is very much of a lottery, even though it be true, as it seems to be, that the sand-

---

[a] Tarr, R. S., Artesian well sections at Ithaca, N. Y.: Jour. Geol., vol. 12, 1904, pp. 69–82.

gravel series beneath the clay is very extensive and everywhere water bearing.

The beds in which the wells are located are a part of the series of ice-front deposits associated with the stand of the glacier in this region. The source of the water is unquestionably the moraine series that occupies the bottom of the valley of the Cayuga Inlet, farther south. This being the case, the water supply must be great. That the water has long been underground is indicated by the analyses (Nos. 9 and 10 in the table)[a] and by its temperature, which in both August and December was 52° F. The Ithaca experiment will be watched with interest, for if only two-thirds of the estimated capacity of the wells is realized it will show a set of conditions in glacial deposits favoring, in a very limited region, the development of an extensive water supply.

---

[a] The analyses of the water of the Illston well doubtless represent approximately the condition of the water in the neighboring wells of which analyses are not yet available.

WATER RESOURCES OF CENTRAL AND SOUTHWEST-
ERN HIGHLANDS OF NEW JERSEY.

By Laurence La Forge.

INTRODUCTION.

Location.—This region, that part of the Highlands south of Andover
and Pompton, about two-thirds of the Highland area of the State,
includes portions of the Morristown, Lake Hopatcong, Hackettstown,
Delaware Water Gap, Somerville, High Bridge, and Easton quadrangles, and comprises parts of Morris, Warren, Hunterdon, Somerset,
and Sussex counties, covering an area of about 685 square miles. The
limestone valleys lying within the Highlands are usually considered
a part of the area and will be so regarded in this paper, together with
the partially inclosed valleys, such as the Pohatcong and Pequest valleys, which extend between the Highland ridges for some distance, but
open out into the Great Valley.

Population.—The population of the area is a little over 75,000.
This includes the entire population of such places as Morristown,
Boonton, and Bernardsville, which lie partly within the Highlands,
but not the population of places like Clinton, Pompton, or Morris
Plains, which are just outside the base of the Highlands, nor of places
like Philipsburg and Belvidere, which are more properly comprised
in the Kittatinny Valley. Of these 75,000 people, somewhat over
two-thirds live in towns and villages, the strictly rural population
averaging less than 35 per square mile. There are 20 towns and villages having 500 or more inhabitants, and about 50 more having from
100 to 500 inhabitants. The rate of increase for the area as a whole
is about 1 per cent a year, but the actual increase is confined to the
larger manufacturing towns in the valleys and to the parts of the
southeastern border which are becoming popular suburbs of New
York. In all of the upland portion and throughout the larger part
of the valleys of the northern and central parts of the area the population is decreasing, there being a steady drift away from these locali-

141

ties. This, however, when the whole area is considered, is slightly more than offset by the influx from outside to the manufacturing towns and popular residence localities.

Industries.—The agricultural industry of the region is not so important as formerly, being now largely confined to the valleys and the peach-growing districts on the less rugged uplands of the southern portion. Lumbering is conducted on a small scale. Iron mining, formerly carried on in a small way at many places in the area, has become concentrated at a few important points and the mines elsewhere have been abandoned. There is a considerable and growing amount of manufacturing in the larger towns, mainly connected with the iron industry. The eastern slope of the Highlands, notably about Morristown and Bernardsville, has become a popular suburban region and is being rather thickly settled, while certain of the wilder and more beautiful of the upland districts, such as those on Mine and Schooley mountains and about Budd, Cranberry, and Hopatcong lakes, are favorite outing places, and of late years have a considerable summer population.

Climate, soil, etc.—The average temperature of the Highland region for the year is about 47°, the average for the winter being about 27 , and the ground and streams are usually frozen during the greater part of the winter. The average annual rainfall ranges from 38 to 48 inches in various parts of the region, and from 30 to 55 inches in different years. Usually there is somewhat more precipitation in summer than in winter. The average annual snowfall is about 5 feet, which is equivalent to about 6 inches of rain, so that about 12 to 15 per cent of the precipitation falls as snow.

North of the terminal moraine the soil is largely glacial material, entirely so along the moraine and in the filled valleys. South of the moraine the bulk of the soil in the valleys is also of glacial derivation, consisting of outwash plains in front of the moraine and of stream-washed drift lower down the valleys, but there are also areas of clay soil resulting from the weathering of the limestone and shale. On the uplands the soil is derived from the disintegration of the gneiss and is as a rule stony and not very fertile.

Forests cover a large proportion of the area, a proportion ranging from 90 per cent and more at the north and northeast to 30 per cent or less at the south and southeast. A greater percentage of the upland than of the valleys is forested, and a greater proportion of the glaciated area than of that lying outside the moraine. Except in parts of the limestone valleys and the plains of washed drift the soil is not very fertile, and the region, on this account and because of the ruggedness of a great part of it, is not well adapted for cultivation. The forest-covered area is increasing, as much of the land that was formerly cultivated is being abandoned.

TOPOGRAPHY.

The Highland region is a dissected plateau, lying about 500 feet above the lower country on either side, and bounded on the southeast by the Piedmont Plain and on the northwest by the Great Appalachian Valley, here known as the Kittatinny Valley. Northeastward the Highlands continue into New York, and to the southwest they extend past the Delaware for some distance into Pennsylvania.

The belt trends northeast to southwest and consists of several roughly parallel ridges and valleys which have nearly the same trend, but are not without local irregularities. A considerable portion of the original plateau surface remains as the summits of the broader ridges, reaching elevations of 1,200 to 1,300 feet in the northwestern part and declining gently both southeastward and southwestward. The highest point in the area, standing at an altitude of 1,333 feet above sea level, is in the town of Jefferson, northeast of Lake Hopatcong. The descent of about 500 feet from the plateau to the lower regions on each side is rather abrupt, and the longitudinal valleys within the highlands have been cut down about the same amount toward their lower ends.

DRAINAGE.

Lakes and ponds.—As there is abundant rainfall in the region and much comparatively level upland surface, with deep soil on most of the unglaciated portion, and natural storage basins everywhere in the glaciated portion, the entire area is well watered. On the slopes there are abundant springs, none very large, but many which never fail, and which can be depended on as sources of supply. Since much of the area is forested and other conditions are favorable to the absorption of water by the rocks, a large proportion of the rainfall becomes ground water, and a large part of the remainder is stored for a time in natural or artificial reservoirs.

Lakes and ponds are abundant in the glaciated area, the principal ones being enumerated in the following list. Of these several are entirely artificial, and nearly all the rest have been enlarged by raising the outlet level. Five are supply reservoirs for the Morris Canal and most of the others are or have been mill or forge ponds.

Name.	Area of drainage basin.		Eleva-tion.	Drainage system.
	Acres.	Square miles.	Feet.	
Lake Hopatcong...................	2,443	25.4	928	Musconetcong.[a]
Budd Lake........................	475	4.5	933	Raritan.
Green Pond (Morris County)......	460	1.7	1,045	Rockaway.
Stanhope reservoir................	339	4.9	859	Musconetcong.
Split Rock Pond..................	315	5.3	815	Rockaway.
Denmark Pond....................	172	4.5	818	Do.
Cranberry Lake	154	3.0	771	Musconetcong.
Green Pond (Warren County).....	117	5.2	399	Pequest.
Stickle Pond	110	1.7	783	Pequanac.
Forge Pond	96	10.1	775	Rockaway.
Shongum Pond....................	70	2.9	698	Do.
Waterloo Pond	68	(?)	640	Musconetcong.
Allamuchy Pond	56	1.8	775	Pequest.
Durham Pond.....................	47	(?)	880	Rockaway.
Panther Pond	41	0.5	766	Pequest.
Bear ponds.......................	38	0.6	977	Musconetcong.
Dixon Pond	35	3.5	560	Rockaway.
Wright Pond	31	3.4	743	Musconetcong.
Stag Pond	23	0.3	820	Do.

a Lake Hopatcong is naturally drained down Musconetcong River, but it supplies water for the summit level of the Morris Canal, so that a considerable part of its drainage actually passes down the Rockaway.

In addition to the lakes and ponds named above, there are about 100 small mill ponds and reservoirs, all artificial and mostly outside the moraine, and about 30 glacial kettle holes, which contain ponds each several acres in extent.

Swamps are numerous within or immediately in front of the glacial boundary, and largely increase the amount of storage on some of the streams. The largest, but also the lowest in elevation, are the Great Meadows of the Pequest and the long swamp on the upper course of the Black River. The rugged forest-covered area in Sussex and northern Morris counties contains scores of small mountain swamps. caused by glacial choking of old drainage channels, many of which lie at the very heads of the streams, at elevations of 1,000 feet or more.

Streams.—The region is drained partly by streams flowing along longitudinal valleys to the Delaware, and partly by streams flowing in both longitudinal and transverse valleys and reaching New York Bay. The principal rivers are the Pequest, Musconetcong. Raritan. and Passaic, the two last named being the largest and most important streams of the State. The entire region, except areas traversed by Pohatcong Creek and some minor streams that flow directly into the Delaware, is drained by these four rivers.

Pequest River is not properly a highland stream, its source and most of its headwaters being in the Kittatinny Valley, but in the lower part of its course it flows in a valley which separates Jenny Jump and Mohepinoke Mountains from the rest of the Highlands, and throughout its length it receives a part of the Highland drainage. At the head of Great Meadows, where it enters the Highlands, it is at an elevation of 532 feet. From here to Townsbury, where it escapes from the terminal moraine, it falls but 22 feet in 10 miles. This part of its course is through the bed of a temporary lake which was formed behind the moraine after the retreat of the ice, and until a few years ago, when extensive drainage works were completed, the meadows were an impassable morass. From Townsbury to Buttzville, where the river emerges again from the Highlands, it falls 130 feet in 6 miles and considerable water power is developed along this stretch.

Pohatcong Creek rises near Mount Bethel at an elevation of 1,000 feet and falls 400 feet in 5 miles to the valley at Karrville, along which it flows for 18 miles farther, reaching the Delaware at 135 feet elevation. Small water powers are developed at several points.

The ultimate source of the Musconetcong is in the town of Sparta, north of the area under consideration, at an elevation of 1,200 feet. It flows through Lake Hopatcong at 928 feet, Stanhope reservoir at 859 feet, and Waterloo Pond at 640 feet, emerging from the moraine at Hackettstown, 12 miles from its source, at 560 feet. From here to the Delaware, 31 miles below, it falls 430 feet. Lubber Run, its principal tributary, falls 450 feet in 10 miles from its source to where it joins the main stream near Waterloo Pond. Abundant water power is developed at a number of points on both streams.

Raritan River, the largest in the State, drains the central and southern parts of the area through three principal tributaries—the North and South branches and Black River—and empties into Raritan Bay, an arm of New York Bay. South Branch, the principal stream, rises in Budd Lake at 933 feet and falls 270 feet in 3 miles to Bartley, where it reaches the low-lying German Valley, along which it turns and runs 11 miles to Califon, falling 200 feet on the way. Here it plunges through a narrow gorge 6 miles long and emerges from the Highlands at High Bridge at an elevation of 220 feet. It carries a large volume of water and furnishes abundant power at many places. North Branch is not so large; its ultimate source is near Calais at an elevation of 1,000 feet. It falls 600 feet in the first 6 miles, but is here hardly more than a brook. From Ralston it cuts through Mine Mountain in a gorge of uncommonly picturesque beauty and emerges upon the lower plateau at only 180 feet. Though of small volume, it has a swift current and is utilized for power at several points. Black River, the third branch, rises in the moraine a little east of the southern end of Lake Hopatcong, crosses the Succasunna Plains, falling only 200 feet in 11 miles, and then drops 440 feet through a gorge 6 miles long in the Eastern Range, leaving the Highlands at Pottersville Falls at 240 feet.

The remainder of the area is drained by Passaic River, which empties into Newark Bay, another arm of New York Bay. The main stream is of little importance in its Highland portion. It falls only 289 feet in 5 miles from its source near Mendham to where it leaves the gorge at 280 feet elevation. The Whippany, another small branch, rises near Mount Freedom and falls 500 feet in 3 miles, but is here a mere brook. For the next 3 miles along Washington Valley to Morristown it is sluggish and flows through swamps most of the way. It leaves the Highlands at 310 feet and lower down becomes an important power stream. The Pequanac, another tributary of the Passaic lying altogether outside this area, drains a part of the northeastern corner.

The largest branch of the Passaic, and the most important in this area, is the Rockaway. This river rises high on the plateau to the north of the area, enters it in the Longwood Valley at about 750 feet, flows southwestward nearly to Wharton, where, at 660 feet, it turns eastward and cuts obliquely through the Eastern Range, emerging below Boonton at 240 feet. In the last 1½ miles of its gorge it falls 250 feet, furnishing one of the fine water powers of the State. It has a large flow and is an important stream both for power and for water supply.

GEOLOGY.

The rocks of the Highlands proper are granitic and hornblendic gneisses, hornblende and biotite-schists with lenses of magnetite, highly metamorphosed limestone and serpentine, and narrow bands of conglomerate and quartzite, also considerably metamorphosed. The origin of the gneisses and schists is uncertain, but a part at least were certainly igneous, and some of the remainder were probably of sedimentary origin. Throughout much of the area they are arranged in bands differing more or less in character, and the strike of these, as well as of the lenses of serpentine and iron ore and of the strips of sedimentary rocks, usually differs but little from the northeast-southwest trend of the ridges and valleys.

These rocks are for the most part very resistant to weathering, and give a bold, rugged character to the topography occupying the higher portions of the area, but the longitudinal valleys throughout most of their length are cut in much less resistant limestones and shales. These valley rocks and the strips of sediments on the ridges are of Cambrian, Ordovician, and Silurian age, while the gneisses and schists are pre-Cambrian. The age of the metamorphic white limestones is not yet settled. There are a few dikes of diabase and amphibolite in certain areas, but as a rule intrusive rocks are wanting.

The terminal moraine of the Wisconsin ice sheet crosses the area from Denville to Buttzville, and in the northern third of the region considerable areas of rock surface are found on the ridges, while the valleys are filled with sand and gravel to a depth of 200 feet in places. Glacial modifications of drainage are numerous, and in this part of the area a number of swamps drain in several directions.

WATER RESOURCES.

Importance of region.—This region, together with its northeastward extension into New York, is perhaps the most important in the United States from the point of view of water supply, because of its proximity to the great metropolitan area about New York City and its natural advantages as a collecting ground. The population of the portion of New Jersey east of the Highlands and north of Somerville and New Brunswick is over 1,100,000, and of that part of New York south of the Highlands and exclusive of Long Island, nearly 2,300,000. While the cities and towns of a considerable part of this area are still supplied with water from local sources, much of the region is rapidly becoming so thickly settled that in a short time it will be impracticable, except in a few limited cases, to depend longer on the local supply, both because of its inadequacy and of the increasing danger of pollution. The need of further supply for New York City is again becoming pressing, and because of the limited area of the Highlands within the State of New York, from which area the present supply is drawn, it may be necessary to go far up the Hudson or outside the State limits into Connecticut or New Jersey for additional water. Besides the population already mentioned, that of Brooklyn and Queens boroughs and Nassau County on Long Island amounts to 1,300,000. A considerable part, possibly all, of the supply for this territory can probably always be obtained from the deep water-bearing gravels of Long Island, but in the future legal complications may interfere with the obtainment from this source of the quantity needed, and at least a part of the supply may have to be procured from the mainland.

Natural advantages.—The Highlands lie in immediate proximity to this belt of dense population, forming its natural boundary on the northwest. The rainfall is abundant, the region is thinly settled and largely forested, and most of it is poorly adapted to the purposes of agriculture. There are many and capacious natural storage basins, and the whole district is so elevated that water may be delivered by gravity and under considerable head to the more densely peopled district. There is but little local contamination, and the region is so poorly adapted to ordinary uses that the sequestration of large parts of it as gathering and storage grounds for metropolitan water supplies is practicable at relatively small expense. These advantages have long been recognized, and the Highlands have repeatedly been pointed out as the future source of water supply for the cities of northern New Jersey, and the conservation of the Highland waters for that purpose has been often urged.

MINERAL SPRINGS.

There is but one mineral spring of note in the area, the formerly well-known Schooley Mountain Spring, located on the western slope of Schooley Mountain, about 2 miles south of Hackettstown. The result of an analysis of its water as given by Peale[a] is shown below:

Analysis of water of Schooley Mountain Spring.

	Parts per 1,000,000.
Calcium sulphate	28.7
Calcium carbonate	24.3
Magnesium carbonate	27.4
Iron carbonate	9.9
Sodium carbonate	9.9
Sodium chloride	7.4
Silica	12.7
Alumina	2.4
Ammonia	Trace.
Carbon dioxide (dissolved gas)	Undet.

This spring was formerly a popular resort; buildings were erected about it and the water was sold to visitors, but it is now abandoned and the water is no longer used.

WATER POWER.

The climatic and topographic conditions being favorable, the amount of available water power along the streams of the Highland region is considerable and it has been largely utilized. Of late years a notable change has been made in the use of this power, as the number of plants making use of water power has diminished, but the total amount utilized remains about the same. This is due to the fact that whereas there were formerly numerous small sawmills, flour mills, iron forges, and the like, each using a little power, many of these have been abandoned, but there are in place of them several large establishments which have been erected at favorable points and which use a considerable amount of power.

From a census of water powers of the State,[b] made by the geological survey of New Jersey in 1890, it appears that in the region discussed in this paper there was a total of about 11,500 horsepower used, 3,000 horsepower being used by mills on the Musconetcong and branches, 2,200 on Rockaway River and branches, and 4,500 on the two branches of the Raritan and their tributaries, the remainder being developed along various small streams, the Pequest standing at the head of these.

---

[a] Peale, A. C., Mineral springs of the United States. Bull. U. S. Geol. Survey No. 32, p. 43.
[b] Vermeule, C. C., Final Rept. Geol. Survey New Jersey, vol. 3; appendix 1, water powers.

The largest development is in the gorge at Boonton, where there is a fall of 250 feet from the pond at Powerville to Old Boonton, with an estimated available horsepower of over 2,000, of which about 1,600 was utilized. At Highbridge, on the Raritan, about 1,200 horsepower is developed, and considerable amounts are used at Hughesville and Riegelsville, on the lower Musconetcong.

Though a much larger amount of power than has yet been utilized could be made available on these streams by proper conservation of the water in ponds, and the amount of manufacturing thus considerably increased, the fact must be recognized that the uses of water for power and for city supply conflict, and that one must necessarily largely preclude the other. If, as seems highly probable, the larger part of the stream and pond water in this region will before many years have passed be sequestrated for municipal supplies, the developed water powers of the Highland region are not likely to be further increased, but will probably be considerably diminished, at least on the Rockaway and Raritan and their branches.

THE MORRIS CANAL.

One of the largest users of water in the Highlands at the present time, and one which controls the water rights of several of the largest and most important storage reservoirs, is the Morris Canal, which traverses the region from Philipsburg to Boonton, passing through Washington and Hackettstown, crossing the low divide near Lake Hopatcong, thence, via Dover and Rockaway, issuing from the Highlands through the gorge at Boonton. It rises from 156 feet at Delaware River to 913 feet at the summit level, and then descends to 288 feet on the level below Boonton and to tide water at Jersey City. At Powerville and at Dover the boats are locked into and out of the Rockaway; therefore the whole flow of the stream is available at these points if needed for the purposes of the canal. Again, along the Musconetcong, from Stanhope to Saxton Falls, the waters of the canal and river mingle at several points.

Lake Hopatcong, Cranberry and Stanhope reservoirs, and Bear and Waterloo ponds are controlled by the canal corporation, and the water of these bodies is used for the canal supply. All except Lake Hopatcong supply the western slope of the canal, and the water thus reaches the Delaware. The Hopatcong feeder enters the canal at the summit level and furnishes the whole supply for the part of the eastern slope within the Highlands, except the water taken from Rockaway River. Nearly all the water drawn from Lake Hopatcong, therefore, passes down the eastern slope and reaches New York Bay, instead of following the original course of drainage from the lake to the Delaware.

Gagings have shown [a] that the canal appears to use very nearly the

a Vermeule, C. C., Final Rept. Geol. Survey New Jersey, vol. 3, Appendix 1, Water Power, p. 196.

whole supply which can be obtained from the reservoirs in their present condition—in fact, the whole available supply in a dry season—as well as an unknown amount from the streams at the points where the waters are mingled; but by raising the dams and increasing the storage the reservoirs could be made to supply all the water needed by the canal in the driest seasons, as the rainfall on the catchment basin is ample when carried over by storage. It would seem, then, that the continued use of the canal would preclude the collecting of any surplus water from the upper Musconetcong and Lubber Run watersheds. Nearly all the water used by the canal, however, goes to make good the leakage, which is large on account of the hillside location of long stretches of the canal and the nature of the soil. From measurements made by Messrs. J. J. R. Croes and G. W. Howell this leakage has been estimated at 1.74 cubic feet per second per mile, and nearly the whole of this is returned to the streams, hence it again becomes available for storage in reservoirs on the lower courses of the streams and is not entirely lost as a source of municipal supply.

The use of water for the ordinary purposes of the canal impairs to a very considerable extent its quality for domestic water supply, though, indeed, the water of the canal is used to some extent for that purpose in some of the towns along its course. Unfortunately, in one or two of the larger towns the canal receives the discharge of sewers, and throughout its length it is the receptacle of more or less refuse of all sorts, so that the use of its water for household purposes ought to be discontinued. The bulk of the leakage occurs by percolation through the soil, and hence the escaped water is to some extent purified by the filtration it undergoes in the process; but much of it escapes as overflow at the spillways, and the filtration of much of the rest is probably imperfect, as it is likely that long-continued leakage along certain lines has established little channels in the soil through which the water percolates with little or no filtration.

The water, then, of valley streams which derive a large part of their flow from the leakage from the canal is still of doubtful purity for domestic use, but, in view of the comparatively small pollution which the canal water must suffer and the large dilution which the escaped water undergoes by mingling with stream water derived from other sources, as well as the purification which would be brought about by its storage in large reservoirs on the lower courses of the streams, it is probable that the escaped water is not unfit, after such dilution and purification, to be stored for a metropolitan supply, especially if the reservoirs are large enough and the dilution is ample. It should be noted in this connection, however, that for some time there has been a movement to bring about the abandonment of the canal. Should this be accomplished, it would set free for other uses the large quantities of water at present required for the purposes of the canal, and would remove that particular source of pollution.

Requirements.—There are about 20 towns and villages in or upon the immediate border of the part of the Highlands discussed in this paper that are large enough to need water systems, and 10 of them have such systems. With the exception of Morristown and Dover these places are all of less than 4,000 population, and the amount of water needed for their supply is comparatively small. The total population of those towns that have a water supply is about 32,000, and of those towns that are likely to need such a supply within the next ten years is about 8,000, so that the total number of people for whom local supplies of any magnitude must be reserved is not great.

Present sources.—As will be seen from the accompanying table, those towns that have public water-supply systems procure water from mountain springs, except Dover and Morristown, which procure a part of their supplies from drilled wells. It will also be noticed that municipal ownership is rare, only Dover, Hackettstown, and High Bridge owning their systems. Most of the systems are owned by companies, and in one or two cases the ownership is private and only part of the village is supplied by the system. The present sources of supply are ample for all present needs as well as for the near future, while in the cases of all but the largest towns it is probable that an ample supply for all future needs can always be obtained from sources now available.

The data incorporated in the table were procured by correspondence with officers of the various towns or of the water companies. No report was received from Gladstone, and the data given are derived from the best information available for that village.

List of municipal water systems.

Town.	Population in 1900.	Source of supply.	Kind of system.	Ownership.	Sufficiency of supply.	Quality.
Morristown	11,267	Springs and drilled wells	Gravity and pumping.	Private	Ample	Good.
Dover	5,986	.....do		Town	....do	Do.
Boonton	3,901	Springs	Gravity	Company.	Not reported.	Not reported.
Washington	3,580	Springs and mountain stream.	.....do	....do	Fair	Good.
Hackettstown	2,474	Mountain stream	.....do	Town	Abundant.	Very good.
High Bridge	1,377	Spring	.....do	....do	Ample.	Good.
Bernardsville	1,300±	Springs	.....do	Company.	....do	Do.
Netcong	941	Brook	Not reported.	Private	Not reported.	Not reported.
Clinton	816	Springs	Gravity.	....do	Plenty.	Good.
Gladstone	300±	Spring	.....do	....do	(?)	Slightly hard.

The water, being taken directly from mountain springs or from drilled wells in favorable locations, is in all cases pure and wholesome, and there is slight danger of contamination. Morristown and Hackettstown report analyses that show water of exceptional purity. For all except the larger towns the springs furnish an ample supply, and hence drilled wells have been resorted to only where the population is so large that the supply from springs has been insufficient.

SURPLUS SUPPLIES.

Since the local population that requires a public water supply is not large, and since the supply is in all cases drawn from wells or springs and not from streams or large ponds, the effects which the needs of this population will have on the available supply for metropolitan purposes is negligible. On the other hand, since for most of those towns an ample supply can in the future be obtained from the present sources, their available supply is not likely to be affected injuriously by the utilization of the lake and stream water for the needs of the large cities. Many of the towns along the eastern bases of the Highlands seem destined to have a large increase of population in the future as metropolitan suburbs, and provision must be made for a greater supply for these, but they are fortunate in being located near small streams that flow from the hills, which are of sufficient volume to supply all their needs, but which run at too low an elevation or furnish too small a quantity of water to be used to advantage for the supply of cities at a distance. The growing manufacturing centers at Washington and about Dover and Wharton may also require larger amounts of water in the future than can be obtained from springs and drilled wells, but the needs of these places may be met without seriously diminishing the supply needed for city purposes. Washington especially, being located on Pohatcong Creek, which flows to the Delaware and is not so situated as to be used to advantage by the cities lying east of the Highlands, will probably always have a sufficient supply for local needs. Shongum Pond and Den Brook may easily be made available for a supply for Dover, and might well be reserved for that purpose.

Needs of the metropolitan area.—The need of water for the metropolitan area has already been briefly mentioned. Leaving out of consideration the 3,400,000 inhabitants of the New York part of the metropolitan district, there are over 1,100,000 in the New Jersey portion—about 60 per cent of the population of the State. The population of this congested area is increasing at the rate of 40 per cent in a decade, and should the increase continue at the same rate in fifty years the population of the New Jersey portion of the area alone will be a little less than 6,000,000. To keep pace with this increase there must necessarily be a gradual abandonment of the local sources of water supply and a turning to more distant regions where the supply will be both larger and in less danger of contamination. Newark has already taken this step, and Jersey City is at present providing for a supply from the Rockaway, at Old Boonton. Other large cities of the district must soon take similar measures.

It has long been known that there is no better region to which to turn for this purpose than the Highlands, where, not more than 25 miles from any of the cities of the district, a pure and abundant supply of water is available. Some of the cities of the metropolitan area are at present consuming about 95 gallons of water a day per capita, and though this amount seems excessive and implies waste of the water, still, to provide for possible contingencies, a source of supply which will be capable of furnishing 100 gallons a day per capita if necessary should be secured. On this basis, if the population of the New Jersey portion of the metropolitan district continues to increase at the present rate, in fifty years the inhabitants will require a supply of not less than 500,000,000 gallons a day.

Amount of supply.—According to the estimates of C. C. Vermeule,[a] the amount of water which can be collected on each of the watersheds of this region with proper storage is as follows:

Rockaway River (above Boonton), 78 million gallons a day; Raritan River (Highland branches), 92 millions; Musconetcong (above Hampton), 82 millions; a total of 252 million gallons a day. An equal or greater amount could be collected from the Ramapo, Wanaque, and Pequannock rivers in the northeastern Highlands, so that the amount of water in the Highlands which could be used for metropolitan supply is upwards of 500 million gallons a day, an amount very much in excess of the present needs of the New Jersey portion of the metropolitan area, and sufficient to meet the wants of this area for a long time. This estimate takes no account of the waters of the Highland portions of the Passaic and Whippany, which are small in amount and issue from the hills at so low an elevation as not to be advantageously used to supply places at a distance; nor of the Pohatcong and Pequest, which flow to the Delaware, and the waters of which could not be carried to the eastern side of the Highlands without considerable expense. Furthermore, these last streams drain limestone areas, and their waters are not so desirable for use as those of the more strictly Highland streams.

The water of the four important Highland streams which flow eastward, the Rockaway and Black rivers and the North and South branches of the Raritan, can be collected in storage basins at or near the points where they issue from the Highlands and delivered by gravity, making them easily available, and at relatively small expense.

---

[a] Vermeule, C. C., Water supply: Geol. Survey of New Jersey, vol. 3, pp. 145, et seq. The writer desires here to acknowledge his indebtedness to this book, which is a mine of valuable information regarding the water resources of New Jersey and has been used freely in the preparation of this paper.

The North Branch of Rockaway Creek, in Hunterdon County, could also be used to advantage by the construction of a reservoir near Mountainville, and, it is estimated, would furnish more than 8 million gallons a day. The waters of the Musconetcong could be made available by the construction of a storage basin at Hampton and a tunnel through the hill to Glen Gardner, and thence to the South Branch of the Raritan; or a part of the Musconetcong could be utilized by the construction of a reservoir in the upper valley and a tunnel through Schooley Mountain. Furthermore, Lake Hopatcong, which drains naturally to the Musconetcong, lies so close to the divide that its waters could be diverted to the Rockaway and thus about 17 million gallons a day could be added to the flow of that stream.

There is a variation of over 50 per cent between the least and greatest rainfall in the Highland region, and thus a very great variation in the amount of water collectible in dry and wet years, but the estimates of Mr. Vermeule are conservative and are based on the computed flow for the driest eighteen consecutive months on record, and the amount of water given as available for the different streams may be relied upon under all conditions.

Character of the water.—The water of the Highlands in its natural state is of the very best quality, and a more satisfactory supply from this point of view could not be desired. Under present conditions. however, the water of the larger streams and of one or two of the lakes suffers contamination from several sources, and this must be guarded against before the water will be entirely fit for use. The Rockaway receives, above the reservoir now being constructed for the Jersey City supply, the sewage of Boonton, Rockaway, Dover, and Wharton. with an aggregate population of 14,000. This sewage, however, is all to be diverted before the reservoir is put into service, and the topography of the region is such that this can be done with little trouble. Hackettstown, with 2,500 inhabitants, is situated on the middle course of the Musconetcong, and there are increasing summer colonies on the shores of Hopatcong and Budd lakes, while there are still one or two important iron-mining localities in the region, besides a large blast furnace at Wharton. All these are to a considerable extent sources of contamination, but in nearly all cases the drainage can be conducted out to points below the sites where storage basins could be most advantageously constructed.

Though the village population is rather large in some other parts of the region, as along the stretch of German Valley, where the villages from Kenville to Califon have an aggregate population of over 3,000. it is as a whole not increasing, and the rural population nearly everywhere is decreasing, so that the danger of contamination other than from the large towns and manufacturing centers seems likely to decrease rather than increase in the future.

Availability.—Here, in immediate proximity to a large and rapidly growing urban district, for which there will be need in the not very distant future of an abundant water supply, is a region which is especially fitted by its natural advantages to be the gathering ground for that water supply—a very nearly ideal gathering ground. It seems certain, if the growth of the metropolitan population continues at its present rate, that in a few decades it will be necessary to utilize very nearly all the available water obtainable in the Highlands for the needs of that population and to conserve the supply by the construction of storage reservoirs. For this purpose it will therefore be necessary to set apart considerable portions of the region, both to provide room for the storage basins and to prevent contamination within their catchment areas.

This seems inevitable, and, like all work of the kind, it can be done at less expense and at greater advantage now than later, and, as a further motive for early action, it should be noted that the present need is great. When there are also taken into consideration the manifest adaptability of this region for use as a State park and the desirability of preserving its natural beauty for the enjoyment of the people of the densely settled area to the east, together with the fact that this use of the region would not conflict with its use as a source of water supply, the argument for the sequestration of a large part of the Highland area by State action, with the conservation of its natural beauty as well as its waters for the use of over one-half of the people of the State, becomes more potent. Indeed, it may well be a subject for interstate action, since, as has been shown, the needs of the population on the New York side of the Hudson are likely soon to be greater that can well be supplied from watersheds within the bounds of New York State, while since the advantages of the Highlands as a recreation ground would be as easily available to the people of New York as to those of New Jersey, the former State could well afford to share with the latter the expense of sequestration of the region, especially since a considerable part of the New York Highlands could advantageously be included within the area which it is desirable to reserve for this purpose.

WATER RESOURCES OF THE CHAMBERS-BURG AND MERCERSBURG QUADRANGLES. PENNSYLVANIA.

By GEORGE W. STOSE.

The Chambersburg and Mercersburg quadrangles are located in southern Pennsylvania, about midway between the eastern and western limits of the State, their southern boundaries lying within 2 miles of the Pennsylvania-Maryland line. They include a portion of the Cumberland Valley, extending from South Mountain at Waynesboro. Franklin County, to Tuscarora Mountain on the west, and a small portion of Fulton County about McConnellsburg.

The principal streams in this area are the Conococheague. flowing from South Mountain westward across the area; the West Branch of the Conococheague, flowing from north to south and joining the Conococheague at the southern boundary of the area; Licking Creek and Back Creek, tributaries of the Conococheague; and Little Antietam Creek, issuing from South Mountain, in the eastern portion of the area. All of these streams have a plentiful flow of water throughout the year. They are dammed at many places for water power to run grist, saw, and woolen mills. Many of the minor streams are similarly utilized on a smaller scale. At the mill above the village of Markes an electric plant has been erected, which supplies Mercersburg and some of the smaller neighboring towns with light. Electric railways are being constructed in the area, and the mountain streams, especially Buck Run at Cove Gap, are contemplated as a source of power. The mountain streams have not been utilized to great extent for this purpose, but they promise to be a fruitful source of power by reason of their constant supply of water, the feasibility of damming, and the large fall which may be obtained.

All the streams in the mountain valleys and ravines furnish good-sized flows of pure, limpid water. On reaching the open limestone valley the water rapidly sinks into the mountain wash and disappears, ultimately finding its way into subterranean fissures in the limestone. There are many large springs in the limestone area, but no system of underground channels has been observed. The water from these lime-stone springs is also pure and sparkling as it flows from its rock caverns, but this is due to its high content of lime carbonate, which is subsequently precipitated as mechanical sediment. On this account it is not so desirable for domestic uses as the pure mountain water, and most of the larger towns pipe their water from the mountains.

Chambersburg, the county seat of Franklin County and the railroad center of the area, obtains its water from Conococheague Creek, from which it is raised 100 feet to reservoirs on the shale hills opposite the town. This creek emerges from the mountains 12 miles east of Chambersburg, where it is fed by many mountain streams. It is in general a large stream, but at this distance from the mountains it is often muddy and at times low, and is subject to contamination by waste from small towns and mills along its course. The water is therefore not of the best quality. A purer and more desirable supply could be obtained from Rocky Spring or Falling Spring, about 4 miles distant, the waters of both of which, however, are hard; or, better still, from Crawford Springs, which issue from the sandstones of South Mountain, 7 miles east of the town.

Waynesboro obtains its water from a small mountain stream in South Mountain 4 miles to the east, where a receiving reservoir has been constructed 250 feet above the town. The distributing reservoir is located on a shale hill to the north of the town. Fayetteville, although it depends chiefly on well water, pipes part of its supply from springs in South Mountain.

The State Soldiers' Orphans' Industrial School at Scotland has a water system of its own, utilizing some small springs on the premises and pumping the water to an elevated tank. Greencastle takes its water from several limestone springs that supply a reservoir situated 2 miles east of the town and 150 feet above it.

Fort Loudon has incased a spring on the east side of Cove Mountain, 1 mile west of the town, and the water is made to flow continually through the old public pumps on the main street, from which the inhabitants help themselves. This supply is unusually pure and cool on account of its continual flow and because it does not stand in an open reservoir. A similar flowing public pump is located at Foltz, at which place the waters of Buck Run are taken where it issues from the mountains. A company has been formed to pipe this water from a point farther back in the mountains to the town of Mercersburg (4 miles southeastward), which at present has only private wells.

McConnellsburg, west of the mountains, is supplied from a spring and reservoir on the mountain slope 200 feet above the town and 1½ miles distant. The other smaller villages in the area depend on private wells and springs for their supply.

In South Mountain there are several noted springs. Mont Alto Park, formerly belonging to the Mont Alto Furnace Company, but recently purchased by the State as a timber reserve, is located in a gap through which flows a stream that is supplied by a number of fine springs, which issue from the sandstone. Pearl and Tarburner springs are the best known. The mountains at the gap have a dense growth of large pine and fir, preserved by the Mont Alto Company, and the park which that company established here, with its pure mountain stream and springs, and outlooks on the heights above, is one of the most attractive resorts in South Mountain. All that it lacks to make it complete is a hotel.

At Crawford Springs, in South Mountain, 2 miles north of Fayetteville, Doctor Crawford once conducted a health resort. A hotel was located here, and baths were built over the springs. The Tarburner Spring was, and is still, similarly utilized. Springs are plentiful in Tuscarora Mountain also, but none have been developed for special purposes.

There are several large limestone springs in the area, marked by beds of water cress and by stately weeping willows. At Falling Spring and Aqua, 3 miles east of Chambersburg, there are numerous large springs, which join to form a good-sized stream. These issue from sandy beds in the limestone series on the sides of an anticline.

Rocky Spring, 4 miles north of Chambersburg, flows from the limestone at a fault contact with overlying shales. This large spring is one of the chief sources of Back Creek.

Blue Spring, 3 miles southwest of Mercersburg, issues in a large pool from the center of a flat anticline in the limestone, but again flows through a cavern before it finally emerges to form one of the chief sources of Licking Creek.

Mount Holly Springs, about 25 miles northeast of this area, along South Mountain, is a very attractive summer resort. The springs issue from the sandstone cut by the gap, and a pretty park has been built around them. Two hotels located here entertain a large number of summer visitors. A paper mill in the gap uses the pure mountain water directly from the springs and makes a high grade of bond paper, which is used by the Government. Another paper mill uses the town water, which is piped from a reservoir at Cold Springs, 3 miles to the southwest. This is not so fresh nor so clear as the water at the upper mill, and consequently the paper is not of so high a grade.

WATER RESOURCES OF THE CURWENSVILLE, PATTON, EBENSBURG, AND BARNESBORO QUADRANGLES, PENNSYLVANIA.

By FREDERICK G. CLAPP.

These quadrangles are situated near the eastern edge of the bituminous coal field in the west-central part of Pennsylvania. The first three named cover a north-south belt lying mostly in Clearfield and Cambria counties, but include small portions of Bedford and Blair; the Barnesboro lies west of the Patton, in Cambria and Indiana counties. Within the area itself there are no large towns; but Clearfield, with 5,000; Altoona, with 29,000; and Johnstown, with 36,000 inhabitants, lie, respectively, 4, 5, and 10 miles outside. Throughout the region are scattered a number of flourishing mining and other towns having a population of less than 3,000, and some of the better parts of the basins support a scanty farming population. About one-half of it consists of extensive barren tracts, either forested or burnt over, which are very sparsely inhabited.

. The Ebensburg quadrangle contains some of the highest land in the State, the anticline along the crest of Allegheny Mountain forming the water parting between the Conemaugh and Little Juniata rivers. A second belt of high land enters the area along the Viaduct and Laurel Hill anticlinal axes; a third, less continuous, along the Nolo axis, in the Barnesboro quadrangle; and a fourth, a broad, high wilderness along the Driftwood axis, cuts across the northwest corner of the Curwensville quadrangle. With a few exceptions the principal streams flow in a northeast course, following in a general way the anticlinal axes. The largest river is the West Branch of the Susquehanna, which rises on the west side of Laurel Hill near Carrolltown and flows 40 miles before finally leaving the area. Chest Creek, a tributary to the Susquehanna at Mahaffey, has a total length of 31 miles. Clearfield Creek, flowing northeastward from Cresson along the western slope of Allegheny Mountain, has 22 miles of its length within the area. Other less important streams are the Conemaugh River, Two Lick,

Black Lick, Anderson, and Little Clearfield creeks. The grade of the streams is usually light, and none of them are used for power except locally. In a few instances, the most important of which are at Patton and Curwensville, the water supply for towns is taken from neighboring creeks.

With the exception of small tracts of Pottsville sandstone, and of an area of lower Carboniferous and Devonian rocks in the southeastern part of the Ebensburg quadrangle, the region is covered entirely by the Lower Productive and Lower Barren Coal Measures, and, as the dips are very gentle, rarely exceeding 200 feet in a mile, springs are numerous throughout, furnishing the farming population with abundant drinking water of excellent quality. Several of the smaller towns obtain their water supply from this source. The principal exceptions to the general abundance of springs are on the barren sandstone flats covering the crests of many of the ridges and in some of the shales of the barren measures, where they form hilltops along the Wilmore and Johnstown basins. The water of the region is nearly always soft.

At Cresson station is situated the Cresson Springs Hotel, now closed, but formerly a summer resort of the Pennsylvania Railroad. which made much of the water from a "magnesia spring" coming from a shale bed some distance above the Mahoning sandstone. A mile southeast of the town are an "iron spring" and a "sulphur spring," located at the horizon of the Upper Freeport coal. Several hotels at Cresson and Altoona are reported to serve water from these springs, and it is also sold extensively in Pittsburg. The following analyses of the waters were made by Prof. F. A. Genth:[a]

Analyses of water in springs near Cresson station, Pa.

[In parts per 1,000,000.]

	Iron spring.	Alum spring.	Magnesia spring.
Sulphate of ferric oxide	Trace.	571.0	....
Sulphate of alumina	27.4	362.6	....
Sulphate of ferrous oxide	401.5	278.0	....
Sulphate of magnesia	386.1	473.6	....
Sulphate of lime	836.5	687.4	1.9
Sulphate of lithia	Trace.	.8	....
Sulphate of soda	28.1	12.0	....
Sulphate of potash	5.5	7.3	....
Chloride of sodium	.7	.4	21.0

[a] Second Geol. Survey Pennsylvania, Rept. HH, p. 36.

Analyses of water in springs near Cresson station, Pa.—Continued.

	Iron spring.	Alum spring.	Magnesia spring.
Bicarbonate of iron..........................	86. 1	64. 1	0. 3
Bicarbonate of manganese.	Trace.	....	Trace.
Bicarbonate of lime	60. 3	....	. 4
Phosphate of lime............................	. 5	Trace.	. 1
Silicic acid.................................	20. 7	32. 0	15. 6
Chloride of magnesium	....	....	9. 6
Chloride of calcium	....	....	22. 3
Bicarbonate of magnesia......................	....	....	7. 1
Bicarbonate of soda	....	....	24. 4
Bicarbonate of potash	....	....	3. 5
Alumina.....................................	....	....	. 1
Nitrous acid	....	....	Trace.
Carbonic acid (free).........................	....	....	11. 3
	1, 853. 4	2, 489. 2	117. 6

The drinking and cooking supply of the towns is obtained from various sources, a tabulated list of which is given below. Several of these are worthy of especial note. Curwensville obtains its principal supply from mountain springs situated 2 to 3 miles from the town, which have been analyzed and found to be exceedingly pure. In the middle of the summer these springs sometimes run low, and at such times the water of Anderson Creek is used, being pumped to a system of sand beds through which it is filtered. Whenever a change is made to creek water the company gives notice to consumers that it should be filtered or boiled before using. The most extensive use of creek water is at Patton, where the water of Chest Creek is used almost entirely, being pumped by the Patton Water Company unto reservoirs. This water is of very good quality. In addition to the appended list Altoona and Holidaysburg must be mentioned, as they obtain their water supply principally from within the Patton and Ebensburg quadrangles. The former place uses water from Burgoons Gap Run, below Delaney; the latter from Blairs Gap Run. The Pennsylvania Railroad supply in the vicinity of Cresson comes from a reservoir in Bear Rock Run, several miles distant.

Sources of supply of the principal towns.

Name.	Population.	Source of supply.	Quality of water.
Gallitzin	2,700	Pumped to reservoir from mountain stream 1½ miles north of town.	Reported fairly good.
Patton *a*	2,600	Pumped to reservoirs from Chest Creek.	Good.
Curwensville ...	1,900	Mountain springs .... Anderson Creek when springs run low.	Excellent. Must be filtered.
Cresson	1,700	Reservoir on hillside east of town.	
Hastings	1,600	Reservoir in ravine above Stirling No. 8 mine.	
Ebensburg......	1,600	Artesian wells.....	Good.
Spangler *a*	1,600	Reservoir in ravine below Benedict.	Contaminated by drainage.
Barnesboro *a*....	1,500		
Lilly	1,300	Small stream 2 miles east of town fed by mountain springs.	Supposed to be good.
Coalport *a*	900	Wells	Good.
Portage.........	800	....	
Carrolltown	800	Pumped from spring	Reported to be best quality.
Mahaffey	700	Piped to town from wells and springs.	
Irvona	700	No waterworks; supply obtained from wells.	
Grampian	600	Two private water lines from springs within the borough limits supply about one-third of town; rest of supply from wells.	Wells are good; water from private lines has been questioned, as it is in danger of contamination.

a Discussed in the text.

The poorest source of supply is probably that of Barnesboro and
Spangler. These two towns have granted a franchise to a private
company by which they are furnished with water from a reservoir in
the ravine southeast of Spangler. Above the reservoir has been built
the mining town of Benedict, the drainage of which dangerously con-
taminates the Barnesboro and Spangler supply. Several alternative
sources have been suggested, the most practicable plan proposed probably being to use the water from Lancashire No. 8 mine, which is supposed
not to be seriously polluted, and which is now pumped to a private reservoir. The use of this water or that from some still less contaminated
ravine in the vicinity might afford relief, but it is doubtful whether
the improvement would be more than temporary. Barnesboro and
Spangler, together with the neighboring towns of Moss Creek, Cym-
bria, Bakerton, Hastings, and Benedict, all within a radius of 3 miles,

comprise a growing mining community which, if it increases in population at the present rate, will in a short time contaminate all the water in the vicinity. It is improbable that any of the ravines containing outcrops of workable coal can escape contamination within the next few years.

Conditions similar to those in the vicinity of Barnesboro exist on a smaller scale at many other mining towns, both within and without the area under consideration. While many of these towns now have satisfactory water supplies, a large number have not, and it seems probable that the question of water supply will soon become serious. Since only temporary relief can be afforded by change from one surface source to another, it will be necessary to prospect below ground.

Heretofore the abundance of good springs in the region has made it unnecessary to sink many wells, although shallow wells are frequent along the valleys of the larger streams and in most of the towns, and deep wells have been sunk at a number of places. The most abundant of these are in the vicinity of Ebensburg and Chest Springs. At Ebensburg they furnish the water supply of the town, which is consequently of the best quality. On the hills immediately east of Chest Springs a number of wells have been drilled in the barren measures and sometimes reach to depths exceeding 100 feet before striking water. Isolated wells were observed at farmhouses on the ridges southeast of Curwensville, south of Mahaffey, and in the vicinity of Marron. At a point directly south of Lumber City a tank has been erected on a hill to contain water for irrigating an orchard. This is the only case of irrigation known in the region. At the tanneries in Curwensville, on the flood plain, several wells have been drilled, which supply abundant water from horizons 100 to 200 feet below the top of the Pottsville formation. At the Westover tannery there is a well 182 feet in depth. The only flowing well observed was at Wilmore. This has a strong flow, due to a head of 1,000 feet, caused by the anticlinal structure of Allegheny Mountain on the east and indicates that if from any cause water should ever become scarce an abundant supply could probably be obtained by sinking artesian wells along the Wilmore basin. Similar conditions, but with a smaller head, must prevail in the other basins. It would seem, therefore, that the only true and permanent remedy for poor water supply in the coal-mining towns will be to drill artesian wells along the deeper basins. In view of the far-reaching importance of pure drinking water, it can not be urged too strongly that test wells be put down at well-chosen points in the several basins.[a] This is the only way to determine with absolute certainty whether or not such a supply is available.

---

[a] Since this report was written the pure-water problem has been successfully solved by the borough of Coalport, which has had two wells drilled into the Pottsville sandstone. These wells are reported to find little water in the sandstone itself, but to obtain a good supply from the fire clay directly below the Brookville or "A" coal. The water has a very pleasing taste and exists in sufficient quantity to supply the entire town. A good head is obtained by pumping to a reservoir on the hill.

WATER RESOURCES OF THE ELDERS RIDGE QUADRANGLE, PENNSYLVANIA.

By RALPH W. STONE.

The Elders Ridge quadrangle is located in west central Pennsylvania, between the valleys of Cowanshannock Creek and Conemaugh River. and covers an area of about 225 square miles. The boundary line between Armstrong and Indiana counties extends from the northeast corner of the quadrangle to Kiskiminitas River, in the southwest corner.

The topography of the quadrangle is hilly. The extremes of altitude range from 825 feet on Kiskiminitas River, near Salina, to 1,625 feet, the elevation of the top of Watt Hill, the highest point in this vicinity. Although the difference in altitude between the highest and lowest points is therefore about 800 feet, the average distance between the valley bottoms and the uplands is not more than 300 feet. A few small areas, each comprising less than a square mile, are approximately level. These are the terrace and flood-plain deposits along the larger streams. Some of the ridges appear comparatively flat topped when viewed from an elevation. The valleys are narrow and, as a general rule, without notable flood plains.

The drainage of this quadrangle is westward into Allegheny River. The largest stream is Kiskiminitas River, which crosses the southwest corner for a few miles. It is shallow, has a number of rifts, and can not be navigated except by rowboats, and even by them only for short distances. The main tributary of the Kiskiminitas in this quadrangle is Blacklegs Creek, the largest stream in the southern half of this territory. It rises in the vicinity of Parkwood and West Lebanon and flows southwestward to its mouth at Saltsburg. This stream is so small that it can be forded at a number of places, and its grade is so gentle that a dam near its mouth backs the water up for a considerable distance.

Crooked Creek flows across the northern half of the quadrangle from east to west, and receives the waters of Plum Creek and Cherry Run. East of Girty the valley of Crooked Creek is comparatively broad, but from that point westward the stream flows through a deep and narrow gorge, with bluffs often 200 feet high. The stream can be forded at but few places in this quadrangle, and carries a sufficient volume of water, even during the low summer stages, to furnish power for a number of grist mills. Dams have been built across it at Cochran Mills, South Bend, and Idaho. It falls 130 feet from Shelocta to Cochran Mills, a distance of nearly 18 miles as the stream flows, and furnishes sufficient head for water power at frequent intervals. Cherry Run, a tributary to Crooked Creek at Cochran Mills, drains the northern portion of the quadrangle. At least one mill derives its power from this small stream. Plum Creek, formed by the junction of its north and south branches, joins Crooked Creek above Idaho and brings in a considerable amount of water. It drains the northeastern corner of the quadrangle and flows across a broad alluvial plain. Its grade is very gentle and it has not yet been utilized at any point for water power. All of the streams in this quadrangle have much steeper grades toward their sources, and in many places are doing rapid cutting in the neighborhood of the divides.

So much of the country has been cleared of timber that, although the rainfall is moderate, after heavy storms the streams rise suddenly and at that time carry enormous volumes of water. The excessive supply, however, runs off almost as quickly as it appears, and damage is done only to such property and crops as are located on the flood plains.

The largest settlement in the Elders Ridge quadrangle is Avonmore, on Kiskiminitas River, which, according to the census of 1900, has a population of 630. The water supply of this village is derived from wells sunk into the sands and gravels of the river terrace on which it is built. Wells dug but a few feet into this loose material obtain a sufficient supply of good water.

The next largest place is Elderton, with a population of 300. It is situated in the northern part of the quadrangle, on a small plateau, at an elevation of 250 feet above Plum Creek. The village water supply is obtained by private wells of the suction and chain type.

This quadrangle is distinctly a rural district, devoted to agriculture, and, as is usual in such cases, depends almost entirely on wells and springs for its water supply. On a few farms windmills are in use for elevating water to private tanks. Roadside springs with water troughs are common, and water can be obtained at almost any point by sinking wells a very short distance beneath the surface. The rocks that are commonly known as good water bearers are the Mahoning sandstone, which lies immediately above the Upper Freeport coal and outcrops over the greater part of this quadrangle, and the Pittsburg sandstone, which overlies the Pittsburg coal and is found in the southern portion of the quadrangle west of Blacklegs Creek.

WATER RESOURCES OF THE WAYNESBURG QUADRANGLE, PENNSYLVANIA.

By RALPH W. STONE.

The Waynesburg quadrangle is located in the southwestern corner of the State of Pennsylvania, in the eastern half of Greene County, its southern boundary being 2 miles north of the West Virginia State line and its western boundary about 15 miles east of the western line of the State. Its dimensions are about 13 by 17 miles, and it comprises about 229 square miles.

The topography of this quadrangle is uniformly hilly. The difference in elevation between the bottoms of the valleys and the crests of the ridges does not exceed 500 feet. The ridges in a general way trend northwest-southeast, but can not be said to have any conspicuous features. This part of Greene County is distinctly an agricultural district, and is reached by Monongahela River and by a narrow-gauge railroad from Washington County.

The drainage of the quadrangle is eastward to Monongahela River. The Monongahela itself crosses the northeast corner of the quadrangle for about 2 miles. Owing to slack-water conditions, this stream is navigable throughout the greater part of the year and affords an outlet for the products of the country about its headwaters at all times, except when it is choked with ice. Without its dams and locks the Monongahela would be little better than a broad creek. The principal tributaries of the river in this quadrangle are the north and south forks of Tenmile Creek, and Muddy, Whiteley, and Dunkard creeks. The north fork of Tenmile enters from Washington County and joins the south fork at the village of Clarksville, a short distance from the Monongahela. The south fork of Tenmile pursues the longest course of any stream in the quadrangle, having its rise in the west-central part of Greene County and extending entirely across this territory in a winding northeast direction. Muddy and Whiteley creeks have low grades in much of their courses. In fact, the streams in this quadrangle have no high grades, except near their headwaters. They are subject to flood and to seasons of slight flow, and in summer are likely to diminish to such an extent that the water stands along their courses

166

in pools, with only a very small volume of running water. It seems that there is not sufficient supply for water power at any one point.

The village of Waynesburg is the largest settlement in the district and has a population of about 3,500. The next largest village is Jefferson, which has a population of 310, and there are a number of smaller hamlets. These villages are located for the most part on the stream courses, where travel is easier than along the uneven crests of the ridges.

Waynesburg derives its water from the south fork of Tenmile Creek, at the western end of the village. It is pumped to a reservoir on the hill north of the village at an elevation of 250 feet above the main street. As the creek carries considerable silt after every heavy rain, the supply is often muddy, although it passes through a sand filter before reaching the reservoir. For days at a time the water drawn from faucets is so heavily charged with sediment as to be almost useless. There seems, however, to be no other adequate supply immediately available. The proposition to drill deep wells has been considered but never tried. Many of the people in Waynesburg—in fact, most of them—use well water for drinking purposes. In most places a well sunk from 17 to 30 feet will reach bed rock and furnish a sufficient amount of fairly pure hard water. The system of waterworks at Waynesburg is the only one in the quadrangle.

The village of Jefferson is located on a terrace deposit of clays and gravels, and obtains its water supply from wells sunk from 20 to 60 feet through this material to bed rock. The supply is sufficient, and but few of the wells have been known to go dry, except during a protracted drought. The water is hard. In all of the other villages in the Waynesburg quadrangle the water supply is obtained from private wells, which are from 15 to 50 feet deep.

Springs are comparatively abundant in this country, and the water comes from various formations. The Upper Washington limestone is a frequent water producer. Springs from this stratum are numerous in Franklin and Washington townships. It is believed that the Waynesburg sandstone, which overlies the Waynesburg coal, and often has a thickness of 40 feet, is usually a water-bearing rock. Wells sunk into it yield an excellent quality of water, but care has to be taken not to penetrate to the coal.

The Waynesburg Cold Storage Company drilled an 8-inch well 134 feet deep at its plant in the village in March, 1901. This well struck water in the Waynesburg sandstone and yields a supply which the pump has never been able to exhaust. Although the pump raises 75 barrels an hour, day and night, there is always about 90 feet of water in the hole. The water is soft and is used for the ice plant.

WATER RESOURCES OF THE ACCIDENT AND GRANTSVILLE QUADRANGLES, MARYLAND.

By G. C. MARTIN.

INTRODUCTION.

Geography.—The Accident and Grantsville quadrangles are located in the "Handle" at the extreme western extension of Maryland. A strip about 2 miles wide, belonging to Pennsylvania, is included in the northern portion of each quadrangle, while a smaller and narrower strip of territory belonging to West Virginia is included in the western portion of the Accident quadrangle. Of the territory in Maryland all but a small area in the southeastern corner of the Grantsville quadrangle lies in Garrett County. Each quadrangle measures approximately 17¼ miles from north to south, 13¼ miles from east to west, and contains about 235 square miles. Friendsville, Md., in the Accident quadrangle, and Elklick, Pa., and Barton, Md., in the Grantsville, are the largest villages.

Relief.—The topography of the Accident quadrangle is mainly that of a plateau which has been deeply cut by streams. The surface of the upland lies in general between 2,500 and 3,000 feet, the highest parts being in the southern and eastern portions. There are no well-defined ridges rising noticeably above the general level of the uplands. The valleys, especially along the main drainage lines, are of the nature of canyons, whose bottoms are frequently from 500 to 1,000 feet below the plateau. In the Grantsville area there are three ridges rising nearly or quite to 3,000 feet, crossing the quadrangle with a northeast-southwest trend, between which plateaus, similar to that of the Accident quadrangle, are developed at an altitude of, between, 2,500 to 2,800 feet.

Drainage.—The drainage of the Accident quadrange is northward by the Youghiogheny to the Monongahela. In the Grantsville area the drainage is in part north by Castleman River to the Youghiogheny in Pennsylvania, and in part southward by Savage River to the Potomac.

168

WATER RESOURCES.

STREAM SUPPLIES.

Youghiogheny River.—The Youghiogheny and its tributaries drain almost the entire area of the Accident quadrangle. It is a large, pure stream whose capacity is far in excess of any probable demand. The only contamination of the main stream comes from the villages of Sang Run, Krug, Friendsville, and Selbysport, and from a few sawmills. The tributaries are all very pure.

Castleman River.—This stream drains the northwest half of the Grantsville quadrangle. It is a large, uncontaminated stream, but there is no demand for its waters in the agricultural region through which it flows in this quadrangle.

Savage River.—This stream drains the central part of the Grantsville quadrangle. It is a large, pure stream and furnishes the water supply for the towns of Piedmont, W. Va., and Westernport, Md.

Georges Creek.—This stream drains the southeast corner of the Grantsville quadrangle. The main stream and the lower courses of its tributaries are so polluted by sewage and mine water as to be totally unfit for any purpose. The headwaters of the tributaries are pure and would furnish good supplies of pure water for the many mining villages in the Georges Creek valley which are annually ravaged by typhoid fever.

SPRING WATER.

There are a great many large, pure springs along the belts of outcrop of the Greenbrier limestone. These belts extend (1) along the western foot of Big Savage Mountain; (2) along the eastern front of Meadow Mountain; (3) along the western front of Negro Mountain; (4) along the eastern front of Winding Ridge; (5) through the valleys of Deep Creek and Marsh Run from Thayerville to McHenry, thence westward to Sang Run and along Youghiogheny River for a distance of 2 miles north and south of Sang Run; and (6) along the northern and eastern edge of the Cranesville valley.

These springs are similar both in geologic relations and in the properties of their water to the group of springs from which the celebrated Deer Park spring water is obtained. The Deer Park springs are about 6 miles south of the southern limits of this folio, and are situated along the direct continuation of the line of springs at the western foot of Big Savage Mountain.

ARTESIAN WATER.

The possibility of obtaining artesian water has never been properly tested in this region. It is, however, probable that the synclines that underlie the valleys of Georges Creek, Castleman River, and Youghiogheny River are artesian basins, and would yield plenty of good artesian water from various horizons.

Several bore holes made in the coal-bearing portions of the synclines have yielded flows of water. This water came from the Coal Measures, and was therefore strongly impregnated with sulphur and iron. It is probable that deeper holes would yield better water from the purer porous sandstones which underlie the Coal Measures. There is, however, no present demand in the region for artesian wells, for the numerous pure streams and springs yield a supply of water that is sufficient for all needs.

WATER RESOURCES OF THE FROSTBURG AND FLINTSTONE QUADRANGLES, MARYLAND AND WEST VIRGINIA.

By G. C. MARTIN.

INTRODUCTION.

Geography.—These quadrangles are mostly in western Maryland, lying just east of the Grantsville quadrangle. Like the latter they cover a narrow strip of Pennsylvania along their northern borders, and include on the south a considerable area lying south of Potomac River and belonging to West Virginia. The portion in Maryland, except a small area in the northeastern part in Garrett County, falls in Allegany County. The West Virginia area is divided between Mineral County on the west and Hampshire County on the east. Cumberland and Frostburg, Md., are in the Frostburg quadrangle, but there are no large towns in the West Virginia portion of the area or in that part of Maryland that is included in the Flintstone quadrangle.

Relief.—The southwestern half of the Frostburg and nearly all of the Flintstone quadrangle is crossed at intervals of a few miles by ridges varying in altitude from 3,000 feet on the west to 1,500 feet or less on the east. Between these ridges, plateaus having altitudes varying from about 2,700 feet on the west to 900 feet on the east are developed. The plateaus, however, exhibit very little of their original level surface, being cut by numerous streams to depths of many hundred feet. The ridges are due to upturned hard rocks of Silurian, Devonian, and Carboniferous age, while the plateaus are composed mainly of Devonian shales and softer coal-bearing rocks of the Carboniferous.

Drainage.—The quadrangles are drained by the Potomac River, which enters the Frostburg area at its southern boundary, flows northeastward to Cumberland, and thence southeastward across the Flintstone area. The minor streams follow in general the trend of the ridges, those of the north flowing southwestward to the Potomac and those on the south northeastward to the same river.

171

North Branch of Potomac River.—The North Branch of the Potomac flows through the southern part of this area for a distance of 35 miles. It furnishes the public water supply for the city of Cumberland. The quantity of water is far in excess of the amount needed, but the quality is extremely bad. The water is, in fact, so polluted that it is entirely unsuitable for domestic or industrial use. The polluting matter consists of the refuse from a number of sawmills and tanneries, the drainage from a large number of coal mines, the chemicals from paper mills, dye works, woolen mills, and gas plants, and the sewage of Piedmont, Westernport, Keyser, Cumberland, and other towns. There is great need of some purer supply.

South Branch of Potomac River.—The South Branch of the Potomac is a large stream, which flows through this area for a distance of about 6 miles. It contains at all seasons a large amount of pure water. It is not used at present, and there is not likely to be any future demand for it. It is, however, very important, as it serves to dilute the impurity of the main stream of the Potomac, and thus improves the quality of the water supply of the city of Washington.

Georges Creek, Braddock Run, and Jennings Run.—These streams flow through the thickly populated mining regions in the western part of this area, and are so badly polluted by sewage and mine water as to be entirely worthless.

Smaller streams.—The headwater streams of the entire region, situated as they are largely in forested areas, are unpolluted and would furnish pure water supplies for the smaller towns. Evitts Creek, Patterson Creek, and Town Creek are the largest of the unpolluted streams, and all contain pure water for the entire length. Their water is not used at present.

The largest springs in this region are on the belts of outcrop of the limestone formations. The lines of upper contacts of the Greenbrier limestone and the Helderberg limestone are marked by a great many springs of large, constant flow and great purity. One of these springs, at the contact of the Mauch Chunk shales and the Greenbrier limestone, at the western foot of Big Savage Mountain, furnishes a large part of the water supply of the town of Frostburg. This supply could be greatly increased by the development of other springs in neighboring localities. The town of Lonaconing, which is at present very poorly supplied with contaminated water, could get a similar supply on the western slope of Big Savage Mountain. The celebrated Deer Park spring water comes from a series of large springs not many miles southwest of this region in this same belt of outcrop of the Greenbrier limestone. A series of similar springs might be developed along the eastern part of the Dans-Piney-Little Allegheny Mountain range which would furnish a partial or even a complete supply of pure water for the city of Cumberland.

The line of contact between the Oriskany sandstone and Helderberg limestone and a large part of the areas of those formations contain a great many large springs which are important for local rural use.

ARTESIAN WATER.

Part of the water supply for the town of Frostburg is obtained from an artesian well 1¼ miles west of that town, on the eastern slope of Big Savage Mountain. Water is said to have been found in sandstones of Carboniferous age at depths of 81, 182, 527, and 1,200 feet. The amount of water procured from the various horizons is not known; nor is it certain whether the water is derived from contaminated horizons near the surface or from deeper pure sources. If this well reached the Mauch Chunk-Greenbrier horizon and the shallower water were cased off an ample pure supply would result. There is an area of about 90 square miles in the northwest central part of the Frostburg quadrangle where, in the Georges Creek syncline, this Mauch Chunk-Greenbrier water horizon could be struck at depths of from 1,000 to 2,400 feet. The water will not rise anywhere in these wells above an elevation of 2,700 feet above sea level, and is probably drained out of the syncline in the vicinity of Braddock Run and Jennings Run to the level of 1,000 feet.

It is highly probable that the Oriskany and Tuscarora sandstones contain artesian water in their synclinal areas. There are areas to the north, northeast, and south of Cumberland where artesian wells, if properly located with regard to the local details of structure, would be almost certain to strike pure flowing water in the Oriskany and Tuscarora sandstones. Similar conditions exist over a large proportion of the Flintstone quadrangle, but except in the vicinity of Cumberland there is no demand for artesian water.

WATER RESOURCES OF COWEE AND PISGAH QUADRANGLES, NORTH CAROLINA.

By Hoyt S. Gale.

The Cowee and Pisgah quadrangles are in western North Carolina, their southernmost limits overlapping into South Carolina. They include parts of Macon, Jackson, Swain, Haywood, Transylvania, Buncombe, and Henderson counties in North Carolina. Their total area is about 1,950 square miles.

Geographically these quadrangles are situated in the heart of the southern Appalachian Mountains, covering an area of comparatively high and roughly dissected country. The headwater valleys of French Broad River, in the Pisgah quadrangle, and of Little Tennessee River, in the Cowee quadrangle, include the greater part of the open country they contain. The smoothly graded débris slopes of these and some other valley bottoms are in strong contrast with the sharp dissection of the territory in general. However, the area contains numerous remnants of old plateaus, recording several distinct periods of ancient peneplanation, and many examples of the original mature topography, as yet untouched by readjusting drainage, are found high up about the uppermost headwaters. Still above these plateau levels rise many residual peaks.

Drainage.—The larger part of the drainage belongs to the Mississippi River system. The Blue Ridge, the main divide between the Mississippi and the Atlantic waters, passes through the southern halves of the two quadrangles. Most of the streams south of this divide have worked their grades back to steep slopes, often escarpments, at their very heads. The Mississippi drainage north of the divide has, however, been less active in its channel cutting, being held up by greater difficulties in its paths and the longer route traversed to reach the sea level. Most of the remnants of old land topographies spoken of above are thus found on the Mississippi side of the Blue Ridge. Subsequent incision of these stream channels has not yet receded upstream far enough to affect grades at the headwaters.

The Cullasaja River is a good example of the streams on the Mississippi side of the Blue Ridge. Heading on the slopes of the residual peaks in the vicinity of Highlands, and also on the Highlands plateau, of about 3,800 feet elevation, it collects its waters from a drainage

174

basin of smooth, well-rounded slopes. About a mile below Highlands it comes to the edge of this plateau and begins to drop off in cascades and falls, and its valley becomes steep sided and deep. The immediate locations of the falls are determined by the harder ledges of rock. In most cases here, as elsewhere in this region, these ledges are granite. In the next 6 miles the river drops 1,200 feet, an average grade of 200 feet in a mile. For the remainder of its course to Franklin its grade again flattens out. Along this stretch is the plateau of Little Tennessee River. From Rabun Gap, the main head of the Little Tennessee, to Franklin this river falls about 100 feet in 25 miles, or 4 feet to the mile. Below Franklin the Little Tennessee comes to the edge of this lower plateau and falls off more rapidly again.

French Broad River is in character very similar to the Little Tennessee. Along both rivers the lower plateaus are at an elevation of about 2,100 feet above sea level, and into them both streams have slightly incised their channels. The French Broad runs a course of about 45 miles from its main forks near Eastatoe Ford to the point where it leaves the Pisgah quadrangle. In this distance it falls from 2,180 feet to 2,000 feet above sea level—180 feet in 45 miles, or 4 feet in a mile. Streams so well graded are very exceptional in this region.

The west fork of Tuckaseegee River gathers into a stream of considerable size the waters of an upper plateau level of about 3,600 feet. Just below Glenville, in Jackson County, it comes to the plateau edge and drops 900 feet in 6 miles, or 150 feet to the mile. The East Fork similarly drops over 100 feet a mile for more than 10 miles.

The heaviest grades are south of the Blue Ridge. The descent is here concentrated into one steep slope from the Blue Ridge plateau down to the Piedmont Plateau. The intermediate levels that occur on the Mississippi streams do not occur here, so that the Blue Ridge is characteristically an escarpment overlooking the Piedmont Plateau. Cæsars Head is a feature of this escarpment. Ten miles west of Cæsars Head the Blue Ridge and the escarpment diverge, and an upper plateau intervenes between the main divide and the fall line of the streams. This plateau extends about 20 miles westward, as far as Whiteside Mountain. Streams heading within this belt, therefore, have flat headwater grades similar to those of Cullasaja River in the Mississippi drainage. However, the drop from the Blue Ridge plateau to the Piedmont Plateau is far greater than any that occurs on the Mississippi side of the divide. For example, the Toxaway River is 3,000 feet at Lake Toxaway, and in 5 miles it falls to 1,300 feet, or at the rate of 340 feet in a mile. It would seem that these streams offer much available power, but the falls are almost always difficult of access and most of them are distant from present lines of transportation.

Springs.—This is on the whole a region sparsely settled and little developed commercially or industrially, and little or nothing is known concerning the underground waters except as they issue in springs. or perhaps through the few existing wells. The country is abundantly supplied with springs, except along the gravel and bowlder deposits, which form the valley bottoms of the larger streams. The water is for the most part very pure. The rocks are typically siliceous or micaceous, containing little readily soluble material either in themselves or in their soils. The soils are loose and coarse, usually sandy rather than clayey, and do not afford much material that will stay in suspension in the water. However, mineralized waters do occur in numerous instances. These can be separated into at least two classes. Besides the granites and micaceous gneisses, a more basic rock, containing much hornblende and other iron minerals, traverses the region in long, narrow bands. Springs issuing along these bands are frequently heavily charged with iron. Their occurrence is rendered noticeable by the deposits of iron hydrates that accumulate about them. A spring that forms an excellent example of this class issue directly from a narrow outcrop of hornblende-gneiss just below the dam at Fairfield Lake, in Jackson County.

Another type of mineralization of spring waters has been noted where formation contacts, and especially zones of faulting, show a development of pyrite, with perhaps other minerals. Weathering of these to a soluble form stains the rocks with copperas and impregnates the water that flows through them. Sulphur and iron waters observed at a number of localities undoubtedly obtain their mineral content in this way. The most accessible and best known sulphur and iron water occurs at Waynesville, at the "Haywood White Sulphur Springs," but the writer is not prepared to say how it originates.

In considering the underground waters it is necessary to distinguish at least three definite types of topography and surface drainage which are conspicuous in this region. The thinly covered ledges of the residual peaks that stand above the old plateau levels shed water rapidly from their steep slopes. Flow within these massive rocks must be slight, except through cracks and fissures. On the plateau levels, however, prolonged weathering has produced a heavy cover of soil which must be comparatively porous. This is well supplied with water from the residuals, and on the other hand well supplied with outlet drains at the plateau edges, and a strong and constant underground flow must result within this cover. In the stream bottoms, at lower levels, are graded plains of wash débris, from which the water does not often come to the surface to form springs. Here the water supply, if not taken from the stream itself, must be drawn from wells or piped from neighboring hillsides. As the larger towns are usually situated in the open valley lands they are least easily supplied with pure water. All of the county seats within the area are so situated.

WATER RESOURCES OF THE MIDDLESBORO-HARLAN REGION OF SOUTHEASTERN KENTUCKY.

By George H. Ashley.

This area, which takes its name from the towns of Middlesboro and Harlan, is a belt of country occupying the region between Pine Mountain on the north and Cumberland Mountain on the south, and is located in Bell and Harlan counties, in southeastern Kentucky. It extends from the headwaters of Yellow Creek, near the Kentucky-Tennessee line, in a course about N. 60° E. to Big Black Mountain, a distance of about 60 miles. At its western end the belt is about 10 miles wide, at the eastern end about 15 miles wide, and its area about 750 square miles.

The topography is that of a basin bounded on the north and south by two high mountain ridges and limited on the east and west by the divides at the heads of the streams draining toward the middle of the basin and passing out by way of the Cumberland River through Pine Mountain at Pineville Gap. The region between the ridges has now been cut down by the rivers until only sharp ridges remain. These commonly vary from 2,500 to 3,400 feet in altitude. The streams are generally from 500 to 2,000 feet below the crest, the Cumberland River having an altitude of 980 feet where it leaves the basin. The slopes of the hillsides are nearly all steep, the only flat lands being narrow belts along the streams. The hills throughout the basin are generally forested.

Geologically the area is a synclinal basin, the axis of which lies to the north of the center, or nearer to Pine Mountain. In the center of the basin the dips are slight, but run up to nearly or quite vertical on the flanks of the bounding ridges. The rocks consist of alternating beds of sandstone, shale, and coal, the sandstones predominating.

Many cabins are scattered over the slopes and crests and in the ravines throughout the hilly region. In fact, the majority of the inhabitants are located among the hills, though the prosperous habitations are nearly all in the towns or along the river bottoms.

Along the bottoms water is obtained either from shallow wells in the river gravels or from near-by springs. In the hilly region springs are

especially abundant and constitute the chief source of supply, even up
to the very crests, although a few shallow wells have been dug on the
hilltops. In times of prolonged drought the springs occasionally fail,
and recourse is had to streams, which, however, are sometimes so low
that water appears only in isolated pools. The waters of the springs
and wells are noncalcareous and are of good quality.

The town of Middlesboro obtains water from Little Yellow Creek,
a stream that flows along the foot of Cumberland Mountain and is fed
from large springs from sandstone. The water is pumped from an
artificial lake made by a concrete dam across the creek to a reservoir
on a near-by hilltop. The pumping station is a mile from the center
of the town. A large spring, which formerly came out a little below
the summit of Cumberland Gap, was diverted by the building of the
railroad tunnel and is now piped to town for the use of the local tannery. A number of deep wells have been bored at Middlesboro for
oil, but, though obtaining no oil, at least one of them—that at the
mouth of Bennett's Fork—gives an abundant flow of water from the
Lee conglomerate. One well is 750 feet deep. Similar artesian water
might be found elsewhere, but no other wells have yet been drilled.

Pineville is supplied from a small tributary of Hagan Mill Branch
of Cumberland River. It is a small stream fed by springs on the
south side of Pine Mountain, 1½ miles east of the gap. The pumping
station is located on the Cumberland River.

Water power on a very small scale is obtained at a number of places
on Cumberland River and on its forks and branches. At only two
points is there promise of considerable power, one of which is on
Shillaly Creek. This stream rises on a broad plateau between Cumberland and Brushy Mountain, and attains a considerable volume before
it descends into the deeper valley. Martins Fork of Cumberland
River heads near Shillaly Creek and descends in a similar but more
gradual manner, and with somewhat larger volume.

SUMMARY OF THE WATER SUPPLY OF THE OZARK REGION IN NORTHERN ARKANSAS.

By GEORGE I. ADAMS.

During the progress of geological work in the Ozark region of northern Arkansas considerable data has been collected in regard to the water resources. This paper is intended as an outline of the conditions of occurrence of the springs, which are numerous and important. They are in many cases the heads of streams, with which the area is well supplied.

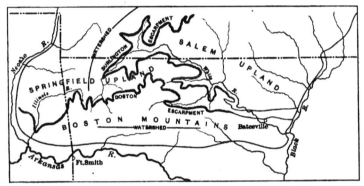

FIG. 32.—Sketch map of Ozark region in northern Arkansas.

The accompanying sketch map of a portion of the State shows the location and extent of the Ozark region. Observations have been largely confined to the western part of the area, but some of the springs in the eastern part have been noted, so that the geologic relations of the ground waters as outlined is thought to be consistent for the whole.

THE OZARK REGION AND ITS DIVISIONS.

The Ozark region in Arkansas includes two distinct types of country, the characters of which are determined by the structure and extent of the rock formations. The southern of these is the Boston Mountains, a dissected highland which extends from the vicinity of Batesville westward to the Indian Territory line. To the south it blends with the Arkansas Valley region. Its northern border is outlined by the Boston

179

escarpment, which is very irregular, being cut by numerous stream valleys, between which are irregular peninsula-like promontories.

North of the Boston Mountains and at a lower level is the more even country that forms a part of the Ozark Plateau, which extends northward into Missouri. The surface features are closely related to the nearly horizontal rock formations. The rocks which outcrop in this area belong principally to two classes. They may be spoken of as the Mississippian limestone and the Ordovician dolomites. The Mississippian limestones are higher geologically and the area in which they outcrop is known as the Springfield upland. The area of the Ordovician dolomites is nearly coextensive with the Salem upland. These two divisions of the plateau are separated by a more or less distinct escarpment, known as the Burlington escarpment, which has a height of approximately 300 feet.

When viewed in a broad way the higher portions of the Springfield and Salem uplands are seen to fall in a slightly warped plain, which is dissected by the streams. The drainage of the western portion of the Ozark region in Arkansas is tributary to Illinois and Elk rivers, which find their way around the western end of the Boston Mountains into Grand River and eventually into the Arkansas. The remaining streams that head in the region are tributary to White River, which flows in an irregular course northward and then southeastward around the eastern end of the Boston Mountains. In the eastern portion of the Salem upland there are a number of streams which enter the State from Missouri. Some of these are tributary to White River within the Ozark region, while others enter Black River, which flows along the eastern border. The larger creeks, and especially the rivers, are perennial. The minor ones flow during most of the year, but during the drier seasons the water in them stands in pools or has a decreased flow, which is contributed by springs. The fact that the country is well watered is largely due to the nature of certain rock formations which hold a large content of water.

SPRINGS OF THE OZARK REGION.

The horizons of the springs are in large measure determined by certain formations which afford easy channels for the underground water, and by others which are impervious and guide the flow along the dip of the rocks to their outcrops. The following classification of the springs of the region is tentative only; further observations and more detailed study will probably enable a classification to be proposed which will more fully cover the conditions existing.

SPRINGS RELATED TO THE LIMESTONES IN THE BOSTON ESCARPMENT.

The rocks that outcrop in the Boston Mountains are principally sandstones and shales. There are, however, two limestone formations which, although usually but from 5 feet to 50 feet thick, are persistent.

The lower is known as the Pitkin limestone and the upper as the Brentwood. The effect of these formations on the topography, as the result of their manner of weathering, is to produce benches. Their outcrops, although often concealed, are frequently seen as conspicuous ledges. The ground water finds an easy path along the bedding planes and joints of these rocks, and because of slight undulations of the formation it converges at certain points and issues as springs. The occurrence of these springs at the headwaters of Illinois River in the vicinity of Prairie Grove, and at many other localities, influenced the early settlers in their selection of sites for homes. A spring of this type is seldom found which is not the site of a settlement, and in some cases they have determined the location of towns.

SPRINGS RELATED TO THE BOONE LIMESTONE AND CHERT.

The principal formation in the Springfield upland is the Boone limestone and chert. Circulating water finds easy channels along its bedding planes and joints, and, as is often the case in such rocks, numerous solution channels, and occasionally caves and sinks, have been formed in it. A large amount of the underground water in the Boone chert is contributed to the streams without appearing as springs. In some places, however, it issues at the heads of valleys or at levels above streams, and forms important springs. A few of the springs are so large that their waters have been dammed or conducted by means of flumes to furnish power for small mills. At many towns they are the source of water supply, and in some instances the locality is utilized as a summer resort, as at Monte Ne, for example.

SPRINGS RELATED TO THE SHALE BED AT THE BASE OF THE BOONE LIMESTONE.

In the western part of the Ozark Plateau a bed of shale ranging from a few feet up to 50 feet in thickness lies underneath the Boone formation. The influence of this impervious bed upon the circulation of the ground water is to carry it along its upper surface with the dip to the outcrop. Many springs issue at the base of the Boone limestone, just above the shale. The best known is Eureka Spring, which has become famed as a health resort. While many of the springs occurring at this horizon do not carry a large volume of water, they are of considerable importance as a source of domestic supply to individual households or neighborhoods.

SPRINGS RELATED TO THE SANDSTONE AT THE TOP OF THE YELLVILLE DOLOMITE.

Descending in the geologic section the next important formation as a source of springs is this sandstone. It is a loosely cemented sandstone, and the waters which issue from it are "softer" than those previously mentioned—that is, they carry but little lime. The water issuing from the sandstone does not usually afford a large supply, but because of its softness it is particularly desirable.

The Yellville dolomite outcrops over nearly the entire area of the Salem upland. Its rôle as a water reservoir is similar to that of the Boone formation. However, there are local beds of shale and argillaceous layers in the Yellville formation, and in some instances they guide the ground water to a point of outcrop. The springs from the dolomite are numerous, but much of the water issues directly into the streams without appearing as springs. The formation as seen in its natural exposures exhibits, especially along the bluffs, so'ution channels showing the point of issuance of ground water before the streams had cut their valleys down to their present positions. Within the area of the Yellville formation there are many sink holes and caverns. The ground water undoubtedly moves in important channels. At Mammoth Springs, a station on the Memphis Railway on the

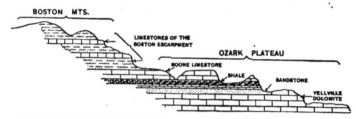

FIG. 33. Diagrammatic section showing principal horizons of springs in northern Arkansas.

northern border of the State, the water issues with such volume that it is often spoken of as a small river. It contributes, as do many of the springs of the region, to the natural beauty of the scenery and the enjoyment of pleasure seekers.

Many of the springs of this region are more or less renowned, and some have been frequented for many years as health resorts. Analyses of the waters from some of them have been made, and certain of them are reputed to have medicinal value and in rare instances do contain unusual minerals in solution. It may be said of the springs of the region in general that they are particularly pure and wholesome. As such they are valuable for domestic purposes, and in some instances as a supply for small towns. A few of them are regarded as possessing therapeutic qualities which render them valuable in specific ailments. None of the springs in this region are hot springs. On the contrary. the temperature of them in summer time is such as to render them cool and potable. The country is at present sparsely settled. No doubt more of the springs which are now allowed to run to waste will. because of their surroundings, become valuable as pleasure and health resorts.

NOTES ON THE HYDROLOGY OF CUBA.

By M. L. FULLER.

INTRODUCTION.

The problems of the water resources of Cuba were brought forcibly to the attention of those interested in the island during the period of occupation by the American army. Notwithstanding the great abundance of water on the island as a whole, it was found that in many of the relatively level strips along the shores the supply was frequently both insufficient and of a poor quality. This led, in a considerable number of instances, to the sinking of wells by the War Department. The attention given, both by the Americans and the natives, to the question of utilizing more fully the splendid natural resources of the island will doubtless, under the present improved conditions, lead to the introduction of pure supplies in many villages.

At many of the mineral springs hotels and baths were constructed and resorts developed by the Spanish. Some of the springs are equal to many of the famous spring resorts of Europe and America in natural beauty of surroundings and in the quality of their waters, and will probably in a few years become popular with many Americans. The caves, though having many natural attractions, are in many instances inaccessible, and have been until recently almost unknown to people outside the regions in which they occur.

The purpose of the present notes is to give a summary of the natural and artificial water resources of the island, including town water supplies and springs and the underground drainage with its resulting caverns.

Acknowledgments.—The facts included in the following notes have been drawn from a number of different sources, but in their presentation it is impossible to give separate credit in each case. The portions relating to the topography and geology are mainly based on the " Report on a Geological Reconnaissance of Cuba," made to Brig. Gen. Leonard Wood by C. W. Hayes, T. W. Vaughan, and A. C. Spencer, published in Cuba in 1901; those relating to water supplies from reports of Assistant Engineer O. Giberga, Maj. H. F. Hodges, and Resident Engineer H. F. Labelle, published in "Civil Report of Brig. Gen. Leonard Wood, January 1 to May 20, 1902," volumes 5 and 6; those relating to mineral water from a report by Mrs. H. C. Brown incorporated in the preceding; those relating to subterranean caves and streams from "Island of Cuba," by Lieut. A. S. Rowan and M. M. Ramsey (Henry Holt, New York, 1896); and those relating to wells in part from published reports of the War Department, but mainly from written memoranda furnished directly by that Department to the United States Geological Survey.

GEOGRAPHY AND TOPOGRAPHY.

The island of Cuba is located south of the State of Florida, the meridian of Washington crossing it about 200 miles from its eastern end. The distance of the nearest point of Cuba from Key West is 85 miles. The length of the island is 730 miles, the breadth from 20 to 90 miles, and the area about 43,000 square miles.

In general terms the island may be said to possess a central highland belt, reaching from Cabo de Maysí on the east, first northwestward and then southwestward to Cabo de San Antonio on the west, attaining an elevation of 2,500 feet and becoming somewhat mountainous in places, but elsewhere dropping nearly to sea level.

The surface may be divided into five topographic provinces, three of which are essentially mountainous while the other two are of low or moderate relief. The easternmost of these topographic divisions coincides approximately with the Province of Santiago. In this province, taken as a whole, there are two principal mountain groups. The southern is an east-west ridge, known as the Sierra Maestra, extending from Cape Cruz to the vicinity of Puerto de Guantanamo, and from there continued by a geologically distinct ridge of the same trend to Cape Maysí. The loftiest mountains of Cuba, some of which are higher than any peaks in the eastern United States, occur in this range. The northern range merges with the southern near the eastern end of the province, but diverges westward, the intervening area constituting the undulating plain of the well-known Cauto Valley, which westward merges with the more extended plains of Puerto Principe. The second topographic division corresponds closely with the Province of Puerto Principe, and is made up of plains or rolling open country broken by occasional hills or low mountains rising above the general level. The third division includes the mountainous portions of the Province of Santa Clara. The island is here crossed by a mountainous belt made up of a number of subordinate groups, the highest point of which is 2,900 feet in altitude. The fourth topographic district comprises the western portion of Santa Clara Province, all of Matanzas and Habana provinces, and the eastern portion of Pinar del Rio. This region, like that of Puerto Principe, is made up of low, flat, or rolling plains, broken by occasional hills several hundred feet in height. The fifth division comprises the greater portion of Pinar del Rio, and is characterized by a prominent range of mountains reaching to a height of 2,000 feet. The coast is bordered at many points, especially on the south side of the island, by marshy belts, while the shore is often fringed with reefs and keys.

GEOLOGY.

The structure of the island in a broad way is anticlinal. The geologic axis, which coincides roughly with the topographic axis, is marked by masses of serpentine, granite, and folded and metamorphic slates and schists, on the flanks of which, dipping gradually away, lie the later sedimentary formations. The dip is strongest to the north. Of the later sedimentary beds the oldest appears to be a semicrystalline blue limestone, which may be as old as the Paleozoic. It is, however, only locally developed, the first persistent beds being the hard gray limestones of the Cretaceous. These are overlain by Eocene limestones and glauconitic sands, sometimes interbedded with volcanic rocks, and by a great thickness of limestone, marls, etc., of Oligocene age. During the accumulation of these rocks the land was deeply submerged, at times possibly completely, except for occasional peaks in southern Santiago Province. In Miocene times the land was uplifted above sea level, where it has remained, except for a possible subsidence of 100 feet in Pliocene times and a number of minor oscillations in Quaternary times, during which the Quaternary shelves of elevated reef rocks were formed. The uplift was greatest along the old axis, the result being the tilting of the beds on both flanks.

The limestones, which are the predominant rocks of the island, constitute nearly its entire surface except along the axis of the island. They are in general distinctly stratified, but their internal structure has been greatly changed by the action of percolating waters, nearly all traces of fossils having been destroyed. Their present thickness is from 800 to 1,000 feet. Since their deposition they have been removed by erosion from portions of the higher lands, while in some of the less elevated lands, although they form the surface, they are frequently cut through by streams.

DRAINAGE.

The highlands of the island are characterized by the presence of abundant springs of considerable volume. While a considerable number of these flow from the metamorphic and igneous rocks, the larger springs are from the limestones which cover a large part of the island. The waters are of that extreme transparency which characterizes the limestone springs of the adjacent mainland at Florida. It is from these springs, or in some instances from underground streams, that the surface streams of the island take their rise. The run-off by these streams is very great, as the high rainfall usually occurs as showers of short duration, the water finding its way quickly down the steep hillsides into the streams. The portion of the rainfall sinking into the ground passes into the porous limestone, through which it finds its way by general seepage or by way of caverns and other underground channels back to the streams. In some instances whole streams disappear into the limestone only to reappear near the margin or beneath the sea.

The arrangement of the streams in Cuba is very simple, their courses being nearly all normal to the coast and their lengths, therefore, very short. The divide between the northward and the southward flowing streams lies near the axis of the island, being generally somewhat nearer the north than the south coast. In the eastern and western provinces the divide corresponds with the mountainous belts, but in the central province it falls on a level plain and is extremely indefinite. In the mountainous regions the valleys, while moderately steep-sided, are fairly wide, but where they cut through the tilted plateau along the northern coast canyon-like valleys have been formed. An exception to the ordinary arrangement of streams is found in Santiago Province, where the Rio Cauto, the largest river of the island, has a trend nearly due west. The length of this stream is 150 miles, and it is navigable for shallow craft for a distance of 50 miles.

Lakes are very rare on the island, although a number occur in the midst of the dense vegetation of the marshes near the coast. Many of these are hardly known to the inhabitants themselves. A very few small lakes are also found in the mountains. The most noted of these is Lake Ariguanabo, which lies 20 miles southwest of Habana. Its surface is 6 square miles in area and it is drained by a stream which, after flowing on the surface a short distance, disappears into a subterranean passage. It is inhabited by fish, which are supposed to have worked their way upward to the lake through this passage.

CLIMATE.

The climate of Cuba is tropical and insular. The year is divided into a rainy and a dry season, the former extending from May to October. Two-thirds of the total precipitation, which amounts to about 50 inches at Habana, falls within this period. The rainfall inland is considerably higher. The average temperature for August is from 89° to 91°, while in December, January, and February it is 10° to 15° lower. The temperature of the north coast is somewhat lowered by the persistent northeast trades. The relative humidity is very great, amounting to an average of 80 per cent.

Notwithstanding the extremely numerous springs and streams of pure water, few public supply systems have been installed, and these only in the largest cities and towns. In the smaller cities and villages the supplies are obtained from a variety of sources. Shallow wells are generally used where water is obtainable, regardless of the fact that they are almost invariably subject to contamination. Cisterns are frequently used, but generally furnish insufficient supplies to meet the needs of the people. To supply this deficiency water is often peddled about the streets in kegs and other receptacles. Even where pure city supplies are at hand the poorer classes, in many instances, are not inclined to connect with the pipe systems because of the expense, but prefer to continue to purchase their supplies from peddlers. In some towns water has been at times so scarce that it has commanded almost fabulous prices.

Habana.—The first water-supply system for the city of Habana was installed at the close of the sixteenth century, when a dam was built across the Almendares River about a mile above Puentes Grandes, and the water brought to the city through an uncovered aqueduct known as the Zanja Real, which reaches the city at a point near the present reservoir. From this point the zanja divides into many branches, the water ultimately making its way into the bay and sea through Matadero, Agua Dulce, and San Lazaro creeks and the city sewers. The water in this aqueduct is now so contaminated as to be unfit for drinking in the built-up portions of the city, and is at present mainly used for irrigation and power. The water rates paid, however, are so low that the expense of keeping the zanja in repair is greater than the income derived from it.

The second system of water supply dates from 1837, when the Almendares River was diverted at a point about 4½ miles above the city. The water was conducted by a 20-inch iron pipe laid to the city, together with a limited number of distributing pipes, the system being known as the aqueduct of Fernando VII. An attempt was made at filtration, but the methods proved defective and in times of heavy rain the supply was turbid from surface wash. The filters, however, although of an antiquated type, were found by the Americans to be in good condition and were cleaned and maintained, in order that they could be utilized to bring in the Almendares water in case of a break in the Vento aqueduct.

The supply from the river proving insufficient a new system was projected in 1858. By this plan a group of 400 springs near Vento, on the banks of the Almendares, about 8 miles above the city, were enclosed in a masonry structure 150 feet in diameter at the base. 25 feet at the top, and 60 feet deep. Masonry dams were built around the top to keep out the surface wash. The water is carried under the river through an inverted siphon consisting of two heavy iron pipes in a masonry tunnel, and thence by gravity through an underground masonry aqueduct to the Palatino reservoir, about 4 miles from the city. From the reservoir the city is supplied by gravity through a system of distributing mains. The construction of the system met with many interruptions, but the works were finally completed about 1893. The supply is about 40,000,000 gallons a day, against 1,333,000 gallons a day of the earlier system.

In 1886, during the construction of the Vento aqueduct and before its completion and the construction of the reservoirs, a branch main, 20 inches in diameter, was run from the Vento aqueduct at a point opposite the filter beds to those beds, and the water of Almendares River was cut off from the beds and the Vento water supplied to the city through the aqueduct of Fernando VII. Since the completion of the Vento system neither this branch main nor that portion of the aqueduct of Fernando VII south of Palatino has been used. They have, however, been maintained in good condition, so that in case of accident to the Vento system below the branch main the city could be supplied with Vento water through this branch and the Fernando VII aqueduct.

Casa Blanca and Regla were formerly dependent upon cisterns or local wells more or less contaminated, or upon water supplied from the Habana mains, which was carried across the harbor in boats and sold at the rate of 2 cents a gallon. In 1899 an iron pipe was laid across the harbor to a pumping station at Cabana Fortress, from which the water is pumped to a 200,000-gallon tank on the hills, whence it flows by gravity to Cabana Fortress, the barracks, and the town of Casa Blanca. Regla is supplied by a gravity main connected with the Habana supply. A pumping station was installed in 1898 near the Palatino reservoirs for the purpose of furnishing supplies for the camp at Quemados, Marianao, and to Camp Columbia, Aldecoa, Principe, and several hospitals, and other places. A project was also started to supply the town of Arroyo Naranjo from springs at Calabazar. Plans were also made for supplying Pirotecnia, Carmelo, and other localities.

The only large sections of Habana now without Vento water are the higher portions of Jesus del Monte and La Vibora and the high sections of Vedado, all of which will have to be supplied by pumping, as they are above the limits of the gravity supply. Detailed plans and estimates were prepared and submitted for the supply of the former places, at an estimated cost of $36,500, but the work was not undertaken by the Americans on account of the lack of funds. Studies were made for the supply of the higher parts of Vedado.

A concession was granted to a private company by the Spanish in 1894 to supply and sell Vento water to the lower portions of Vedado and Carmelo. Some work was done, but the system was found to be incomplete and in bad condition, and the concession was annulled by order of the governor of Habana in 1899. This order was revoked, however, in 1900. The conditions are still very unsatisfactory.

During the Spanish control cast and wrought-iron pipes, principally the latter, were used for house-service connection, and, owing to the rapidity with which they are destroyed by the salts and acids of the soil, breaks were frequent. Many of the distributary pipes were also old and the connections were frequently faulty and complicated. Many of the valves of the old system were found to be almost entirely destroyed. A large amount of repairs, including the replacement of much pipe, the installation of many new valves, and the inspection of 400 fire hydrants and installation of 100 new ones were made by the Americans.

In certain parts of the city leakage was very great and an investigation was made by the Americans, with the result that the daily consumption was decreased from 35,419,342 to 28,760,800 gallons, or from 144 to 117 gallons per capita. The consumption is still high, especially for the hours between 1 and 2 a. m., when the legitimate consumption should be very small. Probably much of this loss is due to leaks that can not be readily located because of the porosity of the soil, which favors rapid absorption of the leakage.

The regulations governing the water supply of the city do not admit of proper control and regulation of the service. In 1902 the installation of water was made compulsory in those portions of the city that are supplied with mains and the installation of meters was required in manufacturing establishments or other institutions using large quantities of water. This order greatly increased the revenues and secured a more just distribution of charges.

During the American occupation it was observed that for a day or two after a freshet a quantity of turbid water entered the main spring or "taza" at Vento. It was at first thought that this was caused by the river water entering the spring through the valves regulating the overflow, which were not designed to resist pressure from the outside, but experiments made later showed that the trouble was due not only to this leakage, but chiefly to underground connections between the outer springs and the main springs. When the river rises beyond the level of the main spring, the head thus formed forces the water through these connections into the spring. Apparently the only way to prevent this is to build a tight wall around the outer springs and carry it above the highest flood level, or else to tightly cover these springs and provide automatic waste valves which will close when the river rises to the danger point.

During the period of the American occupation the **Vento aquedu**:
and the Palatino reservoirs were maintained in good condition, th
reservoirs being thoroughly cleaned at least once a month.   These ar .
however, insufficient in size, holding less than half a day's supply, an .
should be enlarged.   The aqueduct is large enough for present needs.
but a new one is desirable to provide for the possibility of accident t
the present one.   The supply from the springs is ample for present
and future needs, and the water is excellent in quality, being almost
pure organically, but hard from limestone in solution.

Matanzas.—The city of Matanzas has had an abundant supply of
excellent water since 1872.   The source is Bello Springs, from 7 to 1
miles from town.   As in other Cuban cities, many have failed to take
advantage of the public supply but continue to obtain their supply by
purchase or from shallow wells and springs.

During the American occupation the problem of supplying the jail
was investigated.   The water is now obtained from a cistern and
pumped into an elevated 23,000-gallon tank, from which it is distributed to the jail for use in baths and closets.   Since installing the tank
the average daily consumption from the water company's mains has
been reduced from between 5,000 and 10,000 gallons to about 500 gallons, the average charges at the same time falling from $95 to $3.85,
the latter sum being paid for water still obtained from the company
for drinking and cooking purposes.   The total cost of the installation
was $357.

Cárdenas.—Cardenas has been supplied since 1872 with pure wate:
from a subterranean river about a mile distant.   Brackish wells an .
cisterns are still depended upon by some of the people, while other
have until recently purchased their water from street peddlers.

Cienfuegos.—Until a few years ago this city was supplied by cistern-
or by peddlers, but waterworks were being introduced at the time of
the American occupation.

Guantanamo.—This town was formerly supplied by an undergroun.l
stream flowing from a cave, but work on a new supply was begun in
1899 by the Americans, who built a dam across Guaso River, 9 mile-
from Guantanamo.   This dam was completed in 1901 and yields a supply of 955,152 gallons of excellent water daily, or 120 gallons pe:
capita.   The total cost was $210,166.32.   The work was turned ove:
to the municipality in February, 1902.

Puerto Principe.—The water supply of Puerto Principe is derive
chiefly from artesian wells, five of which were drilled during the Amer
ican occupation.   Of these, two are in the northern part of the cit:.
one in the western, one in the central, and one in the southern.   Th
two wells in the northern part of the city are each 486 feet deep, si: .
ated side by side, and supply about 25,000 gallons of water daily '
the public buildings and to the public who call for it.   At this pla-

the water is pumped by steam, the pumps being run day and night. At the other wells—one near Carmen Hospital, where a good and abundant supply was obtained at a depth of 188 feet; one in the market place, 207 feet deep, and one at the Matadero, 154 feet deep—60-foot tower windmills, with 10,000-gallon tanks, have been erected. The tanks are placed 30 feet above the ground.

Santiago.—Santiago has long been supplied with water from Boniato River, a neighboring stream, the water being conducted to the city by an aqueduct. This was cut by the American army during the siege of the city, but was repaired during the American occupation. Prior to this time the supply was only about 9 gallons per capita from June to February, while in March, April, and May the stream was so low that only 2.5 to 5 gallons were obtainable. After certain improvements had been made, measurements showed the supply capable of yielding 800,000 gallons daily, or 18.6 gallons per capita, but the supply was decreased 162,000 gallons daily by the necessity of cutting a part off from the lower part of the city in order to secure sufficient head to supply certain of the higher portions. An additional supply of 30,000 gallons was obtained by cleaning and repairing the old reservoir built in 1860 and connecting it with the city mains. The northern part of the city was supplied from the reservoir, while the southern part was supplied by the direct mains. Pumps were also installed to increase the supply in dry seasons.

As approximately 3,000 people live on the stream within a mile of the intake the water was badly polluted and there was constant danger of epidemics. This, taken in connection with the deficiency of supply, made it imperative to look for a new source. Detailed investigations were accordingly made by the Americans, the country for 30 miles around the city being examined with a view to procuring a new supply. The sources considered were (1) a gravity supply from Baconao River; (2) a gravity supply from Cauto River, and (3) a pump supply from wells in the San Juan Valley.

In the proposed Baconao system the water would be taken from a point about 21 miles above the mouth of the river, from which it would be conducted 19 miles to Daiquirí and 18.5 miles farther to Santiago. Nearly the whole of the line is of easy access, and the location would require very little costly road making. The estimated total cost was $2,844,812 and the annual cost of operation $169,000.

A supply from Cauto River could be brought in by pumping through the Maniel or Rio Frio Pass from a point about 2½ miles below Dos Palmas, a distance of 25 miles from Santiago. The country is favorable to the construction of a conduit, and good locations for dams and power stations are available. The estimated cost was $1,451,127, and the annual cost of operation $105,000.

The preliminary surveys having shown that still better results would be obtained from a well system in the San Juan Valley, a detailed investigation in that region, including the drilling of test wells. was undertaken. The portion of the valley under consideration lies just south of the Sierra Maestra, and drains to the sea through a gorge 500 feet wide at La Laguna. The underflow through the river gravels is very high in some of the branches, and the water is of good quality, notwithstanding the river waters themselves are frequently much polluted. The well sections show: (1) A surface layer of brown clay and vegetable matter from about 4 to 16 feet in depth; (2) a water-bearing stratum of sand, gravel, bowlders, and broken coral rock, 11 to 30 feet thick, with an average of 25 feet; (3) blue, yellow, or brown clay. 5 to 20 feet thick; and (4) a series of sands, gravel, broken coral rock, clay. etc. The coral bed rock has not been reached except along the borders of the valley, where it is near the surface. The following record may be given as typical:

Record of well at Santiago, Cuba.

Strata.	Thickness of strata.	Depth from surface.
Soil	21	21
Fine sand	16	37
Gravel	13	50
Coarse gravel	6	56
Yellow clay	14	70
Blue clay	6	76
Fine sand	17	93
Sandstone	7	100
Coral rock	18	118
Hard coral rock	8	126
Fine sand	9	135
Yellow clay	8	143
Sand and gravel	7	150
Clay	5	155
Sandstone	6	161
Gravel and sand composition	9	170
Total		170

The area of the local underground reservoir to be drawn upon for the supply is estimated as 3½ square miles, and the probable yield computed as from 1,235,000 gallons a day in dry seasons to perhaps 2,434,000 gallons in wet seasons. The average rainfall is estimated at about 60 inches, the run-off as 13.5 per cent, and the total supply of underground storage 23,000,000 gallons a day. If 60 per cent of the 15 feet of storage gravel is utilized the supply is good for 410 days, which, considering the fact that the dry season lasts only four or five months, is amply sufficient. The water of the lower gravels can also be utilized in case of any deficiency of the upper supply. Infiltration galleries, gang-tube wells, and large caisson wells were considered as methods of collecting the water, the latter being finally recommended. San Juan Hill was selected as the best site for the pumping plant. An examination of the water by G. C. Whipple, of Brooklyn, N. Y., showed slight organic matter, low free ammonia nitrites and nitrates, and high chlorine, as well as carbonates and sulphates—especially the former. Storage in the dark, to prevent algous growth, was recommended. The cost of installation was estimated at $408,650 and that of operation as $92,000 annually. Both figures are much lower than either of the other systems proposed.

Manzanillo.—This town formerly had a water system deriving its supply from the river Yara, but it proved unhealthful and was abandoned. Cisterns now furnish the main supply.

SPRINGS.

Owing to the great porosity of the limestone which forms the surface over a large part of Cuba, and which rapidly absorbs and almost as rapidly discharges its waters, springs are very numerous. They are of all sizes, from small rills up to almost river-like streams. Some of the springs seep from the porous limestone, but many of them issue from more or less cavernous orifices. They issue at all altitudes, from the higher portions of the hills down to the lowland border, or even at sea level. It is probable that the marshy tracts bordering the south coast at many points are in part due to the emergence of the underground water through the limestone. Not all of the water comes to the surface of the land as springs, but some passes outward and emerges from the sea bottom along the coast, where in many instances the fresh water can be seen bubbling up through the salt water. Such springs occur in Habana Harbor and at many other points. The fresh water which emerges as copious springs on some of the keys is probably of the same origin, coming from the mainland through subterranean passages in the limestone.

Most of the spring waters of the island are highly calcareous, but are usually not otherwise especially high in mineral matter. They form pure supplies for a number of cities, including Habana, Matanzas, and Cardenas, while at other points not having public supplies, as at Marianao, the water is obtained from adjacent springs and sold

through the town. Although by far the larger number of springs are
of the pure calcareous type, there are a number of mineral springs
upon the island. These are described under "Mineral waters," below.

SUBTERRANEAN STREAMS AND CAVES.

Probably very few regions in the world so abound in caves and
subterranean passages as the island of Cuba. The porous limestone
permits a free entrance of water which, by its passage, dissolves out
caverns of various shapes and sizes. Although the underground passages are very numerous they are not usually large, and the caves
generally lack the great beauty of many of the stalactite-hung caves
of the harder types of limestone.

Many streams disappear from the surface into the limestone. Of
such streams the Rio San Antonio, in the province of Habana, is the
most noted. This stream issues, with a width of several feet and a
strong current, from Lake Ariguanabo, flows a short distance through
the town of San Antonio de Los Baños, and then enters a cavern and
disappears from the surface. In a similar manner the Rio Guanajay flows on the surface for about 12 miles and then disappears into
the limestone near the village of San Andrés. The Rio Jatibonico del
Norte rises in the Sierra de Jatibonico and alternately disappears and
reappears in a succession of cascades. The Rio Mayari, near Santiago,
and the short stream called Moa both pass through caverns in their
courses. Many other streams throughout the island pass at one point
or another in their course under natural rock arches.

Among the larger or more noted of the caves are those of Resol-
ladero Guacanaya, in Guaniguanico; María Belén, in the Sierra de
Añafe; that of Cotilla, near San José de las Lajas, 15 miles southeast
of Habana; the magnificent caves of Bellamar in Matanzas; the cave
of San José de los Remedios; and the caverns of Cubitas, of Gibara,
of Yumurí, of Holguín, and of Bayamo; while north of Guantánamo
are the noted Monte Líbano caverns.

The Bellamar caverns, 3 miles distant from Matanzas, are extensive
and possess a high temperature. The air is supposed to have certain
curative properties, and the caves are resorted to by invalids. At
Baracoa are caverns that have an unusual development of stalactites
and contain animal remains of unknown date.

MINERAL WATERS.

Cuba possesses some true mineral springs, although these are small
in number compared with the common springs. Little use is yet made
of them for drinking purposes, and almost nothing is done in the way
of shipment. On the other hand, however, they are extensively used
for bathing at various watering places and health resorts. The portion

of the following notes relating to the province of Pinar del Rio and to a part of Habana Province was compiled by Mrs. H. C. Brown for the report of Brigadier-General Wood.

Province of Pinar del Rio.—Of the numerous health resorts of Cuba. that of San Diego de los Baños enjoys a particularly large share of popular favor. The town was founded in 1843, and before the numerous wars decimated the population contained 8,007 inhabitants, but at present it counts little more than a quarter of that number. The left bank of the river which flows beside the town bubbles with sulphurous springs, some warm, others cold. The best known of these are called Templado, Tigre, and La Paila. On the authority of Dr. José Miguel Cabarrouy, medical director of the baths, the following analysis of the waters of Templado and Tigre is given:

Analysis of mineral water at San Diego de los Baños, province of Pinar del Rio, Cuba.

	Per cent.
Hydrogen sulphide	0. 152
Carbonic acid	. 062
Sulphate of lime	. 974
Chloride of sodium	. 032
Bicarbonate of magnesia	. 080
Alumina	. 006
Total	1. 306

The density of these waters is given as 1.014. They contain, besides the ingredients noted above, undetermined quantities of silicic acid, carbonate of iron, nitrogen, oxygen, and organic matter. The Templado produces 860,000 liters in twenty-four hours, and the Tigre 240,000, without any variation according to seasons. The water is colorless, has the disagreeable odor of sulphuretted hydrogen, and tastes slightly sulphurous. It has a temperature of about 34° C.

The spring called La Paila shows the following analysis:

Analysis of water from Mineral Spring La Paila, province of Pinar del Rio, Cuba.

	Per cent.
Sulphate of lime	1. 068
Chloride of sodium	. 022
Bicarbonate of magnesia	. 120
Carbonic acid	. 084
Alumina	. 012
Total	1. 306

The waters, which have a temperature from 22° to 25° C., show traces of sulphuretted hydrogen, silica, carbonate of iron, oxygen, nitrogen, and organic matter.

The waters of San Diego de los Baños have proved to be of special value in cases of skin disease, rheumatism, and nervous affection. The place has a wide patronage and makes pretensions as a popular resort.

Springs of mineral water are also reported from the municipal district of Mariel. In the district of San Cristóbal are springs called Soroa.

Province of Habana.—The most prominent springs of Habana Province are found at Guanabacoa, Madruga, and Santa María del Rosario.

The principal use of the mineral water at Guanabacoa is made by the Santa Rita baths, which are constructed of stone and are popular with many of the residents of Habana.

Madruga has warm and sulphur baths, supposed to possess curative properties in cases of skin disease, and also springs of mineral water that is reported to be excellent for stomach troubles. The town is a popular watering place, having several hotels which are extensively patronized from March to October.

The baths of Santa María del Rosario are famous for their medicinal qualities. The principal springs that supply the baths are three in number. The waters of the first, which is called Mina, Templado, or Palmita, issues from an orifice in the rock 3.80 meters below the surface. The water from this spring is clear and colorless, has the odor of sulphuretted hydrogen, and is disagreeable to the taste. It is cold, reaching a temperature of only 22° C. Its reaction is alkaline. The water of the second spring, called Tigre, comes from a well 6.04 meters deep. Although clear and transparent, it has a light yellow color, an odor more pronounced than that of La Mina, and a stronger taste than the first. Its temperature is 21.5° C., and its reaction is freely alkaline. The spring called La Paila is only a meter deep. The water is clear and colorless and has an odor and a taste more pronounced than the water of the other two springs. It is colder, having a temperature of 19.50 C. Analysis of the waters shows appreciable amounts of sulphur and hydrogen sulphide.

Isle of Pines.—There are many natural springs all over the Isle of Pines, and those of Santa Fé have an established reputation for their curative properties, both in Cuba and abroad. The waters are said to be particularly rich in iron and magnesia, and contain also oxygen and carbonic-acid gases, chloride of sodium, sulphate of lime, carbonate of lime, chloride and nitrate of calcium, and silica. The temperature of the waters is generally about 82° F. Some of the larger springs flow a stream of water the size of a man's body.

WELLS SUNK BY THE WAR DEPARTMENT IN CUBA.[a]

Cienfuegos.—A well was sunk by the army authorities at this point to a depth of 111 feet. A good flow of water was obtained. The following is a record of the materials penetrated:

	Feet.
Clay and sand	1- 86
Blue clay	87-101
Blue clay and gravel	102-111

a Compiled from reports and memoranda of the War Department.

Gibara.—Two wells were drilled at this point by the Bacon Air Lift Company, at a cost of $2,500. The first well was started August 8, 1900, but was abandoned at a depth of 212 feet, because of the telescoping of the pipe. A new well was then started. This obtained water at 75 feet and again at 90 feet, but the flow was small. At 560 feet a strong flow of salt water was obtained and the well was then abandoned. The diameter of the casing was 8 inches at the top and 6 inches at the bottom. The failure of this well was seriously felt, as the cistern supply on which the town depended was very insufficient. Several other wells were drilled to a depth of about 600 feet with like results.

Guanajay.—A well was here sunk to a depth of 40 feet, which resulted in a failure.

Holguin.—Most of the supplies at this place are from surface wells and are satisfactory in amount except at times of drought. The water is derived from disintegrated granite and is somewhat liable to surface contamination. It is hard and contains considerable mineral matter. During the American occupation the public wells were deepened and cleaned, and windmills were erected. An artesian well was begun on November 20, 1900. A strong flow of excellent water is reported to have been obtained at 61 feet, but was cased off with the hope of finding more abundant water at a deeper level. Hard water was encountered, however, and the bit broke at 160 feet, at which depth the well ceased. A later well went to a depth of 400 feet, a small amount of water being found at a depth of 125 feet, but none below. A steam pump and tank were put in, and the well gave a supply sufficient for the barracks, but not for the town.

Matanzas.—Wells were drilled by the Government to depths of from 330 to 400 feet, but resulted in failure.

Pasa Caballos.—The wells sunk by the War Department at this point were 78 feet deep and gave 60,000 gallons during a ten-hour pumping test.

Pinar del Rio.—A well begun by the War Department on April 5, 1899, was drilled to a depth of 285 feet, but resulted in a partial failure, the water being very hard and carrying some sulphur. The formation from a few feet of the surface to the bottom of the well is stated to consist of alternate layers of soft and hard rock, parts of which approach clay in character. Water in small quantities was found at 35, 80, and 145 feet, and in smaller amounts between 200 and 285 feet. There are 158 feet of 6-inch casing in the well. The water rises within 95 feet of the surface. The well will not furnish more than one gallon a minute and is regarded as a failure, as the well is pumped down to 150 feet in a half-hour with an ordinary hand pump. The water rises to its former level in about thirty minutes. It cost about $1,800.

Puerto Principe.—A number of wells were drilled by the War Department at this point. The first one of importance appears to have been started on May 1, 1899. It is located on the grounds on which the cavalry barracks were situated in the city of Puerto Principe. and its mouth has an elevation of 337 feet above sea level. The ground in its vicinity is practically level. During the first 30 feet the well passed through decomposed granite, but from 30 feet to the bottom it was drilled through hard crystalline rock of various colors and hardness. The first water of consequence was struck at 95 feet, and rose to within 30 feet of the surface. The boring was continued, however, to a depth of 264 feet without obtaining any decided increase in the amount of water. At this point the drill was once more withdrawn and a pumping test made lasting two days. Five hundred gallons were pumped hourly without lowering the working head of the water more than a few feet. It was, however, decided to push the well to a greater depth. A month later, when the well had reached a depth of 457 feet, the drill was again withdrawn. and the well was tested by a 2-inch pump attached to the walking beam of the well rig and making 30 strokes to the minute. It extended 150 feet down the well, and during the two days an average of 502 gallons of water an hour was raised. At the beginning of the pumping the water stood 30 feet from the surface, but fell rapidly during the first hour, at the end of which it stood at a depth of 43 feet from the surface. During the continuance of the pumping, however, doubtless by reason of the removal of silt from the pores of the adjacent rock, the water slowly rose until, at the end of the fifth hour, it had reached a level of 41 feet from the surface. On standing over night the water returned to its original level of 30 feet, but during the first hour's pumping again fell to a level of 41 feet below the surface. On August 30 the drills were replaced and the well was pushed to a depth of 474 feet. Previous to the sinking of the well, the supply of the barracks had been obtained by hauling water, all of which had to be boiled. from a creek 3 miles away. The test of the well proved that it would supply more than 500 gallons an hour without lowering the water more than 11 feet from its normal level, and it was estimated that from 2,000 to 3,000 gallons an hour could be obtained without lowering the level more than 30 feet. The water is clear, without sediment, and is sparkling and palatable. An analysis of water made at the Military Hospital No. 1 at Habana is given below:

Analysis of water from Puerto Principe.

Odor, none; taste, faint saline; color, no color, very clear; some matter in suspension composed mainly of precipitated salts.

	Parts per 1,000,000.
Free ammonia	None.
Albuminoid ammonia	None.
Chlorine, 0.2	2

	Parts per 1,000,000.
Sulphuric acid as sulphates	35
Sulphureted hydrogen	None.
Total solids	305
Phosphoric acid	Trace.
Silicic acid (silicates)	22
Iron, none; other metals not tested.	
Calcium	69
Magnesium, pyrophosphates	105
Potassium and sodium, somewhat in excess.	

This water is free from harmful materials and is suitable for all domestic purposes.

The water was piped to the barracks and furnished a satisfactory supply. In addition a considerable amount was shipped to Nuevitas by rail.

A second well was begun October 10 and completed December 15, 1900. Its diameter was 8 inches, and it was drilled to a depth of 202 feet. The supply proved ample, and is used in the public buildings of the vicinity. The water stands 135 feet below the surface. Good water was found above 200 feet, but the deeper waters are reported to be brackish. Another well was begun in the market place December 15, 1900. Still others have been described under water supplies.

A well on a proposed site for a military post, presumably near Puerto Principe, is given in the reports of the War Department. It was 8 inches in diameter and 510 feet deep, the water rising to within 8 to 12 feet of the surface. The supply is reported to be of excellent quality.

Santiago.—The army officials report that there is good reason to believe that an abundant supply of water can be obtained from artesian wells along the south coast of Cuba at this point. An examination of the San Juan Valley shows a gathering ground with slight run-off, the remainder reaching the sea by subterranean drainage. It has been proposed to intercept this by tunnels. Some of the shallow wells obtain large flows, but unless carefully located are liable to become contaminated by seepage from the surface. One well, 14 feet in depth, is reported to have so large a flow that it could not be kept down by a pump of 500 gallons capacity.

Triscornia.—An 8-inch well sunk at this point to a depth of 401 feet obtained only a limited supply of water more or less impregnated with salt and unfit for general use.

INDEX.

201

O

[Mount each slip upon a separate card, placing the subject at the top of the second slip. The name of the series should not be repeated on the series card, but the additional numbers should be added, as received, to the first entry.]

U. S. Geological survey.

. . . Contributions to the hydrology of eastern United States, 1904. Myron L. Fuller, geologist in charge. Washington, Gov't print. off., 1905.

Author.

211 p., 1 l. illus., V pl., diagrs. 23^{cm}. (U. S. Geological survey. Water-supply and irrigation paper no. 110)
Subject series: B, Descriptive geology, 46; O, Underground waters, 27.
Contains contributions by various members of the survey.

1. Water, Underground—U. S.

U. S. Geological survey.

. . . Contributions to the hydrology of eastern United States, 1904. Myron L. Fuller, geologist in charge. Washington, Gov't print. off., 1905.

Subject.

211 p., 1 l. illus., V pl., diagrs. 23^{cm}. (U. S. Geological survey. Water-supply and irrigation paper no. 110)
Subject series: B, Descriptive geology, 46; O, Underground waters, 27.
Contains contributions by various members of the survey.

1. Water, Underground—U. S.

U. S. Geological survey.

Water-supply and irrigation papers.

Series.

no. 110. Contributions to the hydrology of eastern United States, 1904. 1905.

U. S. Dept. of the Interior.

see also

Reference.

U. S. Geological survey.

DEPARTMENT OF THE INTERIOR

UNITED STATES GEOLOGICAL SURVEY

CHARLES D. WALCOTT, Director

PRELIMINARY REPORT

ON THE

UNDERGROUND WATERS OF WASHINGTON

BY

HENRY LANDES

WASHINGTON

GOVERNMENT PRINTING OFFICE

1905

CONTENTS.

ILLUSTRATION.

3

LETTER OF TRANSMITTAL.

DEPARTMENT OF THE INTERIOR,
UNITED STATES GEOLOGICAL SURVEY,
Washington, D. C., May 4, 1904.

SIR: I have the honor to transmit herewith, for publication in the series of Water-Supply and Irrigation Papers, a preliminary report descriptive of the underground waters in the State of Washington, prepared by Mr. Henry Landes under the direction of Mr. N. H. Darton, geologist in charge of the western section of hydrology.

It is believed that the report is a valuable contribution to the knowledge of the water resources of the State.

Very respectfully,   F. H. NEWELL,
Hydrographer in charge.

Hon. CHARLES D. WALCOTT,
Director United States Geological Survey.

5

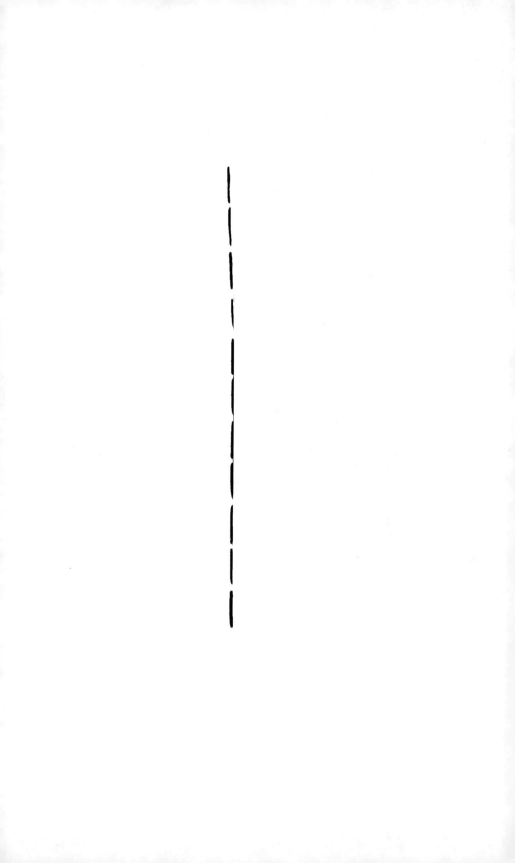

Mean total precipitation

MAP OF W

SHOWING MEAN T

SHINGTON
AL PRECIPITATION
.ANDES

PRELIMINARY REPORT ON THE UNDERGROUND WATERS OF WASHINGTON.

By Henry Landes.

INTRODUCTION.

This report contains a brief account of the water resources of Washington as represented by municipal supplies, deep wells, and springs. For each county a general statement is made, giving the location, rainfall, and most striking features of topography and geology. Following this are detailed statements which give data concerning the municipal systems, deep wells, and springs, and which have been secured entirely by correspondence. The blanks sent to the clerks or other officials of the cities and towns regarding municipal water supplies were practically all filled out and returned, so that this information is complete to the present time. The blanks for the deep wells were not returned as generally as was desired, but almost every section of the State where such wells occur is represented, and those described may be taken as types of their kind in each county. Springs occur so very generally throughout the State that only a small fraction of them may be said to be represented in the blanks filled out and returned, but, as in the case of the wells, those mentioned are typical of their class, and it is only necessary to recall that the number given in every county may be multiplied many times. Every effort has been made to eliminate inaccuracies.

No general statements are made concerning the rainfall, topography, and geology of the State as a whole, since these are given for each county. A rainfall map is included in order to show at a glance what the precipitation is in any section, also the contrasts between one part of the State and another.

ADAMS COUNTY.

General statement.—Adams County lies south and east of the central part of the State, on the line of the Northern Pacific Railway, between Columbia River and Spokane. The topography is that of a gently sloping plain, descending from a height of 1,900 feet in the northeast corner to 700 or 800 feet in the southwestern part as Columbia River is approached. The plain-like character of the surface is modified by occasional valleys locally known as "coulees." The drainage is to the south and the southwest, toward Snake and Columbia rivers. Cow Creek is the principal stream; the other streams as a rule are intermittent and are active only in the late winter and spring months.

This county lies within the dry region of central Washington, where the precipitation is light and is confined to the winter and spring months. In the western half of the county the average yearly rainfall is 10 inches or a little less, while in the eastern half it is from 10 to 15 inches. There is a regular increase in the amount of rainfall with the increase in elevation, and hence the precipitation is greater in the northeastern part of the county than in the southwestern portion, and there is more rain in uplands than in the valleys.

The Columbia River lava, or basalt, is the principal rock in Adams County. It forms the bed rock everywhere except in the southwest corner, where the Ellensburg beds, composed of sands and clays, appear. As a rule the rocks are covered with a thick coating of soil, the exceptions being the valley sides and occasionally the valley bottoms. The soil is fine grained, of good quality, and very productive when supplied with the proper amount of moisture. Grazing and wheat raising are carried on very successfully in nearly all parts of the county.

Despite the small rainfall, there is little or no difficulty in securing a good supply of water for domestic purposes. At depths of from 300 to 500 feet wells yield large quantities of good water. The water-bearing strata are usually layers of very porous or cellular basalt. Wherever such layers outcrop along the border of a valley, springs are commonly found.

Municipal systems.—Ritzville is the only town in the county reported as having a water-supply system. The water is obtained from a well having a depth of 385 feet. In this way a supply of good water is secured. The water level in the well does not vary during the day or year and is not affected by pumping. The water is pumped into reservoirs and is distributed by gravity throughout the town. Besides serving as a domestic supply, the water is used in irrigating the lawns and gardens.

Deep wells.—Besides the well which supplies Ritzville, another deep well has been drilled at the same place by the Northern Pacific Railway Company. This well has a diameter of 8 inches and a depth of 355 feet. The water column stands at a height of about 240 feet. There is no variation in the water level during the day or year, and it is affected slightly by pumping. It is used by the railway to obtain water for locomotives.

At Cunningham a well has been drilled by Thomas and James O'Hair. This well has a diameter of 6 inches and a depth of 426 feet. The water column stands at a height of 356 feet and remains stationary throughout the year. This well was drilled at a cost of $1,066. A wind pump was erected at a cost of $290. The above wells serve as a type for those of Adams County.

Springs.—As a type of the springs found in Adams County, the one owned by G. W. Bassett, near Washtucna, may be noted. The water issues as a stream from the honeycombed or porous basalt, and varies but little in quantity during the year. It is sufficient to supply the town of Washtucna, and some of it is used for irrigating purposes as well. The water is soft, carries no sediment, and has no unpleasant taste.

Along Cow Creek, a tributary of Palouse River, in the eastern part of the county, there are a number of good springs.

ASOTIN COUNTY.

General statement.—Asotin County lies in the extreme southeast corner of the State. It is in the main a plateau region, deeply dissected by canyons. The higher parts of the plateau have an elevation of about 5,000 feet above sea. The canyon of the Snake, along the eastern and a part of the northern boundaries of the county, has a depth of 3,000 to 4,000 feet. Leading into the larger canyon from the southwest are the smaller canyons of Grande Ronde River and Asotin Creek.

The average rainfall is 25 inches and is sufficient to afford an ample water supply. On the higher parts of the plateau the precipitation is sufficient to produce a forest growth of firs, pines, and other coniferous trees. On the lower parts of the plateau bunch grass grows luxuriantly and it is only in the deep valleys that irrigation is necessary.

The bed rock is basalt, except in occasional instances where the streams have cut through the lava and exposed the underlying granites and other crystalline rocks. The soil is very thick and, since it is of basaltic origin, is of good quality.

Only in the valleys has the county been settled to any degree. In the deeper valleys, as that of the Snake, irrigation is necessary in order to produce fruit, vegetables, and other products, but for this purpose there is an ample supply of water at hand. Water from Asotin Creek is used to irrigate a large area of flat land at Clarkston, and a sagebrush plain has been converted into an oasis of alfalfa meadows, fruit orchards, and vegetable gardens. In the gravels and sands of the valleys good water may be had from wells that vary in depth from 20 to 50 feet.

Municipal systems.—Asotin, the county seat, is situated on Snake River at the mouth of Asotin Creek. The water supply of the town comes primarily from Asotin Creek, but some dependence is placed on wells and cisterns. The creek rises in the Blue Mountains and carries pure, clear water without any contamination. Besides serving as a domestic supply the water is used for the irrigation of lawns and gardens. The supply is sufficient for the present and in all probability for all future needs.

Clarkston, on Snake River opposite Lewiston, also gets its supply of water from Asotin Creek. The water is carried by a flume 16 miles in length to a point on the hillside immediately above the town and from there is distributed by pipe lines. At present this forms the sole supply for the town, but a deep well is being bored as a possible additional source of supply. An analysis of the water of Asotin Creek made by Professor Fulmer at Pullman on January 29, 1903, showed 6.18 grains of solid matter in each gallon of water. The solid matter was found to be wholly free from objectionable qualities.

Along Snake River, especially in the vicinity of Asotin, are a number of wells which range in depth from 20 to 35 feet. The water is obtained from gravel and sand and is of good quality. This water supply is intimately connected with the river, as the water in the wells and in the river is always at the same level.

CHEHALIS COUNTY.

General statement.—Chehalis County lies on the western border of the State, fronting the Pacific Ocean. The coast line here is irregular. the most marked indentation being Grays Harbor. From the narrow belt of sand dunes along the coast the surface rises by irregular hills until the low mountains of the Coast Range are encountered on the eastern border of the county. Chehalis and Queniult rivers flow in broad valleys from east to west across the county. The region between these streams is drained by Humptulips River, which flows southwest and empties into Grays Harbor.

Along the coast the yearly rainfall averages about 90 inches, but inland it decreases gradually, dropping to 65 inches at the extreme eastern edge of the county. Practically the whole county is covered with a luxuriant growth of vegetation, the only exceptions being the small areas of outwash glacial gravels, commonly known as " prairies," where but few trees have yet begun to grow.

The rocks of Chehalis County as far as known are of Tertiary age. Marine Eocene fossils have been identified at Porter and Elma, in the eastern part of the county, and marine Pliocene fossils have been found at Granville, at the mouth of Queniult River. The Tertiary rocks are chiefly clastic, consisting of beds of clay, sand, and gravel. Occasionally they contain a small igneous dike, usually of basalt. Nearly all of the county north of Grays Harbor and Chehalis River has been glaciated, and the glacial till, with layers of sand and gravel, is very thick.

The very heavy rainfall yields an abundant supply of water for every purpose. The lakes and the deep porous soil are great storage reservoirs, which feed the streams at all seasons and give them an even flow. The only contamination is due to the decay of vegetal matter in the streams and lakes. In the glacial sediments and in the gravels and sands of the river valleys excellent water may be had from wells of moderate depth. It is not likely that deep wells will ever be necessary in order to afford a supply of water in any part of the county. Since the sedimentary rocks above mentioned are known to be folded, it is possible that artesian basins have been formed, within which flowing wells may be secured by drilling.

Municipal systems.—Aberdeen, at the head of Grays Harbor, gets its supply of water from a creek. The quantity obtained in this way is hardly sufficient for present needs, and water from another creek near by will soon be used. The system of waterworks used is that of direct pressure. As a rule the wells do not yield good water, since in digging them marine deposits of mud and sand are penetrated.

A sanitary analysis of the Aberdeen city water was made April 14, 1901, by Prof. H. G. Byers, of the University of Washington, with the following result:

Analysis of water from city well at Aberdeen.

[Parts per million.]

Total solids	75.69
Volatile solids	12.50
Nonvolatile solids	63.19
Nitrogen as nitrites	None.
Nitrogen as nitrates	.090
Nitrogen as free ammonia	None.
Nitrogen as albuminoid ammonia	.021
Oxygen consumed	.966
Chlorine	6.50

The water supply of Cosmopolis is obtained chiefly from wells, but in part from some creeks owned by the Grays Harbor Commercial Company. From the creeks to the town a gravity system of waterworks has been installed. The water is soft, of good quality, and is ample for present needs and probably for all future demands.

Hoquiam, on Grays Harbor, secures its water supply from the headwaters of Little Hoquiam River, which affords a supply of good water sufficient for present and future needs. The water is pumped into a reservoir and then distributed by gravity. Besides serving the domestic demands the water is also used as a boiler supply. Practically no wells are used in the city, especially in that part which is situated on the tide flats.

Montesano, the county seat of Chehalis County, obtains its water supply from some springs north of the city. The supply is sufficient for the present, but some other source may be necessary in the future. From the springs the water flows into a reservoir and is then distributed by gravity. The water is of good quality and there are no sources of contamination. In Montesano a few wells have been dug, usually about 35 feet in depth. The water-bearing stratum is gravel. The water from the wells is good and may be obtained in large quantities. The water level varies only in the dry summer months, when it falls somewhat.

Ocosta, a town of about 300 inhabitants, on the southwest shore of Grays Harbor, obtains a town supply from wells and springs. The water is of good quality and ample for present needs. The wells have an average depth of 50 feet and the water is obtained from beds of sand. The sand is overlain by clay, so that there is no contamination from the surface. The water level in the wells is somewhat lower in summer than in winter.

CHELAN COUNTY.

General statement.—Chelan County lies a little north of the center of the State, and extends from the summit of the Cascade Mountains southeast to Columbia River. The topography is very rugged, since it includes some of the most broken parts of the Cascades. From the northwestern border of the county, which has an altitude of 7,000 to 8,000 feet, the surface slopes southeastward to the Columbia, where the elevation is only about 700 feet above the sea. The streams flow in deep valleys or canyons, and the divides are very sharp and lined with rows of peaks. There are three prominent drainage systems, all having northwest-southeast courses. Beginning at the north these are Stehekin River and Lake Chelan, Entiat River, and Wenatchee River.

The rocks of Chelan County are mainly granites, gneisses, schists, and other crystallines. The principal exception is the sandstone of Eocene age in the southern part of the county. The sandstone is of lacustrine origin and forms a belt reaching from Wenatchee to Leavenworth in an east-west course and from the southern border of the county northward for about 30 miles.

The rainfall is greatest along the extreme western border, or in the region of the highest mountains, where it is from 40 to 60 inches a year. In the descent eastward it decreases rapidly until in the valley of the Columbia it averages but 15 inches yearly. In the mountains much of the precipitation is in the form of snow, and along the summit of the Cascades are many glaciers. The snow-fed mountain streams carry an abundance of the purest water, and in all parts of the county is an excellent water supply. On the alluvial fans and terraces along Columbia and Wenatchee rivers large tracts are now under successful irrigation, the water being supplied by the mountain streams.

Municipal systems.—The town of Chelan has obtained its water in part from springs and in part from Lake Chelan. A system of water-works is now being installed whereby the water from the lake will be pumped into reservoirs and then be distributed throughout the town by gravity. As the water of the lake comes from the snow fields and streams of the high mountains, and is not contaminated, it is exceedingly pure and healthful. Besides serving as a domestic supply, a large amount of water will be used for irrigation within and about the town. Repeated efforts have been made to obtain water from wells, but although depths of 100 feet were reached no successful wells have been dug on the town site.

Lakeside, at the southern end of Lake Chelan and near the town of Chelan, has in the past depended upon wells for its water supply. Like the town of Chelan, it has under construction a system of water-works which will draw the supply from the lake.

Wenatchee, the county seat of Chelan County, situated on Columbia River at the mouth of the Wenatchee, obtains its supply of water from a creek which flows out from the Cascade Mountains. This supply is sufficient for present needs, but must be supplemented by something better in the future. Besides its use as a house supply, the water is extensively used for irrigating lawns, gardens, and orchards. There are no successful wells in the region about Wenatchee.

Springs.—In the eastern part of the county a number of springs are employed for domestic supply and for irrigation. The spring owned by George Brisson, which may be regarded as a typical one, issues as a stream from a porous basaltic rock. The water is of good quality, carries no sediment, remains constant for the most part, but decreases somewhat in flow in the autumn.

CLALLAM COUNTY.

General statement.—Clallam County lies in the extreme north-west corner of the State, and has a frontage on both the Pacific Ocean and the Strait of Juan de Fuca. It includes the northern half of the Olympic Mountains, the highest peak of which, Mount Olympus, stands on the southern border of the county. The Olympics are rugged, deeply dissected mountains, and reach outward almost to the coast, leaving a narrow belt of hills immediately along the shore. This belt is much wider along the Pacific coast than along the strait. The highest mountain peaks are from 6,000 to 8,000 feet above the sea.

The greatest rainfall in the State is at Cape Flattery, where the average for the year reaches 100 inches. From this point eastward the precipitation decreases regularly, until at the eastern border of the county it is from 20 to 30 inches per year. In the highest mountains much of the precipitation is in the form of snow, and the snow fields and glaciers serve as reservoirs from which the streams are fed during the summer months. The great abundance of pure mountain water will not only serve as an ample supply for all municipal and domestic requirements, but will afford excellent water power. All parts of the county except the highest mountains are densely forest clad, and the forests are eminently helpful in retarding the run-off and equalizing the flow of the streams from season to season.

In a belt varying from 10 to 20 miles in width, extending along both the north and west coasts, the rocks are of Tertiary age and represent marine sediments. Eocene or Oligocene fossils have been identified along the Strait of Juan de Fuca between Twin River and Gettysburg, and fossils that are probably of upper Miocene age have been found near the mouth of the Quillayute River on the west coast. The Tertiary rocks are all sedimentary, consisting chiefly of conglomerates, sandstones, and shales. Within these rocks a good supply of water may be had from comparatively shallow wells.

To the south and east of the belt of rocks above mentioned, within the higher Olympics, the rocks consist of schists, slates, and other metamorphics, with great intrusions and extrusions of igneous rocks. The age of these rocks and their relations to the sedimentaries above described have never been determined. Along the northern border of the county the glacial sediments are commonly very heavy, and from these it is usually easy to secure good water by means of wells from 20 to 40 feet in depth. There are many large lakes within the county, such as Ozette and Crescent, which are reservoirs of the purest water. Lake Crescent is a very deep lake, at the border line of the high mountains, with water of extraordinary blueness and purity.

Municipal systems.—The water supply of Port Angeles, the county seat of Clallam County, comes in part from wells, but in the main is derived from Frazer Creek, which is a small stream rising in the Olympic foothills back of the town and flowing through the town site. About one-half of the water of the stream is taken out in pipes and distributed by gravity. As the water is often subject to contamination, and is of insufficient quantity for the future, it affords a somewhat unsatisfactory supply. A plan is on foot to secure water from Little River, which is within 6 miles of the city. This stream rises in the snow fields of the Olympic Mountains and carries a large volume of excellent water. The wells that have been dug about Port Angeles vary in depth from 12 to 40 feet. The water-bearing strata of sand and gravel are overlain by clay, which prevents contamination from the surface. The water level is lowered somewhat in the summer months, but for domestic uses the wells are never exhausted.

Port Crescent, located on the Strait of Juan de Fuca, a few miles west of Port Angeles, secures its water supply mainly from wells, but to a limited extent from a small lake. The wells range in depth from 16 to 20 feet. Water is obtained from sandstone, which is overlain by a thin layer of soil. The water rises in the wells to within a few few feet of the surface, and the level is scarcely affected by pumping.

CLARKE COUNTY.

General statement.—Clarke County is situated on the southwestern border line of the State, having Columbia River on its southern and western sides, Cowlitz County on the north, and Skamania County on the east. From a low plain along the Columbia the surface rises gradually to the foothills of the Cascades until the northeastern part of the county is reached, which presents a very broken appearance. The principal streams are the North and South forks of Lewis River, Salmon River, and Washougal River.

The yearly rainfall of Clarke County varies in passing from west to east, but it may be said to average about 50 inches. The precipitation is nearly all in the form of rain, the elevation above sea not being sufficient to produce snow to any marked degree.

The bed rock is mainly basalt, except in the northeastern portion of the county, where the metamorphic and granitic rocks of the Cascades prevail. The weathered basalt affords an excellent soil, and agriculture has come to be a very important industry. The soil has formed to such a depth that it contains a good water supply, and surface wells are therefore in common use. Along Columbia River are broad gravel terraces from which large quantities of water are obtained by means of springs and wells. The springs are common along the terrace bases and often yield large amounts of water. The wells are from 25 to 75 feet in depth, and the water-bearing strata are usually overlain by clay, so that surface contamination is at a minimum.

Municipal systems.—Vancouver, the county seat, secures its water supply from 3 springs and 2 wells. The amount thus obtained is sufficient for present needs, and doubtless will be ample for some time to come. There are no sources of contamination, and the water is of excellent quality. A gravity system of waterworks is used. The wells about Vancouver range in depth from 30 to 75 feet. The water is found in a stratum of coarse gravel, which is overlain by clay. It rises to within 25 feet of the surface, and very little change in the level is noted from season to season. The water level is affected only to a slight degree by pumping. Besides its use for domestic purposes, the water system is drawn upon for fire protection, boiler supply, etc. An analysis of the water showed the total solids per gallon of water to be 10.05; organic matter per gallon, 0.60; hardness, 11.30.

Springs.—Five miles east of Vancouver are 3 springs owned by the Vancouver Waterworks Company. The water flows out in streams from a bed of gravel which outcrops upon a hillside. The flow varies slightly at different times, but is about 1,750,000 gallons daily. The water is soft, has a pleasant taste, and carries no sediment.

COLUMBIA COUNTY.

General statement.—Columbia County lies between Wallawalla County, on the west, and Garfield County, on the east, and reaches from the Oregon line northward to Snake River. The southern part of the county lies within the broad plateau of the Blue Mountains, and has a height of from 4,000 to 5,000 feet above the sea. The northern part is a region of high, rolling hills and deep ravines. Along the northern border Snake River flows in a canyon that has a depth of 1,000 to 1,500 feet.

The rainfall varies from 15 inches a year along Snake River to 25 inches in the central and southern parts. In the Blue Mountains it is sufficient to produce a forest covering. In the northern part of the county the bunch grass which once grew so luxuriantly is fast giving way to wheat fields. . In agriculture irrigation is necessary only along the benches or terraces of the Snake, where there are many fine fruit and alfalfa ranches.

As far as known, there are no other rocks in the county save basalt. Some parts of the basalt are very vesicular or porous, and are commonly water bearing. In digging wells an ample supply of good water is generally found within 50 feet of the surface. On the hillsides, wherever the porous basalt outcrops, springs may occur, and they are in common use as domestic water supplies.

Municipal systems.—Dayton, the principal town and the county seat of Columbia County, obtains its water supply from springs. These afford a supply which will doubtless be ample for all time to come. A gravity system of waterworks is used. In a few instances water is obtained from private wells, which range in depth from 25 to 50 feet. The water is found in very porous layers of basalt. In the wells the water level varies slightly with the seasons, but is not affected to any appreciable extent by pumping.

Springs.—Near Dayton are the springs from which the city water supply is derived. The water seeps out from beds of gravel at the base of a hill. The amount of water varies considerably with the season, the flow being reduced in the dry summer months. The water is soft, and no deposits of mineral matter or sediments are found about the springs.

COWLITZ COUNTY.

General statement.—Cowlitz County is situated on the southwestern boundary of the State, north of Clarke County and Columbia River.

The western half of the county is composed of low irregular hills and broad stream valleys; the eastern half is very rough and broken, since it contains the foothills and outlying spurs of the Cascades. The principal streams are Cowlitz, Toutle, and Kalama rivers.

The annual rainfall varies from 65 inches on the western margin to 60 inches in the eastern part. Every part of the county is therefore well watered and heavily forested. There is sufficient water in the streams to supply any possible municipal demands. On the terraces or benches of the larger streams good water is secured from wells at depths between 15 and 50 feet.

In the northern and western parts of the county the prevailing rocks are shales and sandstones of Eocene age. Along Cowlitz River these rocks are coal bearing, and in the neighborhood of Castle Rock and Kelso coal mines have been in operation for some time. The soft coal measures have readily decomposed at the surface, and upon them is a residual soil many feet in thickness. From the soil or the porous rock below water in large quantities is readily obtained by wells of moderate depth. In the southeastern part of the county, in the vicinity of Kalama, the usual rock is basalt, and from it water is secured in wells averaging 25 feet in depth.

Municipal systems.—Castle Rock obtains its water supply from a stream which rises in the foothills east of the town. This supply is sufficient for the present, but must be supplemented in the future by water from some other source. The water is brought from the stream in pipes and distributed by a gravity system. The wells in the vicinity of Castle Rock vary in depth from 15 to 20 feet. The water-bearing material is gravel. The height of the water column depends upon the stage of water in Cowlitz River near by, rising and falling with the river level.

Kalama obtains its water by a gravity system from two creeks. The supply will doubtless have to be increased in the future. The water is very good, the only contamination being from a small amount of decaying vegetation. A few wells are used about Kalama, having an average depth of 25 feet. They are sunk in basalt and, as contamination may be entirely prevented, the water is very good.

DOUGLAS COUNTY.

General statement.—Douglas County is situated immediately east of the central part of the State, within what is commonly known as the "Big Bend" of the Columbia. Its surface is that of a plateau, broken by the Badger Mountains in the western part and by the Saddle Mountains in the southern part, and sloping gently from north to south. Along its northern margin the plateau has an elevation of about 3,000 feet above sea level, which decreases to less than 1,000 feet along the southern border of the county. Within the plateau are three notable canyons or deep clefts, namely, the present valley of the Columbia, and Moses and Grand coulees, the latter doubtless representing a former course of Columbia River. Several small creeks which enter Moses Coulee sink in the sand and do not form a main stream. In Grand Coulee is a chain of lakes, the largest being Blue and Moses lakes. Some of the lakes are fresh, but the majority are alkaline.

Douglas County lies within the region of lowest rainfall in the State. In the northwestern part the average precipitation for the year is 15 inches; in the remainder of the county it is 10 inches or less. The precipitation is largely in the form of snow, especially where the elevation is greatest. There is no forest growth, and prairie conditions wholly prevail. In the highest portions of the county the bunch grass grows abundantly, and cattle raising is an important industry. Here also wheat may be grown successfully, and it has already come to be an important product. In the southern part of the county, where the rainfall is least, irrigation must be practiced in order to carry on agriculture.

The bed rock is practically all basalt except a narrow fringe of granite which is exposed in the canyon walls of the Columbia, and some granite outcrops in the northern end of Grand Coulee. A small area of early Tertiary sandstone underlies the basalt along Columbia River opposite Wenatchee, and late Tertiary lake beds covering a considerable area of the basalt in the southern part of the county should also be noted. From the basalt a very thick and fine-grained soil has been formed, which retains water to an unusual degree. In this way enough water is held within the soil after the winter precipitation to grow and mature the wheat during the coming spring and summer. Within the soil or the porous portions of the basalt below it is usually not difficult to secure a sufficient supply of water for house and farm purposes. Springs are more or less common throughout the county, more particularly at the bases of the cliffs along the Columbia and the coulees. Springs are largely the sources of the lakes mentioned above.

Municipal systems.—The town of Wilsoncreek depends in the main upon wells for its supply of water, but a small stream is also used to some extent. The supply is barely sufficient for present needs, and another source must be sought in the future. While some of the wells have soft water, the others are slightly alkaline. They range in depth from 12 to 54 feet. The more shallow wells obtain water from beds of gravel, while the deeper wells enter the bed rock. The water level in the wells scarcely varies from season to season and is not ordinarily affected by pumping.

Grand Coulee, which has a northeast-southwest direction across the county, was in a former time occupied by Columbia River. Along the course of the coulee there is now a chain of shallow lakes, some of which are fresh, but the most of them are alkaline. The following descriptive matter concerning two of them is taken from Vol. I, of the Annual Report, Washington Geological Survey, for 1901.

Moses Lake, which lies about 12 miles southeast of Ephrata, on the Great Northern Railway, is about 18 miles long and a mile wide, and is very shallow. The average depth is approximately 20 feet. It lies in a shallow basin with low banks, so that a rise of but a few feet would inundate a large section of country. The water is unfit for drinking purposes, but is not strongly alkaline and could probably be used in irrigation. The section of country in which these lakes are located is of course very dry, and supports only a scanty vegetation. Where there is water, however, the soil is very fertile. The lake drains a large area through upper Crab Creek. It has no outlet, but across its foot lies a low range of sand hills through which the water seeps into the sources of lower Crab Creek, which occupies the bed of the canyon below. Along this canyon lie numerous shallow ponds which dry up in summer. The deposits left by these are not of any considerable value, though they contain an appreciable quantity of borax.

An interesting feature of Moses Lake is the fact that it is gradually rising, having risen about 10 feet in the last seven years. If it continues to rise through a few more feet it will break through a clear course into lower Crab Creek and empty into the Columbia.

The analysis of the water of Moses Lake is as follows. The analysis is by H. G. Knight:

[Parts per thousand.]

Total solids	0.32357
Volatile solids	.10095
Nonvolatile solids	.22262
Silica	.01502
Alumina and iron oxide	.00331
Calcium carbonate	.06235
Magnesium carbonate	.07525
Sodium sulphate	.01258
Sodium chloride	.01895
Sodium carbonate	.10914

More interesting is the so-called Soap Lake, or Sanitarium Lake, situated about 6 miles north of Ephrata. This lake is so called because it is so strongly alkaline as to be soapy to the touch, and when a strong wind blows across it the water along the shore is beaten into great rolls of foam. Fish can not live in the water, nor is there any vegetation in this as in Moses Lake. The water is used for bathing, but to those not accustomed to its use the water has a slightly caustic or irritating effect. It is also claimed that it is useful medicinally. There is much of peculiar interest about the lake. It is about 2¼ by three-fourths miles in extent and is very deep in places and probably averages about 40 feet. It drains only a very small area of country and has neither inlet nor outlet in the form of streams. It is located in a deep basin walled to the height of 100 feet or more on the east and west by cliffs of black basalt. The land to the north and south rises slowly; on the south to nearly the height of the cliffs, but on the north the rise is so slight that should the lake rise 15 feet it would empty into the next of the chain of lakes to the north. The source of the water of the lake is said to be a spring in the center. The Indians of the neighborhood assert that only a few years since the lake was very small and was fed by this strongly alkaline spring. Fresh water is, however, continually seeping in from the shores, as is shown by the fact that fresh-water wells may be sunk even but a few feet from the shore, and that the cattle, disliking the strongly alkaline water, face the shore to obtain the sweeter seepage.

The analysis of the water is as follows:

[Parts per thousand.]

Total solids	28. 2669	Potassium carbonate	.5117
Volatile solids	.62503	Lithium sulphate	Trace.
Nonvolatile solids	27. 64186	Phosphorus pentoxide	.1201s
Silica	.12816	Carbon dioxide (semicombined)	1.3708-1
Alumina and iron oxide	Trace.		
Calcium sulphate	Trace.	Borax	None.
Calcium carbonate	Trace.	Iodine	None.
Magnesium sulphate	.39099	Free ammonia	.0340)
Sodium sulphate	6. 34872	Albuminoid ammonia	1.1060
Sodium chloride	5. 81384	The specific gravity	1.0260
Sodium carbonate	14. 08901		

FRANKLIN COUNTY.

General statement.—Franklin County lies between Wallawalla County on the south and Adams County on the north, with Columbia River as its western boundary. The surface is that of a plain, sloping gently toward Columbia and Snake rivers. In the northwest corner of the county the plain is about 300 feet above the surface of the Columbia, giving rise to the cliffs along the stream known as the White Bluffs. In the eastern part of the county the plain rises to a height of 300 to 400 feet above the level of the Snake. Near the confluence of the two rivers the plain decreases in elevation and the banks of the streams are but a few feet in height.

The average rainfall is about 10 inches a year. This will not admit of any forest growth and scarcely permits of the growth of bunch grass. Agriculture may be carried on only where irrigation is possible. Irrigated tracts along the rivers yield very fine returns in alfalfa and fruit.

In the western part of Franklin County the outcropping rocks are thin beds of sand and clay, with layers of volcanic dust, which represent Miocene lacustrine sediments. Such deposits at one time may have covered the entire county, but in the eastern portion basalt alone now appears. At a number of places successful wells have been drilled in the basalt, the water being found in the more porous or vesicular parts of the rock. The depth of the wells varies from about 200 to nearly 700 feet. It is evident that, in advance of drilling, the depth at which water will be found is conjectural. In the western part of the county a well is now being drilled in the sedimentary rocks noted above.

Deep wells.—W. T. Braden, living at Connell, completed a deep well in September, 1902. The well is located in a canyon, and its mouth has an elevation above the sea of 840 feet. It has a diameter of 5 inches and a depth of 676 feet. The well is cased from the surface to solid rock, a distance of 100 feet. The water rises in the well about 36 feet and is brought to the surface by a wind pump. The temperature of the water when brought to the well mouth is 51° F. The water level does not vary during the day or year and is not affected by pumping. The cost of the well, including the pumping machinery, was about $3,000.

There are several firms making a business of drilling wells in Franklin County, the most prominent of which is the Reinbolt Well Drilling Company. A few of the wells drilled by this company, with their depths, are here given: H. W. Brummond, 410 feet; John Finkbiner, 485 feet; John L. Wordheim, 212 feet; Connell Land and Improvement Company, 243 feet; William Fisch, 265 feet; William Elgin, 323 feet; Charles Schelley, 487 feet.

Springs.—A number of springs are known to exist in the eastern part of Franklin County, although but a few of these have been reported. These springs are oftentimes the sources of supply for small lakes which have no outlets.

JEFFERSON COUNTY.

General statement.—Jefferson County is situated in the northwestern part of the State, between Clallam County on the north and west and Chehalis and Mason counties on the south, and extends from Puget Sound to the Pacific Ocean. Topographically the surface presents a great diversity. The county contains the most rugged portions of the Olympic Mountains, the only areas that are comparatively level being in the northeastern part, within a few miles of the shores of Puget Sound. This is the only part of the county inhabited to any degree, there being no inhabitants within the Olympics and but very few west of the mountains along the Pacific coast.

The rainfall along the western coast averages over 85 inches a year. It decreases steadily eastward and is less than 20 inches in the vicinity of Port Townsend. The district east of the Olympics has a low rainfall because the high mountains divert to some degree the rain-laden westerly winds. Within the Olympics are several glaciers and snow fields which feed the streams during the rainless months of summer.

The band of Tertiary rocks which encircles the Olympic Mountains outcrops in both the eastern and the western parts of the county. The rock series is composed chiefly of conglomerates, sandstones, and shales, and without doubt contains large quantities of water which can be secured by means of wells. The bed rock just mentioned is often deeply buried by glacial sediments, which are important reservoirs.

Municipal systems.—Port Townsend obtains its water supply from driven and dug wells owned by the Spring Valley Water Company. From these wells 250,000 gallons per day are obtained by pumping. This supply is hardly sufficient for present needs, and a gravity system from some stream coming out of the Olympic Mountains is in contemplation. The water in use at present is of good quality. The wells do not enter bed rock, the water-bearing materials being sand and gravel. A heavy bed of clay which lies above the sand and gravel prevents contamination. An analysis of the Port Townsend water is given below:

Analysis of water from Port Townsend, Wash.

[Grains per gallon.]

Silica	1.255
Alumina and iron oxide	.146
Calcium carbonate	16.731
Magnesium carbonate	6.987
Magnesium chloride	11.200
Calcium sulphate	1.826
Sodium and potassium chlorides	29.236
Sodium and potassium carbonates	1.531
Total solids	68.912

Port Ludlow obtains its water supply from a small creek. The water is soft, of good quality, and sufficient for all probable future needs. A gravity system of waterworks is used. The water is drawn upon for a boiler supply for the Port Ludlow mills as well as the domestic supply for the town.

KING COUNTY.

General statement.—King County is situated west of the center of the State and extends from the summit of the Cascades westward to Puget Sound. It lies between Snohomish County on the north and Pierce County on the south. Along the shores of the sound the surface is that of a plain, rising to a height of about 300 feet above the sea; east of the plain are hills and ridges, the latter having in general a north-south direction; immediately east of this belt are the mountains, rising from a height of about 3,000 feet along their western border to an average height of 6,000 feet, when the summit line is reached. The mountains are deeply dissected by the several forks of Snoqualmie, Cedar, Green, White, and other rivers.

The annual rainfall varies from 40 inches along the shores of the sound to more than 60 inches within the mountains. About the sound practically all of the precipitation is in the form of rain, but with increase in elevation the snowfall becomes important, and in the higher mountains the amount of snow falling each year is very large.

The presence of snow and glaciers has an important bearing on the run-off of the streams from season to season. Apart from its value as sources of municipal supply, the movement of so much water from higher to lower levels is productive of great power. Already the water power of the cataracts and falls is being harnessed for the use of man. At Snoqualmie Falls, 25 miles east of Seattle, a plant has been installed which develops a total of 10,000 horsepower. Snoqualmie River at this point has a vertical drop of 270 feet, with a flow of about 1,000 second-feet during the driest season and about ten times as much during the periods of high water.

The strata which outcrop about the sound and eastward for a distance of about 25 miles are mainly clastic rocks of early Tertiary age. At many places they are coal bearing, and important coal mines have been developed at Black Diamond, Franklin, Palmer, Renton, Newcastle, Issaquah, and elsewhere. East of the Tertiary sediments are the metamorphic and igneous rocks of the Cascades, which in King County at least are virtually unstudied. In the vicinity of the sound bed rock outcrops but rarely, being covered by a heavy mantle of glacial sediments. At many places this mantle is known to be more than 500 feet in thickness. The glacial sediments comprise beds of till, with stratified sand, gravel, and clay. From the layers of sand and gravel which are interstratified with the till an abundant supply of good water is generally obtained. The wells are usually shallow, it being rarely necessary to go deeper than 40 feet. Springs are very common about the bases of the hills or upon the hillsides where the water-bearing gravels and sands outcrop.

Municipal systems.—The town of Auburn obtains its water supply chiefly from wells, which vary in depth from 40 to 50 feet. The wells are sunk in the alluvium of the White River Valley, the water-bearing materials being sand and gravel. In no instance has bed rock been reached. In some instances shallow wells have been used for a little time, but in these the water is not good. The water from the deeper wells is soft, is not contaminated, and is obtained in ample quantity, in some instances rising to the surface. No other supply of water for drinking purposes is contemplated, but it is likely that water for fire protection will be obtained from White River.

The city of Ballard obtains about 300,000 gallons of water daily from springs and deep wells, about half from each source. The wells are the more satisfactory. While this amount is ample for present needs, it will not be sufficient in the course of time if the city continues its rapid growth. It is very likely that in the near future the supply for Ballard will be secured from the Cedar River system owned by the city of Seattle. The surface wells in Ballard have depths of 12 to 20 feet, but the wells used for the municipal supply average about 160 feet in depth. For the most part they penetrate glacial till or hardpan, the water-bearing strata being beds of sand and gravel. The system of waterworks is direct pressure.

A sanitary analysis of water from one of the deep wells in Ballard, made by Prof. H. G. Byers, of the University of Washington, is as follows:

Sanitary analysis of water from deep well at Ballard.

[Parts per million.]

Total solids	165.66
Oxygen consumed	2.89
Chlorine	5.50
Nitrogen as free ammonia	.874
Nitrogen as albuminoid ammonia	.100
Nitrogen as nitrites	None.
Nitrogen as nitrates	.320

The municipal supply of water for Columbia City is obtained from Seattle. A few private wells are in use having an average depth of 30 feet. They penetrate glacial deposits only and from them water of good quality is obtained.

Water for the town of Enumclaw is furnished from the system of the White River Lumber Company. The water comes from streams flowing out from the Cascade Mountains, and as there are no sources of contamination it is of excellent quality. A gravity system of waterworks has been installed. It is possible that in the future the town supply may be taken from some springs located about 4 miles away. The springs are at a height of about 270 feet above that of the town. A few wells are in use about Enumclaw, which have an average depth of 30 feet. The wells are dug in glacial sediments, a till or hardpan lying above, with sand and gravel below. The wells afford excellent water, the flow rising to the surface in the wet season.

The water supply for Issaquah is furnished by the Gilman Water Company, which owns large springs near the town. The water is of good quality, there being no sources of contamination except possibly a little decaying vegetable matter. The water is conducted through the town by a gravity system. The few private wells in Issaquah range in depth from 20 to 30 feet, the water coming from a stratum of loose gravel. A large quantity of water may be obtained in this way, since it rises almost to the surface and is not affected by pumping. North of the town, toward Lake Sammamish, wells from 60 to 90 feet in depth have been driven, from which there is a continuous flow. The water rises from 3 to 10 feet above the surface.

The town of Kent gets its supply from springs which yield water of excellent quality. While the springs now drawn upon afford a supply for present needs, the product of other springs near by will be drawn upon in the future. From the springs the water is carried throughout the town by a gravity system. In the region about Kent good water may be obtained by means of driven wells. The wells when driven to a depth of 200 feet are artesian in character, the water rising about 6 feet above the surface.

Water for Renton is obtained from a spring situated near the town limits and at a height of 320 feet above the level of the town. The spring belongs to the Seattle Electric Company, but the town has a lease upon it for fifty years. The spring supplies 120,000 gallons of water daily, and should this not be sufficient for future needs water may be obtained from other springs or from Cedar River, which flows through the town. The water, besides being used for domestic purposes, affords fire protection and furnishes the necessary boiler supply for a coal mine and a brickyard. The private wells in use in Renton are mostly shallow, ranging in depth from 10 to 25 feet. Only one enters rock, the others obtaining water chiefly from the gravels of the Cedar River flood plain.

The city of Seattle obtains its supply of water from Cedar River and Cedar Lake. The source of supply is in the Cascade Mountains, the water coming to the reservoir directly from the snow fields. The water is, therefore, soft, clear, and of superior quality. The city owns Cedar Lake and a large portion of Cedar River. The available water supply averages about 600,000,000 gallons per day. There are now piped to the city 22,500,000 gallons daily. Other supply mains will be constructed whenever they are found to be necessary.

A sanitary analysis of the city water of Seattle, made by Prof. H. G. Byers, of the University of Washington, on April 27, 1901, gave the following results:

Sanitary analysis of city water of Seattle.

[Parts per million.]

Total solids	36. 49
Oxygen consumed	1. 26
Chlorine	1. 50
Nitrogen as free ammonia	. 008
Nitrogen as albuminoid ammonia	Trace.
Nitrogen as nitrites	None.
Nitrogen as nitrates	None.

West Seattle obtains water from private wells and from some springs owned by the West Seattle Land and Improvement Company. The water is pumped into tanks, from which it is distributed by gravity. The supply even at the present time can hardly be said to be sufficient, and it is likely that arrangements will soon be made whereby water may be obtained from the Seattle system. The wells about West Seattle range in depth from 30 to 75 feet, water being most commonly found at about 50 feet. The wells are dug entirely in glacial material, mostly sand and gravel.

Springs.—Near the town of Berlin, in the northeastern part of the county, there is a mineral spring owned by the Everett Bottling Works. The water flows out as a stream from the base of a mountain of granitic rock. The flow is uniform from season to season. The quantity flowing has never been measured. The water has a taste of soda. No improvements of any character have been made at the spring, and so far no use has been made of the water. It is probable that in the near future a hotel will be built at the spring and other improvements made.

An analysis of water from the Berlin springs, made by H. G. Knight, of the University of Washington, is as follows:

Analysis of water from Berlin springs.

[Parts per thousand.]

Solids, nonvolatile	0. 5473
Silica	. 0078
Alumina and iron oxide	. 0150
Calcium sulphate	. 0529
Calcium carbonate	. 5627
Magnesium chloride	. 1693
Magnesium sulphate	. 0935
Sodium sulphate	. 9331
Potassium chloride	. 0267
Carbon dioxide	1. 4720

Near the town of Issaquah, at the head of a short, deep valley, is a large spring from which the water supply for the town of Issaquah is taken. The flow has not been measured, but there is sufficient water for a town of 1,000 inhabitants. The water is very clear, cold, and has a pleasant taste. It issues as a stream from a bed of gravel. No improvements have been made at the spring, and none are contemplated. The spring is owned by the Gilman Water Company.

Along the valley side, at the base of a steep hill near Kent, is a large spring from which the town supply of water is largely taken. The daily flow varies from 500,000 gallons in winter to 350,000 gallons in summer. The water is not appreciably charged with minerals, and is very clear and cold when it leaves the spring. It issues as a stream from a bed of gravel which is a part of the glacial sediments. At the spring a reservoir is now being built for storage purposes.

The Great Northern Hot Springs are located near Madison, in the northeast corner of the county. Near the Great Northern Railway. a mile from the springs, the Hot Springs Hotel has been built, with accommodations for 50 guests. The water is piped to the hotel, where it is used for drinking purposes and for baths. The water has been found to be very helpful for rheumatism and for kidney diseases. The water seeps out from the talus rock, and has a temperature of 122° F.

The following analysis of water from the Great Northern Hot Springs was made by C. Osseward, chemist for the Stewart & Holmes Drug Company, Seattle:

Analysis of water from the Great Northern Hot Springs.

[Grains per gallon]

Total solids	9. 9
Chlorine	. 87
Iron	. 76
Lime	2. 33
Magnesia	1. 1
Silica	1. 34
Sodium	1. 63
Potassium	. 34
Sulphuric anhydride	. 52
Ammonia	. 00058

On a hillside near Renton there is a spring issuing from a bed of gravel. The flow is about 120,000 gallons daily. The quantity varies with the season, being about one-fourth less in a very dry season than in a wet one. The water does not carry any sediment, is very clear, and has a pleasant taste. It is used as a town supply by Renton.

KITSAP COUNTY.

General statement.—Kitsap County is situated east of Jefferson and Mason counties, west of King County, and north of Pierce County. It is almost surrounded by the arms or inlets of Puget Sound. With the exception of a range of hills in the southwestern part of the county, along Hood Canal, the surface is that of a plain lying but little above the sea. The coast line is very deeply indented and irregular, abounding in fine bays and harbors.

The rainfall gradually decreases from 60 inches per year in the southwestern part to 30 inches in the northeastern part. It is sufficient to give rise to the very many small streams which are to be found throughout the county and also to produce a very dense forest growth.

Over the major portion of Kitsap County the mantle of glacial sediments is very heavy, so that the bed rock does not often appear at the surface. The wells for the most part penetrate the glacial materials only, and from the latter a satisfactory supply of good water is obtained. As elsewhere within the glaciated area, springs are very common and are often utilized.

Municipal systems.—Bremerton obtains its water supply to a limited degree from wells, but for the most part from a stream which is fed by springs. In this way is obtained a supply of very good water, which is free from any contamination. The quantity will be ample for a long time to come. Gravity is the system employed. The private wells, as a rule, are shallow and are sunk altogether in glacial till, the water flowing from intercalated beds of sand and gravel.

An analysis of water from a spring near Bremerton, made by H. G. Knight, of the University of Washington, is as follows:

Analysis of water from spring near Bremerton.

[Parts per thousand.]

Nonvolatile solids	0. 45194
Silica	. 01334
Alumina and iron oxides	. 04764
Calcium sulphate	. 046385
Magnesium chloride	. 04008
Magnesium sulphate	. 07790
Sodium sulphate	. 23686
Lithium sulphate	. 02128

The source of supply for Charleston is found in some springs and creeks west and south of the town. From these the water is carried to the town in wooden pipes. It is of good quality, and a quantity ample for present and future needs is easily obtained. The private wells about Charleston range in depth from 20 to 150 feet. These wells have been put down in glacial sediments and the water level rises and falls with the tide in the inlet near by.

The water supply for Port Blakeley comes primarily from a stream in the adjoining hills. A reservoir has been made, and from this, by means of gravity, the water is conducted to the town. There is a certain amount of decaying vegetable matter in the reservoir, and this causes some contamination of the water; otherwise the water is good. and sufficient in quantity to serve the town probably for a long time in the future. Besides this method of obtaining water, cisterns and wells are used to a small degree. These wells have a depth ordinarily of about 70 feet. Most of them are wholly in the glacial till, but some of them penetrate rock altogether, the water-bearing stratum in the latter case being a conglomerate.

Some small streams near Port Gamble are drawn upon for the water supply of that town. From reservoirs along these streams the water is distributed to the town by a gravity system. About 150,000 gallons of water are obtained daily, an amount sufficient for the present and the future. The water is soft and of good quality, no sources of contamination being present. Besides its use for domestic purposes. it is employed to a large extent as a boiler supply.

KITTITAS COUNTY.

General statement.—Kittitas County lies near the center of the State, between Chelan County on the north and Yakima County on the south, and extends from the summit of the Cascades eastward to

Columbia River. From the Cascade divide, which has a general height of about 6,000 feet above the sea, the country descends southeastward to a minimum elevation of 500 feet. In the northwestern part of the county the surface exhibits the usual ruggedness of the higher and wilder parts of the Cascades. The highest peak here, Mount Stuart, has an elevation of 9,470 feet.

The rainfall in Kittitas County varies widely because of the great difference in elevation from point to point. In the high mountains the precipitation is from 50 to 60 inches a year. It decreases eastward until it becomes less than 15 inches at Ellensburg and less than 10 inches in the vicinity of Columbia River. Naturally, the character of the vegetation is greatly modified by the wide variation in the amount of rainfall. The higher mountains as a rule are heavily forested, while the plateaus are sparsely covered with trees. In lower altitudes the trees give way to the bunch grass, which is in turn superseded by sagebrush on the still lower levels. The abundant rainfall of the mountains gives rise to many fine streams, which afford water alike for the use of cities on the plains below and for the irrigation of orchards and alfalfa fields.

The southeastern part of the county is largely covered by Miocene basalts. In the neighborhood of Ellensburg are remnants of a sandstone formation which represents the sediments of a middle Tertiary lake of unknown extent. The geology is somewhat complex, as both sedimentary and volcanic rocks of Eocene and Miocene age are found. The sedimentary beds are lacustrine, and consist chiefly of sandstone and shale. In one of the Eocene sedimentary formations round about the towns of Roslyn and Clealum, seams of coal of great economic value have been found, and here are the largest coal mines in the State. In the vicinity of Mount Stuart, near the southern border line of the county, there are large areas of pre-Tertiary rocks of a complex nature geologically, in which there are granites, granodiorites, serpentines, slates, and schists.

Municipal systems.—The water supply for Clealum comes from mountain springs about 3 miles southwest of town. The springs are 180 feet above the town, thus giving a good head and making it easy to distribute the water throughout the town by a gravity system. For the first mile from the springs a 10-inch pipe was laid, for the next mile an 8-inch pipe, and for the third mile a 6-inch pipe. The water pressure in the town is from 90 to 100 pounds per square inch. The amount of water obtained is believed to be ample for a city of 20,000 inhabitants. Should this supply ever fail, water in great abundance may be obtained from Yakima River, a clear mountain stream which flows through a part of the town. Clealum owns 160 acres of land about the springs, thus effectually preventing contamination. Some time ago an effort was made to obtain artesian water. A well was drilled in Clealum to a depth of 800 feet without striking watei. The water supply of the town is used for irrigating as well as domestic purposes.

Ellensburg obtains water from streams which flow out from the foothills of the Cascades. The water obtained in this way, however, is not of the best quality and is barely sufficient for present needs. An unusual demand is made upon the supply, for a large amount of water is used for irrigation as well as for domestic purposes. The private wells which are in use range in depth from 10 to 20 feet. They are essentially surface wells, not entering the rock at all, but obtaining water from beds of gravel. The well water, as a rule, is hard and contains alkali. The water level in the wells varies, the water rising to the surface when the soil is thoroughly saturated as the result of excessive irrigation.

For its town supply Roslyn obtains water from Clealum River, a mountain stream yielding water of excellent quality. This supply is so satisfactory that no dependence is placed upon wells, cisterns, or other sources.

KLICKITAT COUNTY.

General statement.—Klickitat County is situated on the southern border of the State, with Columbia River on the south, Yakima County on the north, and Skamania County on the west. The western part of the county is within the Cascades, and is therefore very rugged. The topography of the central part is that of a broad plateau sloping from the north toward Columbia River. The topography of the eastern end of the county is that of a plain rising but little above the level of the river.

In the western portion of the county the annual rainfall is from 30 to 35 inches, in the vicinity of Goldendale it is 15 or 20 inches, while in the eastern end of the county it is only about 10 inches. As a consequence of the unequal distribution of rainfall, the western portion of the county is a region of forests and streams; the central part has a very sparse forest growth and few streams, but at the same time is one of the best wheat-growing districts in the State; and the eastern part is treeless and virtually without streams, so that agriculture can be carried on only by means of irrigation.

With the possible exception of the extreme western end, all of the county may be said to be covered by Columbia River lava. The bed rock is basalt. As far as known, no deep wells have been drilled into it, as in other parts of the State, to determine whether or not it is water bearing. That it is water bearing is largely proved by the fact that in a number of places springs issue from the outcrops of porous basaltic rock. The springs in some instances have a temperature considerably above that of the surrounding atmosphere. In some cases also they may be classed as mineral springs, and are believed to have medicinal qualities as well.

Municipal systems.—The water for Goldendale is piped from a spring on the southern slope of Simcoe Mountain, 12 miles to the north. The supply is sufficient for present needs, but will probably have to be increased in the future. Other good springs are conveniently near, and the water from these can be easily utilized. The water is soft and uncontaminated. The spring is located at a sufficient height above the city to give a good head of water. A few wells are used, ranging in depth from 12 to 75 feet. Most of the wells are shallow, water being most commonly found at 20 feet in beds of gravel. From the wells a large amount of water may be obtained. The water level shows but little change from season to season, and is affected by pumping only in a slight degree. In Goldendale a large amount of water is used in irrigation.

Springs.—On Government land in T. 6 N., R. 13 E., there is a large spring, or rather a group of springs. At these springs the water flows out in a stream from basaltic rock. The springs are located at the base of a bluff rising from the valley of Klickitat River. The flow is large, there being no perceptible variation from season to season. The water is clear and odorless, but it has an unpleasant taste. It has a temperature of 76° F. Bubbles of gas are constantly escaping from the water. It has been used to a limited degree for medicinal purposes.

On Big Klickitat River in T. 4 N., R. 14 E., is a spring which is considered to have special medicinal properties. By the internal use of the water, stomach and kidney diseases are benefited, and rheumatism is helped by bathing in the water. The spring issues as a stream from basaltic rock. It has a temperature of 100° F. There is a slight film on its surface and it carries a little sediment. It has a strong mineral taste and is charged with gas, iron, and sulphur. Deposits of iron oxide are made along the stream as it flows away from the spring. It is planned to make a health resort at this spring, and a hotel and bath house are in process of construction. An analysis of the water, made by Prof. H. G. Byers of the University of Washington, is as follows:

Analysis of spring water from near Big Klickitat River.

[Parts per million.]

Total solids	758.7
Nonvolatile solids	569.2
Volatile solids	189.5
Silica	85.5
Ferric oxide and alumina	104.9
Calcium carbonate	129.6

Magnesium carbonate 184.8
Calcium sulphate.. 4.5
Sodium chloride 10.1
Potassium chloride 22.3
Potassium sulphate... 27.5
Free carbon dioxide... 700.0

LEWIS COUNTY.

General statement.—Lewis County is in the southwestern part of the State and extends from the summit of the Cascades to Pacific County on the west, with Chehalis, Thurston, and Pierce counties on its northern border, and Wahkiakum, Cowlitz, and Skamania counties on the south. The county has a very diversified topography. The western end of it lies within the Coast Range, where the highest hills or mountains are from 2,000 to 3,000 feet in height. East of the Coast Range is a broad north-south valley which is in reality a part of the Puget Sound basin. East of this great valley rise the mountains of the Cascades, reaching heights of 5,000 or 6,000 feet on the main divide. The mountains have been greatly dissected by the streams, and some of the valleys, notably that of Cowlitz River, are very broad and deep.

The rainfall is heaviest in the western part of the county, where it averages 70 inches per year. It decreases eastward until in the neighborhood of Chehalis and Centralia it is 50 inches. In the mountains of the eastern part of the county the average yearly rainfall is 60 inches. This heavy rainfall gives rise to a very large number of streams and produces a luxuriant forest growth which is equal to any found elsewhere in the State.

The eastern part of the county, or that within the Cascades, is practically unknown geologically. There are a few small areas of coal-bearing sandstones and shales lying along the upper Cowlitz River and in the region about Cowlitz Pass. The geology of the western two-thirds of the county has been studied to some extent, and as far as known the rocks are of Eocene age. Fossils of marine Eocene types occur in great abundance along the Cowlitz in the vicinity of Little Falls. The sedimentary rocks are mainly sandstones and shales and are coal bearing at a number of places, notably about Chehalis and Centralia, in the vicinity of Alpha and Cinebar, and near Morton. Upon the sedimentary rocks in most places a deep soil has formed, and within this water is obtained with ease by means of shallow wells. Springs are common upon the hillsides, and they may be regarded as important sources of good water. The amount of surface water to be had is so very large that it is not probable that recourse to deep wells will ever be necessary. The structure of the sedimentary rocks is such that they undoubtedly contain artesian basins, which may be tapped whenever the need arises.

Municipal systems.—The water for Centralia is obtained partly from wells and partly from Skookumchuck River. From neither of these is very good water obtained, and some other source must be sought out. The wells range in depth from 20 to 30 feet. Water is most commonly found at about 18 feet, in beds of gravel, but as these beds are not overlain by clay or other impervious material, contamination from the surface is not prevented. From these wells a large supply of water may be obtained. The water level varies but little during the year, being somewhat lower in the dry season.

The water supply for Pe Ell is obtained mostly from wells, although a few families derive their supply from springs. The spring water is satisfactory, but the same can hardly be said of the wells. The latter are very shallow and contamination from the surface is very easy. The well water comes from gravel and rises to the surface in some cases. A large supply of water is had, the water level scarcely varying during the year.

LINCOLN COUNTY.

General statement.—Lincoln County is in the eastern part of the State and is bounded on the east by Spokane County, on the west by Douglas County, on the south by Adams County, and on the north by Columbia and Spokane rivers. The surface is that of a plateau having a general height of about 1,500 feet above sea level. On the north the plateau slopes very abruptly to the canyon of the Columbia, while on the south there is a gentle slope toward Crab Creek. Low, rolling hills occur here and there.

The yearly rainfall varies from 10 inches in the western half to 15 inches in the eastern portion. This is too small to permit of forest growth and hence the county was formerly a prairie region clothed with grasses. The bunch grass now, however, has largely given way to wheat fields, and the county produces each year an increasing amount of grain. The streams are in the main of the intermittent type, there being but very few that are of a permanent character. The small streams are active in the runways during the winter and spring months only and disappear with the coming of summer.

With the exception of the outcrops of granite in the vicinity of Columbia and Spokane rivers the bed rock is altogether basalt. The basalt has within it porous layers which become filled with water, thus forming important reservoirs. In the ravines and "coulees" the porous basalt often outcrops and gives origin to springs which afford an ample quantity of water for general house and farm use. As a rule, the springs are active all the year, and in some instances have a flow sufficient for use as a municipal supply. In cases where springs have not been available wells have been dug in the basalt, and whenever the porous rock was encountered ample supplies of water were obtained. Ordinarily the wells vary in depth from 25 to 50 feet.

Municipal systems.—Water for Davenport is obtained from springs near the town. The supply is sufficient for present needs, but it is probable that it will have to be increased. The water is of good quality and there are no sources of contamination. A gravity system of waterworks is used. The wells in the region vary in depth from 20 to 60 feet, the water being most commonly found at 30 feet. The wells enter the rock, and a layer of clay at the surface prevents any contamination from that direction. No flowing wells have been found, but a large supply of water is obtained from the common wells. The water level does not vary during the year and is not affected by pumping. Much water is used in irrigation as well as for domestic purposes.

Harrington obtains its water from wells, although springs are used to some extent. The wells vary in depth from 30 to 100 feet. Water is most commonly found at about 35 feet. All of the wells enter basalt. The water usually rises to within 20 feet of the surface, where it stands constant the year round. It is of good quality, but will doubtless be insufficient for future demands. The water is pumped into a standpipe and distributed by gravity.

The supply for Sprague is obtained in part from a spring and in part from wells. At the spring the water issues from basaltic rock and varies much in quantity from season to season. During the dry season the supply from the spring becomes insufficient and the wells are relied upon. The supply from the wells seems to be inexhaustible and constant pumping makes no impression on the water level. The water from the spring is distributed by gravity, while the well water is obtained by pumping. From these sources excellent water is obtained in quantities believed to be sufficient for all future needs. The private wells of the region vary in depth from 20 to 40 feet. Water is commonly found at a depth of 20 feet. These wells do not enter rock, but are wholly in surface material. The water-bearing materials are sand and gravel. The wells, although shallow, afford a large quantity of water, the level being lowered only in a long, dry season.

Water for Wilbur is obtained from Goose Creek, which has its source in two large springs, and also from some wells. The water is of good quality, and ample in quantity for present and future needs. A gravity system of waterworks has been installed. The wells are all very shallow, the usual depth being about 12 feet. They enter the basaltic rock which underlies the town. Contamination from the surface is prevented by a layer of clay. The wells, although shallow, are relied upon for the domestic supply, and the water from the springs is used largely in irrigation.

Springs.—In sec. 21, T. 25 N., R. 37 E., near the town of Davenport, there is a large spring from which about 360,000 gallons of water flow daily. This supply is fairly constant, although slightly less in summer. The water is clear and cold and has a pleasant taste. It issues as a stream from the rock. It is used as a water supply for the town of Davenport.

Near Davenport, on the farm of J. E Ludy, there is a spring which is located in the bottom of a deep draw or gulch. The water comes out as a stream, and in quantity sufficient to supply about 15 families. It is slightly alkaline, but there are no deposits of mineral matter or sediment about the spring. The water has a temperature of about 45° F.

Near Harrington, in sec. 34, T. 23 N., R. 36 E., there is a spring owned by J. L. Ball. The water issues from the rock at the base of a bluff. It is clear and of a very good quality. It is used for a general farm supply.

L. T. Luper owns 2 springs near Harrington, in secs. 1 and 9, T. 23 N., R. 36 E. The spring in section 1 flows about 23,000 gallons per day, and the one in section 9 about 28,000 gallons per day. Neither spring shows any variation in flow from season to season, and in both the water is clear, of good quality, and apparently not mineral bearing. In each case the water issues as a stream. The water is used as a house supply and in watering stock.

Near Sherman, in sec. 25, T. 29 N., R. 33 E., H. B. Fletcher owns a spring which appears as a stream. The spring is located in a valley. The flow is very large and is constant from season to season. The water as it issues is cold, clear, and of a superior quality. It is used by the farmers for their stock, but it has been proposed to pipe the water to Wilbur as a supply for that town.

MASON COUNTY.

General statement.—Mason County lies in the western part of the State, being separated from the Pacific Ocean by Chehalis County, which borders it on the west. It is bounded on the north and northeast by Jefferson and Kitsap counties, and on the southeast and south by Pierce and Thurston counties. The northwestern portion of the county extends into the Olympic Mountains, and hence is very rugged and broken in character. Most of the remaining part of the county is very hilly, the only portions that are level to any degree being those immediately about Hood Canal and other inlets of Puget Sound. The streams which come from the mountains flow in canyons, and the inter-stream divides are very sharp.

The yearly rainfall is very heavy, ranging from 85 inches on the western border to 60 inches in the extreme eastern portion. As one result of the copious rainfall there is a heavy forest growth, the forests of Mason County being regarded as among the best in the State. In the southeastern part of the county, the only portion which is settled to any extent, water for domestic purposes is obtained very largely from the glacial deposits, which are here very thick. As a rule the wells are shallow, water being obtained at depths varying from 10 to 25 feet. With the abundant rainfall and the heavy mantle of unconsolidated surface materials it is not surprising that springs are very common. The water issues from the springs in the form of streams, and usually in sufficient quantities to afford town supplies.

Municipal systems.—The water supply for Shelton, a town of about 1,000 people, is obtained from some large springs, and will doubtless prove inadequate when the population of the town is doubled, but there are other large springs located conveniently near which may be utilized. The water is soft and pure and very satisfactory for domestic purposes. A gravity system of waterworks is used. In the region about Shelton there are a good many wells from which excellent water is obtained. In some instances the water rises to the surface, and the water level is not distinctly lowered except in September, the month of the least rainfall. The wells are as a rule driven wells, the materials penetrated being glacial till, sand, and gravel. The wells are shallow, varying from 10 to 25 feet in depth. The water-bearing material is overlain by hardpan, so that there is no contamination from the surface.

Springs.—As noted above, the town of Shelton obtains its water supply from two springs located in sec. 12, T. 20 N., R. 4 W. The flow has not been measured, but it is ample for all present demands. The springs issue as streams from gravel beds which outcrop on a hillside above the town. The water contains a small amount of iron, but there are no deposits of mineral matter about the springs.

At other places about Shelton, notably in sec. 18, T. 20 N., R. 3 W., and in sec. 13, T. 20 N., R. 4 W., are other springs which supply large quantities of water. They issue from gravel beds at the base of a bluff about 150 feet high. From the top of the bluff a gravel terrace extends for several miles. Lying upon this terrace, near the springs, there are several lakes, and it has been suggested that the lakes are the sources of the spring water.

OKANOGAN COUNTY.

General statement.—Okanogan County extends from the British Columbia line to Columbia River, and from Ferry County on the east to Chelan and Whatcom counties on the west. This large county presents a great diversity of surface, and within it are rugged mountains, rolling hills, and broad plateaus which merge into plains. The western portion of the county is a region of high mountains and deep valleys. Between Methow and Okanogan rivers is a bold mountain ridge which has a north-south course. Along the eastern border line of the county there is likewise a mountain ridge, with a north-south course and a general height of about 4,000 feet above the sea. Along the southern border of the county the Columbia flows in a deep canyon, but the Okanogan, its chief tributary, is a meandering stream flowing in a broad and comparatively shallow valley. The Methow Valley is likewise very broad and open, except for the last few miles of its course, when it assumes a canyon aspect.

The rainfall of Okanogan County is light in comparison with other parts of Washington. In the mountainous area the annual precipitation is from 15 to 20 inches. Over much of the remaining part of the county it ranges from 10 to 15 inches, and over a small portion it is even less than 10 inches. The forest grows exclusively upon the highlands, chiefly in the Cascades, in the mountains along the eastern border, and on the mountain ridge between Okanogan and Methow rivers. Over those parts of the county where the trees can not grow bunch grass abounds, only the lowest plains along Okanogan River being given over to sagebrush. From the mountains good streams of water flow to the plains and make possible irrigation on a large scale in the valleys of Methow, Okanogan, and Columbia rivers.

As far as known the rocks of Okanogan County are chiefly gneisses, schists, slates, and crystalline limestones. Here and there are small remnants of more or less extensive sediments of Tertiary lakes. Over large areas intrusive and extrusive igneous rocks of various kinds prevail. Over much of the surface there is a thick mantle of soil, and within this in most places water may be obtained by means of wells. Springs commonly occur, and they are relied upon to a large degree by the settlers. Surface water is usually sufficiently abundant to supply the herds of cattle and sheep with their requirements.

Municipal systems.—Loomis obtains its water from Sinlahekin Creek and from a spring owned by J. M. Judd. Neither source is very satisfactory, and it is planned to secure a supply from Toats Coulee Creek, a mountain stream of pure water which flows near the town. A few wells have been dug in the district about Loomis. They are all shallow, ranging in depth from 20 to 30 feet. They do not reach bed rock, but lie wholly in the mantle rock or soil.

PACIFIC COUNTY.

General statement.—Pacific County lies at the southwest corner of the State. The ocean border of the county is exceedingly irregular, due in part to a submergence of the coast and in part to the building out from the headlands of long sand spits. The surface rises gradually from tide water to the summit of the Coast Range, where a maximum height of 3,000 feet above the sea is attained. Much of the coast is low, and alluvial valleys extend for some distance up the streams. The annual rainfall is very heavy, varying from 65 inches at the mouth of the Columbia to 85 inches and over along the north-ern border. In this county the change in amount of precipitation is in a north-south rather than an east-west direction as is commonly the case. As elsewhere the large rainfall produces many streams and an extraordinary forest growth. The larger trees are closely surrounded by a very dense growth of minor forms, so that the forest as a whole is almost impenetrable.

With few exceptions all the rocks of the county are shales, sand-stones, and conglomerates of middle Tertiary age. All of these disin-tegrate readily when exposed to the atmosphere, so that there is a heavy mantle of incoherent porous material many feet in depth. Within this soil there is generally an abundance of water, so that shallow wells and springs form the main sources of domestic supply.

Municipal systems.—Springs afford the water supply for Ilwaco, but they are not very satisfactory and in the near future some other source must be sought. The present supply is not of the best quality and will soon be inadequate. Contamination is due chiefly to decaying vegetable matter. A gravity system of waterworks is used. Ilwaco is situated on a plain which rises but little above high tide. Wells are not used at all, because the water within them is unfit for drinking purposes.

The supply of water for Southbend is obtained from springs. In this way water of excellent quality is had, free from contamination, and in quantity believed to be sufficient for all future needs. The water is carried from the springs and distributed about the town by a gravity system. Very few wells are used about Southbend, and in none of them is the water satisfactory. The wells vary in depth from 10 to 60 feet.

PIERCE COUNTY.

General statement.—Pierce County extends from the summit of the Cascades on the east to Puget Sound on the west, and from King and Kitsap counties on the north to Lewis and Thurston counties on the south. In the vicinity of the sound the surface is that of a plain with a general height of 300 to 400 feet above sea level. Eastward the plain gives way to hills, which in turn soon merge into the high mountains forming the eastern third of the county. Mount Rainier lies in Pierce County, and immediately about it and to the northeast high peaks and deep valleys are very conspicuous.

The lowest annual rainfall is in the neighborhood of Tacoma, where it averages 45 inches. It increases eastward and in the mountains is about 60 inches. Upon the sides of Mount Rainier and the neighbor-ing high mountains the precipitation is principally in the form of snow, and large snow fields and glaciers are found. By the precipitous descent of many of the large streams waterfalls of much economic importance have been produced. Some of the waterfalls are now being utilized for power purposes. The heavy rainfall of the mountains insures an ample supply of the purest water for the cities about the sound or upon the plains.

In the region about Burnett, Wilkeson, Carbonado, and Fairfax the outcropping rocks are chiefly clastic sediments of Eocene age. They are coal bearing, and at the places above mentioned coal mines are now in operation. The extent of the sedimentary series is unknown, since eastward it passes under the lavas of Mount Rainier and westward it disappears beneath the heavy mantle of glacial sediments which covers the western third of the county. The glacial deposits for the most part are made up of coarse gravels and sands, into which the water sinks after each rainfall, so that the drainage is chiefly underground. In the western part of the county there are but few streams save those which have their origin in the hills and mountains. Ordinarily a large quantity of good water is obtained in wells having a depth of 40 to 50 feet. Springs are very abundant and it is from this source that the principal towns of the county derive their water supplies. The springs are commonly found at the bases of the hills, and from them the water usually issues in the form of streams. Not only are the springs numerous, but the quantity of water that flows from some of them is phenomenal.

Municipal systems.—A well situated on an island in White River, near the town, supplies Buckley with water. From this source, by pumping, a very large supply may be obtained, ample for present and future needs. The water is of good quality, there being no known sources of contamination. In the region about Buckley the wells vary in depth from 30 to 40 feet, the water-bearing materials being beds of sand and gravel. The latter are overlain by glacial till or hardpan, so that the wells suffer no contamination from the surface.

The water for Carbonado is obtained from springs located in some gravel beds overlying the coal-bearing rocks. The supply is sufficient for the present and doubtless will be enough for all future needs. A gravity system of waterworks is in use. The water is used for boilers as well as for domestic purposes. There are no wells in Carbonado.

Springs furnish a supply of water for Orting. The water is not contaminated and is very satisfactory. The supply is ample for all the demands made upon it. Other springs near by may be utilized if necessary. A gravity system of waterworks is used. In the region about Orting are a few wells, their depths varying from 14 to 18 feet. The wells are wholly in glacial materials, not reaching the bed rock in any case. The water in the wells is excellent and abundant, usually rising to the surface in the winter months.

The water used in Puyallup comes from Maplewood Springs.

These springs lie within the city limits and have a flow of 20,000,000 gallons daily. The water is of good quality. It is distributed through the town by a gravity system. Puyallup is located on the flood plain of Puyallup River near where it enters the sea. The wells that have been dug in the alluvium of the valley do not yield good water and hence are rarely or never used.

The water system in South Tacoma is owned by Calvin Phillips & Co., who obtain a daily supply of 150,000 gallons from wells. The water is free from any contamination and is quite satisfactory. The usual depth of wells about South Tacoma is 40 to 50 feet. The wells penetrate beds of gravel and sand and in no instance reach bed rock.

For Sumner the water is obtained from a spring located on a hillside 1 mile east of the town. The spring is sufficiently above the level of the town to give a good head to the water, so that it may be distributed by gravity. The supply is ample for present needs, and there are other good springs near by which may be utilized in the future should the needs of the town so demand. No wells are used in the valley about Sumner, all of the water for domestic purposes being piped from the springs on the hillsides which border the valley.

The city of Tacoma obtains its water supply from Clover Creek and some springs. From the former the major portion of the water used in the city is obtained, but the spring water is more satisfactory. The quantity obtained is hardly sufficient for the needs of the city and, as additional sources of supply, three deep wells are being drilled. From the creek and springs water is pumped into reservoirs, from which it is distributed by gravity throughout the city. Tacoma is located upon a plain composed of glacial sediments, till, clay, sand, and gravel, which yield an abundant supply of good water. The depth of the wells varies from 30 to 100 feet. In general the water-bearing strata are overlain by clay or till, so that surface contamination is reduced to a minimum.

A sanitary analysis of water from Clover Creek and springs, made February 23, 1901, by Prof. H. G. Byers, of the University of Washington, resulted as follows:

Analysis of water from Clover Creek and springs.

[Parts per million.]

Total solids	47.20
Nonvolatile solids	25.30
Volatile solids	22.00
Oxygen consumed	6.34
Chlorine	4.40
Nitrogen as free ammonia	.005
Nitrogen as albuminoid ammonia	.080
Nitrogen as nitrites	None
Nitrogen as nitrates	None

Springs.—Near the town of Carbonado, in sec. 4, T. 18 N., R. 6 E., the water issues as a stream from some gravel beds which overlie the coal measures at this place. The spring yields sufficient water to supply the town of Carbonado and the boilers of the Carbon Hill Coal Company.

The Orting Light and Water Company own three springs in sec. 29, T. 19 N., R. 5 E. These springs are small, but afford enough water to supply the town of Orting. They are located upon a hillside, where the water issues as small streams from beds of sand and gravel: The water carries a little sand and is somewhat muddy after hard rains. The flow is said to be a little stronger in summer than in winter.

Maplewood Springs are located near Puyallup, in sec. 32, T. 20 N., R. 4 E. The water issues in streams from beds of gravel at the base of a bluff. The flow is estimated at 20,000,000 gallons per day. The town of Puyallup and a portion of Tacoma are supplied with water from this spring. The spring is owned by the city of Tacoma.

SAN JUAN COUNTY.

General statement.—San Juan County comprises a group of islands lying between Whatcom and Skagit counties in Washington and the island of Vancouver. These islands have rocky, irregular, deeply indented shores and represent the tops of submerged mountains. The surface of some of the islands is that of a plain, but in most cases there are high hills or semimountains. The principal hills are on Orcas Island, the highest point, Mount Constitution, rising 2,200 feet above the sea.

Since the San Juan Islands lie to the leeward of the mountains of Vancouver Island, they have a rainfall that is less than the average of western Washington. The yearly rainfall is between 30 and 35 inches. This is enough to support but a moderate forest growth, so that the trees are not so large as in some parts of the State and there is but a small amount of undergrowth. The rainfall is ample for agricultural purposes and excellent crops are raised.

The rocks of the northernmost islands of the group are of upper Cretaceous age and are a part of the coal-bearing series of the islands of British Columbia. South of the sedimentary rocks just mentioned, and forming the major portion of the islands, are metamorphic and igneous rocks of undetermined age. All of the islands have been glaciated, and oftentimes the rocks are deeply furrowed. The glaciers laid bare the rocks in some instances, and over much of the surface but a thin soil was left. The area of tillable land is thereby largely reduced and limited to the lower valleys. On a few of the islands the water supply is scant and the problem of securing water for domestic purposes is a difficult one. Only a meager supply of surface water is obtained because of the scant rainfall. The slight depth of soil makes it necessary to dig the wells in bed rock in most places. In some instances, however, as about Friday Harbor, the soil is of sufficient depth to yield a water supply by means of shallow wells.

Municipal systems.—The water supply for Friday Harbor comes altogether from wells. These yield water of good quality, which thus far has been quite satisfactory. The wells are all shallow, not often reaching a depth exceeding 18 feet. The water is obtained from beds of gravel. These are overlain by clay, thus preventing surface drainage from entering the wells. A good water supply for the town of Friday Harbor can be had in Sportsmans Lake, which is 3½ miles away and lies at an elevation of about 75 feet above the town. The lake covers 90 acres and is fed by springs.

SKAGIT COUNTY.

General statement.—Skagit County is in the northwestern part of the State and lies between Whatcom County on the north and Snohomish County on the south. It extends from Puget Sound on the west to the summit of the Cascades on the east. The coast is very irregular. The topography of the western part of the county is that of a plain lying but little above the sea. The eastern half is very rough and broken and embraces some of the most rugged portions of the Cascades. A prominent feature of the surface is the great valley of Skagit River, which has an east-west course through the county and upon the broad delta of which the principal towns are located. Dikes have been built along the coast and the lower reaches of the river to keep the alluvial plain from being overflowed. The tide runs up Skagit River for about 20 miles.

The yearly rainfall varies from 30 inches along the coast to an average of 50 inches in the mountains. Practically all of the region was once forest covered, and the cedars of the Skagit Valley are among the largest and best to be found in the State. The timber of the broad alluvial plain along the lower course of the river has been largely removed and excellent farms have taken its place. The most valuable farm land in the State is found here, the amount of rainfall and the excellent quality of the soil being conducive to very high agricultural returns.

In the region south of Hamilton and immediately about Cokedale are outcropping Tertiary sandstones and shales which are coal bearing and in which coal mines have been opened. The extent of these sedimentary areas is unknown, since they are largely concealed by the alluvial deposits of the river, glacial sediments, etc. East of

Hamilton is a north-south belt of slates and schists containing small bodies of iron ore. Among the islands at the western end of the county are frequent outcrops of slates, schists, and other metamorphic rocks. Along the borders of the valley of the lower Skagit are thick deposits of glacial sediments, from which water is often obtained by means of wells and springs. The towns located in the valley of the Skagit may secure supplies of water either from the springs on the adjacent hillsides or from shallow wells that have been sunk into the sands and gravels of the river plain.

Municipal systems.—The town of Anacortes obtains its water from Lake Heart, which is fed by springs. The lake is about 250 feet above the town, so a gravity system of waterworks is used. Lake Heart, although now supplying enough water for the use of the town, will in time become inadequate. Other desirable sources of supply may easily be found in Cranberry and Whistler lakes, which lie near the town. The water now used is of good quality and is quite satisfactory. Some of the wells of the region are shallow, but the best water is obtained at a depth of about 80 feet. Not even the deepest wells enter bed rock, the materials passed through being clay above, with beds of gravel below.

Hamilton secures its water supply exclusively from wells. These are shallow, since the town is located on the flood plain of Skagit River. The wells range in depth from 10 to 25 feet. The height of water in the wells depends upon the stage of water in the river. The water in the wells all comes from the river by seeping through the coarser materials of the flood plain. In each well a lateral movement of the water in the same direction as the flow of the river may be readily seen.

The water that is used at Laconner is brought from a spring by a gravity system. The water is of good quality, and is ample in amount for present demands, but may need to be supplemented in the future.

Water for the use of the town of Mount Vernon is obtained from a large spring. From the spring water is pumped into a reservoir located on a hill above the town and is then distributed by gravity. The water is very pure, there being no sources of contamination, and it is sufficient in amount for present and future needs. The wells of the region range in depth from 10 to 14 feet. They do not enter bed rock, but are altogether in the alluvium of the valley. In some instances the well water is not of good quality.

SKAMANIA COUNTY.

General statement.—Skamania County lies on the southern border of the State, south of Lewis County, between Yakima and Klickitat counties on the east and Cowlitz and Clarke counties on the west. The proportion of low plain to high mountain is less than in any other county in the State. With the exception of a very narrow plain along Columbia River the entire county consists of rugged mountains which have a general height of about 5,000 feet above sea level. From east to west the county covers very nearly the entire width of the higher Cascades in this part of the State. The highest point is Mount St. Helens, with an elevation of about 10,000 feet, standing near the northern border line. St. Helens is a volcano that is said to have shown slight signs of activity since white settlers first came to Washington. It has suffered but little erosion and is symmetrical in outline.

The rainfall ranges from 60 inches or a little more in the northwestern part to 40 inches on the southeastern border. The elevation is such that the rainfall is heavy. It should be noted, however, that not all of the precipitation is in the form of rain, but that some of it, especially in the northwestern part of the county, is in the form of snow. With the exception of a few mountain peaks the surface is entirely forested, and over one-half of the county is within the limits of the Mount Rainier Forest Reserve.

Little is known of the geology of the county. Nothing but basalt, presumably belonging to the Columbia River lava, appears along Columbia River. About 15 miles north of the river, northeast of Skye, outcrops of granite occur. These outcrops are believed to be at the southern end of a belt of granite which runs northward through the county, passing between Mount St. Helens on the west and Mount Adams on the east. The warm springs which occur at Cascades seem to indicate that there are igneous rocks at moderate depths in that vicinity which are not yet entirely cooled. Very little information is at hand regarding the details of the water resources of the county, since it is very sparsely peopled. The only settlements are located in the immediate vicinity of Columbia River.

Springs.—Near Cascades post-office, on Columbia River, in sec. 16, T. 2 N., R. 7 E., is a mineral spring owned by Thomas Moffett. The water seeps out at the base of a hill, and flows 25,000 gallons daily, with no variation from season to season. The water has a temperature of $96°$ F. Gas is constantly escaping and the water is known to carry iron and sulphur. At the spring a bath house and a hotel of 25 rooms were erected some years ago. Plans have been drawn for a new hotel of 100 rooms, new bath houses, and a swimming tank 60 feet long, 20 feet wide, and 7 feet deep. The water is of medicinal value.

been sold in 1902.

SNOHOMISH COUNTY.

General statement.—Snohomish County is located in the northwestern part of the State, between Skagit County on the north and King County on the south, and extends from the summit of the Cascades to the shores of Puget Sound. The surface of the western third of the county is that of a low plain either level or covered with low hills. The eastern two-thirds has a mountainous topography, varying from the low foothills on the west to the lofty snow- and ice-covered peaks on the main divide. The mountains have been trenched by Skykomish River and the North and South forks of the Stilaguamish, all of which flow in deep valleys. These rivers have in general a westerly course, and the watersheds are separated by bold spurs which extend outward from the main mountain mass.

There is a considerable difference between the rainfall of the lowlands along the coast and that in the high mountains. Along the western border of the county the annual rainfall averages 35 inches; at the foot-hills of the mountains it is 50 inches; while within the mountains, and especially toward the eastern summit, it reaches 60 inches and over. In the vicinity of Monte Cristo, about Glacier Peak, and elsewhere in the high mountains the snowfall is very heavy and great fields of snow and many glaciers are the result. With the exceptions of the high mountains that are covered with snow or ice, or are barren of soil, almost all parts of the county are heavily forested. The best timber is found on the plains and the low hills in the western part of the county and in the larger valleys of the mountains. The northeastern part, embracing over a third of the county, is included in the Washington Forest Reserve.

In the western end of the county the bed rock appears at very few places because of the great thickness of the glacial sediments. There are a few outcrops of Tertiary sandstones and shales, which are not coal bearing, as far as observed. The mountainous portion is composed of a great complex of metamorphic rocks with many varieties of extrusive and intrusive igneous rocks. The water supply of the county is obtained in part from the mountain streams and in part from the glacial deposits, in the latter case either from springs or by means of wells. The several sources afford a very satisfactory supply, and there is no part of the county that is not abundantly provided with water of the finest quality.

Municipal systems.—The town of Arlington secures water from springs, wells, and Stilaguamish River. Both springs and wells are very satisfactory, a large supply of excellent water being thus obtained. The wells vary in depth from 20 to 30 feet. They do not enter the rock, but have been dug wholly in clay, sand, and gravel.

Edmunds secures its town supply of water almost wholly from springs, only a few wells being in use. The springs yield a supply that is satisfactory in both quality and quantity. A gravity system of waterworks is used.

Everett obtains a supply of water from 2 creeks, one flowing within a mile and the other within 2½ miles of the city. From these streams the water is conveyed through pipes of 22 and 8 inches in diameter to a reservoir which has the capacity of 1,200,000 gallons. From the reservoir the water is distributed by a gravity system throughout the city. The water is free from contamination and highly satisfactory in every way. The wells in the region about Everett vary in depth from 15 to 115 feet, water being usually found at about 100 feet. The materials penetrated are surface soil, glacial till or hardpan, and finally sand or gravel, which yield the water. Bed rock is not reached even in the deepest wells. A large supply of good water may be obtained from the wells, which is ample in quantity for domestic demands. The following sanitary analysis of the city water of Everett was made by Prof. H. G. Byers, of the University of Washington, on January 10, 1901:

Analysis of city water of Everett.

[Parts per million.]

Total solids	300. 80
Nonvolatile solids	252. 80
Volatile solids	48. 00
Oxygen consumed	7. 27
Chlorine	4. 80
Nitrogen as free ammonia	0. 020
Nitrogen as albuminoid ammonia	0. 050
Nitrogen as nitrites	None.
Nitrogen as nitrates	None.

In Marysville the water supply comes exclusively from wells. The wells are all shallow, since the town is located on the flood plain of Snohomish River. The water is soft, of good quality, and satisfactory for the present, at least. It is planned to use at some future time the water of Lake Stephens, which is located conveniently near.

The water supply for Monroe is obtained from two small creeks. In this way very pure water is had, and in quantity sufficient for present and future needs. A gravity system of waterworks is in operation. The wells are from 18 to 50 feet in depth. None of them enter rock, all being wholly within the alluvial materials of the river valley. An analysis of water from the Monroe supply was made by G. L. Tanzer, of Seattle, on August 3, 1903. He found that there was 0.049 grain of solid matter in 1 liter (33.81 fluid ounces) of water. A qualitative analysis of the solid matter showed it to contain magnesium and sodium chlorides, with a trace of calcium carbonate. No ammonia, nitrogeneous matter, iron, or silicates were found.

Monte Cristo obtains a water supply from a mountain stream, using a gravity system. The water comes directly from the snow fields near by and is very cold and pure. The quantity is greater in autumn and spring than in other seasons, but probably will always be equal to every demand. Besides its uses as a domestic and boiler supply it is also used in an ore concentrator.

The water supply of Silverton comes from mountain streams which head in the fields of snow and ice near the town. The water is therefore of exceptional purity and may be had in quantity sufficient for all possible needs. It is used as a domestic supply, for water power, and in concentrating ore.

Snohomish secures its supply from Pilchuck River, a stream heading to the east in the Cascade Mountains. The water is pumped into a reservoir having a capacity of 500,000 gallons and is then distributed by gravity. There are no sources of contamination and the water is of good quality. The quantity is sufficient for all future needs. The wells about Snohomish range in depth from 15 to 55 feet. In digging the wells glacial till or hardpan is first passed through, and the water is obtained from the beds of gravel that lie below. The following sanitary analysis of water from the city supply was made by Prof. H. G. Byers, of the University of Washington:

Analysis of city water from Snohomish.

[Parts per million.]

Total solids	41.50
Nonvolatile solids	23.22
Volatile solids	18.28
Oxygen consumed	3.75
Chlorine	4.00
Nitrogen as free ammonia	.010
Nitrogen as albuminoid ammonia	.093
Nitrogen as nitrites	None.
Nitrogen as nitrates	.400

SPOKANE COUNTY.

General statement.—Spokane County lies in the extreme eastern part of the State, with Stevens County on the north, Lincoln County on the west, and Whitman County on the south. In the main the surface is that of a plateau, with an average height of about 2,000 feet above the sea. The only exceptions to the plateau aspect to be noted are the mountains in the northeast corner, of which Mount Carlton is the highest, and the high hills or low mountains in the southeastern part along the Idaho boundary. The principal stream, Spokane River, flows in a broad, shallow valley from the eastern border of the county to the city of Spokane; at this point it enters a canyon, of very moderate depth at first, but gradually deepening as the river flows northwestward to join the Columbia.

The average rainfall is about 20 inches per year. The precipitation on the western border is somewhat less than this, while on the eastern side it is a little more. The result is that in a general way the eastern half of the county is forested, while the western half is a prairie. The grassy plains have been largely replaced by wheat fields, since the rainfall is always ample to secure a crop of this cereal. The forest growth is rather sparse, the trees are of only moderate size, and there is little undergrowth.

Spokane County lies partly within the Columbia River lava field, the bed rock of the southwestern half being basalt. As this is the border line of the lava field the basalt is comparatively thin and the underlying rocks occasionally appear through it. The northwestern part of the county is chiefly a granite region, with occasional areas of gneiss and schist. Among these rocks basalt often appears in the form of narrow tongues which have extended outward from the main mass of lava. The valley of Spokane River in Glacial time was almost entirely filled by a gravel train, only a minor portion of which has been removed. At Spokane Falls is a jutting ledge of basalt, and the river drops 130 feet. Within the gravels and sands of the river valleys good water is obtained by means of deep wells. Upon the lava plateau domestic supplies of water are secured from wells ranging from 30 to 135 feet in depth. Some of the basalt is porous, and where such rock outcrops along the bases of hills springs are often found.

Municipal systems.—Cheney obtains a water supply from a lake and from wells. The lake supply is the more commonly used, but water from the wells is more satisfactory. An ample amount is secured for the present and doubtless for all future needs. A direct pressure system of waterworks is in use. The wells vary from 30 to 50 feet in depth, water being most commonly found at 40 feet. They all enter basalt after passing through a layer of clay at the surface. Very good water is secured. The water level scarcely varies from season to season, and is not affected by pumping.

The town of Medical Lake secures water from Clear Lake, located 3¼ miles to the south. The water is hard, but quite satisfactory otherwise. A supply sufficient for all possible demands may be obtained. In the wells about Medical Lake water is usually found at a depth of 15 feet; but the range of depth of the wells is from 10 to 40 feet. The wells are all in basalt, the porous or scoriaceous layers of the rock being water bearing.

An analysis of the water of Medical Lake, made by G. A. Mariner, of Chicago, is as follows:

Analysis of water of Medical Lake.

[Parts per thousand.]

Silica 0.1825
Alumina and iron oxide .... .0120
Calcium carbonate .... .0031
Magnesium carbonate .... .0040
Sodium chloride .... .2869
Potassium chloride .... .1616
Sodium carbonate .... .1089
Potassium carbonate .... Trace.
Lithium carbonate .... Trace.
Borax .... Trace.

Hillyard obtains a supply of water from deep wells. These are commonly from 190 to 200 feet in depth and are wholly in gravel. The water is soft and of the best quality. The supply at present is sufficient for all needs, but it is doubtful if it will be ample for the future.

Latah uses wells exclusively as a source of water supply. Later on water may be taken from a large spring above the town. The wells are shallow, as a rule, varying in depth from 25 to 40 feet. They enter rock, and a layer of clay at the surface tends to prevent contamination from above. The water level varies but little throughout the year. One well sunk to a depth of 135 feet flows constantly.

The city of Spokane obtains its supply of water from Spokane River. The pumping plant is on the river bank about 5 miles above the city. A direct pressure system of waterworks is used. There are no sources of contamination and the water is of a high degree of purity, as well as ample in quantity for all probable future needs.

The following sanitary analysis of city water from Spokane was made on April 23, 1901, by Prof. H. G. Byers, of the University of Washington:

Sanitary analysis of city water from Spokane.

[Parts per million.]

Total solids 50.24
Nonvolatile solids 34.43
Volatile solids 15.81
Oxygen consumed 2.13
Chlorine 1.50
Nitrogen as free ammonia .... .004
Nitrogen as albuminoid ammonia .... .0363
Nitrogen as nitrites .... None.
Nitrogen as nitrates .... .018

Deep wells.—The Hillyard Town Site Company has a well which is used to supply the town of Hillyard with water. It has a diameter of

44 inches and a depth of 200 feet. It was dug in 1900 at a cost of $500. The amount of water obtained is about 50,000 gallons per day. The water level in the well rises and falls with the water level of Spokane River, which flows not far away.

Springs.—William Forthman owns a spring near Latah, in sec. 26, T. 21 N., R. 45 E. The water issues as a small stream at the foot of a hill. The spring has a minimum flow of 720 gallons per day and is much stronger in the springtime. The water is soft, clear, and of excellent quality. It is used for general farm purposes. In the vicinity of Latah are a number of large springs.

STEVENS COUNTY.

General statement.—Stevens County lies in the northeast corner of the State, bordering on Idaho and British Columbia. The surface is characterized by three conspicuous mountain ridges and three north-south valleys. The ridges lie between Columbia and Colville rivers, between Colville and Pend Oreille rivers, and east of the latter stream along the county boundary. Of the three ridges the first one mentioned is the lowest, with an average height of about 4,500 feet. The second and third ridges have a maximum elevation of about 7,000 feet. Columbia River flows in a deep valley that is usually bordered by glacial terraces. The valley of Colville River is wide and the stream has an extensive flood plain, which overflows at certain seasons of the year. For the first 50 miles of its course within the county Pend Oreille River flows very slowly, in a broad valley that is bordered with much agricultural land; farther down the stream crosses a belt of harder rocks, in which it flows in a long, tortuous canyon.

The rainfall averages about 20 inches per year. This is sufficient to permit of agriculture without irrigation. Practically the entire county was once forested, although the region seems to be near the border land of forest and prairie. The trees do not grow very near together and the undergrowth is very scant. As a rule the grasses grow everywhere among the trees. The forest is being removed at a rapid rate, the demands of the lumbermen on one hand and of the farmers on the other tending toward the deforesting of the county in a comparatively short time.

Stevens County is largely a region of crystalline rocks. The north-south ridges between the principal streams are composed chiefly of granite. Flanking the granites and within the valleys are large areas of marbles, quartzites, slates, and other metamorphic rocks. The marbles occur very generally throughout the county and are of economic importance both as an ornamental stone and for the manufacture of quicklime. At several places are outcrops of coal-bearing sandstones and shales, which represent remnants of lacustrine sediments of

Tertiary time. The basalt of the Columbia River lava extends a little way into the county along the southern boundary. The towns depend for the most part upon streams for their water supplies. The water in virtually every stream is free from contamination and is of excellent quality. In general, the soil is of sufficient thickness to contain enough water for the domestic supplies needed in the smaller towns and through the country. The wells for the most part are shallow and do not often exceed 40 or 50 feet in depth.

Municipal systems.—Bossburg secures water from Columbia River, beside which the town is situated. The water is pumped into a reservoir, from which it is distributed about the town by gravity. The supply is satisfactory in every way. There are no wells about Bossburg, but there are some springs which are utilized.

Colville depends mainly upon springs, but in part upon two streams, for its water supply. The water from all of these is of fine quality, there being no sources of contamination. A gravity system of waterworks has been installed. The wells about Colville range in depth from 10 to 30 feet, the usual depth being 20 feet. The water comes from beds of sand, which are overlain by clay.

Water for Marcus is taken from Columbia River. The water is satisfactory from all standpoints. The wells vary from 18 to 60 feet in depth. Beds of clay are first penetrated, then sand, and finally gravel.

Northport obtains its water supply from Deep Creek, a small stream emptying into Columbia River. There are no sources of contamination, and the water is satisfactory, although very hard. A large amount may be secured, sufficient for all future needs. The water is first pumped into tanks which stand upon a hill above the town and is then distributed by gravity. Besides its domestic use it is also used as a boiler supply by the Northport Smelting and Refining Company.

The water for Springdale is taken from a mountain stream. The spring flows for some distance over limestone, hence the water contains some lime. The supply is believed to be sufficient for all future needs. A gravity system of waterworks is in use. The wells about Springdale range from 30 to 100 feet in depth, water being commonly found at depths of 40 to 50 feet. The more shallow wells are wholly in the soil or mantle rock, but the deeper wells all enter bed rock.

Deep wells.—In sec. 9, T. 35 N., R. 39 E., near Colville, the Pacific States Oil Company drilled a well in 1901 when prospecting for oil. The total depth reached was about 700 feet. At the top the well has a diameter of 10 inches, at the bottom 6 inches. A water-bearing stratum was reached at 400 feet, and water now flows from the well mouth at the rate of 3 or 4 gallons per minute. The cost of the well was $3,000. In drilling the well, beds of limestone, sandstone, limestone, shale, and sandstone were passed through in succession.

THURSTON COUNTY.

General statement.—Thurston County is located in the southwestern portion of the State, at the head of Puget Sound, with Chehalis County lying between it and the Pacific Ocean. The county lies entirely within the basin of Puget Sound and the surface is essentially that of a low plain. For the most part the plain is quite level, with here and there occasional low hills. In the western part of the county the Black Hills rise 300 or 400 feet above the general level. In the southeastern part of the county there are a few low hills which mark the extreme outliers of the Cascades. The Puget Sound shore is very irregular. abounding in indentations and bordered by numerous islands.

In the central part of the county the annual rainfall averages 50 inches. In the eastern and western parts it is 60 inches or a little more. The precipitation is practically altogether in the form of rain, since falls of snow occur very rarely.

The bed rock is not often exposed in Thurston County. In the neighborhood of Tenino and Bucoda there are occasional outcrops of sandstone, which are of economic importance, as they afford a good grade of building stone, and also contain coal, which has been mined to some extent. In the northwestern part of the county, at Gate, there are several outcrops of basalt, but how far north, within the Black Hills, this rock extends is not known. Glacial sediments abound over nearly the entire county. These sediments are usually coarse gravels which represent outwash plains. Oftentimes the gravel plains are soil covered to such a slight degree that they are almost barren. Upon the plains the forest growth when present is very sparse and prairie conditions often prevail. The rainwater sinks away quickly into the gravels and the drainage is largely underground. The streams are few, only the larger ones flowing persistently throughout the year. Occasionally a good soil covers the gravel plains and good farming land is found. The hills noted above are heavily forested, and from them several streams flow out upon the plains. An ample supply of water is easily obtained throughout the county. Within the region of the gravel plains water is often obtained from springs and may always be had by means of wells, which rarely exceed 50 feet in depth. From the beds of sand and gravel in the vicinity of Olympia excellent water has been obtained from wells that vary from 125 to 175 feet in depth, the water sometimes outflowing at the surface.

Municipal systems.—The water supply for Olympia is obtained from springs. The water is pumped into a reservoir, from which it is distributed by a gravity system. At the present time about 3,000,000 gallons per day are obtained. This will doubtless prove sufficient for all future needs.

The Tenino water supply comes entirely from wells, which have an average depth of about 35 feet and are wholly in sand and gravel. The supply of water, while ample for present needs, will doubtless have to be superseded by a larger supply in the future.

Deep wells.—On the capitol grounds in Olympia some wells have recently been drilled in order to secure a water supply for the capitol building. No rock was penetrated in drilling the wells, the water-bearing material being sand. Two of these wells have depths of 152 feet each, and the third has a depth of 138 feet. Each well has a diameter of 2 inches. They are all flowing wells, the water rising about 2 feet above the ground. From the deeper wells flows of 6 and 4 gallons per minute are obtained, while from the third well a flow of 2 gallons per minute is had. The wells are located about one-fourth mile from the shore of Puget Sound and their mouths are 25 or 30 feet above mean tide. The rate of flow diminishes at low tide and increases at high tide. The wells have been cased throughout with 2-inch casing. The cost of drilling the wells was $1 per foot.

WALLAWALLA COUNTY.

General statement.—Wallawalla County is located in the southeastern part of the State, along the Oregon boundary, and east of Columbia and Spokane rivers. The region along Columbia River is very low, being but 300 or 400 feet above the sea. From this low plain there is a gradual ascent eastward in the direction of the Blue Mountains. The eastern part of the county is a region of high, rolling hills, with deep ravines or valleys. The hills in outline show the influence of the wind, and many of them are essentially of eolian origin. The prevailing winds are from the southwest, and hence the hills have their more moderate slopes upon their southwestern sides, while their steeper slopes are to the northeast.

The rainfall shows a very close relationship to elevation. In the lowest part, along Columbia River, the annual rainfall amounts to 10 inches or a little less. In a north-south belt through the center of the county it is 15 inches, and in the eastern part it reaches 20 or 25 inches. This is sufficient to afford a tree growth which, though sparse at first, becomes of considerable importance when the summit of the Blue Mountains is reached. Through the central part of the county trees do not grow naturally, and prairie conditions prevail. At the western end of the county, where the rainfall is least, grasses give way to sagebrush. In this part of the county irrigation is necessary in order that agriculture may be carried on successfully. The rolling prairies of the major portion of the county are famous wheat producers, and in the neighborhood of Walla Walla very fine fruit is raised.

In the northwestern part of the county the rocks are in the main
sandstones and shales, which were deposited in a lake in middle Tertiary time.   Along the Snake and the Columbia these rocks are largely
covered by alluvial deposits.   In the remaining part of the county the
bed rock, as far as known, is altogether basalt.   Usually the rock is
deeply buried by the heavy mantle of soil, and outcrops but rarely.
Springs often occur along the bases of the hillsides, wherever porous
basalt appears at the surface.   The wells are usually shallow and dug
entirely within the soil.   The deep well which has recently been drilled
near Walla Walla is of great importance, since it has established the
fact that an artesian basin exists here, and that flowing water may be
secured at depths of from 500 to 600 feet.

Municipal systems.—The water supply of Walla Walla is secured
chiefly from springs, but in part from infiltration ditches.   The amount
of water is barely sufficient in the dry season, and a larger supply is
now being developed at a point about 3 miles above the city.

A sanitary analyses of the Walla Walla city water, made on April
12, 1901, by Prof. H. G. Byers, of the University of Washington,
resulted as follows:

Sanitary analysis of city water from Walla Walla.

[Parts per million.]

Total solids	81. 67
Nonvolatile solids	62. 86
Volatile solids	18. 81
Oxygen consumed	2. 89
Chlorine	1. 50
Nitrogen as free amonia	. 008
Nitrogen as albuminoid ammonia	. 0267
Nitrogen as nitrites	None.
Nitrogen as nitrates	. 452

Waitsburg obtains a water supply from springs in autumn, winter.
and spring, and from a creek in summer.   The springs afford excellent
water, and it is planned to replace the creek water by water from deep
wells.   A gravity system of waterworks is in use.   The wells about
Waitsburg vary in depth, the shallow ones being wholly in soil and
broken rock, while the deeper ones enter the bed rock.   In wells of
the first type water is usually found at 15 to 25 feet, while in the
second instance the usual depth is 40 feet.

Deep wells.—On the farm of the Blalock Fruit Company, near Walla
Walla, a well has recently been drilled (completed May 1, 1903).   It
has a diameter of 6 inches and a depth of 564 feet.   The first rock
encountered was basalt, at a depth of 540 feet.   The well is a flowing
one, the rate of flow being 130 gallons per minute.   The temperature
of the water at the well mouth is 67° F.   The well has been cased to a
depth of 540 feet.   The water is used in irrigation.   The cost of drilling the well was $1,800.

WHATCOM COUNTY.

General statement.—Whatcom County lies in the northwestern part of the State, adjoining British Columbia, and extends from the summit of the Cascades to the Strait of Georgia. Topographically the surface presents two distinct divisions, the plain of the western half and the high mountains of the eastern part. The average height of the plain is 200 or 300 feet above the sea. In the southeastern part of the county are a number of high hills, which give to the plain a broken character. The mountains begin somewhat abruptly, the transition from the plain to the real mountains being quickly made. The highest peak, Mount Baker, stands at the western front of the Cascades, and from this point to the divide on the east the whole area is very rugged. The high mountains have been deeply dissected, and this region is regarded as one of the most difficult to penetrate in the entire Cascades.

The maximum rainfall is 45 inches per year. This occurs in a north-south belt in the vicinity of Mount Baker and also in the northwest corner of the county. From Mount Baker eastward there is a gradual decrease in the rainfall, so that the amount is about 30 inches on the border of the county. Southwest of Mount Baker, also, the rainfall decreases to 35 inches in the vicinity of Bellingham Bay. Mount Baker and all of the highest mountains are covered with snow fields and glaciers. These serve as reservoirs and assure a constant flow to the streams throughout the year. The abundance of precipitation and the great elevations of the region give rise to waterfalls, which will be of great usefulness for power purposes. The falls of the Nooksack, with an effective head of water of 179 feet, are now being developed.

What is now the western part of the county was a large lake in Tertiary (probably Oligocene) time. The deposits made in this lake were mainly sandstones, with some conglomerate and a little shale. These rocks outcrop very commonly in the southeastern part of the county, but in the northwestern portion they are almost entirely covered with glacial sediments. The lacustrine sediments aggregate many thousands of feet in thickness, their extent in this direction being as yet unknown. They are coal bearing near the base. The principal mines for the mining of this coal are located at Blue Canyon, on the shore of Lake Whatcom. The mountainous part of the county is virtually unknown geologically, but the rocks are chiefly metamorphics and igneous intrusions, with great lava flows in the vicinity of Mount Baker.

Whatcom County is abundantly supplied with water. The mountain streams furnish a large amount of excellent water for the use of the towns upon the plains. There are also a number of glacial lakes, such as Whatcom, Padden, and Samish, which are natural reservoirs of good water. From the glacial sediments of the northwestern part of the county large quantities of water are obtained from springs or by means of wells. The glacial sediments are usually made up of alternating beds of till and water-bearing gravel and sand, and surface contamination is thereby eliminated.

Municipal systems.—The water supply for Blaine comes from springs. The water flows into a reservoir, from which it is distributed by gravity. The quality of water is very good, and the quantity sufficient for all future needs. The wells in use about Blaine are shallow, ordinarily not more than 14 feet in depth. Water is found in gravel, usually at a depth of 10 feet. The town supply is used in the boilers of sawmills, and in the salmon canneries, and for general domestic purposes.

Fairhaven obtains a water supply from Lake Padden, a glacial lake south of the city. A gravity system of waterworks is used. A large amount of water may be obtained from this lake, but if the supply should ever be insufficient water may also be taken from the South Fork of Nooksak River or from Lake Samish. The water from Lake Padden is very pure, as is shown by the following analysis, made by the Deakbof Drug and Chemical Company, of Chicago:

Analysis of water from Lake Padden.

[Grains per gallon.]

Silica	0. 467
Calcium sulphate	. 329
Magnesium carbonate	. 129
Sodium and potassium sulphates	. 606
Sodium and potassium chlorides	. 401
Iron oxide	None.
Total solids	1. 932

In Sumas wells are depended upon, primarily, as a source of water supply, although some small streams are used to a slight degree. Water obtained in this way is of good quality, although the quantity is hardly sufficient for present needs, and some other source must be provided in the future. It has been suggested that a deep well be bored as a future source of supply. Wells in and about the town of Sumas are driven wells, for the most part, and have a range in depth of 40 to 90 feet. Water is most commonly found between 40 and 60 feet. The glacial sediments about Sumas are thick, and the wells have been driven entirely in these. Beds of clay were usually encountered first, with beds of sand and gravel below, from which the water was obtained. In a few instances the water in the wells rises to the surface. The water level varies but very little from season to season.

The city of Whatcom obtains its water supply from Lake Whatcom. The lake is several square miles in area and is located in the hills southwest of the city. It lies at a sufficient height above the level of the town to give a good head to the water, so that a gravity system of waterworks may be used. The water is of excellent quality, the only sources of contamination being sawdust and other refuse from the sawmills on the shores of the lake. The catchment basin of the lake is of ample size and insures a large storage of water in the lake basin, so that the water supply will be sufficient for all the future needs of the city. The water is used for domestic purposes and in the many manufacturing plants which are located about Whatcom. The following sanitary analysis of water from Lake Whatcom was made March 16, 1901, by Prof. H. G. Byers, of the University of Washington:

Analysis of water of Lake Whatcom.

[Parts per million.]

Total solids	158.86
Nonvolatile solids	116.52
Volatile solids	42.34
Oxygen consumed	2.03
Chlorine	3.80
Nitrogen as free ammonia	.004
Nitrogen as albuminoid ammonia	.033
Nitrogen as nitrites	None.
Nitrogen as nitrates	None.

WHITMAN COUNTY.

General statement.—Whitman County lies along the Idaho boundary, south of Spokane County and east of Adams County. The surface is in general that of a plateau, with an average height of about 2,000 feet above the sea. Rising above the plateau, especially along the eastern border of the county, are several high hills which represent the extreme outliers of the mountains of Idaho. Of such hills Steptoe Butte is a good example, rising about 700 feet above the plateau. The plateau is generally covered with low hills of a marked eolian type. From the profiles of these hills it is seen that the prevailing winds which fashioned them came from the southwest. Along the southern border of the county Snake River flows in a canyon which has a depth of from 2,000 to 3,000 feet below the level of the plateau. The principal tributaries of the Snake within the county likewise flow in canyons, and hence indicate that the dissection of the plateau is under way.

The average rainfall is about 20 inches. This decreases to 15 inches along the western border and increases to 25 inches in the higher portion along the eastern boundary. With the exception of a slight forest growth on the eastern margin and of a few trees which grow in the valleys along the streams, the entire county was once clothed with grasses. The rainfall is of sufficient amount to make the growth of wheat possible, so that wheat fields now cover almost every part of the plateau. Irrigation is employed only in the canyons of the larger streams, such as the Snake. Upon the terraces or benches that border the latter river large fruit ranches are found, the water for irrigation being supplied by the small tributary streams.

In the plateau above mentioned basalt forms the bed rock almost exclusively. The few buttes or hills along the eastern boundary, which project above the plateau, are composed of granites, gneisses, schists, quartzites, and similar rocks, which indicate the character of the floor upon which the basalt outflowed. Along the canyon of Snake River in two or three places granite appears, showing that the river in trenching the basalt is bringing to light the rocks beneath it. The walls of the canyon of Snake River show the basalt in a series of 8 or 10 practically horizontal flows. The plateau is covered with a deep soil and the underlying rock rarely appears. Because of its eolian nature the soil is of a very fine grain and has the property of retaining moisture to a very marked degree. Oftentimes wells dug only into the soil afford a good supply of water. The towns usually depend upon deeper wells drilled into the basalt for their water supplies. As a general thing these wells range in depth from 100 to 300 feet. In some instances, as at Pullman, water is obtained from beds of sand and gravel between the layers of basalt. In other cases very porous layers of basalt often afford a plentiful supply of water. In some cases flowing wells are secured, this being especially true when the wells are drilled in the bottoms of the canyons or in deep valleys. Springs are very common along the valley sides wherever porous basalt appears at the surface.

Municipal systems.—Colfax obtains its city supply from Palouse River. In this way very good water is obtained; but it is possible that in the future deep wells may be substituted as a source of supply. The water is pumped into a reservoir located on a hill above the city, from which it is distributed by gravity. The wells about Colfax range from 12 to 120 feet in depth. Water is usually found at a depth of about 40 feet. All the deeper wells enter the rock, which is basalt. No flowing wells have been found; in every instance the water must be pumped.

Oakesdale depends chiefly upon wells, although a few springs and cisterns are used. At some places the water is soft, but at other places it is hard. Generally speaking, there are no sources of contamination and the water is quite satisfactory. The wells on the hills about Oakesdale have been sunk to depths of from 45 to 60 feet in order to obtain water. On the plain at the foot of the hills they are shallow, ranging from 10 to 15 feet in depth. In the shallow wells a clay hardpan with sand and gravel below are the only materials penetrated, while all the aeeper wells enter the basaltic rock. The supply of water in the wells is stationary, neither increasing nor decreasing as far as known.

Pullman obtains a water supply from artesian wells. The wells vary in depth from 100 to 130 feet, water being most commonly found at 100 feet. The wells are in basaltic rock, the water coming from beds of sand interstratified with the layers of basalt. About 150,000 gallons of water are obtained daily. The capacity of the wells, however, is such that this supply could be doubled if it were necessary. It is noted that the wells do not flow as strongly as they did when they were first drilled. A gravity system of waterworks is used, the water being first lifted by steam pumps from the wells to a reservoir. An analysis of the city water, made by Prof. Elton Fulmer, of the Washington Agricultural College and School of Science, shows the following composition:

Analysis of city water from Pullman.

[Grains per gallon.]

Silica (SiO_2)	3.49
Potash (K_2O)	.50
Soda (Na_2O)	1.75
Lime (CaO)	1.86
Magnesia (MgO)	1.31
Iron oxide (Fe_2O_3)	.06
Alumina (Al_2O_3)	.03
Sulphuric acid (SO_3)	.08
Carbonic acid (CO_2)	3.73
Chlorine	.15
Total solids	12.96

Tekoa depends entirely upon an artesian well for its water supply. From this well water of a good quality is obtained, there being no contamination as far as known. The supply is at present sufficient, but will not be large enough in the future. In the region about Tekoa water is most commonly found at a depth of about 100 feet, although some of the wells reach depths of 175 feet. All of the wells enter rock. In most instances the water rises to the surface, and in some cases reaches a height of 10 feet above the mouth of the well.

Uniontown obtains its water supply from a deep well. The water is soft and very satisfactory in every way, and the quantity is ample for present needs at least. The deepest wells about Uniontown have a depth of about 200 feet. These wells enter the rock and obtain water from beds of sand and gravel intercalated with basalt. There are some surface wells which do not enter the rock and which are from 7 to 10 feet in depth. In none of the wells does the water rise to the surface. The water level does not vary during the day or year and is not affected by pumping.

Deep wells.—About Palouse a number of wells have been drilled. In general they have a diameter of 6 inches and a depth of from 100

to 300 feet. The depth to the principal source of water is, on the
average, 150 feet. The water is found in beds of sand which lie
below a capping of basalt. From some of the wells the water flows,
while in others the water level stands below the surface and pumping
must be resorted to. The level does not vary during the day or year,
the supply being constant from season to season. The water as a rule
is hard, generally containing magnesia and iron. The usual temperature of the water at the well mouth is 50° F. The wells are mostly
owned by farmers, and the water is used for general farm supply.
The average cost of a well is $300, while the pumping machinery costs
from $100 to $200. As a rule the wells are only partially cased, the
length of casing varying from 20 to 120 feet.

In and about the city of Pullman are a number of artesian wells, of
which the two owned by the city and used as a source of municipal
water supply may be taken as types. One of these wells was drilled
in 1890 and the other in 1899. They are located near the bed of a
stream which flows through the center of the city. The mouths of the
wells are at an elevation of 2,341 feet above the sea. Each well has a
depth of 110 feet, with a diameter of 6 inches. The wells flow and the
water rises to a height of 19 feet above the surface. The temperature
of the water at the well mouth is about 60° F. The supply of water
has decreased since the wells were drilled. Both wells are cased
throughout with 6-inch heavy wrought-iron pipe. The cost of each
well was about $450, and the cost of the pumping machinery necessary
to lift the water to the reservoir was about $3,000.

In Tekoa there is an artesian well that was drilled in 1892 to serve
as a source of water supply for the town. The well was drilled on a
stream bed. It has a depth of 176 feet and a diameter of 6 inches. Only
one water-bearing bed was found, that at the bottom of the well. The
water rises to a height of 8 feet above the surface, and has a temperature of 76° F. at the well mouth. The water level has shown no sea-
sonal variation, and the supply has been constant. The cost of the
well was $750, and the cost of the pumping machinery was $1,000.
The casing has been placed in the well only from the surface to solid
rock. The Oregon Railroad and Navigation Company owns a deep
well at this place, from which is obtained a water supply for locomotives.

YAKIMA COUNTY.

General statement.--Yakima County is located in the south-central
part of Washington, between Columbia River and the summit of the
Cascade Mountains. The surface presents a great diversity topographically. Columbia and Yakima rivers are bordered by broad, low
plains. Between these streams are a number of ridges or low mountains which rise a few hundred feet above the general level of the

plain. Southwest and west of Yakima River the plain merges into an
even-topped plateau, which in the course of 20 or 30 miles gives way
to the foothills of the Cascades. The mountains are comparatively
low, with a general height of 4,000 to 5,000 feet, except in the
vicinity of Mount Adams, in the southwestern part of the county,
where they are much higher, reaching a maximum of 12,400 feet.

As the county lies partly within the Cascade Mountains and partly
within the Columbia Plains, it has geologic features common to both
provinces. The formations of those portions of the Cascades within
the county are scarcely known at all, but are believed to consist mainly
of metamorphic and igneous rocks typical of the interior of the
range. Immediately east of Cowlitz Pass are limited areas of Tertiary
shales and sandstones which contain seams of coal. The coal is a semi-
anthracite, or a true anthracite in some cases. Along the eastern flank
of the mountains are occasional flows of andesite, most of which are
of a very late date.

The eastern part of the county lies within the Columbia Plains, and
here the rocks belong almost wholly to two divisions, viz, the Yakima
basalt and the Ellensburg formation. The Yakima basalt, according
to Smith, has a vertical thickness in the canyon of Yakima River of
more than 2,500 feet, and represents ten or more separate flows. For
the most part the basalt is compact and heavy, but occasionally it is
cellular or scoriaceous. In color it is black, except on weathered surfaces, where it usually has become brown. In a few instances the
so-called ash beds, consisting of fine and coarse tuffs, are found interbedded with the dense compact basalt. The latter has usually a pris-
matic or columnar structure, the result of the contraction or shrink-
age of the lava in the process of cooling. The Ellensburg formation
lies directly on the Yakima basalt, and includes shales, sandstones, and
conglomerates that, in the main, were deposited immediately after the
last flows of lava. In some places there was a final outflow of lava a
little while after the beginning of the deposition of the clastic rocks.
The sediments of the Ellensburg formation accumulated to a thickness
of 1,500 feet or more, and on the evidence of fossil leaves are known
to be of Miocene age. These rocks are but partially consolidated, and
have suffered considerable erosion, with the result that they have been
wholly or in large part removed from much of the area that they once
covered. This formation is of special interest. It contains water, and
within it the artesian wells of the county have been drilled.

The Yakima basalt and the Ellensburg formation have been folded
into a series of arches and troughs, or anticlines and synclines, that
have, in general, an east-west direction. The general structure of
the region is well shown along the course of Yakima River, which has
cut directly or obliquely across the ridges and alternating troughs.
As a rule the arches or ridges are comparatively long and narrow, and

rise from 800 to 1,500 feet above the intervening valleys. The arches are almost invariably capped by basalt, the rocks of the Ellensburg formation being limited to the valleys and the lower flanks of the ridges.

The rainfall of Yakima County shows as great a variation in amount as is to be found in any area of similar size within the State. In the eastern portion the annual rainfall averages 10 inches or a little less. From the vicinity of North Yakima westward it increases regularly toward the summit of the mountains. Upon the plateau it ranges from 15 to 25 inches. In the mountains it varies from 30 inches in the foothills to a maximum of 50 inches along the summit. The effect of the rainfall upon the vegetation is of interest. The western end of the county is so well forested that it is included within the Mount Rainier Forest Reserve. The plateaus and hills of the central and eastern parts are bare of trees, but are abundantly covered with grasses. Along the low plains of Yakima and Columbia rivers the grasses largely disappear and sagebrush takes their place as the principal vegetal covering. Except for the growing of wheat upon the plateaus, no attempt is made to carry on agriculture without recourse to irrigation. The water for irrigation is obtained chiefly from the streams, but to some extent from artesian wells. The streams coming from the mountains, where the rainfall is heavy, carry a large amount of water. Even without the use of impounding reservoirs, enough water may be had to irrigate a large part of the plains. In the vicinity of North Yakima a number of deep wells have been drilled, which furnish sufficient water to irrigate large tracts of land. The character and extent of this artesian basin have been set forth in Water-Supply and Irrigation Paper of the United States Geological Survey, No. 55. The towns, as a rule, depend upon streams for their supply of water, although surface wells are frequently used. Upon the plateaus and about the foothills springs are often found, some of which might be classed as mineral springs.

Municipal systems. – North Yakima obtains its supply of water from Naches River. The quantity obtained is sufficient for the present and probably for all future needs. The Naches rises in the foothills of the Cascades, and as there are no sources of contamination the quality of water is good. A few wells are used about North Yakima. They range in depth from 18 to 25 feet and are wholly in sand and gravel. The city water supply is used to a very large degree for irrigating, besides serving as a domestic supply.

A sanitary analysis of water from the North Yakima city supply. made on April 8, 1901, by Prof. H. G. Byers, of the University of Washington, resulted as follows:

Sanitary analysis of city water at North Yakima.

[Parts per million.]

Total solids	169.53
Nonvolatile solids	114.85
Volatile solids	54.68
Oxygen consumed	1.79
Chlorine	2.75
Nitrogen as free ammonia	.008
Nitrogen as albuminoid ammonia	.0177
Nitrogen as nitrites	None.
Nitrogen as nitrates	None.

The town of Prosser obtains a water supply from Yakima River. The river gets very low in autumn and may not afford sufficient water to meet the future needs of the town. The water is pumped into a reservoir and is distributed by a gravity system. A few wells have been dug in the region, and these have a depth ordinarily of about 40 feet. They do not enter bed rock, but are wholly in gravel.

Deep wells.—In sec. 4, T. 20 N., R. 12 E., is a deep well owned by F. E. Deeringhoff. It is located on a gentle slope at an elevation of about 1,100 feet above the sea. It was completed in April, 1899. The well was drilled to a depth of 275 feet and then bored for the remainder of the way, a distance of 350 feet. The drilled portion has a diameter of 5⅝ inches, while the bored part has a diameter of 3 inches. Three water-bearing beds were found, the first at a depth of 200 feet, the second at 400 feet, and the third and principal one at 625 feet. The principal water-bearing material is sand. When the well was completed the water rose to a height of 40 feet above the surface, but at present it rises to a height of but 1 foot above the well mouth. The temperature of the water at the surface is 74° F. The water is soft and is sulphur bearing to a slight degree. In casing the well 40 feet of 5-inch, 120 feet of 4-inch, and 80 feet of 3-inch pipe were used. The water is used entirely for irrigation.

In sec. 8, T. 12 N., R. 20 E., J. H. Gano has an artesian well located on a plain. The well was drilled, and has a depth of 826 feet. In the upper portion the diameter is 4 inches, while in the lower portion it is only 2½ inches. Water-bearing beds were found at depths of 300 and 400 feet, besides the principal bed at the bottom. The temperature of the water at the well mouth is 78° F. The water rises 40 feet above the surface. The supply has not decreased since the well was completed. The cost of the well was $1,000. The water is used for irrigating purposes.

In sec. 9, T. 12 N., R. 20 E., E. S. Hill has a deep well, which was completed in 1900. The well was drilled on the slope of a hill, and has a depth of 626 feet. From 200 feet downward several water-bearing beds were found, the principal one being a stratum of sand at the bottom. The well flows, and the supply has not increased or decreased since the well was completed. The temperature of the water at the well mouth is 74° F. The cost was $900. In the well 490 feet of casing has been placed, the upper portion of this having a diameter of 4½ inches and the lower portion a diameter of 3½ inches. The water is used exclusively for irrigation.

In sec. 6, T. 12 N., R. 20 E., is an artesian well owned by J. W. Peck. The well is located in a valley, at a height of about 800 feet above sea. It was completed in 1901 at a cost of $1,200. It has a diameter of 6 inches and a depth of 828 feet. The principal flow of water was found in a sandstone, but other minor water-bearing beds were encountered in drilling. The well is a flowing one, the water rising 4 feet above the surface. The water flows a little less strongly in the summer than in the winter, but on the whole the amount of water remains fairly constant. The temperature of the water at the well mouth is 74° F. The water carries a little iron, magnesia, and sulphur.

In sec. 10, T. 12 N., R. 20 E., Robert Rein has a deep well, which was completed in 1900. The depth of the well is 570 feet, the diameter at the top is 6¼ inches and at the bottom 2¾ inches. The well is located on a plain at a height of 1,500 feet above the sea. The water rises about 20 feet above the surface, and the supply has been constant since the completion of the well. The temperature of the water at the well mouth is 60° F. The cost of the well was $700. The water is used for irrigating purposes.

In sec. 8, T. 12 N., R. 20 E., is an artesian well owned by Julius Sauve. It is located on a plain at an elevation of 1,155 feet above sea. The diameter of the well in the first portion is 4¾ inches and in the bottom portion 2 inches. A number of different flows of water were encountered at depths of 790, 861, 876, 890, 907, and finally at the bottom at 1,020 feet. The water rises 80 feet above the surface. The quantity of water has not changed since the well was first drilled. The water at the well mouth has a temperature of 75.2° F. The cost of the well was $1 per foot, or $1,020.

In sec. 8, T. 12 N., R. 20 E., David Walters drilled a well which was completed July 12, 1902. It has a depth of 1,200 feet and a diameter of 5¼ inches. The well is a flowing one. The temperature of the water at the well mouth is 81° F. The cost of the well was $1,600. In casing the well 400 feet of pipe with a diameter of 4 inches and 620 feet of pipe with a diameter of 3½ inches were used.

Springs.—In T. 9 N., R. 12 E., on Government land, spring water issues as a stream from the base of a bluff of basaltic rock. The water is very cold, free from any odor, and colorless. From the water bubbles of gas are constantly escaping. An analysis of the water made by Prof. H. G. Byers, of the University of Washington, resulted as follows:

Analysis of water from spring in T. 9 N., R. 12 E.

[Parts per million.]

Volatile solids	363. 5
Nonvolatile solids	774. 5
Total solids	1, 138. 0
Silica	109. 3
Ferric oxide and alumina	82. 0
Calcium carbonate	266. 4
Magnesium carbonate	177. 7
Sodium chloride	213. 4
Potassium chloride	38. 7
Calcium sulphate	None.
Potassium sulphate	None.

In sec. 9, T. 11 N., R. 15 E., on tribal land belonging to the Yakima Indians, are some warm and cold springs. The water issues as small streams from soil and gravel. About the springs are deposits of reddish matter, presumably iron oxide. The cold springs have a soda taste, and from them all bubbles of gas are constantly escaping. The warm springs are used by the Indians for bathing, and the water is believed by them to possess medicinal value. It is used more for rheumatism than for any other disease. From all of the springs the quantity of flow is constant, no variation being appreciable from season to season.

TABLES OF DEEP WELLS, MUNICIPAL WATER SUPPLIES, AND REPRESENTATIVE SPRINGS.

Deep wells in Washington.

County and post-office.	T.	R.	S.	Name of owner.	Topographic position.	Elevation above sea.
Adams County:						*Feet.*
Cunningham	16	30	22	Thomas and James O'Hair.	Hill	1,278
Ritzville	16	32	33	Northern Pacific Rwy	Valley	1,157
Franklin County:						
Connell	13	32	28	W. T. Braden	Valley	840
Island County:						
Coupeville				E. J. Hancock	Hill	125
Spokane County:						
Hillyard				Hillyard Town Site Co	Plain	2,000
Stevens County:						
Colville	35	39	9	L. J. Walford	Base of hill	1,590
Thurston County:						
Olympia				State of Washington	Plain	30

Deep wells in Washington—Continued.

County and post-office.	T.	R.	S.	Name of owner.	Topographic position.	Elevation above sea.
						Feet
Wallawalla County:						
Walla Walla............	7	35	27	Blalock Fruit Co...........	Plain..............	825
Whitman County:						
Palouse....................				F. P. Egan....................	Valley.............	2,000
Pullman..................	14	45	5	City of Pullman.............	Stream bed........	2,341
Tekoa....................				Town of Tekoa..............	....do..............	
Yakima County:						
North Yakima...........	20	12	4	F. E. Deeringhoff..........	Slope..............	1,100
Do..................	12	20	8	James H. Gano.............	Plain..............	1,100
Do..................	12	20	9	E. S. Hill....................	Slope..............	
Do..................	12	20	6	J. W. Peck	Valley.............	900
Do..................	12	20	10	Robert Rein.................	....do.............	1,500
Do..................	12	20	8	J. Saure	Plain..............	1,155
Do..................	12	20	8	David Walters..............	Slope..............	

County and post-office.	Date when completed.	Kind of well.	Diameter.	Depth.	Depth to principal source of water.	Water-bearing material.
			Inches.	*Feet.*	*Feet.*	
Adams County:						
Cunningham	1902	Drilled	6	426	426	Rock.
Ritzville.................	1901	....do......	8	355	300	Very porous rock.
Franklin County:						
Connell.................	1902	....do......	5	676	660	Porous rock.
Island County:						
Coupeville..............	1889	Dug........	48	125	120	Sand.
Spokane County:						
Hillyard	1900	....do......	44	200	192	Gravel.
Stevens County:						
Colville.................	1901	Drilled .... {	10 / 6	700	400	Sandstone.
Thurston County:						
Olympia	1908	....do......	2	152	152	Sand.
Wallawalla County:						
Walla Walla	1908	....do......	6	564	540	Do.
Whitman County:						
Palouse...................		....do......	6	200	150	Do.
Pullman	1890	....do......	6	110	110	Do.
Tekoa....................	1892	....do......	6	176	176	Porous rock.
Yakima County:						
North Yakima	1899	Drilled and bored. {	5½ / 3	625	625	Sand.
Do		Drilled .... {	4 / 2½	826	820	Do.
Do	1900	....do......		626	620	Do.
Do	1901	....do......	6	828	828	Do.
Do	1900	....do...... {	6½ / 2½	570	570	Do.
Do	1900	....do......	4½	1,020	1,020	Do.
Do	1902	....do......	3½	1,200	1,200	Do.

Deep wells in Washington—Continued.

County and post-office.	Other water-bearing beds found.	Distance water rises above surface.	Distance of water from surface.	Temperature of water at well mouth.	Amount of water obtained daily.	Increase or decrease of supply.
		Feet.	*Feet.*	° *F.*	*Gallons.*	
Adams County:						
Cunningham	None ...	....	356	....	....	Increase.
Ritzville	None ...	....	235	....	....	Stationary.
Franklin County:						
Connell	None ...	....	640	51	....	Stationary.
Island County:						
Coupeville	Yes	....	120	....	....	Do.
Spokane County:						
Hillyard	None .	....	185	....	50,000	
Stevens County:						
Colville	Yes ...	....			....	
Thurston County:						
Olympia	Yes	2	....	....	8,640	Increasing.
Wallawalla County:						
Walla Walla	....	....		67	187,000	Stationary.
Whitman County:						
Palouse	None ...	Several.	....	50	....	Do.
Pullman	None ...	19	....	60	180,000	Decreasing.
Tekoa	None ...	8	....	76	....	Stationary.
Yakima County:						
North Yakima	Two ...	1	....	74	....	Decreasing.
Do	Two	40	....	78	....	Stationary.
Do	Two	Several.	....	74	....	Do.
Do	Yes	4	....	74	....	Do.
Do	Yes	20	....	60	....	Do.
Do	Yes	80	....	75	....	Do.
Do	....	Several.	....	81	....	

County and post-office.	Variation in water level.	Effect of pumping on level of water.	Quality of water.	How water is obtained at surface.
Adams County:				
Cunningham	None	None	Soft	Pumping.
Ritzville	....do	Lowers level 10 feet.	....do	Do.
Franklin County:				
Connell	....do	None	....do	Do.
Island County:				
Coupeville	....	Lowers it slightly.	Hard	Do.
Spokane County:				
Hillyard	Varies during year	Lowers level slightly.	Soft............	Do.
Stevens County:				
Colville	None	....	Iron bearing..	
Thurston County:				
Olympia	Varies with tide..	....		Well flows.
Wallawalla County:				
Walla Walla	None			Do.
Whitman County:				
Palouse	....do	Lowers level.......	Hard	Do.
Pullman	....do		....do	Do.
Tekoa	....do	None	Soft............	Do.

Deep wells in Washington—Continued.

County and post-office.	Variation in water level.	Effect of pumping on level of water.	Quality of water.	How water is obtained at surface.
Yakima County:				
North Yakima	Varies during year		Sulphur bearing.	Well flows.
Do	None		Soft	Do.
Do	....do	None	....do	Do.
Do	Varies during year		....do	. Do.
Do	None	None	....do	Do.
Do	Varies during year		....do	Do.
Do			Sulphur bearing.	Do.

County and post-office.	Cost.	Cost of pumping machinery.	Size and length of casing.	Use made of the water.
Adams County:				
Cunningham	$1,066	$290	Diameter, 2 inches; length, 423 feet.	Farm supply.
Ritzville	2,500		Length, 240 feet	Locomotives.
Franklin County:				
Connell	3,000		Diameter, 5 inches; length, 100 feet.	Farm supply
Island County:				
Coupeville	350	150		Do.
Spokane County:				
Hillyard	500	2,500		Town supply.
Stevens County:				
Colville	3,000		Diameter, 10 inches; length, 40 feet.	None
Thurston County:				
Olympia	152		Diameter, 2 inches; length, 152 feet.	Supply for capitol building.
Wallawalla County:				
Walla Walla	1,800		Length, 540 feet	Irrigation.
Whitman County:				
Palouse	800		Diameter, 5½ inches; length, 150 feet.	Farm supply.
Pullman	450	3,000	Diameter, 6 inches; length, 110 feet.	City water supply.
Tekoa	750	1,000	Diameter, 6 inches	Town supply.
Yakima County:				
North Yakima			Diameter, 5 inches; length, 40 inches. Diameter, 4 inches; length, 120 inches. Diameter, 3 inches; length, 80 inches.	Irrigation.
Do	1,000		Diameter, 4, 3, and 2½ inches.	Do.
Do	900		Diameter, 4½ inches; length, 497 feet.	Do.
Do	1,200			Do.
Do	700			Do.
Do	1,020			Do.
Do	1,600		Diameter, 4 inches; length, 400 feet. Diameter, 3½ inches; length, 620 feet.	Do.

Municipal water supplies in Washington.

Location.	Water-supply system.	Principal source of water.	Other sources.	Sufficient supply for present needs.	Sufficient supply for future needs.
Adams County:					
Ritzville	Yes....	Deep wells	None	No....	No.
Asotin County:					
Asotin	No.....	Asotin Creek	Wells and cisterns....	Yes...	Yes.
Clarkston	Yes....	....do	None	Yes...	Yes.
Chehalis County:					
Aberdeen	Yes....	Creek		No....	No.
Cosmopolis	Yes....	Wells	Creek	Yes...	Yes.
Hoquiam	Yes....	Hoquiam River	None	Yes...	Yes.
Montesano	Yes....	Springs	....do	Yes...	Yes.
Ocosta	No.....	Wells	Springs		
Chelan County:					
Chelan	Yes....	Springs	Lake Chelan	No....	No.
Lakeside	No.....	Lake Chelan	Wells	Yes...	Yes.
Wenatchee	Yes....	Creek	None	Yes...	No.
Clallam County:					
Port Angeles	Yes....	Frazer Creek	Wells and springs	No....	No.
Port Crescent	No.....	Wells	Pond	Yes...	No.
Clarke County:					
Vancouver	Yes....	Springs and deep wells.		Yes...	Yes.
Columbia County:					
Dayton	Yes....	Springs	None	Yes...	Yes.
Cowlitz County:					
Castlerock	Yes....	Creek....	Wells	Yes...	No
Kalama	Yes....	Creeks	None	Yes...	No.
Douglas County:					
Wilsoncreek	No.....	Wells	Streams	No....	No.
Jefferson County:					
Port Ludlow	Yes....	Creek	None	Yes...	Yes.
Port Townsend	Yes....	Deep wells	Wells and cisterns.	No....	No.
King County:					
Auburn	No.....	Wells	White River	Yes...	Yes.
Ballard	Yes....	Springs and deep wells.		Yes...	No.
Columbia City	Yes....	Cedar River	None	Yes...	Yes.
Enumclaw	Yes....	Mountain streams....	Wells	Yes...	Yes.
Issaquah	Yes....	Springs	....do	Yes...	Yes.
Kent	Yes....	....do	None	Yes...	No.
Renton	Yes....	Spring	Wells	Yes...	No.
Seattle	Yes....	Cedar River and Cedar Lake.	None	Yes...	Yes.
West Seattle	Yes....	Springs	....do	No....	No.
Kitsap County:					
Bremerton	Yes....	....do	Wells	Yes...	Yes.
Charleston	Yes....	Springs and creeks	....do	Yes...	Yes.
Port Blakeley	Yes....	Creek		Yes...	Yes.
Port Gamble	Yes....	Creeks		Yes...	Yes.
Kittitas County:					
Clealum	Yes....	Springs	Wells	Yes...	Yes.
Ellensburg	Yes....	Mountain streams	....do	No....	No.
Roslyn	Yes....	Springs and Clealum River.		Yes...	No.

Municipal water supplies in Washington—Continued.

Location.	Water-supply system.	Principal source of water.	Other sources.	Sufficient supply for present needs.	Sufficient supply for future needs.
Klickitat County:					
Goldendale	Yes	Spring	None	Yes	No.
Lewis County:					
Centralia	Yes	Wells	Skookum Chuck River.	No	No
Pe Ell	No	...do	Springs	No	No.
Lincoln County:					
Davenport	Yes	Springs	None	Yes	No.
Harrington	Yes	Wells	Springs	Yes	No.
Sprague	Yes	Springs	Wells	Yes	Yes.
Wilbur	Yes	Creek and wells	Springs	Yes	Yes.
Mason County:					
Shelton	Yes	Spring	None	Yes	No.
Okanogan County:					
Loomis	Yes	Sinlahekin Creek	Spring	Yes	Yes.
Pacific County:					
Ilwaco	Yes	Spring		Yes	No.
Southbend	Yes	...do	None	Yes	Yes.
Pierce County:					
Buckley	Yes	Well	White River	Yes	Yes.
Carbonado	Yes	Springs	None	Yes	Yes.
Orting	Yes	...do	Wells	Yes	No.
Puyallup	Yes	...do	None	Yes	Yes.
South Tacoma	Yes	Wells	...do	Yes	Yes.
Sumner	Yes	Springs	...do	Yes	Yes.
Tacoma	Yes	Clover Creek and springs.	Three wells	No	No.
San Juan County:					
Friday Harbor	No	Wells		Yes	No.
Skagit County:					
Anacortes	Yes	Lake Heart	None	Yes	No.
Hamilton	No	Wells	...do	Yes	Yes.
La Conner	Yes	Spring		Yes	
Mount Vernon	Yes	...do	None	Yes	Yes.
Snohomish County:					
Arlington	No	Springs and wells	Stilaguamish River	Yes	Yes.
Edmonds	Yes	Springs	Wells	Yes	
Everett	Yes	Wood's Creek		Yes	Yes.
Marysville	No	Wells	None		
Monroe	Yes	Creeks	Wells	Yes	Yes.
Monte Cristo	Yes	Mountain stream	None	Yes	Yes.
Silverton	Yes	...do	...do	Yes	Yes.
Snohomish	Yes	Pilchuck River	...do	Yes	Yes.
Spokane County:					
Cheney	Yes	Lake and wells		Yes	Yes.
Hillyard	Yes	Deep wells	None	Yes	No.
Latah	No	Wells			
Medical Lake	Yes	Clear Lake	Wells	Yes	Yes.
Stevens County:					
Bossburg	Yes	Columbia River	None	Yes	Yes.
Colville	Yes	Spring	Creeks and wells	Yes	No.
Marcus	No	Columbia River	None		
Northport	Yes	Deep Creek		Yes	Yes.
Springdale	Yes	Creek	Wells	Yes	Yes.

Municipal water supplies in Washington—Continued.

Location.	Water-supply system.	Principal source of water.	Other sources.	Sufficient supply for present needs.	Sufficient supply for future needs.
Thurston County:					
Olympia	Yes....	Springs	None	Yes...	Yes.
Tenino	No.....	Wells	Spring	Yes...	No.
Wallawalla County:					
Waitsburg	Yes....	Springs	Creeks	Yes...	No.
Walla Walla	Yes....	....do	Creek	Yes...	No.
Whatcom County:					
Blaine	Yes....	....do		Yes...	Yes.
Fairhaven	Yes...	Lake Padden	Three creeks	Yes...	Yes.
Sumas	No.....	Wells	Streams	No....	No.
Whatcom	Yes....	Lake Whatcom	None	Yes...	Yes.
Whitman County:					
Colfax	Yes....	Palouse River	....do	Yes...	Yes.
Oakesdale	No.....	Wells	Springs		
Pullman	Yes....	Artesian wells		Yes...	Yes.
Tekoa	Yes....	....do	None	Yes...	No.
Uniontown	Yes....	Deep well		Yes...	Yes.
Yakima County:					
North Yakima	Yes....	Naches River	None	Yes...	Yes.
Prosser	Yes....	Yakima River	....do	No....	No.

Location.	Quality of water.	Effect of water on the health.	Sources of contamination.	Other sources of supply in contemplation.
Adams County:				
Ritzville	Hard	Good	None	Additional wells.
Asotin County:				
Asotin	Soft	....do	....do	None.
Clarkston	....do	....do	....do	Artesian wells.
Chehalis County:				
Aberdeen	....do	....do		Other creeks.
Cosmopolis	....do	....do	None	None.
Hoquiam	....do	....do	....do	Do.
Montesano	....do	....do	....do	Do.
Ocosta	....do	....do		Do.
Chelan County:				
Chelan	Hard		Surface drainage.	Lake Chelan.
Lakeside	Soft	Good	None	None.
Wenatchee	Contains iron	....do	Surface drainage.	Do.
Clallam County:				
Port Angeles	Soft	Some ill effects in summer.	Decaying vegetation.	Little River.
Port Crescent	Somewhat salty		None	None,
Clarke County:				
Vancouver	Soft	Good	....do	Do.
Columbia County:				
Dayton	....do	....do	....do	Do.
Cowlitz County:				
Castlerock	....do	....do	....do	Do.
Kalama	Hard	....do	....do	Do.

Municipal water supplies in Washington—Continued.

Location.	Quality of water.	Effect of water on the health.	Sources of contamination.	Other sources of supply in contemplation.
Douglas County:				
Wilsoncreek	Alkaline	Good	None	None.
Jefferson County:				
Port Ludlow	Soft	do		
Port Townsend	do	do	None	Mountain streams.
King County:				
Auburn	do	do		White River
Ballard	do	do		Seattle water system.
Columbia City	do	do		None.
Enumclaw	do	do	None	Springs.
Issaquah	do	do	do	None.
Kent	do	do	do	Additional springs.
Renton	do	do	do	Cedar River and other springs.
Seattle	do	do	do	None.
West Seattle	do	do	do	Cedar River.
Kitsap County:				
Bremerton	do	do	do	None.
Charleston	do	do	do	Do.
Port Blakeley	do	do	Decaying vegetation.	Do.
Port Gamble	do	do	None	Do.
Kittitas County:				
Clealum	do	do	do	Yakima River.
Ellensburg	Hard	do	do	None.
Roslyn	Soft	do	do	Do.
Klickitat County:				
Goldendale	do	do	do	Additiona spring.
Lewis County:				
Centralia	do			None.
Pe Ell	do	Some ill effects.	Sewage	Do.
Lincoln County:				
Davenport	do	Good	None	Do.
Harrington	do	do		
Sprague	do	do	None	Do.
Wilbur	do	do	do	Do.
Mason County:				
Shelton	do	do	do	Other springs.
Okanogan County:				
Loomis	do	do	do	Other creeks.
Pacific County:				
Ilwaco	do	Not good	Decaying vegetation.	None.
Southbend	do	Good	None	
Pierce County:				
Buckley	do	do	do	Do.
Carbonado	do	do	Decaying vegetation.	Do.
Orting	do	do	None	Additional springs.
Puyallup	do	do	do	None.
South Tacoma	Hard	do	do	Do.
Sumner	Soft	do	do	Do.
Tacoma	do	do	do	Additional wells.

Municipal water supplies in Washington—Continued.

Location.	Quality of water.	Effect of water on the health.	Sources of contamination.	Other sources of supply in contemplation.
San Juan County:				
Friday Harbor	Soft	Good	None	
Skagit County:				
Anacortes	do	do	do	Other lakes.
Hamilton	do	do	do	None.
Laconner		do		
Mount Vernon	Soft	do	None	Do.
Snohomish County:				
Arlington	do	do	do	
Edmonds	do	do		
Everett	do	do	None	Do.
Marysville	do	do	do	Lake Stephens.
Monroe	do	do	do	Skykomish River.
Monte Cristo	do	do	do	None.
Silverton	do	do	do	
Snohomish	do	do	do	Do.
Spokane County:				
Cheney	do	do		
Hillyard	do	do		Do.
Latah	do	do	None	Springs.
Medical Lake	Hard	do	do	None.
Stevens County:				
Bossburg	Soft	do	do	Do.
Colville	do	do	do	Do.
Marcus	do	do	do	Do.
Northport	Hard	do	do	Do.
Springdale	Slightly alkaline	do	Surface drainage.	Do.
Thurston County:				
Olympia	Soft	do	None	Do.
Tenino	do	do	do	Do.
Wallawalla County:				
Waitsburg	do	do		Deep wells.
Walla Walla			None	
Whatcom County:				
Blaine		Good		
Fairhaven	Soft	do	None	None.
Sumas	do	do	Surface drainage	Deep well.
Whatcom	do	do	Lumber mills	None.
Whitman County:				
Colfax	do	do		Deep wells.
Oakesdale	do	do	None	None.
Pullman	Hard	do	do	Do.
Tekoa	Soft	do	do	Do.
Uniontown	do	do	do	Do.
Yakima County:				
North Yakima		do	do	Do.
Prosser	Soft	do	do	

Municipal water supplies in Washington—Continued.

Location.	System of waterworks used.	Depth of the wells.	Depth at which water is commonly found.	Character of the water-bearing material.	Depth of water from surface.
Adams County:		*Feet.*	*Feet.*		*Feet.*
Ritzville	Pumping and gravity.	a 385 b 200	c 50	Porous	20
Asotin County:					
Asotin		20-35	25	Gravel and sand	10
Clarkston	Gravity				
Chehalis County:					
Aberdeen	Direct pressure	20-40		Sand	
Cosmopolis	Gravity	10	10	....do	d.
Hoquiam	Pumping and gravity				
Montesano	Gravity	35	30	Cement and gravel.	25.
Ocosta		50	40-50	Sand	
Chelan County:					
Chelan	Gravity	Deep.	40-100		e
Lakeside		20-40	Level of lake.	Sand	f.
Wenatchee	Gravity				
Clallam County:					
Port Angeles	....do	12-40	10-20	Sand and gravel	20
Port Crescent		16-20	10-12	Sandstone	e
Clarke County:					
Vancouver	Gravity	30-75	25	Gravel	.
Columbia County:					
Dayton	....do		25-50	Basalt	
Cowlitz County:					
Castlerock	....do		15	Gravel	2.
Kalama	....do		10	Basalt	10
Douglas County:					
Wilsoncreek		12-54	48	Gravel	
Jefferson County:					
Port Ludlow	Gravity				
Port Townsend	Direct pressure	22-100		Sand and gravel	
King County:					
Auburn		40-50	40	...do	
Ballard	Direct pressure	20-160	12-150	....do	
Columbia City	Gravity	12-50	10-30	....do	
Enumclaw	....do		10-30	....do	
Issaquah	....do	20-30	20	Gravel	2-2
Kent	....do		(d)	Sand and gravel	
Renton	....do	10-45	10-25	Gravel	6-
Seattle	....do	15-50	12-40	Sand and gravel	10-2
West Seattle	Direct pressure	30-75	50	....do	25-7
Kitsap County:					
Bremerton	Gravity		10	Sand	
Charleston	....do	20-150	20-150.		
Port Blakeley	....do		70	Gravel	.
Port Gamble	....do				
Kittitas County:					
Clealum	....do	10-16	8-12	Sand and gravel	2-
Ellensburg	....do	10-20	10-20	Gravel	4
Roslyn	Direct pressure	20-60	12-50	Sandstone	1
Klickitat County:					
Goldendale	Gravity	12-75	20	Gravel	

a City well.   c Varies on lower ground.   e No appreciable rise.
b Average well.   d Nearly to surface.   f Lake level.

Municipal water supplies in Washington—Continued.

Location.	System of waterworks used.	Depth of the wells.	Depth at which water is commonly found.	Character of the water-bearing material.	Depth of water from surface.
Lewis County:		*Feet.*	*Feet.*		*Feet.*
Centralia	Direct pressure	20-30	18	Gravel	6
Pe Ell			8	....do	
Lincoln County:					
Davenport	Gravity	20-60	20-40	Basalt	10
Harrington	....do	30-100	35	....do	20
Sprague	....do	20-40	20	Gravel and sand	4-20
Wilbur	....do		10-15	Basalt	6-8
Mason County:					
Shelton	....do	10-50	10	Gravel	
Okanogan County:					
Loomis	....do	20-30	20		
Pacific County:					
Ilwaco	....do				
Southbend	....do	10-60			
Pierce County:					
Buckley	Direct pressure	20-60	30	Sand	10
Carbonado	Gravity				
Orting	....do	6-20	10	Gravel	
Puyallup	Direct pressure	10-20	6-10	Sand	
South Tacoma	....do	40-50	40	....do	40
Sumner	Gravity				
Tacoma	....do	30-100	50	Gravel	30-100
San Juan County:					
Friday Harbor		13-18	10	....do	6-8
Skagit County:					
Anacortes	Gravity	10-80	16	....do	
Hamilton		10-25	18	Sand and gravel	
Laconner	Gravity			Sand	
Mount Vernon	....do	12-14			
Snohomish County:					
Arlington		20-30		Sand and gravel	
Edmonds	Gravity			....do	
Everett	....do	15-115	100	....do	
Marysville			14	Sand	8
Monroe	Gravity	18-50	25	Gravel	8-40
Monte Cristo	....do				
Silverton	....do				
Snohomish	....do		15-55	Gravel	40-45
Spokane County:					
Cheney	Direct pressure	30-50	40	Basalt	
Hillyard	....do	190-200	190	Gravel	185
Latah		25-40	35	Rock	
Medical Lake		10-40	15	Basalt	
Stevens County:					
Bossburg	Gravity				
Colville	....do	10-30	20	Sand	5-8
Marcus			18-60	Gravel	
Northport	Gravity				
Springdale	....do	30-100	40-50	Sand	15-75
Thurston County:					
Olympia	....do				
Tenino		25-40	35	Gravel	

Municipal water supplies in Washington—Continued.

Location.	System of water works used.	Depth of the wells.	Depth at which water is commonly found.	Character of the water-bearing material.	Depth of water from surface
		Feet.	*Feet.*		*Feet.*
Wallawalla County:					
Waitsburg	Gravity	40–60	16–40		12–16
Walla Walla	...do	10–20		Gravel	
Whatcom County:					
Blaine	...do	10–20	10	....do	
Fairhaven	...do				
Sumas		40–90	40–60	Sand and gravel	
Whatcom	Gravity				
Whitman County:					
Colfax	...do	12–120	40	Basalt	
Oakesdale		12–60	10–45	Sand and gravel	10–30
Pullman	Gravity	100–130	100	....do	(*a*)
Tekoa		100–176	100	Basalt	*b* 10
Uniontown		175–200		....do	
Yakima County:					
North Yakima	Gravity	18–25		Gravel	
Prosser	....do	30–40	35	....do	35

Location.	Amount of water obtained daily.	Increase or decrease in supply.	Variation in water level.	Effect of pumping on water level.	Other uses besides domestic supply.
Adams County:	*Gallons.*				
Ritzville		Stationary	No....	No......	Lawns and gardens.
Asotin County:					
Asotin		...do	No....	Little...	Do.
Clarkston					Irrigation.
Chehalis County:					
Aberdeen					
Cosmopolis		Stationary	No....	No....	Boiler supply.
Hoquiam					Manufacturing.
Montesano		Stationary	Yes...	Yes.....	
Ocosta		...do	Yes...		
Chelan County:					
Chelan		...do	Yes...	Yes.....	Irrigation.
Lakeside		...do	Yes...	No.....	Do.
Wenatchee					Lawns and boilers.
Clallam County:					
Port Angeles		Stationary	Yes...	Yes.....	Fire protection and power.
Port Crescent		...do	Yes...	No......	
Clarke County:					
Vancouver		...do	Little.	Little...	Fire protection, power, etc.
Columbia County:					
Dayton		...do	Yes...	No......	None.
Cowlitz County:					
Castlerock			Yes...	Yes.....	Power.
Kalama		Stationary	Yes...	Yes.....	Irrigation.

a A few feet above.   *b* Above.

Municipal water supplies in Washington—Continued.

Location.	Amount of water obtained daily.	Increase or decrease in supply.	Varia-tion in water level.	Effect of pumping on water level.	Other uses besides domestic supply.
Douglas County:	*Gallons.*				
Wilsoncreek			No....	No......	None.
Jefferson County:					
Port Ludlow					Boiler supply.
Port Townsend	250,000	Stationary	Yes...	Yes.....	None.
King County:					
Auburn					None.
Ballard	300,000	Stationary	No....	No......	Fire protection.
Columbia City			Yes...	No......	Irrigation.
Enumclaw		Stationary	Yes...	Little...	None.
Issaquah		do	Yes...	No.....	Do.
Kent					Boiler supply.
Renton	120,000	Stationary	No....	No......	Fire protection and boiler supply.
Seattle	22,500.000	do	Yes...		Fire, irrigation, etc.
West Seattle			No....	No......	None.
Kitsap County:					
Bremerton			Yes...	Yes.....	Fire protection and boiler supply.
Charleston		Decreasing	Yes...	No.....	None.
Port Blakeley					Do.
Port Gamble	150,000	Stationary	No....	No......	Fire protection and boiler supply.
Kittitas County:					
Clealum		do	Yes...	No...	Irrigation, fire protection and boiler supply.
Ellensburg		do	Yes...		Irrigation and boiler supply.
Roslyn		Decreasing	Yes...	Yes....	Fire protection and boiler purposes.
Klickitat County:					
Goldendale		Stationary	No....	No......	Irrigation and fire protection.
Lewis County:					
Centralia			Yes...	No......	None.
Pe Ell		Stationary	Yes...	No.....	Do.
Lincoln County:					
Davenport		do	No....	No......	Irrigation and fire protection.
Harrington			No....	No......	Do.
Sprague		Stationary	Yes...	No....	Do.
Wilbur		do	No....	No......	Irrigation.
Mason County:					
Shelton		do	Yes...		Fire protection.
Okanogan County:					
Loomis					Boiler supply.
Pacific County:					
Ilwaco					None.
Southbend					Boiler supply.
Pierce County:					
Buckley	1,200,000	Stationary	No....	No......	Fire protection.
Carbonado	20,000				Boiler supply.
Orting			Yes...	Very little.	None.
Puyallup			Yes..	No......	Do.
South Tacoma	150,000	Increasing	Yes..	Yes.....	Boiler supply.
Sumner					
Tacoma	8,000,000	Stationary	Yes...	No......	Boiler supply, fire protection, etc.

Municipal water supplies in Washington—Continued.

Location.	Amount of water obtained daily.	Increase or decrease in supply.	Variation in water level.	Effect of pumping on water level.	Other uses besides domestic supply.
San Juan County:	*Gallons.*				
Friday Harbor			Yes		Boiler supply.
Skagit County:					
Anacortes		Stationary	Yes	No	Do.
Hamilton		...do	Yes	No	None.
Laconner					Do.
Mount Vernon		Increasing	No	No	Fire protection.
Snohomish County:					
Arlington			No	No	None.
Edmonds					
Everett			Yes	Yes	Fire protection, boiler supply, etc.
Marysville			Yes	No	None.
Monroe			Yes	Yes	Do.
Monte Cristo					Boiler supply and concentrator.
Silverton					Concentrator.
Snohomish			No		Boiler supply.
Spokane County:					
Cheney			No	No	Fire protection.
Hillyard		Stationary	Yes	Yes	Irrigation.
Latah					
Medical Lake					
Stevens County:					
Bossburg					
Colville		Stationary	No	No	None.
Marcus					
Northport					
Springdale			No	No	Boiler supply.
Thurston County:					
Olympia	3,000,000				Fire protection, boiler supply, etc.
Tenino		Stationary	Yes	Yes	None.
Wallawalla County:					
Waitsburg					Water motor.
Walla Walla			Yes		Fire protection.
Whatcom County:					
Blaine		Increasing	Yes		Boiler supply.
Fairhaven					Boiler supply and fire protection.
Sumas		Stationary	No	No	None.
Whatcom					Boiler supply and fire protection.
Whitman County:					
Colfax					Do.
Oakesdale		Stationary			Fire protection.
Pullman	300,000	Decreasing	No	Yes	Boiler supply and fire protection.
Tekoa		Stationary	No	No	None.
Uniontown		...do	No	No	Street sprinkling.
Yakima County:					
North Yakima					Irrigation.
Prosser			No	No	Do.

Representative springs in Washington.

County and post-office.	T.	R.	S.	Owner of spring.	Topographic position.	Quantity of flow.
Adams County:						*Gallons.*
Washtucna	15	36	{28 {33	G. W. Bassett	Base of bluff	a 40
Chelan County:						
Wenatchee	22	19	24	George Brisson	do	a 8
Clark County:						
Vancouver	1	2	3, 4	Vancouver Waterworks Co.	Hillside	b 1,750,000
Columbia County:						
Dayton	9	39	3	City of Dayton	do	
King County:						
Berlin		11		Everett Bottling Works.	do	a 3
Issaquah	24	6	27	Gilman Water Co.	Head of canyon	
Kent	22	5	19	Town of Kent	Base of steep hill.	b 500,000
McCain	13	26	29	T. G. McCain	Foot of mountain.	a 3
Renton	28	5	20	Seattle Electric Co.	Hillside	b 183,013
Klickitat County:						
Glenwood	9	12		United States	Valley	
Do	6	13		do	Base of bluff	
Goldendale	4	14			In canyon	
Lincoln County:						
Davenport	25	37	21	Town of Davenport.	Base of bluff	b 360,000
Harrington	23	36	34	J. L. Ball		a 2
Do	23	36	14	J. E. Ludy	Deep valley	
Do	23	36	9	L. T. Luper	Valley	b 28,800
Sherman	29	33	25	H. B. Fletcher	do	
Mason County:						
Shelton	20	4	12	Town of Shelton	Hillside	
Pierce County:						
Carbonado	18	6	4	Carbon Hill Coal Co.	do	
Orting	19	5	29	Orting Light and Water Co.	do	
Puyallup	20	4	82	City of Tacoma	Base of bluff	b 18,000,000
Skamania County:						
Cascades	2	7	16	Thomas Moffatt	Hillside	b 25,000
Spokane County:						
Latah	21	45	26	William Forthman.	Base of bluff	b 720
Stevens County:						
Colville				J. U. Hofeteta	Valley	
Wallawalla County:						
Walla Walla				City of Walla Walla.	do	
Yakima County:						
Fort Simcoe	11	15	9	Yakima Indians	Narrow valley	b 57,600

a Miner's inches   b Daily.

Representative springs in Washington—Continued.

County and post-office.	Variation in flow.	Taste.	Temperature.	Quality of water.	Kind of rock
			°F.		
Adams County:					
Washtucna	None	Pleasant...	40	Soft	Basalt.
Chelan County:					
Wenatchee	Decreases in flow in autumn.	...do	55	Hard	Do.
Clarke County:					
Vancouver	Varies slightly	...do		Soft	
Columbia County:					
Dayton	Decreases in summer...	...do		...do	
King County:					
Berlin	None	Soda		Alkaline..	Syenite.
Issaquah	...do.	Pleasant...		Soft	
Kent	Decreases in summer...	...do		...do	
McCain	None	Sulphur ...	122	Sulphur ..	Granite.
Renton	Diminishes in autumn .	Pleasant...		Soft	
Klickitat County.					
Glenwood	None	...do			Basalt.
Do	...do	Unpleasant	76	Charged with gas.	Do.
Goldendale	...do	Mineral ...	100	...do	Do.
Lincoln County:					
Davenport	Decreases in summer...	Pleasant...		Soft	Granite.
Harrington	None			...do	
Do	Decreases in summer...	Pleasant...	45	...do	Basalt.
Do	None	...do		...do	
Sherman	...do	...do		...do	
Mason County:					
Shelton	...do	...do		...do	
Pierce County:					
Carbonado	Decreases in summer...	...do	54	...do	
Orting	None	...do		...do	
Puyallup	...do	...do		...do	
Skamania County:					
Cascades	...do	Sulphur ...	96	Sulphur, iron, etc.	
Spokane County:					
Latah	Maximum in spring....	Pleasant...		Soft	
Stevens County:					
Colville	None	...do		...do	
Wallawalla County:					
Walla Walla	...do	...do		...do	
Yakima County:					
Fort Simcoe	...do	Soda	Warm ..	Charged with gas.	

Representative springs in Washington—Continued.

County and post-office.	Seeps or stream.	Deposits of mineral matter about spring.	Use of water.	Improvements at spring.	Improvements contemplated.
Adams County:					
Washtucna	Stream	None	Town supply and irrigation.	None	None.
Chelan County:					
Wenatchee	....do	....do	Domestic and irrigation.	.....do	Do.
Clarke County:					
Vancouver	....do	....do	City water supply.	.....do	Do.
Columbia County:					
Dayton	Seeps out	....do	....do	....do	Do.
King County:					
Berlin	Stream	Iron	None so far	....do	Hotel.
Issaquah	....do	None	Town supply	....do	None.
Kent	....do	....do	....do	....do	Storage reservoir.
McCain	Seeps out	....do	Medicinal and bathing.	Hotel	Addition to hotel.
Renton	....do	....do	Town supply	None	None.
Klickitat County:					
Glenwood	Stream	Iron	None	....do	Do.
Do	....do	Yes	....do	....do	Do.
Goldendale	....do	Iron	....do	....do	Hotel and a bath house.
Lincoln County:					
Davenport	....do	None	City supply	....do	None.
Harrington		....do	Domestic		
Do	Stream	....do	....do	None	Do.
Do	....do	....do	....do	....do	Do.
Sherman	....do	....do	....do	....do	Do.
Mason County:					
Shelton	....do	....do	Town supply	....do	Do.
Pierce County:					
Carbonado	....do	....do	....do	.do	Do.
Orting	....do	....do	Domestic	....do	Do.
Puyallup	....do	....do	City supply	....do	Do.
Skamania County:					
Cascades	Seeps out	....do	Medicinal, domestic	Hotel and bath house.	New hotel and bath house.
Spokane County:					
Latah	Stream		Domestic	None	None.
Stevens County:					
Colville	Seeps out	....do	Town supply	....do	Do.
Wallawalla County:					
Walla Walla	Stream	....do	City supply	....do	Do.
Yakima County:					
Fort Simcoe	Seeps out	Iron	Bathing	....do	Do.

INDEX.

83

O

Landes, Henry, 1867—

... Preliminary report on the underground waters of Washington, by Henry Landes. Washington, Gov't print. off., 1905.

85 p., 1 l. pl. (chart) 23ᶜᵐ. (U. S. Geological survey. Water-supply and irrigation paper no. 111.)
Subject series: O, Underground Waters, 29.
1. Water, Underground—Washington.

Landes, Henry, 1867—

... Preliminary report on the underground waters of Washington, by Henry Landes. Washington, Gov't print. off., 1905.

85 p., 1 l. pl. (chart) 23ᶜᵐ. (U. S. Geological survey. Water-supply and irrigation paper no. 111.)
Subject series: O, Underground Waters, 29.
1. Water, Underground—Washington.

U. S. Geological survey.

Water-supply and irrigation papers.

no. 111. Landes, H. Preliminary report on the underground waters of Washington. 1905.

U. S. Dept. of the Interior.

see also

U. S. Geological survey.

870

Water-Supply and Irrigation Paper No. 112   Series O, Underground Waters, 30

DEPARTMENT OF THE INTERIOR
UNITED STATES GEOLOGICAL SURVEY
CHARLES D. WALCOTT, DIRECTOR

UNDERFLOW TESTS

IN THE

DRAINAGE BASIN OF LOS ANGELES RIVER

BY

HOMER HAMLIN

WASHINGTON
GOVERNMENT PRINTING OFFICE
1905

CONTENTS.

ILLUSTRATIONS.

LETTER OF TRANSMITTAL.

DEPARTMENT OF THE INTERIOR,
UNITED STATES GEOLOGICAL SURVEY,
RECLAMATION SERVICE,
Washington, D. C., March 28, 1904.

SIR: I have the honor to transmit, for publication in the series of Water-Supply and Irrigation Papers, a manuscript entitled "Underflow Tests in the Drainage Basin of Los Angeles River," by Mr. Homer Hamlin.

The report describes the conditions under which ground water usually occurs in arid regions and the fluctuations in the water level due to rainfall and other causes. The methods used in obtaining this information are of interest, and the paper will contribute valuable material to the important subject of underground waters and their use in the arid regions.

Very respectfully,   F. H. NEWELL,
                                                            Chief Engineer.

Hon. CHARLES D. WALCOTT,
Director United States Geological Survey.

UNDERFLOW TESTS IN THE DRAINAGE BASIN OF LOS ANGELES RIVER.

By HOMER HAMLIN.

INTRODUCTION.

The purpose of this report is to assemble in one publication the results of a series of underflow tests made in the drainage basin of Los Angeles River in 1902 and 1903 by the United States Geological Survey.

This report briefly describes the conditions under which ground water usually occurs, especially in arid regions, and the fluctuations in its water level due to rainfall and to sinking flood waters.

The method of testing used was invented by Prof. Charles S. Slichter, and is fully described in "The Motions of Underground Waters." [a] Up to the present few investigators have used this method.

An attempt has been made to describe briefly the method of sinking test wells and the machinery designed and used during this investigation. The various devices used in testing and the arrangement of the instruments, the methods of testing found most satisfactory, the results obtained at each of the testing stations, and the amount of underflow supposed to pass the Huron street section, are fully described.

GROUND WATER.

By ground water is meant water percolating beneath the surface of the earth. The original source of all ground water is rainfall. Part of the rain soaks into the ground and percolates downward until it reaches a level where the interstices of the rocks, sand, and gravel are already saturated; another part is absorbed by growing plants or evaporated from the surface of the ground; and still another part runs off in surface streams.

---

[a] Slichter, Charles S., The motions of underground waters: Water-Sup. and Irr. Paper No. 67, U. S. Geol. Survey, 1902.

WATER TABLE.

The upper surface of the water-soaked zone in pervious ground is called the water table or water plane. From the surface of the ground to the water table the ground is damp but not saturated, and all contributions of surface water continually tend to sink downward. while below this level the ground is completely saturated; that is. all of the open spaces between the rock particles are full of water. The water table is the level at which water is struck in a well. In artesian belts the water is under pressure, being confined by overlying impervious strata, and when such strata are pierced it rises to the level of the outcrop of the water-bearing formation. The water table is not a level plane like the surface of a lake, except rarely in inclosed basins, but usually has a slope toward the main drainage lines of the region, to which the ground water slowly flows. The slopes usually follow in a general way the surface topography, but are much flatter.

FLUCTUATIONS OF WATER TABLE.

As the ground below the water table is completely saturated, any contributions of water from the surface must result either in raising the water table or in a lateral movement of the ground water toward some outlet. Usually both phenomena occur, the lateral flow being a result of and following the rise in the water table. In humid regions the supply from rainfall is so nearly constant throughout the seasons that the position of the water table is practically fixed, and the rate of lateral movement does not change appreciably from year to year. In arid regions much of the ground water under the plains and valleys is supplied by the streams which pour down from adjacent mountain ranges. When such streams leave the mountain canyons they soon sink in their débris cones of sand, gravel, and bowlders; in such localities the water table is sometimes 200 or 300 feet below the surface. When in flood such streams discharge enormous quantities of water, which sinks and is added to the ground water of the region. The immediate result is the raising of the water table, often 25 to 50 feet, over the region where the sinking occurs. The ground water afterwards flows outward in a generally horizontal direction from these regions and the water table is gradually lowered. In arid regions the water table rises sometimes many feet during rainy seasons and gradually sinks in dry years or during the dry season.

VELOCITY OF UNDERFLOW.

The slowly moving ground water beneath a stream bed or valley is usually designated "the underflow." In order to estimate accurately the amount of ground water passing a given section, it is necessary to know, among other factors, the rate of movement or velocity of the underflow through the pervious beds below. This is conveniently measured by the method invented by Professor Slichter.[a] Briefly the method is as follows:

A group of four or five wells arranged as shown in figs. 9 and 10 is sunk at the locality where the underflow is to be tested. Well A is placed upstream, or in the direction from which the underflow is supposed to come, and the wells B, C, D, and E are spaced downstream at a uniform distance from well A. All of these wells are of small diameter (drive pipe, 1¼ to 2 inches) and have from 4 to 8 feet of perforated screen, usually ordinary well points, at the lower end. These perforations allow the percolating water to pass through the pipe driven into the pervious stratum to be tested. The upper well, A, is charged with a chemical, usually sal ammoniac (NH_4Cl), which dissolves in the water passing through the well and is carried along by the underflow to one or more of the lower wells—which ones depends upon the direction of the underflow. The arrival at one of the lower wells of water containing sal ammoniac in solution is detected by means of electrical instruments.

METHODS AND APPARATUS USED IN UNDERFLOW TESTS.

In September, 1902, the writer was placed in charge of experiments to determine, if possible, the amount of underflow passing through the narrows of Los Angeles River at Huron street, Los Angeles, Cal.

Velocity measurements were begun under direction of Prof. Charles S. Slichter, with the apparatus invented by him. As the work progressed and tests at greater depths were made it was found necessary to modify this apparatus to suit local conditions.

LOCATION OF TEST WELLS.

The first step in testing the velocity of an underflow is to determine, approximately, the direction in which the underflow moves, and the second is to drill the test wells for measuring it. If the locality chosen is in a narrow valley or canyon bounded by rock walls, it is obvious that the underflow must be down the general trend of the valley. If, however, the tests are to be made in a wide valley or plain, it is advisable to ascertain the extent and slope of the water table in the vicinity.

If there are wells near, they should be located by some of the ordinary methods of surveying, and levels referred to some permanent bench mark should be run and the surface elevation at each well determined. The depth to water should be measured and the elevation of the water table computed. It is sometimes necessary to sink additional test wells, which should be located as just described. The data

---

a Water-Sup. and Irr. Paper No. 67, U. S. Geol. Survey, pp. 48-51.

should be platted on a map and contours, or lines of equal elevation, of the water table drawn, and the slope of the water table determined. The movement of the ground water in general is down the slope of the water table.

Fig. 21 (p. 42) is a map of the narrows of Los Angeles River at Huron street, Los Angeles, Cal., showing the extent and slope of the water table at that locality. The broken contour lines show the form of the water table in December, 1902; the dotted contour lines the form in June, 1903. The rise in the water table was due to the sinking flood waters of the Arroyo Seco, which enters Los Angeles River a short distance below. Its channel is usually a dry wash of sand and bowlders, but was in flood during March and April, 1902, and the map shows plainly the sudden rise in the ground water due to the sinking of the storm waters.

A practical conclusion to be drawn from this map is that in order to avoid the disturbing influence of the fluctuations of a tributary stream the test wells should be located a considerable distance either above or below the mouth of such a stream.

When the slope of the water table and the approximate direction of the underflow have been ascertained, wells for testing the velocity should be sunk in groups arranged as shown in figs. 9 and 10 (pp. 24 and 25). They may be sunk by any of several methods, depending on the locality, depth to the water table, coarseness of material, etc.

HAND DRIVING.

For shallow wells in unconsolidated material ordinary well points, as shown in figs. 2 and 3 of Pl. I, *A*, may be used. These are screwed on a length of standard or extra strong wrought-iron pipe and simply driven into the porous strata. Two men using a heavy wood maul, as shown in Pl. I, *B*, can drive such wells to a considerable depth, and, in such cases, this is a rapid and economical method. It is not possible, however, to collect samples of the material penetrated, and the porosity must be estimated from the amount of water yielded when the wells are pumped. This method was used in the first velocity test made in Los Angeles River, but was abandoned as unsatisfactory when deeper tests were attempted.

A portable hand rig is sometimes used to drive well points. Ordinarily two men can drive as many feet a day with a maul as with the hand rig in unconsolidated sands, unless the wells are deep. When used in combination with the hydraulic-jet method of sinking wells, such a rig will sink a pipe to a considerable depth, depending on the material penetrated, but when the material is coarse a point is soon reached beyond which it is impossible to drive any style of small pipe, as it is telescoped or battered by the hammer.

B. DRIVING WELL POINTS.

A. WELL POINTS

WELL-DRILLING RIG.

When wells are to be sunk to a considerable depth, 100 to 150 feet for instance, in very coarse material, some kind of machinery is necessary. A rapid and economical method is a modification of the hydraulic-jet process of deep-well drilling by combination with driving. This method consists of three distinct processes—(1) driving the pipe; (2) chopping up and washing out the sand and gravel that enters the bottom of the pipe as it is driven downward; (3) turning or rotating the pipe to keep the hole straight and to reduce the frictional resistance to driving. A rig equipped for this work, with the machinery assembled, is shown on Pl. II. It consists of the derrick for hoisting the heavy drivepipe, a hammer for driving the pipe, wash rods and chopping bit for hydraulicking, a pump for forcing water down the wash rods under pressure, and a rotator for turning the pipe. The whole is driven by a gasoline engine with a hoisting attachment. This machinery was designed to meet the local conditions of sinking wells to a depth of 100 to 150 feet in very coarse sharp sand, gravel, and bowlders. It has proved economical and satisfactory in actual use for several months.

MACHINE DRILLING.

When starting a well, a hole is first dug to a depth of 4 or 5 feet and a 10-foot length of drivepipe with a shoe firmly screwed on is inserted vertically, passing up through the rotator (a), as shown on Pl. III, *A*. It must be plumbed carefully, for if it is not driven straight the wells will not be properly spaced at the bottom, and this will introduce errors into the velocity measurements. About 13 to 15 feet of wash rod, with a chopping bit attached, is then inserted in the drivepipe, projecting about 3 feet above the top. The drive head (c) is then screwed on the drivepipe with the wash rods projecting up through the hole in the center. The buffer block (e) and hammer (f) should then be slipped over the wash rods which serve as a guide for the hammer, and the water swivel (g), with the hose attached, should be screwed to the top of the wash rods. The ropes for hoisting and driving and the sprocket chain for driving the rotator, etc., are attached as shown on Pl. II, *A*, and Pl. III, *A*.

The rig is operated as follows: Drive the pipe a few feet by tapping lightly with the hammer, then start the pump and force water down the wash rods and through the chopping bit. The rods should be churned up and down like a churn drill until the drivepipe is clear of sand to the bottom of the shoe. The chopping bit can be operated with, or independent of, the rest of the driving apparatus by a line passing over the middle sheave of the hoisting block. When the well is shallow the chopping can be done by hand, but when deep it is necessary to use the winch or winding drum on the engine. The water swivel is made with a projecting base larger than the hole through the hammer (see fig. 7). By means of this device it is possible to drill with the chopping bit and at the same time drive the pipe with the hammer. The length of wash rods is so adjusted that the water swivel is 6 inches to 1 foot above the top of the hammer, when it rests on the drive head. As the ordinary stroke of the hammer is 1 to 2 feet, the wash rods will be raised from 6 inches to 1 foot at each stroke. When the hammer is dropped the rods drop with it until the chopping bit strikes the sand in the hole, while the hammer still falls and almost instantly strikes the drive head. This is done automatically. the only attention required from the driller being the gradual lowering of the wash rods as the pipe is driven downward. The water which is forced out in jets through the chopping bit washes the pulverized sand and gravel up through the drivepipe to the surface. The drivepipe should be turned from three to five times per minute by an ordinary rotator such as is used with hydraulic-well rigs. The rotator should be run from the engine by a sprocket chain passing over sprocket wheels, as shown on Pl. II, B, and Pl. III, A. When the wells are shallow two men with chain tongs can usually turn the pipe.

This method is particularly adapted to sinking wells in coarse sand and bowlders. The drivepipe should be double, extra-strong, steel pipe to secure the strength and stiffness necessary to prevent bending when forced through coarse materials, and its internal diameter should be at least 2¼ inches. It is sometimes very difficult to pull a long string of drivepipe, especially to start it. The only practicable way is to use jackscrews, as shown on Pl. III, B, or use block and tackle with the engine at the same time turning the pipe with the rotator. Below is a more detailed description of machinery used.

APPARATUS.

ENGINE.

In the selection of an engine, consideration must be given to the cost of fuel, transportation facilities, weight of machinery, etc. When fuel is expensive and water scarce, a gasoline engine will be found economical. For ordinary drilling when the depth of the wells does not exceed 150 or 200 feet, a 6 to 8 horsepower engine will furnish the necessary power. A very convenient type of engine is shown on Pl. II, B. It is a vertical gasoline engine geared to a hoisting drum. which is controlled by means of a friction clutch and brakes. The drum is used to raise the hammer and wash rods, for hoisting pipe, etc. From ten to twenty blows per minute can be struck with the hammer in driving pipe. The small winch on the engine is used for pulling

A   WELL-DRILLING RIG

B   WELL-DRILLING MACHINERY

pipe, moving heavy weights, moving the engine, etc. The engine is belted to a force pump, which supplies water for hydraulicking, and is connected to the rotator by a sprocket chain. The engine shown in the cut is but 3 horsepower and is too light for deep wells, but has proved very economical in fuel consumption, using but 3 to 4 quarts of gasoline per day in ordinary drilling; in addition it has required but little attendance, is light, and easily moved.

PUMP.

The pump should be a suction force pump, capable of delivering water under a pressure as high as 150 pounds per square inch, with a discharge of 20 to 50 gallons per minute. It should be provided with a safety valve to prevent excessive pressure. The pump must be able to draw a supply from one to two driven wells, when the water table is within reach. In other cases it will be necessary to haul water for hydraulicking—an expensive and troublesome method.

Connection between the pump and the wash pipe, as shown in Pl. II, A and B, and Pl. III, A, is made by means of 25 to 30 feet of flexible hose of good quality to withstand high pressure. Ten to 15 feet of 2-inch suction hose will be needed to connect the pump with the drive wells from which the wash water will usually be pumped.

LIGHT DRIVEPIPE.

For shallow wells driven by hand in unconsolidated material the ordinary standard or extra-strong pipe will answer for drivepipe. It can usually be pulled and used several times. The outside couplings of such pipe greatly increase the frictional resistance in driving and pulling, and a considerable percentage of pipe is sure to be left in the ground. Extra-strong pipe is sometimes fitted with flush-joint couplings, but so much metal is cut away at the joint that they are frail and will not withstand hard driving, breaking at the joint when the pipe does not go down straight.

HEAVY DRIVEPIPE.

For deep wells the drivepipe should be what is known to the trade as double-extra-strong steel pipe with joints flush inside and out. The inside diameter of the drivepipe must be at least 2¼ inches. The nearest commercial size to this is known as 3-inch pipe; its actual diameter is 3½ inches outside and 2.284 inches inside, the thickness of its shell 0.608 inch and its weight 18.56 pounds per linear foot. Rings of tough Norway iron of the same diameter and thickness as the pipe, and 6 or 8 inches long, should be welded on both

ends of each length of pipe, in which the flush-joint coupling should
be turned. A taper thread with butt joints, as shown in fig. 1. has
been found satisfactory and stands much hard usage. This form of
thread has the advantage of a firm bearing throughout its entire length
when it is screwed up tight. The pipe should be finished in lengths
of exactly 10 feet with a few additional short
lengths of 5 feet. A 1½ to 2 foot length (*b* or
Pl. III, *A*.) should also be provided with a 1¼-
inch hole drilled through it about midway of
its length to allow the wash water and sand to
run out; the drive head should be screwed on
the upper end of this short
length. Such a drivepipe
is stiff and heavy enough to
stand repeated blows from
the hammer. The fact that
there are no outside couplings make it far easier to
drive and pull than the ordinary pipe.

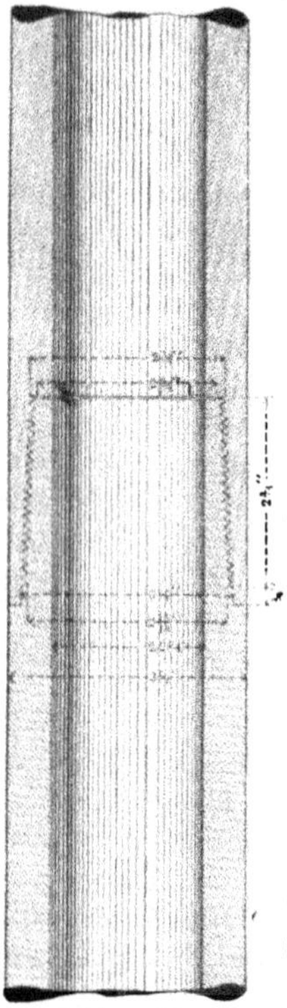

SHOE.

The lower end of the
drivepipe must be protected
from injury by a shoe. The
form shown in fig. 2 has
been found satisfactory. It
should be turned out of
tough steel, the outside diameter being a trifle greater
than the outside of the drivepipe, and the inside diame-

FIG. 1. Heavy drivepipe.

ter the same. If it is to be used with a rotator
the cutting edge should be notched and well tempered, as it will then cut its way downward
more readily. The threads should be the same
as on the drivepipe, so that it will fit on any
length.

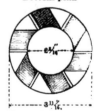

8 threads to 1 inch
6 teeth 1 inch long

Bottom plan

FIG. 2.—Drive shoe.

DRIVE HEAD.

The drive head, as shown in fig. 3, should be made of steel, accurately turned to fit the drive pipe on which it is screwed, and strong
enough to withstand long-continued driving. The wash rods pass up
through the hole in the center of the head, serving as a guide for the
hammer (see *c* on Pl. III, *A*). The four holes drilled in the side are

B. JACKSCREWS.

A. ROTATOR AND DRIVING MACHINERY.

sockets for rods used in screwing on the head or in handling the pipe. A bail for lifting can be sprung into these holes, if desirable, but an eyebolt passing through the hole in the center and loosely fastened with a nut and washer inside the drive head will act as a swivel and be found stronger and more convenient when pulling pipe, hoisting, etc.

BUFFER BLOCK.

A wooden block should be placed between the hammer and the drive head to prevent the destruction of both.   A convenient form is shown

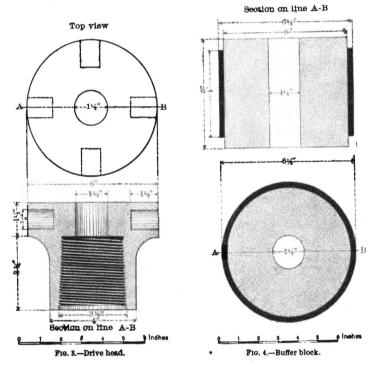

Fig. 3.—Drive head. • Fig. 4.—Buffer block.

in fig. 4.   It is simply an oak or other hardwood block with the grain of the wood vertical, securely bound with an iron band.   A hole should be bored through the middle and the block may then be slipped over the wash rods and rest on the drive head, as shown at e on Pl. III, A.

HAMMER.

A hammer of the form shown in fig. 5 requires no guides, but slides up and down on the wash rods.   It may be made of cast iron with projecting ears for attaching the hoisting ropes.   When it is necessary to add a length of drive pipe, the hammer can be raised off the wash rods, swung to one side, and lowered to the ground.   It is raised and lowered independently of the wash rods by means of two ropes, which pass up through the outside sheaves of the hoisting block at the apex of the derrick and thence to the winding drum of the engine, as shown on Pl. II, *A*, and at f on Pl. III, *A*.

Section on line A-B

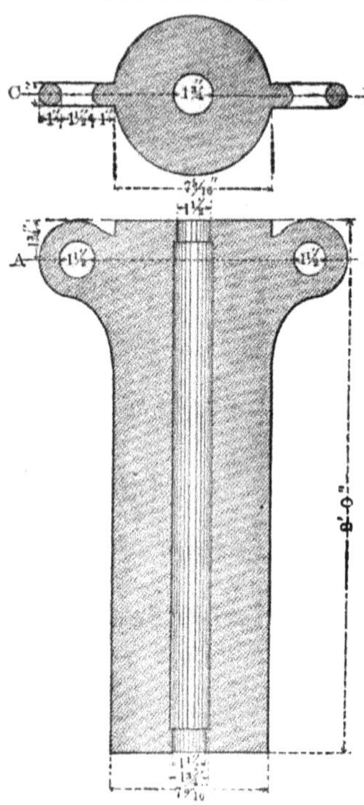

WASH RODS.

The wash rods should be made of what is known as double-extra-strong 1-inch pipe, the diameter being 1.315 inches outside and 0.587 inch inside, the thickness of the shell 0.364 inch, and the weight 3.65 pounds per linear foot.   The rods should have rings of Norway iron on each end and the same style of taper thread as the large drive pipe shown in fig. 1.   When finished, the rods should be exactly 10 feet in length, but several 5-foot and 2-foot pieces should also be on hand.   If deep wells are to be drilled, heavier wash rods will be necessary.   This will require larger holes than those figured in the drive head, buffer block, and hammer.

Section on line C-D
Fig. 5.—Hammer.

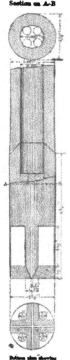

Section on A-B

Bottom plan showing beveled edges of chopping bit.

Fig. 6.—Chopping bit.

CHOPPING BIT.

The chopping bit should be made of best tool steel.   A drawing of the bit in use is shown in fig. 6.   It should be securely welded to a 5-foot length of wash rod instead of being screwed to it, otherwise when drilling in coarse material or bowlders it will probably be broken off and lost in the well.   The cut shows the bit welded to a length of double-extra-strong 1¼-inch pipe, commercial size.

The edges of the star bit should not be drawn down too thin, and must be well tempered to withstand hard usage, for when drilling the impact of the heavy string of wash rods, when dropped several feet, must be borne by the chopping bit. The holes through the bit should be of ample size to deliver enough water to raise the pulverized sand and gravel to the surface and still must be small enough to give the issuing jet a high velocity.

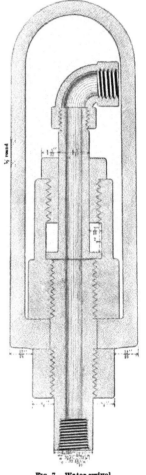

WATER SWIVEL.

A swivel connection between the hose and the wash rods is necessary to permit turning the chopping bit, making connections, etc. The form shown in fig. 7 has been found convenient. It should be strong and well made, as the hammer engages with the lower part of the swivel in raising the wash rod. As sand and gravel in the drive pipe sometimes pack around the drill and rods, the pull on the swivel may be considerable, perhaps enough to stop the engine, in which case the wash rods must be jarred loose by upward blows with the hammer.

DERRICK.

The derrick may be made of three timbers, about 4 by 6 inches, and 30 to 36 feet long, as shown in Pl. II, A. Holes are bored through the top and the timbers loosely connected by a 1-inch bolt, from which is suspended a long clevis, to which the hoisting blocks are attached. Such a tripod can be quickly raised, lowered, or adjusted over a well. Slats nailed on the timbers serve as a ladder for reaching the top.

BLOCKS.

FIG. 7.—Water swivel.

The lines from the hammer should be run up over the outside sheaves of a triple block, and then to the winding drum of the engine, as shown in Pl. II, A. The line from the wash rods is run over the center sheave and then fastened to a cleat on one leg of the tripod. By this arrangement the rods can be easily raised or lowered by hand, or suspended at any height.

A double block should be provided for use with the triple block in pulling the drive pipe, moving the engine, etc.

WELL POINTS.

When wells are shallow and the material to be penetrated is soil or unconsolidated sand, it is economical to drive ordinary 1¼ or 2 inch points. These are usually in 4-foot lengths and are made by perforating standard or extra-strong wrought-iron pipe with seven rows of oblong holes three-eighths by one-half inch. Around the outside of the perforated pipe is wrapped a fine brass screen of No. 35 wire with about 2,500 meshes per square inch; over the gauze is wrapped a perforated brass screen with 28 one-eighth inch holes per square inch, the whole being securely soldered to the perforated pipe along the vertical seam. The gauze strains out the fine sand and silt, but allows the water to enter the well, and the brass screen is put on to protect the gauze. Fig. 1 of Pl. I, A, shows such a well point with the screen and gauze partly removed. This style of point serves for both a well and for part of the testing apparatus as described on page 26.

When the material is compact, or contains cobblestones or bowlders, these well points fail, either by bending, by breaking, or by stripping off the outside screen and gauze. If the material contains stones large enough to deflect the pipe, it will bend and either break in driving or in pulling; in fact, it is almost impossible to pull a crooked pipe, as it will break at one of the couplings. When the material is compact and coarse the frictional resistance on the screen will often strip it entirely off of the pipe, leaving it as shown in figs. 1 or 3 of Pl. I, A. To overcome these difficulties the experiment of placing a fine screen on the inside of the perforated drive pipe was tried and found quite satisfactory. Perforated sheet brass, known to the trade as No. 26 gage, No. 1, with 400 holes, about 0.02-inch in diameter per square inch, should be rolled into a tube slightly smaller than the inside diameter of the perforated pipe (fig. 5 of Pl. I, A), securely soldered along the vertical seam and shoved inside the perforated pipe, where it is held in place by the couplings above and below. It is largely protected from injury and can be withdrawn and cleaned if necessary. Such a well point is shown in fig. 4 of Pl. I, A. The only objection to this style of point is that the holes in the outside tend to fill with sand and clay when the pipe is pulled.

JACKSCREWS.

The jackscrews for pulling pipe should be strong, of at least 32 tons capacity each. A clamp of heavy flat iron, a pulling ring, and steel wedges or dogs with notched edges should be provided for hold-

ELECTRODES AND WELL.

ing the ring in place when the pipe is being pulled.   The arrangement
of the jackscrews, etc., is shown on Pl. III, *B*.

PLACING WELL POINTS FOR ÚNDERFLOW TESTS.

As noted above, ordinary well points are driven into the pervious
strata where it is proposed to test the underflow.

When the heavy drivepipe is used the procedure is somewhat different.   The pipe is driven by machinery to bed rock, or to the desired
depth, and thoroughly cleaned from sand, silt, etc.   A length of well
screen similar to that shown in fig. 4 of Pl. I, *A*, is screwed to a
sufficient length of 1¼-inch standard pipe and lowered to the bottom
of the hole.   The drivepipe is then pulled up, leaving the well screen
and 1¼-inch pipe in the ground with the screen at the proper depth.
The drivepipe is used to sink another well, in which is placed another
screen, etc., until all the well screens of the group are in place.   The
well screens and small pipe, whether set by the first or the second
method, now form part of the testing apparatus, being a portion of
the electric circuit.

TESTING APPARATUS.

The various instruments at first used in testing the velocity of
underflow were made from designs furnished by Prof. Charles S.
Slichter.   The modifications of the original apparatus and the
methods developed are the result of several months' experimenting
by the writer, and may be of value to investigators in the future.

WIRE.

The wire should be a good quality of rubber-insulated copper, about
No. 14.   Single wires are more convenient than a cable or a twisted
lamp cord.   Care must be used and the wire examined occasionally to
see that the insulation is not worn through, for if it is there will
probably be a short circuit between the well casing and the wire.   It
should be kept rolled into open coils, and not kinked or twisted, as
such usage tends to break the insulation.

ELECTRODES.

The electrodes are brass rods from one-fourth to five-sixteenths inch
in diameter and about 4 feet long, usually nickel plated (Pl IV).
They should be heavy, of solid metal throughout, as it is difficult to
sink a light electrode to the bottom of a deep well.   The wire should
be attached to the electrode by some such device as that shown in fig. 8,
this being necessary in deep wells.   It should not be bent or kinked
where attached to the electrode, as it may break and leave the elec-

trode in the bottom of the well. The electrodes are insulated from the well casing by means of wooden blocks (see fig. 8), which should be boiled in paraffin to make them impervious to water. The elec-trodes will need frequent plating, as the action of the electric current and of the sal ammoniac quickly cor-rodes the nickel coating.

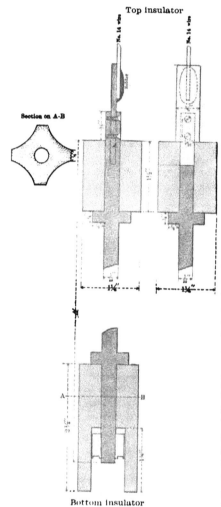

Fig. 8.—Electrodes.

SWITCH CLOCK.

A switch clock found satis-factory in actual use is shown in Pl. V. It is an ordinary eight-day clock with electri-cal devices added. The min-ute hand of the clock carries a steel spring, tipped with platinum, which, at five-min-ute intervals, is brought into contact with platinum strips, each of which is connected with a separate insulated wire passing down through the clock case and thence to the proper binding post. Each contact closes a circuit through one of the test wells. The platinum strips are laid in slits cut in lugs of hard rubber. The strips, the pro-jection of which beyond the edge of the lug is adjustable, are held in place by screws which pass through narrow slots. The hard-rubber lugs are permanently fastened to the clock dial by screws from the back. They are placed at each hour on the dial or at five-minute intervals as meas-ured by the minute hand.

In operating, the spring on the minute hand is carried over the hard-rubber lug, which is so placed that the spring is slightly raised. When the tip of the spring reaches the sharp edge of the lug it suddenly snaps down on the platinum strip and at once closes the circuit. It is

SWITCH CLOCK.

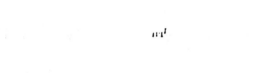

pulled across the platinum strip by the motion of the ı nd and when it reaches the edge snaps away from the strip, suddenly opening the circuit. The time the circuit is closed can be regulated by adjusting the width of the strip. Five to ten seconds is about the proper interval. It is important that the circuit be opened and closed suddenly, otherwise the recording pen, full of ink, may move back and forth on the chart several times and probably obliterate the record.

The current from the positive pole of the battery enters the clock at one of the binding posts, B, C, D, E, etc., and passes through the insulated wire to the minute hand, thence to the clock frame and out to the binding post marked A. By the movement of the minute hand the circuits to the various wells are progressively closed. The twelve switches on the dial permit of thirty hourly records with four lower wells and a thirty-minute interval between records, as shown on Pls. V, A, and VI, B, or of records at intervals of twenty minutes with three lower wells and a ten-minute interval between records. It is also evident that in the first case two groups of wells, or in the second case three groups of wells, can be tested hourly at the same time, with a ten-minute interval between tests.

RECORDING AMPERE METER.

The recording ampere meter (Pl. VII, A) was made by the Bristol Company from designs furnished by Prof. Charles S. Slichter. The instrument keeps a continuous record for twenty-four hours, after which the chart must be changed. The chart is moved at a uniform rate by clockwork. The recording pen is actuated by the electric current sent through the wells at fixed intervals by the switch clock. The electric apparatus is very simple, consisting of a stationary solenoid through which the current passes. A thin disk of soft iron is attached to a nonmagnetic shaft which passes through the solenoid and is supported at its opposite ends by steel knife-edge springs. The recording pen is secured directly to the steel springs and partakes of its angular motion as the disk is attracted by the solenoid. The result is a record such as shown on Pl. VI, A and B.

If an ordinary ampere meter is used the switching must be done by hand, as described on page 27.

BATTERY.

The electric battery should be of ample capacity. Dry cells known as large-size Columbia, $3\frac{1}{4}$ inches in diameter and 8 inches high, have been found satisfactory. The plan and wiring of the battery for use with a recording ampere meter only are shown in fig. 9, and for use with either a recording or ordinary ampere meter are shown in fig. 10.

The battery consists of 12 large-sized dry cells connected in sets of two in parallel. These parallel sets are connected up in series, and by means of the double-throw switches the electromotive force of the battery can be varied from 1.5 to 9 volts, assuming the electromotive force of each cell to be 1½ volts when new. . One, two, three, or more sets of two cells connected in parallel can be used in one series, as the case requires, or if necessary either side of the battery can be cut out by the opening of one of the single-throw switches marked " A. A'." An ordinary ampere meter should, for convenience, be placed in the battery box. Fig. 10 shows such a battery connected up to a group of four wells and arranged for use with an ordinary ampere meter. The switches are set for measuring the current from casing of well B to the electrode of well B, with two sets of cells in series.

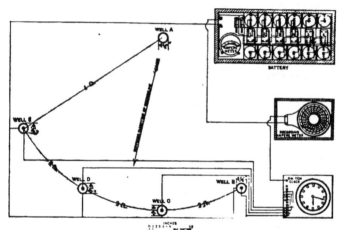

Fig. 9. Plan of test wells and apparatus.

METHOD OF MEASURING THE VELOCITY OF UNDERFLOW.

An outline of the method invented by Prof. Charles S. Slichter has been described on page 11. The following is a description more in detail of the writer's experience during several months' work at Los Angeles, Cal. An attempt has been made to describe the difficulties encountered and the methods developed during the progress of the work.

PUMPING.

After all the test wells have been sunk, as described on page 11, it is necessary before beginning a test to take a small sand pump and thoroughly clean the inside of the well points from sand and silt, and then pump water from each well until it clears. A small hand pump which can be readily moved from well to well is convenient for this

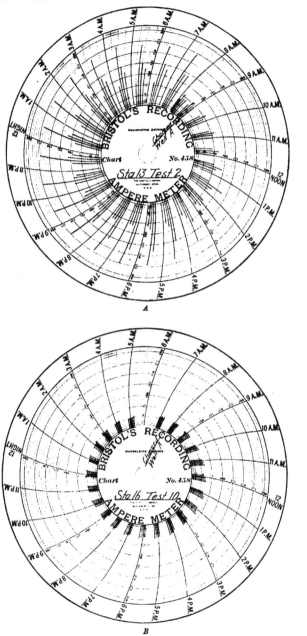

CHARTS USED WITH RECORDING AMPERE METER.

purpose. Pumping is often a tedious job, as the screens become
clogged with silt and mud. In such cases a few buckets of clear
water poured down the well will often open up the screen. Another
way is to occasionally break the vacuum in the pump and let the water
in the pipe fall back suddenly. When these methods fail, success may
be attained by rotating the pipe with a pair of tongs and at the same
time pouring clear water down the well.

If some of the wells pump freely and the rest will not start, it is
almost certain that the screens are clogged, and if all else fails they
may be allowed to stand a day or two. If none of the wells can be
started, or after starting
continue to pump fine
sand and mud, the test
may as well be aban-
doned, for in such cases
the velocity will be too
low to measure by this
method. Testing should
not be begun for a few
hours after pumping, in
order to allow the water
plane to rise to its normal
level.

WIRING.

The next step is to
wire the wells. There
are several ways in which
this may be done, de-
pending on the kind of
electrical instruments
used. If a recording
ampere meter is used,
the wiring should be as
shown in fig. 9. A wire

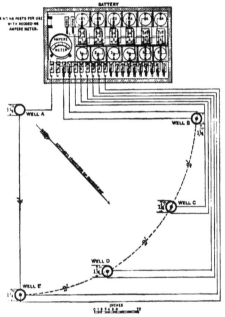

FIG. 10. Plan of test wells and apparatus.

is run from the positive pole of the battery to the casings of lower
wells B, C, D, E, etc., all of which may be connected by soldering
the wire to the galvanized casing of each, thus insuring good connections. A wire is connected to the electrode for each well and
run to the proper binding post on the switch clock; the electrodes
should then be lowered to the bottoms. From the positive binding
post of the switch clock a wire is run to the recording ampere meter,
and from the ampere meter to the negative pole of the battery.
An ordinary ampere meter (in the battery box) is put in the circuit to
check the record of the recording instrument.

The current recorded is that which passes through the water between the casing of the wells and the electrode at the bottom of the well.

The construction of the clock, recording ampere meter, battery, etc.. is described on pages 21 to 24.

If an ordinary ampere meter is used the wiring will be quite different, as all switching must be done by hand. A battery arranged for such a testing and the wiring of a group of wells is shown in fig. 1·. A wire should be run from the proper binding posts, marked A. B. C. D. and E. of the battery to the casing of the corresponding wells, and another from the binding posts, marked B', C', D', E', of the battery to the electrodes of the corresponding wells. The battery is so wired that when the switches A and B are closed, the ampere meter indicates the amount of current flowing through the ground from the casing of well B to the casing of well A. When the switches B and B' (fig. 1·· are closed, the ampere meter indicates the amount of current flowing through the water from the casing of well B to the electrode of well B. After all the wiring is done, the correctness of the various connections should be tested. If the circuit through one of the wells is closed and the electrode lifted out of the well, the ampere-meter needle should fall back to zero, for raising the electrode out of the well opens the circuit. On the other hand, if the casing of the well is touched with the electrode, a strong deflection of the needle should be noted. By these tests an error in wiring may be detected.

CHARGING WELL A.

After all the connections have been made and tested, a new chart should be placed in the recording ampere meter, and the apparatus allowed to run for an hour or so, to get a record of the wells before well A is charged.

Various methods of charging wells have been tried, and the following has finally been adopted. The electrolyte used is sal ammoniac (NH$_4$Cl). This should be placed in a small wooden or paper tub (it rusts iron quickly) and enough water poured on to cover it. A bucket of perforated sheet brass, similar to that used for the inside screen of the well points, about 4 feet long and small enough to readily slip down the inside of the well pipes, should be provided. This should be placed in the tub of wet sal ammoniac and filled with water and the salt by means of a funnel and dipper. The water will drain away into the tub and leave the wet sal ammonia filling the bucket, which should at once be lowered to the bottom of the well.

It is difficult to get a bucket of dry sal ammoniac down a deep well on account of the contained air. In addition, the escaping air causes the formation on the water of a sticky foam, which is carried down by

A. RECORDING AMPERE METER.

B. BATTERY

the next bucket and clogs the screen. The experiment of pouring a hot saturated solution of sal ammoniac down the well pipe was tried, but was condemned. It will either crystallize out when the water cools and leave the pipe filled solidly with the salt, or, if the strata be pervious, will be forced through to the lower wells almost immediately. In any case, undissolved sal ammoniac should not be poured down a well, as it is almost impossible to clean it out for the next test.

The following rule is now observed in charging wells: Put down one bucket of wet sal ammoniac every fifteen minutes for two hours, then one bucket an hour for six hours. Of course, if at any time the electrical instruments show that the sal ammoniac has reached the lower wells, charging should be stopped at once.

DETERMINATION OF VELOCITY OF UNDERFLOW.

When recording instruments are used, it is only necessary to examine them occasionally to ascertain when the sal ammoniac reaches the lower wells. Its arrival is indicated by an increased deflection of the recording pen, as shown by the chart in Pl. VI, *A*. When the velocity is great the deflection is sudden, but when it is low the increase is not so marked. When there is no underflow the deflection of the pen remains the same throughout the test, as shown in Pl. VI, *B*.

When an ordinary ampere meter is used, the circuits to the various wells must be closed by hand at frequent intervals, varying from ten minutes to two hours, depending on the velocity of underflow. When it is probable that the velocity is great, readings must be made often; but if no flow is detected for two or three hours, the time between observations may be longer. The ampere meter must be read and the time and readings recorded in a notebook. From the notes an ampere curve should be platted (fig. 11), from which an estimate of the velocity of flow can be made. The sudden rise in the curve indicates the arrival of the sal ammoniac at the lower well. This rise is primarily due to the fact that the solution of water and sal ammoniac is a better conductor of electricity than is water alone, and when this solution flows into and mingles with the pure water in the well, the rise in the current, due to lessened resistance, can at once be noted.

The time of passage of the solution from well A to one of the lower wells is usually calculated from the hour when charging was begun to the hour when the rising ampere curve reverses, or, in other words, at the point of the maximum rate of increase in the ampere curve. This method corrects for the diffusion or spread of the sal ammoniac in the water and for the error due to the fact that the water in the upper well can not be charged quickly.

The distance between the wells being known and the time of passage of the solution having been noted, it is a simple matter to compute the velocity or rate of flow. This may be expressed in any unit of measure desired. In this report the rate of flow is expressed in feet per day.

When wells have been sunk with the heavy drivepipe, as described on page 21, and the well screens inserted, the first test will evidently be at bed rock, or at the bottom of the hole. After the completion of

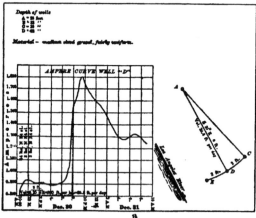

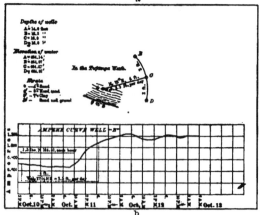

Fig. 11.—Ampere curves.

the first test, the pipe may be pulled up the length of the screens and another test made. This pulling up and testing may be continued to the level of the water table if desirable.

When well points are driven by hand, it is best to drive into the first stratum below the water table, test it, and then drive deeper, continuing the process as far as necessary.

COMPUTATIONS OF DISCHARGE.

In a given section wide differences in the velocity of underflow are to be expected, due to the various factors on which flow depends, such as slope of the water table, porosity, relative size of the sand grains, temperature, etc. It is evident that if the actual percentage of the open spaces or voids between the particles of sand and gravel of each square foot of a measured section was known, and the average velocity at which water flows through the ground had been measured, it would be possible to compute the amount of water passing a given line during a fixed interval of time; moreover, no further data would be necessary.

The difficulties met with when an attempt is made to compute the discharge from a section should be clearly set forth.

Natural sands, as dug from pits or the bed of a stream, consist of grains of different sizes and shapes. Such sand weighs more per unit of volume, as per cubic foot, than any of the various grades of sand sifted from the same bed and having grains of nearly uniform size. This proves that the natural sands contain a less percentage of voids, or, in other words, the porosity is less, than that of the sorted sands.

Experiments made by Prof. F. H. King[a] show that the porosity of unsorted sands from the Tujunga washes in San Fernando Valley, Los Angeles County (see fig. 12), varies from 31.42 to 42.28 per cent. The effective diameter of the sand grains was ascertained to range between 0.05542 and 0.4289 millimeters. The same writer says:

It appears to be generally true that well-rounded grains of nearly uniform diameter tend to give a pore space which lies between 32 and 40 per cent. The mean theoretical pore space for spherical grains of a single size is 36.795 per cent, and this is very close to the mean observed limit for the more simple sands of rounded grains.

A sample of coarse sand from Arkansas River Valley was tested by G. W. Stose in the following manner:

After being thoroughly dried it was saturated with water, and the quantity of water received was determined by weighing. The vessel containing it was then punctured at the bottom, so as to permit the water to drain away, and after a lapse of several days the loss of water was determined by another weighing. It was then found that the sand had received 29 per cent of its volume of water, but afterwards parted with only one-third of the water received, equivalent to 10 per cent of the whole volume of sand.[b]

Both of these samples had been removed from the original beds. The experiments show the probable range of porosity in clean sand of moderately uniform size.

Prof. Charles S. Slichter states that "the porosity of quartz sand will usually vary between 30 and 40 per cent, and that of clay loams

a King, F. H., Nineteenth Ann. Rept. U. S. Geol. Survey, pt. 2, 1899, p. 212.
b Gilbert, G. K., Seventeenth Ann. Rept. U. S. Geol. Survey, pt. 2, 1896, p. 600.

between 40 and 50 per cent, depending upon the variety of sizes in the mixture and the manner of packing the particles."[a]

Very little is known regarding the porosity of naturally deposited and undisturbed beds composed of coarse gravels, coarse and fine sand. and silt. Theory indicates that such beds, having been deposited by the action of running water, or by the waves and currents of the ocean, are far less porous than sands of uniform size, the sorting action of the water tending to pack the material as closely as possible.

It is usual to assume that the porosity of sand is about 33 per cent. and if the grains are rounded, of nearly uniform size, and free from silt, the assumption is, without doubt, about correct.

The actual porosity in each case must be known before any reliable computation of discharge can be made. Concerning this factor, Professor Slichter says:

If two samples of the same sand are packed, one sample so that its porosity is 26 per cent and the other sample so that its porosity is 47 per cent, the flow through the latter sample will be more than seven times the flow through the former sample. If the two samples of the same sand had been packed so that their porosities had been 30 and 40 per cent, respectively, the flow through the latter sample would have been 2.6 times the flow through the former sample. These facts should make clear the enormous influence of porosity on flow, and the inadequacy of a formula of flow which does not take it into account.[b]

Evidently beds which consist of a mixture of bowlders, cobblestones, fine sand, and silt will have a less percentage of voids than graded sand. No flow can occur through bowlders or cobblestones, and if they make up 50 per cent of the volume, and the intervening spaces are filled with sand and silt of graded sizes, there is certainly far less than 33 per cent of voids in the mass. How much less must be determined by experiment.

It is exceedingly difficult in underflow tests to come to any satisfactory conclusion regarding the porosity of the various beds. This is due to the almost insurmountable difficulties in collecting representative samples of the material penetrated. If the testing is done with a small-size drive pipe, 2 to 4 inches, all the coarse material must be excluded, because it is impossible to get it into the pipe or to the surface. When the material is very coarse, it is obviously necessary to dig a well or shaft, or put down a large bore well, 12 inches or more in diameter. Both methods have decided limitations.

It is impracticable, by any reasonable expenditure of money, to dig a well much below the level of the water table. If a large-bore well is sunk, material below the water table must be brought to the surface in a sand bucket. The continual churning of the sand bucket under water washes the fine silt and mud out of the sand and mixes it with

---

[a] Slichter, Charles S., The motions of ground water: Water-Sup. and Irr. Paper No. 67, U. S. Geol. Survey, 1902, p. 17.
[b] Slichter, Charles S., Nineteenth Annual Report U. S. Geol. Survey, pt. 2, 1899, p. 322.

the water in the well. This mixture of mud and water is partly bailed out with each bucket partly filled with sand. How much silt is washed out by the water at any particular point is, of course, impossible to say, but if one will catch the muddy water from the sand bucket and let it settle he will be convinced that it is usually no small amount. No one can tell exactly at what depth or how much of the sand contains silt. It is entirely possible that a bed of clay or silt a foot or more thick may be entirely mixed with the water in the well and no evidence of its presence be shown unless the driller notes that the water is more muddy than usual.

The writer has seen samples of sand taken from wells, carefully washed, cleaned from all silt, clay, etc., and then bottled up and exhibited as water sand. Any estimate of discharge based on such samples is utterly misleading.

Samples are often collected by catching the sand, etc., brought up by a hydraulic rig. Of course these samples are thoroughly washed. In porous gravel the wash water used in hydraulicking continually tends to flow away from the well, being under great pressure. That it carries the fine material near the hole away with it is proved by the fact that the sinking of the wash water when drilling in gravel can often be prevented by pouring clay down the well, which soon fills the voids in the surrounding gravel.

The hydraulic and drive-pipe method of sinking test wells has one great advantage over the sand bucket, in that the presence of a bed of clay or silt is shown at once by the muddy water which rises to the surface through the drive pipe. The depth at which such beds occur can by this means be ascertained.

Another uncertain factor is the unknown horizontal extent of the pervious beds. It can only be assumed that they extend from station to station, and in regions where the pervious beds have been deposited by torrential streams this assumption may lead to error, such gravel beds often being in the form of long trains of coarse material, the buried channel of the stream, and not in a wide extended sheet as supposed by many. To correct for this uncertainty, it is obviously necessary to sink test wells at close intervals along the cross section.

There is much uncertainty regarding the weight to be given velocity measurements. As a rule the velocities recorded are the maximums. Suppose test wells are set 4 feet in a pervious stratum, the upper two feet of which is composed of coarse sand 2 millimeters in diameter, the lower two feet composed of fine sand 0.4 millimeter in diameter, with a pressure gradient of 10 feet per mile. As computed from tables prepared by Prof. Charles S. Slichter,[a] the velocity of flow in the upper bed will be at the rate of 5,386 feet per year, and in the

[a] Slichter, Charles S., The motion of underground water: Water-Sup. and Irr. Paper No. 67, U. S. Geol. Survey, 1902, p. 29.

lower bed at the rate of 216 feet per year. In the first instance this is at the rate of 14.75 feet per day, and in the second at the rate of 0.59 foot per day.

If the test wells are spaced 4 feet apart the chemicals used in charging the upper wells should pass through the coarse stratum and reach the lower well in about six and one-half hours, while it will take about one hundred and sixty-two and five-tenths hours for the same chemical charge to reach the lower well through the fine stratum. In such a case, the salt passing from the upper bed would enter the lower well and cause a sudden rise in the current, as recorded by the instruments. The sal ammoniac in solution, being much heavier than the ground water alone, will settle and fill the lower well. There is no doubt but that the result is correct for the upper 2 feet, but, as the lower well is full of the solution of sal ammoniac, there will be no means of ascertaining when the charge from the lower stratum reached the well, even if the test should be continued for six and three-fourths days, which is highly improbable. It is in fact impossible to know, in testing material naturally deposited, whether the flow is through the whole section tested or through one or two thin strata. The only practical method is to use short well points and test the ground at each few feet in depth.

The most promising field for this method of testing underflow is in regions where the pervious formations are uniform in composition and texture throughout wide areas, and where the water table is not subject to great fluctuations in level. If testing in mountainous regions or in the débris cones of torrential streams is attempted, the testing stations should be close together and tests made at many points and at various depths, to eliminate as much as possible the inaccuracies due to varying texture and porosity of the formations.

BASIN OF LOS ANGELES RIVER.

The drainage basin of Los Angeles River has an area of approximately 502.5 square miles.

The principal eastern tributaries of this stream rise on the west slopes of San Gabriel Mountains and flow down to San Fernando Valley in deep rocky canyons which have been eroded in the granitic rocks of the range (see fig. 12). These streams are torrents during the rainy season, but dwindle to mere rivulets in the summer. When in flood they transport a vast amount of detritus, sand, gravel, and bowlders to the plain below, and have buried the old drainage lines across the east end of San Fernando Valley beneath extensive detritus cones, into which the surface streams sink except in times of extraordinary floods.

The western tributaries of Los Angeles River rise in Santa Susanna Mountains, which bound San Fernando Valley on the north, and in

Santa Monica Mountains on the south. These streams are small, rarely reaching the eastern end of the valley. They flow over sandstones, shales, clays, etc., and are strongly impregnated with alkaline salts, in strong contrast to the pure water from the granitic range to the east.

After flowing southerly in a broad underground channel beneath the detritus in the east end of San Fernando Valley, the water of Los Angeles River is deflected easterly by the impervious rocks of Santa Monica Mountains. This barrier, taken together with the contraction of the underground channel, so obstructs the free percolation below that much of the ground water rises and flows as a surface stream again.

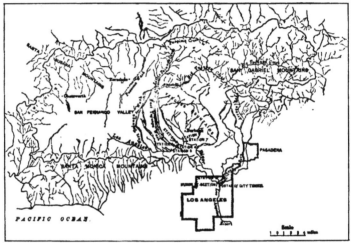

Fig. 12.—Drainage basin of Los Angeles River.

After rounding the eastern end of Santa Monica Mountains the river flows through a region of low, rolling hills in a valley which is about 2¼ miles across at the widest place and narrows to half a mile at Los Angeles, 6 miles below. This valley was formerly deeper than at present, and, in common with San Fernando Valley above, has been partially filled with river detritus, which near the mountains is very coarse but becomes fine and contains beds of sand and silt toward the south. The gradual change from coarse to fine material in connection with the contraction of the underground channel without doubt causes the rise of ground water along this 6-mile section of the river. At the mouth of the Arroyo Seco a marked change occurs in the character of the stream detritus. This stream is steep-graded and torrential, heading on the precipitous slopes of San Gabriel Mountains (see fig. 12). After leaving the mountains it flows in a canyon cut across the hill country to its junction with Los Angeles River. It carries bowlders, coarse gravels, and sand into the river valley, and its wide-spreading débris cone at the mouth of the canyon has forced Los Angeles River to the southwest side of the old valley. There is evidence that this débris cone was once more extensive than now, and it without doubt then acted as a barrier to the flow of the river. thus perhaps accounting in a measure for the fine material in the valley immediately above. About 2 miles below Huron street the river flows out on an alluvial plain and the surface stream soon sinks.

Lower Los Angeles River, extending from Huron street 12 miles northwesterly, being supplied by underground water, is remarkably

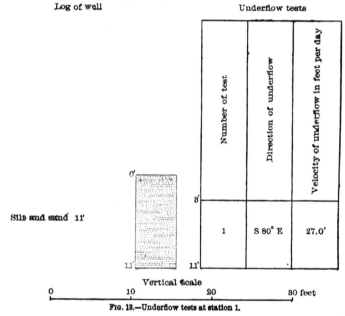

Fig. 18.—Underflow tests at station 1.

uniform in its discharge, little difference being noted throughout the year. The entire surface and part of the underflow is diverted by the city of Los Angeles for domestic use and irrigation.

TESTS IN 1902.

The following preliminary tests were made in Los Angeles River valley during September and October, 1902. The tests made at stations 1 and 2 were mainly for the purpose of adapting the apparatus to local conditions.

STATIONS AND RESULTS.

Station 1 is in Los Angeles River bed, at the Los Felis road bridge, near Huron street (see fig. 21). A group of well points was driven

by hand to a depth of 11 feet. The velocity of the underflow was 27
feet per day between 3 feet and 11 feet, and the movement was S. 80° E.,
or away from the surface stream, which was actually losing water at
this point. Later the water table was found to slope in the direction
indicated by the underflow. The order of the strata, record of the
wells, and the velocity measurements at this station are shown in fig.
13. Sixty-five feet to the east of this station another group of well
points was driven to a depth of 11 feet and tested between 3 feet and 11
feet, but no underflow was detected. They were then driven to a
depth of 19 feet, and tested between 11 feet and 19 feet, with the same

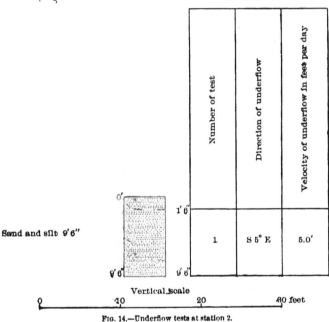

Log of well   Underflow tests

Sand and silt 9' 6"

Number of test	Direction of underflow	Velocity of underflow in feet per day
1	S 5° E	5.0'

Vertical scale

0   10   20   30 feet

FIG. 14.—Underflow tests at station 2.

result. As these were driven wells no samples were collected, but the
material appeared to be fine silt except near the surface, where it was
somewhat sandy, and it is probable that the underflow at station 1
was through this top stratum.

In August and September, 1903, the Southern Pacific Railroad Company dug pits for bridge piers in the river bed about 100 feet east of
this point. The material penetrated was mainly mud and silt, with a
few irregular seams of sand. Only the sand beds were water bearing;
they are from 2 to 4 inches thick.

Station 2 is on the west side of the river, 1 mile above the Huron
street section, as shown in fig. 12. A group of well points was driven

by hand to a depth of $9\frac{1}{4}$ feet, and a test made between $1\frac{1}{4}$ feet and $9\frac{1}{4}$ feet, showing the direction of the underflow to be S. 5° E., velocity 5 feet per day. No samples of the strata pierced were collected, but the amount of water pumped indicated that the ground was not porous. The order of the strata, record of the wells, and the velocity measurements are shown in fig. 14.

The testing apparatus was then moved upstream to San Fernando Valley, where the conditions of underflow and the extent of the ground

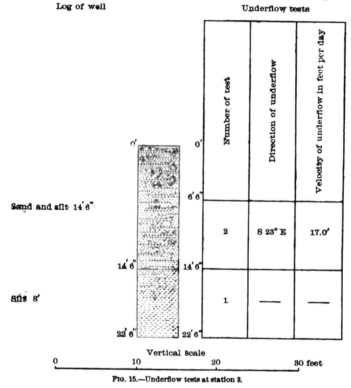

Log of well   Underflow tests

FIG. 15.—Underflow tests at station 3.

water have been thoroughly studied. The object of these tests was to determine, if possible, the relation of the slope of the water table to the direction and velocity of the underflow.

Station 3 is on the east bank of Tujunga Wash, near its junction with Los Angeles River, as shown in fig. 12. A group of well points was driven by hand to a depth of $14\frac{1}{4}$ feet, and a test made between $6\frac{1}{4}$ feet and $14\frac{1}{4}$ feet, where the direction of the underflow was S. 23 E., velocity 17 feet per day. The wells were then driven deeper and a test made between $14\frac{1}{4}$ feet and $22\frac{1}{4}$ feet, but no underflow could be detected. No samples were collected. The underflow appeared to be along the gravel beds in the bottom of the wash.

The order of the strata, record of the wells, and the velocity measurements are shown in fig. 15.

Station 4 is on the east bank of Tujunga Wash, as shown in fig. 12. A group of well points was here driven by hand, and a test made between 6½ feet and 14½ feet, where the direction of the underflow was S. 13° E., velocity 4½ feet per day. These wells were driven

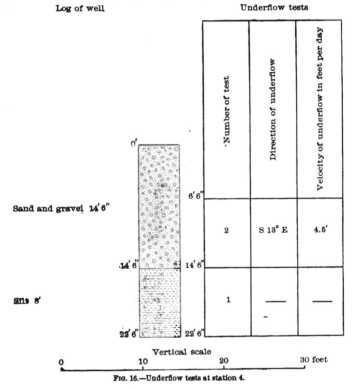

Fig. 16.—Underflow tests at station 4.

deeper, and a test was made as at station 3, but no underflow could be detected. The order of the strata, record of the wells, and the velocity measurements are shown in fig. 16.

Station 5 is on Los Angeles River, near the road from Burbank to Cahuenga Pass, as shown in fig. 12. A group of well points was here driven by hand to a depth of 34½ feet, and a test made between 26½ feet and 34½ feet, where the direction of the underflow was S. 40° E., velocity 48 feet per day. The water from this depth rose to a point 6 feet above the surface, and it is possible that the high velocity was caused by leakage upward around the casing, which drew the charged water toward the well. Borings were made with a 2-inch auger at this station. The order of the strata, record of the wells, and the velocity measurements are shown in fig. 17.

Station 6 is in Tujunga Wash, south of the road from Burbank to Cahuenga Pass, as shown in fig. 12. This group of wells was driven to a depth of 16 feet, and a test made between 8 feet and 16 feet, where the underflow was N. 75° E., velocity 3½ feet per day. Borings were made with a 2-inch auger at this station. The order of the strata,

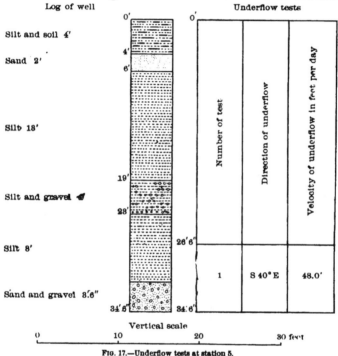

FIG. 17.—Underflow tests at station 5.

record of the wells, and the velocity measurements are shown in fig. 18.

Station 7 is on the left bank of Los Angeles River, 1 mile south of Burbank, at the mouth of a broad sandy wash (see fig. 12). This group of wells was driven by hand to a depth of 8 feet, and a test made between 5 feet and 8 feet, where the direction of the underflow was due east, velocity 2½ feet per day. The underflow was away from the surface stream, and the slope of the water table was afterwards found by leveling to be in the direction indicated by the underflow. The order of the strata, record of the wells, and the velocity measurements are shown in fig. 19.

Station 8 is on the left bank of Los Angeles River, 1 mile south of Burbank, at the mouth of a sandy wash (see fig. 12). This group of wells was driven by hand to a depth of 9 feet, and a test made between 1 foot and 9 feet, where the underflow was found to be N. 60° E., velocity 6.4 feet der day. The order of the strata, record of the wells, and the direction of the underflow are shown in fig. 20.

These tests indicate that in general the underflow is in the direction of the greatest slope of the water table, although there are marked variations. No relation could be traced between the slope of the water table and the velocity of the underflow. The conclusion drawn from the few tests made was that the underflow tends to follow the most

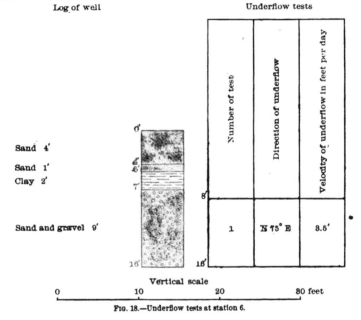

Fig. 18.—Underflow tests at station 6.

pervious strata, even if the extension of these beds varies considerably from the general slope of the water table. The maximum velocities are also found in such strata.

TESTS AT HURON STREET IN 1903.

Extensive tests were made at the narrows of the Los Angeles River along Huron street, Los Angeles, Cal., in 1903 (see fig. 21). This locality will be spoken of as the Huron street section. It was selected for the following reasons:

(1) In 1896 the city drilled several test wells to bed rock, furnishing valuable data regarding the cross section of the pervious gravel beds.

(2) The city water commissioners are now driving a tunnel in bed rock beneath the river channel, beginning about half a mile south of Huron street, to extend entirely across the valley (see fig. 12). It is the intention in the future to drill wells through the pervious gravels and the bed rock above the tunnel and by perforating them draw off the percolating ground water through the tunnel and pump it to the city reservoirs. It was believed that the amount of water intercepted by the tunnel would give an approximate check on the accuracy of the ground-water measurements.

(3) The locality is near shops at which tools and machinery could be repaired, or manufactured if necessary.

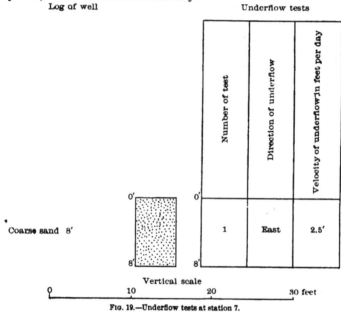

Fig. 19.—Underflow tests at station 7.

The surface profile of Huron street and the cross section of the pervious gravels below are shown in fig. 22. The points marked station 1 and stations 9 to 16 on both map and profile, figs. 21 and 22, represent groups of test wells sunk by the United States Geological Survey. The points numbered 1–7 represent test wells sunk by the city of Los Angeles in 1896, and the wells numbered 8–22 represent house wells in the vicinity. The underground cross section was platted from the information supplied by the various test wells and house wells in the vicinity. This representation of the underground conditions is necessarily incomplete, but no additional information is obtainable at present.

Attention is called to the fact that the pervious beds of sand and gravel in this section (fig. 22) fall into two distinct groups, the lower and most westerly filling the bottom of the old channel of Los Angeles River. The easterly group is without doubt the thinning edge of the Arroyo Seco débris cone. This is proved by the fact that the coarse débris from the Arroyo Seco can be traced directly to this place. The terrace at the east end of the section is also composed of the same material.

The tests of underflow point to the conclusion outlined above; in the lower gravel bed the direction of the underflow is down the general

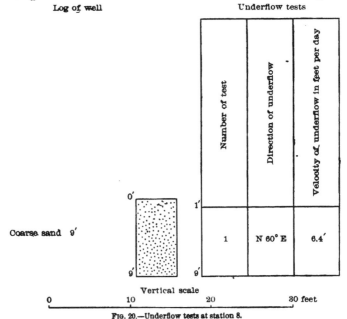

FIG. 20.—Underflow tests at station 8.

trend of the river valley, as shown by the arrows on the map at stations 1 and 9–11; while at stations 12, 13, 14, 15, and 16 the divergence from the trend of the valley is very marked—in fact, at station 15 the flow at present is up Los Angeles River Valley.

It is probable, however, that this reversion of the flow is due to the rise of the water table in the Arroyo Seco débris cone. Normally the underflow is probably from Los Angeles River down the slope of the water table, as indicated by the broken contours in fig. 21, which show the slope of the water table in December, 1902, while the dotted contours show the slope in June, 1903, the change being due, as above noted, to the floods of the Arroyo Seco. The portions of the cross section where measurable velocities were observed are diagonally cross hatched. The rest of the section appears to be practically impervious.

The depths at which the various tests were made and the direction

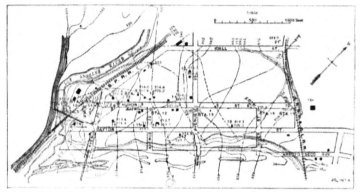

Fig. 21.—Map of Huron street and vicinity. Unbroken lines are surface contours showing topography; broken lines show form of water table in December, 1902; dotted lines show form of water table in June, 1903.

and velocity of the underflow measured are shown in figs. 23–30.

Attention is called to the general agreement in the slope of the water table and direction of the underflow.

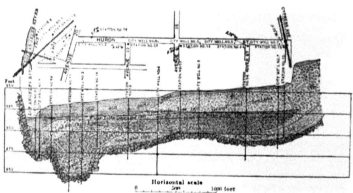

Fig. 22.—Huron street cross section.

STATIONS AND RESULTS.

The first test made was at station 9 (fig. 21), just above station 1. A group of well points was driven with a hand rig and tested between 15 feet and 23 feet. The direction of the underflow was S. 30° E., parallel with the surface stream, velocity 77 feet per day. The well

points were then driven deeper and a test made between 22 feet and
30 feet, where the direction of the underflow was S. 30° E., velocity
20.8 feet per day. The order of the strata, record of the wells, and
the velocity measurements are shown in fig. 23.

The ampere curve of the second test at this station is shown in fig.
11, a. It illustrates what form of curve may be expected with high

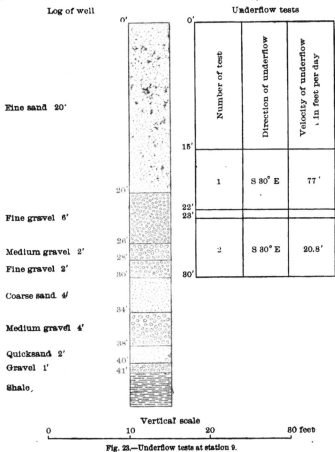

Fig. 23.—Underflow tests at station 9.

velocities and shows that observations at intervals of a few minutes
must be taken to correctly delineate such a curve. This is an extreme
case, however. A test well was sunk by the hydraulic-jetting method
here and the record of the wells is based on the material washed out
of the drivepipe. The pipes broke when it was attempted to drive
them beyond about 30 feet and no deeper tests were made.

Station 10 is on the left bank of the river, as shown in fig. 21. This group of wells was sunk with a hand rig, using the heavy drive-

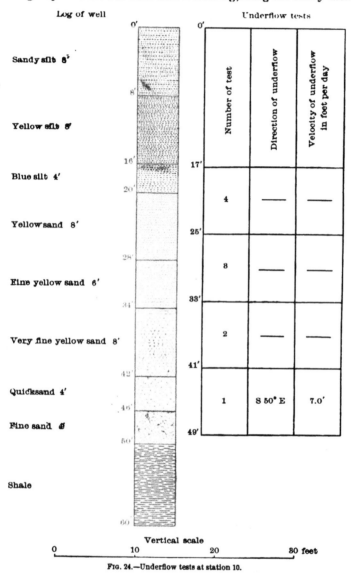

Fig. 24.—Underflow tests at station 10.

pipe described on page 15. The first test was made between 41 feet and 49 feet. The direction of the underflow was S. 50° E., velocity 7 feet per day.　It should be noted that the flow was probably through the stratum between 46 feet and 50 feet.　Tests were afterwards made betweeen 33 feet and 41 feet, between 25 feet and 33 feet, and between 17 feet and 25 feet. but no underflow was detected.　The

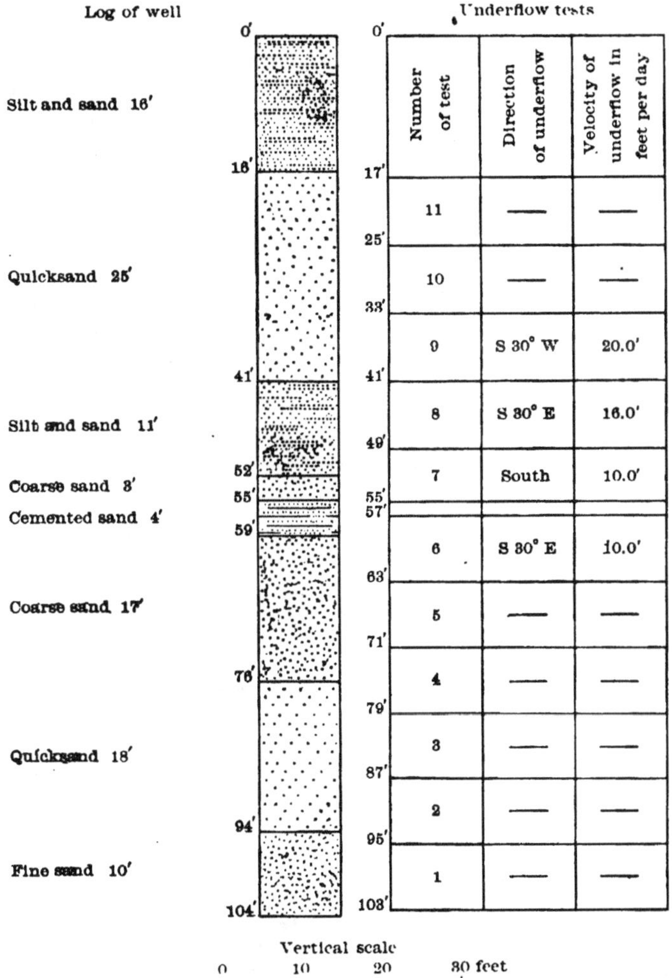

Log of well			Underflow tests			
				Number of test	Direction of underflow	Velocity of underflow in feet per day
Silt and sand 16'	0'	0'				
	16'	17'				
				11	—	—
					25'	
Quicksand 25'				10	—	·
					33'	
				9	S 30° W	20.0'
	41'	41'				
Silt and sand 11'				8	S 30° E	16.0'
		49'				
Coarse sand 8'	52'			7	South	10.0'
Cemented sand 4'	55'	55'	57'			
	59'			6	S 30° E	10.0'
		63'				
Coarse sand 17'				5	—	—
		71'				
	76'			4	—	—
		79'				
Quicksand 18'				3	—	—
		87'				
				2	—	—
	94'	95'				
Fine sand 10'				1	—	—
	104'	103'				

Vertical scale
0　　10　　20　　30 feet

FIG. 25.—Underflow tests at station 11.

material washed out of the drivepipe also indicated that the ground was practically impervious.　The order of the strata, record of the wells, and the velocity measurements at this station are shown in fig. 24.

Station 11 is between the Southern Pacific Railroad and Avenue 20, as shown in fig. 21.   One well of this group was sunk with the hand-rig and the rest by the well-driving machinery, as described on page 12.   The first test was made between 95 feet and 103 feet, but the result was uncertain, the final conclusion being that there was no underflow at this depth.   Other tests were made between 87 feet and 95 feet, between 79 feet and 87 feet, between 71 feet and 79 feet, and between 63 feet and 71 feet, but no underflow could be detected.   The water from 63 feet to 104 feet was heavily charged with sulphureted hydrogen.   A test was then made between 55 feet and 63 feet.

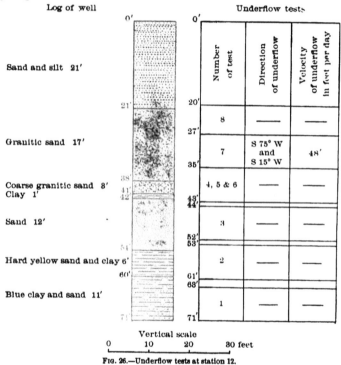

FIG. 26.—Underflow tests at station 12.

The direction of the underflow was S. 30° E., velocity 10 feet per day.   Another test was made between 49 feet and 57 feet.   The direction of the underflow was due south and the velocity 10 feet per day.

This was followed by a test between 41 feet and 49 feet, where the direction of the underflow was S. 30° E., velocity 16 feet per day.   The next test was between 33 feet and 41 feet, where the direction of the underflow was S. 30° W., velocity 20 feet per day.   Two more tests were made, one between 25 feet and 33 feet and another between 17 feet and 25 feet, but no underflow was detected.   The order

of the strata, record of the wells, and the velocity measurements at
this station are shown in fig. 25. The wide divergence in the direction of the underflow at this station, amounting to 60°, indicates
that the water follows the most open strata.

Station 12 is at the intersection of Huron street and Avenue 22, as
shown in fig. 21. Tests were made between 63 feet and 71 feet, between
53 feet and 61 feet, between 44 feet and 52 feet, and between 35 feet
and 43 feet, but no underflow was detected. The next test was made
between 27 feet and 35 feet, where the direction of the underflow was
S. 15° W., velocity 48 feet per day, and also S. 75° W., velocity 48
feet per day, the underflow here apparently being through two porous
strata in which the directions of movement diverge 60°. The last

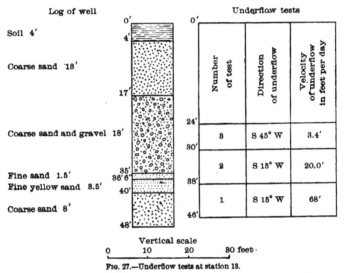

Fig. 27.—Underflow tests at station 18.

test was made between 20 feet and 27 feet, where no underflow was
detected. The order of the strata, record of the wells, and the velocity
measurements are shown in fig. 26. Below 50 feet the water was
impregnated with sulphureted-hydrogen gas.

Station 13 is at the corner of Huron street and Avenue 26, as shown
in fig. 21. A test was made between 38 feet and 46 feet, where the
direction of the underflow was S. 15° W., velocity 68 feet per day.
The next test was between 30 feet and 38 feet, where the direction of
the underflow was S. 15° W., and the velocity 20 feet per day. The
last test was between 24 feet and 30 feet, where the direction of the
underflow was S. 45° W., velocity 3.4 feet per day. The order of
the strata, record of the wells, and the velocity measurements are
shown in fig. 27. Attention is called to the westerly direction of the

underflow, due, without doubt, to the influence of the rising water table.

Station 14 is at the crossing of Huron street and River street, as shown in fig. 21. The first test was between 44 feet and 52 feet, where the direction of the underflow was S. 75° W., velocity 48 feet per day. The next test was between 36 feet and 44 feet, where the direction of the underflow was S. 75° W., velocity 32 feet per day. The

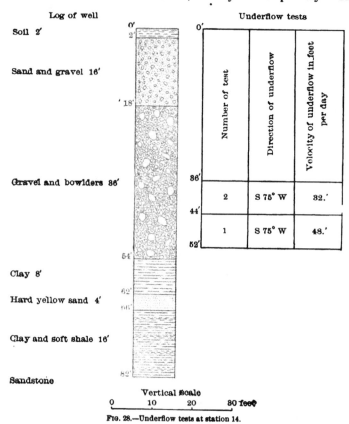

FIG. 28.—Underflow tests at station 14.

top of the well screen being at about the level of the water plane before the rise due to the floods of the Arroyo Seco, it was decided to leave the pipes in the ground and test the direction and velocity of the underflow as the water table gradually falls. That it will fall is proved by the fact that the wells above are now lowering at about 1 foot per week. The order of the strata, record of the wells, and the velocity measurements are shown in fig. 28.

Station 15 is at the eastern termination of Huron street, as shown in fig. 21. The first test was made between 45 feet and 47 feet, but

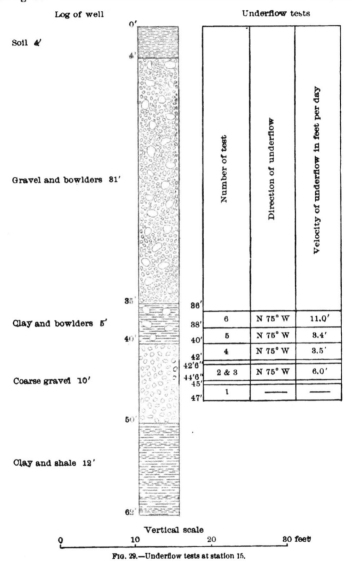

Log of well   Underflow tests

Soil 4'

Gravel and bowlders 31'

Clay and bowlders 5'

Coarse gravel 10'

Clay and shale 12'

Number of test	Direction of underflow	Velocity of underflow in feet per day
6	N 75° W	11.0'
5	N 75° W	8.4'
4	N 75° W	3.5'
2 & 3	N 75° W	6.0'
1	———	———

Vertical scale

0   10   20   30 feet

Fig. 29.—Underflow tests at station 15.

no underflow was detected. A test was then made between 42.5 feet and 44.5 feet, where the direction of the underflow was N. 75° W., velocity 6 feet per day. The next test was between 40 feet and 42 feet, where the direction of the underflow was N. 75° W., velocity

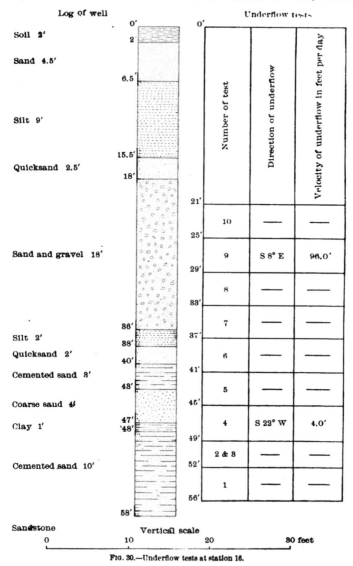

Log of well			Underflow tests		
			Number of test	Direction of underflow	Velocity of underflow in feet per day
Soil **2'**					
Sand **4.5'**					
Silt **9'**					
Quicksand **2.5'**					
			10	—	—
Sand and gravel **18'**			9	S 8° E	96.0'
			8	—	—
			7	—	—
Silt **2'**					
Quicksand **2'**			6	—	—
Cemented sand **8'**					
			5	—	—
Coarse sand **4'**					
Clay **1'**			4	S 22° W	4.0'
			2 & 3	—	—
Cemented sand **10'**			1	—	—
Sandstone					

Vertical scale

0   10   20   30 feet

FIG. 30.—Underflow tests at station 16.

3.5 feet per day. Another test was made between 38 feet and 40 feet, where the direction of the underflow was N. 75° W., velocity 3.4 feet per day. The last test was made between 36 feet and 38 feet, where

the direction of the underflow was N. 75° W., velocity 11 feet per day.
The order of the strata, record of the wells, and the velocity measurements are shown in fig. 29.

Station 16 is north of Huron street, between Avenue 20 and Avenue 22, as shown in fig. 21. Tests were made between 52 feet and 56 feet and between 49 feet and 52 feet, but no underflow was detected. The next test was made between 45 feet and 49 feet, where the underflow was S. 22° W., velocity 4 feet per day. Tests were then made between 41 feet and 45 feet, between 37 feet and 41 feet, between 33 feet and 37 feet, and between 29 feet and 33 feet, but no underflow was detected. A test was then made between 25 feet and 29 feet, where the direction of the underflow was S. 8° E., velocity 96 feet per day. The last test was made between 21 feet and 25 feet, but no underflow was noted here. The order of the strata, record of the wells, and the velocity measurements are shown in fig. 30.

Attention is directed to the probability of this well having penetrated both pervious areas in the Huron street cross section, as shown in fig. 22. All the ground tested here, except the two beds in which underflow was detected, seemed to be close-textured and practically impervious.

DISCHARGE AT HURON STREET.

Owing to the reversal in the normal direction of underflow, by the floods of the Arroyo Seco, it is not possible to estimate, except very approximately, the quantity of ground water passing the Huron street section. The observed facts are here presented, from which engineers may draw their own conclusions regarding the tests and value of the computations submitted.

The western and lower gravel bed, shown on the cross section, fig. 22, has a cross-sectional area of approximately 17,000 square feet, and the upper and eastern gravel bed has a cross-sectional area of about 30,700 square feet below the water plane of 1896. These figures are based on the results of the underflow tests, and on the assumption that all the cross section where flow was not detected has a velocity of underflow of 1 foot per day. The outline of the pervious sections was sketched from station to station.

The average of all the velocities observed in the western gravel bed is 20.6 feet per day, and the porosity is assumed to be 25 per cent. Using this data, the discharge from this portion of the section should be $17,000 \times 20.6 \times 0.25 = 87,550$ cubic feet per day. This discharge should be reduced by about 20 per cent to correct for the obliquity of flow across the section, $87,550 \times 0.8 = 70,040$ cubic feet, or 524,000 gallons per day, in round numbers.

The average of all the velocities observed in the eastern gravel bed is 30.9 feet per day. If it is assumed that under normal conditions the height of the water table is represented by the line marked 1896 in fig. 22, the area of the cross section through which flow takes place will be 30,700 square feet. The slope of the water table was not radically different in 1896 (the assumed normal condition) from what it was in 1903, when the underflow tests were made, but the direction of the flow in the eastern end of the section was deflected nearly or quite 90 degrees from the normal. It is assumed that since the degree of slope of the water table at these two periods is approximately the same, the velocity of underflow was not radically different. Under normal conditions, the direction of flow makes an angle of about 30 degrees with the line of Huron street. The computed discharge has been multiplied by 0.50, the sine of 30 degrees, to correct for the obliquity in the direction of the flow. The porosity is here assumed to be 25 per cent, as before. The discharge from this portion of the section should then be $30,700 \times 30.9 \times 0.25 \times 0.5 = 118,578$ cubic feet, or 889,000 gallons per day, in round numbers.

Although no flow was detected in the rest of the cross section, between the water table and bed rock, it can not be entirely impervious, and it is assumed that the average velocity is 1 foot per day. This is certainly a maximum, as many tests were continued long enough to detect as low a velocity as this.

If the average porosity is assumed to be 25 per cent as above, and the average direction of underflow is at an angle of 30 degrees with Huron street, the discharge from this section should be $60,000 \times 1 \times 0.25 \times 0.50 = 7,500$ cubic feet per day, or 56,000 gallons, in round numbers.

The total computed discharge from the Huron street section will then be 196,118 cubic feet per day, or 2.27 cubic feet per second.

The city tunnel has now been driven 1,163 feet under the river bed, one-fourth mile below. This tunnel yields about 7 second-feet when pumped for a short time at intervals of a few days. The supply from the tunnel has so far been drawn from the water stored in the gravels in the immediate vicinity, pumping not having as yet been continued long enough to demonstrate what the gravel beds will yield.

COST OF TESTS AT HURON STREET.

The following table shows the average cost of testing underflow at Los Angeles, Cal., for a period of two months, using the well-drilling machinery and testing apparatus described in this report:

Cost of underflow tests at Huron street, Los Angeles, Cal., from May 14 to July 14, 1903.

	Total expenses for 2 months.	Driving pipe.	Pulling pipe.	Testing.
Well driller	$150	$96	$22	$32
Laborers	100	64	14	22
Cook	70	40	10	20
Observer	150			150
Subsistence	120	50	20	50
Repairs	40	28	6	6
Chemicals	30			30
Fuel	8	6	2	
Supervision	150	60	30	60
Total	818	344	104	370

Pipe driven, 989 feet; cost per foot, $0.348.
Pipe pulled, 929 feet; cost per foot, $0.111.
Total number of underflow tests, 16; cost per test, $23.125.
Total average cost per test, including driving, pulling, etc., $51.125.

The well-drilling machinery described was constructed in Los Angeles from designs furnished the foundrymen and from machinery purchased at local supply houses.

Cost of well-drilling machinery.

Cost of drilling rig in Los Angeles $870.00
Two hundred linear feet of 3-inch, double, extra-heavy drive pipe, with taper-joint couplings, at $4 per foot 800.00

      Total 1,670.00

SUMMARY.

The following suggestions are based on the experience gained during the work at Huron street and in the San Fernando Valley:

(1) The location of the section where it is proposed to test the underflow should be carefully studied. It should be, if possible, in a straight stretch of the valley, and at some distance, either up or down, from large tributary streams.

(2) The form and slope of the water table should be ascertained and the line of test stations placed most advantageously.

(3) In order to secure accurate results, the testing stations should be close together along the line of the section.

(4) The well screens should be short and the ground should be tested at each 2 to 4 feet in depth, down to bed rock when possible.

(5) If possible, the porosity of the pervious beds should be determined.

(6) When making deep tests some form of drive pipe and screen such as described on the preceding pages should be used.

(7) Recording ampere meter and switch clocks should be used.

The discharge from a given section will undoubtedly be far less than anticipated, the popular tendency being to greatly overestimate the amount of underflow. Even if the results attained by this method of testing are not as accurate as desired, they are, nevertheless, of great value, as they enable investigators to compute, approximately, what could only be roughly estimated before.

INDEX.

55

O

[Mount each slip upon a separate card, placing the subject at the top of the second slip. The name of the series should not be repeated on the series card, but the additional numbers should be added, as received, to the first entry.]

Hamlin, Homer, 1864—

Author.

. . . Underflow tests in the drainage basin of Los Angeles River, by Homer Hamlin. Washington, Gov't print. off., 1905.

55 p., 1 l. illus., VII pl., diagrs. 23^{cm}. (U. S. Geological survey. Water-supply and irrigation paper no. 112.)
Subject series: O, Underground waters, 30.

1. Water, Underground—California. 2. Los Angeles River.

Hamlin, Homer, 1864—

Subject.

. . . Underflow tests in the drainage basin of Los Angeles River, by Homer Hamlin. Washington, Gov't print. off., 1905.

55 p., 1 l. illus., VII pl., diagrs. 23^{cm}. (U. S. Geological survey. Water-supply and irrigation paper no. 112.)
Subject series: O, Underground waters, 30.

1. Water, Underground—California. 2. Los Angeles River.

U. S. Geological survey.

Series.

Water-supply and irrigation papers.
no. 112. Hamlin, H. Underflow tests in the drainage basin of Los Angeles River. 1905.

U. S. Dept. of the Interior.

Reference.

see also
U. S. Geological survey.